Lecture Notes in Networks and Systems

Volume 231

The series "Lecture Notes in Networks and Systems" publishes the latest developments in Networks and Systems—quickly, informally and with high quality. Original research reported in proceedings and post-proceedings represents the core of LNNS.

Volumes published in LNNS embrace all aspects and subfields of, as well as new challenges in, Networks and Systems.

The series contains proceedings and edited volumes in systems and networks, spanning the areas of Cyber-Physical Systems, Autonomous Systems, Sensor Networks, Control Systems, Energy Systems, Automotive Systems, Biological Systems, Vehicular Networking and Connected Vehicles, Aerospace Systems, Automation, Manufacturing, Smart Grids, Nonlinear Systems, Power Systems, Robotics, Social Systems, Economic Systems and other. Of particular value to both the contributors and the readership are the short publication timeframe and the world-wide distribution and exposure which enable both a wide and rapid dissemination of research output.

The series covers the theory, applications, and perspectives on the state of the art and future developments relevant to systems and networks, decision making, control, complex processes and related areas, as embedded in the fields of interdisciplinary and applied sciences, engineering, computer science, physics, economics, social, and life sciences, as well as the paradigms and methodologies behind them.

Indexed by SCOPUS, INSPEC, WTI Frankfurt eG, zbMATH, SCImago.

All books published in the series are submitted for consideration in Web of Science.

More information about this series at http://www.springer.com/series/15179

Radek Silhavy · Petr Silhavy ·
Zdenka Prokopova
Editors

Data Science and Intelligent Systems

Proceedings of 5th Computational Methods in Systems and Software 2021, Vol. 2

Springer

Editors
Radek Silhavy
Faculty of Applied Informatics
Tomas Bata University in Zlín
Zlín, Czech Republic

Petr Silhavy
Faculty of Applied Informatics
Tomas Bata University in Zlín
Zlín, Czech Republic

Zdenka Prokopova
Faculty of Applied Informatics
Tomas Bata University in Zlín
Zlín, Czech Republic

ISSN 2367-3370 ISSN 2367-3389 (electronic)
Lecture Notes in Networks and Systems
ISBN 978-3-030-90320-6 ISBN 978-3-030-90321-3 (eBook)
https://doi.org/10.1007/978-3-030-90321-3

This Springer imprint is published by the registered company Springer Nature Switzerland AG
The registered company address is: Gewerbestrasse 11, 6330 Cham, Switzerland

Preface

This book constitutes the refereed proceedings of the Computational Methods in Systems and Software 2021 (CoMeSySo 2021), held in October 2021.

CoMeSySo 2021 conference intends to provide an international forum for the discussion of the latest high-quality research results in all areas related to systems, software, and networks in intelligent systems.

The addressed topics are the theoretical aspects and applications of software engineering and computational methods in engineering and science or artificial intelligence. The papers address topics as software engineering, cybernetics and automation control theory, econometrics, mathematical statistics or artificial.

CoMeSySo 2021 has received (all sections) 394 submissions and 165 of them were accepted for publication.

The volume Data Science and Intelligent Systems brings papers in which new approaches and methods related to data science and software engineering are applied to real-world problems. Furthermore, the basic research' papers that describe novel approaches in the data science, in algorithm design and in systems and software engineering are included.

The editors believe that readers will find the following proceedings interesting and useful for their research work.

September 2021

Radek Silhavy
Petr Silhavy
Zdenka Prokopova

Organization

Program Committee

Program Committee Chairs

Petr Silhavy, Ph.D., Department of Computers and Communication Systems, Faculty of Applied Informatics, Tomas Bata University in Zlin, Czech Republic

Radek Silhavy, Ph.D., Department of Computers and Communication Systems, Faculty of Applied Informatics, Tomas Bata University in Zlin, Czech Republic

Zdenka Prokopova, Ph.D., Associate Professor at Department of Computers and Communication Systems, Tomas Bata University in Zlin, Czech Republic

Krzysztof Okarma, Ph.D., DSc, Faculty of Electrical Engineering, West Pomeranian University of Technology, Szczecin, Poland

Prof. Roman Prokop, Ph.D., Professor at Department of Mathematics, Tomas Bata University in Zlin, Czech Republic

Prof. Viacheslav Zelentsov, Doctor of Engineering Sciences, Chief Researcher of St.Petersburg Institute for Informatics and Automation of Russian Academy of Sciences (SPIIRAS), Russian Federation

Prof. Lipo Wang, School of Electrical and Electronic Engineering, Nanyang Technological University, Singapore

Silvie Belaskova, Head of Biostatistics, St. Anne's University Hospital Brno, International Clinical Research Center, Czech Republic

Roman Tsarev, Ph.D., Associate Professor, Department of Informatics, Siberian Federal University, Krasnoyarsk, Russia

International Program Committee Members

Organizing Committee Chair

Radek Silhavy, Ph.D., Tomas Bata University in Zlin, Faculty of Applied Informatics, email: comesyso@openpublish.eu

Conference Organizer (Production)

Silhavy s.r.o.
Web: http://comesyso.openpublish.eu
Email: comesyso@openpublish.eu

Conference Website, Call for Paper

http://comesyso.openpublish.eu

Contents

Multiple OLAP Reports and the Best Compromise in the Decision-Making Process

Jan Tyrychtr[1](✉), Martin Pelikán[2], and Ivan Vrana[2]

[1] Faculty of Economics and Management, Czech University of Life Sciences in Prague, Prague, Czech Republic
tyrychtr@pef.czu.cz

[2] Department of Information Engineering, Czech University of Life Sciences in Prague, Prague, Czech Republic

Abstract. It is increasingly difficult for many businesses and organizations to make timely decisions. The main users of OLAP applications are decision-makers (managers, executives, etc.) who are under pressure because of frequent business or organizational tasks. Decision requirements are often associated with many uncertain circumstances. Therefore, managers' decision-making procedures are often based not only on the resulting OLAP report, but also on the opinions of experts who consider a wide range of circumstances. This paper presents the application of the BeCoMe method in managers' decision-making over OLAP outputs. The use of the method is shown in case study with OLAP reports on revenue from selected products sold. The presented OLAP outputs were made up of 10 decision-makers from businesses (managers, experts, or executives). The decision-makers expressed their views (each from their own point of view). Since their views on the issue were inconsistent or even conflicting, we applied the BeCoMe method to achieve a mathematically accurate compromise.

Keywords: Business Intelligence · OLAP · Reports · Decision support · BeCoMe method · Agreement

1 Introduction

1.1 OLAP

OLAP (On-Line Analytical Processing) is a computing technology for multidimensional data analysis that allows end users to perform complex calculations to support planning, budgeting, financial reporting, and forecasting which contribute to effective decision-making in business. OLAP uses a multidimensional view of data where dimensions represent data categories (e.g., product, customer, employee, time, etc.) and fact tables that store measurement calculations as indicators relevant to analysis (profit, revenue, number of sales, etc.). OLAP allows users to query data and analyze it from different perspectives. This technology is a great helper for analysts who need to perform preliminary aggregations and data calculations in the shortest possible time. Theoretical foundations of OLAP are presented in studies [1–4, 7, 8].

R. Silhavy et al. (Eds.): CoMeSySo 2021, LNNS 231, pp. 1–7, 2021.
https://doi.org/10.1007/978-3-030-90321-3_1

1.2 OLAP Decision-Making Process

The main reason for using OLAP as part of Business Intelligence in organizations is to support and improve decision-making processes. Some aspects of decision-making have changed over the last decade. The biggest change is that organizations need to make decisions faster and faster. There is less and less time for adopting decisions, which are often crucial. It is increasingly difficult for many businesses and organizations to make decisions timely. The main users of OLAP applications are therefore decision-makers (managers, executives, etc.) who are under greater pressure because of frequent business or organizational tasks. Decision requirements are often associated with increased periodicity (repeatability) of decisions. The need for managers leads to increased demands on the operational and technological provision of activities by the controlling department and the IT support department in the organization, who prepare the underlying data and their analytical processing.

1.3 Problem Statement

Although OLAP reports are a very useful tool in decision support, they are only data/information sources. The specific decision is always up to the relevant decision maker. In the common situations of medium and large companies and organizations, there are few decisions based on the opinion of a single decision maker. The meetings are attended by members of the annual general meeting, board of directors, the shareholders of the organization and other members or stakeholders. All these important members can receive regular reports from OLAP systems to support their qualified decisions. It often happens that even though all stakeholders receive the same quality messages, their decisions are not uniform, but often different or contradictory (see Fig. 1).

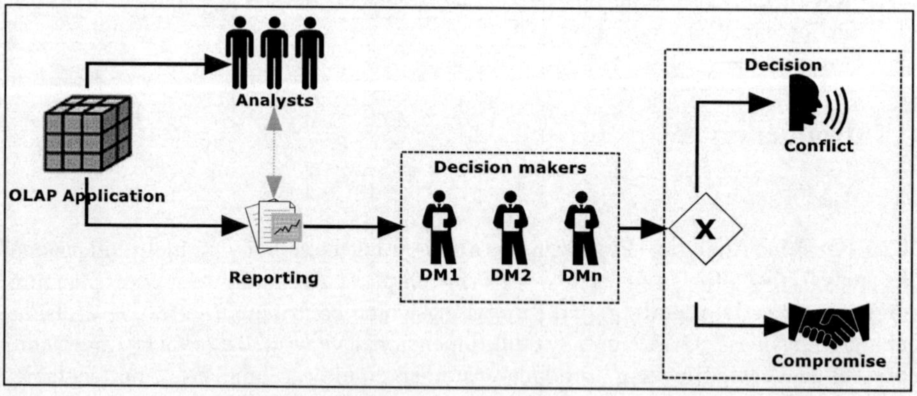

Fig. 1. OLAP reports and a conflict in the decision-making process between multiple decision makers.

The situation (from Fig. 1) often leads to conflicts in the decision-making process or to the adoption of an incorrect (enforced) decision.

The purpose of this paper is to present an existing unique BeCoMe method for quickly finding a compromise between stakeholders' judgements. In the next part of the paper, we will introduce the BeCoMe method and apply it to the case study when making decisions based on OLAP reports.

2 Methods

To create our unique decision-making process for managers and executives who have multiple reports from the OLAP system, we use the BeCoMe method, introduced in [9]. The BeCoMe (best-compromise-mean) method provides a best-compromise solution for multi-expert decision-making under fuzzy conditions. Based on the Shannon theory of entropy, it presents a metric for the level of agreement between experts' judgments.

The BeCoMe method and its instruments provide a solution to both crucial aspects: speed and reliability of decision making in ambiguous conditions. The method can assess experts' answers a certain quantitative parameter or qualitative aspect of the proposed solution (e.g., number of days of leave for employees under benefits). Experts can express their standpoints in three ways: 1) as a crisp number; 2) as a fuzzy number in a format: Preferable value, Lower limit, Upper limit; and 3) as a quantity in the Likert scale: "Strongly disagree", "Rather disagree", "Neutral", "Rather agree", "Strongly agree" or a similar linguistic scale.

The result of BeCoMe: $(\pi\varphi\xi)$ is calculated in the following steps, according to [9]:

1. Collect the triangular experts' judgments, $A_iC_iB_i$.
2. Calculate arithmetic mean Γ: $(\alpha\gamma\beta)$.
3. Calculate center of mass Gx (centroid) for each triangle $A_iC_iB_i$.
4. Rank experts with respect to their centroids.
5. Select median Ω: $(\rho\omega\sigma)$ as the triangle that is in the middle of all triangles $A_iC_iB_i$.
6. Calculate $\Gamma\Omega$Mean: $(\pi\varphi\xi)$.
7. Calculate maximum error Δmax.

The BeCoMe software tool can be used to facilitate calculations. BeCoMe Tool can be downloaded from https://covid19-become.pef.czu.cz/en/. It is simple Excel spreadsheet software for accomplishing all calculations in the case study.

The individual experts' judgments, their distribution along the *x* axis with highlighted arithmetic mean Γ, median Ω, and the final decision $\Gamma\Omega$Mean are visually expressed in Fig. 3.

3 Results

In this section, we present the OLAP decision-making process using the BeCoMe method.

Through the OLAP case study, we shall illustrate how the BeCoMe method can be used for decision-making by managers and executives under the conditions of multiple views on one issue. The aim of this case study was not to describe this management decision-making. Instead, by applying it to typical management situations that have been observed in one company in the Czech Republic, we shall demonstrate the potential of the BeCoMe method as a means of searching for the best-compromise solutions when we have multiple OLAP reports or their variations according to dimensions, and multiple decision-makers with different preferences and opinions.

3.1 OLAP Case Study

Revenue Growth is a key performance indicator (KPI) used to measure how sales increase or decrease over time. Thanks to the OLAP system, which is implemented in the company, the Revenue Growth indicator is divided into meaningful dimensions. By analyzing this KPI (through slice and dice operations), the company can better address the areas that contribute to revenue.

For example, by examining revenue growth by product category, a company can find out which product categories have constantly increasing its revenue, determine which product categories are consistent, or identify which ones are slowing down.

In real business systems, it is very difficult to consider all the circumstances in forecasting, budgetary models. Therefore, when making decisions, it is appropriate to consider not only the data from OLAP systems, but also the opinions of individual decision-makers who have various information, knowledge, and experience in the given field.

In this case study, we submitted multiple OLAP reports to ten selected members of the General Meeting of company. These reports were simulated on data from the Adventure Works database. The group of these experts (decision-makers) received several variants of reports on the company's revenue, for example views according to customers, product categories, or time. Experts also have data about sales volume and price by product, and competitive prices for similar products. One example of an OLAP report is illustrated in Fig. 2.

For the monitoring and at the same time the achievement of the set goal to be transparent, it is necessary to set three "bands" of target values, the so-called red (target not met), yellow (acceptable) and green (target is met). Based on the available data from OLAP, the group of experts was to set a new interval for meeting the target.

The group of experts with diverse expertise was appointed. The expert group engaged in a secret-ballot voting to preserve anonymity of individual actors in the general meeting of the selected company.

Now we shall demonstrate the application of the BeCoMe method. The group of experts was to answer the following issue:

For the next strategic period, propose the value C of the low-revenue interval (red bands) for the Components product category.

Express your proposal in interval in the range of \$ 0 – \$ 900 million and indicate the recommended upper B and lower a limits of your proposal.

Each member of the expert group got acquainted in detail with all financial reports from OLAP systems and then expressed his/her opinion. The suggested values are in the form of fuzzy numbers in Table 1.

In Table 1, the experts' triangular membership functions are expressed by vertex coordinates A, B, and C. G_x is a vertex of the respective triangle.

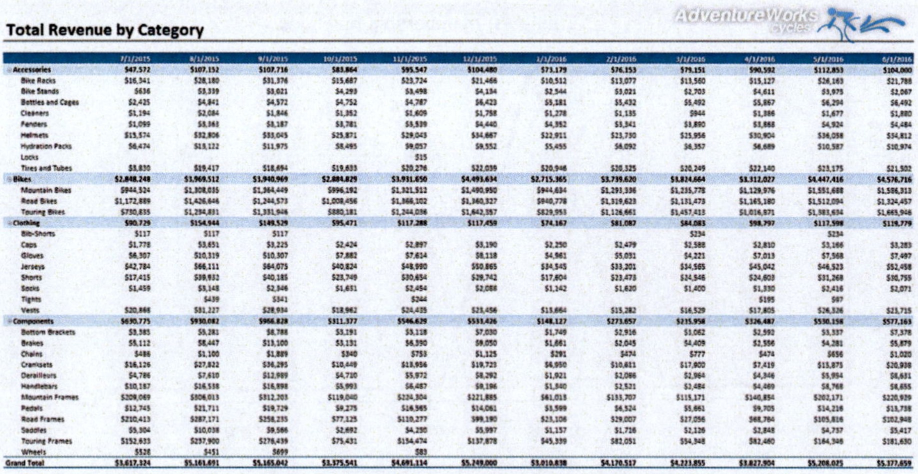

Fig. 2. OLAP report – Total Revenue by product category (according to [6]).

Table 1. Experts' proposals with calculated centroids Gx and median.

Experts	C	A	B	Gx
Chairman of the board	320	280	350	316,67
Member of the board	430	365	430	408,33
Member of the board	**228**	**190**	**275**	**231,00**
Chairman of the supervisory board	220	180	235	211,67
Member of the Supervisory board	200	190	225	205,00
Member of the supervisory board	185	140	205	176,67
Member of the general meeting	167	148	175	163,33
Member of the general meeting	405	310	438	384,33
Financial manager	235	155	310	233,33
Managing director	248	156	340	248,00

1. We ranked the experts with respect to the values of their centroids G_x.
2. As a median Ω: (ρωσ), we selected the triangle, which is in the middle of all triangles, $A_iC_iB_i$. Thus, $\rho = 190$, $\sigma = 275$, and $\omega = 228$ (as expressed in bold letters in Table 1).
3. We calculated the final decision $\Gamma\Omega$Mean: (πφξ), with the following results: $\pi = 191.95$, $\xi = 295.4$, and $\varphi = 247.65$.
4. We calculated that the maximum error $\Delta_{max} = 12.83$ and the value of the best compromise 245 (Fig. 3).

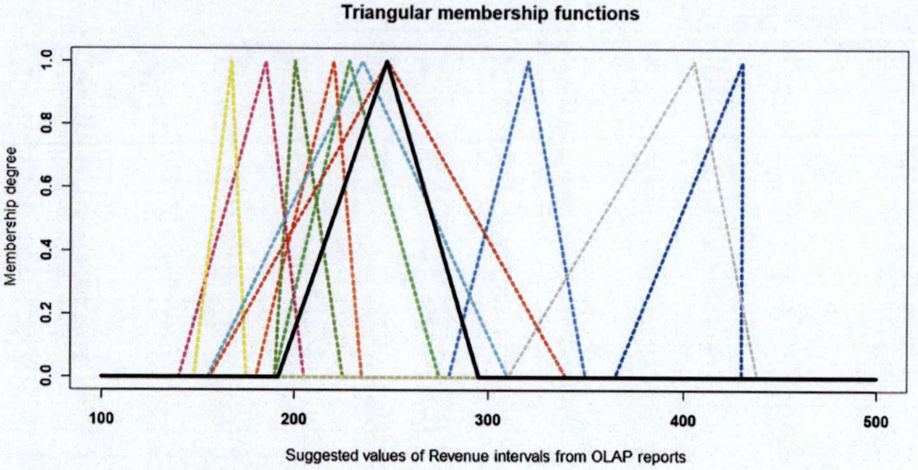

Fig. 3. Graphical expression of the decision situation. The resulting final decision (compromise) is expressed by a solid black line.

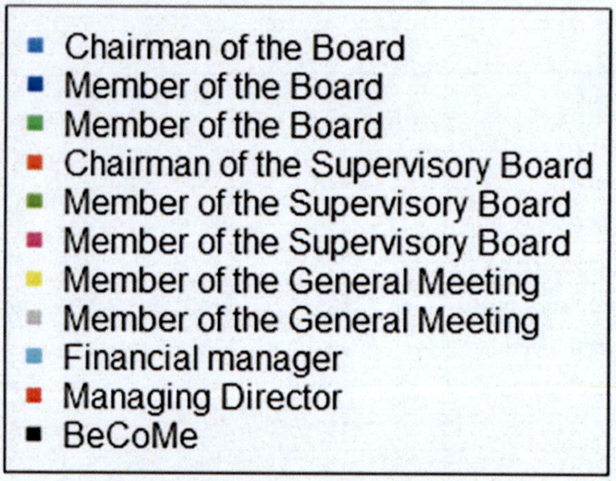

Fig. 4. Legend to the previous Fig. 3.

The company's management now has a mathematically accurate calculation of the compromise value from the various opinions of experts. The KPIs for low-income measure in OLAP can now be set to a compromise value of $ 245 million (revenues below this value will be in the red band) or at the interval with best proposal 247 (e.g., light red band), low limit 191 (e.g., bright red band) and upper limit 295 (e.g., orange band). The accuracy $\Delta_{\max} = 12.83$.

4 Conclusion

Executives, managers and other decision-makers in today's medium and large companies or organizations very often use Business Intelligence tools for their managerial activities. OLAP applications have been used to support the analytical processing of large amounts of data in companies or organizations. Reports, dashboards, or other outputs from OLAP are used by managers for their qualified decisions. This paper presented a unique approach on how to use the OLAP outputs in decision making under conditions of uncertainty. For these purposes, we recommend using the BeCoMe method as a unique and especially easy and fast support to find a compromise between proposals of multiple decision makers.

In the article, we presented the method on a simple and illustrative case study with ten decision-makers who had to agree on a threshold value (or interval) for setting the KPI in OLAP systems.

Acknowledgements. This work was conducted within the project Ambient intelligence in decision-making problems in uncertainty conditions (2019B0008) funded through the IGA foundation of the Faculty of Economics and Management, Czech University of Life Sciences in Prague.

References

1. Abelló, A., Romero, O.: On-line analytical processing. In: Liu, L., Özsu, M.T. (eds.) Encyclopedia of Database Systems, pp. 1949–1954. Springer, Heidelberg (2009)
2. Berson, A., Smith, S.J.: Data Warehousing, Data Mining, and OLAP. McGraw-Hill Inc., New York (1997)
3. Celko, J.: Joe Celko's Analytics and OLAP in SQL. Elsevier, New York (2010)
4. Chaudhuri, S., Dayal, U.: An overview of data warehousing and OLAP technology. ACM SIGMOD Rec. **26**(1), 65–74 (1997)
5. Han, J., Pei, J., Kamber, M.: Data Mining: Concepts and Techniques. Elsevier, New York (2011)
6. Su, E.: Top Finance Expert in Park City, UT: Ellen Su | Toptal https://www.toptal.com/finance/resume/ellen-su. Accessed 19 Aug 2021
7. Thomsen, E.: OLAP Solutions: Building Multidimensional Information Systems. John Wiley & Sons, New York (2002)
8. Tyrychtr, J., Pelikán, M., Štiková, H., Vrana, I.: EM-OLAP Framework. Bus. Inf. Syst. Eng. **60**(6), 543–562 (2018). https://doi.org/10.1007/s12599-018-0533-5
9. Vrana, I., Tyrychtr, J., Pelikán, M.: BeCoMe: Easy-to-implement optimized method for best-compromise group decision making: Flood-prevention and COVID-19 case studies. Environ. Model. Softw. **136**, 10495 (2021)

One Step Deep Learning Approach to Grasp Detection in Robotics

Petr Dolezel(✉), Dominik Stursa, and Daniel Honc

Faculty of Electrical Engineering and Informatics, University of Pardubice, 532 10 Pardubice, Czech Republic
petr.dolezel@upce.cz
https://fei.upce.cz/en/fei/research-team-of-petr-dolezel

Abstract. Grasp point detection is a necessary ability to handle for industrial robots. In recent years, various deep learning-based techniques for robotic grasping have been introduced. To follow this trend, we introduce a convolutional neural network-based approach for model-free one step method for grasp point detection. This method provides all feasible grasp points suitable for parallel grippers, based on a single RGB image of the scene. A case study, which shows the outstanding accuracy of the presented approach as well as its acceptable response time, is presented at the end of this contribution.

Keywords: Grasp point · U-Net · Computer vision · Deep learning

1 Introduction

Grasping is a necessary ability to handle for industrial robots, in order to be beneficial in modern-day production plants or laboratories. Robots, especially robotic arms, should be able to implement various perception capabilities to visually detect near objects, classify them, and finally, locate an optimal grasp point for each object. A grasp point represents not only a point on the body of the manipulated object. It also provides the explicit information, how a robotic End-of-Arm Tooling (EOAT) should be oriented to robustly grab an object in order to securely manipulate it. Apparently, this information depends on the situation and a scene arrangement around a robot, but also on the type of an EOAT.

Traditionally, grasp detection techniques are still developed analytically, or ad-hoc, based on expert knowledge of the application background in combination with numerical computations. Qujiang Lei et al. [12] provided an extensive survey of fast grasp point algorithms, mostly based on the processing of the 3D model of the scene. However, there were also presented several studies of grasp point detection based on RGB-D images [1,15] or simple 2D image of the scene

The work has been supported by the SGS grant no. SGS_2021_019 at the University of Pardubice. This support is very gratefully acknowledged.

R. Silhavy et al. (Eds.): CoMeSySo 2021, LNNS 231, pp. 8–17, 2021.
https://doi.org/10.1007/978-3-030-90321-3_2

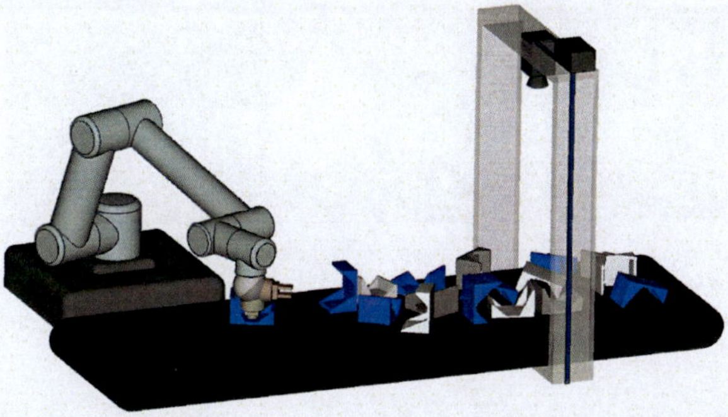

Fig. 1. Robotic grasping using grasp detection system.

[6,11,16]. Even though the mentioned approaches work efficiently and provide acceptable results, they suffer from several limitations, including insufficient generalization and difficult or time-consuming process of the adaptation to different conditions.

Since deep learning methods have provided a very successful results in various computer vision applications [14], many research teams have started to use this approach to grasp point detection. The survey, which aims to serve as a comparison to various grasp point detection frameworks based on deep learning, is available in [7].

In this article, we introduce a novel grasp point detection framework based on an RGB image transformation using a fully convolutional neural network. The aim of the framework is to efficiently detect and locate all feasible grasping points for parallel gripper-based EOATs in cluttered scenes. Since an RGB image is considered as the input information, the framework targets on objects scattered in single-layer manner and for 3 degree-of-freedom EOATs, i.e. EOATs with variable position and fixed pose.

The key part of the proposed framework is a specific application of a fully convolutional neural network. The network transforms the RGB scene image with the objects in one step into a schematic gray-scale image, where the grasp points are highlighted with the gradient geometric shapes. In addition, the full information about the grasp point necessary for the parallel gripper is explicitly coded in those gradient shapes.

We demonstrate the capabilities of the framework using an example with specifically prepared objects. These objects provide different types of edges suitable for parallel gripper - see Fig. 2. In addition, various random clutters of these objects can generate an infinite number of scene arrangements, where the feasibility of the grasp points is affected by mutual positions of the objects. We believe that this example sufficiently demonstrates the capabilities of the proposed framework.

Fig. 2. The object for manipulation (color variations). Each object provides various grasping points for manipulation, depending on its position and pose.

2 Materials and Methods

We introduce an RGB data-driven framework which detects grasp points based on a single RGB image of the scene, and provides all the necessary information for the robotic arm. The main features of the study presented below are as follows.

- The method detects the necessary information about each feasible grasp point in the scene. Unfeasible grasp points are omitted.
- A parallel gripper (see Fig. 3) is considered in this study as a robotic EOAT. Therefore, the necessary information for grasping includes the required position and rotation angle of the EOAT (i.e. pose).
- The introduced framework aims on objects scattered in single-layer manner and for 3 degree-of-freedom EOATs, i.e. EOATs with variable position and fixed pose, because only a single RGB image is considered as the input source of information. The information about depth cannot be appropriately reconstructed from one RGB image.

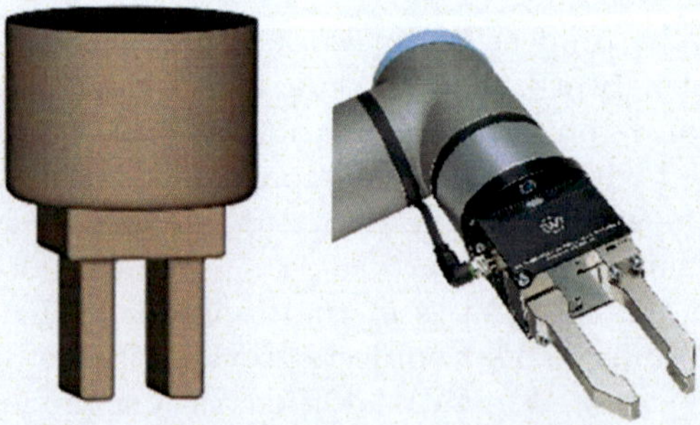

Fig. 3. Paraller gripper (schematic representation and real example).

The proposed framework is intended to be applied not only for separated objects, but also for dense clutters of them. Clearly, various arrangements of the objects in the scene occur, and each individual setting provides a particular group of possible grasp points. The feasibility of the grasp points is affected by close contact and even by overlapping of the objects. Some typical scene settings are figured in Fig. 4. In these figures, the grasp points are highlighted with a schematic position of the parallel gripper.

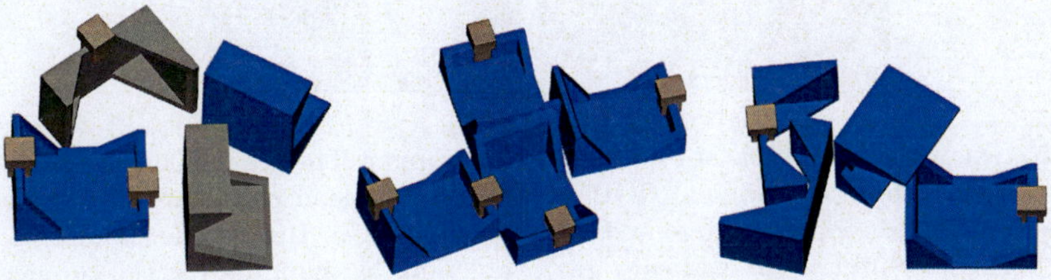

Fig. 4. Grasp points of the randomly oriented objects. Grasp points are highlighted with a symbol of a parallel gripper. Note that some grasp points are not feasible due to the touch positions and overlapping with other objects.

2.1 Fully Convolutional Neural Network Design

In recent years, various convolutional networks have proven to be successful in a semantic segmentation process [5,9]. In simple words, the semantic segmentation is a task, where specific regions in an input data are labeled according to some defined characteristics. We believe, that using the semantic segmentation capability of selected fully convolutional neural networks is able to transform an RGB image of the scene into a schematic gray-scale image, where the necessary information about the grasp points is coded into specific shapes.

Grasp Point Depiction. Our aim is to detect the grasp points for parallel gripper via semantic segmentation-like approach. Specifically, we develop a fully convolutional neural network to label each pixel in an RGB image of the scene. The label goes continuously from 1, which means the optimal, to 0, which is the most unsuited grasp point position. The gradient of the label defines the required spatial orientation of the parallel gripper. Specifically, the slightest gradient descent goes along the edge, while the steepest descent goes perpendicularly from the edge. The situation is depicted in Fig. 5.

Hence, the fully convolutional neural network should work as seen in Fig. 6.

Fully Convolutional Neural Network Topology. We need to select a fully convolutional neural network topology capable of transformation as defined above. To accomplish this goal, the U-Net topology is chosen as an initial architecture for consideration [13]. The U-Net is a U-shaped convolutional network introduced initially for biological and medical image segmentation problems. However, it has then been successfully adopted to many other applications. It follows a typical encoder-decoder architecture with a bottleneck. However, it also

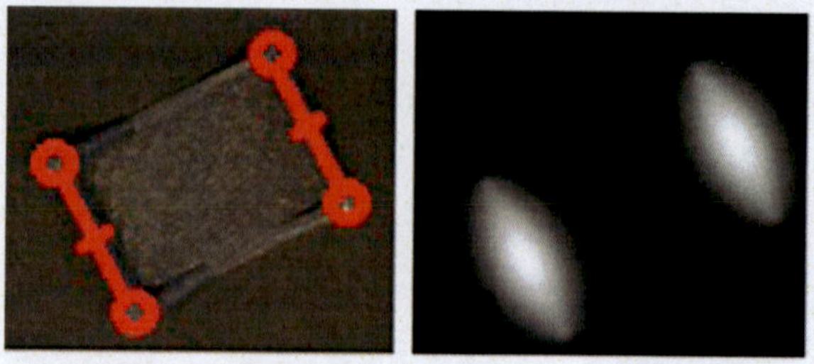

Fig. 5. Grasp point representation for a parallel gripper. The grasp point is represented by a red abscissa on the original RGB image (left); the end points are highlighted by circles and a midpoint is represented by a small mark in the middle. Labels of each pixel are represented by intensities (right); the higher is the intensity, the closer is the label to 1, black color is equal to 0.

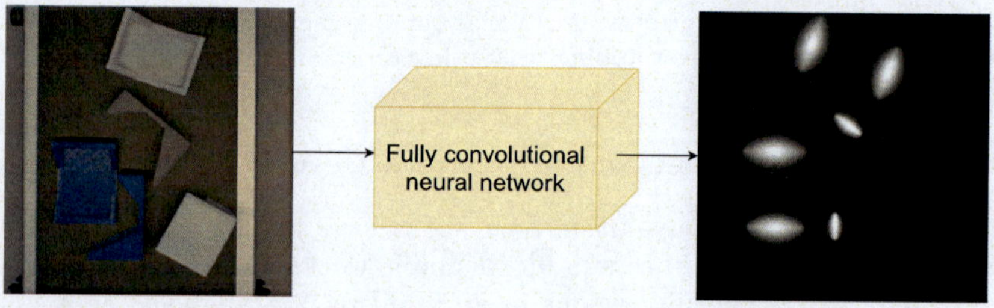

Fig. 6. The neural network is intended to transform the original RGB image into a schematic image, where the positions of the grasping points are highlighted as gradient shapes.

includes direct signals from the encoder part to the decoder part which allow the network to propagate context information to higher resolution layers. U-Net has proven itself to be a very successful in semantic image segmentation, especially in cases where only small training datasets are available.

In addition to U-Net, we select also its derivative called Attention U-Net [10]. Attention U-Net includes attention gates to suppress irrelevant regions in the source image and to highlight salient features useful for a specific task. We believe, that this feature can be useful in our case, since the grasp points take only little space in the source image and the majority of the output remains equal to zero.

As a last considered topology, we include Squeeze U-Net for testing. Squeeze U-Net is a lightweight version of U-Net and according to Beheshti et al. [2], it is much faster for both training and inference than U-Net with the same accuracy. This feature could be useful especially for the deployment on edge-computing devices.

Dataset Acquisition. In order to collect training and testing data, we prepare a robotic stand as seen in Fig. 7.

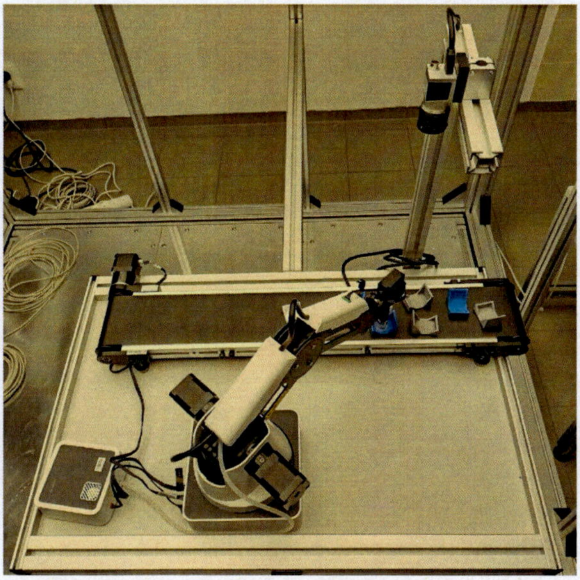

Fig. 7. A robotic stand for data collection. Objects of interest are scattered on a conveyor belt. An industrial camera situated above the scene is used for RGB data collecting.

Initially, in total 716 images are captured for a training set. The collection of images includes from 0 to 9 objects of three colors (blue, white, gray) with various positions and spatial arrangements. Resolution of the images is 288×288 RGB pixels with 8-bit depth. All images are labeled with a custom labeling application in order to get target outputs.

Consequently, we perform data augmentation. Specifically, each image is randomly rotated by angle α between $< -10, 10 >$ degrees and, subsequently, each image is randomly shifted between $< -10, 10 >$ px in both horizontal and vertical direction. Therefore, the original dataset of 716 images is enhanced to 2148 images.

Besides, the testing dataset is also acquired. We acquire 54 special images with the difficult and troublesome arrangements of the objects, such as various contact positions and difficult attainability of the grasp points.

The dataset is publicly available at [4].

Fully Convolutional Neural Network Training. Each selected topology is trained using the ADAM algorithm [8]. Based on our previous experience and based on [3], this algorithm provides good results in most of the cases. We train the networks from scratch with random initial weights (Gaussian distribution). We implement binary cross entropy as the loss function. 30% of the dataset is used as the validation set. We repeat the experiments five times for each topology. The best training session, considering the final values of the loss function over the validation set, is used for further evaluation.

Evaluation Metric. From the accuracy point of view, we adapt well-known IoU metric (Intersection over Union) for evaluation. However, this metric evaluates a detector based on ground-truth bounding boxes and predicted bounding boxes, which is not explicitly provided by our framework. Hence, we define generalized IoU (gIoU) metric as follows.

$$\text{Area of Overlap} = \min(\mathbf{T}, \mathbf{Y}), \tag{1}$$

$$\text{Area of Union} = \max(\mathbf{T}, \mathbf{Y}), \tag{2}$$

$$\text{gIoU} = \frac{\sum \text{Area of Overlap}}{\sum \text{Area of Union}}. \tag{3}$$

In equations above, $\mathbf{Y}$ means the 2D output array from the convolutional neural network, $\mathbf{T}$ means its expected values, min() and max() functions determine the minimal and maximal values at the corresponding array positions and $\sum()$ computes the sum of all values in the input array. IoU $= 1$ means ideal result, 0 means the worst result.

Beside the gIoU metric, the response times of every considered topology are also evaluated. We use the following hardware and software specification: Processor Intel Core i5-8600K, internal memory 16 GB DDR4 (2666 MHz), video card NVIDIA PNY Quadro P5000 16 GB GDDR5x PCIe 3.0, SSD SATA M.2 512 GB, Windows 10 Professional 64bit, Python 3.6 and TensorFlow 2.0.

3 Results

All selected topologies are trained and evaluated according to the procedure described above. The best instances are summarized in Table 1.

Table 1. Results

Topology	Loss	Mean gIoU	T, ms
U-Net	5.73×10^{-2}	0.849	49.9
Attention U-Net	5.73×10^{-2}	0.840	49.8
Squeeze U-Net	5.66×10^{-2}	0.805	29.4

In Table above, Loss means the final value of the loss function evaluated over the validation set. Mean gIoU is the mean value of gIoU evaluated over the testing set and T means the mean response time across the testing set in milliseconds.

From the resulting values in Table 1, we can observe that the U-Net topology provides the best overall gIoU values over the testing set. On the other hand, squeeze U-Net provides the worst results. This is in a contrast with the results during training, where squeeze U-Net reaches the lowest final value of the loss function. This finding could indicate bad generalization capabilities of squeeze U-Net. Considering the response time, squeeze U-Net provides the best results, while the other two topologies are on the same level.

We can say, that for our specific task, U-Net provides the best results with the acceptable response time. In the discussed task, there are no benefits to use

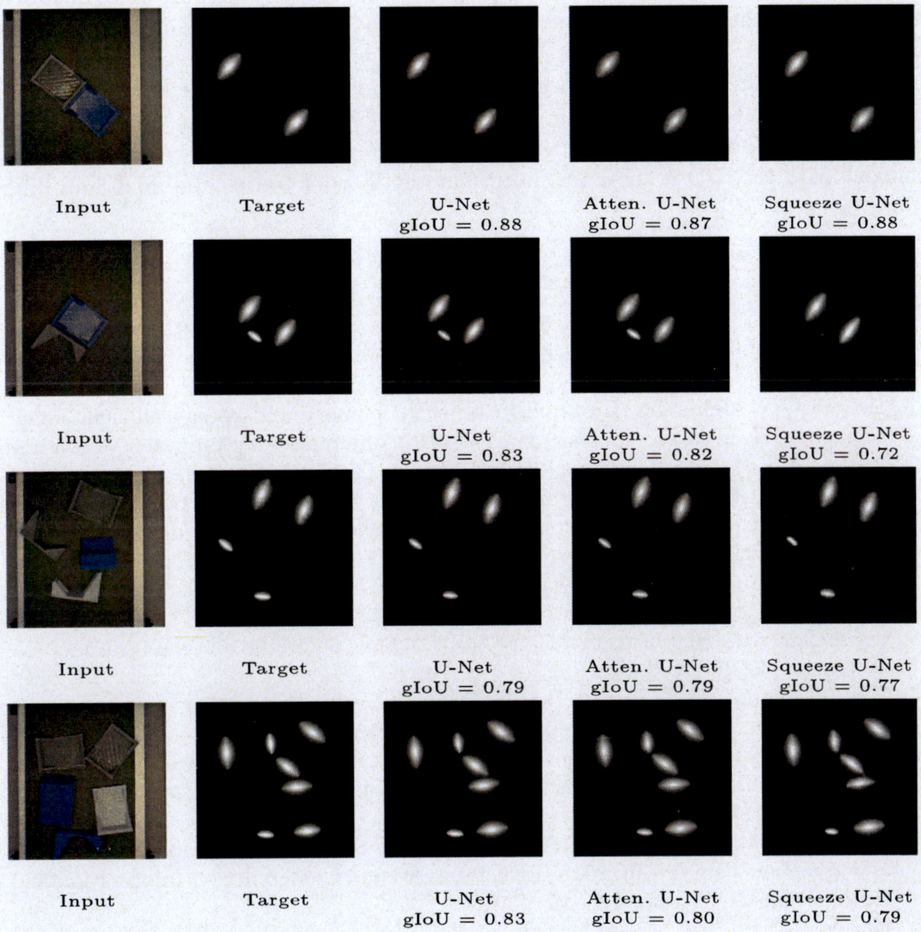

Fig. 8. Responses of the U-Net, Attention U-Net and Squeeze U-Net topologies to the selected input images.

attention gates implemented in attention U-Net. Squeeze U-Net can be implemented in edge-computing applications, where response time is a key element. However, the GIoU decrease is apparent.

4 Conclusions

We introduced a novel approach for grasp point detection suitable for parallel grippers. The key element of the approach is a convolutional network-based transformation of the scene into a schematic image, where all feasible grasp points are highlighted as gradient shapes. The necessary information for a parallel gripper, i.e. the position and the required pose, is coded into the mentioned shape in a very efficient way. Some tests with several convolutional network topologies indicated, that U-Net topology can be suitably implemented for this approach.

Critically, the proposed approach is based on an RGB input only. Hence, it can be applied to scenes with a single layer of objects because of the lacking information about depth. In future work, we are going to improve the approach to work with RGB-D images and point clouds in order to be capable of handling more layers of objects.

References

1. Asif, U., Bennamoun, M., Sohel, F.A.: RGB-D object recognition and grasp detection using hierarchical cascaded forests. IEEE Trans. Robot. **33**(3), 547–564 (2017)
2. Beheshti, N., Johnsson, L.: Squeeze u-net: a memory and energy efficient image segmentation network. In: 2020 IEEE/CVF Conference on Computer Vision and Pattern Recognition Workshops (CVPRW), pp. 1495–1504 (2020)
3. Dogo, E.M., Afolabi, O.J., Nwulu, N.I., Twala, B., Aigbavboa, C.O.: A comparative analysis of gradient descent-based optimization algorithms on convolutional neural networks. In: 2018 International Conference on Computational Techniques, Electronics and Mechanical Systems (CTEMS), pp. 92–99 (2018)
4. Dolezel, P., Stursa, D.: Grasp points dataset for asp u-net (2021). https://www.researchgate.net/publication/350410726_Grasping_points_for_parallel_gripper_and_vacuum_cup_RGB_data
5. Han, C., Duan, Y., Tao, X., Lu, J.: Dense convolutional networks for semantic segmentation. IEEE Access **7**, 43369–43382 (2019)
6. Jabalameli, A., Behal, A.: From single 2D depth image to gripper 6d pose estimation: a fast and robust algorithm for grabbing objects in cluttered scenes. Robotics **8**(3), 69 (2019)
7. Jia, Q., Cai, J., Cao, Z., Wu, Y., Zhao, X., Yu, J.: Deep learning for object detection and grasping: a survey. In: 2018 IEEE International Conference on Information and Automation (ICIA), pp. 427–432 (2018)
8. Kingma, D.P., Ba, J.: Adam: A method for stochastic optimization. CoRR abs/1412.6980 http://arxiv.org/abs/1412.6980 (2014)
9. Long, J., Shelhamer, E., Darrell, T.: Fully convolutional networks for semantic segmentation. In: Proceedings of the IEEE conference on computer vision and pattern recognition, 07-12-June-2015, pp. 431–440 (2015)

10. Oktay, O., et al.: Attention u-net: Learning where to look for the pancreas (2018)
11. Peng, H., Li, B., Ling, H., Hu, W., Xiong, W., Maybank, S.J.: Salient object detection via structured matrix decomposition. IEEE Trans. Pattern Anal. Mach. Intell. **39**(4), 818–832 (2017)
12. Qujiang Lei, Meijer, J., Wisse, M.: A survey of unknown object grasping and our fast grasping algorithm-c shape grasping. In: 2017 3rd International Conference on Control, Automation and Robotics (ICCAR), pp. 150–157 (2017)
13. Ronneberger, O., Fischer, P., Brox, T.: U-net: convolutional networks for biomedical image segmentation. CoRR abs/1505.04597 http://arxiv.org/abs/1505.04597 (2015)
14. Sharma, P., Singh, A.: Era of deep neural networks: a review. In: 8th International Conference on Computing, Communications and Networking Technologies ICCCNT 2017 (2017)
15. Yu, S., Zhai, D.H., Wu, H., Yang, H., Xia, Y.: Object recognition and robot grasping technology based on RGB-D data. In: 2020 39th Chinese Control Conference (CCC), pp. 3869–3874 (2020)
16. Zhang, Q., Gao, G.: Grasping point detection of randomly placed fruit cluster using adaptive morphology segmentation and principal component classification of multiple features. IEEE Access **7**, 158035–158050 (2019)

Value Set-Based Numerical Analysis of Robust Stability for Fractional-Order Retarded Quasi-Polynomials with Uncertain Parameters and Uncertain Fractional Orders

Radek Matušů[1] (✉), Bilal Senol[2], Baris Baykant Alagoz[2], and Abdullah Ates[2]

[1] Centre for Security, Information and Advanced Technologies (CEBIA–Tech), Faculty of Applied Informatics, Tomas Bata University in Zlín, Nám. T. G. Masaryka 5555, 76001 Zlín, Czech Republic
rmatusu@utb.cz

[2] Department of Computer Engineering, Faculty of Engineering, Inonu University, 44280 Malatya, Turkey

Abstract. This example-oriented contribution deals with the value set-based numerical analysis of robust stability for the family of fractional-order retarded quasi-polynomials with both uncertain parameters and uncertain fractional orders. The specific investigated feedback control system consists of the fractional-order PID controller and the controlled plant, represented by a heat transfer process described by the linear time-invariant fractional-order time-delay model with parametric uncertainty (with three uncertain parameters, namely, gain, fractional time constant, and fractional time-delay term, and furthermore two fractional orders). The graphical robust stability analysis is based on the numerical calculation of the value sets and the application of the zero exclusion principle.

Keywords: Robust stability · Fractional-order systems · Parametric uncertainty · Retarded quasi-polynomials · Uncertain fractional orders · Value set · Zero exclusion principle

1 Introduction

Stability is the most important property of all control systems, and its preservation represents a crucial task in the control system design practice. However, stability needs to be ensured not only for the idealized case with a nominal mathematical model of a controlled process but also for the more realistic scenario, i.e., under conditions of uncertainty. Thus, we are frequently interested in robust stability rather than just in (nominal) stability. From the variety of possible uncertain models, the systems with parametric uncertainty belong among the most popular and comprehensible ones [1, 2], especially under the assumption of a box-shaped uncertainty bounding set.

The application of fractional calculus in the field of control engineering has gained the increased attention of many researchers during approximately the last two decades

R. Silhavy et al. (Eds.): CoMeSySo 2021, LNNS 231, pp. 18–23, 2021.
https://doi.org/10.1007/978-3-030-90321-3_3

[3–5] because of the improved performance of fractional-order control systems in comparison with the conventional integer-order ones. Not surprisingly, the robust stability of fractional-order systems under various uncertainty conditions is in line with this positive trend [6, 7]. The investigation of robust stability for systems with parametric uncertainty becomes a complicated task when increasing the complexity of uncertainty structure in the analyzed family of (quasi-)polynomials, even for the integer-order case. Moreover, the fractional orders further increase the intricacy of the robust stability analysis problem [6, 8]. It was already shown in the literature that the value set concept in combination with the zero exclusion principle represents a powerful numerical tool for the investigation of robust stability, especially for linear time-invariant (LTI) fractional-order systems with complicated uncertainty structures [8]. Furthermore, the same graphical test is also applicable to families of retarded quasi-polynomials [8, 9]. Besides, an interesting problem occurs if the systems have crisply known fractional orders [10].

This example-oriented contribution deals with the analysis of robust stability for the family of fractional-order retarded quasi-polynomials with both uncertain parameters and uncertain fractional orders. The feedback control system consists of the fractional-order PID controller and the controlled plant, represented by a heat transfer process described by the LTI fractional-order time-delay model with parametric uncertainty (with three uncertain parameters, to wit, gain, fractional time constant [11], and fractional time-delay term [12], and additionally two fractional orders [10]). This combination leads to the family of closed-loop characteristic fractional-order retarded quasi-polynomials with uncertain parameters and uncertain fractional orders. The robust stability test itself is based on the numerical calculation of the value sets and the application of the zero exclusion principle [2, 8]. The value sets are obtained via sampling (gridding) the uncertain parameters and fractional orders and subsequent direct calculation of related partial points of the value sets for a supposed frequency range.

2 Mathematical Description of a Heat Transfer Process

Suppose a heat transfer process for the heat flux loss case, which was mathematically described in [12, 13] by means of the fractional-order partial differential equation. The same works [12, 13] also derived the transfer function describing the temperature at the point λ_1 with respect to the heat flux at the beginning of the beam. This transfer function has the form [12, 13]:

$$G(\lambda_1, s) = \frac{T(\lambda_1, s)}{H(0, s)} = \frac{K}{T_1 s^{0.5} + 1} e^{-\Theta s^{0.5}} \tag{1}$$

where the experimentally obtained parameters from [12] are $K = 3.2$, $T_1 = 209.27$, and $\Theta = 2.97$.

In [14], it was assumed that three parameters K, T_1, and Θ might vary within $\pm 10\%$ of their nominal values from [12]. In this contribution, we will suppose that not only these parameters but also fractional orders can vary within given bounds that correspond to $\pm 10\%$ of the nominal values. That means that the heat transfer process is described by the parametrically uncertain fractional-order time-delay model:

$$G(\lambda_1, s, K, T_1, \Theta, \alpha, \beta) = \frac{T(\lambda_1, s)}{H(0, s)} = \frac{K}{T_1 s^{\alpha} + 1} e^{-\Theta s^{\beta}} \tag{2}$$

where:

$$
\begin{aligned}
& K \in [2.88,\ 3.52] \\
& T_1 \in [188.343,\ 230.197] \\
& \Theta \in [2.673,\ 3.267] \\
& \alpha \in [0.45,\ 0.55] \\
& \beta \in [0.45,\ 0.55]
\end{aligned}
\tag{3}
$$

The structure of the model (2) and the uncertainty bounding set (3) form the family of controlled plants that will be used in the following analysis.

3 Fractional-Order PID Controller

In [12], several integer-order and fractional-order PID controllers were designed for the nominal system (1) and compared mutually. The fractional-order PID controller:

$$
C(s) = 259.45\left(1 + \frac{1}{67.74s^{0.5}} + 0.18s^{0.5}\right) \tag{4}
$$

with fixed integration and differentiation orders 0.5 and with the parameters obtained by minimization of ITAE criterion seems to represent a very good compromise between the performance indicators and minor steady-state oscillations [12].

4 Value Set-Based Robust Stability Analysis

The family of closed-loop characteristic fractional-order retarded quasi-polynomials, which corresponds to the feedback control system with the plant family (2), (3) and the controller (4), is:

$$
\begin{aligned}
p_{CL}(s, K, T_1, \Theta, \alpha, \beta) &= 67.74T_1 s^{\alpha} s^{0.5} + 67.74s^{0.5} + \cdots \\
\cdots &+ 3163.52574Ks^{0.5}s^{0.5}e^{-\Theta s^{\beta}} + 17575.143Ks^{0.5}e^{-\Theta s^{\beta}} + 259.45Ke^{-\Theta s^{\beta}}
\end{aligned}
\tag{5}
$$

where the parameters K, T_1, Θ, and fractional orders α and β may vary in accordance with (3). For the remainder of this section, they all will be called simply "uncertain parameters".

In some specific cases, the value sets of the fractional-order families can be constructed on the basis of analysis of vertices and exposed edges [6, 15–17]. Nevertheless, in this contribution, the value sets for the family (5) with relatively complicated uncertainty structure, including uncertain fractional orders, are plotted by means of sampling (gridding) the uncertain parameters and subsequent direct calculation of related partial points of the value sets for a supposed frequency range.

The value sets of the family (5) for the frequencies from 0 to 10 with step 0.2 are plotted in Fig. 1. All 5 uncertain parameters are sampled by 5 equidistant values, i.e., each value set for a fixed frequency consists of $5^5 = 3125$ points. The full view from Fig. 1 provides an overview of value sets' shape. Then, a closer view of the situation in

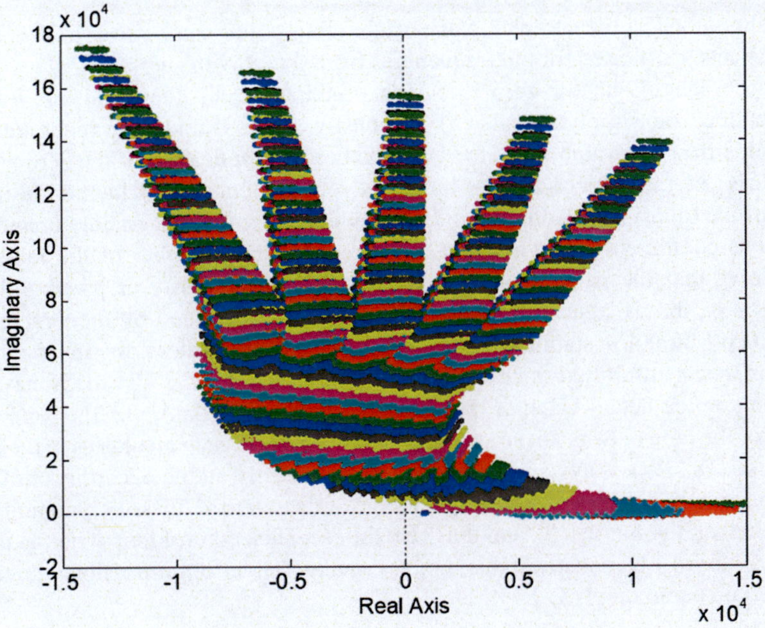

Fig. 1. The value sets of the family (5) for the frequencies from 0 to 10.

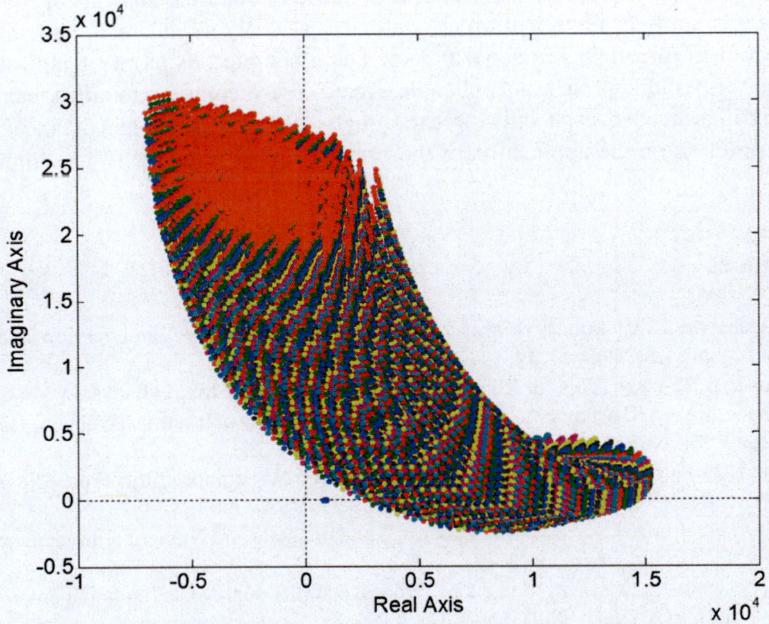

Fig. 2. The value sets of the family (5) for the frequencies from 0 to 2.

the neighborhood of the complex plane origin (zero point) can be seen in Fig. 2, where the value sets are plotted for the frequencies from 0 to 2 with step 0.02.

In accordance with the zero exclusion principle [2, 8], the family of fractional-order retarded quasi-polynomials (5) is robustly stable if and only if the zero point is excluded from the value sets. However, the correct application of the zero exclusion principle is generally preconditioned by several requirements, which include the invariant degree of the (quasi-)polynomial family, pathwise connected uncertainty bounding set, continuous coefficient functions, and at least one stable member of the family. One can observe that the value sets are excluded, and the majority of preconditions are fulfilled (e.g., the existence of a stable member can be verified by the generalization of modified Mikhailov stability criterion from [9]). Nevertheless, the invariant degree condition seems unfulfilled since the degree of the term $67.74T_1s^{\alpha}s^{0.5}$ in (5) may change depending on the actual value of the fractional order $\alpha \in [0.45,\ 0.55]$. (Note that the validity of the index law is not so straightforward in fractional calculus as in the integer-order case [18, 19]). Anyway, the family (5) is robustly stable according to the zero exclusion principle despite the possible change of the retarded quasi-polynomial degree.

Altogether, the family (5), and thus also the feedback control loop with the plant (2) and controller (4), is robustly stable, i.e., it remains stable for all possible variations of the uncertain parameters (3).

5 Conclusion

This contribution was focused on the value set-based numerical analysis of robust stability for the family of fractional-order retarded quasi-polynomials with both uncertain parameters and uncertain fractional orders. The elaborated experiment indicates that the value set-based approach is applicable even to these complex families, but further research will be needed, especially more thorough research on the impact of the invariant degree condition on the application of the zero exclusion principle to fractional-order systems.

References

1. Bhattacharyya, S.P.: Robust control under parametric uncertainty: an overview and recent results. Annu. Rev. Control. **44**, 45–77 (2017)
2. Barmish, B.R.: New Tools for Robustness of Linear Systems. Macmillan, New York (1994)
3. Machado, J.A.T., Kiryakova, V., Mainardi, F.: Recent history of fractional calculus. Commun. Nonlinear Sci. Numer. Simul. **16**(3), 1140–1153 (2011)
4. Chen, Y.Q., Petráš, I., Xue, D.: Fractional order control – a tutorial. In: Proceedings of the 2009 American Control Conference, St. Louis, MO, USA (2009)
5. Xue, D.: Fractional-Order Control Systems: Fundamentals and Numerical Implementations. De Gruyter, Berlin, Germany (2017)
6. Tan, N., Özgüven, Ö.F., Özyetkin, M.M.: Robust stability analysis of fractional order interval polynomials. ISA Trans. **48**(2), 166–172 (2009)
7. Jiao, Z., Zhong, Y.: Robust stability for fractional-order systems with structured and unstructured uncertainties. Comput. Math. Appl. **64**(10), 3258–3266 (2012)

8. Matušů, R., Şenol, B., Pekař, L.: Robust stability of fractional order polynomials with complicated uncertainty structure. PLOS ONE **12**(6), e0180274 (2017)
9. Busłowicz, M.: Stability of linear continuous-time fractional order systems with delays of the retarded type. Bulletin of the Polish Academy of Sciences: Technical Sciences **56**(4), 319–324 (2008)
10. Petráš, I., Chen, Y.Q., Vinagre, B.M., Podlubný, I.: Stability of linear time invariant systems with interval fractional orders and interval coefficients. In: Second IEEE International Conference on Computational Cybernetics, Vienna, Austria (2004)
11. Gómez-Aguilar, J.F., Razo-Hernández, R., Granados-Lieberman, D.: A physical interpretation of fractional calculus in observables terms: analysis of the fractional time constant and the transitory response. Rev. Mex. Fís. **60**(1), 32–38 (2014)
12. Macias, M., Sierociuk, D.: Fractional order calculus for modeling and fractional PID control of the heating process. In: Proceedings of the 13th International Carpathian Control Conference, High Tatras, Podbanské, Slovakia (2012)
13. Sierociuk, D., Dzieliński, A., Sarwas, G., Petráš, I., Podlubný, I., Škovránek, T.: Modelling heat transfer in heterogeneous media using fractional calculus. Philos. Trans. R. Soc. A, **371**(1990), 20120146 (2013)
14. Matušů, R., Pekař, L.: Robust stability of thermal control systems with uncertain parameters: the graphical analysis examples. Appl. Therm. Eng. **125**, 1157–1163 (2017)
15. Yeroğlu, C., Özyetkin, M.M., Tan, N.: Frequency response computation of fractional order interval transfer functions. Int. J. Control Autom. Syst. **8**(5), 1009–1017 (2010)
16. Şenol, B., Yeroğlu, C.: Computation of the value set of fractional order uncertain polynomials: a 2q convex parpolygonal approach. In: Proceedings of the 2012 IEEE International Conference on Control Applications, Dubrovnik, Croatia (2012)
17. Yeroğlu, C., Şenol, B.: Investigation of robust stability of fractional order multilinear affine systems: 2q-convex parpolygon approach. Syst. Control Lett. **62**(10), 845–855 (2013)
18. Labora, D.C., Nieto, J.J., Rodríguez-López, R.: Is it possible to construct a fractional derivative such that the index law holds? Prog. Fract. Differ. Appl. **4**(1), 1–3 (2018)
19. Bhalekar, S., Patil, M.: Can we split fractional derivative while analyzing fractional differential equations? Commun. Nonlinear Sci. Numer. Simul. **76**, 12–24 (2019)

Crisis Risk Management in Organizational Systems

V. N. Tsygichko(✉)

Federal Research Center Computer Science and Control of Russian Academy of Sciences, Vavilova 44, Building 2, Moscow 119333, Russia
vtsygichko@inbox.ru

Abstract. The article discusses choosing strategic decisions' problem in organizational systems (OS) that can cause a crisis (CS). Solving this problem concept is presented. An OS evolution conceptual model under the strategic decisions influence has been built. A choosing a strategic solution problem formal formulation to minimize the CS occurrence risk in the OS is given.

Keywords: Risk management · Organizational system · Crisis · Crisis situation · Strategic decisions · Strategic decisions negative consequences · Conceptual model · System state · Uncertainty · Scenarios

1 Introduction

Crises and the crisis situations preceding them are an OS life all components' evolution constant companion. The CS foresight in OS management processes and the measures' adoption to prevent crises or mitigate their consequences have been and remain in our time the most pressing problems.

One of the most promising directions for solving these problems is a methodological base and methodological tools development for managing CS risks in the OS. The risk management implementation in OS management systems is possible only if there are reliable tools, forecasting and assessing the strategic decisions possible negative consequences risks, potentially leading to the CS emergence [1–6].

Forecasting crisis situations in the management processes of political, economic, social and other OS, in which decision-making takes place in uncertainty a high degree conditions, is a decision-making theory and practice fundamental problem in organizational systems.

An existing publications' analysis shows that at present there are no works that would define the fundamental principles, conditions and possible directions for solving the predicting the CS problem in the OS when making strategic decisions. The article presents a possible direction for solving this problem.

R. Silhavy et al. (Eds.): CoMeSySo 2021, LNNS 231, pp. 24–33, 2021.
https://doi.org/10.1007/978-3-030-90321-3_4

2 Subject Area

Let us introduce the subject area basic concepts - a CS emergence risk management as a strategic decisions implementation consequence in the OS.

By the organizational system (OS) we mean society socio-economic organization any level and any element from the state as a whole to a separate socio-economic unit, for example, a separate enterprise or a part of it.

OS, as a rule, is divided into several parts, which can be considered as independent formations. This division is usually functional. System parts, which internal structure is considered at the analysis selected level, will be called **subsystems of**. For example, industries can be considered as the state economic system subsystems. By **OS elements** we mean its smallest structural units, which internal structure is not considered at the analysis selected level.

All operating systems combine system-wide properties, but each has its specifics, which must be taken into account in a systemic study [7].

Any OS is a wider system part, which is **an external environment for it**.

Each OS and its structural components exist in the given political, legal, economic, natural and other living conditions, within which it functions, without changing its qualitative definition.

The basic concepts one in the subject area we are considering is the OS state concept and its structural components. The term the OS state, its subsystems and elements is understood as parameters values set characterizing them, fixed at some point in time.

Our research subject is crisis situations (CS), which may arise when making strategic decisions in the OS. The term crisis has a wide semantic scale, i.e. it's meaning and content change depending on the context and its application object, which is its different interpretation source.

We believe that this concept most complete content is given on Wikipedia, which defines a crisis (ancient Greekcrisis - decision; turning point) - a coup, a transitional state time, a turning point, a state in which the achieving goals' existing means become inadequate, resulting in unpredictable situations. The crisis reveals latent conflicts and imbalances.

It should be emphasized that OS life each sphere has its crisis concept, reflecting its specificity and existence conditions [1].

Under the OS crisis we mean the parameters value exit describing its state beyond the limits in which the system passes into a new quality or ceases to exist.

Each crisis in any OS life area is preceded by a crisis situation, by which we will understand the OS state combination formed at a certain point in time external and internal factors and its existing conditions, potentially leading to the system crisis. CS is an OS unstable state when a minor event in an external environment or in the system itself can lead to its crisis.

An important concept in the subject area in question is the Strategic Solutions which refers to OS management decisions that change the OS state and conditions, its subsystems, elements and the external environment.

The strategic solutions most important feature in the OS is their uniqueness, as they are always made in conditions unparalleled in the past. The situations' coincidence in the political, social or economic fields is an unlikely event [2]. This excludes using regular forecasting methods possibility to predict the negative consequences and critical situations arising in the strategic solutions' implementation, as there is no statistical and factual information on the current situation.

The only tool known to date to solve this problem is the forecasting scenario method [2], which in practice has proven its effectiveness [8–10].

In the CS case being resolved by the crisis, it is usually, if not too late, the organization introduces measures set to eliminate the crisis and minimize the damage from its consequences.

It is necessary to distinguish between the problems internal and external causes facing the OS, which inadequate solution can lead to a CS.

Internal reasons include:

- Mistaken targeting and inadequate strategic decisions leading to conflict within the OS or with the external environment;
- The OS incompetent management;
- The inability to resolve internal contradictions between the system elements;
- Inadequate response to changing living conditions;
- Resources lack to adapt to new living conditions;
- The OS lifecycle end.

It should be noted that crisis situations most often arise in the OS conflicting decisions implementation case itself and the external environment.

The conflict strategic solutions' implementation tends to cause an actors negative reaction whose interests it affects. By negative reaction we mean the actors actions that negatively affect the OS state and its relationship with the OS external environment.

For example, the decision to raise taxes in the region may entail the enterprises' activities closure or reduction or entire industries in the region and the foreign partners' further cooperation refusal for whom the new living conditions are unprofitable.

Problems external causes that can lead to a crisis situation:

- Changes in living conditions, for example, an unfavourable political situation emergence in the region and the world, a military conflict, a sharp drop in business activity in the commodity and financial markets, etc.;
- Crises in the environment (global, regional, sectoral, etc.);
- Crises associated with the OS evolution cyclical nature vital activity component;
- Insoluble contradictions with the external environment elements;
- Environmental events.

In many cases, CSs arise for the above objective reasons, OS management independent. Such CSs are not this study subject.

Many critical situations are quite obvious, for example, in the ecology field. However, in most cases, crisis situations require recognition.

Historical experience suggests that most economic, political, social and other crises arise unexpectedly for all its actors, although the crises causes and genesis subsequent analysis suggests that they were preceded by a crisis situation, usually caused by inadequate strategic decisions. The 2008 latest global financial crisis is this prime example.

3 OS Evolution Conceptual Model

The strategic decisions choice in the OS can be presented in the following sequence:

- The problem situation assessment and the strategic goals' formulation for the OS in the current conditions;
- Many possible alternatives determination to achieve the set goals.

For each alternative solution:

- The OS many subsystems and elements allocation, which vital activity will be associated with its implementation process;
- The OS subsystems and elements possible reactions' assessment to these reactions strategic decision and possible negative consequences implementation;
- The external environment OS allocation, whose interests may be affected as a making a strategic decision result;
- The external environment OS possible reactions' assessment to the strategic decision implementation and these reactions negative consequences;
- Forecasting a strategic decision negative consequences, potentially leading to a CS emergence, taking into account all actors negative reactions;
- A CS occurrence risk assessment when implementing a strategic decision;
- A CS occurrence preventing or reducing the risk possibilities assessment and determining the resources required for this.

Based on the results obtained, an acceptable choice to the solution is carried out on a CS occurrence minimizing the risks' basis.

A CS emergence problem systematic study in the strategic decisions' implementation requires an OS evolution conceptual model development under this decision influence.

Let us take as our research the abstract object a certain hypothetical OS, in which all internal connections to the variety and connections with the external environment OS inherent in real socio-economic objects, we will represent links on information $r^{I}_{i\mu}$, management $r^{U}_{i\mu}$, material exchange $r^{M}_{i\mu}$, financial exchange $r^{F}_{i\mu}$ and social-political influence $r^{C}_{i\mu}$.

The sets $\{r_{i\mu}\}$ and $\{r_{\mu i}\}$ for the system all elements under consideration determine its structural state S. These sets form a multidimensional array, which dimension is determined by the elements number and the vector components number $r_{i\mu}$.

$$S = \begin{vmatrix} 0 r_{12} \cdots r_{1\mu} \cdots r_{1I} \\ \vdots 0 \\ r_{i1} 0 \\ \vdots 0 \\ r_{\mu 1} 0 \\ \cdots\cdots r_{i\mu} \cdots\cdots 0 \end{vmatrix} \quad (1)$$

The element links μ to the element i are defined by the vector $r_{\mu i}$, which can be interpreted as a vector of feedbacks with respect to $r_{i\mu}$. For example, the information transfer by a control higher level to a lower level is interpreted as a direct link, and the information transfer from the bottom up through the control hierarchy is interpreted as a reverse link.

The matrix S describes the system all structures totality under study. Any structure can be extracted from S by specifying the binary relations corresponding components. For example, the control system is defined by the binary relations submatrix $r^U_{i\mu}$, and the financing system is defined by the submatrix $r^F_{i\mu}$, etc. For hierarchical structures, such as management structures, the $r^U_{i\mu}$ sub-matrix reflects the system elements subordination hierarchy.

A system most important characteristic is its elements internal state. By the element internal state i element $P_i = \{P^J_i\}$, defined on the describing this item parameters possible values' region P^J_i at the analysis selected level, where $j = 1{-}J$ is the parameter number and J - the parameters' number describing the element internal state.

The column vector P. determines the system internal state matrix elements.

$$P = \begin{vmatrix} P_1 \\ \vdots \\ P_i \\ \vdots \\ P_I \end{vmatrix} \tag{2}$$

The elements internal state matrix P implements the lowest, OS description most detailed level.

At the OS representation the second level - P^2 is described as a structure consisting of subsystems, which are characterized by their integral properties P_k determined by the elements internal state parameters aggregate values belonging to each subsystem $\{P^l_k\}$.

$$P^2 = \begin{vmatrix} P_1 \\ \vdots \\ P_k \\ \vdots \\ P_K \end{vmatrix} \tag{3}$$

$$P^2 = \left\{P^l_k\right\} = \overline{P_k}\{P_{ki}\}$$

where $k = 1{-}K$ is the subsystem number, K is the subsystems number, $l = 1{-}L$ - is the parameter number and L - is the parameters' number describing the subsystem internal state;

$\overline{P_k}$ - operator for forming internal state indicators k- of this subsystem. Such operators examples are national economy current state calculating macro-indicators standardized international and domestic methods or the state economy as a whole.

At the generalization' the highest level, the OS internal state as integrity is represented by the macro indicators *m = 1–M*, which generalize its subsystems internal state using the operator

$$P^3 = \begin{vmatrix} P_1 \\ \vdots \\ P_m \\ \vdots \\ P_M \end{vmatrix} \tag{4}$$

$$P^3 = \{P_m\} = \overline{P}\{P_k\}$$

One of the OS state parameters is the characteristic of the legal, economic, political and social environment, which together constitutes its life conditions. The OS environment state vector, represented by the matrix *W*

$$W = \begin{vmatrix} W_1 \\ \vdots \\ W_i \\ \vdots \\ W_I \end{vmatrix} \tag{5}$$

where the element environment is *I*.

It should be noted that the element living conditions include the OS living conditions as a whole and the subsystem living conditions to which the element belongs.

Variables *S, P, W 0669* form the system's states phase space. A point in this space *N* = *{S, P, W}* is called the OS state.

$$H = \begin{vmatrix} H_1 \\ \vdots \\ H_i \\ \vdots \\ H_I \end{vmatrix} \tag{6}$$

where is the element state.

The vector components are the element communication vector with the system other elements and the external environment, the internal state vector, and the living conditions in which there is an element:

$$h_i = \{r_i, W_i, P_i\} \tag{7}$$

where is the row (forward links) and is the column (backlinks) of the matrix *S* (1).

Expressions (6 and 7) are a system state formal description at an arbitrary point in time for any level in the accepted description hierarchy.

The system targeted behaviour is determined by the control vector. In our case, this is a strategic decision U_{st}:

$$U_{st} = |U_i|$$

where is the system element control parameter value implementing the strategic decision.

Strategic decisions U_{st} change the structural state S, the system internal state and both the OS itself life conditions and the external environment OS, with which it is connected by relations determined by the structural state matrix S.

These changes cause the OS subsystems elements and the external environment OS – G to react to the strategic decision.

$$\Gamma = \begin{vmatrix} \Gamma_1 \\ \vdots \\ \Gamma_i \\ \vdots \\ \Gamma_I \end{vmatrix}$$

Strategic decision U_{st}, the external environment reaction and the OS itself elements reaction - G and the new living conditions W determine the OS state - H, which developed after the strategic decision implementation

$$\Gamma = \begin{vmatrix} \Gamma_1 \\ \vdots \\ \Gamma_i \\ \vdots \\ \Gamma_I \end{vmatrix} \tag{8}$$

The implementing the strategic decision process U_{st} is represented by the operator transformations following sequence: $Q = \{Q_w, Q_G, Q_s, Q_p\}$ over the original state OC $H_0 = (W_0, S_0, P_0)$.

$$\begin{aligned} Q_w &: (U_{st}, W_0, S_0, P_0) - W \\ Q_G &: (U_{st}, W, S_0, P_0) - G \\ Q_s &: (U_{st}, W, S_0, P_0, G) - S \\ Q_p &: (U_{st}, W, S, P_0, G) - P \end{aligned} \tag{9}$$

The display system (9) is an OS state evolution conceptual model H_0 when implementing the strategic decision U_{st}.

It should be noted that the mapping operators Q, as a rule, are informal, and their content is determined by an expert. The OS conceptual model (9) allows interpreting and analyse in its terms any processes associated with the strategic decisions' implementation in the OS and to formulate this study problem general formal statement.

4 The Problem Formal Statement

This study problem the most general formulation can be formulated as follows.

Let the parameters' values describing the OS current state at the strategic decision-making procedure start time be known $H_0 = \{S_0, P_0, W_0\}$.

For all parameters characterizing the OS state at representation all levels, their permissible values are determined, which form the OS permissible states multidimensional space boundaries - H_{add}:

$$H_{add} = \{S_{add}, P_{add}, W_{add}\} \tag{10}$$

At least one parameter exit beyond the OS permissible state - H_{add} is a crisis situation and will require an immediate response from the OS management.

Suppose that some problematic situation has developed in the OS that requires a strategic decision to be made by the top management level.

Alternative strategic decisions limited number are known that can be applied in the current situation $U^{st} = \{U_n^{st}\}$, where $n = 1{-}N$ is the alternative decision number and N - is the alternatives number.

It is assumed that these strategic decisions $U^{st} = \{U_n^{st}\}$, with effectiveness varying degrees, can solve the problems that have arisen in the OS.

Each strategic decision purpose $U_n^{st} \in U^{st}$ is to obtain expected benefits C_z^n, which can be both tangible and intangible. We assume that the benefit value is a non-decreasing cost function C^n for decision implementation $C_z^n(C^n)$.

Each decision has its benefit. We will assume that any benefit to our abstract OS top management can be expressed in value terms.

Under the conceptual model (9), each alternative decision $U_n^{st} \in U^{st}$, as its implementation result, changes the OS state of all components - H_0 and causes a reaction G from the system itself elements and the external environment OS. These reactions can have an undesirable effect on the OS state, fraught with negative consequences leading to the CS onset.

A strategic decision negative consequences mean the OS internal state at least one parameter value possible exit and its components beyond the permissible space and material losses in this situation event, i.e. the situation in which

$$P^n \notin P_{add}, P^{2n} \notin P^2_{add}, P^{3n} \notin P^3_{add} \tag{11}$$

Suppose that for each decision $U_n^{st} \in U^{st}$ the influence mechanisms G on the OS state and its components are known, i.e. there are functions a lot $F = \{f_w^n(P), f_w^{2n}(P^2), f_w^{3n}(P^3)\}$ and elements interaction functions a lot with each other $F_{vz} = \{f_{\mu i}, f_{i\mu}\}$ allowing to predict the OS internal state parameters values after each alternative decision implementation $U_n^{st} \in U^{st}$.

It is assumed that the functions F and F_{vz} do not change during the decision implementation.

For each decision, using the functions$F$ and F_{vz} from the structural state matrix S (1), the elements submatrices are highlighted

$$S^{\Gamma} = \{r^{\Gamma}_{i\mu}\} \subseteq S$$

which may be subject to direct or indirect (through other related elements) negative impact G determined by changes in the external environment $W^n = \{W_i^n\}$ as a strategic decision implementation result $U_n^{st} \in U^{st}$.

We will also consider the known costs C^n for each decision implementation $U_n^{st} \in U^{st}$ and the damage C_i^n in value terms from their implementation possible negative consequences for the elements, subsystems and the OS as a whole $C_I^n(H)$.

We assume that the damage $C_I^n(H)$ is a function of F_C^n from the OS elements state P^n:

$$C_I^n(H) = F_C^n(P^n) \tag{12}$$

Within the above assumptions' framework, the OS state evolution under a strategic decision influence seems to be the only value H^n, which allows each decision $U_n^{st} \in U^{st}$ to compare the damage C_I^n (12) from its implementation negative consequences.

In this case, provided that $C^n \leq C_{add}$ and $C^n < C_i^n$ then, the choice will be unambiguous and the only solution will be $optU_n^{st} \in U^{st}$ providing the maximum net benefit $max\Delta C_z^n$:

$$optU_n^{st} \Longleftrightarrow max\Delta C_z^n = max\left[C_z^n(C_{add}) - C_{add} - C_I^n\right] \tag{13}$$

If we introduce a natural assumption that the probability (a CS occurrence risk R^n) is a damage non-decreasing function C_i^n, then a strategic decision choice to minimize damage will correspond to this decision choice to minimize a CS - min risk R^n.

A CS occurrence assessing the risk problem decision in the strategic decisions' implementation is possible only by expert means. The experts' group should construct possible scenarios for the OS evolution in each strategic decision implementation for the parameters' various values describing its state. The constructed scenarios' analysis will make it possible to construct the OS state acceptable values area.

Based on the results obtained, the OS management, as a rule, is forced to make a compromise decision between a CS risk magnitude and its elimination or reduction costs. The search for this compromise is a CS emergence managing the risks process in the OS.

5 Conclusion

Choosing a strategic decision presented a methodological concept to minimize the CS occurrence risk can serve as the basis for creating methodological tools in decision support systems in the OS.

Acknowledgements. The article was prepared with the support of the Russian Foundation for Basic Research (project 19–07-00522).

References

1. Urentsov, O.V.: On the forecasting crises possibility in the financial market. Syst. Res. Method. prob. **35**, 248 (2010)

2. Tsygichko, V.N.: Forecasting socio-economic processes. Preface. In: Gvishiani, D.M. (Ed.) 3rd. revised and add. Book house LIBERKOM, p. 240, Moscow, Russia (2009)
3. Tsygichko, V.N., Chereshkin, D.S., Smolyan, G.L.: Risk management in organizational systems. In: Lambert Academic Press, RU. Beau Bassin, p. 90 (2018)
4. Shannon, K.: Works on information theory and cybernetics. In: Publishing house Foreign. literature, p. 830, Moscow, Russia (1963)
5. Sholomnitsky, A.G.: Risk Theory. In: Selection under uncertainty and risk modelling. Ed. House of the State University Higher School of Economics, p. 400, Moscow, Russia (2005)
6. Tsygichko, V.N.: Strategic decision-making and risk-management in complex systems. In: Silhavy, R. (ed.) CSOC 2020. AISC, vol. 1226, pp. 415–432. Springer, Cham (2020). https://doi.org/10.1007/978-3-030-51974-2_40
7. Sadovskiy, V. N.: General systems' theory foundations: logical and methodological analysis. p. 279, Nauka, Moscow, Russia (1974)
8. Tsygichko, V.N., Smolyan, G.L., Khan-Magomedov, D.D.: Internet in Russia. Development prospects. In: Chereshkina, D.S. Editorial URSS, p. 200, Moscow, Russia (2004)
9. Tsygichko, V.N.: Region socio-economic development forecasting scenario method//The region forecasting the socio-economic development. In: Chereshneva, V.A., Tatarkina, A.I., Glazyev, S., Ekaterinburg, Y. (Eds.) Institute of Economics, Ural Branch of the Russian Academy of Sciences, pp. 90–126 (2011)
10. Tsygichko, V.N., Chereshkin, D.S., Smolyan, G.L.: Strategic decisions negative consequences analysis and assessment in organizational systems. The ISA RAS Proc. **68**(1), 3–23 (2018)
11. Frege, G.: Selected works. Intellectual House, Moscow, Russia (1997).

Fashion Trend Forecasting Using Machine Learning Techniques: A Review

Audrey Aurelia Chang, Cynthia, Devita, Jihad Fahri Ramadhan, Zainina Kyla Shafira Adnan, Bayu Kanigoro(✉), and Edy Irwansyah

School of Computer Science, Bina Nusantara University, Jakarta, Indonesia
{bkanigoro,eirwansyah}@binus.edu

Abstract. Fashion is a fast-paced industry where patterns last up to eight months, being an incredibly competitive industry where trends are constantly evolving. When it comes to companies, forecasting is a significant factor, knowing the trends and making right forecasts in line with the ever-changing trends can make a big difference in the success a company will have. It is important to investigate the various strategies introduced to achieve a more efficient way of predicting demands and consumer needs in order to better understand which techniques are to be used for the best possible outcome. The primary aim of forecasting is to better understand the patterns that impact markets and consumer needs. This paper analyzed the state-of-the-art forecasting methods used to predict market patterns in the fashion industry that could be implemented. The methods implemented to forecast, the models and algorithms used within the methods, the datasets used, and also the trade-offs for each system and model used are described in this analysis. In particular, this analysis will further explore Machine Learning models that are widely applied to support the forecasting efforts carried out in the fashion industry.

Keywords: Fashion · Forecasting · Trend forecasting · Fashion forecasting · Machine learning

1 Introduction

Fashion is a clothing style or behavior that's common at a specific time. It is important as a mode of speech, representing the ever-changing style of clothes worn globally by different cultures. Fashion could be described as a movement involving esthetic expressions using clothing and accessories. The fashion industry includes clothing as well as other items such as belts, shoes, and accessories. Fashion market patterns are continuously changing, and a variety of factors contribute to this situation [13].

If it's the current season (Spring, Summer, Fall, and Winter), the new aesthetic (Goth, Retro, Preppy, etc.), a trend may also be sparked by an artist or social media influencer because of what apparel they're wearing or what kind of accessory they're wearing. That is, fashion is fast-paced. Fashion depends on

R. Silhavy et al. (Eds.): CoMeSySo 2021, LNNS 231, pp. 34–44, 2021.
https://doi.org/10.1007/978-3-030-90321-3_5

the latest trends and models, which conclude that trends play a major role and influence fashion business numbers and data. Trends in the fashion world are fickle, one-piece could be trendy today, and could be considered unfashionable the very next day. Fashion patterns that began decades ago could make a comeback when people least expect it. This makes fashion forecasting an incredibly important thing for businesses to make sure they pursue the right trends.

More precise prediction of demand for fast-moving consumer products is a strategic factor for producers and retailers such as fashion [37]. Fashion forecasting became important for bringing out new and latest trends and sustaining the fashion industry. Furthermore, forecasting also leads to the advancement of strategies, brand image, the extension of product lines, and the securing of new business, and the use of customer buying power. As it is different from other conventional products, the fashion products featured as short-lived period and demand are highly unpredictable with the evolving customer favorite and dynamic fashion trends, making sales forecasts in the fashion industry more challenging [29].

Since forecasts in physical sciences can achieve unprecedented levels of precision, this is not the case in social contexts where virtually all predictions are unpredictable and unambiguously incorrect [21]. Moreover, people have their own preferences in product visual appearances from different categories. The new trend in fashion affects many of them [19]. This matter adds to the awareness that the fashion and apparel industry is unpredictable, and businesses need to respond rapidly and adjust to trend shifts while constantly meeting customer demands. The industry must also become customer-centric and follow sustainable manufacturing practices to alleviate waste production and management. Fashion products' lifecycles are very short, but there is room for implementing different strategies. For example, using large quantities of historical data collected and stored in company databases, as stated by [20]. There are two issues to address in this paper regarding trend forecasting, particularly in the fashion industry. Next, study the methods most widely used in Trend Forecasting.

Second, a comprehensive analysis is taken into account, explicitly addressing the Machine Learning strategies that are appropriate for Fashion Trend and Trend Forecasting.

2 Theoretical Background

According to literature, forecasting itself was first believed to be a form of religious, superstitious or even supernatural belief and has only recently moved from those beliefs. In scientific forecasting however, there are two certainties that we know of, (1) No one possesses prophetic powers and (2) All predictions are uncertain and the only thing that varies is the extent of said uncertainty [22]. Forecasting itself is an extremely hard field to master especially when it comes to merchandising as mentioned in the paper written by [32]. The reason for this is that there are short-term as well as long term forecasting depending on which field the forecasting method will be implemented to numerous methods and models have been developed in order to aid humans in predicting/forecasting with as little error as possible, yet it is still impossible to achieve a forecasting method that is 100% accurate.

One literature mentions Machine Learning as an alternative method to time-series forecasting as a means of trend forecasting method [22], a number of other methods are also available to be implemented. Some of the other methods that are available for implementation for forecasting include Big Data, Data Mining, and Greedy Method. Though each method that has been previously implied are widely implemented and could be implemented independently, some researches combine these methods in order to predict trends more accurately. As previously mentioned, trend forecasting is not an easy feat, when it comes to choosing the forecasting method that one wants to implement there are a number of aspects that need to be considered. As such, each of the aforementioned methods has its own benefits when implemented. Nowadays the two most widely implemented methods are that of Machine Learning as well as Big Data as the implementation of these two methods especially when combined allows businesses and companies alike to achieve a highly accurate prediction that has been based on a vast amount of data.

3 Methodology

This literature review uses PRISMA checklist methodology as the model of systematic review. The authors have chosen the PRISMA method as the PRISMA method allows the authors to report the number of reviews and meta-analyses that the authors had done in the process of writing this review paper and also because the PRISMA method covers concepts and also topics that are generally covered within other systematic reviews [25]. Figure 1 provides a brief overview of the number of journals, papers and articles that have been reviewed for the usage of writing this paper.

The criteria for inclusion of the papers in this review are that the papers are written in the English Language and has Fashion Trend Prediction, Trend Prediction and or Machine Learning as the topic and published within the year of 2016 up to 2020. The inclusion and exclusion of papers depended on how well the papers fit with the topic that is being reviewed, other than that reason for exclusion of papers and articles were i.e. existence of duplicates, non-English articles, studies that were unverified by professionals and articles that were non full-text. The process of inclusion and exclusion within the figure were used to determine which paper will be used as the main foundation of the review paper. The record selection process consists of four steps which are:

1. Identification, which was done by searching online database using platforms such as Science Direct, IEEE, MDPI, and Springer Link as well as other open access platforms such as PQDT Open, Emerald, and informational websites. Searches were done by including keywords such as (1) Fashion Trend Predictions, (2) Trend Forecasting, (3) Trend Prediction and Analysis, (4) Machine Learning. Searches could also be done by combining the keyword mentioned above or combining other words with the keywords.

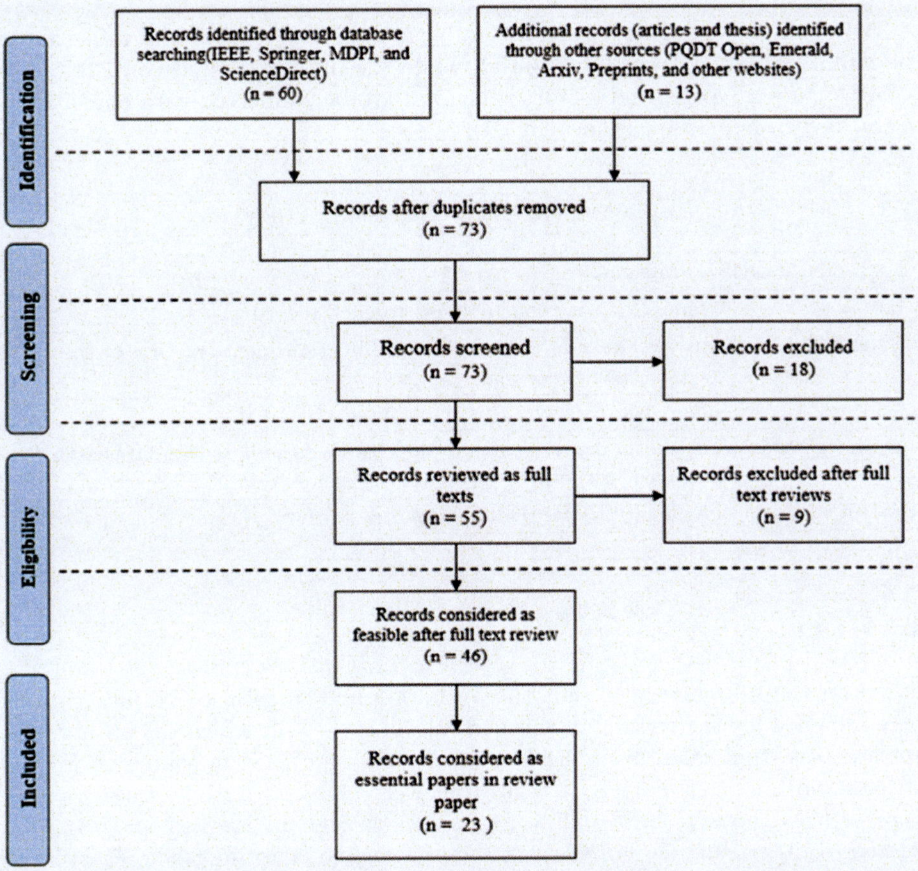

Fig. 1. PRISMA checklist methodology.

2. Screening, in this step duplicate records are removed and the remaining records are screened, records that do not fit with the topic of Trend Forecasting were excluded.
3. Eligibility, records were then further inspected by reading the introduction or abstraction section, records that were found irrelevant to the research questions elicited were eliminated.
4. Included, in this stage full-paper review were conducted to determine whether the articles were key articles or supportive articles for the writing of the review paper.

The data that are extracted and analyzed are obtained by reading the full-text records that were deemed relevant for the purpose of the writing of this paper. Then following items were extracted and analyzed from each record that is of relevance: ID, Reference, Context, Methodologies, and Factor. Table 1 provide a description on the items that were extracted.

Based on the information that has been described within the table above the main context is Trend Forecasting while the data extracted consists of the record's methodologies and factor (which have to be connected with one or more of the keywords previously mentioned).

Table 1. Data Extraction of Each Study.

Extracted data	Description
ID	Unique Identity that belongs to each record
Reference	Authors, year of publication, publication location, title of the record that is reviewed
Context	The context of why this paper was written
Methodologies	The methods, models and algorithms as well as the dataset that was used in the process of writing the records
Factor	The topic that is related to paper review

4 Result

Based on the search results that the authors found, a total of 73 publications were screened for the writing of this systematic review, in which the authors did not find any duplicates from the databases that were used in searching for the articles and publications. The articles that were included in this research paper were articles that (1) have online forms, (2) written in English and (3) were related to either Fashion, Forecasting and or Machine Learning for Forecasting or a combination of the three topics. After ensuring that the publications screened met the conditions, the authors then proceeded to assess the relevancy of these publications with the research questions. Through this assessment, the authors found that a total of 46 publications were deemed relevant to answering the research questions asked. The following table shows the number of publications that were deemed relevant to the research questions.

Table 2. Relevant Publications according to their database platforms.

Electronic database	Papers found	Candidate considered	Publication relevant to research question
IEEE	8	4	2
Link Springer	11	7	3
MDPI	5	5	5
Science Direct	36	28	27
Others	13	11	9
Total	73	55	46

The authors conducted data extraction after completing the main study for this systematic review. The articles and publications that had underwent data extraction were deemed relevant in answering the research questions that the authors had elicited prior to the completion of this paper. The extraction process was done in order to answer the research questions elicited based on findings that were obtained from the publications and articles that the authors have reviewed as full papers.

4.1 RQ1: What Techniques Are Most Commonly Used in Trend Forecasting?

There are a number of different techniques that have been implemented for Trend Forecasting. Some of the techniques that were used in the reviewed papers are Big Data, Data Mining, Time Series, Greedy Method, and Machine Learning.

Big Data was used in research conducted by [1,4,12,30]. Through the paper authored by Atwah et al. [1], we know that the usage of Big Data and Predictive Analysis has a significant impact to increase a company capability in the betterment to manage a crisis situation as well as recover from it. The study conducted by Blekanov et al. [4], provides insight to the importance of using Big Data for forecasting purposes as it explains about the importance of analyzing Social Media and customer feedback and its impact on customer preference and also its impact in dealing with crisis in the clothing industry, as data that obtained from social media is extremely large in quantity it could only be processed by using Big Data. Data obtained from Social Media may be considered a good source for obtaining customer feedback. Izhar et al. [12], described an ontology approach which provides explicit and formal specifications of knowledge. By incorporating the data, the ontology made the process of identifying relevant data much easier to accomplish.

Vincent et al. [39] and Kaur et al. [14], used Data Mining as one of the main methods of their research articles. Vincent et al. [39], used data from 1000 customer records as their research dataset in order to obtain the most common attributes in clothing that are sought out by customers. Kaur et al. [14], implemented data mining on a Bakery Store's sale data in order to obtain a better understanding of the store's customer preference. The greedy method was used by Li et al. [18], in their article on using the Greedy Method as a means of aggregating and decomposing data to determine the intermittent demand of fashion forecasting in the world of retail fashion.

A study about forecasting in Social Settings or social science-based by Makridakis et al. [21], is relevant to this fast-paced sector with its many uncertainties. This paper provides an understanding of the M4 (Makridakis) Forecasting Competition results. The authors also explored the performance of machine learning methods on Time Series forecasting tasks and the difficulties to predict in non-stable environments. While Avdeevaa et al. [3], presents a combined approach to forecast the manufacturing system target indicators. The authors' approach is focused on the formation of long-term forecasts of production targets and improving the accuracy of these forecasts by identifying significant influencing events. Time Series Forecasting Analytics were included in the research of Gómez et al. [10], which analyze the data from Amazon marketplace. Sinha et al. [33], studied Map-reduce based algorithm for Time Series Forecasting.

Machine Learning was used by a number of different authors as a method of trend forecasting and is by far the most used technique for trend forecasting as shown in Table 3. Around 26 articles that were reviewed for the writing of this paper used machine learning or one of its methods as a means of trend forecasting. It could be seen as the most preferred method for trend forecasting.

Table 3. Relevant publication towards the most commonly used techniques for trend forecasting.

Tecnique used	Number of papers	Paper reference
Big Data	4	[1,4,12,30]
Data Mining	2	[14,39]
Time Series	5	[3,6,10,21,33]
Greedy Method	1	[18]
Machine Learning	26	[6,13,19,20,23,30,32,37]

4.2 RQ2: What Are the Machine Learning Techniques that Could Be Used for Fashion Trend and Trend Forecasting?

After reviewing numerous articles and papers with Fashion Trend Forecasting and Trend Forecasting in general, the most commonly used technique for Trend Forecasting is Machine Learning. Therefore a more detailed explanation on the methods, models and algorithms which are used by Machine Learning will be given. There are numerous machine learning techniques that are known in computer science. The most frequently used machine learning techniques in Fashion Forecasting are: Deep Learning, Supervised Machine Learning, Neural Networks, Regression, and Ensemble Learning.

Various researches and studies used Deep Learning techniques [5–7,11,13,15, 32,37], proposed a Sketch-Product Fashion Retrieval Model to develop fashion recommendation system based on Deep Learning, Generative Adversarial Network, and Convolutional Neural Network. While Singh et al. [32], proposed a derived technique from Deep Learning and Tree-based algorithm, concerning the age of style, trend and seasonality, and cannibalization marketing scenario, to forecast fashion demands for new items. Tarallo et al. [37], researched about the Deep Learning and Autoregressive Integrated Moving Average (ARIMA) Model to predict demand in fast-moving customer goods, which includes fashion goods. Brunelli et al. [5], experiment applied Long Short Term Memory (LSTM) Networks and Temporal Convolutional Networks (TCN) to forecast the production by analyzing the production equipment state. Carletti et al. [6], studied Deep Learning based approaches to detect anomalies in refrigerators, namely Convolutional Neural Network and Decision Support System. Chen et al. [7] focused on crude oil price forecasting with a deep learning-based model that included the Deep Belief Network and Long Short Term Memory Network (LSTM).

Nevertheless, Neural Network techniques were widely used regarding forecasting [2,9,11,13,15,19,20,30,35,38]. Guan et al. [11] researched the apparel-based deep learning system design to construct a recommendation system for apparel style. It used Deep Learning methods, Convolutional Neural Network, and SVM. Arslan et al. [2], discussed several methods to enhance the multimodal sequential fashion attribute prediction, using CNN, RNN, and Sequential Prediction. Sun et al. [35], proposed a model that combines semantic and visual embedding, Long Short Term Memory, and CNN to learn fashion compatibility. Vhatkar et al. [38], studied about the models that are applicable to forecast oral-care goods sales, using Neural Network and Supply Chain.

Regression [17,23,30,36] approaches were also adopted to analyze their effectiveness in forecasting related to fashion trend and trend in general. Le et al. [17] applied regression analysis and sentiment analysis to identify the effects of applying big data analytics to the customer's responses through e-commerce market. Parlar et al. [27] analyzed the methods for sentiment analysis using supervised Machine Learning. While Rathor et al. [28] used Supervised Learning to analyze 3000 user reviews on Amazon products. This lead to classifying the reviews based on the positive, neutral, or negative comments, using unigrams and weighted unigrams.

The research by Kraus et al. [15], used Lasso and Ridge Regression models, Random Forest algorithms, Support Vector Machines, Single-Layer, Convolutional, and Recurrent Neural Network. The research applied proprietary business data from operations management, inventory management, and risk management to predict sales, human resources needed, and customer analysis. Loureiro et al. [20] explored several methods including the Deep Neural Networks for sales forecasting in fashion retail. From the study, Deep Learning was found to perform better in predicting sales. He tested different techniques, which include the Random Forest method to forecast sales in fashion retail.

Table 4. Relevant publication towards Machine Learning techniques for fashion trend and trend forecasting.

Tecnique used	Number of papers	Paper reference
Deep Learning	10	[5–8,11,13,15,35–37]
Supervised Machine Learning	2	[27,28]
Ensemble Learning	3	[15,20,26]
Neural Networks	10	[2,9,11,15,16,19,20,24,34,35,38]
Regression	5	[17,23,30,31,36]

Based on the information obtained from the Table 3. We know that most papers that were reviewed during this study implemented one of the five methods that were mentioned within the table, some of the papers even implemented two or more of the aforementioned methods within its experimentations. Though there are multiple methods that could be implemented when it comes to forecasting, based on the results seen from the table it is obvious that the most widely implemented of these methods is that of Machine Learning with a total 68.42% of the papers on trend forecasting used Machine Learning methods as a means of forecasting.

According to the results that were obtained from Table 4., there were 5 main Machine Learning techniques/methods that were implemented within the research papers that were reviewed. The results that were obtained were as follows:

1. Deep Learning methods were implemented in 10 out of the 26 papers studied.
2. Supervised Machine Learning was implemented in 2 papers out of the 26 studied.
3. Methods categorized as Neural Networks were found in 9 out of the 26 papers studied.
4. Regression methods were found in 5 out of the 26 papers studied.
5. Ensemble Learning was found in 3 out of the 26 papers studied.

During the review conducted for the writing of this paper, the authors found that a number of 4 papers reviewed implemented more than one Machine Learning method, these papers were papers [11,15,20,36].

5 Conclusion

This systematic literature review observed and identifies methods and models that were used while forecasting, though the authors' main goal was to observe and identify forecasting methods that were implemented in the field of fashion industry that belongs to the Machine Learning family, the authors have found that related articles and publication on Forecasting in the Fashion Industry extremely scant. Therefore, the authors had to look for implementation in forecasting methods in other fields that have similarities with the fashion industry so that the authors have further understanding as to which of the techniques implemented would be most suitable with the fashion industry. In order to accomplish this, the authors had reviewed a total of 46 publications, the results show that 56.52% of the publication reviewed had topics on implementing Machine Learning for forecasting whilst the other 43.48% consisted of other methods for forecasting as well as some publications that were related with the fashion industry.

According to the information acquired through this systematic review, it is clear that the implementation of Machine Learning as a method of forecasting trend is possible and even beneficial, some of the Machine Learning methods that the authors believe could be implemented in the fashion industry are Deep Learning, Neural Networks as well as Ensemble Learning.

References

1. AL-Ma'aitah, M.A.: Utilizing of big data and predictive analytics capability in crisis management. J. Comput. Sci. **16**(3), 295–304 (2020)
2. Arslan, H.S., Sirts, K., Fishel, M., Anbarjafari, G.: Multimodal sequential fashion attribute prediction. Information **10**(10), 308 (2019)
3. Avdeeva, Z., Grebenyuk, E., Kovriga, S.: Combined approach to forecasting of manufacturing system target indicators in a changing external environment. Procedia Comput. Sci. **159**, 943–952 (2019)
4. Blekanov, I., Krylatov, A., Ivanov, D., Bubnova, Y.: Big data analysis in social networks for managing risks in clothing industry. IFAC-PapersOnLine **52**(13), 1710–1714 (2019)
5. Brunelli, L., Masiero, C., Tosato, D., Beghi, A., Susto, G.A.: Deep learning-based production forecasting in manufacturing: a packaging equipment case study. Procedia Manuf. **38**, 248–255 (2019)
6. Carletti, M., Masiero, C., Beghi, A., Susto, G.A.: A deep learning approach for anomaly detection with industrial time series data: a refrigerators manufacturing case study. Procedia Manuf. **38**, 233–240 (2019)
7. Chen, Y., He, K., Tso, G.K.: Forecasting crude oil prices: a deep learning based model. Procedia Comput. Sci. **122**, 300–307 (2017)
8. Djellali, C., et al.: A new deep learning model for sequential pattern mining using ensemble learning and models selection taking mobile activity recognition as a case. Procedia Comput. Sci. **155**, 129–136 (2019)
9. Donati, L., Iotti, E., Mordonini, G., Prati, A.: Fashion product classification through deep learning and computer vision. Appl. Sci. **9**(7), 1385 (2019)
10. Gómez-Losada, Á., Duch-Brown, N.: Time series forecasting by recommendation: an empirical analysis on amazon marketplace. In: Abramowicz, W., Corchuelo, R. (eds.) BIS 2019. LNBIP, vol. 353, pp. 45–54. Springer, Cham (2019). https://doi.org/10.1007/978-3-030-20485-3_4
11. Guan, C., Qin, S., Long, Y.: Apparel-based deep learning system design for apparel style recommendation. Int. J. Clothing Sci. Technol. **31**, 376–389 (2019)
12. Izhar, T.A.T., Torabi, T., Bhatti, M.I.: Using ontology to incorporate social media data and organizational data for efficient decision-making. Int. J. Comput. Inf. Syst. Ind. Manage. Appl. **9**(2017), 9–22 (2017)
13. Jo, J., Lee, S., Lee, C., Lee, D., Lim, H.: Development of fashion product retrieval and recommendations model based on deep learning. Electronics **9**(3), 508 (2020)
14. Kaur, M., Kang, S.: Market basket analysis: identify the changing trends of market data using association rule mining. Procedia Comput. Sci. **85**, 78–85 (2016)
15. Kraus, M., Feuerriegel, S., Oztekin, A.: Deep learning in business analytics and operations research: models, applications and managerial implications. Eur. J. Oper. Res. **281**(3), 628–641 (2020)
16. Kumar, J., Goomer, R., Singh, A.K.: Long short term memory recurrent neural network (LSTM-RNN) based workload forecasting model for cloud datacenters. Procedia Comput. Sci. **125**, 676–682 (2018)
17. Le, T.M., Liaw, S.Y.: Effects of pros and cons of applying big data analytics to consumers' responses in an e-commerce context. Sustainability **9**(5), 798 (2017)
18. Li, C., Lim, A.: A greedy aggregation-decomposition method for intermittent demand forecasting in fashion retailing. Eur. J. Oper. Res. **269**(3), 860–869 (2018)
19. Liu, Y., Shen, Y.: Personal tastes vs. fashion trends: predicting ratings based on visual appearances and reviews. IEEE Access **6**, 16655–16664 (2018)

20. Loureiro, A.L., Miguéis, V.L., da Silva, L.F.: Exploring the use of deep neural networks for sales forecasting in fashion retail. Decis. Support Syst. **114**, 81–93 (2018)
21. Makridakis, S., Hyndman, R.J., Petropoulos, F.: Forecasting in social settings: the state of the art. Int. J. Forecast. **36**(1), 15–28 (2020)
22. Makridakis, S., Spiliotis, E., Assimakopoulos, V.: Statistical and machine learning forecasting methods: concerns and ways forward. PLOS ONE **13**(3), e0194889 (2018)
23. Merkuryeva, G., Valberga, A., Smirnov, A.: Demand forecasting in pharmaceutical supply chains: a case study. Procedia Comput. Sci. **149**, 3–10 (2019)
24. Mezzogori, D., Zammori, F.: An entity embeddings deep learning approach for demand forecast of highly differentiated products. Procedia Manuf. **39**, 1793–1800 (2019)
25. Moher, D., Liberati, A., Tetzlaff, J., Altman, D.G., et al.: Preferred reporting items for systematic reviews and meta-analyses: the PRISMA statement. Int. J. Surg. **8**(5), 336–341 (2010)
26. Mosavi, A., Faizollahzadeh Ardabili, S., Shamshirband, S.: Demand prediction with machine learning models: state of the art and a systematic review of advances (2019)
27. Parlar, T., Ozel, S., Song, F.: Analysis of data pre-processing methods for sentiment analysis of reviews. Comput. Sci. **20**, 123–141 (2019)
28. Rathor, A.S., Agarwal, A., Dimri, P.: Comparative study of machine learning approaches for Amazon reviews. Procedia Comput. Sci. **132**, 1552–1561 (2018)
29. Ren, S., Patrick Hui, C., Jason Choi, T.: AI-based fashion sales forecasting methods in Big Data era. In: Thomassey, S., Zeng, X. (eds.) Artificial Intelligence for Fashion Industry in the Big Data Era. SSFB, pp. 9–26. Springer, Singapore (2018). https://doi.org/10.1007/978-981-13-0080-6_2
30. Silva, E.S., Hassani, H., Madsen, D.Ø., Gee, L.: Googling fashion: forecasting fashion consumer behaviour using google trends. Soc. Sci. **8**(4), 111 (2019)
31. Singh, D., Leavline, E.J., Muthukrishnan, S., Yuvaraj, R.: Machine learning based business forecasting. IJ Inf. Eng. Electron. Bus. **6**, 40–51 (2018)
32. Singh, P.K., Gupta, Y., Jha, N., Rajan, A.: Fashion retail: forecasting demand for new items. arXiv preprint arXiv:1907.01960 (2019)
33. Sinha, A., Jana, P.K.: MRF: MapReduce based forecasting algorithm for time series data. Procedia Comput. Sci. **132**, 92–102 (2018)
34. Sumaiya, P.: Enhancing user experience using machine learning. Int. J. Eng. Res. Technol. (IJERT) **7**(2), 353–358 (2018)
35. Sun, G.L., He, J.Y., Wu, X., Zhao, B., Peng, Q.: Learning fashion compatibility across categories with deep multimodal neural networks. Neurocomputing **395**, 237–246 (2020)
36. Tanizaki, T., Hoshino, T., Shimmura, T., Takenaka, T.: Demand forecasting in restaurants using machine learning and statistical analysis. Procedia CIRP **79**, 679–683 (2019)
37. Tarallo, E., Akabane, G.K., Shimabukuro, C.I., Mello, J., Amancio, D.: Machine learning in predicting demand for fast-moving consumer goods: an exploratory research. IFAC-PapersOnLine **52**(13), 737–742 (2019)
38. Vhatkar, S., Dias, J.: Oral-care goods sales forecasting using artificial neural network model. Procedia Comput. Sci. **79**, 238–243 (2016)
39. Vincent, O., Makinde, A., Salako, O., Oluwafemi, O.: A self-adaptive k-means classifier for business incentive in a fashion design environment. Appl. Comput. Inf. **14**(1), 88–97 (2018)

Development of the Index Paradigm in Remote Sensing of Soil-Vegetable Cover

Ilya Mikhailenko(✉) and Valery Timoshin

Agrophysical Research Institute of the Russian Academy of Sciences, St. Petersburg, Russia

Abstract. The aim of the work is a systematic analysis and generalization of the generally accepted index paradigm of using Earth remote sensing (ERS) data for assessing the state of the soil and vegetation cover. It has been established that the scalar form and the lack of a mathematical basis do not allow the use of generally accepted vegetation and similar indices for evaluating the vectors of quantitative indicators of the soil and vegetation cover. At the same time, for making many types of management decisions in agriculture, it is important to construct index images reflecting such non-quantitative indicators as types of cultivated and weed plants, the presence of plant diseases, damage to crops and soils, physical and chemical stresses. In terms of its information content, the assessment of such non-quantitative (countable) states is a procedure for recognizing patterns or classes of objects of the soil-plant complex. An inventive empirical approach to the choice of the spectral composition of the indices of their combinations does not currently allow ensuring sufficient reliability of such procedures. Therefore, the purpose of the study in this work is to formalize the process, which allows us to completely exclude the empirical approach of constructing indices and automate the entire procedure for their formation for any number and types of recognizable objects. The basis of formalization is the algorithms for evaluating and selecting the informativeness of features, followed by the construction of index models, which are linear decision rules (LDR) for class recognition. In this case, the features of the classes are spectral subranges, into which the entire spectrum of remote sensing data is divided, and the number of informative features is selected from the condition of ensuring the required reliability of recognition of all observed objects (classes).

Keywords: Index models · Pattern recognition · Informative features · Algorithms · Informativeness of features · Identification of models

1 Introduction

Currently, to work with spectral information in Earth remote sensing (ERS), so-called "index" images are often used. Based on the combination of brightness parameters in certain channels, informative for the selection of the object under study, and the calculation of these values of the "spectral index" of the object, an image is built corresponding to the value of the index in each pixel, which allows you to select the object under study

R. Silhavy et al. (Eds.): CoMeSySo 2021, LNNS 231, pp. 45–56, 2021.
https://doi.org/10.1007/978-3-030-90321-3_6

or evaluate its state. Spectral indices used to study and assess the state of vegetation are commonly referred to as vegetation indices (VI).

The vegetation index is a dimensionless scalar index representing a combination of reflection parameters in different spectral ranges of remote sensing data, reflecting the state of vegetation on a given surface area and in a given image pixel. Currently, there are over 160 variants of vegetation indices. They are selected experimentally (empirically) based on the known features of the curves of the spectral reflectivity of vegetation and soils.

The calculation of most of the vegetation indices is based on the two most stable (independent of other factors) sections of the spectral reflectance curve of plants. The red zone of the spectrum (620–750 nm) has the maximum absorption of solar radiation by chlorophyll, and the near infrared zone (750–1300 nm) has the maximum reflection of energy by the cellular structure of the leaf. High photo-synthetic activity (associated, as a rule, with a large biomass of vegetation) leads to lower reflectances in the red region of the spectrum and higher values in the near infrared. The ratio of these indicators to each other makes it possible to clearly distinguish vegetation from other natural objects.

When processing satellite images in the ENVI software package, there is a special ENVI vegetation index calculator (ENVI VI calculator), which allows you to calculate 27 vegetation indices used to assess the state of vegetation, the content of pigments, nitrogen, carbon, water. For images with only red and near infrared spectral zones, the program can calculate two indices – NDVI and RVI. When working with a hyper spectral image, the calculation of all 27 indices will be offered.

Each of the listed groups of indices is designed to assess some of the properties of the vegetation cover and contains several indices. For specific natural conditions and various tasks, some indices from the group can give more accurate results than others. Comparing the results of calculating indices with field data, you can choose an index that most accurately reflects the property under study. Thus, the accuracy of the results in subsequent processing is significantly increased.

Informational analysis of vegetation and other spectral indices, the list of problems to be solved based on them, as well as the rapidly growing number of such indices shows that at present there is no single method for their formation and the spectral composition of the ERS data used. The absence of such a technique leads to a lack of justification for the introduction of new indices and their spectral composition, and also does not allow evaluating their effectiveness in solving problems in which these indices are used.

This paper proposes a method for the formation of vegetation and similar indices and an assessment of their effectiveness.

2 Materials and Methods

The scalar form of the indices, regardless of the combinations of the used ERS spectral channels, does not allow solving the problems of assessing the quantitative indicators of the investigated properties of the probed plant and soil covers. To evaluate these indicators in accordance with modern information theory, mathematical models of the dependencies of these indicators among themselves and on the main influencing factors, as well as on the relationship between the parameters of reflection of the means of remote sensing and the estimated indicators [19] are needed. In the presence of such models, for the estimation of vectors of quantitative indicators, it is necessary to have ERS data vectors of the corresponding dimension. Any sets or groups of scalar indexes do not meet these requirements.

At the same time, such indices are indicators of non-quantitative conditions of vegetation and soil cover, such as stresses of various natures, the presence of diseases and pests, types of cultivated plants and weeds, and many others. It is in the assessment of such conditions that the most significant results were achieved using various indices. The presence of spatial pictures of estimates of such states is important monitoring information that allows making preliminary control decisions of the beginning and sequence of technological controls, sowing and harvesting, etc. Therefore, increasing the reliability and reliability of such estimates is an urgent problem. It can only be solved if there is a general methodology for the formation of indices, through which it is possible to assess non-quantitative indicators (states) of the vegetation and soil cover.

From the point of view of modern information theory, the assessment of non-quantitative indicators (states) refers to the problem of pattern recognition, and each of such non-quantitative states can be considered as an image or class. As soon as we reduce the estimation procedure based on indices to the problem of pattern recognition, then, according to the theory of recognition, the problem of selecting informative signs on the basis of which this problem is solved immediately arises. When using remote sensing data, the signs are separate gradations (intervals) of the reflection spectrum of the remote sensing means. The feature selection procedure itself practically does not depend on the method of measuring information content. It is only important that this method is the same for all features (groups of features) included in their original set and participating in the selection procedure.

One of the simplest and most effective methods for evaluating the informativeness of features is the Shannon method [27]. He proposes to evaluate the information content as a weighted average of the amount of information falling on different gradations of the attribute. In information theory, information is understood as the value of the eliminated entropy.

The informative value according to the Shannon method is determined as follows

$$I(x_i) = P_i * \sum_{k=1}^{K} P_{i,k} * \log_k P_{i,k} \quad (1)$$

where $I(x_i)$ is the informativeness of the feature x_i, G is the number of intervals (gradations) of the feature, P_i is the probability of the i-th interval (gradation) of the feature, K is the number of feature classes, $P_{i,k}$ is the probability of the i-th feature interval in k-th class.

The probability of the i-th feature interval P_i was calculated by the formula:

$$P_i = \frac{\sum_{k=1}^{K} m_{i,k}}{N} \quad (2)$$

where is the frequency of occurrence of the i-th interval in the k-th class, K is the number of feature classes, N is the total number of objects.

The probability of occurrence of the *i*-th feature interval in the *k*-th class $P_{i,k}$ was calculated by the formula:

$$P_{i,k} = \frac{m_{i,k}}{\sum_{k=1}^{K} m_{i,k}}, \tag{3}$$

where $m_{i,k}$ is the frequency of occurrence of the *i*-th interval in the *k*-th class, K is the number of classes of the feature.

After evaluating the informativeness of all gradations of the spectral range of the used ERS system, to facilitate further selection, they are arranged in descending order of the informativeness indicator. To select the most effective informative features, we introduce a mathematical model of the index

$$I_k = \sum_{i=1}^{n_k} b_{ik} x_{ik}^*, \tag{4}$$

or in vector-matrix form

$$I_k = \mathrm{B}_k^{\mathrm{T}} \mathrm{X}_k^*, \tag{5}$$

where b_{ik} are the parameters of the index model for the *k*-th class, combined into a row matrix $\mathrm{B}_k^{\mathrm{T}}$, x_{ik}^* are the most informative features for the *k*-th class, denoted by the index *i*, combined into vectors X_k^*, I_k are the recognizable patterns denoted sequentially by the numbers 1, 2,3… K, serving as the output quantities of the model.

The parameters of the index model (4), (5) are estimated from the training sample using the recursive least squares method of RLS [7]

$$\begin{aligned} G_k(j) &= R_k(j-1)X_k(j), \\ z_k(j) &= W_k^{\mathrm{T}}(j)G_k(j), \\ e_k(j) &= I_k(j+1) - B_k^{\mathrm{T}}(j-1)X_k(j), \\ B_k(j) &= B_k(j-1) + \tfrac{1}{1+z_k(j)}G_k(j)e_k(j), \\ R_k(j) &= R_k(j-1) - \tfrac{1}{1+z_k(k)}G_k(j)G_k^{\mathrm{T}}(j), \end{aligned} \tag{6}$$

here: $j = 1, 2, 3 \ldots J$ are the numbers of records in the training set, is the current learning, $e_k(j) = I_k(j+1) - B_k^{\mathrm{T}}(j-1)X_k(j)$-error at the *j*-th step of the procedure, $G_k(j)$, $z_k(j)$, $R_k(j)$ are the vector, scalar and matrix parameters of the algorithm, which ensure its convergence.

In model (4), (5) and in the learning algorithm (6), it remains unclear how many most informative features should be selected for each recognized class, i.e., what is the dimension of the n_k vector X_k^*.

To do this, we first give a decision rule, through which a decision is made to assign the current vector of features $X_k^*(j)$ to a specific class $k*$

$$I_k(j) = I_k^*, \quad \text{если} \; \left| I_k^* - \mathrm{B}_k^{\mathrm{T}} \mathrm{X}_k^*(j) \right| \le \delta, \tag{7}$$

where: δ – the specified tolerance for the recognition error, as a rule – $\delta = 0,05I_k^*$.

Taking the frequency of falling of solutions (7) into the tolerance as an estimate of the probability of correct decisions

$$\widehat{P}_k = \frac{n_k}{\sum_{k=1}^{K} n_k}, \tag{8}$$

it is possible sequentially, starting from the first 2 features, to enter in descending order the most informative features until the probability estimate (8) reaches a given value $\widehat{P}_k^*$.

3 Discussion of the Results

Tables 1 and 2 show fragments of the data of the reflection spectra of various plants for the southern and northern latitudes of the Russian Federation. They differ significantly in their spectral patterns, where for plants of northern latitudes the values of reflection parameters are distributed more evenly over the studied spectral ranges (features). Due to the fact that the informativeness determines the sequence of entering features in the model of indices (4), (5), then one should choose compromise options in which there are no zero informative values and they are quite distinguishable for individual features. In accordance with these considerations, a threshold value of 2% was chosen for plants in the southern latitudes (Fig. 1), and 6% for plants in the northern latitudes (Fig. 2).

The selection of informative features for constructing index models in accordance with the given algorithm begins with the first two most informative channels. For plants of southern latitudes, these are channels x_4 – (680–740 nm), x_5 – (800–1100 nm). For plants of northern latitudes, these are channels x_2 – (530–610 nm), x_4 – (680–740 nm). Table 3 shows the results of evaluating the probability of correct recognition of classes for plants of southern latitudes, depending on the dimension of the vector of used characters in the model of indices (5) for a tolerance field of ±10% of the class number.

As can be seen from Table 3, the probability of correct class recognition for the case under consideration does not depend on the number of informative features used. Therefore, for further analysis, we use index models with the first two features. Figure 3, 4 shows the plots of identification of index models for the first plant species (classes) presented in Tables 1, 2. The identification results are presented in Tables 4, 5.

As can be seen from the graphs in Fig. 3, 4 and Tables 4, 5, the selection of the most informative features and setting the parameters of the index models provide high reliability of recognition of plant classes. All currently used vegetation and other indices do not have such capabilities. Preservation of these properties of the recognition system with changes in the spectral characteristics of plants, it is necessary to re-identify the parameters of the index models when changing the phenological phases of plants. This technique can be extended to other objects of soil and vegetation cover with other sets of reflection spectra.

In conclusion, it should be noted that the proposed technique is applicable only in pixel-by-pixel processing of remote sensing data, when it is assumed that an object of the same type prevails in a single pixel.

Table 1. Fragment of spectral data on the state of plants in the southern latitudes of Russia

Name of plants	Class numbers, k	Luminance spectral density (LRD), %				
		450–520 nm	530–610 nm	625–680 nm	680–740 nm	800–1100 nm
		x_1	x_2	x_3	x_5	x_5
Winter wheat (*T. aestivum* L.)	**1**	2,83	4,75	6,19	6,68	6,99
Winter barley (*H. vulgare* L.)	**2**	2,81	4,78	6,26	6,32	6,1
Spring wheat (*T. aestivum* L.)	**3**	2,38	3,94	4,65	4,39	6,07
Spring barley (*H. vulgare* L.)	**4**	2,2	3,76	4,88	5,45	6,03
Corn (*Z. mais* L.)	**5**	1,7	3,42	2,96	4,6	5,25
Sunflower (*H. annuus* L.)	**6**	0,96	1,67	1,11	3,77	5,95
Soybean (*G. max* (L.)	**7**	1,01	1,87	1,42	3,79	5,64
Artemisia ragweed (*Ambrósia artemisiifólia*)	**8**	0,38	0,96	0,45	2,78	4,98
Cocklebur california (*Xánthium californicum*)	**9**	0,76	1,58	0,92	3,61	5,26
Theophrastus (A. *theophrastii Medik.*)	**10**	0,84	2,04	1,1	3,96	4,91
White Mary (*C. album* L.)	**11**	1,02	1,5	1,02	3,14	5,14
Thrownback squid (*A. retroflexus* L.)	**12**	0,89	1,76	1,1	3,46	5,58
Sow thistle (*S. arvensis* L.)	**13**	1,21	1,99	1,74	3,8	5,87
Mousetail foxtail (*Alopecurus myosuroides* Huds.)	**14**	2,01	3,68	4,8	5,96	5,46

Table 2. Fragment of spectral data on the state of plants in northern latitudes of Russia

Name of plants	Class numbers, k	Luminance spectral density (LRD), %				
		450–520 nm	530–610 nm	625–680 nm	680–740 нм	800–1100 нм
		X_1	X_2	X_3	X_5	X_5
Spring wheat (*T. aestivum* L.)	**1**	2,38	3,94	4,65	4,39	6,07
Pikulnik ordinary (*Galeopsis tetrahit*)	**2**	6,67	7,24	6,20	6,00	5,80
White Mary (*C. album* L.)	**3**	5,90	7,25	6,38	6,12	6,00
Field yarok (*Thláspi arvénse*)	**4**	0,96	1,67	1,11	3,77	5,95
Shepherd's bag (*Capsella bursa-pastoris*)	**5**	4,49	6,91	6,0	6,5	6,65
Medium starfish (*Stellaria media* L.)	**6**	5,35	7,72	7,08	7,84	7,60
Spreading highlander (*Persicária lapathifólia*)	**7**	5,67	7,74	7?24	7,15	6,95
Fumitory (*Fumária officinális*)	**8**	7,12	8,96	8,24	8,12	7,42
Bindweed buckwheat (*Fallopia convolvulus*)	**9**	5,2	7,27	6,58	6,15	6,55
Field violet (*Víola arvénsis*)	**10**	5,86	7,49	6,80	7,15	7,45
Purpurea (*Lámium purpúreum*)	**11**	5,35	7,77	7,27	7,10	8,15

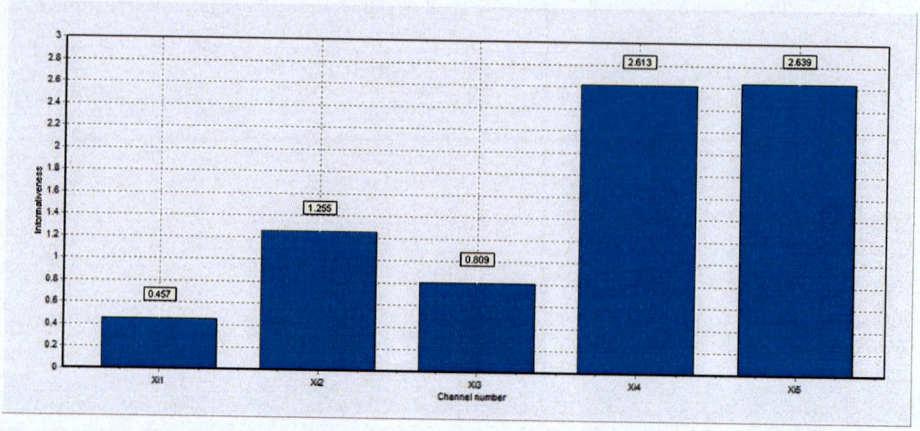

Fig. 1. Diagram of the informativeness of traits for plants of southern latitudes and a sensitivity threshold of 2%

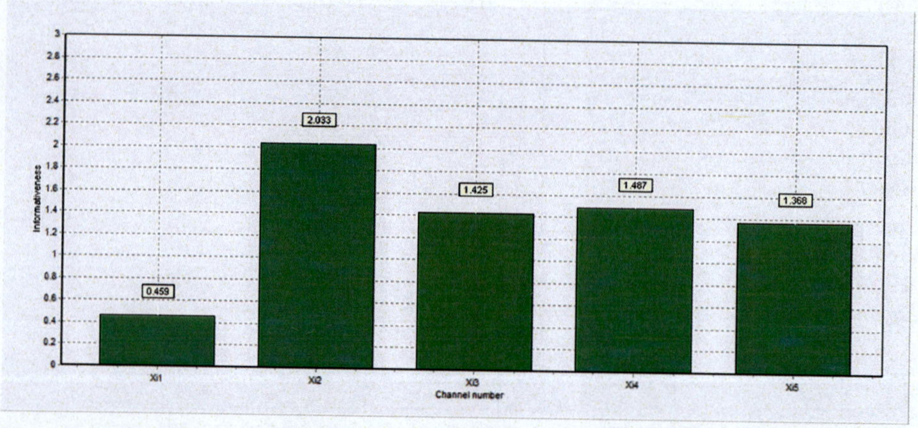

Fig. 2. Diagram of the informativeness of traits for plants of northern latitudes and a sensitivity threshold of 6%

Table 3. Probabilities of correct recognition of classes depending on the number of used features

Numbers classes	Number and composition of features			
	2: x_4, x_4	3: x_4, x_4, x_2	4: x_4, x_4, x_2, x_3	5: x_4, x_4, x_2, x_3, x_1
	Correct classification probabilities			
1	0,95	0,95	0,95	0,95
14	0,95	0,95	0,95	0,95

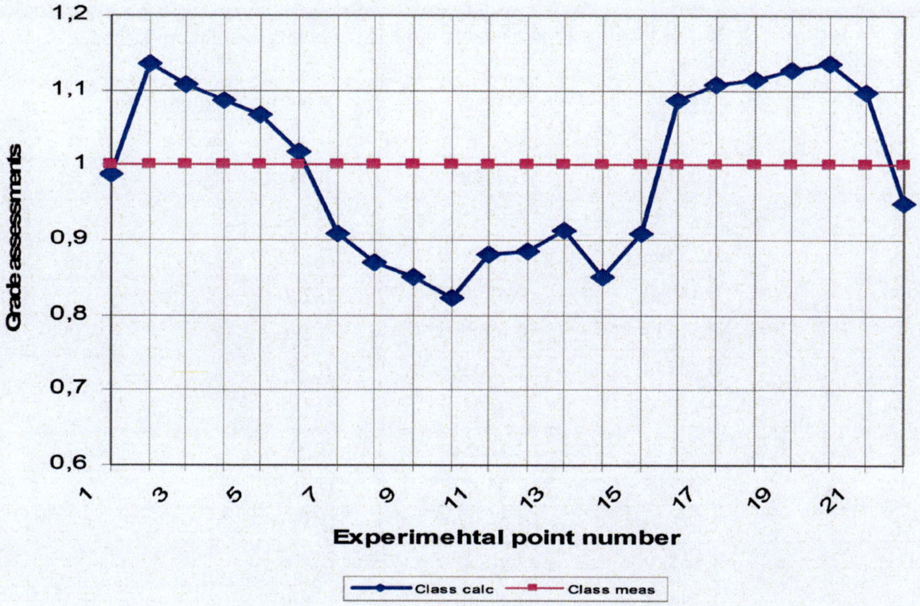

Fig. 3. Identification of the index model of the 1st class of plants in southern latitudes

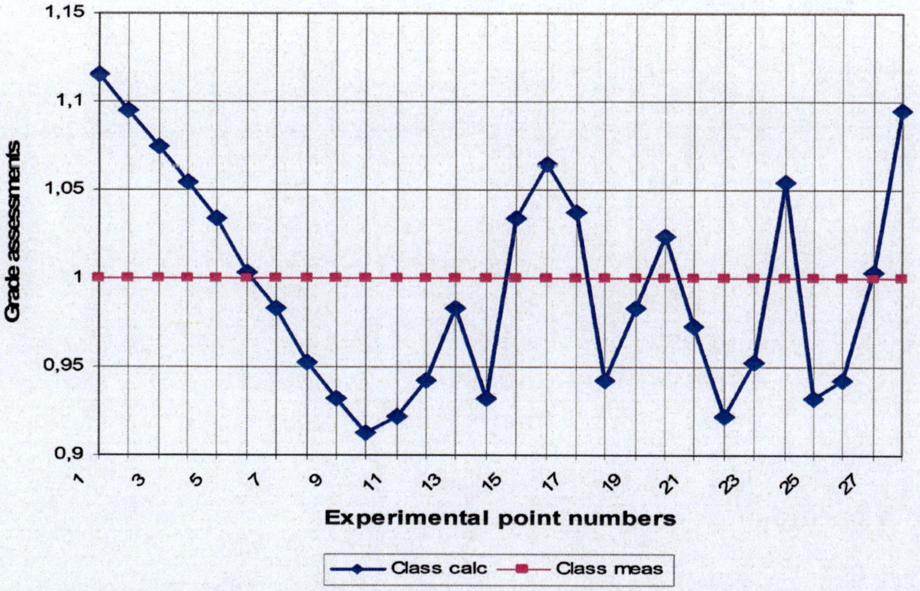

Fig. 4. Identification of the index model of the 1st class of plants in northern latitudes

Table 4. Results of identification of index models for plants of southern latitudes

Class no	Model parameters values		Root mean square error
	b1	b_2	σ
1	0,07175	0,09685	0,037
2	0,15193	0,19765	0,074
3	0,33588	0,37310	0,112
4	0,34083	0,48952	0,149
5	0,51534	0,68969	0,186
6	0,58559	1,30939	0,223
7	0,72659	1,46233	0,261
8	0,87877	2,52582	0,298
9	0,90715	2,05981	0,335
10	1,01760	1,96592	0,372
11	1,34589	2,80647	0,410
12	1,38494	2,71301	0,447
13	1,38991	2,64499	0,484
14	1,03964	1,67757	0,521

Table 5. Results of identification of index models for plants of northern latitudes.

Class no	Model parameters values		Root mean square error
	b1	b_2	σ
1	0,08456	0,08713	0,011
2	0,16562	0,16874	0,023
3	0,22744	0,20778	0,034
4	0,31158	0,30478	0,045
5	0,32315	0,33304	0,057
6	0,42514	0,43709	0,068
7	0,43551	0,47565	0,079
8	0,65740	0,61787	0,090

4 Conclusion

The scalar form, the lack of a model and mathematical basis, as well as substantiation of the spectral composition does not allow the use of modern vegetation and other similar indices for assessing the quantitative parameters of the state of the soil and vegetation cover. At the same time, the assessment of a large number of such non-quantitative conditions as the type of vegetation, plant diseases, physical and chemical stresses, damage, etc. is an extremely urgent problem, the solution of which allows making various management decisions. In terms of its information content, the assessment of non-quantitative (countable) states refers to pattern or class recognition procedures. In turn, the starting point for the implementation of the pattern recognition procedure is the identification of the most informative features, in the space of which the decisive rules of recognition can be built. When using means of optical remote sensing of the Earth (ERS), such signs are separate spectral sub-bands, into which the entire spectral range used by ERS is divided. In this case, the indices are linear decision rules (DM), the number of which is equal to the number of classes, and the DM themselves are a weighted linear combination of the most informative features, the number of which in each DM is selected from the condition of ensuring the required reliability of the recognition procedure. With this approach, the empirical choice of the spectral composition of the indices is completely excluded, and the entire procedure for their formation can be formalized and automated.

References

1. Andreev, G.G., Bocharov, V.P., Sazonov, N.V., Chaban, L.N.: Automation of thematic processing of space images when assessing the state of agricultural crops. Expl. Earth Space 5, 95–102 (1986)
2. Baret, F., Guyot, G., Major, D.: TSAVI: A vegetation index which minimizes soil brightness effects on LAI or APAR estimation. In: 12-th Canadian Symposium on Remote Sensing and IGARSS 1990, Vancouver, Canada, July, pp. 10–14 (1989)
3. Clevers, J.G.P.W.: The derivation of a simplified reflectance model for the estimation of leaf area index. Rem. Sen. of Env. **35**, 53–70 (1988)
4. Crippen, R.E.: Calculating the vegetation index faster. Rem. Sens. of Env. **34**, 71–73 (1990)
5. Crist, E.P., Cicone, R.C.: Application of the tasseled cap concept to simulated thematic mapper data. Photogю Engю Remю Sensю **50**, 343–352 (1984)
6. Elvidge, C.D., Lyon, R.J.P.: Influence of rock-soil spectral variation on the assessment of green biomass. Remю Sensю of Envю **17**, 265–269 (1985)
7. Eikhoff, P.: Fundamentals of identification of control systems. Parameter and State Estimation. – M.: Mir, 681 p. (1975)
8. Huete, A.R., Jackson, R.D., Post, D.F.: Spectral response of a plant canopy with different soil backgrounds. Rem. Sens. of Env. **17**, 37–53 (1985)
9. Huete, A.R.: A soil-adjusted vegetation index (SAVI). Rem Sensю of Envю **25**, 295–309 (1988)
10. Jackson, R.D.: Spectral indices in n-space. Remote Sens. Environ. **13**, 409–421 (1983)
11. Jordan, C.F.: Derivation of leaf area index from quality of light on the forest floor. Ecology **50**, 663–666 (1969)
12. Tu, J., Gonzalez, R.: Principles of pattern recognition. M.: Mir (1978)
13. Kaufman, Y.J., Tanre, D.: Atmospherically resistant vegetation index (ARVI) for EOS-MODIS. In: Proceedings of IEEE International Geoscience and Remote Sensing Symposium, New York, pp. 261–270. IEEE (1992)
14. Kauth, R.J., Thomas, G.S.: The tasseled cap-a graphic description of the spectral-temporal development of agricultural crops as seen by Landsat. In: Proceedings of the Symposium on Machine (1976)

15. Kriegler, F.J., Malila, W.A., Nalepka, R.F., Richardson, W.: Preprocessing transformations and their effects on multispectral recognition. In: Proceedings of the Sixth International Symposium on Remote Sensing of Environment, University of Michigan, Ann Arbor, MI, pp. 97–131 (1969)
16. Kashkin, V.B., Sukhinin, A.I.: Remote sensing of the Earth from space. Digital image processing., M.: Logos, (2001)
17. Leprieur, C., Verstraete, M.M., Pinty, B., Chehbouni, A.: NOAA/AVHRR vegetation indices: suitability for mmonitoring fractional vegetation cover of the terrestrial biosphere. In: Proceedings of Physical Measurements and Signatures in Remote Sensing, ISPRS, pp. 1103–1110 (1994)
18. Lillesand, T.M., Kiefer, R.W.: Remote Sensing and Image Interpretation, 2nd edn., p. 721. John Wiley and Sons, New York (1987)
19. Mikhailenko, I.M.: The main tasks of assessing the state of sowing and soil environment according to space sounding data. Env. Syst. Dev. **8**, 17–25 (2011)
20. Pinty, B., Verstraete, M.M.: GEMI: a non-linear index to monitor global vegetation from satellites. Vegetatio **101**, 15–20 (1991)
21. Qi, J., Chehbouni, A., Huete, A.R., Kerr, Y.H.: Modified Soil Adjusted Vegetation Index (MSAVI). Rem. Sens. of Env. **48**, 119–126 (1994)
22. Richardson, A.J., Everitt, J.H.: Using spectra vegetation indices to estimate rangeland productivity. Geo. Intern. **1**, 63–69 (1992)
23. Richardson, A.J., Wiegand, C.L.: Distinguishing vegetation from soil background information. Phot. Eng. Rem. Sens. **43**, 1541–1552 (1977)
24. Rouse, J.W., Haas, R.H., Schell, J.A., Deering, D.W.: Monitoring vegetation systems in the great plains with ERTS. In: Third ERTS Symposium, NASA SP-351, vol. 1, pp. 309–317 (1973)
25. Duda, R., Hart, P.: Pattern recognition and scene analysis. M.: Mir (1976)
26. Tu, T.-M., Chen, C.-H., Wu, J.-L., Chang, C.-I.: A fast two-stage classification method for high dimensional remote sensing data. IEEE Trans. Geosci. Rem. Sens. **36**(1), 182–191 (1998)
27. Shennon, K.: Works on the Theory of Information and Cybernetics. M.: Foreign Literature Publishing House, 295 p. (1963)

Electric and Hybrid Vehicles: Analysis of Factors Influencing Purchase Intentions of Consumers in Czechia

Tomáš Formánek(✉) and Radek Tahal

University of Economics, Prague, Czech Republic
{formanek,radek.tahal}@vse.cz
http://www.vse.cz

Abstract. Electromobility is a hot topic for discussions held both at the level of governments and in the commercial sphere. Economists and marketers are aware that one of the key factors influencing future consumer behavior is the "purchase intention". This article focuses on the context and factors influencing market potential of electric vehicles in Czechia. Based on primary surveyed data and the research performed, we can say that the ownership of an electric vehicle itself is perceived positively by consumers, but only 17% of consumers see electric vehicles as the dominant source of future mobility. Consumers believe in hybrid engines, yet they do not support major tax breaks or other benefits associated with the ownership and operation of an electric engine car. The only aspect in which state support would be welcomed is the support of infrastructure for charging electric vehicles.

Keywords: Electromobility · Purchase intention · Governmental interventions · Infrastructure

1 Introduction

Electromobility is one of the topics that are currently discussed both at the level of European governments and in the commercial sphere. The EU is working to prevent the effects of climate change by cutting 40% from 1990 levels in greenhouse gas emissions by 2030 [7]. The European Union is setting targets for reducing carbon dioxide (CO2) emissions from new cars and vans. European parliament adopted new rules in its plenary session on 27 March 2019 and emission limits for new passenger cars are expected to fall by 37.5% and for vans by 31%, due in 2030 [8]. Like most other legislative regulations, this legislation has a huge impact on the business environment. The impact of this regulation is substantial because the volume of the automotive sector contributes significantly to GDP and is also a major employer in private sector. The turnover generated by the automotive sector amounts to 7% of EU's GDP. Approximately 13.8 million EU's workers are employed in the automotive industry (directly and indirectly), accounting for 6.1% of all EU jobs [6].

R. Silhavy et al. (Eds.): CoMeSySo 2021, LNNS 231, pp. 57–72, 2021.
https://doi.org/10.1007/978-3-030-90321-3_7

One way of meeting the anti-emission legislature is the massive expansion of electric vehicles (EVs), whose only source of energy is battery electricity. Besides EVs, this category of greener vehicles also includes hybrid electric vehicles (HEVs) that have been commercially produced for a relatively long time and which combine a conventional fossil fuel internal combustion engine (ICE) with electric propulsion, where electricity is produced by surplus power from the ICE or by the recuperation of energy when the vehicle is braking. A more sophisticated alternative to HEVs is the so-called plug-in hybrid electric vehicle (PHEV), in which ICE is combined with a battery that can be recharged from an external source – electric grid.

The key question, however, is which of these technological innovations (and whether at all) will be accepted by the market and what is the adequate role of subsidies and state financial support for reaching sustainable mobility. The corporate world is constantly on the lookout for business frameworks that lead to cost and resource usage efficiencies [15]. Only the market acceptance of a new product can be considered a successful if it is long-term viable and finds enough customers. In a turbulent market environment, producers have to to establish, retain and accumulate adequate resources, as well as frequently review functionality and efficiency of their assets and market knowledge [9].

Purchase Intention

Economists and marketers are aware that one of the key factors influencing future purchases is the *purchase intention* phenomenon. Purchase intention has baffled many researchers for several decades. It has been studied in the context of effectively influencing sales [4]. Marketers have been aware of the importance of the purchase intention for decades and it is still perceived as a core concept in marketing [17].

Products from the automotive sector are among the so-called durable consumer goods. This means that the purchase decision-making process is relatively long and the use of goods lasts for years. Unlike "regular" goods, durable consumer goods have a long product life [24]. Often, marketers are facing a trade-off between increasing purchase likelihood and long-term post-purchase satisfaction [3]. In the context of EV-related marketing, consumers and their co-creating initiatives may play important role in the process of implementing responsible corporate behavior patterns [25].

Research Questions

In this article, we use primary surveyed qualitative (categorical) data and specialized analysis methods to investigate present and future market potential for EVs (and also HEVs). Research questions (RQs) are set as follows:

1. What is the perceived & most likely predominant engine/propulsion type in 20-year time? To what extent has the public accommodated and reconciled with the idea of mobility transitioning to EVs?
2. What are the main perceived features of EVs? What are their perceived strengths and what are the weaknesses and opportunities for improvement?
3. How are EV-owners perceived by the public – in general and in terms of their motives?
4. How does the general public view governmental interventions (a mix of restrictions and subsidies) aimed at promoting EV mobility?

2 Literature Review

For major purchases (from the consumer's point of view), decision-making processes are longer and more thorough. Hence, special interest is given to forecasting purchases of durable goods, consumer buying intentions, expectations, and attitudes [13]. Consumer decision processing algorithms are studied to understand how information is gathered, how it is transformed into opinions and product-choice criteria [18]. Research of purchasing behavior is already the subject of research by sociologists and marketers, but in recent years experts working with information systems and data analyst are also involved. Product involvement and brand attitude effects on purchase decisions are receiving extensive research attention in the fields of quantitative marketing studies [14].

Modern marketing practices are increasingly being applied in the automotive sector. Fundamental restructuring is impacting the automotive industry as opportunities to undertake new strategies and business models are changing the whole market. All major players should take into consideration the increasing need for mobility and social inclusion [21]. Government ingerence and related factors affect both the adoption of cleaner cars and consumers' attitudes toward environmental protection [1]. The propensity to choose greener alternatives is higher among consumers who perceive "green" as a benefit and who regularly buy other ecologically responsible (sustainable) products [16]. Besides sustainability, other considerations have to be addressed, such as prominent price differentials across international borders which are not likely to disappear as long as current car dealership system is in place and significant demand-side differences exist [19].

Politicians in various countries have recognized EVs as the most promising technological alternative to ICE-based cars when it comes to combating global warming. However, considerable barriers have to be removed before EVs can penetrate the general market in mass numbers [2]. Also, the expectations involved with EVs might be just a little bit too high. Electricity used for charging is the largest source of uncertainty in overall environmental impacts. The current energy-power system can face stability issues with rapidly changing levels of EV penetration. However, with proper management and coordination, EVs may be incorporated into the smart grid concept [5].

3 Data and Methodology

This contribution is based on primary data for the Czech Republic, collected during October and November 2019. Hence, our survey predates the onset of the COVID 19 pandemic. Nevertheless, tasks related to analysing factors facilitating the transition towards cleaner mobility remain highly important and relevant. Personal and computer assisted web interviewing (CAWI) data collection methods were used. We have applied a stratified quota sampling approach to obtain a representative dataset based on 402 complete questionnaires. The sampling was performed on the Czech population aged 15 year or more. Table 1 outlines the demographic structure of our population and it shows a very good match between our sample and population proportions. Hence, the data sampled ensure adequate interpretability of our results with respect to the entire Czech population (aged 15+).

Table 1. Czech population structure and the quota sampling used

Category	Strata	Population quotas (%)	Quotas applied (pers.)
Gender	Male	48.72	195
	Female	51.28	207
Age group	15–24 years	11.00	74
	25–34 years	15.68	90
	35–49 years	27.89	107
	50–64 years	22.56	60
	65+ years	22.82	71
Location	Prague	12.01	75
	Bohemia	53.40	186
	Moravia	34.59	141

Planning and execution of the survey was performed by researchers and students at the University of Economics, Prague. A research team consisting of students majoring in marketing research was led by highly experienced university teachers (mostly associate professors). The "Runs" test and related statistical data-validation methods were applied to evaluate the data for potential mishandling [12]. This research is part of a long-term project focusing on consumer behavior with respect to the rapidly advancing technological environment and its impacts [10, 23].

Using the above described interviewing methods, we have collected multiple types of anonymized information from the respondents: data appertaining to socio-demographic classification, various differentiating lifestyle, work and leisure activities and preferences towards important aspects of personal mobility (mostly EV-related). The questionnaire is assembled using different question-types: binary (Yes/No), Likert scale (5-degree scale) and quantitative (interval based: age, earnings, private mileage per year, etc.).

3.1 Polychoric Correlation

To properly analyze and interpret our non-continuous observed variables (binary, Likert scale and interval based), we use polychoric correlation analysis as well as other selected common mathematical and statistical methods. Similar in principle to the standard (Pearson's) correlation, polychoric correlation describes the relationship between two ordered multinomial variables (this includes binary variables as a special case). Polychoric correlation ρ_{pc} is on a scale of $\langle -1, 1\rangle$, it has a "standard" interpretation and we need to bear in mind that correlation does not mean causation.

To calculate polychoric correlation between two multinomial variables (say, A and B), we assume the existence of some corresponding underlying unobserved continuous and normally distributed variables (say, X and Y). Technically, the variables and their relationships may be defined as

$$\begin{aligned} A &= a_i \quad \text{if} \quad \gamma_{i-1} \leq X < \gamma_i, \quad i = 1, 2, \ldots, r \text{ and} \\ B &= b_j \quad \text{if} \quad \tau_{j-1} \leq Y < \tau_j, \quad j = 1, 2, \ldots, s \end{aligned}$$

where γ_i is a threshold corresponding to X (unobserved variable X does not necessarily have an interpretable scale) and a_i is the ith category of A. For B and Y, the notation is analogous. All thresholds and values (categories) of the corresponding multinomial variables follow a strictly increasing order with $\gamma_1 < \gamma_2, \ldots, < \gamma_{r-1}$ and $a_1 < a_2 < \cdots < a_r$. Again, for τ_j and b_j the situation is analogous. Finally, we assume $\gamma_0 = \tau_0 = -\infty$, $\gamma_r = \tau_s = \infty$, and the means and variances for both X and Y variables can be set to $\mu = 0$ and $\sigma^2 = 1$ without any loss of generality. Next, we can use the product-moment correlation between X and Y that is given as: $\rho = \text{cov}(X, Y)/\sigma$ and we may cast the standard normal bi-variate density of X and Y as

$$\phi(x, y; \rho) = \frac{1}{2\pi(1-\rho^2)^{1/2}} \exp\left[\frac{-1}{2(1-\rho^2)}(x^2 - 2\rho xy + y^2)\right], \tag{1}$$

where x and y are observations of variables X and Y. Now, the probability that conditions $A = a_i$ and $B = b_j$ hold simultaneously can be formalized easily [22]. Corresponding likelihood function may be cast and the maximum likelihood estimation method can be used to estimate ρ_{pc} [11].

4 Empirical Results

We start our empirical analysis by focusing on the RQ1 – we evaluate the extent to which Czech consumers have reconciled the general necessity of transitioning personal mobility from fossil fuels to new energy sources. Figure 1 shows the aggregated results: with 51% prevalence, the general public sees HEVs as the predominant mean of transportation in the mid-term future (20 year from now).

While we view the survey results with some precaution, the strong "win" of hybrid cars is well in line with the perceived insufficient range of EVs (as seen in Fig. 3). In a typical setup, hybrid cars feature both the ICE and an electric motor, thus addressing both environmental concerns and the usual battery capacity limitations. Interestingly, all three remaining energy sources shown in Fig. 1 have essentially the same prevalence of roughly 16%. This is especially interesting with respect to the hydrogen fueled cars: ICEs are the most common category in use, EVs are becoming increasingly popular and their sales are rising rapidly – but hydrogen-based cars are still mostly experimental and there is essentially no commercial infrastructure for refueling (theoretically, a Toyota Mirai hydrogen fuel-cell car might be purchased in the Czech Republic).

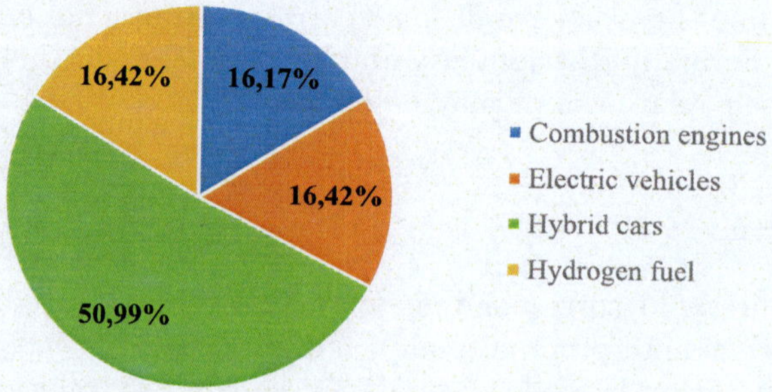

Fig. 1. Perceived main energy sources for personal mobility in 20-year time

To elaborate more thoroughly on the topics related to RQ1, we also focus on the relationship between individual stances (expectations on the predominant energy source for personal mobility shown in Fig. 1) and socio-demographic categorization of respondents. For polychoric correlation analysis, respondents are categorized by different variables: *Gender* (Male = 0/Female = 1), *Age* (age groups as in Table 1), *Mileage* – self-reported annual mileage for private travel and work commuting (0–5 thousand km, 5k–10k, 10k–20k, 20k–40k, 40k+), *Earning* – monthly household earnings (0–20 thousand CZK, 20k–40k, 40k–80k, 80k+; 20k CZK is roughly equivalent to 770 euro) and *Educ.* – achieved education levels (elementary, high school without graduation, high school, university degree). Categories shown in Fig. 1 (combustion vehicles, EVs, etc.) correspond to an unordered multinomial variable (instead of ordered multinomial) and therefore had to be recorded as four separate binary variables for the estimation of polychoric correlations.

All pairwise ρ_{pc} estimates are conveniently summarized in Fig. 2 (a stylized & annotated correlation matrix). Before discussing the actual results relevant for RQ1, we address a few technical aspects of Fig. 2. On the main diagonal, ones represent a perfect match between an observed variable and itself. The correlation matrix is symmetric along its main diagonal – hence $\rho_{pc}(A, B) = \rho_{pc}(B, A)$ for any two variables A and B. Blue colour represents positive polychoric correlations (co-movement of variables), red depicts inverse relationships and white is used for instances of statistically insignificant correlations (insignificant cells also features small-font illegible correlation coefficients).

Figure 2 shows some fundamental socio-demographic relationships, valid in the Czech Republic (as well as in many other countries): we see a strong positive relationship between education and earnings, an even stronger negative relationship between age and education (older individuals report lower education – access to higher education was quite limited before 1989, i.e. under the rule of the communist party). Significant negative correlations among perceived energy sources for personal mobility (see lower right part of Fig. 2) have no relevant interpretation. In fact, this is just a technicality introduced by recording our variables from Fig. 1. For their expectation into to 20-year future, respondents can only choose one of the four options available: EVs, Combustion (ICE vehlices), Hybrid (HEVs, PHEVs) or Hydrogen fuelled cars.

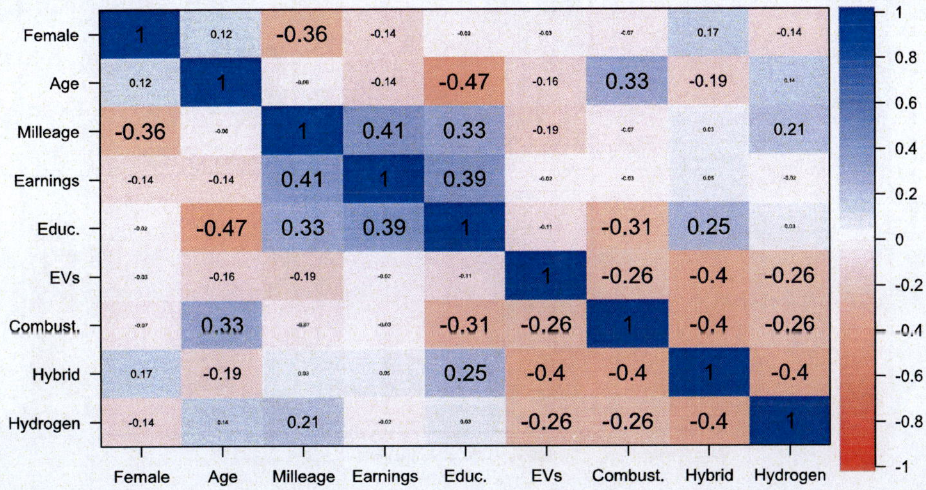

Fig. 2. Perceived energy sources for personal mobility in 20-year time – a socio-demographic analysis

From Fig. 2, we can see that women tend to report significantly less private mileage and work commuting as compared to men. Interestingly, there seems to be no gender-based difference with respect to our RQ1 (i.e. energy sources for private transportation). From the Age (age code) row, we can see that older people tend to pick the classical combustion engines as their favorite/expected power source in 20-year time. Perhaps they have already witnessed multiple bold statements along the lines of "And this is the future of XYZ" that later were disproved by reality. By evaluating private mileage, we can see a positive correlation with earnings and education, which seems reasonable. Interestingly, individuals with higher private mileage seem to champion hydrogen as the future fuel source. This may be due to the short refueling time (on the order of minutes – basically the same as for fossil fuels) – especially if compared to EVs, where recharging may take hours. Earnings seem to have no connection to the surveyed expectations. Similarly, no connection was found for domicile (*Location* as shown in Table 1) and this variable was dropped from Fig. 2 for simplicity. By looking at the education variable, we may see how individuals with higher education clearly turn away from fossil fuels (in a 20-year horizon forecast) and towards hybrids.

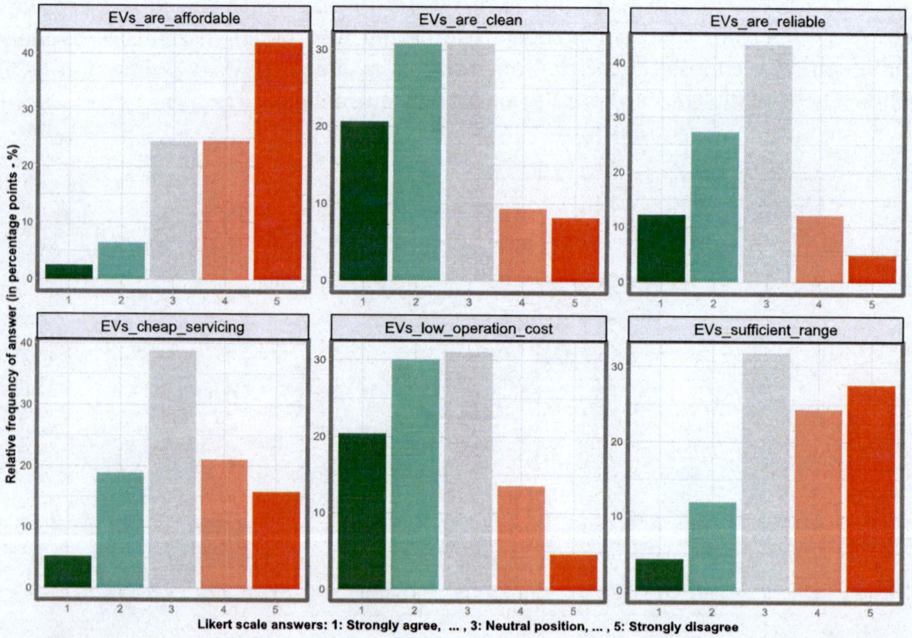

Fig. 3. Main perceived features of EVs: Respondents' stances toward statements (left to right, top first): EVs are affordable, EVs are clean, EVs are reliable, EVs are cheap to service and maintain, EVs have low operation costs, EVs have sufficient range

Next, we turn our attention to RQ2 and discuss selected main features of EVs as perceived by consumers (potential buyers). This analysis is concentrated around six key features of EVs, as summarized in Fig. 3. From the aggregated data shown in Fig. 3, we may observe that EVs are considered expensive in terms of their purchase price. The upper left chart of Fig. 3 relates to the surveyed statement "EVs are affordable" and approximately 66% individuals disagree: 42% strongly disagree and 24.6% mostly disagree with this statement.

Table 2. Polychoric correlations and 95% confidence intervals: significant socio-demographic factors influencing respondent stances for Fig. 3

	Lower empir.	Lower normal	Estimate	Upper normal	Upper empir.	p-value
Age – correlation with:						
EVS are affordable	0.03	0.04	0.14	0.25	0.24	0.01
EVs are clean	0.11	0.11	0.20	0.30	0.28	0.00
EVs are reliable	0.09	0.07	0.17	0.26	0.27	0.00
EVs cheap servicing	0.03	0.03	0.13	0.22	0.22	0.01
EVs low oper. costs	0.13	0.12	0.24	0.34	0.36	0.00
Education – correlation with:						
EVs are reliable	0.01	−0.01	0.10	0.21	0.22	0.07
Gender (Female = 1, Male = 0) – correlation with:						
EVs are clean	−0.34	−0.35	−0.23	−0.09	−0.10	0.00
EVs cheap servicing	0.05	0.05	0.18	0.29	0.27	0.00
Private mileage – correlation with:						
EVs are affordable	0.09	0.08	0.19	0.29	0.28	0.00
EVs sufficient range	0.10	0.09	0.19	0.29	0.29	0.00

Again, polychoric correlations may be used to investigate any relation between socio-demographic factors and the answers summarized in Fig. 3. To save space, the correlation matrix is omitted here as it contains large amounts of duplicated information and statistically insignificant results (this 11×11 matrix contains only a few significant and interpretable correlations). Instead, Table 2 shows selected ρ_{pc} coefficients among key socio-demographic characteristics and questions/statements surveyed. Besides the estimated correlation coefficients, p-values and 95% confidence intervals provide information on statistical significances of the coefficients shown. As shown in previous section, ρ_{pc} estimates are based on normality assumptions regarding the underlying latent variables. Because this assumption may be somewhat limiting, we follow the approach of Revelle [20] and provide both analytical (normality-based) and empirical (bootstrapped, i.e. non-parametric) 95% confidence intervals for all correlation estimates shown in Table 2 (this also applies to Tables 3 and 4).

Please note that the Likert scale answers in Fig. 3 are sorted in ascending order from "1: Strongly agree", to "5: Strongly disagree", which needs to be handled properly while interpreting our results. For the statement "EVs are affordable", we see a positive correlations with age and private mileage variables as shown in Table 2, with ρ_{pc} equal to 0.14 and 0.19 respectively. This means that older individuals and active drivers would be more likely to give higher Likert score answers, i.e. more likely to disagree with EVs being affordable.

As we focus on the second statement: "EVs are clean", there is a general agreement expressed by almost 52% of respondents, accompanied by a relatively strong neutral position – expressed by 30.8% individuals. However, only 17.6% of respondents would disagree with EVs being environmentally clean. A negative correlation $\rho_{pc} = -0.23$ can be observed (Table 2) between the gender variable Female and perceived cleanliness of cars: i.e. women are more likely to consider EVs as clean (they are less likely to disagree with the statement, they tend to record "lower" values at the Likert scale). In contrast, the likelihood of a negative stance increases with age. Other socio-demographic factors are not statistically significant for this EV-related aspect.

Both "EVs are reliable" and "EVs are cheap to service and maintain" statements get the most neutral answers (43.3% and 38.9% respectively) from Fig. 3, presumably due to lack of direct of derived experiences in this field (generally, first-hand experience is not necessary to form an opinion). Besides the neutral position, individuals are generally more inclined towards EVs' reliability than they believe in cheap servicing and maintenance. From the socio-demographic perspective, perceived reliability of EVs decreases with age and with higher levels of education achieved (to a somewhat less extent). From Table 2, we can see that both women and older individuals are less likely to perceive EVs maintenance and servicing as cheap, with ρ_{pc} at 0.18 and 0.13 respectively.

By evaluating consumer stances towards the statement "EVs have cheap operation costs" (which mostly addresses electricity & charging costs but there is some overlap with maintenance and other costs), we see that individuals would mostly agree with this statement, with only 18.4% respondents expressing some form of disagreement. From the socio-demographic perspective, there is not much differentiation related to this statement, with the exception of age ($\rho_{pc} = 0.24$) – older individuals tend to disagree more with the statement.

The so called "range anxiety" phenomenon related to EVs – i.e. the fear of not having sufficient energy (battery capacity) to reach one's destination – is well reflected in respondents' stances towards the "EVs have sufficient range" statement: approximately 52% of individuals disagree, over 30% have a neutral position and only 16.1% would agree. As we look at socio-demographic structuring of answers, the only statistically significant differentiation may be observed in the private mileage variable: the more individuals drive (privately), the more they tend to disagree with the statement – i.e. with increased mileage comes increased range anxiety.

To summarize Fig. 3, purchase price and limited range seem to be the most pronounced bottlenecks that would seem to hold back EV sales. Fortunately, most EV producers seem to be aware of the situation and battery improvements (in terms of reduced costs and/or increased capacity) are announced regularly. Similarly, improving infrastructure (increasing availability of charging stations accessible to the public) may also reduce the range anxiety burden by a significant amount.

Besides the socio-demographic aspects discussed above, there is a relatively high positive correlation between all pairs of answers as per Fig. 3: individuals tend

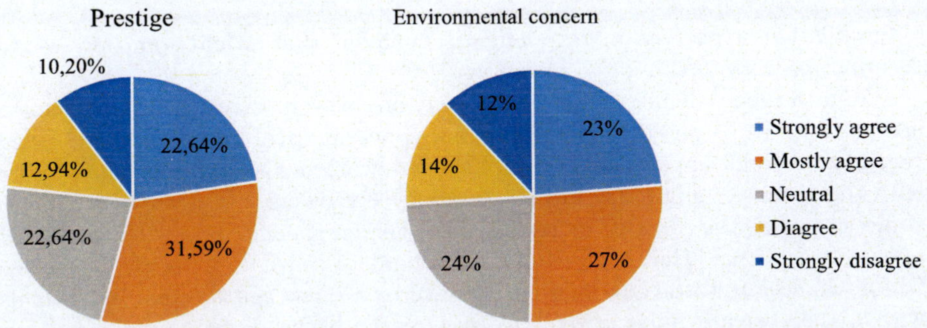

Fig. 4. Perception of EV ownership: respondents' stances towards two statements: EV ownership is a sign of prestige, EV ownership is a sign of environmental concern

to produce similar answers to all questions summarized in Fig. 3 (e.g. with all answer of an individual being either mostly positive, mostly negative or generally neutral).

Table 3. Polychoric correlations and 95% confidence intervals: significant socio-demographic factors influencing respondent stances for Fig. 4

	Lower empir.	Lower normal	Estimate	Upper normal	Upper empir.	*p*-value
Age – correlation with:						
Prestige	–0.23	–0.24	–0.14	–0.04	–0.03	0.01
Earnings – correlation with:						
Prestige	0.01	0.01	0.11	0.20	0.19	0.04
Environmental concern	0.03	0.02	0.14	0.25	0.25	0.03
Gender (Female = 1, Male = 0) – correlation with:						
Environmental concern	–0.38	–0.38	–0.25	–0.13	–0.14	0.00
Private mileage – correlation with:						
Environmental concern	0.14	0.11	0.22	0.32	0.32	0.00

To add complexity and comprehension depth to our analysis, we use RQ3 to evaluate how individuals see EV-ownership motives. Specifically, we have asked our respondent about two distinct (yet not completely disjoint – i.e. not mutually exclusive) buying motives: we focus on buyers'prestige-seeking motives and on their environmental concerns. As shown in Fig. 4, 54.23% agree (strongly or mostly agree) with the environmental motives of EV-owners and slightly less (50.5%) would perceive EV-ownership as motivated by prestige-seeking. The results summarized in the two pie-charts of Fig. 4 may seem very similar, yet polychoric correlation among the two questions is not statistically significant

– there is not a pattern in respondent's behavior that would somehow relate together answers to those two questions.

From the socio-demographic perspective on answers shown in Fig. 4, women are more likely to perceive environmental concerns involved in EV-purchasing (negative correlation in Table 3 implies that women disagree less/agree more with the notion of environmental concerns by EV owners). On the other hand, stances towards the prestige motives are not gender specific. From the age-based perspective, we see that older individuals tend to see EVs more as a prestige factor. Stances towards environmental concerns are not age specific. Individuals with higher private mileage tend to disagree with the environmental concern motive and there is no relation between private mileage and stances towards the prestige motive. Interestingly, higher earning individuals tend to disagree with both motives – yet the relation is not particularly strong at $\rho_{pc} = 0.14$ for environmental concerns and $\rho_{pc} = 0.11$ for prestige. Interestingly, individual stances shown in Fig. 4 are not differentiated with respect to education.

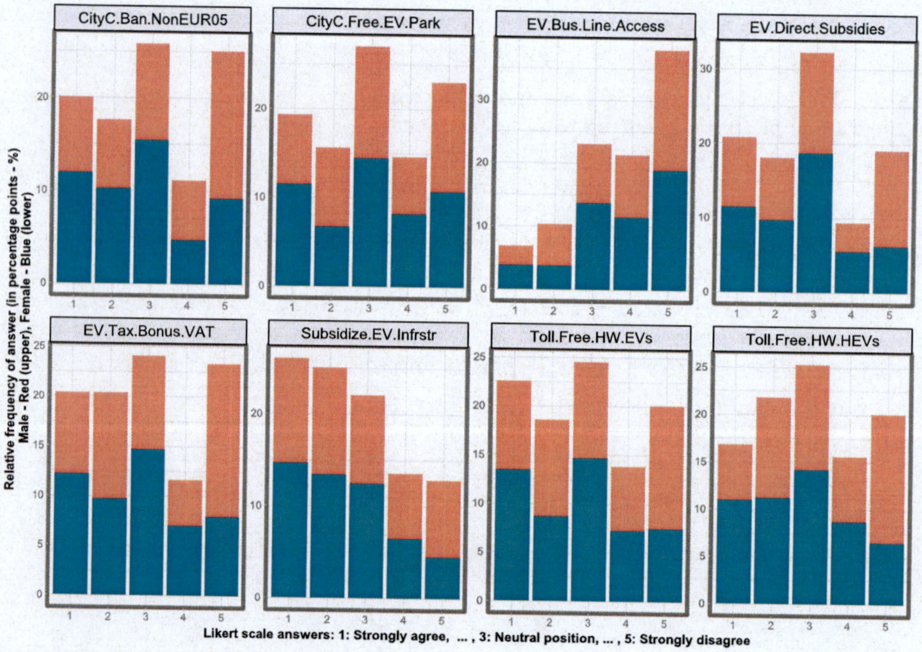

Fig. 5. Perception of possible policy actions by gender (male labelled red, left to right, top first): City-center ban for non-EUR05 cars, City-center free parking for EVS, EV access to bus-lines, EV-purchase direct subsidies, EV tax bonus (e.g. through VAT), Subsidies to EV infrastructure, Toll-free highway access for EVs, Toll-free highway access for EVs & HEVs

Next, we focus on the last main topic analyzed in this contribution: EV-related policy actions are discussed in RQ4. Here, we address and evaluate how respondents perceive different types of governmental interventions aimed at promoting EV mobility. We have selected 8 reasonably distinct areas of policies: seven such policies are aimed at promoting and subsidizing EVs and one policy approach represents restrictive measures banning older ICE-cars (on the grounds of non-compliance with the EUR 05 norm) from city centers. While not applied in Czechia as yet, such measures are already imposed in some cities throughout Germany.

As we inspect the individual elements of Fig. 5, we may see that public opinions towards different policies are not very pronounced: the ratios of individuals supporting and opposing any chosen policy-type seem to be well equalized – with the only exception of respondents mostly disagreeing with allowing EVs into reserved bus-lines in cities (top-row chart, center-right: less than 18% would agree). Polychoric correlations of answers (stances) to different statements show a remarkably strong dependency: all respondents would be very likely to present rather similar answers to all 8 aspects of governmental policies. The highest correlation was identified between toll-free highway access for EVs and toll-free highway access for both EVs and hybrid vehicles (the two bottom-right charts of Fig. 5) at $\rho_{pc} = 0.89$. This would be followed by stances towards direct purchase-subsidies and tax-based (e.g. VAT-based) subsidies for EVs (top-right and bottom-left charts respectively) at $\rho_{pc} = 0.72$. EV access to bus lines is the one aspect where answers are least correlated with other policies surveyed.

Table 4. Polychoric correlations and 95% confidence intervals: significant socio-demographic factors influencing respondent stances for Fig. 5

	Lower empir.	Lower normal	Estimate	Upper normal	Upper empir.	p-value
Age – correlation with:						
Toll free HW for EVs	0.09	0.09	0.18	0.27	0.27	0.00
Toll free HW for EVs & HEVs	0.11	0.11	0.19	0.28	0.28	0.00
City center ban non-EUR05	0.05	0.03	0.14	0.24	0.23	0.01
Subsidize EV infrastructure	0.21	0.19	0.28	0.38	0.38	0.00
City center free EV parking	0.12	0.11	0.19	0.27	0.27	0.00
VAT-based tax bonus for EVs	0.09	0.08	0.17	0.26	0.25	0.00
Direct tax subsidies for EVs	0.10	0.09	0.19	0.28	0.29	0.00
Education – correlation with:						
City center ban non-EUR05	–0.24	–0.24	–0.15	–0.05	–0.07	0.01
Subsidize EV infrastructure	–0.29	–0.30	–0.20	–0.09	–0.09	0.00
City center free EV parking	–0.20	–0.19	–0.09	0.00	–0.02	0.04
VAT-based tax bonus for EVs	–0.23	–0.22	–0.14	–0.03	–0.01	0.01
Private mileage – correlation with:						
Direct tax subsidies for EVs	0.01	0.02	0.13	0.23	0.21	0.02

Correlations between statements shown in Fig. 5 and socio-demographic factors are summarized in Table 4. From this perspective, we may see that female respondents tend to agree more with different types of policies – with the exception of "Free access to bus-lines for EVs", where both genders reject this type of policy at equal rates. Gender-based differences are shown in Fig. 5, other socio-demographic aspects drawn from polychoric correlation analysis are shown in Table 4. With increasing age, individuals tend to disagree more with the policies evaluated (again, with the exception of access to bus-lines, where rejection rates do not depend on age). The most prominent age-based propensity to rejection was observed for "Subsidies to EV infrastructure (i.e. charging stations)" at $\rho_{pc} = 0.28$ (bottom-line, center-left plot in Fig. 5): with increasing age category, individuals are more likely to choose "higher" Likert scale answers, i.e. are more likely to disagree.

Individual stances to government policies are not differentiated with respect to private mileage nor earnings. There is one exception from this general finding: higher private mileage is associated with moderately higher rates of rejection of subsidizing EV infrastructure ($\rho_{pc} = 0.18$). Finally, higher education levels are generally linked to higher acceptance of governmental policies, with two exceptions: education is unrelated to EVs' bus-line access and to EVs' free parking in city centers. Nevertheless, it should be reiterated that despite some socio-demographic differences, there is no predominantly positive or negative stance towards any of the policies discussed (except for the free access to bus-lines, which is generally rejected).

5 Conclusions

In Czechia and in many other countries, automotive industry is a key sector of the economy. It is also a sector whose products directly affect the environment and air quality in cities. This article maps the attitudes of Czech consumers towards different aspects of electromobility – as a preliminary step for predicting their consumer behavior in the future. It is essential for manufacturers in the automotive sector and for the legislature to know the attitude of consumers and be prepared for the market potential in the sale of fully electric vehicles and hybrid cars, while at the same time competently estimating the extent to which government incentives will be needed.

Based on the research performed, we may state that ownership of an electric car itself is taken rather positively by consumers, in particular, consumers declare that these cars are more environmentally friendly. But even so, only 17% of consumers see it as the predominant way of transportation in 20-year time. Consumers see the future more in the combination of an electric motor with a conventional combustion engine – that is, in HEV vehicles. Consumers do not support tax breaks or other benefits for owners of electric vehicles to a significant extent, with the important exception of government support towards EV-charging infrastructure.

Acknowledgement. Institutional research support provided by Faculty of Informatics and Statistics, University of Economics, Prague.

References

1. Adnan, N., Nordin, S.M., Rahman, I., Amini, M.H.: A market modeling review study on predicting Malaysian consumer behavior towards widespread adoption of PHEV/EV. Environ. Sci. Pollut. Res. **24**(22), 17955–17975 (2017). https://doi.org/10.1007/s11356-017-9153-8
2. Ajanovic, A.: The future of electric vehicles: prospects and impediments. Interdiscipl. Rev. Energy Environ. **4**(6), 521–536 (2015). https://doi.org/10.1002/wene.160
3. Aruna, D., Mukherjee, K.: Payment methods and their effect on durable goods replacement. J. Consum. Market. **36**(4), 484–493 (2019). https://doi.org/10.1108/JCM-11-2017-2435
4. Chadha, S., Kansal, P., Goel, S.: The effect of perception of complaint management system on purchase intention: the mediating role of cognitive dissonance. Vis. J. Bus. Perspect. **22**(4), 347–355 (2018). https://doi.org/10.1177/0972262918803473
5. Cox, B., Mutel, C., Bauer, C., Mendoza, A., van Vuuren, D.: Uncertain environmental footprint of current and future battery electric vehicles. Environ. Sci. Technol. **52**(8), 49–89 (2018). https://doi.org/10.1021/acs.est.8b00261
6. European Automobile Manufacturers Association: Facts about the Automobile Industry. https://www.acea.be/automobile-industry/facts-about-the-industry. Accessed 10 Dec 2020
7. European Commission: 2030 climate & energy framework (2018). https://ec.europa.eu/clima/policies/strategies/2030_en. Accessed 10 Dec 2020
8. European Parliament: Emission performance standards for new passenger cars and for new light commercial vehicles (2019). https://www.europarl.europa.eu/doceo/document/TA-8-2019-03-27_EN.html#sdocta10. Accessed 10 Dec 2020
9. Fang, S., Chang, E., Ou, C., Chou, C.: Internal market orientation, market capabilities and learning orientation. Eur. J. Market. **48**(1/2), 170–192 (2014). https://doi.org/10.1108/EJM-06-2010-0353
10. Formánek, T., Tahal, R.: Socio-demographic aspects affecting individual stances towards electric and hybrid vehicles in the Czech Republic. Cent. Eur. Bus. Rev. **9**(2), 78–93 (2020). https://doi.org/10.18267/j.cebr.238
11. Fox, J.: polycor: Polychoric and Polyserial Correlations. R package version 0.7-9. https://CRAN.R-project.org/package=polycor. Accessed 10 Dec 2020
12. Gibbons, J., Chakraborti, S.: Nonparametric Statistical Inference, 4th edn. CRC Press, Berlin (2003)
13. Juster, F.: Prediction and consumer buying intentions. Am. Econ. Rev. **50**(2), 604–617 (1960)
14. Liu, Q., Zhang, X., Huang, S., Zhang, L., Zhao, Y.: Exploring consumers' buying behavior in a large online promotion activity: the role of psychological distance and involvement. J. Theoret. Appl. Electr. Comm. Res. **15**(1), 66–80 (2020). https://doi.org/10.4067/S0718-18762020000100106
15. Maroukian, K., Apostolopoulos, C., Tsaramirsis, G.: Extending model driven engineering aspects to business engineering domain: a model driven business engineering approach. Int. J. Inf. Technol. **9**(1), 49–57 (2017). https://doi.org/10.1007/s41870-017-0009-8

16. Martenson, R.: When is green a purchase motive? Different answers from different selves. Int. J. Retail Distrib. Manag. **46**(1), 21–33 (2018). https://doi.org/10.1108/IJRDM-11-2016-0228
17. Morrison, D.: Purchase intentions and purchase behavior. J. Market. **43**(2), 65–74 (1979)
18. Nisel, R.: Analysis of consumer characteristics which influence the determinants of buying decisions by the logistic regression model. Logist. Inf. Manag. **14**(3), 223–228 (2001). https://doi.org/10.1108/09576050110390257
19. Radoias, V.: Direct and indirect price discrimination in the automotive industries of the European Union. Empir. Econ. **50**(3), 975–990 (2015). https://doi.org/10.1007/s00181-015-0950-y
20. Revelle, W.: psych: procedures for personality and psychological research. https://cran.r-project.org/web/packages/psych/index.html. Accessed 10 Dec 2020
21. Stoean, L.: Trends and top priorities in automotive industry. Qual. Access Success **20**(S2), 624–629 (2019)
22. Tahal, R., Formánek, T.: Reflection of GDPR by the Czech Population. Manag. Market. Challenges Knowl. Soc. **15**(1), 78–94 (2020). https://doi.org/10.2478/mmcks-2020-0005
23. Tahal, R., Formánek, T., Mohelská, H.: Loyalty programs and personal data sharing preferences in the Czech Republic. Econ. Manag. **20**(1), 187–199 (2017). https://doi.org/10.15240/tul/001/2017-1-013
24. Takumi, K.: Loyalty management in durable consumer goods: trends in the influence of recommendation intention on repurchase intention by time after purchase. J. Market. Anal. **7**(2), 76–83 (2019). https://doi.org/10.1057/s41270-019-00050-x
25. Yong, N., Ariffin, S., Ne, G., Wahid, N.: A study of factors influencing consumer's purchase intention toward green vehicles: evidence from Malaysia. Glob. Bus. Manag. Res. **9**(4), 281–297 (2017)

Method of Nonlinear Digital Control Systems Simulation

Eugene Larkin[1], Aleksandr Privalov[2](✉), and Tatiana Akimenko[1]

[1] Tula State University, 300012 Tula, Russia
[2] Tula State Lev Tolstoy Pedagogical University, 300026 Tula, Russia

Abstract. Complex multi-loop objects under digital control are considered. It is shown, that models of such systems have bottlenecks of three types: description of technological unit itself, description of actuator with substantial nonlinearities and model of step-by-step data processing by Von-Neumann type controller, which unfolds in real physical time. Nonlinearities of the first type may be reduced by means of Taylor series expansion at the point of the steady state regime of operation. For simulation of substantial nonlinearities actuator should be described with differential equation, performed in the Cauchy form. When simulation of data processing such properties of controller, as time delays and discrete form of signals processed representation should be taken into account. According to proposed division of system model, the method of the system performance estimation is worked out, due to which part of the system from outputs of actuators till their inputs is considered as linear subsystem, in which Von-Neumann type controller delays are taken into account, while actuators are considered as only units, included the nonlinearities. Theoretical postulates are confirmed by simulation of two-loop digital control system functioning.

Keywords: Object under control · Digital data processing · Time delays · Nonlinearity · Actuator · Cauchy form

1 Introduction

The usage of modern digital technologies in industry supposes wide application of Von Neumann type controllers for managing of continuous processes [1–3]. The feature of controllers of this type is a sequential (operator-by-operator) interpretation of control algorithms [4–6]. As a result, in addition to processing the data, received from the object under control through sensor subsystem, the computer also introduces in managing process time delays, that significantly change the qualitatively characteristics of control, such as an overshooting and a time of reaching the steady state after a step action. To estimate time intervals of delays between transactions, generated by control algorithm, the theory of Markov or semi-Markov processes is used [7–10], however, in the practice of real multi-loop control systems working out, these models are used extremely insufficiently.

R. Silhavy et al. (Eds.): CoMeSySo 2021, LNNS 231, pp. 73–83, 2021.
https://doi.org/10.1007/978-3-030-90321-3_8

Another factor that has a significant impact on the dynamics of the system's behavior is the presence of substantial nonlinearities in actuators that drive the valves of technological units, such as dry friction and limitation on the valve position [11]. To take into account such factors, the classical theory of linear automatic control systems is not enough, however, to describe the linear part of the system, or the part of the system that can be reduced to a linear form, this theory can be used. Thus, the methods of analyzing digital control systems, including the assessment of time delays and significant nonlinearities, are not used sufficiently in engineering practice, which explains the necessity and relevance of the research carried out.

2 Model of Object Under Control

Common structure of the system under investigation is shown on the Fig. 1, and is rather generic [2, 3, 6]. It includes object under control and actuators, which through the interface are linked with digital controller.

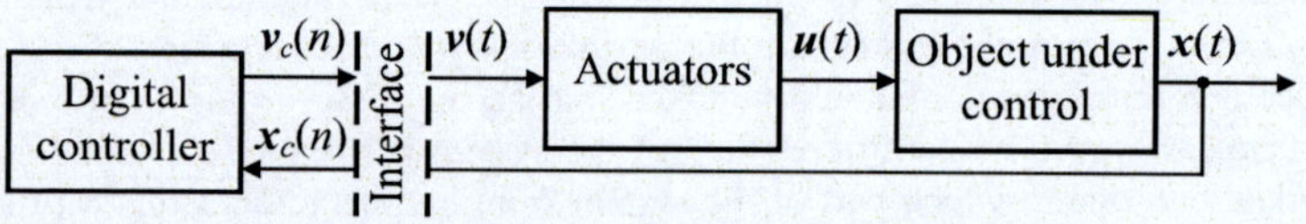

Fig. 1. Flowchart of digital control system

Digital controller processes data vector $\boldsymbol{x}_c(n) = \left[x_{c,1}(n),\ ...,\ x_{c,k}(n),\ ...,\ x_{c,K}(n)\right]^\theta$ and generates action data vector $\boldsymbol{v}_c(n) = \left[v_{c,1}(n),\ ...,\ v_{c,k}(n),\ ...,\ v_{c,K}(n)\right]^\theta$, where θ is the transpose operation; n is the discrete analogue of time t. Vector $\boldsymbol{x}_c(n)$ is formed from the analogue signals state vector $\boldsymbol{x}(t) = [x_1(t),\ ...,\ x_k(t),\ ...,\ x_K(t)]^\theta$, and sequentially element-by element is inputted into controller. Vector $\boldsymbol{v}_c(n)$ is calculated by controller software, element-by-element is transmitted through the interface, transformed into an analogue form $\boldsymbol{v}(t) = [v_1(t),\ ...,\ v_k(t),\ ...,\ v_K(t)]^\theta$ and distributed between actuators. Actuating engines affect at the object under control by means of vector $\boldsymbol{u}(t) = [u_1(t),\ ...,\ u_k(t),\ ...,\ u_K(t)]^\theta$, which is generated at the outputs of engines, and as the result of physical processes in the object under control new state vector $\boldsymbol{x}(t)$ is generated.

In most common case dynamics of physical processes in the object under control is performed with following set of the ordinary differential equations:

$$\begin{gathered}\vartheta_l\left(x_1, \frac{dx_1}{dt}, ..., \frac{d^{m(x,1)}x_1}{dt^{m(x,1)}}, ..., \frac{d^{M(x,1)}x_1}{dt^{M(x,1)}}, ..., x_k, \frac{dx_k}{dt}, ..., \frac{d^{m(x,k)}x_k}{dt^{m(x,k)}}, ..., \frac{d^{M(x,k)}x_k}{dt^{M(x,k)}},\right.\\ \left. ..., x_K, \frac{dx_K}{dt}, ..., \frac{d^{m(x,K)}x_K}{dt^{m(x,K)}}, ..., \frac{d^{M(x,K)}x_K}{dt^{M(x,K)}}, u_l, \frac{du_l}{dt}, ..., \frac{d^{m(u,l)}u_l}{dt^{m(u,l)}}, ..., \frac{d^{M(u,l)}u_l}{dt^{M(u,l)}}\right) = 0\\ 1 \le l \le K,\end{gathered} \tag{1}$$

where $\vartheta_l(...)$ is scalar continual function, differentiable as minimum twice an al parameters included; $x_k(t) \in \boldsymbol{x}(t)$, $u_l(t) = \boldsymbol{u}(t)$; $m(x,k)$, $m(x,k)$ are index functions, which mean m-th degree of state variable x_k derivative; $m(u,l)$ are index functions, which mean m-th degree of actuation variable u_l; $M(x,k)$ $M(u,l)$ are the maximal degree of proper variable derivatives; in common $M(x,k) \neq M(u,l)$.

Taylor series expansion of (1) at the point, defining by non-linear algebraic equation $\vartheta_l(x_1, \ldots, x_k, \ldots, .x_K, u_l) = 0$, $1 \le l \le K$, gives the following linear differential equation system:

$$\boldsymbol{a} \cdot \boldsymbol{x}(t) = \boldsymbol{b} \cdot \boldsymbol{u}(t), \tag{2}$$

where $\boldsymbol{a} = \left[a_{m(x,k,l)}\right]$, $\boldsymbol{b} = \left[b_{m(u,k,l)}\right]$ are $K \times K$ matrices of coefficients; $a_{m(x,k,l)} = \sum_{m(x,k,l)=0}^{M(x,k,l)} \tilde{a}_{m(x,k,l)} \frac{d^{m(x,k,l)}}{dt^{m(x,k,l)}}$; $b_{m(u,k,l)} = \sum_{m(u,k,l)=0}^{M(u,k,l)} \tilde{b}_{m(u,k,l)} \frac{d^{m(u,k,l)}}{dt^{m(u,k,l)}}$. $\tilde{a}_{m(x,k,l)}$ and $\tilde{b}_{m(u,k,l)}$ are constants; $1 \le k, l \le K$.

Direct Laplace transform [12–14] of (2), when $\frac{d^{m(x,k,l)}u_k}{dt^{m(x,k,l)}} = 0$, $\frac{d^{m(u,k,l)}u_k}{dt^{m(u,k,l)}} = 0$, $1 \le k, l \le K, 0 \le m(x,k,l) \le M(x,k,l) - 1, 1 \le k, l \le K, 0 \le m(u,k,l) \le M(u,k,l) - 1$, gives the following system of linear algebraic equations:

$$\boldsymbol{A}(s) \cdot \boldsymbol{X}(s) = \boldsymbol{B}(s) \cdot \boldsymbol{U}(s), \tag{3}$$

where $\boldsymbol{X}(s) = L[\boldsymbol{x}(t)]$; $\boldsymbol{U}(s) = L[\boldsymbol{u}(t)]$; $\boldsymbol{A}(s) = L[\boldsymbol{a}]$; $\boldsymbol{B}(s) = L[\boldsymbol{b}]$; s is the Laplace differentiation operator;

$$\boldsymbol{A}(s) = \left[A_{k,l}(s)\right] = \left[\sum_{m(x,k,l)=0}^{M(x,k,l)} \tilde{a}_{m(x,k,l)} s^{m(x,k,l)}\right];$$

$$\boldsymbol{B}(s) = \left[B_{k,l}(s)\right] = \left[\sum_{m(u,k,l)=0}^{M(u,k,l)} \tilde{b}_{m(u,k,l)} s^{m(u,k,l)}\right].$$

Solution of (3) with respect to $X_k(s)$, $1 \le k \le K$, gives the following expressions:

$$X_k(s) = \sum_{l=1}^{K} U_l(s) \cdot W_{k,l}(s), \quad 1 \le k \le K \tag{4}$$

$$W_{k,l}(s) = \frac{\sum_{i=1}^{K} (-1)^{i+l} \cdot \left|\boldsymbol{A}_{i,l}(s)\right| \cdot \sum_{m(u,i,l)=0}^{M(u,i,l)} \tilde{b}_{m(u,i,l)} s^{m(u,i,l)}}{|\boldsymbol{A}(s)|}. \tag{5}$$

where $\left|\boldsymbol{A}_{i,l}(s)\right|$ is the i-th minor of determinant $|\boldsymbol{A}(s)|$, when expansion it in term of l-th column.

3 Model of Data Processing

In common case digital controller processes data vectors $\boldsymbol{x}_c(n)$ and $\boldsymbol{v}_c(n)$ as follows:

$$\begin{gathered}\vartheta_{c,l}\big(v_{c,1}[n],\ ...,\ v_{c,1}[n+m(v,1)],\ ...,\ v_{c,1}[n+M(v,1)],\ ...,\ v_{c,k}[n],\ ...,\ v_{c,k}[n+m(v,k)],\ ...,\\ v_{c,k}[n+M(v,k)],\ ...,\ v_{c,K}[n],\ ...,\ v_{c,K}[n+m(v,K)],\ ...,\ v_{c,K}[n+M(v,K)],\ f_1[n],\ ...,\\ f_1\big[n+m(f,1)\big],\ ...,\ f_1\big[n+M(f,1)\big],\ ...,\ f_l[n],\ ...,\ f_l\big[n+m(f,l)\big],\ ...,\ f_l\big[n+M(f,l)\big],\\ ...,\ f_K[n],\ ...,\ f_K\big[n+m(f,K)\big],\ ...,\ f_K\big[n+M(f,K)\big],\ x_{c,1}[n],\ ...,\ x_{c,1}[n+m(x,c,1)],\ ...,\\ x_{c,1}[n+M(x,c,1)],\ ...,\ x_{c,l}[n],\ ...,\ x_{c,l}[n+m(x,c,l)],\ ...,x_{c,l}\big[n+M(x,\tilde{n},l)\big],\ ...,\\ x_{c,K}[n],\ ...,\ x_{c,K}[n+m(x,c,K)],\ ...,x_{c,K}[n+M(x,c,K)]\big),\quad 1\le l\le K,\end{gathered} \tag{6}$$

where $\boldsymbol{f}(n) = \big[f_1(n),\ ...,\ f_l(n),\ ...,\ f_K(n)\big]^\theta$ is the desired aim vector, $v_{c,1}[...]$, ..., $v_{c,k}[...]$, ..., $v_{c,K}[...]$, $f_1[...]$, ..., $f_l[...]$, ..., $f_K[...]$, $x_{c,1}[...]$, ..., $x_{c,l}[...]$,..., $x_{c,K}[...]$ are samples of vectors $\boldsymbol{v}_c(n)$, $\boldsymbol{f}(n)$, $\boldsymbol{x}_c(n)$, respectively; $\boldsymbol{f}(n) = \big[f_1(n),\ ...,\ f_l(n),\ ...,\ f_K(n)\big]^\theta$; $M(\xi,\zeta)$ is the maximal degree of ξ-th signal's ζ-th element, finite difference, $\xi \in \{v_c, f, x_c\}$, $1 \le \zeta \le K$.

If in controller only linear operations are realized (f.e., as in digital PID-controller [14]), then (6) looks as follows:

$$\boldsymbol{a}_c\boldsymbol{v}_c(n) = \boldsymbol{b}_c\big[\boldsymbol{f}(n) - \boldsymbol{x}_{\tilde{n}}(n)\big], \tag{7}$$

where $\boldsymbol{a}_{\tilde{n}} = \big[a_{m(v,k,l)}\big]$, $\boldsymbol{b}_c = \big[b_{m(f,x,k,l)}\big]$ are $K \times K$ matrices of coefficients; $a_{m(v,k,l)} = \sum_{m(v,k,l)=0}^{M(v,k,l)} \tilde{a}_{m(v,k,l)}\Delta^{m(v,k,l)}$; $b_{m(f,x,k,l)} = \sum_{m(f,x,k,l)=0}^{M(f,x,k,l)} \tilde{b}_{m(f,x,k,l)}\Delta^{m(f,x,k,l)}$; $1 \le k, l \le K$; $\Delta^{m(v,k,l)}$, $\Delta^{m(f,x,k,l)}$ are signs of discrete limited differences of orders $m(v,k,l)$, $m(f,x,k,l)$, respectively; $\tilde{a}_{m(v,k,l)}$, $\tilde{b}_{m(f,x,k,l)}$ are constants;

$$\Delta^{m(...)}\xi(n) = \sum_{i=0}^{m(...)}(-1)^i \cdot C^i_{m(...)}\xi[n+m(...)-i];\ C^i_{m(...)} = \frac{m(...)\,!}{i\,!\cdot[m(...)-i]\,!}$$

Let samples of discrete vector functions $\boldsymbol{v}_c(n), \boldsymbol{f}(n)\boldsymbol{x}_{\tilde{n}}(n)$ are read out at a time interval τ, i.e. $\boldsymbol{v}_c(n) = \sum_{n=0}^{\infty}\boldsymbol{v}_c(t)\delta(t-n\tau)$, $f(n) = \sum_{n=0}^{\infty}\boldsymbol{f}(t)\delta(t-n\tau)$, $\boldsymbol{x}_c(n) = \sum_{n=0}^{\infty}\boldsymbol{x}_c(t)\times$ $\times\delta(t-n\tau)$ [16]. Then, when $\tau \to 0$, $\boldsymbol{v}_c(n) \to \boldsymbol{v}_c(t)$, $f(n) \to \boldsymbol{f}(t)$, $\boldsymbol{x}_c(n) \to \boldsymbol{x}_c(t)$, and direct Laplace transform may be applied to (7) at initial conditions $\frac{d^{m(v,k,l)}v_k}{dt^{m(v,k,l)}} = 0$, $\frac{d^{m(f,x,k,l)}f_k}{dt^{m(f,x,k,l)}} = 0$, $0 \le m(v,k,l) \le M(v,k,l) - 1$, $0 \le m(f,x,k,l) \le M(f,x,k,l) - 1$, $1 \le k, l \le K$, gives the following system of linear algebraic equations:

$$\boldsymbol{A}_c(s)\cdot\boldsymbol{V}(s) = \boldsymbol{B}_c(s)\cdot[\boldsymbol{F}(s) - \boldsymbol{X}_c(s)], \tag{8}$$

where.

$$\boldsymbol{A}_c(s) = \big[A_{c,k,l}(s)\big] = \left[\sum_{m(v,k,l)=0}^{M(v,k,l)}\tilde{a}_{m(v,k,l)}s^{m(v,k,l)}\right];$$

$$\boldsymbol{B}_c(s) = \left[B_{f,x,k,l}(s)\right] = \left[\sum_{m(f,x,k,l)=0}^{M(f,x,k,l)} b_{m(f,x,k,l)} s^{m(f,x,k,l)}\right];$$

Solution of (8) with respect to $V_k(s),\ 1 \le k \le K$, gives the following expressions:

$$V_k(s) = \sum_{l=1}^{K} \left[F_l(s) - X_{c,l}(s)\right] \cdot W_{c,k,l}(s),\ 1 \le k \le K; \tag{9}$$

$$W_{c,k,l}(s) = \frac{\sum_{i=1}^{K} (-1)^{i+l} \cdot \left|\boldsymbol{A}_{c,i,l}(s)\right| \cdot \sum_{m(f,x,i,l)=0}^{M(f,x,i,l)} \tilde{b}_{m(f,x,i,l)} s^{m(f,x,i,l)}}{|\boldsymbol{A}_c(s)|}. \tag{10}$$

where $\left|\boldsymbol{A}_{c,i,l}(s)\right|$ is the i-th minor of determinant $|\boldsymbol{A}_c(s)|$, when expansion it in term of l-th column.

Both (4) and (9) may be expressed in the matrix form as follows;

$$\boldsymbol{X}(s) = \boldsymbol{W}(s) \cdot \boldsymbol{U}(s); \tag{11}$$

$$V_c(s) = \boldsymbol{W}_c(s) \cdot [\boldsymbol{F}(s) - \boldsymbol{X}_c(s)], \tag{12}$$

where $\boldsymbol{W}(s)$ and $\boldsymbol{W}_c(s)$ are $K \times K$ matrices of transfer functions.

Calculation of action according (7) in Von-Neumann type digital controller is executed according algorithm, which is interpreted operator-by-operator in real physical time. Main destination of the control algorithm is reading data $\mathbf{x}_c(n)$, calculation action $\boldsymbol{v}_c(n)$ and transmitting it to actuator. Algorithm is the cyclic one, but there is no a looping effect in it. In common case algorithm may generate transactions in arbitrary sequence and may have the structure of arbitrary complexity (Fig. 2a, where only transactions operators are shown).

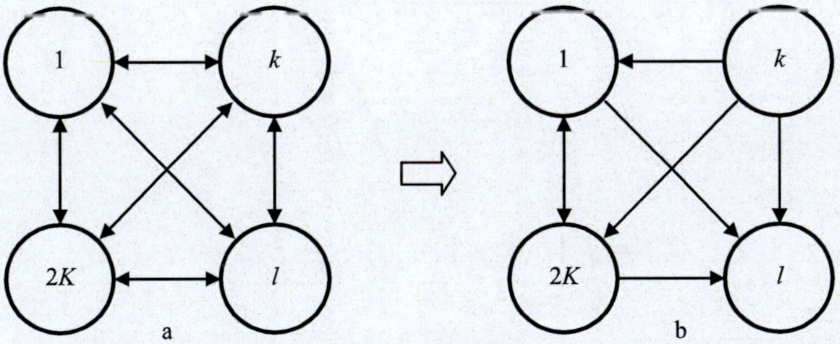

Fig. 2. Common structure of algorithm (a), and the model for time interval estimation (b)

The adequate approach to algorithm simulation is semi-Markov process [7–10], whose states are abstract analogues of transaction operators. Semi-Markov process is represented by the semi-Markov matrix.

$$\boldsymbol{h}(t) = \left[h_{i,j}(t)\right], \tag{13}$$

where $h_{i,j}(t) = p_{i,j} \cdot g_{i,j}(t)$ is the weighted density of execution i-th operator, when after its execution will be executed j-th operator; $p_{i,j}$ is probability of switching to j-th operator after completion execution of the i-th operator; $g_{i,j}(t)$ is pure time density of execution of the i-th operator, when the next operator for execution will be the j-th; $1 \le i, j \le 2K$.

On probabilities $p_{i,j}$ and pure densities $g_{i,j}(t)$ of semi-Markov process following restrictions are imposed:

$$0 < T_{i,j,\min} \le g_{i,j}(t) \le T_{i,j\max} < \infty; \quad \sum_{j=1}^{2K} p_{i,j} = 1; \tag{14}$$

where 2K is the common quantity of transaction operators; $T_{i,j,\min}$ and $T_{kl\max}$ are top and lower bounds of density's $g_{i,j}(t)$ domain.

Due to properties (14) and properties of control algorithm, semi-Markov process (13) is the ergodic one. For estimation of time intervals of wandering through ergodic semi-Markov process from the i-th state till the j-th state it should be transformed to semi-Markov process $\boldsymbol{h}'(t)$, as it is shown on the Fig. 2b. After transforming the i-th state ceases to be ergodic, i-th state became the starting one, and j-th state became the absorbing one. For getting such structure i-th column and j-th row of $\boldsymbol{h}(t)$ should be reset to zeros. The weighted time density of wandering semi-Markov process $\boldsymbol{h}'(t)$ from the i-th first state till the j-th state may be estimated as follows

$$h_{i,j}^{*}(t) = \boldsymbol{I}_i^r L^{-1}\left[\sum_{n=1}^{\infty} \{L[\boldsymbol{h}'(t)]\}^n\right] \boldsymbol{I}_j^c \tag{15}$$

where $\boldsymbol{I}_i^r$ is the row-vector, the i-th element of which is equal to one, and other elements are equal to zeros; $\boldsymbol{I}_j^c$ is the column-vector, j-th element of which is equal to one, and other elements are equal to zeros.

For weighted time density (15) the expectation and the dispersion may be calculated as usual [17]:

$$T_{i,j}^{*} = \int_0^\infty h_{i,j}^{*}(t) \cdot t dt \Bigg/ \int_0^\infty h_{i,j}^{*}(t) dt \tag{16}$$

$$D_{i,j}^{*} = \int_0^\infty \left(t - T_{i,j}^{*}\right)^2 h_{i,j}^{*}(t) dt \Bigg/ \int_0^\infty h_{i,j}^{*}(t) dt \tag{17}$$

Time delays cause deterioration of system performance, hence at the stage of system design one would utilize worst estimations, f.e. with using "three sigma rule" [18], as follows:

$$\tau_{i,j} = T_{i,j}^{*} + 3\sqrt{D_{i,j}^{*}}. \tag{18}$$

For utilizing time intervals achieved, when solving of design problem, one should have a starting point for their counting. Due to the fact, that components of vector

$\mathbf{v}_c(n)$ are calculated on estimation of current object under control state $\boldsymbol{x}_{\tilde{n}}(n)$, below all time intervals are counted from moment of input the first element of $\boldsymbol{x}_{\tilde{n}}(n)$. Moreover, computation of $\boldsymbol{v}_c(n)$ cannot be completed until last element of vector $\boldsymbol{x}_{\tilde{n}}(n)$ will not be inputted into digital controller. So indices of semi-Markov matrix (12) are distributed as follows:

indices $1 \leq i, j \leq K$ belong to abstract analogues of input K elements of vector $\boldsymbol{x}_{\tilde{n}}(n)$ operators of control algorithm;

indices $K+1 \leq i, j \leq 2K$ belong to abstract analogues of output K elements of vector $\mathbf{v}_c(n)$ operators of control algorithm.

Time intervals $\tau_{i,j}$, when $1 \leq i, j \leq K$, describe data skew when inputting feedback signal. Time intervals $\tau_{i,j}$, when $K+1 \leq i, j \leq 2K$, describe data skew when outputting action signal. Time intervals $\tau_{i,j}$, when $1 \leq i \leq K$, $K+1 \leq j \leq 2K$ describe pure lag in feedback contour.

In accordance with the theorem about shifting in the time domain [12, 13] $L[\phi(t-\xi)] = \exp(-\xi s)\Phi(s)$, $\xi > 0$, where ξ is time delay; $\phi(t)$ is a function; $\Phi(s) = L[\phi(t)]$, time delays, affected on a control system performance, may be organized as $K \times K$ delay-matrix:

$$\boldsymbol{D}(s) = \begin{pmatrix} 1 \ldots & \exp(-\tau_{1l}s) & \ldots \exp(-\tau_{1K}s) \\ & \ldots & \\ \exp(-\tau_{k1}s) \ldots & \exp(-\tau_{kl}s) & \ldots \exp(-\tau_{kK}s) \\ & \ldots & \\ \exp(-\tau_{K1}s) \ldots & \exp(-\tau_{Kk}s) & \ldots \exp(-\tau_{KK}s) \end{pmatrix}; \tag{19}$$

where $1 \leq (k = i) \leq K$ are indices, which nominate elements of vector $\boldsymbol{x}_{\tilde{n}}(n)$; $1 \leq (l = j - K) \leq K$ are indices, which nominate elements of vector $\boldsymbol{v}_c(n)$.

When taking into account delay matrix (19), Eq. (12) may be transformed into the following expression:

$$\boldsymbol{V}_c(s) = \boldsymbol{W}_c(s) \cdot \boldsymbol{F}(s) - \tilde{\boldsymbol{W}}_c(s) \cdot \boldsymbol{X}_c(s). \tag{20}$$

where $\tilde{\boldsymbol{W}}_c(s) = \boldsymbol{W}_c(s) \cdot \boldsymbol{D}(s)$ is transfer function matrix of Von-Neumann type digital controller as physical device, which calculates action vector in the real physical time.

When assuming, that time intervals for converting signal vector $\boldsymbol{x}(t)$ into its digital form $\boldsymbol{x}_{\tilde{n}}(n)$, and converting digital action vector $\boldsymbol{v}_c(n)$ into its analogue form $\boldsymbol{v}(t)$ are regarded to software operation, common description of system linear part (from output of actuator till its input is as follows:

$$\boldsymbol{V}(s) = \boldsymbol{W}_c(s) \cdot \boldsymbol{F}(s) - \tilde{\boldsymbol{W}}_c(s) \cdot \boldsymbol{W}(s) \cdot \boldsymbol{U}(s), \tag{21}$$

where $\boldsymbol{V}(s) \cong \boldsymbol{V}_{\tilde{n}}(s)$.

This result is sufficient to solve the problem of modeling the problem of feedback closure through actuators with substantial nonlinearities.

4 Model of Actuator with Nonlinearities

Real actuators, which drive gate valves of object under control, have dry friction at moving mechanical parts, local position feedback, and restriction on valve movement.

Let object under control be homogeneous, and consider model of $u_k(t) \in \boldsymbol{u}(t)$ component formation $u_k(t) \in \boldsymbol{u}(t)$ under action of $v_k(t) \in \boldsymbol{v}(t)$. Foremost, movement actuator valve is restricted, from below by fully closed valve position, and from above by fully opened valve position, so static characteristic of actuator is as follows:

$$u_k = \begin{cases} 0, \text{ when } \tilde{u}_k \le 0; \\ \kappa_1 \tilde{u}_k, \text{ when } 0 \le \tilde{u}_k \le \tilde{u}_{k\max}; \\ \kappa_1 \tilde{u}_{k\max}, \text{ when } \tilde{u}_k > \tilde{u}_{k\max}, \end{cases} \qquad (22)$$

where $\tilde{u}_k$ is the mechanical position of valve; $\tilde{u}_{k\max}$ is fully opened valve position; κ_1 is the transmission ratio of the valve mechanical part.

Cauchy form of differential equation, describing closed loop system with dry friction when $\tilde{u}_k = \tilde{u}_{k,1}$ is as follows:

$$\begin{cases} \frac{d\tilde{u}_{k,1}}{dt} = \tilde{u}_{k,2}; \\ \frac{d\tilde{u}_{k,2}}{dt} = -\kappa_2 \tilde{u}_{k,2} - \kappa_3 \text{sgn} \tilde{u}_{k,2} + \kappa_4 \tilde{v}_k; \end{cases} \qquad (23)$$

$$\tilde{v}_k = v_k - \kappa_5 \tilde{u}_{k,1}; \qquad (24)$$

$$\text{sgn}\left[\tilde{u}_{k,2}(t)\right] = \begin{cases} 1, \text{ when } \tilde{u}_{k,2}(t) > 0; \\ -1, \text{ when } \tilde{u}_{k,2}(t) < 0, \end{cases}$$

where $\kappa_2, \kappa_3, \kappa_4, \kappa_5$ are coefficients; $\text{sgn}\tilde{u}_{k,2}$ is a dry friction, which depends of direction of valve movement; $\tilde{v}_k$ is the feedback error signal.

Expressions (22) and (23) include substantial nonlinearities, so there are known difficulties with application to it direct Laplace transform. So there is the only method of simulation system performance, based on the sampling of the signal $\boldsymbol{v}(t)$ and transforming it to discrete signal $\boldsymbol{v}(n)$:

1) For current, n-th, discrete moment of time and current values of signals $v_1(t)$, ..., $v_k(t)$, ..., $v_K(t)$ computation the following positions of actuator valves, which they take at the $(n+1)$-th discrete moment of time by means of direct integration the system (23). (24).
2) With use static characteristic (22) computation output positions of $\mathbf{u}(t)$, $t = (n+1)$ actuators' valves at the $(n+1)$-th moment, $(n+1)\tau$, where τ is the sampling period.
3) For new output positions of actuators' valves with use of any known method calculation values of signals $v_1(t)$, ..., $v_k(t)$, ..., $v_K(t)$ at the $(n+1)$-th discrete moment of time.
4) Repetition points 1–3 until the integration interval of the system being over.

5 Example of Digital Control System Simulation

As an example, the two-loop digital control system is considered (Fig. 3).

Transfer function, which define OUC dynamics are as follows:

$$W_{11}(s) = W_{21}(s) = \frac{3}{0,5s+1}; \; W_{12}(s) = W_{22}(s) = \frac{1,5}{0,5s+1}. \qquad (23)$$

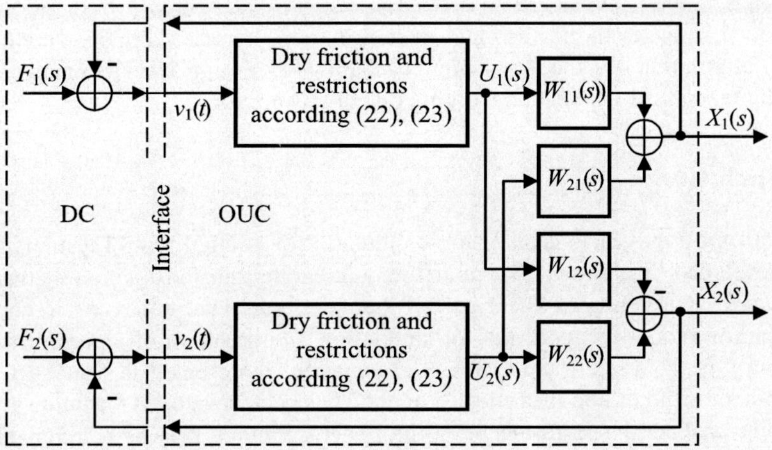

Fig. 3. Two-loop digital control system OUC - Object Under Control, DC - Digital Controller.

In the system proportional feedback is realized. Inputs $F_1(s)$ and $F_2(s)$ are Laplace transform of Heaviside function $L^{-1}[F_1(s)] = 1 \cdot \eta(t)$. Delays of DC are as follows: $\tau_{0,1} = 0$ c; $\tau_{0,2} = 0,02$ c, $\tau_{u,1} = 0,04$ c, $\tau_{u,1} = 0,06$ c.

Figure 4 shows the performance of the system. Plots a and b show the transient process in the system under analogue control, while plots c, and d show the transient process in the system under digital control. Plots a, and c show the pressure in reactor installation, while plots b, and d show the gas concentration in the reactor.

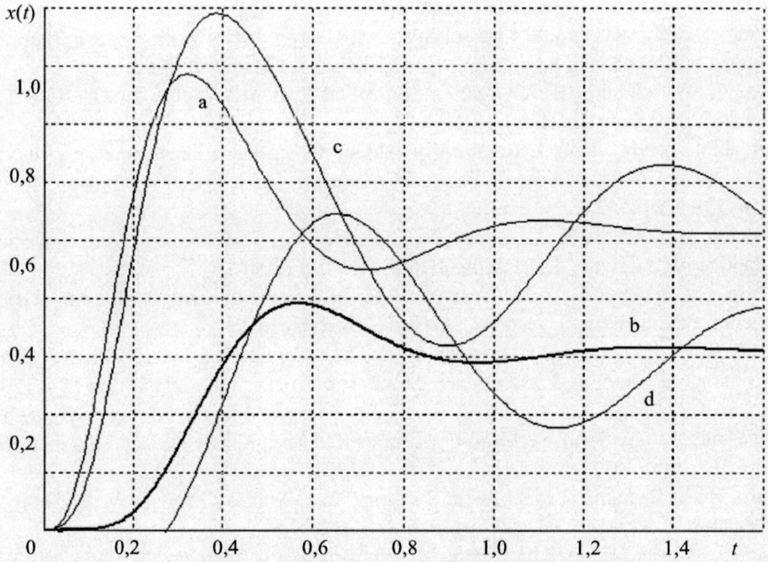

Fig. 4. Performance of the system

As one can see from the plots lags in control loops increase both overshooting, and time of control, that one should take into account when design Von Neumann computer based digital control system and working out a software for it.

6 Conclusion

As a result, the approach to mathematical simulation of a digital control system functioning is developed. It is shown that model may integrate transfer function mathematical apparatus for simulation of linear part of the system an data processing in controller, semi-Markov processes apparatus for simulation interpretation of algorithm by Von-Neumann type controller, and differential equation, represented in Cauchy form for simulation of actuator nonlinearities. A method has been developed that allows to comprehensive analyze the performance of the system. Method may be recommended for use in the engineering practice of designing control systems. Further research in this area can be aimed at developing methods of the digital control algorithms practical synthesis, that are optimal in terms of complexity and quality.

The study was carried out within the framework of the state assignment of the Ministry of Education of Russia on the topic "Innovative approaches to the professional training of future teachers of informatics in the context of digitalization of society", agreement No. 073-03-2021-019/2 of 21.07.2021.

References

1. Roy, R., Stark, R.,Tracht, S., Mon, M.: Continuous maintenance and the future - Foundation and technological challenges. CIRP Ann. **2**(65), 667–668 (2016)
2. Landau, I.D., Zito, G. Digital Control Systems. Design, Identification and Implementation. Springer, London (2006). https://doi.org/10.1007/978-1-84628-056-6
3. Aström, J., Wittenmark, B.: Computer Controlled Systems: Theory and Design. Tsinghua University Press, Prentice Hall (2002)
4. Larkin, E.V., Ivutin, A.N.: Estimation of latency in embedded real-time systems. In: 3-rd Meditteranean Conference on Embedded Computing (MECO-2014), Budva, Montenegro, pp. 236–239 (2014)
5. Erkoyuncu, J.A., Amo, I.F., Ariansyah, D., Bulka, D., Vrabic, R., Roy, R.: A design framework for adaptive digital twins. CIRP Ann. **1**(69), 145–148 (2020)
6. Karnopp, D.C., Margolis, D.L., Rosenberg, R.C.: System dynamics: Modeling, simulation and control of mechatronic systems. Willey, Hoboken (2012)
7. Bielecki, T.R., Jakubowski, J., Niewęgłowski, M.: Conditional Markov chains: properties, construction and structured dependence. Stochastic Proces. Appl. **4**(127), 1125–1170 (2017)
8. Ching, W.K., Huang, X., Ng, M.K., Siu, T.K.: Markov Chains: Models, Algorithms and Applications. International Series in Operations Research & Management Science, vol. 189. Springer, Boston (2013). https://doi.org/10.1007/978-1-4614-6312-2
9. Howard, R.A.: Dynamic Probabilistic Systems. V. 1: Markov Models. V. II: Semi-Markov and Decision Processes. Courier Corporation (2012).
10. Janssen, J., Manca, R.: Applied Semi-Markov processes. Springer, Boston (2006). https://doi.org/10.1007/0-387-29548-8
11. Fadali, M.S., Visioli, A.: Digital Control Engineering: Analysis and Design. Elsevier Inc. (2013)

12. Schiff, J.L.: The Laplace Transform: Theory and Applications. Springer, New York (1991). https://doi.org/10.1007/978-0-387-22757-3
13. Pavlov, A.V.: About the equality of the transform of Laplace to the transform of Fourier. Issues Anal. **5**(23). N.4(76). 21–30 (2016)
14. Li, J., Farquharson, C.G., Hu, X.: Three effective inverse Laplace transform algorithms for computing time-domain electromagnetic responses. Geophysics **2**(81), E75–E90 (2015)
15. Atherton, D.P.: Almost six decades in Control engineering [Hitorical perspectives]. IEEE Control Syst. Mag. **6**(34), 103–110 (2014)
16. Pospišil, M.: Representation of solutions of delayed difference equations with linear parts given by pairwise permutable matrices via Z-transform. Appl. Math. Comput. **294**, 180–194 (2017)
17. Kobayashi, H., Marl, B.L., Turin, W.: Probability, Random Processes and Statistical Analysis. Cambridge University Press, Cambridge (2012)
18. Pukelsheim, F.: The three sigma rule. Am. Stat. **2**(48), 88–91 (1994)

The Oil Production Dynamics Analysis in the World for Half a Century in Value Terms

A. A. Goryachev, R. I. Dzerjinsky(✉), and E. N. Pronina

Applied Mathematics Department, MIREA – Russian Technological University, 78, Vernadsky pros., Moscow 119454, Russia

Abstract. This article examines the oil production dynamics in the world from 1965 to 2020. The linear and exponential production-price supply functions have been investigated, and the world oil production price elasticity has been analysed. The time series fluctuations analysis and trend components are carried out using a shift and autocorrelation function to determine the almost-periods values. The time-critical moments corresponding to phase transitions in the world oil production dynamics are determined. At the initial stage in 1968–2004, the oil resources shortage is latent; it proceeded against the slowly growing production backdrop. However, it was accompanied by an exponential rise in oil prices, which is a similar phenomenon to an expression. The global energy crisis in its essence has become a phase transition from development, which possibilities were practically unlimited, to an oil resources shortage aggravation period. #COMESYSO1120.

Keywords: Almost periodic analysis · Time series · Oil production and cost · Autocorrelation function · Shear function · Trend

1 Introduction

This article is one of an articles series devoted to the world economy development from an economic statistics long-term data almost periodic analysis standpoint prepared by the Department of Applied Mathematics of the IIT RTU MIREA.

Currently, oil is the global world market most demanded part for energy resources, which it will remain in the coming decades. The long-term trends study in the world oil market development, the production volumes reaction to price changes, as well as price variability under the supply changing volumes influence and demand seems to be relevant and decisive.

Affected by political and economic factors, which at first glance is random and always determined by final demand, the oil consumption value is the economic stability an indicator and the new methods' application volume in the energy sector.

The factors same set leads to the need to study the processes' interaction representation using fundamentally different time scales. This is achieved by dividing the initial time series into components that characterize the so-called fast and the process slow components. As a result, the fast movements' analysis makes it possible to determine the time intervals' hierarchy that are closest to periods (near-periods).

R. Silhavy et al. (Eds.): CoMeSySo 2021, LNNS 231, pp. 84–95, 2021.
https://doi.org/10.1007/978-3-030-90321-3_9

Knowing the near-periods allows using them as the initial series smoothing intervals to determine the trends' hierarchy corresponding to the near-periods obtained values and to identify the so-called development cell that determines the prognostic observation period.

The near-proportions problem that can be identified by this study is capable of being reproduced through the near-period proportions [1, 2].

The application of such an analytical ideology makes it possible to analyze markets and systems of different nature [3–5]. The choice of information processing methods is especially relevant when modeling complex technical and socio-economic systems [6, 7].

This article examines the world oil production cost dynamics in the period from 1965 to 2020 [8, 9].

The study aim is to determine the process main characteristics and determine its development phase at a given point in time.

2 Phase Transition, the Relationship Between World Oil Production and Prices

Consider the relationship between world oil production and price, that is, the supply function. The supply function an important characteristic is the demand elasticity, supply or production. Elasticity is understood as one indicator sensitivity measure to changes in another. Elasticity shows how much the first indicator will change when the second changes by one per cent. The demand elasticity reciprocal is price flexibility. It indicates the prices' volatility in the market depending on the supply and demand ratio.

The relationship between price and daily production for the global oil market is shown in Fig. 1.

It can be assumed that until 2005 the world oil market was elastic - the increase in oil prices was accompanied by a proportional increase in production. After 2005, the supply and demand balance is upset, and the growth in oil production does not correspond to price changes. With sharp price fluctuations, insignificant changes in production are observed, of the order of 1%.

If in the linear regression case the straight-line equation approximating the data for the period from 1973 to 2004 has the form, $y = 0.00093x - 42$, then after 2004 the parameters change, $y = 0.0233x - 1820$, where y is the price in dollars per barrel, x - production in barrels per day.

Comparing the slopes in the graph in Fig. 1, we see that their values before and after 2004 differ by approximately 24 times. Thus, 2004 separates the world oil market elastic and inelastic state, a phase transition takes place here.

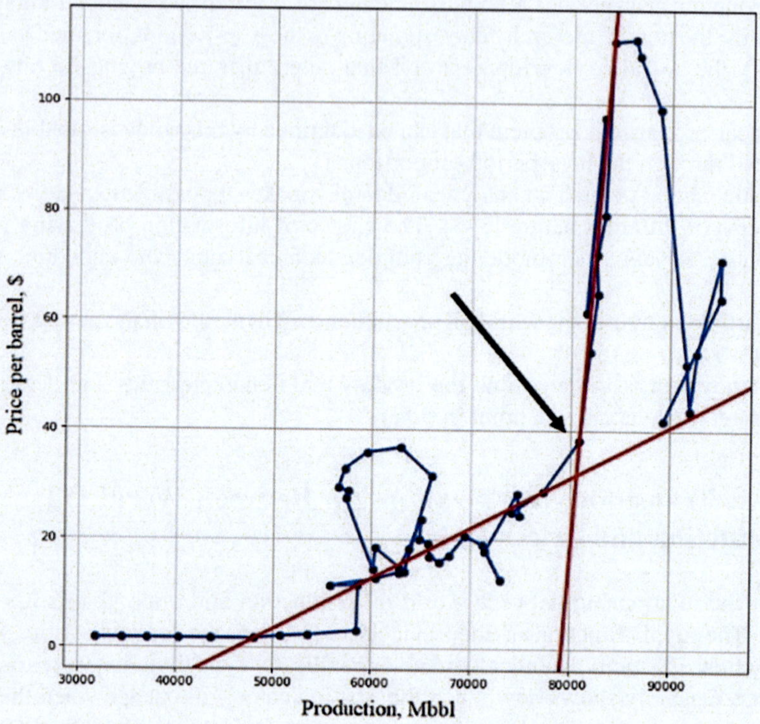

Fig. 1. The relationship between price and daily world oil production. Linear model.

In this case, the slope, defined as the absolute increase in the additional production price per unit, coincides with the marginal production costs calculated per barrel or with the marginal price. Along with price flexibility, this indicator can act as a price sensitivity measure to changes in output. Loop-shaped areas on the graph appear due to non-monotonic changes in the analysed parameters.

3 Exponential Model

Attention is drawn to the angular coefficients increasing nature on the graph (Fig. 1). This process nature corresponds to an increase in prices at an increasing rate. The exponential is the perfect model for developing at an increasing rate. Therefore, we pass to the semilogarithmic coordinate system (Fig. 2).

Exponential development is characterized by the analysed indicator relative growth constancy. Average relative increases in the price of 1 barrel per unit of additional oil production remain constant throughout the entire period, from 1975 to 2020.

As a result, instead of a piecewise linear function glued from two different sections in Fig. 1, one exponential dependence is obtained in the nonlinear model (Fig. 2). The trend line angular coefficient in semilogarithmic is $\gamma \approx 0.0588$. Inverse value, $\varepsilon = \gamma^{-1} \approx 17$. It is a dimensional quantity measured in million barrels per day. The ε value shows the typical production volume: when the daily production increment is 17 million barrels, the price will increase by a factor of *e* times.

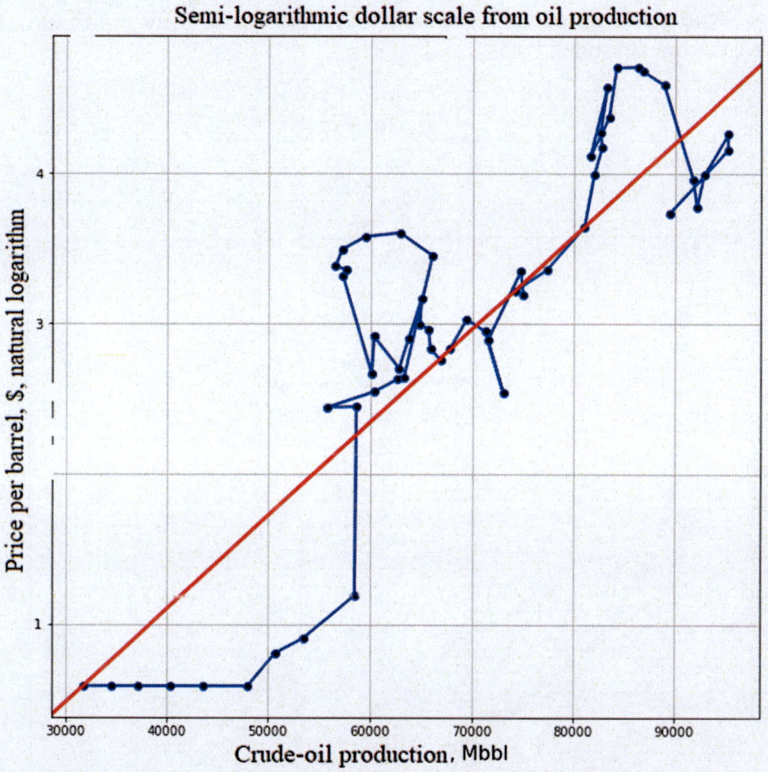

Fig. 2. The relationship between price and daily world oil production. Exponential growth models.

The price doubling requires a daily increase in production by 11.8 million barrels. In other words, the typical time for the oil price doubling is 11.8 million barrels per day.

4 The Trend Component Analysis. Production Dynamics Model with Linearly Decreasing Relative Increments

Critical boundaries are also manifested in the world oil production dynamics. Figure 3 shows the world oil production dynamics in a graph from 1965 to the present, the data source is the British Petroleum website [8]. The first decline in production after a growth long period, shown in the graph in Fig. 3, occurred in the early 70s of the last century. This is the global energy crisis period. The daily production growth rates before and after the crisis were 7% and 1%, respectively.

The accumulated production growth rate dynamics analysis makes it possible to assess the accumulated production inflexion point position. A fairly good approximation is given by the linear model:

$$\frac{dx}{dt} = x(a - bt).$$

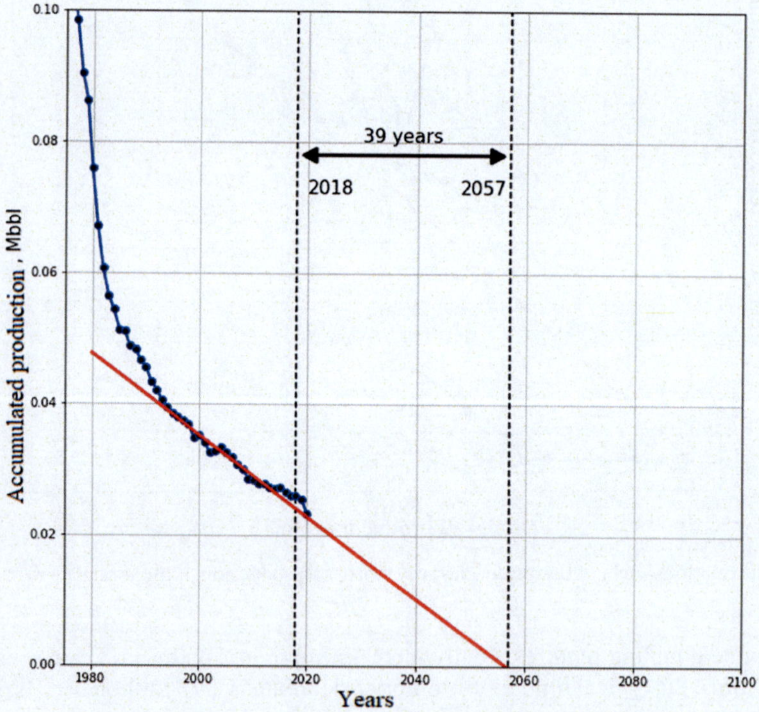

Fig. 3. Growth model with linearly decreasing growth rates.

At the trend intersection point with the abscissa axis $t_0 = \frac{a}{b}$, it falls on 2057, this is the resources' depletion date with the current production trend. The method for determining the inflexion point position is shown in Fig. 3. According to it, the annual production peak falls in 2018. Consequently, the daily oil production is currently in the maximum vicinity. Therefore, in the short and medium-term, there will be production fluctuations around this maximum level.

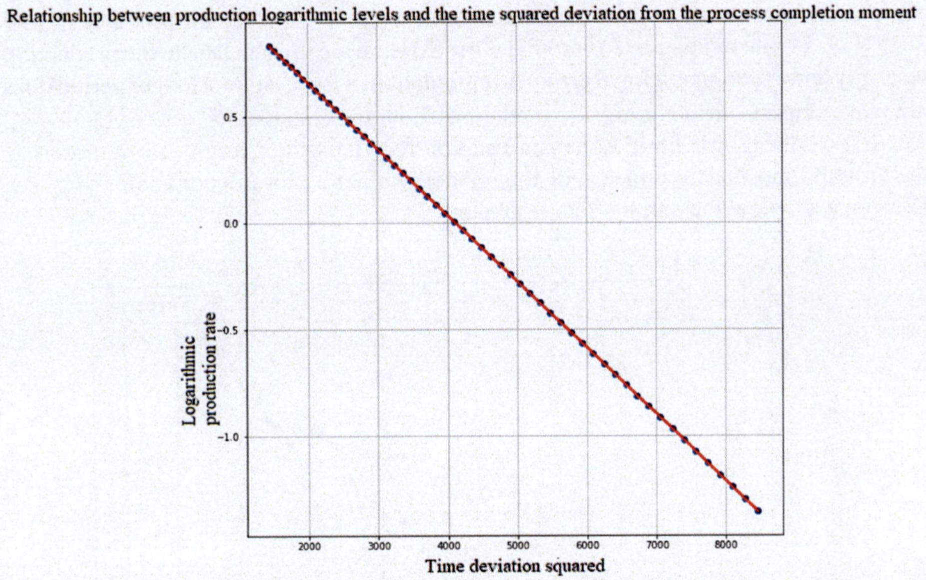

Fig. 4. Anamorphosis.

Integrating the above differential equation, we obtain

$$x(t) = Ce^{-\frac{b}{2}(t_0 - t)^2},$$

this dependence anamorphosis is obtained in the coordinates $(t_0 - t)^2$, ln (x). Its graph is shown in Fig. 4.

The graph intersection point with the y-axis gives recoverable resources forecast estimate equal to 1.314 trillion. barrels. The result obtained coincides with the Institute of Mechanical Engineers data [10].

5 Medium-Term Forecast Based on an Autoregressive Model

For a medium-term forecast, an autoregressive model, ARIMA (0, 1, 1), is used, which is based on exponential smoothing [9]. To filter out the noise and more accurately estimate the local mean, the next observation is predicted using the last few observations mean. To achieve this effect, a simple exponential smoothing model uses past values exponentially weighted moving average. In it, the series' each value is linearly dependent on the previous values. On its basis, a graph was built with the expected oil production until 2025 (Fig. 5), where fluctuations around the maximum are visible. Due to the production volatility and the computational errors' accumulation, a forecast for a longer period does not make sense.Consider a logistic model based on the equation $\frac{dy}{dt} = k * y(1 - \frac{y}{y\infty})$, where $y\infty$ is the growth limit. As a linear relationship intersection result with the abscissa, the growth limit is determined. For this model, we get a plot in coordinates y´/y ~ y. Let's take a look at the current data example:

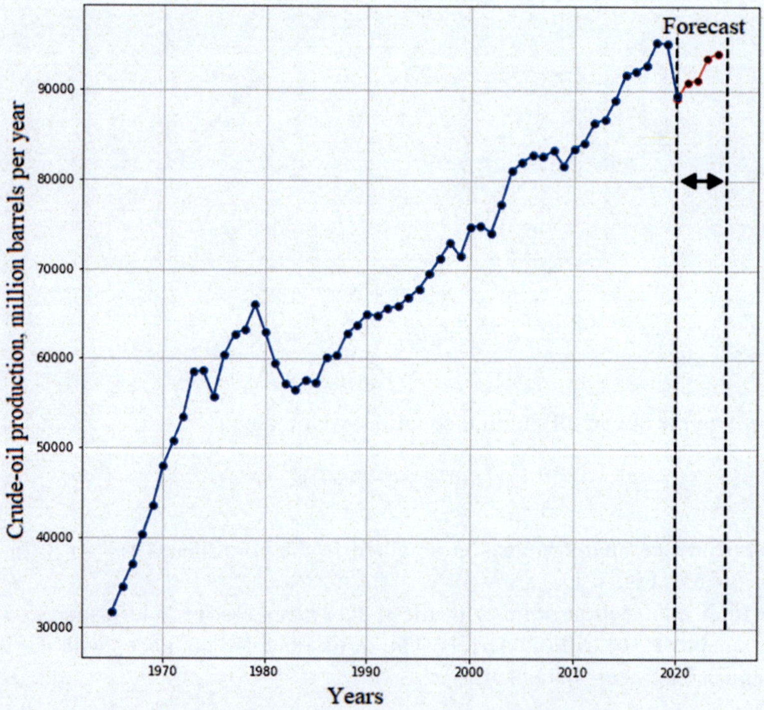

Fig. 5. Oil production medium-term forecast.

6 Fluctuations Analysis

Let us analyse the world oil production time series cyclical component. For this, we construct a shift and autocorrelation function (Figs. 6 and 7).

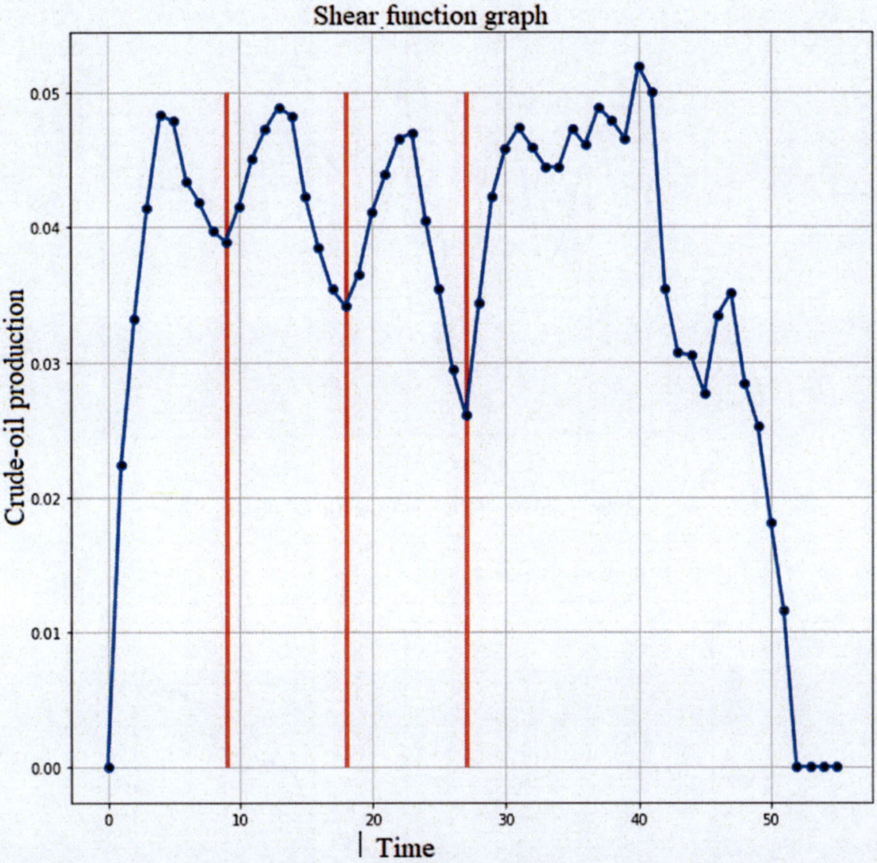

Fig. 6. Shear function graph.

The local minimums and maximums positions indicate an almost period, which duration is equal to 9 years. These short nine-year cycles are marked on the global oil production trend graph in Fig. 8.

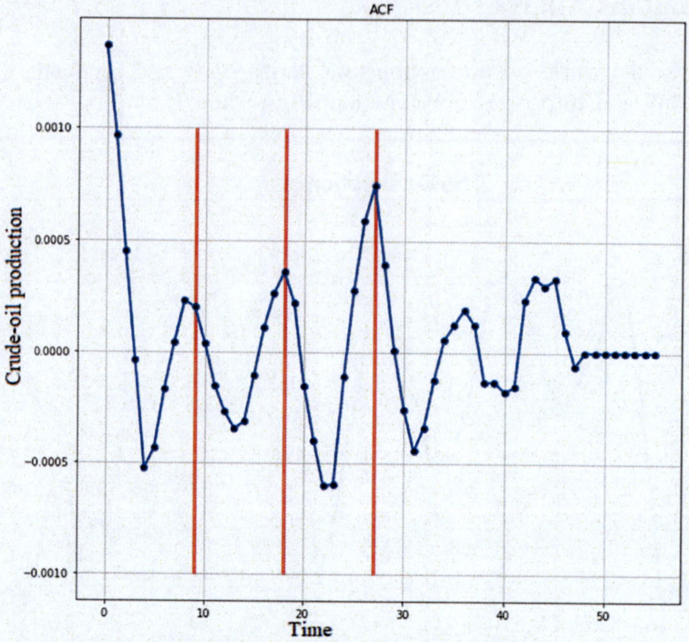

Fig. 7. Autocorrelation function graph.

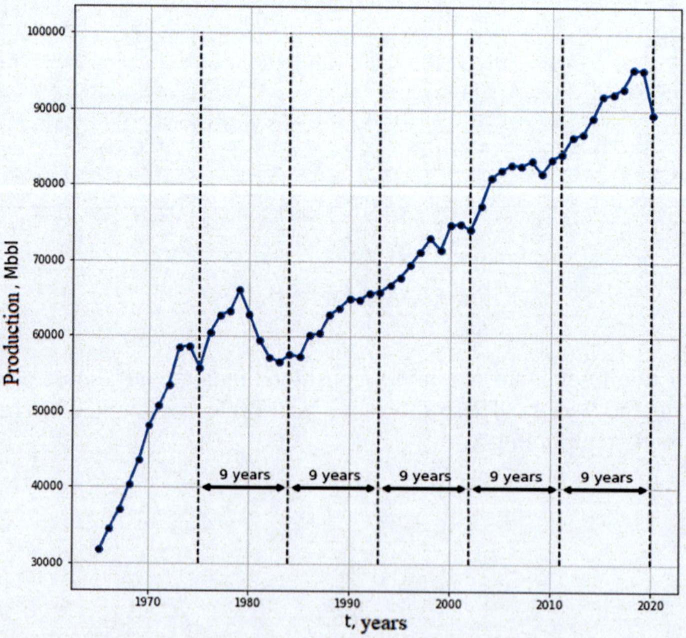

Fig. 8. World oil production dynamics, 1965–2020.

7 Oil Production Dynamics Model in Gold Equivalent

Let's recalculate oil production in volume (barrels) terms, first in dollars, and then in gold grams. To do this, we multiply the daily production level in barrels by the one-barrel current price, and then divide the result by the gold one-gram current price The daily production dynamics in gold equivalent is shown in Fig. 9.

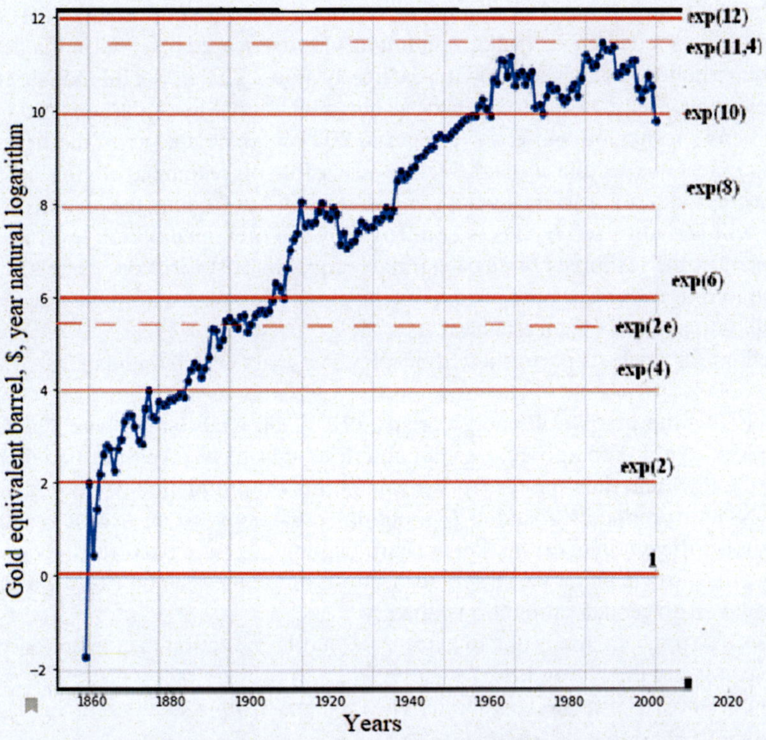

Fig. 9. World oil production dynamics in gold equivalent with a development cell.

Noteworthy is the rhythmic feature manifested in the oil production dynamics in gold equivalent (Fig. 9). Along the vertical, we observe a boundaries uniform sequence spaced from each other on a logarithmic scale by 2 units; in the initial coordinates, it corresponds to a geometric progression with the denominator exp (2). According to [1] development cell, the three uniform measures aggregate forms a complete structure, a development cycle kind. The range from 1 to exp(e) corresponds to a development stable phase, while from exp(e) to exp (3) - to an unstable, restructuring phase. The world oil production dynamics in gold equivalent has entered an unstable growth phase, followed by dramatic changes soon. Further growth in oil production in gold equivalent requires either an increase in production in barrels or an increase in the oil price faster than gold, which is tantamount to ousting gold as a universal equivalent and replacing it with oil.

8 Conclusion

Thus, the global energy crisis in its essence became a phase transition from development, which possibilities were practically unlimited, to an oil resources shortage period. At the initial stage in 1968–2004, the resources' shortage was hidden, it occurred against the slowly growing production background. However, it was accompanied by an exponential increase in oil prices, which is the resource deficit an expression.

At the beginning of the XXI century, the resource deficit nature changes, it becomes obvious. Against a decreased background in newly discovered reserves and a deterioration in their quality, production growth practically stops. The global oil industry is shifting to developing hard-to-recover reserves. Unconventional hydrocarbons are included in the official production and reserve reports. 2004 is associated with the light oil era end. This milestone peculiarity is the easily accessible oil resources obvious shortage.

If the oil production current level is maintained without forcing, the oil barrels proved reserves will last until 2057. This is confirmed by the oil consumption level and based on the recoverable resources forecast estimate, equal to 1.314 trillion. barrels.

In the mid-term forecast, the production level will fluctuate within the maximum, as the production peak has been reached.

Based on the analysis performed, the following years are distinguished:

- 1970–1974 - the first world energy crisis, 1973 - the Arab-Israeli war. The OAPEC introduces an oil embargo against the countries that have taken the Israel side and raises the raw materials' prices. At the same time, oil production decreases by 5%.
- 1985 - Saudi producers abandon production restrictions, as an excess result in the supply over the demand oil market, a sharp decline in prices occurred.
- 2003 - price growth under the supply restrictions influence. The OPEC is forcing other oil producers to reduce production capacity
- 2012 - a sharp rise in prices due to a decrease in oil production in connection with the Arab spring and the war in Libya
- 2020 - coronavirus infection epidemic. The increase in oil production in Saudi Arabia, which led to the dollar per barrel collapse

Thus, oil is no longer a positive factor in economic growth. The future development main task is the new energy production technologies creation.

The proven oil resources depletion date under the current production trend is 2057.

The next expected growth in oil demand after 2021 requires either an increase in physical production or an increase in the oil faster price than gold. This process is tantamount to ousting gold as a universal commodity equivalent and replacing it with oil.

References

1. Veniaminov, S.S.: Revealing Latent structure laws in processes and signals. M., URSS, p. 216 (2014)
2. Kuzmin, V.I., Gadzaov, A.F.: Method of construction of models on empirical data. The manual. M., MGTU MIREA, p. 94 (2012)

3. Pronina, E.N.: Non-linear models of analyzing the long-term tendencies of world oil market. Sci. Rev. **3**, 363–369 (2013)
4. Dzerjinsky, R.I., Pronina, E.N., Dzerzhinskaya, M.R.: The structural analysis of the world gold prices dynamics. In: Silhavy, R. (ed.) CSOC 2020. AISC, vol. 1225, pp. 352–365. Springer, Cham (2020). https://doi.org/10.1007/978-3-030-51971-1_29
5. Paramonov A.A, Kuzmin V.I., Dzerjinsky R.I.: Analysis of almost-periodic and almost-proportional characteristics of a representative sample local minima time series. IOP Conf. Ser. Mater. Sci. Eng. **1047**, 012045 (2020) https://iopscience.iop.org/article/10.1088/1757-899X/1047/1/012045
6. Semenenko, M.G., et al.: How to use neural network and web technologies in modeling complex technical systems. IOP Conf. Ser. Mater. Sci. Eng. **537**, 032095 (2019)
7. Gruzenkin, D.V., et al.: Neural networks to solve modern artificial intelligence tasks. J. Phys. Conf. Ser. **1399**, 033058 (2019)
8. British Petrolium. https://www.bp.com/en/global/corporate/energy-economics/statistical-review-of-world-energy/downloads.html. Accessed 08 Aug 2021
9. Loginom. Autocorrelation function. https://wiki.loginom.ru/articles/acf.html. Accessed 08 Aug 2021
10. Machine learning. http://www.machinelearning.ru/. Accessed 08 Aug 2021

The Earthquake Attributes Disjunctive Form Analysis at the Quickest Form Change Directions

R. I. Dzerjinsky(✉)

Applied Mathematics Department, MIREA – Russian Technological University, 78, Vernadsky pros., Moscow 119454, Russia

Abstract. The as you know, in discrete mathematics there is no limit concept, but there is a function denoted by the term derivative. This is due to the Boolean function expansion in a series similar to the Maclaurin series at the point 00… 0 or the Taylor series at an arbitrary point in space. The article uses the derivatives' form according to Bohman. In practice, the differentiator is used for the Boolean function optimal expansion $f(x_1, x_2, \ldots, x_n)$ into k variables. The criterion for the variables optimal exclusion is to exclude first the variables, upon switching which the Boolean function switches under the conditions maximum number. It is natural to ask which function differential is equal to itself, known for continuous functions as one of Euler's problems. The article analyses a similar formulation concerning differential equations in Boolean variables. A solution is given in a specific Zhegalkin polynomial form.

Keywords: The disjunctive normal form · Differential equations in Boolean variables · The steepest polynomials change · Bohman derivative

1 Introduction

All the mathematical methods used for processing economic information is increasingly reduced to the different dimensions' synthetically generated time series analysis and has both continuous and discrete nature. For continuous argument approximated and analytic functions, (may be given by a grid), the search for extremal and bifurcation regularities has been well studied.

In seismic forecasting cases, such synthetic multidimensional formations behaviour scientifically reliable predictions have not yet been obtained.

Applying bifurcation ideology to the economic characterizations time series allows analyzing different nature markets [1, 2]. The appearance of this discrete variable time series set significantly complicates the search for optimal solutions [3–5]. Meanwhile, the Euler problem general formulation applies to discrete variable functions. This is achieved by the Bohmann derivative formal application to a discrete variable given function. It can be shown that this is true even for a partially deterministic solution.

This work is a naive attempt to find the steepest growth polynomials for Boolean functions.

R. Silhavy et al. (Eds.): CoMeSySo 2021, LNNS 231, pp. 96–101, 2021.
https://doi.org/10.1007/978-3-030-90321-3_10

2 Methodology

We assume, that α_i - information sign of forecasting, which accompanies the coming earthquake, where $\alpha_i = \begin{cases} 0 \\ 1 \end{cases}$.

Among these symptoms we can identify a group of signs «potential constant danger» (including Earthquake regulations zoning), but we should take into account all the signs of another character, having the character of objective documented reality. (Care of water from wells, restless behavior of pets etc.) Also we agree that we indifferent to the duration of manifestation α_i, and is an important of him manifestation prior to the earthquake. We assume, than specific time of manifestation α_i associated with the period of its manifestation which precedes the event. Construct for each information sign of any finite set of earthquakes a n-dimensional vector $\boldsymbol{\alpha_s} = (\alpha_1, \ldots, \alpha_j, \ldots, \alpha_k)$. In general case, for all cases of the earthquake does not match the dimension of number of signs k and presence in it all components j.

Denote by **U** the set $\boldsymbol{\alpha_S}$.

$$\mathbf{U} = \{\boldsymbol{\alpha_1}, \ldots, \boldsymbol{\alpha_m}\}; |\mathbf{U}| = \mathrm{m}$$

The set of all these n-dimensional vectors is called an n-dimensional Boolean cube $\boldsymbol{B^n}$.

Let $S \subset \mathbf{U}$ – arbitrarily, can be $\varnothing$. Compare subset S: $S \rightarrow \boldsymbol{\alpha_S} \in \boldsymbol{B^n}$, where $\boldsymbol{\alpha_S}$ - a vector with the following rule:

$$\boldsymbol{\alpha_S} = (\alpha_1, \alpha_2, \ldots, \alpha_n), \text{ where } n = max(dim\, \boldsymbol{\alpha_m})$$

$$\alpha_i = \begin{cases} 1, & \text{if } \alpha_\mathrm{i} \in S; \\ 0, & \text{if } \alpha_\mathrm{i} \notin S. \end{cases}$$

Then α_S called characteristic for S.

Theorem. Join of subsets of information signs of predicting accompanying the approaching earthquake, meets of the logical sum of their characteristic vectors:

$$\mathrm{S} \cup \mathrm{T} \rightarrow \alpha_s \vee \alpha_T$$

Crossing meets the logical product

$$\mathrm{S} \cap \mathrm{T} \rightarrow \alpha_S \cdot \alpha_T$$

and

$$\overline{S} \rightarrow \overline{\alpha_S}$$

corresponds to the negation of the characteristic vector. The logical sum of two Boolean vectors is a vector obtained from:

$$\alpha = (\alpha_1, \ldots, \alpha_n); \beta = (\beta_1, \ldots, \beta_n);$$

according to the rule:

$$\alpha \vee \beta \overset{\text{def}}{=} (\alpha_1 \vee \beta_1; \alpha_2 \vee \beta_2; \ldots ; \alpha_n \vee \beta_n);$$

Logical multiplication:

$$\alpha\beta \overset{\text{def}}{=} (\alpha_1\beta_1; \alpha_2\beta_2; \ldots ; \alpha_n\beta_n);$$

Logical negation is obtained from the initial coordinatewise negation:

$$\overline{\alpha} = (\overline{\alpha_1}, \ldots, \overline{\alpha_n}).$$

The correspondence between the subsets and characteristic vectors of one-to-one.

Proof. Let's show fidelity of the first compliance. The others are shown similarly Let $\alpha_T = (\beta_1, \beta_2, \ldots, \beta_n)$, $\alpha_s = (\alpha_1, \alpha_2, \ldots, \alpha_n)$;

$$\mathrm{S} \cup \mathrm{T} \to (\alpha_1 \vee \beta_1; \alpha_2 \vee \beta_2; \ldots ; \alpha_n \vee \beta_n).$$

Actually α_S belongs to this set in case it belongs either S either T, i.e. either α_n either β_n is equal 1, but it is also the logical sum.

Theorem: In this operation definition of Boolean cube $\boldsymbol{B^n}$ information signs predicting that accompany the approaching earthquake, it becomes a Boolean algebra. At the same time the role of zero and unit carry out a vectors

$$0 = (0, \ldots, 0); 1 = (1, \ldots, 1);$$

For the proof we must check all axioms of Boolean algebra Such simple reception we can receive Boolean function for each of the considered cases of earthquakes, which represents the conjunctive product $\alpha_1\alpha_2 \ldots \alpha_n$ the information signs of forecasting accompanying the coming specific case of an earthquake.

To all set of cases of cataclysms, there will correspond the disjunctive sum of elementary conjunctions

$$\mathrm{f}(\mathrm{x}_1, \mathrm{x}_2, \ldots, \mathrm{x}_\mathrm{n})|_1 = \vee(\alpha_1\alpha_2 \ldots \alpha_k).$$

One would assume that in each case of earthquake is one and the same set of information signs is essential. However such approach doesn't maintain check by practice. Therefore it is possible to assume about existence of some subset of edges of a Boolean cube $\boldsymbol{B^n}$. His disjunctive sum of elementary conjunction with guarantee accompanies an earthquake.

Let's try to find them.

The derivative of first order $\frac{\partial f}{\partial x_i}$ from a Boolean function f on a variable x_i is the amount on the module of 2 appropriate residual functions:

$$\frac{\partial f}{\partial x_i} = \mathrm{f}(\mathrm{x}_1, \mathrm{x}_2, \ldots, x_{i-1}, 1, \ldots, x_n) \oplus \mathrm{f}(\mathrm{x}_1, \mathrm{x}_2, \ldots, x_{i-1}, 0, \ldots, x_n)$$

where $f(x_1, x_2, \ldots, x_{i-1}, 1, \ldots, x_n)$ – unit residual function;

$f(x_1, x_2, \ldots, x_{i-1}, 0, \ldots, x_n)$ – null residual function; $\oplus$ - the sum modulo 2.

The derivative of first order $\frac{\partial f}{\partial x_i}$ from a Boolean function $f(x_1, x_2, \ldots, x_n)$ determines the conditions under which the function changes the value when switching variable x_i (when you change the value of x_i on the opposite).

The mixed derivative $\frac{\partial^k f}{\partial x_{i_1} \partial x_{i_2} \ldots \partial x_{i_k}}$ from a Boolean function f on a variable x_{i_1}, x_{i_2}, … , x_{i_k} called expression of the form

$$\frac{\partial^k f}{\partial x_{i_1} \partial x_{i_2} \ldots \partial x_{i_{k-1}} \partial x_{i_k}} = \frac{\partial}{\partial x_{i_k}} \left(\frac{\partial^{k-1} f}{\partial x_{i_1} \partial x_{i_2} \ldots \partial x_{i_{k-1}}} \right).$$

The mixed derivative k-th order $\frac{\partial^k f}{\partial x_{i_1} \partial x_{i_2} \ldots \partial x_{i_k}}$ calculated by applying the ratio k times by fixing the variables $x_{i_1}, x_{i_2}, \ldots, x_{i_k}$ (the commit order of the variables does not matter); the number of orderings is equal to $k!$, i.e. the derivative k-th order $\frac{\partial^k f}{\partial(x_{i_1}, x_{i_2}, \ldots, x_{i_k})}$ from Boolean functions $f(x_1, x_2, \ldots, x_n)$ on a variable x_{i_1}, x_{i_2}, … , x_{i_k} equal to the sum of module 2 of all derivatives of the first order, second, third, … , k-th mixed derivatives by fixing the variables $x_{i_1}, x_{i_2}, \ldots, x_{i_k}$:

$$\frac{\partial^k f}{\partial(x_{i_1}, x_{i_2}, \ldots, x_{i_k})} = \sum_i \frac{\partial f}{\partial x_i} \oplus \sum_{i,j,i \neq j} \frac{\partial^2 f}{\partial x_i \partial x_j}$$

$$\oplus \sum_{\substack{i,j,s \\ i \neq j, i \neq s, j \neq s}} \frac{\partial^3 f}{\partial x_i \partial x_j \partial x_s} \oplus$$

$\oplus \ldots \oplus \frac{\partial^k f}{\partial x_{i_1} \partial x_{i_2} \ldots \partial x_{i_k}}, i, j, s, \ldots = i_1, i_2, \ldots, i_k$ [6].

Weight of derivative $P\left(\frac{\partial^k f}{\partial(x_{i_1} x_{i_2} \ldots x_{i_k})}\right)$ from Boolean function called the number of a constituent this derived.

Let's use the differentiation apparatus for optimum decomposition of function $f(x_1, x_2, \ldots, x_n)$ on k variable.

The criterion of an optimum exception of variables consists in an exception at first of variables.

The criterion of optimal elimination of variables.

The optimal criterion for removal of variables is to eliminate the variables first, when you switch the Boolean function which is switched when the maximum number of terms corresponding to the subset of facets of the Boolean cube $\boldsymbol{B}^n$, a disjunctive sum of elementary conjunction guaranteed which accompanies the earthquake. This maximum number is determined by the weight of the derivative.

At an exclusion k variables, find the derivative k-th order from implementable function and search for the maximum value of weight of the derivative $P\left(\frac{\partial^k f}{\partial(x_{i_1} x_{i_2} \ldots x_{i_k})}\right)$, which defines the excluded variables. For receiving residual Boolean functions, we find derivatives again: determined by weight, and the derivative of the considered residual function, having the maximum weight, determines the appropriate variables that are excluded in this step for this residual function, etc. until residual functions will not have simple form.

The research a constituent $P\left(\frac{\partial^k f}{\partial (x_{i_1} x_{i_2} \ldots x_{i_k})}\right)$ from Boolean function inevitably makes us to solve the differential Eq. (1) in a Boolean cube $\boldsymbol{B^n}$

$$\frac{\partial f}{\partial x_i} = f \tag{1}$$

Even

$$\frac{\partial^k f}{\partial \left(x_{i_1}, x_{i_2}, \ldots, x_{i_k}\right)} = f$$

The optimal criterion for removal of variables is heuristic in nature, which is based on the assumption that the greater the weight of the derivative, the more the function f depends on the variables on which it was differentiated.

Transformation, the opposite to differentiation-integration, is interesting from the point of view compositions generalization.

Integral of $\int f(X)\,\mathrm{d}x_i$ is called the set of function $\{F_j\ (\mathbf{X}, x_i)\}$, each of which after differentiation has an appearance $f(X)$:

$$\int f(\mathbf{X})\,\mathrm{d}x_i = \{F_j(\mathbf{X}, x_i)\}, \mathbf{X} = (x_1, x_2, \ldots, x_{i-1}, x_{i+1}, \ldots x_n),$$

$$\frac{\partial F_j(X, x_1)}{\partial x_i} = f(\mathbf{X}).$$

Theorem: The solution of the differential Eq. (1) in a Boolean cube $\boldsymbol{B^n}$ exists and is the carrier of functions with accuracy up to constants, which by analogy with the continuous function argument we will call a Boolean exponent.

$$f := a_0 \oplus x_1 \oplus x_2 \oplus x_3 \oplus \ldots \oplus x_n$$

Proof:

Indeed,

$$\frac{\partial f}{\partial x_i} = f(x_1,\ x_2, \ldots, x_{i-1}, 1, \ldots, x_n) \oplus f(x_1,\ x_2, \ldots, x_{i-1}, 0, \ldots, x_n) = a_0\ = \oplus x_1$$

$$\oplus \ldots \oplus x_{i-1} \oplus 1 \oplus x_{i+1} \oplus \ldots \oplus x_n \oplus a_0 \oplus x_1 \oplus \ldots \oplus x_{i-1} \oplus 0 \oplus x_{i+1} \oplus \ldots \oplus x_n = 1.$$

Note that this result is obtained by inverse iteration, which takes place under the sign of binary addition.

Boolean exponent we will designate by L_e.

Thus existence of the decision follows from the constructive procedure of construction. The decision isn't the only and is defined to within a constant.

$$f = L_e \oplus C \tag{2}$$

Boolean L_e exponent is the only complete polynomial in the class of polynomials Zhegalkin.

Also we have proved.

Theorem: The solution of the Boolean differential Eq. 2 orders in the form of Boolean exponent of a look (2) is

$$f = \overline{\mathbf{C}}$$

3 Conclusion

The Bohman derivative definition formal application to the Euler equation allows us to obtain the fastest change Zhegalkin polynomials a specific form on Boolean variables.

References

1. Dzerjinsky, R.I., Pronina, E.N., Dzerzhinskaya, M.R.: The structural analysis of the world gold prices dynamics. In: Silhavy, R. (ed.) CSOC 2020. AISC, vol. 1225, pp. 352–365. Springer, Cham (2020). https://doi.org/10.1007/978-3-030-51971-1_29
2. Paramonov, A.A., Kuzmin, V.I., Dzerjinsky, R.I.: Analysis of almost-periodic and almost-proportional characteristics of a representative sample local minima time series. IOP Conf. Ser. Mater. Sci. Eng. **1047**, 012045 (2020) https://iopscience.iop.org/article/10.1088/1757-899X/1047/1/012045
3. Veniaminov, S.S.: Revealing Latent Structure Laws in Processes and Signals, p. 216. URSS, Moscow (2014)
4. Kuzmin, V.I., Gadzaov, A.F.: Method of Construction of Models on Empirical Data. The Manual, p. 94. MGTU MIREA, Moscow (2012)
5. Serebrennikov, M.G., Pervozvansky, A.A.: Revealing of the Latent Periodicity, p. 244. Science, Moscow (1965)
6. Gorbatov, A.V., Gorbatova, M.V., Gorbatov, V.A.: Discrete Mathematics, p. 447. AST, Astral, Moscow (2006)

Check for updates

The Changes Dynamics Analysis in the Japanese Stock Exchange Nikkei^225 Index in the Latest Time

R. I. Dzerjinsky(✉), B. A. Krynetsky, and N. V. Chernorizova

Applied Mathematics Department, MIREA – Russian Technological University, 78, Vernadsky pros., Moscow 119454, Russia

Abstract. The exchange rate analysis' the Japan Stock Exchange Nikkei^225 index in the latest time is carried out. This covers data from 1965 to the present. An anamorphosis almost periodic analysis is proposed for the hidden periodicities study in the Nikkei^225 dynamics under the Gompertc models' applicability assumption. The intensive growth periods have been identified. The development' cells found. Critical levels analysis in the system development based on the development' cell predicts characteristic events at the 2022 end; by 2024; in the 2026 spring; in 2037. Development' cells that are not yet finished predict characteristic points in 2023 and 2035. The series structural analysis showed the Japan Stock Exchange Nikkei^225 index stable growth possibility in the period from 2021 to 2035. In the period from 2029–2050 and 2042–2068, we can expect a development unstable phase, known as the perestroika phase.

Keywords: Nikkei^225 index · Intensive growth models · Anamorphosis · Gompertc model · Near-periods · Generalized shear function · Dynamics predictive assessment

1 Introduction

All countries and people's statistician's special adoration subject is the stock exchange price [1, 2]. One of Japan most important stock indices is a synthetic index - Nikkei^225 index. The index is calculated as simple arithmetic mean prices of shares 225 most actively traded companies in the Tokyo Stock Exchange first section [3].

Quotes time series mathematical research for individual companies comes down to presenting them standard technique in fundamentally different time scales. This is achieved by dividing the initial time series into components that characterize the so-called fast and the process slow components. As a result, the fast movements' analysis makes it possible to determine the time intervals' hierarchy that are closest to so-called near-periods.

To determine the trends' hierarchy, understanding near-periods allows you to use them as smoothing intervals for the original series. The initial data presentation in the piecewise linear dependences forms for slow movements, as the main tendencies, makes it possible to analyse directly, without excluding fast movements, based on anamorphosis.

R. Silhavy et al. (Eds.): CoMeSySo 2021, LNNS 231, pp. 102–126, 2021.
https://doi.org/10.1007/978-3-030-90321-3_11

In the study course, almost-proportions are determined, which can be identified both by specialized algorithms means and by analysing the near-periods ratio [4–6].

The consideration' such an ideology application makes it possible to analyse different nature markets [7, 8].

This principle applies to the actual definition of the expected value of Japan's stock exchange activity in recent times was no exception. This paper examines the Nikkei^225 index dynamics in the period from April 1965 to April 2021 [9, 10].

Higher complexity models will require a fuzzy logic apparatus use [10–13].

The study aim is to determine the Japanese stock exchange Nikkei^225 index main characteristics and to determine its development phase at a given moment in time.

2 Main Characterizations

As the analysed data, the Japanese stock index Nikkei^225 values time series was taken - the index is calculated as the Tokyo Stock Exchange first section 225 most actively traded companies shares prices simple arithmetic average - such as Honda Motor Co, Mazda Motor Corp, Mitsubishi Heavy Industries, Toyota Motor, Nippon Soda Co., Nissan Chemical Industries, Canon, Casio Computer Co, Kawasaki Heavy Industries and others. The chart below shows the index price direct values in yen over time (Fig. 1):

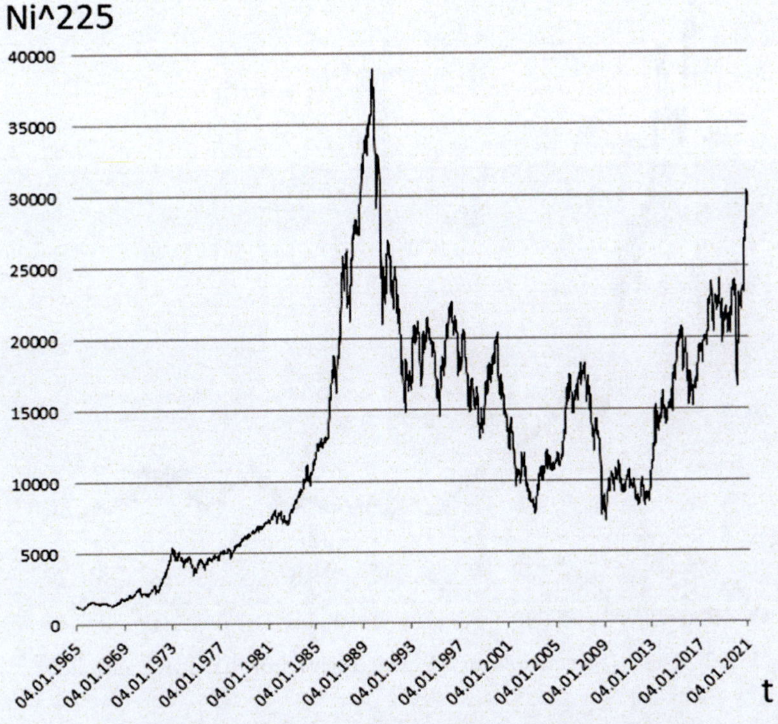

Fig. 1. Data on the index value for the studied time interval.

In the work course, we will consider some models' application for studying time series to identify hidden patterns in the index value dynamics understudy to create predictive estimates.

3 Intensive Growth Areas

In the systems' development process, there are stages at which the reproduction rate increases, that is, for the equation $y' = ky$, the coefficient k increases. To describe the intensive growth processes, the model in coordinates $1/y' \sim t$ is often used. Where y′ are current values and t is time.

This model is a hyperbola and determines the vertical asymptote position at which the function value reaches an infinite value, which is impossible for real systems. Therefore, by the vertical asymptote position, an upper estimate can be made to determine the intensive growth process completion time. To do this, the linearized data in coordinates 1/y′ ~ t must be extended to the intersection with the abscissa axis.

This model with our data will look like this (Fig. 2).

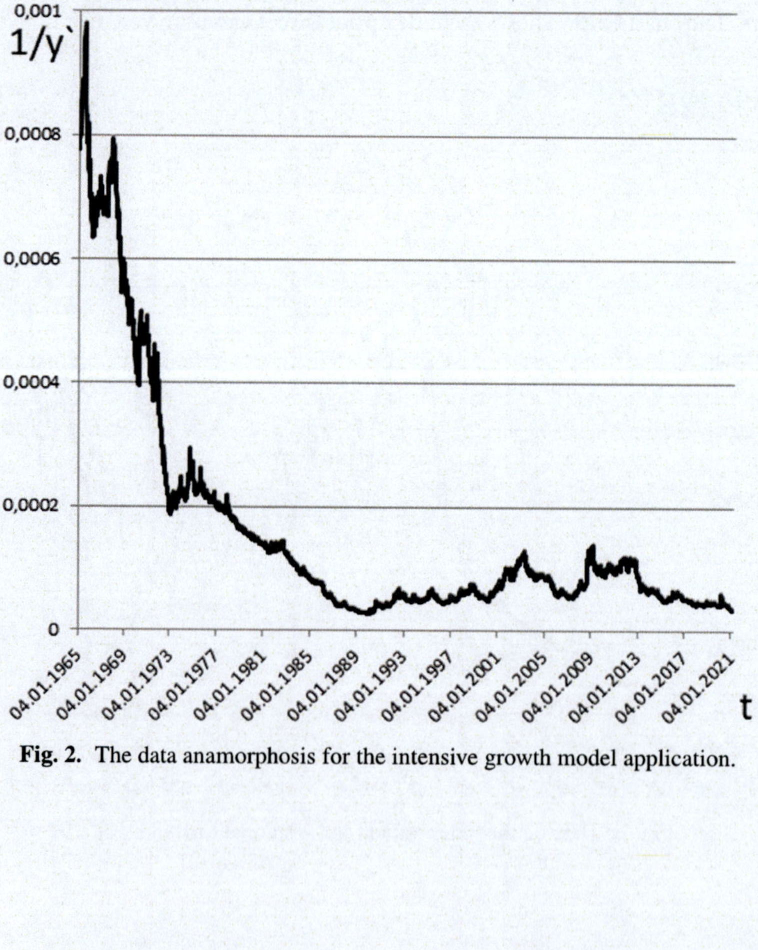

Fig. 2. The data anamorphosis for the intensive growth model application.

Let's draw several obvious lines that characterize the intensive growth process (Fig. 3):

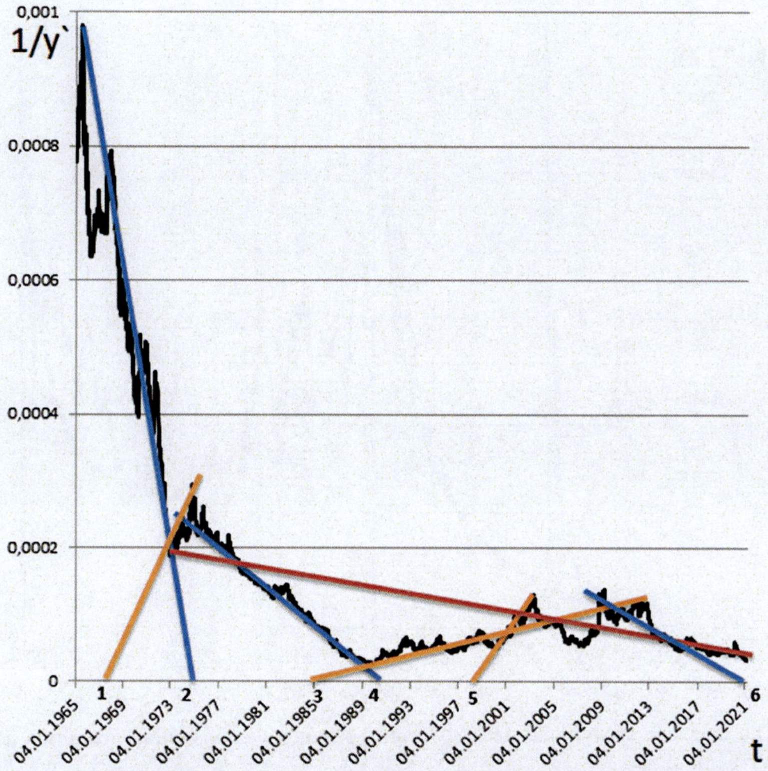

Fig. 3. Intensive growth anamorphosis linearization system.

Blue lines are responsible for intensive growth processes, orange - for intensive decline processes. The red line describes the intensive growth process, which has an asymptote, which value is outside the studied time domain and is predictive. The time axis intersection point will be in 2032 - it is by this time that an intensive increase in the value is expected.

Let us restore the vertical asymptotes position and the intensive growth/decline processes identified during the analysis (Fig. 4):

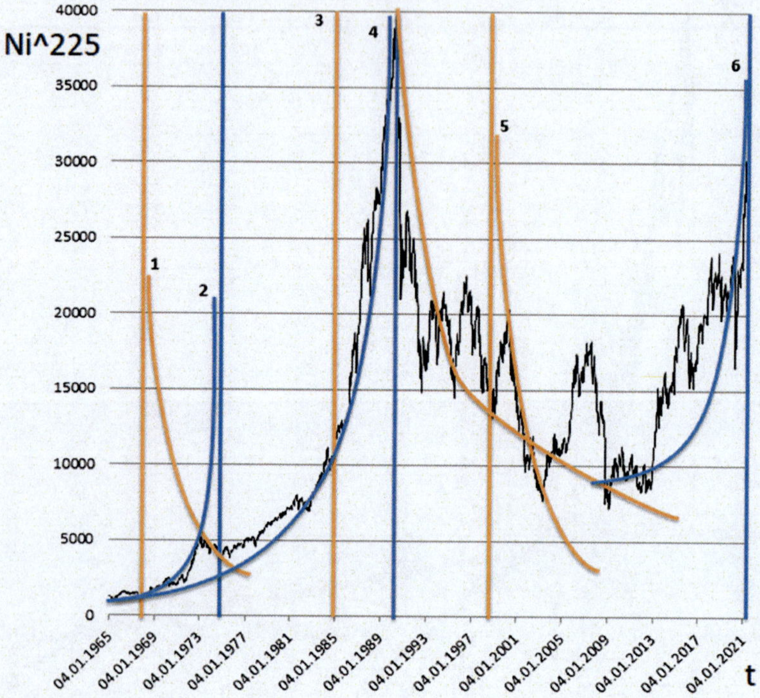

Fig. 4. The intensive growth and decline identified trends demonstration, identified using the model on the initial data.

As a data analysis result using the intensive growth model, we obtained studied index characteristic growth/decline areas, as well as revealed a studied value dynamics predictive estimate.

4 Limited Height, Gompertc's Model

With bounded growth models, you can effectively predict a system growth limit. The only way to effectively identify the model parameters type is to construct empirical data in coordinates that give a linear relationship between the density function and the argument.

The most widespread are the Gompertc model, for which $f(x) = A * e^{-kx}$ and the logistic model obtained at $f(y) = k(y_{\infty} - y)$, where y_{∞} - the growth limit.

For the Gompertc model, let's make construction in the coordinate's ln (y′/y) ~ t. Where y′ is the current values, y is the accumulation, and t is the time (Fig. 5).

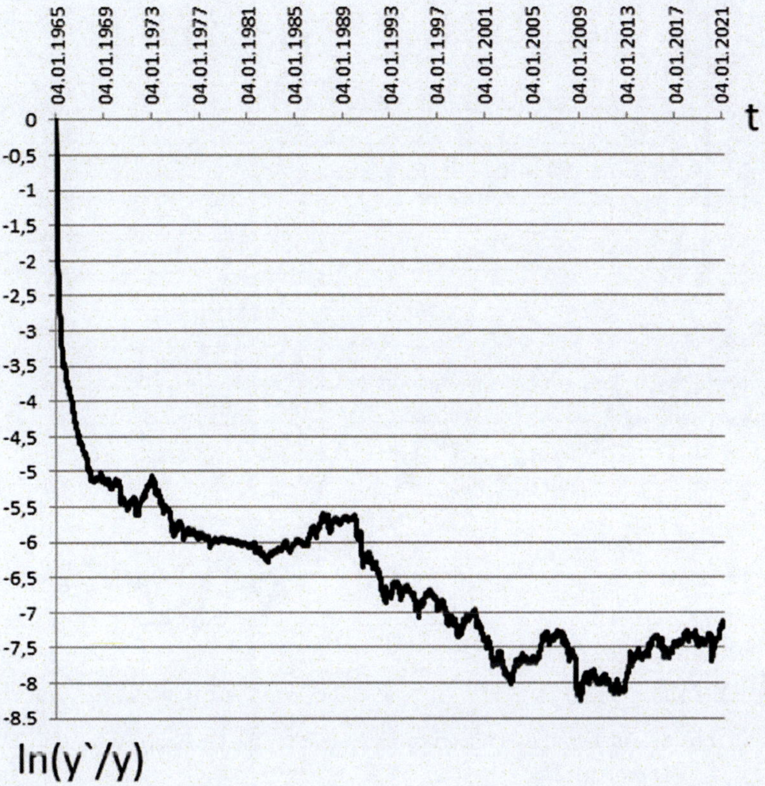

Fig. 5. Data anamorphosis for the restricted stature Gompertz model application.

Let's select the main linear sections (Fig. 6):

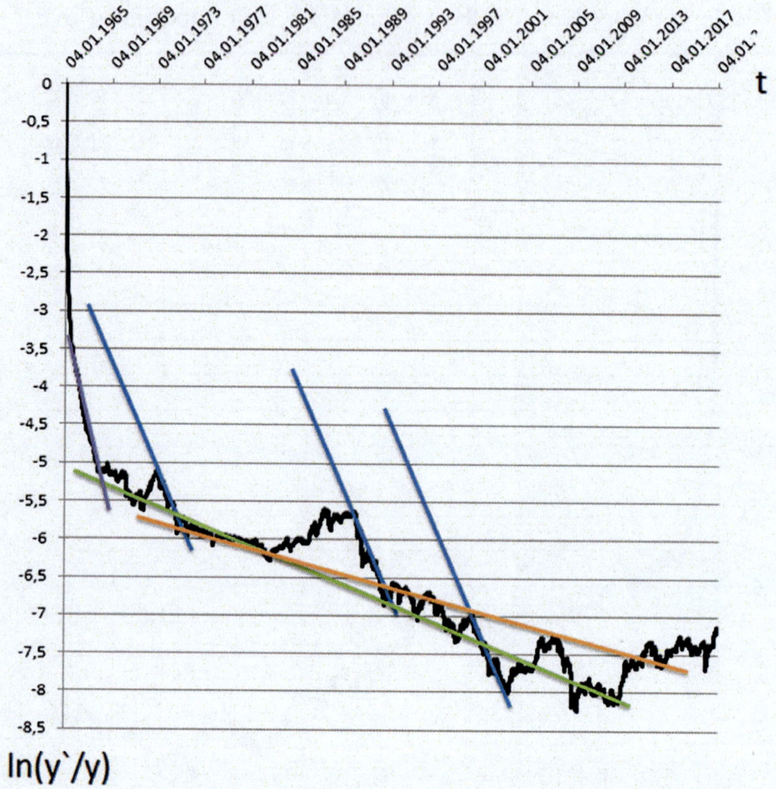

Fig. 6. Anamorphosis linearization system for the Gompertc model.

It should be noted right away that there are several parallel blue straight lines on the chart. With further calculation, these lines slope coefficients will be equal. Having made the appropriate calculations, we obtain the slope coefficient for the violet straight line k1 ≈ 0.016572, for the blue straight lines - k2 $= 0.006097$, for the green - k3 ≈ 0.0012434, for the orange - k4 ≈ 0.0006755.

Further, to obtain a cutoff on the time axis, we take the natural logarithm of each coefficient. ln (k1) ≈ −4.1, ln (k2) ≈ −5.1, ln (k3) ≈ −6.9, ln (k4) ≈ −7.27. Draw horizontal lines on the calculated logarithms value, find their intersections with oblique - these intersections will give t * - inflexion points (Fig. 7).

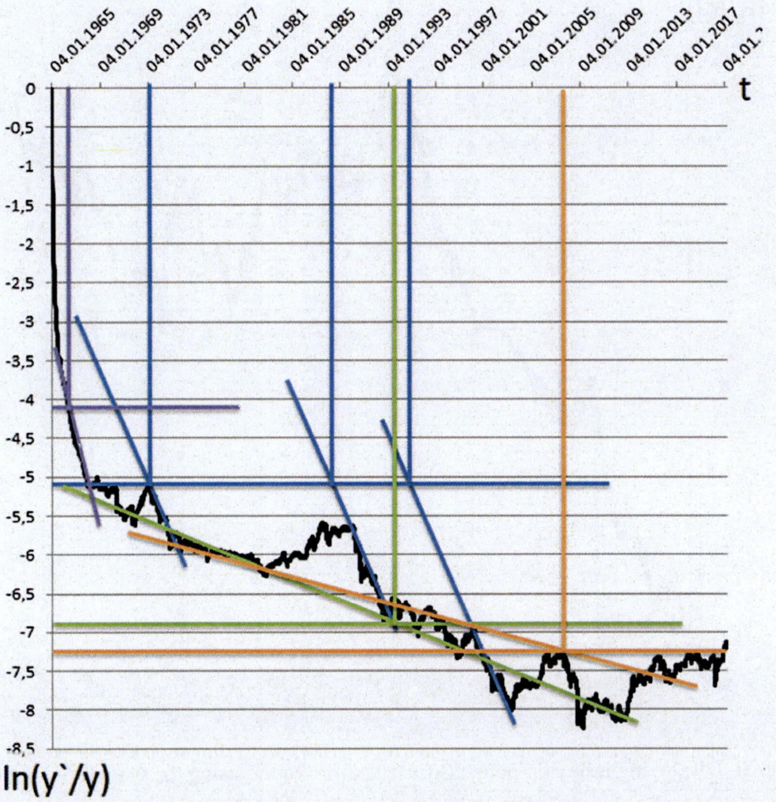

Fig. 7. Restoring the Gompertc model inflexion points position.

Let's transfer the obtained data to a semi-logarithmic graph. Then the maximum points due to the limited growth dynamics will be located at the following points (Fig. 8):

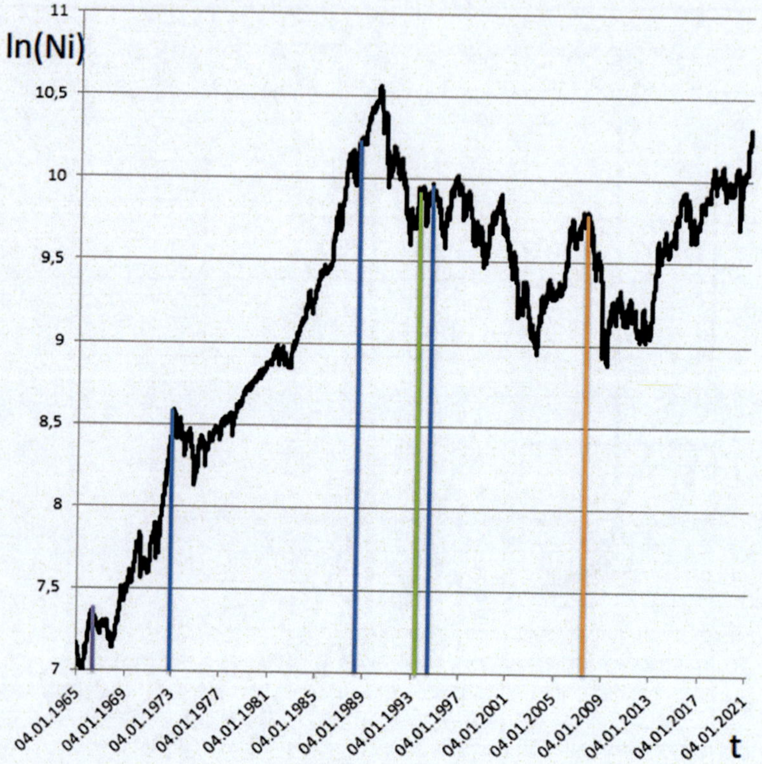

Fig. 8. Displaying inflexion points (local maxima) found using the Gompertc model

5 Limited Growth Logistic Model

Consider a logistic model based on the equation $\frac{dy}{dt} = k * y\left(1 - \frac{y}{y\infty}\right)$, where $y\infty$ is the growth limit. As a linear relationship intersection result with the abscissa, the growth limit is determined. For this model, we get a plot in coordinates y′/y ~ y. Let's take a look at the current data example (Fig. 9):

Fig. 9. Data anamorphosis for the limited growth logistic model application.

Let's draw several simple lines that approximate the logistic model anamorphosis (Fig. 10):

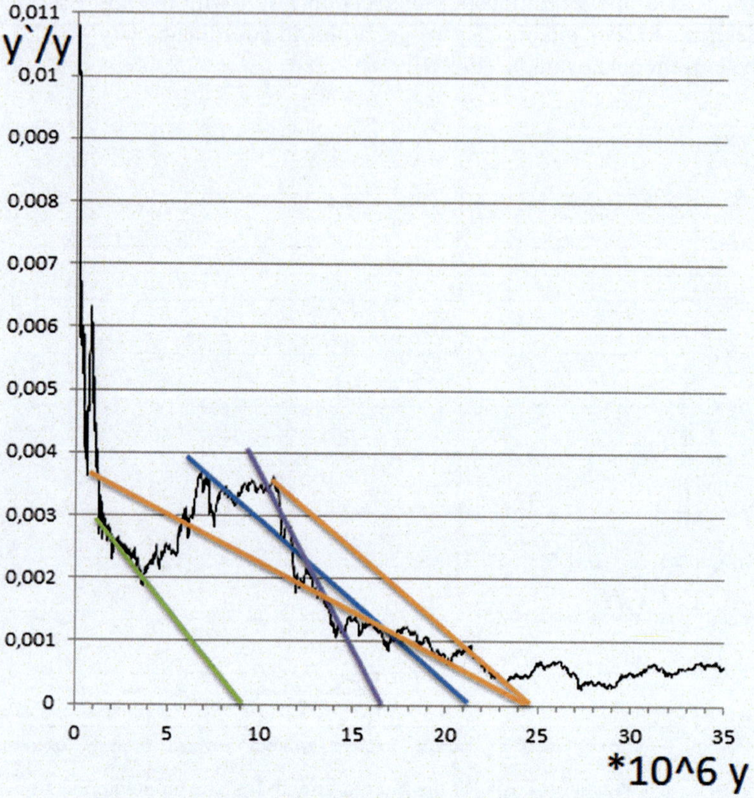

Fig. 10. Linearization system for limited growth logistic model.

The y values are this model growth limits for the selected lines: $9 * 10^6$ for the green line, $16.4 * 10^6$ for the violet, $21 * 10^6$ for the blue and $24.5 * 10^6$ for the orange pair. To restore the inflexion point, divide the growth limits values by 2, and restore the value t^* from y' value for the calculated accumulation value. Let's arrange the obtained points on the semi-logarithmic coordinates graph (Fig. 11):

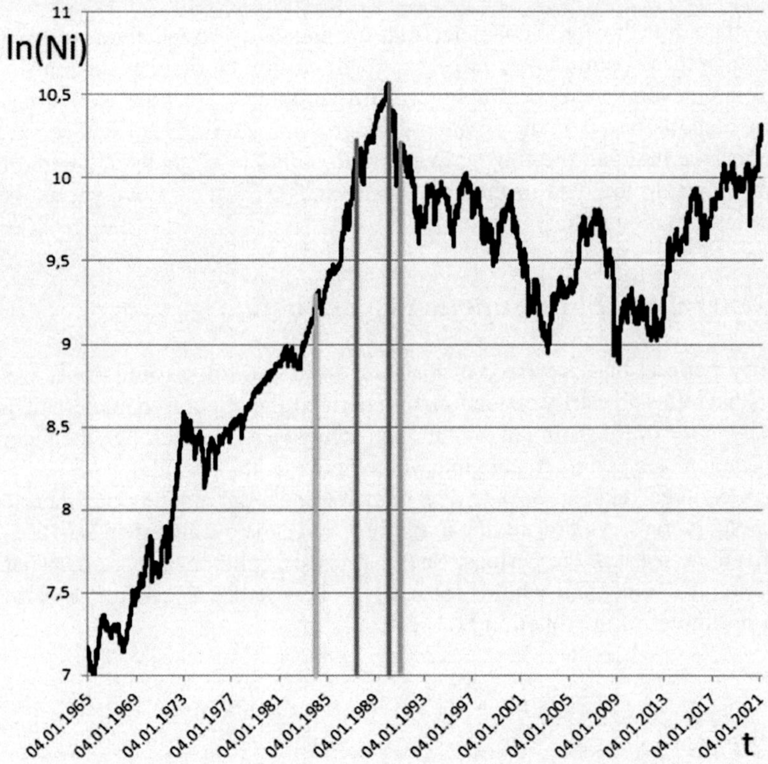

Fig. 11. Displays the local maximum points found using the limited growth logistic model.

Summarizing the data analysis results using limited growth models, it can be noted that both models made it possible to determine with good accuracy the time moments at which the studied index values reached their maximums.

6 The Trend Exclusion

In general, empirical data analysis and processing are often associated with difficulties in identifying fluctuations accompanying trends. Without extracting the trend component, the fluctuations' analysis is difficult to reliably carry out, since the trend distorts the result. There are no regular methods for solving this problem.

We will assume that trend characteristics are encoded through pivot points. The task is to find such these points a position, which will ensure the trend exclusion.

There are several methods for solving this problem. Consider excluding a trend geometric method, based on the geometric progression $y_t = \sqrt{y_{t-\Delta t} * y_{t+\Delta t}}$. Where Δt is a certain time interval, the shift, on which the points $y_{t-\Delta t}$ and $y_{t+\Delta t}$ stand from y_t. Then, transforming the data in the coordinates $\ln \frac{y_{t-\Delta t} * y_{t+\Delta t}}{(y_t)\hat{}2}$ ~ t, we get an exception from the trend sections data.

7 A Generalized Shift Function Construction

To identify periods that are free, if possible, from a priori assumptions, we use an approach that relies primarily on the function period fundamental characteristic property $f(t+\tau) - f(t) = 0$, consisting in the function values repetition through the independent variable equal to the period change interval, where τ is the period.

As a rule, in real data, we have to deal with nonlinear fluctuations, and therefore pure periods are quite rare. As a result, the maximum that can be counted on is to identify the values that are closest to the periods. Such values are called near-periods. That is, τ is a near-period if the inequality holds for it, $|f(t + \tau) - f(t)| < \mathcal{E}$, where >0 is the offset. Then for n samples of the function $f(t)$, we get

$$a(\tau) = \frac{1}{n-\tau} * \sum_{t=1}^{n-\tau} |f(t+\tau) - f(t)|$$

This function is called the shift function. Thus, we are looking for the values closest to zero.

To estimate the near-periods system in experimental data with a trend, it is necessary to expand a shift function concept to include, in addition to the argument τ, the value Δt.

This function minima system investigation in τ and Δt terms will make it possible to determine a near-periods values set corresponding to the initial empirical series. It is possible to identify both individual local minima and τ or Δt such values at which minima channels are formed at variables one fixed value.

To identify such channels, the function is built.

$\Phi(\tau) = \frac{1}{N} * \sum_{\Delta t=1}^{N} a(\tau, \Delta t)$,

where N is the values' number in the corresponding row, and the function $a(\tau, \Delta t)$ is a generalized shift function, which for the excluding a trend geometric method has the form:

$$a(t, \Delta t) = \frac{1}{n - \tau - \Delta t} \sum_{t=1}^{n-\tau-\Delta t} \left| \ln \frac{y_{t-\Delta t+\tau} * y_{t+\Delta t+\tau}}{(y_{t+\tau})^2} - \ln \frac{y_{t-\Delta t} * y_{t+\Delta t}}{(y_t)^2} \right|$$

For the selected data, the generalized shear function level lines graph has the form (Fig. 12):

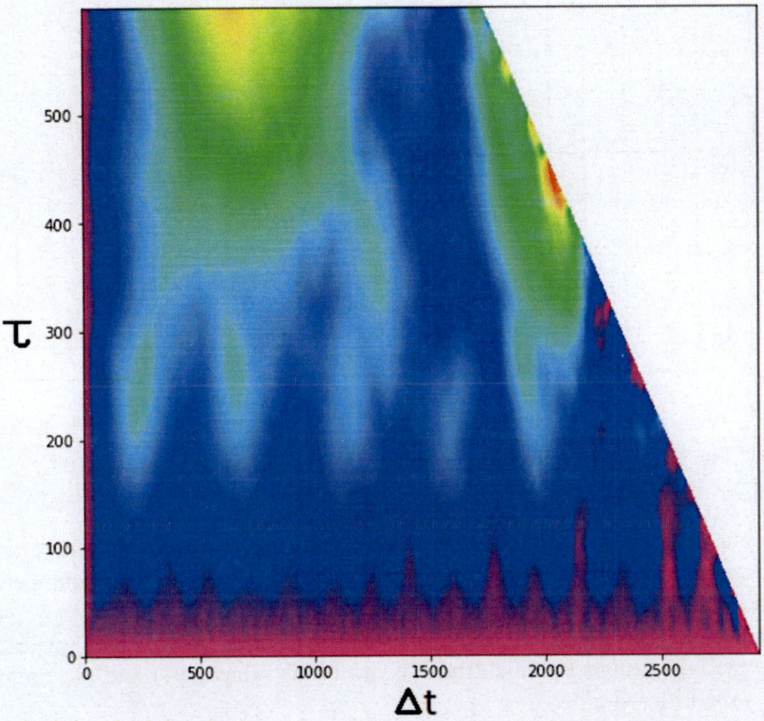

Fig. 12. The generalized shift function level lines graph.

The function section at $\tau = 100$ will be the following graph (Fig. 13):

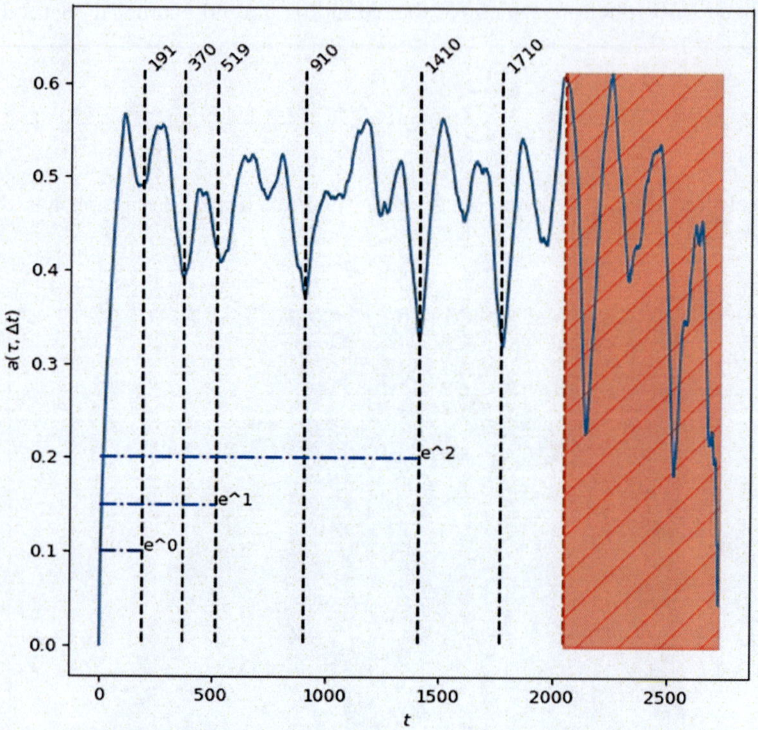

Fig. 13. The generalized shift function cross-section graph at $\tau = 100$.

The data area that is usually not considered in the current analysis is highlighted in red - the last quarter. It is not considered due to the added elements small number, which distorts the result.

The geometric progression in the near-periods values is remarkable: 3 near-periods formed a periods system in which there is an initial value (191) and the progression modulus – the Euler number.

8 Defining the System Development' Cell

When constructing relations scales, the problem is to determine such numbers that singly determine the natural systems' structure. These numbers' one - e = 2.71828 - exponent, Napier's number.

The system development' cell is an arithmetic progression synchronization model with a step of T_k and a geometric progression X_k with a denominator, which has the following form (Fig. 14):

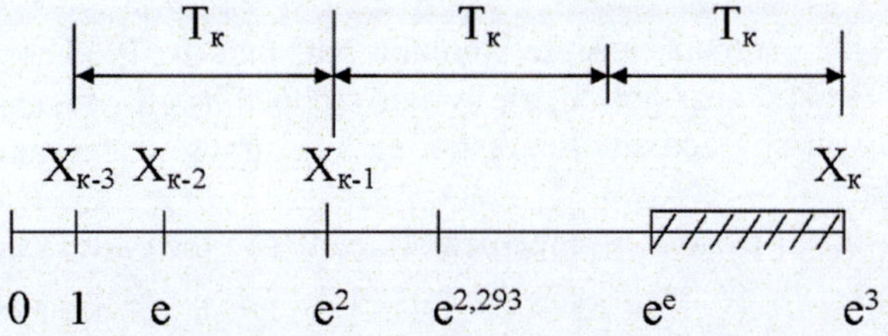

Fig. 14. Arithmetic and geometric progressions' structure synchronization - development' cell.

The relationship between X_k and T_k is done according to the following equation:

$$X_k = \frac{2e}{e-1} T_k \approx 3.164 * T_k$$

In applying the model, special attention is paid to the milestone e^e - from it the so-called restructuring phase begins - an area in which systems undergo unstable fluctuations in preparation for the transition to a new characteristic milestone.

The condition for the progressions' synchronization determines the zero point position and the unit value from which the geometric progression begins, corresponding to the arithmetic progression step T_k.

Consider the system development cell application to the selected data using the calculated near-periods. The first near-period is 191 weeks. Then $X_k \approx 603$ weeks (about 11.5 years), and the system development cell will look like this (Fig. 15):

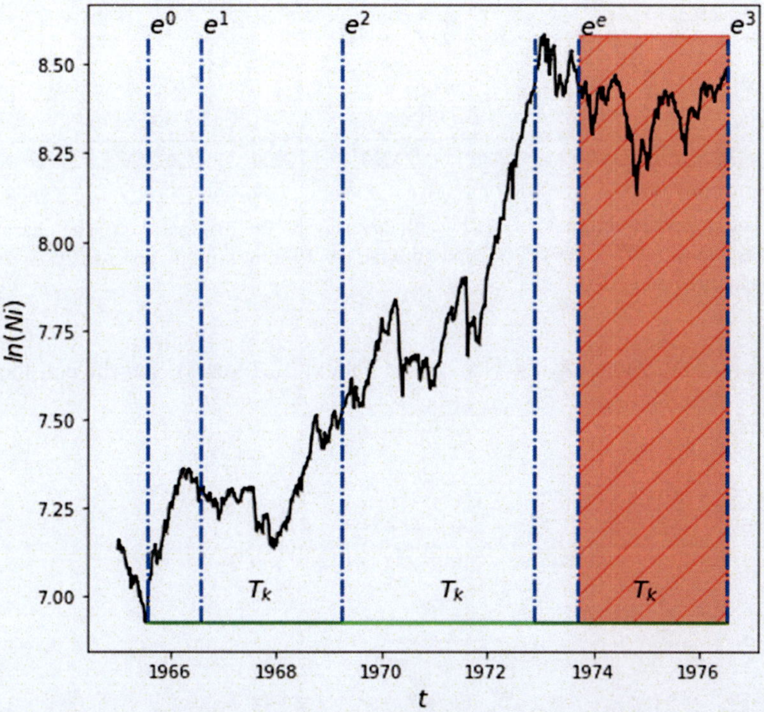

Fig. 15. Development cell for Tk = 191 weeks.

It is seen that the geometric progression unit falls into the 1965 absolute minimum, and at the end of the 2nd near-period, the system stopped its growth, and in the red zone - the restructuring phase - actively fluctuates, ending almost at the same level at which it entered this phase.

Let's look at the all-periods arrangement $T_k = 191$ in the entire sample (Fig. 16):

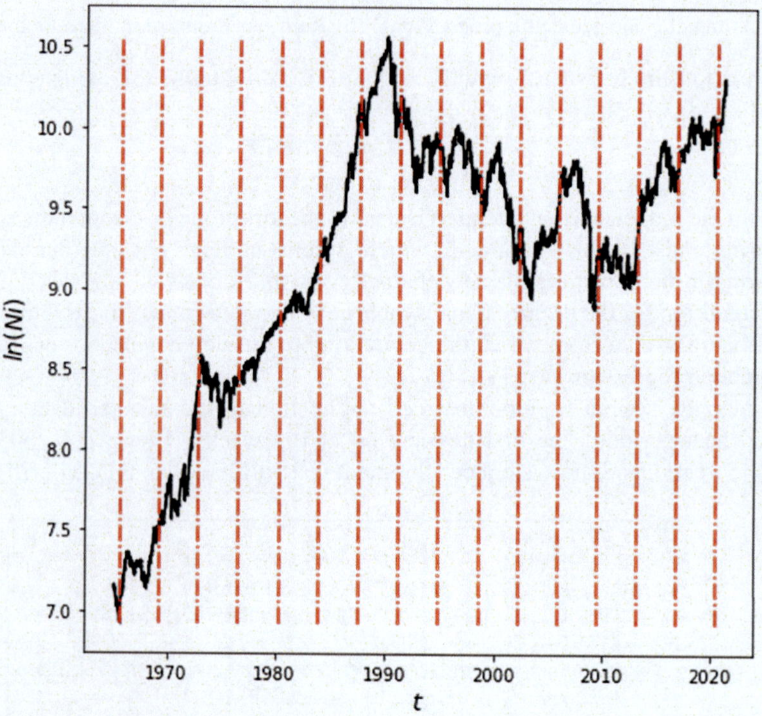

Fig. 16. The period structure Tk = 191 weeks overlay on the original data. The characteristic points are repulsed - 1987, 1991, 2002 maximums, the 1998 minimum, jerks in 2005, 2009 and 2013, dips in 2016 and 2020.

For $T_k = 370$, we get $X_k \approx 1165$ weeks (about 22.3 years), and the cell looks like this (Fig. 17):

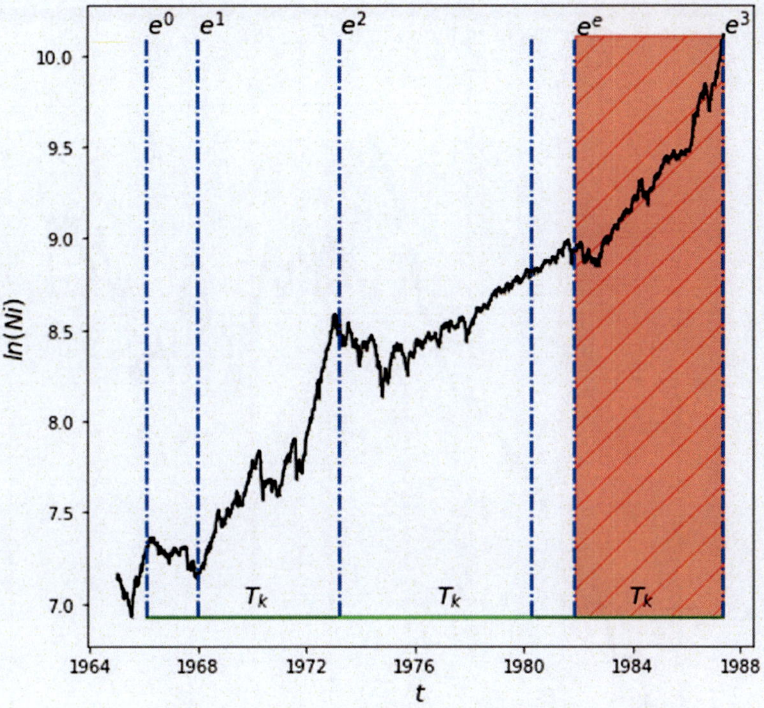

Fig. 17. Development cell for Tk = 370 weeks.

It can be seen that the cell records the 1968 important maximums and minimum positions with good accuracy. It can also be seen that when it enters the restructuring phase, the system has lost its uniform exponential growth (since the graph is presented in semi-logarithmic coordinates, the oblique lines on it represent exponentials), fluctuates slightly and increases the growth rate.

All near-periods arrangement in the sample (Fig. 18):

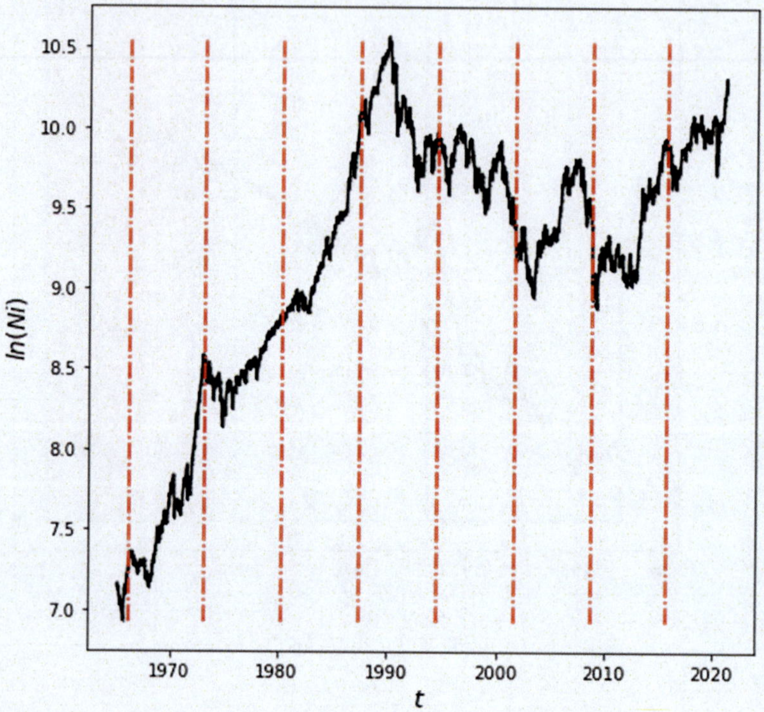

Fig. 18. The period structure Tk = 370 weeks overlay on the original data. In addition to the periods included in the cell, characteristic highs in 1994 and 2015, recessions in 2001 and 2008 were recorded.

For $T_k = 519$, we get $X_k \approx 1627$ weeks (about 31.2 years) (Fig. 19).

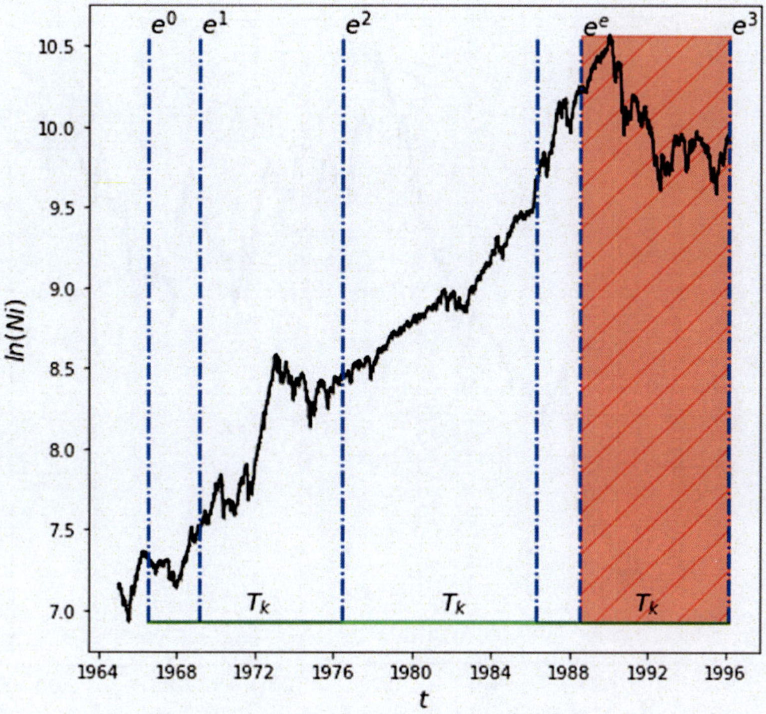

Fig. 19. Development cell for Tk = 519 weeks.

This cell clearly shows how the end of the 2nd near-period ends the local spurt in the growth rate, and in the restructuring phase, the system reaches its absolute maximum and goes into decline.

Near-periods further arrangement (Fig. 20):

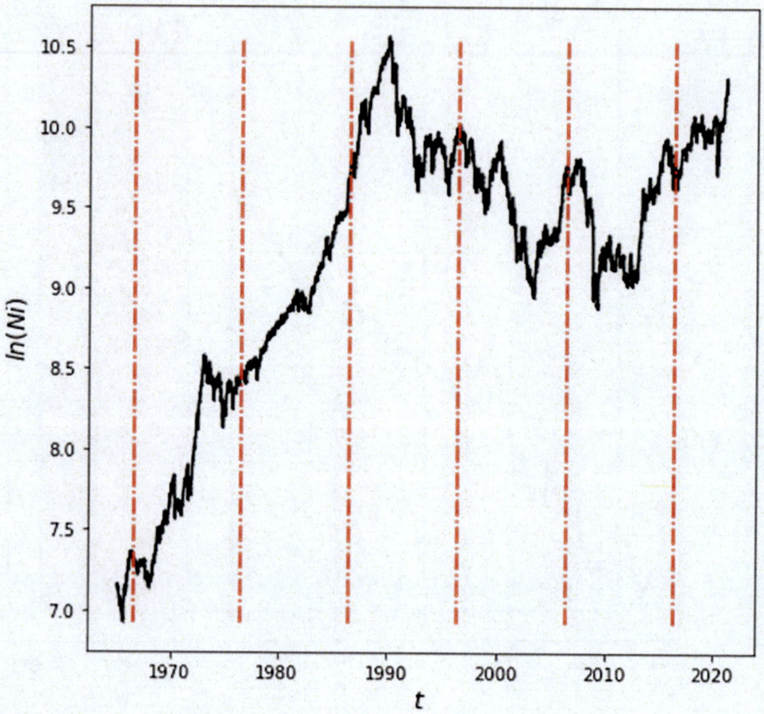

Fig. 20. The period structure Tk = 519 weeks overlay on the original data. After the end of the 3rd Tk period, 2 maximums in 1996 and 2006, and a minimum of 2016 was hit.

The last cell, which completely fits into the available data sample: T_k = 910, X_k ≈ 2852 weeks (about 54.7 years) (Fig. 21).

This cell fixes well the next exponential growth endpoint position in 1985, reveals the maximum position in 2002 and explains the fluctuations from 2006 to 2019-the restructuring phase.

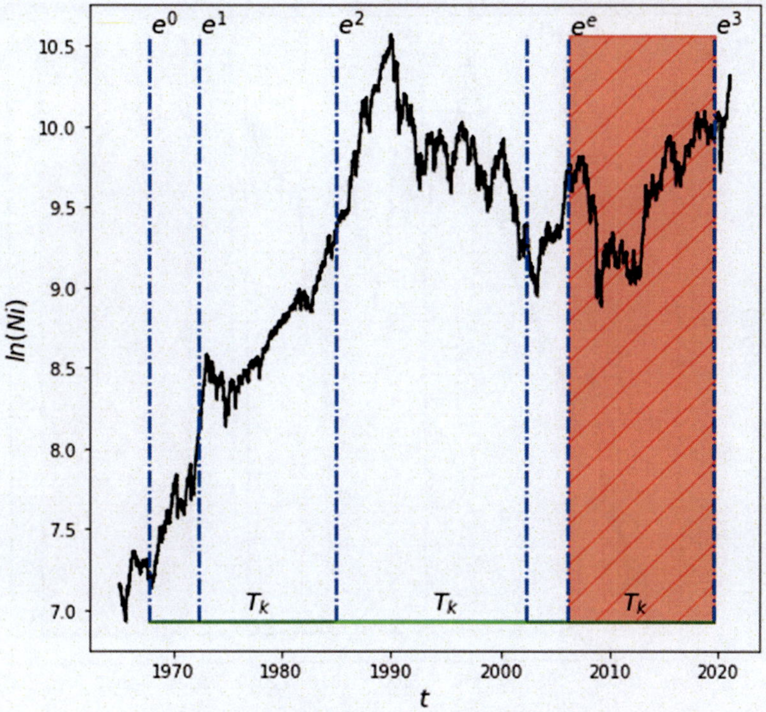

Fig. 21. Development cell for Tk = 910 weeks.

It can be seen that a short time after the development cell end, the system underwent a sharp decline and began to grow. If you look closely at the system behaviour within the intervals, you will notice that this behaviour is also observed at the 1st and 3rd clock cycles beginning.

All the following near-periods are the basis for cells that are larger than the available data sample. So, for $T_k = 1410$, $X_k \approx 4439$ weeks (about 85.1 years) - 2050th year (Fig. 22).

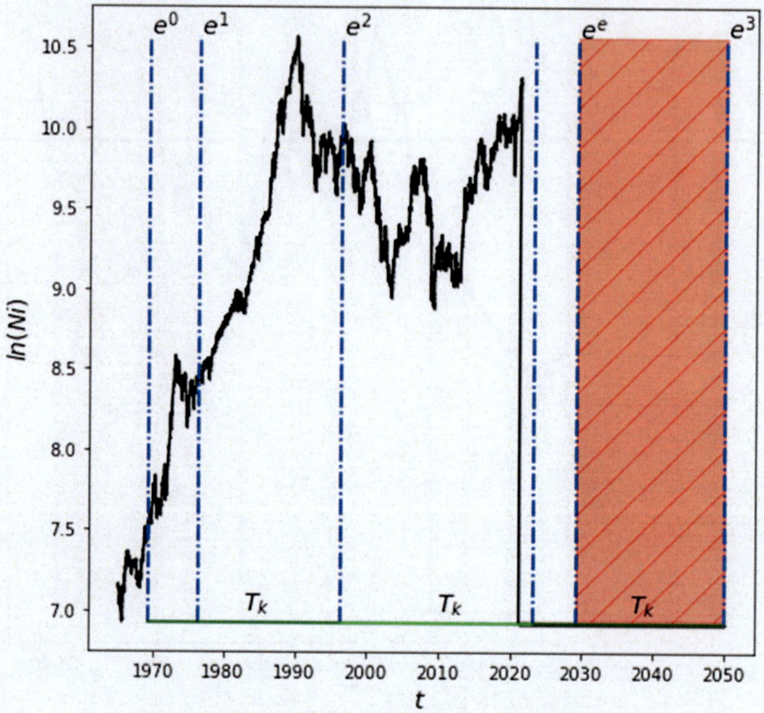

Fig. 22. Development cell for Tk = 1410 weeks.

This cell catches the1996 maximum position and allows predicting the characteristic point around March 2023 and the restructuring area in the period 2029–2050.

The last near-period is $T_k = 1710$, $X_k \approx 5383$ weeks (about 103.2 years) - that is, the cell until 2068 (Fig. 23).

This cell defines 1970 high and the late 2002 low with good accuracy. A characteristic point is assumed in 2035, and a restructuring phase in 2042–2068.

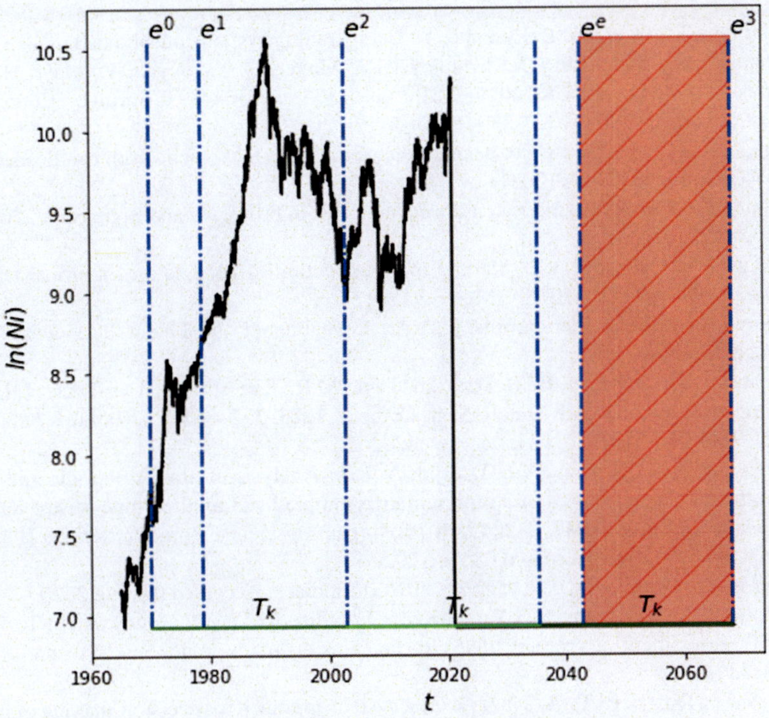

Fig. 23. Development cell for Tk = 1710 weeks.

9 Conclusion

The analysis application made it possible to identify the intensive growth stages in the periods from 1965 to 1973, from 1965 to 1990 and from 2009 to 2019, as well as recessions from 1973 to 1975, from 1990 to 2003. According to the forecast made on the basis, the indicator is expected to grow up to 2032–2035.

Critical levels analysis in the system development based on the development' cell predicts characteristic events at the 2022 end; by 2024; in the 2026 spring; in 2037. The development cells that have not yet finished predict the characteristic points in 2023 and 2035, as well as the restructuring phases in the periods 2029–2050 and 2042–2068.

Summarizing the forecasts, we can say that soon, the system will increase its value, focusing on the 2035 asymptote, along the way, the system will have an extreme points whole queue with 2 years until the current decade middle.

References

1. Inozemtsev, E.: Backstage games of management companies. HSE Econ. J. **21**(2), 283–311 (2017)

2. Pankratova, L., Paientko, T., Lysenko, Y.: Using trading system consolidated models in stock exchange price forecasting. In: Ermolayev, V., Mallet, F., Yakovyna, V., Mayr, H.C., Spivakovsky, A. (eds.) ICTERI 2019. CCIS, vol. 1175, pp. 364–391. Springer, Cham (2020). https://doi.org/10.1007/978-3-030-39459-2_17
3. Hung, C.-C., Chen, Y.-J.: DPP: deep predictor for price movement from candlestick charts. PLoS ONE **16**, e0252404 (2021)
4. Veniaminov, S.S.: Revealing Latent Structure Laws in Processes and Signals, p. 216. URSS, Moscow (2014)
5. Kuzmin, V.I., Gadzaov, A.F.: Method of Construction of Models on Empirical Data. The Manual, p. 94. MGTU MIREA, Moscow (2012)
6. Serebrennikov, M.G., Pervozvansky, A.A.: Revealing of the Latent Periodicity, p. 244. Science, Moscow (1965)
7. Dzerjinsky, R.I., Pronina, E.N., Dzerzhinskaya, M.R.: The structural analysis of the world gold prices dynamics. Adv. Intell. Syst. Comput. **1225**, 352–365 (2020). https://doi.org/10.1007/978-3-030-51971-1_29,lastaccessed2020/08/08
8. Paramonov, A.A., Kuzmin, V.I., Dzerjinsky, R.I.: Analysis of almost-periodic and almost-proportional characteristics of a representative sample local minima time series. IOP Conf. Ser. Mater. Sci. Eng. **1047**, 012045 (2020). https://iopscience.iop.org/article/10.1088/1757-899X/1047/1/012045. Accessed 08 Aug 2020
9. World indeces. https://finance.yahoo.com/world-indices. Accessed 08 Aug 2020
10. Chesalin, A.N., Grodzensky, S.Ya., Pham, V.T., Nilov, M.Yu., Agafonov, A.N.: Technology for risk assessment at product lifecycle stages using fuzzy logic. Russ. Technol. J. **8**(6), 167–183 (2020)
11. Janková, Z., Dostál, P.: Type-2 fuzzy expert system approach for decision-making of financial assets and investing under different uncertainty. Math. Prob. Eng. **2021**, 3839071 (2021)
12. Ahmad, M., Soeparno, H., Napitupulu, T.A.: Stock trading alert: with fuzzy knowledge-based systems and technical analysis. In: Roceedings of 2020 International Conference on Information Technology Systems and Innovation, ICITSI 2020, pp. 155–160. IEEE, Bandung (2020). Art. 9264914
13. Witayakiattilerd, W.: Fuzzy quantitative analysis method for stock selection into portfolio. Chiang Mai J. Sci. **46**(4), 799–811 (2019)

Scalable Architecture of Distributed Control System for Industrial Greenhouse Complexes

Konstantin Krestovnikov(✉), Dmitrii Korshunov, Aleksei Erashov, and Aleksandr Rogozin

St. Petersburg Federal Research Center of the Russian Academy of Sciences (SPC RAS), St. Petersburg Institute for Informatics and Automation of the Russian Academy of Sciences, 39, 14th Line, 199178 St. Petersburg, Russia
{k.krestovnikov,erashov.a}@iias.spb.su

Abstract. The development of greenhouse complexes with soilless technologies and artificial lighting makes it urgent to develop a control system that allows you to automate the growing processes. This paper presents the architecture of a distributed control system for industrial greenhouse complexes. The developed architecture is divided into levels and built on a modular basis. The use of standard modules makes the control system flexible and scalable. Communication between the modules is based on LoRa wireless two-way data transmission technology, which allows the system to be deployed over large areas. The architecture includes 3 types of executive modules, one combined sensor module and a control module. Each of the executive modules operates according to a given algorithm, and its parameters are controlled by a control module, based on a given growing program and information from sensors. This feature allows you to increase the reliability of the system and continue working in the event of a loss of connection with the cloud, as well as to exclude emergencies in the event of a loss of communication between the modules.

Keywords: Automation · Wireless sensor networks · Agriculture in a controlled environment · Internet of things · Vertical farms

1 Introduction

The development of greenhouse complexes near large cities, as well as within their boundaries, for example, the placement of vertical farms in urban development [1], allows you to solve logistical problems and reduce the cost of transporting agricultural products, as well as improve its quality. Due to their multi-level structure and the use of artificial lighting, the complexes of vertical trusses have significant planting areas. Vertical trusses can be classified according to the size and number of levels. However, each system is built on the groundless transfer of essential plant nutrients. The most common option is a hydroponic system based on feeding a nutrient solution to the plant roots. Aeroponic and aquaponic systems are less common. Modern trends in management methods for greenhouse complexes are discussed in [2]. This study provides an overview

R. Silhavy et al. (Eds.): CoMeSySo 2021, LNNS 231, pp. 127–132, 2021.
https://doi.org/10.1007/978-3-030-90321-3_12

of modern greenhouses, controlled environment agriculture and their derivatives, and highlights some of the advances in environmental monitoring, control and optimization. The authors emphasize that for the effective operation of vertical farms, it is necessary to control a significant number of environmental parameters. To solve this problem, as well as many others, typical for industrial greenhouses, it is possible to use distributed control systems that are widely used in industry and robotics [3]. An important property of such systems is scalability, which determines the ability of the system to adapt to different areas of the complex and the number of controlled parameters. The development of such systems is an urgent task, since the automation of the processes of agro-industrial complexes will increase their productivity and product quality [4].

The exchange of information between modules of distributed control systems for greenhouse complexes can be organized using wired or wireless interfaces. The first method was used in work [5], which presents an automatic control system for a greenhouse. The use of this system allows to achieve a reduction in water consumption when watering plants. Data transmission using wired technologies with a large number of actuators and sensors degrades the scalability of the control system and increases the cost of its deployment. Wireless technologies offer scalability advantages and make the system more flexible. These features were used by the authors of [6], and considered the use of various technologies of wireless data transmission in the development of a monitoring system for vertical farms. Wi-Fi, Bluetooth and ZigBee have a number of restrictions on the range of devices relative to each other. The use of LPWAN modulation using LoRa allows to increase the radius of the arrangement of devices in the system, to ensure low power consumption of devices and a low cost of implementation of the system.

The systems that exist at the moment have a number of disadvantages, among which are low adaptability to scalability or the absence of two-way communication between the devices of the system. Of the considered solutions, systems based on wireless data transmission technologies have the best scaling capabilities, the use of which with a modular principle will create a flexible control system for greenhouse complexes. This paper proposes a scalable architecture of a distributed control system for automating the functioning of industrial greenhouse complexes.

2 Architecture of Distributed Control System for Industrial Vertical Farms

Based on the analysis carried out, to form a scalable system with the possibility of reconfiguration in the developed architectural solution, wireless data transmission using Lora-WAN technology is used. This technology allows to ensure the energy efficiency of the system and to reduce the overall performance of the system devices. Based on the data published in articles [7, 8], the use of this technology allows the system to be deployed within a radius of 3 km. This technology is characterized by low cost of radio modules, wide frequency range and high security of communication channel protocols.

The proposed control system architecture, shown in Fig. 1, has three layers. Information transfer between modules of different levels occurs via bidirectional communication channels.

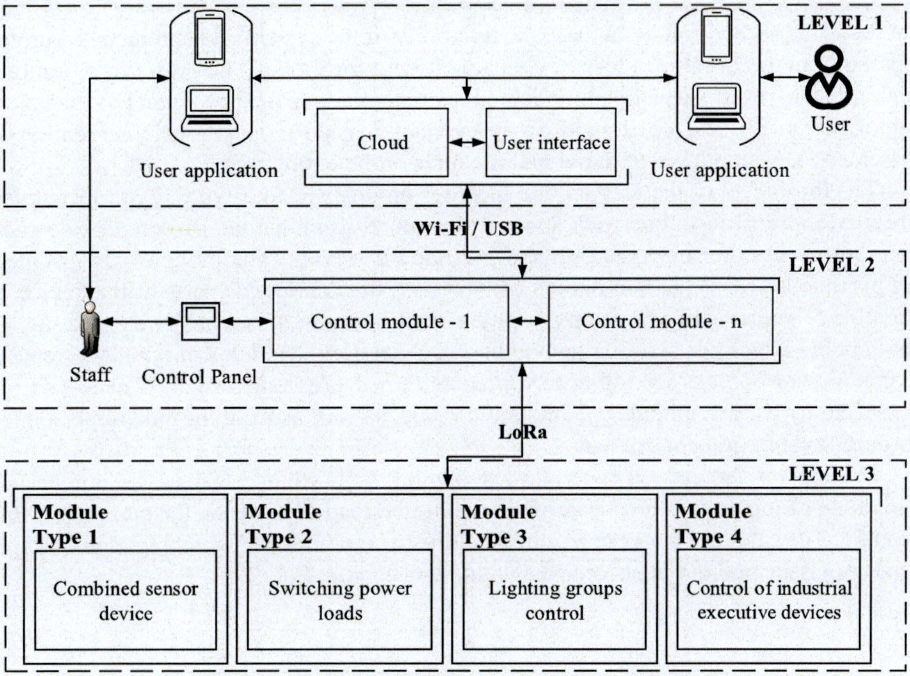

Fig. 1. General architecture of the control system.

The access to the first level of the system through the cloud service [9, 10] has the maintenance personnel and technologists of the greenhouse complex, other interested persons may have access with limited rights. Production technologists have the ability to change all parameters of the system and have access to statistical data and information from sensors. Employees with appropriate access rights can control the operating parameters of the greenhouse complex and participate in the manual control of technological processes.

The second level of the system is represented by control modules. The number of control modules is calculated according to the following Eq. (1) with the resulting result rounded up:

$$M_C = \frac{A_{T1} + A_{T2} + A_{T3} + A_{T4}}{64}, \quad (1)$$

where $A_{T1}, A_{T2}, A_{T3}, A_{T4}$ are the number of modules of the corresponding types.

Control modules allow you to decentralize the control of the entire system, as well as change the type and number of executive modules, adapting the system to the characteristics of each greenhouse complex. The maximum number of modules connected to one control module is limited to 64. This is due to the fact that LoRa technology has a low data transfer rate, which in turn leads to an increase in control time delays with an increase in the number of controlled modules. Control modules have the ability to operate autonomously according to a pre-programmed program and correct the operation of third-level modules by sending messages about changing setpoints or operating modes. This allows you to increase the reliability of the system and continue its correct operation in the event of a loss of connection with the cloud. The presence of buttons and a display in the developed module allows the maintenance personnel to manually control the processes, for example, in the event of a need for operational intervention or for emergency shutdown of individual modules or the entire system.

The third level of the architecture includes modules of four types. Type 1 modules are used to transmit information about temperature, illumination, carbon dioxide concentration and other information from greenhouse sensors. The modules are installed on greenhouse racks, so that human contact with the conductive parts of the device is possible. Therefore, in order to reduce the risk of injury to personnel by electric shock, the module is powered from a low-voltage DC network. Modules of type 2–4 control actuators and operate according to a predetermined program, which, if necessary, is corrected by control modules in automatic mode, as well as from the side of personnel through a cloud or a control panel.

Modules of the second type allow switching high-voltage loads. The number of switching channels and the power of the connected load depends on the circuitry of the module. The calculation of the required number of modules for switching high-voltage loads M_k is carried out in accordance with the following Eq. (2):

$$M_k = \frac{N_{hl}}{n_{ch}}, \tag{2}$$

where N_{hl} is the number of switched load units and n_{ch} is the number of switching channels of each module.

The optimal number of switching channels is selected based on the features of the module application and permissible dimensions, and if it is necessary to switch a large number of loads in one place, it is possible to use several typical modules. The use of these modules is relevant for switching the relief valves of high-pressure pumps of the aeroponic system, drainage pumps and duct ventilation drives.

Modern greenhouses use LED lamps with controlled power supplies. Such power supplies allow you to adjust the power to set the required brightness for each lighting group in accordance with the requirements for growing a certain type of agricultural products. In addition to providing the required growing conditions, brightness control can reduce the power consumption of lighting elements. The purpose of the modules of the third type is to control the drivers of LED lighting lamps. The number of required modules of the third type in the greenhouse complex is calculated by the following Eq. (3), the result is rounded up to the nearest whole:

$$M_L = \frac{S_{dr}}{n_{ch} \cdot N_{dr}}, \tag{3}$$

where S_{dr} is the number of lighting lamp drivers, N_{dr} is the number of drivers in one lighting group, and n_{ch} is the number of module control channels.

To maintain the correct operating modes of the greenhouse complex, it is necessary to control the operation of high-pressure pumps, circulation ventilation and an air conditioner. Typically, industrial solutions are based on asynchronous three-phase motors controlled by a frequency converter (FC). To control the inverter and other industrial actuators, modules of the fourth type are provided, which include several groups of potential-free contacts and an output with an analog signal in the range of 0–10 V with a low level of ripple.

Crop programs are a set of parameters that are stored in a cloud service. When the user selects a specific program, a structured dataset is sent to each of the control modules. The modules of the second and third levels are independent of each other, and can simultaneously implement different growing programs. The data set received by the control module contains the parameters of the microclimate, lighting and irrigation modes, and others related to maintaining correct operation. Further, the control module processes the data and forms a set of functional setpoints for the executive modules. Functional setpoints are values that are used without conversion by level 3 modules. For example, for lighting modules, they will represent the time intervals in which the lamps should work and the intensity of their luminescence. In this case, the low-level algorithm of the third-level modules always remains unchanged, and the obtained setpoints are valid until new ones are received, which does not require constant communication with the control modules and increases the system's reliability. The control modules receive feedback on the monitored parameters from the combined sensor modules. In case of deviations in the feedback values, the control modules correct the settings of the executive modules. If the correction of the parameters is impossible or has not changed the controlled parameter, the control module sends a notification about this to the cloud service. Further, the cloud service notifies users and service personnel about an emergency situation that requires intervention.

The division of the specialization of the modules makes it easy to adapt the system in accordance with the peculiarities of the greenhouse complex. Also, if necessary, it is possible to develop and implement modules with special functionality, while it is not necessary to change the system architecture, control methods and other modules. The developed architecture of the system provides a number of advantages over the already existing options for control systems for industrial greenhouse complexes: scalability, decentralization, large coverage areas without the need to lay information lines.

3 Conclusion

The paper presents the architecture of a distributed process control system for industrial greenhouse complexes. The proposed architecture consists of three levels, which include modules designed for automatic control of the greenhouse climate, lighting and plant nutrition. The paper presents the basic design ratios to determine the required number of modules for each level of architecture. The use of the modular principle makes it possible to implement a decentralized control system and adapt it to the specifics of the greenhouse. The typical structure of the modules makes it easy to expand the functionality of the system by developing new specialized solutions. Increasing the reliability of the system is achieved by applying the principle of autonomous operation of each executive module according to some basic algorithm, the parameters of which are adjusted by control modules. Thus, in case of loss of communication with the control module, the executive module will continue its work. LoRa technology, which is the basis for data exchange in the developed architecture, allows deploying a system with a range of up to 3 km using one control module.

Further research will focus on developing a module structure and software architecture for a control system with the proposed architecture.

References

1. Birkby, J.: Vertical farming. ATTRA Sustainable Agriculture. National Centre for Appropriate Technology (2016)
2. Shamshiri, R., et al.: Advances in greenhouse automation and controlled environment agriculture: a transition to plant factories and urban agriculture (2018)
3. Andreev, V.P., Pletenev, P.F.: Method of information interaction for distributed control systems of robots with modular architecture. In: Trudy SPIIRAN, vol. 57, pp. 134–160 (2018)
4. D'Autilia, R., D'Ambrosi, I.: Is there enough fertile soil to feed a planet of growing cities? Physica A **419**, 668–674 (2015)
5. Sivagami, A., Hareeshvare, U., Maheshwar, S., Venkatachalapathy, V.S.K.: Automated irrigation system for greenhouse monitoring. J. Inst. Eng. (India) Ser. A **99**(2), 183–191 (2018)
6. Wu, J., Tang, X., Lei, J.: Design of wireless monitoring system for greenhouse based on LoRa. J. Phys. Conf. Ser. **1650**(2), 022075 (2020)
7. Petajajarvi, J., Mikhaylov, K., Roivainen, A., Hanninen, T., Pettissalo, M.: On the coverage of LPWANs: range evaluation and channel attenuation model for LoRa technology. In: 14th International Conference on ITS Telecommunications (ITST), pp. 55–59. IEEE (2015)
8. Denisov, A., Shabanova, A., Sivchenko, O.: Data exchange method for wireless UAV-aided communication in sensor systems and robotic devices. In: Ronzhin, A., Rigoll, G., Meshcheryakov, R. (eds.) ICR 2020. LNCS (LNAI), vol. 12336, pp. 45–54. Springer, Cham (2020). https://doi.org/10.1007/978-3-030-60337-3_5
9. Davydov, D.S., Kashevnik, A.M., Kosicin, D.P., Shabaev, A.I., Shabalina, I.M.: Development of the platform for production planning using cloud computing technology. SPIIRAS Proc. **4**(23), 416–430 (2012). (in Russian)
10. Levonevsky, D.K., Vatamanyuk, I.V., Saveliev, A.I.: Multimodal information and navigation cloud system MINOS for corporate cyber-physical intelligent space. Softw. Eng. **8**(3), 120–128 (2017). (in Russian)

Image Segmentation of Acidity of Agricultural Lands in Eastern Siberia

D. A. Perfilyev[1], T. V. Avdyukova[2], I. S. Masich[1], P. A. Zakharov[1], and A. V. Raskina[1](✉)

[1] Siberian Federal University, Svobodny pr., 79, Krasnoyarsk, Russia
[2] Station of Agrochemical Service 'Solyanskaya', Pervomayskaya Street, 19, vil. Novaya Solyanka, Krasnoyarsk Territory 663953, Russia

Abstract. The paper proposes a mathematical model for segmentation of the acidity dynamics of agricultural lands in Eastern Siberia. The segmentation rule is based on the analysis of the image frequency characteristic acidity, which makes it possible to predict the change in the property of the segmented areas. The mathematical approach to the segmentation of the dynamics of agrophysical parameters of agricultural soils has proves to be effective in managing the quality of the natural energy state on the control plots. It is expected that the experience will be applied to the entire responsibility area of 'SAS' 'Solyanskaya' (912.4 thousand ha). Introduction and distribution of innovative practices and technologies will provide new opportunities for monitoring and management of the natural energy state quality of the East-Siberian territory (23 million ha).

Keywords: Image segmentation · Segmentation rule · Segmentation of the dynamics of agrochemical parameters

1 Introduction

The national food security problem cannot be solved without a profound digital transformation of technologies for monitoring and predicting the dynamics of key agrochemical parameters of soil productivity. New technologies should be focused on the optimal use and preservation of the natural energy reserve of agricultural lands.

The transformation of technology for monitoring and managing the quality of natural energy reserve of lands requires the widespread use of highly automated (smart) productivity quality management technologies. This should be based on mathematical analysis and prediction of the dynamics of key agrophysical parameters of soils.

Technologically, such transformation can be seen from the mathematical segmentation of the image.

The mathematical formulation of the segmentation problem is discussed in [1–6].

It is shown that the solution to the segmentation problem of the function $f(a)$ is to divide the image into $S = \{s_1, s_2, \ldots, s_k\}$, non-empty connected areas for which the LP rule takes true values only if any pair of elements from any subset of $s_i \in S$ satisfies the goal of interpreter.

R. Silhavy et al. (Eds.): CoMeSySo 2021, LNNS 231, pp. 133–144, 2021.
https://doi.org/10.1007/978-3-030-90321-3_13

Under alternative conditions, the goal is limited to estimating the dynamics of the distribution density of the segmented image properties.

In general, segmentation can be considered as:

$$\text{Seg} : f(a) \xrightarrow{Lp} S^{(Sem.)},$$

where: $S^{(Sem.)}$– semantic relations, $s_i^{(Sem.)}$ – the name of the $s_i \in S$ area, $LP(s_i^{Sem})$ – the «neighbourhood» model.

In [19], there are the main forms of syntactic neighbourhood relations that are typical for the image area. For the "Area" model of the image $|s_i| > 1$ [19]:

$$s_{10} = <P_i, \{b_i\}^1 \cup \{b_1 - b_{3,} b_5\}^2>,$$

where:* $b_i{}^1$ – u-nary attribute of the s_i area that includes a subset of the elements of the P_i. area. For example, it is the sign of the convexity of the shape of the area; * $b_1{}^2$–$b_3{}^2$, $b_5{}^2$ - models of possible neighbourhood relations of boundary elements of the area. For example, it is the cardinality of the boundary elements of a segmented area with other areas including the «background» area;

A satisfactory segmentation result for S set that is given is an ideal case and is suitable for the "classical" syntax of deterministic mapping of object properties in the image:

1) the problem representation allows for setting in advance the value of the iconic function $f(a)$ for the «background» area. The "background" area fills the entire image which contains the S areas representing the properties of the object. The "background" area is defined by the most contrasting value of the function relative to the segmented areas, usually close to the maximum (minimum) value;
2) the segmented s_i areas have differences in u-nary features (usually in the gradient: $\nabla f(a)$) and neighbourhood, they are also true for the "background" area. The areas have a boundary in the form of a closed external and possibly internal contour in accordance with the set of neighbourhood models of the boundary elements of the $b_1{}^2$–$b_3{}^2$, $b_5{}^2$ area.

In deterministic conditions, the LP rule is a comprehensive estimation of the $\nabla f(a)$ "distance" (proximity) of the $p_i \in f(a)$ current image element and the neighbourhood (adjacency) of the $p_j \in s_i$ form:

$$|p_i \colon \nabla f(a) - s_i| \to min., \text{ and } p_i \cap p_j = 1, \text{ where } p_j \in s_i. \tag{1}$$

Thus, every current element of the $p_i \in f(a)$ image refers only to one $s_i \in S$ area. Accordingly, segmentation is performed on the basis of the uniformity and connectivity of the $f(a)$ image elements.

Without the ideal conditions for displaying the properties of an object, there are regular physical violations of the shape of segmented areas due to distortions in the form of incomplete, false, thickened contours, "overlapping" areas and other related errors [1, 10, 11].

One of the ways to deal with errors is to modernize the rule (1) of the semantic component, as a configurable “radius” that allows for improving the quality of segmentation. Such a rule has the following form:

$$|p_i\text{: } \nabla f(a) - LP(s_i^{Sem})| < r_i^{Sem} \text{ и } p_i \cap p_j = 1\text{, where } p_j \in s_i, \quad (2)$$

where r_i^{Sem} – the radius of uniformity for s_i^{Sem}.

Thus, the assessment of the uniformity of a class can be changed in the process of segmentation [20, 21], by changing the r_i^{Sem} regarding the analysis of the uniformity of adjacent classes. In addition, an adjustment of the power S can be made on the basis of the analysis of the uniformity of s_i^{Sem} and the power of the “background” area.

The use of the adaptation principle (2) on the one hand improves the quality and on the other hand naturally limits the possibility of applying the $LP(s_i^{Sem})$ rule when using image analysis systems with different properties and in different physical display conditions. Moreover, $LP(s_i^{Sem})$ usually represents a priori knowledge based on iconic information, which limits the practical use of formalization [1–4] in solving the problem of interpretation.

The way out of this situation is to recognize the need to consider the image in a purely mathematical aspect, as continuous and differentiable function $f(z)$, problem-subject area and interpretation of the characteristics of the attributes of segmented image areas [22, 23].

The image of the problem-subject area is based on the given physical attributes of the process (object) under study:

Given $F\{z_1, z_2, \ldots, z_n\}$–the complex of attributes of an object represented by a set of images $f(z_i)$, as well as the rules $Lp(S_i^{Sem})$ that determine the uniformity radius r_i^{Sem} of the i-th attribute value.

The segmentation task can be solved by separating the function $f(z_i)$ into N areas (classes) for which $Lp(S_i^{Sem})$ takes true values that comply with the objectives.

In this case, the segmentation is considered a continuous process of analysing the dynamics of the image $f(z_i)$:

$$\text{Seg:} f(z_i) \xleftrightarrow{Lp(S_i^{Sem})/r_i^{Sem}} n_i^{Sem} \quad (3)$$

where $n_i^{(Sem.)} \in N$ is the semantic relation (class name) defined in the analysis of the uniformity of the current state area $f(z_i)$, limited by the radius $r_i^{(Sem.)}$.

The complexity of segmentation is determined by a number of conceptual aspects:

- interpretation objective –(object phenomenology);
- adequacy of imaging –(of phenomenological properties or essential attributes of the adopted model: $f(z)$).

In the general case, segmentation of the function $f(z)$ satisfies the evaluator when any pair of elements $[z_i – z_{i-1}]$ from any subset $z_i \in f(z)$ compiles the contour description.

$$\oint_C f(z)dz = \lim_{\max[z_i - z_{i-1}] \to 0} \sum_{i=1}^{n} f'(s_i)(z_i - z_{i-1})$$

where $f'(s_i)$ describes a change in the physical attribute (semantic relationship, area name–S_i^{Sem});

$Lp\ [z_i - z_{i-1}]$ is the reference to the 'neighbourhood' model, defined by the $r_i^{(Sem.)}$ radius of the uniformity of the physical attribute.

The mathematical 'neighbourhood' model for any area of the function $f(z)$ of the closed loop is described by the Cauchy integral formula.

$$f^{(i)}(z) = \frac{i!}{2\pi i} \oint\limits_{CS_i} \frac{f(z)}{(z_0 - z)^{i+1}} dz \tag{4}$$

If the path in the positive direction (counterclockwise) is defined by a smooth parametric representation $z = z(t)$ $(\alpha \le t \le \beta)$, where $z(\alpha) = a$ and $z(\beta) = b$ (a, b are initial arc integration points), then the calculation is carried out according to the formula.

$$\begin{aligned} f(z)dz &= \oint\limits_{\beta\alpha} f(z(t))z'(t)dt \\ &= \int\limits_{\alpha}^{\beta} \mathrm{Re}[f(z(t))z'(t)dt + i \int\limits_{\alpha}^{\beta} \mathrm{Im}[f(z(t))z'(t)dt \end{aligned} \tag{5}$$

Formula (4) allows for segmenting the contour of area S_i on the image $f(z)$ and (5) describes the contour properties.

In case of a multiple connection (i.e. when bypassing the complement in the negative direction), the segmentation and description of the contour are slightly complicated.

$$f^{(i)}(z) = \frac{i!}{2\pi i} \oint\limits_{CS_i} \frac{f(z_0) - f(z)}{(z_0 - z)^{i+1}} dz. \tag{6}$$

Accordingly, the calculation is carried out by the formula:

$$\begin{aligned} f(z)dz &= \int\limits_{\alpha}^{\beta} \mathrm{Re}[f(z(t))z'(t)dt + i \int\limits_{\alpha}^{\beta} \mathrm{Im}[f(z(t))z'(t)dt \\ &+ \sum_{s=1}^{n} \int\limits_{s_i} \mathrm{Re}[f(z(t))z'(t)dt + i \int\limits_{s_i} \mathrm{Im}[f(z(t))z'(t)dt \end{aligned} \tag{7}$$

It is known that the critical derivation point z_0 serves as the centre of the semantical radius $r_i^{(Sem.)}$. Usually, the value of the critical point is set empirically based on expert knowledge. For example, in the field of plant cultivation, the critical points represent the centres of four groups of cultivated plants with corresponding acidity values.

Conceptually, the value of the critical point is defined on the section of the segmented area boundary and the background where the integral equals to zero at any size and position of the common boundary.

For this reason, it is proposed to connect the position of the critical point with the dynamics of the image frequency characteristic. In this case, the bracketing $r_i^{(Sem.)}$ will ensure the uniformity of the frequency characteristic of the segmented image area.

2 Method of Analysis of the Key Soil Productivity Parameters

The important parameters determining the energy state of agricultural lands in the eastern territories of Krasnoyarsk Territory are the following: acidity, moisture, content of humus, active forms of phosphorus and potassium on slightly acidic and neutral soils in the conditions of sharply continental climate of Siberia [14–17].

An example of the segmentation of images of weighted averages of acidity, active phosphorus and uncomplexed potassium for groups of plants on plots of agricultural land in the area of Stepnyaki village, Brazhensky village council, Kansk district, Krasnoyarsk Territory, is shown in Figs. 1, 2, 3, 4, 5 and 6.

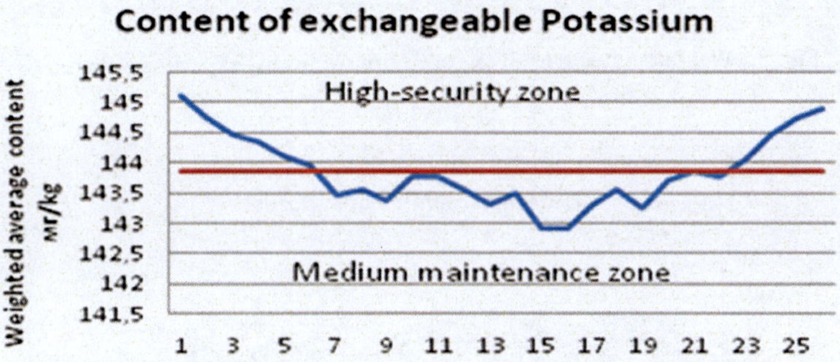

Fig. 1. Weighted averages of the exchange potassium, experimental area

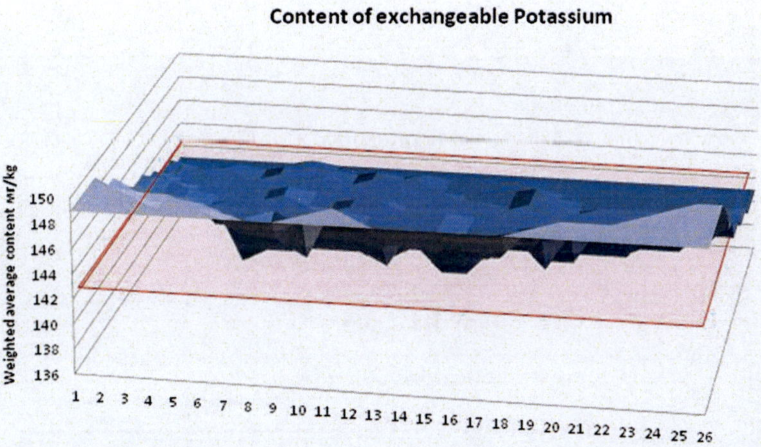

Fig. 2. Image segmentation of the distribution of the weighted averages of the exchange potassium

Usually, the interpretation of the object phenomenology is based on the analysis of the frequency, amplitude and phase characteristics of the segmented areas of the image $f(z_i)$.

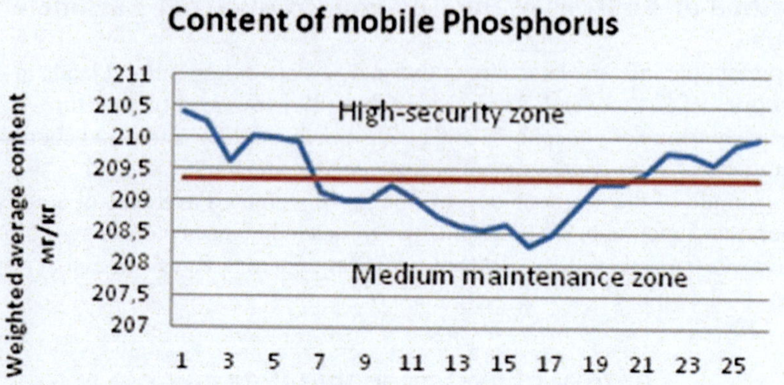

Fig. 3. Weighted averages of the active phosphorus, experimental area

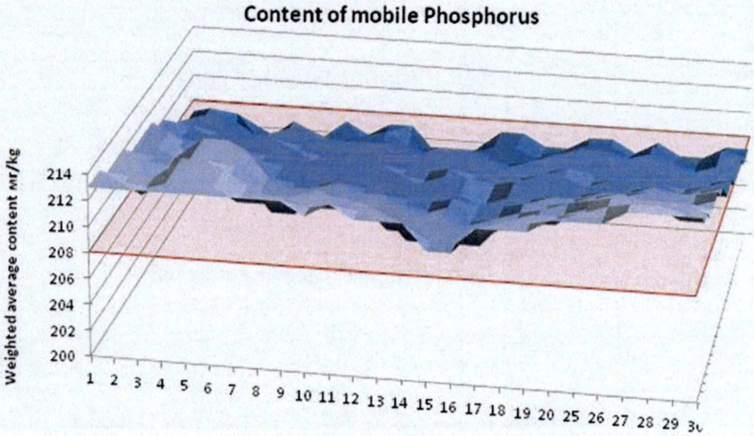

Fig. 4. Image segmentation of the distribution of the weighted averages of the active phosphorus

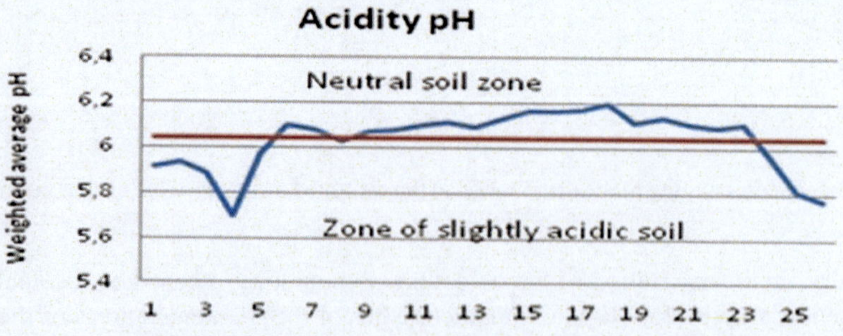

Fig. 5. Weighted averages of acidity, experimental area

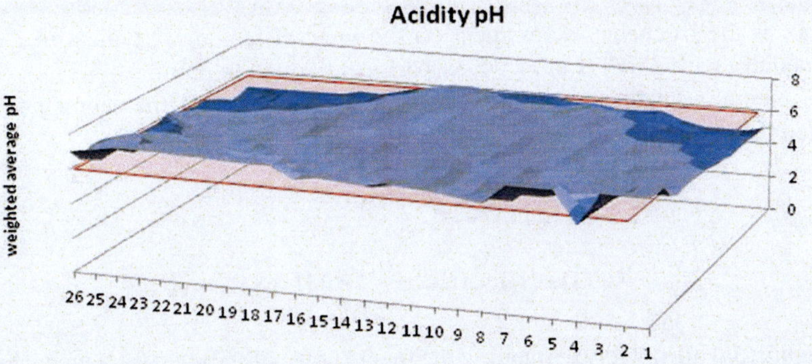

Fig. 6. Image segmentation of the distribution of the weighted averages of acidity

The segmentation starts from the calculation of the centre line of the image graph $f(z_i)$:

$$\mu_{cp} = \frac{\sum_{i=1}^{k} \mu_i}{k},$$

where k is the number of samples.

The uniformity boundaries are calculated using the *Lp* rule as follows:

$$Lp_U = \mu_{cp} + \frac{r_i^{(Sem)} \sigma}{\sqrt{n}},$$

where n is the number of measurements.

Deviation value σ is calculated as follows:

$$\sigma = \sqrt{\frac{\sum_{i=1}^{k} \sum_{j=1}^{n} [\mu_{i,j} - \mu_{cp}]^2}{nk}}$$

The segmentation and contour description are carried out in a similar way in case of a multiple connection, i.e. with the integration over the upper and lower boundaries of the image area.

$$L_{pUL} = \mu_{cp} \pm \frac{r_i^{(Sem)} \sigma}{\sqrt{n}}.$$

Apparently, the segmented areas (see Fig. 1 on the left) by the L_{pUL} rule should be considered relatively uniform since the segmentation is performed on the basis of a rather abstract parameter like a weighted average.

It is convenient to improve the quality of area segmentation and thus obtain a more accurate solution to the task based on the analysis of their frequency characteristic [18–20]. In this case, the segmentation based on the analysis of the frequency dynamics of the segmented image area $f(\omega_r)$ is described by an integral formula of the type [19, 20].

$$f^{(i)}(\omega_0) = \frac{i!}{2\pi i} \oint\limits_{S_i} \frac{f(\omega_r)}{(\omega_r - \omega_0)^{i+1}} d\omega_r. \tag{4a}$$

where: ω_r is the frequency assessment on the interval $[\omega_i - \omega_{i-1}]$ typical of a slightly acidic medium with a radius of r_i^{Sem}; ω_0- critical derivation point.

The integral characterising the change in the frequency of the segmented area is calculated using the formula.

$$
\begin{aligned}
f(\omega_r)d\omega &= \oint_{\beta\alpha} f(\omega_r(t))\omega_r'(t)dt \\
&= \int_{\alpha}^{\beta} \mathrm{Re}[f(\omega_r(t))\omega_r'(t)dt + i\int_{\alpha}^{\beta} \mathrm{Im}[f(\omega_r(t))\omega_r'(t)dt
\end{aligned} \tag{5a}
$$

Formula (5a) allows for segmenting the image area with a given frequency characteristic. The analysis of the frequency dynamics in conjunction with the amplitude analysis of dynamics, phase and image area parameters makes it possible to predict changes in the qualitative uniformity of physical parameters under known seasonality conditions and taking into account the quality of the input data.

Presumably, areas with a relatively high frequency index tend to transfer to the more active soil class and vice versa.

3 Result Analysis and Discussion

The data on the distribution of the agrophysical parameters on the basis of long-term research conducted by the staff of FSUE SAS 'Solyanskaya' provide a general image of the dynamics of agricultural land productivity in Kansk district of Krasnoyarsk Territory, in particular agricultural lands of Stepnyaki village, Brazhensky village council, Kansk district, Krasnoyarsk Territory.

Figure 7 shows an interpretative map with the main parameters of the test plot of Stepnyaki agricultural land.

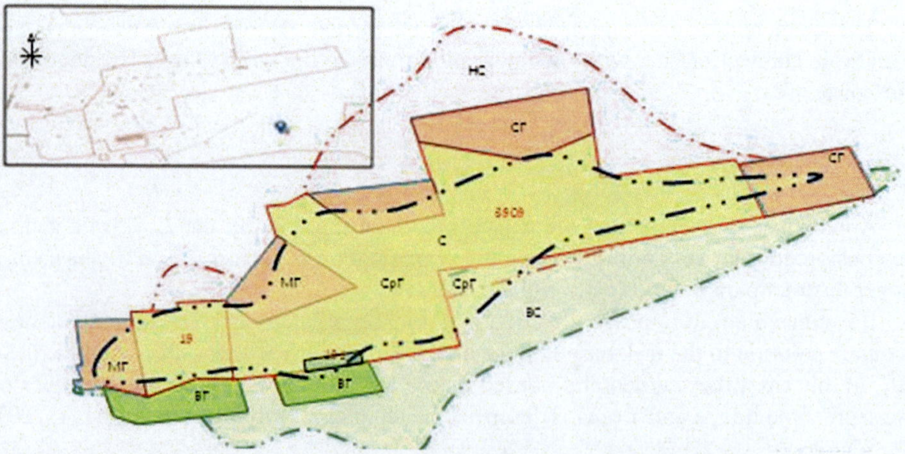

Fig. 7. Interpretative map of agricultural lands of Stepnyaki village

In Fig. 7: LH - low-humus areas, MH – medium-humus areas, HH- high-humus areas, BA - the border of the zone with below-average yield indicators (dark gray and gray forest podzolic soils), A - the border of the zone with average yield indicators (ordinary chernozems, slightly podzolic soils), AA – the border of the zone with above-average yield indicators (leached and ordinary chernozems, slightly podzolic soils).

The calculation of the uniformity radius r_i^{Sem} for the *i*-th class is carried out from the critical point ω_0 as follows:

$$\omega_{0i} = \mu_{cp} \pm \frac{r_i^{(Sem)}\sigma}{\sqrt{n}}.$$

It should be noted that the value of the 'radius' r_i^{Sem} is described by a graph $f(\omega_r)$. As a result, the characteristic $f(\omega_r)$ makes it possible to form the classes of image areas with the radius r_i^{Sem}, which reflect the uniform object properties (5a).

The calculation of the contour area of the segmented areas is carried out according to (5a).

Calculation of the contour frequency:

$$f(\omega_r) = P(\omega) + jQ(\omega) = f(\omega)e^{j\phi(\omega)} = f(\omega)[\cos\phi(\omega) + j\sin\phi(\omega)],$$

where:

$$P(\omega) = \mathrm{Re}f(j\omega);$$

$$Q(\omega) = \mathrm{Im}f(j\omega) = f(\omega)\sin\phi(\omega)];$$

$$\phi(\omega) = arctg\frac{P(\omega)}{Q(\omega)}.$$

Figure 8 shows the result of the image area segmentation that makes it possible to analyse the dynamics of three acidity classes of soils (slightly acidic, circumneutral and neutral) of the Stepnyaki village agricultural land plot.

The central segmented area covers the entire study area of agricultural land in Stepnyaki village. This area belongs to the 'circumneutral' soil class (pH 5.9), which is relatively to the north of more acidic soils and to the south of neutral soils.

According to further analysis of the central plot in relation to the density of image frequency characteristic $f(\omega_r)$, the northern part of the plot, with its relatively high frequency parameter, tends to lower the acidity of the medium to a slightly acidic soil (pH 5.6), while the southern part accordingly tends to increase the acidity to a neutral medium (pH 6.2) (as shown in Fig. 3 on the right). This is also confirmed by the evaluation of the density of the image frequency characteristic $f(\omega_r)$ of the central plot (Fig. 8 on the left).

Overall, the prognosis based on the analysis of the image frequency dynamics shows a relatively high qualitative uniformity of the central plot; the robustness of the plot in maintaining a 'circumneutral' environmental response serves as a good complement to standardised methods of analysis of soil condition of agricultural lands.

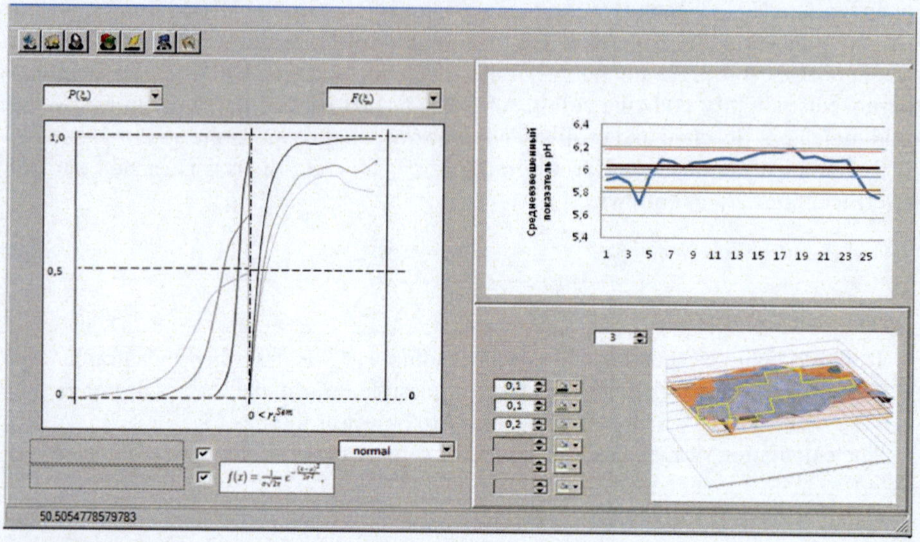

Fig. 8. Segmentation the image frequency properties and evaluation of the uniformity of segmented area classes

4 Conclusion

The proposed mathematical model is harmonised with the accepted methods used to analyse the main agrochemical parameters [21–25] and greatly enhances the possibilities for more accurate analysis, interpretation and prediction of agrophysical characteristics of soil condition.

The segmentation mathematical model based on more complex 'neighbourhood' relations defined by the uniformity radius riSem of the image area frequency characteristic $f(\omega_r)$ allows for performing the fragmentation of the characteristic into N classes under causal conditions, for which Lp(SiSem) takes true values that comply with the objectives.

In this case, object segmentation is a combined strategy:

$$\text{Seg:}f(\omega_r) \overset{Lp(S_i^{Sem})/r_i^{Sem}}{\longleftrightarrow} n_i(\text{Sem.}),$$

where $r_{\text{i}}(\text{Sem.}) \in N$ is the semantic relation (object name) determined during the analysis of the frequency uniformity of the segmented area boundaries.

The mathematical approach to the segmentation of the dynamics of agrophysical parameters of agricultural soils has proved to be effective in managing the quality of the natural energy state on the control plots. It is expected that the experience will be applied to the entire responsibility area of 'SAS' 'Solyanskaya' (912.4 thousand ha).

In the future, it is expected to create a 'Digital atlas' reflecting the dynamics of changes in the main agrochemical parameters of soils as part of the Project 'Strategy for the Development of the Agroindustrial Complex of Krasnoyarsk Territory (5 million ha) for the period until 2030 (http://www.krskstate.ru/2030/plan).

Introduction and distribution of innovative practices and technologies will provide new opportunities for monitoring and management of the natural energy state quality of the East-Siberian territory (23 million ha).

References

1. Denisov, D.A.: Computer aided video analysis methods: monograph, Krasnoyarsk State Technical University Krasnoyarsk (1993)
2. Chochia, P.A.: Pyramidal algorithm of image segmentation. Information processes. Pub. by Moscow State Technical University named after N. E. Bauman. vol. 10, no. 1, pp. 23–35 (2010)
3. Minaee, S., Boykov, Y.Y., Porikli, F., Plaza, A.J., Kehtarnavaz, N., Terzopoulos, D.: Image segmentation using deep learning: a survey. IEEE Trans. Pattern Anal. Mach. Intell. (2021)
4. Brand, M., Kettnaker, V.: Discovery and segmentation of activities in video. IEEE Trans. Pattern Anal. Mach. Intell. **22**(8), 844–851 (2000)
5. Belim, S.V., Larionov, S.B.: Segmentation of images on the basis of algorithm of communities' separation on the graph. Math. Struct. Model. **3**(39), 74–85 (2016)
6. Khanykov, G.: Classification of image segmentation algorithms. The news universities. Instrumentation **61**(11), 978–987 (2018)
7. Avdyushkina, L.S.: Two-stage method nature texture classification. Manage. Process. Sustain. **3**(1), 332–336 (2016)
8. Gonçalves, D.N., et al.: Carcass image segmentation using CNN-based methods. Inf. Process. Agric. (2020)
9. Srikanth, R., Bikshalu, K.: Multilevel thresholding image segmentation based on energy curve with harmony Search Algorithm. Ain Shams Eng. J. **12**(1), 1–20 (2021)
10. Houssein, E.H., El-dinHelmy, B., Oliva, D., Elngar, A.A., Shaban, H.: A novel Black Widow Optimization algorithm for multilevel thresholding image segmentation. Expert Syst. Appl. **167** (2021)
11. Purswani, P., Karpyn, Z.T., Enab, K., Xue, Y., Huang, X.: Evaluation of image segmentation techniques for image-based rock property estimation. J. Pet. Sci. Eng. **195** (2020)
12. Wang, T., et al.: Label group diffusion for image and image pair segmentation. Pattern Recogn. (2020)
13. DeCost, B.L., Holm, E.A.: A computer vision approach for automated analysis and classification of microstructural image data. Comput. Mater. Sci. **110**, 126–133 (2015)
14. Ladhake, K.S.: Semantic image analysis for intelligent image retrieval international conference on intelligent. In: Computing, Communication & Convergence (ICCC-2014) Institute of Management and Technology, India, pp. 193–197 (2014)
15. Banerjee, S., Ghosh, S.K., Datta, S., Saha, S.K.: Segmentation of dual phase steel micrograph: an automated approach. Measurement 2435–2440 (2013)
16. Avdyukova, T.V., Hrutsky, S.A.: Fertility state of arable lands in the east zone of Krasnoyarsk Krai. Agriculture **1**, 25–29 (2017)
17. Milovskih, T.A., Vyatkina, T.A., Smirnova, T.N.: Productivity of agricultural land in Sakhalin Oblast conditions. Sci. Tech. Achievements AIC **33**(4), 53–56 (2019)
18. Chuprova, V.V., Belousov, A.A., Belousova, E.N., Yu.V.: Gorbunova evaluation of agroecological condition of the soils involved in the development of sand and gravel quarries of Kansk district of Krasnoyarsk region. J. Siberian Federal Univ. Agronomic **3**, 16–21 (2019)
19. Tandelov, Yu.P.: Productivity of acidic soils in the agricultural area of Krasnoyarsk Territory. Krasnoyarsk 161 (2012)

20. Perfil'ev, D.A.: Segmentation object strategy on digital image. J. Siberian Federal Univ. Eng. Technol. **11**(2), 213–220 (2018)
21. Perfilyev, D.A.: Object segmentation on a digital image. Regional issues of earth remote sensing; materials of international scient. In: Conference Krasnoyarsk, Siberian Federal University, 366 (2018)
22. Perfilyev, D.A.: Strategy of object segmentation on a digital image. Regional issues of earth remote sensing; materials of VI intern. scient. In: Conference Krasnoyarsk. Siberian Federal University, 368 (2019)
23. Perfilyev, D.A.: Segmentation of frequency nonuniformity of the image. Regional issues of earth remote sensing; materials of VII intern. In: Scientific Conference Krasnoyarsk. Siberian Federal University, 328 (2020)
24. GOST 26269-91: Soils. General Requirements for the Fulfilment of Analyses. Standard Publishing (1992)
25. GOST 26204-91: Soils. Determination of Mobile Compounds of Phosphorus and Potassium by Chiricov Method Modified by CINAO. Standard Publishing (1992)
26. GOST 26205-91: Soils. Determination of Mobile Compounds of Phosphorus and Potassium by Machigin Method Modified by CINAO. Standard Publishing (1992)
27. GOST 26207-91: Soils. Determination of Mobile Compounds of Phosphorus and Potassium by Kirsanov Method Modified by CINAO. Standard Publishing (1992)
28. GOST 26483-85: Soils. Preparation of Salt Extract and Determination of Its pH by CINAO Method. Standard Publishing (1985)

Management of Documented Information to Ensure Traceability of Radioelectronic Equipment Production Using Software

Sergey Fursov and Alexcey Kirov(✉)

RTU (MIREA), Moscow, Russia

Abstract. The article discusses the possibility of using software to ensure the traceability of the production of electronic equipment. The development and production of radio-electronic equipment is a high-tech direction of modern industry and is characterized by a high level of knowledge intensity. These products are subject to high quality assurance requirements. The traceability system is an integral part of the life cycle management of electronic products, which has a significant impact on their quality. Currently, traceability is ensured through the use of accompanying sheets and technological passports. The use of documented information on paper to ensure traceability increases the burden on the quality service and makes it difficult to record changes and evaluate the effectiveness of the QMS. The results of using the software tools proposed in the article will provide the convenience of information processing and create an analytical platform for analyzing information, further testing and subsequent evaluation of the stability of technological processes for the production of electronic equipment and the effectiveness of the QMS.

Keywords: Quality management system · Product life cycle · Traceability · Electronic documentation · Software

1 Introduction

The functioning of the processes of the quality management system (QMS) should be ensured using documented information. Storage and subsequent processing of documented information, which is a confirmation of the correctness of the processes, becomes an urgent task for the functioning of the QMS. This is due to the large amount of information received during the operation of the QMS.

The task of managing large amounts of information to solve without the use of software tools is currently becoming irrelevant and can lead to a loss of competitiveness caused by high costs for quality management.

The most difficult task for the organization of the management of documented information is the use of documented information in the processes of development, production and evaluation of performance. This is due to ensuring the traceability of products at various stages of the life cycle and the specifics of the production of radio-electronic equipment (REA). The solution to the problem of ensuring traceability can be the use of

R. Silhavy et al. (Eds.): CoMeSySo 2021, LNNS 231, pp. 145–152, 2021.
https://doi.org/10.1007/978-3-030-90321-3_14

electronic accompanying documentation. The main advantage of using software tools in the design of accompanying documentation is the subsequent ease of use, since data in electronic form is more convenient to process and carry out their further analysis.

2 Documented Information in the Quality Management System

Modern radio-electronic equipment (REA) contains a large number of components, an electronic component base and multi-level software.

To ensure the quality of REA, traceability of components and components is necessary at all stages of production, including post-production stages.

Documented information is information necessary to ensure the functioning of the quality management system (QMS) [1–3].

The amount of information depends on the complexity of the REA, the specifics of the organization's activities, as well as the specifics of the processes of a particular QMS. At the same time, the procedure for handling documented information should be strictly regulated, which should be given special attention [4].

Figure 1 shows the typical data (information) used in the production of REA and subject to documentation. Given the large volume of data, there is a need for their accounting and processing in electronic form.

Presentation and processing of documented information in electronic form will ensure the traceability of REA products in production processes and during further operation. Traceability is an important factor in ensuring the quality of REA [5] (Fig. 3).

Taking into account the requirements for documented information in the QMS, the following advantages of using software tools to ensure the traceability of REA should be noted:

- transition to "paperless" technologies;
- efficiency of processing and receiving information during the functioning of QMS processes;
- formation of accompanying documentation containing extended data about the REA product;
- providing an initial level of integration of information support and automation of QMS processes;
- creating an environment for the use of CALS technologies;
- reducing the burden on the staff of the quality service by excluding paper accounting logs from the documented information, etc. [6, 7].

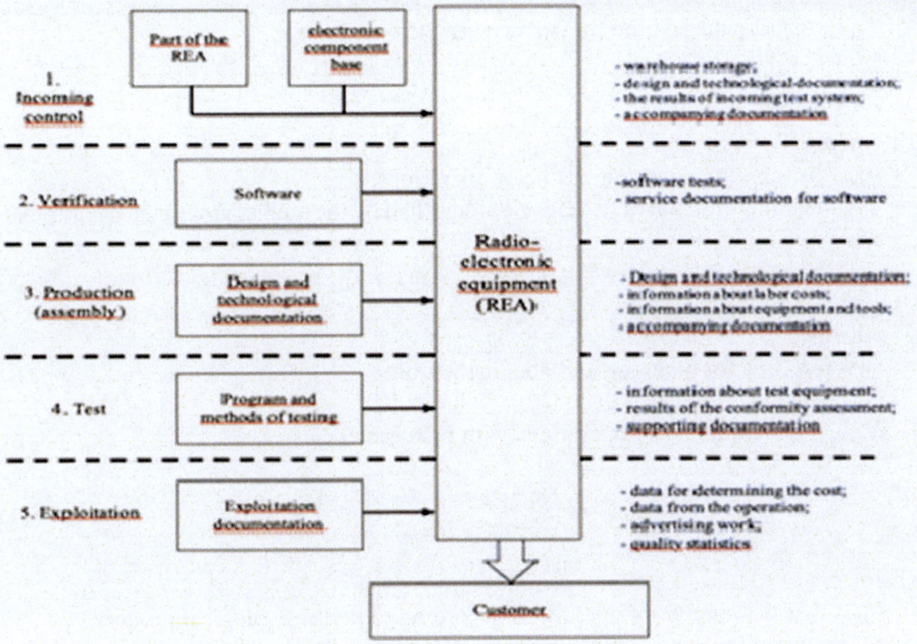

Fig. 1. Typical data of documented information used in the production of REA.

3 Ensuring Traceability Using Software Tools

A set of directions can be implemented using software tools to ensure traceability.

One of the primary tasks facing manufacturers and developers of REA is to ensure traceability and identification at all stages of the life cycle.

Traceability information should be documented and maintained. In addition to saving, the most convenient processing is important, which expands the possibilities of using documented information in QMS processes and ensuring the quality of REA.

Currently, traceability is ensured through the use of accompanying sheets, technological passports, etc. The use of documented information on paper to ensure traceability requires its storage at all stages of the REA life cycle. This leads to an "overgrowth" of the documentation archive, increases the burden on the quality service and makes it difficult to account for changes and assess the effectiveness of the QMS. This is not possible without the use of databases, as maintaining paper records requires large labor costs [8, 9].

The results of changes in labor costs can be estimated in relation to:

$$R_{trace} = \frac{\Delta T}{T_{norm}} \tag{1}$$

where:

ΔT is the time after the implementation of tracking using software tools;

T_{norm} – the standard time for performing the operation.

$$\Delta T = T_{norm} - \left(t_{fact} + t_i\right) \quad (2)$$

where:
t_{fact} – is the actual execution time of operations;
t_i – time for correction of deficiencies identified by the quality assurance department.

$$t_i = \sum\nolimits_{N=0}^{N_{oper.d}} t_{t.d} + t_{pl} \quad (3)$$

where:
t_{pl} – is a time for planning and documentation;
$t_{t.d}$ – time to eliminate defects;
$N_{oper.d}$ – the number of operations with inconsistencies;

$$T_{norm} = \left(\sum\nolimits_{i=0}^{N_{oper}} \frac{t_{oper}}{N_{emp} * K_{sm}}\right) + t_i \quad (4)$$

where:
N_{oper} – is the number of operations according to the technological process;
t_{oper} – operation execution time according to the technological process;
N_{emp} – number of employees involved;
K_{sm} the shift factor.

The assessment can be carried out both on the basis of planned and actual indicators of the technological process. With $R_{trace} \rightarrow 1$, we determine the effectiveness of measures to implement traceability using software tools.

Taking into account the above, it is possible to identify the main directions of ensuring the traceability of REA using software tools. The directions of ensuring the traceability of REA production are shown in Fig. 2.

One of the most important areas of ensuring traceability is the construction of accompanying documentation in electronic form. To build the accompanying documentation in electronic form, it is necessary to determine the information that will be obtained during the production of REA.

The information for constructing the cover sheet using software tools is presented in Table 1.

The received information is recorded using software developed in the Python programming language in files that are convenient for handling and processing, we have chosen the.txt format.

Given the ease of use of software written in Python, as well as the wide possibilities for further processing of information, it is possible to generate documented information that provides traceability (cover sheets) directly at the workplace using portable devices. This will reduce the time of rewriting the results of quality control by employees of the quality service on paper, and the use of the format.txt makes it possible to open the received files on any device for their use in QMS processes.

For further processing of information and obtaining data for quality management, the Python programming language is used with the connection of the following libraries:

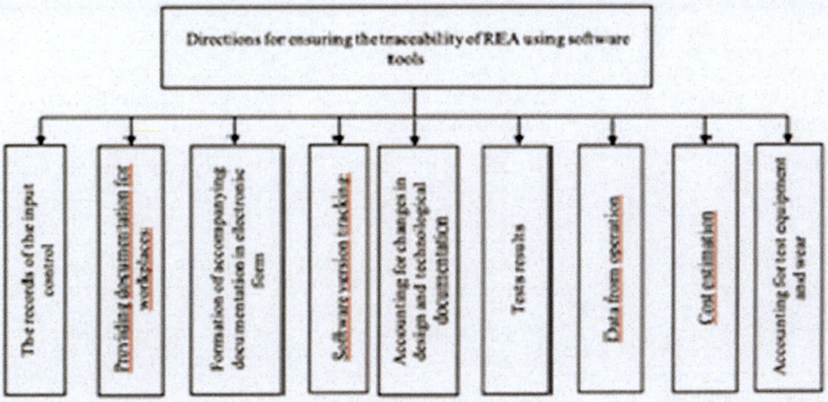

Fig. 2. The main directions of ensuring the traceability of REA using software tools.

1. NumPy;
2. SciPy;
3. Pandas;
4. Mathplotlib, Seaborn, Bokeh, Ploty - for data visualization [10].

The use of software tools for information processing allows you to create an "electronic passport" of the product.

The available software tools are mainly aimed at automating business processes and activities of the enterprise and are closely related to ERP systems that determine the management of enterprise resources [11–13].

The main differences from the available software tools used in the QMS of REA manufacturers are the ability to create databases of information about products even before they are evaluated by the quality service and documented information is processed, as well as further automation.

The use of software tools based on Python scripts will ensure the simplicity of information processing due to the relationship with the database management system and the use of BigData technology algorithms.

4 Conclusion

The software tools used allow you to create a powerful analytical platform for analyzing information, further tests (tests) and subsequent assessment of the stability of the technological processes of REA production and the effectiveness of the QMS.

Possible integration with ERP systems, as well as the use of "paperless" technologies will significantly reduce labor costs for information processing. The information obtained in electronic form can be used to form the electronic case of the product and used for integrated logistics support of the REA life cycle.

Table 1. Information for the cover sheet using software tools.

Name of the transmitted information	Brief description of the information received	Sources of information
Product name	The name of the item of electronic equipment and its modification	• design documentation; • CAD
Name of the technical documentation	Information about technical documentation	• links to the database; • "cloud" storage
Software version	Information about use software	• links to the database; • "cloud" storage
Production conditions	Information about the process operation environment	• IIOT; • electronic logs of microclimate monitoring
Factory (serial) number	Numbering information according to the specification	• assignment in accordance with accounting documents
Information about the input control	Information about the materials used, components, and components that have passed the input control	• information from the input control protocols
Information about operations in the production of	Production and assembly route information	• extracting information from process documentation
Information about tests	Information about tests results	• program and methods of testing; • test reports
Time performance indicators of the technological process	Formation of information about the performance of technological process operations by employees	• marks identifying employees in the technological process; • marks of the quality service on the results of control
Information about the results of control during production	Information about the control of the most important operations and special processes	
Getting data out of operation	Information about the resource, reliability and failures in operation. Conducting the complaint work	• CALS; • CRM-systems; • certificates of compliance
Information about the faults		
Data about performers	Information about the employees involved in the operations (qualification, capacity and etc.)	• accompanying documentation, issued in electronic form; • RFID - identification

```
print('Input test result')
print('Input number")
name = input()
print('Input quantity')
sample_len = int(input())
samples = list()
number = list()
for i in range(sample_len):
    number = input("Input number :')
    sample = input("Input test results " + str(i+1) + ": ")
    samples.append("Part " + str(i+1)+ ' number - ' + number + ": " + sample + "\n")
with open('/Volumes/DOC/CODE/Traceability/file.txt','w', encoding='utf-8') as file:
    file.write(name + '\n")
with open('/Volumes/DOC/CODE/Traceability/file.txt','a', encoding='utf-8') as file:
    for sample in samples:
        file.write(sample)
file.close()
print('View test results?")
x = str(input())
if (x == 'YES'):
    sample_read = open('/Volumes/DOC/CODE/file.txt','r', encoding='utf-8')
    print(*sample_read)
else:
    print('Complete")
```

Fig. 3. Program code for translating information into a text file.

References

1. Manturov, D.V., Kalachanov, V.D., Stateva, G.A.: Organization of production based on the introduction of quality management using CALS technologies (on the example of aviation rocket engineering). Production Organizer 2 (2012). https://cyberleninka.ru/article/n/organizatsiya-proizvodstva-na-osnove-vnedreniya-upravleniya-kachestvom-s-ispolzovaniem-cals-tehnologiy-na-primere-aviatsionnogo. Accessed4 August 2020
2. Michenko, S.V., Ponomarev, S.V., Trofimov, A.V., Pekina, S.S.: Methodology of internal audit of processes of the quality management system of a machine-building enterprise. Bulletin TGTU 1 (2007). https://cyberleninka.ru/article/n/metodika-provedeniya-vnutrennego-audita-protsessov-sistemy-menedzhmenta-kachestva-mashinostroitelnogo-predpriyatiya. Accessed 4 August 2020
3. Michaylovskiy, I.A., Gun, E.I.: Development of a methodology for assessing the significance of product parameters and technological operations of its production in terms of impact on quality. The century of quality 3 (2014)
4. GOST R ISO 9001–2015 Quality management systems. Requirements. Moscow, Standartinform, p. 23 (2015)
5. Grodzinskiy, S.Y., Fursov, S.A., Kirov, A.V.: The use of technology traceability at different stages of the life cycle of electronic products. Inf. Innov. Technol. **1**, 229–231 (2017)
6. Efremov, A.J.: Using the CALS methodology in the quality management system of a particular enterprise. The century of quality, 4 (2011)
7. Fursov, S.A.: Methods of automation monitoring of processes the quality management system of enterprises-producers of electronic component base. Computer science and technology. Innovative technologies in industry and informatics (RSTC FTI-2018), pp. 360–366 (2018)

8. Kirov, A.V., Fursov, S.A.: Use of information technologies as one of the main directions of improving the quality of radio electronics products. Fundam. Prob. Radio-Electron. Instrum. **15**(1), 201–203 (2015)
9. Grodzensky, J.S., Egorov, I.S.: Quality management and CALS-technologies. Innov. Inf. Commun. Technol. **1**, 212–215 (2016)
10. Python software libraries. https://www.python.org. Accessed 10 August 2020
11. Andieva, E.J., Gurieva, E.M., Michailov, V.A.: Analysis of the problems of provision of the ecosystem for developing industrial automation solutions. Omsk Sci. Bull. **3**(159), 126–132 (2018)
12. Orlov, A.B., Gaeva, N.I.: Use of electronic information models at the stage of exploitation of mechanical engineering products. Izvestia TugGU. Ser. Technol. Syst. Eng. **11**, 45–51 (2006)
13. Programs for Automation of Management Activities of the Organization Electronic Resource. https://sci.house/ekonomika-predpriyatiy-scibook/programmyi-avtomatizatsiiupravlencheskoy.html. Accessed 25 April 2019

Comparing Different Supervised Classification Algorithms to Detect Arrhythmia in an ECG

Adrián Silva Méndez(✉) and José Gerardo Tamez Peña

Tecnológico de Monterrey Campus Monterrey, 64000 Monterrey, NL, México

Abstract. In this article, we will review the step process for creating an accurate prediction model that classifies the heartbeat rhythm into two categories: normal and abnormal. This model will be based on the information provided by the UCI Arrhythmia data set consisting of ECG information alongside other attributes such as age, sex, weight, classification (Normal/Abnormal), etc. In the end, we will compare the performance of different types of classification algorithms (on optimal circumstances) and determine which one excels for specific scenarios.

Keywords: Machine learning · CVD · Arrhythmia · Classification · Data processing · Data cleaning · Feature selection · Normalization · PCA · Dimensionality reduction

1 Introduction

There is not a single classification algorithm that excels on all problems. One must choose the algorithm that best fits the problems' characteristics. To determine which algorithm works better for a specific problem, different models should be compared to select the best among all. In this article, we addressed a common supervised classification problem by trying different classification algorithms and choosing the best for each specific situation. In order to compare these models in a "fair" way, we performed some data transformations to achieve the fittest individual model possible before doing any comparison. To compare these models we divided the data set into Training and Testing Samples. Some of these transformations and testing strategies include: Dimensionality Reduction, Feature Selection, and Standardizing/Normalizing Data. The aim of using these techniques in this article is to answer the following research questions:

1) Does Principal Component Analysis (Dimensionality Reduction) fits well for the given problem?
2) Is feature selection an improvement for the models' performance?
3) Is there an optimal amount of selected features?

Note: *All the algorithmic and mathematical steps were made in Python using auxiliary libraries like Sklearn. At the end of the article, you can find a link to the author's project code* [11].

R. Silhavy et al. (Eds.): CoMeSySo 2021, LNNS 231, pp. 153–165, 2021.
https://doi.org/10.1007/978-3-030-90321-3_15

1.1 Problem Background

The classification problem addressed in this article consists of a data set containing ECG information alongside other attributes such as age, sex, weight, etc. The electrocardiogram (ECG) is the record of variation of bio-electric potential with respect to time as the human heartbeats. It provides valuable information about the functional aspects of the heart and cardiovascular system [5]. One additional feature inside the data set is its corresponding classification. This labeling focuses on describing the subject's heart's ability to beat in a normal rhythm. Its values range from normal to different labeled kinds of irregular heartbeat rhythms.

According to the World Health Organization, Cardiovascular diseases (CVD) contribute to $^1/_3$ of deaths worldwide [1]. Sudden Cardiac Deaths coming from abnormal heartbeat rhythms contribute to almost 25% of all those CVD [2]. Unlike the popular opinion that Sudden Cardiac Arrest (SCA) is the same as heart attacks, SCA is not a heart attack (myocardial infarction). Heart attacks happen when there is a blockage in at least one of the coronary arteries, preventing the heart from receiving enough blood. SCA on the other hand occurs when there is a malfunction in the electrical system of the heart making it beat irregularly. The definition of arrhythmia is an irregular heartbeat rhythm [5]. As stated before, these irregular heartbeat rhythms are threatening a big proportion of the population's lives. Early prediction of heart diseases alongside proper medical treatment can save lives in case of sudden cardiac deaths [6].

Therefore, early detection of abnormalities is crucial to identify the people that are at risk and minimize the negative outcome of these anomalies. Nowadays, different Machine Learning approaches aim to correctly predict and classify the corresponding heartbeat abnormality. Lyon et al. (2017) explained and compared some of the most accurate models used to classify different types of Arrhythmia (heart rhythm abnormalities) [3]. These models' accuracy ranges between 95%–99% despite the different classification algorithms used to train them. The key to having competitive models is to shape each of them accordingly and compare the best version of themselves.

1.2 Article's Structure

The article is organized as follows: In Sect. 2 we will be covering all the relevant information regarding the data used to train the classification models. This information includes, but is not limited to, data description, data understanding, data pre-processing, data transformation, and data manipulation. We will also explain step-by-step the methodology followed to select and train each of the classificators. Afterward, in Sect. 3, we will cover the results obtained from the models. We will then proceed to discuss some initial findings and similar articles in Sect. 4. Finally, we will conclude with the results obtained in Sect. 5.

2 Method and Data

2.1 About the Data Set

The data set used and mentioned throughout this article is the UCI - Arrhythmia Data Set [8]. This data set contains information that is described in Table 1.

Table 1. Information about arrhythmia data set [8].

Data set characteristics	Multivariate
Attribute characteristics	Categorical, Integer, Real
Number of instances	452
Number of attributes	279
Missing values	Yes

Some important extra notes about the data set are that unknown values in an attribute are represented by the special character '?' and there is an "additional attribute" that represents the instance classification. The classes are distributed as shown in Table 2.

Table 2. Class distribution of arrhythmia dataset.

Code	Class	No. instances
01	Normal	245
02	Ischemic changes (Coronary Artery Disease)	44
03	Old Anterior Myocardial Infarction	15
04	Old Inferior Myocardial Infarction	15
05	Sinus tachycardy	13
06	Sinus bradycardy	25
07	Ventricular Premature Contraction (PVC)	3
08	Supraventricular Premature Contraction	2
09	Left bundle branch block	9
10	Right bundle branch block	50
11	1st degree AtrioVentricular block	0
12	2nd degree AV block	0
13	3rd degree AV block	0
14	Left Ventricule hypertrophy	4
15	Atrial Fibrillation or Flutter	5
16	Other	22

2.2 Methodology

In this article, the research questions are going to guide our methodology. We are interested in knowing whether the PCA helps any model or not, the same way we want to know if Feature Selection is necessary and if the number of features selected matters or not. Knowing this, the first step is to select the different classification algorithms that are going to be compared throughout this article. The selected algorithms are:

1) Multi-Layer Perceptron Neural Network
2) Random Forest
3) X-Gradient Boosting
4) Adaboosting
5) Logistic Regression
6) SVM (Linear)
7) SVM (RBF)
8) KNN
9) Naive Bayes

In order to compare these algorithms in a "fair" way, we need to have the best version of each of them. To achieve this, we need to perform some data cleaning and data transformations. Some of these transformations include Feature Selection. In order to determine whether Feature Selection is necessary or not, the algorithms that do not naturally need a manual feature selection (this will be explained in detail in Sect. 2.3) will be trained with the original training set and the feature selected training set. Furthermore, the number of selected features will be varied from 10 to 100 in a 10 step interval.

Note: *Since there are many parameters to test and many algorithms to cover in this article, the observations and results will be kept as simple as possible.*

To get the best number of Neighbours in KNN and the best number of estimators in Random Forest, we will perform a test where we increment these numbers 1 by 1 until we find the index with the highest score.

Once we have the best version of each model, we will compare each of them by dividing the data set into training and testing samples and applying distinct metrics such as accuracy, precision, and recall (sensitivity).

2.3 Data Pre-processing

In most of the classification problems, the most time-consuming (and most important) step is the processing and transformation of the data which can take up to 70% of the project's time [9]. It is very uncommon to find a data set that works perfectly as it is for the given problem. There are many times where the data has either empty values, unrealistic values, or unstandardized values. In this Subsection, we will perform all the necessary modifications and transformations to the data to get a more reliable data set to improve the classification models' score.

In this article, we will only focus on binary classification. Since the data set is represented with multiple classes, the first step before moving forward is to abstract the classes into two single classes: Normal Heartbeat (0) and Abnormal Heartbeat (1). To accomplish this, we simply joined all different types of Arrhythmia (classes 2–16) into one single class.

Cleaning Data Set

The next step involves cleaning the data set. After a careful review of each feature and row of the data set, we concluded three things: there are zero empty values, there are special characters ('?') and values range from integers to floats, from positive to negative. Concerning the special characters, we found out that there are multiple rows (almost 90% of the instances) with this character. It is non-viable to delete 90% of the data set since we only have 452 instances. After a deeper analysis, there is one feature that contributes to 82% of the special characters. We can get rid of this attribute on all instances and delete the remaining instances (8%–34 rows) with special characters.

After the last step, we proceeded to get some statistics on the data. We calculated the following information for each attribute as shown in Table 3.

Table 3. Statistics from age and height columns respectively.

Count	452	452
Mean	46.5	166.2
Std	16.5	37.2
Min	0.0	105.0
25%	36.0	160.0
50%	47.0	164.0
75%	58.0	170.0
Max	83.0	780.0

These statistics helped us realize that there were some unrealistic values in some features. For example, in the "Height" column, there were people with a 7 m height, which is highly unlikely. We can also appreciate that the lowest age of the dataset is 0 while the smallest height is 1.05 m. To avoid these outliers, a filter was done to exclude all people higher than 2.3 m and younger than 5 years.

In the end, the data dimensions were reduced from (452, 279) to (418, 278).

Balancing Classes. Before moving on to Feature Selection and standardizing the data set, we need to ensure that the classes are balanced. After counting the instances of each class, we found that there are 237 instances for class "Normal" (0) and 181 instances for the "Abnormal" class (1). Since the data set is very old (1998) it is practically impossible under the current circumstances to get more "natural" data with the same characteristics as the current data. Although there are many data augmentation (Oversampling) techniques to obtain extra artificial data, it is not the aim of this article to explore these techniques. To balance these two classes, a simple Under-sampling was done decreasing the number of records in the majority class by dropping some random instances to match the minority class [10]. The final data dimensions are (362,278) where 181 corresponds to the "Normal" class (0) and 181 corresponds to the "Abnormal" class (1).

Feature Selection. Once both classes are balanced, we can continue with our next step, Feature Selection. Whenever we have a data set with a huge number of features, we need to consider the possibility of Feature Selection. Depending on the algorithm used to generate a classification model, Feature Selection could be a must. Some supervised classification algorithms have their "own Feature Selection" internally. For example, in a Multi-Layer Perceptron Neural Network, each connection between a feature and the next Layer have an assigned weight. This weight controls how important a feature is to determine the output layer (or classification label). As it might be inferred, this algorithm performs feature selection by assigning bigger or smaller weights to a perceptron. Hence, there is no need to do Feature Selection on these types of algorithms, although, it is still a possibility. Some of the classification algorithms that perform their "own Feature Selection" are:

1) Multi-Layer Perceptron Neural Network
2) Random Forest
3) X-Gradient Boosting
4) Adaboosting

Some of the classification algorithms that need Feature Selection to avoid underfitting or overfitting are:

1) Logistic Regression
2) SVM (Linear)
3) SVM (RBF)
4) KNN
5) Naive Bayes

As explained earlier in the article, all of the models will be trained using the Feature Selected training sample. However, the "own Feature Selection" algorithms will also be trained using the original training set. To manually do Feature Selection we will be using the ANOVA (Analisis of Variance) F-Value along with the K-Best Selector from the Sklearn library.

Data Normalization. Whenever we are working with a multi-dimensional problem, we need to consider Standardizing the data. This is important since some algorithms like SVM and Logistic Regression are susceptible to the distance between data points. If for some reason we have an instance that has an outlier in one of its features, this can be classified erroneously due to the distance restriction mentioned before. We will perform Standardization in:

1) Logistic Regression
2) SVM (Linear)
3) SVM (RBF)
4) Naive Bayes

We are avoiding using Standardization in KNN because distance is a crucial factor in determining the classification of an instance. If we normalize the data we are taking away the essence of KNN. On the other hand, the rest of the algorithms perform their own Standardization internally. For standardizing the data we are using the StandardScaler from the Sklearn Library.

Dimensionality Reduction. Principal Component Analysis (PCA) is used as a Dimensionality Reduction technique. It is a statistical process that allows us to summarize the information contained in wide (multiple features) data sets by "summary indices" in smaller components (dimensions) that can be more easily visualized and analyzed. In order to determine whether the resulting PCA helps to improve the performance of a given model by using it as a training set, we selected Naive Bayes as the Guinea Pig model. We will be performing PCA on a SCALED and FEATURE SELECTED (For K = 10) data set and selecting 2 as the number of components. The number of components will be varied between 2 and 16 using 1 as an interval step. The PCA will be performed using the Sklearn library.

3 Results

After applying the steps mentioned before, we obtained the number of neighbors and estimators as shown in Figs. 1 and 2.

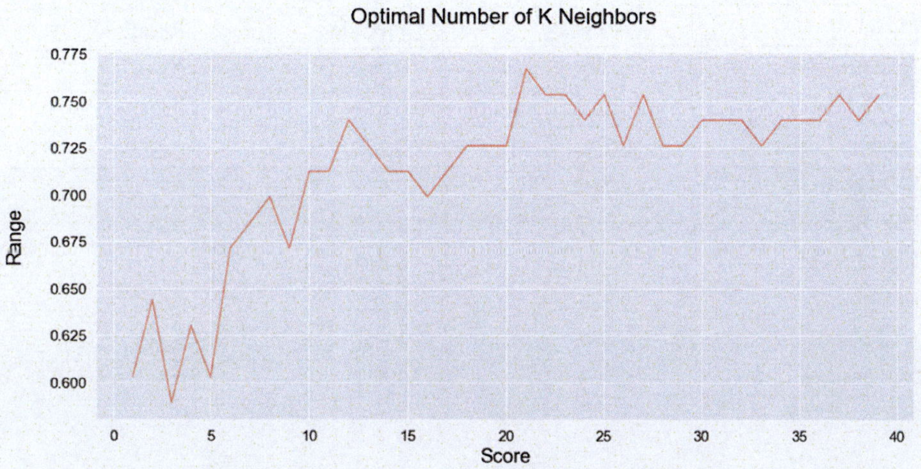

Fig. 1. Test to find the optimal number of neighbors.

Once we have each model under the best shape, the best performance obtained by each of them is given in Table 4. Additionally, in Table 5 the best models for every metric are presented.

Note: *All the algorithms with their "own Feature Selection" were also trained using a manual Feature Selected data set. However, the performance dropped considerably. The best version of these models was trained using the raw data set.*

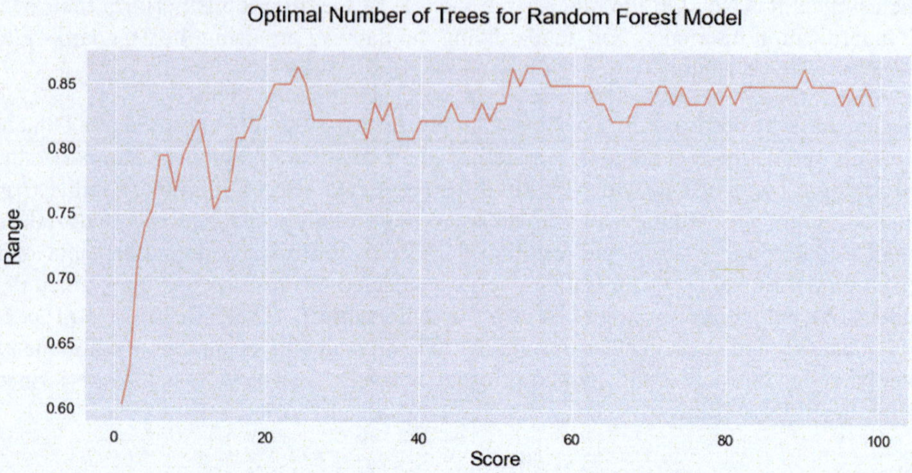

Fig. 2. Test to find the optimal number of estimators.

Table 4. Performance of each model under different metrics.

Model	Accuracy	Precision	Recall
Adaboost	0.795	0.727	0.914
X-Gradient	0.877	0.861	0.886
Random Forest	0.795	0.763	0.829
SVM (Linear)	0.836	0.871	0.771
Log. Regression	0.808	0.818	0.771
Kernel SVM	0.808	0.818	0.771
KNN	0.836	0.861	0.743
MLPNN	0.671	0.641	0.714
Naïve Bayes	0.767	0.821	0.657

Table 5. Best model per metric.

Model	Metric	Value
Adaboost	Recall (Sensitivity)	91.43%
X-Gradient	Accuracy	87.67%
SVM (Linear)	Precision	87.10%

We can now observe the heatmap - confusion matrix of each individual model that outperformed the other models in a specific category in Figs. 3, 4, and 5.

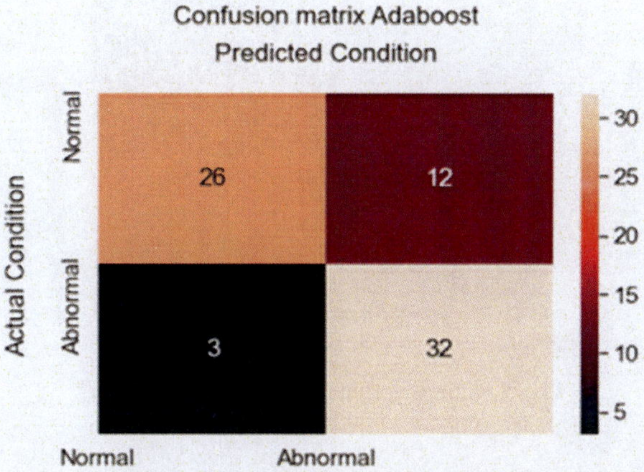

Fig. 3. Confusion matrix for Adaboost.

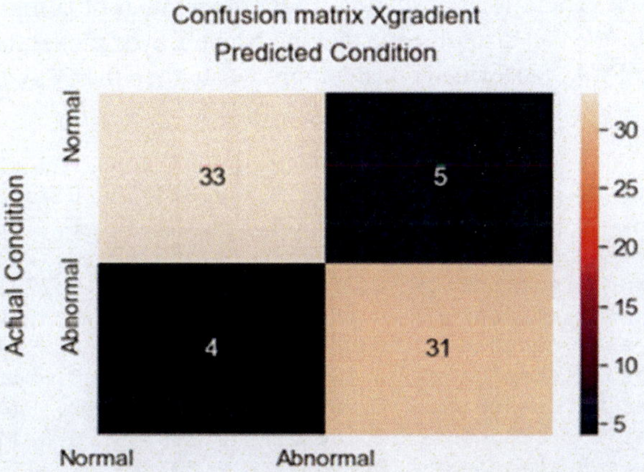

Fig. 4. Confusion matrix for X-Gradient Boosting.

Whenever we are talking about a problem where not identifying positive labels can determine if a person lives or dies, we want to sacrifice some accuracy to increase the True Positives, even if this means getting more false positives. In a real-life scenario, one might look for a balance between increasing true positives and reducing costs (mostly generated by false positives). The Adaboost model is the best example for this scenario. We gained 1 more true positive by sacrificing 7 false positives (when compared to the X-Gradient Boosting model). This might not be ideal, as the cost for treating 7 false-positive patients could be too high. In this scenario, the best model is the X-Gradient Boosting, as it controls a balance between Accuracy, Precision, and Recall without

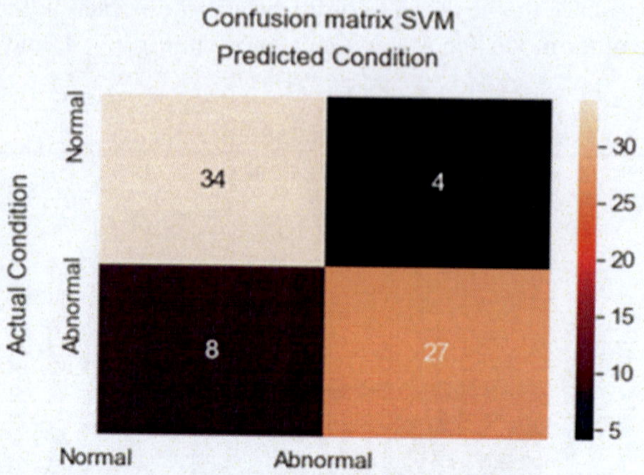

Fig. 5. Confusion matrix for SVM (Linear).

sacrificing much of any of them. However, the best model will depend completely on the problem addressed and its circumstances.

We can now observe in Fig. 6 the results for Naive Bayes varying the number of components in PCA. There is a correlation where the number of components increases, so does the Recall. We can also observe that the Naive Bayes model that used raw data (not processed by PCA) performed considerably worse than the PCA-trained models.

	Model	Accuracy	Precision	Recall	F1 Score
15	Naive Bayes 16 components	0.808219	0.843750	0.750000	0.794118
9	Naive Bayes 10 components	0.821918	0.896552	0.722222	0.800000
11	Naive Bayes 12 components	0.808219	0.866667	0.722222	0.787879
12	Naive Bayes 13 components	0.794521	0.838710	0.722222	0.776119
4	Naive Bayes 5 components	0.808219	0.892857	0.694444	0.781250
5	Naive Bayes 6 components	0.808219	0.892857	0.694444	0.781250
10	Naive Bayes 11 components	0.808219	0.892857	0.694444	0.781250
13	Naive Bayes 14 components	0.780822	0.833333	0.694444	0.757576
14	Naive Bayes 15 components	0.780822	0.833333	0.694444	0.757576
8	Naive Bayes 9 components	0.794521	0.888889	0.666667	0.761905
6	Naive Bayes 7 components	0.780822	0.857143	0.666667	0.750000
3	Naive Bayes 4 components	0.780822	0.884615	0.638889	0.741935
7	Naive Bayes 8 components	0.780822	0.884615	0.638889	0.741935
2	Naive Bayes 3 components	0.698630	0.733333	0.611111	0.666667
1	Naive Bayes 2 components	0.684932	0.709677	0.611111	0.656716
0	Naive Bayes no PCA	0.698630	0.818182	0.500000	0.620690

Fig. 6. Naïve Bayes results from varying numbers of components in PCA.

After gradually increasing the number of Features, we got to a point where the models were falling in overfitting. After experimenting with 100 selected features, the scores for each metric for each classificator fell in terms of accuracy and recall as shown in Table 6.

Table 6. Performance of the 100 feature model using the testing set as validation.

Model	Accuracy	Precision	Recall
X-Gradient	0.808	0.886	0.756
Adaboost	0.781	0.857	0.732
Random Forest	0.795	0.922	0.683
Log. Regression	0.685	0.821	0.561
MLPNN	0.616	0.697	0.561
Kernel SVM	0.699	0.913	0.512
SVM (Linear)	0.685	0.875	0.512
Naive Bayes	0.685	0.875	0.512
KNN	0.589	0.704	0.463

However, when using the training set as a validation, these scores were hitting 100% in most models. This last case is shown in Table 7.

Table 7. Performance of the 100 feature model using the training set as validation.

Model	Accuracy	Precision	Recall
KNN	1.000	1.000	1.000
Adaboost	1.000	1.000	1.000
X-Gradient	1.000	1.000	1.000
Random Forest	0.997	1.000	0.993
MLPNN	0.972	0.978	0.964
SVM (Linear)	0.900	0.937	0.850
Log. Regression	0.879	0.907	0.836
Kernel SVM	0.872	0.964	0.764
Naive Bayes	0.702	0.787	0.529

4 Discussion

In this article, we obtained three different models that have good scores for a specific task. In general terms, the best model was the X-Gradient Boosting as it had a good balance between all three most important metrics: Accuracy, Precision, and Recall (Sensitivity) with 87.67%, 86.11%, and 88.57% respectively. This model has the potential to be used as an assistant tool for classifying ECG based on the model's metrics.

We found that Standardization improved most of the classification models drastically.

We also found that PCA helped improve the performance of the selected classification algorithm (Naive Bayes) considerably making it a good fit for the given problem and circumstances.

Finally, the Feature Selection was very helpful in improving some of the classification models (the ones where their algorithms do not contain an internal feature selection). Although, as we increased the number of features, the generalization properties of the models dropped giving space to overfitting. Since the data set we had for this project was relatively small (<400), we were restricted to the number of features we can use if we wanted to avoid overfitting. If we get a bigger data set of around 4k instances, we will be able to use a higher amount of features, giving space to the models for possible improvement.

Other related articles used the UCI - Arrhythmia Data set and shared their resulting models. M. Mitra and R. Samanta explain how they obtained a great model with high Sensitivity using Correlation Based for Feature Selection (CFS) altogether with the Levenberg-Marquardt (LM) Algorithm as a Classificator. After 100 simulations, the average recall was 86.72% [7].

5 Conclusions

After going through the whole paper, here are some conclusions made that could improve a model's performance:

- Depending on the problem and the circumstances, different models could excel in their own categories. It depends on the data scientist to give the algorithms the necessary tools (by performing all the steps we mentioned in this paper) to take benefit from them.
- PCA (and most algorithms in general) performs better when scaling is done beforehand.
- ALL data transformations should be done in the training set only. A common mistake for data scientists is that they transform data (standardization, feature selection, PCA, etc.) on the whole data set. In reality, we do not have access to the testing set. This means that all testing values (including statistics such as mean, max, outliers, etc.) should be excluded from any transformation. All testing transformations should be done AFTER getting the models (standardization, feature selection, PCA, etc.) fitted with the training set. If we ignore this important point, we could be facing a model that behaves very well in both training and testing samples but fails abruptly in the real world.
- Some classifiers perform their own "feature selection" inside their algorithm. If we do not have the necessary amount of data to try feature selection with a big amount of features, the best decision will be to let these algorithms decide which features to use from the raw data, since they have access to more information which they can take advantage of.
- When comparing classification algorithms, all of them should be in their finest shape. On the contrary, we will be missing potential candidate models due to controllable factors.

For future work, we will try other algorithms such as different Neural Networks. It is also a good idea to try data augmentation techniques to increase the data set and try a higher amount of feature selection. The adequate functioning of neural networks and other classification algorithms depends on the sizes of the training set and test set [4]. Therefore, data augmentation could naturally increase some models' performance. Another possible variable that can be tested is the feature selection method. Instead of using the ANOVA F-value, we can use the Chi-square value or the Maximum Relevance Minimum Redundancy (MRMR) algorithm. Finally, we will find other ECG data sets that contain information spread through time to be able to use Time Series Analysis techniques.

References

1. World Health Organization: Cardiovascular diseases (CVDs) - Key Facts. WHO (2017)
2. Srinivasan, N.T., Schilling, R.J.: Sudden cardiac death and Arrhythmias. Arrhythmia Electrophysiol. Rev. **7**, 45–51 (2018)
3. Lyon, A., Minchole, A., Martinez, J.P., Laguna, P., Rodriguez, B.: Computational techniques for ECG analysis and interpretation in light of their contribution to medical advances. R. Soc. **15**(138), 20170821 (2018)
4. Übeyli, E.D.: ECG beats classification using multiclass support vector machines with error-correcting output codes. Digit. Sig. Process. **17**(3), 675–684 (2007)
5. Cleveland Clinic: Sudden Cardiac Death (Sudden Cardiac Arrest). Cleveland Clinic (2019)
6. Ozbay, Y., Karlik, B.: A recognition of ECG Arrhythmia using artificial neural networks. In: Proceedings 23rd Annual Conference. IEEE/EMBS, Istanbul, Turkey (2001)
7. Mitra, M., Samanta, R.: Cardiac Arrhythmia classification using neural networks with selected features. Procedia Technol. **10**, 76 (2013)
8. Guvenir, H.A., Acar, B., Muderrisoglu, H.: Arrhythmia Data Set. The University of California, School of Information and Computer Science, Irvine, CA (1998)
9. Balatsko, M.: All you want to know about preprocessing: Data preparation. Towards Data Science (2019)
10. Yun-Chung Liu, A.: The Effect of Oversampling and Undersampling on Classifying Imbalanced Text Datasets. Faculty of the Graduate School of the University of Texas at Austin: Thesis (2004)
11. Silva, A.: Arrhythmia UCI Article Repo, Tecnologico de Monterrey Campus Monterrey (2021). https://github.com/adriansilva/Arrhythmia-UCI-Article

Modeling the Effects of Politics Based on a Sociological Reference Scheme for Self-organizing Systems

Harm Hofman and Jan Treur(✉)

Social AI Group, Department of Computer Science, Vrije Universiteit Amsterdam, Amsterdam, The Netherlands

harm@h-hofman.nl, j.treur@vu.nl

Abstract. Many publications discuss how some region became the smartest one in the world. One of the underlying issues is how certain sociological mechanisms lead to the emergence of patterns according to what sometimes are called self-organizing systems. To analyse this, this paper exploits computational modeling based on (multi-order) adaptive network models and inspired by fields addressing self-organizing systems.

1 Introduction

There are many opinions on what an appropriate policy is for the development of a certain region. While some blame local politics for failures, others believe that global development and business is the cause of all evil. A clear answer has not yet been found, even though substantial research has been done on the topic. Questioning the influence of politics may not deserve a yes-or-no type of answer, but rather an answer that involves more aspects. In some way, different types aspects and influences all come together and originate in the region. There are many publications on how some region became the smartest one in the world (Hajer and Dassen 2014; Nooteboom 2004). One of the issues to be addressed is how certain sociological mechanisms that play a role lead to the emergence of patterns according to what sometimes are called self-organizing systems. The aim of this paper is to contribute to this by exploiting computational modeling based on (multi-order) adaptive network models and inspired by fields addressing self-organizing systems in the diverse literature of different disciplines.

First, methods based on a sociological reference scheme and (adaptive) network-oriented modeling (Treur 2020) are briefly introduced. The next step is a distinction between instrument- and target-variables from macroeconomics and governing principles for business administration, used in self-organizing. This is important because government administrators do not only need a model description but also prognoses and scenarios for a policy. This provides the basic ingredients for a computational model that has been developed. For the results, first a historical case is simulated to check the model. Next, a second simulation shows a forecast in a situation where other parameters and variables are required due to contingency (Mintzberg 1979). In the conclusion and discussion section, some potentials and limitations of the introduced computational model are discussed.

R. Silhavy et al. (Eds.): CoMeSySo 2021, LNNS 231, pp. 166–182, 2021.
https://doi.org/10.1007/978-3-030-90321-3_16

2 Methods

The methods used are based on the following:

- A sociological reference scheme
- Macroeconomics and models from other fields
- Levels of Control, timeframe, and self-organizing task groups
- Computational modeling based on multi-order adaptive networks

2.1 A Sociological Reference Scheme

The sociologist (and economist) Parsons developed a reference scheme (Parsons et al. 1953), applied it in practice (Parsons and Platt 1973) and tried to elaborate it using systems theory (Adriaansens 1976); see Fig. 1. In his scheme he used insights from biology and economics (Moss and Sauchenko 2006; Turner 1999). However, systems theory was not yet developed well, and sociologists are not always attracted to mathematical models. So, this by itself useful reference scheme (Kerkhoff 2007; Jong de 1997) was somewhat forgotten and the theory called an important 'high way out of order' (Castellani and Hafferty 2009). Later, computer technology and modeling made great progress; e.g., (Kauffman 2000). Economic science (Moss and Sauchenko 2006) made the best use of this and became dominant over the fragmented sociological science of which it in fact forms a subdomain.

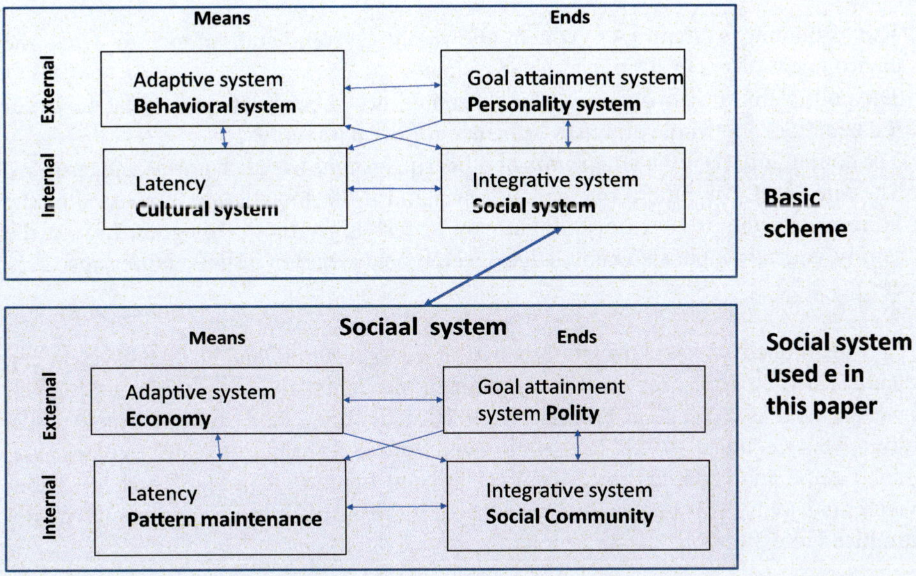

Fig. 1. Parsons reference scheme

The AGIL scheme, as it is called, is a reference scheme within which environmental influences and social changes can be interpreted. It is not a full blueprint but it at least provides recognizable anchors. A system usually tends to reach an equilibrium, but over time it will adapt to changes in the environment to survive. It always differentiates into a number of processes or subsystems, each with its own function (Parsons and Platt 1973; The American University 1973).

These are:

- **Adaptation** that is constantly needed in interaction with the environment because resources are needed from the environment to survive.
- **Goal** attainment establishes priorities that are good for the subsystem.
- **Integration** promotes solidarity in the social system.
- **Latency** or pattern maintenance is necessary to maintain the underlying values of variables. This stems from history or evolution (Parsons and Toby 1977) with learning and education as an important factor in the social system.

In addition, these functions and sub-systems influence each other, for which Parsons indicates media. In this paper we refer to this as a medium 'influence'; that is the core function of all media. Whether it concerns money, power, or value involvement, it is always a question of whether an element or subsystem has the influence to achieve something. Only, sometimes it is called buying, the other time exercising power or keeping on the standards.

In this paper the social system as described above is addressed, which in terms of AGIL means:

- The economic system (A), which allows the system and *adapts* to what the environment offers to survive as a subsystem.
- The politicians and *government* (G) that sometimes-subordinate individual needs for the benefit of the whole and uses influence for the community.
- The community (I) is the *integration* of other subsystems based on norms and customs.
- The *latency* (L) or pattern maintenance system for the social system, also called the fiduciary system, is necessary to maintain underlying values. This goes through the family, education, but also knowledge development is part of it (universities, research, etc.)

As mentioned, the social system is part of a larger whole and each subsystem can be subdivided again, as for example in politics into legislative, executive and judicial power, as well as the federal law (Kerkhoff 2007; Hofman et al. 2000; Hofman 2018; Hofman and Leeuwen 1998). These subsystems are defined in this paper as black box, which can be analyzed, and the social system can function as a black box in a more overarching study. This basic reference scheme is the basis for the computational model introduced in paragraph 2.4 for regions.

2.2 Macroeconomic Models

To make computational analysis useful for policy making or other decisions it needs targets, instruments, and a scope. Economic models therefore perform various functions such as making explanations, analysis, making predictions and designing policy alternatives. This paper introduces a new computational model and tests it on the usability. Within the model, the variables are divided into four main groups.

- Exogenous variables that are not influenced by other model variables with on the one hand *data* that are not under government control and on the other hand the *instruments* to influence the direction of the system.
- Endogenous variables are determined by the model and can be distinguished into *target variables* and variables that are residual, the *irrelevant variables.*

In this paper, the influence of politics is instrument variable (G in AGIL) and the influence of the economic system is the target variable (A in AGIL). The outcomes of the model will show that parameters that are difficult to change may be important for the outcome. This is in line with the contingency theory (Mintzberg 1994) that claims that the course of action depends upon the internal and external situation and that there is no best way to organize. What is a good decision is in one region can be a disaster in another situation and circumstances-related parameters determine to a high degree the output.

What also can be learned from macroeconomics is the scope they declare in their studies:

- The ceteris paribus assumption that 'all other things being equal' or 'other things held constant' meaning that the prediction, although usually accurate in expected conditions, can fail or the relation can be abolished by intervening factors.
- Economic models are based on theories that are neoclassical or Keynesian. The neoclassicals assume price flexibility. Prices and wages quickly find a balance and ensure smooth running of markets and economy. Margret Thatcher in England took up battles with miners on that basis and reorganized the steel industry. Keynes paid more attention to long-term deviations from equilibrium in the markets. Both involve macroeconomic assumptions and not regional differences (see Porter below).

With global disruptions such as the IT bubble, the banking crisis and covid, the predictions no longer hold true. When building computational models, it is important to get awareness of such assumptions. Tools such as network analysis and reference schemes can be general. Policy decisions are made in a context, what is part of the study.

2.3 Models of Business Administration

De Leeuw (Leeuw de 2000, 2003) describes the possibilities and limitations of a model based on systems theory: 'A model is a system, which is (during a period) a representation of aspects of another system, which is used in a certain situation and whose similarity refers precisely to the aspects that are relevant, given the purpose of use.'. That therefore requires a demarcation or system boundary. The system boundary is the distinction made by the viewer between those objects that are considered part of the system and the other objects that are associated with it. This is known as the boundary or demarcation decision. The aim of research is important in this respect, but also the so-called the Real

Life System (RLS) or the world view of the researcher; that is the whole of assumptions and views that together form his or her (obviously subjective) reality. With a variation of this, one can also speak of the RLS of a profession. So there is a certain steering here, consciously or unconsciously. That defines the limits of the study. This applies to the elements that are included in a study environment, but especially also the time frame that is the subject of study. A biologist who studies the origin of humans has a different time frame than the cardiac surgeon. What originated in biology in an environment is the material for the surgeon who takes variables and parameters that arise in biology as a given. There are relationships, but they can hardly be included in one study.

When establishing system boundaries for research, De Leeuw (Leeuw de 2000) distinguishes between a controlled system (BS) and a governing body (BO) that act together as a system in an environment (see Fig. 2). A wide diversity of mechanisms within the networks interact with each other depending on the relations (Leeuw de 2000). Within network oriented modeling this will translate into the inclusion of different formulas or parameters in the relationships. Learning either adapting to or influencing the environment is part of it (Kemp and Weehuizen 2005), but also within the system between BS from BO and vice versa. The same takes place in the internal in subsystems. The governing body (BO) in the basis has more influence on the structure of the controlled system, but depends on the situation and culture, a BS is more self-directed (Endenburg 1992). In organizations this is known as self-managing teams versus top down management (Hofman and Huijsmans 1995; Hofman 1997). This is determined by the relationships, variables and parameters in and between the nodes in network oriented modelling for adaptive networks.

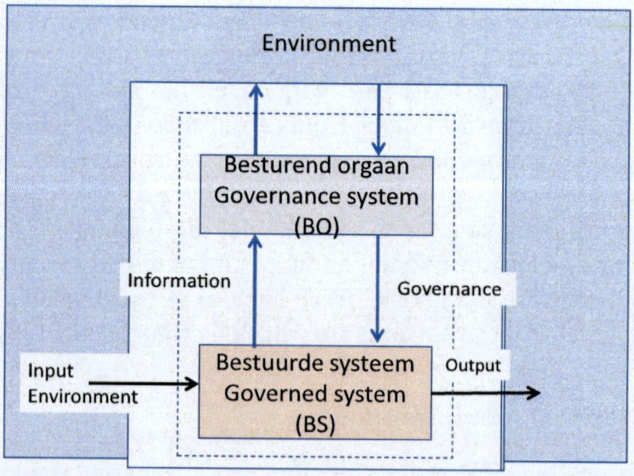

Fig. 2. Levels in social systems (Leeuw de 2000) which is also used for types of learning (Swieringa and Wierdsma 1996; Stacey 2001).

In this context there are many disciplines where good use can be made of, for example:

- (Porter 1990) made an in-depth study of the competitiveness of countries but also of regions. There is an enormous number of studies and publications around his model. Cluster theories are included.
- (Rogers 2003) describes the diffusion of innovations with a lot of empirical material
- Cultures that largely determine the speed of change, i.e. the parameters in models, have been extensively studied by (Hofstede 2001), who also partly used and tested Parsons' work as a basis.
- An example of a context description of the energy transition in the Netherlands is (Verbong et al. 2001; Verbong and Loorbach 2012; Grin et al. 2010; Hofman et al. 2000) integrated many of these aspects in the Parsons reference scheme and this is elaborated for society in (Hofman 2018).

2.4 Network Oriented Modelling for Adaptive Networks Used as a Vehicle

The elements described in the previous sections were addressed by using as a vehicle the modelling approach based on multi-order adaptive complex network models described in (Treur 2020). For this study we consider the social subsystem in a nearly closed system. The political subsystem is initially strongly stimulated as an instrument variable to adjust the target variable to strengthen the economic output. That depends on the mutual connections and the way in which those network-nodes respond to the influences they experience from outside. Finally, first- and second-order adaptation is addressed, fitting to de Leeuw (Leeuw de 2000) and learning theories in different disciplines (Argyris et al. 1985; Hofman and Leeuwen 1998; Kessels and Poell 2001; Nonaka and Takeuchi 1995; Poell 1998; Stacey 2001; Swieringa and Wierdsma 1996; Hofman 1997).

For the basis of the model to be built, Parsons' reference scheme for the social subsystem (the demarcation of the research) was used. For the first-order adaptation level, Hebbian learning is used and for the second-order adaptation level, the speed of learning is influenced by the underlying culture (loosely inspired by how plasticity and metaplasticity are analysed as first-order and second-order adaptation within neuroscience (Abraham and Bear 1996; Hebb 1949)). This is for the two basic nodes I and A, where the parameters have been kept the same for the sake of clarity. The connectivity of the network model is given in Fig. 3 and in Table 1 its states are briefly explained.

In this section, the network-oriented modelling approach used from (Treur 2020) used here is briefly introduced. Following this approach, a temporal-causal network model is characterized by (here X and Y denote nodes of the network, also called *states* with activation values $X(t)$ and $Y(t)$ over time t):

- **Connectivity characteristics** Connections from a state X to a state Y and their weights $\boldsymbol{\omega}_{X,Y}$
- **Aggregation characteristics** For any state Y, some combination function $\mathbf{c}_Y(..)$ (usually with some parameters) defines the aggregation that is applied to the impacts $\boldsymbol{\omega}_{X,Y}X(t)$ on Y from its incoming connections from states X
- **Timing characteristics** Each state Y has a speed factor $\boldsymbol{\eta}_Y$ defining how fast it changes for given impact.

The following difference (or differential) equations that are used for simulation purposes and also for analysis of temporal-causal networks incorporate these network characteristics $\boldsymbol{\omega}_{X,Y}$, $\mathbf{c}_Y(..)$, $\boldsymbol{\eta}_Y$ in a standard numerical format:

$$Y(t + \Delta t) = Y(t) + \boldsymbol{\eta}_Y[\mathbf{c}_Y(\boldsymbol{\omega}_{X_1,Y}X_1(t), \ldots, \boldsymbol{\omega}_{X_k,Y}X_k(t)) - Y(t)]\Delta t \tag{1}$$

for any state Y and where X_1 to X_k are the states from which Y gets its incoming connections. Here the overall combination function $\mathbf{c}_Y(..)$ for state Y is the weighted average of available basic combination functions $\mathbf{c}_j(..)$ by specified weights $\boldsymbol{\gamma}_{j,Y}$ (and parameters $\boldsymbol{\pi}_{1,j,Y}$, $\boldsymbol{\pi}_{2,j,Y}$ of $\mathbf{c}_j(..)$) for Y:

$$\mathbf{c}_Y(V_1, \ldots, V_k) = \frac{\boldsymbol{\gamma}_{1,Y}\ \mathbf{c}_1(V_1, \ldots, V_k) + \ldots + \boldsymbol{\gamma}_{m,Y}\ \mathbf{c}_m(V_1, \ldots, V_k)}{\boldsymbol{\gamma}_{1,Y} + \ldots + \boldsymbol{\gamma}_{m,Y}} \tag{2}$$

Equation (1) and (2) are hidden in the dedicated software environment; see (Treur 2020), Ch 9. As part of this software environment, currently around 45 useful basic combination functions are included in a combination function library; see Table 1 for the two of them used here. The above concepts enable to design network models and their dynamics in a declarative manner, based on mathematically defined functions and relations.

Note that there is a crucial distinction for network models between network characteristics and network states. Network *states* have values (their activation levels) and are explicit representations that may be accessible for network states by connections to and from them and can be handled or manipulated in that way. They can be considered to provide an informational view on the network; usually the states are assumed to have a certain informational content. In contrast, network *characteristics* (such as connection weights and excitability thresholds) have values (their strengths) and determine (e.g., cognitive) processes and behaviour in an implicit, automatic manner. They can be considered to provide an embodiment view on the network. In principle, these characteristics by themselves are not directly accessible nor observable for network states; you can make connections between states but you cannot make connections between network characteristics or between states and characteristics.

Table 1. Examples of basic combination functions from the library used in the current paper

	Notation	Specification	Parameters
Advanced logistic combination function	$\mathbf{alogistic}_{\sigma,\tau}(V_1, \ldots, V_k)$	$\left[\frac{1}{1+e^{-\sigma(V_1+\ldots+V_k-\tau)}} - \frac{1}{1+e^{\sigma\tau}}\right](1 + e^{-\sigma\tau})$	Steepness $\boldsymbol{\sigma} > 0$ Excitability threshold $\boldsymbol{\tau} > 0$
Hebbian learning combination function	$\mathbf{hebb}_{\mu}(V_1, V_2, W)$	$V_1V_2\,(1 - W) + \boldsymbol{\mu}W$	Persistence factor $\boldsymbol{\mu}$

As indicated above, 'network characteristics' and 'network states' are two distinct concepts for a network. Self-modeling is a way to relate these distinct concepts to each other in an interesting and useful way. A *self-model* is making the network characteristics (such as connection weights and excitability thresholds) explicit in the form of adding states (called *self-model states*) for these characteristics and also connections and other network characteristics for these additional states. Thus, the network gets an internal self-model of part of its network structure: it explicitly represents information about its own structure. In this way, by iteration different self-modeling levels can be created where network characteristics from one level relate to network states at a next level.

More specifically, adding a self-model for a temporal-causal base network is done in the way that for some of the states Y of the base network and some of the network structure characteristics for connectivity, aggregation and/or timing (i.e., some from $\boldsymbol{\omega}_{X,Y}$, $\boldsymbol{\gamma}_{j,Y}$, $\boldsymbol{\pi}_{i,j,Y}$, $\boldsymbol{\eta}_Y$), additional network states $\mathbf{W}_{X,Y}$, $\mathbf{C}_{j,Y}$, $\mathbf{P}_{i,j,Y}$, $\mathbf{H}_Y$ (*self-model states*, also called *reification states*) are introduced and connected to other states:

a) **Connectivity self-model**

- Self-model states $\mathbf{W}_{X,Y}$ are added representing connectivity characteristics, in particular connection weights $\boldsymbol{\omega}_{X,Y}$

b) **Aggregation self-model**

- Self-model states $\mathbf{C}_{j,Y}$ are added representing aggregation characteristics, in particular combination function weights $\boldsymbol{\gamma}_{j,Y}$
- Self-model states $\mathbf{P}_{i,j,Y}$ are added representing aggregation characteristics, in particular combination function parameters $\boldsymbol{\pi}_{i,j,Y}$

c) **Timing self-model**

- Self-model states $\mathbf{H}_Y$ are added representing timing characteristics, in particular speed factors $\boldsymbol{\eta}_Y$

The notations $\mathbf{W}_{X,Y}$, $\mathbf{C}_{j,Y}$, $\mathbf{P}_{i,j,Y}$, $\mathbf{H}_Y$ for the self-model states indicate the referencing relation with respect to the characteristics $\boldsymbol{\omega}_{X,Y}$, $\boldsymbol{\gamma}_{j,Y}$, $\boldsymbol{\pi}_{i,j,Y}$, $\boldsymbol{\eta}_Y$: here **W** refers to $\boldsymbol{\omega}$, **C** refers to $\boldsymbol{\gamma}$, **P** refers to $\boldsymbol{\pi}$, and **H** refers to $\boldsymbol{\eta}$, respectively. For the processing, these self-model states define the dynamics of any state Y in a canonical manner according to Eq. (1) and (2) whereby the values of $\boldsymbol{\omega}_{X,Y}$, $\boldsymbol{\gamma}_{j,Y}$, $\boldsymbol{\pi}_{i,j,Y}$, $\boldsymbol{\eta}_Y$ are replaced by the state values of $\mathbf{W}_{X,Y}$, $\mathbf{C}_{j,Y}$, $\mathbf{P}_{i,j,Y}$, $\mathbf{H}_Y$ at time t, respectively.

The network constructed by the addition of a self-model to a base network is called a *self-modeling network* or a *reified network* for this base network. This constructed network is also a temporal-causal network model itself, as has been shown in (Treur 2020), Ch. 10; for this reason, this construction can easily be applied iteratively to obtain multiple levels or orders of self-models, in which case the resulting network is called a *multi-order* or *higher-order self-modeling network*.

The adaptive network model introduced here is described by the states in Table 2 and the connectivity depicted in Fig. 3.

Table 2. State names of the network model including the states A, G, I, L from Parsons at the base level, and the **W**-states and **H**-states for learning at the first- and second-order adaptation level.

State nr	State name	Name	Explanation	Level
X_1	A	Adaptation	Economic target state	Base
X_2	G	Goals	Political instrument state	
X_3	I	Integration	Network state	
X_4	L	Legacy	Pattern maintenance or 'cultural state'	
X_5	$\mathbf{W}_{IA}$		First-order self-model state for Hebbian learning for connection from I to A	First-order adaptation
X_6	$\mathbf{W}_{GI}$		First-order self-model state for Hebbian learning for connection from G to I	
X_7	$\mathbf{HW}_{IA}$		Second-order self-model state for speed factor (learning rate) for $\mathbf{W}_{IA}$	Second-order adaptation
X_8	$\mathbf{HW}_{GI}$		Second-order self-model state for speed factor (learning rate) for $\mathbf{W}_{GI}$	

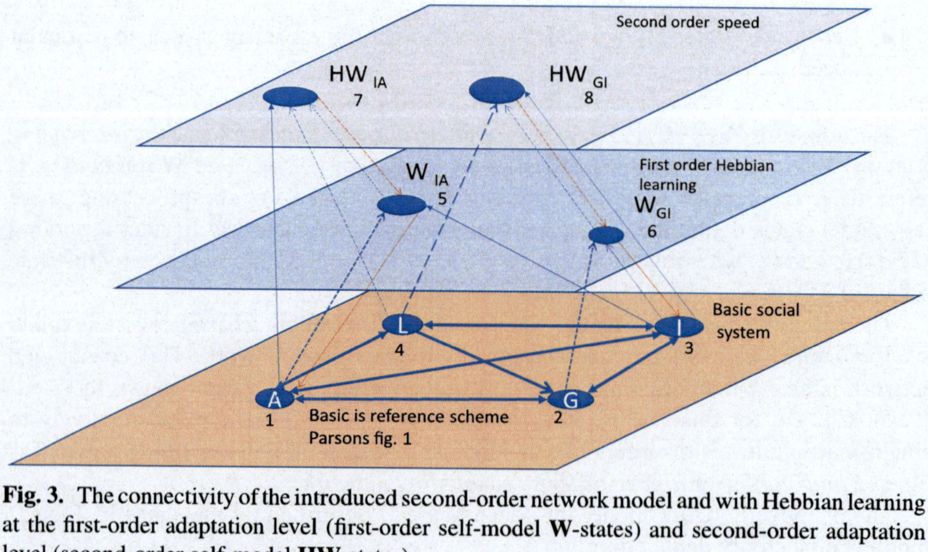

Fig. 3. The connectivity of the introduced second-order network model and with Hebbian learning at the first-order adaptation level (first-order self-model **W**-states) and second-order adaptation level (second-order self-model **HW**-states)

At the basic social level, the core is formed by the states (depicted as ovals for network nodes) and relationships (depicted as arrows for network connections) from the Parsons reference scheme. Here the influence that one state *X* has on another state *Y* depends on the value of that state *X*, the weight $\mathbf{\omega}_{X,Y}$ of the connection from *X* to *Y*. These impacts are aggregated based on a logistic combination function (see Table 1) with parameters and processed for the next stage (for time step Δt). While people work together in co-makership, they learn from that by developing stronger mutual connections. This learning (also called plasticity) is modeled at the first-order adaptation level based on a combination function (see Table 1) for Hebbian learning (Hebb 1949).

- From the government activity (G) and the network activity (I) upward to $\mathbf{W}_{G,I}$ which concerns the weight of the influence of G on I. From $\mathbf{W}_{G,I}$ downward (red arrow) to I: cooperation between government and networks has a positive effect on the network
- And from the network activity (I) and the economy activity (A) upward to $\mathbf{W}_{I,A}$ which concerns the weight of the influence of I on A. From $\mathbf{W}_{I,A}$ downward (red arrow) with influence on companies (A): cooperation in network and companies has a positive effect on companies.

What is important is the extent to which these forms of collaboration are possible and lead to learning. Since it concerns new processes in which one cannot always fall back on old routines and an old pattern can limit learning: the speed of learning is less in the beginning (the second-order self-model states **HW** regulate the learning speed). This process is based on an effect from the relevant nodes (A to $\mathbf{HW}_{IA}$ and I to $\mathbf{HW}_{GI}$) and the pattern enforcement (L, for culture) to the learning speed. The activation of these **HW**-states is done via a logistic function (see Table 1), in which the parameters depend on the culture and activities in the region. The full specification of the model can be found at Appendix (2021).

3 Results

The results are presented here from two perspectives, a first test of the model and a next step towards using the model as a prediction method. This is done with:

- The Eindhoven case: as a test
- The steel industry in North Holland: a simulation as a forecast.

3.1 The Eindhoven Case: As a Test

Eindhoven has an industrial tradition. In the beginning cheap labour played a role with the expansion of Philips and DAF. Philips with historically strong ties between the Philips family and the social community largely withdrew from 1980, but the knowledge centers remained. Supply companies and independent parts of Philips created employment in the field of high-quality technology (including ASML and the VDL Groep, family business). The Brainport foundation in the region is a partnership between companies, knowledge institutions and governments (Triple Helix); (Hofman 2005). There is a strong industry, processing materials into new products, also known as discrete industry. Traditionally, the knowledge environment has been tailored to this.

Philips and its Natlab laboratory was a leading and in the past driving force for the current technical university TUE. The universities of applied sciences of Tilburg, from Catholic courses and the technical college of Eindhoven, merged. Even though Catholicism has lost influence due to apostasy, Noord Brabant still has its own recognizable character. This mix was the breeding ground for today‘s modern High Tech Campus, where research by Philips and IBM is carried out, but more importantly with a multitude of diverse companies that work together in an open manner on innovations for the future. This did not come naturally. The politicians and especially mayor Welchen was one of the initiators in the 1990s. In 2011, the Eindhoven region was 'smartest region in the world'.

The situation can be described as follows:

- Industrial activity has is rapidly declined and was low due to the departure of Philips (production to low labour cost countries; head office to Amsterdam) and problems at DAF. So, the influence of the economic activities (A) is low. But the influence of the politics is high (G) and quickly growing due to mayor Welchen
- The knowledge for discrete production is relatively widely applicable. The knowledge institutes are of high quality and work well together through the influence of Welchen in what is called the triple helix (Hofman 2005); (Hofman et al. 2000). In the beginning there is scepticism among families and the culture used to working from cradle to grave at a social Philips must be converted to a mentality entrepreneurial (L) and will be successful if Welchen's politics shows success quickly (momentum).
- Rabobank points out four essential drivers on business success: innovation, entrepreneurial climate, human capital and management. 'The average quality of management differs between regions. In the Southern Netherlands, the management practices of industrial companies are the best; they perform comparable to companies from top countries such as Sweden and Germany'. So the learning capacity.
- The culture is basically high R&D minded (technically) with cohesion to work in communities.
- Due to the lower demand from Philips, there is unused capacity at the suppliers that can be used more widely than within heavy duty processes. So, the potential of A to grow rapidly is present.

Summarizing: Only relevant data which is changed in this case is mentioned here. As mentioned, economy (A) is initially low due to the crisis, but that of politics (partly due to the mayor) is high and quickly rising to a higher level. Discrete production lends itself to small innovative networks and co-makership (with high learning). In the pattern maintenance (L), the knowledge is available, but there will be some scepticism and new routines have yet to arise. This initially inhibits learning (HW), whereby the culture of management is open to a quick adjustment. Economics (A) is initially low due to the crisis but can quickly rise again due to the structure with appropriate techniques in the environment (τ and δ in a logistic function). These factors have been valued and used as characteristics for the model, after which simulation patterns were created. These patterns align with what has occurred globally in reality; see Figs. 4 and 5.

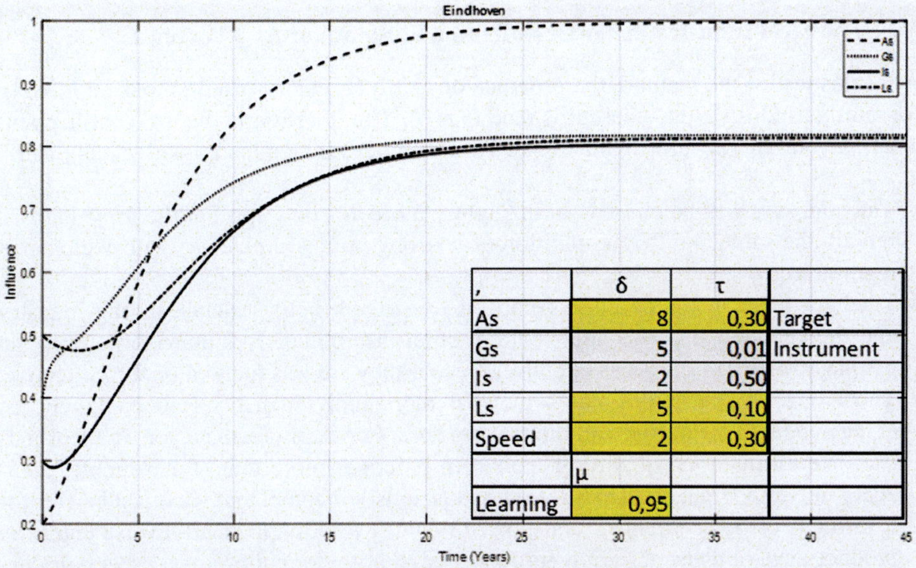

Fig. 4. Simulation with main parameters for Eindhoven. First politics is rising steep and economy and the integration network is following very quick. Pattern maintenance takes a little more time because learning takes time and people are sceptic at the start. However, they take up quite quickly too and the influence of politics, pattern maintenance and the social integration are growing to a nice balance.

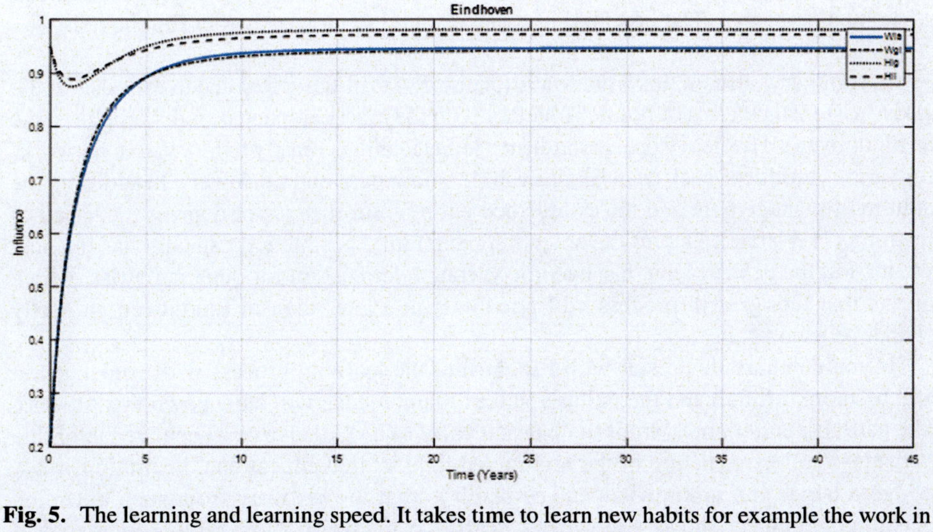

Fig. 5. The learning and learning speed. It takes time to learn new habits for example the work in small groups for innovations (start at 0.2). At the start this is influenced negatively by the speed of change (start at 0.95).

3.2 The Steel Industry in North Holland a Simulation: As a Forecast

After this test of the method, the province of North Holland is used to make a forecast for another industry, the steel mills (see Fig. 6). The question is then: 'What happens when they leave'. Several characteristics are then relevant compared with Eindhoven.

- The company will be phased out in slightly more drastic steps, but the principle will remain the same and the assumption of a strong policy in place as with Welchen is maintained.
- The knowledge institutes are less embedded in a triple helix structure for this industry and the knowledge is less applicable to other disciplines. It is more difficult to get people involved in a process of convergence if they are less focused on the collective.
- The North Holland culture is somewhat more individualistic. Rabobank reports: 'There are also major regional differences here; For example, managers in the Randstad overestimate the quality of their own practices more than, for example, managers in North Brabant'. They seem unconsciously incompetent what is a barrier for learning. Customer-oriented (consumer-oriented) working in small diverse teams on product development of various products is alien to the culture of a process industry, strongly focused on improving the process around one product; the one that will disappear.
- The culture is basically a focus on large scale operations with a specific R&D goal for the process. This process is unique and changes to a basically other field gives a higher resistance.
- The production capacity of suppliers is mainly specific and for maintenance of heavy industry, which has an underutilisation. This has a potential for other products. But with special equipment this takes more efforts and the routes to new appropriate R&D products is scarce. The potential of these resources is lower than in Eindhoven.

In summary: also, in this situation A (economy) is initially low due to the crisis and it is also assumed that that of politics together with the reaction speed is high (kept the same as Eindhoven). The heavy process industry lends itself less easily to innovative networks for various products, such as in Eindhoven (learning parameter μ lower). In addition, the culture of management and the established patterns for that sector depress the speed of learning (**HW**-state). Due to the stagnating economy, production capacity has become vacant, but the capacity that has become vacant is less useful for other activities, which means that this growth process will be slower and later than in Eindhoven an partly cannot be used again.

If you compare this result with Eindhoven, the learning process with comakership will be more difficult and the culture/pattern maintenance will slow down this process. The pattern-maintaining part of the internal culture (L) will also be a negative factor due to the less collective nature of the region and the fact that educational institutions must bridge a larger gap in knowledge. The result is that the recovery from such a decline of a large company shows a slower or nearly no recovery of the region compared to Eindhoven even with a strong policy effort.

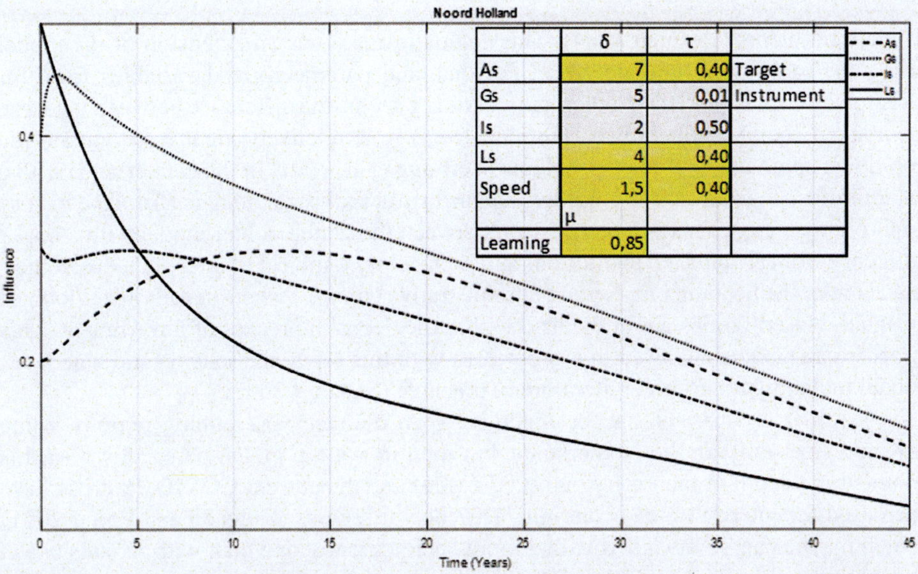

Fig. 6. Noord Holland Steel mills. Here the same model with same polity used, but with adjusted parameters due to other circumstances for North Holland relating to one industry branch, the steel factories (the contingency factors). Policy gets to little momentum. Especially in pattern maintenance where retraining takes more time followed with less successes and more sepsis from the people. This shows the importance of contingency factors expressed in the parameter which are every important.

The given example even shows a downward trend. This has to do with the assumptions that are made in the way as in macroeconomics.

- It is a restricted closed system where all other elements are assumed to remain the same (ceteris paribus). Other developments on the world market have not been considered. This also applies to macroeconomic forecasts for the Dutch economy, which predict less well during the banking crisis and covid.
- Only one industry is assumed here. It is possible that loss will be absorbed by other industries already existing in the region. In fact, the macroeconomic balance using prices as means to get an equilibrium, worked out badly in regions with mines and steel industry in England. The downward trend in this figure may sometimes be more realistic for these regions.

The question from policy makers will be 'How to solve this'. That requires additional studies. This paper is a limited study with a closed regional system. Policy questions require comparisons with fore example foreign countries and involvement of other sectors in the region. What could for example be learned from wind turbine development. Holland is known for its historic windmills, but the modern wind turbines are mainly imported from Denmark. Only the offshore industry benefits somewhat, but in general Holland failed to get competitive advantage out of the extra demand due to solving climate problems (Hofman 1984). This evaluation requires an extension of the model using the same basic modules from Parsons, and parameters and variables from for example, Denmark and the Netherlands. Striking seems the difference between the large number of energy cooperatives (Hofman 1984) with relatively great freedom and the top-down approach to climate problems (Verbong et al. 2001) in the Netherlands with a major influence from energy suppliers and infrastructural stakeholders (from the history of Groningen gas). This is a comparison between the Dutch polder model with a major influence from established institutions and is currently a subject of interest in the Netherlands under the heading of a 'wrong administrative culture', where stakeholders/lobbies in small, closed groups are in the lead within the decision process of government. This process can also be analysed in a subsystem with this reference scheme and a network model including appropriate literature in this area (Sabatier 2007).

A similar comparison can be made between the steel and mining regions in the VK and steel mills in the Netherlands, but than in a more global perspective. At this moment the depth of the recession (2007–2009) and the current COVID-19 crisis have increased the interest in the economic resilience of regions (Delgado and Porter 2021) which maybe can be modelled with a more sociological scope than with models use at this moment (Porter et al. 2014). The disadvantage of including national and regional situations with more industries and levels together in one model is that it becomes complex. The researcher needs to determine whether it makes sense to include this in one model with a good system boundary for research or whether it is better to conduct sub-studies.

4 Conclusion and Discussion

This paper addressed the influence of politics on regional development and the influence of the economy. To set up a computational model based on a second-order adaptive network structure, a reference scheme from sociology was used, which established a relationship between the adaptive networks in reality and mainstream sociology and economics. This was tested against field studies from those disciplines. The contingency theory known from business administration (Mintzberg 1979); where for each department is a suitable culture, working method and professionalization, seems to apply here. This contingency can for example be translated in society in the pattern-maintaining component. The speed at which changes take place depend on:

- The number of times people work together on solutions, represented by Hebbian learning at the first-order adaptation level.
- The speed by which this adaptation takes place, modeled as second-order adaptation.

The introduced second-order adaptive network model provides possibilities to investigate the consequences of changes of characteristics for the situation.

Determining suitable network characteristics for the network relations is essential but difficult. Parsons' reference scheme has many more subsystems that connect to each other. Many variants are then conceivable that can be worked out with the introduced model; Parsons' scheme is not a Grand Theory. It is a reference scheme to research society on a structured way. It is an exceptionally large complex system with many variables and parameters, each interacting with their own deviations.

References

Abraham, W.C., Bear, M.F.: Metaplasticity: the plasticity of synaptic plasticity. Trends Neurosci. **19**(4), 126–130 (1996)

Adriaansens, H.: Talcott Parsons and the conceptuele dilemma. Van Loghum Slaterus, Deventer (1976). (in Dutch)

Appendix (2021). https://www.researchgate.net/publication/352356517

Argyris, C., Putnam, R., McLain Smith, D.: Action Science. Jossey- Bass Inc Publishers, San Francisco (1985)

Castellani, B., Hafferty, F.: Sociology and Complex Science, a New Field of Inquiry. Springer, Heidelberg (2009). https://doi.org/10.1007/978-3-540-88462-0

Delgado, M., Porter, M.: Clusters and the Great Recession (2021). SSRN: https://ssrn.com/abstract=3819293

Delgado, M., Porter, M., Stern, S.: Clusters, convergence, and economic performance. Res. Policy **43**(10), 1785–1799 (2014)

Endenburg, G.: Sociology as social design. Eburon, Delft (1992). (in Dutch)

Grin, J., Rotmans, J., Schot, J.: Transitions to Sustainable Development. Routledge, New York (2010)

Hajer, J., Dassen, T.: Smart Cities. Rotterdam: nai010/PBL (2014)

Hebb, D.O.: The organization of behavior: a neuropsychological theory. Wiley (1949)

Hofman, H.: Evolutionary Regionomics: The evolution of Parsons model in the triple helix, (p. CD). Turijn (2005)

Hofman, H.: Kant's Enlightenment and Eternal Peace. Bravenewbooks.nl (2018). (in Dutch)

Hofman, H.: Danish wind energy has become more businesslike. PT Actueel (1984)

Hofman, H.: Succes or failure working in a product group oriented manner. In: The Second World Congress on Intelligent Manufacturing Processes & Systems. Springer, Budapest (1997)

Hofman, H., Huijsmans, K.: Business process re-engineering in industrieen hbo. van Gorcum, Assen (1995)

Hofman, H., Leeuwen, P.: What can education and trainers learn from each other. Bedrijfskundig vakblad, pp. 26–32 (1998). (in Dutch)

Hofman, H., Foks, O., Kokhuis, J.: Knowledge environment scenarios. van Gorcum, Assen (2000). (in Dutch)

Hofstede, G.: Culture Consequences, Comparing Values, Behaviours Institutions and Organizations Across Nations. Sage Publications, Thousand Oaks (2001)

Jong de, M.: Grandmasters of Sociology. Boom, Amsterdam Meppel (1997). (in Dutch)

Kauffman, S.: Investigation. Oxford University Press, New York (2000)

Kemp, R., Weehuizen, R.: Policy Learning, what does it mean and how can we study it. NIFU STEP (Publin project), Oslo (2005)

Kerkhoff, A.: The society in schemas. De sameleving in schemas. Damon, Budel (2007). (in Dutch)

Kessels, J., Poell, R.: Human resource development, organising the learning. Samson, Groningen (2001)

Leeuw de, A.: Business management, primary process and organization. Van Gorcum, Assen (2000). (in Dutch)

Leeuw de, A.: Business methodology, research management. van Gorcum, Assen (2003). (in Dutch)

Mintzberg, H.: The structuring of organizations. Prentice Hall, Englewood Cliffs (1979)
Mintzberg, H.: Rise and Fall of Strategic Planning. Prentice Hall/FT, London (1994)
Moss, L., Sauchenko, A.: Talcott Parsons, economic sociologist of the 20th century. Blackwell Publishing, Malden (2006)
Nonaka, I., Takeuchi, H.: The knowledge creating company, how Japanese companies create the dynamics of innovation. Oxford University Press, New York Oxford (1995)
Nooteboom, B.: Inter-firm Collaboration, Learning and Networks, an Integrated Approach. Routledge, London and New York (2004)
Parsons, T., Platt, G.: The American University. Harvard University Press, Cambridge (1973)
Parsons, T., Toby, J.: Evolution of Societies. Prentice Hall, London (1977)
Parsons, T., Bales, R., Shils, E.: Working papers in the theory of action. Free Press, Connecticut (1953)
Poell, R.: Organizing work related learning projects, a network approach. Ph.D. Katholieke Universiteit, Nijmegen (1998)
Porter, M.: The Competitive Advantage of Nations. The Macmillan Press Ltd, London (1990)
Rogers, E.: Diffusion of Innovations. Free Press, New York (2003)
Sabatier, P.: Theories of the Policy Process. Westview Press, Colorado (2007)
Stacey, R.: Complex Responsive Processes in Organizations, Learning and Knowledge Creation. Routledge, London (2001)
Swieringa, J., Wierdsma, A.: Towards a learning organization. Wolters Noordhof, Groningen (1996). (in Dutch)
Treur, J.: Network-Oriented Modeling for Adaptive Networks: Designing Higher-Order Adaptive Biological, Mental and Social Network Models. SSDC, vol. 251. Springer, Cham (2020). https://doi.org/10.1007/978-3-030-31445-3
Turner, B.: The Talcott Parsons Reader. Blackwell Publishers, Oxford (1999)
Verbong, G., Loorbach, D.: Governing the Energy Transition. Routledge, New York (2012)
Verbong, G., Selm van, A., Knoppers, R., Raven, R.: A long-term issue, the history of sustainable energy in the Netherlands. Aeneas, Boxtel (2001). (in Dutch)

Selecting a Multi-criteria Decision Analysis Method

Nikita Gorodilov(✉), Maria Dolzhenkova, and Gennadiy Chistyakov

Department of Computers, Vyatka State University, Moskovskaya, 36, 610000 Kirov, Russia

Abstract. Multi-criteria decision analysis (MCDA) methods are widely used in various fields and disciplines. A large number of studies have been focused on the development of new methods and the selection of the most appropriate decision making for a particular task. The purpose of this study is to analyze the existing approaches to selecting MCDA methods, to determine the most appropriate method for solving multi-criteria problems and to provide guidelines how to choose MCDA methods. The paper investigates the existing ways to select a MCDA method. We described the qualitative selection criteria and conducted experiments to determine the most appropriate method using various quantitative indicators. The result of the study is a set of recommendations for selecting a MCDA method.

Keywords: Multi-criteria-decision-analysis · Multi-criteria-decision-making-methods · Multi-criteria-method-selection · Quantitative-analysis-of-multi-criteria-methods · Sensitivity-analysis

1 MCDA Foundations

The main task of multi-criteria decision analysis (MCDA) is to support a decision maker in choosing the most preferable variant from many possible options, taking into account a multitude of criteria characterizing acceptability of individual decision variants (alternatives) [1].

As a rule, the solution of such problems is needed if it is impossible to evaluate different options. Tasks of this kind are associated with uncertainty which can occur in the following situations [2]:

1. Ambiguity occurs when the input data have more than one interpretation;
2. Stochasticity occurs when the input data can be interpreted as random;
3. Partial information is given when part of the input information is missing.

It can also be noted that in the decision making process uncertainty may occur due to the input data or the subjectivity of a decision maker who provides the input data and, in some cases, participates in the decision making process. The types of uncertainty are as follows:

R. Silhavy et al. (Eds.): CoMeSySo 2021, LNNS 231, pp. 183–198, 2021.
https://doi.org/10.1007/978-3-030-90321-3_17

1. Input data uncertainty. This includes ambiguity and partial information, as well as randomness;
2. Subjective uncertainty which occurs as a result of the actions of a decision maker, e.g., the prioritization of criteria or the uncertainty of the information received from a decision maker while a multi-criteria method is being applied.

There are four types of problems that arise in the context of multi-criteria decision making: the choice (selection) problem, the sorting (classification) problem, the ordering (ranking) problem and the description problem [3].

The main aim of the choice problem is to determine the best possible alternative, or to reduce a group of alternatives to a subset of equivalent or inconsistent alternatives. The aim of the sorting problem is to sort alternatives into predefined categories. Sorting can be used to reduce the number of alternatives to consider later. The aim of the ordering problem is to order the alternatives in descending order. In this case, the order may be complete or partial if disparate actions are considered. The aim of the description problem is to describe actions and their consequences (e.g., visualization of the GAIA method).

The literature review provides a range of works on various classifications of multi-criteria problems. In [1,5], the authors divide multi-criteria problems into two types, those that are solved by methods based on the utility or value function and methods of advanced analysis (outranking). There are also other classifications presented by various authors [6–9].

The main task of American school is to try to find the function that reflects the utility of a particular alternative, resulting in a number that reflects the preference. American school uses two types of aggregation to calculate utility: additive and multiplicative [10].

Outranking methods (European school of decision making) uses many rules to find the best alternative rather than a single function, as in American school of decision making. Basically, the solution in such methods is based on paired comparisons of alternatives.

In the multi-criteria analysis, the evaluation of options based on different criteria is aggregated to obtain an overall assessment. In this case, the variety of methods is divided into compensatory and non-compensatory [31]. In compensatory methods, low performance on some criteria can be compensated for by high efficiency on other criteria. In non-compensatory methods, low performance on a number of criteria cannot be compensated for by high performance on other criteria.

Multi-criteria methods are characterized by different types of input data. There are quantitative and qualitative data, which can be expressed in a quantitative and qualitative scale [1,2,11]. Qualitative data are not statistical and are not measured using numbers, instead they are classified based on properties, attributes, or other identifiers. In contrast to qualitative data, quantitative data are statistical. They are measured using numbers and values, which makes them a better option for data analysis.

Qualitative data can be divided into two types: nominal and cardinal [2]. Cardinal data include natural ordered categories, but the distance between the categories is unknown, and nominal data have no concept of order and are only an enumeration of categories.

Multi-criteria methods are characterized by the following data:

1. Alternatives: the list of pre-defined, restricted, and independent alternatives, denoted as: $A_i, i = 1, ..., m$;
2. Criteria: the list of performance indicators required to evaluate alternatives, denoted as: $C_j, j = 1, ..., n$;
3. Decision matrix: the value matrix where the cells indicate the value of alternatives for each of the criteria, denoted as:

$$D_{m,n} = \begin{matrix} \\ A_1 \\ A_2 \\ \vdots \\ A_m \end{matrix} \begin{matrix} C_1 & C_2 & \cdots & C_n \\ \left(\begin{matrix} r_{11} & r_{12} & \cdots & r_{1n} \\ r_{21} & r_{22} & \cdots & r_{2n} \\ \vdots & \vdots & \ddots & \vdots \\ r_{m1} & r_{m2} & \cdots & r_{mn} \end{matrix} \right) \end{matrix}, \quad (1)$$

   where r_{ij}, is the value of i of the alternative and j of the criterion;
4. Criterion weight: indicates the importance of the criterion and may not be used in some methods, it is denoted as: $w_j, j = 1, ..., n$, where w_j corresponds to the weight of c_j of the criterion;
5. Direction of change in criteria: shows which of the values, the highest or lowest, is the priority for a decision maker;
   (a) Positive criterion: from the point of view of a decision maker, the highest value of the criterion is more favorable, it is denoted as C_j^+ (priority is given to the highest value of the j criterion);
   (b) Negative criterion: from the point of view of a decision maker, the lowest value of the criterion is more favorable, it is denoted as: C_j^- (priority is given to the lowest value of the j criterion);
6. Additional parameters: depending on the method, there may be additional parameters which are necessary for the method to perform, e.g., ELECTRE methods use the parameters v (veto threshold), p (preference threshold) and q (indifference threshold) [20].

2 MCDA Method Selection Problem

To date, there have been a large number of multi-criteria methods [1]. This raises the problem of choosing a MCDA method to solve a particular problem. Numerous studies on this problem confirm it. A large number of criteria for selecting a multi-criteria method, as well as an extensive list of evaluation methods are proposed in [1,14,15]. In [12], one of the ways to choose a MCDA method based on the use of the SIMUS method and the Kendall's rank correlation is presented. In [16], the authors choose the most appropriate method to solve the problem of renewable energy planning. There are other studies on selecting MCDA methods.

To solve this problem, it is necessary to determine which of the methods is appropriate. To date, there have been a qualitative and quantitative analyses of methods with the following approaches to evaluation [13]:

1. Qualitative analysis;
   (a) Qualitative approach: methods are compared from qualitative points of view, with no quantitative estimates usually available;
2. Quantitative analysis;
   (a) Approach based on comparison in the context of a real-world case: the procedure for evaluating methods can be described as follows: considering the real-world problem, applying the methods, and discussion and inference based on the data obtained, which will allow selecting the most appropriate method for a particular field;
   (b) Simulation-based approach that generates random data and then analyzes the results.

Each of the approaches is considered below.

3 Qualitative Approach

This approach evaluates methods from the qualitative point of view. For this purpose, various criteria are used, on the basis of which the most appropriate MCDA method is selected.

Currently, there are many studies devoted to choosing MCDA methods, which are focused on conducting qualitative analysis. An approach with a large number of criteria for evaluating MCDA methods, which can be used as a basis for choosing a MCDA method, is proposed [1]. In [13,19], the authors present a set of criteria used to compare different MCDA methods. Exogenous and endogenous criteria, which are used for comparison, are distinguished [15]. The criteria for evaluating the quality of MCDA methods are described [18]. Based on the studies considered, we compiled a generalized list of qualitative criteria for evaluating MCDA methods. The list will be described below.

The analysis revealed that some criteria describe the same concepts, and some are not important. For instance, the analysis of MCDA methods did not reveal a problem with calculations related to a large number of alternatives or the slow operation of the method itself when searching for a solution. Therefore, the "Rate of growth in large problems" [13], "Number of evaluation elements" [15], "Requested time to collect and process data" [15], "Need time period" [19] criteria are excluded from the generalized list. The criteria related to the qualitative evaluation of the method itself, such as "Logic" [13] and "Simplicity" [13], are also excluded since they are too ambiguous and at the same time there are criteria that produce a narrowly focused assessment of the internal structure of MCDA methods. The "Indicator Characteristics" criterion (evaluation of methods in a geographical context) is also excluded [18] since the direction of its evaluation goes beyond the analysis of MCDA methods with the input data and the MCDA method itself available.

The generalized list consists of the following parameters: criterion weight, criterion weight type, uncertainty, type of MCDA problem, input data model, input data type, criteria interdependence, results processing transparency, problem decomposition, group decision making, alternative comparison scale, solution approach, software availability for the method, specialists' role in decision making, compensation effect. The "Compensation Effect" parameter was added to this list, it determines whether the method is compensatory or non-compensatory.

One of the evaluation parameters of the MCDA methods is the need to determine the weight of the criteria. Some methods require a decision matrix to be calculated, and some methods require criteria weights to be determined. There are several ways to evaluate criteria weights: quantitative (ELECTRE), qualitative (ARGUS), and relative (AHP) [1]. In addition to the criteria weights and the decision matrix, many methods require additional input parameters or a specific input data model to be determined, which should be taken into account when selecting a MCDA method [13,15]. Values in the decision matrix can be quantitative or qualitative, but some methods make it possible to work with both types of data simultaneously [15]. Another important evaluation parameter is the possibility to take into account criteria interdependence (AHP, ANP) [18], as well as to decompose the problem (AHP) [19].

Uncertainty is frequently analyzed when evaluating MCDA methods [1,15,18, 19]. By uncertainty, we mean qualitative values of the criteria and the subjective estimates of a decision maker used while the method is being applied, such as the determined criteria weights, preference and indifference thresholds. The parameter defining the type of a MCDA problem is also considered, among them are the problem of choosing the best alternative, classification of alternatives, the problem of partial or complete ranking [1,15].

When analyzing the internal operation of methods, the focus is on the alternative comparison scale and the approach used to solve the problem. In [1,15], three scales for comparing alternatives are described: qualitative (ELECTRE 1), quantitative (TOPSIS), and relative (AHP). In [15], the authors propose three approaches to solving MCDA problems: the approach using the utility function, the outranking and the goal approaches.

When evaluating MCDA methods, the question of the role of specialists in decision making and the question of group decision making are also considered [15,19], which determines the possibility to use the method to develop a solution by a group of people. In [15], the following roles of specialists in decision making, in addition to decision makers, are described: a technical specialist is needed to solve the problem, an assistant is needed to understand the decision making process, a technical specialist and an assistant are needed, specialists, in addition to decision makers, are not required. The transparency in the processing of the results obtained for the problem analysis is also important [18,19].

Along with the above-described parameters for evaluating MCDA methods, the question often arises if the software that can be used to solve the problem is available [1].

The described set of parameters allows giving a qualitative assessment of various MCDA methods in order to determine the most appropriate one for solving a particular problem.

4 Approach Based on Comparison in the Context of a Real-World Case

To determine the most appropriate MCDA method, this approach uses real-world examples of multi-criteria problems solved, i.e., problems for which the result is known.

The analyzed methods return a ranked list of alternatives as a result, where the final rank is known for each alternative. Thus, for the analysis, a list of expected results and a ranked list of alternatives obtained after the method is applied on a specific test case are defined.

To compare the results obtained and determine how close they are, it is possible to use the correlation coefficient. In this paper, the Pearson, Spearman, and Kendall's rank correlations are used.

The Pearson's correlation coefficient is the ratio between the co-variance of two variables and the product of their standard deviations. The range of values of the correlation coefficient is the range from −1.0 to 1.0. The value of 1.0 means that we have a linear relationship describing the relationship between X and Y, with all data points lying on the line for which Y increases with X. The coefficient value of −1.0 means that all data points are on the line, for which the value of the Y variable decreases with the increase in the value of the X variable. The value of 0.0 means that there is no relationship between the variables. The Pearson's correlation coefficient applied to sample data is usually as follows:

$$r_{xy} = \frac{N\sum x_i y_i - \sum x_i \sum y_i}{\sqrt{N\sum x_i^2 - (\sum x_i)^2}\sqrt{N\sum y_i^2 - (\sum y_i)^2}}, \qquad (2)$$

where N is the sample size, x_i and y_i are the individual sample items indexed with i.

The Spearman's correlation coefficient. For a sample of size N, the values of rank x_i and y_i are defined as 3. With this approach, the positions at the top of both rankings are more important. The weight of significance is calculated for each comparison.

$$r_w = 1 - \frac{6\sum_{i=1}^{N}(x_i - y_i)^2((N - x_i + 1) + (N - y_i + 1))}{N^4 + N^3 - N^2 - N}. \qquad (3)$$

The Kendall's correlation coefficient is an alternative to the Spearman's correlation method. It is designed to determine the relationship between two rank variables. The interpretation of the results is defined as the difference between the probabilities of coincidence and inversion in the ranks. For the same values of the variables, the Spearman's correlation coefficient will always be slightly larger than the Kendall's correlation value. The Kendall's correlation coefficient is calculated using the formula:

$$\tau = 1 - \frac{4}{n(n-1)} R \,, \tag{4}$$

$$R = 1 - \sum_{i=1}^{n-1} \sum_{j=i+1}^{n} [[x_i < x_j] \neq [y_i < y_j]] \,, \tag{5}$$

where R is the number of inversions formed by the y_i values, arranged in ascending order of the corresponding x_i.

The quantitative analysis is conducted as follows: for each method, the solution of each of the test cases is calculated and the Pearson, Spearman, and Kendall's correlations are calculated, and then the average correlation value is evaluated. Below is the pseudocode of the calculation algorithm in the context of a real case:

1. We get a list of test cases
2. We get a list of test methods
3. For each of the test cases, it is necessary to:
   (a) Load the test data
   (b) Get a list of expected results
   (c) For each of the test methods, it is needed to:
      i. Perform the calculation using the method for the test case
      ii. Calculate the Pearson, Spearman, and Kendall's correlations for the expected results and those obtained by using the method
      iii. Record the received data for the test case and the test method
4. Calculate the necessary average correlation value from the obtained data.

To calculate the correlation, we use a list of values of alternative positions in the expected list and a list of values of alternative positions obtained by the method solution for a specific test case. Here is one of the examples how to calculate the correlation:

1. We have a test case, and we know what the results should be. The expected result: A_4 (position 1), A_2 (position 2), A_3 (position 3), A_1 (position 4) - $\{1, 2, 3, 4\}$;
2. As a result of calculating some method on this test case, the result was obtained: A_4 (position 2), A_2 (position 1), A_3 (position 3), A_1 (position 4) - $\{2, 1, 3, 4\}$;
3. We get the Pearson's correlation coefficient: 0.8.

5 Simulation-Based Approach

To select the most appropriate MCDA method, this approach uses generated examples of multi-criteria problems. To generate data, it is possible to use sensitivity analysis with changes in the criteria weights in the test cases [21–30].

Now we will explain why it is important to analyze the sensitivity of MCDA methods by changing the weights of the criteria.

Uncertainty emerges due to the inability of decision makers to make accurate judgments about the importance of decision making criteria [32]. One of the conclusions is that in additive models, weights matter, i.e., with a small change in the weights, the optimal alternative may change [33].

It should be noted that, in general, a small change in the criteria weights may change the ranking. The uncertainty associated with choosing the criteria weight causes the need to search for "stable" and "reliable" solutions. The sensitivity analysis is often used to eliminate this uncertainty and assess the reliability of the method. Through sensitivity analysis, various "what ifs" can be analyzed. This helps to observe the impact of the changed criteria on the final ranking result. The "what if" sensitivity analysis is recommended as a means to verify if the results in relation to the subjectivity of expert judgments are stable [17].

The literature review provides three different approaches to the sensitivity analysis [4]. One approach is to give all the criteria equal weight [18,20], another is to set one or more criteria equal to zero [18], and the third approach implies changing the criteria weights by a given interval [21].

In this paper, the sensitivity analysis is used to determine the stability of various multi-criteria methods. To determine the stability, the procedure for changing the weights of the criteria from 0 to 1 is used.

The calculation procedure is carried out as follows: for each method, the sensitivity analysis is performed when solving all test cases, and then the value are calculated based on the results obtained: the percentage of the best alternative in the first position (stability). In this case, the best alternative is the one that has been in the first position the most times after performing all the sensitivity analysis calculations for the selected test case.

When all the methods on all the cases are performed, the average values for the above values are calculated.

The pseudocode of the computation algorithm for the simulation-based approach is as follows:

1. We get a list of test cases
2. We get a list of test methods
3. For each of the test cases, it is necessary to:
   (a) Load the test data
   (b) For each of the test methods, it is needed to:
      i. Perform the sensitivity analysis of the method (for each test case criterion, change the criterion weight from 0 to 1)
      ii. Record the received data for the test case and test method
4. Calculate the required values from the received data.

The sensitivity analysis may be conducted as follows: we load the test data, change the criterion weight from 0 to 1 with a given step (e.g., 0.0001), while changing the weights of other criteria so that the sum is equal to 1, record the result and continue changing the weight until all the criteria are taken.

The pseudocode of the algorithm for changing the criterion weight:

1. We get the initial values of the criteria weights
2. We set the value of the new criterion weight to 0 – 0.0001 (step of weight change)
3. Until the value of the new criterion weight is more than 1, we perform:
   (a) We add 0.0001 to the new weight
   (b) We get the difference between the new value and the initial value of the criterion being changed (this will be the value that will need to be distributed among the rest of the criteria weights, it can be negative or positive)
   (c) We change the weight of all criteria to new values, so that the sum is equal to 1. We carry out until we change the weight of all the criteria:
      i. If this is the criterion for which we change the weight, we assign the previously calculated value (0 at the first iteration) and start the cycle from the beginning for the next criterion, otherwise:
      ii. If the criterion is the last, we subtract the sum of all new values of the criteria weights from 1 and assign a value to this criterion (it is necessary for the sum of the weights to be equal to 1) and end the cycle, otherwise:
      iii. We calculate the value of the criterion weight: from the initial weight of the current criterion, we subtract the difference between the new value and the initial one, multiplied by the initial weight of the current criterion, divided by the sum of the criteria weights without the value of the weight of the criterion being changed (see the numerical example below)
   (d) We record the obtained values of the criteria weights
4. We complete the algorithm since all the values of the criteria weights were obtained, while the weight of one of the criteria changed from 0 to 1 with a step of 0.0001.

An example of calculating weights for 4 criteria with values $\{0.2; 0.15; 0.4; 0.25\}$.

For the first iteration and changing the weight of the first criterion with a value of 0.2:

1. Initial criteria weights $\{0.2; 0.15; 0.4; 0.25\}$
2. We set the value of the weight of the first criterion to 0
3. We get the difference between the new value and the initial one: $0 - 0.2 = -0.2$
4. The sum of the criteria weights without the criterion being changed: 0.8
5. New values of criteria weights $\{0; 0.1875; 0.5; 0.3125\}$:
   (a) 0 - we set ourselves and change with a step of 0.0001
   (b) $0.1875 = 0.15 - (-0.2 * (0.15/0.8))$
   (c) $0.5 = 0.4 - (-0.2 * (0.4/0.8))$
   (d) $0.3125 = 1 - (0 + 0.1875 + 0.5)$.

For the value of the weight of the modified criterion 0.5 (the initial weight of the criterion is 0.2):

1. Initial criteria weights $\{0.2; 0.15; 0.4; 0.25\}$
2. We set the value of the weight of the first criterion to 0.5
3. We get the difference between the new value and the initial one: $0.5-0.2 = 0.3$
4. The sum of the criteria weights without the criterion being changed: 0.8
5. New values of criteria weights $\{0; 0.09375; 0.25; 0.15625\}$:
   (a) 0.5 - we set by ourselves and change with a step of 0.0001
   (b) $0.09375 = 0.15 - (0.3 * (0.15/0.8))$
   (c) $0.25 = 0.4 - (0.3 * (0.4/0.8))$
   (d) $0.15625 = 1 - (0.5 + 0.09375 + 0.25)$.

For the last iteration (the initial criterion weight is 0.2):

1. Initial criteria weights $\{0.2; 0.15; 0.4; 0.25\}$
2. We set 1 to the value of the first criterion weight
3. We get the difference between the new value and the initial one: $1-0.2 = 0.8$
4. The sum of the criteria weights without the criterion being changed: 0.8
5. New values of criteria weights $\{1; 0; 0; 0\}$:
   (a) 1 - we set by ourselves and change with a step of 0.0001
   (b) $0 = 0.15 - (0.8 * (0.15/0.8))$
   (c) $0 = 0.4 - (0.8 * (0.4/0.8))$
   (d) $0 = 1 - (1 + 0 + 0)$.

6 Conducting the Experiment

To conduct the experiment, 20 automated multi-criteria methods were selected: Weighted product model (WPM) [34], WASPAS [35], VIKOR [35], TODIM [35], Weighted sum model (WSM) [34], MOOSRA [36], MOORA (ratio system) [36], MOORA (reference point) [36], Full multiplicative form [36], MULTIMOORA [36], MAUT [35], MABAC [35], ELECTRE 1 [35], EDAS [35], COPRAS [35], ARAS [35], EVAMIX [35], MACKBETH [35], REGIME [35], TOPSIS [37].

Among the presented methods, there are 17 compensatory and 3 non-compensatory ones (MOORA (reference point), ELECTRE 1, REGIME). 3 methods can operate with quantitative and qualitative data without processing (EVAMIX, MACKBETH, REGIME), and 17 methods work with quantitative data only. 17 methods give a solution with full ranking of alternatives, and 3 methods give a solution with partial ranking (WPM, ELECTRE 1, REGIME).

To automate the method calculations that produce a result with partial ranking, we determined the alternative rank based on the number of paired comparisons won.

For the quantitative analysis, 10 examples of solving real multi-criteria problems were selected. The solution to these problems is presented in [21–30].

The results of the quantitative analysis for the approach based on comparison in the context of a real-world case were sorted by descending mean value of the Pearson's correlation and are presented in Figs. 1, 2, 3 and 4.

To conduct the sensitivity analysis, the criteria weight was changed from 0 to 1 with a step of 0.0001. So, for each method, 720,072 measurements were

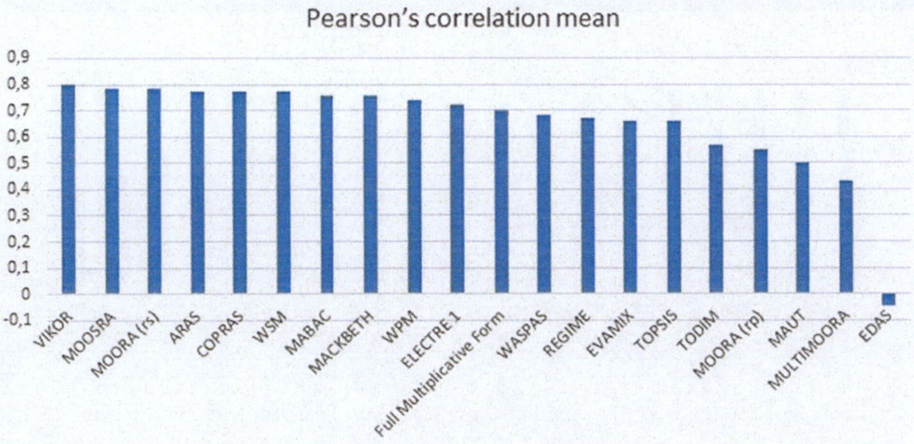

Fig. 1. Pearson's correlation

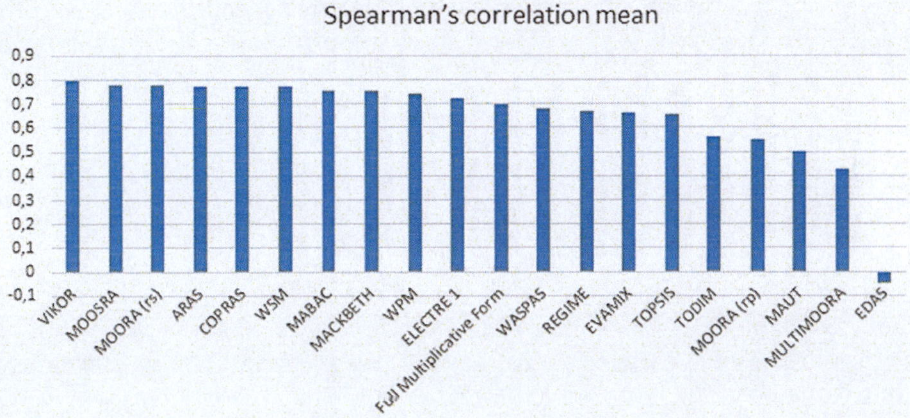

Fig. 2. Spearman's correlation

made on ten test cases. In total, 14, 401, 440 measurements were carried out. The quantitative analysis results for the modeling approach were sorted by the mean of the percentage that the alternative was in the first place in the final ranking. The analysis results are shown in Fig. 5.

The graphs show that the Pearson and Spearman's correlation values are similar, and the Kendall's correlation values are always lower. The method positions for Kendall's correlation differ from sorting by Pearson's correlation values. But, in general, the Kendall's correlation values are decreasing.

Based on the quantitative analysis on real data, VIKOR is the best method to use because, on average, it is better than other methods to give results similar to the original, and the worst option is EDAS, the results of which do not correlate with the expected data.

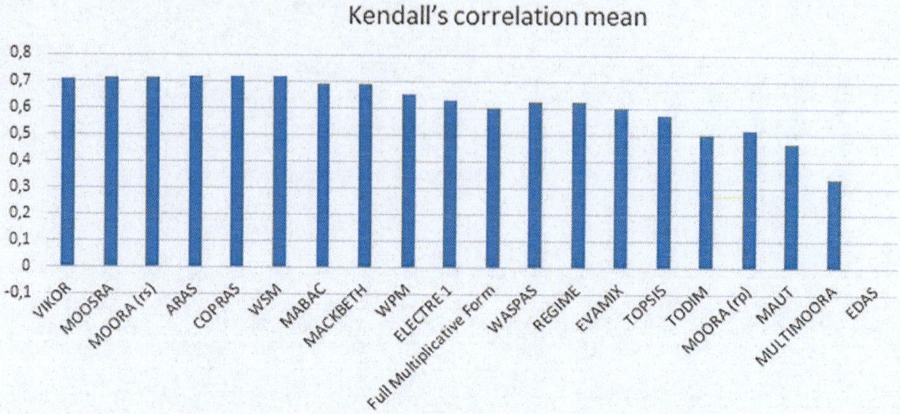

Fig. 3. Kendall's correlation

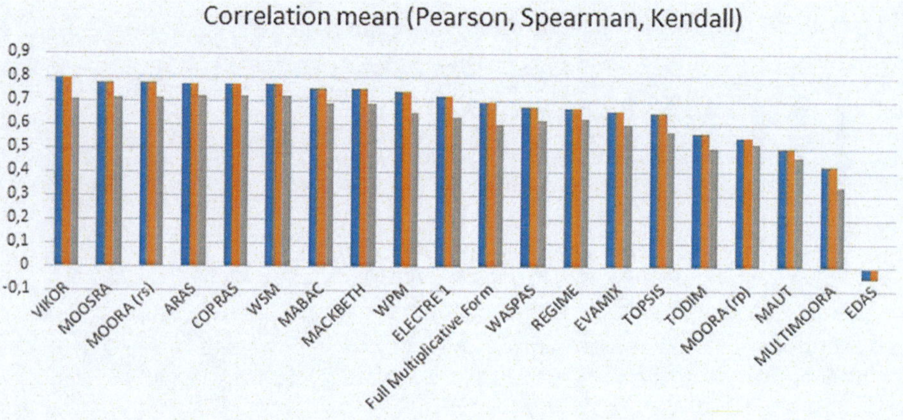

Fig. 4. Pearson, Spearman, Kendall's correlation

Based on the presented results of the sensitivity analysis, TODIM is considered to be the best, which showed the highest indicator of stability (most stable method), and the worst option is the MOORA method (reference point), which showed the lowest indicator of stability of the results to the changes in the criteria weight. It should also be noted that non-compensatory methods (MOORA (rp), ELECTRE 1, REGIME) gave poor results although the stability of these methods was expected to be high, and the EVAMIX, MACKBETH, REGIME methods gave mediocre results, but it should be taken into account that the given methods can work with qualitative data without converting them to quantitative data.

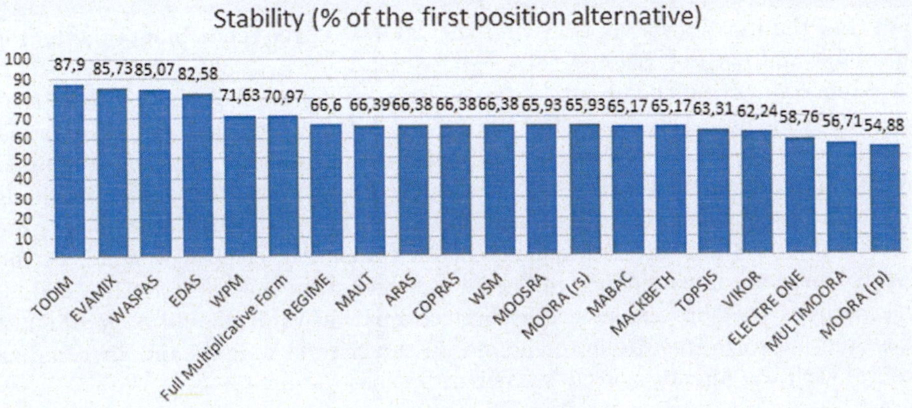

Fig. 5. Sensitivity analysis

7 Conclusion

This paper describes the problem of selecting a MCDA method. We identified and described three approaches to evaluating the MCDA methods: a qualitative approach based on comparison of methods from the qualitative point of view, an approach based on comparison in the context of a real-world case, in which methods are compared in relation to solving a real-world problem, and a simulation-based approach in which data generation is used to compare methods.

When analyzing the qualitative approach in other papers, 40 criteria for evaluating MCDA methods were found, which proves that multi-criteria computing is developing. Many of the criteria described the same concepts, and some were not important to us, so a generalized list of 14 most important criteria for the qualitative assessment of MCDA methods was determined.

We performed the quantitative analysis. For this purpose, 20 automated methods and 10 examples of solving MCDA problems were selected. Each method was implemented using the software.

In the course of the quantitative analysis with the approach based on comparison in the context of a real-world case, the expected and obtained results were evaluated after using a specific method. Pearson, Spearman and Kendall's correlation values were used to perform an assessment. It was found that when performing such an analysis, the VIKOR method is the best option, which gave the best correlation result, and the EDAS method is considered to be the worst option which gave the worst correlation result.

When performing the quantitative analysis with the simulation-based approach, the sensitivity analysis was used, in which all criteria weight in the test cases changed from 0 to 1 with a step of 0.0001. To evaluate the methods, the average value of the maximum percentage of the alternative which is in the final ranking in the 1st position the greatest number of times after all the calculations were made. It was found that using this approach, the TODIM method is the best and the most stable option, and the MOORA (reference point) method is the worst and least stable one.

After conducting the quantitative analysis, we found that the methods that received a higher level of stability received a lower correlation value compared to other methods, and vice versa, the higher the correlation value was obtained by the method, the less stable it is. It should also be noted that stable methods should be used when there is no confidence in the established criteria weights since stable methods are less dependent on changes in criteria weights. In the case when there is confidence in the established values of the criteria weights, the methods that have received the best correlation value should be used since these methods are less stable to changes in the criteria weights and give similar results with the already solved MCDA problems.

References

1. Wątróbski, J., Jankowski, J., Ziemba, P., Karczmarczyk, A., Zioło, M.: Generalised framework for multi-criteria method selection. Omega **86**, 107–124 (2019). https://doi.org/10.1016/j.omega.2018.07.004
2. Pelissari, R., Oliveira, M.C., Abackerli, A.J., Ben-Amor, S., Assumpção, M.R.P.: Techniques to model uncertain input data of multi-criteria decision-making problems: a literature review. Int. Trans. Oper. Res. **28**(2), 523–559 (2018). https://doi.org/10.1111/itor.12598
3. Roy, B.: The optimisation problem formulation: criticism and overstepping. J. Oper. Res. Soc. **32**(6), 427–436 (1981)
4. Höfer, T., Sunak, Y., Siddique, H., Madlener, R.: Wind farm siting using a spatial Analytic Hierarchy Process approach: a case study of the Städteregion Aachen. Appl. Energy **163**, 222–243 (2016). https://doi.org/10.1016/j.apenergy.2015.10.138
5. Whaiduzzaman, M., Gani, A., Anuar, N.B., Shiraz, M., Haque, M.N., Haque, I.T.: Cloud service selection using multicriteria decision analysis. Sci. World J. **2014**, 1–10 (2014). https://doi.org/10.1155/2014/459375
6. Eason, T., Meyer, D., Curran, M., Upadhyayula, V.K.: Guidance to Facilitate Decisions for Sustainable Nanotechnology (2011)
7. Aspen, D.M., Sparrevik, M., Fet, A.M.: Review of methods for sustainability appraisals in ship acquisition. Environ. Syst. Decis. **35**(3), 323–333 (2015). https://doi.org/10.1007/s10669-015-9561-6
8. Roy, B.: The outranking approach and the foundations of ELECTRE methods. Theory Decis. **31**, 49–73 (1991)
9. Bana e Costa, C.A., Vincke, P.: Multiple criteria decision aid: an overview. In: Bana e Costa, C.A. (ed.) Readings in Multiple Criteria Decision Aid, pp. 3–14. Springer, Heidelberg (1990). https://doi.org/10.1007/978-3-642-75935-2_1
10. Choo, E.U., Wedley, W.C.: Comparing fundamentals of additive and multiplicative aggregation in ratio scale multi-criteria decision making. Open J. Oper. Res. Soc. **2**(1), 1–7 (2008). https://doi.org/10.2174/1874243200802010001
11. Mabin, V., Beattie, M.: A practical guide to multi-criteria decision analysis. Victoria University of Wellington (2006)
12. Munier, N., Hontoria, E., Jiménez-Sáez, F.: Analysis of lack of agreement between MCDM methods related to the solution of a problem: proposing a methodology

for comparing methods to a reference. In: Strategic Approach in Multi-Criteria Decision Making, vol. 275, pp. 203–219. Springer, Cham (2019). https://doi.org/10.1007/978-3-030-02726-1_11

13. Asgharizadeh, E., Taghizadeh Yazdi, M., Mohammadi Balani, A.: An output-oriented classification of multiple attribute decision-making techniques based on fuzzy c -means clustering method. Int. Trans. Oper. Res. **26**(6), 2476–2493 (2017). https://doi.org/10.1111/itor.12449
14. Watróbski, J., Jankowski, J.: Guideline for MCDA method selection in production management area. In: Różewski, P., Novikov, D., Bakhtadze, N., Zaikin, O. (eds.) New Frontiers in Information and Production Systems Modelling and Analysis, vol. 98, pp. 119–138. Springer, Cham (2015). https://doi.org/10.1007/978-3-319-23338-3_6
15. Guarini, M., Battisti, F., Chiovitti, A.: A methodology for the selection of multi-criteria decision analysis methods in real estate and land management processes. Sustainability **10**(2), 507 (2018). https://doi.org/10.3390/su10020507
16. Polatidis, H., Haralambopoulos, D.A., Munda, G., Vreeker, R.: Selecting an appropriate multi-criteria decision analysis technique for renewable energy planning. Energy Sources Part B **1**(2), 181–193 (2006). https://doi.org/10.1080/009083190881607
17. Meszaros, Cs., Rapcsák, T.: On sensitivity analysis for a class of decision systems. Decis. Support Syst. **16**, 231–240 (1996)
18. De Toro, P., et al.: Criteria for quality assessment of MCDA methods (2002)
19. Ivlev, I., Kneppo, P., Bartak, M.: Multicriteria decision analysis: a multifaceted approach to medical equipment management. Technol. Econ. Dev. Econ. **20**(3), 576–589 (2014). https://doi.org/10.3846/20294913.2014.943333
20. Roy, B.: Classement et choix en présence de points de vue multiples. Revue Française d'informatique et de Recherche Opérationnelle **2**(8), 57–75 (1968). https://doi.org/10.1051/ro/196802v100571
21. Kumar, P., Singh, R.K.: A fuzzy AHP and TOPSIS methodology to evaluate 3PL in a supply chain. J. Model. Manag. **7**(3), 287–303 (2012). https://doi.org/10.1108/17465661211283287
22. Żak, J., Węgliński, S.: The selection of the logistics center location based on MCDM/A methodology. Transp. Res. Proc. **3**, 555–564 (2014). https://doi.org/10.1016/j.trpro.2014.10.034
23. Patel, M., Bhatt, B., Vashi, M.: SMART-multi-criteria decision-making technique for use in planning activities (2017)
24. Jaini, N., Utyuzhnikov, S.: Trade-off ranking method for multi-criteria decision analysis. J. Multi-Criteria Decis. Anal. **24**(3–4) (2016). https://doi.org/10.1002/mcda.1600
25. Adhikary, P., Kundu, S.: MCDA or MCDM based selection of transmission line conductor: small hydropower project planning and development. IJERA **4**, 2248–9622357 (2014)
26. Ishizaka, A., Pearman, C., Nemery, P.: AHPSort: an AHP-based method for sorting problems. Int. J. Prod. Res. **50**(17), 4767–4784 (2012). https://doi.org/10.1080/00207543.2012.657966
27. Vinodh, S., Jeya Girubha, R.: PROMETHEE based sustainable concept selection. Appl. Math. Model. **36**(11), 5301–5308 (2012). https://doi.org/10.1016/j.apm.2011.12.030
28. Joshi, R., Banwet, D.K., Shankar, R.: A Delphi-AHP-TOPSIS based benchmarking framework for performance improvement of a cold chain. Expert Syst. Appl. **38**(8), 10170–10182 (2011). https://doi.org/10.1016/j.eswa.2011.02.072

29. Fierek, S., Zak, J.: Planning of an integrated urban transportation system based on macro - simulation and MCDM/A methods. Procedia. Soc. Behav. Sci. **54**, 567–579 (2012). https://doi.org/10.1016/j.sbspro.2012.09.774
30. Ferrari, P.: A method for choosing from among alternative transportation projects. Eur. J. Oper. Res. **150**(1), 194–203 (2003). https://doi.org/10.1016/s0377-2217(02)00463-0
31. Jeffreys, I.: The use of compensatory and non-compensatory multi-criteria analysis for small-scale forestry. Small-Scale Forest Econ. Manage. Policy **3**(1), 99–117 (2004). https://doi.org/10.1007/s11842-004-0007-0
32. Malczewski, J.: GIS and Multicriteria Decision Analysis. Wiley, New York (1999)
33. Triantaphyllou, E., Sánchez, A.: A sensitivity analysis approach for some deterministic multi-criteria decision-making methods. Decis. Sci. **28**(1), 151–194 (1997). https://doi.org/10.1111/j.1540-5915.1997.tb01306.x
34. Triantaphyllou, E., Mann, S.H.: An examination of the effectiveness of multi-dimensional decision-making methods: a decision-making paradox. Decis. Support Syst. **5**(3), 303–312 (1989). https://doi.org/10.1016/0167-9236(89)90037-7
35. Alinezhad, A., Khalili, J.: New Methods and Applications in Multiple Attribute Decision Making (MADM). International Series in Operations Research & Management Science. Springer, Cham (2019). https://doi.org/10.1007/978-3-030-15009-9
36. Aytaç Adalı, E., Tuş Işık, A.: The multi-objective decision making methods based on MULTIMOORA and MOOSRA for the laptop selection problem. J. Ind. Eng. Int. **13**(2), 229–237 (2016). https://doi.org/10.1007/s40092-016-0175-5
37. Pavić, Z., Novoselac, V.: Notes on TOPSIS method. Int. J. Res. Eng. Sci. **1**, 5–12 (2013)

Tomato Leaf Diseases Detection Using Deep Learning

Ammar Alhaj Ali, Bronislav Chramcov, Roman Jasek, Rasin Katta(✉), Said Krayem, and Emad Awwama

Faculty of Applied Informatics, Tomas Bata University in Zlin, Zlin, Czech Republic
{alhaj_ali,chramcov,jasek,Katta,krayem,awwama}@utb.cz

Abstract. Early detection and diagnosis of plant leaf diseases is a major necessity in a growing agricultural economy, plant leaf diseases detection is considered crucial at a very early stage as it allows adopting predictive mechanisms that helps avoiding losses to the agri-based economy. Tomato is one of the most important crops that is produced in large quantities with high commercial value and contributes to food chain and security and a lucrative business for many farmers.

This study uses a deep convolutional neural network (CNN) to predict tomato leaf disease by transfer learning. InceptionV3 was used as the backbone of the CNN, the highest accuracy of 99.8% for identifying tomato leaf diseases is achieved on the PlantVillage dataset. And the final model was deployed as a web application on the cloud to be available.

The experimental results show that the proposed model is effective in identifying tomato leaf disease and could be generalized to identify other plant diseases.

Keywords: Deep learning · CNN · Classification · Transfer learning · Plant disease

1 Introduction

The world population is expected to increase by 2 billion people in the next 30 years, from 7.9 billion currently to 9.9 billion in 2050, developing methods for early detection of plant diseases serve the double purpose of increasing crop yield and reducing pesticide use [1]. In some countries of Asia and Africa over 50% of the population depends on agriculture production for their daily living [2]. According to FAO (Food and Agriculture Organization of the United Nations) pests and diseases are responsible for the loss of 40% of the global food production, which poses a threat to food security [3]. In 2019, around 20% of overall global crop productions were lost due to pests and diseases [4].

Tomato is a widely planted crop around the world, it contains rich nutrition and health effects, hence it plays an important role in global production and trade all over the world [5]. According to FAO, tomato is the second most important vegetable crop next to potato [6]. Given the importance of tomatoes in the economic context, thus it is necessary to maximize productivity and improve product quality by using available and evolved

R. Silhavy et al. (Eds.): CoMeSySo 2021, LNNS 231, pp. 199–208, 2021.
https://doi.org/10.1007/978-3-030-90321-3_18

techniques. According to the last statistics, there are as many as 10 types of tomato leaf diseases that have sorely affected Tomato yield from both qualitative and quantitative aspects, this led to massive economic losses, consequently crop production has turned to be a nightmare for farmers due to such kind of diseases. Therefore, early diagnosis and treatment of tomato leaf diseases play an extremely important role in tomato production. Generally, it is possible for a person with high professional knowledge to accurately diagnose plant diseases, but smallholder farmers are not highly educated and do not have the required professional knowledge, so they usually have a low judgment rate of plant diseases.

In the last years, computer vision has provided an interesting and innovative approach for the accurate diagnosis of plant leaf diseases and paved the way for smartphone-assisted disease diagnosis [7, 8]. Various techniques and applications have been made to reduce crop losses due to diseases, some methods have been proposed to identify plant diseases such as Convolutional Neural Networks CNN [9]. The successful identification of plant diseases is fateful to improve agricultural production and reduce the undesirable use of chemical treatment such as fungicide/herbicide. Therefore, disease detection is an important research field to advanced agricultural automation [10].

In the last few decades, Deep Learning has proved its ability of handling large amounts of data to recognize patterns, one of the most popular deep neural networks is the CNN. As CNN represents one of the most powerful techniques for modeling processes and performing pattern recognition in images [11]. Recently several convolutional neural network models have been proposed for plant diseases identifying, and some state-of-the-art architectures such as VGG, Inception-v3, ResNet, and DenseNet have shown promising results.

Few people go to train an entire convolutional neural network from scratch while the alternative approach and most common is using pre-trained models that are already trained on existing large datasets. One of the most common datasets is ImageNet that contains 1.2 million images with 1000 categories, this technique is called transfer learning, it helps to build accurate models in an efficient way [12].

The contributions of this study can be summarized as follows:

1. Presenting a deep learning model to identify tomato leaf diseases at a very early stage to avoid losses to the agri-based economy.
2. Presenting transfer learning as a significant technique to accelerate the training task and increase the accuracy and stability of the classification models.
3. Demonstrating how to improve model accuracy by hyper-parameters tuning and applying data augmentation.
4. Deploying the trained model on the web and make it available as a free app for smallholder farmers.

The rest of the paper is organized as follows: Sect. 2 analyses of the dataset used. Section 3 presents materials and methods like network architecture, transfer learning, data augmentation, and evaluation metrics. Section 4 provides and discusses the experimental results. Section 5 presents the deployment of the final model on the web. Finally, Sect. 6 summarizes the study and draws a conclusion.

2 Dataset

Generally, to develop accurate image classifiers for diagnosing plant disease, we need a large dataset of images for both diseased and healthy plants leaves. This type of dataset was not available short time ago. The PlantVillage project solved this issue as it addressed this problem and started to collect thousands of images from both healthy and diseased crop plants and made them open and free for public use [13]. From this dataset, we extracted only tomato leaf images that are classified in 10 classes. Figure 1 shows one example per each sample class. The total number of images in our dataset is 18,160, each having animage size of 256 × 256 × 3. a summary of our dataset is presented in Table 1.

Fig. 1. Sample images tomato dataset: (a) Bacterial Spot, (b) Early Blight, (c) Late Blight, (d) Leaf Mold, (e) Septoria Leaf Spot, (f) Spider Mites, (g) Target Spot, (h) Yellow Leaf Curl Virus, (i) Mosaic Virus, and (j) Healthy

Table 1. Distribution of classes in Dataset

Class name	Number of images
Bacterial spot	2127
Early blight	1000
Late blight	1909
Leaf mold	952
Septoria leaf spot	1771
Spider mites	1676
Target spot	1404
Yellow leaf curl virus	5357
Mosaic virus	373
Healthy	1591

3 Materials and Methods

This study concentrates on identifying tomato leaf disease using CNN. This section is based on five sub sections, first one displays deep learning-based proposed architecture for identifying tomato leaf disease. In the second and third ones we briefly introduced transfer learning and augmentation. After that training specification is presented. Finally, we explained in the last one the evaluation metrics used in this study.

3.1 Deep Learning Model

In this study, we use Inception-v3 architecture, which is a convolutional neural network architecture from the Inception family that is widely used for classification tasks. This architecture was originally introduced during the ImageNet Recognition Challenge [14]. The Inception V3 network has multiple symmetric and asymmetric building blocks, where each block has several branches of convolution layers, average pooling, max-pooling, concatenated, dropouts, fully-connected layers, and softmax [15]. Figure 2 represents the architecture of the Inception-V3 network for 256 × 256 × 3 image size and 10 classes.

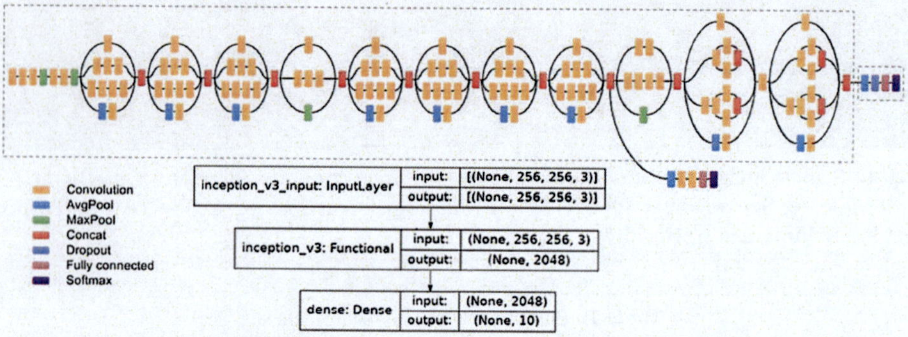

Fig. 2. Architecture of the Inception-v3 neural network

3.2 Transfer Learning

Transfer learning technique is a process that takes features learned on one problem, and leveraging them on another, problem. Transfer learning is usually used when you have small dataset to train a model from scratch [16]. Transfer learning enables us to train models with fewer size data. It can reduce training time and computing resources, see Fig. 3. Using transfer learning (pre-trained models) can boost accuracy without taking much time to converge compared to a model trained from scratch. Sometimes if models are trained from scratch without proper normalization, it can produce misleading results, which might mean that training from scratch is not optimal at all.

In this study, we use the Inception-V3 network already trained on more than a million images from the ImageNet (ImageNet is one of the most common datasets that contains 1.2 million images with 1000 categories).

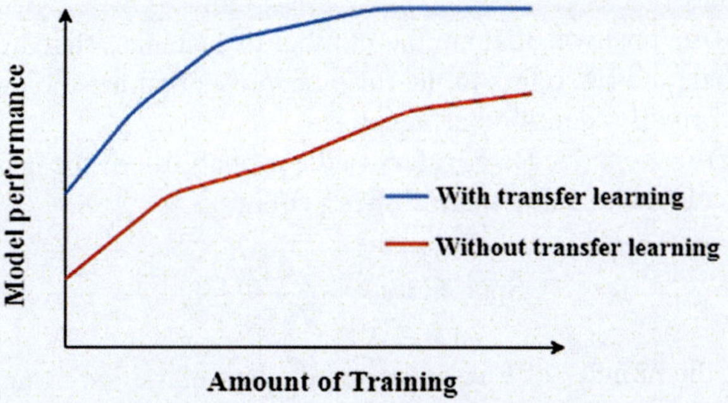

Fig. 3. The benefit of transfer learning

3.3 Data Augmentation

Usually, deep learning models need a lot of data for training, in general, the more the data, the better the performance of the model. But sometimes acquiring enormous amounts of data is considered a major challenge in deep learning. The problem of shortage of data is that the deep learning model might not learn the pattern from the dataset and hence it might not give a good performance.

basically, Data Augmentation (DA) technique is used to increase the amount of data by adding slightly modified copies of already existing data. It is one of the most practical approaches to enhance model performance without spending days manually collecting data [17]. In this study, some data augmentation techniques were proposed, such as shear, zoom, rotation, vertical flip, and horizontal flip to boost the training set for improving the model accuracy.

3.4 Training Specifications

In this study, the experiments are implemented using the Keras framework, and classification was done on Inception-V3 network. Transfer learning is used to increase the accuracy and stability of the model. Our model was first pre-trained on the ImageNet dataset then the model training was done in 30 epochs with a batch size of 16, the model was optimized via Adam optimizer [18] with a learning rate of 0.001, and Categorical CrossEntropy was used as a loss function [19]. To achieve higher accuracy, the learning rate has been reduced whenever a loss metric had stopped improving. The dataset was split into two sets: a training set and a validation set with a ratio of 80:20 respectively.

3.5 Evaluation Metrics

The performance of the model is generally evaluated by different metrics to analyse how well the model performs on test data. Sensitivity, Specificity, Accuracy, Precision, and F1-score metrics are used for evaluating our model.

Sensitivity (True Positive Rate) also known as recall refers to how many positives are correctly identified; this can be calculated by the following equation.

$$\text{Sensitivity(Recall)} = \frac{\text{TP}}{\text{TP} + \text{FN}}$$

Where TP is the true positives that are the number of instances that are positive and are correctly identified, and FN refers to the false negatives that are the number of positive cases that are incorrectly identified as negative.

Specificity (True Negative Rate) refers to the probability of the negative label being true, this can be calculated using the following equation.

$$\text{Specificity} = \frac{\text{TN}}{\text{TN} + \text{FP}}$$

Where TN is the number of true negatives that are classified as negative, and FP is the number of false positives that refers to the negative instances that are incorrectly classified as positive cases.

Accuracy is the most commonly used metric to evaluate classification models. This metric calculates the proportion of samples that are correctly classified, the following equation is used to calculate accuracy.

$$\text{Accuracy} = \frac{\text{TP} + \text{TN}}{\text{TP} + \text{TN} + \text{FP} + \text{FN}}$$

Precision (Positive Predictive Value) is defined as the number of true positives divided by the number of true positives plus the number of false positives, is given by the following equation.

$$\text{Precision} = \frac{\text{TP}}{\text{TP} + \text{FP}}$$

F-score is defined as the harmonic mean precision and recall, as shown in the following equation.

$$\text{F} - \text{score} = 2 * \frac{\text{Precision} * \text{Recall}}{\text{Precision} + \text{Recall}}$$

4 Results and Discussion

The objective of this research was presenting a CNN model to classify tomato plant leaf diseases with high accuracy, our model showed statistically significant performance. Table 2 shows sensitivity, specificity, accuracy, precision, and F-score of Inception-V3.

Table 2. Performance results for Inception-v3

Model	Sensitivity	Specificity	Accuracy	Precision	F-score
Inception-v3	0.9969	0.9998	0.9978	0.9977	0.9978

And Fig. 4 shows the Accuracy/loss curve for Inception-v3 model.

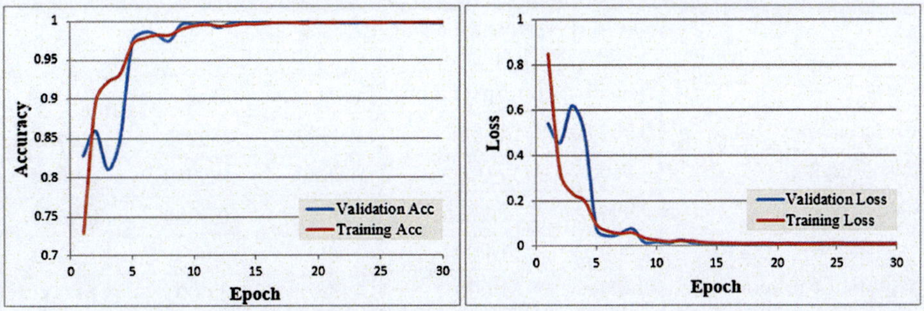

Fig. 4. Accuracy/loss curves of Inception-v3 model

Table 3 shows the confusion matrix of the model; it is possible to visually evaluate the performance of the model. The rows are related to the output classes. The diagonal cells are associated with the observations that are correctly identified, and the off-diagonal cells correspond to the incorrectly identified observations.

Table 3. Confusion matrix based on Inception-V3 model. (1) Bacterial pot; (2) Early Blight; (3) Late Blight; (4) Leaf Mold; (5) Septoria Leaf Spot; (6) Spider Mites; (7) Target Spot; (8) Yellow Leaf Curl Virus; (9) Mosaic Virus; and (10) Healthy

		Predicted									
		1	2	3	4	5	6	7	8	9	10
Actual	1	425	0	1	0	0	0	0	0	0	0
	2	0	196	4	0	0	0	0	0	0	0
	3	1	0	380	0	0	0	0	0	0	2
	4	0	0	0	190	0	0	0	0	0	0
	5	0	0	0	0	354	0	0	0	0	0
	6	0	0	0	0	0	335	0	0	0	0
	7	0	0	0	0	0	0	281	0	0	0
	8	0	0	0	0	0	0	0	1071	0	0
	9	0	0	0	0	0	0	0	0	75	0
	10	0	0	0	0	0	0	1	0	0	317

Among the ten classes, five classes produced 100% correct classification results since those diseases have distinctive features and appearance, which are Leaf Mold, Septoria Leaf Spot, Spider Mites, Yellow Leaf Curl Virus, and Mosaic Virus. The performance measures were calculated for each class and the obtained results are shown in Table 4.

Table 4. Performance results for each class

Class	Sensitivity	Specificity	Accuracy	Precision	F-score
Bacterial Spot	0.9977	1.000	0.9997	0.9997	0.9987
Early Blight	0.9800	1.000	0.9989	0.9989	0.9894
Late Blight	0.9948	0.9985	0.9981	0.9980	0.9964
Leaf Mold	1.000	1.000	1.000	1.000	1.0000
Septoria Leaf Spot	1.000	1.000	1.000	1.000	1.0000
Spider Mites	1.000	1.000	1.000	1.000	1.0000
Target Spot	1.000	0.9997	0.9997	0.9997	0.9998
Yellow Leaf Curl Virus	1.000	1.000	1.000	1.000	1.0000
Mosaic Virus	1.000	1.000	1.000	1.000	1.0000
Healthy	0.9969	0.9994	0.9992	0.9992	0.9980

5 Deployment Final Model on the Web

When a machine learning engineer develops a machine learning model using TensorFlow, PyTorch, Keras, etc. The definitive goal is to make the model available for production. Often when we work on a machine learning project, we focus a lot on Exploratory Data Analysis (EDA), preparing data, feature engineering, build model, hyper-parameters, etc. But we forget that our major goal is to extract important values from the model predictions. Deployment of machine learning models means making models available to the end-users or systems. However, models deployment is sometimes considered tricky [20].

In this study, we build a web application from two parts: Front-end and Back-end. In the front-end, we used HTML and CSS to build a basic front-end with an input form for uploading images and build a back-end of the web application using a Flask framework, Flask is a web application framework used to develop web applications [21]. then we deployed the web application in DigitalOcean cloud [22], after deployment, the model became publicly available and can be accessed via web URL.

6 Conclusions

Plant diseases had become a major concern in agriculture for years. Precision agriculture can help improving both environmental and economic issues, it reduces pesticides usage and obtain the disease information through computer vision effectively at a very early stage to avoid losses to the agri-based economy.

Deep learning has been successfully used in various fields, it showed promising results, and large potential. In agriculture, the role of deep learning has become inevitable due to its powerful deployment capabilities in disease detection, as disease diagnosis using deep learning can enhance the quickness of taking proper decisions to avoid crop losses due to diseases and pests. In addition, the small-holder farmers need such kind of reliable, efficient and cheap technique in their fight against plant disease.

Transfer learning reduces the time of execution drastically and provides various advantages like improving the accuracy of output and requires lesser training data. Although like every other technique, there are few limits to transfer learning, such as overfitting and negative transfer. The inception-V3 model proved to be very effective in image classification and showed great potential in identifying tomato leaf diseases through achieving high classification accuracy.

The main goal of this study was to improve the efficiency of automatic plant disease detection. Experimental results showed that the proposed model can successfully detect and classify tomato leaf disease with an accuracy of 99.8%. In future work, we will extend to focus on detecting diseases in various locations of the plant and at different phases of the disease.

References

1. Sharma, P., Berwal, Y.P.S., Ghai, W.: Performance analysis of deep learning CNN models for disease detection in plants using image segmentation. Inf. Process. Agric. **7**(4), 566–574 (2020)
2. Dhingra, G., Kumar, V., Joshi, H.D.: Study of digital image processing techniques for leaf disease detection and classification. Multimed. Tools Appl. **77**(15), 19951–20000 (2017). https://doi.org/10.1007/s11042-017-5445-8
3. Boulent, J., Foucher, S., Théau, J., St-Charles, P.L.: Convolutional neural networks for the automatic identification of plant diseases. Frontiers in plant science 10 (2019)
4. Keh, S.S.: Semi-supervised noisy student pre-training on efficientnet architectures for plant pathology classification. arXiv preprint arXiv:2012.00332 (2020)
5. Zhang, K., Wu, Q., Liu, A., Meng, X. Can deep learning identify tomato leaf disease? Advances in Multimedia (2018)
6. Tomato. http://www.fao.org/land-water/databases-and-software/crop-information/tomato/en/ Accessed 16 May 2021
7. Wang, Q., Qi, F., Sun, M., Qu, J., Xue, J.: Identification of tomato disease types and detection of infected areas based on deep convolutional neural networks and object detection techniques. Computational intelligence and neuroscience (2019)
8. Mohanty, S.P., Hughes, D.P., Salathé, M.: Using deep learning for image-based plant disease detection. Front. Plant sci. **7**, 1419 (2016)
9. Elhassouny, A., Smarandache, F.: Smart mobile application to recognize tomato leaf diseases using Convolutional Neural Networks. In: 2019 International Conference of Computer Science and Renewable Energies (ICCSRE), pp. 1–4. IEEE, July 2019
10. Ali, A.A., Chramcov, B., Jasek, R., Katta, R., Krayem, S.: Classification of plant diseases using convolutional neural networks. In: Silhavy, R. (ed.) Computer Science On-line Conference, pp. 268–275. Springer, Cham. (2021). Doi: https://doi.org/10.1007/978-3-030-77445-5_24
11. Ferentinos, K.P.: Deep learning models for plant disease detection and diagnosis. Comput. Electron. Agric. **145,** 311–318 (2018)
12. Rawat, W., Wang, Z.: Deep convolutional neural networks for image classi-fication: a comprehensive review. Neural Comput. **29**(9), 2352–2449 (2017)
13. Hughes, D., Salathé, M.: An open access repository of images on plant health to enable the development of mobile disease diagnostics. arXiv preprint arXiv:1511.08060 (2015)

14. Szegedy, C., Vanhoucke, V., Ioffe, S., Shlens, J., Wojna, Z.: Rethinking the inception architecture for computer vision. In: Proceedings of the IEEE Conference on Computer Vision and Pattern Recognition, pp. 2818–2826 (2016)
15. Maeda-Gutierrez, V., et al.: Comparison of convolutional neural network architectures for classification of tomato plant diseases. Appl. Sci. **10**(4), 1245 (2020)
16. Transfer learning & fine-tuning. https://keras.io/guides/transfer_learning/. Accessed 27 May 2021
17. Ali, A.A., Chramcov, B., Jasek, R., Katta, R., Krayem, S.: Classification of plant diseases using convolutional neural networks. In: Computer Science On-line Conference, pp. 268–275. Springer, Cham (2021)
18. Yoo, J., Ahn, N., Sohn, K.-A.: Rethinking data augmentation for image super-resolution: A comprehensive analysis and a new strategy. In: Proceedings of the IEEE/CVF Conference on Computer Vision and Pattern Recognition (2020)
19. Kingma, D.P., Ba, J.: Adam: a method for stochastic optimization. arXiv preprint arXiv:1412.6980 (2014)
20. Gordon-Rodriguez, E., Loaiza-Ganem, G., Pleiss, G., Cunningham, J.P.: Uses and abuses of the cross-entropy loss: case studies in modern deep learning. arXiv preprint arXiv:2011.05231 (2020)
21. How to Easily Deploy Machine Learning Models Using Flask. https://towardsdatascience.com/how-to-easily-deploy-machine-learning-models-using-flask-b95af8fe34d4. Accessed 20 May 2021
22. Flask, web development. https://flask.palletsprojects.com/en/2.0.x/. Accessed 20 May 2021
23. DigitalOcean, The developer cloud. https://www.digitalocean.com/. Accessed 05 May 2021

A Hybrid Approach and Implementation on Root Cause Analysis Logic Model for Enterprise Software Application

Hoo Meng Wong[1(✉)], Sagaya Sabestinal Amalathas[2], and Sayan K. Ray[3]

[1] Taylor's University, Subang Jaya, Selangor, Malaysia
[2] University of Southampton, Nusajaya, Johor, Malaysia
[3] Auckland, New Zealand

Abstract. Despite the Information Technology (IT) era keeps evolving rapidly, many business operations are heavily relying on software application service to process the daily business transactions today. Therefore, software application service is unacceptable if it is unavailable during business hours. However, there are many possibilities can cause the software application service irresponsive. This is because the root cause can be possible to arise mostly in either within the software application layer, or other factors which are falling outside the software application layer. Under such complex situation, it is time consuming to identify the root cause and it is always unavoidable. The objective of Prescriptive Analytical Logic Model (PAL) is not focusing on only solving such problem. Indeed, it aims to minimize the duration spent on root cause analysis activities, and to decide the preferred resolution for the identified root cause. Hence, the algorithm inside the PAL has incorporated with Supervised Learning approach to identify the possible errors of the root cause, and adopted Analytic Hierarchy Process (AHP) to decide the preferred resolution based on the given priority to the errors.

Keywords: Analytic hierarchy process · Log file analysis · Application debugging · Machine learning · Supervised learning

1 Introduction

Business organizations utilize the software application as the IT enabler to strengthen its business operation. By lower down the risk of the software application error and improve the reliability of utilizing the software application will bring business advantages to compete in today's the business industry. Under such circumstance, software application plays an important role to provide service to business operations. More importantly this service is required to be available 24 hours a day, and 7 days a week. It indicates that the expectation of such service availability is high. The toleration of software application service downtime is getting narrower. Give such high expectation of the enterprise software application service, the root cause analysis activity must not be prolonged. Indeed, high accuracy to identify the root cause which has cause the service downtime is crucial. However, dealing with an enterprise software application which

R. Silhavy et al. (Eds.): CoMeSySo 2021, LNNS 231, pp. 209–245, 2021.
https://doi.org/10.1007/978-3-030-90321-3_19

is involving multiple tiers, the root cause analysis will be difficult. This is because the multiple tiers are Front-end tier, Web Server tier, Application Server tier, and Database Server tier. The software application running in each tier has its log file. Especially in the Application Server tier, there can be more than one middleware (software application that is having integration to another software application) running. Each middleware has at least one log file. According to Management Logic (2012), more evidences are required when conducting root cause analysis. Evidences are crucial to identify the root cause. This is very time consuming to collect the evidences especially the evidences are the involved log files across multiple tiers. These logs have to be read through event by event to identify and filter out the alert and warning. Then finally it is still depending on the person who is experience enough to make a judgment on the ultimate error which causes the root cause at the end of analysis activity (Fig. 1).

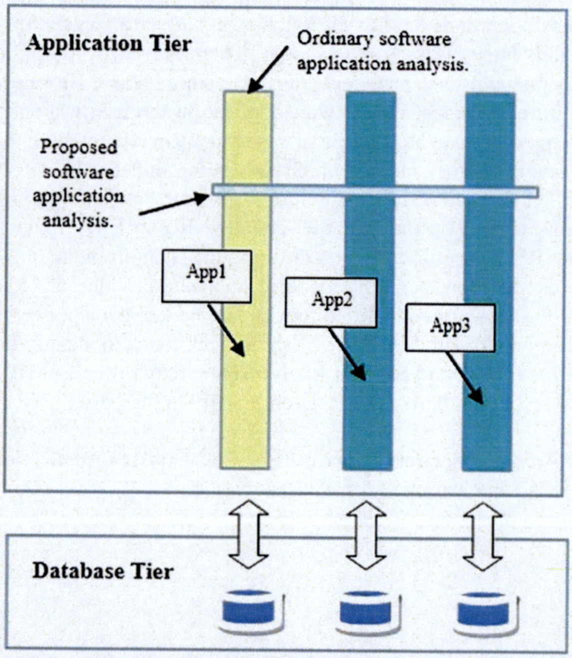

Fig. 1. The algorithm of PAL analyses multiple software application log files.

2 Motive of the Studies

The are several reasons have caused the motivation of this proposed research. Mainly there are four of them. First of all, the enterprise software application itself has more than one log file spanned across multiple tiers. However, with all the enterprise software application error log files, it may not be adequate for the root cause analysis. This is because what if the actual cause is falling outside the enterprise software application boundary. On the next reason, which is the time consuming to read through all the enterprise software application log files manually during root cause analysis activity. Thirdly, due to the root cause analysis activity is conducted by human, it is very subjective to the person who makes judgment and point out the actual error that causes the failure to enterprise software application. Lastly, whenever the operation cost to run the enterprise software application support operation is increasing, it is always facing high expectation from the company's management to have slim down-time on the enterprise software application.

3 Statement of the Problem

To conduct the root cause analysis activity on enterprise software application error, it is time consuming to identify the actual error as it involves multiple tiers and different layers

There are several events and behavior leading to the above statement of problem should be stated out as follows. At the same time, please note that the different layers here are referring to Enterprise Software Application, Operating System, Networking, Virtual Hardware/Physical Hardware, and connectivity to the Database. The enterprise software application behaves in an unexpected manner due to error occurred within or even beyond enterprise software application layer. This can be possibly that enterprise software application defect escapes from software application testing, and it is successfully released to production environment. On the other hand, whenever the root cause analysis activity is conducted manually, human would overlook the important clues and these clues are crucial to identify the actual error. With that, if it is in the situation that the human is lack of analysis experience. This would even waste much more time and cause the entire root analysis activity prolong. Furthermore, due to the root cause analysis activity is conducted by human, it is still very subjective to the human who can really make a good judgment on the actual error. Under such circumstance, the business users are facing difficulty to continue their daily tasks as the enterprise software application is unavailable or malfunction. The human who conducts the root analysis activity will be under pressure on how rapid to rectify the actual root cause.

4 Statement of the Objectives

There are four objectives. It is important to understand the mapping for research questions and research objectives. Here are the table and the diagram to show the mapping between research questions and research objectives (Table 1).

Table 1. Research questions imply the research objectives are required to be fulfilled.

The research question	The research objective
1. What approach can shorten the duration of root cause analysis activity?	1. To mitigate human involvement that can possibly cause prolonging time when conducting root cause analysis activity in the live environment
2. How to identify the error pattern of software application that leads to improvement of accuracy?	2. To improve accuracy on identifying the root cause whenever error event is occurred
3. What method can mitigate software application error and re-occurrence error which human may overlook?	3. To evaluate the proposed algorithm in the logic model can effectively reduce human mistakes during the duration of conducting root cause analysis activity
4. What method can evaluate the effectiveness of the proposed algorithm steps under the logic model?	4. To improve decision-making process on extracting valid error events from the involved log files from the enterprise level of software application

5 Literature Review

According to to Valdman (2001), there are some features under log file analysis presented in more detail while others are just intentions or suggestions. On the other hand, Vaidya and Kumar (2004) had reviewed Analytic Hierarchy Process (AHP) with 150 application papers in total, and there is no knowledge in applying AHP on identifying valid software application error. This means a research gap in implementing a logic model for log file analysis. This will allow to contribute the area of proposed research to incorporate the log data searching technique (introduced in the past research) into this research of prescriptive analytical logic. Apart of adopting techniques of previous reaches, this research introduces AHP to carry out decision making actions such as shortlisting the valid software application error among multiple software application errors are found, and shortlisting the final resolution to the software application error identified. With the AHP is applied, it can help the proposed logic model to be more accurate to evaluate and identify the valid software application error, and of course using AHP again to be more accurate to evaluate and identify the preferred resolution to the software application error.

In addition, by referring to Neves and Camanho (2015), it was a case study proposed to assist in Information Technology (IT) project prioritization by adopting AHP. This case study has two aims:-

i. To explore the expectations of a company's senior management and to identify the challenges during the implementation of the selected method that is decided by the company.

ii. To present a case study for the adoption of AHP along with multi-criteria method (or so-called multi-criteria decision analysis of operation research) that explicitly evaluates multiple conflicting criteria in decision making by a large Brazilian multinational company in the oil and gas industry.

Both aims will derive a well-defined structured process of decision-making analysis to assist on IT project prioritization which would contribute to those companies that are in the Oil and Gas industry. In the above case study, projects were classified into three categories, which are Innovation (technology projects), Growth Productivity (business projects), and Continuity (projects that are for improvements and maintenance on systems). The case study evaluated both the domestic and foreign markets in the oil and gas industry, and based on that defined the criteria for the prioritization of portfolio projects under the consideration of the company's investments. Then, based on the given categories, the criteria are prioritized under the configuration of the multi-criteria tree in a hierarchy was applied as a support tool. After the configuration of the multi-criteria tree was built, the weights were assigned to the criteria and all the sub sub-criteria in the tree. After all, AHP plays an important role to estimate the relative magnitudes of factors through pair-wise comparisons. With that, it allowed the company to obtain a micro view of decision-making as a full picture that came into effect for the departments within the company, to address the strategic directions on their IT projects. With the given case study, it is learned that AHP can actually be applied on analyzing the software application errors, as well as those errors related to the incident of software application failure. Whenever the root cause of software application error is identified, by adopting AHP again to evaluate the possible resolutions, the preferred resolution will help to resolve the software application error under a structured technique.

With the comparison of the proposed logic model to Machine Learning, it has certain similarities based on Klass (2018). However, this proposed logic model is not completely under the Machine Learning approach. One of the similarities is that the proposed logic model has to be trained with training data which consist of inputs paired with the correct outputs. This means the algorithm of proposed logic model will search for the defined patterns in all required log files that correlate with the desired outputs. It contains a little similarity to Machine Learning's Supervised Learning which is applied to predict the correct label for newly present input data, such as $O = f(i)$. O is the predicted output which is determined by a mapping label, and i is the input value that is assigned to a class. Wilson (2019) explained that the Machine Learning's Supervised Learning predicts the output based on the predefined features as label applied on the newly present input data under Classification subcategory. On the other hand, Regression subcategory is a predictive statistical process which attempts to find out the relationship between dependent and independent variables. However, the proposed logic model cannot predict the resolution if the new found error has no straight forward resolution defined in the configuration file.

The reason behind is that when the proposed logic model scan through all the required log files, it is based on the keywords as the label defined in the configuration file. It will require to put the newly present input data into the category of classification (such as Required-for-Instruction) in its configuration file, if the data is not matched with the existing defined label found in the configuration file. This is because in a situation that if the error is from the software application layer, and the resolution is required the software development cycle to do the fix. There is no immediate resolution can be applied during the software application execution. Furthermore, of course there are situations that are having immediate resolution such as restarting the database connectivity or software application process. These resolutions will be stored into the configuration file of proposed logic model. Indeed, there are multiple layers such as CPU, Memory and Storage (i.e., Hard Disk), OS, Network Connectivity, and Database can possibly contribute to the actual root cause. The first level of decision making is to identify which layer is having the cause of error, followed by the second level of decision making is to identify the real root cause within the layer. By understanding this circumstance, the AHP is at the suitable position to conduct the weight evaluation and decision making for the valid error of the root cause.

Supervised Learning has a subcategory, that is Linear Regression. According to Brownlee (2019), it is required to identify well on the relationship between input variable and output variable, and both variables must be linear. A good example provided by Milan (2019) is that in a tropical climate location, if the humidity increases, then the air is holding much more water, and eventually it leads to raining possibility is high. However, in the scenario of software application running in a multiple-tier environment, there is difficult to have a linear relationship. This is because (by example) if end-users accessing to the software application via the Internet and receive a failure page in response, this doesn't always mean that the root cause is because of the software application is failing itself. It can be possibly that:-

i. The server resources on the Web server are running low,
ii. The messaging queue (MQ) is piling up,
iii. The backend agents are not responsive to the requests sent from the Web container,
iv. The network connectivity to the database is intermittent, and etc.

Linear regression is expecting a straight forward outcome which means if it is not in one case then it will be another case. There is no third case in the outcome under such prediction. Therefore, the proposed logic model can adopt this approach under the Regression subcategory for straight forward expectation of the outcome if the linear relationship is clear. Otherwise, AHP will be applied to close the gap of decision making. This also means, there is possibility to design and to develop an algorithm of proposed logic model to be a hybrid of Linear Regression under Supervised Learning alongside with AHP for the decision making.

6 Proposed Logic Model

The Analytic Hierarchy Process (AHP) was developed by Thomas L. Saaty stated in Saaty (1987). The PAL is proposed to consist of algorithm that is having a hybrid approach of both AHP and Supervised Learning to fulfill the statement of the objectives. This proposed Prescriptive Analytical Logic Model does not perform any predictive analytic, but it provides speedy root cause analysis activity by the pre-configured relationship among the related software applications within the enterprise level. Then it performs the analysis based on the exact time of software application error with cross-check the log events of all the related software applications at the time of software application error occurred. The proposed algorithm proposes the following features which highly differentiate from the market products:-

1. Using the "time" of software application error event when it occurred as the key to extract out log events from all the related logs. The logs can be from software applications, Operating System, Network, and even Database to form a dataset for root cause analysis. This would leave no foot print and no interference to all the existing software applications and database operation.
2. Extract only the exact amount of input information is important. It is unnecessary to extract excess information and to waste time on root cause analysis. The log extraction will be 5 min before the time of software application error event when it occurred, and 5 min after the time of software application error event when it occurred. Then, scan the error only from the extracted amount of log rows. This will allow the root cause analysis to understand any symptom or pattern before and after the software application error event occurs. The rule of 5-min-before-and-after log extraction is not only applied on software application log itself. It is applied across all the error log events extracted from the related various software applications and database. This would allow to understand any external factor at other layers could possibly impact to the software application layer.
3. Standard error events are easily classified based on keyword, except for those unknown error events. As far as the error detection concern, this area can be applied Machine Learning's supervised learning to conduct the pattern searching activities on those collected log events. Then for these unknown error events, the proposed logic model will produce a report consists of the compilation of all the related log events from related software applications and database. The log events are before and after 5 min of the point of software application error was detected. Although this report is still required human interference to make judgment and decision to put the final action into the configurations of complex analysis module. With the improvement of the configurations of complex analysis module, the proposed model's intelligent level is increasing until the judgment and decision making can be solely handled by the purposed model itself. Crucial factor is the specific error log events extracted at the right timing for analysis and the compilation of the error symptom and pattern. Unlike the market products which only allow to set trigger for specific action based on the total number of same error events or specific error event occurred.
4. The proposed logic model's standard analysis module does the specific action based on the specific error event occurred. Under the proposed design, it will not allow the same resolution action to be carried out if the same resolution action had been carried in the past. This is because when the same error event re-occurs, but the knowledge-based database of the proposed model identifies the same resolution action had been carried in the past, the proposed model will trigger complex analysis module to carry out further analysis action.

5. Situation can be happened such as multiple software application errors are detected at the same timing. The extracted software application error events can possibility be either false alarm or warning, low impact error or high impact error. The proposed logic model's analysis activity is based on the configurations of standard and complex analysis modules together with the studies of past analysis. The analysis module (based on which module is selected) will then apply the resolution retrieved from the configuration files. However, for the decision-making activity to identify the chosen software application error, this activity will leave it to AHP. This means for each error event is identified by the configuration of the analysis module (in this situation, it is the complex analysis module), and which is the crucial software application error to be fixed first, this is all on the AHP to find out the highest weight assigned to the software application error. AHP is introduced in the proposed model and it is applied to evaluate and to identify the software application error which is carrying the highest weight. Note that the standard analysis module is having straight forward resolution steps, and the complex analysis module will perform deep dive analysis activity by analyzing the log files.
6. Another situation can be happened such as more than one possible resolution is applicable to resolve the identified root cause. Decision making is required. Hence, AHP is in the great position to evaluate and to identify the preferred resolution against the identified root cause.
7. All the current analysis activities and resolution steps will be stored into a knowledge-based database for the future reference. As well as it is for the input information of the future analysis.

The PAL is designed to have two configuration files, one is designed to record down the common errors and room to cater for new found error. Then, another one is designed to record down the resolution steps for the common errors as well as room to cater for human to provide new resolution steps based on the new found error. These preferred resolution actions under the resolution are manual, semi-automatic, and fully automatic. The PAL must obey the configurations in order to carry out the resolution under the preferred actions. Especially for the "Complex Analysis", it is required to store the analysis result and the resolution action into a knowledge-based database which is associated to the PAL for future reference. Regardless the analysis activity is under simple analysis or complex analysis, the knowledge based is catered for upcoming analysis activity and fine tuning the logic. It is mainly for improving the decision making on selecting the final resolution to be applied to the new found error in the near future.

PAL is required to be trained. This is because the PAL cannot be intelligent by itself without the initial training. Human interaction will still involve at the beginning stage to handle the fine tuning of the configurations of PAL. Since training is crucial, it is required real data as the input of the training. With all the important features stated above, the proposed algorithm inside the PAL is derived and shown as follows (Table 2).

Table 2. Proposed algorithm inside the Prescriptive Analytical Logic Model.

AHP	Supervised learning	Algorithm
		1. Collect or integrate various log files obtained from different software application databases with/without a predefined database schema
		2. Identify whether the newly reported software application error is first time occurrence or re-occurrence by cross-checking the database which is associated to the prescriptive analytical logic
	✓	3. Identify possible log data and select the necessary log data (by error patterns) for analysis under the defined software application error classification
✓		4. Allocate weight to each possible software application error based on *Analytic hierarchy process (*AHP)
✓		5. Shortlist the software application error under the highest weight
		6. Analyze the selected log data for shortlisted software application errors and define possible resolution option
✓		7. Allocate weight to each possible resolution option based on AHP
✓		8. Shortlist the preferred resolution option under the highest weight
		9. Deploy the preferred resolution option to fix the software application error under the predefined condition if the resolution does not involve SDLC. Otherwise, produce the Analysis Report
		10. Store the analysis result and resolution action into a database which is associated to the prescriptive analytical logic for future reference and knowledge base activities

7 Design, Development and Implementation

7.1 System Design

The complete picture of the PAL proposed algorithm is derived into the following process flow diagram. The objective is to provide a better explanation about the complete flow for the proposed logic model's process design (Fig. 2).

The Process Flow Design explains the use of the configuration files (first and second) during the execution of the logic model. First configuration file is for error cross-reference and the second configuration file is for carrying out the preferred action based on the recognized error. The entire flow is explained as follows:-

Process 1
It is required to look for the option to use any existing market product to detect Software Application error. Unless it is not possible to use any existing market product to

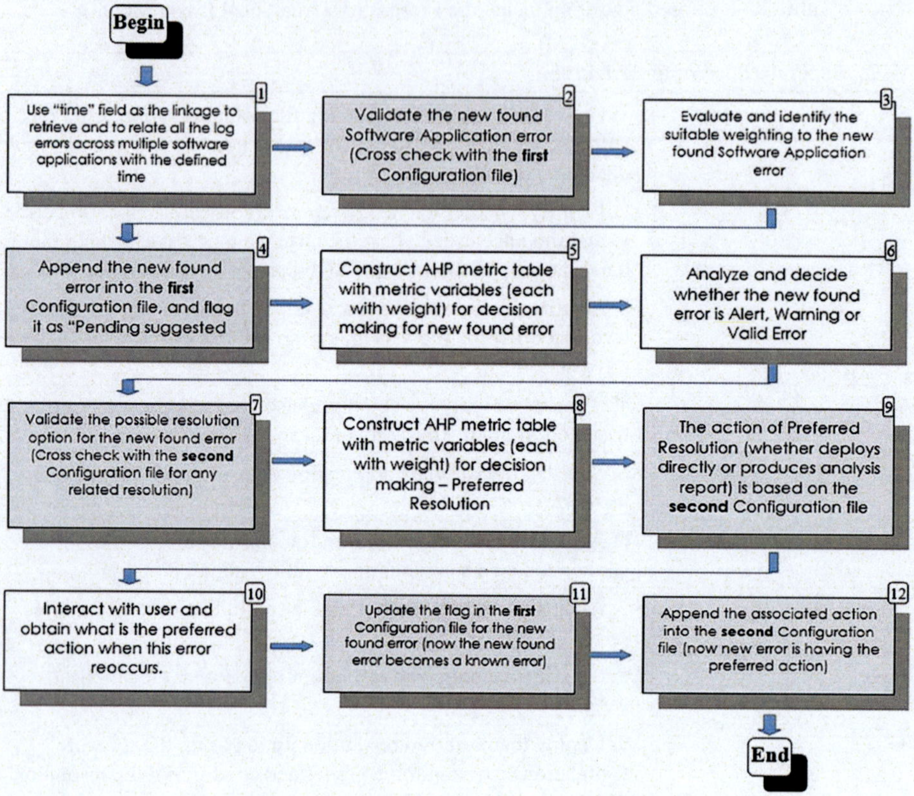

Fig. 2. The complete process flow of PAL.

detect Software Application error, then this process is required to code by programming approach. This is because the PAL should not re-invent the wheel. It should utilize the existing available approach from the market for software application detection if possible.

Process 2
Validate the new found Software Application error (Cross check with the **first** Configuration file). The first configuration file contains the information regarding to the involved software application, databases, and the knowledge-based database.

i. Based on the date and time of the new found Software Application error, PAL extracts the events from the involved software application and databases to form a dataset.
ii. Then, based on the new found software application error as well as any error found in the dataset as per the given date and time, PAL can cross-reference from the first configuration file to identify out whether it is a new error or known error. Which means, PAL will require to map out any error defined in the first configuration file.

iii. Based on the defined error to identify the new found error is actually a known error in either an Alert, Warning or Valid Error category, or it is unidentified and really a new found error.
iv. If it is a new found error, retrieve the last index number stored in the first configuration file, and load into the variable identification_number. Then, increase the index number count by 1 in the variable identification_number, to prepare for the unique identification number for the new found error. Later this identification number will be appended into first and second configuration files for linkage used for information fetching.
v. If this error is identified as known error, then a value (−1) to represent known error stored in a variable. At the same time, addition validation is to ensure that whether it is a re-occurrence error after validating from the knowledge based.
vi. Another value to represent re-occurrence error (−2) will be stored into another variable.

Process 3
Evaluate and identify the suitable weighting to the new found Software Application error. In this process, only a weight value is assigned to a known error based on the error type, and no weight value is assigned to a new found error.

Process 4
Append the new found error into the **first** Configuration file, and flag it as "Pending AHP Process". The PAL needs to recognize the new found error to become a known error however at this stage the AHP process has not been conducted yet. Therefore, the new found error must be flagged as "Pending AHP Process" at this moment.

Process 5
Construct AHP metric table with metric variables (each with weight) for decision making for new found error. In this process, a metric table under AHP will be constructed. The detail on how to construct the AHP table will be explained later in this chapter.

Process 6
Analyze and decide whether the new found error is either Alert, Warning or Valid Error. In this process, it is required to decide which is the final valid error with the highest priority should be fixed. For those Alerts and Warnings are safe to be ignored, it will be assigned with a lower weight.

Process 7
Validate the possible resolution option for the new found error (Cross check with the **second** Configuration file for any related resolution). Which means, the second configuration file will be referenced to retrieve the predefined resolution after the final valid error is decided by the previous process.

i. In the situation that the predefined resolution is clear cut (which has the only option), the next process will be bypassed.
ii. However, in the situation that more than one options of the predefined resolution are available, the next process is crucial.

Process 8
Construct AHP metric table with metric variables (each with weight) for decision making – Preferred Resolution. This process is to construct a metric table under AHP if the predefined resolution is more than one option.

i. The weight for each option is already predefined during the time of the second configuration file is built. Hence, the highest weight of the predefined resolution will be picked up.
ii. In the situation that both options are having the same weight, then the decision making of the preferred resolution is to produce an analysis report to include both preferred resolutions.
iii. An advice statement is required to revise the second configuration file stated on the analysis report, if there are more than one option is carrying the same weight.

Process 9
The action of Preferred Resolution (whether deploys directly or produces analysis report) is based on the **second** Configuration file.

i. In this process, if the only option of the preferred resolution is available, proceed to deploy the resolution steps (whether deploys the resolution directly or produces analysis report).
ii. Otherwise, proceed to produce an analysis report to include both preferred resolutions. An advice statement is required to revise the second configuration file.

Process 10
Interact with user and obtain what is the preferred action when this error reoccurs.

i. In this process, a dialog box will be popped up to ask for proceed or not to proceed the preferred resolution if the variable is having the value (-2).
ii. However, if the variable has no (-2) value but another variable is having (-1) value. Then another dialog box will be popped up, and user needs to fill up the fields to make the new found error become known error under the correct error category, as well as the suggested action for this new found error.

Process 11
Update the flag in the **first** Configuration file for the new found error (now the new found error becomes a known error).

i. In this process, the target is to make the new found error to become a known error. Therefore, the first configuration is updated with the identification_number and the required information for the new found error.
ii. Update the flag "Pending AHP Process" to become "Known Error" in the first Configuration file.
iii. Then, proceed to store the value of the variable identification_number in the **first** Configuration file for the use of next complete process flow.

Process 12
Append the associated action into the **second** Configuration file (now new error is having the preferred action). In this process, the second configuration is updated with the identification_number and the associated preferred action for the new found error.

With that process flow design, the PAL can recognize this specific error (new found error) on the next round of error cross-reference.

7.2 The Analytic Hierarchy Process (AHP) Design

To identify the software application error along may not be straight forward if it is other factors forming a chain reaction to cause the software application fails. The root cause can be hard to determine if the log files from the involved software applications, Database, Operating System, and Networking are not factored into the root cause analysis process. The primary objective is to reduce the duration on conducting the root cause analysis for software application error.

As with the algorithm inside the PAL model, this algorithm consists the Analytic Hierarchy Process (AHP) process for decision making on both identifying the root cause and applying the preferred resolution action. This AHP process plays an important role in the PAL, and this AHP process was introduced by Thomas L. Saaty. The source can be referenced in Saaty (1987).

The design of AHP hierarchy will depend on the software application errors and other involved errors found from different layers. These errors become the participants. The participants can be grouped further based on the error categories. Once the hierarchy is constructed, the participants analyze it through a series of pairwise comparisons that derive numerical scales of measurement for the nodes. Then, the priorities are associated with the nodes, and the nodes carry weights of the criteria or alternatives. Based on the priority, the valid error and the preferred resolution can be decided.

In short, AHP is a decision-making process as a tool, which can be utilized to weight the both the possible root causes and possible resolutions under a hierarchical structure. Then, narrow down to the final root cause, and decide the best resolution among the shortlisted resolutions. Ideally, the AHP will be utilized at the following two (highlighted) specific proposed activity under the PAL, which are at the row 4, 5, 7 and 8 in the following Table 3.

By referring to the proposed algorithm that are number 4, 5, 7 and 8 (with the Note *) in the above table. These proposed activities in the algorithm play an important role of AHP process. The activity details are explained respectively as follows:-

- The activity number 4 helps to identify the possible software application errors and to filter out all the false alarms. On those possible software application errors, each error will be further identified by the its error characteristic and categorized into specific error category under a predefined software application error category list.
- The activity number 5 will perform assigning the weight to each possible software application error based on the error category the impact level of each error within the category. With the assigned weight on each error, it would easily shortlist the highest priority of software application error as the crucial error to be fixed.

Table 3. The proposed activities must be incorporated into the above proposed algorithm.

No.	AHP	Proposed algorithm
1		With the granted software application access, use "time" field as the linkage to retrieve and to relate all the log errors across multiple software applications with the defined time
2		Identify whether the newly reported software application error is first time occurrence or re-occurrence by cross-checking the database which is associated to the PAL
3		Identify possible log data (by error patterns) and select the necessary log data for analysis under the defined software application error classification
4	✓	**Allocate weight to each possible software application error based on *Analytic hierarchy process* (AHP)**
5	✓	**Shortlist the software application error under the highest weight**
6		Analyze the selected log data for shortlisted software application errors and define possible resolution option
7	✓	**Allocate weight to each possible resolution option based on AHP**
8	✓	**Shortlist the preferred resolution option under the highest weight**
9		Deploy the preferred resolution option to fix the software application error under the predefined condition
10		Store the analysis result and resolution action into a database which is associated to the PAL for future reference and knowledge base activities

- The activity number 7 handles the process of assigning weight on each possible resolution after the analysis activities has been conducted at Sect. 6. This is because possible circumstance can be happened when two similar resolutions are selected, but we need to identify which is the most suitable resolution can be applied to resolve the software application error.
- The activity number 8 helps to identify the final preferred resolution to be applied for the crucial error after evaluating the weight, and this action will isolate multiple resolution actions to be applied to the crucial error to be fixed.

With AHP process, the proposed PAL under the complex analysis area will conduct the important actions on filtering errors and identify which is the crucial error to be fixed, and later it also helps to identify which is the preferred resolution action to be applied on the crucial error to improve effectiveness on resolving the software application error.

7.3 Part I – Decision Making on the Valid Error

From the previous sections, the proposed model is incorporated with AHP. As for the proposed AHP design, there are two modules which are required to apply AHP process. The two modules are (i) AHP Error Evaluation and (ii) AHP Resolution Evaluation. On the first model, it is going to discuss about the module design and the error evaluation.

The goal of this AHP module design is applied is to identify the software application error which carries the highest priority among those detected software application errors in the main software application for root cause analysis. The importance on the initiation of the software application error detection in the proposed model is always began at the main software application boundary first. Then, spanning the error detection to other related boundaries under the relationship between the main software application and other software applications. Only the possible factors are identified, then we can assign the correct weight to each factor. Factors can only be found is all depending on the flow of software application error detection.

The possible factors to be considered will be based on the following flow of software application error detection:-

- Detect software application error based on keyword.
- Identify software application error into category based on predefined classification.
- With the given time of the software application error occurred, extract other related logs and discover any error occurred at the same/similar time, and identify the relationship of the errors.
- Retrieve past error event from knowledge-based database and identify whether any past error event had been occurred.
- Calculate the total number of past error event had been occurred.

Based on the given flow above, the following factors will be the possible factors and each factor will be required to assign a suitable weight:-

- Error based on keyword.
- Error based on category or type.
- Error based on the relationship with other error.
- Past error event recorded.
- Total number of past error event recorded.

All the above considerations will be factored into the design of the module, which must be included into AHP Error Evaluation module.

Next is to determine the attributes (or criteria) in AHP. Since the errors are involving multiple layers and involved components, therefore the attributes (or criteria) in first AHP table (to identify the valid error) are:-

- Server Resources
- Networking
- Operating System
- Database
- Middleware
- Software Application

As per Wettergreen (2014)'s video, it is a straight forward approach to assign weight to each attribute. Please note that the attributes are referring to the involved software applications in the environment. As for the weight on each attribute, it is either a value 1 (means more important than) and a value 0 (means less important than). The evaluation will start from the left row against the vertical column, from the sequence of left to right. The complete first AHP table (also known as the Pairwise Comparison Chart) is shown as follows (Table 4):-

Table 4. The first AHP table for the required attributes assigned with the respective weight.

	Server Resources	Networking	Operating System	Database	Middleware (Message Queue)	Middleware (Backend Agent)	Software Application	Total
Server Resources	-	0	1	1	1	1	1	5
Networking	1	-	1	1	1	1	1	6
Operating System	0	0	-	1	1	1	1	4
Database	0	0	0	-	1	1	1	3
Middleware (Message Queue)	0	0	0	0	-	1	1	2
Backend Agent	0	0	0	0	0	-	1	1
Software Application	0	0	0	0	0	0	-	0

By referring to the value in the "Total" column on the first AHP table, the ranking of the attributes reflects the importance on each error and its priority sequence.

Ranking of the attributes are:-

i. Networking (Most Important)
ii. Server Resources
iii. Operating System
iv. Database
v. Middleware
vi. Backend Agent
vii. Software Application (Least Important)

For the attributes, please note that:-

i. Server Resources are referring to CPU, Memory and Hard Disk Space.
ii. Networking is referring to Local Area Network (LAN) and Wide Area Network (WAN).
iii. Operating System is referring to the Operating System layer where Software Application, Database, Message Queue and Backend Agent are running onto it.
iv. Middleware is referring to Messaging Queue (MQ), in short, refers as Message Queue (MQ).
v. Backend Agent is referring to the small software application that handles data selection from MQ, data extraction (uncompressing the compressed data file), data processing, and data loading (to upload data into the Database after processing).

As for each attribute within it, the log events in a log file will be scanned with sequential direction. The known alerts and warnings will be filtered by cross referencing to the first configuration file's **Configuration Section**. Leaving the next action is to identify the first error. This first encountered error in the log events will be picked, and the date and time of this first encountered error can be served as a reference to retrieve the log events from all the related log files before and after five minutes based on the given date and time. These log events will be served as the data set for further analysis and investigation. The challenging point is that which is the first log file to be scanned before the first AHP table is required to trigger for decision making of valid error. This is depending on the decision of the user.

In the data set, if two errors found from two different log files, then the assigned weight on the contrast of two attributes depend on the decision of the user. Different user will have different judgment on which attribute is more or less important in the comparison. From the first AHP table, the most important attribute is Networking and the least important attribute is software application. With this decision-making information, any error event found in the collected log files, based on the combination of the attributes, if the issue is involved in Networking, it must be the first priority to be resolved.

7.4 Part II – Decision Making on the Preferred Resolution

On the other hand, based on the identified valid error, a preferred resolution is suggested by the PAL. This suggested resolution must be initially predefined by the user before the PAL can be trained. However, there are possibility happened to have more than one preferred resolution toward to solve the same error (Table 5).

Table 5. The relationship table for error and resolution.

Error	Preferred resolution	Relationship
1	1	1:1
1	2	1:M

In the 1:M relationship, the second AHP table (also known as the Pairwise Comparison Chart) is required to identify the ultimate preferred resolution for the identified error. Generally, there are three attributes should be considered for the identified valid error that is found in the data set.

For the attributes, please note that:-

- "Time stamp" for the identified valid error is required as it can help to understand when was the time the error was occurred. For example, a backup process for the server executes at midnight time, and the identified valid error occurred at the time when backup process is still running. Therefore, with the error time stamp, it provides crucial information for the decision making of the preferred resolution.
- "Frequency" for the identified valid error is required as it can help to understand how many times the same error was occurred in the data set.
- "Reoccurred" for the identified valid error is required as it can help to understand whether this error had been occurred in the past and it had been resolved before. This information can be stored and retrieved (the past analysis result) from Knowledge based Database under the PAL design.

The complete second AHP table (also known as the Pairwise Comparison Chart) is shown as follows (Table 6):-

Table 6. The second AHP table for the required attributes assigned with the respective weight.

	Time Stamp	Frequency	Reoccurred	Total
Time Stamp	-	1	0	1
Frequency	0	-	0	0
Reoccurred	1	1	-	2

By referring to the value in the "Total" column on the second AHP table, the ranking of the attributes reflects the importance on each error and its priority sequence.

Ranking of the attributes:-

1. Reoccurred (Most Important)
2. Time Stamp
3. Frequency (Least Important)

7.5 Pseudo Code

The entire PAL is developed based on the following Pseudo Code, and the Pseudo Code of PAL is shown as follows:-

- Start **AHP Log Analysis Run.**

- Scan and setup run configuration by grabbing **configuration file** (File: *prototype.properties*).
  1. Configurable properties include Logs Path, Related Log Range, Log Date Format, etc.

- Retrieve **Logs** from config folder based on given **Log File Paths** (Log Types: WASv5, MQ, RPA, PPA, AgentDP)

- Open and Read each **Log** (Repeat Steps 5~8 for each Log Type).

- Scan each **Log** for **Error** messages based on given **Error Keywords** (Config File: *error_keywords.csv*) parsed against ***X* words** length (defaulted to 5-word error messages).
  1. Parse each line in Log against given datetime format to scan only **Timestamps** in given range (Properties: *log.time.start* & *log.time.end*)
  2. Each Line parsed against each **Error Keyword** for matching String.

- Parse and Process each **Error** found.
  1. Group any dynamic errors found according to given **REGEX Pattern List** (Config File: *error_keywords_pattern.csv*).
     i.e. [Error found in hostname *'server.host.com'*] and [Error found in hostname *'another.host.com'*] can be grouped into [Error found in hostname *[HOSTNAME]*]

2. Ignore any false positive or blacklisted errors found according to given **Blacklisted Word List** (Config File: *error_keywords_blacklist.csv*)
   i.e. [NewErrorLog created] can be ignored when keyword Error is picked up by specifying *NewErrorLog*.

- Save each processed **Error** found into **Found Error List.**
  1. Save detailed information on **Error**. i.e., Timestamp, Source Log, etc.

- Process and Save each **Error** found against **Error Table File** (File: *error_table.csv*)
  1. For **Known Errors**, update Number of Occurrences, Last Occurrence, Last Log Found, etc.
  2. For **Unknown Errors**, add new entry with default Solution (No Solution), number of duplicate occurrences in current Run, and other default information.

- Processed all found **Error Log Types** and retrieve **Primary Error(s).**
  1. Apply **TimSort Algorithm** and reverse to retrieve **Primary Error(s)** with highest **Weight** according to given **AHP Log Priority Scores** (Config File: *ahp_log_priority.csv*).
  2. Error Log Type(s) with **Highest Weight/Priority Score** will be marked as Primary Error(s).

- Retrieve **Related Error(s)** based on found Primary Error(s).
  1. For each Primary Error found, **Compare** each element in **Found Errors List**. Retrieve any Errors within given **Log Relation Range** (Properties: *log.related.minutes*).
  2. Other Log Types that are within the **Log Relation Range** are marked as Related Error(s) against the Primary Error compared.

- Generate **Report Contents.**
  1. Map all **Found Errors** (Primary, Related, Others) with matching **Solution** from given **Error Solutions Table** (Config File: *solutions_table.csv*)
  2. Apply detailed information. i.e., Log Found, Log Type, etc.

- Write and generate **Report File.**
    1. Print **Header Template** (Config File: *report-header.txt*)
    2. Print **Primary Error(s)**, Matching Solutions, Detailed Error Info.
    3. Print **Related Error(s),** Matching Solutions, Detailed Error Info.
    4. Print **Other Error(s),** Matching Solutions, Detailed Error Info.
    5. Print **Footer Template** (Config File: *report-footer.txt*)
    6. Save **Report File** with **Timestamp.**

- End **AHP Log Analysis Run.**

As far as the prototype development concern, there are certain modules which are required more time to develop. These GUI modules are treated as "future works". These modules are:-

i. Graphical User Interface (GUI) for the AHP table, which allows user to set the priority to each involved software application.
ii. Graphical User Interface (GUI) for the AHP table, which allows user to set the priority to each resolution if there is more than one resolution involved to solve one valid error.

Meanwhile, to achieve the objective of setting the priority (for the involved software applications, as well as for the resolutions), the Comma Separated Values (CSV) file is used and the settings are kept in the configuration files as of now. Remember, the frequency of updating the priority to each involved software application in the AHP table, it is only once which is at the initial stage of PAL execution under the training phase. As for the frequency of updating the priority to each resolution, it only requires when the scenarios happen such as either there is no resolution is sat or the resolution is required to improve.

7.6 System Development

In the log scanning part, there are two options for prototype development of PAL. The first option is to look for any existing market product to handle the log scanning process, and the second option is to develop the scanning process as a module from scratch. Between the two options, only one option is required and the option is solely depending on the environment where the prototype executes. Either option can still deliver the same objective.

7.7 Log Scanning

Option 1: By using tool such as Filebeat, it is an easy way to collect log files from all the involved software applications. Then, apply necessary modification on the collected logs using Logstash. After all, ship the logs to ElasticSearch for indexing. Indeed, these log events are indexed in the ElasticSearch based on given patterns. Even on this option, the trigger point to activate PAL is still be depended by the error found from the collected logs if the error is found as valid error during log scanning.

Option 2: By using programming technique as the approach to collect logs hourly, the PAL will scan the collected logs and to identify any valid error can be found. With that, the initiation point is that PAL will be triggered if the valid error is found during the log scanning process.

In both options, once the valid error is identified, the PAL will be triggered. The error event along the event date and time are supplied to the PAL. PAL uses this input information as the keyword to pull the log event from all the involved software applications under the enterprise.

Note:

i. In either option, PAL is still a standalone software application to be triggered after the valid error is found from the base log file. Please understand that this so-called valid error is not the ultimate error decided for preferred resolution decision. The error event date and time of this valid error will be utilized to obtain the log events from the remaining involved software applications. With all the involved log events (from the respective involved software application) put together in a table under AHP process, then the decision on deciding the root cause will be more realistic.
ii. The end users are accessing the Web application to perform their daily works in the business operation. The Web application is hosted in a WAS container. In this case, the WAS log is the main log event to be the base for scanning.

In this research, the option 2 is selected due to the organization which provides the environment for this research has yet to purchase the paid package from Elastic. The required programming processes will be discussed on the following sections in this paper.

7.8 Training the PAL Logic

The PAL is required a proper training to understand the differentiation between Alerts, Errors, and Warnings. Although Gogus A. (2012) has explained the Learning Cycles and it is more for human beings. In the actual of fact, there is no software application can have the intelligent by itself without any training cycle in the training phase (Fig. 3).

It is important to lay down the detail steps on how to train the PAL Logic before it can handle real life errors in the actual environment (i.e., live environment). These steps are crucial and it is also known as learning cycle for the PAL Logic. Therefore, training the PAL Logic is very important and the steps are shown as follows.

1. Scan all the involved log files and only scan for the line of logged events that are not scanned within the same day. Whenever error event is found (more than one error event) in the all the involved log files, utilize the event date and time of the

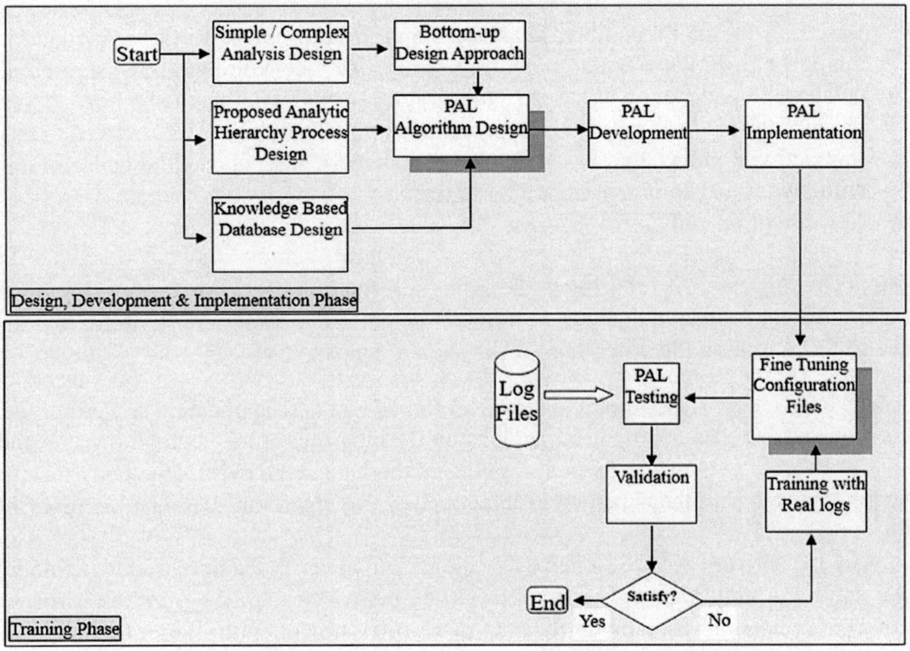

Fig. 3. Training phase of the PAL will make the logic to learn.

error events (as keyword) to relate these error events. However, if only one error event is found during the scan, then this is the only error at this moment which is required to fix.

2. Filter the collected error events (from all the involved log files) based on the first configuration file to avoid fixing alerts and warnings.
3. If this error is not found in the first configuration file, tag this error as "Unknown". Generate an index number as for the reference. Append this error into the first configuration file. Put a flag "on" to indicate a pending update to the second configuration file.
4. If this error is found in the first configuration file, then go to 9.
5. If the error is more than one, then construct the AHP array to decide which is the most priority error to get it fixed.
6. Once the final error is identified, refer to the second configuration file to obtain either suggestion/fix (steps to fix).
7. If the suggestion/fix (steps to fix) is more than one, then construct the AHP array to decide which is the most priority suggestion/fix (steps to fix) to apply.
8. Update the first configuration file if the error is tagged as "Unknown" to be "Known". Update the flag to "off". Then update the second configuration file with the suggestion/fix (steps to fix) associated to this error based on the index number stored in first configuration file. Interact with knowledge-based database, update the flag and update the table column with predefined comment regarding to the error. Go to 10.

9. Based on the index number, fetch the suggestion/fix (steps to fix) associated to this error from the second configuration file. Interact with the knowledge-based database to understand was this suggestion/fix (steps to fix) applied before. If yes, produce the analysis report. If no, proceed to apply the suggestion/fix (steps to fix). Interact with knowledge-based database, update the flag, update the table column with predefined comment regarding to the error.
10. The execution ends.

Note: The suggestion is referring to the gather information (from the involved logs) and it is waiting for human interaction to provide the preferred resolution steps. So that the second configuration file will record down the fix (steps to fix).

The PAL's logic is incorporated into a developed software application as a prototype. It is required to be trained by using the customized log files at the beginning stage, and it has to be trained by using the real log files at the later stage eventually. Only then, it can be deployed into the actual environment (i.e., live environment) to be the proof of concept (POC).

With the delivery of POC, it helps to validate and to verify the proposed logic model PAL can accurately identify the real error (i.e., improve accuracy) from the involved software applications running on the enterprise environment, with shorter time manner (i.e., reduce time duration on root cause analysis). Hence, by improving accuracy and reducing time duration on root cause analysis, it aligns with the research objective of this research.

7.9 System Testing

Software application testing (in short, testing) is a process that cannot be ignored. According to Pan (1999), testing is to evaluate and verify whether the software application meets the expectations. It is one of the crucial processes during the development cycle of PAL. Hence, it is required to have a Test Plan for PAL. The Test Plan consists of necessary test cases to ensure the functionality of PAL to work as expected. Each test case will define the specific input to the PAL, and will validate the output after processing is done by the PAL.

In the PAL prototype, it is important to understand what are the configuration settings. These configuration settings are important to be tested. If the settings are on different sections within a configuration file, or on separate configuration files. Each separation can be tested in different Test Cases, to validate whether the actual result can meet the expected result. The following Test Cases are defined and it is required to test and validate. Each Test Case has a table to list down items for validation. In the common term, this table is served as "Checklist".

7.10 Test Case 1: To Validate All the Configuration Files Can Be Accessed During PAL Execution

There are seven configuration files in PAL. Each configuration file carries specific property or a set of properties which help to define the behavior and expectations of PAL. The most important action to achieve in this test case, is to validate all the required configuration files can be accessed during PAL execution without any error occurs (Tables 7, 8 and 9).

Table 7. The purpose of configuration file: prototype.properties. **File Path:** ahp-log-analysis_1.0.0/resources/

Item number	Item to be validated	Purpose
a	prototype.properties	Main configuration file of PAL to define the properties for log file scanning

Table 8. The respective purpose for the rest of configuration files. **File Path:** ahp-log-analysis_1.0.0/resources/config/

Item number	Item to be validated	Purpose
b	ahp_log_priority.csv	To define the priority of involved software applications
c	error_keywords.csv	To define the keywords of all valid errors
d	error_keywords_blacklist.csv	To define the keywords of all possible alerts and warnings
e	error_keywords_pattern.csv	To define the valid error patterns can be found in the logging events
f	error_table.csv	To record the valid error found and number of occurrences during the given duration
g	solutions_table.csv	To define the preferred resolution for each valid error

Table 9. The test case 1 checklist.

Item number	Item to be validated	Unit testing	Integration testing
a	prototype.properties	Validated and Pass	Validated and Pass
b	ahp_log_priority.csv	Validated and Pass	Validated and Pass
c	error_keywords.csv	Validated and Pass	Validated and Pass
d	error_keywords_blacklist.csv	Validated and Pass	Validated and Pass
e	error_keywords_pattern.csv	Validated and Pass	Validated and Pass
f	error_table.csv	Validated and Pass	Validated and Pass
g	solutions_table.csv	Validated and Pass	Validated and Pass

Under Unit Testing, each configuration file is required to test individually. This is to ensure that the column or section of property from the configuration file is fetched and defined in either constant or variable of PAL correctly.

Under Integration Testing, all the configuration files are accessed by the PAL during the execution. The input of PAL is the customized log files and the output are validated to ensure that the results can meet all expectations.

7.11 Test Case 2: To Validate Other Properties Are Accurately Accessed During PAL Execution

There are other (remaining) properties of PAL which are fairly important and it is required to test and validate (Table 10).

Table 10. The test case 2 checklist.

Item number	Item to be validated	Validation result
a	To validate the interval of log events is correct for all involved log files	Validated and Pass
b	To validate the duration when the primary error is found among the involved log files	Validated and Pass
c	To validate all valid errors are found based on the defined valid errors in the configuration file	Validated and Pass

- To validate the interval of log events is correct for all involved log files. The defined interval is the start time and end time of the log events which PAL will scan on each involved log file.
- To validate the duration when the primary error is found among the involved log files. The defined duration is to scan the log events when the first valid error is found. This means whenever the first valid error is found (after scanning across all the involved log files), PAL will look for log events of each involved log file based on the defined duration before and after of the timing which the first valid error is found. For example, if the duration is defined at 10 min. When the first valid error is found in one of the involved log files, PAL will scan log events on each involved log file at 10 min before and 10 min after, based on the timing of first valid error is occurred.
- To validate all valid errors are found based on the defined valid errors in the configuration file. This is archived by cross-checking with the defined valid errors in the configuration file: error_keywords.csv. As for the alerts and warnings in the involved log files, it must be safely ignored based on the defined alerts and warnings in the configuration file: error_keywords_blacklist.csv.

7.12 Test Case 3: To Validate the Analysis Report of PAL is Shown as Expected

The last Test Case is required to test and validate the Analysis Report of PAL. This Analysis Report must contain information that is simple and easy to understand (Table 11).

Table 11. The test case 3 checklist.

Item number	Item to be validated	Validation result
a	To validate the header of Analysis Report	Validated and Pass
b	To validate the valid error, number of occurrences, and relationship between valid errors found	Validated and Pass
c	To validate the footer of Analysis Report	Validated and Pass

- To validate the header of Analysis Report. On every Analysis Report, it will be plugged with the defined header. The location of the header file is defined in configuration file: prototype.properties. The actual location of header file is located at ../ahp-log-analysis_1.0.0/resources/reports/templates/report-header.txt. This is required to test and to validate.
- To validate the valid error, number of occurrences, and relationship between valid errors found. The listing and the output arrangement for the valid error, number of occurrences, and relationship between valid errors found must be validated. It must be displayed as per expectation.
- To validate the footer of Analysis Report. On every Analysis Report, it will be plugged with the defined footer. The location of the footer file is defined in configuration file: prototype.properties. The actual location of footer file is located at ../ahp-log-analysis_1.0.0/resources/reports/templates/report-footer.txt. This is required to test and to validate.

8 Validation and Verification

The verification is carried by providing actual log files as input to the PAL, and the examination is based on the output of PAL whether the expectations are met (Table 12).

Table 12. The actual log files are used for prototype verification.

Software application name	Log file name
WebSphere Application Server	SrmDailyInternal.log
Repository Processor Agent	Banco_Itau.ansible.ibm.com.srmdata.summary.20201203.03.00.56.log
Push Processor Agent	pushproc0401.20201203.log
Message Queue	AMQERR01.LOG
Agent Data Processing	bld04vc01.2020.12.03.log

The activities carried out under prototype verification is crucial as it can verify the algorithm of PAL in the proposed logic model whether it is actually working if it is developed and implemented. Here are the major activities carried out for prototype verification.

i. Indeed, the verification involves a lot of manual works because the number of valid errors must be tallied with the result of the prototype. In addition, the relationship between valid errors must be same result comparing with the results after the prototype execution. The result is successful.
ii. As for the platform independence, PAL has been tested on AIX, Linux and Windows platforms. These platforms are installed and setup with Java Runtime Environment (JRE). The result is successful, this is because PAL is written with Java Programming language.

8.1 The Duration of PAL Execution

Before PAL execution, the time is captured at Mon Feb 1 09:35:27 +08 2021.

```
[wonghm@oc7774266761 ahp-log-analysis_1.0.0]$ date
Mon Feb  1 09:35:27 +08 2021
[wonghm@oc7774266761 ahp-log-analysis_1.0.0]$
```

As soon as PAL execution is finished, the time is captured at Mon Feb 1 09:36:26 + 08 2021.

```
[wonghm@oc7774266761 ahp-log-analysis_1.0.0]$ date
Mon Feb  1 09:36:26 +08 2021
[wonghm@oc7774266761 ahp-log-analysis_1.0.0]$
```

The total time taken of PAL execution is based on the given actual log files collected from the live (real) environment, it is **01:01 min.**

8.2 The Identification of Valid Error Types

In order to ensure the valid error types can be extracted out correctly by PAL, a manual work to extract out the valid error types in the involved log files is unavoidable. Only have the exact number of valid error types found in the involved log files, it can be compared with the result of the PAL execution. By manual check through the involved log files, there are 26 types of valid error in total.

These valid error types are shown as follows (Table 13).

After PAL execution, the configuration file: **error_table.csv** has recorded the exact number of valid error type, which is 26 types of valid errors in total.

8.3 The Relationship Between the Valid Errors

In PAL, as long as the first valid error is found, based on the given interval, it will extract the valid errors within the given interval. During the training phase, PAL has no idea about the relationship between valid errors at the beginning. It is required to train in term of improving the intelligent level. It is similar to human being. No normal human being

Table 13. The valid error types that are found from the involved log files.

Error no.	Error message	Log type
1	Error on receive from host	MQ
2	PushProcessorException: RegistrationResponse RC 1001 Account:	PPA
3	IBMExceptional logged	PPA
4	java.lang.NullPointerException	PPA
5	PushProcessorException: Registration:evaluate null EntityId:	PPA
6	IBMExceptional	PPA
7	SQLException in processConfigDataType:	PPA
8	Error: SQLCODE = -545 SQLSTATE = 23513 SQLERRMC = PERFORM.SERVER.SQL031006171548640 DRIVER = 4.19.66	PPA
9	com.ibm.msg.client.jms.DetailedJMSException: JMSWMQ2002: Failed to get	PPA
10	DetailedIllegalStateException in sun.reflect.NativeConstructorAccessorImpl.newInstance0(Native Method):	PPA
11	com.ibm.msg.client.jms.DetailedIllegalStateException: JMSCC0020: This session is	PPA
12	PushProcessorException: Unable to link this	PPA
13	DetailedIllegalStateException in sun.reflect.GeneratedConstructorAccessor43.newInstance(Unknown Source):	PPA
14	DetailedIllegalStateException in com.ibm.msg.client.wmq.common.internal.Reason.reasonToException(Reason.java:496): JMSWMQ0018:	PPA
15	MQException in com.ibm.msg.client.wmq.common.internal.Reason.createException(Reason.java:223): JMSCMQ0001:	PPA
16	PushProcessorException: RegistrationResponse RC 1001 Cannot	PPA
17	PushProcessorException: Insufficient data in winconfig	PPA
18	Invalid value thrown out. [HOSTNAME/ID]	AgentDP
19	"Invalid value thrown out." [HOSTNAME/ID]	AgentDP
20	ERRORCODE = -4229 SQLSTATE = null: [jcc][t4][102][10040][4.19.66] Batch	AgentDP
21	ERRORCODE = -4229 SQLSTATE = null	AgentDP
22	Error DataType = diskvcw	AgentDP
23	Error	AgentDP
24	Error DataType = netstatvc	AgentDP
25	Error: SQLCODE = -911 SQLSTATE = 40001 SQLERRMC = 2 DRIVER = 4.19.66	AgentDP
26	Error DataType = hcvc	AgentDP

can speak at the time of new born. The human being has to go through learning cycle to pick up the speaking skill.

```
==========================================================
Log Analysis Report
==========================================================
Listed below are the errors found during the last Log Analysis Cycle.
Details:
Date of Report: 2021.02.01_09.36.20

==== PRIMARY ERRORS ====
Primary ERROR #1: Log Type: MQ | Line #: 8914 | Error Found: Error on receive from host | In Log: AMQERR01.LOG | Occurred 17 times in this log. | Error Time: 2020.12.03_02.04.03 |
Suggested Solution for ERROR #1: No known solution. Please update solutions_table.csv if known. |

==== RELATED ERRORS ====
== ERRORS RELATED TO ERROR #1 ==
Related ERROR #1-1: Log Type: PPA | Line #: 726727 | Error Found: PushProcessorException: RegistrationResponse RC 1001 Account: | In Log: pushproc0401.20201203.log | No known solution. Please update solutions_table.csv if known. | Occurred 28 times in this log. | Error Time: 2020.12.03_02.10.01 |
Related ERROR #1-2: Log Type: PPA | Line #: 726728 | Error Found: IBMExceptional logged | In Log: pushproc0401.20201203.log | No known solution. Please update solutions_table.csv if known. | Occurred 29 times in this log. | Error Time: 2020.12.03_02.10.01 |
```

The preferred resolution or advice can be updated directly into the configuration file: **solutions_table.csv**. In the configuration file: **solutions_table.csv**, each preferred resolution or advice is indexed by **#ID (#)** column. On the other hand, the configuration file: **error_table.csv** has the same index column, but it is named **SOLUTION ID (#)**. Therefore, both **error_table.csv** and **solutions_table.csv** have the linkage to provide resolution based on the valid error.

8.4 The Regression Testing for Additional Module

Any future work on the development of additional module for PAL, regression testing can be conducted to ensure new code changes have no side effect on existing functionalities of PAL. As per Guru99 (2020) has stated that, regression testing is applied to confirm that any code change or new module integration has no affect or impact to the existing features or functionalities of the software application.

9 Results and Discussion

The prototype of PAL is not a software application solution or engineering solution. It is a prototype which is developed under the proposed logic model. The prototype is to prove the algorithm inside the proposed logic model (i.e., PAL) whether it is working as per expectation. What it actually does matter is the algorithm inside the logic model. Indeed, the results of the prototype are important as it is the evidence to prove the algorithm. These results will help to identify the contributions of the research in order to align with the research objectives.

9.1 Shorten the Duration of Reading Log Files and Improve Accuracy

For the log file analysis conducted by human, it is all depending on an individual may read faster or slower. However, according to Capitalize My Title (2020), an adult typically would read at about 125 words per minute when it is required for more focus and attention to detail. In a mathematics calculation, it can easily calculate how much time to spend on reading through word by word manually on a log file (Fig. 4).

```
[wonghm@oc7774266761 Logs]$ wc bld04vc01.2020.12.03.log
  6766922  41923120 548743991 bld04vc01.2020.12.03.log
[wonghm@oc7774266761 Logs]$ █
```

Fig. 4. A log file size can be very huge depending on the detail level of the logged event.

By given a log file that contains 41923120 words, on average an individual would need 41923120/125 = 335,384.96 min to finish reading it. The more time spends on reading, the higher chance overlooks crucial information in the logged event. This will definitely lead to prolong the analysis activity easily if the root cause is still not identified. The actual of fact is that not only one log file will be read through manually. It is a necessity to ensure that no error is left out during the reading of multiple log files, and at the same time this duration which is spent on reading the log files must be reduced (Table 14).

Total number of words = 1651531 + 235 + 85694590 + 491336 + 41923120 = 129,760,812.

Total number of minutes (on average) spent on reading = 129760812/125 = 1,038,086.496.

However, PAL uses **01.01 min** (which is less than 1 h) to complete the entire root cause analysis activity (Table 15 and Fig. 5).

By having PAL, the duration of scanning through all the involved log files is now less than 2 h. All the valid errors can be collected based on the given interval. As far as the valid errors are found, seeing from each valid error alone, it is hard to identify the relationship. However, by using time to relate all the valid errors, it can clearly see which are those valid errors occurred within the same interval. Based on the time as the relationship, it can easily reflect the chain reaction between errors. Of course, further clarification and additional judgment will be provided by human during the training phase. It is to ensure

Table 14. The word count of each involved log file.

Software application name	Log file name	Word count	Minutes on reading
WebSphere Application Server	SrmDailyInternal.log	1651531	13212.248
Repository Processor Agent	Banco_Itau.ansible.ibm.com.srmdata.summary.20201203.03.00.56.log	235	1.88
Push Processor Agent	pushproc0401.20201203.log	85694590	685556.72
Message Queue	AMQERR01.LOG	491336	3930.688
Agent Data Processing	bld04vc01.2020.12.03.log	41923120	335384.96
Total number of minutes (on average) spent on reading			**1,038,086.496**

Table 15. The hours spent on each involved log file for root cause analysis.

Log File	**Hours Spent Manually**	**Hours Spent by PAL**
SrmDailyInternal.log	220.2041333	0.016833333
Banco_Itau.ansible.ibm.com.srmdata.summary.20201203.03.00.56.log	0.031333333	
pushproc0401.20201203.log	11425.94533	
AMQERR01.LOG	65.51146667	
bld04vc01.2020.12.03.log	5589.749333	

Log File	**Hours Spent Manually**	**Hours Spent by PAL**
SrmDailyInternal.log	220	< 1
Banco_Itau.ansible.ibm.com.srmdata.summary.20201203.03.00.56.log	0	< 1
pushproc0401.20201203.log	11426	< 1
AMQERR01.LOG	66	< 1
bld04vc01.2020.12.03.log	5590	< 1

Note: 0.01 indicates that it is less than 1 hour.

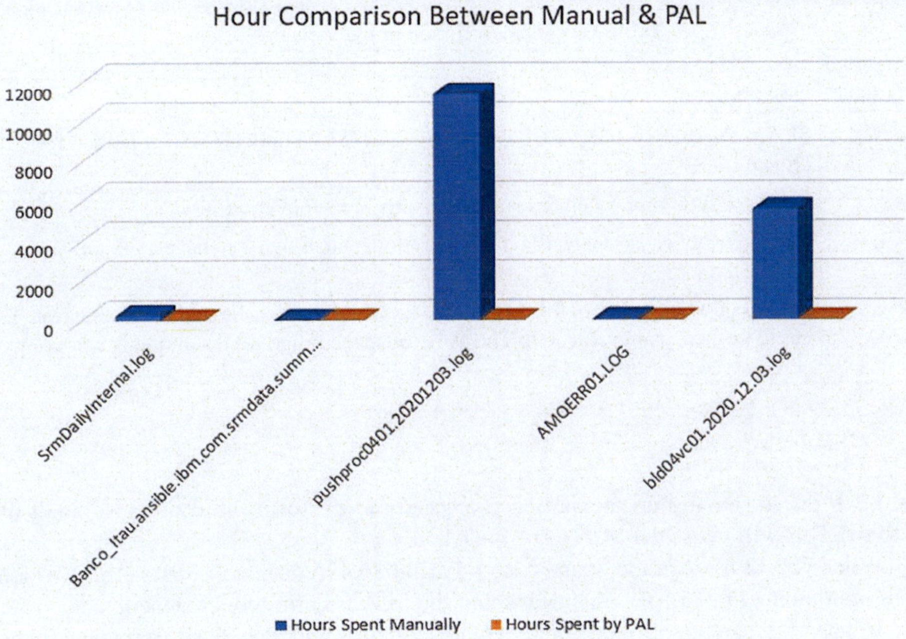

Fig. 5. The hour comparison between manual and PAL spent on root cause analysis.

that the intelligent of the PAL can further understand the relationship between valid errors. Therefore, PAL is in the position to reduce the duration which is spent on reading the log files, and improve the accuracy on identifying valid errors and the relationship between valid errors.

9.2 Shorten the Duration of Root Cause Analysis Activity

In any root cause analysis activity, normally it is to search through and to collect all the errors which are found in a log file, these errors are required to be filtered (safely ignore the alerts and warnings). That will help to narrow down the number of found errors. For those remaining errors after the filtering process, these errors are so called identified as valid errors. Next is to find out the relationship between the errors if there are more than one error is found. Followed by, identify these errors whether there is any relationship with it, or these errors are actually related and occurred in a ripper effect. With that, the root cause can be possibly identified. However, the resolution of the root cause is always depended by how fast the root cause analysis activity can be completed. In the early chapter, the duration of root cause activity can be shortened by the explanation of using a simple mathematics approach (Table 16).

Now, if $\mathbf{W} = \mathbf{X} + \mathbf{Y}$, then $W = (2 + 6) = 8$ h.

X may not be always 2 h as it is depending on how fast the root cause is identified manually. The accuracy to identify the valid error manually is still depending on an engineer's experience, knowledge, and skill. In the worst-case scenario, X can be more

Table 16. The description of the variables.

Variable	Description
SLA	SLA = Assume that Service Level Agreement (SLA) only allows 8 h in total for critical issue
W	Let W = Total time taken to resume the software application service
X	Let X = Root cause analysis activity duration, and assign it to have an initial value, 2 h
Y	Let Y = Development of fix, Testing the fix, and Deployment and regression test. Y has to be a constant value, 6 h. This is because the proposed resolution is not fixing here
p	Let p = Prescriptive Analytical Logic Model

than 2 h due to having an inexperience engineer to perform the root cause analysis activity. This will cause the SLA to be breached.

When **X > 2** h ← In the actual case [(1,038,086.496/60) = 17,301.4416 h (which can round up to 17,301 h)]. This means that the SLA is definitely breached.

Hence, if **X** can be shortened or kept below 2 h, this will have positive impact on the hours in total for W. Which means, p as the Prescriptive Analytical Logic Model (PAL) to reduce the time spent on root cause analysis activity and keep X to be exact 2 h or even lesser.

When **X < = 2** only if X-p, then.

W = (X-p) + Y, W < SLA ← With that, W can be kept below 8 h. Which means **X-p** is after deploying PAL. In the actual case, PAL only uses 01:01 min to complete the entire root cause analysis activity.

In the overall total time taken during a software application is having downtime, it is from the point of time software application encounters error, until the point of time software application error is resolved. The first activity is conducting root cause analysis to identify the actual software application error. There are activities which are having dependency on the root cause activity. For these activities, it must be conducted only after the root course analysis activity is completed. These activities are such as the duration of developing, the duration of testing and the duration of deploying the resolution. All these activities are all in a row (sequence). Each activity has the dependency of previous activity. Which means it is depending on the completion of previous activity before it can start. Indeed, how fast and accurately the root cause of the software application error can be identified. It leads to how fast the next activity can be carried out. With that, it can kick started the activity of resolving the identified error by performing the required predefined steps, or by advising the development of the resolution under the SDLC based on the situation. By focusing on the duration of root cause activity, and with the duration for the rest of activities remain constant, the mature PAL is proven that it helps to shorten the duration of root cause analysis activity. No doubt, it can definitely lead to shorten the entire software application downtime as the final result.

10 Conclusion

The enterprise software applications come along with the event logging feature for record trial and analysis purposes. This contributes to the proposed methodology on identifying the root cause analysis if there is any possibility no interruption or footprint during the execution of these enterprise software applications. Although there are third party tools available in the market which are utilized for analysing the enterprise software application log files. These tools are mainly conducting analysis based on the collected log files from the enterprise software applications. Such an analysis approach is reasonable if the hardware resources of the system are under healthy stage while the enterprise software application is still running in a good condition. However, it is unavoidable that whenever the hardware memory is running low, or the central processing unit (CPU) resource is saturated, or even hard disk space is running low. All these causes would lead to the software application crashes or unstable stage. Indeed, in the event that such circumstance happens, the enterprise software applications will never able to log the exact error event into their log files. Hence, if the root cause analysis activity is solely depending on software application error log files, it is insufficient to identify the actual error accurately. PAL contains algorithm that focus heavily on root cause analysis to identify the real root cause correctly. Therefore, it is reactive and it does not perform any data prediction. The aim is to resolve the valid root cause and prevent the same issue to be reoccurred in the future (i.e., continuous improvement). Any learning gains from root cause analysis activities is served as "knowledge" or "intelligent" to improve the error detection and resolution provision of PAL. With that, PAL can fulfill the objectives to reduce the root cause analysis duration, and increase the accuracy of root cause identification. In conclusion, this is a good opportunity and positive potential to recommend PAL as the brand-new logic model for the root cause analysis of the enterprise software application.

References

da Silva Neves, A.J., Camanho, R.: The use of AHP for IT project priorization – a case study for oil & gas company. Procedia Comput. Sci. **55**, 1097–1105 (2015). ScienceDirect. Elsevier. https://www.sciencedirect.com/science/article/pii/S1877050915015513. Accessed 20 Apr 2020

Gogus, A.: Learning cycles. In: Seel, N.M. (ed.) Encyclopedia of the Sciences of Learning. Springer, Boston (2012). https://doi.org/10.1007/978-1-4419-1428-6_143

Guru99: Unsupervised Machine Learning: What is, Algorithms, Example (2020). www.guru99.com. https://www.guru99.com/unsupervised-machine-learning.html. Accessed 24 Feb 2020

Hoo Meng, W., Amalathas, S.S.: A new approach towards developing a prescriptive analytical logic model for software application error analysis. In: Silhavy, R., Silhavy, P., Prokopova, Z. (eds.) CoMeSySo 2019 2019. AISC, vol. 1046, pp. 256–274. Springer, Cham (2019). ISBN:978-3-030-30328-0. https://doi.org/10.1007/978-3-030-30329-7_24

Wong, H.M., Amalathas, S.S.: An approach towards developing an algorithm for software application error analysis. David Publishing Company (2019). https://doi.org/10.17265/2328-2185/2019.04.006

Wong, H.M., Amalathas, S.S., Zitkova, T.: A business operation stability by improving the speed of restoration on software application service. David Publishing Company (2020). https://doi.org/10.17265/2328-2185/2020.01.008

Wong, H.M., Amalathas, S.S., Zitkova, T.: An approach towards designing a prescriptive analytical logic model for software application root cause analysis. Int. J. Adv. Res. Publ. (IJARP) (2019). http://www.ijarp.org/published-research-papers/nov2019/An-Approach-Towards-Designing-A-Prescriptive-Analytical-Logic-Model-For-Software-Application-Root-Cause-Analysis.pdf

Valdman, J.: Log File Analysis (2001). https://www.kiv.zcu.cz/site/documents/verejne/vyzkum/publikace/technicke-zpravy/2001/tr-2001-04.pdf. Accessed 18 Jan 2017

Pan, J.: Software Testing. Carnegie Mellon University (1999). http://www.sci.brooklyn.cuny.edu/~sklar/teaching/s08/cis20.2/papers/software-testing.pdf

Murínová, J.: Application Log Analysis (2015). http://is.muni.cz/th/374567/fi_m/thesis_murinova.pdf. Accessed 18 Jan 2017

Brownlee, J.: Linear Regression for Machine Learning. Machine Learning Mastery Pty. Ltd (2019). https://machinelearningmastery.com/linear-regression-for-machine-learning/. Accessed 24 Feb 2020

Klass, L.: Machine learning - definition and application examples. MM International (2018). https://www.spotlightmetal.com/machine-learning--definition-and-application-examples-a-746226/?cmp=go-aw-art-trf-SLM_DSA-20180820&gclid=Cj0KCQjwxYLoBRCxARIsAEf16-vBblTDXGp3-EuK0AsxQu90ckOqGbTyXEvjQBWSA3D3MHB51TxDhqUaAlPDEALw_wcB. Accessed 21 Feb 2020

Cmielowski, L.: Adoption of machine learning to software failure prediction. IBM Watson Data (2017). https://medium.com/ibm-data-science-experience/adoption-of-machine-learning-to-software-failure-prediction-e8d85ed0338f. Accessed 2 Apr 2019

Management Logic: Root-Cause Analysis (2012). http://www.management-logic.com/toolbox/sales/Root-Cause%20Analysis/Index.html. Accessed 28 Nov 2015

Wettergreen, M.: Pairwise Comparison Charts 2: Setting Up and Running Them [Video]. YouTube, 7 August 2014. https://www.youtube.com/watch?v=orrQFHKlocs&list=PLGVNLF9na6k7P2c-NQ_8CBEBbLw4pjzG9&index=7&t=0s. Accessed 28 Aug 2020

Vaidya, O.S., Kumar, S.: Analytic hierarchy process: an overview of applications. Department of Mechanical Engineering, Army Institute of Technology, Pune, National Institute of Industrial Engineering (NITIE), Vihar Lake, Mumbai (2004). http://ac.els-cdn.com/S0377221704003054/1-s2.0-S0377221704003054-main.pdf?_tid=39850adc-5d78-11e7-b5bf-00000aab0f26&acdnat=1498815840_e0c9a10c99c46ad30db8da4ef17e817b. Accessed 12 Sept 2018

Roman, V.: Unsupervised machine learning: clustering analysis. Towards Data Science. towardsdatascience.com (2019). https://towardsdatascience.com/unsupervised-machine-learning-clustering-analysis-d40f2b34ae7e. Accessed 9 Mar 2020

Saaty, R.W.: The analytic hierarchy process - what it is and how it is used. Pergamon Journals Ltd, Great Britain (1987). http://ac.els-cdn.com/0270025587904738/1-s2.0-0270025587904738-main.pdf?_tid=55956250-6b90-11e7-b75e-00000aab0f02&acdnat=1500365511_5a54a3ecf035ad3e45ab72b909b7632e. Accessed 12 Sept 2018

Saaty, T.L.: Decision making with the analytic hierarchy process. Int. J. Serv. Sci. **1**(1) (2008). http://citeseerx.ist.psu.edu/viewdoc/download?doi=10.1.1.409.3124&rep=rep1&type=pdf. Accessed 12 Sept 2018

Peng, W.W., Wallace, D.R.: Software Error Analysis. NIST Special Publication (1993). http://www.geocities.ws/itopsmat/SoftwareErrorAnalysis.pdf. Accessed 12 Dec 2015

Wilson, A.: A brief introduction to supervised learning. Towards Data Science (2019). https://towardsdatascience.com/a-brief-introduction-to-supervised-learning-54a3e3932590. Accessed 21 Feb 2020

Wong, H.M., Amalathas, S.S.: The error analysis for enterprise software application using analytic hierarchy process and supervised learning: a hybrid approach on root cause analysis. In: Silhavy, R., Silhavy, P., Prokopova, Z. (eds.) CoMeSySo 2020. AISC, vol. 1295, pp. 633–643. Springer, Cham (2020). https://doi.org/10.1007/978-3-030-63319-6_59

Wong, H.M., Amalathas, S.: A new analysis approach incorporated with analytic hierarchy process for the software application in a multiple tiers environment. Eur. J. Electr. Eng. Comput. Sci. **3**(6) (2019). https://doi.org/10.24018/ejece.2019.3.6.160

Wong, H.M., Amalathas, S.: The root cause analysis algorithm design incorporated with analytic hierarchy process for software application error. Eur. J. Electr. Eng. Comput. Sci. **4**(1) (2020). https://doi.org/10.24018/ejece.2020.4.1.166

Wong, H.M., Amalathas, S., Zitkova, T.: A prescriptive logic model for software application root cause analysis. Eur. J. Electr. Eng. Comput. Sci. **3**(5) (2019). https://doi.org/10.24018/ejece.2019.3.5.133

Teaching and Learning with Technology: The Factors that Affect ICT Integration in Gauteng Public Schools

Freddy Chireka(✉) and Tranos Zuva

Vaal University of Technology, Vaal, South Africa
{218283865,tranosz}@vut.ac.za

Abstract. The purpose of the study was to determine the selected factors that affect ICT in Gauteng public schools. This study used quantitative approach in which inferential statistical techniques were utilized to answer the research objective. A survey instrument was used to address the research. This study allowed for the determination of associated factors and a five-factor questionnaire was used to evaluate the data obtained from the teachers in Gauteng public schools. The results of the multiple regression equation showed that expertise, access, and support were the significant variables in explaining the determinants of ICT Integration among the public schools. The statistical results also showed that there was no positive correlation between teachers' attitudes and teachers' socio-economic profiles. The study therefore recommends that the South African Government through the Department of Education is encouraged to increase the level of professional development, access, and readiness of ICT tools for all and continued support for teachers using ICTs for teaching and learning.

Keywords: Information · Communication and technology · ICT integration · Gauteng provincial department of education and public schools

1 Introduction

Information Communication and Technology (ICT) has been the buzz word of the 21st century, playing a key role in all areas of modern life in education, health care, sports, commerce, and other activities. The way the human race thinks, works and lives has been transformed by the integration of technology through various innovations Hanson [1], Agar [2]. Information and Communication Technology (ICT) has advanced tremendously in the past few decades Agar [2]. No sphere of life has not been affected by ICT. The advancement in ICT has inevitably produced advancements and unparalleled improvements in other subsets of life.

The appropriate use of technology to aid or support the process of teaching and learning is referred to as integration of ICT. Students are mostly familiar with technology and as such they will learn much better within a technology-enabled environment. This is so because pedagogical aspects are addressed through the use of ICT for effective teaching and learning Finger et al. [3]. Education leaders worldwide have invested billions in

R. Silhavy et al. (Eds.): CoMeSySo 2021, LNNS 231, pp. 246–257, 2021.
https://doi.org/10.1007/978-3-030-90321-3_20

educational technologies to improve teaching and learning Bebell et al. [4]. Educational technology has been deployed in the Gauteng Department of Education and yet there is limited research focused on measuring the levels of effectiveness of the ICT integration into teaching and learning in the Gauteng schools and the perceptions of the teachers.

The Organization of Economic Co-operation and Development [5] (OECD) published its report in 2015 entitled, "Students, Computers and Learning" that found out that resources invested in ICT for education are not linked to improved student achievement in reading, mathematics, or science. Technology is viewed as a distraction to the valuable human engagement needed to build deep, conceptual understanding and high order thinking skills. The report argues that we have not yet become good enough at the kind of pedagogies that make most of the technology, and teachers are diluting the effectiveness of teaching by applying 20th century teaching practices to 21st century technologies [5] (OECD). The report further argues that technology can amplify great teaching, but great technology cannot replace poor teaching. The promise of technology will not be achieved, however, if the usage of ICT remains limited.

ICT Integration assists educators and learners in learning more about their different subjects. Technology inspired teaching and learning offer diverse ways of learning such as animated ideas, instant feedback on assessments, game-based learning, simulations, cloud storage of data, manipulation of electronic databases and other ancillary uses that will make the teaching and learning process more engaging and beneficial Finger et al. [3]. Previous researchers noted that the integration of ICT in teaching and learning significantly improves the students' active learning capabilities Ghavifekr and Rosdy [6].

Koehler and Mishra [7] identified the complex challenges of technology integration into the teaching and learning process. At the heart of good teaching with technology are three core components: content, pedagogy, and technology, plus the relationships among and between them. The interactions between and among the three components, playing out differently across diverse contexts, account for the wide variations seen in the extent and quality of educational technology integration. These three knowledge bases (content, pedagogy, and technology) form the core of the technology, pedagogy, and content knowledge framework. An overview of the framework is provided in the following section, though more detailed descriptions may be found elsewhere Koehler and Mishra [7]. This perspective is consistent with that of other researchers and approaches that have attempted to extend Shulman [8] idea of pedagogical content knowledge to include educational technology.

Teachers need more pedagogically professional development programmes to enable them to understand the pedagogical implications of using ICT in order to utilize the resources properly Hennessy and Patterson [9]. Professional development programmes play a critical role in the integration of ICT. Educators who successfully attend such courses are evidently more competent than those who do not undergo professional development Winzenried et al. [10]. It was proven that teachers either avoid or deliberately substitute the usage of ICT if they did not receive adequate training. An institution in Ireland found out that low confidence among teachers was a reason for avoiding technology use. Some educators in Canada also admitted low confidence as a main reason for not adopting technology. In some instances it was found that students actually know better than their teachers when it comes to technology Hennessy and Patterson [9].

The South African schooling community is still very uneven with mostly urban learners being more advantaged that their counterparts in poorer rural and township areas Mlitwa and Koranteng [11]. The Government of South Africa has adopted the e-Education policy of 2004 in order to equip schools with ICT resources that can assists the schools achieve the promise of quality education in all schools Eom et al. [12]. The purpose of the e-Education Policy of 2004 is to enable the deployment of technology both hardware and software as well the training of teachers to be able to utilize those resources. The Gauteng Province is one such province that has taken a giant leap in bringing ICTs into the township schools through its paperless classrooms launched in 2014 Mlitwa and Koranteng [11].

There has been an increased interest from the National Education Department to integrate technology into the teaching and learning spaces. The Gauteng Department of Education has launched its Classroom of the Future with technology being the basis towards a paperless classroom. This study seeks to find out the factors that affect integration of technology and teachers' perceptions in selected schools in the Gauteng Province of South Africa.

Despite some researchers who have looked at this aspect there is limited research done in determining the factors that affect ICT integration into the education process and teachers' perceptions in the Gauteng Province.

The question which remains to be addressed is which factors effect ICT integration in public schools in Gauteng Province in South Africa are. This study motivated by the intention to assess the factors that affect ICT integration in the teaching and learning processes in public schools in South Africa.

This paper is arranged as follows: Sect. 1 Introduction, Sect. 2 Background, Sect. 3 Literature Review, Sect. 4 Methodology, Sect. 5 Results, Sect. 6 Conclusions and Recommendations.

1.1 Background

The Gauteng Department of Education has heavily invested in ICTs in education through its Classroom for the Future initiative in a total of 375 public schools. In the South African context, the 2004 White Paper on e-Education: Transforming Learning and Teaching through Information and Communication Technologies (ICTs) states:

> "Enriching the learning environment through the use of ICTs is a continuum; it is a process that takes learners and teachers through learning about ICTs (exploring what can be done with ICTs), learning with ICTs (using ICTs to supplement normal processes or resources), and learning through the use of ICTs (using ICTs to support new ways of teaching and learning)".

Despite this, there is little empirical evidence as to the actual factors that influence ICT integration of these technologies into teaching and learning. Furthermore, there is little empirical evidence on the teachers' perceptions towards the ICTs for teaching and learning in the Gauteng Province of South Africa. This study may help the Gauteng

Education Department policymakers to have insight into the factors that affect technology integration and the relationship between the factors. This will help the policymakers in making decisions on the investment of resources that positively impact greater technology integration by the teachers. The study aims to gain an insight into the selected factors that affect ICT integration in Gauteng public schools.

2 Literature Review

Integration of ICT in education refers to the use of computer-based communication that is incorporated into daily classroom instructional processes. In line with preparing students for the 21st century world, teachers are the key players in using ICT in their daily classrooms. This is due to the capability of ICT for providing a dynamic and proactive teaching-learning environment Hatlevik and Arnseth [13]. The integration of ICT improves the quality of the delivery of teaching and learning to learners while also connecting them to the broader global community Albirini [14]. The adoption of ICT is an ongoing and continuous process that seeks to support teaching and learning comprehensively Mindess et al. [15].

The appropriate use of technology to aid or support the process of teaching and learning is referred to as integration of ICT. Students are mostly familiar with technology and as such they will learn much better within a technology-enabled environment. This is so because pedagogical aspects are addressed through the use of ICT for effective teaching and learning Finger et al. [3]. Education leaders worldwide have invested billions in educational technologies to improve teaching and learning Bebell et al. [4].

2.1 Empirical Literature Review

Integration of ICT involves the appropriate use of technology to aid teaching and learning. This was identified as the first main stage of integration, followed by enhancement which involves the modification of the use of the ICT. This was identified as the second stage. The complementary stage involves the use of ICT to complement existing content and improve learning outcomes. This was identified as the third stage Hermans et al. [16].

Integration of ICT in education refers to the use of computer-based communication that is incorporated into daily classroom instructional processes. In conjunction with preparing students for the 21st century world, teachers are the key players in using ICT in their daily classrooms. This is due to the capability of ICT for providing a dynamic and proactive teaching-learning environment Hatlevik and Arnseth [13]. The integration of ICT improves the quality of the delivery of teaching and learning to learners while also connecting them to the broader global community Albirini [14]. The adoption of ICT is an ongoing and continuous process that seeks to support teaching and learning comprehensively Mindess et al. [15].

In Queensland (Australia) an instrument to quantitatively measure the level of Information and Communication Technologies (ICTs) in the teaching and learning process was developed in 2003. This was because of the lack of a methodology that effectively measured student outcomes because of ICT integration Jamieson-Proctor et al. [17]. Since its development in 2003, the instrument has been extensively tested and evaluated for Education Queensland. This comprehensive evaluation of the instrument comprised three major components: A statistical analysis of the data obtained when the instrument was completed by 929 Queensland state schoolteachers; a peer review of the instrument involving a 15-member expert Peer Review Team; and interviews 170 Computers in the Schools with 42 teachers from six purposely selected schools across Queensland who had recently used the instrument. The revised instrument is currently used by Education Queensland as part of its ICT census that all schools must complete annually. System-wide data are collected with the instrument on the systemic key ICT driver: learning, teaching, and the curriculum Jamieson-Proctor et al. [17].

There have been similar studies done in this area. In another study the researchers Ghavifekr and Rosdy [6] used the theory of diffusion Rogers [22], the Technology Acceptance Model (TAM), Venkatesh et al. [18] and the Conceptual Framework of Study Venkatesh et al. [18]. The study focused on looking at the component of teachers' attitudes on the integration of ICT for teaching and learning. The results indicate that ICT integration has great effectiveness for both teachers and the students. The results further indicate that teachers' well-equipped preparation with ICT tools and facilities is one of the main factors in the success of technology-based teaching and learning. It was also found that professional development training programmes for teachers also played a key role in enhancing students' quality learning Ghavifekr and Rosdy [6]. The study above utilized the outdated TAM model with a sample of 101 educators randomly drawn from 10 schools in Malaysia.

Another South African study was conducted entitled "Integration of ICT into Curricula in Western Cape Schools: The Activity Theory Perspective" by Mlitwa and Koranteng [11]. Here researchers looked at the dynamics of ICT integration into the teaching and learning process and the structure of the curriculum in marginalized areas of Western Cape. The activity theory Hardman [19] was utilized to investigate how electronic learning policy could be rolled out together with the deployment of both hardware and software. The overall goal was to ensure that ICT integration is at the centre of the curriculum in South Africa. The study found that most schools were not yet prepared for effective ICT integration into the curriculum.

In the case of Jordanian schools, teachers had low levels of ICT usage for teaching and learning. It was also found that a significant positive correlation exists between teachers' level of ICT Integration and their attitudes Al-Zaidiyeen et al. [20]. Their paper suggests that integration of ICT into teaching must be given greater consideration as teachers already possess positive attitudes towards the promise of technology. A similar study was done in Queensland State schools which sought to answer the question of whether ICT integration programmes were contributing significantly to effective teaching and learning Jamieson-Proctor [17]. The study found out that a significant population of females was less confident than their male colleagues and overall findings indicate and agree with the OECD report that says that the promise of technology is yet to be realized. The current programmes are not yet making great progress in the education space. This study shows a pattern that teacher attitudes are directly linked to the usage of ICT resources.

3 Methodology

This study used quantitative approach in which inferential statistical techniques were utilized to answer the research objectives and predict the factors that affect the level of ICT integration. This study allowed for the determination of associated factors and a five-factor questionnaire was used to evaluate the data obtained from the teachers in Gauteng public schools. To determine the relationship between the level of ICT integration, canonical correlation was used. Multiple regression was utilized to interrogate the degree to which the variables were predictive to the level of ICT integration in the teaching and learning process.

Teachers who were practising in both primary and secondary public schools were considered for the study. These participants had to be voluntarily willing to complete the survey instrument. Teachers who teach in private schools in Gauteng Province were excluded from the study together with teachers who teach in public schools but who were unwilling to participate in the study. Teachers who teach in other provinces were also not included in the study.

The sample size N was calculated using the Dobson formula in (1) Ghavifekr and Rosdy [6].

$$N = \frac{Z^2 P(1 - P)}{(D)^2} \tag{1}$$

where P = 0.91 the estimated proportion of teachers with good levels of ICT integration a design effect of 1, Z = 1.96 for an alpha level of 0.025 in each tail.

The sample size was based on a 95% level of confidence and the allowable margin of error was set at 0.005.

The proposed sample size using the Dobson formula was calculated to be 128. Participants were drawn from 50 schools using purposeful sampling. The schools were purposefully sampled from public schools from the Gauteng Department of Education using the criteria of those who benefitted from the e-learning rollouts. Participants were then randomly selected from these schools.

A total of 166 questionnaires were distributed. Out of the 166 questionnaires, 133 usable questionnaires returned to the researcher. The sample was comprised of 90 males and 43 females.

3.1 A Subsection Sample

The Questionnaire was adopted from Isleem [21], and it consisted of six sections. The sections are clearly outlined below.

Section A: Integration of ICT Integration for teaching and learning: The section includes items that relate to the level of ICT integration in teaching, for example preparation of lessons, delivery of lessons, pre- and post-evaluation, and all related administrative reporting.

Section B: Level expertise in ICT Integration: The section includes items that relate to the level of ICT expertise in teaching, for example preparation of lessons, delivery of lessons, pre- and post-evaluation, and all related administrative reporting.

Sections C: Access to computers: The section includes items that relate to the level of access to ICT for teaching, for example preparation of lessons, delivery of lessons, pre- and post-evaluation, and all related administrative reporting.

Section D: Attitudes towards ICT: This construct relates to the attitudes and beliefs about ICT for teaching and learning.

Section E: Level of support: This construct relates to the support for ICT Integration for teaching and learning.

Section F: Selected characteristics of the teacher (teacher profile): This construct relates to the selected characteristics of the teacher.

The specified objectives and the study purpose were clearly outlined on the cover page as well as the role of the participants. Also, the questionnaire gave room for anonymous and confidential inquiry from either the researcher or the supervisors. Likert scale format was used for the questionnaire. For this study, the following model was proposed:

Table 1. Questionnaire layout

Section	Number of questions	Construct	Purpose
A	18	Integration of ICT Integration for instructional purposes	The section includes items that relate to the level of ICT integration in teaching for example preparation of lessons, delivery of lessons, pre- and post-evaluation, and all related administrative reporting
B	18	Expertise in ICT Integration	Level of expertise in ICT Integration: The section includes items that relate to the level of ICT expertise in teaching, for example preparation of lessons, delivery of lessons, pre- and post-evaluation, and all related administrative reporting
C	16	Access to computers	Access to computers: The section includes items that relate to the level of access to ICT for teaching, for example preparation of lessons, delivery of lessons, pre- and post-evaluation, and all related administrative reporting

(*continued*)

Table 1. (*continued*)

Section	Number of questions	Construct	Purpose
D	20	Attitudes towards ICT	Attitudes towards ICT: This construct relates to the attitudes and beliefs about ICT for teaching and learning
E	20	Support	Level of support: This construct relates to the support for ICT Integration for teaching and learning
F	9	Selected characteristics of the teacher (teacher profile)	This construct helps to provide the selected characteristics of the teacher

Source: Author's analysis

The Diffusion of Innovations theory was identified as the theoretical framework Rogers [22]. Diffusion can be defined as the process through which innovation is transcended through various channels over time among members of a social system Rogers [22]. In attempting to measure the factors that affect ICT integration in instructional use by teachers in public schools in Gauteng Province, the Diffusion of Innovations theory provided the most relevant tools (Fig. 1).

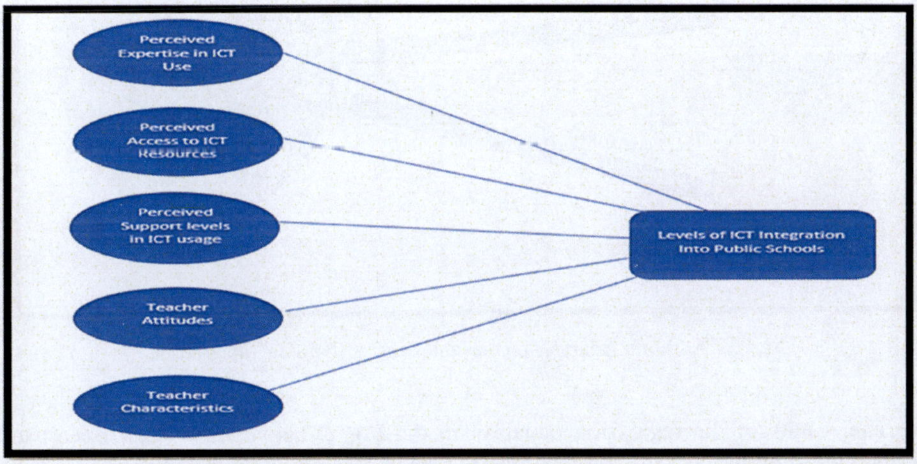

Fig. 1. Proposed model for the factors affecting ICT integration in teaching. Source: Isleem [21]

4 Results

To measure the validity of the study, a pilot study was utilized. Such a study was undertaken like the actual study but just focusing on a smaller sample size. The Cronbach alpha coefficient value was found to be above 0.7 and were deemed to be acceptable Ursachie et al. [23] .

4.1 Multiple Regression Analysis Results

Multiple regression was run to predict levels of influence of the variables in the model. The significant variables are summarised in the model below: expertise, access, and support.

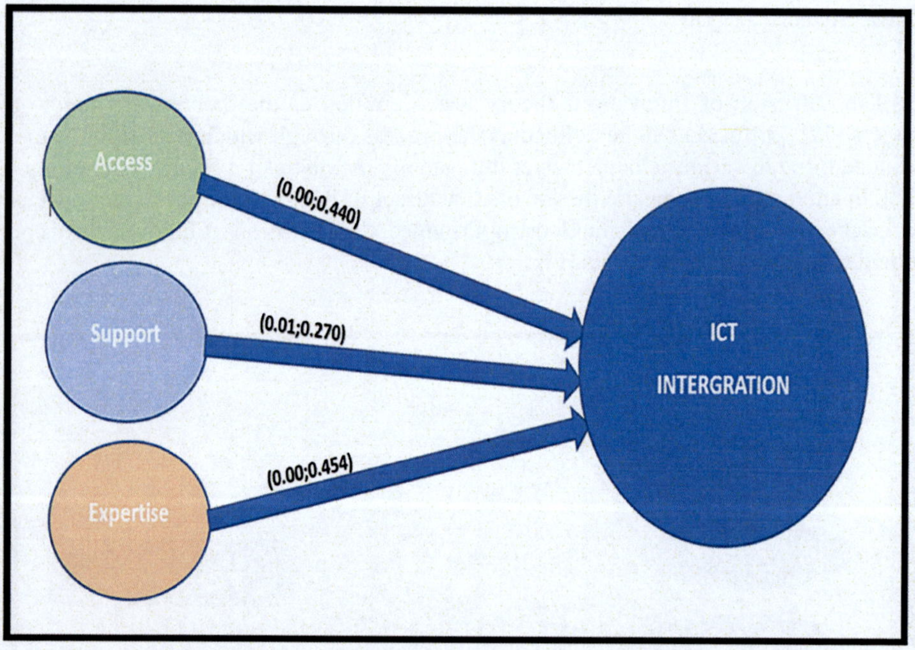

Fig. 2. Author's Analysis on variables that affect ICT Integration.

The results of the regression equation in the Fig. 2 above shows that expertise, access, and support were the significant variables in explaining the determinants of ICT integration among the public schools. The variables were all significant at the 1% level of significance. The results further indicate that expertise was the most significant at 1% level of significance (P-value 0.000) with a positive influence on ICT Integration. In this way a unit change in the expertise of the teachers is associated with a 0.454 change in the level of ICT Integration. These results were supported by various research studies in literature Aslan and Zhu [24] (Gichovi 2013).

5 Conclusion, Recommendation and Future Work

The rapid worldwide technological introductions and development of ICTs (Information, Communication and Technology) has impacted the process of teaching and learning globally. In education, Integration of Information, Communication, and Technology (ICT) refers to the use of technology in the day-to-day classroom teaching and learning processes. Teachers have emerged as the key player in the integration of ICT. This study sought to investigate the factors that influence the ICT integration in Gauteng Province of South Africa among teachers in public schools. The results of the multiple regression equation showed that expertise, access, and support were the only significant variables in explaining the determinants of ICT Integration among the public schools. The statistical results also showed that there was no positive correlation between teachers' attitudes (−0.073;0.493) and their socio-economic profiles (−0.425;0.841).

Therefore, the study recommends that firstly, the government of South Africa should intensify support for ICT for instructional purposes. ICT support refers to diverse services that ensure that ICT resources are always in good working order. The Government should ensure that the necessary mechanisms and infrastructure are available to the instructors and the learners to increase the speed at which ICT integration progresses in the province and the country in general. The reality of the fourth industrial revolution should act as a drive to the government to ensure that support is intensified so that the instructors and the learners have access to ICT tools.

The other important aspect that should be considered by the government of South Africa is the issue of professional development by making that the instructors are equipped with the necessary ICT expertise. ICT expertise refers to a high-level understanding and applying a range of computer programmes, software, and other applications. It has been shown that expertise is an important determinant of ICT integration in Gauteng Province. The availing of the necessary infrastructure in the Department of Education will ensure that the expertise of the instructors will advance the integration of ICT in South Africa. For instance, the government must ensure that all the schools in both rural areas and those in urban areas have the important tools that will enable the instructors and the learners to have access to those materials which will improve their expertise.

Access to ICT tools has also been found to be important to ensure that ICT integration can improve in the province. Access to ICT refers to the availability of ICT tools for use and application. The Department of Education, therefore, must ensure that schools have access to ICT tools. When schools have access to the necessary ICT infrastructure, this will ensure that instructors and learners will have access to ICT tools which will improve the integration of ICT in Gauteng Province and South Africa in general.

5.1 Future Work

Future studies should look at measuring the levels of ICT integration in the schools used in this study.

References

1. Hanson, J.: 24/7: How cell phones and the internet change the way we live, work, and play. Greenwood Publishing Group (2007)
2. Agar, J.: Constant Touch: A Global History of the Mobile Phone, Cambridge. Icon Books. City and publishers for all book items? City: Icon Books? But no city required if you are using APA, 7th edn. (2003a)
3. Finger, G., et al.: Teaching teachers for the future (TTF) project TPACK survey: summary of the key findings. Aust. Educ. Comput. **37**, 13–25 (2013)
4. Bebell, D., O'Dwyer, L.M., Russell, M., Hoffmann, T.: Concerns, considerations, and new ideas for data collection and research in educational technology studies. J. Res. Technol. Educ. **43**, 29–52 (2010)
5. OECD Students, Computers and Learning. OECD Publishing
6. Ghavifekr, S., Rosdy, W.A.W.: Teaching and learning with technology: effectiveness of ICT integration in schools. Int. J. Res. Educ. Sci. **1,** 175–191 (2015). Make sure that you use capital letters consistently for your book and journal article titles. Check and correct throughout. Compare this item with the EOM item above. Has the same method been used?
7. Koehler, M., Mishra, P.: What is technological pedagogical content knowledge (TPACK)? Contemp. Issues Technol. Teacher Educ. **9**, 60–70 (2009)
8. Shulman, L., Sparks, D.: Merging content knowledge and pedagogy: an interview with lee shulman. J. Staff Dev. **13**(1), 14–16 (1992)
9. Hennessy, J.L., Patterson, D.A.: Computer architecture: A quantitative approach. Elsevier (2011)
10. Winzenried, A., Dalgarno, B., Tinkler, J.: The interactive whiteboard: a transitional technology supporting diverse teaching practices. Australas. J. Educ. Technol. **26**, 534–552 (2010)
11. Mlitwa, N.W., Koranteng, K.: Integration of ICT onto curricula in the western cape schools: the activity theory perspective. J. Commun. Inf. **9** (2013)
12. Eom, S.B., Wen, H.J., Ashill, N.: The determinants of students' perceived learning outcomes and satisfaction in university online education: an empirical investigation. Decis. Sci. J. Innov. Educ. **4**, 215–235 (2006)
13. Hatlevik, O.E., Arnseth, H.C.: ICT, teaching and leadership: how do teachers experience the importance of ICT-supportive school leaders? Nordic J. Digital Literacy **7,** 55–69 (2012). Capital letters again. Compare this item with the one above.
14. ALBIRINI, A.: Teachers' attitudes toward information and communication technologies: the case of Syrian EFL teachers. Comput. Educ. **47**, 373–398 (2006)
15. Mindess, S., Young, J.F., Darwin, D.: Concrete: Prentice Hall. Is Concrete the city? (2003)
16. Hermans, R., Tondeur, J., van Braak, J., Valcke, M.: The impact of primary school teachers' educational beliefs on the classroom use of computers. Comput. Educ. **51**, 1499 (2008)
17. Jamieson-Proctor, R., Watson, G., Finger, G., Grimbeek, P., Burnett, P.C.: Measuring the use of information and communication technologies (ICTs) in the classroom. Comput. Schools **24**, 167–184 (2007)
18. Venkatesh, V., Morris, M.G., Davis, G.B., Davis, F.D.: User acceptance of information technology: toward a unified view. MIS Q. **27,** 425–478 (2003)
19. Hardman, J.: Activity theory as a framework for understanding teachers' perceptions of computer usage at a primary school level in South Africa. South Afr. J. Educ. **25**, 258–265 (2005)
20. Al-Zaidiyeen, N.J., Mei, L.L., Fook, F.S.: Teachers' attitudes and levels of technology use in classrooms: the case of Jordan schools. Int. Educ. Stud. **3**, 211 (2010)
21. Isleem, M.I.: Relationships of selected factors and the level of computer use for instructional purposes by technology education teachers in Ohio public schools: a statewide survey. The Ohio State University, Ohio (2003)

22. Rogers, E.M.: Diffusion of innovations. Simon and Schuster. Schumpeter, E. B. 1961. English overseas trade statistics, pp. 1697–1808. Clarendon Press (2010)
23. Ursachi, G., Horodnic, I.A., Zait, A.: How reliable are measurement scales? external factors with indirect influence on reliability estimators. Procedia Econ. Financ. **20**, 679–686 (2015)
24. Aslan, A., Zhu, C.: Influencing factors and integration of ICT into teaching practices of pre-service and starting teachers. Int. J. Res. Educ. Sci. **2,** 359–370 (2016). Here you do have capital letters for the journal titles so be consistent throughout.

Analysis of Machine Learning Classification Techniques for Anomaly Detection with NSL-KDD Data Set

Ana Cholakoska(✉), Martina Shushlevska, Zdravko Todorov, and Danijela Efnusheva

Computer Technologies and Engineering Department, Faculty of Electrical Engineering and Information Technologies, Ss. Cyril and Methodius University in Skopje, Skopje, Macedonia
{acholak,danijela}@feit.ukim.edu.mk

Abstract. Along with the high-speed growth of the Internet, cyber-attacks are becoming even more frequent, so detecting network intrusions is essential for keeping network connections under control. However, in the modern big data environment, traditional methods do not meet the network requirements regarding adaptability and efficiency. Therefore, intelligent intrusion detection systems can only be built if an adequate data set is used. This way, a data set with a sizable amount of quality data can mimic real-time network traffic to train and test an intrusion detection system (IDS). The NSL-KDD data set is a refined version of its predecessor KDD"99 data set. In this paper, the NSL-KDD data set is analyzed and used to study the effectiveness of various classification algorithms in detecting anomalies in network traffic patterns. The results show that the Random Forest algorithm provides the best results with accuracy, precision, recall and an F1 score of 99%.

Keywords: Machine learning · NSL-KDD data set · Anomaly detection · Network security · Intrusion detection and prevention system

1 Introduction

With the fast-paced growth of computer networks and the number of connected devices, and the increasing number of running applications, many systems are being targeted by malicious attacks. It is shown that computer systems suffer from security vulnerabilities that are technically difficult and costly for manufacturers to address. Therefore, the role of intrusion detection systems [1], as special-purpose devices for detecting anomalies and attacks in the network, is becoming increasingly important in the implementation of security in computer and communication systems.

Research in intrusion detection methods has been generally based on two possible methods: detecting an anomaly or a deviation from normal traffic and recognizing patterns of misbehavior by detecting only intrusions that contain known patterns of attack. The first approach, anomaly detection, is usually used in academic research as a more robust method because of its theoretical potential for addressing unauthorized attacks.

R. Silhavy et al. (Eds.): CoMeSySo 2021, LNNS 231, pp. 258–267, 2021.
https://doi.org/10.1007/978-3-030-90321-3_21

The latter approach is favored in commercial products due to its predictability and high accuracy [2].

Conducting a thorough analysis of recent anomaly detection research reveals the application of machine learning methods, which have a very high detection rate of 98% and a false prediction rate below 2% [3]. Machine learning techniques are used and applied in many studies [4–9], where they have been shown to provide solid results in detecting intrusions. The applicability of machine learning for intrusion detection systems is due to the technologies on which it is based (identification, extraction, classification, regression and prediction) and the accessibility to well-known data sets consisting of real-time network traffic with a large number of features.

Machine learning (ML) uses past behavior to identify patterns and build patterns that help predict future behavior and future events without explicit programming [10]. The learning process begins with observations (or data) and direct experience (or instruction) in order to find patterns in the data and make better decisions in the future, based on the given examples [11].

IDS is a powerful technology used to detect both known and unknown malicious attacks. However, among the state-of-the-art solutions for intrusion detection systems and commercial tools, few products use anomaly detection approaches, and practitioners still think it is not a mature technology. In order to explain the usage of ML algorithms for solving a problem with unauthorized attacks in network traffic, this paper provides an analysis of several machine learning algorithms over the NSL-KDD data set.

Intrusion detection is broadly considered to be a classification problem. The main issue in standard classification problems lies in minimizing error probability while making the classification decision. Hence, the critical point is how to choose an effective classification approach to build accurate intrusion detection systems in terms of high detection rate while keeping a low false alarm rate. In complex classification domains of intrusion detection, dataset attributes may contain false correlations, which hamper the process of detecting intrusions. Some attributes of a dataset may be redundant because the information they add is contained in other attributes. Some extra attributes can increase computation time, having an impact on the detection accuracy. In IDS, effective attributes selection improves the classification accuracy by searching for the subset of attributes. Also, some data in the dataset may not be helpful for intrusion detection and thus can be eliminated before training. Therefore, for the preprocessing and feature selection of the NSL-KDD data set RFE (Recursive Feature Elimination) method has been used [12].

The rest of the paper is organized as follows: Sect. 2 presents some related work based on intrusion detection. Section 3 gives a brief description of the NSL-KDD dataset and provides details about the machine learning model built with reduced feature selection. Section 4 provides analysis of the dataset with various classification techniques, including Naïve Bayes (NB), Decision Tree (DT), Support Vector Machine (SVM) and Random Forest (RF). Section 5 deals with a conclusion and future work.

2 State of the Art

Intrusion detection systems can be software or hardware based, with a purpose of monitoring network packets or systems to detect malicious activity and take certain measures.

In general, if some malicious activity is detected in the network, the source IP address is blocked for a certain period of time and all packets coming from that address in the future are rejected [1]. There are several types of intrusion detection systems, and these are discussed below.

A Network Intrusion Detection System (NIDS) detects malicious activity by monitoring and analyzing network traffic. This type of IDS is usually activated when packets enter a particular network and its job is to monitor and accept/reject packets from Internet and pass them on to the local network. An example of such a network system is Snort [13].

Host-based intrusion detection system (HIDS) detects malicious activity by monitoring and analyzing system calls, application logs, access control lists, etc. Host-based intrusion detection systems help organizations to monitor processes and applications running on devices such as servers and workstations. Examples of such systems are: Tripwire, OSSEC [14].

Wireless Intrusion Detection System (WIDS) monitors wireless networks to detect malicious behavior (analysis of the number of packets sent during a single time window, too many de-authentication packets, too many broadcast requests, and etc.). Most often such systems operate on access points - AP (Access Point) and do not allow certain users to connect to them if any harmful activity is detected. Examples of such systems are Kismet and NetStumbler [15].

Network behavior analysis (NBA) passively monitors network traffic in order to detect unknown and unusual patterns that could cause damage [16]. It is recommended to be used together with a firewall, as well as with other types of IDS systems.

An Intrusion Prevention System (IPS) upgrades an intrusion detection system. The IDS is used to detect and log an attack, and the IPS is used to detect, block and log the attack. Actually, IPS systems can prevent certain attacks while they occur. There are several types of IPS systems, including: NIPS, HIPS, WIPS, NDA [17]. In doing so, it is necessary to distinguish that the firewall is used to leak or block traffic, while IPS systems analyze whether the traffic that passes through the firewall is harmful. It follows that IDS/IPS is not a replacement for a firewall or vice versa, but that these two tools must be used together in order to implement a higher level of protection.

Given that the previously discussed systems work with a large amount of data, lately the application of machine learning in IDS anomaly detection systems is becoming more relevant. There are several such studies that have yielded good results [4–9]. For example, the authors of [4] present implementation of Fuzzy ARTMAP neural network as an intrusion prevention system.

The research given in [5] shows an update to the Snort IDS in order to target the problem that Snort faces when it cannot determine the outcome of a particular event. For this purpose, a special data mining algorithm (frequent item sets mining) is applied, which is based on MapReduce for data processing. Similarly, the authors of [6] propose an intrusion detection model that uses MapReduce to process large structured and unstructured data placed in key/value pairs. The way MapReduce generates key pairs/values for attack detection relies on using a combination of Fuzzy CMeans (FCM) and Support Vector Machine for classification.

Another research [7], proposes a hybrid attack detection system based on Support Vector Machine and C5.0. Using such a combination of algorithms improves the accuracy of attack detection, compared to being used separately. A hybrid system has also been proposed in [8], where two algorithms (K-means and NB) are used to group and classify data. Moreover, the authors of [9] present a comparative study of various deep learning approaches for intrusion detection and also discuss the application of different data sets for that purpose. Specifically, in this paper the NSL-KDD data set is used in order to develop machine learning model based on classification algorithms, applied as anomaly detection IDS. More details are given in the next chapter.

3 Methodology for Building ML Models for Anomaly Detection

3.1 Overview of NSL-KDD Data Set

The NSL-KDD data set solves some of the inherent problems of the KDD'99 data set, which is considered as standard benchmark for intrusion detection evaluation. The training dataset of NSL-KDD is similar to KDD' 99, as a predictive model capable of distinguishing between "bad" connections, called intrusions or attacks, and "good" normal connections [18]. Each connection is labeled as either normal or attack type. It contains 43 features per record, with 41 of the features referring to the traffic input itself and the last two are labels (whether it is a normal or attack) and Score (the severity of the traffic input itself). In total there are 37 number of types of attacks, and these attacks are grouped in 4 different classes of attacks: Denial of Service (DoS), Probe, User to Root (U2R), and Remote to Local (R2L).

Although these attacks exist in the data set, the distribution is heavily skewed. A breakdown of the record distribution can be seen in Table 1, where three different data sets are observed: 20% subset of the NSL-KDD Train data set (KDDTrain + 20%), full NSL-KDD train data set (KDDTrain +), and full NSL-KDD test data set (KDDTest +). Essentially, more than half of the records that exist in each data set are normal traffic, and the distribution of U2R and R2L attacks is extremely low. Although this is low, this is an accurate representation of the distribution of modern-day internet traffic attacks, where the most common type of attack is DoS, while U2R and R2L are hardly ever seen.

Table 1. Number of records and percent of attacks in NSL-KDD dataset.

Dataset	Number of records	Normal	DOS	Probe	U2R	R2L
KDDTrain + 20%	25192	53%	37%	9.16%	0.04%	0.8%
KDDTrain +	125973	53%	37%	9.11%	0.04%	0.85%
KDDTest +	22544	43%	33%	11%	0.9%	12.1%

The features in a traffic record of the NSL-KDD data set provide the information about the encounter with the traffic input by the IDS and can be broken down into four categories: Intrinsic, Content, Host-based, and Time-based. Below is a description of the different feature categories:

- Intrinsic features can be derived from the header of the packet without looking into the payload itself, and hold the basic information about the packet. This category contains features labeled 1–9, named as: duration, protocol_type, service, flag, src_bytes, dst_bytes, land, wrong_fragment and urgent, accordingly.
- Content features hold information about the original packets, as they are sent in multiple pieces rather than one. With this information, the system can access the payload. This category contains features labeled 10–22, named as: hot, num_failed_logins, logged_in, num_compromised, root_shell, su_attempted, num_root, num_file_creations, num_shells, num_access_files, num_outbound_cmds, is_host_login and is_guest_login, accordingly.
- Time-based features hold the analysis of the input traffic over a two-second time window. These features are mostly counts and rates rather than information about the content of the input traffic. This category contains features labeled 23–31, named as: count, srv_count_serror_rate, srv_serror_rate, rerror_rate, srv_rerror_rate, same_srv_rate, diff_srv_rate, srv_diff_host_rate, accordingly.
- Host-based features are similar to time-based features, except instead of analyzing over a 2-s window, the analysis is made over a series of connections (ex. How many requests made to the same host over x-number of connections). These features are designed to extract attacks, which span longer than a two-second window time interval. This category contains features labeled 32–41, named as: dst_host_count, dst_host_srv_count, dst_host_same_srv_rate, dst_host_diff_srv_rate, dst_host_same_src_port_rate, dst_host_srv_diff_host_rate, dst_host_serror_rate, dst_host_srv_serror_rate, dst_host_rerror_rate, dst_host_srv_rerror_rate, accordingly.

Feature selection, a process of selecting a subset of original features according to certain criteria, is an important and frequently used dimensionality reduction technique for data mining. It reduces the number of features, removes irrelevant, redundant, or noisy data, and brings the immediate effects for applications: speeding up data mining algorithm, and improving mining performance such as predictive accuracy and result comprehensibility. Therefore, in the next subsection details are provided about building a ML model based on the NSL-KDD data set with reduced set of features, according to the RFE feature selection method [12].

3.2 Building ML Models from NSL-KDD Data Set

To process the NSL-KDD data set, Python is used in combination with multiple libraries, including: Numpy, Seaborn, Pandas, Sklearn [19]. These libraries are purposed to simplify the evaluation of the data set, and also to support the building of the ML models from the NSL-KDD data set.

Firstly, the data is read and then preprocessed. One-Hot-Encoding (one-of-K) is used to transform all categorical features into binary features. Then, the encoded categorical data frame is joined with the non-categorical data frame and the dataset is split into 4 datasets for every attack category. This is shown in Table 2.

Table 2. Joined encoded categorical data frame with non-categorical data frame

Dataset	Dimensions of DOS	Dimensions of probe	Dimensions of R2L	Dimensions of U2R
Train data set	(113270, 123)	(78999, 123)	(68338,123)	(67395, 123)
Test data set	(17171, 123)	(12132, 123)	(12596, 123)	(9778, 123)

Feature Scaling and Feature Selection. Feature scaling is a method used to normalize the range of independent variables or features of data in a given data set, during data pre-processing. In machine learning and statistics, feature selection also known as variable selection, attribute selection or variable subset selection, is the process of elimination redundant and irrelevant data by selecting a subset of relevant features that fully represents the given problem.

Recursive Feature Elimination, or shortly RFE, is a widely used algorithm for selecting features that are most relevant in predicting the target variable in a predictive model - either regression or classification [12]. RFE applies a backward selection process to find the optimal combination of features. First, it builds a model based on all features and calculates the importance of each feature in the model. Then, it ranks the features and removes the one(s) with the least importance iteratively based on model evaluation metrics (e.g., RMSE, accuracy, and Kappa). Feature importance can be computed based on the model (e.g., the random forest importance criterion) or using a model-independent metric (e.g., ROC curve analysis). This process continues until a smaller subset of features is retained in the model (technically, RFE can still keep all (or most) of the features in the final model). Table 3 shows the features that are selected, for each attack group, by applying the RFE method.

Table 3. Selected features with RFE method

Attack Type group	selected features
DOS	['logged_in', 'count', 'serror_rate', 'srv_serror_rate', 'same_srv_rate', 'dst_host_count', 'dst_host_srv_count', 'dst_host_same_srv_rate', 'dst_host_serror_rate', 'dst_host_srv_serror_rate', 'service_http', 'flag_S0', 'flag_SF']
Probe	['logged_in', 'rerror_rate', 'srv_rerror_rate', 'dst_host_srv_count', 'dst_host_diff_srv_rate', 'dst_host_same_src_port_rate', 'dst_host_srv_diff_host_rate', 'dst_host_rerror_rate', 'dst_host_srv_rerror_rate', 'Protocol_type_icmp', 'service_eco_i', 'service_private', 'flag_SF']
R2L	['src_bytes', 'dst_bytes', 'hot', 'num_failed_logins', 'is_guest_login', 'dst_host_srv_count', 'dst_host_same_src_port_rate', 'dst_host_srv_diff_host_rate', 'service_ftp', 'service_ftp_data', 'service_http', 'service_imap4', 'flag_RSTO']

(*continued*)

Table 3. (*continued*)

Attack Type group	selected features
U2R	['urgent', 'hot', 'root_shell', 'num_file_creations', 'num_shells', 'srv_diff_host_rate', 'dst_host_count', 'dst_host_srv_count', 'dst_host_same_src_port_rate', 'dst_host_srv_diff_host_rate', 'service_ftp_data', 'service_http', 'service_telnet']

Application of ML Classification Algorithms for Anomaly Detection. In this step, four different algorithms have been used (Naïve Bayes, Decision Tree Classifier, SVM Classifier and Random Forest Classifier), for creating four different models. All classifiers are trained for a data set with complete set of features (Basic NSL-KDD) and for a data set with reduced features (Reduced NSL-KDD), for later comparison. Next, test data is used to make predictions of the model. Therefore, Multiple metrics are considered such as: accuracy score, recall, f-measure and precision, confusion matrix. A 10-fold cross-validation is also performed.

4 Analysis of Results for Various ML Classification Algorithms

The performance of IDS is evaluated based on its capability of classifying network traffic into a correct type. In order to avoid the effect of data sampling when assessing the IDS, therefore, we conducted experiments by using repeated k-fold (kf) cross-validation method, and the value of k is considered as 10. Different attribute selection algorithms are applied on the dataset to select the attributes.

In order to evaluate the performance of the proposed IDS, we make a comparison between two different models: one with all features called Complete NSL-KDD and one with RFE-based feature selection, or with reduced features called Reduced NSLKDD. The results can be seen on Table 4 and 5.

As shown in Table 4, reducing the NSL-KDD features improves the accuracy of the Naive Bayes algorithm for DOS and U2R attacks, while for the Probe and R2L attacks,

Table 4. Accuracy and precision results for complete and reduced NSL-KDD data sets for various ML algorithms

Classification algorithm	Class name	Accuracy for complete NSL-KDD	Accuracy for reduced NSL-KDD	Precision for complete NSL-KDD	Precision for reduced NSL-KDD
Naïve Bayes	DOS	0.86733	0.87362	0.98822	0.97240
	Probe	0.97898	0.94238	0.97323	0.94306
	U2R	0.97259	0.98967	0.60157	0.68017

(*continued*)

Table 4. (*continued*)

Classification algorithm	Class name	Accuracy for complete NSL-KDD	Accuracy for reduced NSL-KDD	Precision for complete NSL-KDD	Precision for reduced NSL-KDD
	R2L	0.93562	0.81423	0.89097	0.74817
SVM	DOS	0.99359	0.93122	0.99001	0.98912
	Probe	0.98591	0.95153	0.97568	0.93788
	U2R	0.99550	0.99540	0.99550	0.93108
	R2L	0.96745	0.87703	0.94853	0.91280
Decision tree	DOS	0.99639	0.99738	0.99505	0.99692
	Probe	0.99571	0.99085	0.99392	0.98674
	U2R	0.99652	0.99652	0.86295	0.87538
	R2L	0.97920	0.97459	0.97151	0.96689
Random forest	DOS	0.99819	0.99750	0.99866	0.99799
	Probe	0.99670	0.99374	0.99577	0.99128
	U2R	0.99795	0.99744	0.96125	0.96678
	R2L	0.98182	0.97634	0.97544	0.96865

Table 5. Recall and f-measure results for complete and reduced NSL-KDD data sets for various ML algorithms

Classification algorithm	Class name	Recall for complete NLD-KDD	Recall for reduced NSL-KDD	F-measure for complete NSL-KDD	F-measure for reduced NSL-KDD
Naïve Bayes	DOS	0.70308	0.72989	0.82145	0.83372
	Probe	0.96051	0.87269	0.96654	0.90225
	U2R	0.97911	0.74888	0.66091	0.70212
	R2L	0.95508	0.66767	0.91620	0.69025
SVM	DOS	0.99531	0.85107	0.99265	0.91485
	Probe	0.98050	0.90709	0.97805	0.92134
	U2R	0.70218	0.67257	0.76329	0.73801
	R2L	0.96110	0.73874	0.95456	0.78416
Decision tree	DOS	0.99665	0.99705	0.99585	0.99698
	Probe	0.99267	0.98467	0.99392	0.98566
	U2R	0.90958	0.89540	0.88210	0.87731
	R2L	0.96958	0.96086	0.97051	0.96379
Random forest	DOS	0.99718	0.99625	0.99792	0.99712
	Probe	0.99391	0.98911	0.99483	0.99017
	U2R	0.88784	0.84622	0.91780	0.88683
	R2L	0.97298	0.96419	0.97419	0.96635

still the use of complete NSL-KDD data set with all features gives better accuracy. Meanwhile, the SVM algorithm gives better accuracy and precision for the complete NSL-KDD data set (reducing the features has no impact on the results). The same goes for recall and f-score, as shown in Table 5, so it can be said that the reduction of features does not have a good effect on SVM algorithm. In the last metric (f-measure) it can be seen that reducing the features for Naive Bayes gives better results for DoS and U2R attacks. Using the Decision Tree algorithm, it can be seen that reducing the features gives better accuracy on DoS attacks, but on the Probe and R2L attacks better results are achieved for the complete NSL-KDD data set with all features. For the precision, better results are achieved in the DoS and U2R attacks. For the recall, only the DoS attacks has a better recall for the complete NSL-KDD data set. As it can be seen in Table 4, when the Random Forest algorithm is used, reducing the features contributes to higher precision only on the U2R attacks. The Random Forest algorithm also shows the highest accuracy compared to the rest of the algorithms for the reduced and the complete NSL-KDD data set. For the recall and f-measure, Table 5 shows that in all four types of attacks it gives better results for the complete NSL-KDD data set.

5 Conclusion

In this paper an intrusion detection system based on NSL-KDD data set has been implemented. The dataset was trained and tested for four class attack categories: R2L, DoS, U2R and Probe. The training and testing was done using the whole dataset and a reduced one – RFE methodology. The results were evaluated and both showed that the Random Forest algorithm provides the best results with a score of 99%. The Decision tree algorithm had similar results as RF, as well as SVM. Naïve Bayes was the least effective algorithm for intrusion detection purposes. For future research, the NSL-KDD data set can be further expanded with new types of attacks and other machine algorithms may be proposed and analyzed to further improve the intrusion detection.

References

1. Yeo, L.H., Che, X., Lakkaraju, S.: Understanding modern intrusion detection systems: a survey. Eastern Michigan University, USA (2017)
2. Jyothsna, V.V.R.P.V., Prasad, V.R., Prasad, K.M.: A review of anomaly based intrusion detection systems. Int. J. Comput. Appl. **28**(7), 26–35 (2011)
3. Tavallaee, M., Bagheri, E., Lu, W., and Ghorbani, A.A.: A detailed analysis of the KDD CUP 99 data set. In: Second IEEE International Conference on Computational Intelligence for Security and Defense Applications, pp. 53–58 (2009)
4. Vilela, D.W.F.L., Lotufo, A., Santos, C.R.: Fuzzy ARTM AP neural network IDS evaluation applied for real IEEE 802.11w data base. In: IEEE International Joint Conference on Neural Networks, pp. 1–7 (2018)
5. Tong, Z., Ying, H.: Application of frequent item set mining algorithm in IDS based on hadoop framework. In: IEEE Chinese Control and Decision Conference, pp. 1908–1911 (2018)
6. Hajare, S.A.: Detection of network attacks using big data analysis. Int. J. Recent Innov. Trends Comput. Commun. **4**(5), 86–88 (2016)

7. Golman, V.: An efficient hybrid intrusion detection system based on C5.0 and SVM. Int. J. Database Theor. Appl. **7**(2), 59–70 (2014)
8. Tanpure, S.S. et al.: Intrusion detection system in data mining using hybrid approach. Int. J. Comput. Appl., 0975–8887 (2016)
9. Ferrag, M.A., Maglaras, L., Moschoyiannis S., Janicke, H.: Deep learning for cyber security intrusion detection: approaches, datasets, and comparative study. J. Inf. Secur. Appl. **50**(1) (2020)
10. James, D.: Introduction to Machine Learning with Python: a Guide for Beginners in Data Science, 1st edn. CreateSpace Independent Publishing Platform, USA (2018)
11. Joshi, A.V.: Machine Learning and Artificial Intelligence. Springer (2020)
12. Chen, X., Jeong, J.C.: Enhanced recursive feature elimination. In: Machine Learning and Applications Conference, pp 429–435 (2007)
13. Caswell, B., Beale, J., Baker, A.: Snort Intrusion Detection and Prevention Toolkit. Syngress. MA, Burlingthon (2007)
14. Ambati, S.B., Vidyarthi, D.: A brief study and comparison of open source intrusion detection system and tools. Int. J. Adv. Comput. Eng. Netw. **1**(10), 26–32 (2013)
15. Hutchison, K.: Wireless Intrusion Detection Systems, SANS Institute. White Paper (2005)
16. Stallings, W.: Network Security Essentials: Applications and Standards, 6th edn. Pearson, USA (2017)
17. Stiawan, D., Shakhatreh, A.I., Idris, M.Y., Bakar, K.K.A., Abdullah, A.H.: Intrusion prevention system: a survey. J. Theor. Appl. Inf. Technol. **40**(1) (2012)
18. Protić, D.D.: Review of KDD CUP ‘99, NSL-KDD and KYOTO 2006+ datasets. Military Tech. Courier **66**(3), 580–596 (2018)
19. Madhavan, S.: Mastering Python for Data Science. Packt Publishing, UK (2015)

Research and Selection of Rational Methods for Obtaining Framework of Schedules for the Parallel Programs Execution

Valery Bakanov(✉)

MIREA (MIREA Russian Technological University), HSE University (National Research University Higher School of Economics), Moscow, Russia
http://vbakanov.ru/left_1.htm

Abstract. This paper deals with the creation of rational methods for developing a schedule (execution plan) of programs for parallel computing systems of a given (homo- and heterogeneous) architecture for arbitrary algorithms. To solve this task, a software system is proposed that allows the implementation of various (heuristic) methods for building up an execution schedule for parallel programs and to quantitatively evaluate and compare the computational complexity of these methods. The quantitative results gained from the application of some methods for constructing rational parallel execution plans for classical data processing algorithms are presented.

Keywords: Analysis of the program information structure · Graph representations of the algorithm · Purposeful reorganization of the stacked-parallel form of the data dependence graph · Rational parameters of parallel programs execution · Parallel program execution framework · Procedure for constructing a rational parallel program execution plan

1 Introduction

With Currently, the data processing parallelization is used mainly to reduce computational time. Data are simultaneously processed in parts on many different computing devices with the subsequent combination of results. Parallel execution allows evading the fundamental law formulated by Lord Rayleigh in 1871. According to this law (as applied to the CPU heat dissipation), their heat dissipation power is proportional to the fourth power of CPU clock speed (speed doubling increases the heat dissipation by 16 times) and, in fact, to replace it with a linear one in the number of parallel computers (while maintaining the CPU clock speed). Nothing is given for free. The task of identifying (usually hidden from the uninitiated observer) the parallelism potential in algorithms is plain to see, and even more so the efficiency of its (parallelism) use.

To obtain a plan (schedule) for the parallel program execution (hereinafter PPE) becomes a problem when organizing parallel computations. The fact is that this sequential program can be represented in parallel form in many ways (while maintaining the

R. Silhavy et al. (Eds.): CoMeSySo 2021, LNNS 231, pp. 268–280, 2021.
https://doi.org/10.1007/978-3-030-90321-3_22

invariant in the form of a set of operations), and each way will have a different execution efficiency on a computer of a given architecture.

The starting point for work is algorithms. The interest in their intrinsic properties has recently been growing again [1].

2 Methods

A specialized software system of the instrumental level was created to solve the issue of determining the methods for calculating the rational PPE plans. The generalized chart and data flow diagram of this system are shown in the following figure.

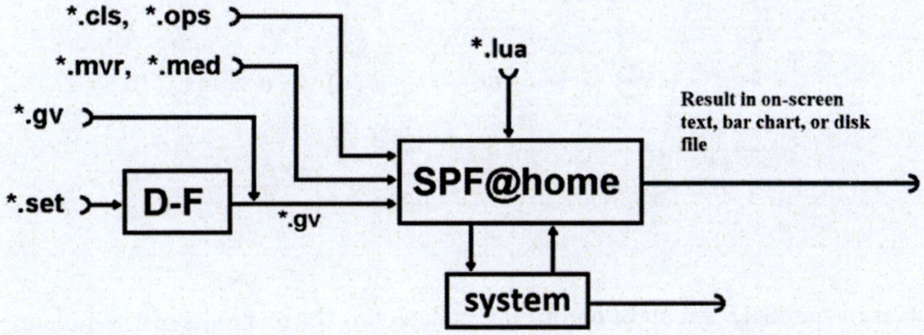

Fig. 1. Instrument complex diagram (*.set and *.gv are the program file and the analyzed program information graph file respectively, *.mvr, *.med are the files of metrics for the algorithm graph vertices and arcs respectively, *.cls, *.ops are the files of computer and program operator parameters respectively,*.lua is a text file in the Lua language that contains the methods of SPF reorganization).

The complex input (diagram in Fig. 1) receives a description of the analyzed algorithm in the form of a traditional program or a formal description in the form of an oriented acyclic information graph of the algorithm (hereinafter IGA) – an "operators - operands" dependence. The graph vertices are associated with program operators (groups of operators), and arcs are associated with data transmission lines.

The actor model imitation (Data-Flow module) and construction of IGA special sections in the form of its stacked parallel form (hereinafter SPF, SPF@home module) are implemented to reveal and analyse the internal logical parallelism in algorithms, [2]. Both of the mentioned modules are developed using the C/C++ in the GUI style for the Win'32 model (the command line mode is additionally implemented for mass calculations). Both modules are completely OpenSource and can be downloaded for free use from the following addresses: http://vbakanov.ru/dataflow/content//install_df.exe and http://vbakanov.ru/spf@home/content/install_spf.exe (installation file format). The D-F module builds up the schedules according to the asynchronism model both at the beginning and at the end of the operator execution. SPF@home style schedules correspond more to synchronism at the beginning of execution (Fig. 2).

The SPF@home background is the IGA file itself (always present, because it is built by the translator when analyzing the program source code) and additional files

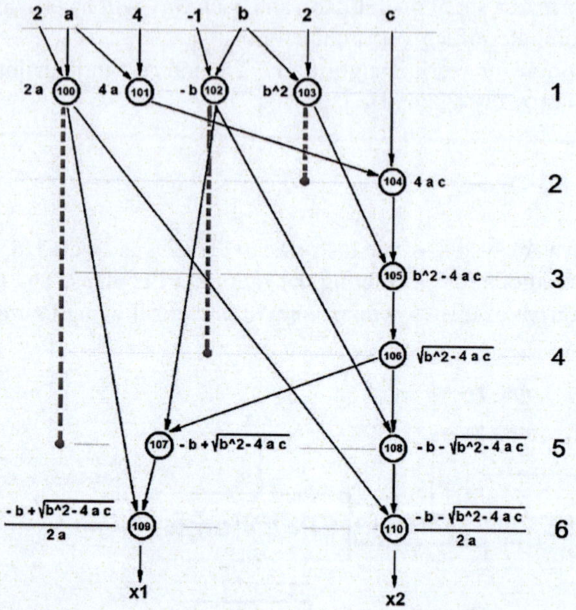

Fig. 2. A stacked-parallel form of the algorithm for detecting the real roots of a complete quadratic equation in the upper version (the tier numbers are shown on the right, the dotted line shows the valid positions of operators along the SPF tiers).

that determine the measures of graph arcs and vertices (for taking into account, for example, the temporal characteristics of transfers and operator execution) and parameters of individual computers of a parallel (including heterogeneous) computational field. The SPF@home module also allows to determine the local data lifetime between SPF tiers, which is necessary for determining/optimizing the parameters of temporary data storage devices (high-speed shared memory, for example, general-purpose registers).

Due to *NP*-completeness of the scheduling problem [7], it can be said that the only rational methods for its solution, which involves the use of heuristic methods for SPF transformation is developed. The actual transform methods are described using the Lua scripting language [8]. Lua calls serve as a wrapper for API functions of SPF@home; calculation of various graph parameters and careful modeling process registration are provided. Herewith an iterative approach to gradually improve the quality of heuristic methods while moving towards the stated goal is assumed.

The system is mainly aimed at analyzing programs developed using the high-level programming languages without explicitly specifying parallelism and in systems with the ILP (*Instruction-Level Parallelism*) concept [5]. Although the SPF@home module capabilities make it possible to use indivisible blocks of a commands sequence of any size. The concerned system builds up exactly the *framework* of programs for parallel execution. A specific parallel programming technology, methods to call for subroutines, interrupts processing, cache usage, etc. are solved for each specific case separately.

Taking into consideration the long-known fact about the movement of an active (where at a given period of time there is a vigorous machine instructions execution) operators domain along the front of the program being executed, it is advisable to process in blocks (within individual files, subroutines) sequentially from the beginning of the program to its end. In this case, the output vertices of the previous block are input for the next one.

The internal data implementation, of course, is not obliged at all to provide for the explicit building of the SPF in a 2D array. It can be any convenient for computer implementation one. For example, in the naive case, it establishes a one-to-one correspondence between the IGA in the form of a set of directed arcs $\{k, l\}$ (adjacency matrix) and twos of numbers of the vertices i_k, j_k and i_l, j_l, where i, j are the numbers of rows and columns in the SPF.

3 Results

The results of research carried out using the SPF@home module as one characterized by more flexibility to achieve the set goals are given below. Experiments on the computation intensity management using the D-F module were published earlier (see, for example, paper [4]).

Three typical subtasks were defined reasoning from the fact that the SPF is already in fact (possibly requiring improvement) a plan for parallel program execution (as blocks of statements are executed sequentially in tiers from top to bottom along the SPF). In real cases, it is often required to execute several of these subtasks simultaneously. These three subtasks are:

- Balancing the number of operators over all the tiers of a given SPF without increasing its height. Balancing means obtaining the maximum evenness of operator distribution over the tiers while maximizing the use of hardware capabilities and minimizing the number of parallel computers required to solve the problem.
- Obtaining a program execution schedule on a given number of parallel computers with a possible increase in the SPF height (program execution time).
- Obtaining a program execution schedule on a heterogeneous field of parallel computers.

For the SPF@home module, the actual sequence of obtaining a rational schedule for executing a parallel task will be as follows (each and many other actions are performed by corresponding Lua calls):

- Obtaining the original SPF (which have no restrictions on the width of tiers).
- This SPF modification in the needed direction by purposeful rearrangement of operators from tier to tier while preserving the original connections in the algorithm information graph.

1. Programs execution schedule on the minimum number of parallel computers while maintaining the SPF height

Such an SPF organization allows obtaining a schedule with the maximum code density. Computers of the VLIW architecture (*Very Long Instruction Word*, machine instruction with a very long word) often lack such density, [9]. It should be understood that this task has more of a methodological value, because the number of physical parallel computers is usually much less than the width of SPF for real tasks.

The 1-01_bulldozer empirical method is aimed at the evenest operators distribution over the SPF tiers without increasing in its height (saving the program execution time). The operators are moved down only (initially, the SPF is built in the upper version). To do this, the method tries to move the operators from the tiers of higher than average width to the tiers with the smallest width.

The 1-02_bulldozer method is a modification of the previous one with adaptation. We will call the range of possible operators' placement over the SPF tiers without changing the information connections in the algorithm graph as the variability maximum. For an operator with a variability maximum, the upper and lower limits of its possible placement along the tiers are calculated within the tier.

In many cases, it is possible to significantly (up to 1.5–2 times) reduce the width of the SPF but almost never to the minimum value (the arithmetic mean of the widths of the tiers). On the whole, the 1-02_bulldozer method is somewhat more efficient but loses in terms of computational complexity. In most cases, an increase in the processed data size increases the efficiency of balancing. Obviously, this is connected to an increase in the number of SPF degrees of freedom.

2. Programs execution schedules on a fixed number of parallel computers with the possibility of increasing the SPF height

Methods with an increase in the SPF height are of practical interest. Two methods were used with the metaphorical names - Dichotomy and WidthByWidth. At the same time, we have to accept the increase in program execution time (previously described features typical to the VLIW architecture are also present here). The finite width of the transformed SPF was set during the computational experiments. The coefficient of variation characterizing the non-evenness of the distribution of tier widths (the value that inversely characterizes the code density) was calculated as $CV = \sigma/\overline{W}$, where σ is the mean-square deviation of the number of operators for all SPF tiers, $\overline{W}$ is the arithmetic mean number of operators in the tiers.

Dichotomy method assumes the transfer of half of the operators to a newly developed tier below the current one to unload the unnecessarily wide tiers. WidthByWidth method implements a gradual transfer of operators to the newly developed SPF tiers. In most cases, WidthByWidth method leads to better results than Dichotomy. For example, the transformed SPF height is significantly smaller and this corresponds to a reduction in the parallel program execution time with a smaller number of operator transfers).

3. Program execution schedule on a fixed number of heterogeneous parallel computers

Modern multi-core processors are increasingly being developed with computing cores of various capabilities. Therefore, it is practically useful to be able to build a parallel program execution schedule for such systems (with a heterogeneous field of parallel computers).

SPF@home module supports this feature by comparing the information from two files - for operators and computers (*.ops and *.cls, respectively). It is possible to set a match for a variety of freely assignable attributes in relation to any range of operators/computers. The condition for this operator to be satisfied on a given calculator is $minVal\_1 \leq Val\_1 \leq maxVal\_1$ for the same parameter. *Val_1, minVal_1, maxVal_1* are the numerical values of this parameter for the operator and the computer, respectively.

Development of a schedule for a program executing on a heterogeneous field of parallel computers is a more complicated procedure than the ones described above, and the emphasis here is on programming in Lua (API functions of the SPF@home system provide the minimally necessary support). As the operators that require different computers can be on the same SPF tier, the concept of splitting the SPF tiers into families of sub-tiers can be useful. Each family corresponds to a block of calculators with certain capabilities. Since all given operators of the tier have the GKV-property, the sequence of their execution within the tier/sub-tier in the first approximation is arbitrary. The diagram below shows the splitting of operators on one of the SPF tiers in the case of six parallel computers of three types (Fig. 3).

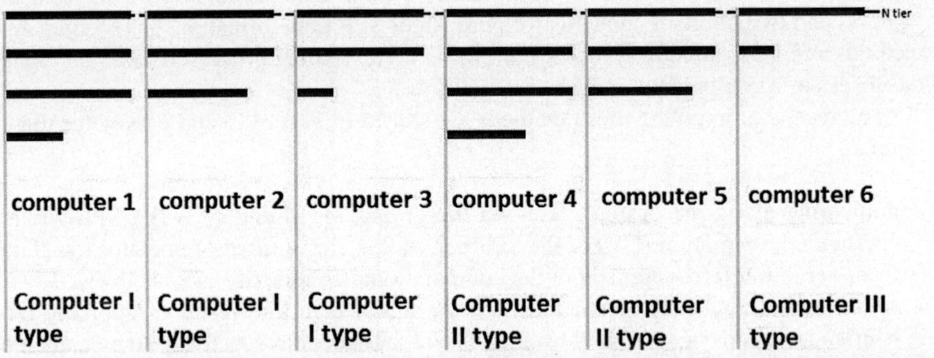

Fig. 3. The scheme for SPF tiers splitting into families of sub-tiers when solving the problem of determining the schedule for heterogeneous field of parallel computers.

In this case, the total problem solving time T is determined by the sum over all the tiers of maximum operator execution time values on the sub-tiers of the given tier:

$$T = \sum_j \left(\max_{k_j} \sum_i t_{ik} \right),$$

where j is the number of tiers, i is the number of sub-tiers on a given tier, k_j are computer types on j-th tier, t_{ik} is the execution time of i type operator on a k type computer.

If the maximum performance is a task, it is quite possible to determine the number of computers of a particular type that minimizes T. The task to minimize the total T solution time becomes more complicated if it is possible to execute each operator on several computers due to the t_{ik} ambiguity in the expression given above. The additional balancing by sub-tiers is required here.

As the developed techniques are supposed to be used as part of parallelizing translators (possibly virtual machines) generating programs for execution in the EPIC style

(*Explicitly Parallel Instruction Computing*, a set of instructions with explicit parallelism) [9] so the issue of performance is important. More precisely, the balance between the performance of the resulting program and the complexity of work on developing a plan for its implementation. As it was said, the computational complexity of SPF transformation will be estimated in units of operator transferring from level to level, and the quality – in the statistical parameters of width evenness for the SPF tiers. When obtaining the results presented below, the ability to build a schedule with a minimum amount of local data was not used to minimize the amount of temporary data storage during computations.

Maintaining the original SFP height is equal to executing the algorithm (program) in the shortest possible time. The evenest load on all the parallel system computers will be with the same width of all SPF tiers (arithmetic mean). In the SPF of real algorithms and the span of processed data, the arithmetic mean is quite large (almost always this number is much larger than the number of individual computers in a parallel system). So the concerned case is rare in practice but still needs to be analyzed.

We selected algorithms that are commonly used in data processing (mainly, linear algebra) and two heuristic methods for purposeful SPF transformation, 1-01_bulldozer methods and 1-02_bulldozer, for the comparison. The span of processed data (put along the abscissa) was a parameter in the calculations.

The results of applying these methods are shown in Figs. 4, 5 and 6. Key for these figures:

- numbering along the ordinate axes on the graphs a), b) and c) is the SPF width, coefficient of variation (CV) of the SPF tier widths, the number of operator transfers from tier to tier (characteristic of the computational complexity), respectively;
- solid, dotted, and dash-dotted lines are the initial data, the result of applying the heuristic methods of 1-01_bulldozer and 1-02_bulldozer, respectively (they coincide in Fig. 6a).

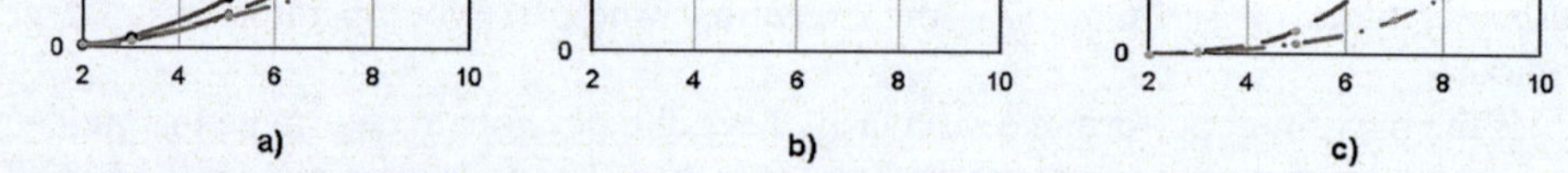

Fig. 4. Parameters of the parallel execution plan while maintaining the SPF height for the algorithm of classical multiplication of square matrices of the 2, 3, 5, 7, 10th orders (corresponds to the numbering along the abscissa).

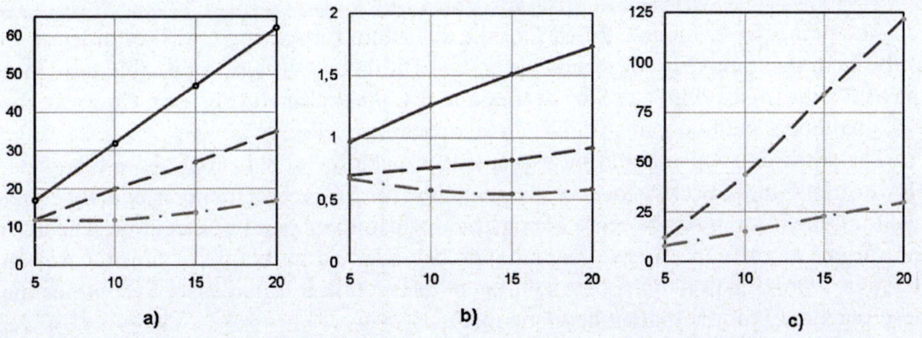

Fig. 5. Parameters of the parallel execution plan while maintaining the height of SPF for the algorithm for computing the pair correlation coefficient by 5, 10, 15, 20 points (corresponds to the numbering along the abscissa axes).

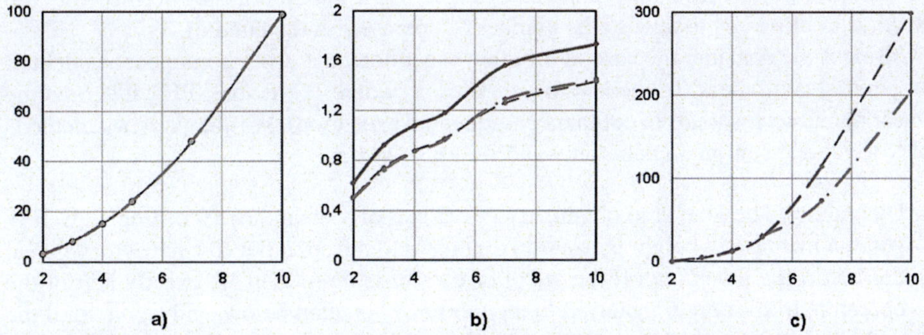

Fig. 6. Parameters of the parallel execution plan while maintaining the height of SPF for the algorithm for solving the linear algebraic equations system (LAES) for the 2, 3, 4, 5, 7, 10th order (corresponds to the numbering along the abscissa axes) by the direct (non-iterative) Gauss algorithm

Figures 4, 5 and 6 show that in many cases it is possible to get closer to the specified goal. For example, Fig. 4a) illustrates a decrease in the SPF width up to 1.7 times (1-01_bulldozer method) and up to 3 times (1-02_bulldozer method) when multiplying matrices of the 10th order.

Coefficient of variation of the SPF tier widths (Fig. 4b) approximates to 0.3 (the data set homogeneity border) when using the 1-02_bulldozer empiric and, which is important, is stable over the entire range of data dimensions.

The complexity of achieving the result (Fig. 4c) with 1-02_bulldozer is significantly lower (up to 3.7 times with the 10 order of matrices) than with 1-01_bulldozer. It is important that the method efficiency increases with the growth of the processed data dimension.

The 1-02_bulldozer method proved to be no less effective on the algorithm for calculating the pair correlation coefficient (Fig. 5).

An attempt to reorganize the SPF of the algorithm for solving LAES of order up to 10 by both methods (Fig. 6) turned out to be of little use. It was impossible to reduce the SPF width at all (Fig. 4a). The decrease in CV is very small (Fig. 6b). However, the 1-02_bulldozer method gains slightly in labor intensity (Fig. 6c).

The research scope should be expanded to make the conclusions more complete. However, it is already clear that not every method for SPF reorganizing is equally effective for transforming (in the sense of constructing a rational parallel execution schedule) the various algorithms. As the general approach adopted in this paper presupposes an iterative approximation to the best solution of the set task it is necessary to continue the development of more efficient heuristic methods.

The formulation of the task in the form of programs' execution on a given number of parallel computers is as close as possible to the real case of scheduling a VLIW machine with a given number of processor instructions in a machine word (bundle size) of a very long instruction word). At this, the task is posed of obtaining a schedule with the minimum execution time (minimum SPF height) at the maximum code density (the SPF width is as close as possible to the number of operators in the bundle).

Below, we consider the case of program execution on a given homogeneous field of W parallel computers (from $W = W_0$ to $W = 1$, where W_0 is the SPF width, and the lower bound corresponds to completely sequential execution). We compare two methods of SPF reorganization – Dichotomy and WidthByWidtn:

- Dichotomy. The goal is to obtain an SPF variant of a width not exceeding a given W with an increase in height by transferring operators from a tier to a newly developed tier below the given one. If the width of the tier is more than W, exactly half of the operators from it are transferred to the tier newly created below, and so on, until the width is no higher than the specified W. The method works very quickly, but roughly. The SPF height is clearly excessive and the unevenness of the tiers widths is high).
- WidthByWidtn. Only those operators of the tier with the number of operators higher than the specified $N > W$ are subject to transfer by creating under this tier the M tiers with the number equal to:

$$M = \frac{\begin{array}{l} Quot = N/W \text{ for integer division (on each of the lower tiers} \\ \text{there will be exactly } Quot \text{operators.} \end{array}}{\begin{array}{l} N/W + 1 \text{ in the case of } N/W \text{ division with the remainder (then} \\ \text{on the last tier we have exactly } N\%W \text{ operators)} \end{array}}$$

An important feature of this method is the rule for selecting the specific operators to be transferred to the newly created tiers.

Figures 7, 8 show the results of applying these methods to two common linear algebra algorithms – multiplication of square matrices by the classical method and the solution of linear system algebraic equations by the direct (non-iterative) Gauss algorithm. Curves 1 and 2 in these and the following figures correspond to the heuristic methods of WidthByWidtn and Dichotomy, respectively. The reformed SPF width here corresponds to the number of instructions in the bundle of a very long machine word.

As can be seen from Figs. 7 and 8, both methods on the indicated algorithms lead to similar results (from considerations of representing the SPF by a flat table and

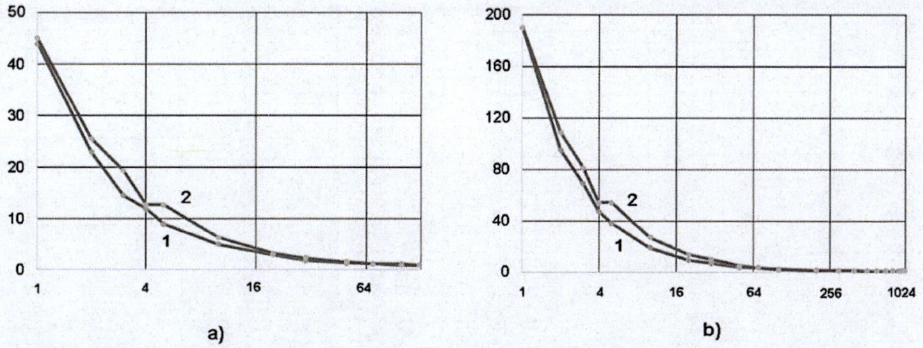

Fig. 7. Increase in height (ordinate) when limiting the SPF width (abscissa), times; the algorithm for multiplying square matrices by classical method of the 5th and 10th orders – Fig. a) and b), respectively

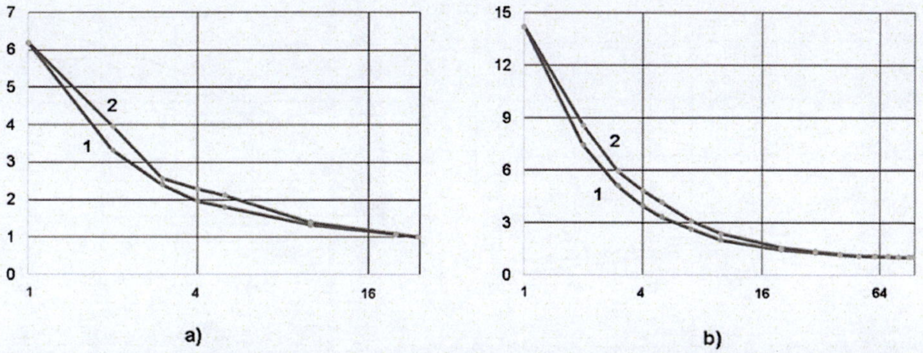

Fig. 8. Increase in height (ordinate) when limiting the SPF width (abscissa), times; an algorithm for solving a linear system algebraic equations by the direct (non-iterative) Gauss algorithm of the 5th and 10th orders – Fig. a) and b), respectively.

the invariance of the total number of operators in the algorithm, this is, of course, a hyperbola!).

However, when comparing, the curves corresponding to WidthByWidtn are located *lower* than those in the Dichotomy method. This corresponds to higher running speed. The results obtained by WidthByWidtn practically coincide with the ideal SPF height equal to $N_{tot.}/W_{av.}$, where $N_{tot.}$ is the total number of operators, $W_{av.}$ is the arithmetic mean number of operators over the SPF tiers at a given width.

The results analysis shown in Figs. 9 and 10 is of greater interest (at least because it has a purely practical interest - computational complexity of the SPF transformation). As can be seen from Figs. 9 and 10, for the cases considered, WidthByWidtn has a lower (approximately 3–4 times) computational complexity (in units of the number of operators transferring from tier to tier) compared to Dichotomy (although the opposite is expected at first glance). However, WidthByWidtn has more complex internal logic compared to Dichotomy (in the latter case, it is primitive). It is logical to use the Dichotomy method

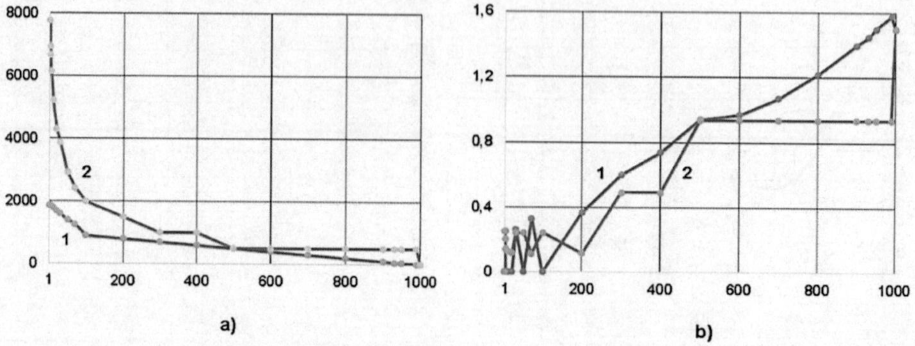

Fig. 9. The number of operator transfers between the tiers - a) and the CV coefficient of variation - b) with decreasing the SPF width for the algorithm of multiplying square matrices of the 10th order by the classical method (abscissa axis is the SPF width after reforming).

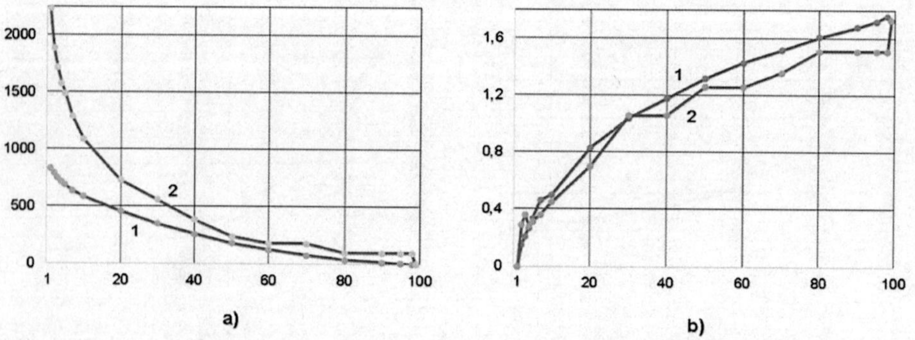

Fig. 10. The number of operator transfers between the tiers - a) and the CV coefficient of variation - b) with decreasing the SPF width for the algorithm for solving the linear system algebraic equations of the 10th order by the direct (non-iterative) Gauss algorithm (abscissa axis is the SPF width after reforming)

in parallelizing translators for fast (but rather rough) construction of parallel program execution plans and the WidthByWidtn method for those construction in optimization mode.

4 Discussion

In general, the developed software package has confirmed its effectiveness in detecting hidden parallelism of algorithms and researching the possibilities of its rational use in data processing. The approach of using a scripting language for the development of heuristic methods of purposeful transformation of the algorithm information graph form has shown the greater flexibility and transparency for the Researcher. At the same time, flexibility is achieved by using an interpreted scripting language, and processing speed is achieved by executable code of the parent application compiled language.

In the course of the research, a significant dispersion it was revealed in properties of the algorithms (represented by information graphs) according to the possibility of forming the frameworks for parallel execution plans with the maximum code density (or, which is practically the same - with the maximum use of the resources of a parallel computing system).

Moreover, different algorithms require different methods for their effective transformation. It seems to be an important task of a priori (even before the SPF reorganization, at the time of its receipt) determination of algorithm parameters for the selection of effective procedures for their purposeful reformation. Creation of algorithms classifying system on some parameters that determine the effective methods of their transformation should be a tool here. According to the author of this paper, the use of artificial intelligence formal methods for solving this problem may be promising.

The limited amount of processed data caused exclusively by the complexity of manual algorithms compilation may seem to be a certain drawback of the study. Indeed few people are practically interested in the tasks of solving the LAES or multiplying matrices of the 10th order. However, all experiments (including those carried out on algorithms artificially synthesized in accordance with the specified parameters) show that the revealed trends only intensify as the dimension of the processed data increases. And this gives confidence that this trend will continue in the future.

In accordance with the iterative principle of gradual approximation to the best solution of the problem under consideration, inherent in the heuristic approach, the author is confident in the possibility of a quantitative improvement (with respect to the above parameters) in the methods for determining program execution schedules on a given field of parallel computers.

The target consumers of the developed methods for generating parallel program execution schedules are, first of all, the translators and virtual machines developers, researchers of the algorithms' properties (in the direction of finding and using the potential of their latent parallelism). The developed system can be recommended to teachers and those who study the basics of system programming for parallel computing architectures.

References

1. AlgoWiki. Open Encyclopedia of Algorithm Properties. http://algowiki-project.org. Accessed 30 June 2021
2. Fedotov, I.E.: Parallel programming. Models and techniques. M.: SOLON-Press, p. 390 (2018)
3. Dennis, J.B., Misunas, D.P.: A preliminary architecture for a basic data-flow processor. In: Proceedings of Second Annual Symp. Computer Architecture, Houston, Texas, pp. 126–132, January 1975
4. Bakanov, V.M.: Controlling the dynamics of computations in processors of streaming architecture for various types of algorithms. Softw. Eng. J. **9**, 20–24 (2015)
5. Tanenbaum, E., Austin, T.: Computer Architecture. SPb.: Peter Publishing House, p. 816 (2019)
6. Bakanov, V.M.: Software complex for modeling and optimization of program implementation on parallel calculation systems. Open Comput. Sci. **8**(1), 228–234, ISSN (Online) 2299-1093. https://doi.org/10.1515/comp-2018-0019
7. Gehry, M., Johnson, D.: Computing machines and intractable tasks. Mir, Book on Demand, p. 420 (2012)

8. Ierusalimschy, R.: Programming in Lua. 3rd edn. PUC-Rio, Brasil, Rio de Janeiro, p. 348 (2013)
9. Fisher, J.A.: Very long instruction word architectures and the ELI-512. In: International Symposium on Computer Architecture. Proceedings of the 10th Annual International Symposium on Computer Architecture, pp. 140–150. Association for Computing Machinery (ACM), New York (1983)

Application of Cluster Analysis for the Study of Factors Affecting the Rating of Schools in Moscow

T. M. Bosenko(✉), E. S. Konopelko, E. V. Lavrenova, and Yu. V. Frolov

Institute of Digital Education, Moscow City Pedagogical University, 28, Sheremetyevskaya Street, Moscow 129594, Russia

Abstract. The article presents the results of the study, during which groups of schools in Moscow were studied for some indicators that affect their rating. The process of clustering schools was carried out on the basis of eleven indicators characterizing not only educational results, school activity in the Olympiad movement, but also the financial and economic activities of general education organizations. The study applied cluster analysis using the k-secondary method and other methods that made it possible to determine the estimated number of groups of schools in Moscow with similar features. Based on the results of the study, factors that significantly affect the rating of schools on the studied dataset were determined.

Keywords: Secondary schools · School rating · Database · Cluster analysis methods · Study result analytics

1 Introduction

With Quality management system in general education includes, among other things, the procedure for determining and comparative analysis of school ratings. Schools ranking is based, as a rule, on key figures that cover the main areas of activities on education and upbringing of children. Thus, in Moscow they use the key figure blocks, which contain, in particular, criteria describing the activities of schools in such areas as the educational process quality (unified state examination and diagnostics of students' knowledge); participation of students in the Moscow and All-Russian Olympiads for schoolchildren; working with children needed the special education; school activities on the use of the municipal social and cultural resources; work on the development of professional skills in students, for example, within the WorldSkillsRussia; family satisfaction activities and other key figures.

This study was aimed at determination of the number of clusters and their composition on the data set of schools provided by the Moscow Center of Quality of Education, Moscow Department of Education and Science and Accounts Chamber of the Russian Federation.

The scientific problem of this paper was to apply data analysis revealing significant factors that affect the formation of clusters and school ratings. Thus, a permanent analytical system that guides schools to achieve high key figures within the education quality management system can be designed in the future.

R. Silhavy et al. (Eds.): CoMeSySo 2021, LNNS 231, pp. 281–295, 2021.
https://doi.org/10.1007/978-3-030-90321-3_23

The topicality of the research question is due to the need to design a n education quality management system based on data and supporting proactive management decisions.

2 Review of Literature

Recently, the use of analytical methods in the social sphere has been expanding in monitoring, forecasting its development, and making proactive management decisions by federal and regional authorities [1–5]. In this regard, the target figure for business analysts and the purpose of analytical tools is to ensure the proper quality of materials, based on which the decisions on social sphere development are prepared.

Education is one of the key sectors of the social sphere. It contributes to the formation of human capital assets [6]. In the papers [4, 5, 7–11], various directions of educational data analytics were studied. The advantages of analytical tools were also presented using examples of forecasting the number of teachers in the regional education system, assessing the need of the Russian economy for digital specialists, as well as clustering the users who generate and consume the educational content on the Moscow Electronic School platform.

One of the important education issues is the creation of such a quality management system that would be focused on meeting the expectations of all stakeholder groups: consumers (children, parents), education authorities, teaching staff, and other social partners [6]. Since the quality as a target function depends on many factors, its management cannot be imagined without the analytics of data generated both inside the educational institution (as an open system) and outside it.

3 Methods

Herein the cluster analysis was applied to group Moscow schools being rated according to 11 factors presented in Table 1.

Table 1. Figures of schools used in cluster analysis.

No	Predicate name	Q-ty
1	Secondary School Ranking	340
2	Number of wins in Moscow Olympiads	340
3	Number of wins in All-Russian Olympiads	340
4	Number of runners-up in Moscow Olympiads	340
5	Number of runners-up in All-Russian Olympiads	340
6	Average school score_State Final Examination	340
7	Average school score_Unified State Examination	340

(*continued*)

Table 1. (*continued*)

No	Predicate name	Q-ty
8	Amount of funds invested in the school development (under contracts)	340
9	Teachers' number dynamics	340
10	Average accrued salary of teachers	340
11	Average monthly salary fund for teachers at school	340

As known, k-means algorithm operation can be conditionally divided into four main stages [10–14]:

- determination of the cluster k centers;
- definition of objects belonging to clusters;
- k cluster centroid calculation;
- comparison of clusters centers and centroids.

Clusters centers can be determined both randomly and on the basis of some criteria (for example, based on a comparison of the mean Euclidean distance for objects). Centroids values (cluster centers) are calculated by the average values of objects features (figures) included in this cluster.

At the final algorithm stage, the centers (centroids) identified at previous iteration are compared with the centroids obtained after redefining (at the next iteration stage) of objects belonging to clusters. If they match, the clustering algorithm terminates. In other cases, it returns to the second stage, at which the objects belonging to clusters is determined on the earlier found centroids. In this study, computational experiments on clustering and the calculation of statistical key figures were performed within the IBM Statistica package [15].

The k-means method complexity lies in the absence of recommendations and/or observable features for choosing the number of clusters. In this regard, an important methodological issue when conducting cluster analysis is to predict the optimal number of clusters for the data set under study in terms of their compliance with statistical characteristics. One approach is based on analysis of variance helping to determine the significance of clustering features. In this case, the number of clusters is determined by F-statistic value. The higher total F-statistic value for the cluster as a whole and the greater the contribution of clustering features to clustering, the more justified the existence of a particular cluster is.

Estimation of the clusters number with the subsequent determination of their composition was carried out by computational experiments in k-means using the following two methods:

- silhouette analysis [12, 14, 16]
- by calculating the F-criterion values followed by assessment of cluster occupancy with objects (schools) and final cluster centers (average values of factors characterizing schools in the cluster).

Silhouette analysis allows to interpret and check the consistency of objects within clusters and provides a graphical representation of the expected number of valid clusters in the form of a silhouette on study data samples.

A silhouette profile indicates how similar ("cohesion") or different ("separation") the clusters (objects that make up a cluster) are from each other. If clusters have a relatively high value (greater than zero), then we can conclude that there are valid clusters on the examined sample of data in Table 1. If the objects included in the silhouette are represented on the visualization in a zone less than zero, then, based on the results of data clustering, the absence of stable results can be expected, namely, there may be too many or too few clusters.

4 Problem Setting

Clustering of schools included in the Moscow ratings for 2018–2019 was identified as the study objective. The base of factors affecting the rating included the key figures shown in Table 1.

In the course of the study, the hypothesis was tested about the possible influence of the funding volume for schools on the effectiveness of their activities. The same time, efficiency was understood as the ratio of the results achieved by the school (for example, the value of the rating or the average USE score), and the costs - investments in the development of the school infrastructure and/or the amount of funds allocated to pay teachers.

5 Results

Using the silhouette analysis the proposed number and profiles of valid clusters were estimated. The study was carried out by sequentially increasing the expected number of clusters formed using the k-means method and analyzing the silhouettes (profiles) of all the clusters obtained. Figures 1, 2, 3, 4, 5, 6, 7, 8, 9, 10 and 11 illustrate the study results.

At the concluding study phase, the final cluster centers were assessed. For illustrative purposes, Table 2 and Table 3 show the values of factors characterizing schools for two variants of dataset grouping into six and eight clusters.

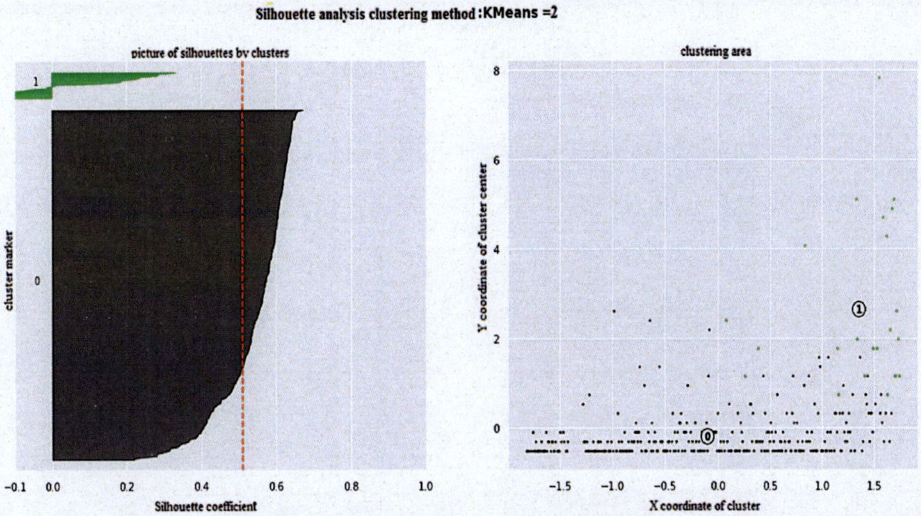

Fig. 1. K-means cluster silhouettes (2 clusters).

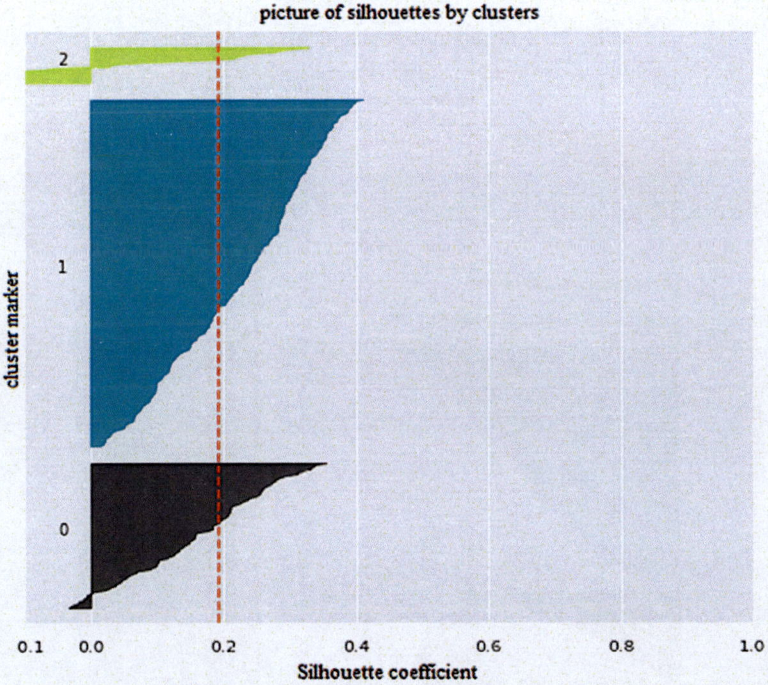

Fig. 2. K-means cluster silhouettes (3 clusters).

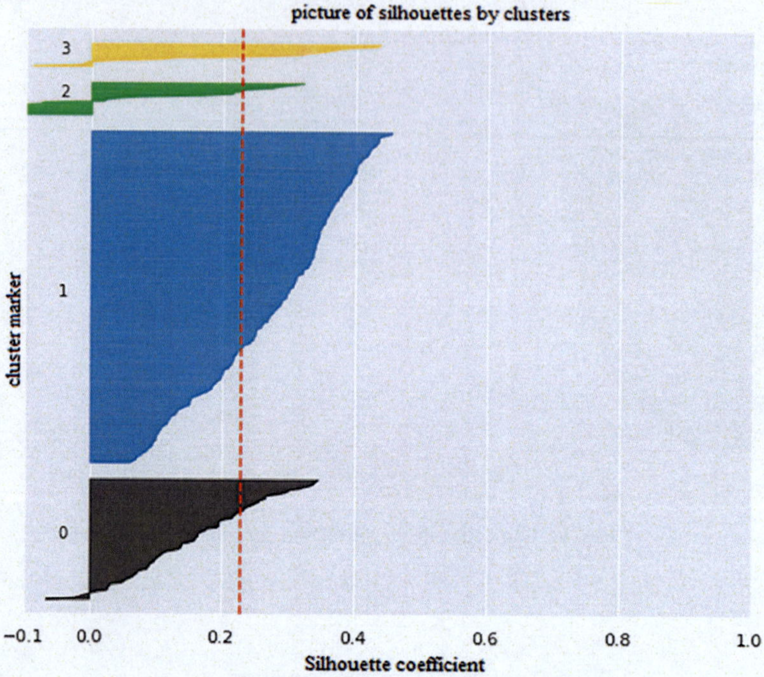

Fig. 3. K-means cluster silhouettes (4 clusters).

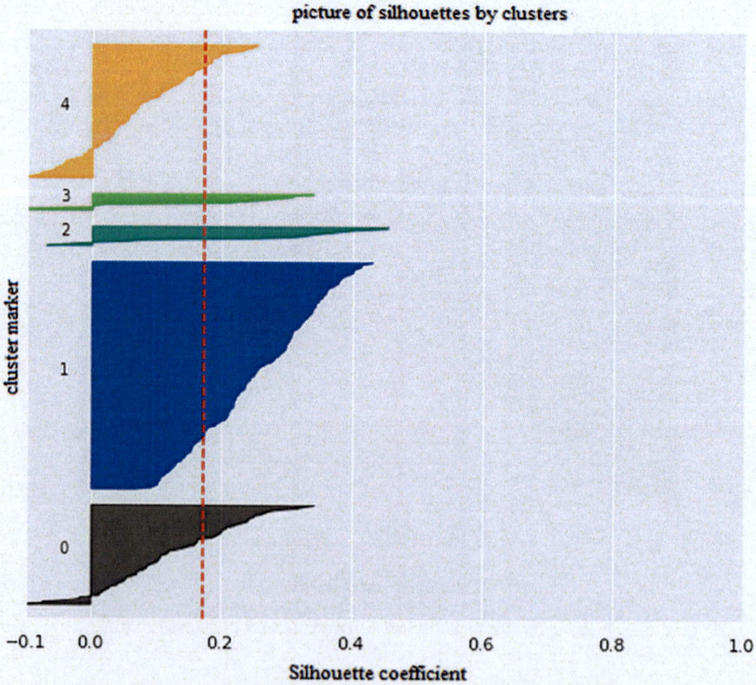

Fig. 4. K-means cluster silhouettes (5 clusters).

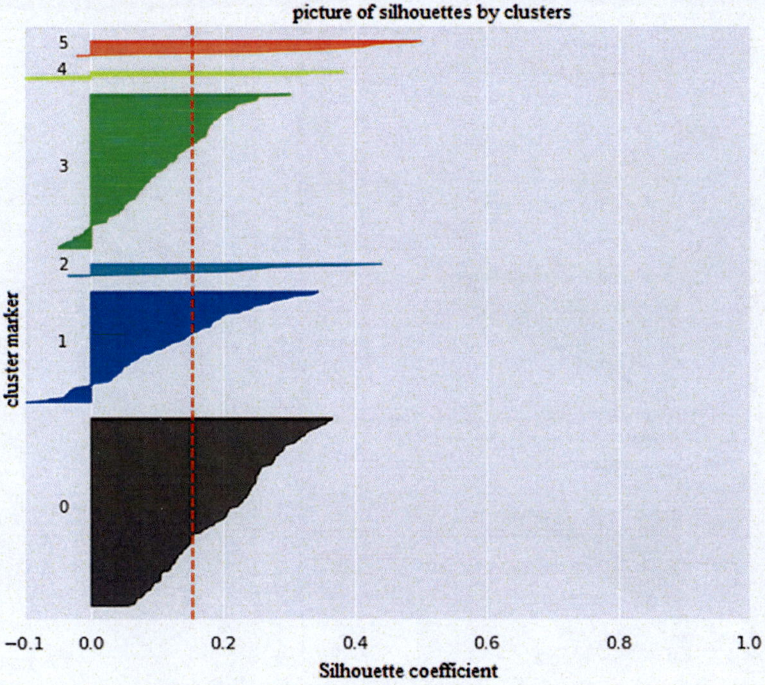

Fig. 5. K-means cluster silhouettes (6 clusters).

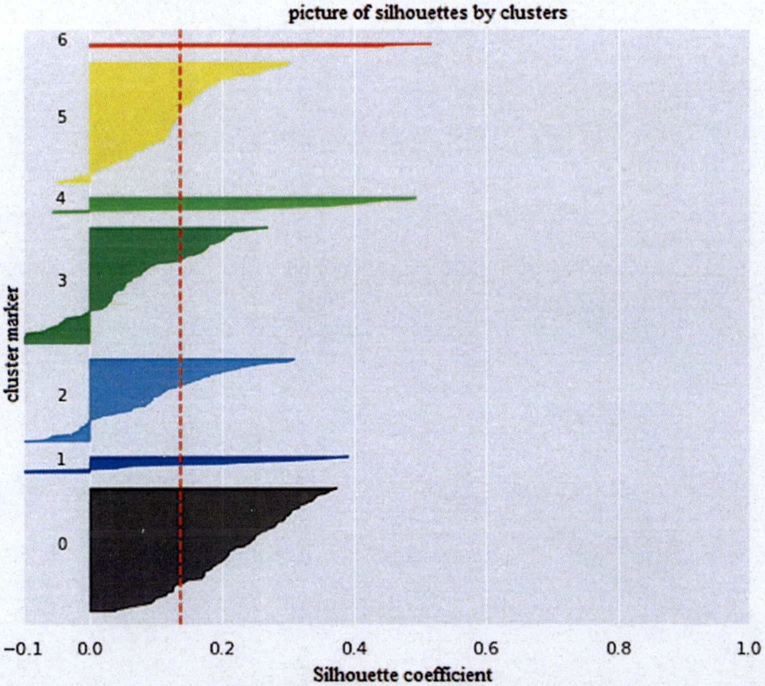

Fig. 6. K-means cluster silhouettes (7 clusters).

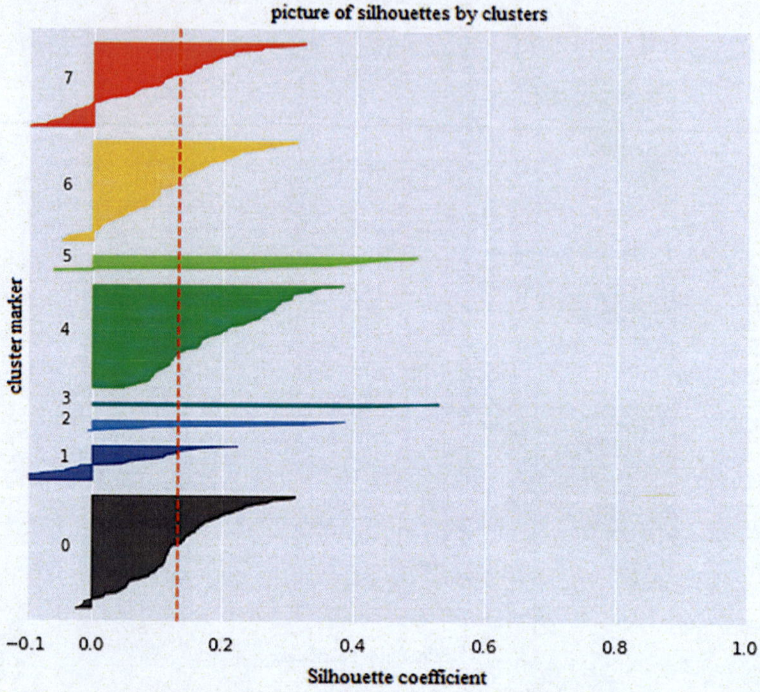

Fig. 7. K-means cluster silhouettes (8 clusters).

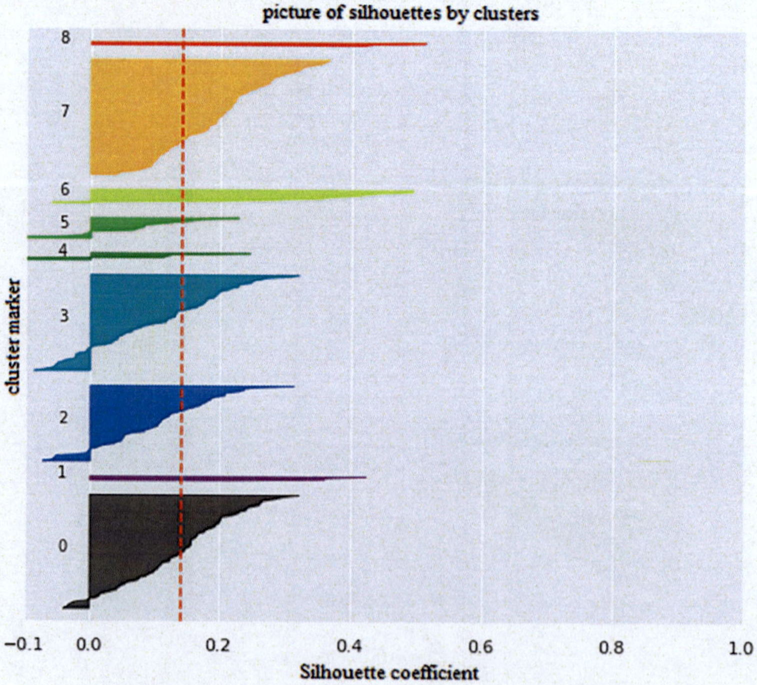

Fig. 8. K-means cluster silhouettes (9 clusters).

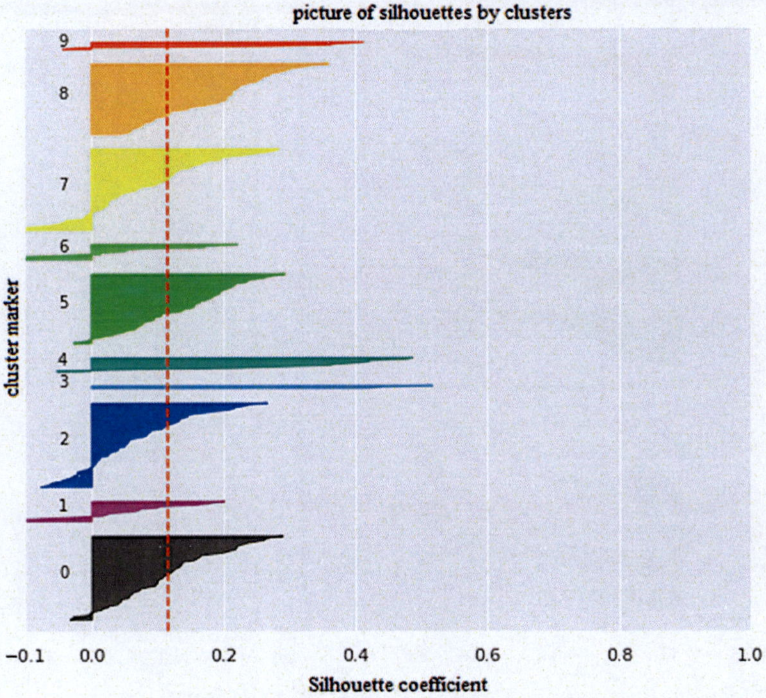

Fig. 9. K-means cluster silhouettes (10 clusters).

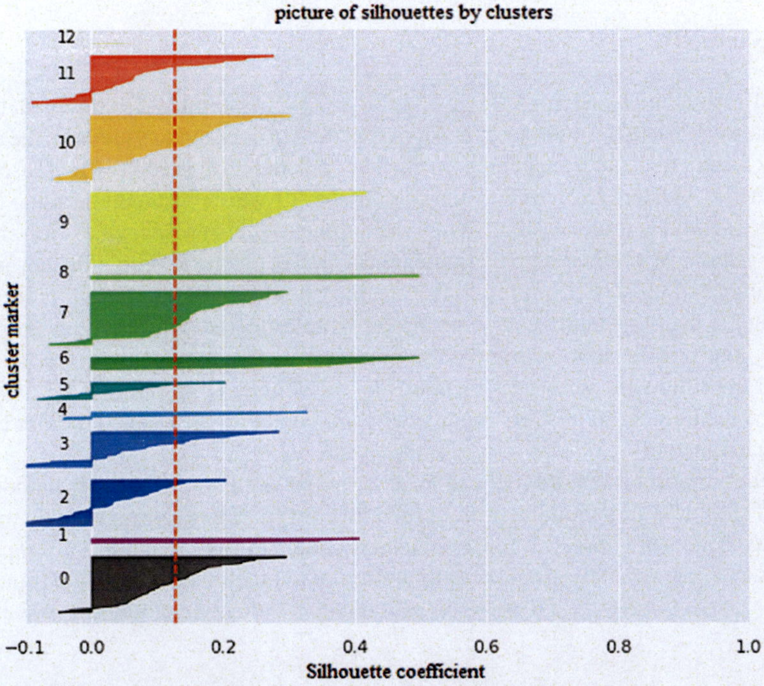

Fig. 10. K-means cluster silhouettes (13 clusters).

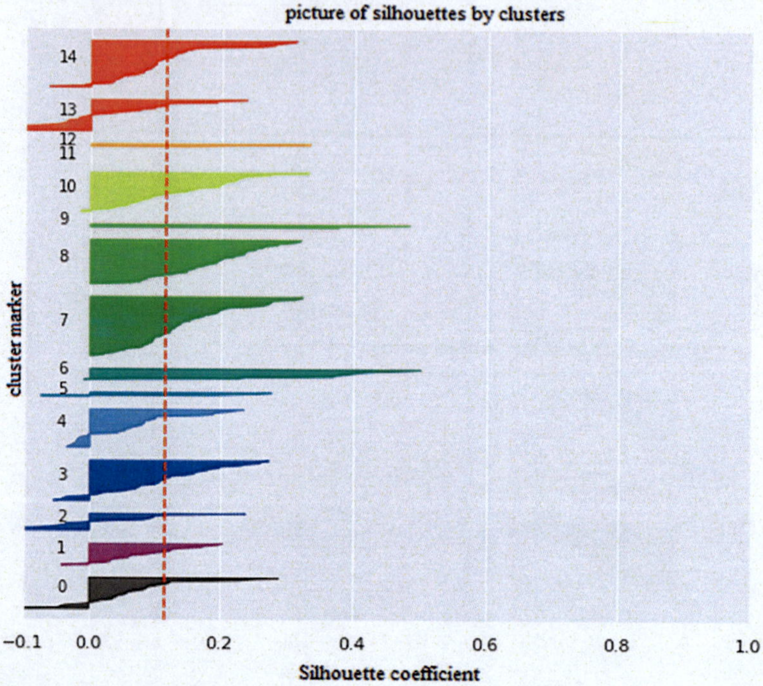

Fig. 11. K-means cluster silhouettes (15 clusters).

6 Discussion

Obtained results analysis (Figs. 1, 2, 3, 4, 5, 6, 7, 8, 9, 10 and 11) shows that with a sequential increase in the number of clusters specified in the k-means method, the number of valid clusters that can be accepted for further consideration gradually increases. With the number of clusters from five to six, the number of valid clusters determined by the silhouette method is three clusters. When the number of clusters increases from seven to nine, the silhouette method signals there are four valid clusters. With a further increase in the number of clusters from ten to fifteen, the number of valid clusters predicted by the silhouette method stabilizes at five. At the same time, the number of clusters with a relatively low number of objects and with "branches" extending into the zone of negative values of the silhouette parameter increases on the silhouette graphs.

Hence the number of clusters predicted by silhouette on the studied dataset is either four or five clusters.

Another important feature that can be determined using the silhouette method is the degree of similarity of silhouette profiles of valid clusters. Thus, approximately the same silhouette profiles are observed for four clusters shown in Figs. 7 and 8. This apparently indicates an approximately similar configuration of the factors that form the found valid clusters.

Table 2. Average values of factors characterizing schools (6 clusters).

End cluster centers						
	Cluster					
Cluster number	1	2	3	4	5	6
Number of observations per cluster	172	46	1	21	92	8
Secondary School Ranking	201	266	248	306	147	336
Number of wins in Moscow Olympiads	2	3	11	3	3	3
Number of wins in all Russian Olympiads	2	2	10	3	2	3
Number of runners-up in Moscow Olympiads	7	10	17	11	11	15
Number of runners-up in all-Russian Olympiads	15	14	69	19	14	16
Average for school State Final Examination	4	4	4	4	4	4
Average for school Unified State Examination	64	63	70	64	65	65
Amount under contracts	217591582	316252406	832737549	415046730	128563938	543858446
Teachers' number dynamics	1	1	1	1	1	1
Average salary of teachers	117381	117674	98901	119906	116200	119760
Payroll for teachers	1228286	1614144	5632652	2031571	877542	1700875

Clusters Nos. 3 and 6 were excluded from the analysis due to the low number of objects therein.

With an increase in the number of clusters generated by the k-means method, the process of "fragmentation" is observed, i.e. the number of silhouettes (clusters) increases, their width decreases, which signals the number of objects (schools) included in the cluster. Moreover, as the number of clusters grows, the number of silhouettes with branches located in the negative zone of the silhouette parameter increases. This evidence a decrease in the reliability (validity) of the analysis results for determining the number of clusters.

Table 3. Average values of factors characterizing schools (8 clusters).

End cluster centers								
	Cluster							
Cluster number	1	2	3	4	5	6	7	8
Number of observations per cluster	1	16	126	8	62	41	81	5
Secondary School Ranking	248	318	210	281	148	268	168	353
Number of wins in Moscow Olympiads	11	3	2	2	3	3	2	5
Number of wins in all-Russian Olympiads	10	3	3	1	3	2	1	3
Number of runners-up in Moscow Olympiads	17	12	8	8	11	11	7	19
Number of runners-up in all-Russian Olympiads	69	23	17	9	16	13	9	18
Average for school State Final Examination	4	4	4	4	4	4	4	4
Average for school Unified State Examination	70	65	64	64	66	63	64	62
Amount under contracts	832737549	402115361	231758162	473509873	113323444	321502043	177679993	568984827
Teachers' number dynamics	1	1	1	1	1	1	1	1
Average salary of teachers	98901	119150	117625	122441	116044	117574	116752	118035
Payroll for teachers	5632652	2040828	1268014	1897076	835745	1610000	1094490	1688030

Clusters Nos. 1, 4, 8 were excluded from the analysis due to the low number of objects therein

At the next stage of the study, the F-criterion was assessed as a whole for the studied clusters (from two to fifteen), and for each factor included in the study. Statistical key figures was calculated within the IBM Statistica package [15].

It was assumed, firstly, that the maximum total (calculated for all factors from the cluster) values of the F-criterion indicate those clusters that can be considered valid with the possibility of performing regression analysis on statistically significant factors – school key figures (Fig. 12). Secondly, the higher the value of the F-criterion for a particular factor characterizing the school is, the greater significance it has in the formed cluster (Table 4). This allows the found significant factors to be included for research in the regression analysis.

Table 4. Significance of studied factors.

Predicate	F	Significance, $p < 0.05$
Amount to be invested under contracts	2091.997	0.000
Payroll for teachers	28.858	0.000
Secondary school ranking	7.867	0.000
Number of wins in Moscow Olympiads	2.195	0.014
Number of runners-up in Moscow Olympiads	1.842	0.046
Number of runners-up in All-Russian Olympiads	1.552	0.112
Average school score for State Final Examination	1.455	0.147
Number of wins in All-Russian Olympiads	1.440	0.153
Teachers' number dynamics	1.310	0.217
Average school score for Unified State Examination	1.292	0.227
Average salary of teachers	1.136	0.332

Figure 12 illustrates the change in the total value of the F-criterion for clusters obtained using the k-means method.

As seen on Fig. 12, starting with a set of clusters numbering 12, F-criterion begins to decrease, and, therefore, further division into clusters is inexpedient.

The above results show that, in order of decreasing significance in terms of influence on the formation of clusters, the factors can be arranged in the following row: Amount to be invested under contracts; Payroll for teachers; Secondary school rating; number of wins in the Moscow Olympiads; number of runners-up in Moscow Olympiads. The rest of the factors do not have a significant effect on the formation of clusters in the studied data sample.

Taking in consideration the results presented in Tables 3 and 4, indicates, in particular, that factors such as the number of victories in Moscow Olympiads, the number of runners-up in the All-Russian Olympiads, the average school State Final Examination score, the number of victories in the All-Russian Olympiads, the dynamics of the number of teachers, the school average value of the Unified State Examination, the average salary of teachers does not have a significant impact on the formation of clusters in the studied

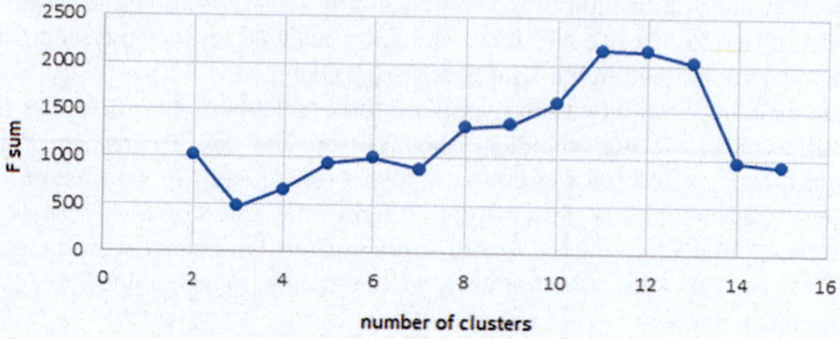

Fig. 12. F-criterion values for the study clusters.

data sample. The same time, such key figures of school activities as the amount to be invested under contracts, payroll for teachers, the secondary schools rating have values that differ when moving from cluster to cluster and, therefore, largely determine the formation of clusters that include Moscow schools. At the same time, the influence of economic key figures (investment volume and pay-roll fund) on the school rating is seen.

7 Conclusion

In this study, the k-means clustering method is investigated as an intelligence and reconnaissance analytics tool for data characterizing Moscow secondary schools. Statistical studies have been carried out. Thus, the silhouette method was applied to determine the probable number of clusters on the study dataset. The optimal number of clusters with a volume of four-five clusters was determined by silhouette method. F-criterion made it possible to identify the most significant factors (Amount to be invested under contracts, Payroll for teachers, secondary school rating) that determine the process of cluster formation. An empirical relation has been determined between the school rating, on the one hand, and the economic performance of schools such as the amount to be invested under contracts and the payroll for teachers on the other hand.

As stated, the research should be continued on a sample of data that includes a large set of factors that effect the rating of Moscow schools.

The performed studies, according to the authors, create the preconditions for future design of the education quality management system at the regional and federal levels.

References

1. On the Monitoring System for the Implementation of National Projects. https://d-russia.ru/o-sisteme-monitoringa-realizatsii-natsproektov.html. Accessed 03 Mar 2021
2. Methodology for Ranking the Constituent Entities of the Russian Federation for Evaluating the Effectiveness of Measures Being Implemented According to Federal Projects that are Part of the National Demography, Healthcare, Education and Science Projects. https://is.cminp.ru/. Accessed 03 Mar 2021
3. Klyachko, T.L., Sinelnikov-Murylev, S.G.: Strategy for Russia: Education. Scientific Papers: Education, p. 118. Delo Publishing House, RANEPA, Moscow (2018)

4. Frolov, Y., Ofitserov, V.P., Ofitserov, M.V.: Forecasting the key figures of need for teaching staff at forming the regional education system strategy. Educ. Econ. **3**, 50–64 (2012)
5. Ryabov, V.V., Pishchulin, N.P., Frolov, Yu.V.: Education Strategy: Formation Foundations, Methods of Assessment and Forecasting, p. 206. MGPU, Moscow (2003). Ed. Ryabova, V.V.
6. Frolov, Yu.V.: Knowledge Management, 2nd edn. Rev. and enlarged. Textbook for Bachelor's and Master's Degree Programs, p. 324. Yurayt, Moscow (2021)
7. Frolov, Yu.V., Bosenko, T.M.: Training of personnel for the development of innovative entrepreneurship. Acad. Entrepr. J. **26**(1)(327), 1–6 (2020)
8. Frolov, Y., Bosenko, T.M.: Pre-processing of data on the behavior of users of the Moscow Electronic School service. J. Phys. Conf. Ser. **1691**, 012224 (2020). https://doi.org/10.1088/1742-6596/1691/1/012224
9. Lavrenova, E.V., Frolov, Y.V., Bosenko, T.M.: Analytical studies of behavior of users of the Moscow electronic school service. In: Silhavy, R. (ed.) CSOC 2021. LNNS, vol. 229, pp. 121–132. Springer, Cham (2021). https://doi.org/10.1007/978-3-030-77445-5_11
10. Skobliakova, I.V., Semenova, E.M., Zakharov, A.V., Tokmakova, E.N., Morozova, O.I.: Training of innovative personnel in the digital economy. In: Shakirova, O.G., Bashkov, O.V., Khusainov, A.A. (eds.) Current Problems and Ways of Industry Development: Equipment and Technologies. LNNS, vol. 200, pp. 683–692. Springer, Cham (2021). https://doi.org/10.1007/978-3-030-69421-0_73
11. Guangrui, T., Yong, Z., Yan, C.: Effectiveness evaluation of five chain integration of state grid corporation of china based on TOPSIS method. In: E3S Web of Conferences 2021, vol. 236 (2021). https://doi.org/10.1051/e3sconf/202123604016
12. Peter, J.: Rousseeuw Silhouettes: a graphical aid to the interpretation and validation of cluster analysis. J. Comput. Appl. Math. **20**, 53–65 (1987). https://doi.org/10.1016/0377-0427(87)90125-7
13. Mirkin, B.G.: Clustering for Data Mining. A Data Recovery Approach. Taylor & Francis Group, Boca Raton (2005)
14. Duda, R.O., Hart, P.R., Stork, D.G.: Pattern Classification, p. 738. Willey, New York (2001)
15. Software platform for statistical analysis IBM SPSS STATISTICS 26. https://www.ibm.com/ru-ru/products/spss-statistics
16. de Amorim, R.C., Hennig, C.: Recovering the number of clusters in data sets with noise features using feature rescaling factors. Inf. Sci. **324**, 126–145 (2015). https://doi.org/10.1016/j.ins.2015.06.039. ISSN 0020-0255

BRICS-Eurasian Credit Markets: Problems and Prospects

Irina Yarygina[1], Inna Lukashenko[1], Ernesto Lee[2], Patrick Washington[3], and Roman Tsarev[4,5](✉)

[1] Financial University, Moscow, Russia
[2] Broward College, Davie, FL, USA
[3] Morehouse College, Atlanta, GA, USA
[4] V.A. Trapeznikov Institute of Control Sciences of Russian Academy of Sciences, Moscow, Russia
[5] International Academy of Science and Technologies, Moscow, Russia

Abstract. Strengthening financial cooperation between BRICS-Eurasian member-countries is the best solution to mitigate trade wars, unfair competition and worsening of international economic relations. In this paper, we analyze and determine cooperation between BRICS and Eurasian union (EAEU) credit markets, taking into consideration their reasonable interaction. The aim of the article is to introduce recommendations aimed at optimizing financial cooperation of developing countries as members of the intergovernmental alliances. The existing data is processed using combination of Bloomberg and Excel tools. Testing of initial hypothesis proved interdependence of credit markets. It is mostly developed and presented in data. The analysis of the research concludes the main industries financed by governments or institutional investors vary depending on the strategic orientation of each country's economy. For example, for the Russian Federation, South Africa, India, Brazil, these sectors are Materials, energy, innovations with domination of the Materials sector as an essential for building mutual value-chains ecosystems. In China, the key industry is the financial sector, which indicates the predominance of the service sector. As for the EAEU countries, most of the loans are directed to state needs, covering external or internal debt (Belarus, Armenia, etc.) and are forced to resort to borrowed funds to cover their needs. Kazakhstan and Russia use credit facility for developing national economies in the interest of the population. Anyway, the research founds out, that the indexes of the emerging credit markets still rely on the production of raw materials and depends to the certain extent on the price movements of commodities. We consider reasonable to take into consideration the international experience in risk management as well as approach conveying better access to loans for commercial entities; improving inter government grants to new industries and taxing relief for manufactured exported goods. In order to break the economic deadlock of some BRICS - EAEU credit markets it is worth to lead harmonized financial policies that would pave the way for mutual development. #COMESYSO1120

Keywords: Intensive BRICS · Eurasian Economic Union · Credit market · Indices · Economic cooperation

R. Silhavy et al. (Eds.): CoMeSySo 2021, LNNS 231, pp. 296–318, 2021.
https://doi.org/10.1007/978-3-030-90321-3_24

1 Introduction

All Trade and economic cooperation of BRICS (Brazil, Russia, India, China and South Africa) is one of the priority areas of their Economic Partnership Strategy. In the list of key objectives of this Strategy, an important role plays the promotion of trade cooperation, expanding access to the financial markets of member countries and diversifying investment cooperation. It was noted that financing trade cooperation, along with other areas, is designed to strengthen balanced and inclusive economic growth, as well as increase the level of international competitiveness of the BRICS economies, which account for 17,3% of world trade in goods, 12,7% of world trade in services, 21% of global gross domestic product (GDP) [1].

The expansion of national currencies is the key aspect in ensuring BRICS economic relations as well as cross-border relations with other partner - countries, such as Mongolia, Vietnam, the countries of the Eurasian Economic Union, etc. Areas of interaction are energy, oil products, wood processing, agriculture, machinery and equipment, production of construction materials, medicine and innovations. Good prospects for financing in national currencies remain in the sphere of export – import operations, infrastructure development, transport, new production technologies and environmental protection. Interstate support for commercial, industrial centers, aimed at expanding trade ties, can open broad prospects, including in the framework of implementation of investment projects, joint ventures, small and medium enterprises. The basis of the interaction of the parties is a legal framework of BRICS and the agreement between the central banks of partner countries (reached in 2007) on the increasing use of national currencies, including development of the service sector.

Establishing Eurasian Economic Union, the member-states aspired for a very simple and noble cause. They joined forces, pooled knowledge and capabilities in order to protect their sovereignty and independence more efficiently, to make their voices better heard in international and global affairs, and to foster their internal economic, political and social development based on mutual respect and recognition of their specificities. These are entirely legitimate tasks, fair and logical, compatible in all their aspects with the aims and principles of contemporary international law. Member – countries have shown that achieving these goals is feasible and attainable. Coordination of efforts, activities and policies of the member-states on a wide variety of issues produced positive impact on the international economy and finance. Nevertheless, results could have been more impressive if the cooperation among members of the Eurasian Economic Union was stronger and deeper. The Eurasian Economic Union is an economic union consisting of Armenia, Belarus, Kazakhstan, Kyrgyzstan and Russia. The treaty of the Union came into force on the 1st of January 2015. This Union provides a single market for the 183 million inhabitants of the countries. This region has a gross domestic product of $4 trillion according to The World Bank [2]. The single market provides free movement of goods, capital, services and people. It also allows each member to use common policies in for their macroeconomic sectors such as, transport, industry, technical regulation, agricultural, energy, foreign trade and investment and competition.

Provisions for a single currency and greater integration are envisioned for the nearest future. The day-up to-day activity of the Eurasian Union is performed by the Eurasian Economic Commission, which is similar to the European commission. The main purpose of the union is to establish and open economic development between the countries. Great attention is paid to the Credit markets' facilities and its possible alliance.

It is worthwhile mentioning, that the infrastructure of the central banks of BRICS-EAEU member countries, financial markets, banks for development is currently operating. It is important to note the positive dynamics of the interaction of the BRICS New Development Bank (NDB), Eurasian Development Bank with participants in the economic relations of member - countries. In order to accelerate the development of financial relations it is important to use the levers of agreed benefits and preferences, as the main incentive for entrepreneurs is economic feasibility of business.

2 Methods

The data collected consists of current data of the indices of the financial markets of BRICS-EAEU and includes an external study of the stock indices. The data ranges in December 2020. Bloomberg is widely used to collect the data. The data is processed true a combination of Bloomberg and Excel in order to produce the most accurate and correct results.

An informal analysis provides a basic outlook and information of the data set. It helps to begin the formation of a starting point for the presented search. This is an analysis that will include descriptive statistics and correlograms. These will form an understanding of the properties, which generally benefit a researcher with further testing. Included are standard econometric testing used to prevent spurious results. The first step after acquiring the data is to analyze and interpret the indices to assist in supporting the research question. Additional information presents trends and behaviors of the selected variables over a given time. The information received is used to compare variables or series which provides statistical quantities, which may have some prediction qualities.

In order to identify problems and determine the prospects for lending operations of the countries under consideration, as well as assess the proximity of their strategic positions on certain issues, for example, trade policy promoting interaction between the parties, the study uses cluster analysis. Such approach contributes to the formation of relatively homogeneous groups of countries based on quantitative data analysis. Key country comparison indicators are used for this purpose. The first indicator is GDP per capita in purchasing power parity (PPP), playing an important role in the implementation of cross-country analysis. Since it reflects not only output, but also the level of population income and prices in the country. The second indicator is the share of high-tech exports (in%), characterizing the development level of industries producing industrial goods and technologies. Based on the data obtained, it seems possible to determine the intentions of countries to increase the share of production of goods with high added value or trade in other goods of mutual interest. Let us consider the method of quantitative assessment used in the study on the specific example. Table 1 presents quantitative data for the BRICS countries used for the implementation of cluster analysis.

Table 1. Input data for cluster analysis

Country	Brazil	Russia	India	China	South Africa
GDP PPP (per capita. $)	15553.4	25763.3	7166.2	16842.4	13526.2
Share of high-tech exports (%)	12.3	11.5	7.0	23.8	4.6

Source: World Bank Group [2].

To reach the goals of the study we also implement cluster analysis by using an agglomerative hierarchical classification algorithm. We apply the principle of "nearest neighbor", choosing the usual Euclidean distance as the distance between objects:

$$p(x_i x_j) = \sqrt{\sum_{l}^{k} (x_{il} - x_{jl})^2},$$

where l - features, k - number of features.

The following example is presented hereunder: features X1 - GDP in PPP (per capita, $), feature X2 - the share of high-tech exports, %), countries: 1 - Brazil, 2 - Russia, 3 - India; 4 - China, 5 - South Africa. The search for the smallest distance is performed, and the closest objects are combined into one relatively homogeneous cluster. The result of the conducted iterative process is presented in Table 2.

Table 2. Cluster analysis result - least distance matrix.

№	1,4	2,3	5
1,4	0	8920.908	2027.215
2,3	8920.908	0	12237.1
5	2027.215	12237.1	0

Source: compiled by the authors.

The table contains data determining the smallest distances between objects. As a result of the analysis, we obtain 3 clusters: S (1,4), S (2,3), S (5). Consequently, according to the selected criteria, X1 and X2, Brazil and China, Russia and India are the closest to each other in pairs. South Africa falls into a separate cluster. Generally, the results obtained correspond with the results of the analysis for the BRICS bilateral ties: China is Brazil's main partner in import and export; Brazil is China's main import partner. The data are also valid for India and Russia: India is the second import and export partner for the Russian Federation after China; Russia is India's most important import partner (after China).

3 Results

The following features of the BRICS credit market are revealed hereunder.

For the Russian Federation, the largest sector in terms of lending is the industrial sector, which accounts for 36.1% of total loans and amounts to 37 billion USD. Russian banks with foreign participation of ING Group (Netherlands), Commerzbank Group of German banks, Société Générale (France), dominate on the credit market for production of goods [Table 3, 4 and 5].

Table 3. Credits by sectors of the Russian economy.

Sectors of the economy	The volume of credits ($ mln)	Deals	Participants
Materials	37021.71	68	27
Financials	16614.45	15	7
Industrials	15331.99	11	6
Energy	14651	29	9
Communications	13919.53	15	2
Utilities	2267.41	37	5
Consumer Staples	1936.09	2	2
Government	500	1	1
Consumer Discretionary	117.54	2	1

Source: Bloomberg, date of access 12.01.2021

Table 4. Largest borrowers in Russia.

Sectors of the economy	Credit market participants (borrowers)	The volume of credit ($ mln)	Deals (as on 31.12.2020)
Technology	Invest holding USM	2403.81	9
Financials	Russian credit bank	1582.84	4
Industrials	OAO Sovcomflot	991.50	5
Energy	PJSC Gazprom	7987.21	12
Communications	PJSC AFK Sistema	12347.56	13
Utilities	PJSC RusHydro	1629.94	23
Consumer Staples	JSC Agromir	1745.65	1
Government	Government of the Russian Federation	500.00	1
Consumer Discretionary	PJSC Detsky Mir	117.54	2

Source: Bloomberg, date of access 12.01.2021

It is not surprising that the main corporate borrowers in the Russian economy are energy and extractive industries. In Russia, which has embarked on the development of technologies, the raw materials industries still dominate, as illustrated in Fig. 1.

Table 5. Largest creditors of the Russian economy.

Sectors of the economy	Credit market participants (creditors)	The volume of credits ($ mln)	Deals (as on 31.12.2020)
Technology	ING Group	2462.41	21
Financials	ING Group	482.33	4
Industrials	ING Group	181.01	5
Energy	Commerzbank	598.59	4
Utilities	Société Générale	24.66	1

Source: Bloomberg, date of access.12.01.2021

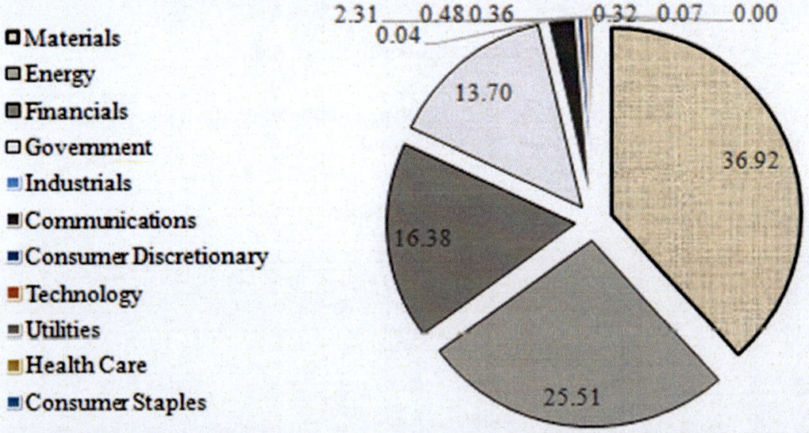

Fig. 1. Volume of credits to the key sectors of the Russian economy (%). *Source*: Bloomberg, date of access 12.01.2021

In its turn, the energy sector is the leader in terms of lending in India. It accounts for 21.18% (US $ 36 billion) of loans issued at the end of 2020. Banks of Great Britain, Japan and others are the lenders. This geographical coverage indicates the attractiveness of lending in the Indian market. The analysis showed that the country has a favorable investment climate with low credit risks and relatively high profitability of operations compared to resident countries of foreign banks. India's credit market data are presented in Tables 6, 7 and 8.

If the leading industry-borrower in Russia is one and a half times larger than the industry following it, in India the second largest borrower is not far behind the leader (Fig. 2).

Table 6. Credits by sectors of the Indian economy.

Sector s of the economy	The volume of credits ($ mln)	Deals	Participants
Energy	36618.39	99	56
Financials	31579.32	141	75
Utilities	21828.46	102	59
Government	21483.52	65	7
Materials	20321.87	66	31
Industrials	20280.62	112	68
Communications	8522.14	20	7
Consumer Discretionary	7780.05	48	36
Technology	2665.6	6	5
Health Care	1604.6	20	18
Consumer Staples	205.09	8	8

Source: Bloomberg, date of access 12.01.2021

Table 7. Largest borrowers in India.

Sectors of the economy	Credit market participants (borrowers)	The volume of credits ($ mln)	Deals (as on 31.12.2020)
Energy	Reliance Industries Ltd	13859.6	15
Financials	Power Finance Corp Ltd	5932.95	23
Utilities	ReNew Power Pvt Ltd	960.09	11
Government	Republic of India	18898.93	55
Materials	Tata Steel Ltd	7989.1	9
Industrials	Macquarie Group Ltd	1061.88	9
Communications	Reliance Jio Infocomm Ltd	6324.3	9
Consumer Discretionary	Air India Ltd/India	1860	7
Technology	Tata Sons Pvt Ltd	2000	2
Health Care	Glenmark Pharmaceuticals Ltd	282.5	2
Consumer Staples	Slmg Beverages Pvt Ltd	59.44	1

Source: Bloomberg, date of access 12.01.2021

Table 8. Largest creditors to the Indian economy.

Sectors of the economy	Credit market participants (creditors)	The volume of credits ($ mln)	Deals (as on 31.12.2020)
Energy	L&T Financial Services	1358.35	15
Financials	Standard Chartered Bank	1642.82	20
Utilities	Yes Bank	2100.29	17
Government	Sumitomo Mitsui Financial	896.36	5
Materials	Standard Chartered Bank	1244.04	14
Industrials	Yes Bank	2172.56	20
Communications	Mitsubishi UFJ Financial Group Inc	561.47	6
Consumer Discretionary	Indusind Bank Ltd	489.58	10
Technology	Standard Chartered Bank	215.03	2
Health Care	Yes Bank	258.15	5
Consumer Staples	ICICI	94.37	3

Source: Bloomberg, date of access 12.01.2021

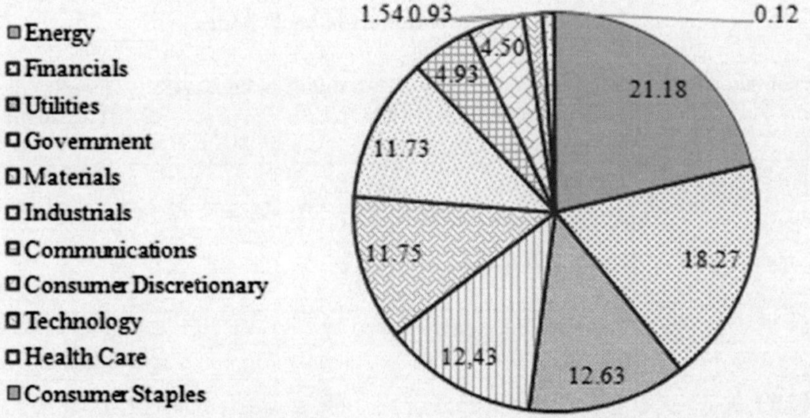

Fig. 2. Volume of credits to the key sectors of the Indian economy (%). *Source:* Bloomberg, date of access 12.01.2021

At the end of 2020, in terms of the credits issued in the Republic of South Africa (South Africa), the industrial production sector is leading with a total volume of 18 billion US dollars. Second place belongs to financing the activities of credit institutions (US $ 8.7 billion). It is important to note the largest banks in Europe (Nedbank Ltd, Standard Chartered Bank), Japan (Sumitomo Mitsui Trust Holdings Inc) and Australia (Commonwealth Bank Australia) are the creditors of various sectors of the South African economy. This picture also indicates an attractive investment climate in South Africa (Tables 9, 10 and 11).

Table 9. Credits by sectors of the South African economy.

Sectors of the economy	The volume of credits ($ mln)	Deals	Participants
Materials	18822.78	33	19
Financials	8723.1	29	15
Government	7121.93	3	1
Utilities	5240.86	8	3
Health Care	4874.61	4	4
Consumer Staples	3685.02	13	10
Communications	2204.68	8	5
Consumer Discretionary	1948.61	14	10
Energy	1834.98	11	10
Industrials	1395.51	17	8
Technology	110.12	4	2

Source: Bloomberg, date of access 12.01.2021

Table 10. Largest borrowers in South Africa.

Sectors of the economy	Credit market participants (borrowers)	The volume of credits ($ mln)	Deals (as on 31.12.2020)
Materials	Gold Fields Ltd	2419.29	5
Financials	Investec PLC	2165.00	6
Government	Republic of South Africa	7121.93	3
Utilities	Eskom Holdings SOC Ltd	4990.37	6
Health Care	Aspen Pharmacare Holdings Ltd	4037.24	1

(*continued*)

Table 10. (*continued*)

Sectors of the economy	Credit market participants (borrowers)	The volume of credits ($ mln)	Deals (as on 31.12.2020)
Consumer Staples	Steinhoff International Holdings NV	1254.54	2
Communications	MTN Group	1792.62	4
Consumer Discretionary	Truworths International Ltd	196.91	2
Energy	Mainstream Renewable Power Ltd/South Africa	400.56	2
Industrial	Waco International Ltd/South Africa	181.21	5
Technology	Community Investment Ventures Holdings Pty	103.30	3

Source: Bloomberg, date of access 12.12.2020

Table 11. Largest creditors by sectors of the South African economy.

Sectors of the economy	Credit market participants (creditors)	The volume of credits ($ mln)	Deals (as on 31.12.2020)
Materials	Nedbank Ltd	640.20	4
Financials	Standard Chartered Bank	467.85	11
Utilities	African Export-Import Bank	449.68	2
Health Care	FirstRand Ltd	780.03	2
Consumer Staples	FirstRand Ltd	1271.37	3
Communications	Sumitomo Mitsui Trust Holdings Inc	312.50	1
Consumer Discretionary	FirstRand Ltd	482.60	4
Energy		0.00	
Industrials	Commonwealth Bank Australia	211.02	1

Source: Bloomberg, date of access 12.01.2021

In South Africa, as well as in Russia, credits for the Materials dominates, and just like in India, Financials are the second in borrowing. Considering the consumer industries, crediting for everyday goods tends to be in the bottom line of borrowers, behind Consumer Discretionary. For Africa and Belarus in our clusters, Consumer Staples industry is above the Consumer Discretionary sector in the list of borrowers. The African structure of industrial credits is shown in Fig. 3.

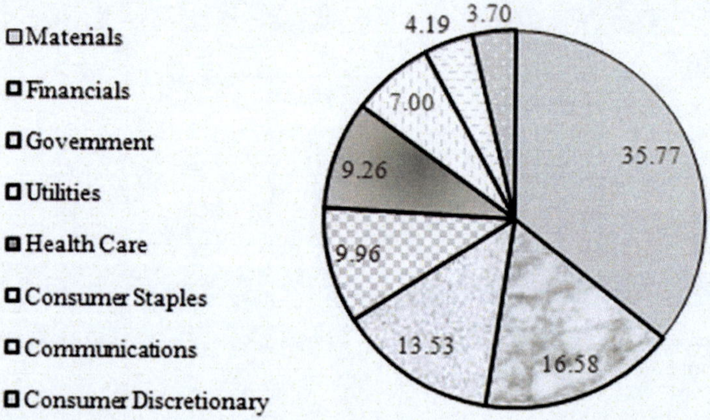

Fig. 3. Volume of credits to key sectors of the South African economy (%). *Source:* Bloomberg, date of access 12.01.2021

The energy sector is in the lead in Brazil with USD 22 billion (35%) in terms of the total volume of credits, issued at the end of 2020. The industrials are in second place. The largest world financial institutions are represented on the country's credit services market: CITI, ING Grogg, etc. The largest Brazilian bank, Banco Itaú BBA, is one of the participants in the credit market. It provides a business entity with universal services at favorable prices. The Brazilian government is creating an attractive investment climate and reducing the level of financial risks in banking. The data characterizing the features of the Brazilian credit market are presented in Tables 12, 13 and 14.

Table 12. Credits by sectors of the Brazilian economy.

Sectors of the economy	The volume of credits ($ mln)	Deals	Participants
Energy	22195.86	30	12
Materials	22006.68	29	12
Financials	3996.11	5	4
Industrials	3240.14	11	5
Consumer Discretionary	3157.12	5	3
Utilities	2875.42	24	8

(*continued*)

Table 12. *(continued)*

Sectors of the economy	The volume of credits ($ mln)	Deals	Participants
Government	2421.46	6	2
Communications	1918.94	7	3
Consumer Staples	1557.64	9	7

Source: Bloomberg, date of access 12.01.2021

Table 13. Largest borrowers in Brazil.

Sectors of the economy	Credit market participants (borrowers)	The volume of credits ($ mln)	Deals (as on 31.12.2020)
Energy	Petroleo Brasileiro SA	18271.24	13
Materials	Votorantim SA	1160.00	7
Financials	Notre Dame Intermedica Participacoes SA	679.79	2
Industrials	Rumo SA	1457.61	5
Consumer Discretionary	Sogefi Filtration do Brasil Ltda	2917.12	3
Utilities	Enel SpA	421.04	10
Government	Federative Republic of Brazil	2400.00	5
Communications	TIM SA/Brazil	723.94	4
Consumer Staples	BRF SA	630.59	2

Source: Bloomberg, date of access 12.01.2021

Table 14. Largest Creditors to the Brazilian economy.

Sectors of the economy	Credit market participants (creditors)	The volume of credits ($ mln)	Deals (as on 31.12.2020)
Energy	Citi	2420.00	4
Industrials	Scotiabank	3811.31	10
Financials	ING Group	404.95	2
Materials	Citi	200.00	1

(continued)

Table 14. (*continued*)

Sectors of the economy	Credit market participants (creditors)	The volume of credits ($ mln)	Deals (as on 31.12.2020)
Consumer Discretionary	Banco Itau BBA	150.00	1
Government		0.00	
Communications	ING Group	456.67	3
Consumer Staples	Banco Itau BBA	306.13	6

Source: Bloomberg, date of access 12.01.2021

Brazil and Russia have the same top three, with a slight difference in the order of their location. Unlike Russia, where there is a clear leader, and besides him, three more sectors are large borrowers, Brazil lends to the two largest industries almost equally, while crediting of the remaining industries does not go slightly higher than 6% of the total number of credits, as you can see at picture below (Fig. 4).

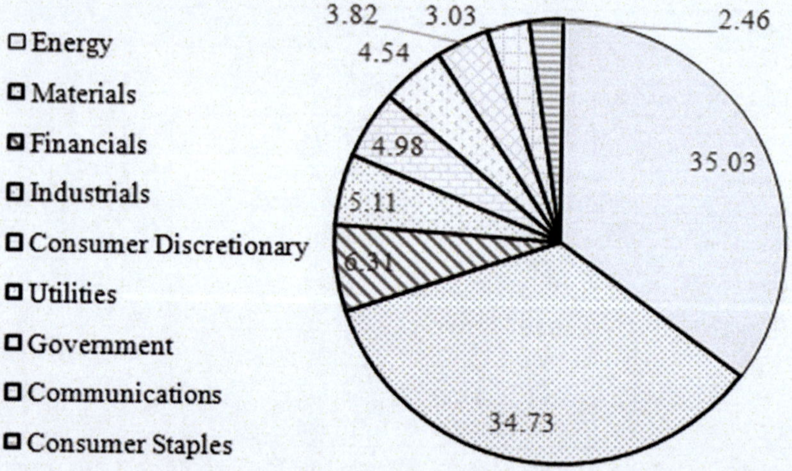

Fig. 4. Credits to key sectors of Brazil (%). *Source:* Bloomberg, date of access 12.01.2021

Analysis of the data of the Chinese economy showed the dominance of the financial sector in the credit resources market. The total volume of lending to this sector of the economy amounted to 4 billion US dollars. It constitutes 45% of the total lending to various sectors of the PRC economy. The country has a sufficiently developed financial infrastructure, a stable and tightly regulated financial market rather attractive to foreign companies. It is important to note the Bank of China is the leading participant in the Chinese credit market, coordinating the implementation of the country's credit policy.

International credit institutions have the limited license. Such policy excludes the possibility of competition between Chinese banks and foreign ones. This approach seems to be very reasonable in the conditions of the planned economy of the PRC. It is worth noting that the promotion of market transformations in the Chinese economy assumes a long process excluding “shock therapy”. the last characterizes several Central and Eastern European countries and associated with a sharp decline in the living standards of the population in the countries of this region. PRC, the largest country - a participant in world economic relations, with a key reserve currency, does not allow market experiments. China is aimed at improving financial relations based on sustainable development. PRC transformed into the exporting economy. Tables 15, 16 and 17 present data on the Chinese credit market.

Table 15. Credits by sectors of the Chinese economy.

Sectors of the economy	The volume of crediting ($ mln)	Deals	Participants
Financials	254829.26	729	465
Industrials	85005.48	248	221
Consumer Discretionary	62904.44	223	170
Materials	45709.73	156	106
Energy	35672.76	62	48
Technology	26687.8	70	47
Utilities	21925.9	61	44
Communications	13453.29	10	7
Consumer Staples	11325	46	35
Health Care	5762.08	43	38
Government	2745.56	8	2

Source: Bloomberg, date of access 12.01.2021

Table 16. Largest borrowers in China.

Sectors of the economy	Credit market participants (borrowers)	The volume of lending ($ mln)	Deals (as on 31.12.2020)
Financials	CIF Hoings Group C Ltd	2671.14	16
Industrials	China COSCO Shipping Corp Ltd	6502.3	3
Consumer Discretionary	Fosun International Holdings Ltd	2637.47	6
Materials	HBIS Group Co Ltd	10835.28	12

(*continued*)

Table 16. *(continued)*

Sectors of the economy	Credit market participants (borrowers)	The volume of lending ($ mln)	Deals (as on 31.12.2020)
Energy	Shandon Qingyuan Group Co Ltd	1723.89	5
Technology	Semiconductor Manufacturing International Corp	2169.35	16
Utilities	Beijing Enterprises Group Co Ltd	1749.95	5
Communications	Naspers Ltd	4780	2
Consumer Staples	China Mengniu Dairy Co Ltd	1791.83	4
Health Care	Fosun International Holdings Ltd	146.74	3
Government	People's Republic of China	2615.56	7

Source: Bloomberg, date of access 12.01.2021

Table 17. Major creditors to the Chinese economy.

Sectors of the economy	Credit market participants (creditors)	The volume of crediting ($ mln)	Deals (as on 31.12.2020)
Financials	Bank of China	61027.00	298
Industrials	Bank of China	27991.64	144
Consumer Discretionary	Bank of China	12355.59	98
Materials	Bank of China	17087.99	84
Energy	Bank of China	10148.69	28
Technology	Bank of China	4535.39	14
Utilities	Bank of China	3979.58	25
Communications	Bank of China	904.09	4
Consumer Staples	Bank of China	1324.00	24
Health Care	Bank of China	1931.16	25

Source: Bloomberg, date of access 12.01.2021

Figure 5 shows, that the largest borrowers in China are financial institutions, that significantly outperform other sectors of the economy in terms of borrowing. From the previous table, you can see that these are mostly domestic credits.

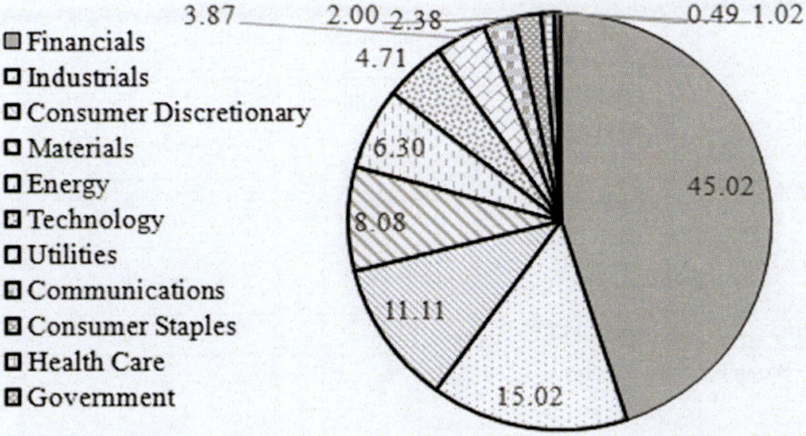

Fig. 5. Volume of credits to the key sectors of Chinese economy (%). *Source:* Bloomberg, date of access 12.01.2021

The analysis of the data obtained, considering the cluster analysis, showed that the strategic interests of the BRICS coincide within the framework of financing of the activities of economic entities in the following sectors of the economy: production, finance, energy, technology, interstate loans, financing of trade in Industrials, Consumer Staples, and Consumer Discretionary.

The data obtained made it possible to identify the peculiarities of the other developing credit market – the market of the EAEU.

The government is the main participant of the credit market in Armenia with the volume of borrowings USD 586 million. According to the rating agencies, the country currently has a low credit rating with a high probability of default. Financial resources provided by state institutions are mainly used to cover external and internal debt. The Bank of Armenia, Armeria Group Cy Ltd, Spayka LLC are the main creditors of economic entities in Armenia, confirming the presence of national banks in the country's credit market [3]. Information characterizing the credit market in Armenia is presented in Tables 18, 19 and 20.

Table 18. Credits by sectors of the Armenian economy.

Sectors of the economy	The volume of credits ($ mln)	Deals	Participants
Government	568.56	7	n/a
Financials	69.04	3	n/a
Materials	43972	1	n/a

Source: Bloomberg, date of access 12.01.2021

Most of the new credits in Armenia are aimed at refinancing public debts, rather than funding real expanded production (Fig. 6).

Table 19. Largest borrowers in Armenia.

Sectors of the economy	Credit market participants (borrowers)	The volume of credits ($ mln)	Deals (as on 31.12.202)
Government	Republic of Armenia	560.23	6
Financials	Ameria Group Cy Ltd	46.20	1
Industrials	Spayka LLC	21.50	1

Source: Bloomberg, date of access 12.01.2021

Table 20. Largest creditors of the Armenian economy.

Sectors of the economy	Credit market participants (borrowers)	The volume of credits ($ mln)	Deals (as on 31.12.2020)	Shares of the market (%)
Government	Republic of Armenia	560.23	6	98.53
Financials	Ameria Group Cy Ltd	46.20	1	66.92
Industrials	Spayka LLC	21.50	1	100.00

Source: Bloomberg, date of access 12.12.2020

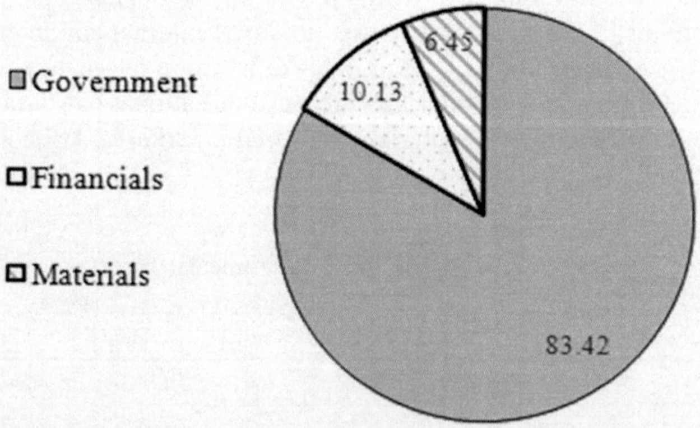

Fig. 6. Breakdown by market size in Armenia (%). *Source:* Bloomberg, date of access 12.01.2021

As in Armenia, the public sector is the main borrower in Belarusian economy with a total volume of debts in USD 1.1 billion at the end of 2020. The financial sector ranks second with a total volume of credits in 532 million USD. According to the available information, the government of the country does not have sufficient funds to solve the problem of external debt. High credit risks do not contribute to the expansion international banking and its participants in Belarus. Data, characterizing credit market in Belarus is presented in Tables 21, 22 and 23.

Table 21. Credits by sectors of the Belarusian economy.

Sectors of the economy	The volume of credits ($ mln)	Deals	Participants
Government	1108.6	10	n/a
Financials	532.82	12	2
Utilities	191.12	2	n/a
Consumer Staples	82.05	3	n/a
Materials	64.06	2	n/a
Industrials	41.56	2	n/a
Consumer Discretionary	12724	2	n/a

Source: Bloomberg, date of access 12.01.2021

Table 22. Largest borrowers in Belarus.

Sectors of the economy	Credit market participants (borrowers)	The volume of credits ($ mln)	Deals (as on 31.12.2020)
Government		0	
Financials	Alfa-bank	240.83	3

Source: Bloomberg, date of access 12.01.2021

Table 23. Largest creditors of the Belarusian economy.

Sectors of the economy	Credit market participants (borrowers)	The volume of credits ($ mln)	Deals (as on 31.12.2020)	Share of the market (%)
Government	Republic of Belarus	1108.60	10	100.00
Financials	Belarusbank	336.86	3	63.22
Utilities	Minskvodokanal UE	191.12	2	100.00
Consumer Staples	Eurotorg Holding PLC	75.83	2	92.42
Materials	VMG Industry	64.06	2	100.00

(*continued*)

Table 23. (*continued*)

Sectors of the economy	Credit market participants (borrowers)	The volume of credits ($ mln)	Deals (as on 31.12.2020)	Share of the market (%)
Industrials	Stadler Rail AG	22.62	1	54.43
Consumer Discretionary	Bel Parfum Distribution Group	44021	1	80.00

Source: Bloomberg, date of access 12.01.2021

Figure 5 shows that in Belarus as in Armenia the government is the greatest, creditor of the national economy (Fig. 7).

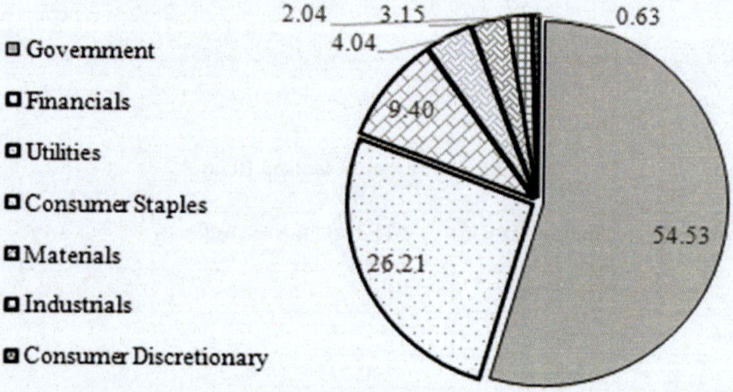

Fig. 7. Breakdown by market size in Belarus (%). *Source:* Bloomberg, date of access 12.01.2021

Production of goods is leading in Kazakhstan in terms of the total borrowing volume at the end of 2020. The debt amounts to USD 2.3 billion (30% of issued loans). The country is also experiencing problems with paying off its external debt. Banks with foreign participation are represented on the credit market of Kazakhstan: Credit Suisse, Japanese financial conglomerates Mizuho Financial, Mitsubishi UFJ Financial Group Inc. Despite stable economic relations, Russian banks refrain from expanding their activities and lending to Kazakhstani business entities. Information characterizing the credit market in Kazakhstan is presented in Tables 24, 25 and 26.

Table 24. Credits by sectors of the Kazakh economy.

Sectors of the economy	The volume of crediting ($ mln)	Deals	Participants
Materials	2371.09	6	4
Government	1750	2	1

(*continued*)

Table 24. (*continued*)

Sectors of the economy	The volume of crediting ($ mln)	Deals	Participants
Energy	1457.53	8	5
Financials	972.01	8	7
Industrials	634	3	3
Communications	368.33	5	1
Utilities	212.13	5	5
Consumer Discretionary	107.63	2	2
Consumer Staples	6,86	1	1

Source: Bloomberg, date of access 12.01.2021

Table 25. Largest borrowers in Kazakhstan.

Sectors of the economy	Credit market participants (borrowers)	The volume of credits ($ mln)	Deals (as on 31.12.2020)
Materials	KAZ Minerals PLC	1825	3
Government	Republic of Kazakhstan	1750	2
Energy	Beineu-Shymkent Gas Pipeline LLP	720	1
Financials	NATIONAL WELL-BEING FUND	600	1
Industrials	BAKAD Investment & Operation LLP	585	1
Communications	NATIONAL WELL-BEING FUND	368.33	5
Utilities	CAPEC Green Energy LLP	105.72	1
Consumer Discretionary	Aktau Tourism City Ltd	98.31	1
Consumer Staples	Resmi Group Ltd	6,86	1

Source: Bloomberg, date of access 12.01.2021

It is obvious that in a resource-producing country like Kazakhstan, the materials sector is primarily credited by national and international financial institutions (Fig. 8).

Data on the credit market in Kyrgyzstan are not sufficiently presented. In November 2020, Moody's agency assigned a B2 credit rating to Kyrgyzstan. The outlook is negative, which is due to the difficulty in the development of the credit market and the reduction in the presence of international credit institutions with foreign participation in Kyrgyzstan.

Table 26. Largest creditors of the Kazakh economy.

Sectors of the economy	Credit market participants (creditors)	The volume of credits ($ mln)	Deals (as on 31.12.2020)
Materials	Credit Suisse	225.00	1
Energy	Mizuho Financial	40.00	1
Financials	Mitsubishi UFJ Financial Group Inc	600.00	1

Source: Bloomberg, date of access 12.01.2021

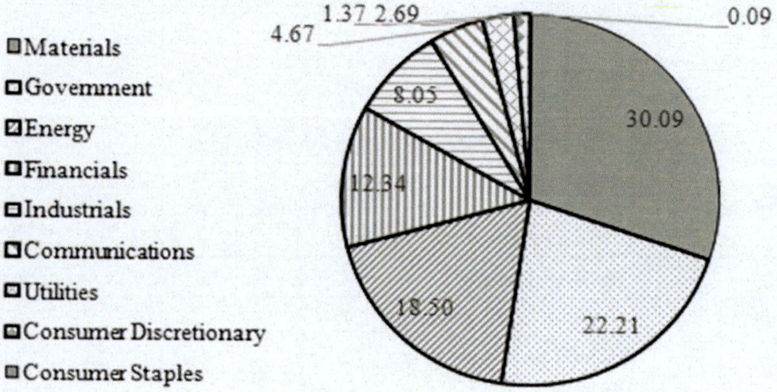

Fig. 8. Breakdown by market size in Kazakhstan (%). *Source:* Bloomberg, date of access 12.01.2021

In general, it seems possible to draw a conclusion about the favorable development prospects of the countries of the Eurasian Economic Union in the field of production crediting and interstate financing the development of economic processes, based on the analysis of the EAEU credit market. It seems to be reasonable to consider the historically and geographically justified attraction of the considered clusters.

4 Discussions

The analysis of the BRICS and EAEU credit markets made it possible to identify problems and reveal the prospects for cooperation between the countries [4–10].

Considered above developing countries - members of interstate associations have common development goals. The achievement of them is associated with ensuring sustainable growth of national economies and improving the living standards of the population. In this regard, each country determines short-term and long-term development goals, and determines the priorities for financing certain sectors of the economy. For example, production, energy, communications are the key credited sectors of the economy in Brazil, Russia, India, South Africa. Borrowed funds provide the development of these sectors with the necessary resources. In China, the service sector including finance continues to develop. It is important to note that the Chinese economy continues to develop. Financing reproduction processes is sustainable, despite the difficult epidemiological situation caused by the COVID19 pandemic. In general, the sustainable development of the economies of the BRICS countries creates a favorable investment climate and attracts foreign financial institutions to participate in the BRICS credit markets. In turn, most of the EAEU countries are experiencing the burden of debt problems, the attracted resources are used to cover the obligations of members-countries. However, attracting foreign investors to participate in multilateral financing on flexible terms for lenders and borrowers has favorable prospects.

The results allow us also to stress on the necessity of optimizing the intergovernmental BRICS-EAEU projects and programs of trade and economic cooperation in the difficult period of the global epidemiological threat and of overcoming its consequences. Flexible conditions for interaction between the parties, the provision of preferences for BRICS-EAEU business entities will create conditions for expanding instruments for direct and indirect financing the key sectors of the BRICS-EAEU economies and the development of trade.

5 Conclusion

The conducted analysis confirmed the need in the formation of short-and long-term trade and investment strategy for BRICS-EAEU economic partnership. it should be noted that the priorities of long-term strategy for BRICS-EAEU is strengthening the position of countries in the global economy. Evaluation of financial markets indices of the BRICS-EAEU showed contribution of a coherent long-term strategy for BRICS-EAEU economic partnership to the accumulation of the resources necessary to address important strategic objectives and achievement of sustainable and inclusive development. In turn, the short-term strategy promotes the implementation of mechanisms of joint action in the agreed period. Short-term strategy considers the volatility of the world economy and finance, the strategic interests of partner countries soon. The proposed inter-state support of BRICS-EAEU business entities creates favorable conditions for the effective solution of problems and reaching the strategic goals. Exhibitions, trade fairs on a regular basis contribute to the strengthening of partnerships in promoting development, transfer of competences and broadening the cooperation between the parties through dialogue and exchange of experience in issues of multilateral cooperation. It seems reasonable to accelerate participation of development banks BRICS-EAEU participation and institutional investors of all forms of ownership in the processes of acceleration business processes. Harmonization of trade, financial and other types of economic contacts between BRICS-EAEU will also expand cooperation between the parties. In order to manage the risks of economic and financial contacts it seems appropriate to speed up the process of harmonization of regulation of member countries and to implement ecosystems approach to develop value chains cooperation for the benefit of member countries.

References

1. Arapova, E., Chkoniya, A.E.: BRICS trade potential: problems and prospects. Natl. Interests Priorities Secur. **7**, 152–164 (2016)
2. Bank of Russia, Meeting of finance ministers and Central Bank governors of the BRICS countries (2019). https://www.cbr.ru/today/ms/inf_2019
3. Gusarova, S.A.: The development of electronic commerce in the BRICS countries. Econ. Sci. **11**(156), 34–42 (2017)
4. International Trade Center, Trade Map (2018). https://www.trademap.org
5. Martínez-Zarzoso, I.: The Log of Gravity Revisited. Appl. Econ. Taylor Francis J. **45**(3), 311–327 (2013)
6. Mathur, S., Dasgupta, M.: BRICS: Trade policies, institutions, and areas of deepening cooperation. Centre for WTO studies Working paper (2013)
7. Mogilat, A.N., Sal'nikov, V.A.: Trade effects estimation for the case of eurasian economic space countries: application of regional gravity model. J. New Econ. Assoc. **3**(27), 80–108 (2015)
8. Tinbergen, J.: Shaping the World Economy: Suggestions for an International Economic Policy. Twentieth Century Fund, New York (1962)
9. UNCTAD, Publications Library (2018). https://unctad.org/en/PublicationsLibrary/tdr2018overview_ru.pdf, Accessed 10 Apr 2019
10. Yarygina, I.Z.: Public-private partnership of the BRICS member countries as a risk reduction factor for the global economy. Econ. Taxes Law **4**, 56–59 (2013)

Updated Methodology for Task Knowledge Based Development of Parallel Programs

Alexander V. Ilyin[1] and Vladimir D. Ilyin[2(✉)]

[1] State Research Institute of Aviation Systems, 7, Viktorenko Street, 125319 Moscow, Russia
[2] A. A. Dorodnicyn Computing Center of the Federal Research Center "Computer Science and Control" of the Russian Academy of Sciences, 44/2, Vavilova Street, 119333 Moscow, Russia

Abstract. The updated methodology for *task knowledge based* development (*tkb*-development) of parallel programs includes our updated methods for *tkb*-construction of *non-parallel program systems* (*nps*) and automatic parallelizing. *TKB*-system of *parallel program systems* construction (*pps*-construction) consists of five components: the *pps*-construction task knowledge base, the *tkb*-constructor of parallel programs with automatic parallelizer embedded, the *tkb*-constructor of runtime support and graphic tools. *PPS*-construction system produces *pps* to be run in a specified runtime environment. The foundations of updated methodology for *pps*-construction, task knowledge representation language *OBRAZ,* automatic parallelizing method *MPM* and *cooperative problem solver* (*CPS*) based on digital twins are presented in short.

Keywords: Task knowledge based development of parallel programs · Task knowledge representation language · Automatic parallelizing method · Cooperative problem solver based on digital twins

1 Introduction

Since programming has existed, programmers have been looking for ways to improve their productivity [1–8]. Modern instrumental programming systems, as a rule, have built-in mechanisms for applying templates of developed products and ready-made solutions for individual components. Such systems contain tools for editing program texts, translators, tools for debugging, testing and optimizing programs, graphical tools for interface design, text templates for programs of various types, program libraries, etc. Traditional tools for developing parallel programs include languages [9], translators [10, 11] and environments for parallel programs [12, 13].

Currently, there are a huge number of solvable tasks, programmed in various languages and implemented as part of system and application complexes. As the number of programmed tasks grew, so did the need to reuse proven programs. Already at the initial stages of the development of programming technologies, together with languages and translators, software developers began to create various program libraries [14, 15].

What should be the system of knowledge about programmable tasks to serve as the basis for effective automation of program development? The answers to this question are

R. Silhavy et al. (Eds.): CoMeSySo 2021, LNNS 231, pp. 319–328, 2021.
https://doi.org/10.1007/978-3-030-90321-3_25

constantly relevant. Along with other answers to similar questions in [4, 5] an approach to program construction was proposed that incorporated an interactive synthesis of task constructive objects. In the system of automated construction of programs (called the system of generating programs), the system of knowledge about programmable tasks is presented in the form of structures built from task constructive objects. In [16], this approach was developed for the construction of parallel programs.

It is advisable to create systems of knowledge about programmable problems, using which it would be possible to find systematized data on problem statements and algorithms, programs and test examples. In such systems, each problem must be represented as a respectively specified constructive object. Software development instrumental systems should be linked to regularly updated knowledge systems about programmable tasks.

1.1 Markuping Text Fragments and Writing Formulas

For markuping text fragments, the s-modeling language *TSM* (*Textual Symbolic Modeling*) is used:

□<text fragment>□ means a definition;

∆<text fragment>∆ means an example;

◊<text fragment>◊ means a note.

Italics are used to highlight the concepts and sentences, to which the authors want to attract the attention, and also for variables names [5].

2 Symbolic Modeling (*s-Modeling*) of Programmable Tasks (*s-Tasks*): Basic Concepts

□ *A system Sc of notions* is {a set S^{Sc} of notions, a family *rel* (S^{Sc}) relations defined on S^{Sc}}. □

□ *S-task* is {*Formul, Rulsys, Alg, Prog*}, where *Formul* is the statement of the task; *Rulsys* is the set of systems of *mandatory* and *orienting* rules for solving the task, *Alg* is the union of sets of algorithms, each of which corresponds to one element from *Rulsys*; *Prog* is the union of sets of programs, each of which is set in accordance to one of the elements of *Alg*.

A task formulation *Formul* is {*Mem, Rel*}, where *Mem* is a set of task notions on that the partitioning is given [$Mem = Inp \vee Out$ ($Inp \wedge Out = 0$)], and a set *Rel* of relations between the notions that defines the binary relation $Rel < Inp * Out$. A set of *Mem* is called the *task memory*, and *Inp* and *Out* are its *input* and *output*, the values of which are supposed to be set and searched, respectively.

Each element from *Rulsys, Alg* and *Prog* has *a usage specification.* □

◊ In the general case, the sets *Rulsys, Alg,* and *Prog* can be empty: the number of their elements depends on the degree of study of the problem. ◊

The *Rulsys* element usage specification includes the specification of the *task solver* type [autonomous *s-machine* (desktop, laptop, etc.), s-machines cluster, etc.]; requirement for data security, etc. The specification for the use of elements from *Alg* includes data on the permissible modes of operation of the task solver (automatic local, automatic distributed, interactive local, etc.), on the requirements for the obtained result, etc. The specification for program usage includes data on programming languages, operating systems, etc.

The notion of *task constructive object* (*tco*) is introduced along with the number of other notions derived from it. They form the complex of notions used in *s*-modeling of the process of transition from the problem to the programmed problem. This complex should allow the software developer to formulate his ideas in *tco*-terms and to advance step by step towards the producing of the specified program system, keeping in mind the meaning of every transition [4, 5].

3 PPS-Construction

The *pps*-development in *pps*-construction environment is considered as *tco*-constructions building. The *tco*-concept is used to support the «stage to stage» transition from a task formulation (*what*-representation of task is being programmed) to a program (final *how*-representation) [4].

TCO is presented by the finite tree, where task formulation corresponds to the tree root and programs correspond to the leaves. The intermediate (on the way from formulation to program templates) are necessary for taking into account the implementation specification. The process of *pps*-construction consists of other processes which are performed at different stages of working with *tco* in the world of programmable tasks. This world is understood as the aggregate of *tco*-spaces.

□ *PPS*-construction is a formalized process based upon both the set of *tco* and the family of *tco*-construction rules. The representation of *tco* is a special graph called a *task relation graph* (*g-graph*). The family of *tco*-construction rules is based on both the set of *g*-graph operations and the set of inference procedures which search subgraphs of *g*-graphs called *solving structures* [5]. *PPS*-construction is realized in spaces of *tco* and consists of such processes as *tco*-(*creation, concretization, specialization, generalization* and *notion shell replacement*). From the formal view point, *pps*-construction may be considered as a process of inductive *tco*-set construction. □

□ The construction processes of *nps* and *pps* have common components. The first of them is creation of *basic tco* (*b-tco*) and it is called *p-creation.* □

□ *The specialization of b-tco and the derived ones* (*p-specialization*) is the process of setting the correspondence of some *tco* and the non-empty set of derived *tco.* A derived *tco* is designed to solve a problem which is a partial case of the problem represented by its base *tco.* □

□ *TCO notion shell replacement* (*p-replacement*) assigns names to certain problem domain notions in accordance with a given lexicon. □

□ Making *tco-constructions* is the process of linking *tco* by means of special communication functions. □

□ The process of *tco-concretization* called *p-concretization* defines the correspondence between the problem formulation and the method of its solution, program template and program in accordance with the implementation specification. □

There are some additional components of *pps*-construction that will be mentioned later (Sect. 4).

4 OBRAZ Language for Task Knowledge Representation

The conception and formalization of task knowledge representation are developed to allow the building of the task knowledge base of the system of computer aided program construction.

OBRAZ is designed to describe notion systems, which contain knowledge that is necessary for supporting the processes of *tco-(definition, specialization, generalization, construction* and *concretization*) [4, 5]. The language implements an idea to replace the development of problem specification by the development of the specification of application domain notions system. Interpreting of the notions system specification produces specifications of tasks that are necessary for solving a given problem. These tasks specifications and the specification of *task solving environment* (computational resources, data sizes, etc.) provide all the data necessary to build the program system.

The central element of OBRAZ is a *notion* defined by its name and specification. A notion specification is a set of other subordinate notions (attributes) and the relations between them. A language statement usually contains a notion definition including such relations as affiliation and inheritance. Language statements can also describe restrictions for a notion value set and equivalence of notions as a special case of restriction.

The notion can have several definitions that express various points of view. The language interpreter considers different definitions as separate notions. A user can set up his own point of view to the problem area, selecting a single definition among multiple possible ones.

OBRAZ language is developed to be a tool that supports the representation of knowledge, which is necessary to build a target program system producing values of some notions (*output*) by means of processing of values of other notions (*input*). This goal has determined that the notion model should be based on notions represented by programmable tasks and called *task relations*. Other than task relations, the *notion specification* can contain some relations on notion value set that are called *constraints* and are used to define suitable conditions when application of this specification is possible.

□ The task that formulates a relation on a set of notions is also a notion with special meaning. *Task specification* contains two non-intersecting subsets of attributes called *input* and *output*. The aggregation of *input* and *output* is called *task memory*. □

The program construction environment has no means of evaluating task relations stored in its database. This determines the absence of variables in OBRAZ, because there is no temporary data to store. ◊ The *notion specification* is close to the concept of *object class* in object-oriented languages. ◊

To be brief, we'll sometimes skip in subsequent paragraphs words «specification» or «description», so one should consider terms «notion» and «task» as «notion specification» and «task specification». Text enclosed in /* and */ (multiple lines) is a comment. Text between // and end of line is also a comment.

4.1 A Notion Representation

□ The basic structural unit in OBRAZ language is a *notion definition*. It is represented by its name and specification that in turn is a collection of definitions of other notions and relations between them. Specification can be formed from other specifications using inheritance. □

Some notions in OBRAZ have built-in specifications. There are the most general notions such as «text», «number», «set» for which OBRAZ language supports constants and special syntax extensions that increase readability and compactness of notation. There are also notions that are used to build the task based models, such as «task relation», «task memory element», «input», «output», «task graph», «query». All these notions are called predefined.

□ Every notion in OBRAZ is defined by means of specifications of some other notions. If notion*A* is defined inside notion *B,* then we'll say that *B* is the owner of *A* and *A* is the attribute of *B*. This creates an affiliation relations hierarchy that covers the whole notion system. Every notion that can be specified using OBRAZ is the owner of something or an attribute of something or both. Affiliation relations are transitive. For any notion, the owner of its owner is also its owner, and the attribute of its attribute is also its attribute. □

4.2 Operations on Notion Specifications

□ *Notion specifications* can be built from other notions specifications by applying to them operations of *concatenation, intersection* and *exclusion.*

General form of notion definition is

<name> : <expression>;

where <expression> contains operators: «+» – for concatenation, «–» – for exclusion and «*» – for intersection.

Operands in an expression are specifications represented by an explicit set of definitions enclosed in curly braces «{}» or by names of notions possessing them. □

□ The notion is called *base notion* if its specification is an operand in an expression defining another notion that is called a *derived* one. □

The concatenation operator «+» adds to its left-side operand all attributes, relations and constraints that are contained in the right-side specification. A notion specification defined by concatenation of some base notions contains a specification of its base. Those attributes that appear in a result specification from specifications of other notions are called *inherited* ones, unlike own attributes that were defined inside curly braces.

The exclusion operator «–» selects only those attributes and relations from the left-side specification that are not defined in the right-side specification. The latter can contain some notions that are unknown in the left-side specification – these notions are ignored. A specification to be excluded can be represented by a notion name or as an explicit list of attributes and relations enclosed in curly braces.

The group of two or more specifications separated by intersection operator «*» defines a new specification containing only those attributes and relations that are defined in every member of this group.

Relations defined using operators «=» and «<>» are more than restrictions on the value sets. The operator «=» also means synonymy, mutual concatenation of specifications and establishing of links between memory elements of *tco*-construction. The operator «<>» prohibits the concatenation and equalizing of the notions it connects and similarly for their derivatives.

4.3 A Task Relation

□ *A task relation* is a notion whose specification is formed similar to a specification of any other notion. It is predefined in OBRAZ with the following description:

```
trel {
        mem;
        // generalization of the notion of «memory element»;
        input, output : mem;
        // «input» and «output» notions are derived from it
    }
```

This specification is interpreted as follows: there is the «task relation» notion (*trel*) that has the attribute «task memory» (*trel.mem*) and attributes «task input» (*trel.input*) and «task output» (*trel.output*) derived from *trel.mem.* □

To specify some notion x as a task relation one should derive it from *trel*:

x : trel + {...};

Concatenation of x attributes {...} with *trel* specification gives to x inherited attributes *x.mem*, *x.input* and *x.output*.

The t*ask relation* notion has special built-in derivatives: *function, equation, program.*

□ The *function* notion differs from the inherited *trel* notion with only the output element being implicitly equivalent with the function itself. □

□ An *equation* is a task relation where all memory elements are derived from a *mem* notion (not from *input* or *output*). The language interpreter dynamically redefines the equation's memory according to its current needs during the search of a problem solution so that one of the memory elements becomes output and all the others form the input. So an equation can be considered as a compact notation defining task relations in one specification. □

□ The *program* notion is a task relation containing some attributes that allow for the description of the process of generating source or object program code. The attribute set structure depends on the software and hardware environment used to build the final program system and execute it. □

◊ The built-in OBRAZ program notion is a template and cannot specify any program code generation process. ◊

There are some notions in the construction system knowledge base that are derived from «program» and define proper attributes to describe construction in various special environments.

4.4 A *Task Relation Graph* (*Tr-Graph*)

□ Every notion can be considered as a *tr*-graph if it contains some task relations or other *tr*-graphs that are *tr-graph vertices.* The union set of memory elements of all tasks in *tr*-graph forms *tr-graph memory.*

Memory element equivalence of *tr*-graph vertices forms vertices memory intersections that are called *tr-graph edges.* Every group of equivalenced notions with at least one task memory element in the group, forms one *tr-graph memory element*. All members of such a group are synonyms. Every task memory element that isn't connected with any other notion also forms one *tr*-graph memory element. □

5 MPM: The Automatic Parallelizing Method

MPM is developed to support construction of safe and efficient *pps*. MPM-parallelizer is designed to transform a non-parallel algorithm to a parallel one.

□ *MPM-parallelizer* is the abstract machine for the automatic parallelizing of *tco*-represented algorithms. Its input is a *nps*-algorithm (*A*) and a runtime environment specification. Its output is an *pps*-algorithm (*B*) and a runtime environment configuration. □

5.1 A *Task Constructive Automaton* (*TCA*)

TCA is designed for the representation of an algorithm to be parallelized. □ Algorithm *A* is considered as operations set *S* on which order relation *R* is defined ($A = <S, R>$). If an algorithm $PA = <SPA, RPA>$ is derived from an algorithm $NA = <SNA, RNA>$ by parallelizing then $SPA = SNA$ and either $RPA \neq RNA$ or $RPA = RNA$. If $RPA = RNA$ an algorithm *NA* can not be parallelized. □

□ *TCA* is the task constructive object to be executed in a parallel environment. It has hierarchical unlimited memory (*TCA*-memory). Input (*TCA*-input) and output (*TCA*-output) subsets are defined on *TCA*-memory. The intersection (*TCA*-input * *TCA*-output) = 0. If *TCA*-input is available (*TCA*-input[av]) *TCA* is ready to be executed (*TCA*[re]). When *TCA*[re] obtains computational resource it becomes executable (*TCA*[ex]). Then output elements of *TCA*[ex] become available (*TCA*-out[av]). □

There are two types of *TCA*: *TCA*-unit and *TCA*-block. A *TCA*-unit has one-level unlimited memory (Δ a single operation of algorithm may be represented by *TCA*-unit Δ). A *TCA*-block consists of *TCA*-units that are linked by memory relations. A *TCA*-block memory elements set is union of sets of the memory elements of the *TCA*-units included. *TCA*-block may consist of other *TCA*-blocks whose behavior is the same as the behavior of *TCA*-units.

□ Communication between *TCA*-block memory and memory of *TCA*-unit obeys the following assertions.

- If the memory state condition of *TCA*-block output is «true» (all elements are computed) then memory access of *TCA*-unit to *TCA*-block is allowed to read from *TCA*-block and write to *TCA*-unit memory
- If the memory state condition of *TCA*-unit output is «true» then memory access of *TCA*-block to *TCA*-unit is allowed to read from *TCA*-unit and write to *TCA*-block memory. As a result *TCA*-block memory state is altered. □

5.2 MPM-Parallelizing Stages

There are two situations of MPM-parallelizing: the first is outside the *pps*-construction system and the second is inside it. In the first situation the non-parallel algorithm is represented by non-parallel source code and a parallel algorithm – by parallel source code. In the second situation the non-parallel algorithm is produced from problem specification and an algorithm representation is a non-parallel *TCA*-block. The first situation differs from the second one in that the automatic parallelizing process includes two additional components: transformation from non-parallel source code to non-parallel *TCA*-block and transformation from parallel *TCA*-block to parallel source code.

□ Thus *MPM-parallelizing stages* are transformations from:

(a) non-parallel source code to non-parallel *TCA*-block;
(b) non-parallel *TCA*-block to maximum parallel *TCA*-block;
(c) the maximum parallel *TCA*-block to parallel *TCA*-block implemented in accordance with computational resources specification;
(d) parallel *TCA*-block to parallel source code.

The stage (b) is the main one. This transformation is a step by step process that is implemented in accordance with the assumption that the computational resources of the maximum parallel *TCA*-block are unlimited.

Stage (b) obeys the following rules:

1. If on parallelizing the step*k* of non-parallel *TCA*-block there are one or more *TCA*-units that are *TCA*[ex] then the transformator includes them in the box *k* of parallel *TCA*-block ($k = 1, 2, \ldots$).
2. The parallel *TCA*-block is a sequence of boxes ($k = 1, 2, \ldots, n$). □

6 Cooperative Problem Solver (CPS) Based on Digital Twins

CPS may be used in *e-services for tkb-development of pps* (*cps*-services).

The elements of *cps-cooperation* (called *dt-inf*s) are considered as learnable *s-machines* [5] for solving problems. A *dt*-inf is an *s*-machine with a finite set of states, having a *given goal* understood as the set of tasks solvable on the *tr*-graph. In the processes of solving problems, *dt*-infs interact with each other and environmental elements (*enels*) through the exchange of unified messages.

The behavior of *dt*-infs is determined by the following requirements:

- to receive and send messages, interacting with other *dt*-infs and *enels*;
- to interpret messages, giving out information about errors if they occur and allowing correction of received messages;
- to perform the work requested in the message, allowing dialogue, pauses and cancellation;
- to formalize the result of the completed work for sending to the correspondent;
- to participate in cooperations of *dt*-infs.

Table 1. *DT*-inf as communication s-machine

The name of *dt*-inf component	Work performed by the *dt*-inf component
BEFORE	Initialization, startup, training
START	Receiving, buffering, and interpreting messages
STEP	Step in the process of completing the job requested in the message
END	Shutting down for the current message
AFTER	Shutting down *dt*-inf

Each component of the *dt*-inf can be in one of the states belonging to the set $State = \{head, tail, dial, intr, paus, canc, resl, fail, null\}$, where

head is the ready state;
tail – the job is finished;
dial – interaction with *enels*;
intr – interaction with other *dt*-inf;
paus – pause;
canc – cancel;
resl – result received;
fail – failed (the result could not be obtained);
null – a state that requires unintended intervention by the sender of the message (*enel* or *dt*-inf) [17].

7 Conclusions

The results of the experimental implementation of the first version of *the methodology for tkb-development of pps* has confirmed its effectiveness [5, 16].

The updated methodology focused on implementation in the form of e-services based on *cps* (*cps*-services) includes a model of a *cooperative problem solver* (*cps*) based on digital twins [17].

There is reason to believe that the use of *cps*-services for software development will significantly increase the efficiency and reliability of designed software systems.

References

1. Dijkstra, E.: Structured programming. Software engineering techniques. NATO Science Committee, (edited by J. N. Burton and B. Randell), pp. 89–93 (1969)
2. Wirth, N.: Program development by stepwise refinement. Commun. ACM **14**(4), 221–227 (1971). https://dl.acm.org/doi/10.1145/362575.362577. Accessed 15 June 2021
3. Luker, P., Burns, A.: Program generators and generation software. Comp. J. **29**, 315–321 (1986)

4. Ilyin, V.: The program generating system. Nauka, Moscow (1989). https://www.elibrary.ru/item.asp?id=24889586. Accessed 15 June 2021
5. Ilyin, A., Ilyin, V.: S-modeling of tasks and construction of programs. Institute of Informatics Problems, Russian Academy of Sciences, Moscow (2012). https://www.elibrary.ru/item.asp?id=25816802. Accessed 15 June 2021
6. Imam, S., Sarkar, V.: Integrating task parallelism with actors. In: The ACM International Conference on Object Oriented Programming Systems Languages and Applications, OOPSLA 2012, pp. 753–772. ACM, New York (2012)
7. Grasso, I., Pellegrini, S., Cosenza, B. et al.: A uniform approach for programming distributed heterogeneous computing systems. J. Parallel Distrib. Comput. **74**(12), 3228–3239 (2014). https://www.sciencedirect.com/science/article/pii/S0743731514001440. Accessed 15 June 2021
8. Meil, D.: The search for unlimited productivity. Commun. ACM **64**(6), 10–11 (2021). https://cacm.acm.org/magazines/2021/6/252830-the-search-for-unlimited-productivity/fulltext. Accessed 15 June 2021
9. Specialized parallel languages and extensions of existing languages. Parallel.ru (2021). https://parallel.ru/tech/tech_dev/par_lang.html. Accessed 15 June 2021
10. Jo, C., George, K., Teague, K.: Parallelizing translator for an object-oriented parallel programming language. In: The Tenth Annual International Phoenix Conference on Computers and Communications, Scottsdale, AZ, USA, 27–30 March 1991, pp. 265–271. IEEE (1991). https://ieeexplore.ieee.org/document/113821. Accessed 15 June 2021
11. Takehiko, N., Suda, R.: APTCC : auto parallelizing translator from C to CUDA. In: International Conference on Computational Science, ICCS 2011, Procedia Computer Science, vol. 4, pp. 352–361. Elsevier (2011). https://www.sciencedirect.com/science/article/pii/S1877050911000950?via%3Dihub. Accessed 15 June 2021
12. Butler, R., Gropp, W., Lusk, E.: A scalable process-management environment for parallel programs. In: Dongarra, J., Kacsuk, P., Podhorszki, N. (eds.) Recent Advances in Parallel Virtual Machine and Message Passing Interface, EuroPVM/MPI 2000. LNCS, vol. 1908, pp. 168–175. Springer, Heidelberg (2000). https://doi.org/10.1007/3-540-45255-9_25, https://link.springer.com/chapter/10.1007/3-540-45255-9_25. Accessed 15 June 2021
13. Ikeda, T., Yuen, S.: A reversible runtime environment for parallel programs. In: Lanese, I., Rawski, M. (eds.) Reversible Computation, RC 2020. LNCS, vol. 12227, pp. 272–279. Springer, Cham (2020). https://doi.org/10.1007/978-3-030-52482-1_18, https://link.springer.com/content/pdf/10.1007%2F978-3-030-52482-1_18.pdf. Accessed 15 June 2021
14. Sweger, K.: 1130 Scientific subroutine package (1130-CM-02X): Programmer's manual. IBM (1967). http://media.ibm1130.org/1130-006-ocr.pdf. Accessed 15 June 2021
15. Arushanyan, O., Volchenskova, N.: Providing internet access to the numerical analysis library of the research computer center of Moscow State University. Comput. Methods Program. **1**(2), 1–8 (2000). https://www.elibrary.ru/item.asp?id=1271758. Accessed 15 June 2021
16. Ilyin, V.: A methodology for knowledge based engineering of parallel program systems. In: Forsyth, G., Moonis, A. (eds.) The Eighth International Conference «Industrial and Engineering Applications of Artificial Intelligence and Expert Systems», IEA/AIE 95, Melbourne, Australia, 6–8 June, 1995, pp. 805–809. Gordon and Breach (1995)
17. Ilyin, V.: The model of the cooperative problem solver based on digital twins. Syst. Means Inform. **29**(2), 172–179 (2019). https://www.elibrary.ru/item.asp?id=41273564. Accessed 15 June 2021

The Support Vectors and Random Forest Methods Analysis in the Forecasting Customer Churn Problem in Banking Services

Roman I. Dzerzhinsky[1](✉), M. D. Trifonov[2], and E. V. Ledovskaya[2]

[1] The Applied Mathematics Department, MIREA – The Russian Technological University, The Information Technologies Institute, Vernadsky pros., 78, 119454 Moscow, Russia

[2] Applied Mathematics Department, MIREA – Russian Technological University, Vernadsky pros., 78, 119454 Moscow, Russia

Abstract. Based on the typical data analysis from the Kaggle website, the bank's clients probable predictive behaviour was investigated. Customer behaviour signs probability distributions families have been constructed, their mutual correlation has been checked by calculating the Pearson and Spearman coefficients, and characterizations classes have been constructed. The classes balancing is carried out using the completion synthetic method. Categorical features replacement was made using the principal components method. The resulting dataset was used to train support vector machine algorithms and a random forest and then used to predict customer churn. To check the models quality, the corresponding estimates are calculated and constructed. Thus, algorithms have been obtained that improve their work due to the experience - machine learning models accumulation. #COMESYSO1120.

Keywords: Random forest · Support vector machine · Principal components · Unbalanced classes · Distributions · Machine learning

1 Introduction

With the information technology development, various companies have been able to store information huge amounts about their customers. Such data is of some value since from them, it is possible to identify user behaviour in various patterns and try to predict their actions.

One of the banking sector forecasting tasks is forecasting customer churn. This task is quite relevant because attracting new customers is much more expensive than retaining old ones.

An effective solution to such problems is the machine learning methods use. The outflow problem is a binary classification, for which solution two methods are used in this work. The first is the support vector machine, which is linear classifiers representative but retains its versatility even in the data nonlinearity case. The second is the random forest, which is a non-linear classifier. In most tasks, it shows the best results, but it has its drawbacks.

R. Silhavy et al. (Eds.): CoMeSySo 2021, LNNS 231, pp. 329–336, 2021.
https://doi.org/10.1007/978-3-030-90321-3_26

This paper examines the data on an anonymous bank' clients taken from the Kaggle platform [1].

This work purpose is to compare the support vector machine [2] and the random forest model [3] in the predicting customer churn problem.

2 General Data Analysis and Visualization

For this work, we took information about the bank's clients, presented in the table in the csv format. This table contains 10127 examples and 21 features, including 8 categorical, 12 quantitative and one feature - customer identification number.

For classifiers quality training, the classes' ratio must be approximately the same. The leaving and remaining customers' ratio in the original dataset is shown in Fig. 1.

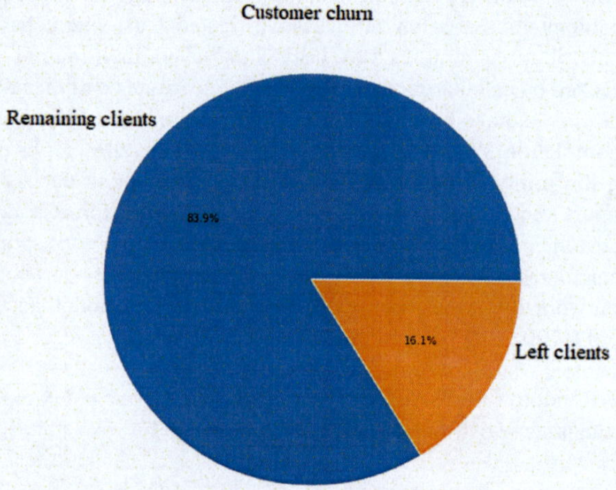

Fig. 1. Left and remained customers ratio.

Thus, the class ratio is approximately 5.25: 1, which is a poor indicator due to the non-probabilistic classification models sensitivity to unequal class sizes [4]. The imbalance can be corrected by various methods, in this work the synthetic minority addition method will be considered and used, or SMOTE [5].

The clients' age distribution is an almost normal distribution (see Fig. 2) with truncated tails due to the restrictions on the bank's clients age. This allows the confidence intervals construction to be used for various hypotheses statistical evaluation.

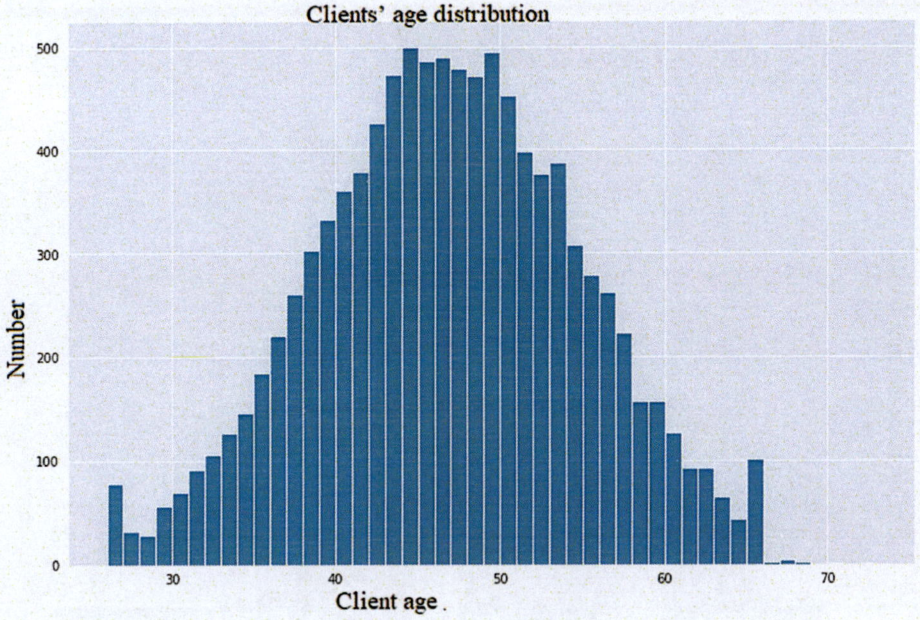

Fig. 2. Clients' age distribution.

3 Data Preprocessing

For the classifiers correct training, it is necessary to carry out the following operations: remove gaps, standardize quantitative features, and transform categorical features. There are no missing values in the original dataset, and categorical features are processed using One-Hot-Encoding. Because of this transformation, out of 8 categorical features, 22 binary are obtained.

The class imbalance problem can be solved using the synthetic minority complement method. This algorithm finds the nearest neighbours for a smaller class' each representative in the feature space. Then a random one is selected from them, and a new synthetic object is randomly generated on the interval between them. Thus, a smaller class is supplemented to a larger one or a given amount. This method is recommended to be used with an initial reduction in a larger class objects number, and then a minority addition. This allows not to generate synthetic objects a huge number that can distort existing patterns.

Numerous features slow down the models, so the resulting features are converted using the principal component method [6]. It allows reducing the features number while preserving the information maximum amount from them. The explained component variances plot is shown in Fig. 3:

For this work, it was found by the enumeration method that the optimal amount is 4 new principal components that explain the variance' 55%. For initial binary features such as a number, this is a sufficient result. A further increase in the principal components' number worsens the models quality final assessment.

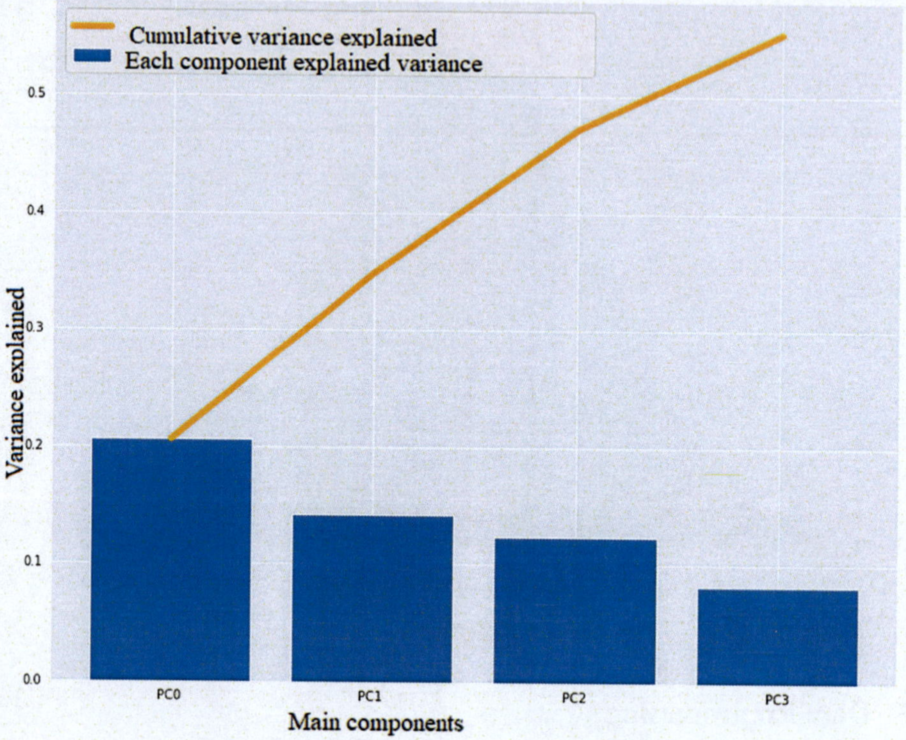

Fig. 3. Principal components explained variances.

To help in the features' selection to be used within the model, a Pearson correlation matrix [7] is built with the main components (Fig. 4):

Among the features pairs, which correlation exceeds the 0.5 value, pair one is automatically discarded.

Thus, using the correlation and enumeration matrix to achieve the best forecasting result, the following features were selected: the transactions total number for the last year, the transactions total number ratio in the fourth quarter to the first quarter, the purchased bank products and 4 new components total number.

4 Churn Forecast and Quality Assessment

To check the models overall quality and relevance with the resulting data, a cross-validation check is performed [8]. At each of the training sample sections, the F1-score of the classifier [9] trained on the remaining sections is checked. In this work, a division into 5 sections is made, the assessment for them each is shown in Fig. 5:

The following features of the BRICS credit market are revealed hereunder.

For the Russian Federation, the largest sector in terms of lending is the industrial sector, which accounts for 36.1% of total loans and amounts to 37 billion USD. Russian banks with foreign participation of ING Group (Netherlands), Commerzbank Group of

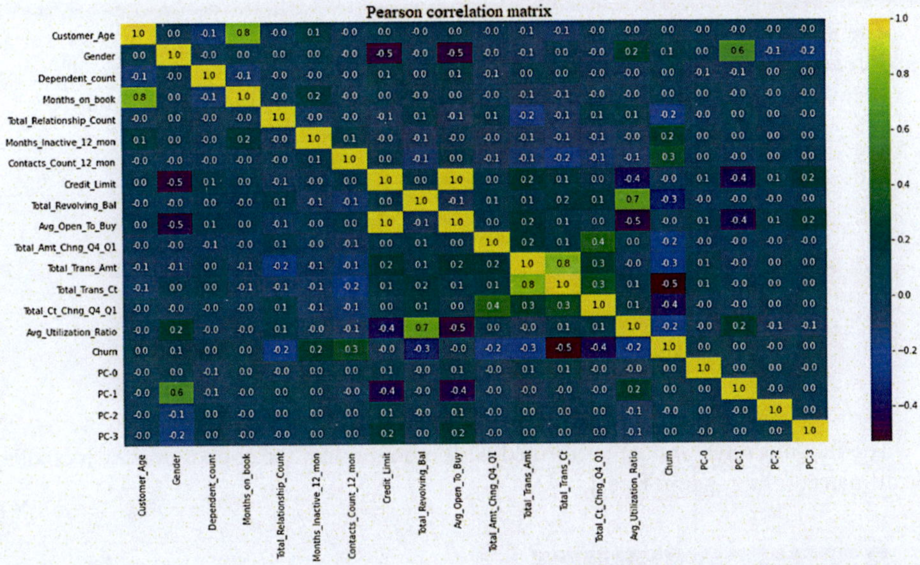

Fig. 4. Pearson correlation matrix.

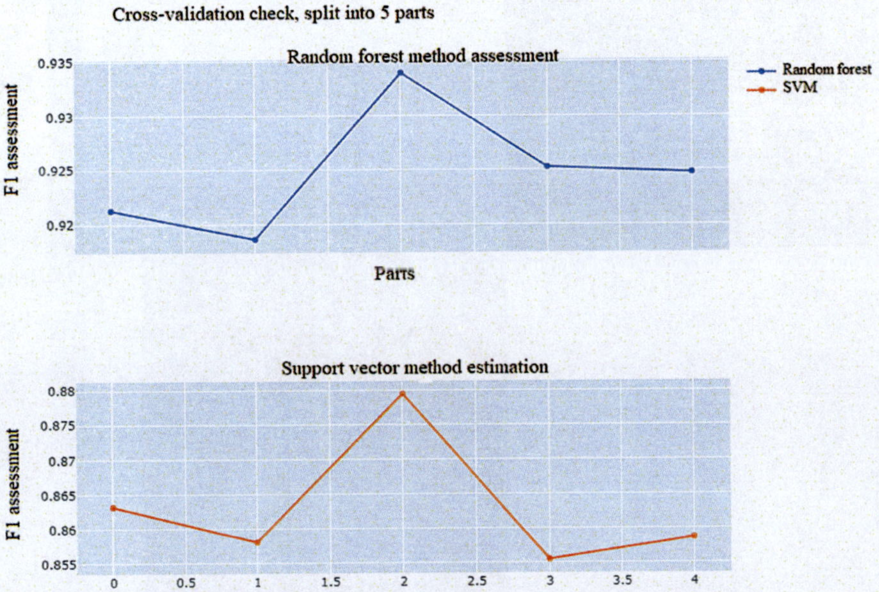

Fig. 5. Cross-validation check.

German banks, Société Générale (France), dominate on the credit market for production of goods (Table 3–5).

The F1-score ranges from (0.91; 0.935) for a random forest and (0.855; 0.88) for a support vector method. Thus, these methods relevance using has been confirmed.

Below are the classifiers work estimates on the test sample for both methods:

Table 1. Models estimation on augmented data.

Model	F1-estimate on the test sample
Random forest	0.92
Support vector method	0.87

The random forest model performed a little better than the support vector method. Error matrix for a random forest:

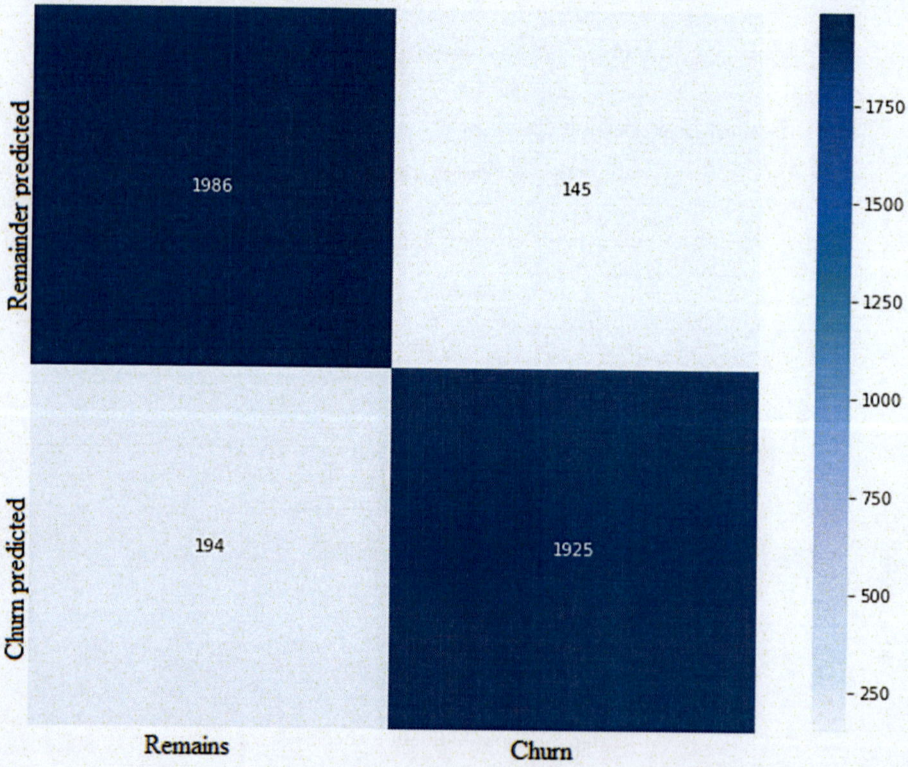

Fig. 6. The random forest model errors matrix on augmented data.

The using the synthetic minority complement method relevance is confirmed by training the same algorithms, but on the initial data, and then predicting the churn:

Table 2. Models on the initial data evaluation.

Model	F1-score on a test sample (no padding)
Random forest	0.6
Support vector method	0.52

As you can see, the new algorithms estimates are worse by about 0.2, which means classification quality a serious loss.

Error matrix for a random forest model trained on the original data:

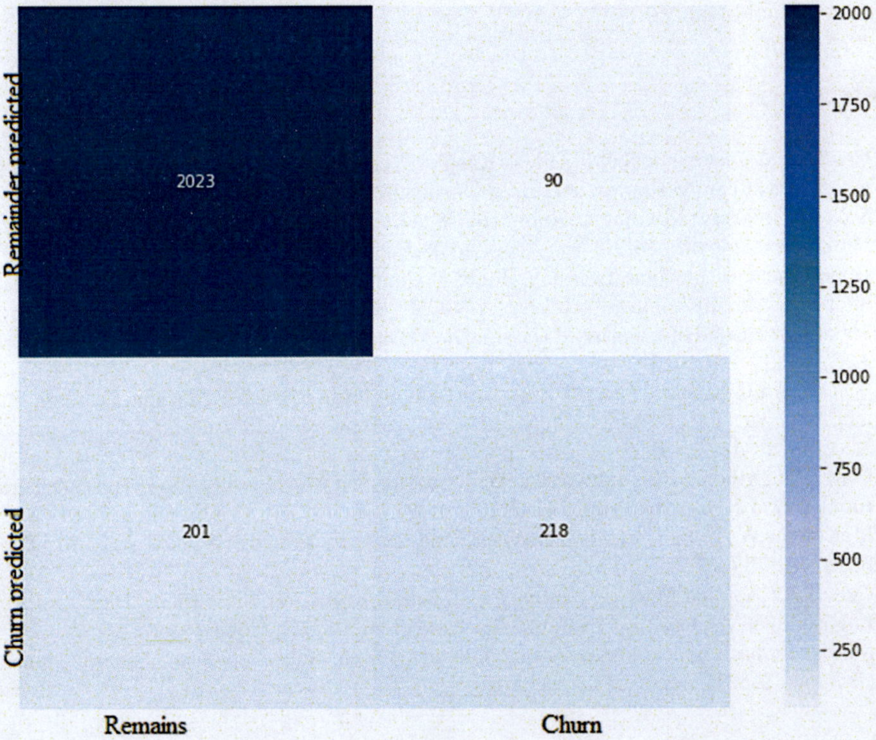

Fig. 7. Error matrix for a random forest model trained on the original data.

Table 1 shows that the random forest model allows for a more accurate outflow forecast than the support vector method. Figure 6 indicates to us correctly predicted values a high proportion, equal to 8.7%, as well as erroneous predictions a small proportion for the classes each. This indicates the SMOTE method correct application. Without its use, the model performs worse, as can be seen in Fig. 7, where the erroneous predictions' number for false positive and false negative values is comparable to the correct positive churn predictions. Also, from Table 2, it can be seen that both models F1-score fall significantly in comparison with the models trained on augmented data (see Table 1).

5 Conclusion

The In this paper, a data set about the bank's clients were analysed, and their churn a forecast was built using linear and nonlinear classifiers, and their comparison was obtained.

Thus, the random forest model proved to be more efficient. Also, this model is interpretable, since the criteria by which the splitting is carried out, highlight the priority features and their threshold values. But even an information large amount about the study objects does not allow building a good classification model if the classes in the original dataset are not balanced.

References

1. Credit Card Customers, Sakshi Goyal: Kaggle-Access mode. https://www.kaggle.com/sakshigoyal7/credit-card-customers, Accessed 08 Aug 2020
2. Sokolov, Evgeniy: Machine Learning course at the Higher School of Economics. https://github.com/esokolov/ml-course-hse, Accessed 08 Aug 2020
3. Open Course in Machine Learning. Topic 5. Compositions: bagging, random forest, Habr. https://habr.com/ru/company/ods/blog/324402/#2-sluchaynyy-les, Accessed 08 Aug 2020
4. Should it be afraid of unbalanced classes. David Dale. Habr. https://habr.com/ru/post/349078, Accessed 08 Aug 2020
5. SMOTE for Imbalanced Classification with Python. Jason Brownlee. Machine Learning Mastery. https://machinelearningmastery.com/smote-oversampling-for-imbalanced-classification, Accessed 08 Aug 2020
6. Plas, J.V.: Python for complex tasks. Data Science and Machine Learning. Translated from English - Pal'ty. I., SPb.: Peter, 576: ill (2018)
7. McKinney, W.: Python and data analysis. Translated from English Slinkin, A.A., M.: DMK Press, 482: ill (2015)
8. Geron, A.: Applied Machine Learning with Scikit. Learn and TensorFlow. Translated from English, SPb.: Alpha Kniga LLC, 688: ill. Parallel. tit. English (2018)
9. Metrics in machine learning problems. Labintsev, Egor. Habr. https://habr.com/ru/company/ods/blog/328372/, Accessed 08 Aug 2020

Algorithm for Determining Target Point of Manipulator for Grasping an Object Using Combined Sensing Means

Polina Kozyr(✉), Aleksei Erashov, and Anton Saveliev

St. Petersburg Federal Research Center of the Russian Academy of Sciences (SPC RAS), St. Petersburg Institute for Informatics and Automation of the Russian Academy of Sciences, 39, 14th Line, 199178 St. Petersburg, Russia
{erashov.a,saveliev}@iias.spb.su

Abstract. In this paper, we propose an algorithm for determining the coordinates of the center of the target object, which are necessary to complete the task of grasping this object with the end-effector of the manipulator. To obtain information about the position of the object, data on the position of the center of the ArUco marker and data from the matrix of pressure sensors of a capacitive type installed in the working area of the manipulator are used. Using the sensor array monitors the position of the target. Combining the data obtained with the matrix and the computer vision system allows obtaining a three-dimensional representation of the position of the object relative to the base frame of the manipulator. The article proposes design ratios that allow to calibrate the sensor matrix for more accurate positioning of objects. According to the results of the experiments, the coordinates of the object on the plane of the matrix are determined with the mean root mean square errors of 1.38 mm and 1.34 mm along the axes, which may be sufficient for the grasp of objects by the manipulator. In addition, a number of experiments were carried out to test the performance of the algorithm presented in this article in the problem of grasping an object using a three-link manipulator with three degrees of freedom. According to the results of these experiments, the average success rate of the grasp reached 84% .

Keywords: Grasping point of the manipulator · Target object · Matrix of combined sensors · Capacitive sensor · ArUco-marker

1 Introduction

Modern industrial robots are widely used in solving problems of objects transfer, welding, and assembly. The advantages of such systems are speed, continuous operation and accuracy. Modern manipulators designed to perform such tasks often work with a predetermined position of the target object, or this position is aligned by personnel. At the same time, such systems are practically unable to adapt to changing conditions and possible changes in the locations of target objects. Robotics are currently being developed to eliminate the aforementioned disadvantages. In the overwhelming majority of existing

R. Silhavy et al. (Eds.): CoMeSySo 2021, LNNS 231, pp. 337–350, 2021.
https://doi.org/10.1007/978-3-030-90321-3_27

systems, obtaining information about the location of the target object is implemented on the basis of visual methods [1–3]. For further processing of this information, machine learning methods are often used [4–6], which are characterized by large time costs for collecting data and training models. Thus, the problem of developing systems for determining the position of an object in space for robotic systems that do not require lengthy preprocessing of input data and are characterized by simplicity of work is relevant.

2 Related Works

In recent years, many different studies have been published, the purpose of which was to develop an effective algorithm for determining the position of an object in space with its subsequent capture [1–3]. In particular, the authors of [1] proposed a method for solving the problem of palletizing containers with cargo, which is implemented using computer vision data. The principle of the method is based on the bin-packing algorithm [7]. The results of the experiments showed that the time savings in task execution reached 20% compared to the situation when the workers were palletizing goods without coordination by the algorithm presented in this work.

In work [2], the authors presented a mobile robotic tool for grasping and moving objects. This article proposes a method combining an adaptive Monte Carlo localization algorithm and a triangle matching method based on data from indoor laser reflectors. Object detection is carried out using an SSD (Single Shot MultiBox Detector) neural network [8] based on RGB image data. According to the results of the experiments, the mobile manipulator had 90% success in the localization task and 95% in the capture of the object.

In order to increase the accuracy of determining the position of objects in space and expand the working area, it was proposed in [3] to combine data received from several cameras, the position of objects is estimated using ArUco markers. The presented principle of marker position detection is based on the use of the Kalman filter [9], which processes noisy data. The accuracy of determining the coordinates of the center of the object by the proposed method was about 10–15 μm along the X and Y axes and 30 μm along the Z.

To obtain information about the features of an object based on the data of depth maps in [10], the SURF algorithm developed by the authors is used. To improve the accuracy of determining the center of an object the RANSAC method (RANdom SAmple Consensus) [11] is also in operation. According to the results of the experiments, the error in tracking the position of the object by this method was, on average, ~ 2.5 mm. In [12], a system for packing objects into a container using a manipulator is considered. Information about the size and orientation of an object is obtained using a single RGB-D sensor. The optimal position for an object in a container is calculated by recognizing the available space of the container with another RGB-D sensor. According to the results of the experiments, the average error in determining the coordinates of the center of the object along the X axis was 4.3 mm, along the Y axis – 4.1 mm.

In the study [4], a method that includes a lightweight neural network for detection and segmentation (LNNDS), developed by the authors of the work, as well as a neural network PointNet [13], for autonomous harvesting of apples based on image data from an RGB-D camera is presented. The developed robotic harvesting system achieves a harvesting success rate of 0.8. The article [5] presents a system for harvesting kiwi, consisting of four manipulators with end-effectors. The computer vision system uses a full convolutional neural network [14] to perform semantic image segmentation. The position of each kiwi fruit is determined by comparing stereo point matching [15] obtained from the left and right cameras under natural lighting conditions. The test results of system performance show that the presented robot is able to successfully harvest 51.0% of the total amount of kiwi in the garden.

In [6], in the task of processing data on the position of an object obtained from depth maps, a 6-DOF GraspNet neural network model was used. When testing this method, a successful grasp with a manipulator was accomplished in 88% of the experiments. In the article [16], the PointNetGPD neural network model was used to solve a similar problem. This model is based on the PointNet neural network. The model was trained on a dataset [17], which included objects of various shapes, sizes, and textures. When testing this method using the UR5 collaborative manipulator, the percentage of successful captures of objects was 82%. The authors of [18] used the residual neural network ResNet [19] to search for the grasp point of the object based on the depth map data. As a result of the experiments, the object was grasped successfully in ~ 82.17% of cases.

Based on the carried out analysis, it is possible to judge a variety of developed methods for searching for grasp points of a target object using data obtained exclusively with the help of technical vision [1–3]. Some of the described works include deep learning methods, the use of which in the problem of finding the gasp point of a target object by a manipulator [4–6] requires the formation of an extensive database, long training of models and the availability of high-performance systems for using models of neural networks. Such approaches claim to be versatile, but require additional investment of time and the use of high-performance computing devices. Many machine learning methods implement the determination of object capture points based on depth maps [6, 16, 18]. However, in situations of insufficient quality overview of the target object, the effectiveness of systems based purely on visual methods decreases.

The use of tactile sensors in conjunction with visual methods will allow avoiding some of the problems described above, simplifying the process of developing such grasp systems, and also reducing the complexity of signal preprocessing. This paper presents an algorithm for determining the grasping point of an object by a manipulator based on data obtained using an ArUco marker placed on the surface of an object and a matrix of pressure sensors on which the target object is directly located. The performance of the algorithm is verified experimentally using a three-link manipulator with three degrees of freedom.

3 Installation Scheme

In this work, we consider the problem of determining the grasp point of the manipulator based on the interpretation of the indicators of a capacitive tactile sensor [20] and data obtained using the ArUco marker. A schematic representation of a setup consisting of a manipulator, an array of pressure sensors, and an object with an ArUco marker is shown in Fig. 1a.

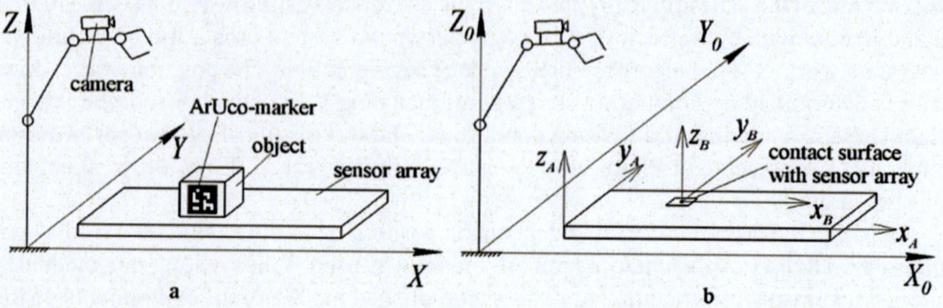

Fig. 1. a – Schematic representation of a setup consisting of a manipulator, a pressure sensor matrix and an object with an ArUco marker; b – Reference systems of the manipulator, a pressure sensor matrix and an object.

The manipulator is a stationary robotic device. In the working area of the manipulator there is a matrix of pressure sensors, on which the object is located. An ArUco marker is installed on the surface of the object. To capture images, further search and obtain the spatial characteristics of the marker, a camera is used, which is located on the grip of the manipulator.

According to the algorithm developed in this work, the coordinates of the position of the center of the object along the X and Y axes of the reference frame of the manipulator are determined based on the signals of the matrix of capacitive sensors. The search for the center of the object along the Z axis is based on the spatial characteristics of the ArUco marker located on the surface of the object, which is recognized using a camera.

The coordinates of the center of the object along the X, Y axes are initially found relative to the frame of reference of the matrix, and the coordinate Z is relative to the frame of reference of the camera (Fig. 1b). All three coordinates must be brought to a single coordinate system of the manipulator, since the control systems for manipulation systems require the description of target points relative to the reference frame of the robot. After transformations, the coordinates are fed to the manipulator control unit to grasp the object.

3.1 Capacitive Sensor Matrix

Capacitive-type sensors have high sensitivity, resistance to temperature fluctuations and a large dynamic range [21, 22]. The prototype of the matrix used in this work consists of 64 sensors of the capacitive type, with which it is possible to estimate the magnitude of the pressure applied to their surface. Each capacitive cell has dimensions of 12 × 12 × 4.2 mm and covers a pressure range from 0 to 245 kPa. The cells in the matrix are arranged in four rows of 16 cells each. The STM32 microcontroller reads the digitized signals via the I2C interface from a 16-bit ADC, which receives signals from the cells of the pressure sensor matrix. The software for the microcontroller is written in C. The readable matrix signals are transferred via USB to a personal computer, which includes an Intel® Core ™ i3-9100F processor. The code for processing data from the cells of the sensor matrix and the manipulator control algorithm on this device are implemented using the Python 3 language. Figure 2 shows the matrix used in this work.

Fig. 2. Matrix of pressure sensors for 64 cells.

To compare the distance and coordinates of the matrix cells, a local reference system was introduced for it. Graphically, the coordinate system and the capacitive matrix are shown in Fig. 3.

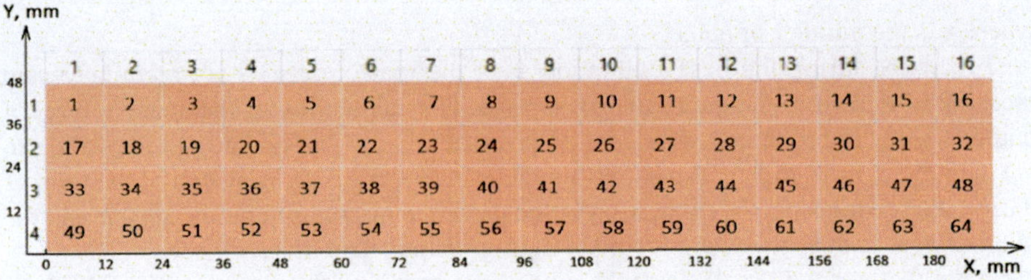

Fig. 3. Coordinate system of the matrix of pressure sensors and the location of the matrix cells.

The origin of the matrix coordinate system corresponds to the left edge of the cell located in row 4 of column 1. The Y axis is defined by rows, and the X axis is defined by columns. Coordinate values are given in millimeters. Since the length and width of each cell is 12 mm, and the distance between cells is 1 mm, the cell in row 4 of column 1 has the coordinate [6 mm, 6 mm], in row 1 of column 2 - [19 mm, 6 mm], in row 2 columns 1 - [6 mm, 19 mm], etc.

3.2 Calibrating Data Acquired by the Matrix of Pressure Sensors

Matrix cells have different sensitivity due to the peculiarities of their design, so their values must be calibrated. Calibration is the process of adjusting the output values of the matrix cells using the proportionality factor between the change in cell values when pressure is applied to it and the total set calibration change. The algorithm for calibrating the output data is presented below.

During the first stage, data is taken using a matrix in the absence of pressure applied to its surface. After that, the average level of output signals for each cell E_{imean_np} is calculated according to (1), provided that $E_{imean} = E_{imean\_np}$, the level of the output signal from each cell of the sensors in the absence of pressure on the sensor E_{inp} corresponds to e, m_i – the number of values that were recorded during the first stage for each cell, i – the cell number.

$$E_{imean} = \frac{\sum_{i=1}^{n} e}{m_i} \tag{1}$$

According to (2), the total average level of output signals for all cells of the matrix is calculated, taking into account that $E_{imean} = E_{imean\_np}$ and E_{av} corresponds to the total average level of output signals E_{av_np} for all cells of the matrix in the absence of pressure on its surface.

$$E_{av} = \frac{\sum_{i=1}^{n} E_{imean}}{n}, \tag{2}$$

where n is the number of cells.

Likewise, during the next calibration step, the matrix outputs are recorded when pressure is applied to the entire matrix surface. The calculation of the average level of output signals for each cell with applied pressure E_{imean_p} is carried out according to (1) with $E_{imean} = E_{imean\_p}$, the output signal level $e = E_{ip}$ from each cell of the sensors during the second stage of calibration in the presence of pressure on the sensor and the number of m_i values that were recorded during second stage for each cell.

After that, using (2) calculates the overall average level for the output signals E_{av_p} under the conditions when the pressure is applied. Moreover, $E_{av} = E_{av\_p}$, $E_{imean} = E_{imean\_p}$.

Then the proportional factor k_i is calculated for each cell:

$$k_i = \frac{E_{av\_p} - E_{av\_np}}{E_{imean\_p} - E_{imean\_np}} \tag{3}$$

After that, each output value of the matrix is transformed as follows:

$$E_{i\_new} = (E_i - E_{imean\_np}) \cdot k_i \tag{4}$$

where E_i is the level of the output signal from the matrix of pressure sensors after the completion of the cell calibration procedure.

After the matrix is calibrated, the manipulator becomes available for capturing objects.

4 Algorithm for Determining the Grasp Point

The flowchart of the algorithm for determining the grasp point of the manipulator based on data from the matrix of pressure sensors and data on the position of the center of the ArUco marker relative to the camera is shown in Fig. 4.

The measurement initialization block starts sampling frames from the video stream and activates measurements on the sensor matrix.

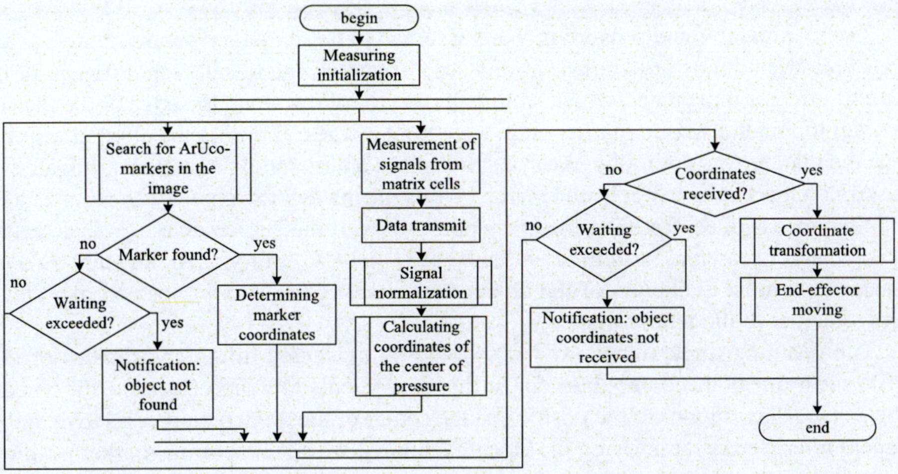

Fig. 4. Algorithm for determining the target point based on data obtained using a matrix of pressure sensors and data on the position of the center of the ArUco-marker.

The received signals from the cells of the sensor matrix are transmitted to the control unit of the manipulator, the values are corrected and the coordinates of the center of pressure are calculated. These coordinates are determined relative to a coordinate system referenced to one of the corners of the sensor array. The search for the (x, y) coordinates of the center of the target is based on pressure data from all cells where the signal level is increasing. When the signal values in a cell exceed a certain threshold level, which indicates the presence of an object on the cell surface, the cell number and the corresponding value are written into a new array. Determination of the coordinate of the center of the object along the X and Y axes is calculated using the (5), (6) based on the data of the new array.

The values of the coordinates (x, y) of the point of the center of pressure will indicate how far along the plane of the matrix it is necessary to move the fingers of the end-effector to further grasp the object.

$$X_{mid} = \frac{\sum_{i=1}^{n} X_i P_i}{\sum_{i=1}^{n} P_i}, \tag{5}$$

$$Y_{mid} = \frac{\sum_{i=1}^{n} Y_i P_i}{\sum_{i=1}^{n} P_i}, \tag{6}$$

where i is the ordinal number of the cell from the new array, n is the number of cells, X_i is the X-coordinate of the cell, Y_i is the Y-coordinate of the cell, P_i is the pressure value in ADC units corresponding to the cell i.

The manipulator and sensor array are assumed to be stationary – their relative positions and orientations are known. In this case, it is possible to set such a trajectory of the end-effector movement, which would allow it to follow along the edge of the matrix in order to find the ArUco marker on the resulting images. If during the allotted waiting time the marker was not found, the system displays a notification that the coordinates of the ArUco marker have not been received – this means that the object was not detected and the process of determining the grasp point stops. If the ArUco marker was detected by the marker search algorithm provided by the OpenCV library, then the marker coordinates determine the height of the object center (along the Z axis), located in a plane perpendicular to the matrix plane.

The resulting three values (x, y, z) should be presented in the coordinates of the reference frame of the manipulator for further movement of its end-effector to the object. The z value corresponds to the position of the center of the ArUco marker relative to the camera reference system. Using the OpenCV library, we can determine marker position and orientation, then convert them to the base frame of the manipulator, provided that the position and orientation of the camera's frame of reference is known relative to it. Using similar transformations, you can also represent the x and y coordinates relative to the manipulator's base frame.

After obtaining the coordinates of the center of the target object, it is possible to move the end-effector to the target point found by the proposed algorithm and to grasp the object. All the necessary actions for moving the links of the manipulator are parts of the "Moving the working body" block and are associated with classical tasks for the process of grasping by a robotic device.

5 Experiments

To test the presented algorithm of finding the center of the target object, a manipulator with three degrees of freedom was used, the kinematic diagram of which is shown in Fig. 5a. The manipulator has three revolute joints. The inverse kinematics is solved analytically due to the classical kinematic scheme of the manipulator. To solve it, the coordinates of the point *a*, located in the geometric center of the end-effector, are set

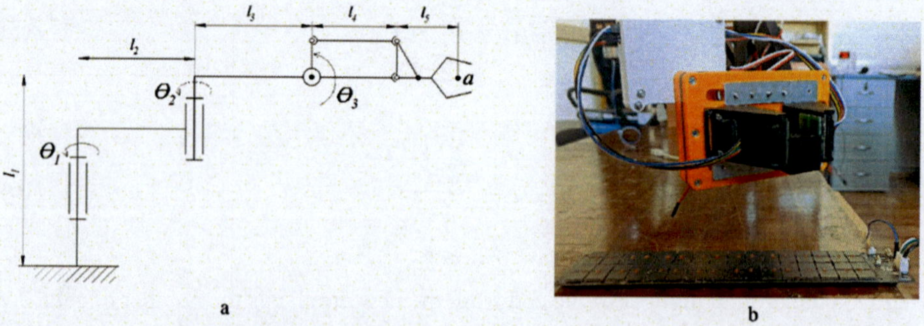

Fig. 5. a – Kinematic diagram of the manipulator; b – Appearance of the two-fingered gripper of the manipulator with an object and a matrix of sensors.

relative to the base frame of the manipulator. Upon completion of the movement of the end-effector, this point must coincide with the specified accuracy with the grasping point.

The end-effector installed in the manipulator, shown with the array of sensors and the object in Fig. 5b, has two fingers, into which combined pressure and proximity sensors of a capacitive type are embedded [23]. With the help of these sensors, it is possible to control the grasping force and the orientation of the gripping elements relative to the object to grasp it without displacement and deformation [24].

To reduce the influence of inertness and smooth regulation during manipulator operation, a trajectory generator should be used [25]. The paths for changing the angle of rotation in the first and second joints can be the same. In this case, the trajectory for the third joint differs from the previous ones in that the duration of the transient process is increased for it, which makes it possible not to collide with the object when the links move. To prevent the collision of the fingers of the end-effector with the object during the movement of the third link, the desired duration of the transient process was increased by ~ 66%, in comparison with the work of the previous links. In the experiments, trajectory of movement of the manipulator links is based on polynomials of the 5th degree.

Before the experiments, the measuring system was calibrated. After calibration, in the absence of pressure applied to the matrix surface, the matrix values are reduced to zero, and when the same pressure is applied to several cells, the latter have comparable values (Fig. 6).

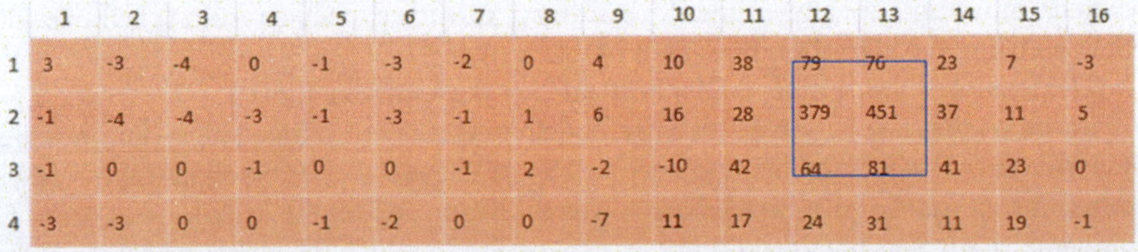

	1	2	3	4	5	6	7	8	9	10	11	12	13	14	15	16
1	3	-3	-4	0	-1	-3	-2	0	4	10	38	79	76	23	7	-3
2	-1	-4	-4	-3	-1	-3	-1	1	6	16	28	379	451	37	11	5
3	-1	0	0	-1	0	0	-1	2	-2	-10	42	64	81	41	23	0
4	-3	-3	0	0	-1	-2	0	0	-7	11	17	24	31	11	19	-1

Fig. 6. Values of the matrix cells: a) in the absence of pressure; b) in the presence of pressure.

The signal from each cell of the matrix is represented in the normalized output values of the ADC. Figure 6 shows a situation when an object, indicated by a blue frame, is located completely on cells in 12, 13 columns in the second row, partially overlapping cells in the first and third rows. It can be seen from the figure that the cells, to the surface of which the same pressure is applied, have output values of approximately the same level. Non-zero readings of the cells in the absence of pressure on its surface in Fig. 7 can be explained by the influence of measurement noise. For this case, the maximum deviation is 0.07% of the maximum ADC output value (2^{14}).

The target object, the gripping point of which is determined by the system, is a cube with dimensions of 24 × 24 × 24 mm. In the experiments carried out, this object was placed on the array of sensors in various positions. To test the operability of the pressure matrix, four groups of experiments were carried out with 10 experiments in each group: the object completely overlaps the cells (Fig. 7a), the object is located on the edge of the matrix (Fig. 7b), several cells are only partially overlapped (Fig. 7c), the object is located at an angle of 45° relative to the matrix (Fig. 7d).

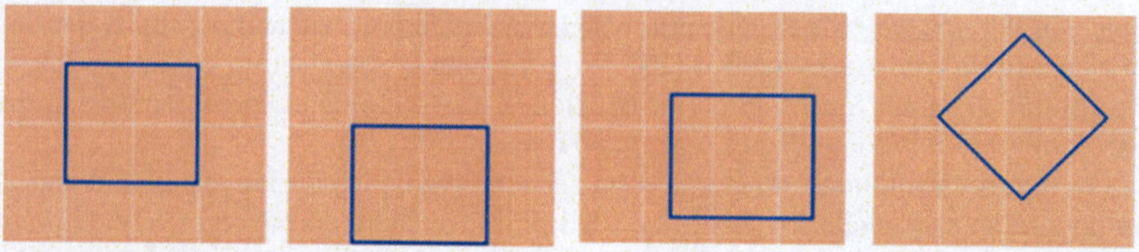

Fig. 7. Position of the object during four experiments: a) the object completely overlaps the cells; b) the object is located at the edge of the matrix; c) several cells are only partially overlapped; d) the object is at an angle of 45°.

As a method for assessing the accuracy of determining the coordinates of the center of an object along the X and Y axes for each group of experiments, the root mean square error (RMSE) is used for 10 experiments related to each type of experiments. It is calculated by (7):

$$RMSE = \sqrt{\frac{1}{b}\sum_{i=1}^{b}(y_i - \hat{y}_i)^2} \tag{7}$$

where b is the number of experiments performed, y_i is the expected result, $\hat{y}_i$ is the result obtained using the algorithm presented in this article.

The results of four groups of experiments are presented in Table 1.

Based on the data in Table 1, we can conclude that the developed algorithm has the highest accuracy when the target object strictly overlaps the cells below it. In this case, the root mean square error on the X axis was 0.81 mm, and on the Y axis – 1.20 mm. A similar accuracy was obtained during an experiment in which the object is located at an angle of 45° relative to the matrix (RMSEx – 1.17 mm, RMSEy – 0.92 mm). The lower accuracy for the group of experiments number 2 when placing the object on the edge of the matrix (RMSEx – 2.32 mm, RMSEy – 2.09 mm) compared to the rest of the groups of experiments can be explained as follows. When the target object is located on the

Table 1. Results of experiments on assessing the accuracy of determining the coordinates of the target object along the X and Y axes by the developed algorithm based on the data of the pressure sensor matrix.

Experiment group	Position of the object on the matrix	RMSE x, mm	RMSE y, mm
1	Object completely overlaps cells	0.81	1.20
2	The object is positioned on the edge of the matrix	2.32	2.09
3	Several cells are only partially overlapped	1.05	1.12
4	The object is located obliquely	1.17	0.92

matrix away from its edges, the system receives useful data not only from the cells on which the object is directly placed, but also from those cells that surround it. If an object placed at the edge of the matrix, data from the side of its border will not be received, as a result of which the values of the coordinates of the position of the center of the object along the X and Y axes may be somewhat distorted.

To evaluate the operation of the algorithm presented in this work, experiments were carried out to grasp the target object using a manipulator at various placements of the object on the sensor array, similar to Fig. 7a-d. Several experiments were carried out for each position of the object (~20). The obtained values of the probability of successful grasp according to the developed algorithm are shown in Table 2.

Table 2. Results of experiments on the capture of objects by the manipulator.

Experiment group	Position of the object on the matrix	Capture probability
1	Object completely overlaps cells	0.94
2	The object is positioned on the edge of the matrix	0.73
3	Several cells are only partially overlapped	0.81
4	The object is located obliquely	0.86

According to the results of experiments with object grasping, the use of the object position search algorithm proposed in the article to determine the target point of the manipulator was successful on average in 84% of the grasps performed. Table 2 shows that the maximum probability of successful grasps was 0.94 and corresponded to experiment 1. The lowest grasp success compared to these experiments is characterized by experiment 2 (grasp probability was 0.73). As can be seen from Tables 1 and 2, the system requires additional refinement of the matrix design, as well as software in order to increase the accuracy of determining the center point of the object when it is located at the edge of the matrix.

The developed algorithm is characterized by the following computational and time capacities. The code for calculating the calibration coefficients for each cell is made 25 s before the data is read from the pressure sensor matrix and uses 0.5% of the CPU and 57.5 MB of RAM. Calibration of the matrix output data using the calculated coefficients is carried out at the time of data collection. When reading data from a matrix of pressure sensors, the processor is used by 0.6%, the RAM – by 57.6 MB. The program for calculating the position of the object relative to the X and Y axes of the matrix is executed in 0.51 s, using 0.3% of the processor's work and 48.1 MB of RAM.

Thus, within the framework of this study, an algorithm for finding the grasp point of the target object was developed and tested on the basis of data obtained from a matrix of pressure sensors and data on the position of the center of the ArUco marker.

Table 3 shows data on the average probability of a successful capture using the developed algorithm and other existing methods.

Neural networks, as usual, are highly accurate, but at the same time, they are characterized by the complexity of implementation and time-consuming preparation of data for

Table 3. Comparison of the results of the proposed algorithm and existing methods.

Methods and algorithms	Probability of a successful grasp
The proposed algorithm	0.84
SSD [2]	0.95
LNNDS and PointNet [4]	0.80
6-DOF GraspNet [6]	0.88
PointNetGPD [16]	0.82
ResNet [18]	~ 0.82

use. The algorithm presented in this study is simpler to implement, requires short data calibration, and uses significantly less computing power compared to other solutions from Table 3, which are based on deep learning. It can be seen from the results that the developed algorithm was able to surpass the methods based on the use of the PointNet neural network and ResNet in the task of capturing an object. This method is inferior to the SSD and 6-DOF GraspNet methods, which can be explained by the imperfection of the manipulator control system, the looseness of the links and the effect of noise on the operation of the pressure sensor matrix. In the future, it is planned to eliminate these disadvantages.

6 Conclusion

The algorithm for solving the problem of determining the position of a target object presented in this work combines the use of a camera in conjunction with an ArUco marker and the use of a matrix of pressure sensors. This algorithm is simple to implement, does not require a long time to prepare the system for work with certain objects and large computing power. Regardless of the visibility conditions, the algorithm will be able to determine the center of the object along the X and Y axes.

To test the operability of the matrix of pressure sensors, four types of experiments were carried out, in which the object was located on the matrix in different positions. According to the results of the experiments, the minimum root-mean-square errors in determining the coordinates of the center of the object along the X and Y axes corresponded to the location option when the object completely overlapped the cells on which it was located. On the X-axis, the root-mean-square error was 0.81 mm, and on the Y-axis – 1.2 mm. The maximum root mean square error corresponded to the location of the object at the edge of the matrix (RMSEx – 2.32 mm, RMSEy – 2.09 mm).

The coordinate of the center of the object along the Z axis was found using a camera and an ArUco marker located on the object. In the course of this study, the developed algorithm for determining the coordinates of the center of the target object was implemented and tested in a three-link manipulator with three degrees of freedom. On average, 84% of all made grasps were successful.

The presented algorithm for determining the target point of movement of the manipulator for grasping an object using a combination of the output signals of the pressure sensor matrix and ArUco-marker data can find its application in production in the tasks of automation of storage, loading and unloading of goods, as well as in problems of human-machine interaction [26, 27].

References

1. Lamon, E., Leonori, M., Kim, W., Ajoudani, A.: Towards an intelligent collaborative robotic system for mixed case palletizing. In: 2020 IEEE International Conference on Robotics and Automation (ICRA), pp. 9128–9134. https://doi.org/10.1109/icra40945.2020.9196850
2. Martello, S., Pisinger, D., Vigo, D.: The three-dimensional bin packing problem. Oper. Res. **48**(2), 256–267 (2000)
3. Zhou, H., Chou, W., Tuo, W., Rong, Y., Xu, S.: Mobile manipulation integrating enhanced AMCL high-precision location and dynamic tracking grasp. Sensors **20**(22), 6697 (2020)
4. Liu, W., et al.: SSD: single shot multibox detector. In: Leibe, B., Matas, J., Sebe, N., Welling, M. (eds.) ECCV 2016. LNCS, vol. 9905, pp. 21–37. Springer, Cham (2016). https://doi.org/10.1007/978-3-319-46448-0_2
5. Popescu, D.C., Dumitrache, I., Caramihai, S.I., Cernaianu, M.O.: High precision positioning with multi-camera setups: adaptive kalman fusion algorithm for fiducial markers. Sensors **20**(9), 2746 (2020). https://doi.org/10.3390/s20092746
6. Kalman, R.E.: A new approach to linear filtering and prediction problems. J. Basic Eng. **82**(1), 35–45 (1960). https://doi.org/10.1115/1.3662552
7. Lin, H.T., Chiang, M.H.: The integration of the image sensor with a 3-DOF pneumatic parallel manipulator. Sensors **16**(7), 1026 (2016)
8. Lee, J.J., Kim, G.: Robust estimation of camera homography using fuzzy RANSAC. In: Gervasi, O., Gavrilova, M.L. (eds.) ICCSA 2007. LNCS, vol. 4705, pp. 992–1002. Springer, Heidelberg (2007). https://doi.org/10.1007/978-3-540-74472-6_81
9. Hong, Y.D., Kim, Y.J., Lee, K.B.: Smart pack: online autonomous object-packing system using RGB-D sensor data. Sensors **20**(16), 4448 (2020). https://doi.org/10.3390/s20164448
10. Kang, H., Zhou, H., Wang, X., Chen, C.: Real-time fruit recognition and grasping estimation for robotic apple harvesting. Sensors **20**(19), 5670 (2020). https://doi.org/10.3390/s20195670
11. Qi, C.R., Su, H., Mo, K., Guibas, L.J.: Pointnet: deep learning on point sets for 3d classification and segmentation. In: Proceedings of the IEEE Conference on Computer Vision and Pattern Recognition, pp. 652–660 (2017)
12. Williams, H.A.M., et al.: Robotic kiwifruit harvesting using machine vision, convolutional neural networks, and robotic arms. Biosys. Eng. **181**, 140–156 (2019). https://doi.org/10.1016/j.biosystemseng.2019.03.007
13. Long, J., Shelhamer, E., Darrell, T.: Fully convolutional networks for semantic segmentation. In: Proceedings of the IEEE Conference on Computer Vision and Pattern Recognition, pp. 3431–3440 (2015)
14. Fielding, G., Kam, M.: Applying the Hungarian method to stereo matching. In: Proceedings of the 36th IEEE Conference on Decision and Control, vol. 2, pp. 1928–1933 (1997). https://doi.org/10.1109/cdc.1997.657875
15. Mousavian, A., Eppner, C., Fox, D.: 6-dof graspnet: variational grasp generation for object manipulation. In: Proceedings of the IEEE International Conference on Computer Vision, pp. 2901–2910 (2019)
16. Liang, H., et al.: Pointnetgpd: detecting grasp configurations from point sets. In: 2019 International Conference on Robotics and Automation (ICRA), pp. 3629–3635. IEEE (2019)
17. Calli, B., Singh, A., Walsman, A., Srinivasa, S., Abbeel, P., Dollar, A.M.: The ycb object and model set: Towards common benchmarks for manipulation research. In: 2015 international conference on advanced robotics (ICAR), pp. 510–517. IEEE (2015)

18. Shao, Q., et al.: Suction grasp region prediction using self-supervised learning for object picking in dense clutter. In: 2019 IEEE 5th International Conference on Mechatronics System and Robots (ICMSR), pp. 7–12. IEEE (2019)
19. Szegedy, C., Ioffe, S., Vanhoucke, V., Alemi, A. A.: Inception-v4, inception-resnet and the impact of residual connections on learning. In: Thirty-first AAAI conference on artificial intelligence, vol. 301, no. 1 (2017)
20. Krestovnikov, K., Erashov, A., Bykov, A.: Development of circuit solution and design of capacitive pressure sensor array for applied robotics. Rob. Tech. Cybern. **8**(4), 296–307 (2020)
21. Chi, C., Sun, X., Xue, N., Li, T., Liu, C.: Recent progress in technologies for tactile sensors. Sensors **18**(4), 948 (2018). https://doi.org/10.3390/s18040948
22. Park, M., Bok, B.G., Ahn, J.H., Kim, M.S.: Recent advances in tactile sensing technology. Micromachines **9**(7), 321 (2018). https://doi.org/10.3390/mi9070321
23. Krestovnikov, K., Cherskikh, E., Zimuldinov, E.: Combined capacitive pressure and proximity sensor for using in robotic systems. In: Ronzhin, A., Shishlakov, V. (eds.) Proceedings of 15th International Conference on Electromechanics and Robotics "Zavalishin's Readings." SIST, vol. 187, pp. 513–523. Springer, Singapore (2021). https://doi.org/10.1007/978-981-15-5580-0_42
24. Erashov, A., Krestovnikov, K.: Algorithm for controlling manipulator with combined array of pressure and proximity sensors in end-effector. In: 16th International Conference on Electromechanics and Robotics "Zavalishin's Readings" (2021)
25. Craig, J.J.: Introduction to Robotics. Pearson Education International, Boston (2005)
26. Galin, R., Meshcheryakov, R., Samoshina, A.: Mathematical modelling and simulation of human-robot collaboration. In 2020 International Russian Automation Conference (RusAutoCon), pp. 1058–1062. IEEE (2020)
27. Kharchenko, S., Meshcheryakov, R., Turovsky, Y., Volf, D.: Implementation of robot–human control bio-interface when highlighting visual-evoked potentials based on multivariate synchronization index. In: Ronzhin, A., Shishlakov, V. (eds.) Proceedings of 15th International Conference on Electromechanics and Robotics "Zavalishin's Readings." SIST, vol. 187, pp. 225–236. Springer, Singapore (2021). https://doi.org/10.1007/978-981-15-5580-0_18

Automated Computer Analysis of Microscopy Images Based on Laplacian and Neuro-Fuzzy Networks

I. G. Shelomentseva[1,2](✉), S. V. Chentsov[2], and N. V. Yakasova[2,3]

[1] Krasnoyarsk State Medical University named after Professor V.F. Voino-Yasenetsky, 1, Partizan Zheleznyak Avenue, Abakan, Krasnoyrsk 660022, Russia

[2] Siberian Federal University, 79, Svobodny Avenue, Abakan, Krasnoyrsk 660041, Russia

[3] Khakassia State University named after N. F. Katanov, 90, Lenin Avenue, Abakan, Republic of Khakassia 655017, Russia

Abstract. Automated medical imaging is at the heart of modern medical diagnostic systems. The authors propose a model for computer analysis of light microscopy images. This model includes filtering and segmentation modules, a feature extraction module, and a classification module. The formation of the classifying function is based on a neuro-fuzzy model, and segmentation was carried out using the Laplacian-Gaussian. The authors propose a modified version of the Laplacian of the Gaussian by adding a non-sharp masking procedure. The authors tested the quality of filtering and segmentation using the Normalised Colour Difference (NCD) metric. The feature extraction module is based on contour and color descriptors. The authors use Fourier descriptors as contour descriptors and the color values of the RGB and HSV model parameters of the object and the area around the object as color descriptors. The resulting model is implemented as a prototype of an automated system and tested on a test set of images.

Keywords: Computer-aided diagnosis · Automated microscopy · Laplacian · Neuro-fuzzy networks

1 Introduction

By now, methods and algorithms for computer analysis of images are firmly included in the functioning of medical information systems. One of the types of biomedical research is the digital scanning of microscopic preparations. As a rule, these images are characterized by a monochromatic background, on which located such objects of interest as mycobacteria, elements of cancer tumors, histopathological objects.

The authors consider automated computer analysis of microscopy images using the example of images of sputum analyzes stained by the Ziehl-Neelsen method. This technique is one of the ways to diagnose tuberculosis. Despite some advances in technology and programs, tuberculosis remains among the deadliest infectious diseases, and without significant advances in diagnosis and treatment, the epidemic cannot be stopped. The authors present a model of an automated system for diagnosing tuberculosis by the Ziehl-Neelsen microscopy method using the Laplacian and neuro-fuzzy networks.

R. Silhavy et al. (Eds.): CoMeSySo 2021, LNNS 231, pp. 351–355, 2021.
https://doi.org/10.1007/978-3-030-90321-3_28

2 Materials and Methods

The functional structure of the model includes an image database, an image preprocessing module, a segmentation and ROI generation module, a feature vector extraction module, and an intelligent processing module (Fig. 1). During the computer analysis, the automated system forms an image database, an ROI database, a feature vector database. The image database and ROI database format is BMP, the feature database format is CSV. An experienced microscopist obtained digital images from various slides of Ziehl-Neelsen stained sputum smears (belonging to patients of a tuberculosis polyclinic). The resolution of the images is 572 × 422, the size is 942 KB and the color model is RGB (red, green, and blue) (Fig. 1).

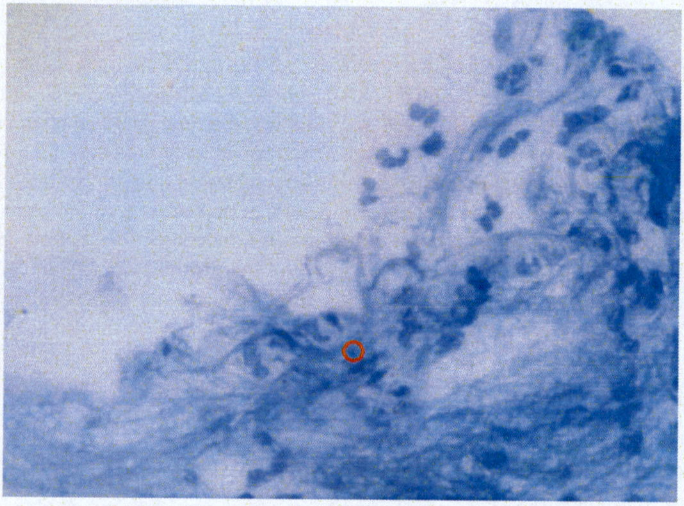

Fig. 1. Ziehl-Neelsen stained microscopy image

The analysis process begins with loading the analyzed image from the image database. The first stage of the algorithm (Fig. 2).

Is the preprocessing stage, which includes the choice of the filtering method - linear or nonlinear. This stage can be skipped if the image is submitted for processing of sufficiently good quality. The process of the automated computer analysis is described by an algorithm represented in Fig. 2.

The next stage of the algorithm is the stage of image processing in the form of segmentation, and this stage is mandatory. The segmentation uses the modified Laplacian of the Gaussian [1]. Since a single analysis can process up to 300 microscopy images, additional filtering will subsequently speed up the classification operation. For this purpose, during the segmentation of the first image, the parameter A is calculated, which is then used to correct the filter kernel during the segmentation of subsequent images [2].

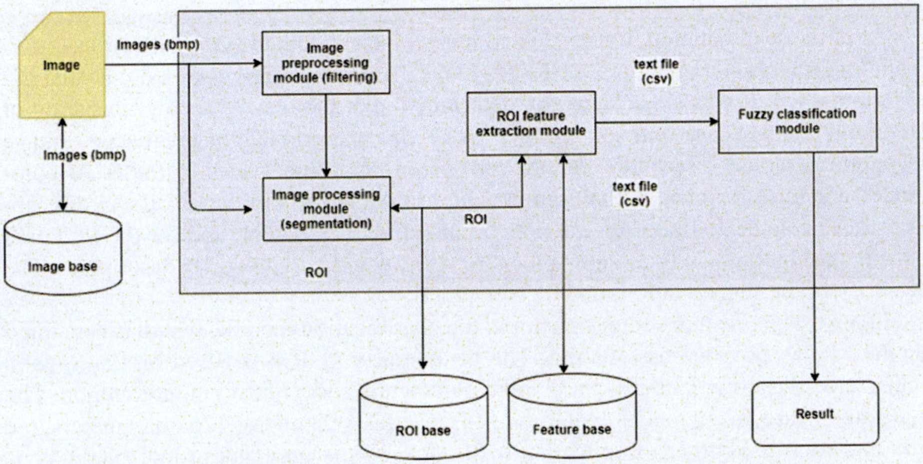

Fig. 2. Functional diagram of automated computer analysis of microscopy images

By the modified Laplacian of the Gaussian, the authors understand the use of the unsharp masking technique for the kernel of the Laplacian of the Gaussian (formula 1–2) [3].

$$f'(x, y) = Af(x, y) - \overline{f(x, y)} \quad (1)$$

$$f'(x, y) = \begin{cases} Af(x, y) - \nabla^2 f(x, y), ifw(0, 0) < 0 \\ Af(x, y) + \nabla^2 f(x, y), ifw(0, 0) > 0 \end{cases} \quad (2)$$

Where
$f'(x, y)$ – the result of high-frequency filtering
$f(x, y)$ – the original image
$\overline{f(x, y)}$ – an unfocused copy of the image
A–non-negative real parameter.

Authors use an NCD criterion to calculate the parameter A. NCD criterion compares the original and the filtered image by numeric value, which depends on the CIE Luv color model (formula 3). As soon as this criterion reaches its minimum value, then the increase in parameter A should be stopped and the kernel of the Laplacian of the Gaussian should be modified [4].

$$NCD = \frac{\sum_{i=1}^{m}\sum_{j=1}^{n}\sqrt{\left(y'_{ijL} - y_{ijL}\right)^2 + \left(y'_{ijU} - y_{ijU}\right)^2 + \left(y'_{ijV} - y_{ijV}\right)^2}}{\sum_{i=1}^{m}\sum_{j=1}^{n}\sqrt{\left(y_{ijL}\right)^2 + \left(y_{ijU}\right)^2 + \left(y_{ijV}\right)^2}} \quad (3)$$

m x n - image dimension
y' - the pixel value of the filtered image
y - the pixel value of the original image
L, U, V - parameters of the CIE Luv color model

After the segmentation of the selected image, the automated system identifies informative morphological features of the ROI, such as color, shape and contour analysis descriptors [5]. It was experimentally determined that the feature vector consisting of color and contour descriptors is optimal for solving the problem of microscopy image recognition. Color descriptors include red, green, and blue values of the RGB color model, and the hue, tone, and saturation values of the HSV color model. Color descriptors are calculated for the ROI, the area bounded by the contour, and for the ROI area outside the contour. Contour descriptors are represented by 10 Fourier descriptors. The accuracy of the segmentation module is 99.98%.

In next stage, feature vector transfer to the classification module, which is generated by the Matlab program as a fis file. The neuro-fuzzy system is based on the Sugeno subtractive clustering model with 32 input parameters and a linear output function. The generated neuro-fuzzy model contains 13 fuzzy rules, 893 nodes, 1261 parameters, the min and prod methods as implication methods, and the max and probor methods as aggregation methods. For deffasification, the weighted average of all the output data of the rules is used [6]. The resulting accuracy of the classification module ranged from 87.5% to 93.4%.

3 Computational Experiment

Authors developed a prototype of a system for automated computer analysis of microscopy images in C++ to conduct a computational experiment. The system modules are managed using control elements, interface windows, and drop-down vertical menus that select the route for image analysis.

The computational experiment was conducted with a test set of images, including 14 images with mycobacteria, 4 images with one mycobacterium (such preparations have indeterminate results and require a retake of analysis) and 18 images without mycobacteria.

Among the preparations that were found to contain acid-resistant mycobacteria using simple light microscopy, a positive result was found in 100% of cases.

Among the preparations with an indeterminate result, in which the presence of one or two acid-fast mycobacteria was established using simple light microscopy (4 preparations), an indeterminate result was obtained only in 50% of cases, in other cases a false-negative result was shown.

Among the preparations on which the absence of acid-resistant mycobacteria was established using simple light microscopy (18 preparations), using the developed prototype, a negative result of the study was obtained in 89.9% of cases (in 16 preparations), in 11.1% (2 preparations) – an invaluable result, on the basis of which it is necessary to recommend repeating the microscopy examination.

4 Conclusion

The presented system of automated computer analysis of microscopy images based on the Laplacian and neuro-fuzzy networks allows successfully segmentation and classification of mycobacterium tuberculosis. The system can be easily expanded to other microscopy images - for this it is enough to change the kernel of the Laplacian-Gaussian, select the parameter A for more optimal segmentation, and import the neuro-fuzzy inference model. The algorithms and models proposed by the authors will provide full automation of the microscopy process, improve its quality and efficiency.

References

1. Narkevich, A., Shelomentseva, I., Vinogradov, K., Sysoev, S.: Comparison of segmentation methods for digital microscopic images of sputum stained by the Ziehl-Neelsen method. Eng. J. Don **4**, 1–11 (2017)
2. Shelomentseva, I.: Results of filtration and segmentation of the images of analysis of sputum stained by Ziehl-Neelsen method. Int. J. Adv. Stud. **7**(4–2), 110–114 (2017)
3. Gonzales, R.C., Woods, R.E.: Digital Image Processing, 2nd edn. Pearson Education, London (2002)
4. Shelomentseva, I.: Parameters of filtering by LOG filter of microscopic images of sputum stained by Ziehl-Neelsen method. Econ. Inf. Technol. **47**(2), 362–371 (2020)
5. Chentsov, S.V., Shelomentseva, I.G., Yakasova, N.V.: Creating of feature dictionary using contour analysis, moments and fourier descriptors for automated microscopy. In: Silhavy, R., Silhavy, P., Prokopova, Z. (eds.) CoMeSySo 2020. AISC, vol. 1295, pp. 397–403. Springer, Cham (2020). https://doi.org/10.1007/978-3-030-63319-6_36
6. Shelomentseva, I.G., Chentsov, S.V., Narkevich, A.N.: Selection of optimal parameters of the neuro-fuzzy ANFIS system for classification of Ziehl-Nielsen stained sputum smear images. Econ. Inf. Technol. **46**(2), 305–310 (2019)

Admonishment System for Human Physical Distancing Violators Using Faster Region-Based Convolutional Neural Network with Detectron2 Library

Glenn Magwili[1], Michael Jherriecho Christiane Ayson[2](✉), and Mae Anne Armas[3]

[1] Mapúa University, Intramuros, Manila, Philippines
gvmagwili@mapua.edu.ph
[2] Mapúa University, Imus, Philippines
mjcvayson@mymail.mapua.edu.ph
[3] Mapúa University, Caloocan, Philippines
maaarmas@mymail.mapua.edu.ph

Abstract. Physical distancing is essential to help prevent the spread of Coronavirus Disease 2019 (COVID-19). However, people often are unaware of their physical distances from each other. This study developed an autonomous system that monitors and admonishes physical distancing violators by spotting lights at them and playing a pre-recorded audio to remind them to keep their distance. It uses a downward-looking Fisheye IP camera to take a picture of the platform every 30 s. It uses Faster R-CNN with Detectron2 Library to detect people. Violators are then identified as individuals who have less than 1 m of computed Euclidean distance between other people. The environment being monitored was divided into four quadrants, each of which is lit with COB LEDs (controlled by an Arduino UNO) if they contain at least one violator. At the same time, a corresponding audio admonishment is being played through a loudspeaker. The performed single-sample t-test proves the accuracy of the Euclidean distance measurement of the system since the probability value of 0.6969, derived from the computed t-statistic of 0.3914, is significantly greater two-tailed significance level of 0.05.

Keywords: Physical distancing · Faster R-CNN · Detectron2

1 Introduction

According to the World Health Organization (WHO), as of the 2nd of September 2020, 25.3 million recorded Coronavirus Disease 2019 (COVID-19) cases worldwide, out of which 848 thousand have died. Oum and Wang (2020) stated that these numbers are just manifestations of communicable and deadly COVID-19 [1]. In an attempt to address this situation, the Philippine Government filed the "New Normal" Bill, which proposes several protocols, including the mandatory 1-m-radius physical distancing, requiring

R. Silhavy et al. (Eds.): CoMeSySo 2021, LNNS 231, pp. 356–371, 2021.
https://doi.org/10.1007/978-3-030-90321-3_29

everyone to be at least 1 m away from other people. Grocery stores and malls have also implemented measures such as manually monitoring the physical distance between people within the establishment. However, despite all the efforts, there are still many violators, as several images uploaded on social media platforms show.

Published studies focusing on the automation of physical distancing protocol implementation, however, are scarce. One is the study conducted by Pouw, Toschi, van Schadewijk, and Corbetta (2020) [2], wherein they used the commercial pedestrian tracking sensors equipped in Utrecht Central railway station in The Netherlands to detect physical distancing violators. Meanwhile, Kazuki, Satoru, Takayuki, and Tetsuo developed a robot in 2019 to admonish pedestrians smoking in public places [3]. Tsaipei, Yun-Yi, Fong-Wen, and Yi-Fu (2019) [4] improved and optimized the Mask R-CNN object detection algorithm to detect humans in a top-view fisheye image. Manually implementing a physical distancing protocol is ineffective and inefficient. Manual implementation calls for an automated approach. However, the aforementioned studies do not address this problem but at least provide valuable insights. No study has been published that focuses on an active automated approach regarding the said protocol to the best of the researchers' knowledge.

This study aims to develop a system that detects and admonishes physical distancing violators. The specific objectives are as follows: 1) to develop a prototype that captures images and admonishes violators by playing a prerecorded audio and spotting lights at them; 2) to use Faster RCNN to detect people and formulate software to detect violators and admonish them; 3) to integrate the software and hardware; and 4) to test the experimental set-up in a controlled environment.

An admonishment system can be helpful and practical for the automated overseeing of the physical distancing protocol since such systems effectively perform various tasks, primarily surveillance [5]. Moreover, the usefulness of the proposed admonishment system will not vanish when the pandemic ends. It is being expected that after the pandemic, maintaining physical distance in public will be part of the new normal, not only for preventing the spread of diseases but also for everyone's security and safety.

This study is limited to developing an admonishment system, with the fisheye IP camera-equipped computer performing surveillance in a closed environment to detect violators of physical distancing protocol. Meanwhile, the lights and speech-playback system receive the violators' coordinates, which will spotlight the concerned area and play a pre-recorded audio, reminding the violators to comply with the physical distancing protocol. It can only detect and admonish the violators within the camera's visual field every 30 s, limiting its current application to relatively small establishments such as barangay halls.

2 Methodology

2.1 Conceptual Framework

Figure 1 presents the conceptual framework of this study. It starts with the computer taking a top-view fisheye picture of the platform every thirty (30) s using the Internet Protocol fisheye camera equipped with it. This fisheye image is then passed to the pre-trained model of Detectron2 Model Zoo designed for human detection. Once the humans

Input	Process	Output
Top-view fisheye image of the platform	Pass the input image to Detectron2 Model Zoo for human detection Measure the Euclidean distance between people Identify and localize the physical distancing protocol violators Interpret the calculated data in terms of which area to send a light alert to and which pre-recorded audio to play	Enclose in red boxes the violators detected in the input image Send light alert to the area containing violator/s Play a pre-recorded audio intended to admonish the violators in the concerned area

Fig. 1. Conceptual framework

are identified, their Euclidean distance is measured, out of which the physical distancing protocol violators are identified: those who are less than 1 m apart. The locations of these violators are then lit with directive Chip-on-Board (COB) LEDs to serve as light alerts. At the same time, a pre-recorded audio intended to admonish the detected violators based on their locations is played through a loudspeaker.

2.2 Software Development

Main Process

Figure 2 presents the main process flowchart of the program. It starts by importing the necessary libraries and performing basic setup, including Numpy, Math, CV2, and Detectron2 Configuration and Weight. Afterward, it retrieves the latest frame captured by the fisheye IP camera and imports it as an image, passed into the pre-trained Detectron2 model for human detection. The resulting data includes the detected humans and their coordinates in the image. Using their coordinates and treating them as points in cartesian coordinates, computing their Euclidean distances becomes simple. Based on their Euclidean distances, violators are then identified if they are less than 1 m apart. These violators are stored in a data frame enclosed in red boxes, while non-violators are enclosed in blue boxes. The violators are then subjected to admonishment by light alert and audio playback. Afterward, the program execution halts if the Keyboard Interrupt is triggered in the computer, i.e., Ctrl + M + I. Otherwise, the process loops back to obtaining the latest frame captured by the IP camera, and the process continues.

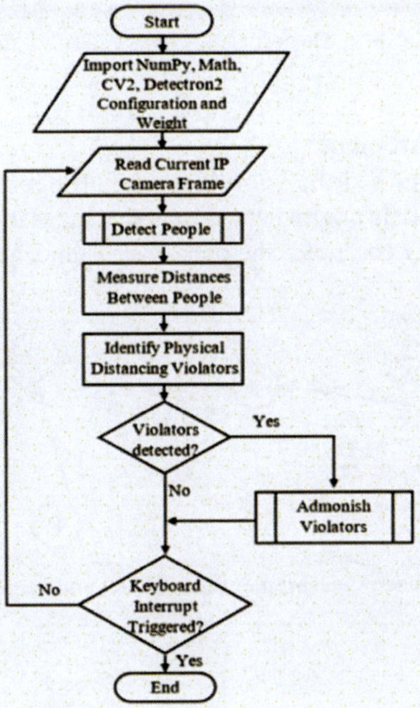

Fig. 2. Main process flowchart

People Detection

Figure 3 shows the system architecture of the Faster RCNN Algorithm for object identification and localization. It starts by taking an image as input and then extracting the feature map using VGG16, from which the Region Proposal Network (RPN) bases its prediction. The RPN generates anchor boxes and then predicts which of these boxes contain the object/s of interest. These anchor boxes are then ranked, and the boxes most likely to have the object/s are proposed. These proposals are then forwarded for RoI pooling, and the object/s are then identified and localized using Detectron2 Library. Only objects identified as a "person" are retrieved for this implementation, and other entities are ignored. However, this complex process was not explicitly implemented in

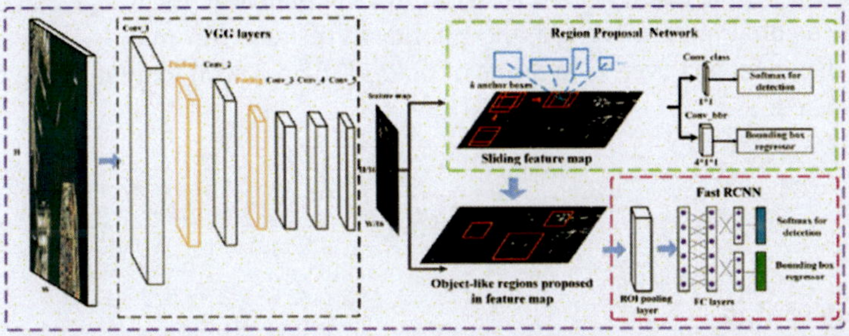

Fig. 3. System architecture of the faster region-based convolutional neural network (Faster R-CNN) algorithm.

this study as it only uses the pre-trained Detectron2 Model Zoo published by Facebook AI Research (FAIR).

Pairwise Distance Measurement

After the predictor has returned the coordinates of all people detected, their pairwise distances are then measured through a nested loop that takes the coordinates of the people as input and then iteratively computes the pairwise distances between all possible pairs.

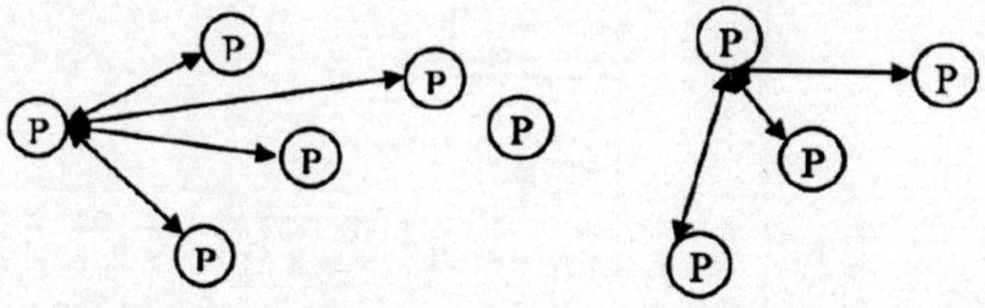

Fig. 4. Pairwise distance measurement first (Left) and second (Right) iterations

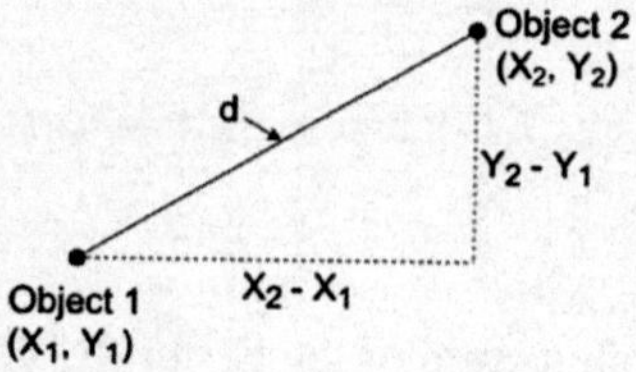

Fig. 5. Euclidean distance between two points

Pairwise distance between two people is computed using the Euclidean distance formula. Say Person 1 located at the coordinates (x1, y1) and person two at (x2, y2), and then their pairwise distance d is defined by the equation:

$$d = \sqrt{(x_2 - x_1)^2 + (y_2 - y_1)^2} \tag{1}$$

Since the image is taken from directly above the platform, the image can be treated as two-dimensional, and, therefore, the Euclidean distance formula is applicable. The calculated distance, however, is in terms of pixels. To convert it to meters, the following equation is used:

$$d_{meters} = \frac{d_{pixels}}{pixels\ per\ meter} \tag{2}$$

Violation Detection

After all possible pairwise distances are measured, people are labeled as violators if they have at least one pairwise distance of less than 1 m with any other person.

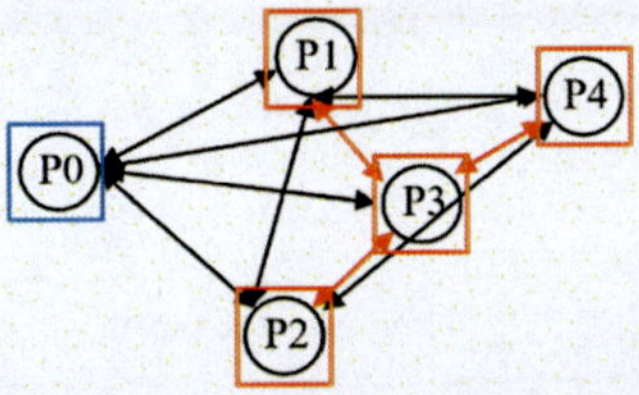

Fig. 6. Physical distancing violation detection

Violator Admonishment

As shown in Fig. 7, the testing platform is divided into four quadrants. A light alert is raised on quadrants containing violators, i.e., by spotting a directive light on that area and playing a pre-recorded audio intended to admonish the individuals present in that quadrant.

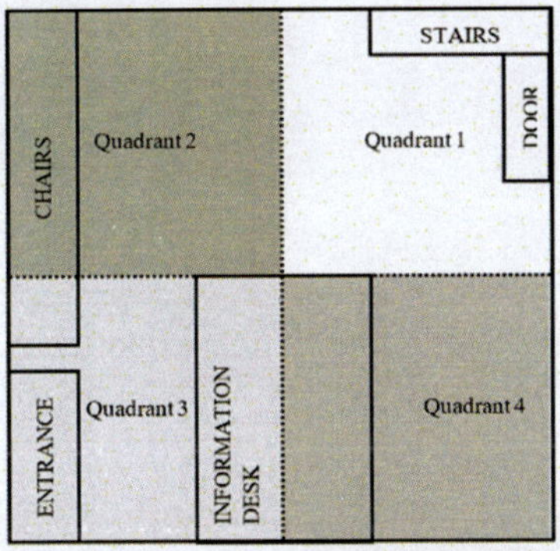

Fig. 7. Violator admonishment setup

Figure 8 shows the process of admonishing the detected violators. First, it retrieves the coordinates of the violators, then compares their coordinates to the set boundaries of each quadrant of the testing site. It checks which quadrants contain the violators, plays the pre-recorded audio corresponding to the quadrants, then forwards the locations of the violators to the ceiling-mounted Arduino UNO via the RF24 module.

Light Alert

The ceiling-mounted Arduino UNO is a standalone device that takes data from the computer through the nRF24L01 module as input and turns the LEDs on, accordingly, as output. Figure 8 shows the process the Arduino UNO performs. First, it takes the quadrant locations of the violators as input from the computer. Next, it checks if quadrant 1 contains violator/s. If yes, it turns LED1 on for 10 s, i.e., the LED pointed towards quadrant 1. It then performs the same process with the other quadrants, checking if they contain violator/s and turns on their corresponding LEDs for the same amount of time, accordingly.

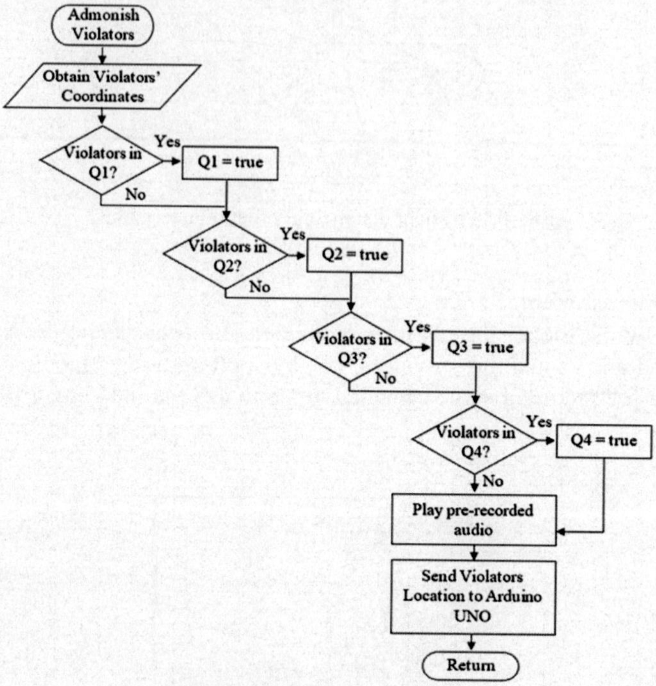

Fig. 8. Admonish violators flowchart

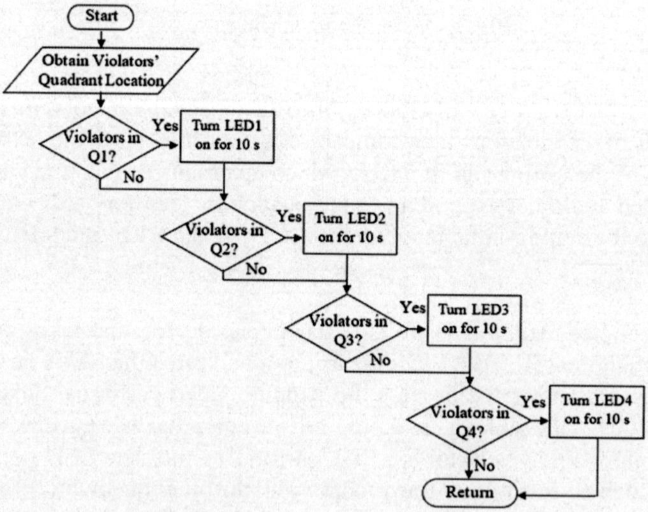

Fig. 9. Arduino UNO light alert flowchart

2.3 Hardware Development

Block Diagram

Figure 10 shows the unscaled positions and connections of the components relative to each other. It also shows how data is transferred throughout the system. The fisheye IP camera is located at the center of the ceiling to give it the maximum vision of the platform. The laptop first obtains a frame from the fisheye IP camera that shows the entire platform, which is then processed to identify the presence and location of physical distancing violators. The areas corresponding to their locations are then passed to an Arduino UNO connected to the laptop via serial connection and interfaced with an nRF24L01 transceiver module. The data received by this Arduino UNO is forwarded via the nRF24L01, which is, in turn, acquired by another nRF24L01module connected to another Arduino UNO that is located near the IP camera. In this instance, the violator's location has already been sent from the laptop to the ceiling-mounted Arduino UNO. As a result, it will turn on the LEDs following the received areas. These Chip-on-Board (COB) LEDs, however, are powered by the mains voltage. Hence, the Arduino UNO excites the relay module, turning on the COB LEDs by completing its connection with the mains voltage. At the same time, the laptop plays through the Bluetooth speaker, which is also mounted near the IP camera, a prerecorded audio intended to admonish the detected violators based on their locations. Thus, the detected violators are successfully admonished through a light alert and a prerecorded audio.

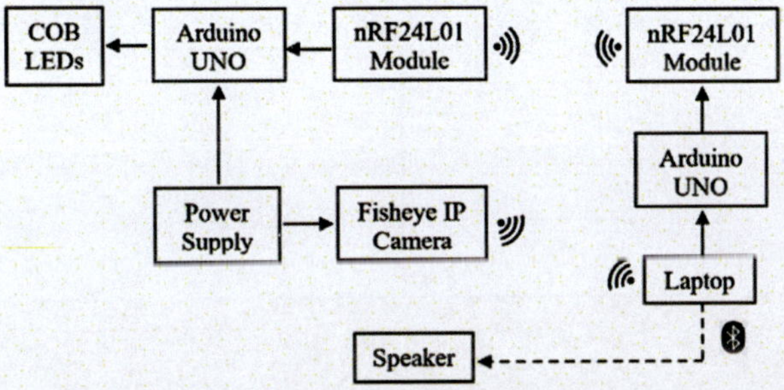

Fig. 10. Block diagram of the setup

Light Alert

Figure 11 shows the schematic diagram of the light alert wherein the nRF24L01 Transceiver is connected to the Arduino Uno to the 4-Channel Relay Module to the Driverless Chip-on-Board (COB) Light Emitting Diode (LED), which is powered by an individual AC power supply. The nRF24L01 Transceiver would acquire the data from the other nRF24L01 Transceiver, which would be analyzed by the microcontroller Arduino Uno by connecting them with their corresponding pinouts. Then, the Arduino Uno supplies 5V to the 4-Channel Relay Module, and its pin 4 to 7 is connected to input 1 to 4. Then, the relay is connected to the COB LED, and the AC power supply, wherein

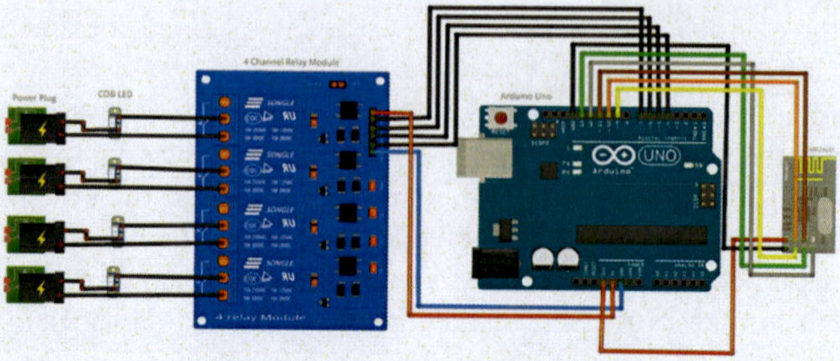

Fig. 11. Light alert schematic diagram

the connection lets the relay act as a switch. Moreover, this would allow yielding the corresponding output based on the data received.

Device Positions

Figure 12 illustrates the positions of the devices relative to the testing site. The white rectangle represents the floor of the testing site at the bottom of the Figure. As shown, the IP camera, represented by the black hemisphere, is mounted on the ceiling. It is positioned at the center of the testing site. Beside it is the Arduino-controlled light alert prototype.

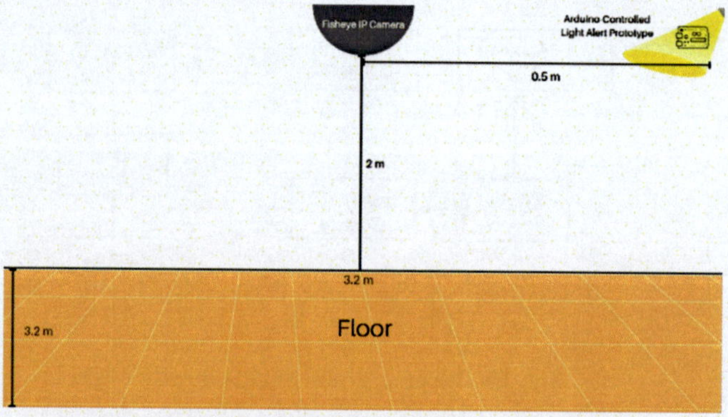

Fig. 12. Device positions relative to the testing site

2.4 Data Gathering

Device Calibration

Calibration was performed to determine the approximate number of pixels per meter. Figure 13 shows a single trial for measuring the pairwise distance, in pixels, between two people. The trial was repeated 30 times, with two people standing in different locations, exactly 1 m apart.

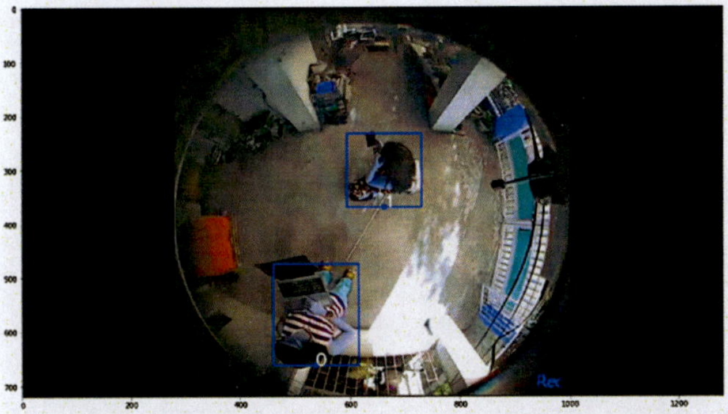

Fig. 13. Sample calibration trial

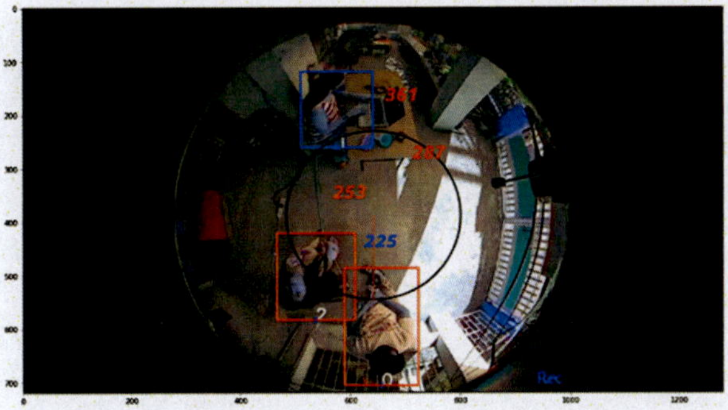

Fig. 14. Calibration results

Figure 14 shows a sample result of the pairwise distance measurement and violator detection. It was found that within a circle of radius 252 pixels from the center, 1 m is approximately equal to 253 pixels. Beyond that circle, 1 m is around 361 pixels. If the measurement is made between a point within the circle and another point outside the circle, 1 m is approximately 287 pixels. The Figure also shows a successful identification of physical distancing violators, where persons 0 and 2 were labeled as violators since they are less than 1 m apart. On the other hand, person 1 is more than a meter away from them.

2.5 Controlled Testing

After the calibration, the completed prototype was subjected to testing in a controlled environment. The testing was supposed to be done at Barangay 28 Hall.

Figure 15 shows the target testing site. It is the barangay hall of Barangay 28 in Caloocan, Metro Manila. Its dimensions are 4.6 m in length, 3.6 m in width, and 2.6 m in height.

Figure 16 shows the garage of Ayson residence, which was used as the site for controlled testing. A table was placed in the testing area, as seen on the upper-left part

Fig. 15. Target testing site

Fig. 16. Actual testing site

of the platform, which functions as a makeshift information desk to mimic the target testing area as much as possible.

In the controlled testing, various scenarios were enacted, and the system's responses were observed and recorded. Also, only three persons were available for testing, as they were the only ones residing in the same house.

3 Results and Discussion

3.1 Data Gathered

Distance Measurement

Data was gathered from controlled testing, which consisted of a total of 30 trials. The first 15 trials consisted of two individuals standing at various testing sites, while the remaining 15 trials involved three individuals.

Figure 17 shows the relationship between the actual distance and the computed distance between the two persons on the meter scale. Hence, it shows that there are only minor discrepancies with the values.

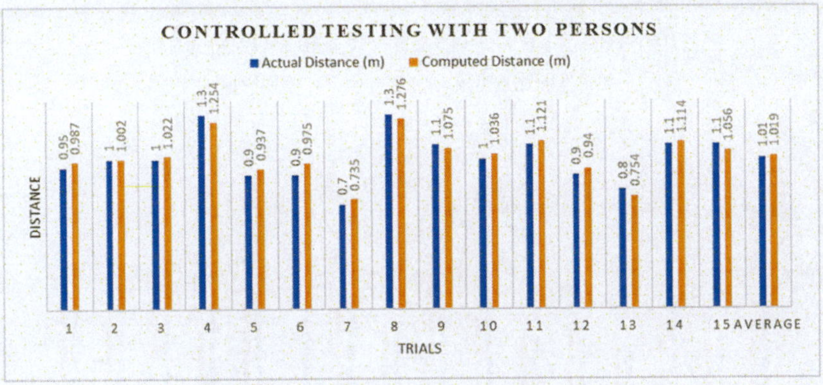

Fig. 17. Controlled testing with two persons

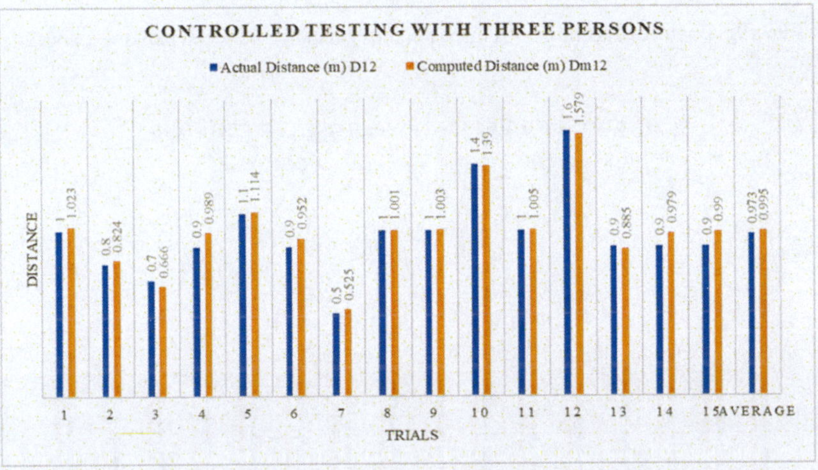

Fig. 18. Controlled testing with three persons (distance between persons 1 and 2)

In the three-person controlled test, Figs. 18–20 illustrate the relationship between the actual and calculated distances between persons 1 and 2, persons 2 and 3, and persons 1 and 3, respectively. As shown in the graph, there are only minor differences due to approximation.

Violator Identification

The image processing algorithm autonomously made the violator identification. Anyone having a pairwise distance of less than one meter with respect to any other person was labeled a violator.

Figures 21 and 22 compare the actual and detected number of violators for the two-person and three-person controlled tests, respectively. Both values are equal for all trials, indicating the accuracy of the system.

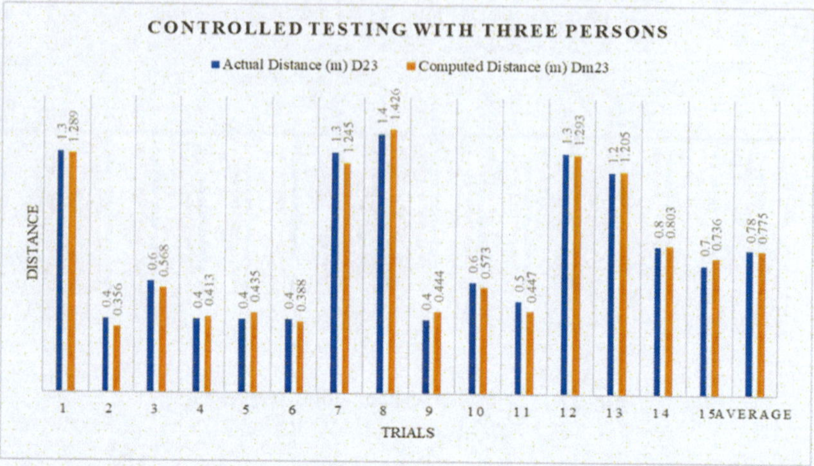

Fig. 19. Controlled testing with three persons (distance between persons 2 and 3)

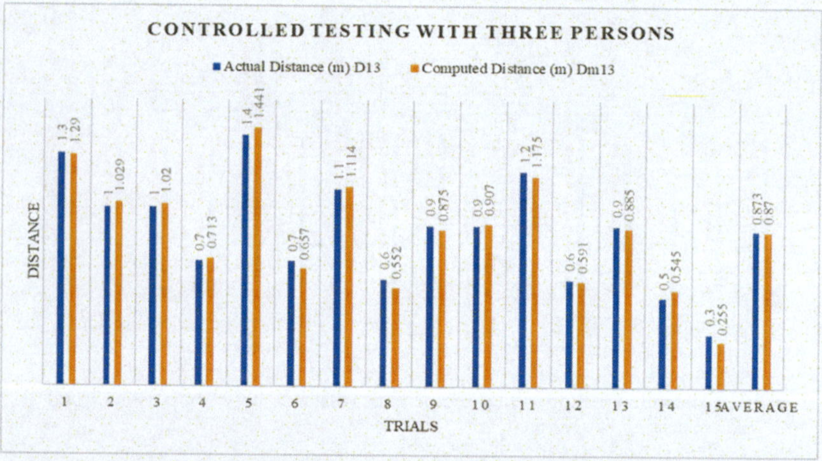

Fig. 20. Controlled testing with three persons (distance between persons 1 and 3)

3.2 Statistical Treatment

A T-test was used to determine whether a significant difference lies between the actual distance (Dactual) and the computed distance (Dcomputed) to evaluate the accuracy of the prototype in measuring distances between people.

Single-Sample T-Statistic is the calculated difference represented in standard error units and is obtained from the Single-Sample T-Test.

Furthermore, probability value was also used, which is the probability measure of finding the observed results when the null hypothesis of a given statistical test is valid. It was used to determine a significant difference, with a confidence level of 95%, between

Fig. 21. Violator identification involving two persons

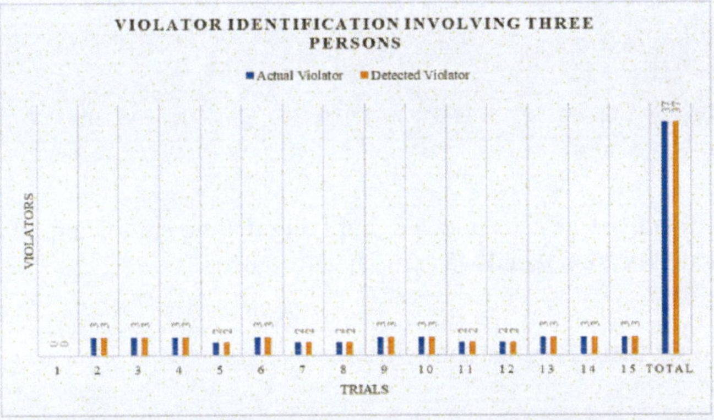

Fig. 22. Violator identification involving three persons

actual and identified violators. The p-value is easily obtained from the t-table, based on the obtained t-statistic and degree of freedom.

Table 1 summarizes the comparison between the actual distances manually measured using a meter scale and the computed distances provided by the image processing algorithm. A two-tailed single-sample t-test was used to determine whether a significant difference exists between the actual and calculated distances. As shown in the table, the actual distances have a mean of 0.909 m, while the computed distances have a mean of 0.915. Their standard deviations are found to be 0.114 and 0.104, respectively. The number sample of 60 was obtained from all the distances measured and computed. Fifteen distances were measured from the controlled testing involving two persons, and three distances were measured per trial in the controlled testing involving three persons with 15 attempts.

From the calculations and based on the gathered data, the t-statistic was found to be $t = 0.3914$, resulting in a p-value of $p = 0.6969$. Since the p-value is significantly higher than the two-tailed significance level, i.e., $p > 0.05$, the null hypothesis was retained,

Table 1. Comparison of actual and computed distances between people

	Dactual	Dcomputed
D	1.01	1.019
D12	0.973	0.995
D23	0.78	0.775
D13	0.873	0.87
Mean ($\overline{x}$)	0.909	0.915
Standard deviation (s)	0.114	0.104
Number sample (n)	60	60
T-Statistic (t)	0.3914	
Significance level (Two-tailed)	0.05	
Degree of freedom (n−1)	59	
p-value	0.6969	

and the alternative hypothesis is rejected, implying that no significant difference exists between the actual distances and the computed distances.

4 Conclusion and Recommendation

4.1 Conclusion

From the findings derived, the researchers conclude that the measurements of the developed admonishment system are accurate, as proven by the gathered data and the conducted single-sample two-tailed t-test. Having a t-statistic of $t = 0.3914$ and a probability value of $p = 0.6969$, it was proven that there is no significant difference between the actual and computed distances between people. Therefore, the assessment of the admonishment system regarding physical distancing violators is accurate and reliable. Hence, the developed admonishment system can effectively monitor and help implement physical distancing between people within a closed environment.

4.2 Recommendation

The admonishment system presented in this paper can accurately measure the distances between people and admonish those who are violating the physical distancing protocol. However, there is a processing time delay, which can be improved using various methods in which can be improved. Moreover, its application is currently limited to relatively small and closed environments such as barangay halls, convenience stores, and many more. Thus, researchers who would like to pursue the same topic are recommended to focus on this limitation. Possible solutions include taking advantage of an objection detection algorithm to improve the processing time and using multiple cameras and integrating them into a single system to monitor a more extensive area effectively. Implementing better camera calibration techniques is also recommended to take advantage of fisheye cameras' broad field of view. Also, using brighter lights for the light admonishment system.

Additionally, due to the restrictions of the community quarantine being implemented as of the time of writing, only controlled testing involving three persons was performed. Thus, if circumstances would allow, it would be better for future researchers to perform uncontrolled testing involving more people to test the system further. Finally, it is also recommended to implement an object-tracking mechanism so that that directive lights would track the detected physical distancing violators.

References

1. Oum, T.H., Wang, K.: Socially optimal lockdown and travel restrictions for fighting communicable virus including COVID-19. Transp. Policy **96**, 94–100 (2020). https://doi.org/10.1016/j.tranpol.2020.07.003
2. Pouw, C.A.S., Toschi, F., van Schadewijk, F., Corbetta, A.: Monitoring physical distancing for crowd management: real-time trajectory and group analysis, July 2020
3. Mizumaru, K., Satake, S., Kanda, T., Ono, T.: Stop doing it! Approaching strategy for a robot to admonish pedestrians. In: 2019 14th ACM/IEEE International Conference Human-Robot Interact, p. 772 (2019)
4. Wang, T., Hsieh, Y.Y., Wong, F.W., Chen, Y.F.: Mask-RCNN Based People Detection Using A Top-View Fisheye Camera (2019). https://doi.org/10.1109/TAAI48200.2019.8959887
5. Din, A., Jabeen, M., Zia, K., Khalid, A., Saini, D.K.: Behavior-based swarm robotic search and rescue using fuzzy controller. Comput. Electr. Eng. **70**, 53–65 (2018). https://doi.org/10.1016/j.compeleceng.2018.06.003

Comparison of the Performance of NewSQL Databases Based on Linux OS

Sarah Myriam Lydia Hahn(✉), Ionela Chereja, and Oliviu Matei

Computers and Information Technology, Technical University Cluj Napoca, Cluj-Napoca, Romania
sarah-hahn@gmx.net, oliviu.matei@holisun.com

Abstract. In the field of cloud technology and big data a new kind of databases arose - NewSQL databases. In the meantime there are many different suppliers and databases available. In this article the most common NewSQL databases in the field of Linux OS are presented and their performance is measured and compared to each other. Therefore a data set with weather data was prepared and used. Furthermore the test setting is explained. The results of the different tests are illustrated and discussed. Recommended actions are given at the end.

Keywords: NewSQL · Performance · Linux · Cloud technology · Test series · Distributed databases · Big data · Latency · Execution time

1 Introduction

"Big data" - a term which is frequently used in the meantime. It is not only a fashion or marketing term but also a topic which is very relevant in sciences [7]. Gartner described this publicly known term with the 3Vs: volume, velocity and variety [5]. An increasing amount of data volume is not only caused by a growth of data sources but also by an increasing file size. The swiftness of data generation is defined by velocity. Caused by a continuously data generation the data cannot be loaded in batches anymore but in real time data streams. Examples for streaming data are in the field of healthcare by monitoring patients or in retail for location-based marketing. The third characteristic variety reflects the varying types of data structures - structured, semi-structured and unstructured. Relational databases or csv files are structured data whereas videos or articles are unstructured documents.

The rise of big data is also possible because of the new prospects of cloud computing. Even there are several definitions for this term [13], we define it as parallel and distributed system based on various, virtualized computers provided by a cloud vendor [8]. Cloud vendors such as Google, Amazon or Microsoft supply hardware, simple access to the system and professional skills. The service is a pay per use model [13]. The advantages of cloud computing are a lesser up-front investment, lower operating costs, higher scalability and elasticity, easy access trough the internet, reduced business risks and maintenance expenses [26].

R. Silhavy et al. (Eds.): CoMeSySo 2021, LNNS 231, pp. 372–384, 2021.
https://doi.org/10.1007/978-3-030-90321-3_30

To handle the growing data volume, multiple accesses and availability scalability is needed. Scalability is possible in two ways - horizontal and vertical. Horizontal scaling means that more hardware resources are added to an existing machine. In the opposite vertical scaling is defined by adding more machines to an existing resource group. Because horizontal scaling is not flexible enough the trend is going to use vertical scaling. Vertical scaling was first implemented by a middleware which seems to be one database application but in the background directs the queries to separate nodes. With the movement to NoSQL databases new database management systems (DBMS) were developed to fulfill these requirements [23].

NoSQL databases derived their wording from "not only SQL". They were not only developed for scaling but also to eliminate the disadvantages of relational database management systems (RDBMS). They wanted to be schemaless, supporting big data, provide fast I/O at low cost respectively be open source [18,24]. Caused by the different behaviour of NoSQL databases and RDBMS a third type of database was developed - NewSQL databases. They have the objective to pair the advantages from both DBMS, like scalability from NoSQL databases and the ACID (Atomic, Consistency, Isolation and Durability) properties from RDBMS [12,23].

Over the course of time many different NewSQL DBMS were developed. There are a few articles which focus on the performance of several NewSQL databases. But none of them measure all of the relevant existing databases [12,19]. Also there are no performance tests for many of the databases. This paper focuses on databases for Linux operating systems (OS). The most common NewSQL databases in the field of Linux OS are presented. Such a market overview is not existing yet and helps in the decision-making to find the right database. In the main part of this article a performance test is being made. Especially the performance of read, write and update transactions are measured and compared. The structure will be the following: the first section "related work and explanations" provides the definition of NewSQL databases as well as some concepts and definitions about performance. In the third chapter "methodology" the approach of the performance test itself and the test setting is described. It is followed by the "performance test results". Last but not least the "conclusion" of the test is shown.

2 Related Work and Explanations

The section explains related work and explanations for the following performance test.

2.1 NewSQL Databases

The term NewSQL databases first arose in 2011 in an article by Matthew Asslet [3]. The objective of this new kind of database is to combine the best parts of

NoSQL databases and RDBMS. The advantage of NoSQL databases is especially the scalability which accompanies high performance and availability. The characteristics of the databases like ACID are adopted from RDBMS [4,19,23].

ACID is an acronym for the four properties atomicity, consistency, isolation, durability. A request which supports these is called a transaction. Atomicity defines that a request as a whole is completely successful or not - there is no state in between. In the case of a failure during request execution the entire request fails and the database is set to the pre-state. Consistency means that the state of the database must always be valid. It results in consistent constraints, cascades and triggers. The third characteristic Isolation defines that each transaction is viewed separately. For the possibility of a parallel execution of separate requests transactions can lock tables temporally. Durability guarantees that the state after a successful execution of a transaction is ensured also after a failure, like power breakdown [15].

Furthermore NewSQL databases allow (partial) structured query language (SQL) queries. SQL is the main query language for applications. A parallel execution of read and write transactions is ensured with a non-locking concurrency control mechanism [4,19]. NewSQL databases supports online transaction processing (OLTP) as well as online analytical processing (OLTP) workloads. Concepts of NoSQL databases such as column-oriented data model, distributed nodes or in-memory processing are used for the implementation [22]. Detailed evaluations between the most common NewSQL databases were made in the past [10,11,16]. All in all these evaluations focus on the properties and not on the performance.

2.2 Performance

Performance concerning time, hardware resources, capacity and modes is an important factor for data integration processes. These factors can be measured as [25]:

- Time efficiency: grade of low response and processing times
- Resource utilization: resources utilized by the process differentiate between the type of resource and the quantity
- Capacity: demand that can be executed without missing the other requirements
- Modes: modes of data processing (batch, real-time etc.).

Especially in the field of NoSQL databases there are many performance tests being made. The tests are made on different databases with different data sets and test cases. All in all it can be reduced to the following types of test cases [2].

- Write latency: the time it takes to insert a number of records into the database
- Update latency: the time it takes to update a number of records into the database
- Read latency: the time it takes to read a number of records from the database.

Latency refers to time efficiency. Caused by a dependency between time efficiency and resource utilization the focus on the performance test will be on the latency. Regarding capacity and modes the standard settings of each database were used.

3 Methodology

3.1 Databases

There are a few surveys which focus on the performance of NewSQL databases but they do not contain all the common DBMS [12,19]. To get an overview about the market and the existing NewSQL DBMS a research of existing scientific work (including, but not only [1,6,9,14,17,19–21]) were made. In these the databases Actor DB, Altibase, Apache Trafodion, Citus, CockroachDB, Drizzle, Faircomdb (former C-TreeACE), FaunaDB, GemFire XD, Google Spanner, GridGain IMDB, Harper DB, MariaDB Xpand (former ClusterixDB), NuoDB, Singlestore (former MemSQL), TIBCO Active Spaces, TiDB, TokuDB, Vitess and VoltDB are mentioned.

Out of these databases the focus is only on databases, which are not abandoned and based on Linux OS. Several projects are not alive anymore like Actor DB, Altibase, Apache Trafodion, Drizzle, GemFire XD and TokuDB. The databases itself must not only support Linux OS but can also support other OS, cloud environment like Amazon Web Services or Microsoft Azure or container technology like Docker. Furthermore they have to fit into the test setting. Following the databases Citus, CockroachDB, Faircomdb, GridGain IMDB, Harper DB, TIBCO Active Spaces, TiDB, Vitess and VoltDB are relevant.

3.2 Data Set

For comparability reasons the different test series on the different databases were made with the same data sets. The data itself is a weather data set which was collected between 2008 and 2020 in different Romanian cities. Depending on the measurement station there are different values like the temperature in different heights, soil temperature, moisture and rainfall. Further there is information about the measurement station itself e.g. if the coupler is attached and the status of the battery. The original files are in .hobo or .dtf format. They have been converted to .csv files with HOBOware. Afterwards out of this data the SQL statements were build. Depending on the database the specific SQL dialect was chosen.

The target table contains the columns in Table 1.

In total the data contains over 12 million of data rows distributed over 13 years.

Table 1. Data set columns

Column name	Data type	Length
id	Integer	
bad_battery	Text	100
batt_v	Decimal	
coupler_attached	Text	100
coupler_detached	Text	100
Date	Text	100
end_of_file	Text	100
event_units	Decimal	
ext__line_event_units	Decimal	
good_battery	Text	100
host_connected	Text	100
Index	Integer	
moisture_m_m	Decimal	
moisture_m_m_2	Decimal	
Rainfall	Decimal	
Stopped	Text	100
Temp_C	Decimal	
Temp_F	Decimal	
Temp_F_1	Decimal	
Temp_F_2	Decimal	
Temp_F_3	Decimal	
Temp_F_4	Decimal	
Temp_F_5	Decimal	
Temp_F_6	Decimal	
Temp_F_soil_1	Decimal	
Temp_F_soil_2	Decimal	
Temp_F_soil_3	Decimal	
Time	Text	100
Timestamp	Text	100
table_name	Text	255
City	Text	255

3.3 Test Settings

To have an identical test setting the virtual machines (VMs) were set up similarly for the series. For the virtualization Oracle VirtualBox 6.1.18 is used. The hardware beneath is the same and physical.

- Processor: AMD Ryzen 3 2200U, 2500 MHz
- CPU: 2 physical cores
- RAM: 8 GB (4 GB to virtual machine)
- Disk: 50 GB to virtual machine.

The configuration of each VM is the same. There is one VM for each database to prevent any conflicts. Each instance is a single node cluster. Caused by the different databases different Linux OS were needed. For comparability reasons as less different systems as possibly were chosen. Table 2 gives an overview about the OS.

Table 2. Linux OS per NewSQL DBMS

NewSQL DBMS	Linux OS
Citus	Ubuntu 20.4 LTS
CockroachDB	Ubuntu 20.4 LTS
Faircomdb	Ubuntu 18.4 LTS
GridGain IMDB	RedHat 8.3
Harper DB	Ubuntu 18.4 LTS
TIBCO Active Spaces	RedHat 8.3
TiDB	RedHat 7.9
Vitess	Ubuntu 18.4 LTS
VoltDB	Ubuntu 18.4 LTS

For each database the standard configuration was used. This is the configuration the manufacturer recommends for first tests. Depending on the use case another configuration could fit better. For the overall performance test there was no special use case but the performance in total was measured. For the execution the free standard tool of each database was used which can be also the terminal.

3.4 KPIs

The different latencies are measured by their execution time with different SQL queries.

- Write latency: Insert a set of x records whereby each record is a separate statement

- Update latency: Update a set of x records whereby each record is a separate statement
- Read latency: Select a set of x records whereby each record is a separate statement.
- Read latency total: Select a set of x records whereby all records are queried in one statement.

Each query is one test and is executed as one transaction. The same test is executed at least five times and the average was taken to eliminate outlier. The number of records ranges from 1.000 to 10.000 (steps of 1.000). Per database this setup translates to 200 tests (4 different queries, 10 different numbers of records and 5 executions).

4 Performance Test Results

By executing the test series some limitations came up. TiDB was not usable for the defined tests. At least it was not possible to execute a sample of 100 records. Following there are no results for this database. This is shown with a data point at zero (no) execution time.

Figure 1 shows the results of the write latency. The databases CockroachDB and FaircomDB are conspicuous. Their execution time is in all test series multiple times higher than the other databases. Since the execution time was that protracted it was not possible to execute a test series with over 4000 records.

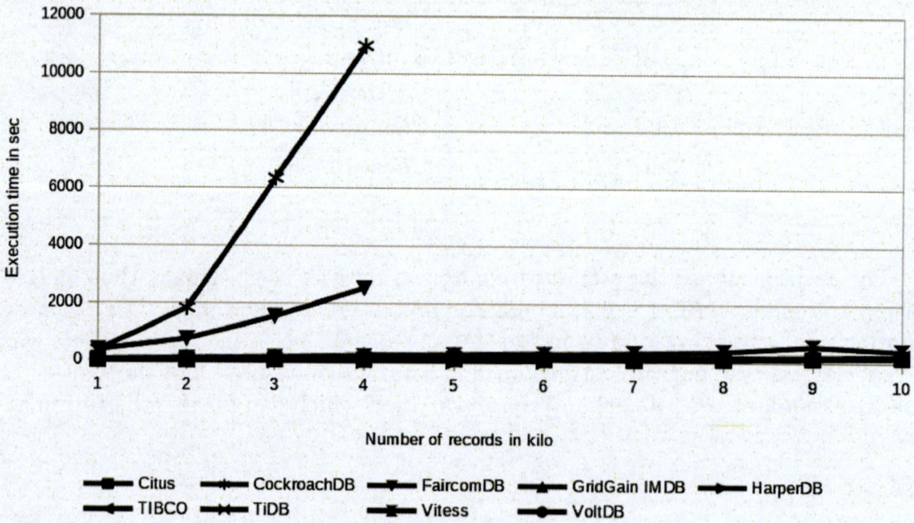

Fig. 1. Write latency

The other databases can be seen in detail in Fig. 2. HarperDB was the database with the best execution times. However the test settings did not admit to execute the statements with more than 6000 records. Also GridGain IMDB did not support the execution of more than 6000 records. GridGain IMDB and Vitess had similar execution times. Compared to the other databases the results of these are in the middle of the pack. Besides HarperDB the databases Citus, TIBCO and VoltDB have a good performance.

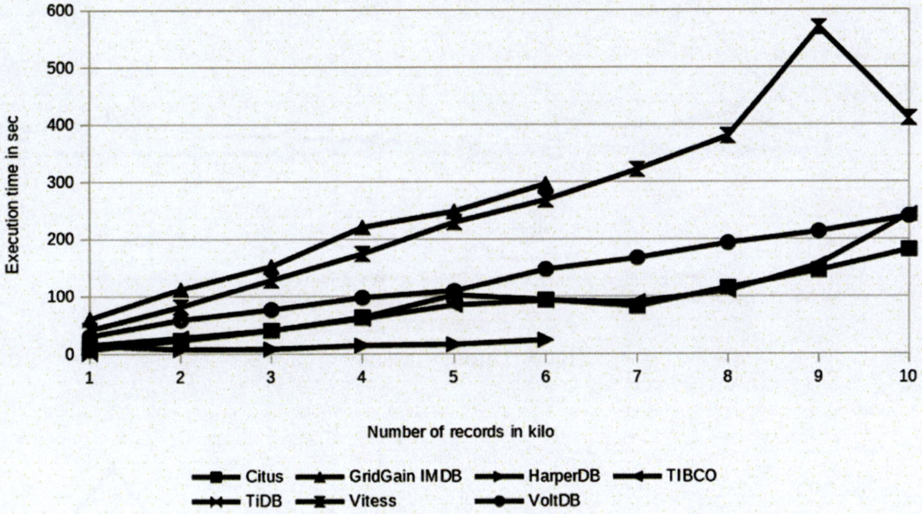

Fig. 2. Write latency in detail

The next figure, Fig. 3, displays the results of the update latency tests. It stands out that the performance of CockroachDB is really bad. The execution time of FaircomDB is not good but it steady independent of the number of records. The test settings did not permit to execute more than 4000 records. So the further trend is not known. HarperDB did not support the execution of updates in the test settings.

In Fig. 4 shows that the performance of the databases Citus, TIBCO and VoltDB is similar. They are the best databases in the performance test regarding the update latency. GridGainn IMDB and Vitess are also similar but execution time is higher.

The results of the read latency tests are pictured in Fig. 5. Again CockroachDB has the worst test series results followed by FaircomDB. HarperDB did not support the execution of a selection of several rows in the test settings.

The results of the other databases are shown in Fig. 6. GridGain IMDB and Vitess are almost identical followed by TIBCO. The databases with the lowest execution time regarding read latency are Citus and VoltDB.

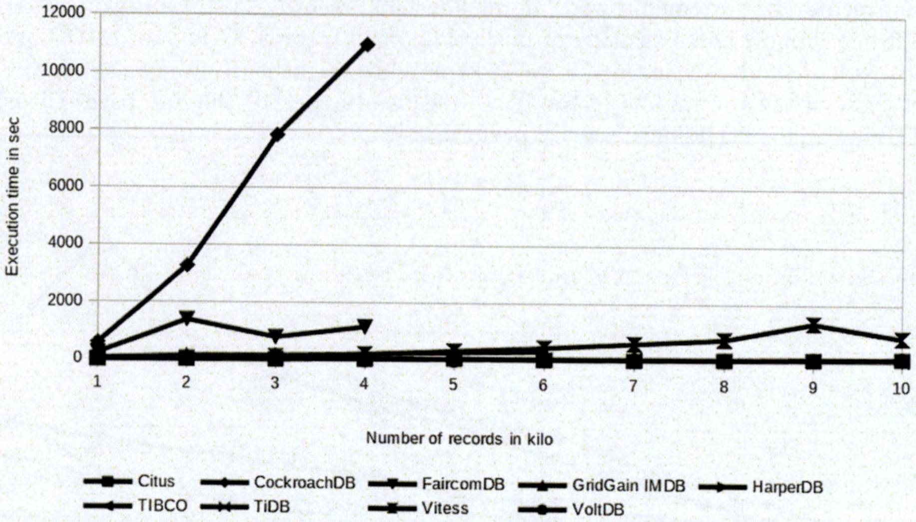

Fig. 3. Update latency

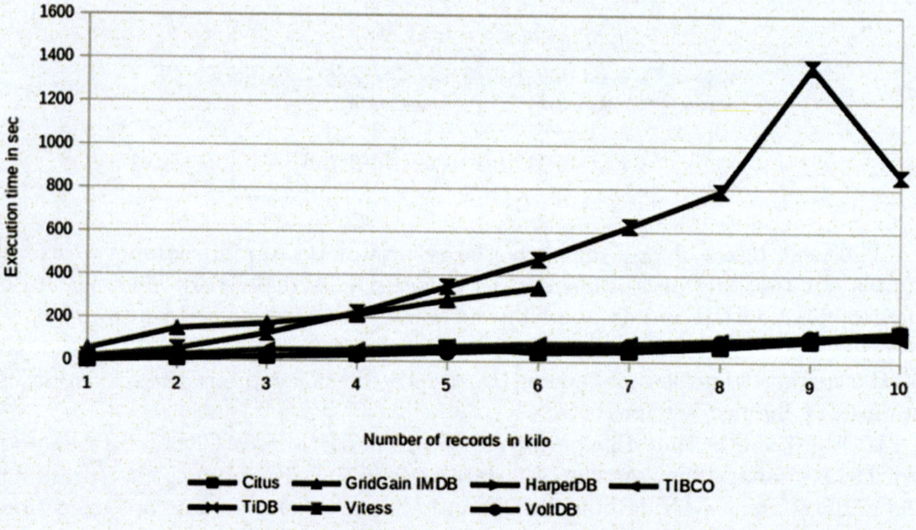

Fig. 4. Update latency in detail

The total read latency is illustrated in Fig. 7. Though there are variations and the execution time is not linear the databases with the highest execution times are GridGain IMDB, FaircomDB, HarperDB and TIBCO. The execution time of Vitess is in the middle. Citus and VoltDB are the fastest databases in the test regarding the total read latency.

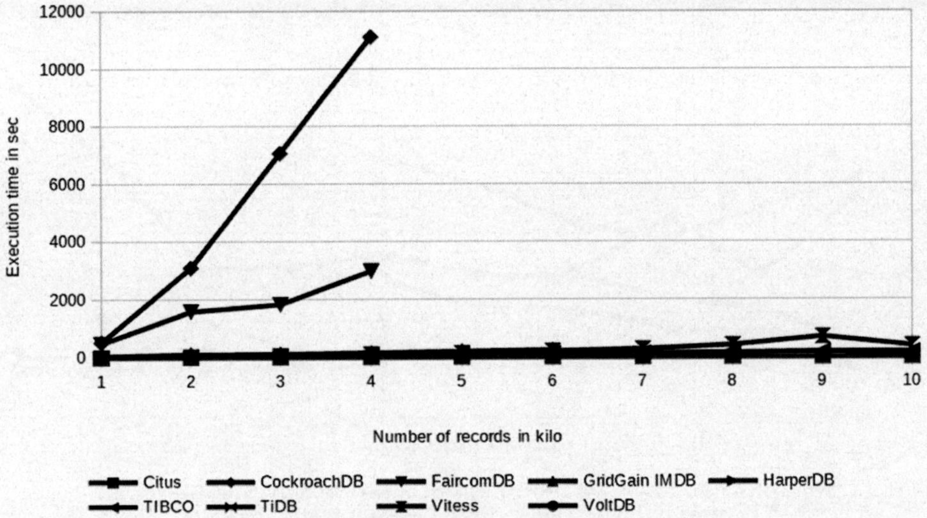

Fig. 5. Select row-by-row latency

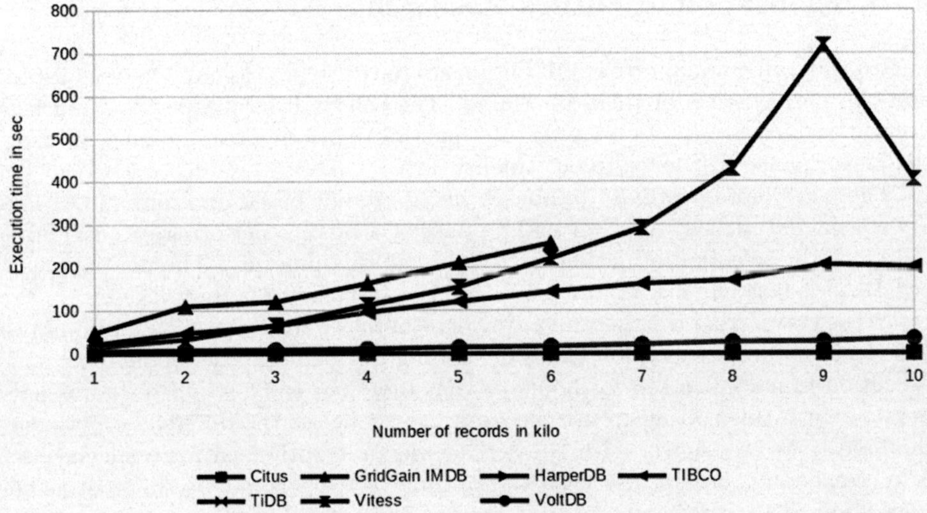

Fig. 6. Select row-by-row latency in detail

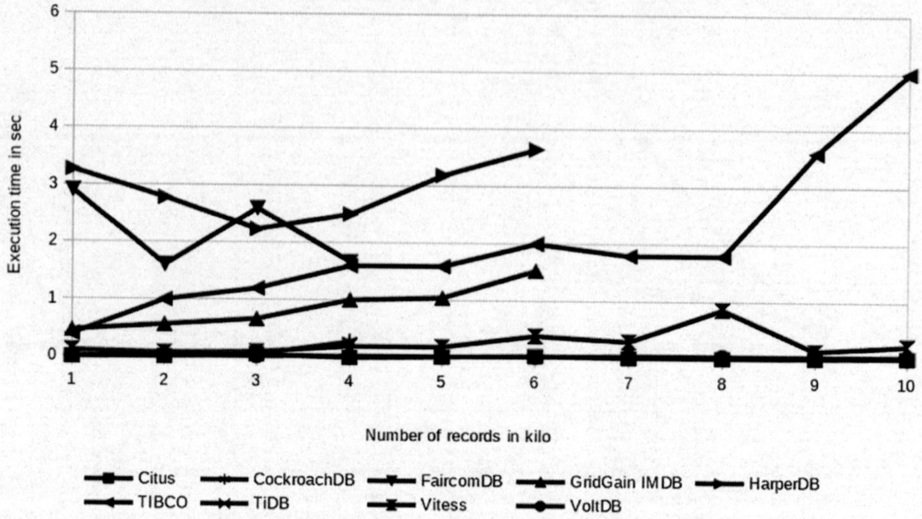

Fig. 7. Select row-by-row latency

5 Conclusion and Future Research

In this work all common NewSQL databases in the field of Linux OS were listed and the performance of them measured. The selected databases are a market overview which can be used for the first step during a decision making. Further the performance test helps to cut the list.

There are more NewSQL databases which are not based on Linux OS. They are considered in this work but will be part of a further performance test. Furthermore the selection of databases is made at this juncture. The development of DBMS is fast moving and in the future the test has to be re-worked.

In the test setting the chosen hardware resources were reduced to a minimum. This leads to lower execution times depending on the database. Some tests had execution times which ran for hours so that they had to be stopped. It was not possible with this setting to use big data. Nevertheless the differences between the databases are shown. Also the setting was not optimized for a special use case. Depending on the use case the optimal setting could be different. The suggestion is to carry out another performance test based on each use case which is relevant. If the database is used for a program and not only for ad hoc statements normally program code is used. The performance of each database can vary depending on the kind of connection, the program language and the chosen software.

The best execution times in all of the tests have the databases Citus and VoltDB. This is the reason why the suggestion is to focus on these databases and have a closer look at them. Regarding to the test results the usage of the databases CockroachDB and FaircomDB is not recommended. They have the worst test results compared to the other tested databases. Nevertheless these are the results with this special test setting. Other settings or further properties have to be evaluated.

Acknowledgments. This work was supported by a grant of the Romanian National Authority for Scientific Research and Innovation, CCCDI - UEFISCDI, project number COFUND-ICT-AGRI-FOOD-MUSHNOMIC, within PNCDI III.

References

1. Almassabi, A., Bawazeer, O., Adam, S.: Top newSQL databases and features classification. Int. J. Database Manage. Syst. **10**, 11–31 (2018)
2. Almootassem, O., Husain, S.H., Parthipan, D., Mahmoud, Q.H.: A cloud-based service for real-time performance evaluation of noSQL databases. arXiv preprint arXiv:1705.08317 (2017)
3. Anslett, M.: NoSQL, newSQL and beyond: the drivers and use cases for database alternatives (2011)
4. Aslett, M.: How will the database incumbents respond to noSQL and newSQL (2011)
5. Beyer, M.A., Laney, D.: The Importance of 'Big Data': A Definition, pp. 2014–2018. Gartner, Stamford (2012)
6. Binani, S., Gutti, A., Upadhyay, S.: SQL vs. NoSQL vs. NewSQL-a comparative study. Database **6**(1), 1–4 (2016)
7. Buhl, H.U., Röglinger, M., Moser, F., Heidemann, J.: Big data (2013)
8. Buyya, R., Yeo, C.S., Venugopal, S., Broberg, J., Brandic, I.: Cloud computing and emerging it platforms: vision, hype, and reality for delivering computing as the 5th utility. Futur. Gener. Comput. Syst. **25**(6), 599–616 (2009)
9. Chandra, U.: A comparative study on: NoSQL, newSQL and polygot persistence. Int. J. Soft Comput. Eng. (IJSE) **7** (2017)
10. Chereja, I., Hahn, S.M.L., Matei, O., Avram, A.: Multidimensional analysis of newSQL database systems. In: Computer Science On-line Conference. Springer (2021, to appear)
11. Chereja, I., Hahn, S.M.L., Matei, O., Avram, A.: Operationalizing analytics with newSQL. In: Computer Science On-line Conference. Springer (2021, to appear)
12. Fatima, H., Wasnik, K.: Comparison of SQL, noSQL and newSQL databases for internet of things. In: 2016 IEEE Bombay Section Symposium (IBSS), pp. 1–6. IEEE (2016)
13. Geelan, J., et al.: Twenty-one experts define cloud computing. Cloud Comput. J. **4**, 1–5 (2009)
14. Gurevich, Y.: Comparative survey of noSQL/newSQL DB systems. The Open University of Israel, Department of Mathematics and Computer Science (2015)
15. Haerder, T., Reuter, A.: Principles of transaction-oriented database recovery. ACM Comput. Surv. (CSUR) **15**(4), 287–317 (1983)
16. Hahn, S.M.L., Chereja, I., Matei, O.: Analysis of transformation tools applicable on newSQL databases. In: Computer Science On-line Conference. Springer (2021, to appear)
17. Hajoui, O., Dehbi, R., Talea, M., Batouta, Z.I.: An advanced comparative study of the most promising noSQL and newSQL databases with a multi-criteria analysis method. J. Theor. Appl. Inf. Technol. **81**(3), 579 (2015)

18. Han, J., Haihong, E., Le, G., Du, J.: Survey on noSQL database. In: 2011 6th International Conference on Pervasive Computing and Applications, pp. 363–366 (2011)
19. Kaur, K., Sachdeva, M.: Performance evaluation of newSQL databases. In: 2017 International Conference on Inventive Systems and Control (ICISC), pp. 1–5. IEEE (2017)
20. Khasawneh, T.N., AL-Sahlee, M.H., Safia, A.A.: SQL, newSQL, and noSQL databases: a comparative survey. In: 2020 11th International Conference on Information and Communication Systems (ICICS), pp. 013–021 (2020)
21. Kumar, R., Gupta, N., Maharwal, H., Charu, S., Yadav, K.: Critical analysis of database management using newSQL. Int. J. Comput. Sci. Mob. Comput. **3**, 434–438 (2014)
22. Moniruzzaman, A.B.M.: NewSQL: towards next-generation scalable RDBMS for online transaction processing (OLTP) for big data management. Int. J. Database Theory Appl. **7** (2014)
23. Pavlo, A., Aslett, M.: What's really new with newSQL? ACM SIGMOD Rec. **45**(2), 45–55 (2016)
24. Sharma, V., Dave, M.: SQL and noSQL databases. Int. J. Adv. Res. Comput. Sci. Softw. Eng. **2**(8) (2012)
25. Theodorou, V., Abelló, A., Lehner, W.: Quality measures for ETL processes. In: Bellatreche, L., Mohania, M.K. (eds.) DaWaK 2014. LNCS, vol. 8646, pp. 9–22. Springer, Cham (2014). https://doi.org/10.1007/978-3-319-10160-6_2
26. Zhang, Q., Cheng, L., Boutaba, R.: Cloud computing: state-of-the-art and research challenges. J. Internet Serv. Appl. **1**(1), 7–18 (2010)

Distance Monitoring with Contact Tracing Logging System Using Received Signal Strength Indicator

Jessie Jaye R. Balbin, Mark Raniel C. Cruz, and Ken Robert D. San Juan(✉)

Mapua University, Intramuros, Manila, Philippines
jjrbalbin@mapua.edu.ph, {mrccruz,krdsanjuan}@mymail.mapua.edu.ph

Abstract. In this study, the researchers developed 2 wearable hardware that utilizes Arduino Nano 33 BLE Sense as it's microcontroller in the implementation of social distancing and contact tracing with an application and database. The application provides an indicator for risk status, notifications for social distancing if the two devices are within 100 cm of each other. The two devices were calibrated at a displacement value of 1 m and yielded a RSSI value of −69 dBm for device 1 and −67 dBm for device 2 respectively. The database also provides a manual override for the risk status that is displayed in the graphical user interface in the instances of human error.

Keywords: Arduino Nano · Wearable device · Contact tracing · Social distancing · RSSI

1 Introduction

In In light of what is happening around the world about the novel coronavirus or COVID-19, the whole world tastes a pandemic that strikes a new era which is commonly called as the "new normal" since it is thought that life will never be the same again. In this "new normal", social distancing and effective contact tracing is integral to the new way of life and this needs to be implemented correctly to further flatten the infection curve of this disease. In the Philippines, after being in strict lockdown for months, the capital, Manila, and its nearby areas are now placed under General Community Quarantine wherein there is more freedom in movement, but social distancing and contact tracing must be implemented. According to Dr. Troy Gepte, a public health consultant for the Senate Committee on Health, the methods of contract tracing in the Philippines needs to be digitalized. In the current situation, the methods of contract tracing are lacking here in the Philippines since it relies on an exhaustive contract tracing and strict military measures instead of public health measures that is using digital and automated contract tracing.

The objective of this study is to develop 2 wearable device that can be used to acquire the distance and location of each other for purposes of social distancing and contact tracing. To be able to do the main objective, the specific objectives of this study

R. Silhavy et al. (Eds.): CoMeSySo 2021, LNNS 231, pp. 385–401, 2021.
https://doi.org/10.1007/978-3-030-90321-3_31

must be done and it is (1) To make an application that will display the location, risk status, RSSI Values for both devices, will provide notifications for social distancing and health status; (2) acquire the distance of device1 and device 2 for social distancing protocol using Received Signal Strength Indicator; (3) create a database that will display the timestamps, location, risk status, and provide manual override for the users of the device. The significance of the study is that since the world now lives in this new era wherein social distancing and effective contact tracing must be implemented, researchers must find ways to effectively implement this new normal in the society. Even after the pandemic, social distancing will be practice and this thesis study if proven effective may help in this cause.

This study will only focus on procurement of two devices in utilizing Arduino Nano 33 BLE Sense in implementation of social distancing and contact tracing monitoring. The hardware component of the device will use Arduino Nano 33 BLE as it's microcontroller and it will be connected to a HM Bluetooth Device and a 5 V DC source. It will utilize PLA Filament as straps since it will be wearable like a watch. The study aims to (1) To make an application that will display the location, risk status, RSSI Values for both devices, will provide notifications for social distancing and health status; (2) acquire the distance of device1 and device 2 for social distancing protocol using Received Signal Strength Indicator; (3) create a database that will display the timestamps, location, risk status, and provide manual override for the users of the device (Fig. 1).

2 Methodology

2.1 Conceptual Framework

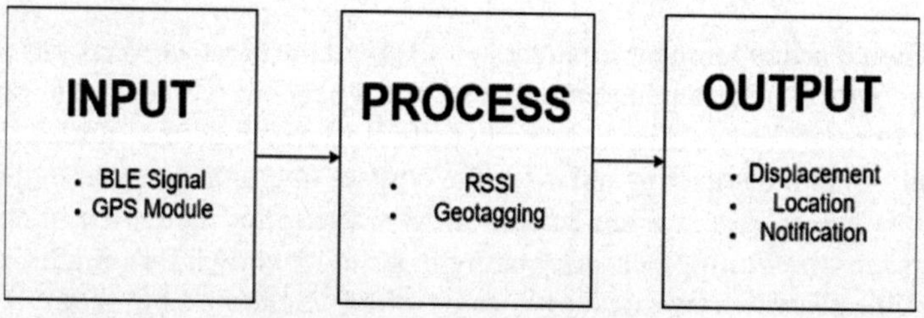

Fig. 1. Conceptual framework

The figure above shows the conceptual framework of this proposal and how it will work. The input of this research will be the data obtained from the user using the BLE/GPS Module in one of the nodes. Two nodes will be created and the BLE module will be used for the acquisition of data from Device 1 to Device 2. The GPS Module will be used for location tracking and to know whether a contact was made to a possible PUI. An application will be developed so that the user of the devices will know whether they have been in contact with another user that has either contracted the virus or to report whether they themselves have had contact with someone. This is for the purpose of proper implementation of social distancing and contract tracing since the person that will be notified that have been in danger can practice self-quarantine to help decrease the spread of the virus (Fig. 2).

2.2 System Flow Chart

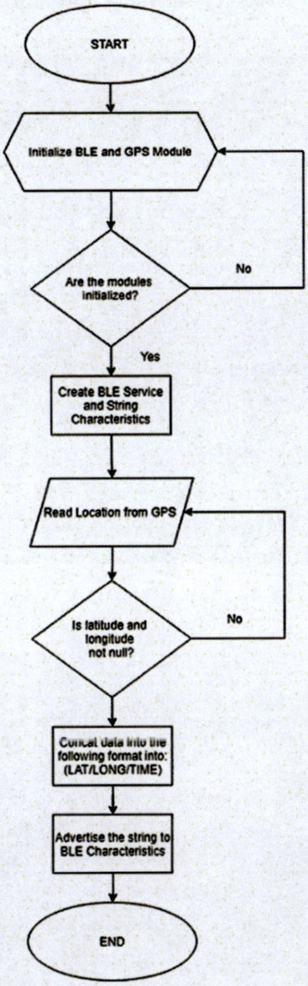

Fig. 2. Process flow chart of the system

The figure above shows the process flow chart for the system. The first thing is for the modules and sensor will now initialize. After initialization, the BLE service and string characteristics will now be shown

2.3 Prototype

In the figure above, it is the actual prototype done by the researchers. Two wearable devices that can notify the users to observe social distancing when they are within 90-cm distance of one another and indicate if you are at risk if the report button would be clicked when the observe social distancing notification would be prompted that would be seen on the Graphical User Interface of the Application (Figs. 3 and 4).

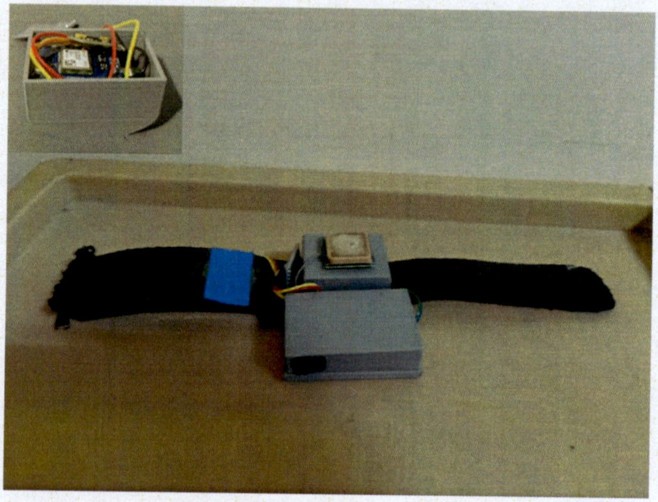

Fig. 4. Parts of the prototype

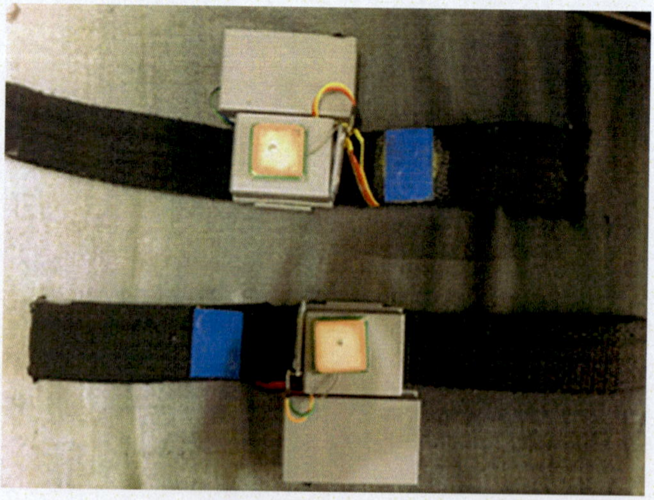

Fig. 3. Actual prototype

In this figure is the parts of the prototype. The prototype is functioned by the Arduino Nano 33 BLE Sense as its microcontroller and Bluetooth module found at the side. There is also the GPS Module, Ublox Neo – 6M GPS Module, that would trace the location of the user and send the coordinate in a longitude latitude format in the Show Contact Graph of the Graphical User Interface. There is also the 5V DC Power Supply used along with the switch to turn on and off the device. The strap used for the prototype is the PLA Filament Strap. In the upper left, you could see the inside view of the prototype, where the modules and microcontroller are found (Figs. 5 and 6).

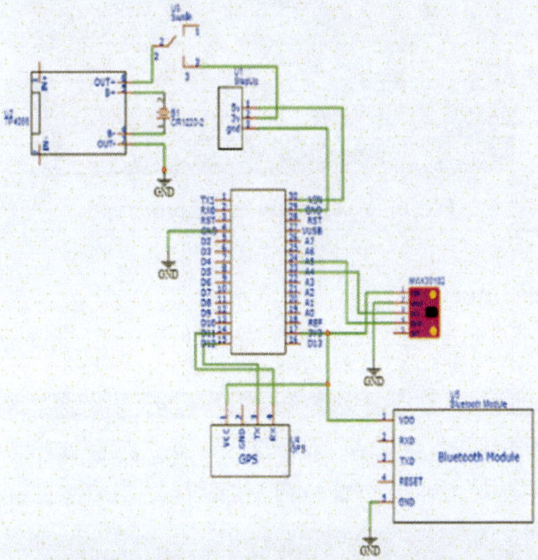

Fig. 5. Schematic diagram of the hardware

For this study, two Arduino Nano 33 BLE Microcontroller-based wearable devices were done by the researchers for the purpose of contact tracing and social distandcing monitoring by using the concept of Received Signal Strength Indicator. In figures above, the parts of the prototype is shown with the hardware is composing of BLE Module, Ublox GPS Module, a 5 Vdc voltage source, and the microcontroller used to program the different modules will be the Arduino Nano 33 BLE Sense. It utilizes a PLA Filament strap for it's wearability. It can also be seen how the device can be worn. There is also the schematic diagram that have been used for the whole study (Fig. 7).

Fig. 6. Wearability of the prototype

2.4 Graphical User Interface

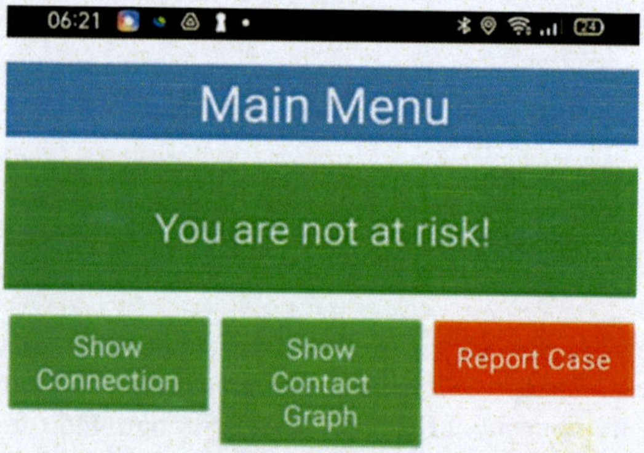

Fig. 7. Actual graphical user interface

It can be accessed using an android phone that will be connected to mobile Wi-Fi or a router. The figure above displays the home part of the application. The graphical user interface also shows the connection of the device, the RSSI values, and the contact graph. As seen from the graphical user interface, the application that will be developed in this research will display messages whether the user of the device is at risk or not. This is crucial in effectively lowering the transmission rate since the user will see in the application that he/she may be a risk and will start to practice self-quarantine (Figs. 8 and 9).

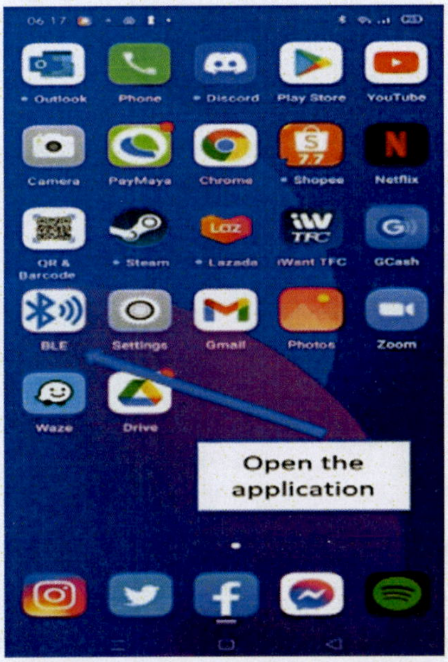

Fig. 8. Accessing the graphical user interface

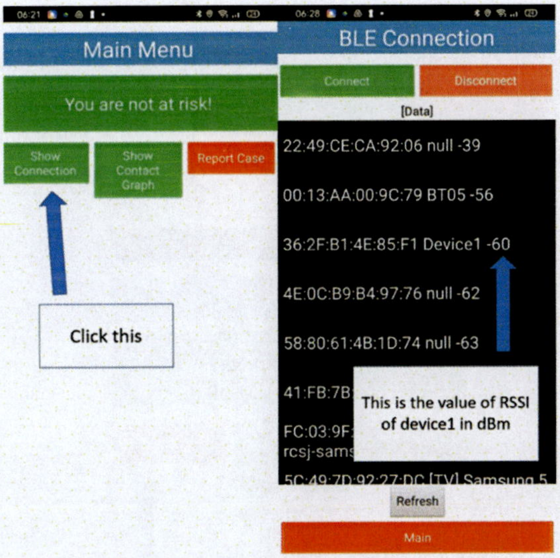

Fig. 9. Accessing the BLE connection tab

In the figure above, it shows the application home interface on the left side and the connection interface on the right side. The researchers will be clicking device 1 to connect the hardware to the GUI. After clicking device 1, the researchers will click on connect and it will redirect back to the home screen. It is also shown in the right side of the screen the mac address and the RSSI value for the device made (Fig. 10).

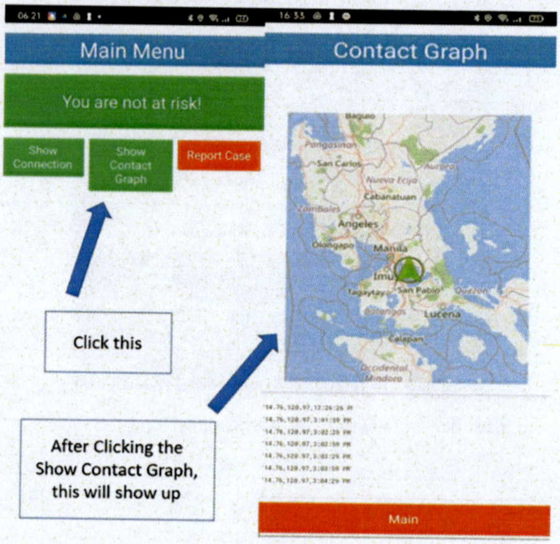

Fig. 10. Accessing the show connection tab

The researchers will click on the show contact graph and this will utlizing the concept of geotagging. The location retrieved from the GUI used the Ublox GPS module which will receive signals from satellite and ground stations to determine the exact location of the user (Fig. 11).

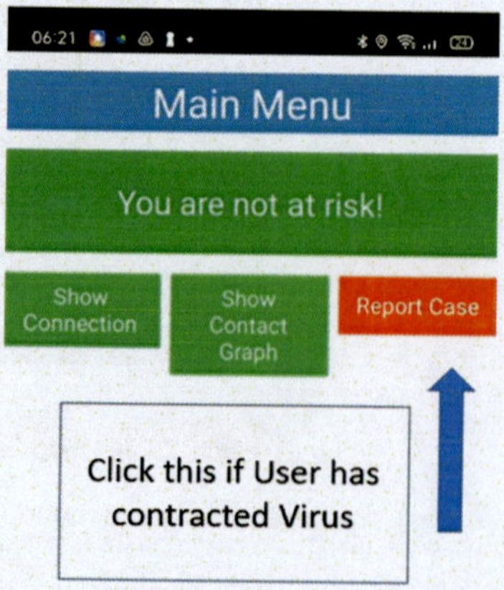

Fig. 11. Accessing the report case button

When the user of either device 1/device 2 has contracted the virus or has encountered a positive case, he/she will click on the report case button to notify the other device that was encountered (Fig. 12).

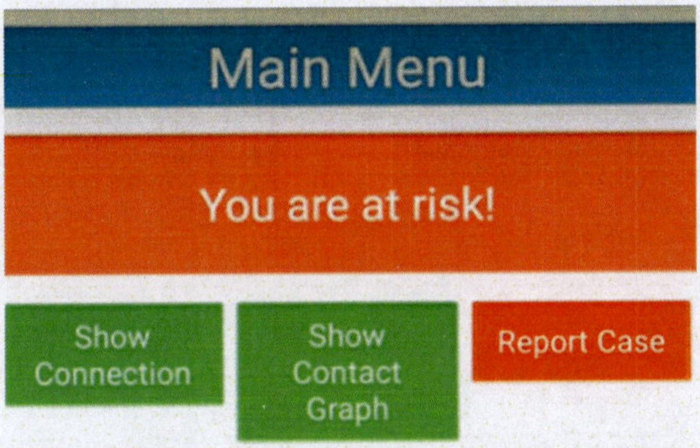

Fig. 12. Actual graphical user interface when risk button is clicked

Once the report case button was pressed, the Indicator in the Graphical User Interface changes it's message from "You are not at risk" to "You are at risk". The color of the indicator also changes from green to red to signal a warning (Fig. 13).

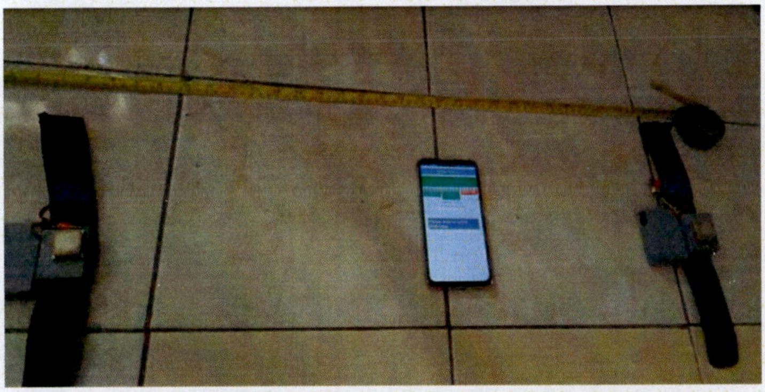

Fig. 13. Social distancing notification in the GUI

Another feature of the Graphical User Interface is a social distancing notification when it detects a breach in protocol. This notification will pop-up when device 1 and device 2 are less than 100 cm from each other wherein 100 cm (1 m) is the standard for social distancing (Fig. 14).

2.5 Calibration and Testing

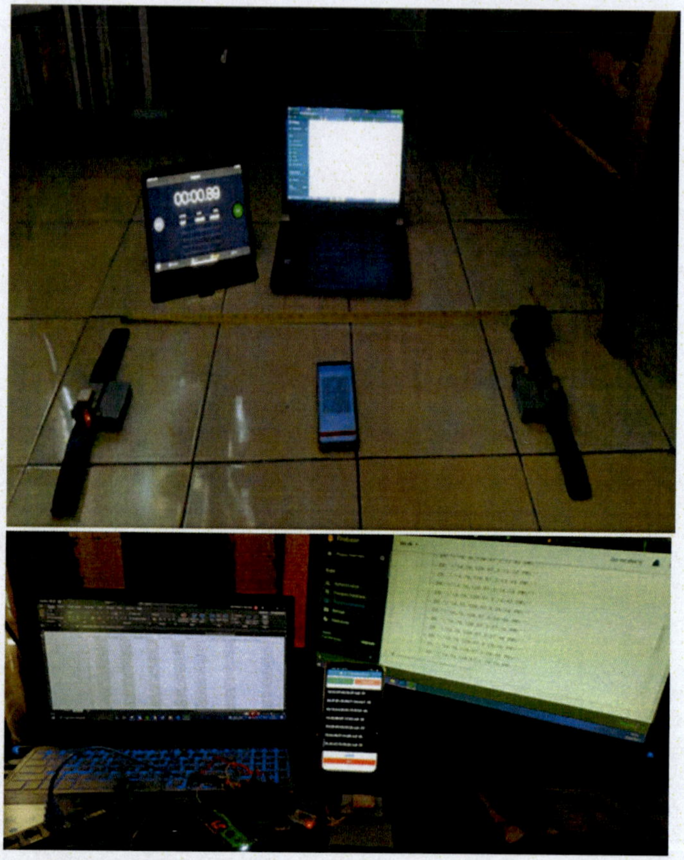

Fig. 14. Calibration for device testing setup

This is the calibration testing setup that the researchers have used for the testing and calibrating. There is a laptop where the database would be observed. There is also a timer along with the laptop for the testing of the time of connection of the devices and graphical user interface. The two devices are displaced with one another and the measurement for the distance is guided by the meter stick. Lastly, the phone is also there to open the application that would prompt the user to be directed to the Graphical User Interface of the device (Table 1).

Table 1. Calibration table

RSSI @ 1m						
Trial #	Device 1			Device 2		
	Min RSSI	Max RSSI	Mean RSSI	Min RSSI	Max RSSI	Mean RSSI
1	−65	−71	−68	−65	−71	−68
2	−65	−68	−67	−65	−68	−67
3	−68	−70	−69	−68	−70	−69
4	−69	−65	−67	−69	−65	−67
5	−69	−71	−70	−65	−71	−68
6	−68	−70	−69	−65	−70	−68
7	−68	−72	−70	−63	−72	−68
8	−66	−68	−67	−66	−68	−67
9	−69	−74	−72	−61	−66	−64
10	−67	−73	−70	−60	−68	−64
11	−68	−72	−70	−66	−68	−67
12	−65	−71	−68	−65	−71	−68
13	−65	−68	−67	−65	−68	−67
14	−68	−70	−69	−68	−70	−69
15	−69	−65	−67	−69	−65	−67
16	−69	−71	−70	−65	−71	−68
17	−68	−70	−69	−65	−70	−68
18	−68	−72	−70	−63	−72	−68
18	−66	−68	−67	−66	−68	−67
20	−69	−74	−72	−61	−66	−64
21	−67	−73	−70	−60	−68	−64
22	−68	−72	−70	−66	−68	−67
23	−65	−71	−68	−65	−71	−68
24	−65	−68	−67	−65	−68	−67
25	−68	−70	−69	−68	−70	−69
26	−69	−65	−67	−69	−65	−67
27	−69	−71	−70	−65	−71	−68
28	−68	−70	−69	−65	−70	−68
29	−68	−72	−70	−63	−72	−68
30	−66	−68	−67	−66	−68	−67
Average device 1			−69	Average device 2		−67

The data gathered showed in the table above is for calibration by repeating trials of measurement of RSSI of both device 1 and device 2. It utilized the sample mean to get the average RSSI of device 1 and device 2 and it was found that the average RSSI of device 1 is −69 dBm while for device 2 the value is −67 dBm. Using the RSSI table below, it can be concluded that both device 1 and device 2 falls in the threshold of acceptable signal strength.

In order to acquire the distance from device1 to device2 for the study, the researchers will be using the formula for RSSI or Received Signal Strength Indicator.

$$d = 10^{\frac{A-RSSI}{10n}} \tag{1}$$

Wherein:
d = Estimated Distance.
A = measured RSSI @ 100 cm.
RSSI = Received signal strength indicator.
n = path loss index.
when d < 100 cm as computed using the formula, the GUI will display a notification in the mobile phone as shown in the previous section of the methodology (Fig. 15).

Fig. 15. Controlled test set-up (open area)

The test for measurement will have a controlled setup of having the two devices be in a distance of 1 m without any obstruction (Fig. 16).

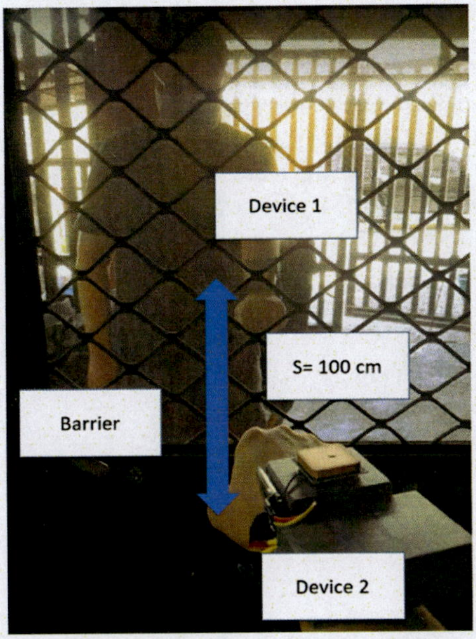

Fig. 16. Uncontrolled test set-up (physical barrier)

The uncontrolled setup will still be in 1 m displacement, but obstruction will be placed (Table 2).

Table 2. Mean square error for RSSI

Uncontrolled testing (Barrier)		Controlled testing (Open Area)	
Device 1	Device 2	Device 1	Device 2
−43	−44	−44	−44
−45	−46	−45	−45
−48	−48	−48	−49
−51	−50	−51	−50
−55	−52	−56	−54
−60	−56	−56	−56
−67	−60	−60	−60
−65	−64	−64	−61
−68	−67	−66	−66
−70	−69	−70	−69
MSE	7.2	MSE	1.6

The RSSI values from both devices will be compared using Mean Squared Error. A good value to get for MSE is near 0 as this means there is little to no error in the performance of the two device. Presented with the data above, the MSE for Controlled Testing is 1.6 while for the 2 uncontrolled testing with obstruction, the value of MSE is 7.2. It can be said that the error between device 1 and device 2 is low because the MSE value is near 0. The formula of MSE is shown below (Figs. 17 and 18):

$$MSE = \frac{1}{N}\sum\nolimits_{i=1}^{N}(Y_i - \hat{y}_i)^2 \quad (2)$$

Wherein:
n = is the number of data points.
Yi = represents observed values.
Yî = represents predicted values.

2.6 Database

Fig. 17. Database showing location

The figure above shows the database at real time. It will display the location, time, parameters, and risk status. The default setting for the risk status is shown as "no" meaning there is no risk. The Graphical User Interface will also display the "no risk" in the indicator part of the application. Should the user encounter/catch the virus while in use of the device, he/she must press the Report Case button in the GUI as shown in the previous section of this paper.

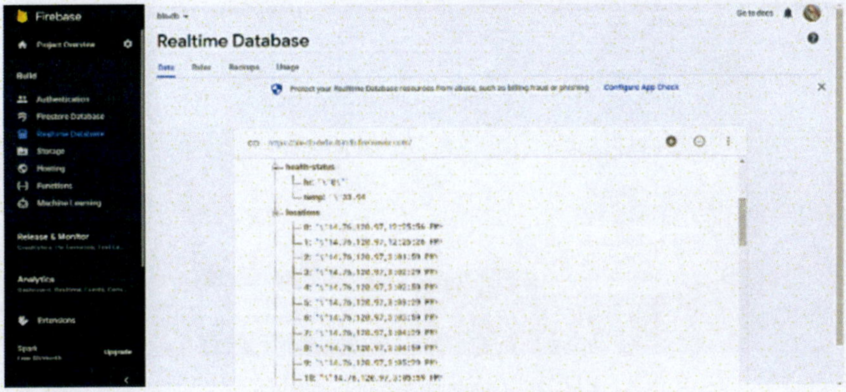

Fig. 18. Database default status

Once the user of either device 1 or 2 has clicked the report case button, the Graphical User Interface will change it's indicator status and this will be reflected in the database for the research as shown below (Fig. 19).

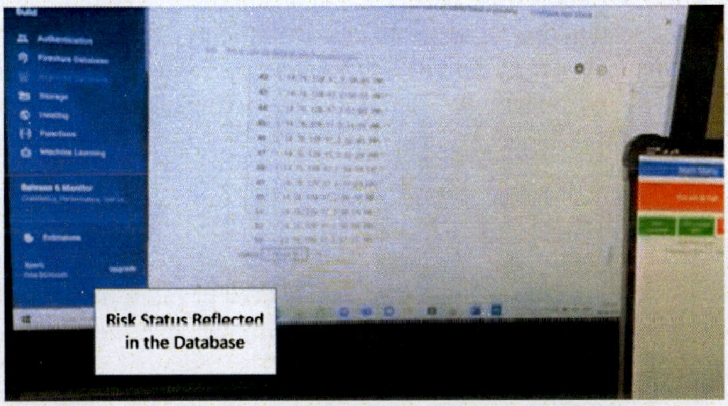

Fig. 19. Database showing risk status with GUI

With the data presented, it can bee seen that the notification system for contact tracing purposes work with GUI and database which is an important part of this research since when a user of one of the devices developed by the researches contracts the device, he or she will be notified immediately through looking at the database or looking at the phone application.

In the case that the report case button was misclicked or any other mishaps that may occur, the researchers put a manual override for the design. This is simply done by issuing a command in the database of "no". This will reset the GUI and will be demonstrated by the image below (Fig. 20).

Fig. 20. Database manual override

3 Conclusion

The first specific objective was to create an application that will display the parameters of the hardware fabricated and provide a Graphical User Interface that can read the RSSI values in the Connections button, the Location is read in the Show Contact Graph Button, and a working Report Case button and indicator displaying the risk status of the user was done. The second objective was to create a database that will store information received from device1, device2, and GUI. The manual override was placed for the case of mishaps in the report case button. Overall, the researchers have met all the main and specific objectives for this study.

4 Recommendations

The researchers would like to recommend three devices or more for setup and testing. The researchers would also recommend a further study wherein the GUI and hardware are already integrated in the device using a screen. The researchers also recommend a further study in getting RSSI values in the outdoor settings as the uncontrolled setting.

References

1. Hernandez-Orallo, E., Manzoni, P., Calafate, C.T., Cano, J.C.: Evaluating how smartphone contact tracing technology can reduce the spread of infectious diseases: the case of COVID-19. IEEE Access 8, 99083–99097 (2020). https://doi.org/10.1109/ACCESS.2020.2998042
2. Centers for Disease Control and Prevention. Coronavirus disease 2019: pregnancy and breastfeeding. https://www.cdc.gov/coronavirus/2019-ncov/prepare/pregnancy-breastfeeding.html.
3. Kajikawa, N., Minami, Y., Kohno, E., Kakuda, Y.: On availability and energy consumption of the fast connection establishment method by using bluetooth classic and bluetooth low energy. In: Proceedings - 2016 4th International Symposium Computing and Networking, CANDAR 2016, pp. 286–290 (2017). https://doi.org/10.1109/CANDAR.2016.76.

4. Dimaunahan, E.D., Cruz, J.C.D., Agasino, F.M.L., Gonda, N.Z.T., Isada, N.G.P., Tolentino, B.C.A.M.: An arduino-based self-sustaining buoy prototype for a non-contact infrared sensor system raw crude oil detector. 2018 IEEE 10th International Conference on Humanoid, Nanotechnology, Information Technology, Communication and Control, Environment and Management, HNICEM 2018, pp. 2–6 (2019). https://doi.org/10.1109/HNICEM.2018.8666263
5. Shang, F., Su, W., Wang, Q., Gao, H., Fu, Q.: A location estimation algorithm based on RSSI vector similarity degree. Int. J. Distrib. Sensor Netw. 2014 (2014). https://doi.org/10.1155/2014/371350
6. Sese, J.T., et al.: Adaptation of the ITU-T E.161 international standard as mapping layout for a wearable text-input device for the blind. In: 2019 IEEE 8th Global Conference on Consumer Electronics, GCCE 2019, 590–591 (2019). https://doi.org/10.1109/GCCE46687.2019.9015434
7. Oguchi, K., Maruta, S., Hanawa, D.: Human positioning estimation method using received signal strength indicator (RSSI) in a wireless sensor network. Procedia Comput. Sci. **34**, 126–132 (2014). https://doi.org/10.1016/j.procs.2014.07.066

A Percentile Whale Algorithm an Application to the Set Covering Problem

Lorena Jorquera[1], Pamela Valenzuela[1], Leonardo Causa[2], Paola Moraga[1(✉)], and José-Miguel Rubio[3]

[1] Escuela de ingeniería en construcción, Pontificia Universidad Católica de Valparaíso, Valparaíso, Chile
{lorena.jorquera,pamela.valenzuela,paola.moraga}@pucv.cl
[2] Facultad de Ingeniería y Negocios, Universidad de le Américas, Santiago, Chile
lcausamo@edu.udla.cl
[3] Facultad de Ingeniería, Ciencia y Tecnología, Universidad Bernardo O'Higgins Santiago, 8370993 Metropolitana, Chile
josemiguel.rubio@ubo.cl

Abstract. The design of robust continuous metaheuristic techniques to apply them to combinatorial problems is a line of research with great applied potential. There are two major challenges in this regard. The first is related to the fact that many of the problems addressed at an industrial level are of the combinatorial type and the second has to do with the fact that no lesser subset of these problems are of the NP-hard type. In this study, a binarization mechanism has been proposed for continuous swarm intelligence metaheuristics which uses the percentile concept to perform binarization. This percentile concept is applied to Whale's optimization algorithm to solve the set coverage problem (SCP). To identify the contribution of the method, experiments were designed which are compared with a baseline. Additionally, the proposal is compared with recently published algorithms, using reference instances. The results indicate that the binary percentile whale (BPWH) algorithm obtains adequate results when it is evaluated with a combinatorial problem such as the SCP.

1 Introduction

In recent years, an area of investigation of algorithms has been developed which are inspired by natural phenomena and have been applied to solve optimization problems. As examples of these algorithms we have Cuckoo Search [1], Bat Algorithm [2] and Whale Algorithm [3] among others. Due to the type of phenomena in which these algorithms are inspired, a large part of these algorithms operate naturally in continuous spaces. On the other hand, the combinatorial problems have a great relevance at the scientific and industrial level. We find them in different areas, such as Civil Engineering, Bio-Informatics [4], Operational Research [5–9], Big Data [10], resource allocation [11–13], programming problems [14–16], routing problems [17–19], civil engineering, [20–23], among others. In particular, if we want to use continuous algorithms to solve combinatorial problems, these

R. Silhavy et al. (Eds.): CoMeSySo 2021, LNNS 231, pp. 402–412, 2021.
https://doi.org/10.1007/978-3-030-90321-3_32

algorithms must be adapted considering not to strongly affect the exploration and exploitation properties of the algorithm since the efficiency of the algorithm may be affected.

Various binarization methods have been developed in order to address this situation. In a literature search, the main binarization methods used correspond to transfer functions, angle modulation, and quantum approach. In this article, we present a binarization method that uses the percentile concept to group solutions and then performs the binarization process. To verify the efficiency of our method, we use the Whale algorithm, which mimics the social behavior of humpback whales.

To check our binary percentile whale optimization algorithm (BPWH), we use the well-known set covering problem. Experiments were developed using a random operator to validate the contribution of the percentile technique in the binarization process of the fireflies algorithm. In addition, to verify our results, the JPSO algorithm developed by [24] and MDBBH developed in [25] were chosen. The results show that the BPWH algorithm obtains competitive results.

2 Set Covering Problem

The set covering problem corresponds to a classical problem in combinatorics, and complexity theory. The problem aims to find subsets of a set. Given a set and its elements, we want to find subsets that completely cover the set at a minimum cost. The SCP is one of the oldest and most studied optimization problems. It is well-known to be NP-hard.

SCP is an active problem because medium and large instances often become intractable and cannot be solved any more using exact algorithms. Additionally, SCP due to its large number of instances, is used to verify the behavior of proposed new algorithms. In recent years, SCP has been approached by various continuous Swarm intelligence metaheuristics and using different methods to binarize metaheuristics. In [26], they used the teaching-learning-based optimisation with a specific binarization scheme to solve medium and large size instances. An SCP in fault diagnosis application was developed by [27]. In this article, the Gravitational Search algorithm was used and transfer functions were applied to perform the binarization. A rail scheduling problem application using SCP was developed in [25]. In [24] a Jumping PSO was used to solve SCP and in [14] CS metaheuristics were binarized using a percentile algorithm.

SCP has many practical applications in engineer, e.g., vehicle routing, facility location, railway, and airline crew scheduling [28–31] problems.

The SCP can be formally defined as follows. Let $A = (a_{ij})$, be a $n \times m$ zero-one matrix, where a column j cover a row i if $a_{ij} = 1$, besides a column j is associated with a non-negative real cost c_j. Let $I = \{1, ..., n\}$ and $J = \{1, ...m\}$, be the row and column set of A, respectively. The SCP consists in searching a minimum cost subset $S \subset J$ for which every row $i \in I$ is covered by at least one column $j \in J$, i.e.,:

$$\text{Minimize } f(x) = \sum_{j=1}^{m} c_j x_j \tag{1}$$

$$\text{Subject to } \sum_{j=1}^{m} a_{ij} x_j \geq 1, \forall i \in I, \text{ and } x_j \in \{0,1\}, \ \forall j \in J \tag{2}$$

where $x_j = 1$ if $j \in S$, $x_j = 0$ otherwise.

3 Binary Percentile Whale Algorithm

The application of statistics and machine learning techniques, are used in many areas such as transports, smart cities, agriculture, and computational intelligence, examples of applications are found in [32–37]. In this article, we explore the percentile concept applied to binarization of continuous metaheuristics. As an algorithm to binarize, the whale algorithm is used and as a problem to solve, SCP. The Proposed BPWH has two modules. The first module corresponds to the initialization of the feasible solutions and is detailed in the Sect. 3.1. Once the initialization of the particles is performed, it is consulted if the maximum of iterations is satisfied. Subsequently, if the criterion is not satisfied, The BPWH is then run in conjunction with the percentile operator, this is detailed in Sect. 3.2). Once the transitions of the different solutions are made, we compare the resulting solutions with the best solution previously obtained. In the event that a superior solution is found, this replaces the previous one. The general algorithm scheme is detailed in Fig. 1.

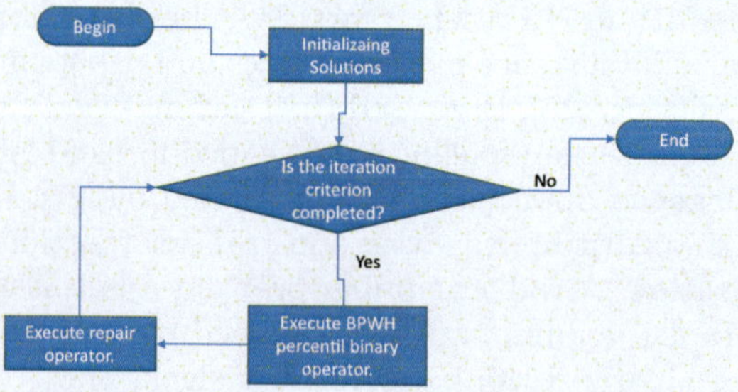

Fig. 1. Flowchart of the BPWH algorithm.

3.1 Initialization

For initialization of a new solution, a column is randomly chosen. It is then queried whether the current solution covers all rows. In the case of the solution does not meet the coverage condition, the heuristic operator is called. This operator aims to select a new column. This heuristic operation is iterated until all rows are covered. Once the coverage condition is met, the solution is optimized. The optimization consists of eliminating columns where all their rows are covered by more than one column. The detail of the procedure is shown in Algorithm 1.

Algorithm 1. Initialization Operator

Function Initialization()
Input
Output Initialized solution S_{out}
$S \leftarrow$ SRandomColumn()
while All row are not covered **do**
    S.append(Heuristic(S))
end while
$S \leftarrow$ dRepeatedItem(S)
$S_{out} \leftarrow S$
return S_{out}

3.2 Percentile Binary Operator

Since our firefly algorithm is continuous swarm intelligence metaheuristic, it works in an iterative way by updating the position and velocity of the particles in each iteration. As firefly is continuous, the update is done in $\mathbb{R}^n$. In Eq. 3, the position update is written in a general way. The $x(t+1)$ variable represents the x position of the particle at time $t + 1$. This position is obtained from the position x at time t plus a Δ function calculated at time $t + 1$. The function Δ is proper to each metaheuristic and produces values in $\mathbb{R}^n$. For example in Cuckoo Search $\Delta(x) = \alpha \oplus Levy(\lambda)(x)$, in Black Hole $\Delta(x) = \text{rand} \times (x_{bh}(t) - x(t))$ and in the Whale, Bat and PSO algorithms Δ can be written in simplified form as $\Delta(x) = v(x)$.

$$x(t+1) = x(t) + \Delta_{t+1}(x(t)) \tag{3}$$

In the percentile binary operator, we considering the movements generated by the algorithm in each dimension for all particles. $\Delta^i(x)$ corresponds to the magnitude of the displacement $\Delta(x)$ in the i-th position for the particle x. Subsequently these displacement are grouped using $\Delta^i(x)$, the magnitude of the displacement. This grouping is done using the percentile list. In our case the percentile list used the values {20, 40, 60, 80, 100}.

The percentile operator has as entry the parameters percentile list (pList) and the list of values (vList). Given an iteration, the list of values corresponds to the magnitude $\Delta^i(x)$ of the displacements of the particles in each dimension. As a first step the operator uses the vList and obtains the values of the percentiles given in the pList. Later, each value in the vList is assigned the group of the smallest percentile to which the value belongs. Finally, the list of the percentile to which each value belongs is returned (pGroupValue). The algorithm is shown in Algorithm 2.

A transition probability through the function P_{tr} is assigned to each element of the vList. This assignment is done using the percentile group assigned to each value. For the case of this study, we particularly use the Step function given in Eq. 4.

$$P_{tr}(x^i) = \begin{cases} 0.1, & \text{if } x^i \in \text{group } \{0,1\} \\ 0.5, & \text{if } x^i \in \text{group } \{2,3,4\} \end{cases} \tag{4}$$

Afterwards the transition of each particle is performed. In the case of whale search the Eq. 5 is used to perform the transition, where $\hat{x}^i$ is the complement of x^i. Finally, each solution is repaired using the heuristic operator.

$$x^i(t+1) := \begin{cases} \hat{x}^i(t), & \text{if } rand < P_{tg}(x^i) \\ x^i(t), & \text{otherwise} \end{cases} \tag{5}$$

Algorithm 2. percentile binary operator

Function percentileBinary(vList, pList)
Input vList, pList
Output pGroupValue
percentileValue = getPValue(vList, pList)
for each value in vList **do**
    pGroupValue = getPGroupValue(pValue,vList)
end for
return pGroupValue

4 Results

4.1 Insight of BPWH Algorithm

In this module, we developed experiments that allowed us to study the contribution of the binary percentile operator with respect to the quality of the solutions obtained and the iterations used when solving SCP instances. To perform this evaluation, we used the instances of Balas and Carrera. In the comparison, a random operator was designed which replaces our binary percentile operator. This random operator instead of executing transitions assigning transition probabilities per group uses a fixed value of 0.5. To compare the distributions of the results of the different experiments we use a violin Chart. The horizontal axis X corresponds to the problems, while Y-axis uses the measure % - Gap.

Furthermore, a non-parametric test, Wilcoxon signed-rank test is carried out to determine if the results of BPWH with respect to the random algorithm have a significant difference or not. The parameter settings and browser ranges are shown in Table 1.

Table 1. Setting of parameters for Binary percentile firefly Algorithm.

Parameters	Description	Value	Range
N	Number of solutions	25	[20, 25, 30]
G	Number of percentiles	5	[4, 5, 6]
Iteration number	Maximum iterations	800	[700, 800, 900]

BPWH corresponds to our percentile algorithm. *random.0.5* is the random variant. The results are shown in Table 2 and Fig. 2. When we compared the Best Values between BPWH and *rnd.0.5* which are shown in Table 2. In the Average comparison, BPWH outperforms *rnd.0.5* in all problems. The comparison of distributions is shown in Fig. 2. We see the dispersion of the *rnd.0.5* distributions are bigger than the dispersions of BPWH. Therefore, the percentile binary operator contributes to the precision of the results. Finally, the BPWH distributions are closer to zero than *rnd.0.5* distributions, indicating that BPWH has consistently better results than *random.0.5*. When we evaluate the behavior of the algorithms through the Wilcoxon test, this test indicates that there is a significant difference between the two algorithms.

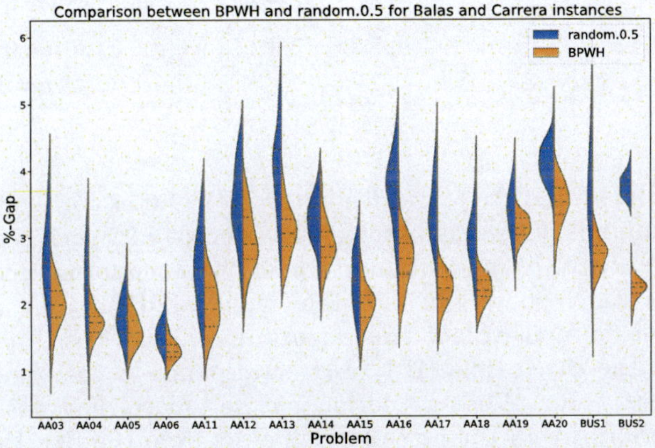

Fig. 2. Evaluation of percentile binary operator.

4.2 BPWH Compared with JPSO and MDBBH

In this section, Table 3, we detail the comparisons made with the objective of evaluating the performance of our BPWH algorithm. For the evaluation, the larger problems of the OR-library were chosen. To develop the comparison, two algorithms were selected. The first one corresponds to a discretization Particle Swarm Optimization (PSO) technique called Jumping PSO (JPSO), [24]. The second one, is a binarization of the Black Hole technique called Multi Dynamic

Table 2. Balas and Carrera instances.

Instance	row	col	Density	Best known	BPWH (avg)	std	*rnd.0.5* (avg)	std std
AA03	106	8661	4.05%	33155	33596.1	129.7	33693.5	243.5
AA04	106	8002	4.05%	34573	34872.6	61.3	34931	234.1
AA05	105	7435	4.05%	31623	31881.3	62.6	31889.2	107.6
AA06	105	6951	4.11%	37464	37653.7	64.3	37651.3	93.5
AA11	271	4413	2.53%	35384	35768.3	138.6	35901.2	199.5
AA12	272	4208	2.52%	30809	31455.5	157.2	31631.3	202.1
AA13	265	4025	2.60%	33211	33996.1	134.8	34201.4	236.7
AA14	266	3868	2.50%	33219	33879.4	121	33967.3	124.2
AA15	267	3701	2.58%	34409	34822.2	117.3	34846.8	168.1
AA16	265	3558	2.63%	32752	33328.7	144.2	33601.3	185.4
AA17	264	3425	2.61%	31612	32006.9	93.9	32386.2	145.7
AA18	271	3314	2.55%	36782	37291.4	79.5	37417.6	146.5
AA19	263	3202	2.63%	32317	33099.8	73.8	33109.1	138.3
AA20	269	3095	2.58%	34912	35843.3	96.3	35963.1	96.6
BUS1	454	2241	1.88%	27947	28484.1	88.4	28756.1	233.4
BUS2	681	9524	0.51%	67760	68548.4	89.8	68678.9	105.3
Average				35495.56	36032.1	109.3	36164.08	165.11
p-value							3.52e−04	

Binary Black Hole (MDBBH) Algorithm, [25]. JPSO uses a discrete PSO based on the frog jump. JPSO works without the concept of velocity, replacing this one with a component of random jump which allows performing the movement in the discrete search space. On the other hand, MDBBH uses a binarization mechanism specific to the Black Hole algorithm. This is based on the concept of closeness to the Black Hole (BH). BH corresponds to the solution that has obtained the best value. When a solution is close to BH, the transitions are less likely to be performed. In the case of being away from the BH, the probability of transition is greater.

For example, E and F, which are medium in size, the results of the MDBBH and BPWH algorithms are similar when comparing their Best Value. When analyzing the average, JPSO obtained better results. For the case of problems G and H, BPWH has similar results to MDBBH in most of the problems when comparing its Best Value. In the case of averages, JPSO was superior in all cases. At this point, we must emphasize that the percentile technique used in BPWH allows binarizing any continuous swarm intelligence algorithm, unlike JPSO which is specific to PSO.

5 Conclusions

In this work, the BPWH algorithm based on the percentile concept was proposed to perform the binarization of whale metaheuristics. We must emphasize that this percentile concept can be applied in the binarization of any continuous metaheuristics of the swarm intelligence type. To evaluate the performance of Our percentile concept the set covering problem was used together with the whale algorithm. Experiments were designed to evaluate the contribution of the binary operator percentile of the algorithm. It was found that the operator contributes significantly to improve the accuracy and quality of the solutions, all these visualized through violin graphics. As future works, we believe that the autonomous search for efficient configuration is a little-explored area and it could help to improve the performance of the binarizations both in the quality of the solutions and in the times of convergence. Additionally, it is interesting to see how this percentile technique works with other metaheuristics and other NP-hard problems. Finally, we believe that using machine learning we can generate indicators or obtain correlations that allow us to understand how the percentile algorithm translates the exploration and exploitation properties from the continuous space to the binary space.

Table 3. OR-Library benchmarks.

Instance	row	col	Density	Best known	JPSO Avg	MDBBH best	Avg	BPWH best	Avg
E.1	500	5000	10%	29	29	29	29	29	31.1
E.2	500	5000	10%	30	30	31	31,6	30	33.6
E.3	500	5000	10%	27	27	27	27,4	27	28.6
E.4	500	5000	10%	28	28	28	29,1	28	30.2
E.5	500	5000	10%	28	28	28	28	28	30.6
F.1	500	5000	20%	14	14	14	14,1	14	15.2
F.2	500	5000	20%	15	15	15	15,3	15	17.6
F.3	500	5000	20%	14	14	14	14,8	14	15.4
F.4	500	5000	20%	14	14	14	14,9	14	15.3
F.5	500	5000	20%	13	13	14	14,1	14	16.1
G.1	1000	10000	2%	176	176	177	178,5	177	179.8
G.2	1000	10000	2%	154	155	157	160,6	156	161.4
G.3	1000	10000	2%	166	167,2	168	170,4	168	171.9
G.4	1000	10000	2%	168	168,2	169	170,9	170	172.4
G.5	1000	10000	2%	168	168	168	169,8	168	172.2
H.1	1000	10000	5%	63	64	64	64,9	65	67.4
H.2	1000	10000	5%	63	63	64	64	65	65.6
H.3	1000	10000	5%	59	59,2	59	60	60	62.3
H.4	1000	10000	5%	58	58,3	59	60,4	59	62.8
H.5	1000	10000	5%	55	55	55	56,4	56	57.8
Average				67.1	67.3	67.7	68.7	67.9	71.7

References

1. Yang, X.-S., Deb, S.: Cuckoo search via lévy flights. In: Nature & Biologically Inspired Computing, 2009. NaBIC 2009. World Congress on, pp. 210–214. IEEE (2009)
2. Yang, X.-S.: A new metaheuristic bat-inspired algorithm. In: Nature Inspired Cooperative Strategies for Optimization (NICSO 2010), pp. 65–74 (2010)
3. Mirjalili, S., Lewis, A.: The whale optimization algorithm. Adv. Eng. Softw. **95**, 51–67 (2016)
4. Altimiras, F., et al.: Altered gut microbiota in a fragile x syndrome mouse model. Front. Neurosci. **15** (2021)
5. Crawford, B., et al.: Investigating the efficiency of swarm algorithms for bridge strengthening by conversion to tied-arch: a numerical case study on San Luis bridge. Iran. J. Sci. Technol. Trans. Civ. Eng. 1–13 (2020)
6. Crawford, B., Soto, R., Astorga, G., García, J.: Constructive metaheuristics for the set covering problem. In: Korošec, P., Melab, N., Talbi, E.G. (eds.) Bioinspired Optimization Methods and Their Applications. International Conference on Bioinspired Methods and Their Applications, pp. 88–99. Springer, Cham (2018). https://doi.org/10.1007/978-3-319-91641-5_8
7. García, J., Peña, A.: Robust optimization: concepts and applications. In: Nature-Inspired Methods for Stochastic, Robust and Dynamic Optimization, IntechOpen (2018)
8. García, J., et al.: A Db-scan binarization algorithm applied to matrix covering problems. Comput. Intell. Neurosci. 2019 (2019)
9. Crawford, B., et al.: Q-learnheuristics: towards data-driven balanced metaheuristics.Mathematics **9**(16) (2021)
10. García, J., Altimiras, F., Peña, A., Astorga, G., Peredo, O.: A binary cuckoo search big data algorithm applied to large-scale crew scheduling problems. Complexity **2018** (2018)
11. García, J., Maureira, C.: A KNN quantum cuckoo search algorithm applied to the multidimensional knapsack problem. Appl. Soft Comput. **102**, 107077 (2021)
12. García, J., Lalla-Ruiz, E., Voß, S., Droguett, E.L.: Enhancing a machine learning binarization framework by perturbation operators: analysis on the multidimensional knapsack problem. Int. J. Mach. Learn. Cybern. **11**(9), 1951–1970 (2020). https://doi.org/10.1007/s13042-020-01085-8
13. García, J., Moraga, P., Valenzuela, M., Pinto, H.: A db-scan hybrid algorithm: an application to the multidimensional knapsack problem. Mathematics **8**(4), 507 (2020)
14. García J., Crawford B., Soto R., Astorga G.: A percentile transition ranking algorithm applied to knapsack problem. In: Silhavy, R., Silhavy, P., Prokopova, Z. (eds.) Applied Computational Intelligence and Mathematical Methods. Proceedings of the Computational Methods in Systems and Software, pp. 126–138, Springer, Cham (2017). https://doi.org/10.1007/978-3-319-67621-0_11
15. García, J., Crawford, B., Soto, R., Astorga, G.: A percentile transition ranking algorithm applied to binarization of continuous swarm intelligence metaheuristics. In: Ghazali, R., Deris, M., Nawi, N., Abawajy, J. (eds.) Recent Advances on Soft Computing and Data Mining. International Conference on Soft Computing and Data Mining, pp. 3–13, Springer, Cham (2018). https://doi.org/10.1007/978-3-319-72550-5_1

16. García, J., Astorga, G., Yepes, V.: An analysis of a KNN perturbation operator: an application to the binarization of continuous metaheuristics. Mathematics **9**(3), 225 (2021)
17. Tapia, D., et al.: A Q-learning hyperheuristic binarization framework to balance exploration and exploitation. In: Florez, H., Misra, S. (eds.) International Conference on Applied Informatics, pp. 14–28. Springer, Cham (2020). https://doi.org/10.1007/978-3-030-61702-8_2
18. García, J., Crawford, B., Soto, R., Astorga, G.: A clustering algorithm applied to the binarization of swarm intelligence continuous metaheuristics. Swarm Evol. Comput. **44**, 646–664 (2019)
19. García, J., Măntoiu, M.: Localization results for zero order pseudodifferential operators. J. Pseudo-Differ. Oper. Appl. **5**(2), 255–276 (2013). https://doi.org/10.1007/s11868-013-0084-y
20. García, J., Yepes, V., Martí, J.V.: A hybrid k-means cuckoo search algorithm applied to the counterfort retaining walls problem. Mathematics **8**(4), 555 (2020)
21. García, J., Martí, J.V., Yepes, V.: The buttressed walls problem: an application of a hybrid clustering particle swarm optimization algorithm. Mathematics **8**(6), 862 (2020)
22. Yepes, V., Martí, J.V., García, J.: Black hole algorithm for sustainable design of counterfort retaining walls. Sustainability **12**(7), 2767 (2020)
23. Martínez-Muñoz, D., Marté, J.V., Garcéa, J., Yepes, V.: Embodied energy optimization of buttressed earth-retaining walls with hybrid simulated annealing. Appl. Sci. **11**(4) (2021)
24. Balaji, S., Revathi, N.: A new approach for solving set covering problem using jumping particle swarm optimization method. Nat. Comput. **15**(3), 503–517 (2015). https://doi.org/10.1007/s11047-015-9509-2
25. García, J., Crawford, B., Soto, R., García, P.: A multi dynamic binary black hole algorithm applied to set covering problem. In: Del Ser, J. (eds.) Harmony Search Algorithm. International Conference on Harmony Search Algorithm, pp. 42–51. Springer, Singapore (2017). https://doi.org/10.1007/978-981-10-3728-3_6
26. Lu, Y., Vasko, F.J.: An or practitioner's solution approach for the set covering problem. Int. J. Appl. Metaheuristic Comput. (IJAMC) **6**(4), 1–13 (2015)
27. Li, Y., Cai, Z.: Gravity-based heuristic for set covering problems and its application in fault diagnosis. J. Syst. Eng. Electron. **23**(3), 391–398 (2012)
28. Crawford, B., Soto, R., Monfroy, E., Astorga, G., García, J., Cortes, E.: A meta-optimization approach for covering problems in facility location. In: Figueroa-Garcia, J., López-Santana, E., Villa-Ramírez, J., Ferro-Escobar, R. (eds.) Workshop on Engineering Applications, pp. 565–578. Springer, Cham (2017). https://doi.org/10.1007/978-3-319-66963-2_50
29. Kasirzadeh, A., Saddoune, M., Soumis, F.: Airline crew scheduling: models, algorithms, and data sets. EURO J. Transp. Logistics **6**(2), 111–137 (2015). https://doi.org/10.1007/s13676-015-0080-x
30. Horváth, M., Kis, T.: Computing strong lower and upper bounds for the integrated multiple-depot vehicle and crew scheduling problem with branch-and-price. Central Eur. J. Oper. Res. **27**(1), 39–67 (2017). https://doi.org/10.1007/s10100-017-0489-4
31. Stojković, M.: The operational flight and multi-crew scheduling problem. Yugoslav J. Oper. Res. **15**(1) (2016)
32. García, J., Crawford, B., Soto, R., Castro, C., Paredes, F.: A k-means binarization framework applied to multidimensional knapsack problem. Appl. Intell. **48**(2), 357–380 (2017). https://doi.org/10.1007/s10489-017-0972-6

33. García, J., Pope, C., Altimiras, F.: A distributed k-means segmentation algorithm applied to Lobesia botrana recognition. Complexity **2017** (2017)
34. Graells-Garrido, E., García, J.: Visual exploration of urban dynamics using mobile data. In: García-Chamizo, J., Fortino, G., Ochoa, S. (eds.) Ubiquitous Computing and Ambient Intelligence. Sensing, Processing, and Using Environmental Information. International Conference on Ubiquitous Computing and Ambient Intelligence, pp. 480–491. Springer, Cham (2015). https://doi.org/10.1007/978-3-319-26401-1_45
35. Graells-Garrido, E., Peredo, O., García, J.: Sensing urban patterns with antenna mappings: the case of Santiago, Chile. Sensors **16**(7), 1098 (2016)
36. Peredo, O.F., García, J.A., Stuven, R., Ortiz, J.M.: Urban dynamic estimation using mobile phone logs and locally varying anisotropy. In: Gómez-Hernández J., Rodrigo-Ilarri J., Rodrigo-Clavero M., Cassiraga E., Vargas-Guzmán J. (eds.) Geostatistics Valencia 2016, pp. 949–964. Springer, Cham (2017). https://doi.org/10.1007/978-3-319-46819-8_66
37. Maureira, C., Pinto, H., Yepes, V., Garcia, J.: Towards an AEC-AI industry optimization algorithmic knowledge mapping an adaptive methodology for macroscopic conceptual analysis. IEEE Access 1 (2021)

A Machine Learning Whale Algorithm Applied to the Matrix Covering Problem

Matias Valenzuela[1], Paola Moraga[1(✉)], Leonardo Causa[2], Hernan Pinto[1], and José-Miguel Rubio[3]

[1] Escuela de ingeniería en construcción, Pontificia Universidad Católica de Valparaíso, Valparaíso, Chile
{matias.valenzuela,paola.moraga}@pucv.cl, hernan.pinto@puv.cl

[2] Facultad de Ingeniería y Negocios, Universidad de las Américas, Santiago, Chile
lcausamo@edu.udla.cl

[3] Facultad de Ingeniería, Ciencia y Tecnología, Universidad Bernardo O'Higgins Santiago, 8370993 Metropolitana, Chile
josemiguel.rubio@ubo.cl

Abstract. In the industry, the need for optimization naturally arises, with which there is a large number of optimization problems, particularly combinatorial and NP-hard type. Therefore, there is an important motivation for the development of algorithms that address these types of problems. IN this line, a large number of metaheuristic algorithms have been developed which work only in continuous spaces. Modifying the latter in order to address combinatorial problems has important applications. In this article, we will study a general binarization mechanism of continuous metaheuristics based on clustering techniques and it will be applied to the whale algorithm. Experiments are designed in order to demonstrate the contribution of the clustering technique in the binarization process. The results indicate that the Ballena binary optimization algorithm (MLWH) obtains adequate results when it is evaluated with a combinatorial problem such as the SCP.

1 Introduction

At the industry level, combinatorial problems are highly relevant, they appear in different areas such as Operational Research [1–4], resource allocation [5–7], problems of programming [8–11], routing problems [12–15], civil engineering [16], Bio-Informatics [17]. On the other hand, in recent years algorithms inspired by natural phenomena have been generated to solve optimization problems. As examples of these algorithms we have Cuckoo Search [18], Black Hole [19], Bat Algorithm [20] and Whale Algorithm [21], among others. Many of these algorithms work naturally in continuous spaces and therefore must be adapted to solve combinatorial problems. In the adaptation process, the algorithm's exploration and exploitation mechanisms can be altered, having consequences on the algorithm's efficiency.

Several binarization techniques have been developed to address this situation. In a literature search, the main binarization methods used correspond to transfer

R. Silhavy et al. (Eds.): CoMeSySo 2021, LNNS 231, pp. 413–422, 2021.
https://doi.org/10.1007/978-3-030-90321-3_33

functions, angle modulation, and quantum approach. In this article, we present a binarization method that uses the unsupervised k-means technique to group the solutions and then perform the binarization process. To verify the efficiency of our method, we used the MLWH. This algorithm was proposed in [21].

To check our MLWH, we use the well-known set covering problem. Experiments were developed using a random operator to validate the contribution of the k-means technique in the binarization process of the whale algorithm.

2 Set Covering Problem

The set covering problem corresponds to a classical problem in combinatorics, and complexity theory. The problem aims to find subsets of a set. Given a set and its elements, we want to find subsets that completely cover the set at a minimum cost. The SCP is one of the oldest and most studied optimization problems. It is well-known to be NP-hard [22].

SCP is an active problem because medium and large instances often become intractable and cannot be solved any more using exact algorithms. Additionally, SCP due to its large number of instances, is used to verify the behavior of proposed new algorithms. In recent years, SCP has been approached by various continuous Swarm intelligence metaheuristics and using different methods to binarize metaheuristics. In [23], they used the teaching-learning-based optimisation with a specific binarization scheme to solve medium and large size instances. An SCP in fault diagnosis application was developed by [24]. In this article, the Gravitational Search algorithm was used and transfer functions were applied to perform the binarization. A rail scheduling problem application using SCP was developed in [25]. In [26] a Jumping PSO was used to solve SCP and in [27] CS metaheuristics were binarized using a k-means algorithm.

SCP has many practical applications in engineer, e.g., vehicle routing, facility location, railway, and airline crew scheduling [1, 28–30] problems.

The SCP can be formally defined as follows. Let $A = (a_{ij})$, be a $n \times m$ zero-one matrix, where a column j cover a row i if $a_{ij} = 1$, besides a column j is associated with a non-negative real cost c_j. Let $I = \{1, ..., n\}$ and $J = \{1, ...m\}$, be the row and column set of A, respectively. The SCP consists in searching a minimum cost subset $S \subset J$ for which every row $i \in I$ is covered by at least one column $j \in J$, i.e.,:

$$\text{Minimize } f(x) = \sum_{j=1}^{m} c_j x_j \tag{1}$$

$$\text{Subject to } \sum_{j=1}^{m} a_{ij} x_j \geq 1, \forall i \in I, \text{ and } x_j \in \{0, 1\}, \ \forall j \in J \tag{2}$$

where $x_j = 1$ if $j \in S$, $x_j = 0$ otherwise.

3 Binary Whale Algorithm

The application of statistics and machine learning techniques, are used in many areas such as transports, smart cities, agriculture, and computational intelligence, examples of applications are found in [31–36]. In this article, we explore the k-means technique applied to binarization of continuous metaheuristics. As an algorithm to binarize, Whale is used and as a problem to solve, SCP. The Proposed MLWH has two modules. The first module corresponds to the initialization of the feasible solutions and is detailed in the Sect. 3.1. Once the initialization of the particles is performed, it is consulted if the maximum of iterations is satisfied. Subsequently, if the criterion is not satisfied, The MLWH is then run in conjunction with the k-means technique, this is detailed in Sect. 3.2). Once the transitions of the different solutions are made, we compare the resulting solutions with the best solution previously obtained. In the event that a superior solution is found, this replaces the previous one. The general algorithm scheme is detailed in Fig. 1.

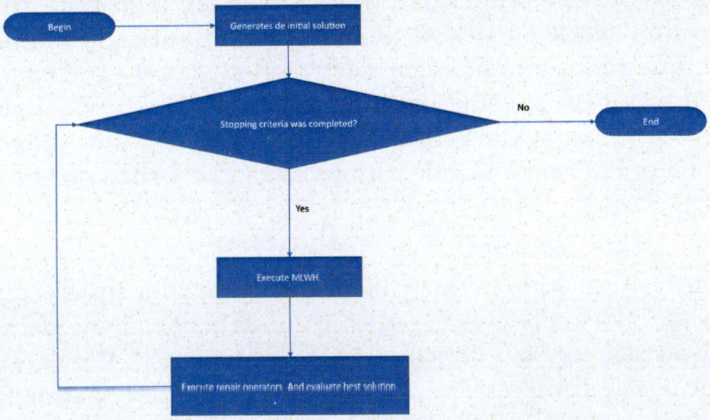

Fig. 1. Flowchart of the MLWH algorithm.

3.1 Initialization

For the initialization of a new solution, a column is randomly chosen. It is then queried whether the current solution covers all rows. In the case the solution does not meet the coverage condition, the heuristic operator is called (Sect. 3.3). This operator aims to select a new column. This heuristic operation is iterated until all rows are covered. Once the coverage condition is met, the solution is optimized. The optimization consists of eliminating columns where all their rows are covered by more than one column. The detail of the procedure is shown in Algorithm 1.

Algorithm 1. Initialization Operator

Function Initialization()
Input
Output Initialized solution S_{out}
$S \leftarrow$ SRandomColumn()
while All row are not covered **do**
  S.append(Heuristic(S))
end while
$S \leftarrow$ dRepeatedItem(S)
$S_{out} \leftarrow S$
return S_{out}

3.2 Mcahine Learning Transition Ranking Operator

Consider our metaheuristic is a continuous and population based. Due to its iterative nature, it needs to update the position of particles at each iteration. When the metaheuristic is continuous, this update is performed in $\mathbb{R}^n$ space. In Eq. 3, the update position is presented in a general manner. The x_{t+1} variable represents the x position of the particle at time t+1. This position is obtained from the position x at time t plus a Δ function calculated at time t+1. The function Δ is proper to each metaheuristic and produces values in $\mathbb{R}^n$. For example in Cuckoo Search $\Delta(x) = \alpha \oplus Levy(\lambda)(x)$, in Black Hole $\Delta(x) =$ rand $\times (x_{bh}(t) - x(t))$ and in the Whale, Bat and PSO algorithms Δ can be written in simplified form as $\Delta(x) = v(x)$.

$$x_{t+1} = x_t + \Delta_{t+1}(x(t)) \tag{3}$$

In the k-means transition ranking operator, a model for transitions in discrete space is proposed. This model is based on considering the movements generated by the metaheuristic in each dimension for all particles. $\Delta^i(x)$ corresponds to the magnitude of the displacement $\Delta(x)$ in the i-th position. Subsequently, these displacements are grouped using the magnitude of the displacement $\Delta^i(x)$. This grouping is done using the k-means technique where k represents the number of clusters used. Finally, a generic P_{tr} function given by the Eq. 4 is proposed to assign a transition probability.

$$P_{tr} : \mathbb{Z}/k\mathbb{Z} \rightarrow [0,1] \tag{4}$$

A transition probability through the function P_{tr} is assigned to each group. Naturally, this P_{tr} function is modeled as a cumulative probability function. For the case of this study, we particularly use the linear function given in Eq. 5. In this equation, N(x^i) indicates the location of the group to which $\Delta^i(x)$ belongs. The α coefficient corresponds to the transition probability and β to the transition separation coefficient. Both parameters are estimated in each metaheuristic. For our particular case, N$(x^i) = 0$ corresponds to elements belonging to the group that has the lowest Δ^i values. N$(x^i) = 7$ corresponds to the group of elements that have the greatest Δ^i values.

$$P_{tr}(x^i) = P_{tr}(N(x^i)) = \alpha + \beta N(x^i)\alpha \tag{5}$$

A transition probability through the function P_{tr} is assigned to each group. Naturally, this P_{tr} function is modeled as a cumulative probability function. For the case of this study, we particularly use the linear function given in Eq. 5. In this equation, $\mathrm{N}(x^i)$ indicates the location of the group to which $\Delta^i(x)$ belongs. The α coefficient corresponds to the transition probability and β to the transition separation coefficient. Both parameters are estimated in each metaheuristic. For our particular case, $\mathrm{N}(x^i) = 0$ corresponds to elements belonging to the group that has the lowest Δ^i values. $\mathrm{N}(x^i) = 7$ corresponds to the group of elements that have the greatest Δ^i values.

$$x^i(t+1) := \begin{cases} \hat{x}^i(t), \text{ if } rand < P_{tg}(x^i) \\ x^i(t), \quad \text{otherwise} \end{cases} \tag{6}$$

3.3 Heuristic Operator

The goal of the Heuristic operator is to select a new column for cases where a solution needs to be built or repaired. As input variables, the heuristic operator considers the incomplete solution S_{in} which must be completed. The operator obtains the columns that belong to S_in, then obtains the rows R that is not covered by the solution to S_in. With the set of rows not covered and using Eq. 7 we obtain in line 4 the best 5 rows to be covered. With this list of rows ($lRows$) on line 5, we obtain the list of the best columns according to the heuristic indicated in Eq. 8. Finally randomly in line 6, we obtain the column to incorporate.

$$WeightRow(i) = \frac{1}{L_i}. \tag{7}$$

Where L_i is the sum of all ones in row i

$$WeightColumn(j) = \frac{c_j}{|R \cap M_j|}. \tag{8}$$

Where M_j is the set of rows covered by Col j

Algorithm 2. Heuristic operator

1: **Function** Heuristic(S_{in})
2: **Input** Input solution S_{in}
3: **Output** The new column $colOut$
4: $lRows \leftarrow$ getBestRows(S_{in}, N=10)
5: $listcolumsnOut \leftarrow$ getBestCols($LRows$, M=5)
6: $colOut \leftarrow$ getCol($lcolsOut$)
7: **return** $colOut$

4 Results

In this section, we detail the behavior of MLWH when it is applied to SCP. The contribution of the binary k-means operator was studied when solving the different SCP instances. Additionally, MLWH was compared with other algorithms that have recently resolved SCP. To solve the different SCP instances, a PC with Windows 10, a core i5 processor, and 16 GB in RAM was used. The program was coded in Python 3.5. For the statistical analysis, the non-parametric Wilcoxon signed-rank test was used in addition to violin charts. The analysis of violin charts is performed by comparing the dispersion, median and interquartile range of the distributions.

4.1 Insight of MLWH Algorithm

In this module, we developed experiments that allowed us to study the contribution of the binary k-means operator with respect to the quality of the solutions obtained and the iterations used when solving SCP instances. To perform this evaluation, we used the instances of Balas and Carrera. In the comparison, a random operator was designed which replaces our binary k-means operator. This random operator instead of executing transitions assigning transition probabilities per group uses a fixed value of 0.5. To compare the distributions of the results of the different experiments we use a violin Chart. The horizontal axis X corresponds to the problems, while Y-axis uses the measure % - Gap defined in Eq. 9.

$$\% - Gap = 100\frac{BestKnown - SolutionValue}{BestKnown} \tag{9}$$

Furthermore, a non-parametric test, Wilcoxon signed-rank test is carried out to determine if the results of MLWH with respect to the random algorithm have significant difference or not. The parameter settings and browser ranges are shown in Table 1.

Table 1. Setting of parameters for Binary percentile firefly Algorithm.

Parameters	Description	Value	Range
N	Number of solutions	25	[20, 25, 30]
K	Number of clusters	5	[4, 5, 6]
Iteration number	Maximum iterations	800	[700, 800, 900]

MLWH corresponds to our standard algorithm. *random.0.5* is the random variant. The results are shown in Table 2 and Fig. 2.

When we compared the Best Values between MLWH and *rnd.0.5* which are shown in Table 2. In the Average comparison, MLWH outperforms *rnd.0.5* in all problems. The comparison of distributions is shown in Fig. 2. We see the dispersion of the *rnd.0.5* distributions are bigger than the dispersions of MLWH. Therefore, the k-means binary operator contributes to the precision of the results. Finally, the MLWH distributions are closer to zero than *rnd.0.5* distributions, indicating that MLWH has consistently better results than *random.0.5*. When we evaluate the behavior of the algorithms through the Wilcoxon test, this test indicates that there is a significant difference between the two algorithms.

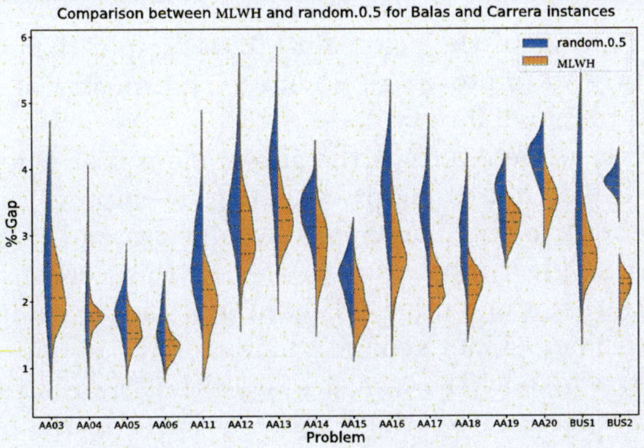

Fig. 2. Evaluation of k-means binary operator.

Table 2. Balas and Carrera instances.

Instance	row	col	Density	Best known	MLWH (avg)	std	*rnd.0.5* (avg)	std std
AA03	106	8661	4.05%	33155	33357.4	125.1	33693.5	247.6
AA04	106	8002	4.05%	34573	34984.8	62.5	34931	233.1
AA05	105	7435	4.05%	31623	31739.2	64.7	31889.2	113.8
AA06	105	6951	4.11%	37464	37672.6	59.5	37651.3	99.4
AA11	271	4413	2.53%	35384	35801.1	132.7	35901.2	196.3
AA12	272	4208	2.52%	30809	31437.9	148.1	31631.3	218.6
AA13	265	4025	2.60%	33211	34022.3	138.2	34201.4	236.8
AA14	266	3868	2.50%	33219	33774.1	126.1	33967.3	127.3
AA15	267	3701	2.58%	34409	35019.6	118.4	34846.8	166.1
AA16	265	3558	2.63%	32752	33206.4	138.3	33601.3	188.9
AA17	264	3425	2.61%	31612	32012.8	91.2	32387.7	152.1
AA18	271	3314	2.55%	36782	37165.3	76.1	37419.3	149.5
AA19	263	3202	2.63%	32317	33212.6	78.4	33108.3	147.8
AA20	269	3095	2.58%	34912	35832.2	101.1	35962.6	102.1
BUS1	454	2241	1.88%	27947	28414.9	80.8	28758.2	187.2
BUS2	681	9524	0.51%	67760	68844.1	79.4	68680.6	121.1
Average				35495.6	36030.6	100.9	36164.4	168.0
p-value							3.1e−04	

5 Conclusions

In this work, the MLWH algorithm based on the k-means technique was proposed to perform binarization of whale metaheuristics. We must emphasize that this k-means technique can be applied in the binarization of any continuous metaheuristics of the swarm intelligence type. To evaluate the performance of Our k-means technique the set covering problem was used together with the whale algorithm. Experiments were designed to evaluate the contribution of the binary operator k-means of the algorithm. It was found that the operator contributes significantly to improve the accuracy and quality of the solutions, all these visualized through violin graphics.

As future works, we believe that the autonomous search for efficient configuration is a little-explored area and it could help to improve the performance of the binarizations both in the quality of the solutions and in the times of convergence. Additionally, it is interesting to see how this k-means technique works with other metaheuristics and other NP-hard problems. Finally, we believe that using machine learning we can generate indicators or obtain correlations that allow us to understand how the k-means algorithm translates the exploration and exploitation properties from the continuous space to the binary space.

References

1. Crawford, B., Soto, R., Monfroy, E., Astorga, G., García, J., Cortes, E.: A meta-optimization approach for covering problems in facility location. In: Figueroa-García, J., Lopez-Santana, E., Villa-Ramírez, J., Ferro-Escobar, R. (eds.) Applied Computer Sciences in Engineering Workshop on Engineering Applications, pp. 565–578. Springer, Cham (2017). https://doi.org/10.1007/978-3-319-66963-2_50
2. García, J., Crawford, B., Soto, R., Astorga, G.: A percentile transition ranking algorithm applied to binarization of continuous swarm intelligence metaheuristics. In: Ghazali, R., Deris, M., Nawi, N., Abawajy, J. (eds.) Recent Advances on Soft Computing and Data Mining International Conference on Soft Computing and Data Mining, pp. 3–13. Springer, Cham (2018). https://doi.org/10.1007/978-3-319-72550-5_1
3. Tapia, D., et al.: A Q-learning hyperheuristic binarization framework to balance exploration and exploitation. In: Florez, H., Misra, S. (eds.) Applied Informatics. International Conference on Applied Informatics, pp. 14–28. Springer, Cham (2020)
4. Crawford, B., et al.: Q-learnheuristics: towards data-driven balanced metaheuristics. Mathematics **9**(16) (2021)
5. García, J., Moraga, P., Valenzuela, M., Pinto, H.: A db-scan hybrid algorithm: an application to the multidimensional knapsack problem. Mathematics **8**(4), 507 (2020)
6. García, J., Altimiras, F., Peña, A., Astorga, G., Peredo, O.: A binary cuckoo search big data algorithm applied to large-scale crew scheduling problems. Complexity **2018** (2018)
7. García, J., Lalla-Ruiz, E., Voß, S., Droguett, E.L.: Enhancing a machine learning binarization framework by perturbation operators: analysis on the multidimensional knapsack problem. Int. J. Mach. Learn. Cybern. **11**(9), 1951–1970 (2020). https://doi.org/10.1007/s13042-020-01085-8
8. García, J., Maureira, C.: A KNN quantum cuckoo search algorithm applied to the multidimensional knapsack problem. Appl. Soft Comput. **102**, 107077 (2021)
9. García, J., Crawford, B., Soto, R., Astorga, G.: A clustering algorithm applied to the binarization of swarm intelligence continuous metaheuristics. Swarm Evol. Comput. **44**, 646–664 (2019)

10. Crawford, B., Soto, R., Astorga, G., García, J.: Constructive metaheuristics for the set covering problem. In: Korošec, P., Melab, N., Talbi, E.G. (eds.) International Conference on Bioinspired Methods and Their Applications, pp. 88–99. Springer, Cham (2018). https://doi.org/10.1007/978-3-319-91641-5_8
11. García, J., Astorga, G., Yepes, V.: An analysis of a KNN perturbation operator: an application to the binarization of continuous metaheuristics. Mathematics **9**(3), 225 (2021)
12. García, J., Peña, A.: Robust optimization: concepts and applications. In: Nature-inspired Methods for Stochastic, Robust and Dynamic Optimization, p. 7 (2018)
13. Astorga, G., Crawford, B., Soto, R., Monfroy, E., García, J., Cortes, E.: A meta-optimization approach to solve the set covering problem. Ingeniería **23**(3) (2018)
14. García, J., et al.: A db-scan binarization algorithm applied to matrix covering problems. Comput. Intell. Neurosci. **2019** (2019)
15. García, J., Măntoiu, M.: Localization results for zero order pseudodifferential operators. J. Pseudo-Differ. Oper. Appl. **5**(2), 255–276 (2013). https://doi.org/10.1007/s11868-013-0084-y
16. Crawford, B., Soto, R., Lemus-Romani, J., Astorga, G., Matus de la Parra, S., Peña-Fritz, A., Paredes, F.: Investigating the efficiency of swarm algorithms for bridge strengthening by conversion to tied-arch: a numerical case study on san Luis bridge. Iran. J. Sci. Technol. Trans. Civ. Eng. **45**(4), 2345–2357 (2021)
17. Altimiras, F., et al.: Altered gut microbiota in a fragile x syndrome mouse model.Front. Neurosci. **15** (2021)
18. Yang, X.-S., Deb, S.: Cuckoo search via lévy flights. In: World Congress on Nature & Biologically Inspired Computing, 2009. NaBIC 2009, pp. 210–214. IEEE (2009)
19. Hatamlou, A.: Black hole: a new heuristic optimization approach for data clustering. Inf. Sci. **222**, 175–184 (2013)
20. Yang, X.-S.: A new metaheuristic bat-inspired algorithm. In: Nature Inspired Cooperative Strategies for Optimization (NICSO 2010), pp. 65–74 (2010)
21. Mirjalili, S., Lewis, A.: The whale optimization algorithm. Adv. Eng. Softw. **95**, 51–67 (2016)
22. Gary, M.R., Johnson, D.S.: Computers and intractability. In: A Guide to the Theory of NP-Completeness (1979)
23. Lu, Y., Vasko, F.J.: An or practitioner's solution approach for the set covering problem. Int. J. Appl. Metaheuristic Comput. (IJAMC) **6**(4), 1–13 (2015)
24. Li, Y., Cai, Z.: Gravity-based heuristic for set covering problems and its application in fault diagnosis. J. Syst. Eng. Electron. **23**(3), 391–398 (2012)
25. García, J., Crawford, B., Soto, R., García, P.: A multi dynamic binary black hole algorithm applied to set covering problem. In: Del Ser, J. (eds.) Harmony Search Algorithm. International Conference on Harmony Search Algorithm, pp. 42–51, Springer, Singapore (2017). https://doi.org/10.1007/978-981-10-3728-3_6
26. Balaji, S., Revathi, N.: A new approach for solving set covering problem using jumping particle swarm optimization method. Nat. Comput. **15**(3), 503–517 (2015). https://doi.org/10.1007/s11047-015-9509-2
27. Garcia, J., Crawford, B., Soto, R., Astorga, G.: A percentile transition ranking algorithm applied to knapsack problem. In: Silhavy, R., Silhavy, P., Prokopova, Z. (eds.) Applied Computational Intelligence and Mathematical Methods. Proceedings of the Computational Methods in Systems and Software, pp. 126–138, Springer, Cham (2018). https://doi.org/10.1007/978-3-319-67621-0_11
28. Kasirzadeh, A., Saddoune, M., Soumis, F.: Airline crew scheduling: models, algorithms, and data sets. EURO J. Transp. Logis. **6**(2), 111–137 (2015). https://doi.org/10.1007/s13676-015-0080-x

29. Horváth, M., Kis, T.: Computing strong lower and upper bounds for the integrated multiple-depot vehicle and crew scheduling problem with branch-and-price. Central Eur. J. Oper. Res. **27**(1), 39–67 (2017). https://doi.org/10.1007/s10100-017-0489-4
30. Stojković, M.: The operational flight and multi-crew scheduling problem. Yugoslav J. Oper. Res. **15**(1) (2016)
31. García, J., Crawford, B., Soto, R., Castro, C., Paredes, F.: A k-means binarization framework applied to multidimensional knapsack problem. Appl. Intell. **48**(2), 357–380 (2017). https://doi.org/10.1007/s10489-017-0972-6
32. García, J., Pope, C., Altimiras, F.: A distributed k-means segmentation algorithm applied to Lobesia botrana recognition. Complexity **2017** (2017)
33. Graells-Garrido, E., García, J.: Visual exploration of urban dynamics using mobile data. In: García-Chamizo, J., Fortino, G., Ochoa, S. (eds.) Ubiquitous Computing and Ambient Intelligence. Sensing, Processing, and Using Environmental Information International Conference on Ubiquitous Computing and Ambient Intelligence, pp. 480–491. Springer, Cham (2015). https://doi.org/10.1007/978-3-319-26401-1_45
34. Graells-Garrido, E., Peredo, O., García, J.: Sensing urban patterns with antenna mappings: the case of Santiago, Chile. Sensors **16**(7), 1098 (2016)
35. Peredo, O.F., García, J.A., Stuven, R., Ortiz, J.M.: Urban dynamic estimation using mobile phone logs and locally varying anisotropy. In: Gondez, J., Rodrigo-Ilarri, J., Rodrigo-Clavero, M., Cassiraga, E., Vargas-Guzman, J. (eds) Geostatistics Valencia 2016, pp. 949–964, Springer, Cham (2017). https://doi.org/10.1007/978-3-319-46819-8_66
36. Maureira, C., Pinto, H., Yepes, V., García, J.: Towards an AEC-AI industry optimization algorithmic knowledge mapping: an adaptive methodology for macroscopic conceptual analysis. IEEE Access **9**, 110842–110879 (2021)

Modelling World Population Dynamics

R. I. Dzerzhinsky(✉), A. S. Zausailov, and A. A. Vorontsov

MIREA - The Russian Technological University, 78, Vernadsky pros., Moscow 119454, Russia

Abstract. The world population long-term dynamics quantitative regularities are analysed and its size long-term forecasts are obtained. The intensive growth models, limited growth, in particular, the logistic model, as well as relative population increases in a model with non-monotonic dynamics, are considered. The all considered models parameters were identified using real statistical data; on their basis, computational algorithms and programs were developed, and the limiting world population predictive estimates were obtained #COMESYSO1120.

Keywords: Population size · Model and growth rates · Logistic model · Logistic curve

1 Introduction

The current demographic situation in different regions and countries is rather heterogeneous. The world's population of 70% is concentrated on an area that makes up 7% of the land. More people live in the Eastern Hemisphere (about 85%) than in the Western Hemisphere, and in the North - more than in the Southern (about 10%). In the planet South, there is a rapid population growth: the population of Africa, Asia and Latin America practically doubles every 25 years. In the North, in Europe, the picture is reversed: constancy or even decline in numbers. So, for example, since the 1990s beginning, Russia's population declines annually by almost 1 million until around 2009. The population increased continuously from 2009 to 2017, but the population has been continuously decreasing over the past three years. By 2026, he predicts an increasing decrease in the Russian population.

The entire demographic problem an important aspect is the heterogeneous population growth in the world different parts. In addition, it is complicated by the unequal demographic trends for different peoples.

It would be more correct to use the fuzzy logic apparatus to describe such processes, by analogy, as is done, for example, in [1].

The site Worldometers [2] was selected as statistical data a source. Population estimates are given as of mid-current 2021.

This work purpose is to analyse the world population long-term dynamics quantitative patterns and obtain its size long-term forecasts.

R. Silhavy et al. (Eds.): CoMeSySo 2021, LNNS 231, pp. 423–432, 2021.
https://doi.org/10.1007/978-3-030-90321-3_34

2 The World Population Intensive Growth. Population Explosion

Throughout mankind history, the world population number has been increasing and only in certain, relatively short historical periods associated with wars, epidemics, natural disasters, is it temporarily decreasing.

The history early stages were characterized by extremely slow population growth, this is characterized by production development at a low level and strong dependence of humans on the environment.

Figure 1 shows the world population dynamics from 200 to the present.

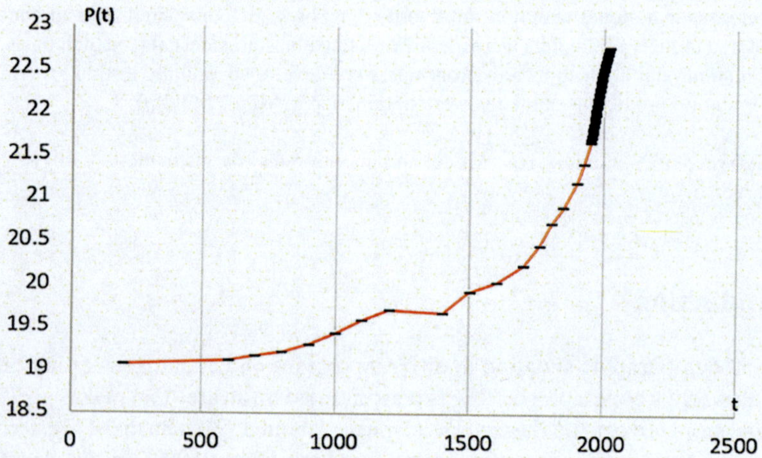

Fig. 1. The world population dynamics in 200–2021, million people Semi-logarithmic coordinate system. The graph is based on data from Worldometers [1].

The graph in Fig. 1 is plotted in a semi-logarithmic coordinate system. A semi-logarithmic coordinate system turns out to be developed, identifying areas naturally with constant relative increments.

Growth rates are calculated using the following formula:

$$r(t) = ln\left[P\frac{t+1}{P(t)}\right], \tag{1}$$

where:

t is the year,

$r(t)$ is the growth rate in the interval from the year t middle to the year $t + 1$ middle,

$P(t)$ - population size in the year t middle,

ln is the natural logarithm.

The world demographic dynamics shows growth at an increasing rate - the slopes on the graph (Fig. 1) are increasing. The world population growth rates dynamics in 1600–2021, shown in Fig. 2.

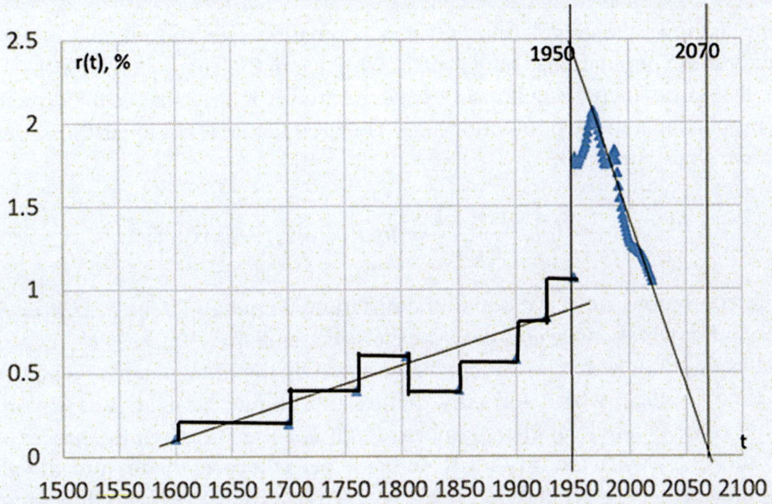

Fig. 2. Relative annual growth dynamics in the world population, 1600–2021. In 1600–1950 relative increments were calculated for 50 years, in some places less often, after 1950 - annual data.

Maintaining a high birth rate results in sharply reduced mortality, primarily in developing countries, is a demographic explosion. This was due to the World Health Organization's program implementation to vaccinate the world population against epidemic diseases and teaching the developing countries' population to hygiene rules for obstetrics and caring for children under 1-year-old. Developed countries, with fertility rates of 11–15 births per thousand population per year, have moved to a constant population level or weak growth, which is observed in European countries.

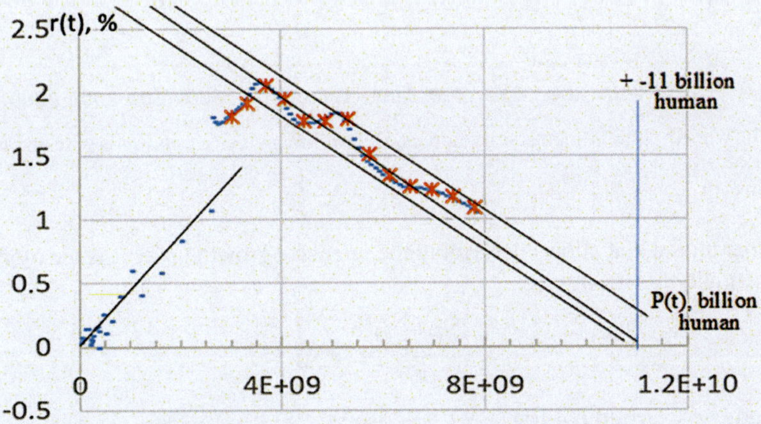

Fig. 3. Dependence connecting the world population size and its relative growth: annual growth is shown in blue, red - averaged over five years in 1600–2021.

Figure 3 shows the correlation between population size and relative growth rates. Consider first the populations' range that took place in the years 1650–1950.

Here there is an increasing linear dependence r(t) = K·P(t), that is, the growth rate is directly proportional to the population size. The growth rate is the logarithmic derivative of a discrete analogue, since

$$r(t) = ln\left[P\frac{t+1}{P(t)}\right] = ln(P(t+1)) - ln(P(t)). \tag{2}$$

Passing to continuous time, we obtain the differential equation $d[\ln(P)]/dt = K \cdot P$ or $dP/dt = K \cdot P^2$, which integral is the hyperbolic dependence $P(t) = 1/K \cdot (t_0 - t)$, with a vertical asymptote, which position is determined by the time instant $t = t_0$.

The time point t_0, which formally defines the infinite population, falls on 2025–2030. This value is a radical change moment, an upper estimate in the process nature. Figure 4 shows a way to identify it [3]. In the hyperbola case, the reciprocal values for population sizes $1/P(t)$ depend on t linearly, $1/P(t) = K \cdot (t_0 - t)$ so the parameter t_0 value corresponds to the straight line $(t_0 - t)$ with the abscissa axis.

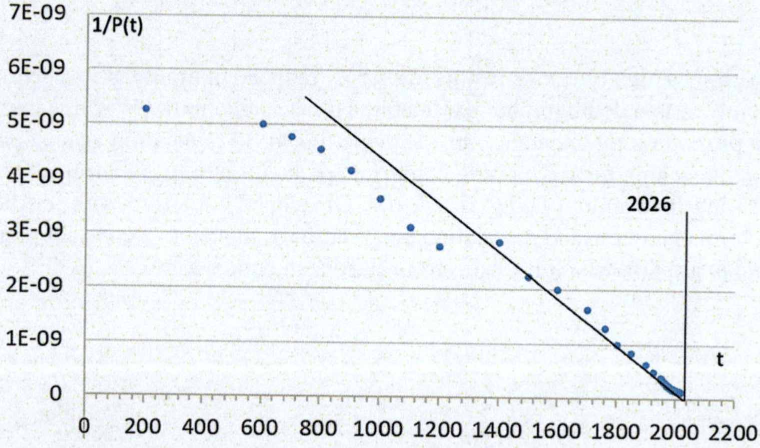

Fig. 4. The world population inverse dynamics. Determining when a trend ends with an increasing pace.

Another model that allows describing the intensive growth process is the model with an exponentially increasing rate

$$\frac{dP}{P \cdot dt} = a \cdot e^{bt}, \tag{3}$$

which study was carried out in [3].

The differential equation integration determines the process trajectory,

$$\ln P = \frac{a}{b} \cdot e^{bt} + \ln C, \tag{4}$$

where

$$\ln\left(\ln\left(\frac{P}{C}\right)\right) = \ln\left(\frac{a}{b}\right) + b \cdot t. \quad (5)$$

For such a process starting point to be at the coordinates' origin, it is required to analyse the data at the scale $(\ln(\ln(P/C \cdot e)), t)$, then at the starting point $\ln \ln e = 0$ (Fig. 5). According to this model, 1950 turns out to be demographic dynamics a critical point - here the dependence parameters change abruptly, which had been preserved for 350 years up to that moment. The exponential relationship b does not remain unchanged: after 1950, its values increase, and then, starting from the 1990s, they decrease (Fig. 5).

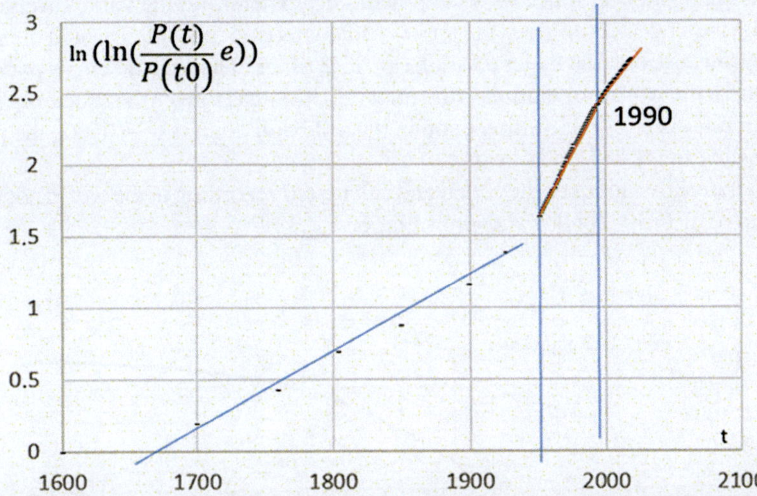

Fig. 5. Straightening transformation for intense growth, a process with exponentially increasing rates. The break in the trend, corresponding to the abrupt change in the parameters, falls in 1950 and 1990.

The Forrester, Meadows, and other global dynamics mathematical models were based on the intensive population growth trend since their parameters were estimated over the time interval 1900–1970. This trend extrapolation automatically led to a crisis, which manifested itself in the resources' depletion, an increase in pollution and a decrease in the birth rate and an increase in mortality caused by these factors.

3 Demographic Transition. Population Growth at a Falling Rate

On the downside, the straight-line dependence of relative increases on numbers (Fig. 3) means that $r(t) = k \cdot (P^{\propto} - P)$ passing to continuous time, we get the differential equation:

$$\frac{dP}{dt} = k \cdot P \cdot \left(P^{\propto} - P\right), \quad (6)$$

which is called the logistic equation [3].

The solution to this equation is a limited growth curve, also called logistic:

$$P(t) = \frac{P^{\propto}}{1 + e^{-k\cdot(t-t^{*})}}, \tag{7}$$

where:

P^{∞} is the growth limit that determines the horizontal asymptote position,

t^{*} - the inflexion point abscissa,

k - the curve growth rate an indicator.

The value $P^{\propto}/P(t) - 1 = P^{\propto} - P(t)/P(t)$, represents the relative potential increase in numbers from the moment t to stabilization at the maximum possible level P^{∞}. For logistic dependence, it decreases exponentially. Therefore, the value inverse to the growth rate index k^{-1}determines the process characteristic time, after which the relative potential growth decreases by a factor of $e \approx 2.72$ times. The logistic curve parameters estimation for the world population in the interval 1968–2021 gives the following results: the growth limit is $P^{\infty} \approx 11$ billion people, the inflexion point is $t^{*} \approx 1992$, the process characteristic time is $k^{-1} \approx 33$ years.

Comparative dynamics of the actual and calculated according to the world population logistic curve in 1950–2050 is shown in Fig. 6.

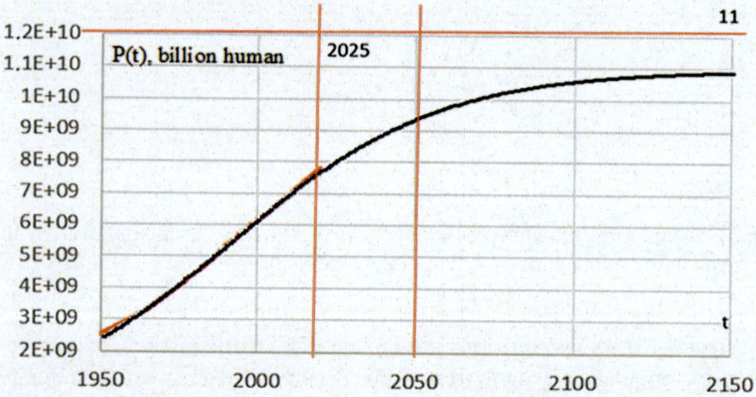

Fig. 6. Comparative dynamics of the actual and calculated according to the world population logistics model in 1950–2050.

A similar comparison in rectifying logistic coordinates is given in Fig. 7.

A certain rhythmic structure is manifested in the world demographic dynamics. In the logistic growth area, the main rhythm duration coincides with the characteristic time of the logistic curve $k^{-1} \approx 33$ years (Fig. 7).

As a result, four critical points can be identified in the population dynamics in the logistic growth area:

- 1959–1960 - demographic transition - a radical change in the structure, the transition from intensive development with increasing rates to development with decreasing rates, the logistic growth beginning (the population reached 3 billion humans);

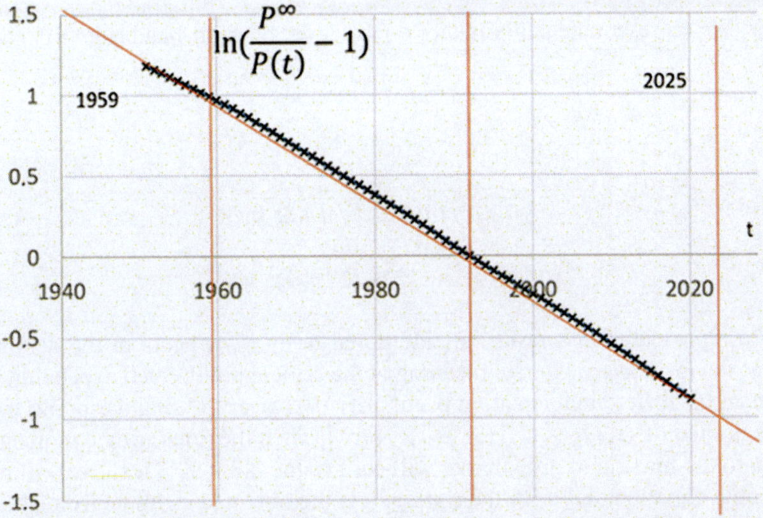

Fig. 7. Logistic dependence in rectifying coordinates $\left(t,\ \ln\left[P^{\infty}/P(t) - 1\right]\right)$. Demographic dynamics rhythmic structure.

- 1992 - inflexion point on the logistic curve of the population $P(t)$ (population - 5.49 billion human);
- 2024–2025 - inflection point on the annual growth curve $\Delta(t) = P(t+1) - P(t)$ (predicted population estimate of 8 billion human);
- 2058 - logistic growth and demographic transition completion, the world population stabilization at the level of 11 billion humans, unless other growth constraints come into effect earlier.

4 Relative Gains Non-monotonic Dynamics

In the S P Kapitsa work [5] proposed the world demographic dynamics a mathematical model, which allows describing in a unified way both the regime with exacerbation before the transition and the transition itself. Express the rate of change in population size as a time function by substituting $dP/dt = K \cdot P^2$ on the equation right side hyperbolic dependence,

$$P(t) = \frac{1}{K \cdot (t_0 - t)}, \tag{8}$$

then,

$$\frac{dP(t)}{dt} = \frac{1}{\left[\sqrt{K} \cdot (t_0 - t)^2\right]}. \tag{9}$$

Near the critical moment t_0, the equation right-hand side (9) increases indefinitely, which leads to absurd results, both for the population size values and for its absolute and relative increases. We represent the denominator on the equation right-hand side (9) in the form:

$$[(t_0 - \tau - t) + \tau]^2, \quad (10)$$

where

$$then\ for\ t \le T_1\ и\ \tau \ge 0, \quad (11)$$

$$[(T_1 - t) + \tau]^2 \ge (T_1 - t)^2 + \tau^2 \quad (12)$$

As such a representation a result, an additional term τ^2, appears in the denominator, which has the inverse growth rate meaning at the critical point. Such a technique allows to eliminate the differential equation peculiarity (9) - near the singular point, the speed becomes limited $dP(t)/dt \le C/\left[(T_1 - t)^2 + \tau^2\right]$. Then the time-critical moment corresponding to the maximum growth rate will shift to the point T_1. Next, we will consider a differential equation, in which the right part is the obtained estimate from above:

$$\frac{dP(t)}{dt} = \frac{C}{\left[(T_1 - t)^2 + \tau^2\right]}, \quad (13)$$

where $C = 1/K$.

The solution to the differential Eq. (13) is the inverse trigonometric function, which in appearance coincides with the Cauchy distribution of mathematical statistics. This function describes the period before the system transition at time T_1, the transition from increasing relative increases in numbers to decreasing ones, and the period in which the system will be after the transition to the new regime.

The inverse trigonometric function is:

$$P(t) = \frac{C}{\tau} \cdot arcctg\left(\frac{T_1 - t}{\tau}\right). \quad (14)$$

If the explosive hyperbolic growth model with a vertical asymptote worked before the transition, then after it – limited growth a model with saturation, with a horizontal asymptote, which is located at the level $\pi \cdot C/\tau$. As a result, the growth of asymptotics changes.

Kapitsa's model (14) approximates the world population dynamics using an inverse trigonometric function. Function (14) contains three parameters: the passing of the inflexion point moment T_1, the time scale parameter τ and the number scale parameter C/τ.

In this work, the timescale parameter τ values determination and the moment T_1 is carried out by constructing an anamorphosis for function (14).

The trigonometric function (14), which describes the $P(t)$ dynamics, is rectified in coordinates $(ctg(\pi P(t)/P\infty); (T_1 - t/\tau))$. To plot a graph in such coordinates, the limiting population P∞ knowledge is required. We will take this estimate from the logistic model and assume that P∞ ≈ 11 billion humans.

Then the anamorphosis slope will give the timescale parameter value $1/\tau \approx 49$, and the graph intersection point with the horizontal axis is the moment $T_1 = 1992$ (Fig. 8).

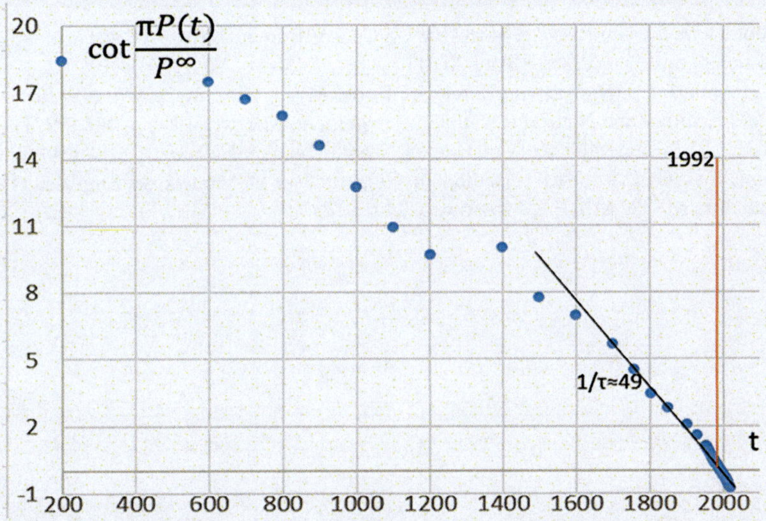

Fig. 8. Anamorphosis. The timescale determination and the inflexion point position on the absolute increments curve.

The abundance actual values at the characteristic points are equal to:

$$\begin{array}{l} P(1955) = 2.8 \ billion \ P(1969) = 3.6 \ billion \\ P(1992) = 5.5 \ billion \ P(2014) = 7.3 \ billion \end{array}. \quad (15)$$

Compared to model ones:

$$\begin{array}{l} P(1955) = 3.2 \ billion \ P(1969) = 3.9 \ billion \\ P(1992) = 5.5 \ billion \ P(2014) = 7.0 \ billion \end{array}. \quad (16)$$

5 Conclusion

In this work, using the methods discussed in more detail in [6], the world population long-term dynamics quantitative patterns were analysed, and its size long-term forecasts were obtained.

The moment in time that determines the hyperbolic dependence vertical asymptote position is critical - it corresponds to 2026.

The current stage in the world civilization development marks growth with falling relative and absolute growths, it will lead to the world population stabilization at the level of 11 billion people if the critical level of 9.1 billion people is successfully passed.

References

1. Chesalin, A.N., Grodzensky, S.Y., Van, T.F., Nilov, M.Y., Agafonov, A.N.: Technology for assessing risks at the product life cycle stages using fuzzy logic. Russian Technol. J. **8**(6), 167–183 (2020)

2. Worldometers, the United Nations and the US Census Bureau, https://www.worldometers.info/world-population/. Accessed 8 Aug 2020
3. World Population. Demographic Reference Book. Mysl', Moscow, p. 478 (1989)
4. Watt, K.: Ecology and Natural Resource Management. Mir, Moscow, p. 464 (1971)
5. Kapitsa, S.P.: A General Theory of Human Growth. Nauka, Moscow, p. 188 (1999)
6. Kuzmin, V.I., Gadzaov, A.F.: Method of Construction of Models on Empirical Data. The Manual. The MSTU MIREA, Moscow, p. 94 (2012)

Data Analysis on Terrorist Activity. Predicting the Terrorist Activity Level

Roman I. Dzerzhinsky(✉), E. V. Ledovskaya, and A. A. Shchenov

The Applied Mathematics Department of the Institute of Information Technologies, MIREA - The Russian Technological University, 78, Vernadsky pros., 119454 Moscow, Russia

Abstract. Using the trend exclusion method, the system development critical levels theory, the least-squares method, the autoregression model and by constructing the Alter-Johnson function, the world terrorism system development almost-periods and phases were determined, clustering was carried out. Models have been selected to predict the terrorist attacks level, and their comparison has been made. #COMESYSO1120

Keywords: Predicting the terrorist attacks level · Excluding trend methods · Alter-Johnson function · Least squares method · Restructuring phase · Autoregressive method

1 Introduction

Over the past few decades, the ongoing terrorist attacks have become more dangerous and deadly. If in the late 19th - early 20th century only people a few or dozens became terrorist attacks victims, then by the end of the 20th - early 21st century the one terrorist victim's number act may be in the thousands. Despite the problem growing global nature, there are still not developed mathematical methods that allow analysing, predicting and identifying patterns of terrorist activity [1]. A huge amount of information that has been accumulating for decades remains unexplored and not fully analysed.

As research an object, a data set on terrorist acts is used, the world terrorism system development almost-period and phase are determined, and models are selected and compared to predict the terrorist activity level [2, 3].

2 Information About the Data Set Under Study and Primary Visualization

As the data on which basis all further research will be carried out, information was taken from the Global Terrorism Base from 1970 to 2017. The dataset includes the following data on terrorist attacks: date, country, region, city, latitude and longitude, the deaths number and the injured number during the attack, the terrorist group that carried out the attack, the target against which the attack was carried out, information about a target type (government sector, military sector, etc.).

R. Silhavy et al. (Eds.): CoMeSySo 2021, LNNS 231, pp. 433–442, 2021.
https://doi.org/10.1007/978-3-030-90321-3_35

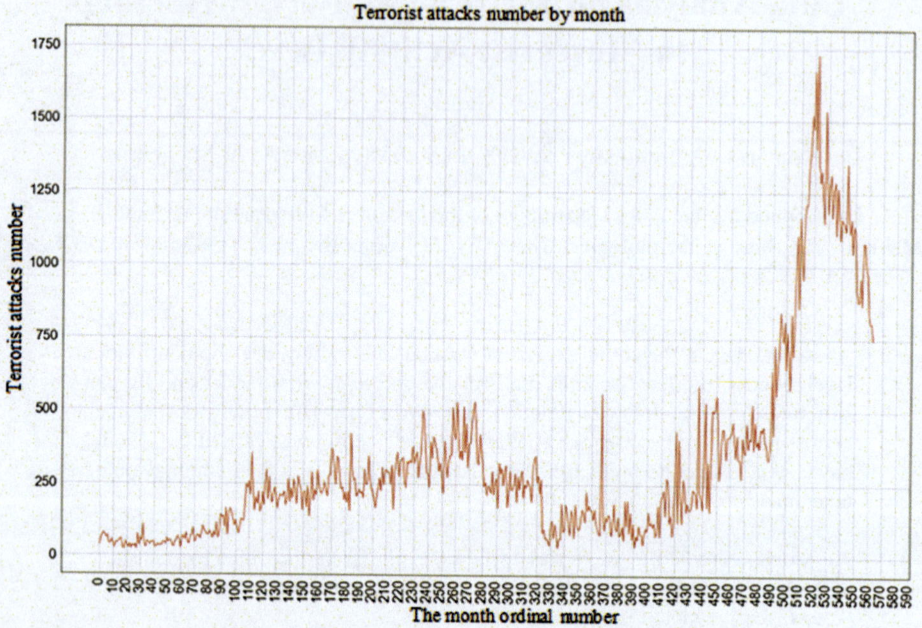

Fig. 1. Terrorist attacks monthly number from 1970 to 2017.

Let's perform the data as an initial visualization, displaying the terrorist attacks monthly number in a graph from 1970 to 2017 (see Fig. 1).

Figure 1 shows that the terrorist attacks monthly number level in the 21st century, compared with the 20th-century end, has grown significantly. If at the 20th-century end the act's monthly number was 500–550 cases, then at the 21st century beginning their number began to grow rapidly, and by 2014 reached the 1750 monthly cases mark.

3 Eliminating a Trend and the System Almost-Periods Defining

Let's eliminate the graph trend component shown in Fig. 1 to make it stationary [4, 5]. Based on the stationary data series, we construct the Alter-Johnson shift function to determine the system almost-periods understudy (see Fig. 2).

In Fig. 2, black vertical lines mark the deepest local minima of the Alter-Johnson function, which are almost periods of the system under study. It can be seen that the function first local minimum is at point 12. All subsequent local minima of the function are approximately at the same distance from each other (12, 24, 36, etc.). From this, we conclude that the almost-period of the studied data is 12 months, and if we take into account that the studied data series starts from 01/01/1970, then we can conclude that the terrorist activity peak falls on each year summer.

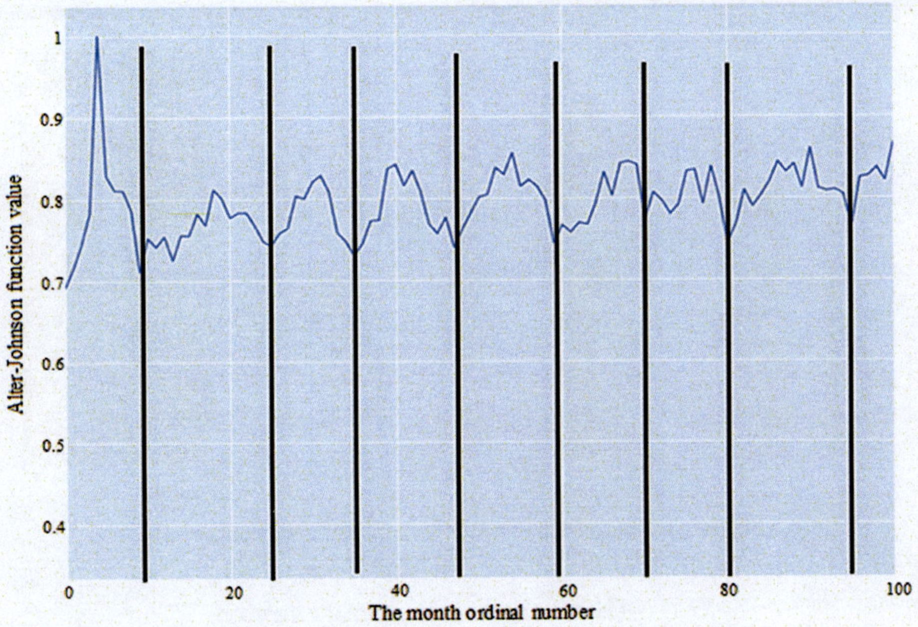

Fig. 2. The considered data almost-periods system.

4 Systems Development Critical Levels. The Restructuring Phases

To study a time series' development and behaviour, it is necessary to determine the system development critical levels [6, 7]. A critical point is a special point, which an example is a catastrophe, crisis, jump, phase transition and others. In the interval between the critical level constants e^e and e^3, there is a zone in which the systems change their structure and, after the restructuring phase end, the system general level either begins to grow intensively or to decline intensively. The noted critical constants and the system restructuring phases are shown in Fig. 3.

In Fig. 3, vertical rectangular sections mark the phase restructuring periods. It can be seen from the figure that in the intervals between the restructuring phases, the system intensively grew each time. Consequently, it can be concluded that once the restructuring current phase ends in mid-2029, which starts in mid-2021, the terrorist activity growth will again begin to increase intensively and reach even greater proportions.

5 Predicting the Terrorist Activity Level

We make a short-term prediction of the terrorist attacks monthly number using a least-squares method and an autoregressive model, then compare the two results [8].

Figure 3 shows the terrorist attacks monthly number in a graph on a semilogarithmic scale, as well as the system under study trend, which is built using the least-squares method. Since the terrorist attack's monthly number logarithm increases at a constant

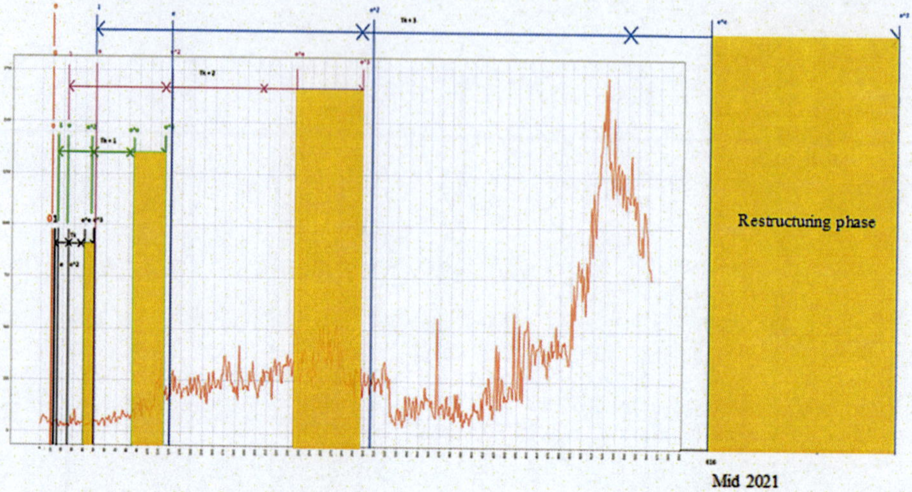

Fig. 3. Critical constants and the system restructuring phases.

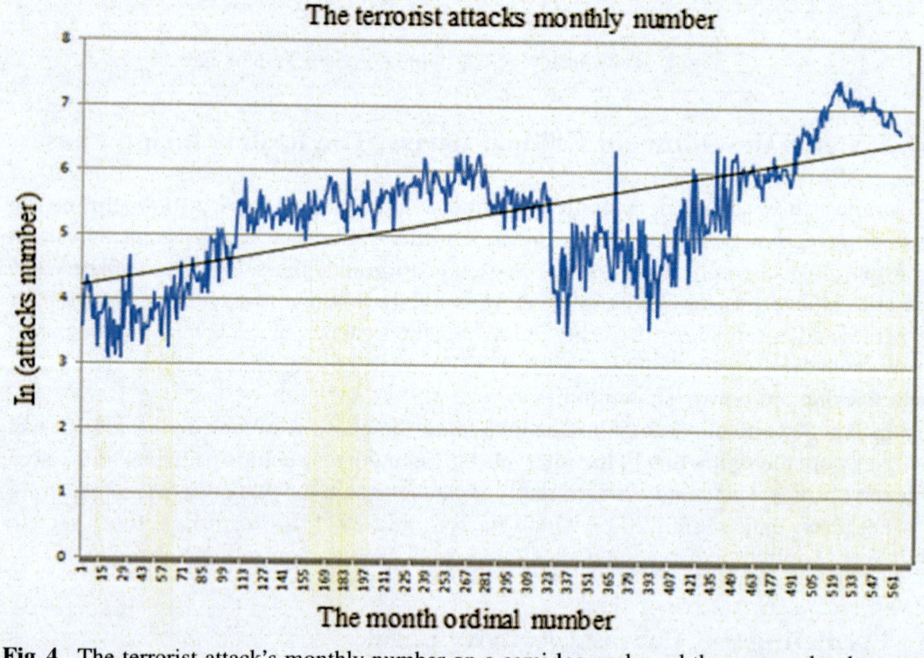

Fig. 4. The terrorist attack's monthly number on a semi-log scale and the system development trend.

rate, that is, it presents a relationship similar to a linear one, a linear trend will best approximate the data by repeating this relationship (Fig. 4).

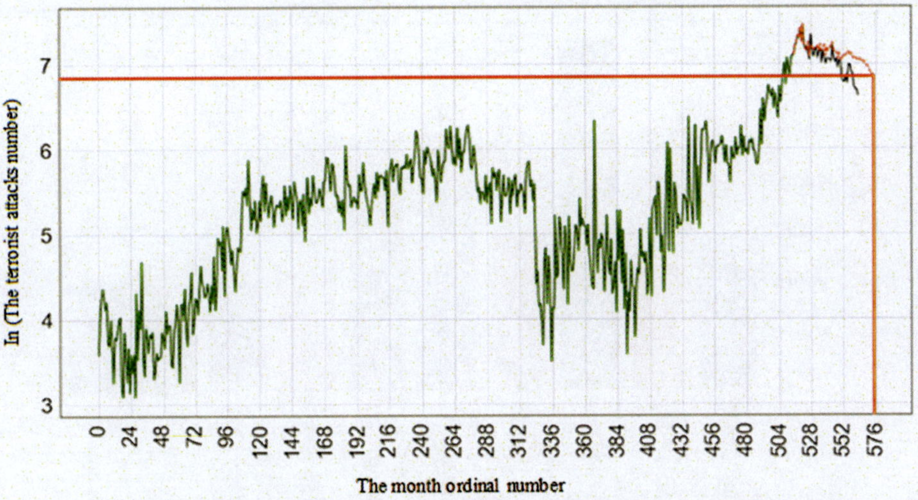

Fig. 5. Predicting the terrorist activity level using the autoregressive model.

In total, 10 months from our data endpoint, the monthly terrorist attacks number is predicted to be around 850. The resulting model approximation coefficient is 0.6716, which is a good indicator for a linear trend.

Now let's predict the terrorist attacks monthly number using the autoregressive model. To do this, we divide the dataset into training and test samples in the ratio: the share of 0.8 of our data is attributed to the training set and the rest to the test set. The forecast result is shown in Fig. 5.

Figure 5 shows the terrorist attacks monthly forecast number in a graph. At the graph branching point: the upper curve is the data obtained using the model, the lower one is the real data. The forecast is made for ten months ahead. By its value, one can judge that the terrorist attacks monthly number will be $e^{6.9} = 990$ cases. The model means square error value was 0.035, which is a very good indicator.

The predictions result from the least-squares method and the autoregressive model are similar. However, it should be borne in mind that the autoregressive method is more accurate, since its forecast in this case is based on 56 previous months, and not on the system's behaviour tendency.

6 Clustering Terrorist Attacks and Identifying Gang Patterns

Let's cluster our data on terrorist groups using the k-means method. Since the data contains information about the ten most dangerous terrorist groups attacks, we set the clusters' number equal to 10. The clustering method helps us to classify a new terrorist act as a cluster of existing acts, thereby helping to identify the terrorist act culprit [9, 10].

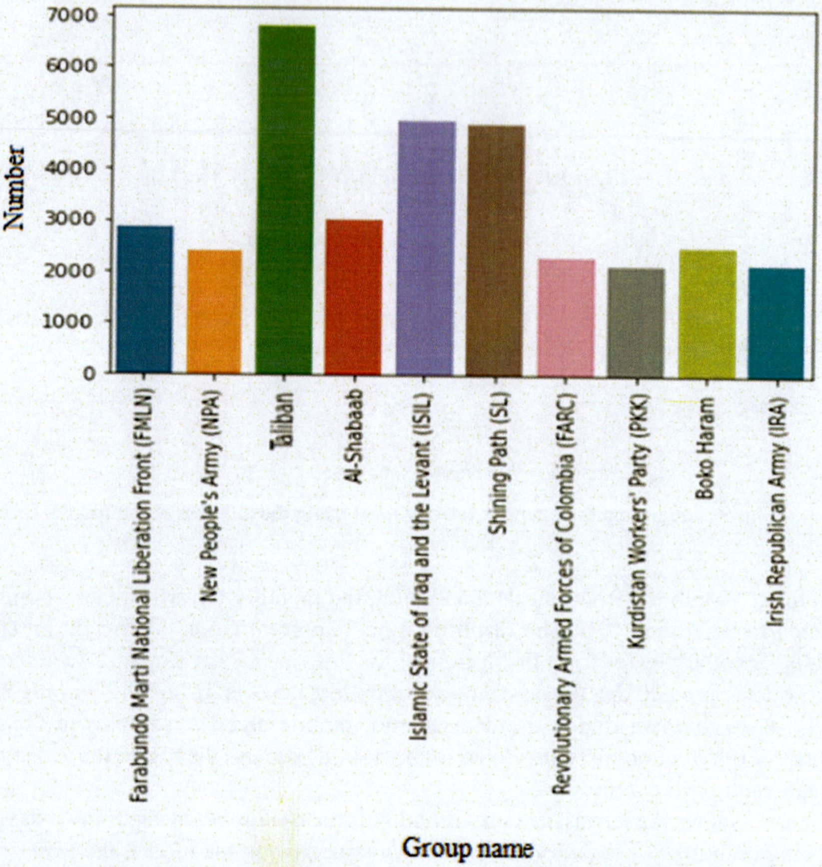

Fig. 6. k-means clustering result.

Applying the k-means method to our dataset, we get Fig. 6:

Figure 6 shows that our data is divided into clusters. However, this graph does not fully reflect how well this separation has happened. Let's apply the principal component analysis (PCA) to downsize our data and display the clustering result on a three-dimensional plane [11].

To display data on a three-dimensional plane (see Fig. 7) we need to select the principal components' number equal to 3.

Figure 7 depicts clustering after data downsizing. Each of the three principal components: x, y, z is a linear combination of all the other features of the data in question. It can also be seen from the figure that the data is rather accurately divided into clusters, however, there are some outliers.

Let us apply the k nearest neighbours method to determine the attacks patterns by terrorist groups, classifying our clusters into 2 groups: outlier and real data. We are faced with the separating outliers task from clusters, which we turn into a full-fledged classification task.

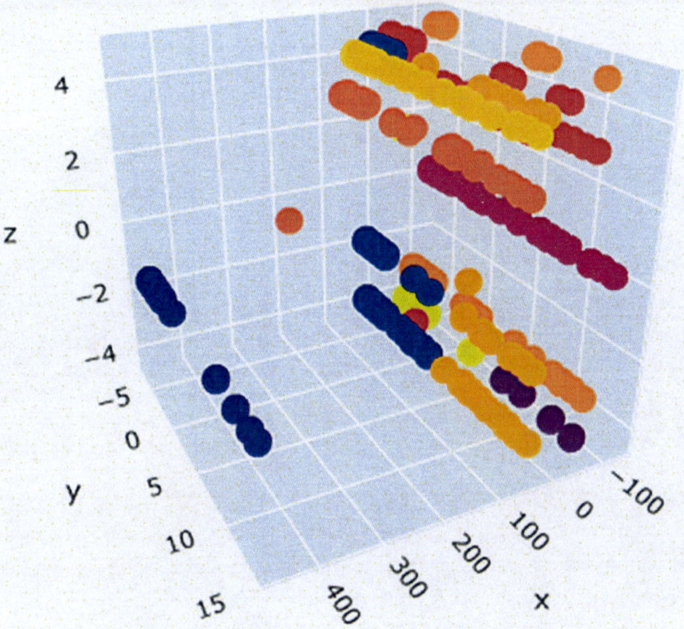

Fig. 7. The clustering result visualization.

Figure 8 shows the result of applying the k-nearest neighbour method to the cluster responsible for the ISIL faction. The blue sector on the diagram right-hand side represents outliers, the orange sector represents the grouping behaviour pattern.

The attacks pattern by the ISIL group is as follows:

- coverage area: Iraq, Syria, Turkey;
- the region regardless, military formations, police, government agencies and businesses are attacked by detonating explosives;
- the region regardless, they attack civilians with firearms.

This method application is considered on a single cluster basis, however, the k-nearest neighbours' method is applied within each cluster to determine each grouping behavioural model.

After applying the method to each cluster and labelling the data into real and outliers, we remove outliers from each cluster. The data rest in the cluster is the terrorist group behavioural model.

7 The Terrorist Groups' Behavioural Models Comparison

After we have removed the outlier data from each cluster, it is necessary to compare the terrorist groups behavioural patterns and determine how similar clusters they share.

Figure 9 shows 10 pie charts representing the clusters that correspond to the terrorist groups in question.

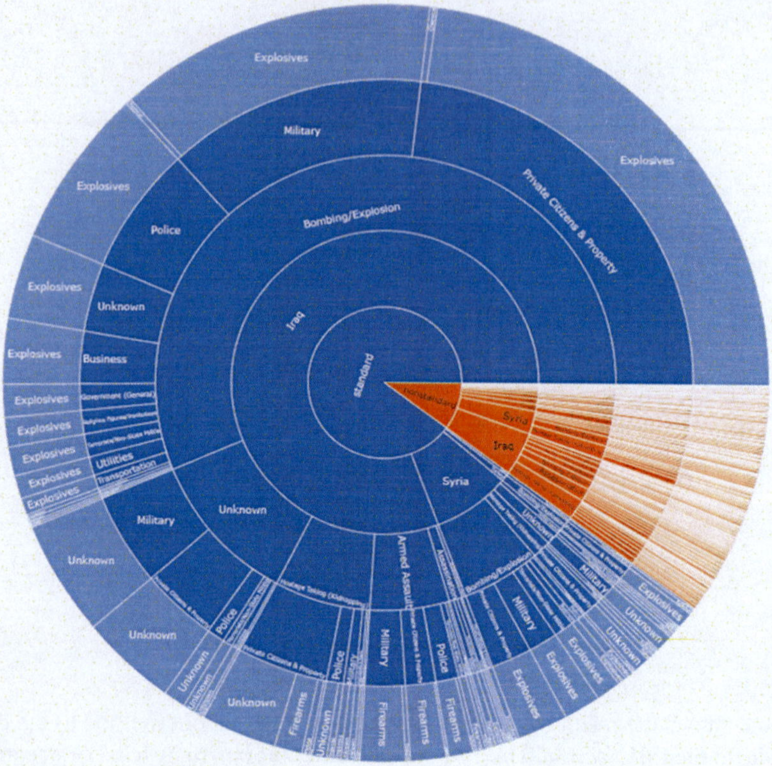

Fig. 8. The result of applying the k-nearest neighbours' method to the ISIL terrorist group cluster (banned in the Russian Federation).

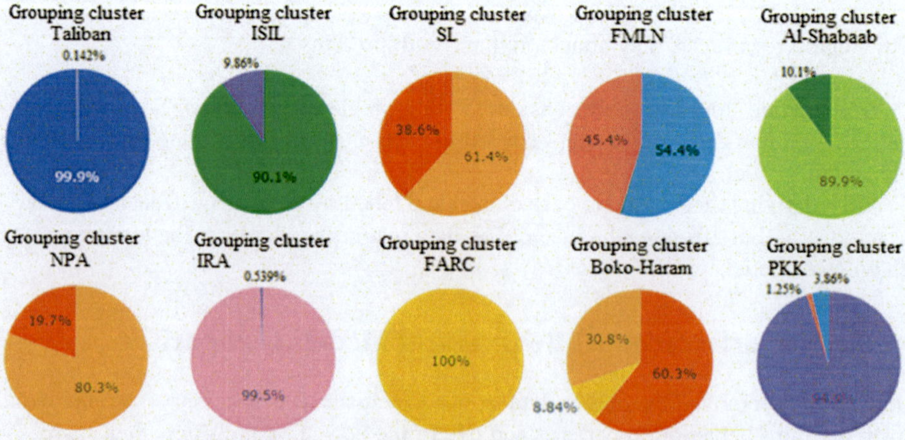

Fig. 9. The terrorist groups' clusters comparis.

Based on the presented diagrams, the following conclusions can be drawn:

- clusters of all groupings except SL, NPA, FMLN and PKK mainly consist of data related to only one grouping. Even if their cluster contains data on other groups, their number is very small. It can be concluded that these groups' responsibility zone is clearly defined and practically does not overlap with other groups zones;
- the SL and NPA groupings share the same cluster. This means that their behaviour patterns and responsibility areas are very similar;
- the FMLN and PKK groupings also share the same cluster. Their pattern and area of responsibility are similar.

8 Conclusion

This article looked at a set of real-world data on terrorist attacks. The system almost period was determined, which is equal to 12 months, and it was also determined that the world terrorism system is in a state of restructuring, after which end, in 2029, an increase in the terrorist attacks monthly number will most likely occur. Using two forecasting methods, the terrorist attacks monthly number a short-term forecast as of October 2018 was realized: for the least-squares method, the figure was 850 monthly attacks, for the autoregressive method it was 990. It was found that the autoregressive model shows a very small error on the data series under study and is better suited for forecasting because it is based on the system development historical data. Also, the data was divided into 10 clusters under the terrorist groups' number represented in the studied data series, the data dimension was reduced to more clearly demonstrate the clustering result. Using the k-nearest neighbours' method, outlier data were identified and eliminated in each cluster, thereby identifying attacks patterns by terrorist groups. The applying this method result was considered on the ISIL group a clustering example (banned in the Russian Federation). All terrorist groups clusters were compared, and a conclusion was drawn about their responsibility areas.

References

1. Couto, G., Pimentel, P., Roias, C., Castanho R.A.: The impact of terrorism on capital markets. Wseas Trans. Bus. Econ. **18**, art. 91, 962–968 (2021)
2. Kuzmin, V.I., Gadzaov, A.F.: methods for constructing models from empirical data. In: Moscow State Institute of Radio Engineering, Electronics and Automation, Technical University, Moscow, p. 94 (2012)
3. Kuzmin, V.I., Gadzaov, A.F.: Technical Analysis. Textbook for the course "Technical Analysis" for students. for ex. 04/09/01 "Informatics and Computer Technology", p. 71. MIREA, MGUPI, Moscow (2015)
4. Kuzmin, V.I., Gadzaov, A.F.: Models and Methods of Scientific and Technical Forecasting. Textbook for students, training. for ex. 04/09/01 "Informatics and Computer Technology", p. 90. MIREA, Moscow (2016)
5. Hasti, T., Tibshirani, R., Friedman, J.: Foundations of Statistical Learning: Data Mining, Inference and Forecasting, 2nd ed. Per. from English LLC "Dialectics", SPb, p. 768 (2020)

6. Spiegelhalter, D.: The Art of Statistics. How to find answers in data. "Mann, Ivanov and Ferber". MYTH Nauchpop (2019)
7. Brunton, S.L., Kutz, J.N.: Data Analysis in Science and Technology. Per. from English Slinkina, A. A, p. 574. DMK Press, Moscow (2021)
8. Autoregressive Models for Forecasting Time Series Using Python. https://www.machinelearningmastery.ru/autoregression-models-time-series-forecasting-python, Accessed 08 Aug 2020
9. Aurelien, G.: Applied Machine Learning with Scikit-Learn and TensorFlow. Per. from English, SPY, OOO "Alfa-kniga", p. 688 (2018)
10. McKinney, W.: Python and Data Analysis, p. 540. DMK Press, Moscow (2020)
11. Reducing the Dimension of Data. https://your-scorpion.ru/log-loss-pca, Accessed 08 Aug 2020

Method for Reconfiguring Kinematic Structure of Modular Robots Using Deep Reinforcement Learning

Dmitry Blinov, Irina Vatamaniuk(✉), and Anton Saveliev

St. Petersburg Federal Research Center of the Russian Academy of Sciences (SPC RAS), St. Petersburg Institute for Informatics and Automation of the Russian Academy of Sciences, 39, 14th Line, 199178 St. Petersburg, Russia
vatamaniuk@iias.spb.su

Abstract. Modular robotic systems are a group of identical robots capable of reconfiguring depending on the assigned tasks. The task of reconfiguring the kinematic structure is the construction of a trajectory connecting the initial and target positions for each module of the system. This paper presents a method for reconfiguring the kinematic structure of a modular robotic system using deep reinforcement learning. This method is based on the deep Q-learning algorithm. In addition, a method for the formation of Q-tables has been developed, which makes it possible to effectively scale the system, as well as to obtain the most complete information about the joint position of the system and transfer it to the neural network, a mechanism for forming the observation vector has been developed. The aim of the training is to build a reward-maximizing control algorithm. To assess the effectiveness of the proposed method, a computer simulation of a modular robotic system consisting of 5, 10 and 15 modules was created. The percentage of test episodes completed without collision was about 94% at 5 modules, 89% at 10 and 82% at 15, and the percentage of successfully completed episodes was about 87%, 64% and 0%, respectively. In addition, in most of the episodes that were completed without a collision, most of the modules were in target positions. In this connection, we can conclude that for the agent was successfully trained to reconfigure 5 and 10 modules, and for 15 a partial solution of the reconfiguration problem was obtained.

Keywords: Modular robotics · Reconfiguration · Kinematic structure · Deep learning · Reinforcement learning · Q-learning

1 Introduction

Modular robotic systems allow to solve many problems using the same set of modules: manipulation, monitoring, research of hard-to-reach and dangerous objects, transportation of goods, construction of maps of the surrounding space [1, 2]. In addition, modular robotic systems have several advantages over classical ones. These are adaptability, increased maintainability, and lower production costs. Thus, failure of one of

R. Silhavy et al. (Eds.): CoMeSySo 2021, LNNS 231, pp. 443–451, 2021.
https://doi.org/10.1007/978-3-030-90321-3_36

the elements of a classical robotic system often entails either partial or complete loss of functionality; in modular robotic systems, if one element is damaged, only the damaged module fails, while the rest of the system continues to work. Reducing the cost of production in comparison with classical robots ensures that there is no need to adjust the production process when changing the tasks for robots [3–5].

A special subspecies of modular robotic systems are self-reconfigurable modular robotic systems. They can independently rebuild depending on the task at hand. For this, each module must have an identical kinematic structure and have a control system that allows the modules to be correctly combined into an integral robotic system [6].

The development of self-configuration methods is a promising direction in the development of modern modular robotics. The ability to independently unite and rebuild into solid robots allows you to abandon human participation completely or partially, which increases the possibilities of using modular robotic systems [7]. However, to date, the reconfiguration of modular robotic systems in non-deterministic environments requires additional research. The main reason for the inefficiency is the low speed of autonomous operation due to the software limitations of the control system of modular robots. Increasing the speed of reconfiguration of the kinematic structure is one of the ways to improve the efficiency of the system. This can be achieved through the development and improvement of methods for reconfiguring the kinematic structure of modular robotic systems, minimizing the search time for a route in the environment and using adaptive self-configuration algorithms [7].

The solution to the problem of reconfiguring the kinematic structure of a modular robotic system can be considered as a classical motion planning problem, considering each module as a separate robot, and applying these planning methods to them [8, 9]. However, this task is complicated by the need to analyze a large amount of data arising from the synchronous operation of many modules. To reduce the amount of data analyzed, you can apply sampling of the environment. Each module in each environment must perform one action at a time [10].

In addition, the problem can be solved by modernizing classical motion planning methods and using machine learning methods. The main methods of machine learning include methods using genetic algorithms, neural network methods and machine learning methods.

Genetic algorithms are based on the application of the mechanisms of natural evolution: mutation, crossing and selection. In [11], the authors used a genetic algorithm to solve the problem of self-configuration of modular robots. Motion planning is viewed as a set of functions that are applied sequentially to modules. In this case, the genetic algorithm determined the order in which these functions should be performed.

Another machine learning method capable of solving the problem of reconfiguring the kinematic structure of modular robotic systems is deep reinforcement learning. Reinforcement learning is a method of machine learning in which the test environment interacts with the environment through feedback, while receiving a reward that shapes the policy of action. The effectiveness of reinforcement learning for robotic control has been demonstrated in [12–14]. With the help of reinforcement learning, the authors have solved the problems of navigation, manipulation, and planning of movement.

Application of Q-learning is also applicable to solving the problem of reconfiguring modular robotic systems. So, in [15], the authors proposed a method for reconfiguring modular robotic systems using Q-learning. In this paper, an algorithm was presented that allows you to effectively train an agent to move a different number of modules. However, the results obtained show that this method is poorly scalable and needs to be modernized. So, when testing it on modular robotic systems consisting of 5, 10 and 15 modules, the training results showed successful occupation of a position according to the target configuration in 704, 32 and 0 cases out of 2000, which indicates poor scalability of this method.

The aim of this work is to improve the efficiency of the reconfiguration of the kinematic structure of modular robotic systems. Even though there are works confirming the effectiveness of reinforcement learning for solving various control problems, its use for solving the problem of planning the movement of modular robotic systems is a little-studied area. In those studies that use reinforcement learning to solve the problem of reconfiguring modular robotic systems, it has shown weak results and poorly poor scalability. Therefore, the goal of this work is also to improve the efficiency of reinforcement learning by using deep Q-learning and developing mechanisms to increase the scalability of the system. This work is organized as follows: the second section describes the general provisions of the Q-learning algorithm; the third section contains a description of the methods used to improve the efficiency of Q-learning; the fourth section shows the results of testing the developed method for reconfiguring a modular robotic system consisting of a different number of modules.

2 Materials and Methods. General Approach for Solving the Problem of Reconfiguring the Kinematic Structure of Modular Robots Using Deep Reinforcement Learning

Reinforcement learning is a method of machine learning, during which some test agent (agent) interacts with some environment (environment) through feedback (Fig. 1). In this case, the agent interacts with the environment through actions (action), and the environment, in turn, sends the agent reactions to these actions - observations (environment states) and reward.

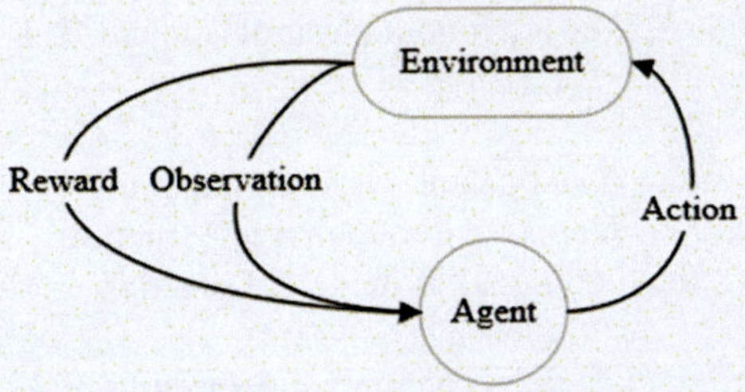

Fig. 1. Interaction of agent and environment in reinforcement learning

Using reinforcement learning, it is necessary to find the optimal strategy for the agent's interaction with the environment so that at an arbitrary time moment t the agent, being in the state $s \in S$, chooses an action $a \in A(s_t)$ that would allow him to receive the maximum reward $R = \sum_{t=1}^{n} r_t$, where r_t is the reward after the action is taken a.

Since changes in the state of the environment, it is necessary to use a discrete action space, the most rational method to use to solve the problem is Q-learning. Q-training is a method based on training an agent to predict his behavior in a specific environment. Q-learning is one of the most effective types of reinforcement learning. Based on the reward received from the environment, the agent forms a Q-function. This function sets the agent's assessment of the reward that he can receive for his action. The learning process consists in refining the value of the Q-function at each step of the agent to obtain the maximum reward.

In deep Q-learning, the role of the Q-function is played by an artificial neural network, which is trained on small data packets during the agent's work. The expediency of using neural networks instead of Q-tables is used in cases where it is impossible to enumerate all possible states of the environment.

To solve the problem of reconfiguring the kinematic structure of modular robotic systems, let us define a set of actions $a_1, a_2, ..., a_n$ for each module of the robotic system, allowing it to move from the point of its initial position (x_0, y_0) to point $(x_k, y_k) \in K$, where K is a given configuration, receiving the maximum possible reward R.

To successfully complete the reconfiguration of a modular robotic system, it is necessary to ensure that the modules reach the target configuration, avoiding collisions with other modules and the boundaries of the workspace. In this work, a discrete set of 5 actions is used as a space of actions, four of which correspond to moving the module one cell to each side, and one to maintaining the current position. Since several elements interact with each other in the workspace, a 1D (one-dimensional vector) of length l is received as an action at each step, where l is the number of elements interacting in the workspace.

The goal of the agent is to maximize the reward function so that, because of training, the modules reach the target position along the shortest distance. In addition, it should be considered that the agent needs to interact with several elements at the same time; when forming the reward function, it is necessary to consider that the reward should be evenly distributed among the elements. In addition, the agent needs to be rewarded for completing global tasks.

To solve these problems, a reward function was developed. This function is calculated as follows:

1. Reward equal to 1 for each step taken without collision
2. Reward equal to 1 for each approach to the target structure
3. Reward equal to 2 if the module takes the target position

The total reward for an episode is calculated as the sum of all rewards received for each step.

To improve the efficiency of the Q-learning algorithm when solving, a new method of forming a Q-table for the possibility of scaling was proposed and a normalized vector of observations was used.

3 Methods to Improve the Effectiveness of Deep Q-Learning

3.1 Normalized Observation Vector

The kinematic structure of modular robotic systems in a certain workspace of width w and length l can be converted into a 2D matrix, in which the module is represented as the number "k" $k = i + 2$, where i is the serial number of the module, the target configuration is represented as the number "1", and the empty space - as "0", moreover, if the module reaches the target configuration, then its representation in the matrix changes as "-k". The transformation of the workspace 2a, length 10, width 10, in which 5 modules interact, is shown in Fig. 2b.

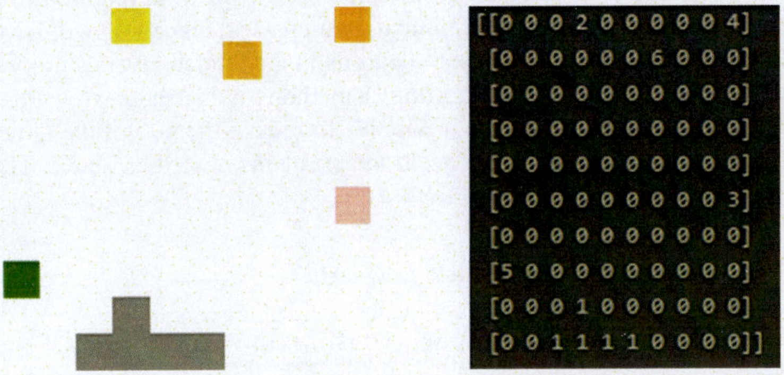

Fig. 2. Workspace (a), converting workspace to matrix (b)

The matrix way of describing the workspace (environment) allows you to determine the vector of observations. To determine the observation vector, it is necessary to transform the 2D matrix 1D vector and normalize it by dividing each of its elements by $n + 1$, where n is the number of modules in the workspace. This method of forming observations performs two important functions: it provides the agent with the most detailed description of the environment and the mutual position of the modules relative to each other, and normalizes the input vector, which has a positive effect on training.

4 A New Method of Forming a Scalable Q-Tables

In this paper, the Q-learning algorithm has been improved to be scalable. This makes it possible to train an agent who can interact with different sets of modules. For this, a scalable Q-table method was developed. During the learning process, the agent predicts what actions he needs to take to achieve the maximum reward. Using a neural network, the algorithm generates a Q-table - a table containing the values of expected rewards, in the process of performing an action. In a standard deep reinforcement learning agent, a Q-table is a vector, where each possible action has its own value max $Q(a)$. However, in this case, this approach is not effective. To control each module of the system, 5 actions are enough. Therefore, it is sufficient to require a vector (table) of actions with a length equal to the number of modules multiplied by the number of actions, in which each action will correspond to the movement of one of the modules, while maintaining their positions of all the others. However, this option for solving the problem is not effective, since it significantly increases the number of steps to complete the task. Another option for using the standard approach to the formation of a Q-table is to use a vector of length 5^l, l is the number of modules, where all possible actions that are possible because of the interaction of modules are combined with each other. This brings several problems: poor scalability, for example, the number of elements in the Q-table will be 5^5, and with $l = 10$ will be 5^{10}, and therefore the total number of calculations increases, which negatively affects the learning rate, especially with an increase in the number of modules. In addition, when using deep learning, in which the vector 5^l is required to predict a given Q-table, the load will also be significantly increased, since you will have to fight the "bottleneck" effect and use at least 5^l neurons in the hidden layers.

To improve the efficiency of the algorithm, a method for distributing rewards between modules was developed, which made it possible to reduce the size of the Q-table to $5 \cdot l$ elements, which made it possible to avoid the problems described above. The idea of constructing a Q-table is presented in Table 1.

Table 1. Q-table

	a1	a2	a3	a4	a5
Module 1	Q_{11}	Q_{12}	Q_{13}	Q_{14}	Q_{15}
Module 2	Q_{21}	Q_{22}	Q_{23}	Q_{24}	Q_{25}
…					
Module *n*	Q_{n1}	Q_{n2}	Q_{n3}	Q_{n4}	Q_{n5}

In this table, rewards can only be added to one element in each row, thereby combining possible actions.

5 Test Results

To test the proposed method, an environment was created that allows visualizing and testing the kinematic structures of modular robotic systems with a different set of modules. So, in each new episode, a certain target area is randomly created, and the modules are given random positions. After that, the agent is allowed to move the modules so that they completely fill the target area. The developed method was tested in such an environment for a configuration of 5, 10 and 15 modules. The software implementation was performed in the python programming language using the tkinter library for visualization and the keras neural network library for training the neural network. Examples of graphical visualization of the environment for each set of modules are shown in Fig. 3.

As a result of training, graphs of the sums of the average remuneration of each module received for every fifty episodes were obtained. These graphs are the main characteristic of the effectiveness of reinforcement learning. The resulting graphs are shown in Fig. 4.

(a) (b) (c)

Fig. 3. An example of the initial position of the operating environment of a modular robotic system, consisting of 5 (a), 10 (b) and 15 (c) modules

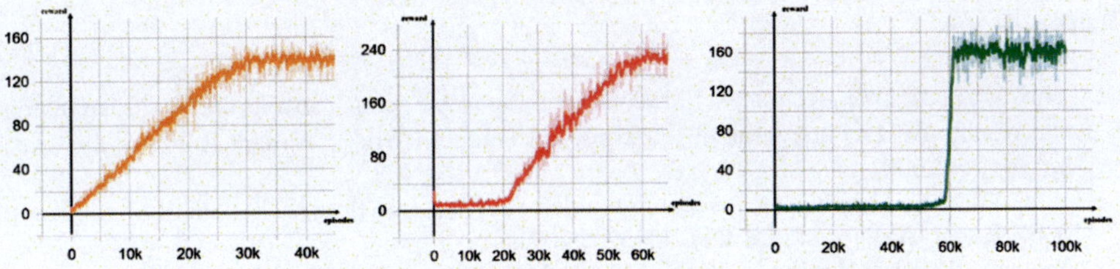

(a) (b) (c)

Fig. 4. Graphs of the sums of the average reward of modules

Based on the graphs obtained, convergence of reward is achieved in each of the environments, which indicates the completion of training. So, in the first experiment, convergence is achieved at 31000 episodes and reaches a value of about 140, which corresponds to the expected maximum reward. In the second experiment, convergence is achieved at 62000 episodes and reaches 230. In the third, convergence is achieved at 61000 episodes and reaches 160, which is less than the expected reward.

In addition, each of the agents after training was tested on a control dataset. The percentage of test episodes completed without collision was about 94% with 5 modules, 89% with 10 and 82% with 15, and the percentage of successfully completed episodes was about 87%, 64% and 0%, respectively. In addition, in most of the episodes that were completed without a collision, most of the modules were in target positions. In this connection, we can conclude that for the agent was successfully trained to reconfigure 5 and 10 modules, and for 15 a partial solution of the reconfiguration problem was obtained.

6 Conclusion

In this paper, deep reinforcement learning has proven to be effective in training an agent to reconfigure 5 and 10 modules. However, when increasing the modules to 15, the agent, although he achieves the convergence of the reward, he fails to achieve such a level of reward, which ensured the fulfillment of the task assigned to him. Presumably, this is due to insufficient memory of the agent, or to the number of neurons in the hidden layers of the artificial neural network. It is possible to fix this problem by adding mechanisms to increase the amount of memory and the number of neurons in hidden layers.

A method for reconfiguring the kinematic structure of modular robotic systems using deep reinforcement learning is presented. Special state and action spaces have been developed that most effectively allow the deep learning agent to interact with the kinematic structure of the modular robotic system. The method was based on the Q-training algorithm, which showed efficiency for training agents in discrete environments. The algorithm is adapted to work with a scalable action space and can receive rewards for several elements of the system. This method has been tested on three environments, consisting of 5, 10 and 15 modules. The results obtained indicate the scalability of the algorithm.

References

1. Yim, M., Duff, D.G., Roufas, K.: Modular reconfigurable robots, an approach to urban search and rescue. In: 1st International Workshop on Human-friendly Welfare Robotics Systems, pp. 69–76 (2000)
2. Yim, M., et al.: Modular self-reconfigurable robot systems [grand challenges of robotics]. IEEE Robot. Autom. Mag. **14**(1), 43–52 (2007)
3. Yim, M., White, P., Park, M., Sastra, J.: Modular self-reconfigurable robots. In: Encyclopedia of Complexity and Systems Science, pp. 5618–5631 (2009)
4. Feczko, J., Manka, M., Krol, P., Giergiel, M., Uhl, T., Pietrzyk, A.: Review of the modular self reconfigurable robotic systems. In: 2015 10th International Workshop on Robot Motion and Control (RoMoCo), pp. 182–187 (2015)
5. Yim, M., Duff, D.G., Roufas, K.D.: PolyBot: a modular reconfigurable robot. In: Proceedings 2000 ICRA. Millennium Conference. IEEE International Conference on Robotics and Automation. Symposia Proceedings, pp. 514–520. IEEE (2000)
6. Mezenceva, O.S., Petrenko, V.I., Zhilina, E., Pavlov, A.S., Apurin, A.A.: Developing a concept of available multi-functional modular robot for education and research. In: CEUR Workshop Proceedings SLET (2019)
7. Stoy, K., Brandt, D., Christensen, D.J., Brandt, D.: Self-reconfigurable Robots: An Introduction. The MIT Press, London (2010)
8. Kovács, G., Yusupova, N., Smetanina, O., Rassadnikova, E.: Methods and algorithms to solve the vehicle routing problem with time windows and further conditions. Pollack Periodica **13**(1), 65–76 (2018)
9. Lynch, K.M., Park, F.C.: Modern Robotics. Cambridge University Press, Cambridge (2017)
10. Liu, J., Zhang, X., Hao, G.: Survey on research and development of reconfigurable modular robots. Adv. Mech. Eng. **8**(8), 1–21 (2016)
11. Ababsa, T., Djedl, N., Duthen, Y.: Genetic programming-based self-reconfiguration planning for metamorphic robot. Int. J. Autom. Comput. **15**(4), 431–442 (2017). https://doi.org/10.1007/s11633-016-1049-4
12. Nguyen, T.L., Nguyen, D.V., Le, T.H.: Reinforcement learning based navigation with semantic knowledge of indoor environments. In: 2019 11th International Conference on Knowledge and Systems Engineering (KSE), pp. 1–7. IEEE (2019)
13. Katyal, K., Wang, I., Burlina, P.: Leveraging deep reinforcement learning for reaching robotic tasks. In: Proceedings of the IEEE Conference on Computer Vision and Pattern Recognition Workshops, pp. 18–19 (2017)

14. Han, Y., Guo, Y., Mi, Z., Yang, Y.: Robot path planning in dynamic environments based on deep reinforcement learning. In: Ning, H. (ed.) CyberDI/CyberLife -2019. CCIS, vol. 1137, pp. 265–283. Springer, Singapore (2019). https://doi.org/10.1007/978-981-15-1922-2_19
15. Petrenko, V.I., Tebuyeva, F.B., Pavlov, A.S., Gurchinskiy, M.M.: Method for reconfiguring the kinematic structure of a mechatronic-modular robot in non-deterministic conditions. Syst. Eng. Inf. Technol. **2**(4), 57–56 (2020). (in Russian)

On the Infinite-Dimensional Model Multiple-Parameter Estimation Using Feedback-Relay Identification Test

Libor Pekař[1,2](✉), František Zezulka[1], and Petr Dostálek[2]

[1] College of Polytechnics Jihlava, Tolstého 1556, 586 01 Jihlava, Czech Republic
libor.pekar@vspj.cz

[2] Tomas Bata University in Zlín, nám. T. G. Masaryka 5555, 76001 Zlín, Czech Republic

Abstract. The objective of this contribution is twofold. First, it demonstrates a case study on applying the standard single-run relay-feedback parameter identification test to a representative of infinite-dimensional systems. Namely, a delayed mathematical model of a circuit heating laboratory appliance process is used. Second, an initial estimation of the model parameters is done via the parameter identification of another – simpler – model. The transition between these two models adopts the idea of dominant spectrum assignment that is solved by using the well-established Levenberg-Marquardt algorithm. Finally, the remaining model parameters are estimated by solving another nonlinear optimization problem in the frequency domains. As transfer function denominator parameters are set independently to the numerator ones, the proposed technique significantly reduces the number of additional relay experiments. Numerical results indicate that the method needs improvements regarding time-response as well as frequency-response accuracy.

Keywords: Relay-based identification · Infinite-dimensional model · Initial estimation · Pole assignment · Levenberg-Marquardt algorithm

1 Introduction

One of the very popular frameworks for process model parameters identification is based on the closed-loop experiment with a relay placed instead of the controller [1]. It was implemented in practice, especially in (but not limited to) the process and chemical industry [2]. This original test used the on/off (or ideal) relay with two-level symmetrical output without hysteresis. As the experiment output, one (ultimate) point of the process model frequency response is estimated via the so-called describing function (DF), which gives rise to the determination of only two unknown parameters. Hence, dozens of enhanced techniques and methods have been developed [3] since the groundbreaking work. For instance, time-domain [4] and frequency-domain [5] data evaluation have been investigated, various relay types have been applied [6, 7], multiple tests have enabled to provide more information about the process dynamics [8], or even these extended data have been extracted from the single test [9].

R. Silhavy et al. (Eds.): CoMeSySo 2021, LNNS 231, pp. 452–464, 2021.
https://doi.org/10.1007/978-3-030-90321-3_37

Infinite-dimensional models (IDMs) are characterized by the infinite number of impulse-response modes, or equivalently, by the infinite number of state-matrix eigenvalues that correspond to the solution of the characteristic function. Let us call these eigenvalues poles for simplicity. IDMs can inherently describe dynamics of complex high-order processes, even if the model dynamics is of a low order (measured by order of derivatives). Time-delay models (TDMs) constitute typical representatives of the IDMs family [10]. They comprise delays in dynamics (i.e., the so-called internal delay) besides the habitual delay in the input-output (IO) relation. Unfortunately, according to the authors' best knowledge, there have been published only a limited number of results on relay-based identification of TDMs. A simple four-parameter model (SFPM) was identified using the standard relay and a priori knowledge of the IO delay [11]. The Fourier transform and a relay with saturation [7] were applied in [12] to identify parameters of a five-parameter TDM representing a simplified model of a circuit heating laboratory process. However, a more precise heating process model (HPM) based on the use of heat and mass balance equations included nine parameters [13] (in a selected single-input single-output relation). The method of moments applied to the series of the SFPM and a high-order finite-dimensional non-periodic subsystem can be found in [14].

Although the SFPM is assumed to be a "universal" model, it cannot estimate a complex system dynamics. However, one can take it as the first-attempt approach to modeling processes with unknown dynamics. Its advantage is all of its four parameters can be detected under a single relay experiment. Two parameters can be obtained from the ultimate cycle data and the corresponding DF, while two remaining ones can be read from the shape of the limit cycles and its numerical integration (or from the process step response). Contrariwise, the HPM or another multi-parameter TDM requires either a multiple relay test (which is time-consumptive) or to solve a computationally demanding nonlinear optimization problem that usually arises from an advanced single-relay test.

This contribution is primarily focused on applying the standard relay test to the nine-parameter HPM under a single run. This goal can be achieved via a priori performance of identifying the SFPM, the dynamics of which is given by its poles. These poles are then attempted to be identical to the poles of the HPM, which yields the initial guess of the characteristic function parameters. To solve this partial task, the Levenberg-Marquardt algorithm (LMA) is used. By adopting the IO delay value estimation technique and the computation of the steady state from the relay test, there are only two remaining HPM parameters that do not depend on model poles. Hence, these remaining parameters are then estimated via the standard test using the knowledge of the DF. This last task can be formulated using a relatively simple nonlinear problem in the frequency domain.

The rest of the paper is organized as follows. Section 2 introduces the problem in question in more detail and provides the reader with the HPM and the standard relay identification test via the DF. In Sect. 3, the SFPM and its basic spectral properties are given first. Then, the LMA – as the crucial numerical tool – is described. The proposed method for the identification of the HPM using the SFPM and the DF is eventually summarized, and its algorithm is provided. A numerical example that performs all the steps of the proposed method is presented in Sect. 4. A comparison of the actual and estimated HPM is given in the time domain and the frequency domain. Finally, Sect. 5 concludes this study.

2 Problem Statement

The framework problem is to apply the standard feedback relay identification test on the nine-parameter HPM under a single run. The estimation of the characteristic quasipolynomial parameters constitutes a particular research question that serves to solve the given problem. The characteristic function solutions are simply the characteristic quasipolynomial zeros, i.e., system poles. This initial guess enables to determine the remaining HPM parameters with a lower effort. Hence, the introduction of HPM and the standard relay [1] test follows.

2.1 Infinite-Dimensional Model of the Thermal Process

The HPM, i.e., a TDM of the laboratory circuit heating process, derived via heat and mass balance laws when considering all the significant process delays, have been published in [13]. A photo of the laboratory appliance is displayed in Fig. 1. The meaning of numbered positions follows.

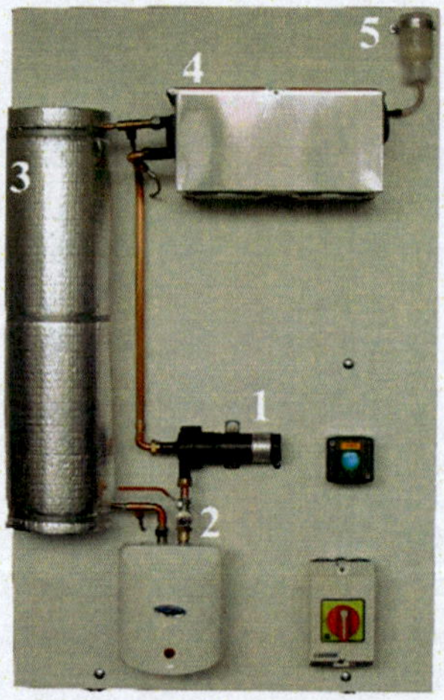

Fig. 1. A photo of the laboratory circuit heating appliance (numbered positions are described in the text body).

The distilled water is driven via a continuously controllable pump {1} through a flow heater {2} to a long insulated coiled pipeline {3}. Then, the heating fluid enters a cooler (air-water plate-and-fin heat exchanger, radiator) {4} with two cooling fans and flows back to the pump. An expansion tank serving for the compensation of the water thermal expansion effect is placed at the top of the appliance {5}.

The manipulated inputs (or measurable disturbances if needed) are the pump input voltage, the heater power, and the input voltage to cooler fans. The outputs are represented by measured outlet temperature from the heater, inlet temperature to the cooler, and the outlet temperature from the cooler.

The eventual model is quite complex, nonlinear, and multivariable. Let us consider the single-input single-output submodel where the heater power is taken as the model input $u(t)$, while the outlet temperature from the heater represents the output $y(t)$. The linearized relation between $u(t)$ and $y(t)$ (in a particular operating point) can be expressed by the delay-differential equation

$$\dddot{y}(t) + a_2\ddot{y}(t) + a_1\dot{y}(t) + a_0 y(t) + a_{0D} y(t-\vartheta) = b_0 u(t-\tau) + b_{0D} u(t-\tau-\tau_0) \tag{1}$$

where $a_0, a_{0D}, a_1, a_2, b_0, b_{0D}$ are real-valued non-delay parameters, ϑ, τ, τ_0 mean the internal (state) delay and delays in the IO relation, respectively, and the dot notation is used for the time derivative. Delay values are naturally non-negative.

Relation (1) is equivalent to the transfer function

$$G_{HPM}(s) = \frac{Y(s)}{U(s)} = \frac{B(s)}{A(s)} = \frac{b_0 + b_{0D}\mathrm{e}^{-\tau_0 s}}{s^3 + a_2 s^2 + a_1 s + a_0 + a_{0D}\mathrm{e}^{-\vartheta s}}\mathrm{e}^{-\tau s} \tag{2}$$

where s is the Laplace transform variable and the denominator quasipolynomial of (2) represents the characteristic quasipolynomial, the roots of which agree with system poles (characteristic values). The number of poles is infinite (i.e., the model is infinite-dimensional) whenever $\vartheta \neq 0$.

A complex two-step parameter identification procedure have been made in [13]. First, non-dynamic parameters have been determined from the measurement of suitable static characteristics. Second, remaining parameters depended on the system dynamics have been identified from step responses. The eventual parameters of (2) have been found as follows.

$$b_0 = -2.50 \cdot 10^{-7}, b_{0D} = 2.27 \cdot 10^{-6}, a_0 = 1.30 \cdot 10^{-4}, a_{0D} = -7.22 \cdot 10^{-5}, a_1 = 8.51 \cdot 10^{-3},$$
$$a_2 = 0.17, \tau = 141, \tau_0 = 1.5, \vartheta = 151 \tag{3}$$

Let us take result (3) are the actual ("true") values for further comparisons.

2.2 Relay-Based Identification Using Describing Function

The relay identification test [1] s the feedback scheme depicted in Fig. 2. If the process is stabilizable by the nonlinear element (relay), the initially excited output reaches sustained oscillations of the constant ultimate amplitude A and period T_u, see Fig. 3 (according to [12]). Process input has the same period yet a different amplitude B. The overall phase lag between $e(t)$ and $y(t)$ is $-\pi$; however, the phase leg between $u(t)$ and $y(t)$ can be within the range $[-\pi, 0)$ (depending on the used relay element).

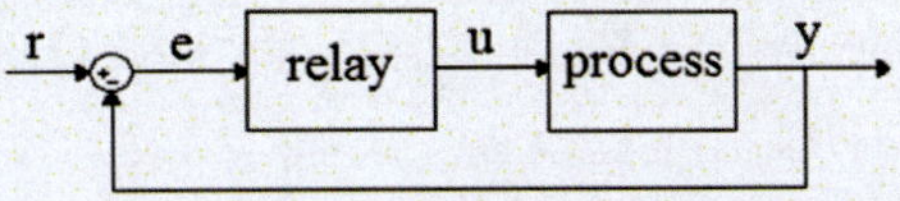

Fig. 2. Relay feedback test scheme.

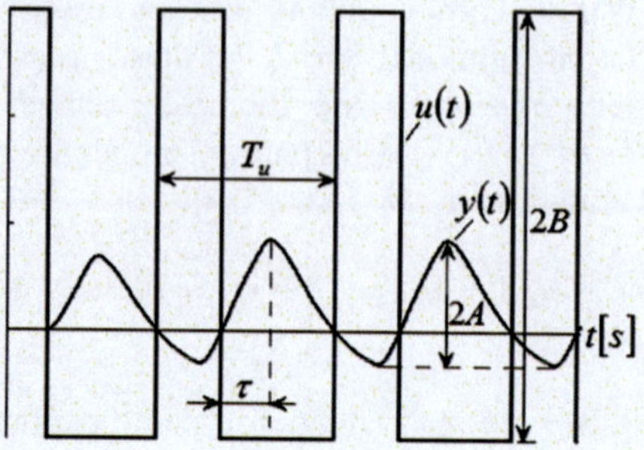

Fig. 3. Sustained oscillations data.

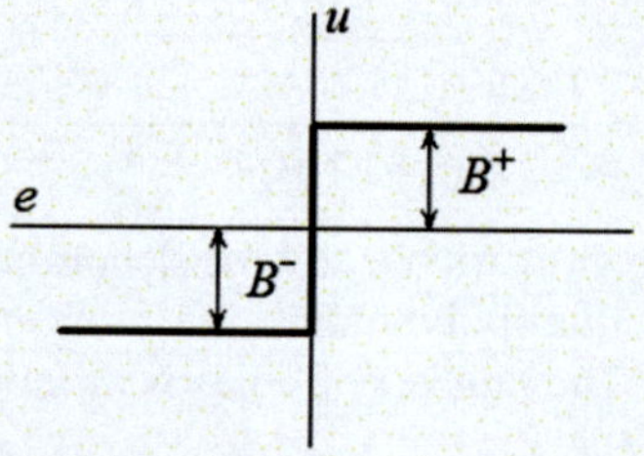

Fig. 4. Asymmetric relay static characteristics [12].

The static characteristics of the asymmetric on/off relay (without hysteresis) can be seen in Fig. 4. The ideal relay satisfies $B = B^+ = B^-$.

The concept of DF utilizes an approximation of the nonlinear relay behavior, usually via truncation of the Fourier series expansion [15]. The DF of the relay specified by Fig. 4 has the form [16]:

$$k_u = \frac{4B}{\pi A}\sqrt{1 - \left(\frac{\delta}{A}\right)^2} \tag{4}$$

where $\delta = 0.5\left|B^+ - B^-\right|$. Hence, for the ideal relay, one has $\delta = 0$. It holds for the sustained oscillations that

$$k_u G(\mathrm{j}\omega_u) = -1 \Leftrightarrow \begin{cases} |k_u G(\mathrm{j}\omega_u)| = 1 \\ \measuredangle(k_u G(\mathrm{j}\omega_u)) = -\pi \end{cases} \tag{5}$$

where the ultimate frequency $\omega_u = 2\pi/T_u$, $G(\mathrm{j}\omega)$ stands for the process (model) frequency transfer function, and $\mathrm{j}^2 = -1$, which identifies the critical point -1 of the open-loop Nyquist plot.

Formula (5) can identify only two process-model parameters. However, the IO delay τ can be obtained graphically from Fig. 4. Besides, the model static gain $k = G(0)$ can be computed from

$$k = \frac{\int_t^{t+T_u} y(\xi)\mathrm{d}\xi}{\int_t^{t+T_u} u(\xi)\mathrm{d}\xi} \tag{6}$$

when using the asymmetric relay (i.e., $\delta \neq 0$), for a sufficiently high t when the oscillations are settled [7, 11]. It is assumed that the measured signal is not affected by dc components, such as static load disturbance. Note that the value of k can also be simply determined from the step response.

To sum up the problem in question, the goal is to use the ideal relay (4) to identify parameters of model (2) via the solution of the condition (5) under a single feedback experiment. Besides, the static gain and the IO delay are estimated using (5) and Fig. 4, respectively. Hence, there remain five parameters to be determined in the model, which gives rise to the main research question.

3 Proposed Technique

3.1 Simple Infinite-Dimensional Model

The leading idea of how to solve the given research question is to identify a simpler TDM (i.e., the SFPM) and then adopt its dynamics (spectrum) to the HPM (2). The SFPM is governed by the transfer function

$$G_{SFPM}(s) = \frac{b_{0,m}}{s + a_{0D,m}e^{-\vartheta_m s}} e^{-\tau_m s} \tag{7}$$

As the static gain of (7) (i.e., $k_m = b_{0,m}/a_{0D,m}$) and τ_m can be obtained as described in Subsect. 2.2, the remaining two parameters can be determined by the single relay test. When applying (5) on model (7), one gets [11]:

$$\vartheta_m = \frac{\pi - \cos^{-1}(k_u k_m \cos(\omega_u \tau_m))}{\omega_u}, \quad a_{0D,m} = \frac{\omega_u}{\sin(\omega_u \vartheta_m) - \tan(\omega_u \tau_m)\cos(\omega_u \vartheta_m)}$$
$$\sin(\omega_u \vartheta_m) - \tan(\omega_u \tau_m)\cos(\omega_u \vartheta_m) \neq 0 \tag{8}$$

Remark 1. A disadvantage of the analytic solution (8) is that it inherently supposes that $k_u k_m \cos(\omega_u \tau_m) \in [-1, 1]$. If this condition does not hold, formulae (8) cannot be used, and a numerical solution of (5) has to be used instead.

The model is asymptotically stable, i.e., its denominator has all its roots in the left-half Gauss plane, if and only if $0 < a_{0D,m}\vartheta_m < \pi/2$. Moreover, the rightmost pole is real whenever $0 < a_{0D,m}\vartheta_m \leq 1/\mathrm{e}$. Otherwise, there is a complex conjugate pair of poles [8, 11, 17].

3.2 Levenberg-Marquardt Algorithm

The LMA can solve a set of n nonlinear algebraic equations by minimizing the corresponding residual sum of squares [18]. Namely, let the set be

$$\mathbf{f}(x_i, \mathbf{p}) = \mathbf{0} \tag{9}$$

where $\mathbf{f} = (f_1, f_2, \ldots f_n)^T$, $\mathbf{0} = (0, 0, \ldots 0)^T$, x_i is a function variable value, and $\mathbf{p} \in \mathbb{R}^m$ stands the parameter set to be found. Superscript T denotes the matrix transpose. Then, the sum

$$\sum_{j=1}^{n} \left(f_j(x_i, \mathbf{p})\right)^2 \tag{10}$$

is minimized via the solution formula

$$\mathbf{p}^{(k+1)} = \mathbf{p}^{(k)} - \left(\left[\mathbf{J}\left(\mathbf{f}\left(x_i, \mathbf{p}^{(k)}\right)\right)\right]^T \mathbf{J}\left(\mathbf{f}\left(x_i, \mathbf{p}^{(k)}\right)\right) + \lambda^{(k)}\mathbf{I}\right)^{-1} \left[\mathbf{J}\left(\mathbf{f}\left(x_i, \mathbf{p}^{(k)}\right)\right)\right]^T \mathbf{f}\left(x_i, \mathbf{p}^{(k)}\right) \tag{11}$$

where the superscript (k) indicates the iterative step, $\mathbf{J}$ stands for the Jacobian with respect to $\mathbf{p}$, and $\mathbf{I}$ is the identity matrix of size $m \times m$. Factor λ is decreased when (10) decreases, and vice versa.

The algorithm is used twice in this study. First, it serves for the computational setting of denominator parameters of (2) such that dominant poles of (7) coincide with those of (2). Second, once the parameters of the characteristic quasipolynomial are set, numerator parameters of (2) are estimated via the solution of (5). Note that the initial setting of $\mathbf{p}^{(0)}$ is discussed in Subsects. 4.2 and 4.3.

3.3 Technique Summary

All steps of the proposed technique for single-run parameter identification of the HPM (or, generally, a multiple-parameter TDM) can be summarized by the following algorithm.

Algorithm 1. (Single-run relay identification test for the HPM using dominant pole placement based on the SFPM).

Input: HPM, SFPM with undetermined parameters.
Output: Parameters estimation of the HPM.

1: Set $0 < \delta \ll 1$, $B > 0$, and perform the feedback relay experiment as in Fig. 2 until sustained oscillations appear. Read T_u, A.
2: Calculate ω_u, k_u (4), and $k_m = k$ (5). Estimate $\tau_m = \tau$ (Fig. 3).
3: Substitute $b_{0,m} = a_{0D,m} k_m$ in (7) and solve (5) for G_{SFPM} (7) using (8) => $a_{0D,m}, \vartheta_m$.
4: If $a_{0D,m}, \vartheta_m \notin \mathbb{R}$, reset $a_{0D,m}, \vartheta_m$ and go to step 5; else, go to step 7.
5: Set $\lambda^{(0)}$ and the initial denominator parameters of (7).
6: Use the LMA (11) to solve (5) for G_{SFPM} (2) => $a_{0D,m}, \vartheta_m$.
7: Set $\lambda^{(0)}$ and the initial denominator parameters of (2).
8: Use the LMA (11) to set poles of (2) identical to those of (7) => $a_2, a_1, a_0, a_{0D}, \vartheta$.
9: Substitute $b_0 = (a_0 + a_{0D})k - b_{0D}$ in (2).
10: Set $\lambda^{(0)}$ and the initial numerator parameters b_{0D}, τ_0 of (2).
11: Use the LMA (11) to solve (5) for G_{HPM} (2) => b_0, b_{0D}, τ_0.

A numerical example demonstrating all the algorithm steps follows.

4 Numerical Results

Consider model (2)–(3) as the exact heat-process model for this example purposes. The example provides the reader with particular numerical results of Algorithm 1.

4.1 Simple Model Parameter Identification

Let $\delta = 0.05, B = 100$ (note that practical issues of this setting are concisely discussed in Subsect. 4.4). The relay-feedback test yields the sustained oscillations with the following ultimate data: $A = 0.963$, $T_u = 363.4$ s, i.e., $\omega_u = 1.73 \cdot 10^{-2}$ rad/s. Then, $k_u = 132.04$ according to (4). It can be estimated from Fig. 2 that $\tau = \tau_m \approx 136.5$ s and the computation of (6) gives $k = k_m = b_{0D,m}/a_{0D,m} = 3.23 \cdot 10^{-2}$.

The attempt to calculate (8) results in $k_u k_m \cos(\omega_u \tau_m) = -3.013$, which (unfortunately) unable to get real-valued SFPM parameters, see Remark 1. Therefore, the LMA is used to determine $a_{0D,m}$, ϑ_m.

Let $\lambda^{(0)} = 1$ and the initial parameters' estimation be $a_{0D,m} = \omega_u = 1.73 \cdot 10^{-2}$, $\vartheta_m = 0.25\pi/a_{0D,m} = 45.43$. It comes from the assumption that $1/a_{0D,m}$ is close to the process (natural) time constant (for the delay-free system), and the selected value of ϑ_m means the "center of stability region". Factor λ is multiplied or divided by 5 in every single iteration step, according to the change of the residual sum (10). The use of the LMA then results in

$$a_{0D,m} = 4.776 \cdot 10^{-3},\ \vartheta_m = 6.828\,\text{s} \tag{12}$$

and (10) for the nonlinear equation set (5) (with $G(s) = G_{SFPM}(s)$) eventually has the error value of 0.2392.

4.2 Thermal Process Model Characteristic Quasipolynomial Estimation

The SFPM rightmost poles read

$$\begin{aligned} &s_1 = -0.00494, s_2 = -0.73822, s_{3,4} = -0.82529 \pm 1.05290\text{j}, \\ &s_{5,6} = -0.89818 \pm 2.00889\text{j}, s_{7,8} = -0.94808 \pm 2.94507\text{j} \end{aligned} \tag{13}$$

The leftmost two real poles and the first conjugate pair are set as poles of (2) by solving the set of equations $A(s_i) = 0, i = 1, 2$, and $\text{Re}A(s_3) = \text{Im}A(s_3) = 0$, where $A(s)$ is the denominator (i.e., the characteristic quasipolynomial) of (2). The set is solved by the LMA again.

By comparing denominators of (2) and (7), the ideal setting reads $a_2 = a_0 \to 0, a_1 \to \infty, a_{0D} \to a_1 \cdot a_{0D,m}, \vartheta = \vartheta_m$ or $a_0 = a_2 \to 0, a_1 \to a_{0D}/a_{0D,m}, a_{0D} \to \infty, \vartheta = \vartheta_m$, which is, however, non-feasible. Among various other initial settings, two interesting results with a low minimum of (10) have been obtained

$$\begin{aligned} &1:\ a_2 = 65.495, a_1 = 2093.747, a_0 = -6.854, a_{0D} = 10, \vartheta = 6.797 \\ &2: a_2 = 0.3363, a_1 = 0.9625, a_0 = 2.386 \cdot 10^{-2}, a_{0D} = 2.267 \cdot 10^{-2}, \vartheta = 8.146 \end{aligned} \tag{14}$$

for initial settings $a_2 = 5, a_1 = 10/a_{0D,m}, a_0 = 5, a_{0D} = 10, \vartheta = \vartheta_m$ and $a_2 = 0.1, a_1 = 0.1, a_0 = 0.1, a_{0D} = 0.1, \vartheta = \vartheta_m$, respectively. Let data vector 1 in (14) be denoted as the denominator of the HPM 1, while the latter parameters set in (14) be that of the HPM 2.

Poles of the HPM 1 are almost identical to (13), while those of the HPM 2 read

$$s_1 = -0.00494, s_2 = -0.73822, s_{3,4} = -0.16760 \pm 0.96169\mathrm{j}, s_{5,6} = -0.82529 \pm 1.0529\mathrm{j} \quad (15)$$

As can be seen, especially the real part of $s_{3,4}$ and the imaginary part of $s_{5,6}$ are pretty far from the required roots (13); however, the HPM 2 provides interesting time and frequency responses (see Subsect. 4.4).

4.3 Remaining Thermal Process Model Parameter Identification

The remaining parameters of HPM 1 and HPM 2, i.e., b_0, b_{0D}, τ_0, can be obtained from the substitution according to step 9 of Algorithm 1 and estimated from the relay test data again.

The initial setting follows the idea that $b = b_{0D}$ and τ_0 is "a small positive number". Hence, since $b_0 + b_{0D} = (a_0 + a_{0D})k$, we do let set $b_{0D} = 0.5(a_0 + a_{0D})k$ and $\tau_0 = 0.1$. Then, the LMA with the identical control parameters set to that in the preceding subsection is applied to solve (5) with $G(s) = G_{HPM}(s)$. The eventual results for HPM 1 and HPM 2 are, respectively

$$\begin{aligned} &1\!: \; b_0 = 0.1728, \;\; b_{0D} = 0.1530, \;\; \tau_0 = 1027.311 \\ &2\!: \; b_0 = 7.420, \;\; b_{0D} = -7.419, \;\; \tau_0 = 4.95 \cdot 10^{-4} \end{aligned} \quad (16)$$

The corresponding residual errors (10) for the HPM 1 are 0.2970 and 6.1736, where the former value holds for the identity of real and imaginary parts in (5), while the latter holds for the identity of the gain and the phase. The errors for the HPM 2 are 0.2998 and 6.1717, respectively.

4.4 Results Evaluation

Let us display step responses and frequency responses (namely, Nyquist plots) of the original (exact) process model and the eventual models SFPM, HPM 1, and HPM 2. The time-domain responses are provided to the reader in Fig. 5. Note that the input step change is $\Delta u = 100$ W and the output data unit is °C. The HPM 1 and HPM 2 responses include non-smooth and high-oscillatory sections, respectively; see details in Fig. 6.

Errors of time-domain responses measured by the Integral Absolute Error (IAE) and the Integral Time Absolute Error (ITAE) from the original response are given in Table 1. As can be seen, the HPM 1 offers the best response in the time domain. However, the non-smooth section indicates high-frequency modes in the model dynamics.

Nyquist plots are displayed in Fig. 7. Note that the frequency range is selected as $\omega \in [0, 0.1]$ rad/s. The corresponding errors measured by the Root Mean Squares (RMS) and the 2-norm of the differences from the original data are given in Table 2. These results require a detailed analysis.

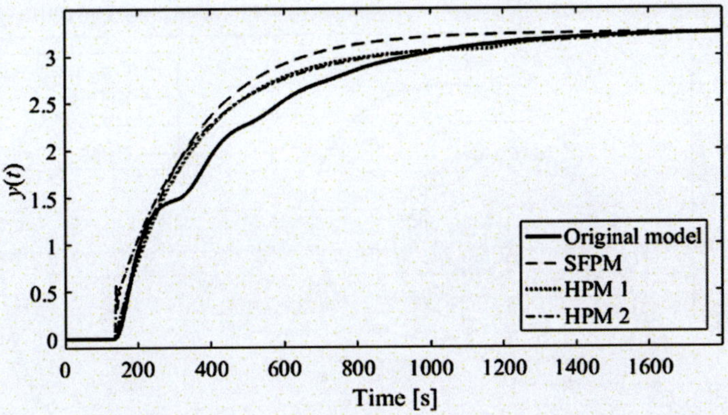

Fig. 5. Step responses of the original process, the SFPM, HPM 1, and HPM 2.

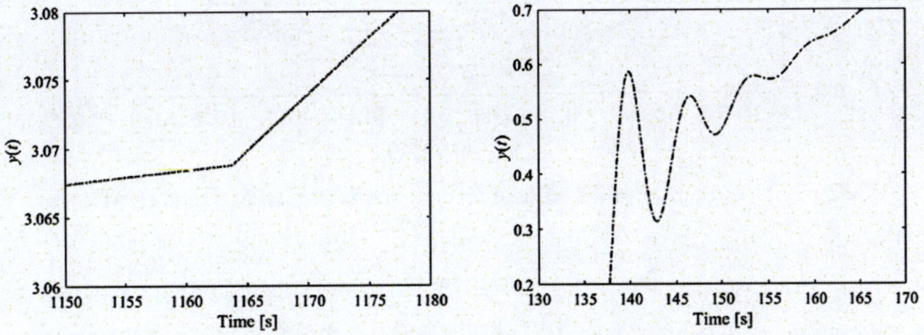

Fig. 6. Selected detail views on HPM 1 (left) and HPM 2 (right) time responses.

Table 1. Error measures of step responses.

Model	IAE	ITAE
SFPM	248.707	$1.533 \cdot 10^5$
HPM 1	164.270	$9.030 \cdot 10^4$
HPM 2	185.553	$9.380 \cdot 10^4$

By comparing data in Fig. 7 and Table 2, it is surprising that the HPM 2 has worse performance than the SFPM as the course of its Nyquist plot is closer to that of the original model. The justification arises from a frequency warping of the HPM 2 so that the corresponding points in Fig. 7 (for a particular frequency value) are farther from the original data than those of the SFPM. The same holds for the HPM 1, where – in addition – an extraordinary shape appears. Its gain and phase vary fast so that these functions are not monotonous and oscillate. However, the course of its Nyquist plot envelops that of the SFPM.

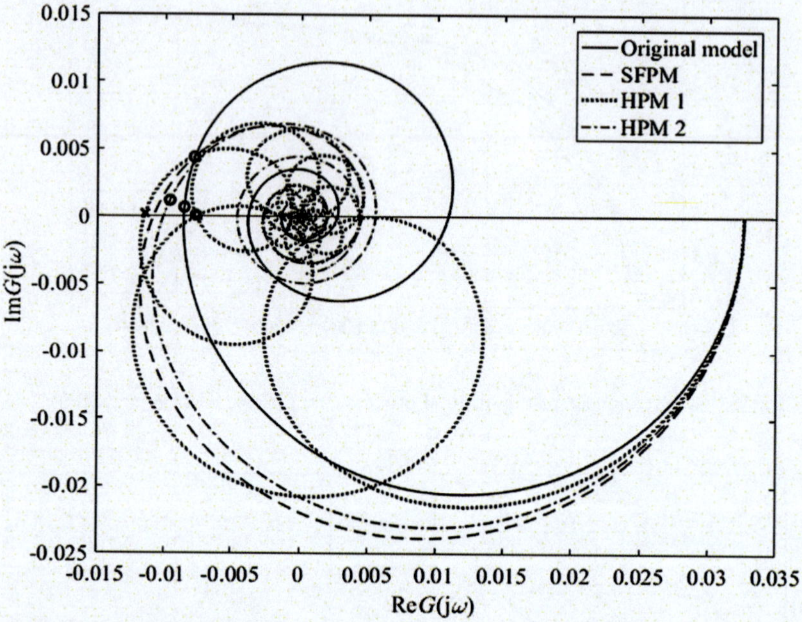

Fig. 7. Nyquist plots of the original process, the SFPM, HPM 1, and HPM 2.

Table 2. Error measures of Nyquist plots.

Model	RMS	2-norm
SFPM	$2.641{\cdot}10^{-2}$	0.118
HPM 1	$4.416{\cdot}10^{-2}$	0.198
HPM 2	$4.218{\cdot}10^{-2}$	0.189

Another exciting issue is positions of $\mathrm{Re}G(\mathrm{j}\omega_u)$ and $\mathrm{Im}G(\mathrm{j}\omega_u)$ (recall that $\omega_u = 1.73 \cdot 10^{-2}$). From the relay test, the theoretical values read $\mathrm{Re}G(\mathrm{j}\omega_u) = -1/k_u = -0.0076$, $\mathrm{Im}G(\mathrm{j}\omega_u) = 0$ (denoted by the square in Fig. 7). This position almost coincides with $\left[\mathrm{Re}G(\mathrm{j}\omega_u), \mathrm{Im}G(\mathrm{j}\omega_u)\right] = \left[-0.0077, 9.88 \cdot 10^{-5}\right]$ for the HPM 1 (a circle symbol). However, this is not the true critical (ultimate) point. The same Nyquist plot position is already reached for $\omega_u = 0.86 \cdot 10^{-2}$. Moreover, the point for which every change of $\mathrm{Re}G(\mathrm{j}\omega)$ implies the transition from stability to instability (and vice versa) of the feedback loop (i.e., the crossing of -1 for the open-loop frequency response) agrees with the leftmost crossing point of the real axis. This true critical point is denoted by the cross symbol and appears at $\omega = 1.27 \cdot 10^{-2}$. Note that $\left[\mathrm{Re}G(\mathrm{j}\omega_u), \mathrm{Im}G(\mathrm{j}\omega_u)\right]$ for the original model, the SFPM, and the HPM 2 are $\left[-0.0085, 6.91 \cdot 10^{-4}\right]$, $[-0.0070, 0.0044]$, $[-0.0096, 0.0012]$, respectively (circles).

To sum up, both the HPMs yield better performance than the SFPM in the time domain, while the opposite result holds in the frequency domain. The poor results for

HPMs are given mainly by the hybrid nature of Algorithm 1. That is, the pole assignment attempts to get the HPM as close to the SFPM as possible. However, the subsequent LMA optimization is designed to utilize relay-feedback test data to the HPM. Hence, the eventual HPMs represent a trade-off between the original process dynamics (i.e., its ultimate frequency data) and the already obtained SFPM.

Finally, let us make a concise note on the practical implementation. The used relay has no hysteresis. However, it is better to set a nonzero dead-zone of the relay static characteristics due to signal noise. We also use a single asymmetric test. It is suitable for the estimation of the process static gain, yet it causes an error in the ultimate data determination. Therefore, it is better to use a double test under one experiment. First, to apply an asymmetry, then to use the ideal relay for more precise ultimate data.

5 Conclusion

An approach to a single-run relay-feedback identification experiment for time-delay models with multiple parameters has been proposed. The method has been based on using a simple model, the parameters of which are estimated first using the standard relay experiment. Then, the eventual dominant simple-model poles have been assigned to the pole loci of the complex infinite-dimensional model of a thermal circuit process. Finally, the remaining parameters of the thermal process model have been determined from the known ultimate data of the original relay experiment. The well-established Levenberg-Marquardt algorithm has been used to solve the particular sets of nonlinear algebraic equations arising in some steps of the algorithm.

The obtained numerical results yield a necessity to enhance the algorithm regarding eventual parameters' estimation since the step responses and Nyquist plots of the models have revealed some unpleasant properties. A multiple relay test (yet with a reduced number of requisite experiments) or more powerful (yet more complex) numerical tools for the solution of given equations have to be used in the future.

Acknowledgements. This research was supported by the College of Polytechnics Jihlava, under Grant no. 1170/10/2137.

References

1. Åström, K.J., Hägglund, T.: Automatic tuning of simple regulators with specifications on phase and amplitude margins. Automatica **20**(5), 645–651 (1984)
2. Dharmalingam, K., Thangavelu, T.: Parameter estimation using relay feedback. Rev. Chem. Eng. **35**(4), 505–529 (2019)
3. Liu, T., Wang, Q.-G., Huang, H.P.: A tutorial review on process identification from step or relay feedback test. J. Process Control **23**(10), 1597–1623 (2013)
4. Wang, Q.-G., Hang, C.C., Zou, B.: Low-order modeling from relay feedback. Ind. Eng. Chem. Res. **36**(2), 375–381 (1997)
5. Ma, M.D., Zhu, X.J.: A simple auto-tuner in frequency domain. Comput. Chem. Eng. **30**(4), 581–586 (2006)

6. Jeon, C.H., Cheon, Y.J., Kim, J.S., Lee, J., Sung, S.W.: Relay feedback methods combining sub-relays to reduce harmonics. J. Process. Control **20**(2), 228–234 (2010)
7. Yu, C.C.: Autotuning of PID Controllers: A Relay Feedback Approach, 2nd edn. Springer, London (2006)
8. Hofreiter, M.: Relay feedback identification with additional integrator. IFAC-PapersOnLine **52**(13), 66–71 (2019)
9. Bi, Q., Wang, Q.-G., Hang, C.C.: Relay-based estimation of multiple points on process frequency response. Automatica **33**(9), 1753–1757 (1997)
10. Hale, J.K., Lunel, S.V.: Introduction to Functional Differential Equations. Springer, New York (1993)
11. Vyhlídal, T., Zítek, P.: Control system design based on a universal first order model with time delays. Acta Polytech. **41**(4–5), 49–53 (2001)
12. Pekař, L., Prokop, R.: Saturation relay vs. relay transient identification tests for a TDS model. In: Proceedings of the 27th European Conference on Modelling and Simulation (ECMS 2013), Alesund, Norway, pp. 446–452 (2013)
13. Pekař, L.: Modeling and identification of a time-delay heat exchanger plant. In: Pekař, L. (ed.) Advanced Analytic and Control Techniques for Thermal Systems with Heat Exchangers, pp. 23–48. Academic Press (Elsevier), Cambridge, MA, USA (2020)
14. Hofreiter, M.: Fitting anisochronic models by method of moments for anisochronic control of time delay systems. Int. J. Math. Mod. Meth Appl. Sci. **10**, 71–79 (2016)
15. Úředníček, Z.: Nonlinear systems-describing functions analysis and using. MATEC Web Conf. **210**, 02021 (2018)
16. Ramana, K.V., Majhi, S., Gogoi, A.K.: Identification of DC–DC buck converter dynamics using relay feedback method with experimental validation. IET Circ. Devices Syst. **12**(6), 777–784 (2018)
17. Pekař, L.: Root locus analysis of a retarded quasipolynomial. WSEAS Trans. Syst. Control **6**(3), 79–91 (2011)
18. Marquardt, D.: An algorithm for least-squares estimation of nonlinear parameters. SIAM J. Appl. Math. **11**(2), 431–441 (1963)

A Binary Whale Algorithm Applied to Knapsack Problem

Paola Moraga[1], Pamela Valenzuela[1], Francisco Altimiras[2(✉)], Luis Lopez[1], and José-Miguel Rubio[3]

[1] Escuela de ingeniería en construcción, Pontificia Universidad Católica de Valparaíso, Valparaíso, Chile
{paola.moraga,pamela.valenzuela,luis.lopez}@pucv.cl
[2] Facultad de Ingeniería y Negocios, Universidad de las Américas, 7500975 Santiago, Chile
faltimiras@udla.cl
[3] Facultad de Ingeniería, Ciencia y Tecnología, Universidad Bernardo O'Higgins, 8370993 Santiago, Chile
josemiguel.rubio@ubo.cl

Abstract. Combinatorial and particularly combinatorial NP-hard problems are of importance and interest in operations research. Therefore, the design of efficient algorithms that address these combinatorial problems are applied research lines with interesting potential. Inspired by the above, this article designs a binarization method so that continuous metaheuristics can solve combinatorial problems. The binarization method uses the concept of percentile. This percentile method is applied to the firefly algorithm. The multidimensional backpack problem (MKP) was used to verify our algorithm.

Keywords: Combinatorial optimization · Multidimensional knapsack problem · Metaheuristics

1 Introduction

Combinatorial problems frequently appear in the industry. For example we found combinatorial problems in civil engineering [1–5], Operations Research [6–10], allocation problems [11–14], vehicle routing problems [15,16], robust optimization, and problems of scheduling [17–19], among others.

In recent years we have witnessed the generation of a large number of nature-inspired algorithms that are capable of solving optimization problems efficiently. Examples of these algorithms are a black hole, gravitational search, and multiverse algorithm, among others. Due to the nature of the phenomenon, we find that many of these algorithms work in continuous search spaces and must be adapted to solve the MKP since the latter is a combinatorial problem. The adaptation process must ensure that the exploitation and exploration mechanisms are preserved so that the efficiency of the algorithm is maintained.

R. Silhavy et al. (Eds.): CoMeSySo 2021, LNNS 231, pp. 465–474, 2021.
https://doi.org/10.1007/978-3-030-90321-3_38

When performing a state of the art, it is found that there are several binarization techniques that allow adapting algorithms that work in continuous spaces based on binary spaces. Among the main binarization methods used are the functions of transfer, angular modulation, and quantum focus; The reader can consult [20] for more details. In this article, we developed a binarization method based on the percentile technique to perform the binarization process.

Whale metaheuristic algorithm was selected to verify the effectiveness of the percentile technique in the binarization process. Whale algorithm has been successfully applied to various engineering and operations research problems. The binary version of the firefly algorithm will be denoted by BPWH and this version will be applied to MKP. Experiments were developed that account for the percentile operator contribution in the binarization process. For this, a random operator was built that serves as a baseline to measure the contribution.

2 KnapSack Problem

The allocation of resources which is modeled by MKP is a frequent problem at the industrial level. The objective of MKP is to identify a subset of objects, which produce the greatest benefit by satisfying a set of restrictions specific to each problem. MKP corresponds to one of the most studied problems that fall in the NP-hard category. When we look at the literature, we see that MKP has been solved using different techniques. For example, in [21] an unsupervised learning technique was applied. A quantum binarization was designed in [22] to address MKP. In [23] a fruit fly algorithm was adapted, and in [24] a differential algorithm was used. Formally MKP is defined as:

$$\text{maximize} \sum_{j=1}^{n} p_j x_j \tag{1}$$

$$\text{subjected to} \sum_{j=1}^{n} c_{ij} x_j \leq b_i, i \in \{1, \ldots, m\}. \tag{2}$$

With $x_j \in \{0, 1\}$, $j \in \{1, \ldots, n\}$. p_j represents the profit of the item j. c_{ij} corresponds a cost related to dimension i and element j. In each dimension, the constraints i are noted by b_i. The solution can be conceptualized using a binary representation, in this way a 0 means absence of the element inside the knapsack.

3 Binary Percentile Whale Algorithm

The areas of data mining and statistics are cross-cutting disciplines and we can find applications of them in different situations, such as transports, agriculture, cities, and mathematics [15,21,25–29]. The percentile technique applied to the binarization of metaheuristics that work in continuous spaces is explored in this section. In Fig. 1, a general diagram of the solution is shown. As a first stage, the solutions must be initialized, then it is determined if the criterion of the maximum number of iterations has been met. In the case that it has not been fulfilled, the firefly algorithm is executed and using the solutions obtained, the percentile algorithm is applied to perform its binarization and subsequently a repair algorithm for the case that there is any restriction that is not complied with. Once the binarization has been carried out through the percentile technique, the solutions obtained are compared with the best. In the case of having a superior, a local search operator is executed. Otherwise, the previous steps are repeated.

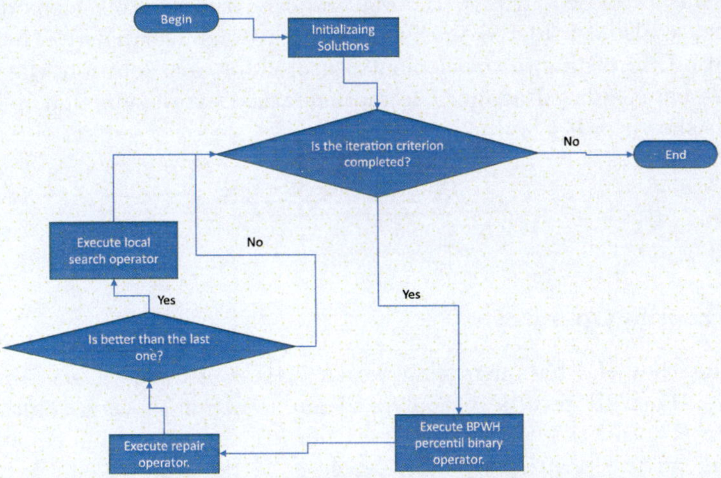

Fig. 1. Flowchart of the binary percentile whale algorithm.

3.1 Initialization and Element Weighting

Since the bat algorithm is a swarm algorithm, to begin the search for the best solution, it is necessary to initialize the list of solutions. The mechanism used in the generation of each solution is as follows: first, a random element is chosen. Then, it should be verified if it is feasible to incorporate new elements, therefore, it is necessary to evaluate the restrictions of the problem. In the selection of the new element, if it is feasible, a list of possible candidates is generated that is consistent with the restrictions. Each item is assigned a previously calculated weight and the one with the best weight is selected. This procedure is repeated until no additional element can be incorporated.

In the calculation of the weight of an element, several heuristics have been developed. For example, in [30] a pseudo-utility was proposed. The equation that allows the calculation is written in Eq. 3. In this equation, the variable w_j corresponds to the surrogate multiplier whose value is between 0 and 1. This multiplier can be interpreted intuitively as a shadow price of the constraint j.

$$\delta_i = \frac{p_i}{\sum_{j=1}^{m} w_j c_{ij}} \tag{3}$$

A more intuitive measure focused on the average resource occupancy was proposed by [31]. It is shown in Eq. 4.

$$\delta_i = \frac{\sum_{j=1}^{m} \frac{c_{ij}}{m b_j}}{p_i} \tag{4}$$

In this article, a variation based on the average occupation will be used. This variation was proposed in [21]. In this proposal, the existing elements in the backpacks are also considered to calculate the average occupancy. Therefore, in each iteration, according to the elements selected for each solution, the measurement must be calculated again. The mathematical expression that defines this measure is shown in Eq. 5.

$$\delta_i = \frac{\sum_{j=1}^{m} \frac{c_{ij}}{m(b_j - \sum_{i \in S} c_{ij})}}{p_i} \tag{5}$$

3.2 Percentile Operator

Considering that the bat algorithm works in a continuous space and also it is iterative. Then, it results in the speed and position of each solution being updated in $\mathbb{R}^n$. In Eq. 6 the above is written in general. In this equation x_{t+1} corresponds to the position of the particle x at the time $t+1$. To calculate the position, the *Delta* function, which is specific to each algorithm, must be considered additionally.

$$x_{t+1} = x_t + \Delta_{t+1}(x(t)) \tag{6}$$

In the case of the bat algorithm, binarization is done through a binary percentile algorithm. This algorithm considers that, given a solution x, we calculate the magnitude of its displacement $\Delta^i(x)$ in the i-th component and then group all the displacements in order to obtain the values for the percentiles {20, 40, 60, 80, 100}. In each percentile, a transition probability is associated that corresponds to the values shown in Eq. 7. Using these transition probabilities together with Eq. 8 the binarization of the solutions is performed. The algorithm is detailed in Algorithm 1.

$$P_{tr}(x^i) = \begin{cases} 0.1, & \text{if } x^i \in \text{group } \{0,1\} \\ 0.5, & \text{if } x^i \in \text{group } \{2,3,4\} \end{cases} \tag{7}$$

$$x^i(t+1) := \begin{cases} \hat{x}^i(t), & \text{if } rand < P_{tg}(x^i) \\ x^i(t), & \text{otherwise} \end{cases} \tag{8}$$

Algorithm 1. Percentile operator

Function pbinary(vLst, pLst)
Input vList, pList
Output pGroupValue
pValue = getPValue(vLst, pLst)
for each value in vList **do**
    pGroupValue = getPGroupValue(pValue,vLst)
end for
return pGroupValue

3.3 Repair Operator

Because the percentile operator and the local search operator can generate solutions that do not satisfy any of the constraints, it is necessary to apply a repair operator. The measure described in Eq. 5 was used to select the elements that perform the repair. In the case that repair is required, the element with the maximum measure is chosen and removed from the solution. This process is iterated until a solution that satisfies all constraints is obtained. The pseudocode is shown in Algorithm 2.

Algorithm 2. Repair Algorithm

Function Rep(Sl_{in})
Input Input solution Sl_{in}
Output The Rep solution Sl_{out}
$S \leftarrow S_{in}$
while needRepair(S) == True **do**
    $s_{max} \leftarrow$ MaxWeight(S)
    $S \leftarrow$ removeElement(S, s_{max})
end while
state ← False
while state == False **do**
    $s_{min} \leftarrow$ MinWeight(S)
    if $s_{min} == Null$ **then**
        state ← True
    else
        $S \leftarrow$ addElement(S, s_{min})
    end if
end while
$S_{out} \leftarrow S$
return S_{out}

4 Results

4.1 Understanding BPWH Contribution

In this section, the experiments are developed to determine the contribution of the percentile binary operator in the binarization process. To determine this contribution, problems cb.5.250 of the OR library were selected. As comparison artifacts, violin plots, tables, and statistical comparisons were used. Particularly, the non-parametric Wilcoxon test was used to perform analyses. In the plots, the X-axis identifies the analyzed instances and the Y-axis the % -Gap defined in Eq. 9. The Wilcoxon test aims to verify whether the difference between the results obtained by BPWH is significant compared to the other algorithms. The exploration of the parameters is detailed in Table 1.

$$\% - Gap = 100\frac{BestKnown - SolutionValue}{BestKnown} \tag{9}$$

Table 1. Setting of parameters for binary percentile bat search algorithm.

Parameters	Description	Value	Range
N	Solutions	30	[20, 25, 30]
G	Percentiles	5	[4, 5, 6]
Iteration number	Maximum number of iterations	1000	[1000]

Percentile Binary Operator Contribution. To understand the contribution of the percentile operator to the binarization process a random operator was constructed to be considered as a baseline. This random operator was configured to perform transitions with a fixed probability of 0.5 without considering in which percentile the speed variable is located. Two configurations were analyzed: the first configuration includes the local search operator and the second configuration does not consider this operator. The above with the objective of decoupling the contribution of the local search operator and the percentile. BPWH corresponds to our standard algorithm. *rd.ls* is the random variant that includes the local search operator. *BPWH.wls* corresponds to the version with a binary percentile operator without a local search operator. Finally, *rd.wls* describes the random algorithm without a local search operator.

The first comparison corresponds to comparing the best values between BPWH and *rd.ls*. These results are shown in Table 2. BPWH exceeds *rd.ls*. Despite being higher, the values are close. Our second comparison corresponds to considering the averages. In this case, BPWH exceeds *rd.ls* in almost all problems bluntly. The comparison of distributions through the violin plots shown in Fig. 2, indicates that the dispersion of the distributions *rd.ls* is greater than the dispersions of BPWH. In particular, this is notorious in problems 1, 2, 4, and 9. Therefore, the percentile and local search operators as a whole contribute to improving and making the results more robust. Finally, BPWH distributions are closer to zero than the *rd.ls* distributions, indicating that BPWH has consistently better results than *random.ls*. When we evaluate the behavior of the algorithms through the Wilcoxon test, this indicates that the difference is significant.

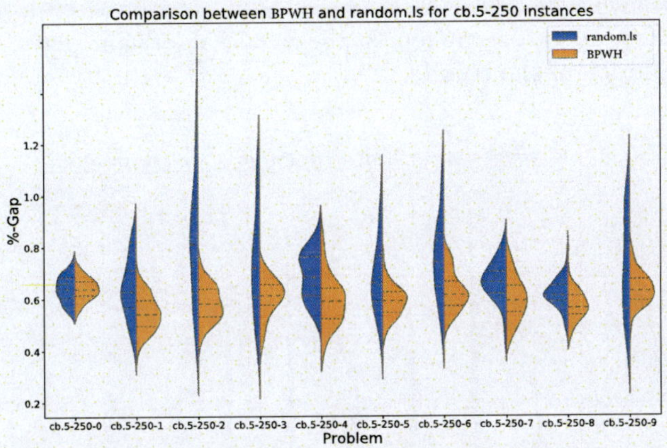

Fig. 2. Evaluation of percentile binary operator with Local Search operator

The following analysis aims to separate the contribution of the local search operator from the binary percentile operator, in order to be able to size the contribution of the percentile operator to the binarization process. For this, the comparison between the *wls* and *rd.wls* algorithms is made.

Table 2. Evaluation of percentile binary operator

Set	Best Known	Best *rd.ls*	Best BPWH	Best *rd.wls*	Best *wls*	Avg *rd.ls*	Avg BPWH	Avg *rd.wls*	Avg *wls*
cb.5.250-0	59312	59211	59225	59158	59175	59124	59122.1	59038.3	59101.7
cb.5.250-1	61472	61435	61435	61409	61409	61276.1	61338.5	61284.9	61344.1
cb.5.250-2	62130	62036	62074	61969	61969	61862.4	61964.1	61771.4	61900.7
cb.5.250-3	59463	59367	59446	59365	59349	59205.5	59257.5	59113.4	59214.5
cb.5.250-4	58951	58914	58951	58883	58930	58671.3	58802.3	58664.8	58747.5
cb.5.250-5	60077	60015	60056	59990	60015	59874.9	59901.9	59823.6	59924.1
cb.5.250-6	60414	60355	60355	60348	60349	60196.1	60286.4	60210.6	60266.2
cb.5.250-7	61472	61436	61472	61407	61407	61253.6	61282.8	61209.3	61341
cb.5.250-8	61885	61829	61885	61790	61782	61686.3	61738.9	61615	61683
cb.5.250-9	58959	58832	58866	58822	58787	58762.7	58671.1	58631.8	58735.5
Average	60413.5	60343.0	60376.5	60314.1	60317.2	60191.3	60233.2	60136.3	60225.8
p-value								1.6 e−04	3.4 e−04

The results are shown in Table 2: when analyzing the best results, we observe that *wpe* gets better results than *05.wpe* in all problems except 1, 6 and 7. However, the results are again very close. When we analyze the average, *wpe* exceeds *05.wpe* in all problems. The Wilcoxon test indicates that the difference is significant. This indicates that *wpe* is consistently better than *05.wpe*. Note that in this case we are not using the local search operator. The violin plots are shown in Fig. 3. This plots show that the dispersion of the solutions for the case of *05.wpe* is much greater than in the case of *wpe*. This indicates that the percentile binary operator applied to the whale algorithm plays an important role in the accuracy of the results.

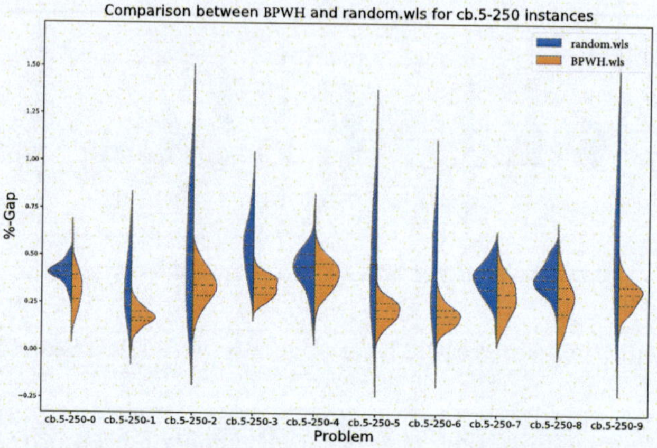

Fig. 3. Evaluation of percentile binary operator without local search operator

5 Conclusions

In this article, the percentile technique was explored as a method to perform the binarization of the bat algorithm. We must emphasize that the percentile technique is a general binarization technique and therefore can be applied in the binarization of any continuous swarm intelligence algorithm. MKP was used to evaluate the performance of the binary algorithm. The contribution of the percentile binary operator was analyzed, and it was concluded that this operator contributes to the accuracy and quality of the solutions obtained.

As a future line of research, it is interesting to use the bat algorithm in the three binarization mechanisms: Percentile, k-means, and transfer functions. In addition, it is interesting to binarize other metaheuristics with the percentile technique along with solving other NP-hard problems. Another line of longer-term research corresponds to exploring adaptive techniques to dynamically change the parameterization of metaheuristics, in this sense it is interesting to explore reinforced learning techniques.

References

1. García, J., Yepes, V., Martí, J.V.: A hybrid k-means cuckoo search algorithm applied to the counterfort retaining walls problem. Mathematics **8**(4), 555 (2020)
2. García, J., Moraga, P., Valenzuela, M., Pinto, H.: A db-scan hybrid algorithm: an application to the multidimensional knapsack problem. Mathematics **8**(4), 507 (2020)

3. García, J., Martí, J.V., Yepes, V.: The buttressed walls problem: an application of a hybrid clustering particle swarm optimization algorithm. Mathematics **8**(6), 862 (2020)
4. Crawford, B., et al.: Investigating the efficiency of swarm algorithms for bridge strengthening by conversion to tied-arch: a numerical case study on San Luis Bridge. Iran. J. Sci. Technol. Trans. Civ. Eng. **45**, 2345–2357 (2020). https://doi.org/10.1007/s40996-020-00531-5
5. Martínez-Muñoz, D., Martí, J.V., García, J., Yepes, V.: Embodied energy optimization of buttressed earth-retaining walls with hybrid simulated annealing. Appl. Sci. **11**(4) (2021). https://www.mdpi.com/2076-3417/11/4/1800
6. Crawford, B., Soto, R., Astorga, G., García, J.: Constructive metaheuristics for the set covering problem. In: Korošec, P., Melab, N., Talbi, E.-G. (eds.) BIOMA 2018. LNCS, vol. 10835, pp. 88–99. Springer, Cham (2018). https://doi.org/10.1007/978-3-319-91641-5_8
7. García, J., Crawford, B., Soto, R., Astorga, G.: A percentile transition ranking algorithm applied to binarization of continuous swarm intelligence metaheuristics. In: Ghazali, R., Deris, M.M., Nawi, N.M., Abawajy, J.H. (eds.) SCDM 2018. AISC, vol. 700, pp. 3–13. Springer, Cham (2018). https://doi.org/10.1007/978-3-319-72550-5_1
8. Crawford, B., Soto, R., Monfroy, E., Astorga, G., García, J., Cortes, E.: A meta-optimization approach for covering problems in facility location. In: Figueroa-García, J.C., López-Santana, E.R., Villa-Ramírez, J.L., Ferro-Escobar, R. (eds.) WEA 2017. CCIS, vol. 742, pp. 565–578. Springer, Cham (2017). https://doi.org/10.1007/978-3-319-66963-2_50
9. García, J., Crawford, B., Soto, R., Astorga, G.: A clustering algorithm applied to the binarization of swarm intelligence continuous metaheuristics. Swarm Evol. Comput. **44**, 646–664 (2019). http://www.sciencedirect.com/science/article/pii/S221065021730528X
10. Crawford, B., et al.: Q-learnheuristics: towards data-driven balanced metaheuristics. Mathematics **9**(16) (2021). https://www.mdpi.com/2227-7390/9/16/1839
11. García, J., Crawford, B., Soto, R., Astorga, G.: A percentile transition ranking algorithm applied to knapsack problem. In: Silhavy, R., Silhavy, P., Prokopova, Z. (eds.) CoMeSySo 2017. AISC, vol. 662, pp. 126–138. Springer, Cham (2018). https://doi.org/10.1007/978-3-319-67621-0_11
12. Astorga, G., Crawford, B., Soto, R., Monfroy, E., García, J., Cortes, E.: A meta-optimization approach to solve the set covering problem. Ingeniería **23**(3), 274–288 (2018)
13. García, J., Lalla-Ruiz, E., Voß, S., Droguett, E.L.: Enhancing a machine learning binarization framework by perturbation operators: analysis on the multidimensional knapsack problem. Int. J. Mach. Learn. Cybern. **11**(9), 1951–1970 (2020). https://doi.org/10.1007/s13042-020-01085-8
14. García, J., Maureira, C.: A KNN quantum cuckoo search algorithm applied to the multidimensional knapsack problem. Appl. Soft Comput. **102**, 107077 (2021)
15. García, J., Peña, A.: Robust optimization: concepts and applications. In: Nature-Inspired Methods for Stochastic, Robust and Dynamic Optimization, p. 7 (2018)
16. García, J., Astorga, G., Yepes, V.: An analysis of a KNN perturbation operator: an application to the binarization of continuous metaheuristics. Mathematics **9**(3), 225 (2021)
17. García, J., Crawford, B., Soto, R., García, P.: A multi dynamic binary black hole algorithm applied to set covering problem. In: Del Ser, J. (ed.) ICHSA 2017. AISC,

vol. 514, pp. 42–51. Springer, Singapore (2017). https://doi.org/10.1007/978-981-10-3728-3_6
18. García, J., Altimiras, F., Peña, A., Astorga, G., Peredo, O.: A binary cuckoo search big data algorithm applied to large-scale crew scheduling problems. Complexity **2018** (2018)
19. Crawford, B., Soto, R., Monfroy, E., Astorga, G., García, J., Cortes, E.: A meta-optimization approach to solve the set covering problem. Ingeniería **23**(3), 274–288 (2018)
20. García, J., et al.: A Db-scan binarization algorithm applied to matrix covering problems. Comput. Intell. Neurosci. **2019** (2019)
21. García, J., Crawford, B., Soto, R., Castro, C., Paredes, F.: A k-means binarization framework applied to multidimensional knapsack problem. Appl. Intell. **48**(2), 357–380 (2017). https://doi.org/10.1007/s10489-017-0972-6
22. Haddar, B., Khemakhem, M., Hanafi, S., Wilbaut, C.: A hybrid quantum particle swarm optimization for the multidimensional knapsack problem. Eng. Appl. Artif. Intell. **55**, 1–13 (2016)
23. Meng, T., Pan, Q.-K.: An improved fruit fly optimization algorithm for solving the multidimensional knapsack problem. Appl. Soft Comput. **50**, 79–93 (2017)
24. Liu, J., Wu, C., Cao, J., Wang, X., Teo, K.L.: A binary differential search algorithm for the 0–1 multidimensional knapsack problem. Appl. Math. Model. **40**(23–24), 9788–9805 (2016)
25. García, J., Pope, C., Altimiras, F.: A distributed k-means segmentation algorithm applied to lobesia botrana recognition. Complexity **2017** (2017)
26. Graells-Garrido, E., García, J.: Visual exploration of urban dynamics using mobile data. In: García-Chamizo, J.M., Fortino, G., Ochoa, S.F. (eds.) UCAmI 2015. LNCS, vol. 9454, pp. 480–491. Springer, Cham (2015). https://doi.org/10.1007/978-3-319-26401-1_45
27. Graells-Garrido, E., Peredo, O., García, J.: Sensing urban patterns with antenna mappings: the case of Santiago, Chile. Sensors **16**(7), 1098 (2016)
28. Peredo, O.F., García, J.A., Stuven, R., Ortiz, J.M.: Urban dynamic estimation using mobile phone logs and locally varying anisotropy. In: Gómez-Hernández, J.J., Rodrigo-Ilarri, J., Rodrigo-Clavero, M.E., Cassiraga, E., Vargas-Guzmán, J.A. (eds.) Geostatistics Valencia 2016. QGG, vol. 19, pp. 949–964. Springer, Cham (2017). https://doi.org/10.1007/978-3-319-46819-8_66
29. Maureira, C., Pinto, H., Yepes, V., Garcia, J.: Towards an AEC-AI industry optimization algorithmic knowledge mapping an adaptive methodology for macroscopic conceptual analysis. IEEE Access **9**, 110842–110879 (2021)
30. Pirkul, H.: A heuristic solution procedure for the multiconstraint zero? One knapsack problem. Nav. Res. Logist. **34**(2), 161–172 (1987)
31. Kong, X., Gao, L., Ouyang, H., Li, S.: Solving large-scale multidimensional knapsack problems with a new binary harmony search algorithm. Comput. Oper. Res. **63**, 7–22 (2015)

How Does Corruption Affect Firm Growth in Russia? A Case Study of the Krasnoyarsk Region

N. N. Tarun Chakravorty[1(✉)], Vladislav Rutskiy[1,2], and Elena Sevastyanova[1]

[1] Siberian Federal University, Krasnoyarsk, Russia
n.n.tarun.chakravorty@bath.edu
[2] HSE University, Moscow, Russia

Abstract. This study investigates how corruption affects firm growth in the manufacturing industrial sector of Russia. Both quantitative and qualitative analyses have been done in the study. An econometric exercise on the impact of the corrupt behaviour of government officials on firm growth has been done using World Bank's Enterprise Survey data. OLS, IV, and GMM techniques have been applied to a panel dataset. The dataset suffers from some missing values which have been imputed using an econometric tool. To overcome the weakness of the quantitative analysis because of the missing values for a bribe and to have a better understanding of the issues, primary data are collected directly from the firm owners and managers from Krasnoyarsk region of Russia, and through Focus Group Discussions (FGD) with firm owners & managers and experts. The analysis of the secondary data shows a positive impact of corruption on firm growth with a highly significant coefficient of the key variable of interest, which is a bribe. The findings support the so-called speed money theory and the argument for this is that when bureaucracy is stuck in corruption, some extra informal payment would help to expedite the process. Most respondents in the primary survey and FGDs have not mentioned bribery as a barrier to growth, which means bribery helps them to speed up the firm growth.

Keywords: Firm growth · Administrative obstacles · Corruption · Bribe · Russia

1 Introduction

Governance issues have been widely investigated as a factor of economic growth in recent decades. The studies such as [1–7] have found economic growth to be correlated with governance and institution. However, investigations on the impact of the governance problem, namely corruption have been conducted predominantly at the macro-level. At the firm level, this kind of investigation is few. [8] is one seminal study done on firm-level, which has investigated the impact of taxation and the corrupt behaviour of the government officials on firm growth in Uganda. [9] has examined the same in Vietnam. A similar study has been done by Chakravorty on Bangladesh [10]. These studies provide mixed results. The motivations for conducting new research on Russia are: small enterprises

R. Silhavy et al. (Eds.): CoMeSySo 2021, LNNS 231, pp. 475–492, 2021.
https://doi.org/10.1007/978-3-030-90321-3_39

in transition economies have received significantly less attention, little work has been done at firm-level so far and Russia differs from other transition economies in that the degree of regulatory penetration by firms is the highest in Russia [11]. Moreover, the literature on the impact of corruption is clearly divided. In the literature the debate on the effect of corruption has led to two opposing strands: efficiency-enhancing strand and efficiency reducing strand. Those who advocate the efficiency-enhancing strand such as [12–15] argue that corruption increases economic growth firstly because the money paid as bribe acts as 'speed money' or 'greasing wheel' which moves the files on and secondly because bureaucrats taking bribe work harder to increase their 'earnings'. Bribes here can circumvent the obstacles created by bureaucrats and act as a trouble-saving device which, in turn, raise efficiency and promote investment and growth [16]. However, [17] suggests that this 'speed money' may help the economy grow only when it is stuck in wide and deep corruption. [18] and [19] are among many others who support the 'speed money' idea.

Efficiency reducing strand advocated by [17, 20, 21] and many others, finds corruption to be reducing economic growth. [17] opposes 'efficiency argument' as she finds that corruption help growth in some areas/sectors for a certain period but corruption would not be confined in those sectors. It will rather spread over those areas/sectors and the whole economy would be stagnant and economic activities would slow down after a point in time. Another cause of such slow-down is that corrupt officials might wait for higher bribes for the same 'job' and delay in moving the files.

In recent studies, the impact of corruption on growth and investment has appeared to be indeterminate. For example, Wedeman [22] do not find any impact of the overall level of corruption on the GDP growth rate of a country. [23]'s re-investigation of [24]'s work using a better instrumental variable ends up with inconclusive results. From these perspectives, causal relations between corruption and growth and other macroeconomic variables are still unclear and it demonstrates the need for further empirical investigations on different countries and different periods.

The impact of corruption on growth or investment has been studied mainly at the macro level through cross-country analyses in the literature. [8] is the first micro-level study which is done on Ugandan firms. Another study is [9] which investigates the impact of corruption on firm growth in some African countries. Chakravorty [10] has examined the impact of corruption on firm growth in Bangladesh. No such work, to the best of our knowledge, has so far been done on Russian firms while Russia is a transition economy. Therefore, it would be interesting to see how corruption as a governance problem affect firm growth in Russia and it was the motivation to conduct a new study.

To this end, in this study, we address the research question: does corruption as a governance problem affect the firm growth in the Krasnoyarsk region and at the national level of Russia?

Based on our literature review and our general wisdom, we build our hypothesis that: corruption negatively affects firm growth.

2 Literature Review

Kaufmann and Wei [25] argue that when bureaucrats enjoy discretionary power in applying/enforcing regulations, officials having intentions of extracting bribes can often customize the nature and amount of harassment on firms meaning that they would segment the market and charge firms with their respective abilities to pay. As a result, in equilibrium, there is no guarantee that firms that paid higher bribes would not face higher red tape. In their model, they find that bribes firms pay and harassment government officials create are positively correlated. The bribe determination process might have been monopolistic where the bribe is endogenously determined and the bureaucrat segments the market and charges a higher bribe to those who appear to have a higher capacity to pay and lower to those who have the lower capacity [26]. Efficient firms are seen to have a greater ability to pay a bribe as well. Therefore, "the firms which have paid more bribes would have larger growth" [10]. The Russian case is a typical example of the existence of this discretionary power of government officials and segmentation of the market as [11] describes Russia as a country full of regulations and an elaborate system of permits and licenses, where bureaucrats enjoy a lot of discretion in enforcement and monitoring creating the scope for the rent-seeking.

Heterogeneity in reported findings on the effects of corruption on growth leads scholars to think that the impact of corruption is inconclusive. Recently many studies including the seminal paper [24], which used econometric analyses to make inferences on a causal relationship between corruption and growth, have been re-investigated and have been seen to be fraught with problems such as reverse causality and omitted variable problems. These studies and many further studies have ended up with inconclusive results [27, 28], which seems to be predominant in the literature lately. This inconclusiveness has also been supported by [29–32]. However, the inconclusiveness might have appeared also due to publication biases. Campos, Dimova, and Saleh [33] and Ugur [34] conducted meta-analyses using 29 and 41 empirical studies respectively and tried to investigate the predominant findings in those studies overcoming the biases and found the negative effect of corruption to be predominant.

Different types of government spending create different types of scope or different degrees of possibilities for bureaucrats or politicians to engage in corrupt practices, extract bribes and go without being caught [17, 21, 35–37].

Owing to the secrecy and limited competition, military spending is prone to a relatively high level of informal contracts creating scope for rent-seeking activities and other corrupt practices, which crowd out productive investment in the private sector exerting a negative impact on growth [38].

d'Agostino, Dunne, and Pieroni [39] investigated how corruption affects the link between government expenditures and growth in Africa. To do that, they used an extended endogenous growth model which allows corruption to affect different types of government spending differently. Their investigation finds corruption to have negative effects on growth and that corruption's negative impact aggravates when there is a higher burden of military expenditure.

Campos et al. [33] in their meta-analysis using 41 empirical studies, find evidence that corruption diminishes the positive effect of institutions and trade openness on growth and thus the final effect of corruption on economic growth is seen to be negative after tackling the biases existing in the studies.

The meta-analysis done by [34] which uses 29 primary studies, confirms corruption's direct negative effect on per-capita GDP growth after controlling for publication selection bias and within-study dependence.

d'Agostino, Dunne, and Pieroni [39] investigate the effect of corruption using an endogenous growth model which accounts for the effect of corruption allowing it to capture the indirect effect of corruption on military spending and government investment expenditure. This strategy helped to capture the heterogeneous effects on per capita growth rates. Their findings confirm that government investment spending positively impacts economic growth and that large military burden, current government spending and corruption have negative effects. This study also finds that the negative effect of the military burden on the growth rate gets stronger in presence of corruption. The extent to which military expenditure affects the corruption-growth link has been found to be much smaller.

Using the bootstrap panel Granger causality approach which addresses both cross-sectional dependence and heterogeneity across countries Huang [40] investigates the impact of corruption on economic growth in 13 Asia-Pacific countries throughout 1997–2013. The study finds significant results only in South Korea and China. In the former corruption is seen to have a positive effect on economic growth and in the latter, the increasing economic growth of the country is seen to give rise to corruption.

Célimène, Dufrénot, Mophou, and N'Guérékata in their paper [41] adopt an open economy stochastic growth model and present a standard portfolio argument which says that evasion outcome for the private sector might rather be conducive to economic growth following the development of private investment if the proceeds from tax evasion were invested in the equity market and that the negative externalities of tax evasion on public spending can be diminished if its productivity is high enough. It is noteworthy that, when there is a high incentive for cheating (because the tax collection system is deficient), the negative externalities of tax evasion on public spending can be diminished if its productivity is high enough. This implies that there may be a trade-off between tax governance and policies enhancing the efficiency of public goods and services on the economic growth [41].

The biggest obstacles to firm growth, as put by most CEOs of SME's, are: i) untimely delivery of supply; ii) lack of finance- especially the lack of external financing and its high cost; iii) high interest rates; (iv) lack of land, office space and building; v) lack of credit finance, which prevented both local sales and export market (around a third of the firms in the sample received a loan from a financial institution over the past three years. Less than one third heard of the availability of special financing programs for SMEs [42, 43]. Pissarides, Singer and Svejnar [42] have done an econometric analysis of the data on the dealings between firms and public officials, firm characteristics and the business environment received from the survey conducted in 1995 in the Russian region of Krasnodar and Bulgaria using a multinomial logit model. The main thrust of this econometric analysis was to find the determinants of the most important constraints that would emerge from characteristics of the entrepreneur, of the firm and of the environment in which firms operate. They find that the external financing constraint exerted the biggest negative effect on SME growth. While comparing the Russian case and Bulgarian case they find that the constraints in the areas of production and finance are all-encompassing in Russia meaning that these constraints affect SME growth irrespective of the characteristics of the CEO, the firm, or the sector environment, which is not the case in Bulgaria. It suggests that the nature of the constraints prevailing in production and financial sectors became more ubiquitous and all-encompassing after the fall of central planning in Russia than in Bulgaria.

Zazdravnykh [44] examined the factors making differences in the degree of development of enterprises in the manufacturing industry among Russian regions with particular attention to European and Siberian territories of the country through an econometric analysis using maximum likelihood estimator. The dependent variable in the model was the share of the employed population. The explanatory variables were the share of the urban population, the share of the employed population in small enterprises in the industry, the share of the urban population, the share of the employed population in industry, the density of infrastructure per 1000 km^2, the share of the employed population with tertiary education in a region, the share of the population aged 20–34. For the analysis, the author used a panel data set over 2005–2011 and 80 regions. The study finds that the existence and a high fraction of young people, specialization of a region and infrastructure make differences in the building of entrepreneurship. Development of cities in this purpose put a contribution in the European part of Russia and is not for Siberia and that the proportion of the population with tertiary education appears to be more important for Siberia. The findings also include that historical and geographical factors play important roles in the development of entrepreneurship.

According to Broadman [45], corruption being pervasive in Russia has been barring new entry into industries. Bribes need to be paid to tax inspectors, customs officers and local bureaucrats to initiate businesses and then to survive. It is possible to get tax payments manipulated by local authorities and have deductions.

The surveys conducted by Transparency International and by the World Bank rank Russia as a very corrupt country. According to the Corruption Perceptions Index (CPI), developed and calculated annually by Transparency International, Russia was 79th out of 91 countries surveyed in 2001. In 2010, Russia was in 154th place, with a score of 2.2, out of 178 countries [46].

According to the study conducted by the All-Russian Center for Research of Public Opinion (VCIOM) in 2008, 30% of the respondents perceived corruption to be very high in Russia, 44% perceived it just high, 19% considered it as mid-level and only 1 percent considered it as low-level. The study ranks different sectors in terms of the level of corruption starting from the highest: traffic police (33%), local authorities (28%t), police (26%), healthcare (16%), education (15%). The study also reveals that 52% had experiences of giving money or gifts to medical professionals, while 36 percent made informal payments to educators [47].

According to estimates presented by Osipian in [48], the number of reported corruption cases in Russia is on the rise. "There were 12,500 cases of bribery reported in 2008. That is a 7.7% increase as compared to 2007. The number of recorded abuses of public office in 2008 has reached 43,500, i.e. an 11.4% increase since 2007. According to the data, presented by the Attorney General, major categories of corrupt civil servants in

Russia are members of college and university faculty, school teachers, policemen, and doctors, but not state bureaucrats" [48].

Osipian in [48] on the basis of his qualitative analysis using studies, surveys and reports, predicts high levels of education corruption to harm total factor productivity in the long run. The channels for this effect, as he suggests, lowering the level of human capital and slowing the pace of its accumulation. What would be more detrimental for the whole economy is that the practices in other professions would depend very much on the ethical standards learned in universities; therefore, corruption in the education sector would negatively affect economic growth because educational institutions would supply corrupt professionals to other sectors.

Broadman in [45] reports that most small enterprises have to pay the Mafia for their survival in the business and this payment ranges from 5% of their profit to above, which the Mafia collect on a monthly basis. Owners, directors and managers are worried also because violent crime might be inflicted on them by the Mafia at any time.

Following the transition in Russia, most enterprises have experienced ownership change but the industrial configuration determined by the then central planning administration is still in place preventing competitive market-based structure, conduct and performance. Thus industrial growth- especially SME growth has been severely limited [45]. Significant restructuring of dominant firms or effective measures to eliminate non-economic barriers to entry were not undertaken in Russia. Horizontal and vertical dominance among existing firms and barriers to entry by new businesses are more predominant in Russia's industrial sector than in many other transition economies. Therefore, the absence of new businesses in Russia is striking [45].

As entry barriers into the industries in Russia have been considered the biggest factor preventing growth [45] lists the following institutionally or administratively determined constraints on entry: "(i) discrimination concerning business licensing and registration; (ii) blocked access to (state-owned) warehousing and distribution channels; (iii) a harsh and uncertain regulatory regime (changing "rules of the game") with weak mechanisms for dispute resolution; (iv) difficulty in obtaining suitable business premises and real estate; (v) corruption and organized crime; (vi) poor access to seed capital and competitively priced bank credit; and (vii) a large tax burden, including high rates, multiplicity and complexity, and lack of stability, predictability and transparency" [45].

3 Methodology

This study has used both quantitative and qualitative methods. Initially, we use World Bank Enterprise Survey data and analyze the impact of the corrupt behaviour of public officials on firm growth. While doing that we discover that data on corruption- or bribe-related questions are missing in large quantities. However, the missing values have been imputed using an econometric tool. An econometric exercise on the impact of the corrupt behaviour of government officials on firm growth has been done. OLS, IV and GMM techniques have been applied to a panel dataset. To overcome the weakness of the quantitative analysis because of the missing values for the bribe and to have a better understanding of the issues, primary data are collected directly from the firm owners and managers from Krasnoyarsk region of Russia, and through Focus Group Discussions

(FGD) with firm owners & managers and experts. Besides investigating the impact of corruption on firm growth, we wanted to investigate the factors affecting firm growth in Krasnoyarsk. For all these reasons, we ventured to survey the firms of the industrial sector of Krasnoyarsk but because of lack of resources and reluctance of the firm owners and managers that venture almost failed. We ended up collecting data from only 8 firms. Then we took the venture to hold focus group discussions to collect qualitative data from firm owners & managers, experts, current bureaucrats, retired bureaucrats, economists and businessmen on the barriers to firm growth with a special reference to the governance problem in the form of corruption of the government officials.

In the regression model of firm growth, several control variables, namely age, size, firms' innovation, infrastructure, have been used with the main variable of interest, a bribe. For econometric analysis, we have used Stata. It was not possible to run regression on the data collected by our research team members directly from firm owners & managers because there are only 8 observations. However, it helped us to compare the World Bank data with this dataset and come to a better understanding of the issues we have addressed in this study. We have analyzed the data collected through Focus Group Discussions in a qualitative method using NVivo.

Inferences on the research question have been made by combining the findings received from quantitative analysis and those from qualitative analysis.

3.1 Sampling Method

The data for the first part of our analysis was picked up from the Enterprise Survey 2012 [49] conducted by the European Bank for Reconstruction and Development and the World Bank under the title "Business environment and enterprise performance survey". For this survey, a combination of stratified random sampling and multi-stage cluster sampling methods has been adopted. The sampling method, sample size, sampling area and data collection method and other related information are available in detail on the survey website.

The data for the second and third parts of our analysis (i.e. quantitative data collected by our enumerators from firm owners or managers and qualitative data collected through Focus Group Discussions) were collected based on convenience sampling i.e. respondents and participants were chosen by personal communications with acquaintances. Although it is not scientifically a random sampling based on the listing of populations it is still random.

3.2 Dataset

As we have already been mentioned, for the quantitative analysis we have used the data from the World Bank's Russian Enterprise Survey which were collected throughout 2009–2012. To supplement this analysis and to have a better understanding of the issues we have collected primary data directly from the firm owners and managers belong to the industrial sector of the Krasnoyarsk region. For the qualitative analysis, we held five Focus Group Discussions (FGD) with experts (lawyers, political scientists, experts on enterprise, management, banking & finance, innovation, governance & institution and so on), current bureaucrats, retired bureaucrats, businessmen and economists. The Principal

Investigator conducted the FGD's while the Research Associate and Project Manager helped him by interpreting from English to Russian and from Russian to English. Each FGD consists of 10 discussants – 2 from each category of people. Each FGD lasted two and a half hours. We held 5 FGD's in total. FGD's were all recorded with the consent of the participants besides the note-taking. Data received from FGD's were first processed by excel and then analyzed by the software (NVivo).

3.3 Model Specification

For our regression analysis we use the following model which is similar to [8]:

$$y_i = \beta_0 + \beta_b b_{ij} + X_{ij}'\delta + \beta_\theta \theta_{ij} + \eta_{ij}. \tag{1}$$

Where i refers to a firm and j refers to the industry, y_i is firm growth, b_{ij} stands for corruption of the public officials in the form of bribes, θ_{ij} is a firm-specific unobservable factor (s) that may affect both rates of bribery and growth of firms. The term $X_{ij}'\delta + \eta_{ij}$ captures a firm's growth potential where X_{ij} is a vector of observable characteristics of firms such as age, size, regulation, competition, involvement in foreign trade etc. b_{ij} consists of two terms: B_j which is location-specific and B_{ij} which is firm-specific. B_j is the average bribe paid at the location j where the firm is situated, which, in turn, represents the underlying characteristics of that particular location, determining the amount of bribes bureaucrats can extract. B_{ij} denotes the idiosyncratic component [8]. The sign of b_{ij} in the above equation was expected to be negative as per the hypothesis built.

3.4 Description of Variables

The firm's growth is the endogenous variable of the model. It is measured by the change in firm sales from 2008 to 2011. The following formula is used to measure growth:

$$(\text{Firm's growth}) = [\log(\text{Sale in 2011}) - \log(\text{Sale in 2008})]/2 \tag{2}$$

We have converted the amounts of monetary values from Russian rubbles to US dollars using the relevant year's exchange rate. To measure bribe i.e. how much had to be paid as an informal payment in a year in 2008, first, we have taken this amount as a fraction of yearly sales as we need to scale down the level of bribe payments while examining the relationship between bribe and growth because bribe payments depend on the size (sale) of the firms. In the regression, we have used the log of bribe. This is because logged transformations of the values bring distributions closer to normal distributions. Also, a linear model in levels may fail to provide exact predictions because it restricts the effects of regressors to be additive [50]. Another advantage of using the logged form of variables is that it reduces unobserved heterogeneity (which is a feature of survey data) to a good extent [51]. Likewise, some other variables have been used in their log forms discussed below.

Since older firms may have better access to finance and are able to develop relations with bureaucrats, the age of a firm is to be correlated with its sales growth. Therefore, the log of the firm's age has been used as a control variable in the regression equation.

The size of a firm might affect its future growth. Therefore, size has been used in the regression equation to control this effect. Size has been measured by the following formula:

$$(\text{Firms}' \text{ size}) = [\text{lnsale08} + \text{lnsale11}]/2 \tag{3}$$

Since firms have to go through regulations made by the state, there is a need to go to government offices to get things. It creates the scope for government officials to ask for bribes. All the firms do not pay the same amount of bribe. How much a firm would pay depends on how many regulations it has to go through, which, in turn, depends on the type of industries they belong to. Therefore, to tackle this, we have used a variable "regulation" which has been measured by the log of the senior management's time spent for settling issues with regulations and government offices.

Innovation plays a very important role in the growth of firms. The more it invests in innovation, the higher would be its productivity. Therefore, innovation has been used in the regression equation to tackle the heterogeneity of innovativeness of firms.

Infrastructure is an important factor for growth. Infrastructure includes power supply, transportation, communications, roads, highways, waterways, rail and so on. Firms running in areas with good infrastructure, benefit from it and can speed up their growth. On the other hand, firms running in areas with poor infrastructure suffer from slow growth. Therefore, a variable "infrastructure" has been used to control for it.

4 Results

4.1 Descriptive Statistics

Mean, standard deviation, minimum and maximum values of the variables used in the regression model could be found in Table 1. Values of the region-wise firm's growth rate and bribe paid could be found in Table 2. Values indicating the quality of the governance could be found in Table 3.

Table 1. Descriptive statistics.

Varibales	Observation	Mean	Std. Dev.	Min	Max
Growth	2992	0.102	0.585	−3.787	6.913
Bribe	460	0.011	0.125	0	2.529
Age	5224	32.097	192.054	1	2021
Size	2992	13.838	1.827	6.824	21.92
Innovation	4220	1.795	1.037	0	2
Infrastructure	279	1038092	14400000	0	240000000

The number of days taken (days_elect) to get an electric connection is seen to vary widely over regions. In Moscow, it takes 730 days while in Kaliningrad Region and

Table 2. Region-wise growth rate and bribe paid.

Regions	Growth rate	Bribe
Belgorod Region	0.058362	1428571
Central Region	0.334922	9012.238
Chelyabinsk Region	0.08671	3571.428
Far East Region	0.308526	–
Irkutsk Region	0.202203	17857.14
Kaliningrad Region	1.1216	2500
Kaluga Region	−0.00943	80357.14
Khabarovsk Territory	−0.04663	2857.143
Lipetsk Region	−0.00467	375
Moscow City	0.106368	17901.79
Moscow Region	0.158852	12053.57
Murmansk Region	−0.22936	54857.14
North-West Region	0.271873	2467.532
Novosibirsk Region	0.972437	17857.14
Omsk Region	0.256336	6326.531
Republic of Mordovia	0.254052	1428.571
Republic of Sakha (Yakutia)	−0.24798	16517.86
Republic of Tatarstan	0.838374	14910.71
Rostov Region	−0.16674	714.2857
Saint Petersburg	−0.40571	3214.286
Samara Region	−0.22839	132142.9
Siberia Region	0.176103	38.61004
South Region	0.312409	542.328
Sverdlovsk Region	−0.18407	7857.143
Tver Region	−0.04677	13035.71
Ulyanovsk Region	−0.00775	8003.247
Ural Region	0.525439	204.0816
Volga Region	0.232845	5.102041
Volgograd Region	0.019382	3571.428
Voronezh Region	0.117254	1785.714
Yaroslavl Region	0.346669	357.1429

Table 3. Some statistics on a few variables indicating the quality of governance.

Region	days_elect	days_water	days_customs	days_input
Russia	157	34	8.5	33.5
Belgorod	15	15	5	30
Central	180	90	3	30
Kaliningrad	1	1	7	22
Kaluga	1	14	5	30
Leningrad	30	30	30	5
Moscow	730	90	3	90
North-West	45	3	7.5	17.5
Volga	365	60	7	60
Krasnoyarsk	47.5	30	1	22

Source: World Bank Enterprise Survey, 2013
Note: days_elect is the number of days taken to get an electric connection, days_water is the number of days taken to get water line, days_customs is the number of days required to get clearance from the customs, days_input is the number of days of inventory of most important input.

Kaluga Region it takes only 1 day. In this region of our special interest i.e. the Krasnoyarsk Region it takes 47.5 days. The number of days taken to get water line (days_water) takes a smaller number of days in comparison to electric connection. Variation in it is again very high. For the waterline, the Moscow region and the Central region, and the Volga Region take more than the national average. The smallest number of days taken by Kaliningrad Region and then by North-West Region – 1 day and 3 days respectively. Krasnoyarsk Region clears customs in the smallest number of days i.e. in 1 day, Central Region and Moscow in 3 days, Belgorod Region and Kaluga Region in 5 days. The highest number of days for this is taken by Leningrad, which is 30 days. Other regions take a similar number of days. Average days of inventory of most important input is another indicator of governance quality (days_input). In this respect, the best performance is shown by Leningrad Region and the worst is by Moscow Region. The region of our particular interest i.e. Krasnoyarsk region is moderate.

4.2 Regression Models Output

The results of building the most relevant regression models explaining a firm's growth dynamics could be found in Table 4 (OLS), Table 5 (2SLS) and Table 6 (GMM).

Table 4. The results of building Regression model 1 (OLS).

Source	SS	df	MS	Number of obs = 372
Model	1.90523598	5	0.381047195	F(5, 5) = 4.63
Residual	0.411863105	5	0.082372621	Prob > F = 0.0591
Total	2.31709908	10	0.231709908	R-squared = 0.8223 Adj R-squared = 0.6445 Root MSE = 0.28701

growth	Coef	Std. Err	t	P > |t|	[95% Conf. Interval]	
bribe	7.868459	2.732379***	2.88	0.035	0.8446543	14.89226
age	−0.0008268	0.0008908	−0.93	0.396	−0.0031167	0.001463
size	0.0177432	0.069358	0.26	0.808	−0.1605472	0.1960335
inov	−0.6213712	0.3127853***	−1.99	0.104	−1.425411	0.182669
infra	−1.44e−07	1.89e−07	−0.76	0.480	−6.31e−07	3.42e−07
_cons	1.165108	1.043342	1.12	0.315	−1.516889	3.847104

Table 5. The results of building Regression model 2 (2SLS).

Source	SS	df	MS	Number of obs = 370
Model Residual	1.92534561 0.391753475	5 5	0.385069122 0.078350695	F(5, 5) = 4.91 Prob > F = 0.0527
Total	2.31709908	10	0.231709908	R-squared = 0.8309 Adj R-squared = 0.6619 Root MSE = 0.27991

growth	Coef	Std. Err	t	P > |t|	[95% Conf. Interval]	
bribe	7.558449	2.727049	2.77	0.039	0.5483465	14.56855
bribe	0	(omitted)				
age	−0.0214786	0.0199213	−1.08	0.330	−0.0726879	0.0297307
size	0.0181031	0.0656488	0.28	0.794	−0.1506524	0.1868586
inov	−0.5892654	0.3094861	−1.90	0.115	−1.384825	0.2062941
infra	−1.60e−07	1.78e−07	−0.90	0.412	−6.17e−07	2.98e−07
_cons	1.208008	0.9755056	1.24	0.271	−1.299609	3.715625

Table 6. The results of building Regression model 3 (GMM).

Instrumental variables (GMM) regression						Number of obs = 370
GMM weight matrix: Robust						Wald chi2(5) = 455.18 Prob > chi2 = 0.0000 R-squared = 0.7658 Root MSE = 0.2221
Growth	Coef	Robust Std. Err	z	P > |z|	[95% Conf. Interval]	
bribe	9.310317	2.348519	3.96	0.000	4.707304	13.91333
age	−0.0269261	0.0114421	−2.35	0.019	−0.0493521	−0.0045001
size	0.0812954	0.0299453	2.71	0.007	0.0226036	0.1399871
inov	−0.4115891	0.2037816	−2.02	0.043	−0.8109937	−0.0121845
infra	−2.10e−07	9.50e−08	−2.21	0.027	−3.96e−07	−2.35e−08
_cons	0	(omitted)				

From the tables above it can be seen that the coefficient of the variable of our interest i.e. bribe has appeared with a 1% significance level (highly significant coefficient) that is, bribe affects firm growth positively. Precisely, if bribe increases by 1%, firm growth increases by 9.3% according to IV with GMM estimation, which should be considered the most accurate. All the control variables in the regression are seen to be highly significant. Size is seen to affect growth positively while age, innovation and infrastructure are seen to affect growth negatively. Results will be explained further.

4.3 Discussion of Results

Regression Results from World Bank Data

The regression results show a positive effect of bribery on firm growth, which goes against our hypothesis. However, the positive impact of bribery supports the ‘speed money’ or ‘greasing the wheel’ theory which was formulated based on the fact that in South Korea, the Philippines, Taiwan, Hong Kong there was a high level of corruption but at the same time those countries achieved high-level growth. It means that when corruption is wide and deep in the bureaucracy bribe helps to speed up the economic activities and thus have a positive impact on firm growth. The general wisdom is corruption is very high in Russia and this general wisdom is supported by the high corruption perception index produced by Transparency International and World Bank. Therefore, in Russia like the countries mentioned above, firms benefit from bribery because otherwise the economic activities would slow down and firm growth would fall.

Quantitative Primary Data Collected Directly from Firms

It is seen that firms are seen reluctant to reveal information about the bribe. It is also revealed that the respondents do not consider bribery a big problem for them. They don’t complain about bribe payments, it means that they benefit from paying the bribe and achieve better growth. They rather tend to mention other obstacles for their growth such as lack of finance, high tax rate, prices of power etc.

Results from Focus Group Discussion
On the question of obstacles to firm growth, the participants of FGD's put forward obstacles that are unique to Russia – may be typical of transition economies that have switched from Socialist centrally planned system. Most respondents mentioned the problem of the existence of the vertical system of decision making and allocation and distribution, which make the firms and businesses located far from the capital city or at the regional level suffer immensely. Monopolization of businesses by big companies patronized and favoured by the government or high-ups of the government in all sectors has appeared to be a big problem. And this Monopolization is created by the companies which are owned by the people connected to the power structure of the government or by the government high officials or ministers themselves. This Monopolization prevents new firms from coming into being. Lack of finance has been marked as another big problem. Even if it is available it costs too high in terms of interest rate to borrow. Participants recall how the banking system was destroyed by allocating state funds to 3 to 3.5 thousand young people related to government high-ups and an extremely inefficient and dishonest system of banking was introduced. Now many of those banks are sold out. A proper banking system that is commercially viable was absent in this country for a long time. Thus, real entrepreneurs are not coming up. The high tax rate and complicated tax system have been held as another big obstacle to firm growth. A complicated tax system encourages tax evasion, which creates the scope for corruption and thus government officials reduce tax figures for businessmen in exchange for a bribe.

Although there is general wisdom inside and outside Russia that corruption is very high in this country, most of the participants in the FGD's do think that corruption is a big problem anymore in this country. However, this response seems to be biased because Russian people traditionally do not tend to talk about corruption and admit the existence of corruption or the payment of a bribe. They have not denied the existence of corruption but not on large scale. They have also suggested corruption exists mainly at the top level which is engaged in offering big contracts like oil and gas sectors, highways, bridges, transportation and so on. At a lower level, corruption is low, which does not create big problems. They also suggested that at a lower level around 10 years ago corruption has been reduced.

On the question of the existence of syndicated corruption of a group consisting of bureaucrats, police, judge, politicians and mafia, the participants admitted that it existed until 2010 but after 2010 the situation has improved although it has not fully gone. Even nowadays when an entrepreneur takes the initiative to build a factory, an existing established and powerful entrepreneur, being heard of it and considering the new entrant a rival, send musclemen to stop it. He manages to stop it with the help of local politicians, bureaucrats, police and the mafia. In such a case new entrepreneur has close down his entrepreneurship in mid-way incurring him a huge loss.

On the question of 'does corruption help' a good number of participants replied in affirmative. They mentioned that 'kick-back' type bribe helps to get things done from government offices or clear goods from customs or cutting the tax figure but in such a case, one has to have or build personal relations with the related government officials.

5 Conclusion

We can conclude that one important and notable feature of Russian businessmen and the citizens of the country is that people do not want to talk about corruption or bribery even if they pay the bribe and this bribe payment affect the growth of their firms or life as a whole.

This may be the legacy of communist rule from which they fear imprisonment. A deep-rooted fear may exist inside the minds of the Russian people that if they talk about the corruption of the public officials, politicians or government, they may have to be in trouble in the form of imprisonment or any other punishment imposed by the government. During Focus Group Discussions (FGD) we experienced two incidents: in one FGD, after the discussion was over, one entrepreneur came to the Principal Investigator and said to him whispering 'of course we need to pay bribe but we do not dare to mention it even in a small gathering like this because the fear of imprisonment haunts us and this fear of imprisonment has been injected into our mind from our parents and grandparents who experienced imprisonment during Stalin's period.' In another FGD when we started asking the question 'do you need to pay a bribe for clearing customs or get things done in government offices?' the only bureaucrat participating in the FGD, who is a lady left the room without telling anybody. In the Krasnoyarsk region, all data for the variable bribe are missing.

The information above tells us that it is difficult to assess the bribery level in Russia. Therefore, assessing the impact of bribery on firm growth quantitatively is also difficult. Therefore, we held FGD's but in FGD's most respondents didn't mention bribery as a barrier to growth, which means bribery may help to speed up their growth while they know that corruption of bureaucrats would create obstacles for them and they must pay bribes and benefit.

The inability to collect some quantitative data on firm growth and bribery was a big limitation of the study. Even in secondary data analysis data on important variables of the interest of the study were hugely missing. However, if resources and funding are available, in future we can conduct further research on a large scale over the whole country, which would enable us to have useful findings for the development of the industrial sector of the country and understand how corruption may be reduced mainly from the top level of the government.

The following policy suggestions may be put forward to the policymakers based on the findings of our research:

1. The structure of the government has to be made horizontal breaking the vertical structure and regional offices have to be empowered. Industrial and business units all over the country must enjoy equal access to opportunities.
2. The monopolization of companies has to be broken. Must take initiative to bring new entrepreneurs in the market providing subsidies, tax holidays, tax release and other facilities.
3. The tax structure has to be made easier and simpler. The tax rate has to be cut substantially.
4. Subsidies and other facilities have to be provided to the business.
5. The price of fuel, electricity and other basic resources need to be reduced.

6. A commercially viable and efficient banking system has to be introduced and low-cost funds for businesses have to be made available.
7. To reduce corruption, the supporting system has to become more transparent by introducing digital services, arranging for lodging complaints against government officials digitally, increasing the salaries of public officials and increasing the punishment of corruption.
8. Reducing the number of documents or licenses for start-ups and all relevant papers should be made available in one office.

The results of our study could be valuable for further studies on obstacles to firm growth in Russia and relevant policies ensuring economic development.

References

1. La Porta, R., Lopez-de-Silanes, F., Shleifer, A., Vishny, R.: The quality of government. J. Law Econ. Organ. **15**(1), 222–279 (1999)
2. North, D.: Growth and structural change. Norton, New York (1981)
3. De Long, J.B., Shleifer, A.: Princes and merchants: European city growth before the industrial revolution (1993)
4. Knack, S., Keefer, P.: Institutions and economic performance: cross-country tests using alternative institutional measures. Econ. Polit. **7**(3), 207–227 (1995)
5. Easterly, W., Levine, R.: Africa's growth tragedy: policies and ethnic divisions. Q. J. Econ. **3**, 1203–1250 (1997)
6. Acemoglu, D., Johnson, S., Robinson, J.A.: Institutions as a fundamental cause of long-run growth. In: Handbook of Economic Growth, vol. 1, pp. 385–472 (2005)
7. Aidt, T., Dutta, J., Sena, V.: Governance regimes, corruption and growth: theory and evidence. J. Comp. Econ. **36**(2), 195–220 (2008)
8. Fisman, R., Svensson, J.: Are corruption and taxation really harmful to growth? Firm level evidence. J. Dev. Econ. **83**(1), 63–75 (2007). https://doi.org/10.1016/j.jdeveco.2005.09.009
9. Nguyen, T.T., Van Dijk, M.A.: Corruption, growth, and governance: private vs. state-owned firms in Vietnam. J. Bank. Financ. **36**(11), 2935–2948 (2012)
10. Chakravorty, N.N.T.: Corruption in Bangladesh: its implications for firm level and macroeconomic growth. Ph.D. thesis, University of Bath, EThOS, BRITISH LIBRARY (2015). http://ethos.bl.uk/OrderDetails.do?uin=uk.bl.ethos.687332EThOS. BRITISH LIBRARY database
11. Safavian, M.S., Graham, D.H., Gonzalez-Vega, C.: Corruption and microenterprises in Russia. World Dev. **29**(7), 1215–1224 (2001)
12. Leff, N.H.: Economic development through bureaucratic corruption. Am. Behav. Sci. **8**(3), 8–14 (1964)
13. Huntington, S.P.: Political Order in Changing Societies. Yale University Press, New Haven (1968)
14. Lui, F.T.: An equilibrium queuing model of bribery. J. Polit. Econ. **93**(4), 760–781 (1985)
15. Leys, C.: What is the problem about corruption? J. Mod. Afr. Stud. **3**(02), 215–230 (1965)
16. Méon, P.-G., Sekkat, K.: Does corruption grease or sand the wheels of growth? Public Choice **122**(1–2), 69–97 (2005)
17. Rose-Ackerman, S.: Corruption: A Study in Political Economy. Academic Press, New York (1978)
18. Khan, M.: Determinants of corruption in developing countries: the limits of conventional economic analysis. In: International Handbook on the Economics of Corruption, pp. 216–244 (2006)

19. Morgan, T.: The theory of error in centrally-directed economic systems. Q. J. Econ. **4**, 395–419 (1964)
20. Myrdal, G.: Asian drama, an inquiry into the poverty of nations (1968)
21. Shleifer, A., Vishny, R.: Corruption. Q. J. Econ. **108**(3), 599–617 (1993)
22. Wedeman, A.: Looters, rent-scrapers, and dividend-collectors: corruption and growth in Zaire, South Korea, and the Philippines. J. Developing Areas **1**, 457–478 (1997)
23. Shaw, P., Katsaiti, M.S., Jurgilas, M.: Corruption and growth under weak identification. Econ. Inq. **49**(1), 264–275 (2011)
24. Mauro, P.: Corruption and growth. Q. J. Econ. **110**(3), 681–712 (1995)
25. Kaufmann, D., Wei, S.-J.: Does "grease money" speed up the wheels of commerce? (1999)
26. Reinikka, R., Svensson, J.: Local capture: evidence from a central government transfer program in Uganda. Q. J. Econ. **3**, 679–705 (2004)
27. Aidt, T.S.: Corruption, institutions, and economic development. Oxford Rev. Econ. Policy **25**(2), 271–291 (2009)
28. Svensson, J.: Eight questions about corruption. J. Econ. Perspect. **19**(3), 19–42 (2005)
29. Li, S., Wu, J.J.: Why China thrives despite corruption. Far Eastern Econ. Rev. **170**(3), 24–37 (2007)
30. Rock, M.T., Bonnett, H.: The comparative politics of corruption: accounting for the East Asian paradox in empirical studies of corruption, growth and investment. World Dev. **32**(6), 999–1017 (2004)
31. Vial, V., Hanoteau, J.: Corruption, manufacturing plant growth, and the Asian paradox: Indonesian evidence. World Dev. **38**(5), 693–705 (2010)
32. Wedeman, A.: Development and corruption: the East Asian paradox. In: Political Business in East Asia, pp. 34–61. Routledge, London(2002)
33. Campos, N.F., Dimova, R.D., Saleh, A.: Whither corruption? A quantitative survey of the literature on corruption and growth (2010)
34. Ugur, M.: Corruption's direct effects on per-capita income growth: a meta-analysis. J. Econ. Surv. **28**(3), 472–490 (2014)
35. Mauro, P.: The effects of corruption in investment and government expenditure: a cross-country analysis. In: Elliot, K.A., (ed.) Corruption and the Global Economy, Institute for International Economics, Washington DC, Chapter 4 (1997)
36. Tanzi, V.: Corruption around the world: causes, consequences, scope, and cures. IMF Staff Pap. 1–24 (1998)
37. De la Croix, D., Delavallade, C.: Growth, public investment and corruption with failing institutions. Econ. Gov. **10**(3), 187–219 (2009)
38. d'Agostino, G., Dunne, J., Pieroni, L.: Corruption and growth in Africa. Eur. J. Polit. Econ. **43**, 71–88 (2012)
39. d'Agostino, G., Dunne, J.P., Pieroni, L.: Government spending, corruption and economic growth. World Dev. **84**, 190–205 (2016)
40. Huang, C.-J.: Is corruption bad for economic growth? Evidence from Asia-Pacific countries. North Am. J. Econ. Financ. **35**, 247–256 (2016)
41. Célimène, F., Dufrénot, G., Mophou, G., N'Guérékata, G.: Tax evasion, tax corruption and stochastic growth. Econ. Model. **52**, 251–258 (2016)
42. Pissarides, F., Singer, M., Svejnar, J.: Objectives and constraints of entrepreneurs: evidence from small and medium size enterprises in Russia and Bulgaria. J. Comp. Econ. **31**(3), 503–531 (2003)
43. Viktorova, T.S., Shesterikov, D.S.: State of development of small and medium entrepreneurship (2015)
44. Zazdravnykh, Y.A.: Why there are more entrepreneurs-manufacturers in one regions and less in others: an empirical evidence. Econ. Reg. **3**, 140–146 (2014)

45. Broadman, H.G.: Reducing structural dominance and entry barriers in Russian industry. Rev. Ind. Organ. **17**(2), 155–175 (2000)
46. Corruption perception index 2010. Transparency International (2010)
47. Leonidov, A.: Bribes in HEIs: money, gifts and sex (2009)
48. Osipian, A.L.: Education corruption, reform, and growth: case of Post-Soviet Russia. J. Eurasian Stud. **3**(1), 20–29 (2012)
49. Enterprise Survey 2012. Russian Federation, 2011–2012. World Bank, European Bank for Reconstruction and Development (2012)
50. Wooldridge, J.: Introduccion a la econometria/Introductory Econometrics: A Modern Approach. Cengage Learning Editores (2009)
51. Cameron, A.C., Trivedi, P.K.: Microeconometrics using stata, vol. 5. Stata Press, College Station (2009)

Conceptual Model for a Shared Cybersecurity Operations Center for ICS

Willian Dimitrov[1](✉), Georgi Dimitrov[1], Kamen Spasov[2], and Svetlana Syarova[1]

[1] University of Libraraian Studies and Information Technologies, 1784 Sofia, Bulgaria
v.dimitrov@unibit.bg
[2] Sofia University, 1000 Sofia, Bulgaria

Abstract. The article offers an analysis of the scope of a conceptual model for a Shared Industrial Control Systems (ICS) Cybersecurity Operations Center (SCOC). It ensures continuous proportional ICS protection against malicious actors and in accordance with regulations and standards. The focus of the research is on the three basic functions of the cybersecurity cycle – proactive functions, security operations management, and cybersecurity incident response. The aim of the study is to increase the effectiveness of cybersecurity for all Shared SOC subscribers, to overcome the problem of staff shortages, and allow each company to focus on its core business. The proposed analysis provides reference points for synthesizing a private solution for multiple clients of Shared SOC, a basis for the concept of operations and preparation for technical design. Shared SOC provides a powerful tool for meeting the requirement for multi-layered cyber protection, which is already a fact in many regulatory measures. Shared SOC creates conditions for all subscribers with ICS to increase their level of maturity in cybersecurity. Our contribution is in determining the meta scope of operations performed by the SCOC team.

Keywords: Cyber · Security · Operations · Center · ICS · Shared · centre

1 Cyber Defense Issues for ICS

Industrial Control Systems (ICS) guarantee basic operations for critical infrastructure from energy, water supply, transport and manufacturing. The topic of sharing of common cybersecurity services for ICS [1,2] is an important direction for research. This is caused by (1) the outstripping growth of the potential of hackers over that of cyber defenders [1–3]. According to publications [4,5], an

This work is supported by the PPNIP-2021-09/12.04.2021 "Analysis and optimization by algorithms for the classification of signals from Smart IoT devices" and National Science Program "Information and Communication Technologies for Unified Digital Market in Science, Education and Security".

R. Silhavy et al. (Eds.): CoMeSySo 2021, LNNS 231, pp. 493–503, 2021.
https://doi.org/10.1007/978-3-030-90321-3_40

attacker who is already in the corporate information system could penetrate the industrial network in 82% of cases. (2) Limitations of existing solutions for Security Operations Centers (SOCs). In a previous article [6] we published a comparative study of the services offered by cybersecurity products and the real needs, in accordance with the legal framework in the field of ICS and SCADA [7–9]. Cybersecurity technologies have a limited scope based on the corporate experience of the respective manufacturer. These proposals lack models for organizing the activities of security teams. The challenges arise from the diversity of the security technologies and services from different providers. Combining them in cyber defense systems leads to a lack of interoperability between different security systems, insufficient visibility from end to end, inadequate resources to reduce risks [10,11]. (3) The convergence of IoT and ICS has led to the notion of Operation Technologies (OT), which has significantly complicated the threat surface and cyber defense efforts [12,13]. Our review of ICS security breaches in the area of critical infrastructure covers the period after 2003 [7,14,15].

2 Methodology of the Research

We introduce a conceptual model for a Shared Cybersecurity Operations Center (SCOC). The model has a holistic nature and presents a common vision for the three main components of cybersecurity governance in relation to an organizational structure and integrated technology solution. The target function of this type of centre is to deliver cybersecurity services to a number of ICS installations. This approach allows organizations to outsource cybersecurity services to ICS and information systems, as well as their communication connectivity protection to an external structure. Its main goal is to increase the effectiveness of cybersecurity for all SCOC subscribers and to allow to overcome the problems of (1) the need to combine the requirements of different standards and frameworks in the management of ICS cybersecurity, (2) staff shortages, and (3) any organization that owns ICS and is a subscriber to SCOC services will be relieved of much of its cybersecurity efforts and focus on its core business. The proposed model provides reference points for synthesizing a shared private solution for multiple clients of SCOC, a basis for conceptual research and drafting of terms of reference and project. We justify the need to integrate an extended set of functionalities in the conceptual model for SCOC. We believe that they are the basis for a proportional response in the cyber defense of the ICS in the current conditions of increasing the capabilities of malicious actors, complicating regulatory requirements and the lack of cyber expertise. Our contribution is in determining the meta scope of operations performed by the SCOC team. This can be a methodology for compiling a framework for constructing a private solution for SCOC.

What is the reason for synthesizing the proposed conceptual model? We believe that the existing security models that are applicable in such ecosystems are partially acceptable in the scope of our preposition, but none of them covers all the organizational and technical aspects of the layered cybersecurity view from individual IT assets to corporate policy. Our vision for laying the model includes creating a methodology for their application and playing the role of umbrella.

The model includes regulatory restrictions, which in individual cases can be considered according to the specifics of the respective industry area and technological environment. In determining the set of operations that are delegated to the SCOC, we take into account existing laws, directives and standards applied in the field of ICS security.

We believe that the operations required for the SCOC determine the functionalities of the organizational solution holistically integrated with the technical design of the architecture, and the elements of the conceptual model are based on the set of functionalities. Center congestion and scalability requirements are topics of requirements for technical architecture that go beyond the scope of this article.

The analyzed model is based on empirical experience from the establishment of crisis management centers for natural disasters and accidents and cyber protection solutions for large ICT systems. What is common here is the notion that the cybersecurity incident handled by the SCOC has the nature and effects of a crisis. The process management requirements in both cases are similar [16]. We use a methodology for the synthesis of functions depending on the needs for implementation of processes (Feature-driven development, FDD), effective at the stages of research and technical assignment [17], a multidisciplinary approach in the design of reliable and robust protection systems [18].

The proposed conceptual model for a shared cybersecurity center is characterized by the completeness of the range of services offered, typical of the umbrella approach [19]. Differences from commercially available cybersecurity technologies and private solutions that are integrated into enterprise cybersecurity systems cover:

1. Anticipatory threat intelligence,
2. Dynamic adaptation of subscriber protection,
3. Applied contextual studies of potential threats and change management in subscriber technology environments,
4. The full cycle of a cyber-incident,
5. Control over the supply chains of ICT assets,
6. Competencies update of the SCOC team and the teams of the client companies.

Our vision for the full cycle of an incident includes all the processes that are initialized after its onset. The client organization of the shared cybersecurity center, various departments of the center itself and many external third parties are involved in these processes. Our empirical experience in security incident management shows that this process involves a broad ecosystem of participating organizations dealing with various aspects of the consequences of the incident. In practice, we consider that the incident is completed only after all possible commitments of the injured organization to the affected third parties and individual citizens have been definitively dropped together with all institutional regulatory obligations regarding the incident.

We accept that the purpose of SCOC is to protect against cyber-attacks a set of ICS, which are or not connected to organizational information systems and contain both modern and outdated automation technologies.

Centralized security features will increase the effectiveness of the technologies installed in the SCOC. Much of the simultaneous effort to manage cybersecurity and supporting assets for every subscriber will be avoided. In our concept these are delegated to SCOC.

To specify the requirements for the model we analyze the function of the SCOC based on the existing legal frameworks, standards and sectoral security requirements of the ICS, published by the leading institutions worldwide. SCOC provides cyber protection for more than one ICS installation, the owner of each ICS installation may be different.

The purpose of the analysis is to identify a set of functionalities that will enhance the effect of SCOC. We take into account many differences between security requirements and the needs of the traditional ICT system and ICS. The common set is listed, for example, in [8]. The ICS cybersecurity ecosystem is now far beyond standards and includes many elements of the digital space related to internet security and the supply chain [20]. At the same time, the vision of cyber defenders for proactive behavior and reuse is evolving.

The model is dimensioned to prevent high-level cybersecurity incidents proportionally. According to [21], these are interruptions that are the most serious and are considered significant. The response to these incidents must be immediate. High-level incidents include life-threatening activities, compromising critical systems or information, compromising privileged accounts, and breaking the law.

Behind the term cybersecurity incident management, we mean a complex process that encompasses preventive activities, detection, analysis, response and recovery from destructive events with subsequent feedback to prevent similar situations. The purpose of incident management is to mitigate the impact of a devastating event. To achieve this goal, processes are established that detect and analyze events to determine if an incident is ongoing, respond, eliminate the consequences, and improve the organization's ability to respond to a future incident.

The resources and efforts required for management depend on the impact of the incident and the organization's operational environment on third-party supplies and product placement. We believe that in the event of a security incident, the most important part of remedial action is clarity at the end. Once the organization has completed its review and reported on the lessons learned, it is important to close the incident. The closure of the incident is an official statement that no further action is required. Incident closure activities include notifying all stakeholders affected by the incident that it has been resolved and that they should not experience additional effects. Important practices for improving incident management capabilities include root cause post-incident analysis [21] and updating cybersecurity policies within change management and knowledge management functionalities.

In building this conceptual model, we took into account the general characteristics of invasion detection processes and the activities for recovery and elimination of the consequences of cybersecurity failures. We share the view that the response to an incident is the most problematic part of the whole cyberdefense management process [22,23].

We did a systematization of the set of functions that ICS SCOC has to perform on the basis of the analysis of cyber incidents known in the last twenty years in critical infrastructure sites. The obtained set of functions provides reduction of the operating costs of SCOC for each subscriber; it overcomes the shortage of cybersecurity staff in the individual organizations managing ICS; allows to focus on the accumulation of experience and knowledge, incorporation of advanced technologies that allow proactive security policies; detection of new threats at early stages of attacks; providing the enterprise with access to a package of services, including remote control of the perimeter.

3 Scope of SCOC Conceptual Model

We assume that the subscribers to SCOC have an approximately similar technical architecture in their respective layers of IMS: sensors, PLCs, non-IP networks, data collection channels, portals, databases, servers and orchestration systems, connections with external systems to suppliers, subcontractors and institutions. The communication connectivity and the main network services DNS, NTP, SMTP for the subscribers are provided through SCOC. The three main pillars of the SCOC are aggregates of proactive functions, operational management and incident response, covering the cross-section of cybersecurity services required for subscribers (Fig. 1).

Proactive features	Operational Security Management	Responce to a Cybersecurity Incident
Connections with External SOC	Asset Management	Operational Measures
Threat Intelligence	Vulnerabilities Management	Gathering Evidences
Supply Chin Control	Updates Management	Recovery
Knowledge Management	Privileged User Control	Investigative Bodies
Change Management	Endpoint Security Management	Supply Chain Disclosure
Regulatory Compliance	Penetration Testing	External Consultants
Standards Compliance	SIEM	Publicity
Research and Development	**Cyber security Incidents Escalation**	Insurance Events
Security incident response planning		Elimination of the Consequences
		Security Incident Management

Fig. 1. Generic SCOC conceptual model

Proactive Features. The features in this stage have the greatest weight in the whole cybersecurity model from the view point of real cyber protection activities. It means governance of the activities according corporate cybersecurity policy aligned with business strategy. Organizations maintain secure connections with third parties and contractors. Any such relationship poses a risk that must be monitored.

Cybersecurity policy management includes compliance with regulatory requirements, security standards, and business policy and organizational objectives. The shared SCOC can manage the core of the policy, which is common to all subscribers, and provide methodological guidance on the specific sections that affect each of them. Sharing the security of basic network services with the SCOC abilities will increase the level of protection for all subscribers [24,25].

Relations with External Organizations. Regulatory measures require organizations to maintain communication with many external organizations during different stages of the cyber security lifecycle: in communications with external providers, in operational management. In dealing with the consequences of a security incident, communications with investigative bodies, external consulting companies, insurance companies, government institutions, the press and others are a must. The escalation of incidents requires communication with government departments and branch organizations, subcontractors for the provision of cybersecurity services, public relations. These communications cover the entire life cycle of a security incident from initial detection to return to steady state. The team is made up of security professionals dedicated to protecting the business.

Research and Development. This Kind of Activities are in the Scope of the Supply Chain Control. The SCOC team can manage the test site with simulation of common ICS and employee training [26]. They research advanced technologies for future implementation in SCOC and controls subcontractors who develop new solutions. Laboratory testing of updates is a prerequisite for avoiding problems when applying security updates to live systems. The SCOC team multiplies the effect as the results are used by all affected subscribers.

Ensuring Security Throughout the Supply Chain. Organizations communicate with multiple contractors in the supply chain, which increases the area for attack. The SCOC implies a protective strategy to reduce the risks of third parties in the supply chain. It must include technical and organizational measures, compliance with standards, regulations and control over the content of contracts that may create preconditions for cyber incidents. This means accepting the principle of zero confidence and the need for constant monitoring of suppliers [27], control over the delivery of products; contract management with cloud service providers and telecommunications operators in the security and SLA (Service Level Agreement) sections. These functionalities reflect the dynamics of the regulatory measures which have been growing in recent years, and ensure compliance with standards. It gives a complete picture of the risk of third parties for each subscriber of SCOC. SCOC provides an opportunity to study advanced technologies and their timely implementation.

SCOC creates conditions for assessing the impact of the proposed setting and tests the correction for non-critical assets in a laboratory environment - operating systems, databases, dashboards, firewalls and other peripheral protection devices. This ensures that operating systems are not adversely affected by updates. SCOC provides security system updates once for all, instead of updates for each subscriber. The same applies to ex-post network segmentation and access control checks. SCOC proactively performs regression testing before implementing updates to operational or critical assets. This leads to possibility to implement learning algorithms with their own proprietary data sets collected from real attacks [28].

Operational security management includes responding to events from real-time monitoring and incident reporting systems; identification of penetration points and removal of vulnerabilities, penetration testing, asset management and communication links, exchange performed by each asset, detection of problems with incorrect configurations and other weaknesses. The exact matching of the vulnerabilities to the controllers is performed on the basis of the firmware version data. The traffic from mirror ports in the switches is submitted to the IDS for analysis. This set of features is already in the appropriate condition and established in the research and market. Usually it includes the next technologies: SIEM, IDS/IPS, Antispam, Encryption, Network Access Control. The important component of this part of the model is organizational solution. It depends on the maturity of corporate management and implemented security policies.

Our Vision for the Operational Engagement of the SCOC Units Covers Activities that Affect the Entire Subscriber Security Ecosystem. The SCOC team identifies the increased risks, removes the false positive results, identifies the security events that have occurred, which are related to already published incidents, and investigates the threats. They rise alarms for detected incidents after analysis of related security events. The SCOC monitors system configurations for invasion attempts, security vulnerabilities, regular updates, and signs of compromise. They acquire filtered normalized log files and perform correlation and analysis of log files with the latest technologies - big data analysis, ML, AI. They do anomaly analysis of indicators of compromise (IoC) to find enemy tools or artifacts; create hypotheses for testing; random search, use of automated tools for signal detection, tests for security vulnerabilities in the changing domain of ICS [29].

Automated Security Activities. SCOC provides a single portal with employees trained in the use of automation tools for security processes. Literature sources describe the effects of fatigue on security experts 24/7 [30,31].

Response to a Cybersecurity Incident. It includes incident registration, prioritization, tracking and providing a complete history of events and activities throughout the life cycle of the incident, managing processes to eliminate the consequences of security incidents.

The prepared in advance Cyber Incident Response Plan is activated (employees must be trained to implement it), the negotiation of commitments of third parties: external lawyer, consultants, investigative bodies, law enforcement and regulatory authorities; public relations management through press conferences and publications on company websites. The main links between operational management and the response to a cybersecurity incident are presented at Fig. 2.

Review of the Reaction After the Incident and Lessons Learned. Forensic research on preserved original data and media has to be done. It includes collection of all appropriate log files: from DNS, firewalls, proxies, system event log, social security communication for additional data, damage assessment. SCOC is an institution responsible for assessing security incidents. This requires combined technical skills and experience in investigation. Suitable indicators can be selected from [32].

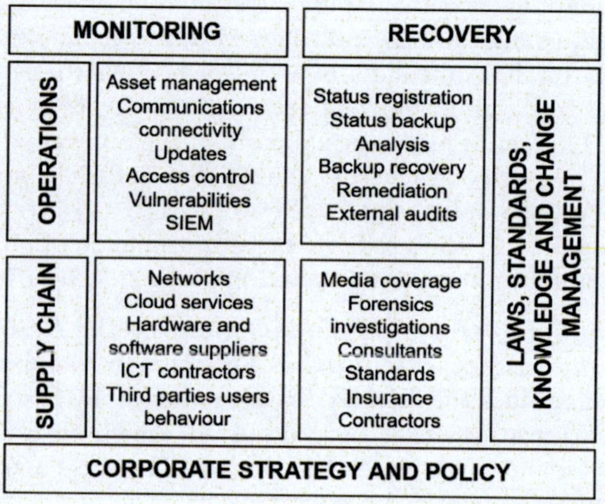

Fig. 2. Complexity of response to a cybersecurity incident

Investigation of Security Incidents. The organization may have suffered successive cyber attacks [33]. The analysis of many incidents shows that (1) they take a long time after all technical measures have been taken in response to the incident, (2) third parties are involved in the subsequent analysis of the incident, (3) third party claims arise, (4) the organization continues to allocate resources to address these long-term effects, (5) needs coordinated management of the processes associated with these effects.

4 Conclusion

The SCOC must address the following gaps in ICS cybersecurity: lack of active investigation and integrity of evidence and data; lack of compatible forensic tools for field devices. SCOC's clients, for historical reasons, also support outdated and advanced technologies. Identifying data sources in SCADA and the many layers of connectivity in a complex architecture is a difficult task in its essence [9].

Today's SOCs will develop radically in the near future as cutting-edge technologies are integrated by the cybersecurity industry. The idea of a SCOC provides a powerful tool for meeting the requirement for multi-layered cyber protection, which is already a fact in many regulatory measures. SCOC creates conditions for all subscribers with ICS to increase their level of maturity in cybersecurity [34]. The conceptual model can be useful for security system architects, system architects and integrators. In accordance with design standards, the logical next step in research is synthesizing of conceptual architecture and organizational solution for management of the individual functionalities of SCOC.

References

1. Tweneboah-Koduah, S., Skouby, K.E., Tadayoni, R.: Cyber security threats to IoT applications and service domains. Wirel. Pers. Commun. **95**(1), 169–185 (2017)
2. Schwab, W., Poujol, M.: The State of Industrial Cybersecurity 2018. Kaspersky ICS Whitepaper.pdf (2018)
3. Livingstone, D., Baylon, C., Brunt, R.: Cyber Security at Civil Nuclear Facilities. Understanding the Risks. Chatham House Report (2015)
4. Reese, E., Miller, S.: A totally tubular treatise on TRITON and TriStation (2018). https://www.fireeye.com/blog/threat-research/2018/06/totally-tubular-treatise-on-triton-and-tristation.html
5. Technologies, P.: Industrial companies: attack vectors (2019). https://www.ptsecurity.com/ww-en/analytics/ics-attacks-2018
6. Dimitrov, W., Syarova, S.: Analysis of the functionalities of a shared ICS security operations center. In: 2019 Big Data, Knowledge and Control Systems Engineering (BdKCSE), pp. 1–6. IEEE (2019)
7. Andersson, G., et al.: Cyber-security of SCADA systems. In: 2012 IEEE PES Innovative Smart Grid Technologies (ISGT). IEEE (2012)
8. Stouffer, K., Falco, J.: Guide to Industrial Control Systems (ICS) Security. Recommendations of the National Institute of Standards and Technology. Special Publication 800–82 (2011)
9. Mittal, S.: The Issues in Cyber-Defence and Cyber-Forensics of the SCADA Systems (2015). https://www.cisoplatform.com/profiles/blogs/the-issues-in-cyber-defence-and-cyber-forensics-of-the-scada
10. Managed Security Services. DDCC-1437/11/13 (2013). https://www.dimensiondata.com/en-us/services/managed-services/managed-security-services
11. Hristov, P., Hristova, T.: Explaining the DLT applications in the context of a customers, facility managements and utility companies relationship. In: 2019 16th Conference on Electrical Machines, Drives and Power Systems (ELMA). IEEE (2019)
12. Conti, M., Dehghantanha, A., Franke, K., Watson, S.: Internet of things security and forensics: challenges and opportunities. https://arxiv.org/pdf/1811.09239.pdf
13. ENISA. Baseline security recommendations for iot in the context of critical information infrastructures (2017)

14. Cusack, B., Mahmoud, A.: Digital forensics investigative framework for control rooms in critical infrastructure. In: Australian Digital Forensics Conference (2018)
15. Fielder, A., Li, T., Hankin, C.: Defense-in-depth vs. critical component defense for industrial control systems. In: 4th International Symposium for ICS & SCADA Cyber Security Research 2016 (ICS-CSR 2016). BCS Learning & Development (2016)
16. ENISA. Mapping of OES Security Requirements to Specific Sectors (2017). https://www.enisa.europa.eu/publications/mapping-of-oes-security-requirements-to-specific-sectors/
17. Nawaz, Z., Aftab, S., Anwer, F.: Simplified FDD process model. Int. J. Mod. Educ. Comput. Sci. **9**(9), 53–59 (2017)
18. Gaidarski, I., Minchev, Z.: Virtual enterprise data protection: framework implementation with practical validation. In: BISEC 2018, Belgrade, Serbia (2018)
19. Lee, S., Levanti, K., Kim, H.S.: Network monitoring: present and future. Comput. Netw. **65**, 84–98 (2014)
20. Ginter, A.: The Top 20 Cyberattacks on Industrial Control Systems. Waterfall Security Solutions (2017)
21. Funded and supported by Department of Homeland Security under ContractNo. FA8721-05-C-0003 with Carnegie Mellon University. Crr supplemental resource guide. volume 5. incident management version 1.1
22. Sabbagh, B.: Cybersecurity Incident Response A Socio-Technical Approach. Department of Computer and Systems Sciences, Stockholm University, Stockholm (2019)
23. Tisdale, S.M.: Cybersecurity: challenges from a systems, complexity, knowledge management and business intelligence perspective. Issues In Information Systems (2015)
24. Cowperthwaite, A., Somayaji, A.: The futility of dnssec. In: Proceedings 5th Annual Symposium Information Assurance (ASIA 10), pp. 2–8 (2010)
25. Gurubaran, S.: BIND DNS Software Flaw Let Remote Attackers to Cause DOS attack (2019). https://gbhackers.com/bind-dns-software-vulnerability-let-remote-attackers-to-cause-a-denial-of-service-condition
26. Pauna, A.: (ENISA) Certification of Cyber Security skills of ICS/SCADA professionals (2014). https://www.enisa.europa.eu/publications/certification-of-cyber-security-skills-of-ics-scada-professionals
27. Nichols, S.: Remember Stuxnet? You'll endure its hated-by-critics sequel if you don't patch your holey Siemens industrial kit (2019). https://www.theregister.co.uk/2019/07/10/siemens_controllers_flaw
28. Teixeira, M., Salman, T., Zolanvari, M., Jain, R., Meskin, N., Samaka, M.: SCADA system testbed for cybersecurity research using machine learning approach. Fut. Internet **10**(8), 76 (2018)
29. Pfrang, S., Meier, D., Friedrich, M., Beyerer, J.: Advancing protocol fuzzing for industrial automation and control systems. In: Proceedings of the 4th International Conference on Information Systems Security and Privacy. SCITEPRESS (2018)
30. Bada, M., Nurse, J.R.: Developing cybersecurity education and awareness programmes for small- and medium-sized enterprises (SMEs). Inf. Comput. Secur. **27**(3), 393–410 (2019)
31. Hadlington, L.: Human factors in cybersecurity examining the link between internet addiction, impulsivity, attitudes towards cybersecurity, and risky cybersecurity behaviours. Heliyon **3**(7), e00346 (2017)
32. CIS. Center for internet security. a measurement companion to the cis critical security controls (2015)

33. Walker, P., Hern, A.: Labour suffers second cyber-attack in two days (2019). https://www.theguardian.com/politics/2019/nov/12/labour-reveals-large-scale-cyber-attack-on-digital-platforms
34. Mattioli, R., Moulinos, K.: Analysis of ICS-SCADA Cyber Security Maturity Levels in Critical Sectors. ENISA (2015)

Using Real-Time Traffic Information for Transportation Planning

Taulant Amiti, Karimi Moahammad Ali, Benjamin Wüthrich, and Thomas Hanne(✉)

University of Applied Sciences and Arts Northwestern Switzerland, Olten, Switzerland
thomas.hanne@fhnw.ch

Abstract. In this research, we investigate the feasibility of integrating real-time traffic information into transportation planning in order to receive a more accurate result. In our model, we optimized the Travelling Salesman Problem (TSP) from the OpenCI optimization framework and combined the traffic information from Google API within this problem (TSP) in order to make a better tool for more accurate transportation planning.

Keywords: Real-time traffic data · Travelling salesman problem · Transport planning

1 Introduction

Transport plays an important role in today's economy and society and has a large impact on growth and employment. The transport industry directly employs around 10 million people and accounts for about 5% of gross domestic product (GDP). Logistics, such as transport and storage, account for 10–15% of the cost of a finished product for European companies (Commission 2016). Effective transport systems are fundamental for companies' ability to compete in the world economy. This has led to planning for transportation as one of the challenging debates for organizations. The planning of transportation services is often quite inaccurate due to imprecise and fluctuating information on road accessibility and driving times impaired by congestion. New services such as Google Maps, BMW, Sygic started to provide real-time information about traffic conditions such as caused by changing densities of traffic, but they provide real-time information from one start point to one destination (from location A to location B). Indeed, traffic information only from a starting point to one destination is not enough for companies with tens of cities to distribute products and goods to customers or branches. It is important for transportation planning to have accessibility to real-time traffic information about all the paths which should be passed to transport commodities to all destinations, and this is just one side of the problem in transportation planning. On the other hand, transportation planners are looking for an optimal routing tour, so they have to calculate the value of all possible paths to visit all different cities to make the most accurate plan. For instance, if a company has ten branches and wants to deliver commodities to them, the company has ten factorial possible paths to choose (~3.6 million possible paths) and planners need

R. Silhavy et al. (Eds.): CoMeSySo 2021, LNNS 231, pp. 504–518, 2021.
https://doi.org/10.1007/978-3-030-90321-3_41

to calculate all the possible path's distances (or equivalently time or costs) in order to choose the shortest or cheapest way. This problem is known as Travelling Salesman Problem, which will be described more in next part of our paper.

1.1 Problem Statement and Research Question

The planning of transportation services is often quite inaccurate due to imprecise and fluctuating information on road accessibility and driving times impaired by congestion and other poorly predictable aspects. New services such as Google Maps started to provide real-time information about traffic conditions such as caused by volatile densities of traffic. In this project, we are going to analyze the feasibility of using real-time traffic information in transportation planning (in the sense of a proof of concept) and looking for a corresponding concept of using such data. In particular, in this paper using the respective information for a routing algorithm for calculating alternative routes in case of changed traffic conditions will be studied. This includes a consideration of available (or necessary) interfaces and integration with a routing algorithm.

Based on this transportation problem "Inaccurate Transportation Planning because of a lack of real-time traffic information and road accessibility", we are going to answer the following question: Is it possible to integrate real-time traffic information into transportation planning and receiving a more accurate result?

This paper is organized as follows. In Sect. 2 we briefly present an overview of the body of knowledge and a short description of the TSP and real-time traffic information providers. In Sect. 3 we present a methodology, how to integrate a TSP solution and routing algorithm with Real-Time Traffic information to find a better solution for transportation planning. In Sect. 4 we present our solution and conclusions.

2 Literature Review

Models like the one we propose appear only rarely, but the literature on the Travelling Salesman Problem (TSP), Real-Time Traffic Information, and more general vehicle routing is vast. We briefly review some salient topics.

According to the research about infrastructure for the TSP (Hornik and Hahsler 2007), the concept of the traveling salesman problem is to seeking a tour of a specified number of cities (visiting each city exactly once and returning to the starting point) where the length of the tour is minimized, with N cities there are (N-1)! possible tours (Hornik and Hahsler 2007). The TSP is a classic combinatorial optimization problem, which is simple to state but difficult to solve. The problem is known to be NP-hard and, thus, there are no exact polynomial time solution algorithms. The TSP is an optimization problem which has various applications such as vehicle routing and scheduling, planning and logistics. There are two ways to solve the TSP problem, exact solutions (e.g. Branch and Bound Algorithm) and heuristics solutions including various metaheuristics.

The OpenCI and its predecessor OpenOpal (Dornberger et al. 2012) is open-source optimization software written in Java. It provides an interactive mode with a graphical user interface (GUI) and freely combinable modules till a full batch mode. It runs on a single processor as well as on multiple ones distributing/balancing the computations via Java remote method (RMI). OpenCI provides algorithms for automatic optimization, design of experiment, and machine learning. OpenCI allows creating a problem, connecting it with one or more problem-solving algorithms, and running the optimization. The object-oriented architecture supports the use and implementation of new problems or optimization/search methods. Additionally, a data viewer is integrated which visualizes the results and thus supports the understanding of the optimization runs and the retrieved results as well. Dornberger et al. (2012) provide different solutions to solve the TSP like a genetic algorithm, ant colony algorithm, Kohonen Neural Networks and more.

On the other hand, it is important to know the impact of Real-Time Traffic on route switching to find an accurate path for transportation. According to the research, the predictive capabilities can be enhanced if real-time traffic information are considered within the model (Jou et al. 2004).

Cheong and White (2012) investigate the value of choosing the next stop based on current traffic conditions. The aim is to minimize the expected total travel time within a multi-stop trip. Cheong and White modeled this problem as a Markov decision process. According to Cheong and White (2012), the value of using current traffic information for dynamic tour determination can be significant.

Another important issue regarding designing an artificial instance for transportation is how to define a realistic instance. Maggioni et al. (2014) provide a standard methodology to extend routing instances from the literature incorporating real data provided by sensors networks.

They considered the multi-path TSP (mpTSPs) with stochastic travel costs, a recently introduced stochastic variant of the TSP. Given a set of nodes, where each pair of nodes is connected by several paths and each path shows a stochastic travel cost with unknown distribution, the mpTSPs aims at finding an expected minimum Hamiltonian tour connecting all nodes. The mpTSPs arises in City Logistics when one has to design tours to provide services such as garbage collection, periodic delivery of goods in urban grocery distribution, and periodic checks of shared resources as in bike sharing services. In these situations, the decision maker must provide tours that will be used for a time horizon, which spans from one to several weeks. In such a case, the different paths connecting pairs of nodes in the city are affected by uncertainty due to different time- dependent travel time distributions of the different paths. Moreover, in many cases, even an approximated knowledge of the travel time distribution is difficult because of the large size of the data involved and the high variability of the travel times.

Maggioni et al. (2014) contributed to extending the transportation literature along different directions. First, they introduced a standard methodology to build realistic instances with time-dependent travel times. The methodology they propose is easy to replicate, not requiring a deep statistical analysis on the historical data and can be adapted to other cases where only a partial information about the road network is available. Second, they analyzed the results obtained by combining the academic instances with real data on the traffic flows taken from the city of Turin in Italy, showing the benefits and the limits of the methods presently available in the literature to solve the mpTSP.

2.1 Real-Time Traffic Data

Currently, there are different companies, which provide real-time traffic information, for instance, Google Maps, Sygic (Sygic 2017) or BMW (BMW 2017).

2.1.1 Google Maps API

Google provides a wide variety of APIs related to their product Google Maps. Some of them also include real-time information on traffic conditions (Google Maps 2017b). The basic usage of the API is free of charge. A detailed technical description of the API is mentioned in Sect. 4.1.

2.1.2 Sygic

Sygic provides also real-time traffic information, based on data from TomTom Traffic (Sygic 2017). This information is collected from more than 400 million drivers and updated every 2 min, so drivers always have the most accurate and up-to-the-minute information. GPS data is collected from connected personal navigation devices (PNDs), commercial fleet GPS devices, mobile phone signals, road sensors, journalistic data, smartphones, and car dashboard systems. Drivers do not have to report anything; the data is collected automatically and anonymously for convenience. All the information is merged and evaluated to generate precise real-time traffic information to help drivers make better decisions on the road.

Drivers will know where the congestion is and how long it will take them to get through it. Knowing of the delay is good, but this feature is far more useful than that. Drivers will also receive automatic faster route suggestions, so they can take an alternative route to avoid the traffic jam.

2.2 Existing Solutions

One existing solution comes from Google itself. The tool offers different services and is specially addressed to enterprise customers. On the website, there are only some previews, and the functionalities are not described in detail (Google Maps 2017a). What is mentioned, is a ride planner-website from Harley-Davidson. On the other hand, it is unclear and not mentioned how the optimal route is calculated.

Another solution for routing and solving Travelling Salesman is called OpenCI which provides an optimization framework. OpenCI is an open-source software with search and optimization methods. It is a modular optimization framework used and further developed at the University of Applied Sciences and Arts Northwestern Switzerland FHNW. OpenCI is used in teaching at Master of Science in Business Information System at FHNW and for the purpose of different student's projects. The software includes search and optimization methods such as the simplex method, genetic algorithms, evolution strategies in various variants, procedures in the field of Swarm Intelligence such as Ant Colony Optimization, methods for multi-objective optimization, several methods in the field of neural networks and visualization technics. It is a software written in Java. It also provides a totally interactive mode with a graphical user interface (GUI) and freely combinable modules, till a full batch mode. OpenCI provides algorithms for automatic optimization, design of experiment, and machine learning. It allows creating a problem, connecting it with one or more problem-solving algorithms, and running the optimization. The object-oriented architecture supports the use and implementation of new problems or optimization/search methods. Additionally, a data viewer is integrated, which visualizes the results and thus supports the understanding of the optimization runs and the retrieved results as well (FHNW University 2017).

2.3 TSP Within OpenCI

For routing and finding an appropriate path, best practice is to use the Travelling Salesman Problem (TSP). In OpenCI a TSP template is available. The TSP focuses on how a salesman may travel through several cities by choosing the shortest way (called a tour), at the end returning to the starting city. As TSP is one of the most intensively studied optimization problems in computational mathematics, it is used for its integration in OpenCI. The aim has been a very generic implementation of TSP in OpenCI leaving the option to adapt the TSP to particular academic as well as real-life applications, such as vehicle routing problems or order picking problems, for instance in automated warehouses. Due to this, a generic instance of the TSP is implemented in OpenCI using cities located in Euclidian space at X-Y coordinates and direct bidirectional routes between them. The extension of the generic TSP can then be done by creating a new instance of the TSP class adding the specifics of the particular problem (e.g., one-directional, weighted routes).

As described before to find the shortest way for a multi-stop tour, OpenCI provides a framework which can be used. To solve the TSP there are three different solvers available in OpenCI: Ant Colony Optimization, Kohonen, and Christofides Heuristic. This paper focuses on Christofides Heuristic because the underlying algorithm is easily adaptable. To find the shortest way for a multi-stop tour in OpenCI, it is required to create a TSP problem as well as a Solver and to connect the two instances. The TSP, as well as the Solvers, offer different adjustments within their options (e.g., number of cities to visit). As the settings are made, the set up can be saved and run to find an accurate path for the whole tour.

For transportation planning finding the shortest route is just one side of the problem. To make accurate plans the information about the real-time traffic on the road is another factor which is very important (In order to have that in this model, the real-time traffic information from Google will be combined with the TSP in OpenCI to find a better solution for transportation planning. More information will be presented in Sect. 4. There are also lots of applications, software, and web tools, which provide real-time traffic information like Google, BMW navigator, Sygic and other tools.

2.4 Tools in General

As mentioned before there are lots of applications, software and other tools which provide real-time traffic information, but the most important difference between our proposed solution (see Sect. 4) and other tools is that our solution focuses on finding the optimal route based on real-time traffic data.

Most of the existing solutions and tools provide real-time traffic information from one starting point to a destination (from location A to location B). There are a few tools which try to provide solutions for a route respectively for a multi-route journey in combination with real-time data. But none of this solution combines the data with known TSP solvers to reach the best possible solution.

3 Methodology

The methodology chosen for our research is considered as design science research. The goal of design science research is about designing and creating an artifact (constructs, models, methods, and instantiations). Hevner and Chatterjee (2010) cite that design science research is a research paradigm in which a designer answers questions relevant to human problems via the creation of innovative artifacts, thereby contributing new knowledge to the body of scientific evidence (Hevner and Chatterjee 2010).

For our study, we selected OpenCI as a modular optimization framework for developing the artifact which already includes various algorithms for the TSP. We used the GitLab repository of OpenCI (Institut für Wirtschaftsinformatik - FHNW 2017). Based on an analysis of existing solutions and concepts, and gaps were identified. Based on these insights, further research was required in the field of using different Application Programming Interfaces (APIs) and, as a result, we came up with API from Google Maps for the purpose of transportation planning and distance measurement. We studied the Google Distance Matrix API and asked Google employees to send us some API specification and possible user instructions. Finally, an effort was put on implementing Google real-time traffic information API with the Travelling Salesman Problem for transportation planning to find an accurate transportation plan to minimize the total cost and time of a travel route.

4 Solution Design

For analyzing the feasibility of using real-time traffic information in transportation planning (proof of concept) and looking for a corresponding concept of using such data, we investigate the problem in two phases. First, accessing real- time traffic data from Google, and second; finding the shortest possible path based on the real-time data. In the first phase, we are going to use real-time traffic data transportation planning and combine it with the TSP.

In the second phase, we have to understand how we can find the shortest possible route for a multi-stop trip in particular for transportation planning. In order to find the shortest path, it is required to calculate the distance or in our case the travel time for all possible ways. We assume that we have to visit N cities and turn back to the first city. If City1 is our starting point, in the end of the trip we will turn back to this city. In this case for N cities, there are (n-1)! possible paths to travel (Hornik and Hahsler 2007), as we deal with an asymmetric TSP. Our TSP is asymmetric because there may be a traffic jam between location A and location B, but the opposite direction runs smooth. Therefore, the back-and-forth edges (in the sense of directed connections of the respective graph) between both locations have different weights respective durations. To find the shortest possible path there are a lot of software and tools available, which implement known algorithms and heuristics. In this project, we are going to use OpenCI optimization framework.

After carefully reviewing the technical requirements, we came up with a solution design. Figure 1 shows a diagram of technical requirements and steps that we follow with the aim to achieve our solution.

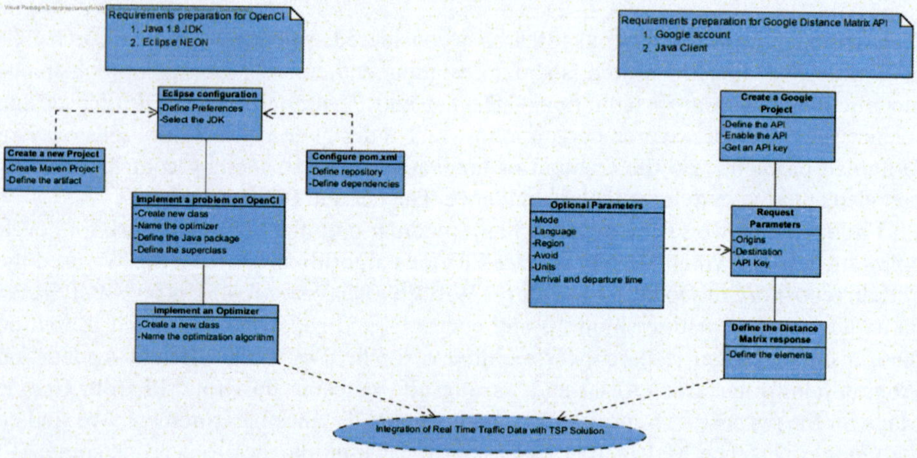

Fig. 1. Solution design diagram.

In this model, we use the OpenCI software to calculate the weight of all possible paths and finally find the shortest path. As OpenCI works with a Distance Matrix, we focused on the Distance Matrix API from Google to combine with the TSP within OpenCI for transportation planning to find an accurate transportation plan to minimize the total cost and time of travel tour. The goal of this paper is to find a new solution and create a new/own TSP problem within OpenCI so that with this TSP solver the best solution can be calculated/solved. In next part, we will describe how to integrate Google API with Travelling Salesman Problem within OpenCI. In other words, we will define a new TSP problem using real-time traffic data for an accurate transportation plan to minimize the total cost and time of travel tour. In next part of this paper, we describe how Travelling Salesman Problem and TSP solver algorithm in OpenCI works and finally how we are going to optimize the TSP within OpenCI and combine traffic data to calculate and find the best possible transportation plan.

Note that for a more realistic transportation planning, variants of the TSP such as the capacitated vehicle routing problem (CVRP) and further variants might be more useful as they allow to include further requirements and restrictions, for instance time windows. In addition, the resulting routes may allow for further optimization potential during dynamic replanning such as exchanges of customer orders between the routes of different vehicles as a result of changing traffic conditions.

In the next section, we will, however, stick to the more simple TSP and the describe Google API and the combination of the TSP and Google Traffic Data.

4.1 Google API

Google provides a set of application programming interfaces (API) which allow communication with Google Services and their integration with other services. Examples of these include Search, Gmail, Translate or Google Maps. Third-party apps can use these APIs to take advantage of or extend the functionality of the existing services.

The APIs provide functionality like analytics, machine learning as a service (the Prediction API) or access to user data (when permission to read the data is given). Another important example is an embedded Google map on a website, which can be achieved using the Static maps API, Places API or Google Earth API.

With Google API it is possible to access to matrix and data about the traffic condition on roads in particular time of the day, which consists of the cities and the distances between them. The travel distance and time for the matrix or origins and destination are provided through the Google Maps Distance Matrix API service. The service is based on the recommended route between the start and the endpoint. The data is defined by an array, whose entry in row i and column j is the distance from city i to city j in miles, as shown below in Fig. 2.

```
self.matrix = [
[   0, 2451,  713, 1018, 1631, 1374, 2408,  213, 2571,  875, 1420, 2145, 1972], # New York
[2451,    0, 1745, 1524,  831, 1240,  959, 2596,  403, 1589, 1374,  357,  579], # Los Angel
[ 713, 1745,    0,  355,  920,  803, 1737,  851, 1858,  262,  940, 1453, 1260], # Chicago
[1018, 1524,  355,    0,  700,  862, 1395, 1123, 1584,  466, 1056, 1280,  987], # Minneapol
[1631,  831,  920,  700,    0,  663, 1021, 1769,  949,  796,  879,  586,  371], # Denver
[1374, 1240,  803,  862,  663,    0, 1681, 1551, 1765,  547,  225,  887,  999], # Dallas
[2408,  959, 1737, 1395, 1021, 1681,    0, 2493,  678, 1724, 1891, 1114,  701], # Seattle
[ 213, 2596,  851, 1123, 1769, 1551, 2493,    0, 2699, 1038, 1605, 2300, 2099], # Boston
[2571,  403, 1858, 1584,  949, 1765,  678, 2699,    0, 1744, 1645,  653,  600], # San Franc
[ 875, 1589,  262,  466,  796,  547, 1724, 1038, 1744,    0,  679, 1272, 1162], # St. Louis
[1420, 1374,  940, 1056,  879,  225, 1891, 1605, 1645,  679,    0, 1017, 1200], # Houston
[2145,  357, 1453, 1280,  586,  887, 1114, 2300,  653, 1272, 1017,    0,  504], # Phoenix
[1972,  579, 1260,  987,  371,  999,  701, 2099,  600, 1162, 1200,  504,    0]] # Salt Lake
```

Fig. 2. Google distance matrix shows the distance between cities.

The Google Maps Distance Matrix API, provides the user with different programming libraries for developers and it is available in different programming interfaces such as Java Client, Python Client, Go Client and Node.js Client for Google Maps Services. To be able to use the Google Maps Distance Matrix API service, you first need to create an account or use an existing Google account and then create a project (Google Maps 2017b). After that, there is a list of different APIs, where the developers can choose more than one API for a specific project purpose and activate the API to get a key for further programming processes.

Two main reasons account for the choice of the Google Maps APIs. Firstly, there is a Java Client available which allows a seamless integration into OpenCI, which is also based on Java. Secondly, the APIs from Google are well documented.

4.2 Solution Architecture

Currently, there are three solvers for TSP within OpenCI (FHNW University 2017), one of them is Christofides Heuristic which solves the TSP based on the distance of randomly generated cities (see Fig. 3).

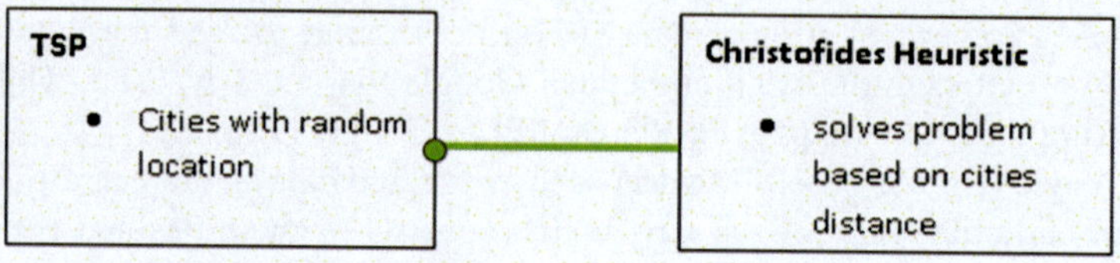

Fig. 3. Functionality of OpenCI: Christofides heuristic solves TSP.

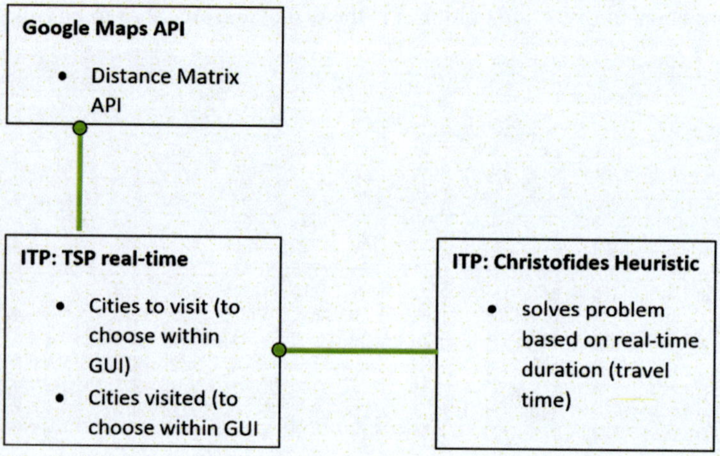

Fig. 4. Functionality of OpenCI: Christofides heuristic solves TSP.

To succeed with the combination described, the existing ITP and solver within OpenCI have to be amended (see Fig. 4). Instead of randomly generated cities, the user should be able to choose cities he wants to visit, and he can state which city he already visited. The cities already visited are deducted before proceeding. The location (Geolocation API), as well as the travel time (Distance Matrix API) between the cities chosen by the user, can then be requested with Google Maps API. As this information arrives, the amended Christofides Heuristic calculates a solution no longer based on the distance but on the travel time between the cities. The location of the cities (Geolocation API) is used for the visualization of the solver's result.

5 Development

The planned solution design (see Sect. 4) has been implemented in the existing OpenCI framework with the help of the integrated development environment Eclipse.

5.1 Implementation

As described, the existing concepts in OpenCI had to be amended according to the solution design to achieve the goal set and to answer the research question. This section explains the most relevant changes within the following three Java classes. The code for these three classes is available from the authors.

5.1.1 TSP.Java

The TSP is the class which contains the problem, and according to the superclass it has to implement the two methods init() and evaluate(). Within the project, most changes affected method init(). The init() method loads the destination cities chosen by the user. In a next step, the cities already visited are loaded (for supporting a dynamic update of tours during execution). As the cities already visited no longer need to be considered within the solution, these cities are removed.

For every destination city, the real-time TSP requests geolocation information, respective the latitude and longitude from Google Geocoding API (Google Maps 2017e). Based on this information, the cities are later displayed on the visual result.

In the next step, the second newly created method calculateDistanceMatrix() within TSP.java is called. This method creates the distance matrix, which is later the base for the solver. This distance matrix is no longer based on static distances but on real-time traffic data which is requested at Google via an API key. Prerequisite for receiving real- time date is to choose the so-called travel mode "DRIVING" as well as to set the departure time to right now (Google Maps 2017d).

It may happen, that the user input regarding the cities to visit is not usable or is not recognized by Google. In this case, the exception is caught and subsequently, it is tried to identify the city which causes the error. If the city is found, this information is printed out to the end user, with the aim that he or she can adjust the input.

5.1.2 ChristofidesHeuristic.Java/CCity.Java

Within the ChristofidesHeuristic class, only limited changes were necessary. The Distance Matrix created in the TSP class is stored in the local double array. This is a prerequisite for accessing the information in the matrix within the CCity class.

The CCity class calculates among other things the distance between two cities. This information is needed to solve the TSP with the Christofides Heuristic (or other the solver methods). Instead of returning a randomly generated number as previously, the method getDistanceTo() was amended to return the real-time traffic data, respective the duration in traffic from Google which has been stored in the distance matrix. The further calculations within ChristofidesHeuristic are made with this data and thus the solution is based on real-time traffic data.

5.1.3 Launcher.Java

In order to access the functionality of the real-time TSP as well as the respective amended Christofides Heuristic, the two classes had to be inserted into the OpenCI Launcher (for the framework GUI). The two modules are now available with the right mouse button in the OpenCI framework.

5.2 Possible Use Case

The aim is that the artifact has a value for transportation or logistic companies delivering goods to different locations i.e. cities. The route can be planned in real-time and cities visited are no longer considered for the rest of the route.

Figure 5 presents a possible use case for the real-time TSP. A delivering truck starts its journey in Olten and follows the planned route to Aarau and Oberentfelden. At every location it visits, the driver unloads some goods and updates the field "Cities visited" in the user interface manually. The rest of the route is then calculated based on the real-time traffic data from Google.

As the truck driver updates the visited locations in Unterkulm, the route changes due to changing traffic conditions. As presented in Fig. 5, the route no longer goes to Schöftland (red line) but to Triengen (green line) and afterward to Schöftland.

With this change of the route, the driver can avoid an accident and saves time compared to the originally planned route, which now takes longer due to the traffic issue. To sum up, the real-time aspects result in an optimized routing, which in return saves time, fuel, and money, and thus leads to more efficient solutions.

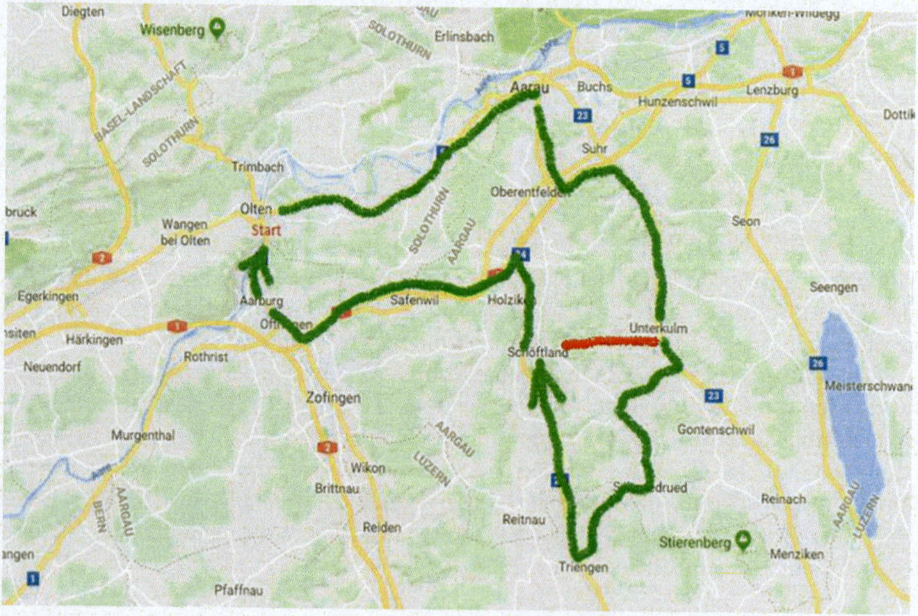

Fig. 5. Use case real-time TSP.

5.3 Limitations

The limitations of our artefact include the Google APIs and the Christofides Heuristic.

The standard usage limits for the Google Maps Distance Matrix API allow using 2'500 free elements per day. The number of elements is counted through the numbers of origins multiplied by the number of destinations. Anyway, there is a possibility to increase the daily limit of elements upon payment (Google Maps 2017c). As our research funds are limited, we preferred to work with the free access account.

Furthermore, there is a maximum of 25 origins and 25 destinations per request (Google Maps 2017c), which counts also upon payment. If the daily limit is reached, an exception is thrown. A respective exception is caught in the code, and the information is printed out to the end-user.

The used Christofides Heuristic has the limitation that the number of cities to visit needs to be at least nine. Otherwise, an error is displayed. This limitation is not specific to the amended Christofides Heuristic Solver, but also to the general one containing the latest version of the OpenCI (FHNW University 2017). As the Christofides Heuristic is only one possible way to solve the TSP, this limitation could easily be removed by implementing another solver for the real-time TSP.

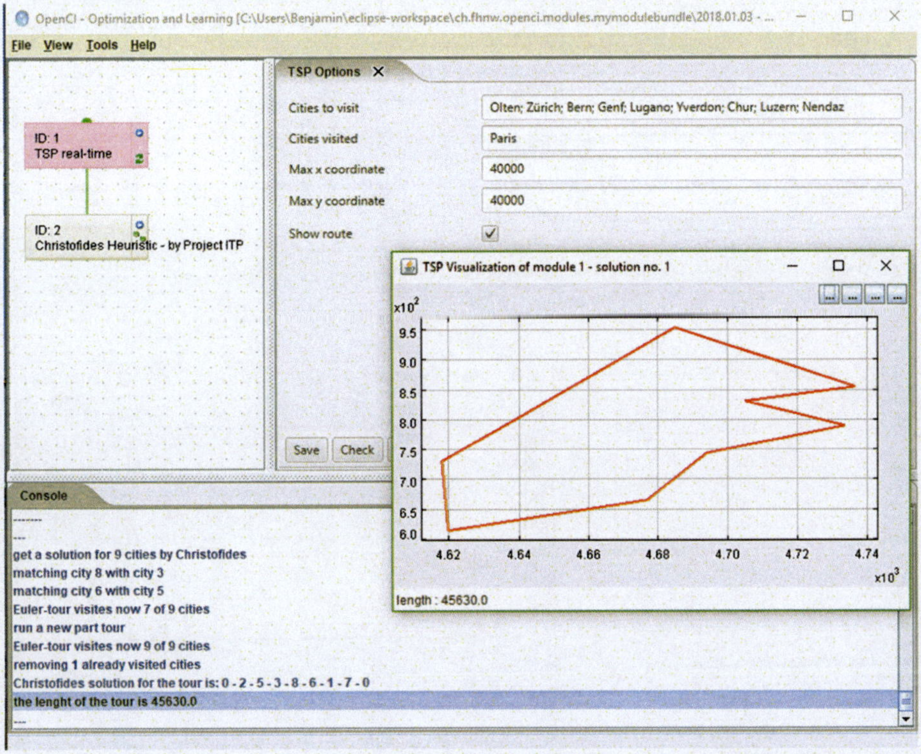

Fig. 6. Real-time tour.

6 Evaluation

As the artifact has been generated, the next step in the design research methodology is to evaluate the result and assess, whether the described problem can be solved, and the research question can be answered accordingly.

6.1 Simulation of Traffic Jam

Assuming to visit following nine cities within Switzerland: Olten, Zürich, Bern, Genf, Lugano, Yverdon, Chur, Luzern, and Nendaz. TSP real-time calculated the following optimal route based on real-time traffic data from Google: 0-2-5-3-8-6-1-7-0 (see Fig. 6).

According to this solution, it takes 3481 s to drive from city 2 (Bern) to city 5 (Yverdon), this information comes from Google Maps API. Assuming there is a traffic jam on A1 motorway between Bern (2) and Yverdon (5), which evolves as the driver is on the route between Olten (0) and Bern (2). As a result, the travelling time between Bern (2) and Yverdon (5) rises to two hours (7200 s). This situation is simulated "manually" by changes directly in program code.

As a result, the previously optimal route changes to 0-2-8-5-3-6-1-7-0 (see Fig. 7). The route no longer goes from Bern (2) to Yverdon (5), but from Bern (2) via Nendaz (8) to Yverdon (5). As the driver is on the way from Olten (0) to Bern (2), he can dynamically

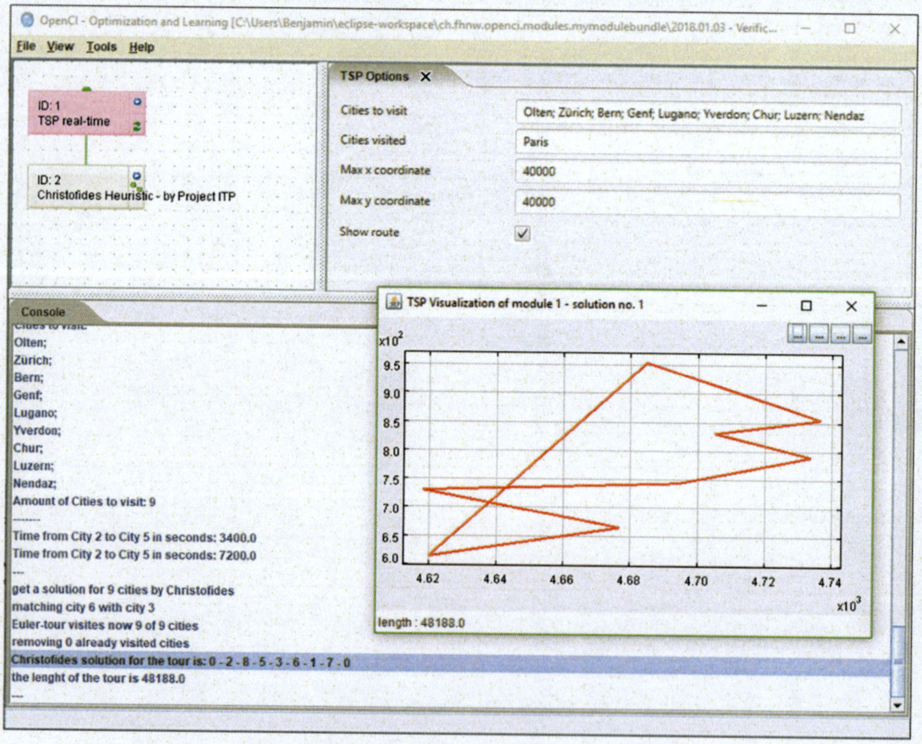

Fig. 7. New tour with different sequence

change the route and proceed with the direction to Nendaz (8) instead of Yverdon (5) (Fig. 7).

Table 1 summarizes the influence of the different behavior to the total travel time. Assuming the driver sticks to the original plan, the route lasts 3719 s longer, as the route between Bern (2) and Yverdon (5) no longer takes 3481 but 7200 s. If the driver adapts the route, the journey only takes 2558 s longer (48'188–45'630 s). As a result, in this example, the driver has a time advantage of 1161 s (3719–2558 s), if he adapts to the new optimal route generated by the Christofides Heuristic.

Table 1. Route is faster if updated real-time.

Route behavior	Time planned originally	Actual time with traffic jam/Total travel time
Stick to plan: 0–2-5–3-8–6-1–7-0	45′630 s	45′630 + 3′719 = 49′349 s
Change during drive from Olten (0) to Bern (2) to new route: 0–2-8–5-3–6-1–7-0	45′630 s	48′188 s

7 Conclusions

Due to a lack of real-time traffic information and road accessibility transportation planning is often inaccurate. The aim of this paper was to investigate the possibility of integrating real-time traffic information into transportation planning in order to receive a more accurate result.

According to research discussed in Sect. 2, it is possible to integrate real-time traffic information into transportation planning, which results in a more accurate transportation planning. But there was no product available which combines the algorithm for the optimal solution with real-time data. Our artefact combines existing solvers for the Travelling Salesman Problem within the OpenCI framework with existing real-time traffic data from Google Maps. As described in Sect. 5.2, the updated traffic information can be used during the execution of transport activities to change the related planning. As shown in Sect. 6, the use of our artefact can have an impact.

As the artefact still needs some manual work, we are aware, that this is not supporting drivers in the best manner. On the other hand, this is an aspect for further improvement and discussed in the next paragraph.

As mentioned in Sect. 5.3 and because of the limitations of the Christofides Heuristic, it would be useful to combine further solvers for the TSP with the real-time data access. With our current solution, the user has to enter the cities visited manually. A new feature could be added to automate the process of removing visited cities from the route. Google Maps, for instance, offers a service called Geolocation API which determines the current location based on the cell id (Google Maps 2018). In our opinion, this feature would improve the value of the existing solution. Additionally, if the user no longer needs to manually add the visited cities, there is also the possibility to automate the re-planning cycle. For instance, an algorithm could recalculate the optimal solution every second minute and compare it to the current optimal route. If a new optimal route is available, the driver would be informed. Moreover, the real-time TSP could also be integrated into other planning software.

References

BMW. Make Progress Instead of Standing Still. Von Real Time Traffic Information 5 Oct 2017. https://www.bmw.ie/en/topics/fascination-bmw/connected-drive/rtti.html

Cheong, T., White, C.C.: Dynamic traveling salesman problem: value of real-time traffic information. IEEE Trans. Intell. Trans. Syst. **13**(2), 619–630 (2012)

Commission, E.: 14 Jul 2016. Von https://ec.europa.eu/jrc/en/research-topic/transport-sector-economic-analysis

Dornberger, R., Hanne, T., Frey, L., Ernst, R.: Solving Optimization Problems by Metaheuristics Using the OpenOpal Framework – Integration of Travelling Salesman Problem and selected Solvers. Olten: Institut für Wirtschaftsinformatik IWI (2012)

FHNW University - Institut für Wirtschaftsinformatik.System Engineering – OpenCI, 9 Dec 2017. Von FHNW https://www.fhnw.ch/de/die-fhnw/hochschulen/hsw/iwi/systems-engineering/open-ci

Google Maps. 10 Nov 2017a. Google Maps. Von Customers: https://enterprise.google.com/intl/ALL_ALL/maps/customers/

Google Maps. 9 Dec 2017b. Google Maps APIs. Von Distance & duration for multiple destinations and transport modes: https://developers.google.com/maps/documentation/distance-matrix/

Google Maps. 9 Dec 2017c. Google Maps APIs. Von Distance Matrix API Usage Limits: https://developers.google.com/maps/documentation/distance-matrix/usage-limits

Google Maps. 9 Dec 2017d. Google Maps APIs. Von Developer's Guide: https://developers.google.com/maps/documentation/distance-matrix/intro

Google Maps. 9 Dec 2017e. Google Maps APIs. Von Geocoding API: https://developers.google.com/maps/documentation/geocoding/start abgerufen

Google Maps. 10 Jan 2018. Google Maps APIs. Von The Google Maps Geolocation API: https://developers.google.com/maps/documentation/geolocation/intro

Hevner, A., Chatterjee, S.: Design research in information system: theory and practice. In: Design Research in Information System: Theory and Practice (2010)

Hornik, K., Hahsler, M.: TSP – Infrastructure for the Traveling Salesperson Problem. Wirtschaftsuniversität Wien, Wien, Austria (2007)

Institut für Wirtschaftsinformatik - FHNW. (2017). IWI - OpenCI. Von gitlab FHNW: https://gitlab.fhnw.ch/IWI-OpenCI/copenci/wikis/home

Jou, R.-C., Lam, S.-H., Weng, M.-C., Chen, C.-C. (2004). Real Time Traffic Information and its Impacts on Route Switching Behavior of Expressway Drivers. University of Science and Technology, Taiwan

Maggioni, F., Perboli, G., Tadei, R.: The Multi-Path Traveling Salesman Problem with Stochastic Travel Costs: Building Realistic Instances for City Logistics Applications. University of Bergamo, Italy (2014)

Sygic. Real-time Traffic Information 5 Oct 2017. Von How Does it Work? https://www.sygic.com/blog/2016/real-time-traffic-information-how-does-it-work

Semantic Artificial Intelligence

Alexander A. Kharlamov[1,2,3,4](✉)

[1] Institute of Higher Nervous Activity and Neurophysiology of RAS, Butlerov Street, 5A, 117485 Moscow, Russia
kharlamov@analyst.ru
[2] Moscow State Linguistic University, Ostozhenka Street, 38, 119034 Moscow, Russia
[3] Higher School of Economics University, Pokrovsky bvd, 11, 109028 Moscow, Russia
[4] Moscow Institute of Physics and Technology, Institutsky Line, 9, 141701 Dolgoprudny, Moscow Region, Russia

Abstract. As it is known, the main difference between humans and animals is the presence of the second signaling system. The introduction of the second signaling system into the artificial intelligence system including the linguistic model of the world that works in conjunction with the extralinguistic model of the world, makes it possible to separate the presentation and processing of specific information into lower-level and upper-level representations. Thus, along with the detailed description, the model of the world also contains a generalized representation, which makes it easy for the user to interact with the system using natural language.

Keywords: Semantic AI · Second signaling system · Model of the world · Language model of the world

1 Introduction

From the very beginning, artificial intelligence systems (AIs) have also been called (in academic circles) knowledge-based systems. Therefore, from the point of view of V.I. Gorodetsky [1], "the main research and development in AI is concentrated around three problems of technology for working with knowledge, namely: (1) acquiring knowledge; (2) representation and transformation of knowledge; and (3) use of knowledge".

The main difference between humans and animals is known; this is the presence of a second signaling system [2]. That is why the main two intellectual functions of humans are (1) the generation of meaningful coherent natural language texts; and (2) comprehension of texts. The second signaling system is based on the first signaling system. That is, in order to understand how to model natural intelligence, it is necessary to understand how both the first and second signaling systems work.

Consider the so-called "Semantic Artificial Intelligence". This concept was proposed by V.V. Borisov in a private conversation; we will correlate it with the model of natural intelligence.

R. Silhavy et al. (Eds.): CoMeSySo 2021, LNNS 231, pp. 519–525, 2021.
https://doi.org/10.1007/978-3-030-90321-3_42

2 Semantic Artificial Intelligence

Definition. Semantic artificial intelligence is an informational model of natural intelligence based on the model of the world. This is a simulation of how natural intelligence processes and manipulates specific information, including, but not limited to the process of purposeful behavior.

Let us consider how the information model of natural intelligence is implemented: how information processing takes place in the human mind.

2.1 Semantic Artificial Intelligence Structure

In the most general terms, the SAI structure includes: (1) sensing and (2) motor skills, which enable the SAI interaction with the environment. It includes (3) a model of the world that can be used in behavior planning.

The model of the world includes, in addition to the extralinguistic part (on which the work of the first signal system is based), the linguistic part as well. Despite the differences in their content, both of these parts in the human brain are represented by the same information processing mechanisms, such that they effectively interact and form a single whole. Differences in their content are determined by the peculiarities in the organization of processing specific information of various modalities, which are different in various sensors.

2.2 Semantic Artificial Intelligence Functions

In contrast to the SAI structure, its functions (although these are structured in accordance with the aforementioned chain of processes "acquiring knowledge – representation and transformation of knowledge – use of knowledge") turn out to be very complex. Knowledge acquisition is a multilevel procedure for constructing a model of the world, which is inextricably linked with the representation of knowledge of both various levels and various modalities (and multimodal ones as well). And the use of knowledge includes the processes of planning and monitoring the implementation of purposeful behavior.

The described SAI architecture as a model of natural intelligence assumes functions that can (or should be) implemented in such a way so that the mentioned SAI could manifest its intelligence. The main function that allows this is (1) purposeful behavior [3]. This is the most complicated and complex type of SAI behavior that includes other less complex types of behavior. Purposeful behavior allows manipulation of the model of the world in the process of achieving a goal.

Simpler types of behavior are reduced to the formation of a model of the world at its various levels. The learning process (education) required for the formation of a model of the world includes (2) the formation in the sensory cortex of dictionaries of images for events of varying complexity and various modalities, on the one hand. On the other hand, (3) the formation of templates of situations based on these dictionaries in the hippocampus as combinations of words of dictionaries in space and time. The generated templates are used for: (4) segmentation of the world when it is perceived; (5) identification of the degree of novelty of input situations; (6) manipulating the representations of situations (in the form of their names in one or another modality) in the anterior cortex when planning and controlling the implementation of purposeful behavior.

Planning and monitoring the implementation of purposeful behavior (7) also become possible in the learning process (now of the anterior cortex), where the hierarchy of its columns is the basis for formation of representations of chains for names of situations of varying complexity that, in fact, represent behavioral plans.

The sensory part of the model of the world is formed on the basis of sensory information obtained in the process of signal processing of information of various modalities. Therefore, it strongly depends on the modality of the information received: a one-dimensional speech wave at the signal level is processed differently than a two-dimensional image of the three-dimensional world.

3 Semantic Artificial Intelligence Implementation

3.1 Separate Independent Parts

Separate independent parts of the architecture that implement certain functions in the complex can be listed as follows: (1) speech analyzer; (2) generator of meaningful texts; (3) visual analyzer; (4) bimodal (multimodal) analyzer; (5) speech-to-text synthesizer; (6) motor control structures. These individual parts, individually and in combination, can be used to implement various intelligent functions. Together, with the addition of missing architectural elements, they can form the core of the SAI system.

Speech Analyzer. Two main intellectual functions inherent in humans (understanding speech and generating meaningful texts) are basically the main functionality of both SAI and AI in general. All other intelligent functions are based on these two. Speech recognition of these two has advanced further at present, therefore, we will talk about the auditory verbal analyzer first.

Signal Peripherals. The peripherals of the auditory verbal analyzer are relatively simple (in comparison, for example, with the visual analyzer). In fact, one can use spectral processing, which provides good results in most automatic speech recognition systems. These are 13 mel-cepstral coefficients, their first and second derivatives.

Symbolic Processing in the Sensory Cortex and Above. Structural processing of the input code sequence leads to the formation in the columns of the sensory cortex of a hierarchy of dictionaries of level-forming elements of varying complexity (from the dictionary of acoustic-phonetic elements to the dictionary of pairwise compatibility of root bases) [4].

In the hippocampus [4], templates of situations are formed, which in the dominant hippocampus include descriptions of situations in terms of language phrases (spoken, written, analysis and synthesis), in this case – a phrase as it is perceived by ear, with the emphasis on the name – that is, the name of the situation for subsequent use.

Generator of Meaningful Texts. The generation of meaningful texts is the "ceiling" of the SAI functions, since purposeful behavior is implemented in this case [5].

Symbolic Processing in the Anterior Cortex and Below. In this case, the mechanism of attention identifies a fragment of the extralinguistic model of the world, which in the process of generation must be explicated in the form of text. It is projected onto the linguistic model of the world, which includes a hierarchy of representations, similar to the hierarchy in the auditory verbal analyzer [6], but working with representations of commands that control articulation. A speech message is formed in parallel with the participation of all levels of the hierarchy as a sequence of phrases describing various parts of this fragment of the extralinguistic model of the world. The sequence is determined by the prehistory of learning purposeful (in this case, speech) behavior; first general, then secondary, or first secondary, then general, or in combination.

Signal Peripherals. After the formation of the control sequence, the articulation control channels form states of the articulators corresponding to the control sequence and activate the corresponding sources of sound generation.

Visual Analyzer. The visual analyzer is much more complicated than the auditory one, since the two-dimensional image of the three-dimensional field of view differs by at least one dimension from the one-dimensional speech wave.

Signal Peripherals. Primary image processing is especially difficult. Breaking down the image by the points of the greatest informational value [7], which is performed by the oculomotor system, allows not only to save resources during processing, but also to arbitrarily enhance processing in places that are difficult to understand. The breaking-down scenario depends on the problem statement and preliminary learning process.

Symbolic Processing in the Visual Cortex and Above. In the visual cortex, as well as in the auditory cortex, a hierarchy of dictionaries of images for events of varying complexity is formed, which are then used to form the representation of situation templates in the hippocampus.

Bimodal (Multimodal) Analyzer. Combining analyzers provides an improvement in the quality of information processing (for example, in the recognition process). Multimodal models of the world perceive the world in different ways; therefore, the uncertainty in one modality can be eliminated by the information of another modality. This works especially well when some fragment of the subject domain is better represented in one of the modalities than in the other.

Motor Control. Various motor skills are currently quite well developed in android robots (for example, the Boston Dynamics laboratories). The upper levels are implemented by the mechanisms of purposeful behavior.

3.2 Combined Action of the Mechanisms

These mechanisms interact with each other in the process of solving current tasks. These tasks include, for example, conducting a dialogue, finding an answer to a question, solving a problem, and other similar tasks.

Let us describe their interaction when solving some typical tasks.

Finding a Solution To a Problem (Planning Purposeful Behavior). The search for a solution to a problem is the most common (almost the only one: all the rest are reduced to it) elementary act of purposeful behavior. Its scheme looks like the following.

There is a shortage of some resources, no matter what – real or mental ones. The purpose of the search is identified. If it is introduced explicitly, it means that it is set, if implicitly, the goal is identified in the process of interaction with the user.

Purposeful behavior consists in planning the achievement of the goal and in monitoring the implementation of the plan. Planning purposeful behavior means building a chain of situations from the current one to the target one. At the same time, the semantic network of situations in the anterior cortex, represented by a list of available pairs of situations, involves searching on it by associative enumeration with the identification of chains of situations to reach the target one, provided that there is a current one.

One can start with a simple case where the target situation is one step further from the current one. There is no choice: one just needs to "enter" it. A difficult case is when a chain of situations cannot be built. In this case, the chain is built as far as possible, and an opposite chain is built from the target situation. Thus, a counter text is looked for, which will ensure the connection of these two chains (the Newton's apple).

Solving a Problem (Monitoring the Implementation of the Plan). The implementation of the plan is monitored at each step of entering the next situation in the chain. If the realities of the environment when using the planning template of the situation coincide with those presented in the template, a decision is made at the current step on the implementation of the plan. If they do not coincide, a new (adjusted) plan is built.

Search for an Answer To a Question. An example explaining the procedure for planning and implementing a plan is solving the problem of finding an answer to a question, as well as the actual answer to a question in the process of dialogue behavior. Finding an answer to a question and answering a question is one step of the dialogue behavior (its passive part).

The question asked contains the user's goal. In answering the question, the SAI plans the purposeful behavior described above. The answer to the question should contain information that is available in the SAI knowledge base. If it is not there, additional detailed targeted behavior may be planned to search for missing information in external sources, for example, on the Internet.

Before the search for an answer to the question begins, some actions are taken to understand the form of the question (the syntactic level of the language model is used); the question is projected onto the SAI extralinguistic model of the world, which makes it possible to understand the content of the question [8].

Answer to the Question. To answer the question, it is required to obtain the information the user needs. This information can be obtained in the form of a text (language) answer. If there is no direct answer, then a template is looked for, where this answer is. But in this case, in addition, natural language descriptions of situations are formulated, presented in the chain of situations that led to the received answer.

Dialogue Behavior. The text thus formed can generate subsequent questions from the user, which leads to a dialogue where the SAI purpose is to explain its point of view.

Thus, passive participation in dialogue is similar to simple answering questions. If we consider conducting a dialogue as purposeful behavior, a goal is needed, which on the SAI part can only be in the case presented above (explanation of the mechanism for obtaining an answer).

4 Possible Applications

A concept is worthless if it is not experimentally confirmed. In other words, it needs to be implemented. Alas, full-scaled implementation is not available at the moment; therefore, one needs to make an attempt at implementing its essential fragment (it is desirable that this one is significantly in demand), which can be further developed to more or less complete implementation.

Such a fragment can be, for example, a model of the agent's behavior, synchronous with the behavior of a real person, which is included in the context of normative behavior. For example, it can be a model of behavior of a person (for example, an employee of a certain enterprise) in terms of participation in the production process, who has to follow certain rules (follow instructions).

Such modeling is relevant, since the business suffers from the unpredictability of employees' behavior, which is also monitored by humans, that is, it is also unpredictable. Eliminating this uncertainty dramatically improves the production situation.

The agent's behavior is modeled by projecting onto two parts of the world model: linguistic and extralinguistic. The projection onto the extralinguistic model of the world makes it possible to take into account the physical conditions of the agent's behavior, while the projection onto the language model takes into account the text context.

At the same time, this bimodality allows us to consider the model as the core of the SAI system with the possibility of its subsequent development in terms of the expansion of sensing, the inclusion of motor skills, semantic representations of various levels, dialogues. All this can be structured in terms of the accepted concept.

A sufficiently long observation of the behavior of a real agent makes it possible to form a model of his/her world in the form of a semantic network: the behavior of the agent in the real world is projected into behavior on the extralinguistic model of the world, which is described by a sequence of sentences, that is, a natural language text. Based on the text, a semantic network of this text can be built [4]. To analyze the dynamics of a situation, it is sufficient to simply have a set of texts describing the situation at various points in time. These texts are easy to find, for example, in current media publications [6].

And one more thing to bear in mind: further hardware simulation of the system is possible, subject to the development of a neurochip and an architecture based on this neurochip, which will allow implementing the details of the concept within the architecture of the model core.

5 Conclusion

The introduction of the second signaling system into the artificial intelligence system including the linguistic model of the world that works in conjunction with the extralinguistic model of the world, makes it possible to separate the presentation and processing of specific information into lower-level and upper-level representations. The lower-level representations implemented mainly in the extralinguistic part of the model of the world detail the description of situations, while the upper-level representations implemented mainly in the linguistic part of the model of the world generalize the description. Thus, along with the detailed description, the model of the world also contains a generalized representation, which makes it easy for the user to interact with the system using natural language.

References

1. Gorodetsky V.I.: Artificial intelligence: scientific contents, trends, opinions and pitfalls. https://www.youtube.com/watch?v=U-Gt4GXEKDw. Accessed 21 July 2021
2. Porshnev, B.F.: On the Beginning of Human History (Problems of Paleopsychology), 2nd edn. Nauka, Moscow (1974)
3. Kharlamov, A.A.: The language model of the world and purposeful human behaviour. J. Brain Behav. Cogn. Sci. **2**(1), 1–5 (2018)
4. Kharlamov, A.A.: Associative Memory As an Environment for the Formation of a Space of Knowledge From Biology To Applications. Palmarium Academic Publishing, Dusseldorf (2017)
5. Luriya, A.R.: Language and conscience. Piter, St.-Pb. (2019)
6. Kharlamov A.A., Pilgun, M.: Analysis of the situation connotation on the example of assessing the reaction of society: social media data. Int. J. Future Gener. Commun. Netw. (IJFGCN) NADIA **3**(13), 37–44 (2020)
7. Zavalishin, N.V., Muchnik, I.B.: Visual perception model and image analysis algorithms. Publisher, Moscow (1974)
8. Weston, J., et al.: Towards AI-complete question answering: a set of prerequisite toy tasks. arXiv:1502.05698v10 [cs.AI], 31 December 2015

Intellectualization of Production Facilities Management in Conditions of Uncertainty

Victor L. Burkovsky, Evgeniy V. Sidorenko, Vladimir Zh. Bocharov(✉), and Stepan L. Dobrynin

Voronezh State Technical University, Voronezh, Russian Federation
bvl@vorstu.ru

Abstract. The article is devoted to the consideration of the intellectualization of the management of production facilities in conditions of uncertainty. The authors emphasize the relevance of the chosen research topic. The classification of intelligent management approaches in conditions of uncertainty is considered: the proactive approach and the reactive approach; the description of the features of these approaches is given. Bayesian networks are described in the context of management under uncertainty. On the basis of Bayesian networks, the problems of managing additive manufacturing processes are shown. And the use of Bayesian neural networks is shown in the light of the consideration of the problems of managing robotics complexes and regional power networks #COMESYSO1120.

Keywords: Management in conditions of uncertainty · Bayesian network · Neural network

1 Introduction

Traditional approaches to the management of production facilities usually consider a completely deterministic environment in which there is no uncertainty. Situations that arise during the operation of production facilities are in most cases unpredictable due to the impact of various factors on the objects, dynamic changes in the external environment, violations in the management system. In addition, traditional approaches use numerical values of control system parameters (operation execution time, process flow rate) to represent the state of a real system. These numerical data are a potential source of incompatibility of models with a physical object, since they may change at the time of recording or be insufficiently known. The compatibility of the model with a physical object is realized by defining and modeling input data that is presented as uncertain, inaccurate or incomplete. Uncertainty is associated with doubts about the reliability of knowledge or with not knowing whether a statement is true or not. Inaccuracy occurs due to problems in the formulation of knowledge. Incompleteness is associated with a lack of knowledge or partial knowledge about some characteristics of the system.

Sources of uncertainty in production facilities can be divided into three categories: natural variability of production processes, information uncertainty due to inadequate, poor-quality, missing or erroneous data, and uncertainty caused by modeling, assumptions and approximations [1].

R. Silhavy et al. (Eds.): CoMeSySo 2021, LNNS 231, pp. 526–536, 2021.
https://doi.org/10.1007/978-3-030-90321-3_43

Taking into account each kind of uncertainty in the management of production facilities is a difficult task. The requirements for production optimization have evolved from the usual traditional efficiency criteria described in terms of static optimality to new efficiency criteria described in terms of reactivity, adaptability and robustness. In this connection, there is a need to classify intelligent management approaches in conditions of uncertainty.

2 Methods

2.1 Classification of Intelligent Management Approaches in Conditions of Uncertainty

The structural classification scheme covers new types of intelligent approaches to managing production facilities under conditions of uncertainty, such as proactive and reactive. It is developed independently of the subject areas and applied problems and can be applied to various types of production facilities (control of robotic complexes, technological processes of additive manufacturing, distributed objects of regional power supply of a nuclear power plant) (Fig. 1).

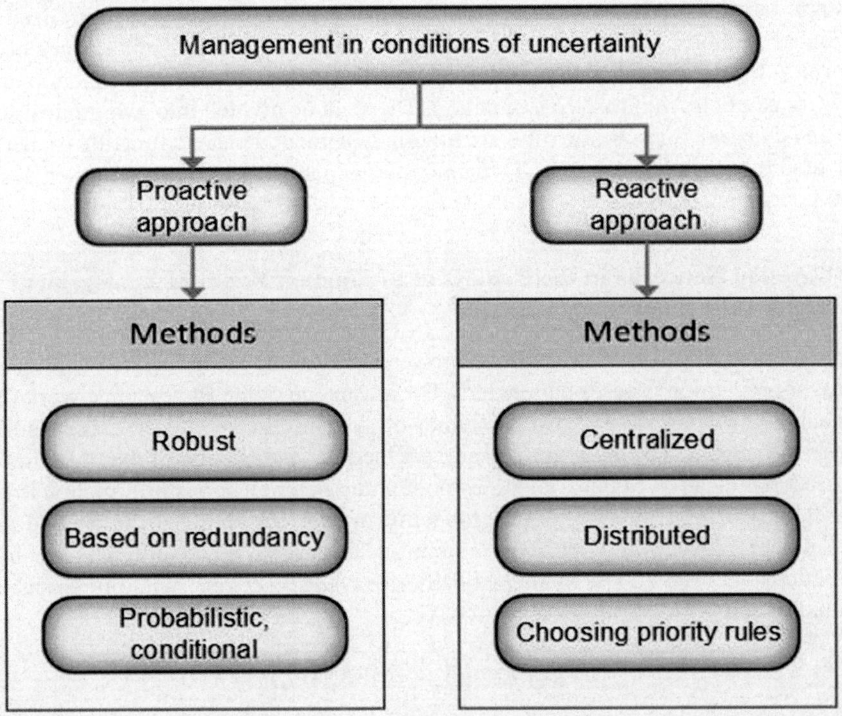

Fig. 1. Block diagram of classification of intelligent management approaches under uncertainty

A proactive approach is an important data management tool to prevent the expected unavoidable violations in the management system during its operation. This approach focuses on the construction of predictive models that predictably meet the performance requirements of objects in a dynamic environment [2]. The proactive approach usually uses redundancy (temporary or resource-oriented), probabilistic and conditional methods (several possible variants of control actions are being developed, between which it is possible to switch adaptively). Taking into account that real-time events in the production process are usually stochastic in nature, the main problem of the proactive approach is the definition of predictability models and their results.

The reactive approach is used in highly perturbed environments, where the uncertainties are random. In such an environment, autonomous planning of control actions becomes almost impossible, and all decisions are made in real time. In this context, a centralized or distributed decision-making process is envisaged. Centralized methods use expert planning systems: for example, using meta-rules (for selecting rules about the rules that will be applied), built on the basis of expert knowledge. The idea of the centralized method is to identify and dynamically update the states and transition probabilities that are used by the expert system to solve the problem of planning and management in real time. Distributed methods are widely used for reactive planning under conditions of uncertainty, mainly using multi-agent systems. Multi-agent systems are suitable for developing flexible and reliable distributed management, but the performance of such a system strongly depends on the coordination mechanism [3]. In both centralized and distributed methods, one of the important issues is the static or dynamic choice of priority rules. In general, the priority rules change depending on the target function under study (for example, minimization or delay). They can be divided into two groups: static rules and dynamic rules. Static rules are time-independent and are a function of the data associated with tasks or resources. Dynamic rules depend on time and on the actual state of the system.

2.2 Bayesian Networks in the Context of Production Facilities Management Under Uncertainty

A large number of production facility management systems are associated with quality control or production process parameters. The advantage of the Bayesian network (BN) approach is that it allows for several studies of system parameters, such as sensitivity analysis, dimensionality reduction, resource allocation, calibration of the object model and direct propagation of uncertainty, by combining heterogeneous information [4].

A Bayesian network is a graph probability model consisting of nodes and arcs, where nodes represent variables in the system, and arcs represent conditional probability distributions between nodes. Mathematically, a Bayesian network is a joint probability distribution of a set of variables $X = (X_1, X_2, \ldots, X_n)$, represented as

$$Pr_B(X) = \prod\nolimits_{i=1}^{n} Pr_B(X_i|\Pi_{X_i}) \tag{1}$$

where Π_{X_i} is the set of parent nodes X_i, and $Pr_B(X_i|\Pi_{X_i})$ represents the conditional probability distribution of X_i taking into account its parent nodes. If X_i has no parent nodes, then $Pr_B(X_i|\Pi_{X_i})$ represents a particular probability distribution of X_i.

Figure 2 shows a Bayesian network that represents the joint probability distribution of three random variables A, B, C.

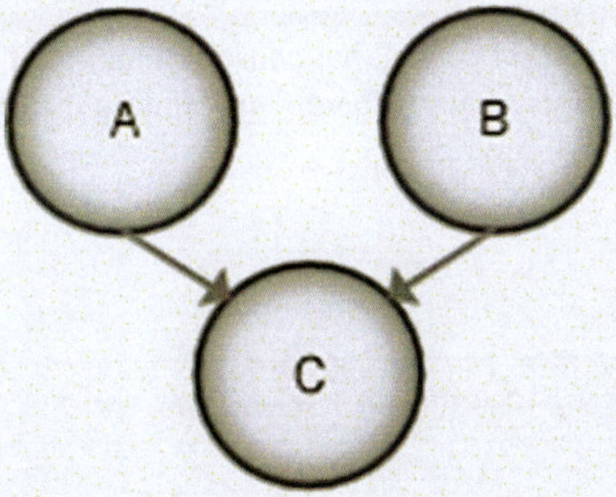

Fig. 2. A Bayesian network with three variables

The joint probability distribution can be represented as

$$Pr_B(A, B, C) = Pr(A)Pr(B)Pr(C|A, B) \tag{2}$$

Since A and B do not have parent nodes, private probability distributions are used for them instead of conditional ones. BN allows two types of analysis: direct and reverse. Direct analysis involves calculating the uncertainty in C, taking into account the uncertainty in the root nodes (nodes without parent nodes) - A and B. As a reverse analysis, model calibration is widely used, which includes updating the limit distributions of A and B when new data appears in the variable C.

In the case where only data for variable B is available, parametric or nonparametric probability distributions may correspond to the available data. Using the probability distribution of the match, you can perform further analysis, direct or reverse.

The conditional probability distribution associated with $Pr(C|A, B)$ can be interpreted as a stochastic model linking variables A, B and C. The presence of model errors leads to uncertainty in the distributions of conditional probabilities. Analyses such as model validation can be performed to quantify model errors, and the revised model, after accounting for all errors, can be used as a conditional probability distribution in a Bayesian network. Thus, the results of the uncertainty of the data and the model can be integrated into the Bayesian network.

3 Results and Discussions

3.1 Intellectualization of Additive Manufacturing Process Management in Conditions of Uncertainty

Additive manufacturing (AP) is the process of creating objects based on the data of a three-dimensional model, usually layer by layer (the finished product is formed by adding the material). Thanks to the layer-by-layer manufacturing, the AP allows you to create objects of complex geometric shape with minimal consumption of materials, such as plastics, polymers and metals [5]. One of the main obstacles to the widespread use of AP technologies is the difference in the quality of manufactured parts. There are various sources in AP technological processes that cause product quality variability due to the spread and aggregation of uncertainty, thereby preventing the re-production of products with constant quality.

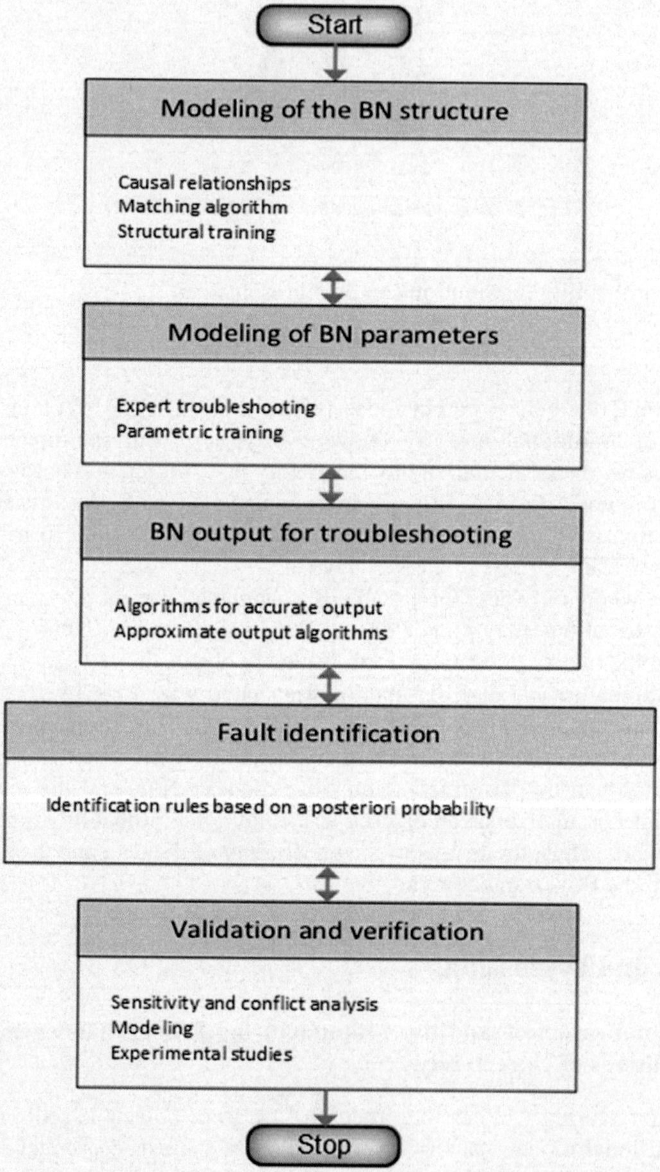

Fig. 3. Block diagram of fault diagnostics in the AP based on BN

The procedures for diagnosing the parameters of technological processes in the AP and the quality of parts using the BN consist of modeling the structure of the BN, modeling the parameters of the BN, output of the BN, fault identification, as well as validation and verification. Figure 3 shows a block diagram of fault diagnostics in an AP based on BN. The BN-based fault diagnosis model includes five stages:

1. A structural BN model can be created using cause-and-effect relationships, matching algorithms, or a structural learning method;
2. A parametric BN model can be created using expert troubleshooting and parametric training;
3. Exact or approximate algorithms can be performed to output BS data;
4. System failures can be identified directly using a posteriori probability or various failure identification rules;
5. Validation and verification of the model can be carried out using various methods, such as sensitivity analysis, conflict analysis, modeling and experimental studies. If the diagnostic results are unsatisfactory, one or more of the first four stages are reviewed or replaced by the specific method used until satisfactory diagnostic results are obtained.

3.2 Intellectualization of Control of Robotic Complexes in Conditions of Uncertainty

A robotic complex (RC) is a set of robotic devices combined into a single system that works according to a certain program, which can change during operation. There are a large number of different implementations of RC control systems, but the most advanced of them are feedback systems.

A large number of disturbing factors act on the control actions in the RC control systems, which lead to the appearance of uncertainty in the system. Such disturbing factors, for example for mobile robotics, can be electromagnetic interference; temperature changes; humidity fluctuations affecting the clock circuits; radioactive radiation, vibration, etc. Thus, it can be argued that the performance of RC control systems depends on the stability of the control system and its insensitivity to disturbances. At the same time, one should not be mistaken in the fact that completely ignoring uncertainties increases the stability of the system.

The use of BN in robotics can significantly increase the independence of the RC control system in decision-making. For example, for the implementation of automatic operation functions: cross-country movement of a mobile RC or horizontal positioning algorithms for robots with a large number of degrees of freedom of actuators whose mobility is limited by the external environment.

In addition to the use of BN in robotics, there is often a question of machine vision, which is used in complex RCs for accurate positioning of manipulated objects and an adequate assessment of the surrounding space. BN is a mechanism for working with probability, which is hardly applicable in this case. Quite often, facet neural networks are used to work with technical vision, which can significantly simplify the process of recognizing visual images.

An artificial neural network (ANN, NN) is a mathematical model built in the image and likeness of the organization of communication of biological neurons. At the same time, the system is self-learning.

Artificial neuron – designed for performing simple calculations and transmitting further over the network. These neurons can be divided into the main types: input, hidden and output. A set of neurons similar in purpose is called a layer. Artificial neurons have two main parameters, namely output and output data.

A synapse is a connection between neurons in the NS. Synapses have a weight that changes the input data when transmitted between neurons. At the same time, data on synapses is transmitted from neurons that have a large weight. Thanks to the weights, the data gets to the inputs from neurons with a large weight.

The state of a neuron can be determined by the formula

$$S = \Sigma(i = 1)^n x_i \cdot w_i \tag{3}$$

where n is the number of neuron inputs, x_i is the value of the i-th neuron input, i is the weight of the i-th synapse.

The value of the axon of a neuron is determined by the formula

$$Y = f(S) \tag{4}$$

where f is the activation function. The activation function is intended for data normalization. The most common activation functions are linear, sigmoid, and hyperbolic tangent.

The most commonly used activation function is sigmoid, which has a range of values from 0 to 1. The activation function is sigmoid:

$$f(x) = \frac{1}{1 + e(-ax)} \tag{5}$$

For values in the range from -1 to 1, the hyperbolic tangent activation function is used:

$$f(x) = \frac{e^2x - 1}{e^2x + 1} \tag{6}$$

Usually in neural networks, connections between neurons are set using weights (numbers), but instead of weights, you can use probabilities calculated using the Bayes formula. Thus, we get a Bayesian neural network (BNN). Systems built on the basis of the BNN have a number of major advantages in applying to the RC:

- With the help of BNN, the control system can work with unknown values of variables by determining dependencies between variables;
- BNN allow management systems to conduct predictive analysis;
- With the help of the BNN, the control system can combine patterns obtained from data from sensors and vision systems, as well as pre-set ones in the control system;
- Control systems with BNN are more resistant to data noise and excessive complication of the overfitting model.

Otherwise, the BNN differ little from the usual NN. This makes it possible to build RC control systems based on BNN, as well as on the basis of NN. As an example of the use of BNN, let's consider the RC ROIN RTP-M control system.

Accidents that occur at oil and gas industry facilities pose a huge threat to human life and health, cause irreparable damage to the environment and cause huge losses. The possibility of liquidation in the shortest possible time and with minimal participation of people will reduce the negative consequences described above and, more importantly, reduce the risk to people's health and life. To solve this problem, the company JSC MGK "Intechros" has developed a robotic complex RTP-M. This robotic complex allows remote guidance of the shut-off wellhead assembly to the gushing wellhead during the elimination of burning gas and oil fountains (the main negative factors are high temperature, high concentrations of explosive volatile compounds, open flame, etc.). Control over the execution of technological operations is carried out by explosion-proof chambers.

The RTP-M robotic complex is equipped with a control system that allows remote control over a radio channel at a distance of up to 1000 m in line-of-sight conditions. At the same time, the operator of the robotic complex has access to various modes of operation of the control system for performing various technological operations.

The RTP-M ROIN control system was developed using BNN. The control system allows the RC to localize accidents with the minimum possible human participation. The developed RC and control system are focused on application in the elimination of accidents at oil and gas industry facilities. For example: for extinguishing and eliminating gas fountains, analyzing blockages in the area of an accident and eliminating other consequences.

Partial control and execution of technological operations were transferred from the operator to the control system, to increase both the overall reliability of the RTC operation and the speed of operations. First of all, the automation of the RC allows you to avoid negative consequences if you lose communication over the radio channel between the robotic complex and the operator's remote-control point. Each technological operation is transferred under the control of its BNN. Thus, the same actuators can be available for the control system regardless of the current technological operation. The set of BNN technological operations forms the layers of the general neural network of the neuro-controller. With this approach, the operator of the robotic complex is only required to indicate which operation needs to be performed, and then control its execution, correcting the control actions if necessary. A block diagram of the control system is formed: each layer is responsible for its own technological operation, after summing up the control actions from the layers of the neural network, the data is transmitted to the main control system, from which the control action is transmitted to the control object. At the same time, the RTC operator monitors the control system (Fig. 4).

This control system is installed on the RTP-M robotic complex, after which a training cycle of the BNN layers is carried out. Then comparative tests of the neural network control and the standard control system are carried out for qualitative and quantitative analysis of changes in the efficiency of the robotic complex.

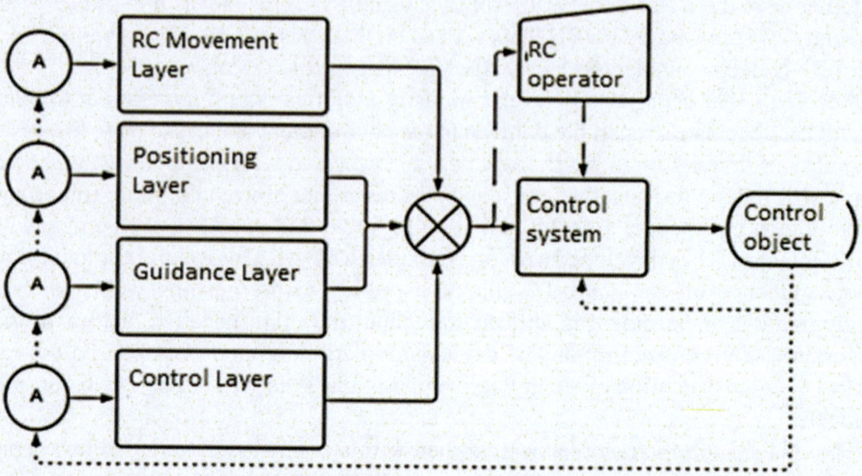

Fig. 4. Block diagram of the control system

3.3 Intellectualization of Regional Power Grid Management in Conditions of Uncertainty

The functioning and development of global and regional energy systems largely depends on the ability of the system to predict the load on the power grid. The more accurate and earlier the forecast is made, the easier it is to maneuver the generating capacities to smooth out the peaks of the highest and lowest load.

There are a large number of factors that affect the load of the energy network: daily fluctuations in consumption, weather conditions, the state of the power grid, etc. The widely used term "forecasting method" is used both for simple calculations and for multi-step expert surveys, in addition, as a method of practical and theoretical action [10]. It is these factors that determine the reliability of the forecast.

As a measure to combat uncertainty in predicting the load of power grids, you can use BNN by dynamically changing the number of input parameters of neurons, that is, changing the structure of the neural network as necessary (Fig. 5).

In this BNN is considered not only as a summing function. Since a particular neuron is not always involved in the accumulation of data by the activation function, it is rational to choose a hyperbolic tangent. Thus, the neuron accepts the following ranges of values:

1. The values from $[-1; 0]$, from $[0; +1]$ and 0 are the total accumulated data and the BNN behaves like an ordinary classical neural network;
2. The value of a neuron from $[+1; +\infty]$ - it is required to add another neuron, change the structural scheme of the network and determine new weight values of the BNN probabilities. Adding a neuron normalizes the network and leads it to 1 state;
3. The value of the neuron from $[-\infty; -1]$ already requires the exclusion of the neuron from the BNN, so it gives redundant data that interferes with the network.

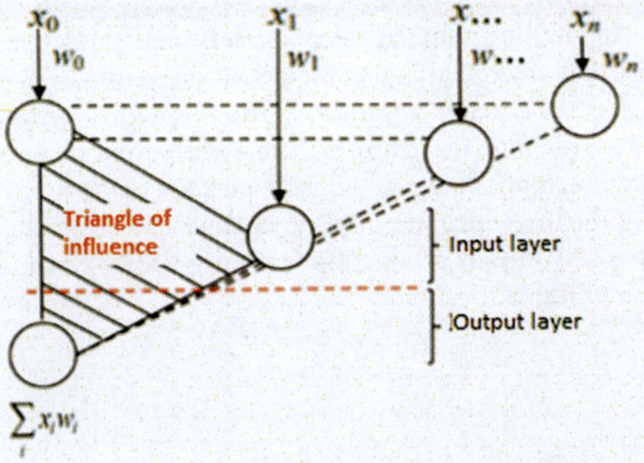

Fig. 5. Block diagram of the BNN of the power grid

The mathematical description of a neuron can be represented as follows [11]:

$$f(x) = n \cdot \tan(x \cdot n) \tag{7}$$

The input data of the BNN are divided into [11]: parameters of the quality of energy supply, parameters of the functioning (consumption) of the subscriber, parameters of emergency cases (weather conditions) or special requirements of subscribers to the quality of electricity. Parameters of the subscriber's feedback (timely payment, presence on the spot in case of an emergency, response to service interruptions), etc.

The number of input neurons and neurons of the inner layers is dynamic and changes during the functioning of the BNN, the number of output layers is unchanged. Some of the output neurons have additional connections with the neurons of the first layer [11].

The use of the BNN allows us to solve the problem of predicting the load on the power grid in the most rational way.

4 Conclusion

Within the framework of this article, the theoretical foundations of the intellectualization of management of the most popular production facilities operating under conditions of uncertainty are considered.

Sources of uncertainty in production facilities can be divided into three categories: natural variability of production processes, information uncertainty due to inadequate, low-quality, missing or erroneous data, and uncertainty caused by modeling, assumptions and approximations.

Taking into account each kind of uncertainty in the management of production facilities is a difficult task. The requirements for production optimization have evolved from the usual traditional efficiency criteria described in terms of static optimality to new efficiency criteria described in terms of reactivity, adaptability and robustness. In this connection, there is a need to classify intelligent management approaches in conditions of uncertainty.

The approaches to intelligent management are reflected: proactive – the construction of predictive models that predictably meet the performance requirements of objects in a dynamic environment; the reactive approach is used in highly perturbed environments where uncertainties are random. In such an environment, autonomous planning of control actions becomes almost impossible, and all decisions are made in real time.

The questions of the theoretical description of Bayesian networks and their application in control systems of various production facilities: additive manufacturing, RC and power systems are highlighted.

References

1. Li, Z., Ierapetritou, M.: Process scheduling under uncertainty: review and challenges. Comput. Chem. Eng. **32**(4–5), 715–727 (2008)
2. Liebrecht, C., et al.: Multi-criteria evaluation of manufacturing systems under uncertainty 4.0. Procedia CIRP **63**, 224–229 (2017)
3. Yong, B., Brintrup, A.: Multi agent system for machine learning under uncertainty in cyber physical manufacturing system. In: Borangiu, T., Trentesaux, D., Leitão, P., Giret Boggino, A., Botti, V. (eds.) SOHOMA 2019. SCI, vol. 853, pp. 244–257. Springer, Cham (2020). https://doi.org/10.1007/978-3-030-27477-1_19
4. Nannapaneni, S., Mahadevan, S., Dubey, A., Lee, Y.-T.: Online monitoring and control of a cyber-physical manufacturing process under uncertainty. J. Intell. Manuf. **32**(5), 1289–1304 (2020). https://doi.org/10.1007/s10845-020-01609-7
5. Dobrynin, S.L., Burkovsky, V.L.: Problems of additive manufacturing management based on industrial internet of things technologies. Bull. Voronezh State Tech. Univ. **17**(2), 7–13 (2021)
6. Bocharov, V., Danilov, A., Burkovsky, V., Gusev, K., Gusev, P.: Analysis of resource availability of production enterprise based on fuzzy neural network. In: Ronzhin, A., Shishlakov, V. (eds.) Proceedings of 15th International Conference on Electromechanics and Robotics "Zavalishin's Readings." SIST, vol. 187, pp. 459–467. Springer, Singapore (2021). https://doi.org/10.1007/978-981-15-5580-0_37
7. Bocharov, V.Zh., Burkovsky, V.L.: Control of a robotic complex in extreme operating conditions based on fuzzy neural networks. Bull. VSTU (6) (2016)
8. Bocharov, V.Zh., Burkovsky, V.L.: A neural control system based on a 3D network in a special-purpose robotic complex. Bull. VSTU (3) (2017)
9. Bocharov, V.Zh., Burkovsky, V.L.: Modular type control system and 3D neural network power controller for roin RTS r-300/04. In: Energy and Resource Conservation-XXI Century: Materials of the XVIII International Scientific and Practical Conference, Orel (2020)
10. Vedernikov, A.S.: Planning of operating modes of electric power systems. Samara State Technical University, pp. 193. SSTU, Samara (2019)
11. Urumbayeva, O.B., Shalaev, T.A., Shikulskaya, O.M.: The concept of intelligent power grid management. Eng.-Build. Bull. Caspian Reg. Sci. Tech. J. **3**(33), 69–74 (2020). Astrakhan State Architectural University. GAOU JSC VO "AGASU", Astrakhan

Thinking Activity Dynamics Evaluation Based on EEG Analysis

Natalya Bodrina, Konstantin Sidorov(✉), and Natalya Filatova

Tver State Technical University, Lenina Avenue 25, Tver, Russia

Abstract. The process of assessing the dynamics of cognitive activity based on the characteristics of the EEG is described. As signs reflecting the level of cognitive activity, estimates of the density of trajectories of the projection of the phase portrait in the vicinity of the origin were used. Six pairs of leads (Cz-A2 and C4-A2; Pz-A1 and P3-A1; P4-A2 and P3-A1; C3-A1 and C4-A2; O1-A1 and O2-A2; Pz-A1 and O1-A1) in which there is a significant change in these characteristics in the process of cognitive activity were identified. Experiments are described. The level of cognitive activity was assessed before and after emotion stimulation of various durations (5 and 10 min). The results show an improvement in the indicators of cognitive activity during the process of sequential emotion stimulation. The average density of trajectories increases after the stimulation. Also, the total time for completing tasks and the number of errors decrease. These indicators improve when stimulation time increases.

Keywords: Cognitive activity · Stimulation of emotions · EEG signal

1 Introduction

Dynamic models describing intellectual processes in the human brain are usually based on the analysis of EEG signals [1–9]. For this purpose, spectral methods are usually used to study mental activity, as well as considering the characteristics of the power spectra for individual leads [10, 11], or the coherence functions for a pair of interconnected leads [12, 13].

In our research, we use the methods of nonlinear dynamics that allow reconstructing an attractor for a time series recorded in one EEG lead. The experiments have shown that the attractor properties change with changing cognitive activity [14].

These changes differ significantly depending on the location of the EEG sensor. Partial data integration is possible when constructing a phase portrait using the lead axes (O1, O2, P3, P5, C3, C4,…), which display the amplitude values from the corresponding EEG sensor signals. Then each point in the phase space has a corresponding value of the brain electrical activity recorded at the moment t_i in each lead.

It is shown that in order to analyze and assess cognitive activity, it is enough to use the projections of the multidimensional phase portrait in several leads.

The trajectories reproduced on the graph display the dynamics of cognitive activity in the Δt interval. Their analysis is based on the methods of nonlinear dynamics,

R. Silhavy et al. (Eds.): CoMeSySo 2021, LNNS 231, pp. 537–543, 2021.
https://doi.org/10.1007/978-3-030-90321-3_44

phase portrait characteristics, which are used when comparing attractors reconstructed by physical signals.

It is necessary to assess the possibility of using these characteristics to build a model of the cognitive activity dynamics. For this purpose, we use an experimental approach and a specialized bioengineering system [15]. The system allows forming various experimental procedures to assess cognitive activity or emotional reactions. This creates conditions for experimental testing of two hypotheses. First, it is possible to assess the effect of the emotiogenic stimulation duration on cognitive activity. Second, it is possible to single out a limited number of EEG leads in order to build a model of cognitive activity dynamics, which simplifies further interpretation of simulation results.

2 Methods

2.1 The Used EEG Characteristics

The main characteristics used when constructing a model of the cognitive activity dynamics are related to density in the vicinity of the origin of the coordinates of the reconstructed attractor trajectories (Fig. 1a) or a multidimensional phase portrait (Fig. 1b) [16, 17].

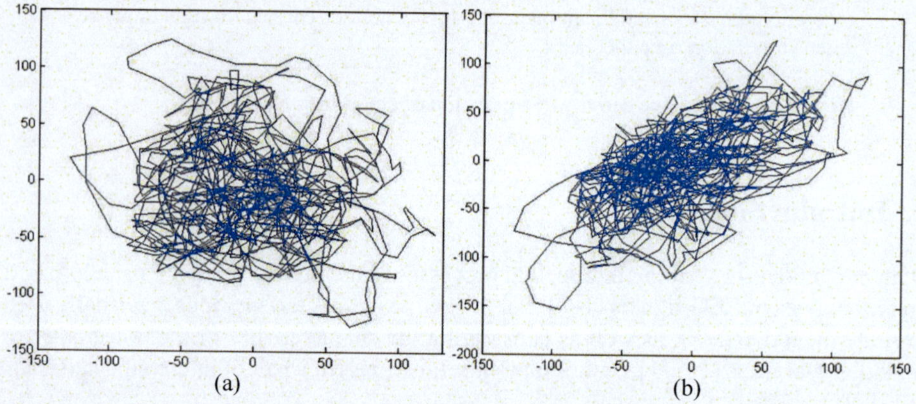

Fig. 1. An attractor reconstructed on P4-A2 lead (a) and a phase portrait projection by P4-Pz leads (b)

This characteristic is defined as the sum of the trajectory points in the four central cells of the grid covering the two-dimensional projection of the multidimensional phase portrait on the plane formed by two leads.

2.2 The Experiment

The experiments involved using the EEG-Speech + multichannel bioengineering system [18]. The system included the Encephalan-131-03 electroencephalograph (Taganrog, Russia). The EEG signal was recorded in 19 leads according to the 10–20 system, the sampling rate was 250 Hz.

The participants in the experiments were healthy males aged 20–25 years. The sessions took place during the day in a quiet room, with the testee seated in a comfortable chair.

According to the experiment scenario (Fig. 2), the blocks of cognitive tasks alternated with periods of emotiogenic stimulation of various duration. A cognitive load [12] was a flow of slides with the same type tasks – multiplying a two-digit number by a one-digit number. Every 10 tasks were combined into groups and presented simultaneously. The difficulty of the tasks was at the same level. The transition to the next task was possible only after a correct answer of a testee. Skipping tasks was prohibited, and the execution time was not limited.

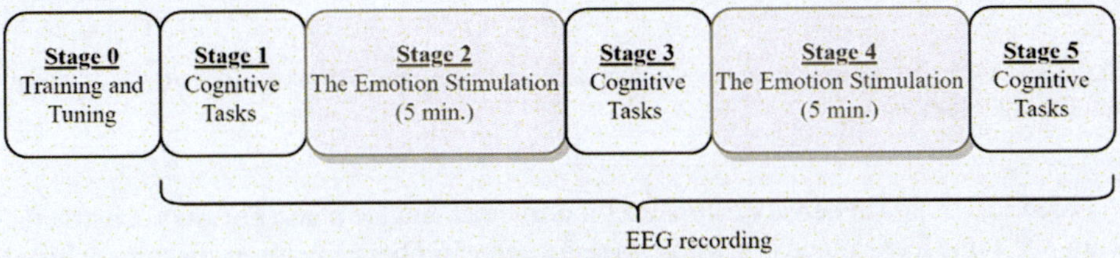

Fig. 2. The experiment scenario

At the stage 0, a testee got acquainted with the tasks and got used to work. Also, the EEG electrodes were installed and the skin impedance was monitored. The stages 1, 3, and 5 contained 100 cognitive tasks each. The stages 2 and 4 involved emotiogenic stimulation. At the stage 2, the emotion stimulation lasted 5 min, at the stage 4 it lasted 10 min. The stimuli were videos without sound. Weak negative emotions were stimulated. For this purpose, the subjects of stimuli were selected individually according to the results of a preliminary survey of the testees.

Each experiment session resulted in an obtained EEG record and a protocol with the recorded time stamps of events and all testee's responses. From each EEG record, we have obtained 100 artifact-free fragments lasting 4 s.

3 Results

The indicator (ν) was calculated for each EEG signal fragment. Figure 3 shows a diagram of this characteristic changes for the leads P4-A2 and C4-A2 during one experiment session. On the average, the density characteristic increases after stimulation, both when assessing it using attractors (according to P4-A2 lead and Pz-A2 lead) and when using the projection of the phase portrait (P4-Pz) reconstructed according to the indicated pair of leads to assess the density.

The experimental results show that the tendency of density changes in the attractor trajectories for a time series recorded in one EEG lead (for example, C4-A2 or Cz-A2) practically coincides with the tendency of a similar characteristic obtained for the projection of a multidimensional phase portrait on the plane formed by these leads (for example, C4-Cz).

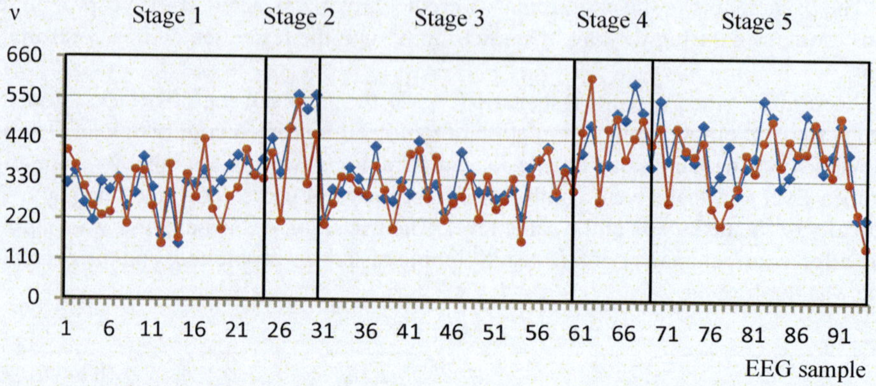

Fig. 3. The density (ν) change in the attractor reconstructed using EEG signals in the P4-A2 (red) and C4-A2 (blue) leads

Identical trends in density estimates for attractors and for phase portraits constructed using the same leads made it possible to reduce the number of features that are included in the model of the cognitive activity dynamics.

We have identified six leads with a significant change in these characteristics during cognitive activity. In accordance with these facts, the model includes the following components:

- f_1 is the central density of the trajectories of the projection of the multidimensional phase portrait on the plane formed by Cz-A2 and C4-A2 leads,
- f_2 is the central density of the trajectories of the projection of the multidimensional phase portrait on the plane formed by the Pz-A1 and P3-A1 leads,
- $f_3, \ldots, f_6$ is the central density of the trajectories of the projection of the multidimensional phase portrait on the plane formed by the leads (P4-A2 and P3-A1), (C3-A1 and C4-A2), (O1-A1 and O2-A2), (Pz-A1 and O1-A1), respectively.

The experiments have confirmed the assumption about the possibility of using estimates of the central density of trajectories of the projections of the multidimensional phase portrait on the plane formed by pairs of leads in order to monitor cognitive activity.

The diagram (Fig. 4) shows a comparison of the average value of the characteristic (ν) for different experiment stages. For comparison, we use the density estimates obtained for attractors reconstructed from the C4-A2 and Cz-A2 leads. The graph for this characteristic obtained for the projection of the phase portrait on the plane formed by the pair of leads C4 and Cz illustrates the same trends.

The diagrams given (Fig. 3, Fig. 4) illustrate the influence of the stimulation duration on the level (ν). The average density at the stage 6 is 23% higher than this estimate after the first stimulation (stage 4), the duration of which is 40% less than the second one.

In addition, we have recorded a decrease in the total time for completing calculating tasks (by almost 40%), as well as in the number of errors made by testees after emotiogenic stimulation (Fig. 5). These indicators improve when stimulation time increases ($\Delta t2 = 2 * \Delta t1$).

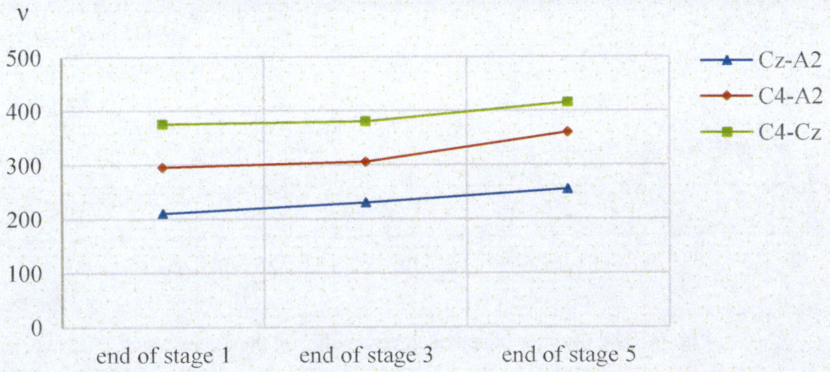

Fig. 4. The change in the central density of the attractor at individual experiment stages

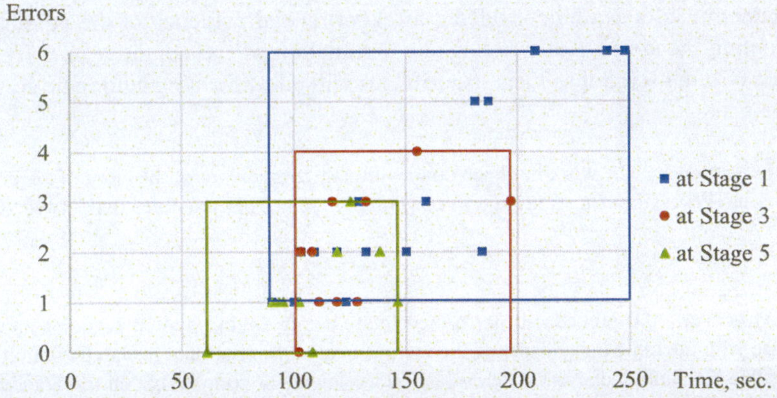

Fig. 5. The results of task performing depending on the time of emotiogenic stimulation

During the experiments, the errors' nature also changed. After emotiogenic stimulation, there is a sharp decrease in repeated errors, and this process progresses with the transition to the second stage of stimulation (Fig. 6).

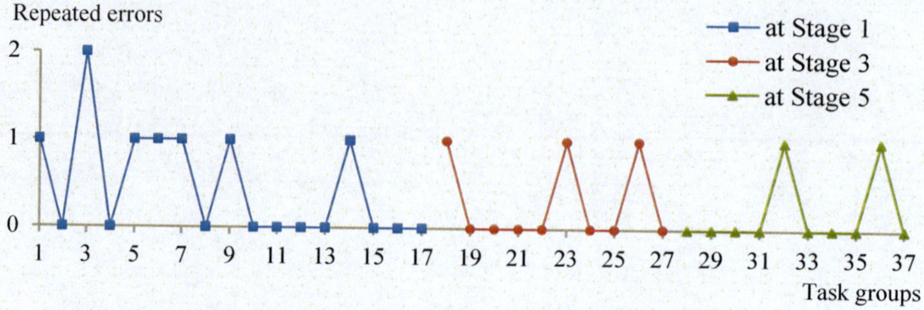

Fig. 6. The number of repeated errors during the experiment

4 Conclusion

The obtained results can be considered as experimental evidence of the possibility of implementing the scheme of emotiogenic stimulation of the cognitive activity of the computer system user. Subsequent experiments will allow forming automatic correction algorithms.

Acknowledgements. The work has been done within the framework of the grant of the President of the Russian Federation for state support of young Russian PhD scientists (МК-1398.2020.9).

Ethical Approval. All procedures performed in studies involving human participants were in accordance with the ethical standards of the institutional and/or national research committee and with the 1964 Helsinki declaration and its later amendments or comparable ethical standards.

References

1. Rabbi, A.F., Ivanca, K., Putnam, A.V., Musa, A., Thaden, C.B., Fazel-Rezai, R.: Human performance evaluation based on EEG signal analysis: a prospective review. In: 2009 Annual International Conference of the IEEE Engineering in Medicine and Biology Society, Minneapolis, pp. 1879–1882. IEEE (2009). https://doi.org/10.1109/IEMBS.2009.5333877
2. Rahman, M., Karwowski, W., Fafrowicz, M., Hancock, P.: Neuroergonomics applications of electroencephalography in physical activities: a systematic review. Front. Hum. Neurosci. **13**, 182 (2019). https://doi.org/10.3389/fnhum.2019.00182
3. Ismail, L.E., Karwowski, W.: Applications of EEG indices for the quantification of human cognitive performance: a systematic review and bibliometric analysis. PLoS One **15**(12), e0242857 (2020). 10.1371journal.pone.0242857
4. Rabinovich, M.I., Muezzinoglu, M.K.: Nonlinear dynamics of the brain: emotion and cognition. Adv. Phys. Sci. **180**(4), 371–387 (2010). https://doi.org/10.3367/UFNr.0180.201004b.0371. (in Russian, Uspekhi Fizicheskikh Nauk)
5. Baars, B.J., Gage, N.M.: Cognition, Brain, and Consciousness: Introduction to Cognitive Neuroscience, 2nd edn., pp. 653. Elsevier Academic Press: Burlington (2010). ISBN: 978-0-12-375070-9

6. Lu, Y., Jaquess, K.J., Hatfield, B.D., Zhou, C., Li, H.: Valence and arousal of emotional stimuli impact cognitive-motor performance in an oddball task. Biol. Psychol. **125**, 105–114 (2017). https://doi.org/10.1016/j.biopsycho.2017.02.010
7. Klimesch, W.: EEG alpha and theta oscillations reflect cognitive and memory performance: a review and analysis. Brain Res. Rev. **29**(2–3), 169–195 (1999). https://doi.org/10.1016/s0165-0173(98)00056-3
8. Basar, E., Basar-Eroglu, C., Karakas, S., Schurmann, M.: Brain oscillations in perception and memory. Int. J. Psychophysiol. **35**(2–3), 95–124 (2000). https://doi.org/10.1016/S0167-8760(99)00047-1
9. Basar, E., Basar-Eroglu, C., Karakas, S., Schurmann, M.: Gamma, alpha, delta, and theta oscillations govern cognitive processes. Int. J. Psychophysiol. **39**(2–3), 241–248 (2001). https://doi.org/10.1016/s0167-8760(00)00145-8
10. Reva, N.V.: Linear and nonlinear characteristics of human EEG with evoked emotions: PhD. 03.00.13. Novosibirsk (2000). (in Russian, Linejnye i nelinejnye harakteristiki EEG cheloveka pri vyzvannyh jemocijah)
11. Sidorov, K.V., Bodrina, N.I., Filatova, N.N.: Monitoring human cognitive activity through biomedical signal analysis. In: Kryzhanovsky, B., Dunin-Barkowski, W., Redko, V., Tiumentsev, Y. (eds.) NEUROINFORMATICS 2020. SCI, vol. 925, pp. 309–315. Springer, Cham (2021). https://doi.org/10.1007/978-3-030-60577-3_37
12. Lukoyanov, M.V., Krilov V.N.: EEG coherence in passive perception of lines with different inclination angles in left-handers and right-handers. Bulletin of the Nizhny Novgorod University named after N.I. Lobachevsky. Biology **5**(1), 135–139 (2012). (in Russian, Vestnik Nizhegorodskogo universiteta im. N.I. Lobachevskogo. Biologija)
13. Sidorov, K.V., Filatova, N.N., Bodrina, N.I., Semaev, P.D.: Analysis of biomedical signals as a way to assess cognitive activity during emotional stimulation. Proc. Southwest State Univ. Ser. Control Comput. Eng. Inf. Sci. Med. Instrum. Eng. **9**(1), 74–85 (2019). (in Russian, Izvestiya Yugo-Zapadnogo gosudarstvennogo universiteta. Serija: Upravlenie, vychislitel'naja tekhnika, informatika. Meditsinskoe priborostroenie)
14. Filatova, N.N., Sidorov, K.V., Bodrina, N.I., Voronkov, M.E., Klyuev, P.S.: Monitoring the level of trainee's cognitive activity by EEG signal fragmentary analysis. Proc. Southwest State Univ. Ser. Control Comput. Eng. Inf. Sci. Med. Instrum. Eng. **9**(4), 8–23 (2019). (in Russian, Izvestiya Yugo-Zapadnogo gosudarstvennogo universiteta. Serija: Upravlenie, vychislitel'naja tekhnika, informatika. Meditsinskoe priborostroenie)
15. Filatova, N., Bodrina, N., Sidorov, K., Shemaev, P., Vinogradov, G.: Biotechnical system for the study of processes of increasing cognitive activity through emotional stimulation. In: Kovalev, S., Tarassov, V., Snasel, V., Sukhanov, A. (eds.) IITI 2019. AISC, vol. 1156, pp. 548–558. Springer, Cham (2020). https://doi.org/10.1007/978-3-030-50097-9_56
16. Sidorov, K.V., Bodrina, N.I.: A software package for monitoring human emotional reactions and cognitive activity by analyzing biomedical signals. In: Silhavy, R., Silhavy, P., Prokopova, Z. (eds.) CoMeSySo 2020. AISC, vol. 1295, pp. 444–459. Springer, Cham (2020). https://doi.org/10.1007/978-3-030-63319-6_41
17. Sidorov, K.V., Bodrina, N.I.: Monitoring the characteristics of human emotional reactions based on the analysis of attractors reconstructed according to EEG patterns. In: Silhavy, R., Silhavy, P., Prokopova, Z. (eds.) CoMeSySo 2020. AISC, vol. 1295, pp. 430–443. Springer, Cham (2020). https://doi.org/10.1007/978-3-030-63319-6_40
18. Filatova, N.N., Bodrina, N.I., Sidorov, K.V., Shemaev, P.D.: Managing cognitive process with emotion stimulation. In: Proceedings of 17th RCAI 2019, vol. 1, pp. 240–248. UlSTU, Ulyanovask (2019)

Survey on Data Mining Tools Using Decision Analysis Resolution Method Applied for Context-Aware Data Mining Scenarios

Anca Avram(✉), Oliviu Matei, and Ionela Chereja

Technical University of Cluj-Napoca, University Center at Baia Mare, Baia Mare, Romania
{anca.avram,ionela.chereja}@ieee.org, oliviu.matei@holisun.com

Abstract. Nowadays there is a large spectrum of tools for implementation of data mining processes. Researchers could be put in a difficult position when having to choose a proper tool for performing the experiments they design. This paper presents an overview of some well known data mining tools and describes the implementation of the Decision Analysis Resolution (DAR) method for choosing the most appropriate one in Context-Aware Data Mining (CADM). DAR method is explained and applied for CADM scenarios.

Keywords: Data mining tools · Decision analysis resolution

1 Introduction

Currently there are multiple tools that allow implementing data mining processes, for educational purposes and not only. Depending on what step of the analysis one is (data manipulation; modeling) or the type of problem to be solved (social networking analysis, text mining), one tool could be more appropriate than another.

The target of this research is not to explore every possible tool, but to choose some of the well known ones and take an informed decision on what is the most appropriate to use for implementing Context-Aware Data Mining (CADM) scenarios. CADM respects the same steps as classical data mining just that it comes with an extra step of integrating context in the process [3]. The main purpose of the experiments revolves around some specific weather stations available data and the goal was predicting the soil moisture on some specific locations by using context-awareness. More details on the specific data and scenarios can be found in [2] and [1].

Section 2 presents an overview of some of the best known tools in the industry and Sect. 3 highlights the main steps of the Decision Analysis Resolution (DAR) method. Following, Sect. 4 depicts the DAR implementation for the CADM scenarios, while conclusions are summarized in Sect. 5.

R. Silhavy et al. (Eds.): CoMeSySo 2021, LNNS 231, pp. 544–554, 2021.
https://doi.org/10.1007/978-3-030-90321-3_45

2 State of Art

Slater et al. [16] provide a brief description of 40 possible tools to be used for educational data mining. However, because there is such a large market of possible tools, they also admit that their research might not cover some tools that might be favoured by other researchers. They present the most used, accessible and powerful ones. Starting from this, the focus of current research was more on the possible tools to be used for modeling.

- **RapidMiner**[1] is a tool developed on an open core model. It provides support for all the phases of the data mining process, from data preparation, applying a large range of machine learning algorithms, model validation and optimization to results visualisation [10]. It was developed in 2001 at the Technical University of Dortmund under the name YALE (Yet Another Learning Environment) and is written in Java [9]. The tool provides a rich graphical user interface that allows the development of complex workflows called "processes". Each process is designed connecting one or more so called "operators" while one operator is designed to perform a single task.
- **Weka**[2] (Waikato Environment for Knowledge Analysis) is a free software developed in New Zealand, at the University of Waikato, currently used in various areas, but especially in the educational field. It has a graphical user interface that allows the user to perform complex data mining tasks for data preprocessing, machine learning and data visualisation [23]. It can also be accessed from a Java API or other terminal applications.
- **KEEL** (Knowledge Extraction based on Evolutionary Learning[3]). It is an open source tool, written in Java, with a simple graphical user interface, that allows performing multiple data mining tasks [19], using a data flow design. It was designed especially for education and research, with extensive support on evolutionary algorithms, but with limited support on other types of algorithms.
- **KNIME** (KoNstanz Information MinEr[4]). It is an open analytics platform that revolves around data pipelines. It started as a research tool, in a software company specialized in pharmaceutical applications, in 2004, in Germany. KNIME allows data visualisation and manipulation [4] by creating data science workflows. It allows manipulation of data sources from different origins from files, to databases, Azure or sources like Twitter or Google Sheets. Also, permits creating different statistics, aggregation and cleaning of data, data visualisation, applying machine learning algorithms and creating models for different purposes (classification, regression, clustering), validating models and sharing insights.
- **Orange**[5] is an open source software that allows the implementation of data mining tasks, from data analysis, preprocessing and visualisation to applying

[1] https://rapidminer.com/.
[2] https://www.cs.waikato.ac.nz/ml/weka/.
[3] http://sci2s.ugr.es/keel/.
[4] https://www.knime.com/.
[5] https://orangedatamining.com/.

machine learning algorithms and evaluate results. The framework revolves around the concept of data analysis workflows. These workflows are visually designed on a canvas interface and the steps in the workflow are performed by widgets. Each widget performs a single functionality, from a wide range of possible actions, for example data input, filtering, data visualisation, classification, regression etc.

- **SPSS** (Statistical Package for the Social Sciences[6]) is a tool from IBM, designed for data mining and text analytics. In 1994, it was originally named Clementine, while from 2010 on, it goes with the name SPSS. One of the biggest goals of this tool was to make prediction models easy to use, without the need of the user to understand into details how the algorithms work behind the scenes, making it easy to use by users of different skill levels.

As Romero and Ventura emphasize in [15], even though usually the design of the data mining tools is more oriented towards performance and flexibility than on usability, a tool to be used in educational purposes has to be more intuitive with an easier to use interface, easy to configure, execute and visualize the results.

Pynam et al. [14] did a study comparing Rapid Miner, Weka, Knime, Orange and R Tool. Their research is mostly focused on the main features available in each tool. They consider Weka and Knime as tools that do not need coding knowledge, while Rapid Miner is more adequate for experts; R is a very good tool for visualisation, while Orange is adequate when the requirements are less complex.

Wahbeh et al. [21] did a comparative study of the performance of a few tools available for performing classification tasks. According to their study, Weka was considered to have the highest applicability, then Orange, Tanagra and KNIME. Also, Weka had the highest improvement of performance when the test mode was changed to Cross Validation.

3 Methodology

To decide upon the most suitable tool to be used further on, we used Decision Analysis and Resolution (DAR). It is a formal technique commonly used for decision making when facing with multiple options and it can be applied in different areas of a system. It assures that the decision is taken in a standardized, objective way, and also properly documented [24].

This process is required to be applied for all decisions taken in companies that implement Capability Maturity Model Integration (CMMI) Level 3, also known as the defined model [6,18]. This level is an evolution from Level 1 (Initial level) and Level 2 (Managed level), in the sense that every step in the product flow has to be properly documented and has to follow a specific process [5].

The steps for DAR are: (1) define significant criteria; (2) identify the possible alternatives; (3) define the methods that will be used to evaluate the alternatives; (4) decide on the best alternative.

[6] http://www.ibm.com/analytics/us/en/technology/spss/.

As starting point for selecting the criteria, were used the software product qualities as specified in the ISO25010 standard. According to the standard, the software product quality relates to eight important areas: functional suitability; performance efficiency, compatibility, usability, reliability, security, maintainability and portability. From the product qualities defined in the ISO25010 standard were chosen the ones that were relevant for implementing CADM processes. Then was decided the list of possible tools to be used for describing the CADM scenarios.

As part of the method, the criteria chosen were split into more levels, depending on the importance of that criteria to be met by the chosen tool. Then was established another scoring of the tool based on how well the criteria is satisfied. A pay of matrix was then defined and the tools were scored based on this matrix. The tool with the best score is the tool of choice.

4 Decision Analysis Resolution (DAR) Implementation

The purpose of this initiative is to analyse the most suitable tool option to be used further on in the experiments for implementing CADM processes. It is important to weight cost-benefits of using the selected tools for implementing data mining processes, including the significant criteria.

4.1 Step 1: Definition of Significant Criteria

Starting from the software product qualities defined in ISO25010, were identified the most relevant attributes needed in designing and implementing CADM processes.

Quality Criteria. The ISO25010 list was analysed and the qualities that have to be met by the chosen tool were determined as follows:

- **Accesibility** - chosen tool must be effective, efficient and easy to use.
- **Debugability** - sufficient instrumentation should be available in order to be able to track an issue occuring in the development.
- **Emotional factors** - choosing an appealing tool should improve the development and the learning curve.
- **Learnability** - since there is no knowledge in other tools used in data mining, the learnability of the new tool is very important. Having a fast learning tool would speed up the implementation and would allow more time to focus on analysing the results and the processes.
- **Operability** - it is very important that the software chosen is one that can be easily operated.
- **Usability** - it is essential that the tool can be used to achieve the specified goal of designing CADM processes in an effective and efficient manner.

Functional Criteria. As an addition to the quality attributes previously enumerated, the following functional aspects were also defined as necessary to be part of the decision criteria.

- **Affordability** - being in the research phase, with no funds in this regard, it is essential that the tool is free and can be used like that.
- **Easy integration with CSV and SQL** - majority of data available for the CADM processes dedicated to soil moisture is in CSV or SQL database format, hence it would be important that the integration with the tool would be as easy as possible.
- **Allow data transformations** - since the data sources might vary and also the quality of data provided as input for the CADM processes, there is a specific need that data transformation is possible and also easy to be achieved.
- **Large range of machine learning algorithms** - it is important that various machine learning algorithms are available for being able to test more behaviours of CADM scenarios.
- **Dedicated support for time series analysis** - data achieved from the weather stations that will be the main data for the thesis is in the time series area, so having specific support for this is a nice to have.
- **Built in support for optimization** - it would a helpful feature to test more setups for the machine learning algorithms and be provided support in choosing the best setup.
- **Suggestion mechanism** - this functionality would be nice to have.

4.2 Step 2: Alternatives Chosen

As mentioned, the purpose of the current research is not to make an exhaustive evaluation of the tools in the market, but to choose one suitable tool for further development of CADM processes, with a special target on predicting the soil moisture.

From the list of possible options presented in Sect. 1, were considered two commercial tools: **RapidMiner** and **KNIME** and two free tools, largely used in research and for scientific purposes: **Weka** and **Orange**. RapidMiner and KNIME were both included by Gartner in the leading tools of its Magic Quadrant for Data Science and Machine Learning Platforms [11]. Weka and Orange have also a more than 20 years history with data mining.

4.3 Step 3: Methods Used to Evaluate the Data Mining Tools

The **importance** of the criteria was rated as follows: **3 - Essential** - if the tool does not fulfill this criteria, it no longer represents an option; **2 - Important** - it is important for the tool to have the criteria, but the tool still can be an option if other workarounds can be found in this matter; **1 - Nice to have** - the specific criteria is not important, but having it fulfilled would ease the further development; **0 - Not necessary**. The chosen criteria from Step 1 were rated as it is presented in Table 1.

Table 1. DAR criteria importance in deciding the software tool to design and develop CADM processes

Criteria	Importance
Accessibility	Essential
Debugability	Important
Emotional factors	Important
Learnability	Important
Operability	Important
Usability	Essential
Affordability	Essential
Easy integration with CSV & SQL	Important
Allow data transformation	Essential
Large range of machine learning algorithms	Important
Dedicated support for time series analysis	Nice to have
Built in support for optimization	Nice to have
Suggestion mechanism	Nice to have

Depending on the level the criteria is satisfied by the tool analysed, more satisfaction scores were defined. These scores are depicted in Table 2.

Table 2. DAR scores for the level of satisfaction on a specific criteria

Satisfaction score	Satisfaction level	Description
4	Exceptional	Exceptional capabilities
3	Good	Good capabilities
2	Limited	Limited capability, difficult to implement/use
1	Minimal	Minimal to no capability or requires adjustments
0	NA	Not Applicable

Finally, the defined pay of matrix that is used in scoring each tool, based on the importance of the criteria versus level of satisfaction, is presented in Table 3.

Table 3. DAR pay of matrix used for scoring a tool on a criteria

	Exceptional	Good	Limited	Minimal	NA
Essential	12	9	6	3	0
Important	8	6	4	2	0
Nice to Have	4	3	2	1	0
Not necessary	0	0	0	0	0

4.4 Step 4: Decision on Best Option

Table 4 presents an overview of the ratings obtained according to the criteria chosen, while the explanation for each criteria is depicted as follows.

Table 4. DAR ratings for the criteria chosen and tools: RapidMiner, Weka, KNIME and Orange

Criteria	RapidMiner	Weka	KNIME	Orange
Accessibility	Exceptional	Good	Exceptional	Limited
Debugability	Exceptional	Limited	Good	Limited
Emotional factors	Exceptional	Limited	Good	Limited
Learnability	Exceptional	Good	Exceptional	Limited
Operability	Exceptional	Good	Exceptional	Good
Usability	Exceptional	Good	Exceptional	Limited
Affordability	Good	Exceptional	Exceptional	Exceptional
Easy integration with CSV & SQL	Exceptional	Exceptional	Exceptional	Exceptional
Allow data transformation	Good	Limited	Limited	Limited
Large range of machine learning algorithms	Exceptional	Exceptional	Exceptional	Exceptional
Dedicated support for time series analysis	Good	Good	Good	Good
Built in support for optimization	Exceptional	Good	Good	Limited
Suggestion mechanism	Exceptional	NA	Limited	Limited

In terms of **accessibility**, the most efficient tools in the selection are RapidMiner and KNIME, because being also commercial tools, a part of the investment is done in this area. Weka and Orange, by being more like research tools do not offer the same level of accesibility. This also has an impact on the **emotional factors, usability and learnability**.

In terms of **usability and learnability**, RapidMiner and KNIME have built in tutorials, detailed documentation, description videos available and also support from the community can be provided when needed. KNIME and RapidMiner also provide a list of recommended operations at each step in the modeling, based on what the community users have used on a similar design. For Weka, Witten et al., in their book "Data Mining: practical machine learning tools and techniques with Java implementations" [22], beside describing the basics of data mining, they also have a few chapters where they exemplify the concepts using Weka. This books and some free online courses can be used as starting point when learning Weka. Orange users can also start the learning from a set of videos available, have a detailed documentation of all the components and also provide access on the blog to different examples. Still, in comparison to RapidMiner and KNIME, where beside the user interface is very suggestive and documentation and demo for each component are provided, the access to the support for each operation is not as intuitive for Weka [17] and Orange, needing either a hover on the operation, either supplementary search in the documentation.

For **debugging**, RapidMiner offers the possibility to add a breakpoint before or after any step in the process. By a right click and select or a combination of key strokes, this functionality is available. In order to do debugging in Weka, one needs to use usual Java debuggers, so it is a bit more difficult to perform and requires specific knowledge. KNIME offers the possibility to add breakpoints in the designed work flows, while Orange does not provide an out of the box method for this operation.

In terms of **operability**, all of the chosen tools allow an intuitive design of the data mining processes. Still, the design of the flows in RapidMiner and KNIME is easier to implement and maintain, because of the visually suggestive hints provided by each operator in the process (input/output parameter hints, support available etc.).

On the **cost, initial and life-cycle cost/affordability**, the ranking is as follows. Weka, KNIME and Orange are free open source tools, widely used for teaching and research. KNIME is free as concerns the Analytics Platform, but has a separate fee if one wants to also have access to the KNIME servers that allow the deploy of the designed workflows into production. RapidMiner is a commercial tool that needs an annual subscription fee for their services but has some special facilities for educational purposes - it offers the possibility to be used for free in a limited version (the main drawback is that the user is limited in this case to 10000 rows of data); beside that it allows one year access to the entire range of functionalities if one creates an educational account for that. So, considering this, RapidMiner is still a suitable option from the cost perspective.

Related to the **easy integration with CSV and SQL** criteria, all the four tools have dedicated support for both integration with Excel and SQL database integration. As concerns the need that the tools provide support for complex **data transformation**, RapidMiner provides a large set of operators for normalization, handling missing values, outlier detection, duplicate removal and dimensionality reduction. Orange, KNIME and Weka offer support for these operations but in a limited and less intuitive manner.

In terms of **range of machine learning algorithms**, Weka provides a large number of implemented algorithms, for classification and regression [8], but with a higher support for classification. In the same time Rapid Miner and KNIME [13] have support for both classification and regression, but are especially powerful in predictive analytics. Orange also offers a large selection of algorithms, while the complicated parts are written in C++, for performance reasons, making it an efficient tool to be used [7].

As Kosorus et al. presented in their research [12], both RapidMiner and Weka can be used successfully for **time series analysis**. RapidMiner specifically provides a supplementary extension for time series. Also KNIME and Orange have specific operations available for time series analysis. Also, all the four tools allow applying the windowing technique [20] to time series elements.

All the four tools have a **built in support for optimization**, in order to find the best possible combinations of parameters. Orange has two classes that are created with tuning purposes: one for fitting a single parameter and one for trying multiple combinations of parameters, using these assumes a deeper understanding of the application behavior. KNIME has a specific extension for optimization with multiple nodes, but it is still in an experimental phase and assumes using a combinations of nodes to perform a check on the best parameter combinations. Weka allows setting different values for the algorithms run, but is a manual process where one has to re-configure the algorithm parameters for as many combinations one wants to test. RapidMiner provides an operator that allows entering different values for all the operators that are placed inside, in a simple and intuitive manner.

Suggestion Mechanism. Orange registers the preferences of the users over time and responds accordingly, while Weka does not offer support in this sense. KNIME also offers some limited support in this matter, while RapidMiner has hints on the user interface, that change depending on the user choices and on the most used operators by the community. Also, suggestions are provided about the values of the parameters and the percentage of users that used those values. More than that, they came up with a framework where you only enter the data and they provide some suggestions on how the data could be modeled and what would be the recommended algorithms.

As a conclusion, after establishing the rating of the criteria for each of the four tools (Table 4), the pay off matrix was computed for each rating based on the initial established importance of the criteria (Table 1) and on the decided scores (Table 3). Table 5 presents the obtained results.

Table 5. DAR final scores for the criteria chosen and tools: RapidMiner, Weka, KNIME and Orange

Criteria	RapidMiner	Weka	KNIME	Orange
Accessibility	12	9	12	6
Debugability	8	4	6	4
Emotional factors	8	4	6	4
Learnability	8	6	9	4
Operability	8	6	9	6
Usability	12	9	12	6
Affordability	9	12	12	12
Easy integration with CSV & SQL	8	8	8	8
Allow data transformation	9	6	6	6
Large range of machine learning algorithms	8	8	8	8
Dedicated support for time series analysis	3	3	3	3
Built in support for optimization	4	3	3	2
Suggestion mechanism	4	0	2	2
Total score	**101**	**78**	**96**	**71**

As it can be seen the best score was obtained by RapidMiner (101 points), followed closely by KNIME (96 points), while Weka and Orange obtained a total score of 78 points, respectively 71 points. In conclusion, after the DAR implementation for the four tools, the tool of choice is RapidMiner.

5 Conclusions

The market of data mining tools is very extensive. For the implementation of the CADM scenarios the most suitable tool was needed. The technique applied for decision making in the described context is Decision Analysis and Resolution. It is based on four main steps that were followed. There were selected the most important criteria necessary for the CADM processes to be successfully applied, then the tools chosen for analysis: RapidMiner, KNIME, Orange and Weka. Further on, an objective scoring was decided based on the importance of each selected criteria. Then, research was performed and the tools were rated for each criteria and finally the scores were computed.

From the four tools, the one that had the best score is RapidMiner, followed closely by KNIME. These tool were evaluated with "Good" and "Excellent" for all the selected criteria. The next one in line is Weka, while Orange is the last one.

The article proved that DAR can be successfully applied for taking objective decision. More than that, the same mechanism can be used in any part of the data mining processes when decision making is needed.

References

1. Avram, A., Matei, O., Pintea, C., Anton, C.: Innovative platform for designing hybrid collaborative and context-aware data mining scenarios. Mathematics **8**(5), 684 (2020)
2. Avram, A., Matei, O., Pintea, C.-M., Pop, P.C.: Context quality impact in context-aware data mining for predicting soil moisture. Cybern. Syst. **51**(7), 668–684 (2020)
3. Avram, A., Matei, O., Pintea, C.-M., Pop, P.C., Anton, C.A.: Context-aware data mining vs classical data mining: case study on predicting soil moisture. In: Martínez Álvarez, F., Troncoso Lora, A., Sáez Muñoz, J.A., Quintián, H., Corchado, E. (eds.) SOCO 2019. AISC, vol. 950, pp. 199–208. Springer, Cham (2020). https://doi.org/10.1007/978-3-030-20055-8_19
4. Berthold, M.R., et al.: Knime-the konstanz information miner: version 2.0 and beyond. ACM SIGKDD Explor. Newsl. **11**(1), 26–31 (2009)
5. Chrissis, M.B., Konrad, M., Shrum, S.: CMMI for Development: Guidelines for Process Integration and Product Improvement. Pearson Education, London (2011)
6. Dadhich, R., Chauhan, U.: Integrating CMMI maturity level-3 in traditional software development process. Int. J. Softw. Eng. Appl. **3**(1), 17 (2012)
7. Demšar, J., Zupan, B., Leban, G., Curk, T.: Orange: from experimental machine learning to interactive data mining. In: Boulicaut, J.-F., Esposito, F., Giannotti, F., Pedreschi, D. (eds.) PKDD 2004. LNCS (LNAI), vol. 3202, pp. 537–539. Springer, Heidelberg (2004). https://doi.org/10.1007/978-3-540-30116-5_58
8. Desai, A., Sunil, R.: Analysis of machine learning algorithms using weka. Int. J. Comput. Appl. **975**, 8887 (2012)
9. Deutsch, G.: Rapidminer from rapid-i at cebit 2010 (2010)
10. Hofmann, M., Klinkenberg, R.: RapidMiner: Data Mining Use Cases and Business Analytics Applications. CRC Press, Boca Raton (2016)
11. Idoine, C., Krensky, P., Brethenoux, E., Linden, A.: Gartner magic quadrant for data science and machine learning platforms, vol. 2019 (2019)

12. Kosorus, H., Honigl, J., Kung, J.: Using r, weka and rapidminer in time series analysis of sensor data for structural health monitoring. In: 2011 22nd International Workshop on Database and Expert Systems Applications, pp. 306–310. IEEE (2011)
13. Naik, A., Samant, L.: Correlation review of classification algorithm using data mining tool: Weka, rapidminer, tanagra, orange and knime. Procedia Comput. Sci. **85**, 662–668 (2016)
14. Pynam, V., Roje Spanadna, R., Srikanth, K.: An extensive study of data analysis tools (rapid miner, weka, r tool, knime, orange). SSRG Int. J. Comput. Sci. Eng. (SSRG–IJCSE)–Volume **5**, 4–11 (2018)
15. Romero, C., Ventura, S.: Educational data mining: a survey from 1995 to 2005. Expert Syst. Appl. **33**(1), 135–146 (2007)
16. Slater, S., Joksimović, S., Kovanovic, V., Baker, R.S., Gasevic, D.: Tools for educational data mining: a review. J. Educ. Behav. Stat. **42**(1), 85–106 (2017)
17. IT Central Station. It central station. Unbiased reviews from the tech community
18. CMMI Product Team. CMMI for systems engineering/software engineering/integrated product and process development/supplier sourcing, version 1.1, continuous representation. CMU/SEI (2002)
19. Triguero, I., et al.: Keel 3.0: an open source software for multi-stage analysis in data mining. Int. J. Comput. Intell. Syst. **10**(1), 1238–1249 (2017)
20. Wagner, N., Michalewicz, Z.: An analysis of adaptive windowing for time series forecasting in dynamic environments: further tests of the DyFor GP model. In: Proceedings of the 10th Annual Conference on Genetic and Evolutionary Computation, pp. 1657–1664 (2008)
21. Wahbeh, A.H., Al-Radaideh, Q.A., Al-Kabi, M.N., Al-Shawakfa, E.M.: A comparison study between data mining tools over some classification methods. Int. J. Adv. Comput. Sci. Appl. **8**(2), 18–26 (2011)
22. Witten, I.H., Frank, E.: Data mining: practical machine learning tools and techniques with java implementations. ACM SIGMOD Rec. **31**(1), 76–77 (2002)
23. Witten, I.H., Frank, E., Hall, M.A., Pal, C.J.: Practical Machine Learning Tools and Techniques. Morgan Kaufmann, Burlington, p. 578 (2005)
24. Zubrow, D.: Measuring software product quality: The ISO 25000 series and CMMI. Technical Report, Carnegie-Mellon University Pittsburgh PA Software Engineering Institution (2004)

The Redundant Distributed Data Labeling Under Conditions of the Decentralized Training Datasets Storing for the Fog- and Edge-Robotic Environment

Anna Klimenko[1](✉), Eduard Melnik[2], and Alexander Kozlovsky[3]

[1] Scientific Research Institute of Multiprocessor Computer Systems of Southern Federal University, 2, Chekhov Street, 347928 Taganrog, Russian Federation

[2] Federal Research Center The Southern Scientific Center of the Russian Academy of Sciences, 41, Chekhov Street, 344006 Rostov-on-Don, Russian Federation

[3] Southern Federal University, 105/42 Bolshaya Sadovaya Street, Rostov-on-Don 344006, Russia

Abstract. The process of learning is typical for the most assistive and autonomous robots. Considering the learning as a complex process, the issue of the data labeling is in the focus of this paper, because of the labeling noise effect on the learning results. Distributed redundant labeling is discussed in recent works, yet, no known publications contain the process description in the focus of the distributed decentralized processes, although the manner of distributed implementation of the various areas of the robotics is relevant now in the angle of fog- and edge-robotics. The current paper is devoted to the question of the redundant distributed labeling under the conditions of the decentralized data storing, which is quite relevant in the fog- and edge- robots environments. Two main approaches of the distributed monitoring of the labeling experts groups are considered and compared, as well as some experimental research of the redundant labeling results has been conducted with the results presented.

Keywords: Assistive robots · Machine learning · Data labeling

1 Introduction

Assistive robots area contains a huge number of examples: wheelchair robots, walking assist robots, communication robots [1–4] as well as assistive robots for autonomous cars [5] and many others.

By the forecast of the assistive robotics market, it was valued at USD 3.1 billion in 2018 and is expected to reach USD 11.2 billion by 2024. The top assistive robotics companies are: Shadow Robot Company, Panasonic, Honda Motor Company, Kinova Robotics, Cyberdyne, Dream Face Technologies and others [6]. Mobile assistive robots are usually controlled by software and use sensors to identify their surroundings. Some of these systems are embedded with artificial intelligence technology, which allows them

R. Silhavy et al. (Eds.): CoMeSySo 2021, LNNS 231, pp. 555–567, 2021.
https://doi.org/10.1007/978-3-030-90321-3_46

to function. These systems are mainly used in applications such as elderly assistance, companionship, and public relations.

According to a recent survey published by the Evans Data Corporation Global Development, machine learning and robotics was at the top of developers' priorities for 2016, with 56.4% of participants stating that they're building robotics apps and 24.7% of all developers indicating the use of machine learning in their projects [7].

So, the problem of machine learning application to the robotics field is quite relevant, while the machine learning is the problem itself, which includes such issues as reinforcement learning of the robots, the update of the training datasets, etc. One of the considerable issues of the machine learning field is the training dataset forming, updating and data labeling processes. Data labeling generates high costs, it can cause the multiple errors through the application functioning in case of dataset low quality. Dataset quality is partially determined by the data labeling quality, which depends on the methods and techniques of the data labeling and eliminating of the labeling noise from the datasets [8].

At the same time, fog- and edge-robotics involve the new, distributed types of the dataset storages, including the fully decentralized and ledger-based [9–13].

However, to the best of our knowledge, the area of the training datasets forming under conditions of the distributed data storage is not considered properly, though, some projects of the decentralized data labeling are presented and function.

This paper considers the data labeling for the robots learning under conditions of the decentralized, ledger-based data storage. It is presupposed that the distributed redundant labeling is made by the distributed decentralized groups of the experts, so, the problem is to organize their cooperation in the angle of data labeling. In the focus of this paper we compared two known approaches of the group monitoring and control, estimated the data exchange volume of the approaches, spotlighted the pros and cons of the approaches considered. Also, the experimental research of the distributed redundant labeling has been made. This has allowed to estimate the level of labeling errors depending on various experimental conditions. The experimental material has been taken from the dataset containing the benthos, which is the perfect content for labeling by the distributed non-qualified users.

2 Machine Learning Application to Robotics and the Issues of the Training Datasets Forming

The following directions of the machine learning field are applied to the robotics [7, 14]:

- Computer vision;
- Imitation learning;
- Self-superwised learning;
- Multi-agent learning.

An increase of data volumes i.e. visual information available on the web (including annotated/labeled photos and videos) has propelled advances in computer vision, which in turn has helped further machine-learning based structured prediction learning techniques, leading to robot vision applications like identification and sorting of objects.

Self-supervised learning approaches enable robots to generate their own training examples in order to improve performance; this includes using a priori training and data captured close range to interpret "long-range ambiguous sensor data." It's been incorporated into robots and optical devices that can detect and reject objects (dust and snow, for example); identify vegetables and obstacles in rough terrain; and in 3D-scene analysis and modeling vehicle dynamics. Watch-Bot uses a 3D sensor, a camera, laptop and laser pointer to detect 'normal human activity', which are patterns that it learns through probabilistic methods. Watch-Bot uses a laser pointer to target the object as a reminder (for example, the milk that was left out of the fridge). In initial tests, the bot was able to successfully remind humans 60% of time (it has no conception of what it's doing or why), and the researchers expanded trials by allowing their robot to learn from online videos (the project called RoboWatch). Other examples of self-supervised learning methods applied in robotics include a road detection algorithm in a front-view monocular camera with a road probabilistic distribution model (RPDM) and fuzzy support vector machines (FSVMs), designed at MIT for autonomous vehicles and other mobile on-road robots. Autonomous learning, which is a variant of self-supervised learning involving deep learning and unsupervised methods, has also been applied to robot and control tasks.

So, most of the mentioned directions need a training stage on the basis of the previously labeled training dataset. Though there is a considerable number of the datasets prepared in the Internet (Imagenet (more than 14 million images), MS-COCO (more than 200000 labeled images), Visual QA (more than 250000 labeled images), The street View House Numbers (630420 images), CIFAR-10 (60000 images in 10 classes) and others.), a set of projects demands the individual labeled data according to the project requirements. The last decade has generated some techniques to label the data, they are presented in the table below.

Table 1. Approaches to the data labeling

Approach	Description	Pros	Cons
In-house labeling	Assignment of tasks to an in-house data science team	Predictable results of a good quality and the ability to track the process of labeling	The duration of the process and possibly high cost
Outsourcing	Temporary employees of freelance platforms, posting vacancies on social media and job search sites	The ability to evaluate applicant's skills	The need to organize the workflow

(*continued*)

Table 1. (*continued*)

Approach	Description	Pros	Cons
Crowdsourcing	Cooperation with the freelancers from crowdsourcing platforms	Cost savings and fast results	Quality of work can suffer
Specialized outsourcing companies	Hiring and external team for a specific project	Assured quality	Higher price compared to crowdsourcing
Synthetic labeling	Generating data with the same attributes of real data	Training data without mismatches and gaps; Cost- and time-effectiveness	High computational power required
Data programming	Using the software that programmatically label data to avoid manual work	Automation; fast results	Dataset of a relatively low quality

Nowadays crowdsourcing approaches are quite widespread and are to be cost-efficient, however, crowdsourcing approach can generate datasets of a poor quality as well. Labeling errors, or labeling noise is a serious issue for every ML project. The picture below (Fig. 1) illustrates the dependency between the recognition errors number and the number of labeling errors.

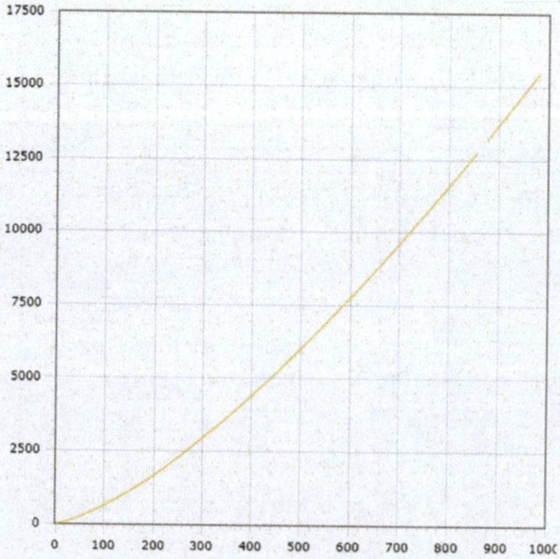

Fig. 1. Dependency between the recognition errors number and the number of labeling errors

The main sources of labeling noise are as follows [8]:

- the information which is provided to the expert may be insufficient to provide reliable labeling;
- the errors can occur in the expert's labeling process itself;
- the labeling task is subjective, as, for example, in medical applications and image data analysis;
- the presence of data encoding errors and communication problems, which is caused by insufficient level of dependability.

At least four major techniques for labeling noise elimination are described in literature:

- label noise-robust methods; [15–17]
- probabilistic label noise-tolerant methods [18, 19];
- data cleansing methods [20, 21];
- model-based label noise-tolerant methods [22–24].

Yet, there is at least one more interesting and prospective method to minimize the data labeling errors: it is the so-called distributed labeling, which presupposes the labeling experts redundancy.

3 Redundant Distributed Data Labeling Approaches

From the angle of cost, the following distributed labeling methods can be outlined:

- with rewards;
- volunteer labeling.

In the scope of this paper we consider the volunteer labeling, which is based on the gameplay or on such psychological human features as a wish to help and to support some new ideas. Nowadays there are some studies devoted to the labeling through the game [25–29] as well as functioning projects, e.g. Ihearu project, which implements the audio data labeling by means of the game.

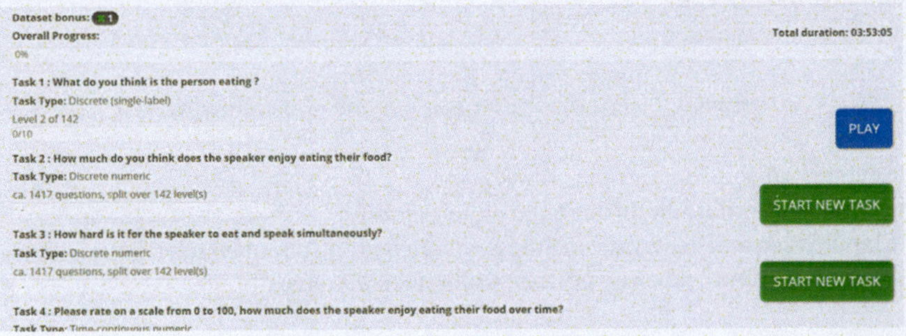

Fig. 2. Ihearu game process example

Also, there are projects with volunteer labeling:

- Functional & Anatomical Imaging & Shape Analysis Lab;
- Project «Zooniverse»;
- Volunteers to for sentiment labeling.

It must be mentioned that, according to the studies [30]: «For all algorithms, the overall accuracy it can achieve under the volunteer context is evidently higher than the paid crowd, without any exception. For the same workflow, the overall accuracy it can achieve in volunteer context is normally around 30% higher than the paid crowd context for workflows with two to three questions.»

Majority voting [30] now is the basic approach in the field of distributed labeling, but, in general the main problem is to reach the consensus on labeled data under conditions of the highly dynamic groups of volunteer experts.

The next important issue is to organize the experts cooperation under conditions of the distributed data storing (e.g., distributed ledger concept-based storing).

The groups of experts, which label the same piece of content, must be monitored somehow to reach a consensus on the labeled content element. Excluding the centralized group monitoring, the two following approaches can be used: fully decentralized and with the distributed leader.

The following section contains the comparison of these approaches from the data volume generation and computation complexity of the general stages of the system functioning.

4 The Comparison of the Monitoring Strategies as Applied to the Users Cooperation in the Groups

Consider the approaches to the experts groups monitoring in the angle of consensus on the labeled data. We presuppose that the data labeling system functions in a semi-automatic decentralized way, such it is described in [31, 32].

As is mentioned above, the monitoring of the experts groups can be implemented in the following ways:

- with a distributed leader;
- in a non-leader manner.

In case of the distributed leader, the leader implements the following tasks:

1. the request of the content;
2. the content distribution through the experts group;
3. the labeled content gathering and the decision making about which labeling is "right";
4. sending the "right" labeling content to the labeled data storage.

Also, if there is a problem with the decision about the "right" labeling, for example, split voting, or the dynamically changing group made its quantity such as no decision can be made (2 members with split voting, for example), the leader must act so as to restore the group possibility to deliver the labeled content.

Besides, there are procedures of the group forming and leader election under conditions of the geographically distributed and non-interconnected participants.

In cases of the changes in the experts group, the consensus on the group composition must be reached. As the leader gathers the labeled data and chooses the most recommended labels, there is no distributed consensus on the labeled content – this is a decision made by the group leader, based on the votes selected.

Another approach to the groups monitoring is a fully decentralized, non-leader one. According to this approach the following stages take place:

- every participant (expert) determines his relation to the current labeling task;
- every participant must find and make a connection with other participants of the same labeling task;
- a consensus on the group must be reached;
- a participant asks the content for labeling;
- a participant labels the content and sends the labeled element to other participants in the group;
- a participant labels the content elements receiving them (labeled) from the other participants;
- all participants exchange the labeled data packages, performing the voting on which labeling is right;
- every participant gathers the labeled data packages and sets the right labeling based on the virtual voting;
- the package of the right labeled data is sent to the ledger.

So, a consensus on data must be achieved in a distributed group, as well as the group composition must be monitored, detected and handled. Thus, the group composition consensus must be achieved as well as the particular procedure for the participant addition. The latter is relevant to the distributed leader-based approach too.

So, comparing two approaches, the following basics of their implementation must be spotlighted.

1. In the distributed leader-based approach the group composition consensus is needed. Then, the leader makes a decision about the "right" labeling and sends the result to the distributed storage. Also, the leader makes a decision to change the group composition if there is no possibility to choose the right labeling.
2. Non-leader approach presupposes the group composition consensus as well as the labeled data one. The task of the group composition change is relevant as well and requires the consensus on which participant can be invited to the group. It must be mentioned that the non-leader approach has the following important disadvantage: every participant sends its results to the distributed storage, generating the data redundancy.

Estimate the group monitoring approaches from the data exchange volume angle, excluding such parameter as the size of the content request:

Consider N as the participants number in the group, Vel as the content element size.

Then for the group with the distributed leader the data volume is as follows:

$$V_l = V_{el} + 2V_{el}(N - 1) \quad (1)$$

For the distributed (fully decentralized) group the data volume is as follows:

$$V_d = 2NV_{el} + 2NV_{el}(N - 1) \quad (2)$$

And the dependency between the data exchange volume and the number of group participants is presented in Fig. 3.

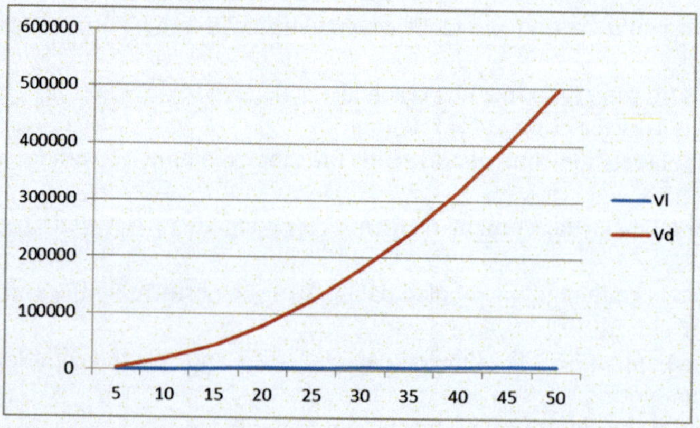

Fig. 3. The data exchange volume dependant on the participants number

The dependency between the data exchange volume and the content element size is shown in Fig. 4.

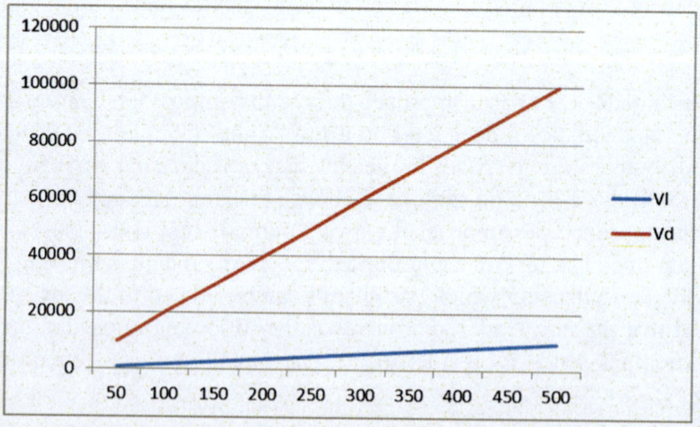

Fig. 4. The dependency between the data exchange volume and the content element size

As to the leader change procedure, which takes place in the case of leader failure (the user disconnects from the group), in case of redundant data labeling it is of low significance, and there is no need to store the full datalog (as is done in [33]). As the procedure of the labeling is without rewords, the new participants can label the same content.

5 The Experiments with the Distributed Redundant Labeling

In this section some selected results of the redundant distributed data labeling are presented. We implemented the process of the distributed data labeling based on the distributed leader approach for three participants (the number 3 is important because it is the smallest number of the participants for the majority voting).

In the figure below the examples of the content to be labeled are presented.

Fig. 5. Benthos examples

First experiment was conducted under the following conditions: every user has labeled 50 content elements at the first stage, and then 50 content elements was labeled with redundancy and verified by the other users. The graph illustrates the quality of the labeling, and one can see that the individual labeling brings up to 24% of the errors in the dataset, while the redundant labeling reduces this share to 12%.

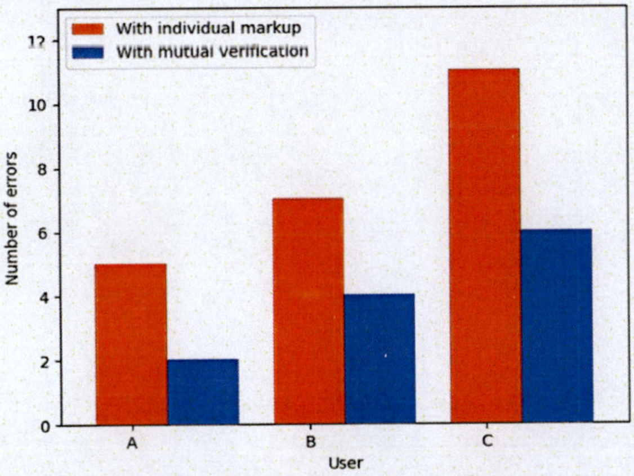

Fig. 6. Users labeling errors diagram

The second experiment was conducted with the same number of users, but for the various dataset sizes: 25, 50, 100 content elements.

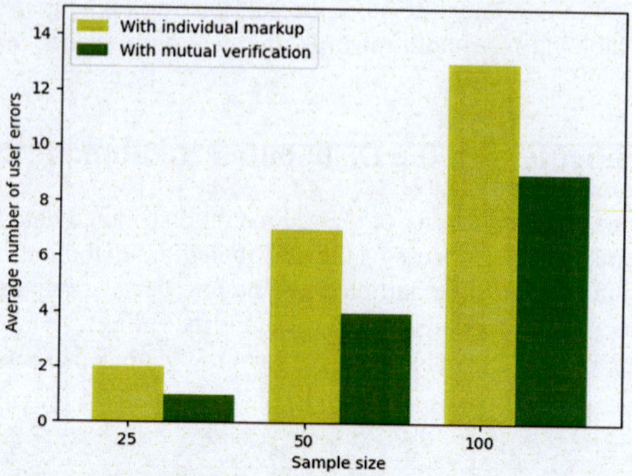

Fig. 7. Labeling of content packages of 25,50,100 elements

The tendency of the labeling error number reduction takes place, yet, it must be mentioned that the dependency between the errors number and the dataset size is not linear. This is caused by the users exhaustion and, in general, is a good reason to use the redundant distributed labeling for the large datasets.

However, the experimental research of the content elements number increase and its effect on the labeled dataset quality, the following dependency was traced.

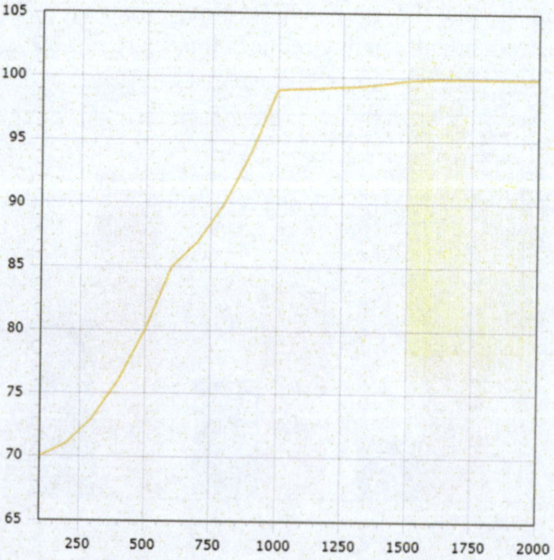

Fig. 8. Content elements number increase and its effect on the labeled dataset quality

So, the number of the content elements can be restricted because of the inexpediency of the large number of labeled elements in the dataset. It causes the possible restrictions on the experts number.

6 Conclusions

In this paper a problem of the redundant distributed labeling is considered. Data labeling is one of the paramount problems of the machine learning application to the robotics area, and labeling errors cause low quality of learning. Some projects propose the distributed labeling without rewards, but none of them can cooperate with the distributed and decentralized storages and services.

In the current paper we consider the questions of the monitoring and control of the labeling users groups, as well as the positive effect of the distributed redundant labeling on the dataset quality.

The comparison of the group monitoring methods presents the superiority of the distributed leader-based monitoring and control within the semi-automatic machine-human interaction system from the angle of data volume generating, data redundancy and the computational complexity of the procedures. The experimental research presents a considerable improvement of the dataset quality in case of mutual verification of the labeling experts, as well as the existence of the upper boundary for the expedient number of the labeling users in the group.

Acknowledgement. This study is granted by the RFBR projects 18-29-22086.

References

1. Levine, S.P., Bell, D.A., Jaros, L.A., Simpson, R.C., Koren, Y., Borenstein, J.: The Navchair assistive wheelchair navigation system. IEEE Trans. Rehabil. Eng. **7**(4), 443 451 (1999)
2. Tsukahara, A., Kawanishi, R., Hasegawa, Y., Sankai, Y.: Sit-tostand and stand-to-sit transfer support for complete paraplegic patients with robot suit HAL. Adv. Robot. **24**(11), 1615–1638 (2010)
3. Wada, K., Shibata, T.: Living with seal robots—its sociopsychological and physiological influences on the elderly at a care house. IEEE Trans. Rob. **23**(5), 972–980 (2007)
4. Komatsu, T., Iwaoka, T., Nambu, M.: Leaving a message with the PaPeRo robot: the effect of interaction experience with real or virtual PaPeRo on impression evaluation. In: Harper, R., Rauterberg, M., Combetto, M. (eds.) ICEC 2006. LNCS, vol. 4161, pp. 37–42. Springer, Heidelberg (2006). https://doi.org/10.1007/11872320_5
5. https://escholarship.org/content/qt1hq115j2/qt1hq115j2_noSplash_6236e018b74cc55d74fb31603fb3fc55.pdf?t=pkr1ps
6. https://www.globenewswire.com/news-release/2019/11/08/1943913/0/en/Global-11-2Bn-Assistive-Robotics-Market-by-Mobility-Type-Application-Geography-Forecast-to-2024.html
7. https://emerj.com/ai-sector-overviews/machine-learning-in-robotics
8. Frénay, B., Verleysen, M.: Classification in the presence of label noise: a survey. IEEE Trans. Neural Netw. Learn. Syst. **25**(5), 845–869 (2014)
9. Blockchain and Fog Computing for Cyber-Physical Systems: The Case of Smart Industry

10. He, Y., Wang, Y., Qiu, C., Lin, Q., Li, J., Ming, Z.: Blockchain-based edge computing resource allocation in IoT: a deep reinforcement learning approach. IEEE Internet Things J. **8**(4), 2226–2237 (2021). https://doi.org/10.1109/JIOT.2020.3035437
11. Chand, M., Ramachandran, N., Stoyanov, D., Lovat, L.: Robotics, artificial intelligence and distributed ledgers in surgery: data is key! Tech. Coloproctol. **22**(9), 645–648 (2018). https://doi.org/10.1007/s10151-018-1847-5. Epub 2018 Sep 21. PMID: 30242534
12. Panesar, S., Cagle, Y., Chander, D., Morey, J., Fernandez-Miranda, J., Kliot, M.: Artificial intelligence and the future of surgical robotics. Ann Surg. **270**(2), 223–226 (2019). https://doi.org/10.1097/SLA.0000000000003262. PMID: 30907754
13. Tran, J., Ramachandran, G., Shah, P., Danilov, C., Santiago, R., Krishnamachari, B.: SwarmDAG: a partition tolerant distributed ledger protocol for swarm robotics. Ledger **4** (2019). https://doi.org/10.5195/ledger.2019.174
14. Wu, C., Zhang, J., Selman, B., Savarese, S., Saxena, A.: Watch-bot: unsupervised learning for reminding humans of forgotten actions. In: 2016 IEEE International Conference on Robotics and Automation (ICRA), pp. 2479–2486 (2016). https://doi.org/10.1109/ICRA.2016.7487401
15. Manwani, N., Sastry, P.S.: Noise tolerance under risk minimization. IEEE Trans. Cybern. **43**(3), 1146–1151 (2013)
16. McDonald, R., Hand, D., Eckley, I.: An empirical comparison of three boosting algorithms on real data sets with artificial class noise. In: Windeatt, Terry, Roli, Fabio (eds.) MCS 2003. LNCS, vol. 2709, pp. 35–44. Springer, Heidelberg (2003). https://doi.org/10.1007/3-540-44938-8_4
17. Abellán, J., Masegosa, A.: Bagging decision trees on data sets with classification noise. In: Link, Sebastian, Prade, Henri (eds.) FoIKS 2010. LNCS, vol. 5956, pp. 248–265. Springer, Heidelberg (2010). https://doi.org/10.1007/978-3-642-11829-6_17
18. Joseph, L., Gyorkos, T.W., Coupal, L.: Bayesian estimation of disease prevalence and the parameters of diagnostic tests in the absence of a gold standard. Am. J. Epidemiol. **141**(3), 263–272 (1995)
19. Perez, C.J., Giron, F.J., Martin, J., Ruiz, M., Rojano, C.: Misclassified multinomial data: a Bayesian approach. Rev. R. Acad. Cien. Serie A. Mat. **101**(1), 71–80 (2007)
20. Brodley, C.E., Friedl, M.A.: Identifying mislabeled training data. J. Artif. Intell. Res. **11**, 131–167 (1999)
21. Gamberger, D., Boskovic, R., Lavrac, N., Groselj, C.: Experiments with noise filtering in a medical domain. In: Proceedings 16th International Conference on Machine Learning, Bled, Slovenia, June 1999, pp. 143–151. Springer, San Francisco (1999)
22. Krauth, W., Mezard, M.: Learning algorithms with optimal stabilityin neural networks. J. Phys. A Gen. Phys. **20**(11), L745 (1987)
23. Clark, P., Niblett, T.: The CN2 induction algorithm. Mach. Learn. **3**(4), 261–283 (1989)
24. Cantador, I., Dorronsoro, J.: Boosting parallel perceptrons for label noise reduction in classification problems. In: Mira, J., Álvarez, J.R. (eds.) IWINAC 2005. LNCS, vol. 3562, pp. 586–593. Springer, Heidelberg (2005). https://doi.org/10.1007/11499305_60
25. Yang, J., Fan, J., Wei, Z., Li, G., Liu, T., Du, X.: A game-based framework for crowdsourced data labeling. VLDB J. **29**(6), 1311–1336 (2020). https://doi.org/10.1007/s00778-020-00613-w
26. von Ahn, L., Dabbish, L.: Labeling images with a computer game. In: Proceedings of the SIGCHI Conference on Human Factors in Computing Systems, pp. 319–326 (2004)
27. Jingru, Y., Fan, J., Wei, Z., Li, G., Liu, T., Du, X.: Cost-effective data annotation using game-based crowdsourcing. Proc. VLDB Endowment **12**(1), 57–70 (2018)
28. Games-With-A-Purpose for Data Labelling. http://eecs.qmul.ac.uk/phd/phd-opportunities/items/games-with-a-purpose-for-data-labelling.html. Датаобращения. Accessed 29 Jan 2021

29. iHEARu-PLAY is a citizen science project focusing on voice and speaker recognition research. https://www.ihearu-play.eu. датаобращения. Accessed 29 Jan 2021
30. Bu, Q., Simperl, E., Chapman, A., Maddalena, E.: Quality assessment in crowdsourced classification tasks. Int. J. Crowd Sci. **3**(3), 222–248 (2019)
31. Melnik, E.V., Klimenko, A.B.: A complex approach to the data labeling efficiency improvement. In: Kovalev, S., Tarassov, V., Snasel, V., Sukhanov, A. (eds.) IITI 2019. AISC, vol. 1156, pp. 41–50. Springer, Cham (2020). https://doi.org/10.1007/978-3-030-50097-9_5
32. Melnik, E.V., Klimenko, A.B., Ivanov, D.Y.: The distributed ledger-based technique of the neuronet training set forming. In: Silhavy, R., Silhavy, P., Prokopova, Z. (eds.) CoMeSySo 2019 2019. AISC, vol. 1047, pp. 7–15. Springer, Cham (2019). https://doi.org/10.1007/978-3-030-31362-3_2
33. Liskov, B., Cowling, J.: Viewstamped Replication Revisited (2012)

Check for updates

Optimized Decision Forest for Website Phishing Detection

Abdullateef O. Balogun[1,2(✉)], Hammed A. Mojeed[1], Kayode S. Adewole[1], Abimbola G. Akintola[1], Shakirat A. Salihu[1], Amos O. Bajeh[1], and Rasheed G. Jimoh[1]

[1] Department of Computer Science, University of Ilorin, Ilorin PMB 1515, Nigeria
{balogun.ao1,mojeed.ha,adewole.ks,akintola.ag,salihu.sa, bajehamos,jimoh_rasheed}@unilorin.edu.ng

[2] Department of Computer and Information Sciences, Universiti Teknologi PETRONAS, Bandar Seri Iskandar, 32610 Perak, Malaysia
abdullateef_16005851@utp.edu.my

Abstract. The development of web and internet technology has resulted in its application in a wide range of services. This has resulted in an increase in the number of cybersecurity issues over the years, the most famous of which is the phishing attack, in which hostile websites impersonate genuine websites to acquire naïve users' data required for illegal access. Current mitigation measures, including anti-phishing software and machine learning (ML) approach, have proven to be successful in identifying phishing operations. Hackers, on the other hand, are coming up with new techniques to get around these counter-measures. Nonetheless, given the dynamism of phishing efforts, there is a constant requirement for novel and efficient website phishing detection solutions. In this study, an optimized decision forest (ODF) method for detecting website phishing is proposed ODF involves the use of a genetic algorithm (GA) for the selection of optimal diverse individual trees in a forest to generate an efficient sub-forest. Specifically, accurate and diverse trees from a decision forest are passed into GA as an initial population to generate a more robust forest with high efficacy. The performance of the proposed ODF is evaluated using three phishing datasets from the UCI repository. Findings from the experimental results revealed that ODF performed better than selected baseline classifiers. Particularly, ODF recorded a high detection accuracy (98.37%), AUC (0.999), f-measure (0.98), MCC (0.967) values with a low false-positive rate (0.016). In addition, ODF outperformed some existing ML-based phishing attack models. Consequently, the proposed ODF method is recommended for dealing with sophisticated phishing attacks.

Keywords: Cybersecurity · Decision forest · Genetic algorithm · Machine learning · Phishing attack

1 Introduction

The upsurge adoption and implementation of Information Technology (IT) has resulted in a greater variety of web-based solutions available on the internet. These operations

R. Silhavy et al. (Eds.): CoMeSySo 2021, LNNS 231, pp. 568–582, 2021.
https://doi.org/10.1007/978-3-030-90321-3_47

span from critical services like financial and health services to important activities like educational services [1, 2]. Studies have shown that financial transactions, online entertainment platforms, and social networking applications are among the most prominent and commonly used web services with many active users. The large number of people who embrace these web services attests to their current advancement. The goal is to make regularly used web services more accessible and available. Conversely, due to the lack of a standard and general internet policy, the limitless accessibility of these web-based services paves way for cyber-attacks [3–5].

Cyber-attacks create serious flaws and threats for web services and their users. Also, cyber-attacks often lead to significant data losses and in some cases huge financial losses. Phishing attacks on the internet are a common type of this cyber-attack. Cyber-criminals are increasingly creating phoney websites to collect personal information from unwary end-users and exploit it illegally [2, 6].

The website phishing attack is a common cybersecurity problem that has choked and rampaged the internet, putting netizens and web services at risk [7, 8]. According to [2], a phishing attack is a frequent scam in which an illegitimate website impersonates a reputable website to obtain data from unsuspecting users. Consequently, phishing attacks represent a significant threat to internet-based services. [9–11]. In 2018, the Anti-Phishing Working Group (APWG) labelled 51,401 phishing websites on the internet. According to RSA, phishing attacks cost worldwide corporations about $9 billion in 2016 [12, 13]. These cases demonstrate how rapidly phishing attacks from illegal websites spread, resulting in considerable monetary losses and inconveniences [9, 11, 14].

Several cybersecurity experts and analysts have formulated anti-phishing solutions for detecting phishing websites [15–17]. To mitigate website phishing attacks, one of these methods is to employ a blacklist strategy. The blacklisting function in web browsers compares the provided universal resource locator (URLs) to previously saved phishing website URLs to evaluate its legitimacy. The inability of blacklist anti-phishing solutions to identify new phishing URLs owing to their reliance on compiling blacklisted phishing URLs is a serious drawback [3, 18]. However, cybercriminals are employing innovative strategies that allow them to quickly avoid the blacklisting procedure. Due to the extreme dynamism of cyber-attacks, Machine Learning (ML)-based technologies are used to assess the trustworthiness of websites to deal with the complicated presence of online phishing attacks on website characteristics [12, 15, 19].

The effectiveness of the ML-based phishing detection method, on the other hand, is dependent on the effectiveness of the chosen ML technique in detecting phishing websites. Several ML algorithms have been employed to detect phishing websites, albeit with low detection accuracy and large false-positive rates [6, 20–22]. This might be related to data quality issues, such as unbalanced datasets, which reduce the performance of ML models [23, 24]. Considering phishing websites are dynamic, more advanced ML approaches are required.

Consequently, an optimized decision forest (ODF) method for identifying phishing websites is proposed. ODF involves the use of a genetic algorithm (GA) for the selection of optimal diverse individual trees in a forest to generate an efficient sub-forest. Specifically, accurate and diverse trees from a decision forest are passed into GA as an initial population to generate a more robust forest with high efficacy. The working principle of the proposed method is based on the notion that GA with a quality initial population typically leads to a good quality final solution. [25, 26].

The remainder of this paper is organized as follows. Section 2 looks at existing related studies. Section 3 depicts the methodology and experimental procedure. Section 4 presents and discusses the experimental results and findings. In the end, Sect. 5 gives the conclusion of the study and proposes possible future works.

2 Related Works

This section explores and examines prominent phishing detection strategies developed using diverse anti-phishing and ML methods.

Mohammad, Thabtah and McCluskey [1] identified phishing websites using a self-structuring neural network. Their methodology is based on a variable adaptive learning rate that is introduced before new neurons and network structures are introduced. The suggested model's accuracy values for the training, testing and validation sets were 94.07%, 92.48%, and 91.12%, respectively. Wang, Zhang, Luo and Zhang [27] developed a URL-based website phishing detection method that is independent of third-party services such as DNS and search engines. To feed the deep learning model and detect phishing assaults, the suggested method used deep learning to extract useful features. To identify website phishing, the proposed approach integrated Recurrent Neural Network (RNN) and the Convolutional Neural Network (CNN). The proposed approach yielded encouraging results. However, the great accuracy of deep learning-based models comes at the expense of a long detection time. PhishDump, a novel interPhishDump, a novel internet phishing detection system for mobile devices, was introduced by Rao, Vaishnavi and Pais [28]. The authors claim that current detection techniques are ineffective on mobile devices with limited processing capacity and little random-access memory (RAM). Furthermore, as the number of smartphones grows, so does the number of phishing attempts targeting smartphones. As a result, the authors proposed a simple detection strategy for mobile-targeted website phishing. Support Vector Machine (SVM) and Long Short Term Memory (LSTM) are combined in this method. Two data sets were used to test the suggested strategy, and the findings were encouraging. The detection time of the [28] technique is a concern, similar to the previously presented method [27], because the complexity of the base estimators is large, resulting in a long detection time.

Vrbančič, Fister Jr and Podgorelec [2] utilised the bat meta-heuristics search method to improve DNN. The proposed technique has a maximum accuracy of 96.9%. These experiments show that neural network models are nearly as successful as traditional classifiers in detecting phishing websites. Rahman, Rafiq, Toma, Hossain and Biplob [11] evaluated the efficacy of different ML and ensemble techniques in identifying website phishing (K-Nearest Neighbour (KNN), Decision Tree (DT), SVM, Random Forest (RF), Extreme Randomized Tree (ERT), and Gradient Boosting Tree (GBT)). Correspondingly, Chandra and Jana [9] investigated the use of meta-classifiers to improve phishing website detection. According to their findings, ensemble techniques beat single classifiers. Alsariera, Elijah and Balogun [12] used Forest Penalizing by Attributes (ForestPA) ensemble variants to detect phishing websites. To develop strong trees, Forest uses weight assignment and an increment method. The suggested meta-learner ForestPA variants are highly excellent at detecting phishing websites, according to their data, with a minimum accuracy of 96.26%. To select optimum functions, Chiew, Tan, Wong, Yong and Tiong [13] developed a Hybrid Ensemble FS (HEFS) technique based on a unique cumulative distribution function gradient (CDF-g) method. HEFS's RF estimate was 94.6% accurate. Also, Aydin and Baykal [14] employed subset-based functionality derived from a website URL for phishing website detection. Alpha-numeric character, keyword, security, domain identity, and rank-based techniques were used to assess the retrieved characteristics. Afterwards, the retrieved features were submitted to NB and Sequential Minimal Optimization (SMO). Precision was 83.96% for NB and 95.39% for SMO, respectively.

Ubing, Jasmi, Abdullah, Jhanjhi and Supramaniam [17] suggested a phishing approach based on feature selection (FS) and the Ensemble Learning Mechanism (ELM). The FS technique was the Random Forest Regressor (RFG), and the ELM was decided by majority vote. Their experimental results indicated that the suggested approaches outperform experimented baseline and ensemble methods.

Based on the above reviews, there is a need for more dependable and efficient solutions, as most current techniques are relatively ineffective. Consequently, an optimized decision forest (ODF) method based on the metaheuristic method (GA) is proposed for identifying phishing websites in this study.

3 Methodology

This section describes the experimental methodology used in this study—specifically, the proposed ODF website phishing detection technique. The phishing datasets used for training and testing, detection performance metrics, and experimental procedures are discussed in this section.

3.1 Optimized Decision Forest (ODF) Method

Optimized decision forest (ODF) is based on the use of a genetic algorithm (GA) for the selection of ideal diverse individual trees in a forest to produce an efficient sub-forest. In particular, quality and diversified trees from a decision forest are transferred into GA as an initial population to produce a more robust and efficacious forest. In other words, ODF enables the incorporation of high-quality trees as the first population for GA that can be referred to as GA-based subforest selection [26]. In other words, ODF allows the infusing of high-quality trees as the initial population for GA. The working principle of the proposed method is based on the notion that GA with a quality initial population typically leads to a good quality final solution. [25, 26]. GA, as deployed in this study is an evolutionary-based metaheuristic algorithm [29, 30]. GA represents a viable solution in a basic data structure known as a chromosome. Generally, the implementation of a GA starts with a collection (population) of randomly selected chromosomes. Then, to generate a new population, GA employs genetics-inspired operators such as crossover and mutation. Then, these chromosomes are assessed periodically, and chromosomes that offer a better answer are retained in the mechanism to be allowed a greater likelihood of generation (reproduction). The result of GA is the chromosome with the optimal method (solution) thus far. Algorithm 1 as shown in Fig. 1 presents the pseudocode for ODF [26].

```
Algorithm 1:
input: A Decision Forest DF.
output: A Subforest Sub_PGA.
Required:
    T ⟵ 100 (Size of DF) ; |P| ⟵ 20 (Size of
    Population);J ⟵ 20 (Number of Iterations);
end
P_Curr ← ∅ ;Step 1: Initial Population Selection
    for (i = 1 to |P|) do
        if (i mod 2 = 0) then Cr_i ← Get_Cr_Random();
        else Cr_i ← Get_Cr_Strata();
        P_Curr ← P_Curr ∪ {Cr_i};
    end
    Cr_SFBest ← Get_Best_Cr(P_Curr);
end
for (j = 1 to J) do
    Cr_CurrBest ← Get_Best_Cr(P_Curr);
    Step 2: Crossover and Elitist Operation
        P_TempCurr ← P_Curr;
        for (k = 1 to |P_Curr|/2) do
            Cr_b ← Get_Best_Cr(P_TempCurr);
            Cr_r ← Get_Roulette_Cr(P_TempCurr);
            /* Cr_b ≠ Cr_r                                   */
            O ← Crossover(Cr_b, Cr_r) ; /* Two Offspring O ←
                {O_k, O_k+1} are generated                  */
            P_TempCurr ← P_TempCurr ∩ {Cr_b} ∩ {Cr_r};
            P_Mod ← P_Mod ∪ {O_k} ∪ {O_k+1};
        end
        Elitist_Operation(Cr_SFBest, Cr_CurrBest, P_Mod);
    end
    Step 3: Mutation and Elitist Operation
        for (k = 1 to |P_Mod|) do
            t ← Gen_Rnd_Num(1, T) ;
            /* Random Number between 1 and T                 */
            P_Mod ← 1-Bit_Flipper(Cr^k_Mod, t);
        end
        Elitist_Operation(Cr_SFBest, Cr_CurrBest, P_Mod);
    end
    Step 4: Chromosome Selection for the Next Iteration
        P_Pool ← P_Curr ∪ P_Mod;
        P_Curr ← Get_Roulette_Pop(P_Pool, J);
    end
end
Step 5: Rectification of So Far Best Chromosome
    for (k = 1 to T) do
        if (Cr^k_SFBest == 0) then Rectify(Cr'_SFBest, k));
    end
    for (k = 1 to T) do
        if (Cr^k_SFBest == 1) then Rectify(Cr'_SFBest, k));
    end
end
Sub ← Cr_SFBest;
return Sub_PGA;
```

Fig. 1. Pseudocode for optimized decision forest [26]

3.2 Website Phishing Datasets

In the experimental phase of this work, three phishing datasets were employed. These datasets are widely available and are often utilized in existing studies. [1, 11–13, 15, 31]. The First dataset (Dataset 1) comprises 1,353 instances (702 phishing, 548 real, and 103 suspicious) with a total of ten attributes [31]. In the second dataset (Dataset 2), there are 11,055 instances (4,898 phishing and 6,157 legitimate instances) and 30 attributes [1]. The third dataset (Dataset 3) consists of 10,000 instances, 5,000 of which are legitimate and 5,000 of which are phishing. Dataset 3 has 48 discrete, continuous, and categorical features [11, 13]. Dataset 1 differs from Datasets 2 and 3 in that it has three class labels. For additional information about the phishing datasets, you can check these studies [1, 11–13, 15, 31].

3.3 Experimental Framework

This section describes the experimental framework utilised in this study, as depicted in Fig. 2. The process is designed to experimentally assess and test the effectiveness of the proposed method for identifying phishing websites. The proposed and other experimented methods are trained and tested using three phishing datasets from the UCI repository. The experimented website phishing detection models are implemented and tested with the K-fold (k = 10) Cross-Validation (CV) approach. The 10-fold CV was chosen for its capacity to construct phishing models while reducing the impact of the class imbalance problem [32–35]. Since the K-fold CV technique allows each instance to

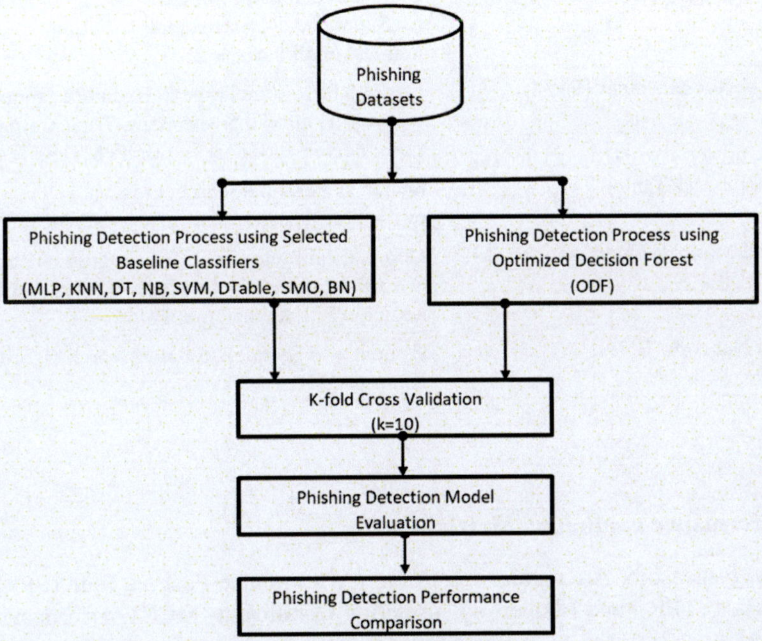

Fig. 2. Experimental framework

be used iteratively for both training and testing [33, 36–39], the proposed method (ODF) and selected baseline classifiers (Multilayer Perceptron (MLP), KNN, DT, Naïve Bayes (NB), SVM, Decision Table (DTable), SMO, Bayesian Network (BN)) were deployed on phishing datasets based on 10-fold CV. The selected baseline classifiers were chosen based on their usage and performance from existing studies [40–43] and Table 1 presents the parameter settings of these classifiers as used in this study. The phishing detection efficiency of the proposed phishing method (ODF) is then evaluated and compared to other experimented and existing phishing detection approaches. All experiments were performed using the WEKA machine learning tool in the same environment [44].

Table 1. Parameter settings of selected baseline classification algorithm

Classification algorithm	Parameter setting
Multilayer Perceptron (MLP)	activationFunction = Approximate Sigmoid; lossFunction = SquaredError; numFucntions = 2; method = BackPropagation
K Nearest Neighbour (KNN)	K = 1; distanceWeighting = NULL; nearestNeighbourSearchAlgorithm = LinearNNSearch; meanSquared = False
Decision Tree (DT)	confidenceFactor = 0.25; minNumObj = 2; subtreeRaising = True; binarySplits = False; useMDLcorrection = True
Naïve Bayes (NB)	useKernelEstimator = False; useSupervisedDiscretozation = False; numDecimalPlaces = 2;
Support Vector Machines (SVM)	Eps = 0.001; kernelType = RadialBasisFunction; loss = 0.1; nu = 0.5; shrinking-True; normalize = false; degree = 3; cost = 1.0; SVMType = C-SVC
Decision Table (DTable)	Search = BestFirstSearch; crossVal = 1; evaluationMeasure = accuracy; useIBK = False
Sequential Minimal Optimization (SMO)	C = 1.0; calibrator = Logistic regression; epsilon = 1.0E-12; filterType = NormalizeTrainingData; kernel = PolyKernel; numFolds = -1
Bayesian Network (BN)	Estimator = SimpleEstimator; searchAlgorithms = K2, useADTree = False

3.4 Performance Evaluation Metrics

Accuracy, F-measure, Area under the Curve (AUC), False-Positive Rate (FPR), True Positive Rate (TPR), and Mathew's Correlation Coefficient (MCC) performance evaluation measures are used to evaluate the detection performance of the experimented phishing methods. The choice for these measures emanates from their broad and continuous usage in previous research for website phishing detection. [11, 12, 17–19, 45, 46].

i. Accuracy is the average degree at which the actual labels of all instances are predicted correctly. It is computed as outlined in Eq. (1):

$$\text{Accuracy} = \frac{\text{TP} + \text{TN}}{\text{TP} + \text{FP} + \text{TN} + \text{FN}} \tag{1}$$

i. F-measure shows the weighted average of the Recall (R) and Precision (P). It stresses a classifier's ability to maximize both precision and recall at the same time. Equation 2 represents the computation of the F-measure.

$$\text{F} - \text{measure} = \frac{2 \times \text{TP}}{2 \times \text{TP} + \text{FP} + \text{FN}} \tag{2}$$

ii. The AUC plots the FP rate on the X-axis and the TP rate on the Y-axis. AUC is not vulnerable to plurality bias and does not overlook the minority samples during its assessment.

iii. The False Positive Rate (FPR) is the proportion of legitimate instances mistakenly reported as phishing attacks.

$$\text{FPR} = \frac{\text{FP}}{\text{FP} + \text{TN}} \times 100 \tag{3}$$

iv. True Positive Rate (TPR) is the rate at which actual phishing website instances are correctly classified as that phishing website.

$$\text{TPR} = \frac{\text{TP}}{\text{TP} + \text{FN}} \times 100 \tag{4}$$

v. The Mathews Correlation Coefficient (MCC) is a statistical rate that provides a high score if the prediction produces good outcomes in all four classes of the confusion matrix (true positives, false negatives, true negatives, and false positives), in proportion to the scale of the positive and negative elements in the dataset. MCC can be computed as shown in Eq. 5.

$$\text{MCC} = \frac{\text{TP} \times \text{TN} - \text{FP} \times \text{FN}}{\sqrt{(\text{TP} + \text{FP}) \times (\text{TP} + \text{FN}) \times (\text{TN} + \text{FP}) \times (\text{TN} + \text{FN})}} \tag{5}$$

4 Results and Discussion

This section covers the experimental findings acquired after implementing, training, and testing the proposed ODF and selected baseline classifiers (See Table 1) on three phishing datasets. In addition, the performance of the proposed ODF is compared with website phishing models from existing studies.

4.1 ODF Performance Comparison of Proposed ODF and Baseline Classifiers

On Dataset 1, the performance of the proposed ODF is compared to that of selected baseline classifiers, as shown in Table 2. For the performance comparison, six performance measures are employed (See Sect. 3.4). In terms of accuracy, ODF outperformed NB (84.10%), SVM (85.66%), SMO (86%), DTable (84.47%), KNN (86.32%), DT (87.58%), MLP (84.77%), and BN (84.33%). Similar performance can be observed in terms of f-measure and AUC values. Specifically, ODF had a superior f-measure and AUC values of 0.895 and 0.968 respectively over KNN (0.863, 0.880), DT (0.891, 0.916), MLP (0.840, 0.927), SVM (0.825, 0.867), NB (0.825, 0.948), SMO (0.846, 0.900), DTable (0.839, 0.954) and BN (0.828, 0.948). In addition, when compared to the selected baseline classifiers, ODF on Dataset 1 had the highest TP-Rate (0.894) and lowest FP-Rate (0.078).

Table 2. Experimental results of ODF and baseline classifiers on dataset 1

	Accuracy (%)	F-Measure	AUC	TP	FP	MCC
ODF	**89.43**	**0.895**	**0.968**	**0.894**	**0.078**	**0.812**
NB	84.10	0.825	0.948	0.841	0.120	0.722
SVM	85.66	0.825	0.867	0.857	0.123	0.734
SMO	86.00	0.846	0.900	0.860	0.109	0.757
DTable	84.47	0.839	0.954	0.845	0.110	0.737
MLP	84.77	0.840	0.927	0.848	0.108	0.742
KNN	86.32	0.863	0.880	0.863	0.104	0.761
DT	87.58	0.891	0.916	0.890	0.082	0.803
BN	84.33	0.828	0.948	0.843	0.118	0.727

On Dataset 2, a similar performance of the proposed ODF method was observed. That is, the proposed ODF recorded a superior performance over the selected baseline classifiers on Dataset 2. Particularly, ODF achieved the highest accuracy value (97.26%), F-Measure value (0.973), AUC value (0.995), TP-Rate value (0.973), MCC value (0.944), and the lowest FP-Rate value (0.029) when compared with the performance of the baseline classifiers as presented in Table 4. Similarly, comparable results were obtained on the performance of ODF on Dataset 3, as shown in Table 5. ODF recorded the best accuracy value (98.34%), F-Measure value (0.984), AUC value (0.999), TP-Rate value (0.984), MCC value (0.967), and the lowest FP-Rate value (0.016).

Based on the experimental results, it was observed that in most cases the proposed ODF outperformed the majority of the experimented selected baseline classifiers. These findings suggest that ODF delivered comparable results (performance) for phishing detection across all three datasets, independent of dataset size. In other words, ODF outperformed baseline classifiers in the identification of website phishing.

Table 3. Experimental results of ODF and baseline classifiers on dataset 2

	Accuracy (%)	F-Measure	AUC	TP	FP	MCC
ODF	**97.26**	**0.973**	**0.995**	**0.973**	**0.029**	**0.944**
NB	90.70	0.907	0.962	0.907	0.098	0.811
SVM	94.60	0.946	0.944	0.946	0.059	0.891
SMO	92.70	0.927	0.925	0.927	0.078	0.852
DTable	93.44	0.934	0.981	0.934	0.073	0.867
MLP	94.76	0.948	0.983	0.948	0.053	0.894
KNN	96.84	0.968	0.967	0.968	0.034	0.936
DT	95.87	0.959	0.984	0.959	0.045	0.916
BN	92.98	0.93	0.981	0.930	0.075	0.858

Table 4. Experimental results of ORF and baseline classifiers on dataset 3

	Accuracy (%)	F-Measure	AUC	TP	FP	MCC
ORF	**98.37**	**0.984**	**0.999**	**0.984**	**0.016**	**0.967**
NB	85.15	0.85	0.949	0.852	0.149	0.715
SVM	91.49	0.915	0.915	0.915	0.085	0.83
SMO	93.87	0.939	0.939	0.939	0.061	0.878
DTable	95.79	0.958	0.982	0.958	0.042	0.916
MLP	95.92	0.959	0.983	0.959	0.041	0.918
KNN	95.53	0.955	0.955	0.955	0.045	0.911
DT	97.31	0.973	0.976	0.973	0.027	0.946
BN	95.79	0.958	0.992	0.958	0.042	0.916

4.2 Performance Comparison of Proposed ODF with Existing Methods.

In this section, the performance of the proposed ODF is further compared with existing state-of-the-art methods for website phishing detection. Table 5 shows the performance comparison of ODF with existing methods on Dataset 1.ODF outperformed existing methods as proposed by Rahman, Rafiq, Toma, Hossain and Biplob [11].

On Dataset 2, Table 6 compared the performance of ODF with existing methods. The experimental findings obtained by Ubing, Jasmi, Abdullah, Jhanjhi and Supramaniam [17], Alsariera, Elijah and Balogun [12], and Vrbančič, Fister Jr and Podgorelec [2] are comparable to that obtained by ODF. However, the performance of ODF was still superior to other existing methods on Dataset 2. Furthermore, Table 7 compared the performance of the ODF with existing methods based on Dataset 3. Specifically, the performance of ODF was superior to methods proposed by Chiew, Tan, Wong, Yong and Tiong [13] and Rahman, Rafiq, Toma, Hossain and Biplob [11] on Dataset 3.

Table 5. Performance evaluation of ODF and existing models on Dataset 1

Phishing models	Accuracy (%)	F-Measure	AUC	TP-rate	FP-rate	MCC
Rahman, Rafiq, Toma, Hossain and Biplob [11] (KNN)	88.00	-	-	-	0.099	-
Rahman, Rafiq, Toma, Hossain and Biplob [11] (SVM)	87.00	-	-	-	0.087	-
*Proposed ODF	**89.43**	**0.895**	**0.968**	**0.894**	**0.078**	**0.812**

Table 6. Performance evaluation of ODF and existing models on Dataset 2

Phishing models	Accuracy (%)	F-Measure	AUC	TP-Rate	FP-Rate	MCC
Aydin and Baykal [14]	95.39	0.938	0.936	-	0.046	-
Dedakia and Mistry [8]	94.29	-	-	-	-	-
Ubing, Jasmi, Abdullah, Jhanjhi and Supramaniam [17]	95.40	0.947	-	-	0.041	-
Hadi, Aburub and Alhawari [10]	92.40	-	-	-	-	-
Chiew, Tan, Wong, Yong and Tiong [13]	93.22	-	-	-	-	-
Rahman, Rafiq, Toma, Hossain and Biplob [11] (KNN)	94.00	-	-	-	0.049	-
Rahman, Rafiq, Toma, Hossain and Biplob [11] (SVM)	95.00	-	-	-	0.039	-
Chandra and Jana [9]	92.72	-	-	-	-	-
Folorunso, Ayo, Abdullah and Ogunyinka [19] (Stacking)	95.97	-	-	-	-	-
Folorunso, Ayo, Abdullah and Ogunyinka [19] (Hybrid NBTree)	94.10	-	-	-	-	-
Al-Ahmadi and Lasloum [47]	96.65	0.965	-	-	-	-

(*continued*)

Table 6. (*continued*)

Phishing models	Accuracy (%)	F-Measure	AUC	TP-Rate	FP-Rate	MCC
Alsariera, Elijah and Balogun [12]	96.26	-	-	-	0.040	-
Ali and Malebary [21]	96.43	-	-	-	-	-
Ferreira, Martiniano, Napolitano, Romero, Gatto, Farias and Sassi [48]	87.61	-	-	-	-	-
Vrbančič, Fister Jr and Podgorelec [2]	96.50	-	-	-	-	-
*Proposed ODF	**97.26**	**0.973**	**0.995**	**0.973**	**0.029**	**0.944**

Table 7. Performance evaluation of ODF and existing models on Dataset 3

Phishing models	Accuracy (%)	F-Measure	AUC	TP-Rate	FP-Rate	MCC
Chiew, Tan, Wong, Yong and Tiong [13]	94.60	-	-	-	-	-
Rahman, Rafiq, Toma, Hossain and Biplob [11] (KNN)	87.00	-	-	-	0.078	-
Rahman, Rafiq, Toma, Hossain and Biplob [11] (SVM)	91.00	-	-	-	0.067	-
*Proposed ODF	**98.37**	**0.984**	**0.999**	**0.984**	**0.016**	**0.967**

Summarily, proposed ODF indeed produced significant performance in phishing website detection as compared with baseline methods such as MLP, KNN, DT, NB, SVM, DTable, SMO and BN with better accuracy and other performance evaluation metrics. This performance is replicated across the three datasets that were considered in this study. Also, the performance of ODF in most cases is superior in terms of accuracy, F-Measure, AUC, TP-Rate, FP-Rate, and MCC values to existing state-of-the-art website detection methods.

5 Conclusion and Future Works

Phishing attacks are among the most serious cyberattacks that have a widespread detrimental impact on web users. A phishing attack on a website may be damaging to online users and internet-based solutions in general. A website phishing attack allows an adversary to get access to victims' personal information, which may subsequently be used to carry out fraudulent transactions or steal users' identities. Nonetheless, detecting phishing websites has proven to be a challenge owing to attackers' sophisticated and dynamic methods. As a result, this study proposed an optimized decision forest (ODF) that involves the use of a genetic algorithm (GA) for the selection of optimal diverse individual trees in a forest to generate an efficient sub-forest to detect phishing websites. ODF demonstrated excellent detection performance, outperforming baseline models such as MLP, SVM, DT, NB, BN, and other state-of-the-art techniques for detecting phishing websites.

In the future, the proposed ODF will be tested on real-time phishing website datasets to evaluate its generalizability in detecting phishing websites. Furthermore, more advanced approaches for building and optimizing tree-based classifiers will be studied.

References

1. Mohammad, R.M., Thabtah, F., McCluskey, L.: Predicting phishing websites based on self-structuring neural network. Neural Comput. Appl. **25**(2), 443–458 (2013). https://doi.org/10.1007/s00521-013-1490-z
2. Vrbančič, G., Fister Jr, I., Podgorelec, V.: Swarm intelligence approaches for parameter setting of deep learning neural network: Case study on phishing websites classification. In: Proceedings of the 8th International Conference on Web Intelligence, Mining and Semantics, pp. 1–8 (2018)
3. Ali, W., Ahmed, A.A.: Hybrid intelligent phishing website prediction using deep neural networks with genetic algorithm-based feature selection and weighting. IET Inf. Secur. **13**, 659–669 (2019)
4. Verma, R., Das, A.: What's in a URL: Fast feature extraction and malicious url detection. In: Proceedings of the 3rd ACM on International Workshop on Security and Privacy Analytics, pp. 55–63 (2017)
5. Azeez, N., Misra, S., Margaret, I.A., Fernandez-Sanz, L.: Adopting automated whitelist approach for detecting phishing attacks. Comput. Secur. **108**, 102328 (2021)
6. Alqahtani, M.: Phishing websites classification using association classification (PWCAC). In: 2019 International Conference On Computer and Information Sciences (ICCIS), pp. 1–6. IEEE (2019)
7. Abdelhamid, N., Ayesh, A., Thabtah, F.: Phishing detection based associative classification data mining. Expert Syst. Appl. **41**, 5948–5959 (2014)
8. Dedakia, M., Mistry, K.: Phishing detection using content based associative classification data mining. J. Eng. Comput. Appl. Sci. **4**, 209–214 (2015)
9. Chandra, Y., Jana, A.: Improvement in phishing websites detection using meta classifiers. In: 6th International Conference on Computing for Sustainable Global Development (INDIACom), pp. 637–641. IEEE (2019)
10. Hadi, W.e., Aburub, F., Alhawari, S.: A new fast associative classification algorithm for detecting phishing websites. Appl. Soft Comput. **48**, 729–734 (2016)
11. Rahman, S.S.M.M., Rafiq, F.B., Toma, T.R., Hossain, S.S., Biplob, K.B.B.: Performance assessment of multiple machine learning classifiers for detecting the phishing URLs. In: Raju, KSrujan, Senkerik, R., Lanka, S.P., Rajagopal, V. (eds.) Data Engineering and Communication Technology. AISC, vol. 1079, pp. 285–296. Springer, Singapore (2020). https://doi.org/10.1007/978-981-15-1097-7_25
12. Alsariera, Y.A., Elijah, A.V., Balogun, A.O.: Phishing website detection: forest by penalizing attributes algorithm and its enhanced variations. Arab. J. Sci. Eng. **45**(12), 10459–10470 (2020). https://doi.org/10.1007/s13369-020-04802-1

13. Chiew, K.L., Tan, C.L., Wong, K., Yong, K.S., Tiong, W.K.: A new hybrid ensemble feature selection framework for machine learning-based phishing detection system. Inf. Sci. **484**, 153–166 (2019)
14. Aydin, M., Baykal, N.: Feature extraction and classification phishing websites based on URL. In: IEEE Conference on Communications and Network Security (CNS), pp. 769–770. IEEE (2015)
15. Adeyemo, V.E., Balogun, A.O., Mojeed, H.A., Akande, N.O., Adewole, K.S.: Ensemble-based logistic model trees for website phishing detection. In: Anbar, M., Abdullah, N., Manickam, S. (eds.) ACeS 2020. CCIS, vol. 1347, pp. 627–641. Springer, Singapore (2021). https://doi.org/10.1007/978-981-33-6835-4_41
16. Pham, B.T., Nguyen, V.-T., Ngo, V.-L., Trinh, P.T., Ngo, H.T.T., Tien Bui, D.: A novel hybrid model of rotation forest based functional trees for landslide susceptibility mapping: a case study at Kon Tum Province, Vietnam. In: Tien Bui, D., Ngoc Do, A., Bui, H.-B., Hoang, N.-D. (eds.) GTER 2017, pp. 186–201. Springer, Cham (2018). https://doi.org/10.1007/978-3-319-68240-2_12
17. Ubing, A.A., Jasmi, S.K.B., Abdullah, A., Jhanjhi, N., Supramaniam, M.: Phishing website detection: an improved accuracy through feature selection and ensemble learning. Int. J. Adv. Comput. Sci. Appl. **10**, 252–257 (2019)
18. Abdulrahaman, M.D., Alhassan, J.K., Adebayo, O.S., Ojeniyi, J.A., Olalere, M.: (2019): Phishing attack detection based on random forest with wrapper feature selection method. Int. J. Inf. Process. Commun. (IJIPC) **7**, 209–224 (2019)
19. Folorunso, S.O., Ayo, F.E., Abdullah, K.-K.A., Ogunyinka, P.I.: Hybrid vs ensemble classification models for phishing websites. Iraqi J. Sci. **61**, 3387–3396 (2020)
20. Alsariera, Y.A., Adeyemo, V.E., Balogun, A.O., Alazzawi, A.K.: AI meta-learners and extra-trees algorithm for the detection of phishing websites. IEEE Access **8**, 142532–142542 (2020)
21. Ali, W., Malebary, S.: Particle swarm optimization-based feature weighting for improving intelligent phishing website detection. IEEE Access **8**, 116766–116780 (2020)
22. Osho, O., Oluyomi, A., Misra, S., Ahuja, R., Damasevicius, R., Maskeliunas, R.: Comparative evaluation of techniques for detection of phishing URLs. In: Florez, H., Leon, M., Diaz-, J.M., Belli, S. (eds.) ICAI 2019. CCIS, vol. 1051, pp. 385–394. Springer, Cham (2019). https://doi.org/10.1007/978-3-030-32475-9_28
23. Balogun, A.O., Basri, S., Abdulkadir, S.J., Adeyemo, V.E., Imam, A.A., Bajeh, A.O.: Software defect prediction: analysis of class imbalance and performance stability. J. Eng. Sci. Technol **14**, 3294–3308 (2019)
24. Yu, Q., Jiang, S., Zhang, Y.: The performance stability of defect prediction models with class imbalance: An empirical study. IEICE Trans. Inf. Syst. **100**, 265–272 (2017)
25. Rahman, M.A., Islam, M.Z.: A hybrid clustering technique combining a novel genetic algorithm with K-Means. Knowl.-Based Syst. **71**, 345–365 (2014)
26. Adnan, M.N., Islam, M.Z.: Optimizing the number of trees in a decision forest to discover a subforest with high ensemble accuracy using a genetic algorithm. Knowl.-Based Syst. **110**, 86–97 (2016)
27. Wang, W., Zhang, F., Luo, X., Zhang, S.: Pdrcnn: precise phishing detection with recurrent convolutional neural networks. Security and Communication Networks 2019 (2019)
28. Rao, R.S., Vaishnavi, T., Pais, A.R.: CatchPhish: detection of phishing websites by inspecting URLs. J. Ambient Intell. Humanized Comput. **11**(2), 813–825 (2019). https://doi.org/10.1007/s12652-019-01311-4
29. Mirjalili, S.: Genetic algorithm. Evolutionary algorithms and neural networks, pp. 43–55. Springer, Cham (2019). Doi: https://doi.org/10.1007/978-3-319-93025-1
30. Oluwagbemiga, B.A., Shuib, B., Abdulkadir, S., Marian, G., Thabeb, A.: A hybrid ant colony tabu search algorithm for solving next release problems. Int. J. Innov. Technol. Exploring Eng. **8**, 191–198 (2019)

31. Balogun, A.O., et al.: Improving the phishing website detection using empirical analysis of function tree and its variants. Heliyon **7**, e07437 (2021)
32. Balogun, A.O., et al.: SMOTE-based homogeneous ensemble methods for software defect prediction. In: Gervasi, O., et al. (eds.) ICCSA 2020. LNCS, vol. 12254, pp. 615–631. Springer, Cham (2020). https://doi.org/10.1007/978-3-030-58817-5_45
33. Yadav, S., Shukla, S.: Analysis of k-fold cross-validation over hold-out validation on colossal datasets for quality classification. In: IEEE 6th International Conference on Advanced Computing (IACC), pp. 78–83. IEEE (2016)
34. Balogun, A.O., Basri, S., Abdulkadir, S.J., Hashim, A.S.: Performance analysis of feature selection methods in software defect prediction: a search method approach. Appl. Sci. **9**, 2764 (2019)
35. Balogun, A.O., et al.: Impact of feature selection methods on the predictive performance of software defect prediction models: an extensive empirical study. Symmetry **12**, 1147 (2020)
36. Arlot, S., Lerasle, M.: Choice of V for V-fold cross-validation in least-squares density estimation. J. Mach. Learn. Res. **17**, 7256–7305 (2016)
37. Balogun, A.O., et al.: Search-based wrapper feature selection methods in software defect prediction: an empirical analysis. In: Silhavy, R. (ed.) CSOC 2020. AISC, vol. 1224, pp. 492–503. Springer, Cham (2020). https://doi.org/10.1007/978-3-030-51965-0_43
38. Basri, S., Almomani, M.A., Imam, A.A., Thangiah, M., Gilal, A.R., Balogun, A.O.: The organisational factors of software process improvement in small software industry: comparative study. In: Saeed, F., Mohammed, F., Gazem, N. (eds.) IRICT 2019. AISC, vol. 1073, pp. 1132–1143. Springer, Cham (2020). https://doi.org/10.1007/978-3-030-33582-3_106
39. Balogun, A.O., Lafenwa, F.B., Mojeed, H.A., Usman-Hamza, F.E., Bajeh, A.O., Adeyemo, V.E., Adewole, K.S., Jimoh, R.G.: Data sampling-based feature selection framework for software defect prediction. In: Abawajy, J.H., Choo, K.-K., Chiroma, H. (eds.) EATI 2020. LNNS, vol. 254, pp. 39–52. Springer, Cham (2021). https://doi.org/10.1007/978-3-030-80216-5_4
40. Ahmad, S.N.W., Ismail, M.A., Sutoyo, E., Kasim, S., Mohamad, M.S.: Comparative performance of machine learning methods for classification on phishing attack detection. Int. J 9 (2020)
41. Jain, A.K., Gupta, B.: Comparative analysis of features based machine learning approaches for phishing detection. In: 2016 3rd International Conference on Computing for Sustainable Global Development (INDIACom), pp. 2125–2130. IEEE (2016)
42. Karabatak, M., Mustafa, T.: Performance comparison of classifiers on reduced phishing website dataset. In: 2018 6th International Symposium on Digital Forensic and Security (ISDFS), pp. 1–5. IEEE (2018)
43. Balogun, A.O., et al.: Empirical analysis of rank aggregation-based multi-filter feature selection methods in software defect prediction. Electronics **10**, 179 (2021)
44. Hall, M., Frank, E., Holmes, G., Pfahringer, B., Reutemann, P., Witten, I.H.: The WEKA data mining software: an update. ACM SIGKDD Expl. Newsl. **11**, 10–18 (2009)
45. Adewole, K.S., Akintola, A.G., Salihu, S.A., Faruk, N., Jimoh, R.G.: Hybrid rule-based model for phishing URLs detection. In: Miraz, M.H., Excell, P.S., Ware, A., Soomro, S., Ali, M. (eds.) iCETiC 2019. LNICSSITE, vol. 285, pp. 119–135. Springer, Cham (2019). https://doi.org/10.1007/978-3-030-23943-5_9
46. AlEroud, A., Karabatis, G.: Bypassing detection of url-based phishing attacks using generative adversarial deep neural networks. In: Proceedings of the 6th International Workshop on Security and Privacy Analytics, pp. 53–60 (2020)
47. Al-Ahmadi, S., Lasloum, T.: PDMLP: phishing detection using multilayer perceptron. Int. J. Network Secur. Appl. **12**, 59–72 (2020)
48. Ferreira, R.P., et al.: Artificial neural network for websites classification with phishing characteristics. Social Networking **7**, 97 (2018)

Development of the Intelligent Method and Software for Prediction of the Risk of Cardiovascular Complications in Patients with Atrial Fibrillation and Heart Failure

Alina Sheina, Larisa Burmistrova, Andrey Kuzmin(✉), Olga Kuznetsova, Nikita Tereshin, and Maxim Burmistrov

Penza State University, Penza, Russia
a.v.kuzmin@pnzgu.ru
http://www.pnzgu.ru

Abstract. The article is devoted to the problem of assessment of the risk of complications in special groups of patients. The most important problems and dangerous consequences are linked with atrial fibrillation and heart failure in patient's history. Score based methods are used for risk assessment in cardiology (CHA2DS2-VASc, HAS-BLED, ATRIA, EHRA etc.). The authors analyze the application of machine learning methods in this field and propose neural network based method and software for the assessment of the risk of complications. The authors' method that includes questionnaire and scale is analyzed. The task is formalized and considered as classification. The architecture of perceptron with one hidden layer is chosen. The authors developed an application for data input, prepared the training and testing sets, formulated the requirements and recommendations for the training process. The results allow concluding that the prediction method based on neural network is technically implementable and potentially effective for this task. Future work includes training and testing the neural network and assessment of the accuracy of the results.

Keywords: Prediction · Neural network · Atrial fibrillation · Heart failure · Software

1 Introduction

Currently, cardiovascular diseases occupy the leading position in morbidity and mortality among the population of the entire planet. Today, among all cardiac arrhythmias, atrial fibrillation (AF) is the most common persistent form of arrhythmia in the elderly, which causes significant complications such as stroke, thromboembolism and decreased quality of life [1]. Chronic heart failure is an acute public health problem. Despite advances in the study of pathogenesis, clinical picture and treatment, the prognosis of chronic heart failure patients remains unfavorable.

According to the Framingham study, the prevalence of chronic heart failure among the population is 0.9% among the 50–59 age group and increases in the 80–89 age group

R. Silhavy et al. (Eds.): CoMeSySo 2021, LNNS 231, pp. 583–592, 2021.
https://doi.org/10.1007/978-3-030-90321-3_48

to 6.9 and 8.2% for men and women, respectively. After the diagnosis of heart failure, the 5- and 10-year patient survival rates are 50 and 10%, respectively [2].

Atrial fibrillation (AF) and heart failure (HF) are diseases that often occur together in patients, and each year the prevalence of these two diseases only progresses. AF combined with HF affects the duration and quality of life of patients, and also undoubtedly has an impact on costs in the health care system [3]. AF often occurs as a cause or consequence of heart failure. The high prevalence of AF in patients with HF increases among patients of different ages, but this is especially true of elderly patients. The pathophysiological changes that occur in patients with HF and AF are complex and poorly understood.

AF often results in ineffective active contraction. Rapid, irregular conduction to the ventricles impairs both systolic and diastolic function of the left ventricle (LV), which subsequently leads to the development of HF [4]. According to various scientific research works, it is known that fibrosis of the left atrium (LA) and ventricle is an important pathogenetic link between AF and HF. In addition, the stratification of the risk of cardiovascular mortality is interrelated with the definition of the HF subtype according to echocardiography with the study of such a parameter as left ventricular ejection fraction (LVEF). LVEF is an indicator that reflects the volume of blood expelled by the left ventricle (LV) at the time of its contraction (systole) into the aortic lumen. EF calculated based on the ratio of the volume of blood ejected into the aorta to the volume of blood in the left ventricle at the time of its relaxation (diastole). The risk of mortality in patients with concomitant AF and HF with reduced LVEF is slightly higher than in patients with concomitant AF and HF with preserved LVEF. On the other hand, the prognostic effect of HF in AF has similar features among patients with decreased LVEF and preserved LVEF. In addition, there is evidence that left atrial size and functional assessment of the LA and LV are important in predicting HF in AF in addition to generally accepted risk factors [6].

Changes in neurohormonal activation, electrophysiological parameters and mechanical factors interact with each other, thereby creating favorable conditions in which HF predisposes to AF and vice versa, AF aggravates HF. Mechanisms of pathogenesis include atrial remodeling and tachycardia-induced cardiomyopathy. Atrial remodeling refers to adverse electrophysiological, cellular and structural changes in atrial myocardial tissue that occur in response to pressure and volume overload or various arrhythmic disorders. The main causes of left atrial remodeling are AF and ventricular/valve disease. However, noncardiac conditions such as sleep apnea, hypertension, diabetes, and obesity are also important factors in LA remodeling. Many conditions can coexist and accelerate the adverse remodeling associated in most cases with AF. In addition to ion channel changes and impaired electrophysiological properties, interstitial fibrosis is a hallmark of AF-induced LA remodeling and is associated with chamber dilation, spherical deformity, and decreased atrial function, further contributing to the development of AF in a vicious circle ("AF gives rise to AF"). The degree of structural tissue remodeling varies greatly in individuals with AF and may also be related to genetic factors/predisposition [7].

Arrhythmias and atrial contractions contribute to the development of ineffective coronary blood flow. Irregular, shortened diastole with altered AF-induced flow reserve is a common cause of ventricular ischemia and type 2 myocardial infarction. Patients with AF have a 3-fold increased risk of developing myocardial infarction, regardless of other risk factors. The coexistence of risk factors common to patients with AF and HF, along with increased inflammation and platelet activation, is likely to increase the risk. Importantly, anticoagulant therapy appears to protect against incident myocardial infarction in the AF and HF patient population [8].

The development of AF in HF seems to increase the number of deaths, since AF changes the pumping function of the heart, decreases its contractility and, as a consequence, develops systolic myocardial dysfunction. However, despite the mechanisms studied, the currently available therapeutic options for AF in patients with HF are diverse, their impact on prognosis remains unknown and is the subject of ongoing clinical trials [5].

2 State of the Art

Different score based methods are used in clinical practice. Most of them include a questionnaire with a set of valuable prediction parameters and a scale for the risk evaluation. Among widely used score methods there are CHADS2 and its modifications (for example CHA2DS2-VASc [9]), HAS-BLED [10], ATRIA [11] and method based on recommendations of European Heart Rhythm Association (EHRA).

Traditionally arrhythmia detection involves record and analysis long of term ECG monitoring data. Wearable ECG devices are used for ECG monitoring [12, 13]. Different signal transform and analysis methods are applied for arrhythmia detection (for example Wavelet [14]). The obtained parameters could be also used for prognosis [15].

The most up-to-date approach to the problem of prediction is machine learning. Methods of this field allow processing a large amount of information that is accumulated in medical institutions. The application of classical classification methods (such as decision trees or naive Bayesian method) was analyzed by the authors.

An approach to the classification of cardiac arrhythmias is considered in work [16]. Three different parameters are considered for the study: cardiac arrhythmia, normal sinus rhythm, and chronic heart failure. The data was pre-processed using a discrete wavelet transform. The authors of this paper used a support vector machine (SVM) for classification. The obtained accuracy was 95.92%. But this method has a significant drawback, it is unstable with respect to noise in the initial data. If the training sample contains outliners, they will significantly distort the parameters of constructed separating hyperplane.

In work [17], an automated method was developed for the selection of informative signs of arrhythmia, which were then used in the k-nearest neighbor classification algorithm. The classification accuracy was 98%. But this method has a drawback: the constructed model cannot be "separated" from the data: all examples must be used to classify a new example. This feature severely limits the use of the algorithm.

Paper [18] describes the models for predicting in-hospital mortality based on the machine learning method using data from patients with random hypothermia. Machine learning methods were used (lasso, random forest and gradient boosting for decision trees). The results of the methods were evaluated using ROC analysis and the area under the curve (AUC): lasso method 0.784, random forest 0.794, gradient boosting for decision trees 0.780. These models can help doctors and patients make decisions. However, the applicability to the clinical practice is still unclear. Thus, further prospective study is required to assess clinical utility.

Work [19] proposed a multi-parameter assessment based on a combination of electrocardiographic and echocardiographic parameters. The random forest method was applied to analyze the significance of the features. The Monte Carlo cross-validation method was used for verification. The results show the average AUC 0.809.

In work [20], the characteristics of groups of heart failure and their relationship with concomitant diseases were investigated. Five clusters of heart failure were identified: low load, metabolic-vascular, anemic, ischemic and metabolic. Latent Class Analysis (LCA) was used to identify these clusters.

Paper [21] explored machine learning methods for predicting hospital admissions and mortality in patients with heart failure: random forests, decision trees, support vector machines, neural networks, and Bayesian methods. A limitation of research using machine learning methods was that most of them were not externally valid, and customization was rarely evaluated. In the only study that was externally valid in a single dataset, the area under the curve was 0.83.

Basing on the obtained analysis of the works, it can be concluded that these methods have a significant drawback, the results of these methods are difficult to interpret.

Other machine learning methods are neural networks that can reveal hidden patterns in the data. For this group of methods, an analysis of the works was also carried out.

In work [22] the classification of electrocardiograms (ECG) is carried out. In this work, a compact deep neural network method is proposed. This method is based on convolutional networks. The ensemble of such networks provided accuracy 99.88%.

The ECG analysis methods were also investigated in [23]. A deep learning approach has been proposed for the automated identification of congestive heart failure and arrhythmias with high accuracy and low computational requirements. This study presents an algorithm that combines a convolutional neural network with a non-stationary Gabor transform with a constant Q. The proposed one showed an accuracy of 98.82%.

In the study [24], coarse-grained entropy was used. The accuracy of this method was 99.23%.

The article [25] proposes a system for detecting atrial fibrillation. The paper considered a neural network using the error backpropagation learning algorithm. The accuracy of this method was 83.55%.

Basing on the obtained analysis of the works, it can be concluded that neural network methods have one significant drawback, they cannot explain the result to the final user. Fuzzy neural networks have a different principle of operation. In the structure of this network, a knowledge base is created - this is the experience of an expert in this field. With the help of neural network training methods and a knowledge base, these networks allow not only obtaining the forecasting result, but also explaining it. For further research, it is advisable to experiment with different architectures of neural and fuzzy networks.

3 Problem Statement

The combination of AF and HF significantly affects the prognosis: prognosis for patients with HF and AF is worse than the prognosis of patients with HF without AF. It is important to predict the risk of cardiovascular complications in patients with AF and HF. Recognizing cardiac arrhythmias and providing medical attention as quickly as possible is essential to prevent sudden and untimely death.

Early identification of a high-risk group of patients can help prevent adverse outcomes, which in turn can improve patient quality of life and lower health care costs (by reducing the cost and duration of treatment). Often, when predicting risk, doctors are based on measuring or determining a small number of indicators (for example, hematocrit, mean erythrocyte volume, fibrinogen and fibrinolytic activity, adrenaline-induced platelet aggregation are taken into account for thromboembolic complications). But, as experience shows, the calculation taking into account a smaller number of parameters does not give a high forecast accuracy.

There are various formulas for calculating the risk (for example, a formula for predicting the time of maintaining sinus rhythm after stopping paroxysm of atrial fibrillation taking into account echocardiography indicators, a formula for predicting the restoration of sinus rhythm in patients with paroxysmal atrial fibrillation, taking into account ECG indicators, etc.). But all these formulas are cumbersome and inconvenient for manual calculations, especially in the practice of medical workers. From the studied literature, it can be concluded that assessment systems and formulas for predicting risk in patients with HF have not yet been developed. Among patients with AF and HF, the risk of complications and adverse events is higher. However, the manifestations of HF are almost not taken into account when assessing the risk of an unfavorable outcome in a patient.

Thus, the development of a risk assessment methodology for patients with AF and HF that is sufficiently reliable and suitable for practical use is urgent.

4 Solution Vision

To solve the problem of predicting the complications in patients with rhythm disturbances and HF, it is proposed to develop a neural network based method for analyzing prognostic factors and implement it in the form of software. The software will allow the doctor to use an additional tool for rapid assessment of the patient's condition and determine the further tactics of treatment.

The system should meet the following requirements:

- Contactless communication between a doctor and patients using the telemedicine and Internet technologies;
- Correlation of the input parameters with the risk of complications;
- Software implementation and digital distribution of the developed method;
- Simple, convenient and intuitive interface that does not require serious training in the field of information technology.

5 Materials and Methods

The medical aspect of the study includes a method for assessing the patient's current state of health. The input data is organized in accordance with a special score-based method proposed by the authors. It takes into account the most significant prognostic factors. These parameters are divided into 2 groups.

1. Factors related to AF:

   (a) form of detected AF,
   (b) duration of the last paroxysm of AF,
   (c) most often used method to stop a paroxysm of AF,
   (d) radiofrequency ablation in history,
   (e) symptoms accompanying the paroxysm of AF (specific and non-specific),
   (f) incidence of AF paroxysms.

2. Parameters for HF assessed by echocardiography:

   (a) systolic dysfunction of the LV (ejection fraction),
   (b) diastolic dysfunction among for the patients with HF (transmitral flows, the ratio of the filling rates of the LV in early diastole and atrial systole),
   (c) LA size for the patients with AF and HF.

All the values of the listed parameters are evaluated in corresponding points (from 1 to 4). Patients are divided into 3 risk groups depending on the total score:

- the patient's condition is at a good compensated level, does not require a doctor's consultation, belongs to the "green zone".
- the state of health is at a satisfactory level, related to the "yellow zone", the condition may worsen in the near future, an outpatient consultation with a specialist is required to correct the treatment and carry out additional diagnostics;
- an unfavorable condition that threatens the patient's life, requiring immediate hospitalization in a hospital, related to the "red zone".

6 Results

6.1 Neural Network Design

Basing on the analysis of the task and input data it can be concluded that for further experiments it is necessary to use neural networks. Neural networks have a different architecture, type, and based on the problem it is necessary to decide what type of network to use, how many layers and how many elements in each layer. Since the number of input and output elements is determined by the properties of the input and output data, the study of the architecture of a multilayer network should begin by determining the size of one hidden layer. Too many neurons in the hidden layer will increase the network training time to an almost unrealistic value. It can also lead to overfitting of the network, which manifests itself in the fact that the network will work fine on the training set, but very badly on the input examples that are not included in it. One of these rules is the geometric pyramid rule [26]. According to this rule, the number of neurons in the hidden layer in a three-layer perceptron is calculated as the square root of the product of the number of neurons in the input layer and the number of neurons in the output layer. In our case the number of neurons in the input layer is 9, the number of neurons in the output layer is 3 and the resulting number of neurons in the hidden layer is 5.

Analyzing the input parameter 1a (see part 5), it is clear that the data are assessed on a point system. Values should be transformed using scaling and coding methods. This indicator is a qualitative unordered variable. To encode the values of such a variable, a binary vector of 4 components is used. The first category is coded as {1000}, the second, respectively – {0100}, the third – {0010} and the fourth – {0001}.

The next variable (1b) is a quality ordered variable. Therefore, it should be encoded as binary vectors with as many components as the variable has states. The states of the qualitative variable are ordered as follows: then the state corresponds to the vector {100}, the state corresponds to the vector {110}, the state corresponds to the vector {111}.

Parameters 1c-1f are qualitative unordered variables. To encode the values of such variables, a binary vector of n components is used.

Parameters of group 2 are quality ordered variables. Therefore, they should be encoded in the form of binary vectors with as many components as there are states in the variable.

After the data has been reduced to one format, a training method should be chosen. The backpropagation algorithm imposes low memory requirements and has a relatively low computational complexity. Therefore, it is decided to use a slower but less resource intensive gradient descent algorithm with the maximum number of training cycles 10000 and the hyperbolic tangent as an activation function.

6.2 Data Sets

The datasets that will be used for learning and testing the proposed neural network are formed on the base of impersonalized clinical records. 3 groups of patients were identified, depending on the stratification of the risk cardiovascular accidents (high, medium and low risk groups). Total number of patients – 147. Among them:

- High risk group – 58;
- Medium risk group – 45;
- Low risk group – 44.

120 datasets will be used for learning (40 for each risk group) and the rest for testing.

It was found that the most common risk factors among the patients with AF and HF are the following:

- the duration of paroxysm is more than 24 h;
- the frequency of occurrence of paroxysms (according to anamnesis, ECG and HM-ECG), frequent attacks that cause the appearance of the pronounced one;

- the method of stopping the paroxysm is medication or with the help of EIT.
- Echocardiographic parameters: reduced LVEF (less than 40%) and LV diastolic dysfunction of 2–3°are found, an altered LA size was more often recorded, which indicates LA dilatation and a high development of thrombosis
- All patients had an elevated atrial natriuretic peptide, which confirms the presence of the diagnosis of HF.

7 Conclusion

The authors proposed to predict the risk of cardiovascular complications in patients with AF and HF using machine learning methods. A special score of predictors is proposed and implemented as software. The authors chose the neural network architecture and prepared sets of clinical data for learning and testing the system.

The risk factors discovered and analyzed in this research allow to develop an instrument for prediction of cardiovascular complications that negatively affect the duration and quality of life of elderly patients. In this category of patients, the presence of risk factors should be taken into account in order to provide timely qualified medical care.

The application of special software and neural network looks promising for prediction of complications in patients with AF and HF. Future work includes training and testing the neural network and assessment of the accuracy of the results.

References

1. Koca, M., et al.: Impact of atrial fibrillation on frailty and functionality in older adults. Irish J. Med. Sci. (1971) **189**(3), 917–924 (2020). https://doi.org/10.1007/s11845-020-02190-x
2. Osmolovskaya Yu.F., Romanova N.V., Zhirov I.V., Tereschenko, S.N.: Epidemiology and management of heart failure patients with atrial fibrillation. Med. Council (Meditsinskiy sovet) **10**, 93–97 (2016). (in Russ.). https://doi.org/10.21518/2079-701X-2016-10-93-97
3. Ling, L.H., Kistler, P.M., Kalman, J.M., Schilling, R.J., Hunter, R.J.: Comorbidity of atrial fibrillation and heart failure. Nat. Rev. Cardiol. **13**(3), 131–147 (2016). https://doi.org/10.1038/nrcardio.2015.191
4. Bisbal, F., Baranchuk, A., Braunwald, E., Bayés de Luna, A., Bayés-Genís, A.: Atrial failure as a clinical entity: JACC Review Topic of the Week. J. Am. Coll. Cardiol. **75**(2), 222–232 (2020). https://doi.org/10.1016/j.jacc.2019.11.013. PMID: 31948652
5. Taniguchi, N., Miyasaka, Y., Suwa, Y., Harada, S., Nakai, E., Shiojima, I.: Heart failure in atrial fibrillation - an update on clinical and echocardiographic implications. Circ. J. **84**(8), 1212–1217 (2020). https://doi.org/10.1253/circj.CJ-20-0258
6. Wilson, B.D., Wasmund, S.L., Sachse, F.B., Kaur, G., Marrouche, N.F., Cannon-Albright, L.A.: Evidence for a heritable contribution to atrial fibrillation associated with fibrosis JACC: Clin. Electrophysiol. **5**(4), 493–500 (2019)
7. Lee, H.Y., Yang, P.S., Kim, T.H., et al.: Atrial fibrillation and the risk of myocardial infarction: a nation-wide propensity-matched study. Sci. Rep. **7**, 12716 (2017). https://doi.org/10.1038/s41598-017-13061-4
8. Ragbaoui Y., Chehbouni C., Hammiri A.E., Habbal, R.: Epidémiologie de l'Association fibrillation atriale et insuffisance cardiaque Epidémiologie de l'association fibrillation atriale et insuffisance cardiaque [Epidemiology of the relationship between atrial fibrillation and heart failure]. Pan Afr. Med. J. **26**, 26–116 (2017). (in French). https://doi.org/10.11604/pamj.2017.26.116.11470

9. Joundi, R.A., Cipriano, L.E., Sposato, L.A., Saposnik, G.: Ischemic stroke risk in patients with atrial fibrillation and CHA2DS2-VASc score of 1: systematic review and meta-analysis. Stroke: J. Cerebral Circulation **47**(5), 1364–1367 (2016). https://doi.org/10.1161/strokeaha.115.012609
10. Pisters, R., Lane, D.A., Nieuwlaat, R., De Vos, C.B., Crijns, H.J., Lip, G.Y.: A novel user-friendly score (HAS-BLED) to assess 1-year risk of major bleeding in patients with atrial fibrillation. Chest. **138** (5), 1093–1100 (2010). https://doi.org/10.1378/chest.10-0134
11. Fang, M.C., et al.: A new risk scheme to predict warfarin-associated hemorrhage: the ATRIA (Anticoagulation and Risk Factors in Atrial Fibrillation) study. J. Am. Coll. Cardiol. **58**(4), 395–401 (2011)
12. Serhani, M.A., El Kassabi, H., Ismail, H., Nujum Navaz, A.: ECG monitoring systems: review, architecture, processes, and key challenges. Sensors **20**(6), 1796 (2020). https://doi.org/10.3390/s20061796
13. Kuzmin, A., Safronov, M., Bodin, O., Baranov, V.: Mobile heart monitoring system prototype. In: Balandin, S., Balandina, E. (eds.) Tools and Technologies for the Development of Cyber-Physical Systems, pp. 153–175. IGI Global (2020). https://doi.org/10.4018/978-1-7998-1974-5.ch006
14. Kuzmin, A., Safronov, M., Bodin, O., Prokhorov, S., Stolbova, A.: Mobile ECG monitoring system prototype and wavelet-based arrhythmia detection. In: 21st Conference of Open Innovations Association FRUCT. pp. 210–216. FRUCT Oy, Helsinki (2017). https://doi.org/10.23919/FRUCT.2017.8250184
15. Mitrokhin, M., et al.: Investigation of models for prognosis of critical values of non-invasive electrophysiological parameters of pregnant women with abnormalities of heart rate In 21st Conference of Open Innovations Association FRUCT, pp. 238–243. FRUCT Oy, Helsinki (2017). https://doi.org/10.23919/FRUCT.2017.8250188
16. Kumari, C.U., Ankita, R., Pavani, T., Vignesh, N.A., Varma, N.T., Manzar, M.A., Reethika, A.: Heart rhythm abnormality detection and classification using machine learning technique. In 4th International Conference on Trends in Electronics and Informatics (ICOEI), pp. 580–584. IEEE (2020). https://doi.org/10.1109/ICOEI48184.2020.9142914
17. Sun, X., Park, J., Kang, K.: Arrhythmia classification using nearest neighbor search with principal component analysis. In 6th ACM Conference on Bioinformatics, Computational Biology and Health Informatics September, pp. 553–555. Association for Computing Machinery, NY (2015). https://doi.org/10.1145/2808719.2811573
18. Okada, Y., Matsuyama, T., Morita, S., et al.: Machine learning-based prediction models for accidental hypothermia patients. J. Intensive Care **9**(1), 6 (2021). https://doi.org/10.1186/s40560-021-00525-z
19. Gallard, A., et al.: Prediction of response to cardiac resynchronization therapy using a multi-feature learning method. Int. J. Cardiovasc. Imaging **37**(3), 989–998 (2020). https://doi.org/10.1007/s10554-020-02083-1
20. Gulea, C., Zakeri, R., Quint, J.K.: Model-based comorbidity clusters in patients with heart failure: association with clinical outcomes and healthcare utilization. BMC Med. **19**, 9 (2021). https://doi.org/10.1186/s12916-020-01881-7
21. Shin, S., et al.: Machine learning vs. conventional statistical models for predicting heart failure readmission and mortality. ESC Heart Failure **8**, 106–115 (2021). https://doi.org/10.1002/ehf2.13073
22. Ihsanto, E., Ramli, K., Sudiana, D., Gunawan, T.S.: An efficient algorithm for cardiac arrhythmia classification using ensemble of depthwise separable convolutional neural networks. Appl. Sci. **10**(2), 483 (2019). https://doi.org/10.3390/app10020483

23. Eltrass, A.S., Tayel, M.B., Ammar, A.I.: A new automated CNN deep learning approach for identification of ECG congestive heart failure and arrhythmia using constant-Q non-stationary Gabor transform. Biomed. Signal Process. Control **65**, 102326 (2021). https://doi.org/10.1016/j.bspc.2020.102326
24. Zhanga, T., et al.: Assessing multi-layered nonlinear characteristics of ECG/EEG signal via adaptive kernel density estimation-based hierarchical entropies. Biomedical Signal Processing and Control 65, 102520 (2021). https://doi.org/10.1016/j.bspc.2021.102520
25. Afdala, A., Nuryani, N., Nugroho, A.S.: Detection of atrial fibrillation using artifical neural network with power spectrum density of RR interval of electrocardiogram. J. Phys.: Conf. Ser. **795**, 012073 (2017). https://doi.org/10.1088/1742-6596/795/1/012073
26. Widrow, B., Lehr, M.A.: 30 years of adaptive neural networks: perceptron, madaline, and back-propagation. Proc. IEEE **78**(9), 1415–1442 (1990). https://doi.org/10.1109/5.58323

Investigation of Manifestations of Aggressive Behavior by Users of Sociocyberphysical Systems on Video

Mikhail Uzdiaev and Irina Vatamaniuk(✉)

St. Petersburg Institute for Informatics and Automation of the Russian Academy of Sciences, St. Petersburg Federal Research Center of the Russian Academy of Sciences (SPC RAS), 39, 14th Line, 199178 St. Petersburg, Russia
uzdyaev.m@iias.spb.su

Abstract. This paper discusses the problem of assessing aggressive behavior, presented on videos that are publicly available on the Internet, of users of the sociocyberphysical systems in various subject situations: a game video stream and a conversational and entertainment video stream. This work presents the technique of preliminary selection of video samples containing manifestations of aggression in various situations. This technique is based on the naturalistic behavior observation approach. It allows to evaluate the presence of aggressive behavior in the video and to identify the type of aggression, as well as additional behavioral, emotional and other manifestations that accompany the aggressive behavior of users. An assessment based on the analysis of contingency tables of the difference in the types of aggression shown in the situations under consideration is made. The assessment of the difference between the accompanying emotional and behavioral manifestations in different types of aggression is performed. The differences in specific aggressive actions for the considered object situations are also highlighted. The results obtained make it possible to clarify the criteria for assessing aggressive behavior on video for various subject situations, which, in turn, makes it possible to clarify the methodology for marking up video datasets for machine learning systems.

Keywords: External observation · Analysis of behavior · Aggressive behavior · Analysis of contingency tables

1 Introduction

In recent years, sociocyberphysical systems, the so-called "smart" environments, and the Internet of Things systems, have become increasingly widespread. Along with this, the number of users of such systems is also growing. However, with an increase in the number of users, the risk of conflict situations increases, which are often accompanied by destructive manifestations such as aggression. In addition, recently there have been several high-profile fatal events [1] and associated with the so-called trash streams. During these streams some participants showed acts of violence against others. These

R. Silhavy et al. (Eds.): CoMeSySo 2021, LNNS 231, pp. 593–604, 2021.
https://doi.org/10.1007/978-3-030-90321-3_49

were verbal insults, acts of physical aggression (hitting, pushing, slapping, throwing objects), other humiliation and bullying. Despite the fact that the very phenomenon of trash streams requires a deep and comprehensive study by various scientific and philosophical disciplines, it is already becoming urgent to identify such behavior in order to quickly respond to it and prevent negative consequences.

Let us first consider the concept of aggression, which we will adhere to in the future. According to [2], aggression is a purposeful destructive and offensive behavior that violates the norms and rules of coexistence of people in society, damaging the objects of attack (animate and inanimate), causing damage or causing them psychological discomfort: negative experiences, states of tension, fear, depression, etc. Based on this definition, it is possible to formulate an initial idea that can be used as the basis for various methods for assessing aggressive behavior.

In modern scientific psychology, there are many different classifications of aggression, based on various criteria for the manifestation of human aggression. One of the most complete classifications was given by A. Buss [3]. He identified three parameters by which the forms of aggressive behavior are characterized: physical - verbal, active - passive, direct - indirect. The combination of these parameters gives eight forms of aggressive behavior, which are described in detail in [4, 5].

Also, the modifications of this classification are presented in many methods for assessing aggressive behavior and aggressive personality traits [6]. In this study, A. Bass's classification was used to determine the preliminary groups into which one or another manifestation of aggression identified in the video can be combined. This grouping makes it possible to significantly facilitate the process of processing empirical material by the fact that during the initial assessment it is quite difficult to group a large number of primary private categories into a small number of general groups that have a common leading feature. In this study, the following manifestations of aggression were selected as the initial categories: physical aggression directed at inanimate objects, physical aggression directed at animate objects, and verbal aggression directed at animate objects.

It is noted that some types of aggression are accompanied by strong emotional manifestations. Some researchers even single out emotional or affective aggression in a separate category [7–9]. Therefore, an important aspect of the study is also the identification of those emotions that are manifested during acts of aggression. In addition, it is advisable to determine the activity accompanying aggressive acts, which is not directly related to offensive destructive actions. Examples of such activity are: general motor exitation, which manifests itself in pronounced movements of the body, head; gestures, manifested in hand movements, including various gestures, including obscene ones; the severity of facial expressions; features of non-verbal speech behavior: the pitch and loudness of the voice. In addition, it is important to identify the features of the aggressive actions themselves: what type of aggression is manifested in a given situation, the features of verbal aggression (the nature of statements, the presence of obscene vocabulary, etc.); features of physical aggression (what actions are most often performed by the aggressor); the orientation of aggression in various situations (at whom or what the aggression is directed at in a particular objective situation), etc.

It is especially worth emphasizing that various subject situations significantly affect the ways people behave in these situations. Aggression, accordingly, in different subject situations should manifest itself in different ways. Within the framework of this study, the manifestation of aggression in two common subject situations is considered - game streams and conversational and entertainment streams.

2 Approaches to Assessing Aggressive Human Behavior

There are various books and practical guides for psychologists, psychiatrists, and social workers [10–12]. Mostly, they are oriented at practical application during work with patients with different disorders or people under specific conditions, such as prisoners in jail, children in orphanages, etc. These techniques are mostly presented in the form of questionnaires and imply a subjective assessment of the situation either by a specialist or by the person him/herself. It is a challenging task to characterize and measure aggressiveness due to its overly complex nature, variative definitions, misinterpretations and lack of conceptual clarity [13]. Aggressive behavior has traditionally been classified into two subtypes: impulsive or premeditated. According to this division, a variety of Impulsive/Premeditated aggression scales (IPAS) have been developed. The most known are the following ones: Buss-Perry Aggression Questionnaire (BPAQ) [14], State-Trait Anger Expression Inventory (STAXI) [15], Lifetime History of Aggression (LHA) [16], Aggression Interview [17]. Various modifications of these scales formed the basis of the modern common and specialized approaches, such as [18–21]. Yet, the task of detecting aggressive behavior in sociocyberphysical systems is still unsolved and requires developing a methodology which would provide a clear and unambiguous way to automatically mark up the large amount of multimodal data [22–24].

3 Methodology for External Naturalistic Observation of Behavior

In this study, we use a method of external naturalistic behavior observation of people in various situations. The situations under investigation were presented in open-access video on the Internet. This completely excludes the influence of researcher on the situation, which ensures the naturalness of behavioral manifestations of individuals. Another obvious advantage of this method is that there is no need to obtain consent to participate in the experiment, since the recordings are access-free in the public domain on public hosting platforms such as YouTube. External naturalistic behavior observation is objective behavior assessment method due to the fact that data on motor reactions, verbal and non-verbal behavior are recorded from the outside and do not depend on the personal characteristics of the subjects. In addition, external observation can be used to assess the behavior presented in the recording, for example, in audio and/or video. The naturalism of observation is provided by more or less free recording of the behavioral manifestations of individuals. However, naturalistic observation implies recording as complete a list of all possible manifestations in a given situation as possible. The analysis of the observation results will make it possible to determine the most common aggressive actions, as well as accompanying manifestations, such as motor excitation, facial expressions, emotions, non-verbal speech behavior in various situations, etc. As a result, this will make it possible to refine and customize the process of data markup for the corresponding machine learning systems.

3.1 Preliminary Search and Selection of Samples Containing Manifestations of Aggression

This study examined two situations that arise in sociocyberphysical systems, in which aggression is often manifested. These are game streams and conversational and entertainment streams. A game stream is characterized by presence in the frame of one or more players of a computer game. Usually, the player sits in front of the webcam in such a way that his head and/or body is visible down to the waist. The frame usually contains furniture, dishes, computer components, such as microphone, headphones, keyboard, mouse, etc. A conversation and entertainment stream, on the other hand, has more variability in the interior, the number of participants and the situations that may occur there.

As part of the study, a dataset was collected containing video samples that show manifestations of aggression in the settings described above. The dataset was collected on the video hosting service youtube.com by the following queries in the search engine of the service (in Russian): "жесть на стриме" ("trash stream"), "драка на стриме" ("fight on stream"), "агрессия на стриме" ("aggression on stream"), "троллинг на стриме" ("trolling on stream"), "оскорбления на стриме" ("insults on the stream","ругань на стриме" ("swear at stream"), as well as requests containing obscene expressions. A total of 200 videos were selected at random, which allegedly contained violent actions. The average duration of video sample was 25.3 s. For objective reasons, it was impossible to determine the exact age of the people in the video. The video from the dataset contains a total of 80 people, of which 69 are men and 11 are women.

3.2 Marking up Procedure

The marking up of a pre-selected dataset to determine aggressive behavior was carried out by 11 annotators aged 21–29 years, 6 of which were men, and 5 were women.

To perform the markup, instructions were drawn up to be followed by the annotators. To begin with, in order to identify the primary framework for the assessment, the annotators were explained in detail the definition of aggression given by S.N. Enikolopov mentioned above. At the same time, such aspects as the purposeful nature of actions, the presence of the object of the attack, or the fact that this object is meant, were especially emphasized. Then the annotators had to watch each video and rate it according to the selected annotation categories. In this study, the following categories of annotation were identified: whether there is aggression in the video being watched; free-form abstract verbal description that reflects what is happening on the video; who or what is the aggression or object of aggression directed at; situation. Table 1 shows the observation protocol and an example of its filling, which was compiled on the basis of the selected assessment categories.

The annotator first evaluates whether the video contains violent behavior. This must be done because the original sample contained a sufficient number of videos that did not show manifestations of aggressive behavior, but, for example, manifestations of anger and strong motor arousal, etc. Further, the annotator briefly describes the situation in free form, highlighting the main actions and events that are present in the viewed video fragment. The lack of a clear formalization of the description is due to the peculiarities

Table 1. Example of observation protocol.

No.	Presence of aggression	Verbal description of behavior (actions) if there is aggression	Who or what is the aggression directed at	Setting	Number of people in the frame
0	+	Anger; targeted swearing; hysterics; hits the chair;	Computer chair	Game stream	1
1	+	Anger; targeted swearing; Hysterics; heart-rending screams; hit the table	Table	Game stream	1

of search observation, which is necessary for the primary collection of information. In the course of further processing, this information will highlight the most significant aggressive actions, emotional states, motor excitation, the nature of non-verbal speech behavior, facial expressions, gestures, and the nature of verbal speech behavior. Next, the annotator must highlight the object of aggression or the victim to which the aggressive actions are directed. At the end, the annotator determines the setting: a game stream or a conversational/entertaining stream and indicates the number of people present in the frame.

As a result of annotations from the whole variety of actions described by annotators in free form, the following groups were distinguished:

1. Types of displayed aggression, which included physical aggression directed at other people, physical aggression directed at inanimate objects, verbal aggression;
2. Emotional and behavioral manifestations accompanying acts of aggression, which include pronounced non-verbal speech activity (loudness, voice pitch), manifestation of strong negative emotions, pronounced body motor excitation, pronounced facial expressions;
3. Methods of manifestation of physical aggression directed at other people, which include punches, kicks, grabs, pushing with hands, knocking down on the floor, hitting with objects, strangulation, and other manifestations;
4. Methods of manifestation of physical aggression directed at inanimate objects, which include hitting objects with hands, throwing objects, biting objects, kicking objects, breaking objects and other acts;
5. Ways of manifestation of verbal aggression, which include obscene insults, non-obscene insults, insults based on ethnicity, rude instruction to go away, orders to shut up, accusations, threats, and other manifestations;

All these selected categories in the course of processing and analysis of the assessment results formed the basis of categorical variables for which it is supposed to evaluate the differences indicated above in the research tasks.

Then the videos with the most confidence of aggression presence were selected. This was done by the overwhelming majority of annotators' votes: the video is considered as aggressive if 9 of 11 annotators (81.8%) marked it up as an aggressive one. As a result, 132 videos were selected out of the initial 200 videos, on the basis of the analysis of which all the interrelationships of interest to us were identified in the future.

3.3 Statistical Processing

In this work, the study of the relationship of certain actions, manifested emotions, facial expressions, body motor excitation with a particular situation is carried out. Thus, we are examining categorical variables. The analysis of contingency tables was chosen as such a method for processing categorical variables. To identify differences in the groups of the studied variables, the χ^2 Pearson was used, on the basis of which the statistical significance of the result was calculated.

4 Experiment Results

4.1 Assessment of the Differences in the Types of Aggression in Various Subject Situations

Let us now designate the types of the studied groups of variables. First, the aim of the study is to identify the differences between the types of aggression shown in different situations (Table 2). The rows of the table indicate situations - a game stream and a conversational/entertaining stream. The following variables are considered as columns: physical aggression directed at other people, physical aggression directed at inanimate objects, verbal aggression.

Table 2. Contingency table between object situations and types of aggression.

	Physical aggression against other people	Physical aggression against inanimate objects	Verbal aggression
Game stream	9	21	41
Conversational/entertaining Stream	71	2	45

The analysis of the contingency table showed a significant difference in the manifestations of the types of aggression ($\chi^2 = 55.69$, $df = 2$, $p < 0.001$). Let us now consider the percentage of each type of aggression in the two considered object situations. Figure 1 shows diagrams illustrating the percentage of aggression types for game streams (a) and conversational and entertaining (b).

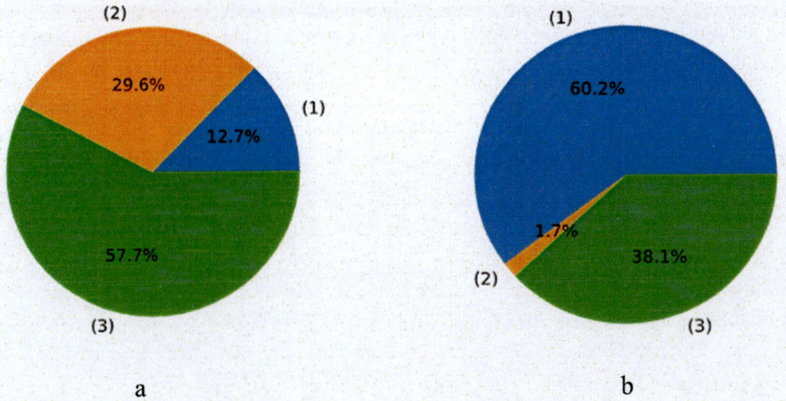

Fig. 1. Percentage ratio of types of aggression for (a) gaming streams, (b) conversational and entertaining streams, where (1) physical aggression directed at people, (2) physical aggression directed at objects, (3) verbal aggression.

Diagram (a) shows that the most common type of aggression is verbal aggression (3), and in second place is physical aggression directed at inanimate objects (2). This ratio can be described by the fact that in the case when one person is in the frame, he cannot directly physically attack another and takes out aggression on virial interlocutors and/or inanimate objects located nearby. Physical aggression directed at other people (1) refers to situations where there is more than one person in the frame. In the case of physical aggression (Fig. 1b), physical aggression directed at people prevails (1). This ratio can be described by the fact that in most videos of a conversational and entertainment nature, more than one person is present in the frame. It is also worth noting a relatively small proportion of physical aggression directed at inanimate objects (2).

4.2 Assessment of the Difference in Accompanying Emotional and Behavioral Manifestations in Various Types of Aggression

Let's designate the types of the studied groups of variables. It is necessary to identify differences in non-verbal speech activity, manifestations of negative emotions, motor excitation, in facial expressions (columns of Table 3) for various types of aggression - physical, aimed at other people, physical, aimed at inanimate objects, verbal aggression (rows of Table 3). Analysis of the contingency table showed a significant difference in the manifestations of types of aggression ($\chi^2 = 19.77$, df = 6, $p < 0.005$).

Let us now consider the percentage of accompanying emotional and behavioral manifestations for each type of aggression. Figure 2 presents a pie chart showing the distribution of associated emotional and behavioral manifestations for physical aggression directed at other people (a), a pie chart showing the distribution of associated emotional and behavioral manifestations for physical aggression directed at inanimate objects (b), circular diagram showing the distribution of accompanying emotional and behavioral manifestations for verbal aggression (c).

Table 3. Table of contingency between types of aggression and accompanying emotional and behavioral manifestations.

	Non-verbal speech activity	Manifestation of negative emotions	Motor excitement	Pronounced facial expressions
Physical aggression against other people	44	56	80	16
Physical aggression against inanimate objects	16	23	23	20
Verbal aggression	78	86	81	44

Analysis of the contingency table showed a significant difference in the manifestations of types of aggression ($\chi^2 = 19.77$, df = 6, $p < 0.005$). Let us now consider the percentage of accompanying emotional and behavioral manifestations for each type of aggression. Figure 2 presents a pie chart showing the distribution of associated emotional and behavioral manifestations for physical aggression directed at other people (a), a pie chart showing the distribution of associated emotional and behavioral manifestations for physical aggression directed at inanimate objects (b), a pie chart showing the distribution of accompanying emotional and behavioral manifestations for verbal aggression (c).

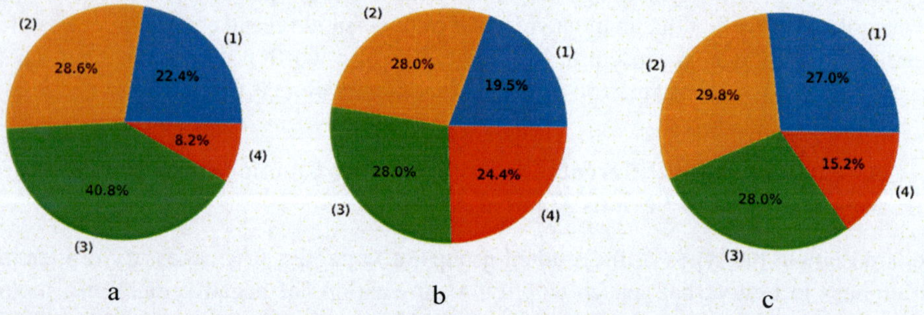

Fig. 2. Percentage of: (a) accompanying emotional and behavioral manifestations for physical aggression directed at other people; (b) accompanying emotional and behavioral manifestations for physical aggression directed at inanimate objects; (c) accompanying emotional and behavioral manifestations for verbal aggression, where (1) is pronounced non-verbal speech behavior; (2) is manifestation of negative emotions; (3) is pronounced motor excitation; and (4) is pronounced facial expressions.

It can be seen from the diagrams that physical aggression (Fig. 2, a) is most often accompanied by pronounced motor excitation (3), while pronounced mimic manifestations are most rare (4). The accompanying emotional (2) and behavioral (3) manifestations are distributed more evenly in the case of physical aggression directed at inanimate objects (Fig. 2, b). In the case of verbal aggression (Fig. 2, c), manifestations of negative emotions (2), pronounced motor excitation (3) and pronounced non-verbal speech behavior (1) have approximately the same frequency of occurrence, while pronounced facial expressions (4) are significantly inferior to them.

4.3 Assessment of the Differences in Specific Aggressive Actions Specific for a Particular Type of Aggression

The studying of the specific acts of the considered types of aggression (physical, directed at other people, physical, directed at inanimate objects, verbal) between game and conversational-entertainment streams using the analysis of contingency tables, shows no significant ($p < 0.05$) difference between the considered situations. However, it will be important to give the distribution of these acts of each type of aggression in the total sample.

In total, 150 different actions attributed to verbal aggression were identified in a selected 132 videos. These include obscene insults (1), not abusive insults (2), rude instruction to go away (3), accusations (4), threats (5), other manifestations (6), which included those manifestations whose share did not exceed 3%. These include wishes for death, humiliation, encouragement of violence, insults based on ethnicity, orders to shut up and other orders. The ratio of their number is shown in the diagram in Fig. 3(a). The analysis of physical aggression directed at other people reveals a total of 155 different actions. These include punches (1), kicks (2), grabs with hands (3), pushing with hands (4), falling to the floor (5), strikes with objects (6), strangulation (7), other manifestations (8), which include dousing with liquid, jumping on a person, leg sweeps, threats with a weapon, twisting the nose, pulling the hair, throwing objects at people, and the share of which did not exceed 3% in the total sample. The ratio of the number of aggressive actions attributed to physical aggression directed at other people is reflected in the diagram in Fig. 3(b). In the case of physical aggression directed at inanimate objects, a total of 41 actions were identified, each of which can be attributed to the following categories: hitting objects with hands (1), throwing objects (2), biting objects (3), hitting objects with feet (4), breakage of objects (5) and other acts (6), which include striking objects with other objects and jumping onto objects. The ratio of the number of aggressive actions attributed to physical aggression directed at inanimate objects is reflected in the diagram in Fig. 3(c).

The most common manifestation of verbal aggression (Fig. 3a) is obscene insults (1). Threats (5) and rude instruction to go out (3) are significantly inferior to abusive insults. The portion of each of the other manifestations does not exceed 10%. The most common manifestations of physical aggression (Fig. 3b) directed at other people are punches (1), kicks (2), hand grabs (3). The most common manifestations of verbal aggression (Fig. 3c) are hitting objects with hands (1), throwing objects (2), biting objects (3). The portion of each of the other manifestations does not exceed 10%.

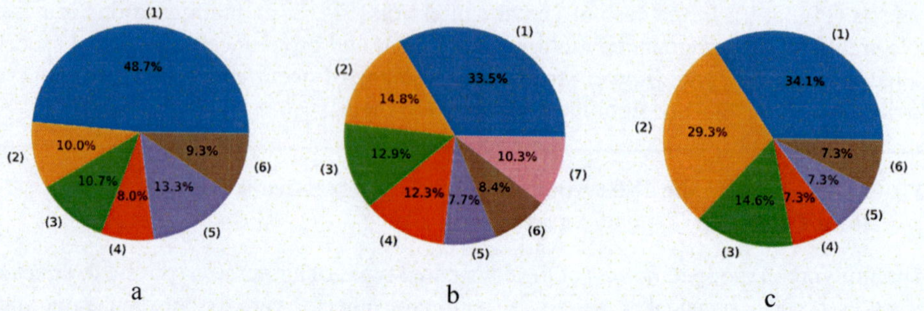

Fig. 3. (a) Percentage of different manifestations of verbal aggression; (b) percentage of different manifestations of physical aggression directed at other people. (c) percentage of different manifestations of physical aggression directed at inanimate objects

5 Discussion

Using the described methodology, a preliminary selection and assessment of the behavior of information space users presented on the video in various subject situations (game and conversational-entertainment streams) was carried out for the presence of aggressive behavior there. All actions, types of aggression, as well as emotional and behavioral manifestations accompanying aggressive behavior were also identified. The obtained material was further grouped into more general semantic groups, which form appropriate categorical variables. According to these variables, the following investigations was carried out: assessment of the difference in the types of aggression shown in the designated object situations; assessment of the difference in the types of aggression shown and the accompanying additional emotional and behavioral manifestations; the difference in specific aggressive actions inherent in various types of aggression in the considered object situations.

Evaluation using contingency tables showed a significant difference for all but the last task. This fact of difference can be used to concretize the video preselection methodology, as well as to develop and test more detailed criteria for the selection of videos containing aggressive behavior. Despite the fact that in various subject situations there were no significant differences in aggressive actions in different types of aggression, identifying the list of these actions and their percentage can also be useful from the point of view of specifying the assessment methodology.

Based on the results of the study, the following specific steps are proposed to improve the methodology. In the observation protocol, the column "Verbal description of behavior (actions), if there is aggression" (Table 1) should be replaced by the categories identified during the study: type of aggression with options, physical aggression directed at other people, physical aggression directed at inanimate objects, verbal aggression; accompanying emotional and behavioral manifestations with options for non-verbal speech activity, manifestation of negative emotions, motor excitation, pronounced facial expressions; actions characteristic of physical aggression directed at other people; actions characteristic of physical aggression directed at inanimate objects; actions characteristic of verbal aggression.

6 Conclusion

The method of preliminary video selection considered in this article followed by a qualitative assessment of the aggressive behavior of users of the sociocyberphysical space allows, firstly, to clarify the characteristic composition of the aggressive behavior presented on the video in various situations: accompanying verbal and non-verbal speech behavior, body motor activity, facial expressions, emotional state, specific acts characteristic of verbal and physical aggression. Secondly, the analysis of the selected videos showed a difference in the characteristic composition for different types of aggression and different situations. All this makes it possible to concretize the protocol for selecting and evaluating videos containing aggressive behavior by introducing specific evaluation criteria into it, as well as specifying these criteria for various object situations and types of aggression shown. In addition, the given methodology can be used for preliminary selection of data for deeper analysis using numerical scales and valid testing methods for assessing behavior.

In the future, it is planned to test this technique in the problem of marking up video datasets containing manifestations of aggression by users, and training machine learning systems that perform single- and multimodal recognition of aggression on the marked data. In addition, the methodology described in the article has prospects of application for assessing other types of behavior presented in the video, for example, the behavior of participants in negotiations, interviews, lectures and lessons, taking place via video conferencing, etc.

Acknowledgements. This research is supported by RFBR Project No.18–29-22061_мк.

References

1. https://www.bbc.com/russian/features-55238100
2. Enikolopov, S.N.: The concept of aggression in modern psychology. Appl. Psychol. **1**, 60–72 (2001)
3. Buss, A.H.: The psychology of aggression. Wiley (1961)
4. Buss, A.H.: The psychology of aggression. - Wiley, 1961.; [Enikolopov, S.N.: The concept of aggression in modern psychology. Appl. Psychol. **1**, 60–72 (2001)
5. Ilyin, E.P.: Psychology of aggressive behavior. SPb.: Peter (2014). T. 368
6. Suris, A., et al.: Measures of aggressive behavior: overview of clinical and research instruments. Aggression Violent Behav. **9**(2), 165–227 (2004)
7. Berkowitz, L.: Affective aggression: the role of stress, pain, and negative affect. In: Human Aggression, pp. 49–72. Academic Press (1998)
8. Geen, R.G.: Processes and personal variables in affective aggression. In: Human Aggression, pp. 1–21. Academic Press (1998)
9. Lindsay, J.J., Anderson, C.A.: From antecedent conditions to violent actions: a general affective aggression model. Personality Soc. Psychol. Bull. **26**(5), 533–547 (2000)
10. Goldstein, A.P., Keller, H.: Aggressive behavior: Assessment and intervention. Pergamon Press (1987)
11. Wehby, J.H.: Issues in the assessment of aggressive behavior. Preventing School Failure: Altern. Educ. Child. Youth **38**(3), 24–28 (1994)

12. Stanford, M.S., Houston, R.J., Mathias, C.W., Villemarette-Pittman, N.R., Helfritz, L.E., Conklin, S.M.: Characterizing aggressive behavior. Assessment **10**(2), 183–190 (2003)
13. Parrott, D.J., Giancola, P.R.: Addressing "The criterion problem" in the assessment of aggressive behavior: development of a new taxonomic system. Aggression Violent Behav. **12**(3), 280–299 (2007)
14. Buss, A.H., Perry, M.: The aggression questionnaire. J. Pers. Soc. Psychol. **63**, 452–459 (1992)
15. Spielberger, C.D.: State-Trait Anger Expression Inventory manual. Odessa, FL: Psychological Assessment Resources (1996)
16. Coccaro, E.F., Berman, M.E., Kavoussi, R.J.: Assessment of life history of aggression: development and psychometric characteristics. Psychiatry Res. **73**, 147–157 (1997)
17. Stanford, M.S., Barratt, E.S.: Procedures for the Classification of Aggressive/Violent Acts, Manual. University of New Orleans, New Orleans (2001)
18. Walters, A.S.: Using safety planning for aggressive behavior in youth: applications to intensive treatment settings. Brown Univ. Child Adolescent Behav. Lett. **37**(1), 8 (2021)
19. Gaynes, B.N., et al.: Preventing and de-escalating aggressive behavior among adult psychiatric patients: a systematic review of the evidence. Psychiatric Serv. **68**(8), 819–831 (2017)
20. Kim, S.C., Berry, B., Young, L.: Aggressive behaviour risk assessment tool for long-term care (ABRAT-L): Validation study. Geriatr. Nurs. **40**(3), 284–289 (2019)
21. Tuente, S.K., Bogaerts, S., Veling, W.: Mapping aggressive behavior of forensic psychiatric inpatients with self-report and structured staff-monitoring. Psychiatry Res. **301**, 113983 (2021)
22. Saveliev, A., Uzdiaev, M., Malov, D.: Aggressive action recognition using 3D CNN architectures. In: 2019 12th International Conference on Developments in eSystems Engineering (DeSE), pp. 890–895. IEEE (2019). https://doi.org/10.1109/DeSE.2019.00165;
23. Uzdiaev, M.: Methods of multimodal data fusion and forming latent representation in the human aggression recognition task. In: 2020 IEEE 10th International Conference on Intelligent Systems (IS), pp. 399–403. IEEE (2020). https://doi.org/10.1109/IS48319.2020.9200112;
24. Mescheriakov, R.V., Iskhakov. A.Y., Evsyutin. O.O.: Analysis of modern methods to ensure data integrity in cyber-physical system management protocols. Inf. Automation **19**(5), 1089–1122 (2020)

Social Media Communities Topic Modeling

Andrey M. Fedorov, Igor O. Datyev(✉), and Andrey L. Shchur

Institute for Informatics and Mathematical Modeling, Kola Science Centre of the Russian Academy of Sciences, 14, Fersman St., Apatity 184209, Russia
{fedorov,datyev,shchur}@iimm.ru

Abstract. Social media is a source of native information generated by user communities. Processing a large amount of text and the related information posted in social media is an urgent task for developers of modern intelligent information systems. Such systems are necessary for information support for decision-making in regional management. Topic modeling is a method of constructing text document collection models, it is used in solving problems of exploratory search, classification and annotation of document collections and news streams. The paper explores the application of different topic modeling methods on social media texts. The specifics of text content are discussed, as well as the methods of assessing the applicability of topic modeling methods in the aspect of social media communities modeling. For different topic modeling methods, the experiment results on calculating UMass metrics with a varying number of topics on datasets created by the authors are presented. Conclusions are made about the preference of the BigARTM library for further work on modeling social media communities.

Keywords: Social media · Topic modeling · Assessment of the various methods' applicability

1 Introduction

Social media is an information technology for the creation and exchange of user information (ideas, opinions, etc.) through online communities. Social media is the pulse of the modern world, a layer of the most relevant operational information. This information is used for standard business tasks related to maintaining brand reputation, increasing sales or providing services, primarily to increase the profit of commercial companies.

In addition to commercial use, there are a number of problems that can be solved using social media. For example, linguists and sociologists pay close attention to the analysis of social networks and the identification of its communities' topic structure, since this opens up wide opportunities for assessing the dynamics of the natural language vocabulary, the evolution of intra-linguistic connections [1, 2]. Today, online social networking technologies find more and more use in state and regional government. One of the most important tasks here is to provide decision-makers with up-to-date information on topics discussed in social media. Social Networking Services (SNS) provide information about almost every settlement updated close to real time, which was unavailable in the previous "non-digital" or "analog" world.

R. Silhavy et al. (Eds.): CoMeSySo 2021, LNNS 231, pp. 605–614, 2021.
https://doi.org/10.1007/978-3-030-90321-3_50

Processing a large array of text data is impossible without automation tools. To date, topic modeling is used to identify the main topics from the text corpora. Topic modeling shows good results on large text collections of coherent texts, one of the best examples of which is the collection of scientific articles. Specific features of social media content imply the use of short messages, emojis and other special characters, multimedia, all of which complicates the use of topic modeling methods without adaptation to these features.

The article continues to develop the concept of a framework for social media analysis proposed by the authors earlier [3]. The main purpose of this study is to assess the applicability of various topic modeling methods to the textual content of social media communities.

2 A Brief Review of Topic Modeling Methods

The basic methods of topic modeling are Latent Semantic Analysis (LSA) [4], Probabilistic Latent Semantic Analysis (PLSA) [5] and Latent Dirichlet Allocation (LDA) [6]. It is also necessary to mention Non-negative Matrix Factorization (NMF) [7, 8], Hierarchical Dirichlet Language Model (HDLM) [9], Bigram model [10], Hierarchical Dirichlet Processes (HDP) [11], Correlated Topic Model (CTM) [12], and Pachinko Allocation model (PAM) [13].

The topic models that consider evolution of topics over time: the Dynamic topic model (DTM) [14], Continuous Time Bayesian Network (CTBN) [15], Dynamic Bayesian networks (DBN) models Topics Over Time (TOT) [16], framework for the discovery of the topical content of a data corpus, and the tracking of its complex structural changes across the temporal dimension (Bayesian mixture modeling, LDA, and HDP) [17], Evolutionary Hierarchical Dirichlet Processes (EvoHDP) [18], Encoder-only Transformer Language Model (ETLM) and weighted Conditional random field regularized Correlated Topic Model (CCTM) [19], TopicFlow [20], Probabilistic Model of Narratives Over Topical Trends in Social Media [21] and others.

One of the distinguishing features of social media texts is the short length. The reduction in effectiveness of short text topic modeling is the result of the very limited set of co-occurrences between words compared to other more traditional document sources. To deal with this issue several approaches have been proposed [22]: directly considering word co-occurrences instead of modeling document identities (biterm topic model - BTM, bursty BTM) [23, 24], considering each short document as arising from one latent topic (mixture of unigrams model - MU, Dirichlet multinomial mixture with collapsed Gibbs sampling (GSDMM), Twitter-LDA) [6, 25–28], and considering heuristic ties between documents in order to string them into larger "pseudo-documents" (Self-Aggregation based topic model (SATM), Pseudo-document-based topic model (PTM)) [29, 30].

It's also worth to mention a separate class of modern methods based on neural networks: Word2Vec, TOP2Vec, Bidirectional Encoder Representations from Transformers (BERT), Generative Pre-trained Transformer 3 (GPT-3) and many others. However, due to the structural and volumetric complexity of neural network methods, the results obtained by these methods practically deny any interpretation.

Also, special attention should be paid to the Additive Regularization for Topic Models (ARTM) framework which enables easy combination of different types of models (multimodal, multilingual, temporal, hierarchical, graph-based, and short-text) to create new models with the desired properties for applications. This modular ARTM framework for topic modeling is implemented in the BigARTM open-source library [31].

3 The Features of Topic Modeling Methods Evaluation in Social Media Dataset

Most research on topic modeling is done on perfect datasets, where data is specially preprocessed and packed in a repository. In real social media texts, everything is different: there are grammar errors, missing punctuation marks, a huge number of rare and unique words, abundance of special characters (hashtags, emojis, etc.) and meta information, multilingualism, an inconvenient way of storing data (different or rare formats, the need for parsing).

But even the problem of having a perfect dataset fades when it comes to comparing topic modeling methods. In this work, we intended to evaluate the methods of topic modeling in conditions close to solving real problems using social media texts. However, we discovered that even with a perfect dataset, it's not that simple to compare the results of different topic modeling techniques.

Of course, it is possible to single out general, so-called qualitative characteristics, by which one can judge the applicability of existing information systems in solving certain problems even before their actual use. The set of characteristics of an ideal topic modeling system is as follows:

- Dynamic: identifies the history of the topics evolution
- Hierarchical: identifies hierarchical relationships between topics
- Interpretable: how every topic is understandable to humans
- Multigram: terms and phrases are inseparable
- Multimodal: authors, links, tags, users
- Multilingual: for cross- and multilingual search (a special case of a multimodal system)
- Segmenting: identifies topics within a document
- Learning: according to assessors' evaluation and user logs
- Determines the number of topics automatically
- Identifies and naming new topics automatically
- Online: processes a collection in one pass
- Parallel, distributed for large collections.

However, practically no topic modeling system possesses all of the aforementioned characteristics simultaneously, and often it's not even required to solve a specific task. Therefore, we are forced to formulate a class of planned tasks: building a model of a social media community, including the possibility of identifying its text topics, and comparing the communities with each other.

Considering this clarification of the class of tasks and the aforementioned specifics of social media content, multimodality seems to be the most preferential criteria, as it will allow taking into account all the parameters of the community's model.

Now let's move on to the complexity of assessing the applicability of topic modeling methods for social media communities, which is directly related to the result of modeling. The main reason for this is the lack of objective universal external metrics that are suitable for all methods. In other words, up to date the comparative (and not only that) quality of models is determined mostly with the help of experts with the main indicator being semantic interpretability, assessed using the intrusion method. The essence of the intrusion method is in addition of extra words to the topic for further detection by an expert. The more additional words the expert has found among all the topics, the higher is the quality of the model.

However, it is necessary to take into account the fact that today topic modeling is no longer only a tool for identifying semantically meaningful topics, but a way to build models in a broader sense - a data embedding and clustering technique, where a dictionary can contain not only words of a natural language, but also special objects (hashtags, emojis, attached files data, like file format, author, etc.) and even actions - used for example, in hypergraph topic models. In such models, the method of determining the quality of models with the involvement of experts to assess the semantic interpretability of topics becomes inapplicable.

There are various intrinsic metrics of the topic models quality such as matrix sparseness, perplexity and others, including those listed in [32]. But they are often incomparable with each other even within the same method (for example, perplexity), not to mention comparing them for topic models obtained by different methods. Among the known automatically calculated, the PMI metric (also called UCI) [32] and the UMass metric [33] has the largest correlation with the human judgements of topic quality. At the current stage of the research, we chose the UMass metric for the experiments, because UMass does not require an external dataset to calculate, unlike PMI.

We should also mention quantitative characteristics that claim to be universal, like time of model execution (which is not always critical) and the number of topics, which can be attributed to initial conditions, with the exception of a small number of methods that determine the number of topics automatically.

Thus, the issue of finding universal automatically calculated quality metrics for topic models created by different methods suitable for comparison with each other is a very acute issue.

4 Experimental Results and Analysis

We have chosen several topic modeling methods. The first of them, Additive Regularization for Topic Models (ARTM), is multimodal for the perspective of taking into account various non-textual parameters when modeling communities of online social networks (BigARTM open-source library) [31]. We used ARTM without the regularization that corresponds to the PLSA model.

The second method, LDA, is the most common method (implemented in the GenSim open-source library). The third method is GSDMM, because the authors of [26] announced the best results specifically for short texts.

It should be noted that source code of all the selected methods is available in the Python programming language and posted on the github site, which simplifies the embedding of these methods into the framework we are developing [3].

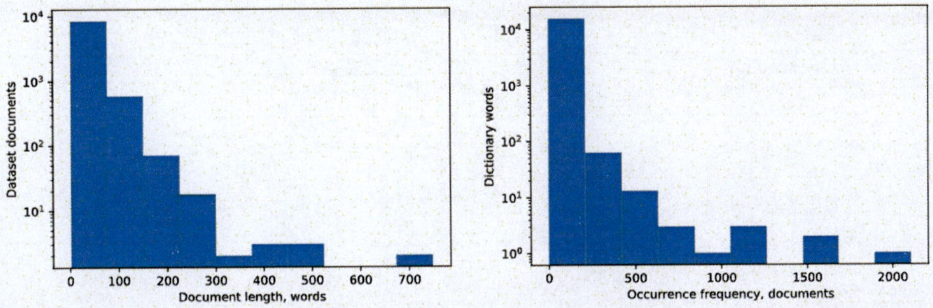

(a)

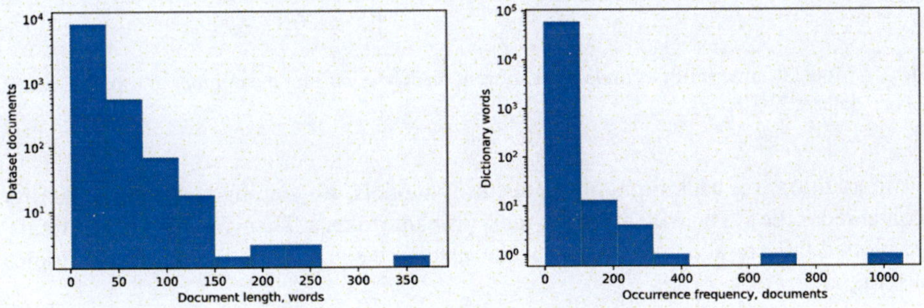

(b)

Fig. 1. The number of documents depending on the length of documents (measured in words), the number of dictionary words depending on the word occurrence frequency (measured in documents quantity) in the text corpus. Dataset 1 (unigram) - (a), dataset 2 (bigram) - (b).

For the experiments, we used a dataset containing all the posts of three local online communities (groups) of the VKontakte social network for 6 months. The dataset was preprocessed by removing special characters and stop words, and further lemmatization. The result was a dataset containing unigrams. In addition, we decided to use bigrams to improve the semantic interpretability of topics. For this, based on the first dataset, we created a second dataset of bigrams, using syntactic templates: "verb+noun", "verb+inf", "verb+adverb", "adverb+verb", "noun+noun", "noun+adverb", "noun+infinitive", "adverb+adjective", "adverb+adverb", "adjective+noun". Thus, in the experiments, two datasets were used, the properties of which are presented in Figs. 1(a) and 1(b), respectively.

Each message (post) of the online community in datasets is presented as a separate document.

For two datasets, we conducted experiments to calculate the model fit time of different methods depending on the number of topics (see Fig. 2).

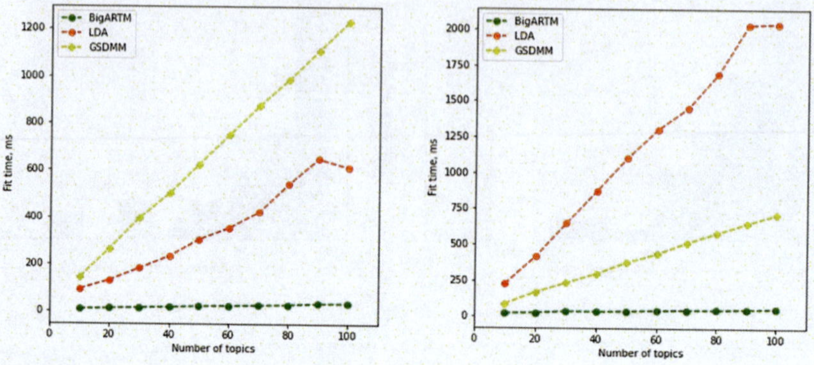

(a) (b)

Fig. 2. Model fit time (milliseconds) for different methods, varying the number of topics. Dataset 1 - (a), dataset 2 - (b).

In addition, for each topic in the resulting models, we summarized the probability (contained in the phi matrix) of the ten most probable tokens. Then, for different methods, we plotted the average and median of these sums depending on the number of topics (see Fig. 3).

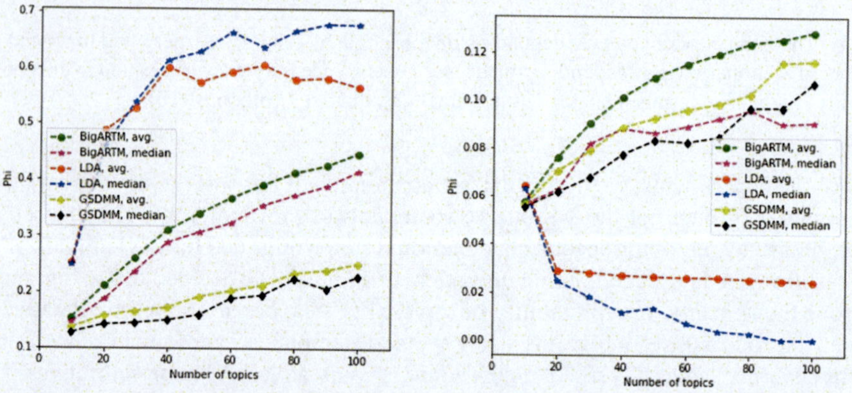

(a) (b)

Fig. 3. Average and median of the probabilities sum of the top ten tokens, varying the number of topics. Dataset 1 - (a), dataset 2 - (b).

In an attempt to assess the semantic interpretability of topic models obtained by different methods, we calculated the average and median UMass depending on the number of topics (see Fig. 4).

The quantitative values of the experimental results based on the dataset of unigrams surpassed the values obtained on the dataset of bigrams. In experiments on calculating the

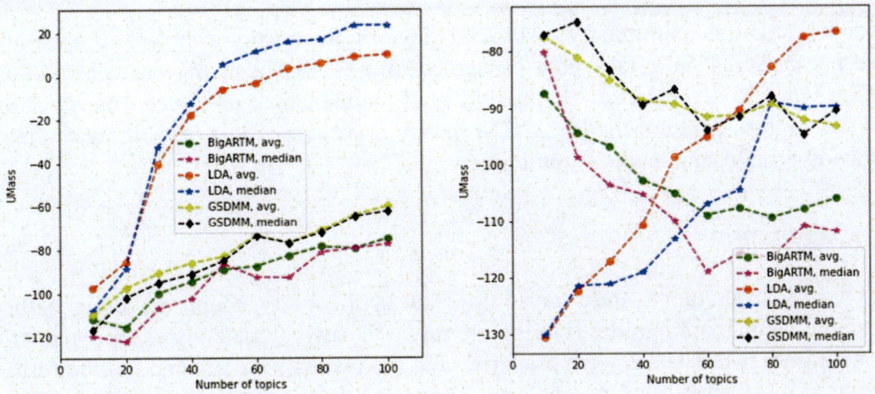

(a) (b)

Fig. 4. Average and median UMass, varying the number of topics. Dataset 1 - (a), dataset 2 - (b).

fit time of the model and the Phi matrix, such a result was expected. This is due to the fact that the probability of the occurrence of a unigram in the text is greater than the probability of the occurrence of a bigram. The results of experiments on calculating the UMass metric can also be explained, since the probability of the joint occurrence of two bigrams in the text is less than the probability of the occurrence of two unigrams. On the one hand, most researchers (including the creators of the UCI and UMass metrics) [32, 33] note the correlation of such metrics with human judgments about the semantic interpretability of the topic. On the other hand, researchers [31] believe that the modeling and formation of topic tokens set based on bigrams can significantly increase semantic interpretability in comparison with themes based on unigrams. Thus, it is almost impossible to compare the results quality of different topic modeling methods, based on the values of existing quantitative metrics in order to identify the best semantically interpreted result. The problem of comparing the values of metrics such as UMass or developing new metrics remains an urgent topic for further research.

Based on the experimental results, we can conclude that all the methods we used need further tuning. BigARTM significantly outperforms other methods in model fitting speed. It should be noted the possibility of using another faster online BigARTM model, the results of which are not presented in this study due to the authors' desire to explore the available methods in the classical form (out of the box), i.e. without fine tuning. However, we are convinced that through such tuning, in particular with the use of so-called regularizers for matrix sparseness and topics decorrelation, BigARTM will demonstrate higher UMass metrics values.

It should be noted that the probabilistic methods of topic modeling were deliberately chosen for the experiments, since the mathematical interpretability of the result and the ability to trace the mathematical inference are very important for understanding and correcting the model.

We emphasize once again that the key advantages of BigARTM that influenced our choice is the ability to create multimodal models that allow us to describe non-text objects within documents and, thus, take into account various parameters when modeling an online community of a social network. Such parameters are authors, time, source, categories, links, tags, phrases, named entities, objects in images, records of user actions, etc. Thus, the speed, the possibility of mathematical inference (interpretation) and the multimodality of the BigARTM library make it the most suitable for modeling and comparing social media communities.

5 Conclusion

This work is an attempt to find a way to assess the applicability of topic modeling methods for social media communities. During the research, several methods of topic modeling were chosen, two datasets were prepared, and the results were obtained in the form of a topic sets. As a result of the analysis of topic models quality metrics, we concluded that UCI and UMass are suitable for assessing and comparing the results obtained by different methods of topic modeling. Based on the experimental results, we found that a direct comparison of the UMass metric values obtained on the unigram dataset and the bigram dataset is impossible. The importance of multimodality and the possibility of mathematical interpretability (inference) of the results should be noted. At the current stage of research, among the tested methods, BigARTM seems to be the most suitable in the framework of our future work for topic modeling of social media communities.

The topic modeling methods are potentially an integral component of decision support information systems at different levels from state- to municipal government. Based on a topic analysis of open data monitoring on social media communities, it is possible to obtain operational reports on the retrospective and current state of society, and forecast its evolution. The combination of topic modeling and the formative artificial intelligence will allow great capabilities to decision-makers in administrative problems solving.

Acknowledgments. The work is supported by the Ministry of Science and Higher Education of the Russian Federation (reg.n. 0226-2019-0036).

References

1. Bodrunova, S., Koltsov, S., Koltsova, O., Nikolenko, S., Shimorina, A.: Interval semi-supervised LDA: classifying needles in a haystack. In: Castro, F., Gelbukh, A., González, M. (eds.) MICAI 2013. LNCS (LNAI), vol. 8265, pp. 265–274. Springer, Heidelberg (2013). https://doi.org/10.1007/978-3-642-45114-0_21
2. Koltsova, O., Maslinsky, K., Koltsov, S.: Protests, elections and their contributions to the topical structure of the Russian blogosphere: a «Big Data Approach». Internet, Politics, Policy 12 2012: Big data, Big Challenges? Oxford Internet University, 20–21 September 2012
3. Datyev, I.O., Fedorov, A.M., Shchur, A.L.: Framework for civic engagement analysis based on open social media data. In: Silhavy, R. (ed.) CSOC 2020. AISC, vol. 1225, pp. 586–597. Springer, Cham (2020). https://doi.org/10.1007/978-3-030-51971-1_48
4. Deerwester, S., Dumais, S.T., Furnas, G.W., Landauer, T.K., Harshman, R.: Indexing by latent semantic analysis. J. Am. Soc. Inf. Sci. **41**(6), 391–407 (1990)
5. Hofmann, T.: Probabilistic latent semantic analysis. In: UAI 1999: Proceedings of the Fifteenth Conference on Uncertainty in Artificial Intelligence, Stockholm, Sweden, 30 July–1 August 1999, pp. 289–296. Morgan Kaufmann (1999). ISBN 1-55860-614-9

6. Blei, D.M., Ng, A.Y., Jordan, M.I.: Latent Dirichlet allocation. J. Mach. Learn. Res. **3**, 993–1022 (2003)
7. Lee, D., Seung, H.: Learning the parts of objects by non-negative matrix factorization. Nature **401**, 788–791 (1999). https://doi.org/10.1038/44565
8. Kuang, D., Choo, J., Park, H.: Nonnegative matrix factorization for interactive topic modeling and document clustering. In: Celebi, M.E. (ed.) Partitional Clustering Algorithms, pp. 215–243. Springer, Cham (2015). https://doi.org/10.1007/978-3-319-09259-1_7
9. MacKay, D.J.C., Peto, L.C.B.: A hierarchical Dirichlet language model. Nat. Lang. Eng. **1**(3) (1995). https://doi.org/10.1017/S1351324900000218
10. Wallach, H.M.: Topic modeling: beyond bag-of-words. In: Machine Learning, Proceedings of the Twenty-Third International Conference (ICML 2006), Pittsburgh, PA, USA pp. 977–984 (2006). https://doi.org/10.1145/1143844.1143967
11. Teh, Y.W., Jordan, M.I., Beal, M.J., Blei, D.M.: Sharing clusters among related groups: hierarchical Dirichlet processes. In: NIPS 2004: Proceedings of the 17th International Conference on Neural Information Processing Systems, pp. 1385–1392. MIT Press, Cambridge (2004)
12. Lafferty, J.D., Blei, D.M.: Correlated topic models. In: Weiss, Y., Schölkopf, B., Platt, J.C. (eds.) Advances in Neural Information Processing Systems, vol. 18, pp. 147–154. MIT Press (2006)
13. Li, W., McCallum, A.: Pachinko allocation: DAG-structured mixture models of topic correlations. In: ICML 2006: Proceedings of the 23rd International Conference on Machine Learning, pp. 577–584. ACM, New York (2006). https://doi.org/10.1145/1143844.1143917
14. Blei, D.M., Lafferty, J.D.: Dynamic topic models. In: ICML 2006: Proceedings of the 23rd International Conference on Machine Learning, pp. 113–120. ACM, New York (2006). https://doi.org/10.1145/1143844.1143859
15. Nodelman, U., Shelton, C.R., Koller, D.: Continuous time Bayesian networks. In: Proceedings of the Eighteenth Conference on Uncertainty in Artificial Intelligence, Alberta, Canada, pp. 378–387 (2002)
16. Wang, X., McCallum, A.: Topics over time: a non-Markov continuous-time model of topical trends. In: Proceedings of the 12th ACM SIGKDD International Conference on Knowledge Discovery and Data Mining, Philadelphia, PA, USA, pp. 424–433 (2006). https://doi.org/10.1145/1150402.1150450
17. Beykikhoshk, A., Arandjelović, O., Phung, D., Venkatesh, S.: Discovering topic structures of a temporally evolving document corpus. Knowl. Inf. Syst. **55**(3), 599–632 (2017). https://doi.org/10.1007/s10115-017-1095-4
18. Zhang, J., Song, Y., Zhang, C., Liu, S.: Evolutionary hierarchical Dirichlet processes for multiple correlated time-varying corpora. In: Proceedings of the 16th ACM SIGKDD International Conference on Knowledge Discovery and Data Mining, Washington, DC, USA, pp. 599–632 (2010). https://doi.org/10.1145/1835804.1835940
19. Gao, W., et al.: Generation of topic evolution graphs from short text streams. Neurocomputing **383**, 282–294 (2020). https://doi.org/10.1016/j.neucom.2019.11.077
20. Malik, S., et al.: TopicFlow: visualizing topic alignment of Twitter data over time. In: Proceedings of the 2013 IEEE/ACM international Conference on Advances in Social Networks Analysis and Mining, pp. 720–726 (2013). https://doi.org/10.1145/2492517.2492639
21. Oghaz, T.A., Mutlu, E.Ç., Jasser, J., Yousefi, N., Garibay, I.: Probabilistic model of narratives over topical trends in social media: a discrete time model. In: Proceedings of the 31st ACM Conference on Hypertext and Social Media (HT 2020), pp. 281–290. Association for Computing Machinery, New York (2020). DOI:https://doi.org/10.1145/3372923.3404790
22. Vayansky, I., Kumar, S.: A review of topic modeling methods. Inf. Syst. **94**, 101582 (2020). https://doi.org/10.1016/j.is.2020.101582

23. Yan, X., Guo, J., Lan, Y., Cheng, X.: A biterm topic model for short texts. In: Proceedings of the 22nd International Conference on World Wide Web, Rio de Janeiro, Brazil, pp. 1445–1455 (2013). https://doi.org/10.1145/2488388.2488514
24. Yan, X., Guo, J., Lan, Y., Xu, J., Cheng, X.: A probabilistic model for bursty topic discovery in microblogs. In: AAAI 2015: Proceedings of the Twenty-Ninth AAAI Conference on Artificial Intelligence, vol. 29, no. 1, pp. 353–359 (2015)
25. Riddell, A.B.: A simple topic model (mixture of unigrams) (2012)
26. Yin, J., Wang, J.: A dirichlet multinomial mixture model-based approach for short text clustering. In: KDD 2014: Proceedings of the 20th ACM SIGKDD International Conference on Knowledge Discovery and Data Mining, pp. 233–242 (2014). https://doi.org/10.1145/2623330.2623715
27. Nigam, K., Mccallum, A.K., Thrun, S., et al.: Text classification from labeled and unlabeled documents using EM. Mach. Learn. **39**, 103–134 (2000). https://doi.org/10.1023/A:1007692713085
28. Zhao, W.X., et al.: Comparing Twitter and traditional media using topic models. In: Clough, P., et al. (eds.) ECIR 2011. LNCS, vol. 6611, pp. 338–349. Springer, Heidelberg (2011). https://doi.org/10.1007/978-3-642-20161-5_34
29. Quan, X., Kit, C., Ge, Y., Pan, S.J.: Short and sparse text topic modeling via self-aggregation. In: Proceedings of the Twenty-Fourth International Joint Conference on Artificial Intelligence (IJCAI 2015), Palo Alto, California, USA, pp. 2270–2276. AAAI Press (2015)
30. Zuo, Y., et al.: Topic modeling of short texts: a pseudo-document view. In: KDD 2016: Proceedings of the 22nd ACM SIGKDD International Conference on Knowledge Discovery and Data Mining, pp. 2105–2114. Association for Computing Machinery, New York (2016). https://doi.org/10.1145/2939672.2939880
31. Kochedykov, D., Apishev, M., Golitsyn, L., Vorontsov, K.: Fast and modular regularized topic modelling. In: 21st Conference of Open Innovations Association (FRUCT), Helsinki, Uusimaa, Finland, pp. 182–193. FRUCT Oy (2017). https://doi.org/10.23919/FRUCT.2017.8250181
32. Newman, D., Lau, J.H., Grieser, K., Baldwin, T.: Automatic evaluation of topic coherence. In: Human Language Technologies: The 2010 Annual Conference of the North American Chapter of the Association for Computational Linguistics (HLT 2010), pp. 100–108. Association for Computational Linguistics, Stroudsburg (2010)
33. Mimno, D., Wallach, H., Talley, Ed., Leenders, M., McCallum, A: Optimizing semantic coherence in topic models. In: Proceedings of the 2011 Conference on Empirical Methods in Natural Language Processing, Edinburgh, Scotland, UK, pp. 262–272. Association of Computational Linguistics (2011)

Modeling Theory Based on the Formal-Constructive Interpretation of the Model

B. Yury Melnikov(✉)

Ural State University of Economics, Yekaterinburg, Russia

Abstract. Modeling is the basis of methods of activity in science, technology, etc. One of the main problems of modeling is to create a method for obtaining the required models. The task of multidimensional multidimensional assessment of the adequacy of the constructed models and improving their adequacy is no less urgent. The author proposed an algebraic approach to building models, which consists in identifying three components: 1) a system of models that can be taken as basic for a given field of activity; 2) systems of typical transformations and typical combinations of models; 3) an approximation mechanism designed to represent or obtain a model as a result of applying typical transformations and typical combinations of basic models. It turned out that traditional interpretations of the concept of "model" do not allow to fully implement the algebraic approach to modeling. This work contains a brief description of the foundations of the author's theory of modeling, based on a formal-constructive interpretation of the concept of "model", within which a model is understood as a system of two components, the first of which, called interface, formalizes the system of information exchange between the prototype and its image, in the second component, called model-content, formalizes the prototype image. The basic components of the author's theory of modeling are briefly described: formal constructive interpretation of the model, "algebra of models", theory of adequacy, theory of strategies.

Keywords: Mathematical modeling · Algebraic approach to modeling · Modeling theory · Model adequacy · Strategy of activity

1 Introduction

The modeling method, in particular, mathematical modeling, is the basis of scientific research, engineering, art and other fields of activity. Paradoxically, a sufficiently universal and constructive method for developing new models has not yet been created; the available recommendations are either very vague and vague, or refer to very specific types of models. To date, a huge number of different models have been developed, but, unfortunately, firstly, the totality of these models is mainly eclectic, it is difficult to call it a system, and secondly, even in the case when there are several models of a certain prototype, none of which individually does not allow achieving the required level of adequacy, the task of obtaining the required model in the form of a combination of existing models is usually quite complex and original. In addition, the problem of assessing

R. Silhavy et al. (Eds.): CoMeSySo 2021, LNNS 231, pp. 615–627, 2021.
https://doi.org/10.1007/978-3-030-90321-3_51

the adequacy of models is nontrivial. The author's theory of modeling is devoted to the solution of these problems, a brief outline of which is given in this article. This theory includes a formal constructive formalization of the concept of "model", a system of typical transformations and typical combinations of models ("model algebra"), the theory of adequacy and the theory of strategies.

2 Methods

The scientific basis of the research is a systematic approach to modeling: consideration of the object and its components as a system, in particular, the representation of the object in the form of a system of models reflecting different aspects of the prototype, and a system of models reflecting the same aspect; the implementation of the principles of integrity, hierarchy, etc. The method of analogy is actively used, in this case it consists mainly in the transfer of some mathematical approaches and methods to "pre-mathematics". In particular, many of the results obtained are based on the axiomatic method of formalizing concepts and the methods of scientific analysis and synthesis based on it.

3 Results

In recent decades, more and more often there is a need to create models that reflect various aspects of the prototype or reflect these aspects in many ways, at different levels [1]. In addition, an important role is played by the ability to describe the same aspect of the prototype using models based on different scientific languages, for example, some economic problems can be described in terms of economic relations, financial science, etc. Vector algebra is described, firstly, in the language of geometry (the "algebra" of directed segments, considered as the result of postponing a vector from a point), and secondly, in a special symbolic language of vector algebra (within which one operates with expressions of the form $\vec{a} = 2\vec{b} - 3\vec{c}$, $\vec{a} \perp \left(\vec{b} - \vec{p}\right)$, etc.), thirdly, in the matrix language on which the coordinate method is based.

3.1 Algebraic Approach to Model Building

By the algebraic approach to building a model, we mean a system of three components: 1) a system of models that are basic for a given class of models; 2) a system of typical transformations and typical combinations of models; 3) an approximation mechanism designed for (generally speaking, approximate) representation of the model in the form of the result of applying typical transformations and typical combinations of basic models.

In this case, "mechanism" is understood in the ideal sense, as a set of states and processes that make up some phenomenon or internal structure, a system of functioning, an apparatus of some kind of activity.

3.2 Formal-Constructive Interpretation of the Model

Typically, the definition of a "model" is based on the requirement that the image is "similar" to the prototype. For example, studying a model should lead to an understanding of processes and phenomena, ways to control them, etc. For example, [2] offers the following interpretation of the mathematical model: "Models are mathematical representations of mechanisms that govern natural phenomena that are not fully recognized, controlled, or understood".

But this creates an almost insurmountable obstacle to the implementation of an algebraic approach to modeling, since one of the models can be "similar" to a prototype in one sense, the other in another, and the result of their combination will not be "similar" to a prototype in any of these senses., i.e. will not be a model in the traditional sense of the word. Therefore, we took the structure of this object as the basis for the definition of the model. First, in our opinion, the model cannot be identified with the prototype image. Indeed, is the equation $p = \sin 2q + p'$ a model, for example, for the development of a company? To answer this question, you need to explain what the symbols *p*, *p*', *q* mean, justify equality, etc. In this case, we need to interpret the results of the analysis of this model in terms used to characterize the prototype. In other words, it is necessary to ensure the exchange of information between the prototype and the image, see Fig. 1.

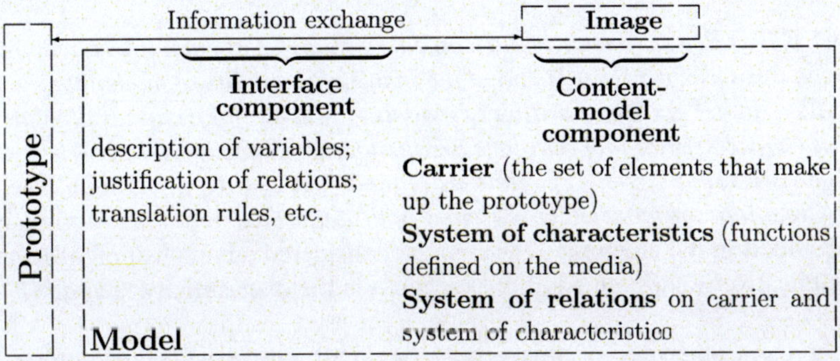

Fig. 1. Illustration for the formal constructive interpretation of the model.

Secondly, modeling an object begins with dividing it into its component parts, i.e. formation of the model carrier, see Fig. 1. Note that the system of constituent parts is defined ambiguously. For example, consider the problem of uniform motion. On the one hand, it can be assumed that the carrier of this model consists of track sections. The system of characteristics initially consists of two quantities, the value of the path section *AB* corresponds to *s*(*AB*) - the length of this section of the path (in some units of measurement), and the second - to the value *t*(*AB*) - the time during which the object has overcome the section *AB*. Then the secondary value $v(AB) = \frac{s(AB)}{t(AB)}$ is entered - the average speed of movement along the section *AB*. Another model can be proposed, in which the elements of the carrier are periods of time [*p*;*q*]. The quantity σ[*p*;*q*] corresponds to this period of time the length of the trajectory that the object has traveled over the time period [*p*;*q*], and the quantity τ[*p*;*q*] - the duration of the period [*p*;*q*], that is, $\tau[p; q] = q - p$.

3.3 Typical Conversions and Typical Combinations of Models

The second component of the algebraic approach to building models is applied in modeling practice, but in an informal way [1, 3]. The formal-constructive interpretation of the model allows one to formalize typical transformations and typical combinations of models. Let's start with the natural relation "to be a submodel": *model **A** is called a submodel of model **B** if all the components of its interface are components of the interface of model **B**, the carrier, system of characteristics and system of relations of model **A**are subsets, respectively, the carrier, system of characteristics and system of relations of model **B***, see Fig. 2.

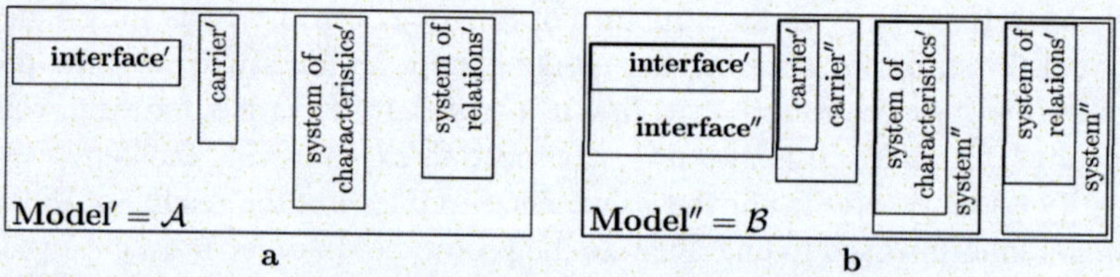

Fig. 2. Иллюстрация к понятию «подмодель».

*Model **B** is called the result of enrichment of model **A** if model **A** is a submodel of model **B***. In this case, *model **A** is called the result of reduction of model **B***.

*We called model **C** a composition of models **A** and **B**, if the carrier of model **C**, its systems of characteristics and relations are the union, respectively, of carriers, systems of characteristics and systems of relations of models **A** and **B**, in addition, usually the system of relations of model **C**is enriched with additional relations*. For example, a field understood in the algebraic sense can be considered a composition of its additive and multiplicative groups, an additional relation is the distributivity law: $a \cdot (b + c) = a \cdot b + a \cdot c$.

Within the framework of this article, we will not provide a complete set of combinations of models.

3.4 Fundamentals of the Theory of Adequacy

The last, third component of the algebraic approach to modeling, which we called the **approximation mechanism**, is often based on the use of various ways to improve the quality of models [4]. The "quality" of various models can be assessed by comparing the consumption of resources for analyzing the model, the level of complexity of the theoretical base, including the complexity of the model language, etc. But the most important characteristic of the "quality of the model" is the level of "similarity" of the image to the prototype. In addition, in addition to the "approximation mechanism", usually one or another interpretation of the "similarity" of the image to the prototype is taken as the basis for the definition of the model. The absence of such a requirement is a characteristic feature in the formal constructive interpretation of the model (see Fig. 1), which is sometimes perceived as a drawback. But in practice, the assessment of "similarity" is a separate task that is solved after the creation of the model and is usually considered as a tool for improving its quality [4–8].

It is usually considered that the adequacy of the model is assessed by comparing the image with the prototype. But in fact, the adequacy of the model is almost always assessed by comparing the calculated values with the results of instrumental measurements, identifying signs of certain properties, relationships, etc. Thus, comparison is almost always carried out by comparison not with a prototype, but with a model taken as a standard, see Fig. 3.

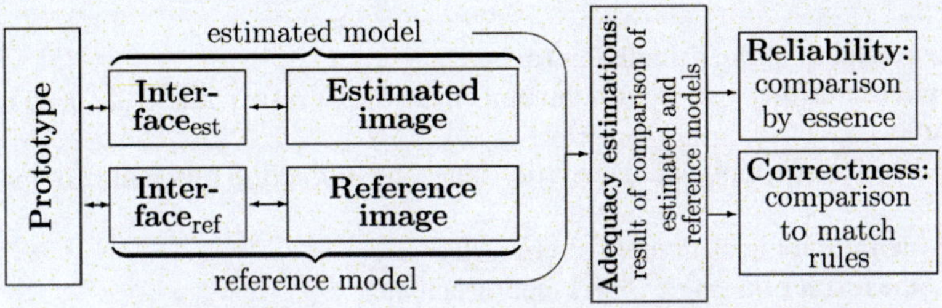

Fig. 3. Illustration for the concept of "model adequacy".

Let's note a few points.

First, the reference model coincides with the prototype, apparently only in the case when the prototype is an ideal object: theory, mathematical construction, scientific concept, etc.

Second, here the term "reference" does not mean "high quality". It only means that we have taken this model as a sample.

Thirdly, the theory of adequacy made it possible to identify the key difference between research and project activities [9]. The design activity differs in that the reference models and adequacy characteristics are focused on assessing the reliability of the result of achieving the goal: the indicators specified in the project assignment must be achieved, the presence of the required relations is recorded, etc. For research activities, it is characteristic that reference models and characteristics of adequacy are focused on assessing correctness: the requirements for conducting an experiment must be met, all rules for conducting a proof must be followed, etc. From all this it follows that it is more correct to talk about the design and research stages of activity.

Evaluation of the value of adequacy is often associated with the comparison of the value of the quantity with some reference value. The presence or absence of a certain relationship between the elements of the carrier or characteristics is also often determined using the values of some quantities (often of a stochastic nature). Typically, the measured or reference value is the result of a direct instrumental measurement. But the instrumental measurement of the value of a quantity is almost always carried out with some error. Assessing the adequacy is complicated when carrying out an indirect measurement of quantities, when the conclusion about the value of the quantities under consideration is made based on the results of measuring other quantities. For example, it makes no sense to obtain the value of the resource of each engine by measuring the time of its operation on the bench until the destruction of this engine. Sometimes there is no certainty that the measured or calculated value characterizes the quantity of interest to us. For example, good values of the specific strength of a material do not guarantee its long-term performance if the material has insufficient resistance to fatigue degradation.

The system of factors influencing the assessment of adequacy, more precisely, the correctness and reliability of the process and the results of measuring the quantities, we called the **metric package**.

Metric package: factors affecting the correctness of the process and measurement results of quantities:

1) the correctness of the calculation rules;
2) compliance of these rules with the objectives of the research and the used research apparatus;
3) a method for obtaining characteristic values, including the following components:

    3.1) definition or assignment of characteristics;
    3.2) the result of a single direct measurement;
    3.3) the result of statistical processing;
    3.4) the result of the calculation based on correlation dependences;
    3.5) the result of a determination using general theories based on accepted assumptions and axioms.

Metric package: factors affecting the reliability of the process and measurement results of quantities:

1) the reliability of the initial data;
2) the adequacy of the models used;
3) the reliability of the theoretical apparatus.

3.5 Axiomatic Definition of Adequacy

Adequacy model*is a model, the carrier of which is a set of models, the domain of definition of adequacy characteristics is included in the set of pairs of models (the components of the pair are the estimated and reference models), and the range of values can be non-numerical, and the following axioms are fulfilled:*

1) *the axiom of determining adequacy as a result of comparing the evaluated and reference models;*
2) *the axiom of representability of the estimated model in the reference model;*
3) *the axiom of measurement using a partially ordered set of values;*
4) *the axiom of modeling adequacy;*
5) *the axiom of estimation based on the metric package.*

The axiomatic definition of adequacy characteristics allowed in the work [4] to determine all possible directions for improving the adequacy of models when carrying out only evolutionary changes in the assessed and reference models, and characteristics of adequacy (comparison method). In this case, evolutionary changes are understood as combinations of the following basic transformations (see Fig. 3): 1) transformation of carriers of the estimated or reference model; 2) transformation of the interface component of the evaluated or reference model; 3) change in the characteristics of adequacy (i.e., the method of obtaining the result of comparing the evaluated and reference models).

3.6 Multi-aspect Modeling. Models-Polyads

Typically, the construction of the model is preceded by the selection of the modeled aspect of the prototype. For example, if the work of an enterprise is modeled, then it is possible to model such aspects as: models of technological processes, models of production relations within the enterprise and relations with other enterprises, models of various aspects of the management process (selection and management of personnel, management structure of the enterprise, incentive system for employees, management, partners, clients), etc. Management of the enterprise requires taking into account several of these models at once. The use of a system of models reflecting several aspects of the prototype is called multi-aspect modeling, see Fig. 4.

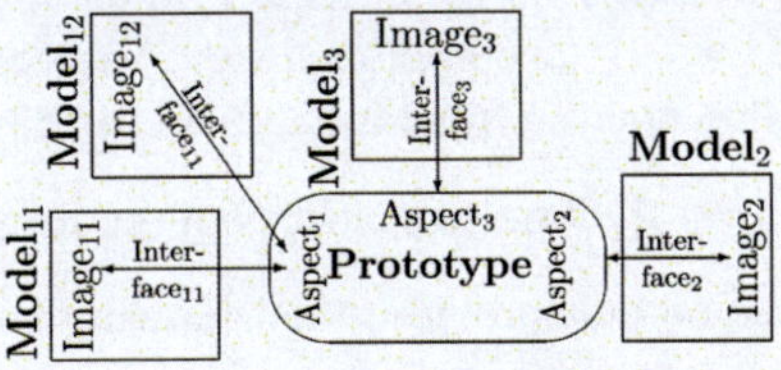

Fig. 4. Multi-aspect modeling.

Multi-aspect modeling is used to achieve the following goals:

1) make a complete idea of the prototype (for example, to make a decision on the feasibility of investing in the activities of the enterprise);
2) highlight the aspects of the prototype that require special attention and correction (for example, a SWOT analysis serves this purpose).

Currently, there is a need to create complex models that simultaneously reflect several aspects, for example, production, environmental, financial, etc.

An "ideological alternative" to multi-aspect modeling is the situation when the same aspect of the prototype (this aspect can be declared a prototype) is reflected by several models, for example, using different languages (for example, one model is formulated in the language of geometry, and the other in the language of matrix algebra), see Fig. 5a. In this case, it becomes necessary to exchange information about the images, for which the corresponding interface is usually formed, see Fig. 5b. Usually, after that, the analysis of the models occurs practically without the participation of the prototype, as a result of which a pair of models with a common interface is obtained, see Fig. 5c. The last system, in which (Interface_1; Image_2) can be considered the model (Model_B) of the

Image_1 object, and (Interface_1; Image_1) can be considered the model (Model_A) of the Image_2 object, we called the **model-dyad**.

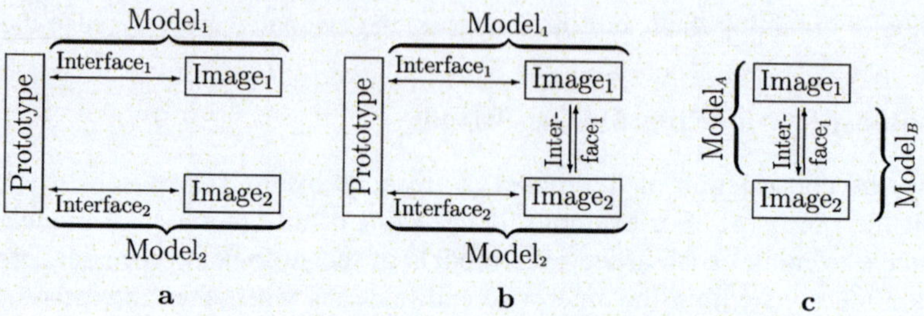

Fig. 5. Illustration for the concept of model-dyad.

The model-tryad, model-quadryad and, in the general case, model-polyad are defined similarly. For example, in mathematics, the algebra of complex numbers is actually presented in textbooks as a quadriad model with components in the form: 1) the algebra of polynomials of degree at most 1 in the variable i (in electronics, the letter j is used) by the additional relation $i^2 = -1$; 2) complex plane; 3) algebras of ordered pairs (a; b) with componentwise addition and multiplication given by the formula $(a; b) \cdot (c; d) = (a \cdot c - b \cdot d; a \cdot d + b \cdot c)$; 4) algebras of matrices of the form $\begin{pmatrix} a & b \\ -b & a \end{pmatrix}$.

Conscious use of polyad models provides the following benefits:

1) the possibility of using the analytical apparatus of several theories;
2) the ability to effectively control the adequacy of reasoning by interpreting the results obtained within the framework of other components of the polyad model;
3) simplification of the search and justification of hypotheses.

We usually give students the following analogy: Imagine a strong, trained athlete near a vertical wall with no protrusions. How high can he climb? Only to the place where it reaches in the jump. Now imagine that next to it, at a convenient distance, there is the same wall, parallel to the original one. Now, pushing off these walls (which simulates the presence of an interface between the images in Fig. 5c, this athlete can climb very high.

3.7 Elements of Strategy Theory

Probably the most requested activity management model is the **activity plan**. The problem of planning activities has at least two main aspects: the creation of a plan and the implementation of the plan (with possible adjustments). We consider an activity strategy as a mechanism for creating a plan. The relationship between strategy, strategy implementation, action plan and plan execution is shown in Fig. 6.

This interpretation of the strategy is not generally accepted. Our analysis showed that almost all interpretations of this term admit an adequate interpretation within the

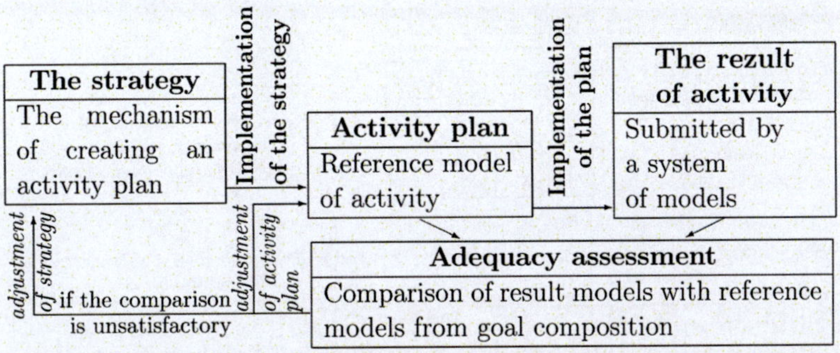

Fig. 6. Strategy, strategy implementation, action plan and plan execution.

framework of the model shown in Fig. 6. In fact, it is our interpretation of the concept of "activity strategy" as a mechanism for creating action plans that is meant in works on engineering [10–12], physics and chemistry [13], medicine [14], computer science [15, 16], radio electronics [17, 18], management [19], education [20], etc.

We have taken as an axiom the assertion that *each point of the activity plan is perceived by any subject of activity as a reference either to the goal of the activity, or to an available algorithm of activity*. Another axiom is the statement that *this perception is subjective*. For example, if a person knows only one algorithm for completing a certain point, he will perceive this point as a link to an algorithm, and if he does not know such a method or knows several ways to accomplish this point of the plan, then he will probably take this point as a reference to a goal. a specific way of achieving which is not specified.

To apply strategies for the purpose of application in information technology, it is necessary to constructively formalize the concept of strategy and its components, in particular, the concept of the goal of activity.

We took it for an axiom that ***the goal of activity****is a model with a carrier from the reference models of the result of activity*. The more fundamentally different reference models are included in a goal, the easier it is to plan for achieving that goal. To control the completeness of the target composition, various classifications of reference models can be used. One of the actual classifications turned out to be the selection of three classes of reference models according to the level of abstraction: 1) model languages; 2) templates of reference models; 3) specific reference samples.

We have identified 3 options for implementing the strategy: local, internal and external, see Fig. 7.

Local implementation of the strategy consists in the fact that the point of the plan, perceived as the goal of the activity, indicated in Fig. 7a through G13 is replaced with a typical plan to achieve this goal. If the resources for choosing a typical plan for achieving the $\mathbf{G}_{13}$ goal are not enough, in order to obtain the necessary resources, it is sometimes enough to complete some of the points of the plan preceding the item under consideration. We called this variant of the strategy implementation internal, see Fig. 7b. Sometimes even the fulfillment of the previous points of the plan may not be enough to obtain the resources necessary to select a plan to achieve the $\mathbf{G}_{13}$ goal. In this case, the necessary

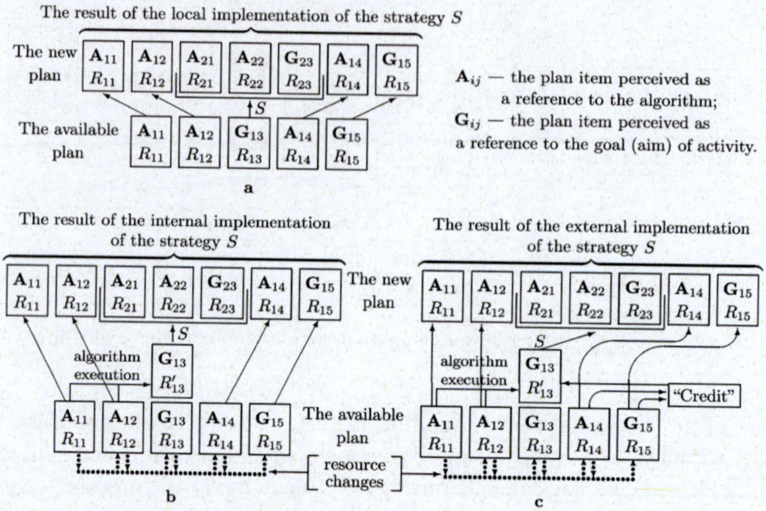

Fig. 7. Local (a), internal (b) and external (c) strategy implementation.

resources will have to be received "on credit", which will have to be paid for in the process of executing the constructed plan, see Fig. 7 **c**.

3.8 Internal Algebraic Representation of an Activity Strategy

In our opinion, the phrase "strategy is a mechanism for creating an activity plan" cannot be taken unambiguously as a definition. In our opinion, it is advisable to formalize such a multifaceted phenomenon as a strategy not deductively, with the help of a definition, but inductively, in this case, with the help of a system of models, including those reflecting different aspects of the strategy. We received one of the strategy models [21] as a result of applying the algebraic approach to obtaining an activity plan, see Fig. 8.

The model shown in Fig. 8, we called the internal algebraic representation of the activity strategy. The term "internal" means that the base models are the elements or components of the original model. In this case, we consider the typical goals of activity and typical algorithms for a specific area of activity as basic models, and we consider them as single-point plans. Typical transformations and typical combinations consist of concatenating plans, replacing a plan item with a subplan for achieving it and vice versa, etc.

The approximation mechanism includes analytical and synthetic methods [22] and other methods.

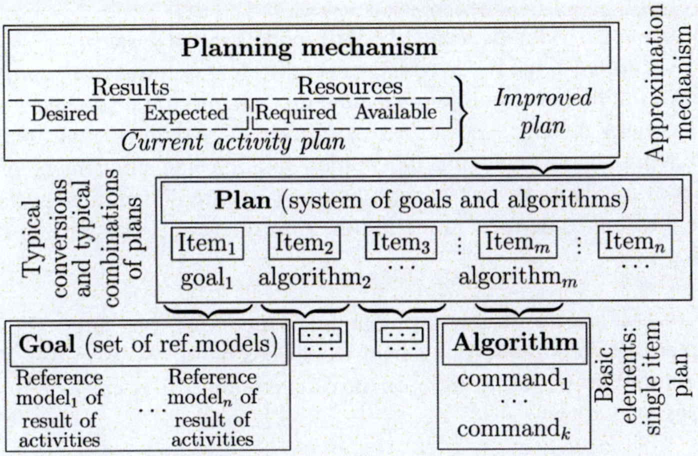

Fig. 8. Internal algebraic representation of the strategy.

3.9 External Algebraic Representation of an Activity Strategy

If in the implementation of the algebraic approach to the description of activity strategies the basic elements are strategies, which we will call **basic strategies**, then we call this implementation of the algebraic approach the external algebraic representation of the strategy. We have built an external algebraic representation of the strategies of routine modeling, routine project activities [23] and routine research activities, which we have defined axiomatically.

4 Discussions and Conclusion

There are several directions for further research:

1) study the algebraic properties of typical transformations and typical combinations of models;
2) to distinguish systems of basic models for sets of known models of different fields of activity (in practice, it turned out that some of the models admit several significantly different formal constructive interpretations in the sense of Fig. 1);
3) to develop the theory of adequacy (at the moment we have identified a complete list of directions for the evolutionary improvement of the adequacy of models), for example, to create an analytical apparatus for assessing the adequacy of models, using our proposed standard transformations and typical combinations of models;
4) to implement an algebraic approach to describing strategies for various types of activities, in particular, to identify systems of basic strategies for them, as we have already done for strategies of routine modeling, routine design and routine research activities.

The systematic use of modeling theory allows:

1) transform a largely chaotic, eclectic set of models into a system;
2) create multidimensional models that reflect different features of the prototype;
3) carry out a multifaceted multidimensional assessment of the adequacy of models;
4) formalize many management processes (forming a system of goals, building plans for achieving them, managing the implementation and adjustment of plans) on the basis of an algebraic approach to building and implementing strategies, create and use uniform tools for automating the management process using information technologies.

This work is a brief overview of the author's theory of modeling. The limitations on the number of references to one's own work does not allow us to provide even the minimum required list of publications on the research topic; they can be found using the elibrary.ru system.

References

1. John, A., Makkar, S., Swaminathan, S., et al.: Factors influencing household pulse consumption in India: a multilevel model analysis. Glob. Food Secur. **29**, 100534 (2021). https://doi.org/10.1016/j.gfs.2021.100534
2. Tedeschi, L.O.: Assessment of the adequacy of mathematical models, pp. 225–247 (2006). https://doi.org/10.1016/j.agsy.2005.11.004.
3. Bredykhin, V., Gurskyi, P., Alfyorov, O., et al.: Improving the mechanical mathematical model of grain mass separation in a fluidized bed. East. Eur. J. Enterp. Technol. **3**, 79–86 (2021). https://doi.org/10.15587/1729-4061.2021.232017
4. Melnikov, Y.B., Onokhina, Y.A., Shitikov, S.A.: Improving the adequacy of economic models. J.Ural State Univ. Econ. **19**(1), 94—106 (2018). https://doi.org/10.29141/2073-1019-2018-19-1-8
5. Reggiani, M., Marchetti, F.: On assessing model adequacy. IEEE Trans. Syst. Man Cybern. **5**, 322–330 (1975). https://doi.org/10.1109/TSMC.1975.5408407
6. Ash, A.: Retrospectively judging the adequacy of a prospective model. Biometrics **40**, 1183 (1984)
7. Kim, S., Lee, S.: Maximum likelihood estimation of probabilistic non-suppression model for OECD NPP electrical fire applying non-negative continuous distribution. Fire Saf. J. **122** (2021). https://www.scopus.com/inward/record.uri?eid=2-s2.0-85104939995&doi=10.1016%2fj.firesaf.2021.103323&partnerID=40&md5=8b487ad1add1e23191d9f81e8dc6e1a6
8. Zhou, L., Rasmy, M., Takeuchi, K., et al.: Adequacy of near real-time satellite precipitation products in driving flood discharge simulation in the Fuji river basin, Japan. Appl. Sci. (Switz.) **11**(3), 1–16 (2021). https://doi.org/10.3390/app11031087
9. Melnikov, Y., Solov'yanov, V.B., Shirpuzhev, S.V.: Research and project activity of students from the position of modeling theory. Pedagogical J. **7**(2A), 36–49 (2017)
10. Shi, C., Yan, F., Ge, G., Wei, Y., Zhai, W.J., Yao,: Significantly enhanced energy storage performances and power density in (1 − x)BCZT-xSBT lead-free ceramics via synergistic optimization strategy. Chem. Eng. J. **426**, 130800 (2021). https://doi.org/10.1016/j.cej.2021.130800
11. Jin, S.H., Lee, B., Kim, J.S., Lee, C.-S.: Improvement strategy of a microfluidic sorter using a pneumatic bilayer valve. Chem. Eng. Sci. **245**, 116834 (2021). https://doi.org/10.1016/j.ces.2021.116834

12. You, D., Shen, X., Liu, G., Wang, G.: Signal anomaly identification strategy based on Bayesian inference for nuclear power machinery. Mech. Syst. Sig. Proces. **161**, 107967 (2021). https://doi.org/10.1016/j.ymssp.2021.107967
13. Zhang, D., et al.: Physical and chemical synergistic strategy: a facile approach to fabricate monovalent ion permselective membranes. Chem. Eng. Sci. **245**, 116873 (2021). https://doi.org/10.1016/j.ces.2021.116873
14. Jiang, W., et al.: Amplified antitumor efficacy by a targeted drug retention and chemosensitization strategy-based "combo" nanoagent together with PD-L1 blockade in reversing multidrug resistance. J. Nanobiotechnol. **19**(1), 200 (2021). https://doi.org/10.1186/s12951-021-00947-9
15. Huang, Z.-B., Fu, G.-T., Fa, T.-H., et al.: High performance ant colony system based on GPU warp specialization with a static-dynamic balanced candidate set strategy. Fut. Gener. Comput. Syst. **125**, 136–150 (2021). https://doi.org/10.1016/j.future.2021.06.041
16. Li, Q., Wang, D., Lu, H., Hong, W., Mu, Q., Xiong, K.: Performance analysis of the collaborative caching strategy in an UAV-RAN. EURASIP J. Wirel. Commun. Netw. **2021**(1), 1–20 (2021). https://doi.org/10.1186/s13638-021-01922-6
17. Zhao, X., Wente, A., Flecha, M., et al.: Culture moderates the relationship between self-control ability and free will beliefs in childhood. Cognition **210** (2021). https://doi.org/10.1016/j.ijepes.2021.107265
18. Chang, P., Song, G., Hou, J., Xu, R.: A novel voltage phase selector for inverter-interfaced system based on two-stage control strategy. Int. J. Electr. Power Energy Syst. **133**, 107299 (2021). https://doi.org/10.1016/j.ijepes.2021.107299
19. Peng, T., He, Q., Zhang, Z., Wang, B., Xu, X.: Industrial internet-enabled resilient manufacturing strategy in the wake of COVID-19 pandemic: a conceptual framework and implementations in China. Chin. J. Mech. Eng. **34**(1), 1–6 (2021). https://doi.org/10.1186/s10033-021-00573-4
20. Perikova, E., Loviagina, A., Byzova, V.: Metacognitive strategies of decision making in educational activities: efficiency in higher education. Sci. Educ. Today. **9**(4), 19–35 (2019). https://doi.org/10.15293/2658-6762.1904.02
21. Melnikov, Y.B., Privalov, S.M.: Internal algebraic representation of strategy as a means of organizing the teaching of mathematical activities. Mod. Educ. **4**, 1–14 (2019). https://doi.org/10.25136/2409-8736.2019.4.31402. https://nbpublish.com/library_read_article.php?id=31402
22. Melnikov, Y., Onokhina, E.A., Shitikov, S.A.: Bottom and top-down analysis as a component of the strategy for making managerial decisions. Theor. Appl. Econ. **4**, 41–49 (2018). https://doi.org/10.25136/2409-8647.2018.4.28207
23. Melnikov, Y., Khripunov, I.V., Chopovda, V.S.: Algebraic approach to the strategies of project activities. Izvestia USUE **2**(53), 115–123 (2014)

A Cognitive Architecture for Mental Processes Involving Mental Models Analysed from a Self-modeling Network Viewpoint

Laila van Ments[1], Raj Bhalwankar[2], and Jan Treur[2(✉)]

[1] AutoLeadStar, Jerusalem, Israel
laila@autoleadstar.com

[2] Social AI Group, Department of Computer Science, Vrije Universiteit Amsterdam, Amsterdam, The Netherlands
r.p.bhalwankar@student.vu.nl, j.treur@vu.nl

Abstract. This paper contributes an analysis of how in mental and social processes, humans often apply specific mental models and learn and adapt them in a controlled manner. It is discussed how controlled adaptation relates to the Plasticity Versus Stability Conundrum in neuroscience. From the analysis an informal three-level cognitive architecture for controlled adaptation was obtained. It is discussed here from a self-modeling network viewpoint how this cognitive architecture can be modeled as a self-modeling network. Making use of the specific network characteristics offered by the self-modeling network structure format, a large number of options for different types of adaptation of mental models and different types of control over adaptation of mental models were obtained. Many of these options were illustrated by a several realistic examples that were formalized by self-modeling networks. Other options that were distinguished from the analysis here, are offered as interesting options for future research.

1 Introduction

The area of mental models within psychology, educational science and other related disciplines addresses how in their mental and social processes, humans often learn, adapt and apply specific mental models as a kind of blueprints, schemas or maps. Most of the many publications on mental models in multiple disciplines are informal and not computational. This may partly be due to the challenging complexity of the different types of processes involved in handling mental models. The main processes that came out of a more detailed analysis of this literature are (Van Ments and Treur 2021c):

(1) *Applying a mental model*
This can be considered a form of internal (mental) simulation. Outcomes of this, affect a person's decisions and actions; e.g., (Craik 1943)

(2) *Developing and maintaining a mental model*
Adaptation of mental models often takes place. This usually involves learning, extinction or forgetting, and revision; e.g., (Piaget 1936; Hebb 1949; Seel 2006)

R. Silhavy et al. (Eds.): CoMeSySo 2021, LNNS 231, pp. 628–647, 2021.
https://doi.org/10.1007/978-3-030-90321-3_52

(3) *Exerting control over a mental model*
In a context-sensitive manner, usually control is exerted over adaptation of a mental model; e.g., (Du Plooy 2016; Darling-Hammond et al. 2008; Hurley 2008; Mahdavi 2014; Pintrich 2000).

To obtain a formalized and computational model of mental processes involving a mental model, these three different types of interacting processes all have to be addressed, which indeed may be a bit challenging.

Some inspiration can be obtained from the wider neuroscientific context. In that context, (2) corresponds to what often is called plasticity and (3) relates to the notion of metaplasticity. In (Sjöström et al. 2008), the latter topic is discussed in relation to what is called the Plasticity Versus Stability Conundrum. More specifically, concerning (2) and (3), within neuroscience it has been found more in general that:

- In the brain *plasticity* can occur in different forms; for example:
  - synaptic neural plasticity; e.g., Hebbian learning (Hebb 1949)
  - nonsynaptic neural plasticity (sometimes called intrinsic plasticity) such as plasticity of excitability thresholds within neurons; e.g., (Chandra and Barkai 2018; Debanne et al. 2019; Sjöström et al. 2008)

- Plasticity turns out not to be constant but can be depend on circumstances; various neural mechanisms have been discovered by which the extent of plasticity varies over different circumstances by being controlled in a context-sensitive manner. This is called *metaplasticity*; e.g., (Abraham and Bear 1996; Magerl et al. 2018; Robinson et al. 2016; Sjöström et al. 2008)

These concepts and the way in which they have been modeled by formalized and computational models in (Treur 2020) provided useful inspiration to obtain formalised computational models for mental processes based on mental models as well.

As a first step, based on the analysis of the different types of interacting processes (1), (2) and (3) involved in mental model handling, an informal global cognitive architecture for mental model handling has been developed taking these processes into account in three different but related levels; see (Van Ments and Treur 2021c). Moreover, using the viewpoint of self-modeling networks, it has been shown how this (informal) cognitive architecture can be formalized in a computational manner in a self-modeling network format. In this self-modeling network, for the abovementioned informal three-level cognitive architecture, by the base level network internal simulation based on a mental model as subnetwork takes place, by a first-order self-model network at the next level, adaptation of this subnetwork representing the mental model and by a second-order self-model network at the second-next level context-sensitive control over this adaptation takes place. This will be discussed in some more detail in Sect. 3. After that, in Sects. 4, 5 and 6 each of the three levels will be discussed in more detail and illustrated by many examples of realistic cases of mental processes involving mental models. First, in Sect. 2 the notion of self-modeling network model is briefly introduced.

2 Self-modeling Network Models

A specific modeling approach addressing dynamics and adaptivity is the network-oriented modeling approach described in (Treur 2020). The current section briefly describes this modeling approach.

2.1 Network Models

According to the network-oriented modeling approach described in (Treur 2020) a network model is characterised by:

- **connectivity characteristics**
  Connections from a node (or state) X to a node Y and their *weights* $\boldsymbol{\omega}_{X,Y}$
- **aggregation characteristics**
  For any node Y, some *combination function* $\mathbf{c}_Y(..)$ defines aggregation that is applied to the single impacts $\boldsymbol{\omega}_{X,Y}X(t)$ on Y through its incoming connections from states X
- **timing characteristics**
  Each node Y has a *speed factor* $\boldsymbol{\eta}_Y$ defining how fast it changes for given (aggregated) impact

The difference (or differential) equations that are useful for simulation purposes and also for analysis of network dynamics incorporate these network characteristics $\boldsymbol{\omega}_{X,Y}$, $\mathbf{c}_Y(..)$, $\boldsymbol{\eta}_Y$: it holds.

$$Y(t+\Delta t) = Y(t) + \boldsymbol{\eta}_Y\big[\mathbf{c}_Y\big(\boldsymbol{\omega}_{X_1,Y}X_1(t), \ldots, \boldsymbol{\omega}_{X_k,Y}X_k(t)\big) - Y(t)\big]\Delta t \tag{1}$$

for any state Y and where $X_1, \ldots, X_k$ are the states from which it gets its incoming connections. Examples of useful combination functions are:

- the simple logistic sum function $\mathbf{slogistic}_{\sigma,\tau}(..)$ defined by:

$$\mathbf{slogistic}_{\sigma,\tau}(V_1, \ldots, V_k) = \frac{1}{1+\mathrm{e}^{-\sigma(V_1+\ldots+V_k-\tau)}} \tag{2}$$

- the advanced logistic sum function $\mathbf{alogistic}_{\sigma,\tau}(..)$ defined by:

$$\mathbf{alogistic}_{\sigma,\tau}(V_1, \ldots, V_k) = \left[\frac{1}{1+\mathrm{e}^{-\sigma(V_1+\ldots+V_k-\tau)}} - \frac{1}{1+\mathrm{e}^{\sigma\tau}}\right]\left(1+\mathrm{e}^{-\sigma\tau}\right) \tag{3}$$

This function is obtained from the simple logistic sum function by subtracting its value for sum 0 from it and rescaling the result for the [0, 1] interval.

The above concepts enable to design network models and their dynamics in a declarative manner, based on mathematically defined functions and relations.

2.2 Modeling Adaptive Networks as Self-modeling Networks

Realistic network models are usually adaptive: their network characteristics often are adapted over time. Therefore, their dynamics is usually an interaction (sometimes called co-evolution) of these two sorts of dynamics: dynamics of the nodes (or states) in the network (dynamics *within* the network) versus dynamics of the characteristics of the network (dynamics *of* the network). Dynamics of the network's nodes are modeled declaratively by declarative mathematical functions and relations. In contrast, the dynamics of the network characteristics traditionally are described in a procedural, algorithmic nondeclarative manner, which then leads to a hybrid type of model. But by using *self-models* within the network, a network-oriented conceptualisation can also be applied to *adaptive* networks to obtain a declarative description using mathematically defined functions and relations; see (Treur 2020). This works through the addition of new nodes to the network (called *self-model states* or *reification states*) which represent (adaptive) network characteristics. Such nodes are depicted at a next level (*self-model level*), where the original network is at a *base level*. These types of characteristics with their self-model states and their roles are shown in Table 1.

This provides an extended network, also called *self-modeling network*. Like for all network models, a self-modeling network model is specified in a (network-oriented) declarative mathematical manner based on nodes and connections. These include interlevel connections relating nodes at one level to nodes on the other.

Table 1. Different network characteristics and self-model states for them

Types of characteristics	Concepts	Notations	Self-model states	Role played by the self-model state
Connectivity characteristics	Connections weights	$\omega_{X,Y}$	$\mathbf{W}_{X,Y}$	Connection weight **W**
Aggregation characteristics	Combination functions and their parameters	$\mathbf{c}_Y(..)$ $\pi_{i,j,Y}$	$\mathbf{C}_{i,Y}$ $\mathbf{P}_{i,j,Y}$	Combination function weight **C** Combination function parameter **P**
Timing characteristics	Speed factors	η_Y	$\mathbf{H}_Y$	Speed factor **H**

The outcome is also a network model (Treur 2020, Chap. 10). This whole construction can be applied iteratively to obtain multiple self-model levels that can provide higher-order adaptive networks, and is quite useful to model, for example, plasticity and metaplasticity in the form of a second-order adaptive network with three levels, one base level and a first- and a second-order self-model level; e.g., (Treur 2020, Chap. 4).

To support the design of network models and simulation of them, for any application from a library predefined basic combination functions $\mathrm{bcf}_i(..)$, $i = 1, ..., m$ are selected by assigning weights $\boldsymbol{\gamma}_{i,Y}$, where the combination function then becomes the weighted average.

$$\mathbf{c}_Y(..) = (\boldsymbol{\gamma}_{1,Y}\mathrm{bcf}_1(..) + ... + \boldsymbol{\gamma}_{m,Y}\mathrm{bcf}_m(..))/(\boldsymbol{\gamma}_{1,Y} + ... + \boldsymbol{\gamma}_{m,Y}) \tag{4}$$

Furthermore, parameters of combination functions are specified, so that $\mathrm{bcf}_i(..) = \mathrm{bcf}_i(\mathbf{p},\mathbf{v})$ where **p** is a list of parameters and **v** is a list of values.

3 Modeling the Cognitive Architecture for Mental Models as a Self-modeling Network

Based on the different processes in which mental models are used as briefly discussed above, a cognitive architecture for handling mental models has been designed covering the three types of processes in an integrated manner as depicted in Fig. 1, left hand side. For more details of this architecture, see (Van Ments and Treur 2021c).

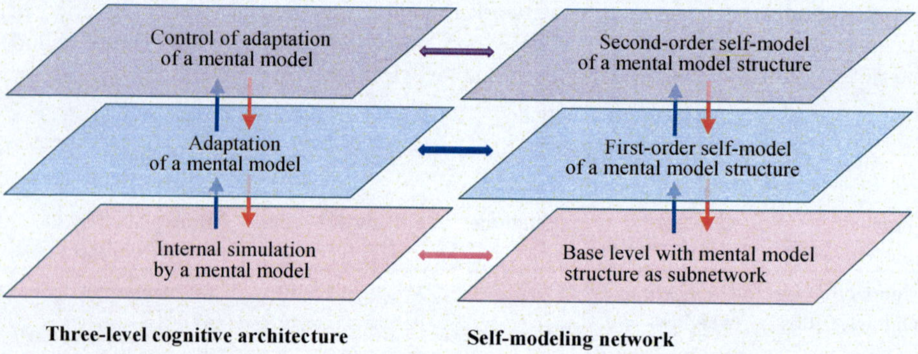

Fig. 1. Modeling the three-level cognitive architecture for mental model handling by a self-modeling network

The mapping from the three levels of the cognitive architecture to a self-modeling network is as follows (see also Fig. 1):

- **Lower level: a mental model as a subnetwork**
  The mental model as a relational structure at the base level within the cognitive architecture is modeled as a (sub)network structure of states (nodes) and connections between them at the base level of the self-modeling network; the dynamics of the states of this subnetwork model internal simulation of the mental model
- **Middle level: a first-order self-model of a mental model representing adaptation of its network structure**
  The level for adaptation of a mental model within the cognitive architecture is modeled as a first-order self-model of the mental model structure as represented at the base level; the dynamics of the states of this first-order self-model model adaptation by making changes in the structure of the mental model

- **Upper level: a second-order self-model of a mental model representing control of adaptation of its network structure**
  The level for control of adaptation of a mental model is modeled as a second-order self-model of the mental model, which is a self-model for the self-model for adaptation of the mental model; the dynamics of the states of this second-order self-model model control of adaptation by making changes in the structure of the first-order self-model that describes the adaptation of the mental model

So, mental models and the way they are handled can be considered as being described through multiple representations: they can be viewed from three levels of representation according to the three planes depicted in Fig. 1, right hand side. At the lower, base level depicted by the lower (pink) plane, a mental model, which in general essentially is considered to be a relational structure, is represented by nodes and connections between these nodes. For internal simulation, the nodes have activation levels that vary over time; based on the relations these activation levels affect each other over time. Next, at the adaptation level depicted in Fig. 1 right-hand side by the middle (blue) plane, it is represented how the mental model relations change over time by some adaptation specification. Finally, at the top level depicted by the upper (purple) plane in Fig. 1 it is indicated how the adaptation at the middle level is controlled. In this way, to model mental processes in which mental models play a role, within the self-modeling network these mental models do not get a single but a three-fold representation by which the different uses and operations on the mental model are distinguished like they are distinguished by the levels in the cognitive architecture.

In the next three sections each of the levels is discussed in some more detail and illustrated by many examples of realistic cases involving mental models in mental processes.

4 How Mental Models Can Be Used

From the viewpoint of the self-modeling network format, mental models are represented as subnetworks at the base level, which have their own internal connections (between their own mental model states), but also connections from and to other mental states that do not belong to the mental model. It is through the latter types of connections that a mental model can affect a person's mental processes in a wider sense and, more specifically, their decisions and actions. It was pointed out, among others already by Craik (1943) that having mental models enables organisms to make better decisions on how to act, as future developments can be predicted by these mental models:

> 'If the organism carries a "small-scale model" of external reality and of its own possible actions within its head, it is able to try out various alternatives, conclude which is the best of them, react to future situations before they arise, utilise the knowledge of past events in dealing with the present and future, and in every way to react in a much fuller, safer, and more competent manner to the emergencies which face it.' (Craik 1943, p. 61)

From an analysis of many cases, it was found that mental models can affect a wide variety of types of other mental states and behavioural actions. Mental states relating to goals and actions, preparation and ownership for actions, and action execution are among the mental states and behaviour that can be affected by mental models. However, also other types of mental states can be affected by mental models, such as emotional responses, awareness states, belief states, and states in other mental models. In Table 1, an overview can be found of several realistic examples of how mental models affect other (base level) mental states which are not part of the mental model; this indicates the way in which the mental models are used in the overall mental processes and behaviour. All of them have been formalized in computational self-modeling network format based on the cognitive architecture shown in Fig. 1; see the references in the table to papers where they are described in much more detail. In particular, the following cases are addressed in Table 2.

Case 1 Multiple mental models
At the first row a reference is made to a visualization case study where states of a geometric mental model affect states of an arithmetical mental model to reveal arithmetical relations.

Case 2 Flashbacks in PTSD
The second row refers to how for PTSD a mental model made of a traumatic event is sometimes triggered (flashback) and then in turn triggers stressful emotions and awareness of the traumatic event.

Case 3 Counterfactual thinking
In the third row a counterfactual thinking case is addressed, where based on mental models of different alternative scenarios, beliefs are revised.

Case 4 Self-interpretation in therapy
The fourth row refers to a realistic therapy case in which self-interpretation based on a mental model of a person's own functioning leads to stronger awareness and revised beliefs.

Case 5 Metaphors for joint decision making
In the fifth row mental models for competitive and cooperative metaphors for joint decision making are considered that affect action ownership states that form the basis of a decision.

Case 6 Mental God-model and empathy
The sixth row refers to a case study of mental God-models, where these mental models affect a number of other mental states concerning the person's own actions, goals and also emotions; for example empathic or disempathic actions are influenced by a mental model of an empathic or disempathic God.

Case 7 Mental attachment model
The seventh row addresses a case for Attachment Theory, where mental models of self and other developed based on a primary caregiver during childhood, affect preparation for actions with respect to significant others later in life.

Case 8 Shared mental model
Finally, in the last row a case study for shared mental models for hospital teamwork is addressed, where, like in row 5, these mental models affect action decisions via their ownership states.

All in all, the influences of mental models on a person's mental processes at the base level can be diverse.

Table 2. Overview of the example mental models and which other mental states and behaviour they affect

		Action preparation	Action ownership	Action execution	Goal	Emotion	Awareness	Belief	Other mental model
Multiple mental models	(Treur 2021a)								+
Flashbacks in PTSD	(Van Ments and Treur 2021b)					+	+		
Counterfactual thinking	(Bhalwankar and Treur 2021c)							+	+
Self-interpretation in therapy	(Treur and Glas 2021)						+	+	
Metaphors for joint decision making	(Van Ments and Treur 2021c)		+						
Mental God-model and empathy	(Van Ments et al. 2022)	+	+	+	+	+			
Mental attachment model	(Hermans et al. 2021)	+							
Shared mental model	(Van Ments et al. 2021)		+						

To illustrate this in some more detail, for Case 2 and 3 the conceptual connectivity view on the network model design is provided in Fig. 2 and 3. Notice the three levels as introduced in Fig. 1: the base level in pink, the first-order self-model level to allow for adaptation in blue and the second-order self-model level to control adaptation in purple. Again, more depth on the models can be found in the references indicated in Table 2.

In Fig. 2 sensing of a traumatic event te consisting of a sequence of phases or steps is modeled by sensor states ss_{te1}, ss_{te2}, ss_{te3}. For example, te1 or traumatic event step 1, is a potentially dangerous situation for a child you observe, the second step te2 is an action from your side with the intention to save the child from that situation and te3 is an unfortunate failure of your action such that the child actually gets hurt. During this traumatic event sequence, sensory representations srs_{te1}, srs_{te2}, srs_{te3} are activated, and by sensory preconditioning (Brogden 1947; Hall 1996) the connections between these sensory representations are learned through a Hebbian learning mechanism (Hebb 1949). By this learning process, the mental model of the traumatic event sequence is formed and represented by base states srs_{te1}, srs_{te2}, srs_{te3} and their connections (see the small pink parallelogram within the base plane in Fig. 1) with first-order self-model states $\mathbf{W}_{srs_{te1},srs_{te2}}$ and $\mathbf{W}_{srs_{te2},srs_{te3}}$. What can be seen in Fig. 2 is that the mental model states have outgoing base level connections to two other mental states that are not part of the mental model: to the awareness state as_{te} of the trauma and to the emotional response preparation state ps_b. This is indeed what is shown in the second row of Table 1.

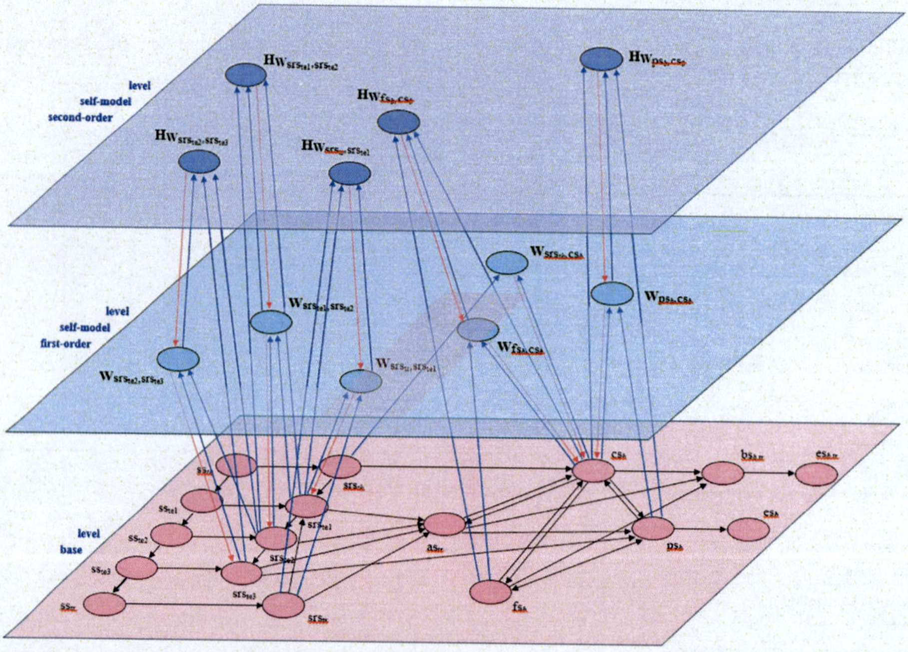

Fig. 2. Connectivity of the second-order adaptive network model of Case 2 described above

In the example for Case 3 shown in Fig. 3 the mental models are depicted within the small outlined areas within the (pink) base level plane. In this case, there is no direct connection within the base plane from mental model states to belief states but a causal pathway through the first-order self-model level (the middle blue plane); in this case (by a liberal interpretation) this is also considered as a causal effect on the belief, which at the same time is part of another mental model. Therefore, in row 3 of Table 1 there are two + indications.

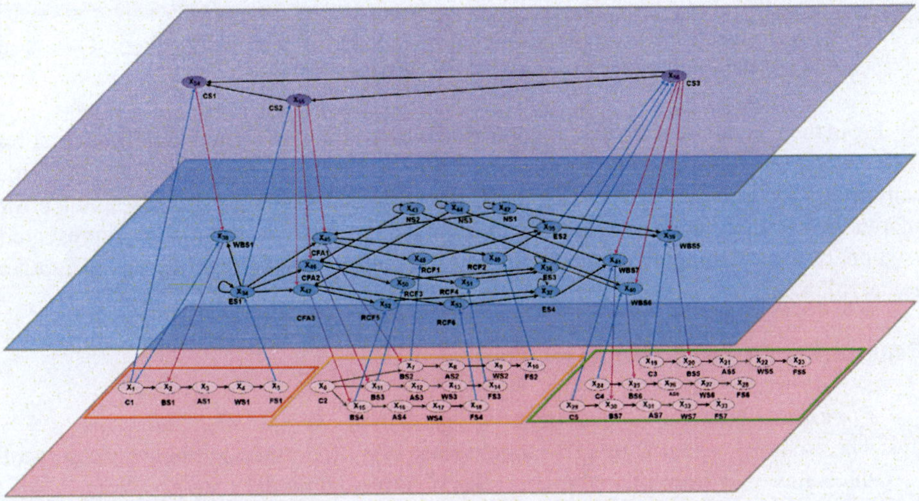

Fig. 3. Conceptual model of case 3 described above including the controlled adaptive network model for counterfactual thinking

5 How Mental Models Can Be Adapted

Different forms of adaptation or learning can be applied to mental models. Examples of individual types of learning are

- Learning of connections of mental models by observation of a process in the real world; for example, based on Hebbian learning (Hebb 1949)
- Learning of excitability of mental model nodes; for example, (Chandra and Barkai, 2018; Debanne, Inglebert, Russier, 2019; Sjöström et al. 2008)
- Learning of connections of mental models by instruction or by being told; for example by instructional learning, e.g., (Hogan and Pressley 1997; Seel 2006)
- Learning of connections of mental models from other, related mental models; for example, learning arithmetical or algebraic relations based on visualization like in Case 1 in Sect. 3; e.g., (Du Plooy 2016; Koedinger and Terao 2002)
- Learning of mental models based on counterfactual thinking like in Case 3 in Sect. 3; e.g., (Van Hoeck et al. 2015)
- Learning of mental models based on Theory of Mind self-interpretation, like in Case 4 in Sect. 3; for example (Treur and Glas 2021)

Also at the non-individual, social level, adaptation of mental models can play an important role. Three examples are

- Bonding based on homophily (or similarity), where the mental models describing the internal representations of connections between persons are adapted over time; e.g., (McPherson et al. 2001; Treur 2021b)
- Development of attachment relations as described by Attachment Theory, like in Case 7 in Sect. 3; e.g., (Bartholomew and Horowitz 1991; Hermans et al. 2021)

- Forgetting connections in a shared mental model during teamwork, like in Case 8 in Sect. 3; e.g., (Burtscher and Manser 2012; Van Ments et al. 2021)

From the viewpoint of the self-modeling network format as formalization, adaptation is described by (first-order) self-models of parts of the (base) network. This automatically offers a number of specific available network characteristics that are suitable for adaptation and that can be used in particular for adaptation of mental models represented as subnetworks. Examples of such networks characteristics that can be made adaptive are as follows.

Connectivity for mental model relations

- For example, mental model connections and their weights
- These can be modeled by self-model states $\mathbf{W}_{X,Y}$ representing the weight $\boldsymbol{\omega}_{X,Y}$ of the connection from mental model state X to mental model state Y

Aggregation for mental model states

- For example, excitability thresholds of mental model states
- These can be modeled by self-model states $\mathbf{T}_Y$ representing the excitability threshold $\boldsymbol{\tau}_Y$ of mental model state Y

Timing for mental model states

- For example, speed factors of mental model states
- These can be modeled by self-model states $\mathbf{H}_Y$ representing the speed factor $\boldsymbol{\eta}_Y$ of mental model state Y.

In Table 3, an overview can be found of such examples of adaptation of mental models that enabled to model a variety of realistic scenarios involving mental models. The first 8 rows correspond to the eight cases discussed in Sect. 3. In this table four new cases are shown:

Case 9 Car driving mental model

This case addresses the learning of a mental model of how a car works and how you can drive it; this takes place through a combination of instructional (by being told) and observational learning (by Hebbian learning).

Case 10 Self-controlled learning

In this case of a car driving mental model, the learner first learns by observation and then takes initiative to ask for (confirmative) instruction.

Case 11 Analysis and support
This case describes in the context of providing support, an adaptive mental model for analysis of possible problems and an adaptive mental model of suitable support actions for such problems.

Case 12 Bonding by homophily
This case addresses how the mental models of the (adaptive) connections between two persons are affected by how similar the (adaptive) mental models of characteristics or states of the two persons are.

For Case 2 depicted in Fig. 2 in row 2 of Table 2 it is indicated that the flashback movie is learned by Hebbian learning; for Case 3 depicted in Fig. 3, it is indicated in row 3 of Table 3 that it addresses learning by counterfactual thinking.

Table 3. Overview of realistic cases of how mental models can be adapted

		Learning by						Bonding by	
		Hebbian learning	Excitability learning	Information source	Counterfactual thinking	Other mental models	Theory of Mind analysis	Homophily	Attachment
Multiple mental models	(Treur 2021a)	+				+			
Flashbacks in PTSD	(Van Ments and Treur 2021a)	+							
Counterfactual thinking	(Bhalwankar and Treur 2021c)				+				
Self-interpretation in therapy	(Treur and Glas 2021)	+	+				+		
Metaphors for joint decision making	(Van Ments and Treur 2021b)	+							
Mental God-model and empathy	(Van Ments et al. 2022)	+							
Mental attachment model	(Hermans et al. 2021)	+							+
Shared mental model	(Van Ments et al. 2021)	+		+					
Car driving mental model	(Bhalwankar and Treur 2021a)	+		+					
Self-controlled learning	(Bhalwankar and Treur 2021b)	+		+					
Analysis and support	(Treur 2021c)	+	+						
Bonding by homophily	(Treur 2021b)							+	

6 How Mental Model Adaptation Can Be Controlled

In this section, different ways are discussed in which control over mental models and their adaptation can be exerted. Again this is done from the viewpoint of the network format provided by self-modeling networks. In (Sjöström et al. 2008), the topic of control of adaptation is discussed in relation to what is called the Plasticity Versus Stability Conundrum. Also within an AI-context, in machine learning examples of this conundrum and controlled adaptation to address it are known, such as the (decreasing) temperature parameter in simulated annealing and the sensitive balancing between exploration and exploitation in reinforcement learning, also called the explore-exploit dilemma (Holland 1975; March 1991; Wilson et al. 2014). This is explained in (Wilson et al. 2014) as follows:

> 'When you go to your favorite restaurant, do you always order the same thing, or do you try something new? Sticking with an old favorite ensures a good meal, but if you are willing to explore you might discover something better. This simple conundrum, deciding between something you know and something you do not, is commonly referred to as the exploration– exploitation dilemma.'

Applied to mental models in particular, this quote illustrates that on the one hand decision making based on known mental models can be very efficient (navigating based on a well-known map), but on the other hand this may prevent someone from learning even better decisions (exploring still unknown territory).

In a controlled adaptive network model for mental models based on the self-modeling network format, adaptation is modelled by a first-order self-model, as discussed in Sect. 4. There are a number of network characteristics involved in the structure of a first-order self-model used for the adaptation. By systematically going through these possible network characteristics, the following examples of network characteristics for adaptation to be controlled can be distinguished and are illustrated by various examples. Recall from Sect. 4 how exactly at the middle level self-model states can be introduced to represent adaptive network characteristics from the lower level. For example, self-model state $\mathbf{W}_{X,Y}$ represents an adaptive connection weight $\boldsymbol{\omega}_{X,Y}$ from the base level, and $\mathbf{H}_Y$ represents speed factor $\boldsymbol{\eta}_Y$ from the lower level, and so on. This can be iterated for the middle and upper level to obtain a second-order self-model. For example:

- second-order self-model state $\mathbf{H}_{\mathbf{W}_{\mathrm{X,Y}}}$ can be used to represent the adaptive adaptation speed of first-order self-model state $\mathbf{W}_{X,Y}$
- second-order self-model state $\mathbf{M}_{\mathbf{W}_{\mathrm{X,Y}}}$ can be used to represent the adaptive persistence parameter $\mu_{\mathbf{W}_{\mathrm{X,Y}}}$ of first-order self-model state $\mathbf{W}_{X,Y}$.
- second-order self-model state $\mathbf{W}_{Z,\mathbf{W}_{\mathrm{X,Y}}}$ can be used to represent the adaptive weight $\boldsymbol{\omega}_{Z,\mathbf{W}_{\mathrm{X,Y}}}$ of the connection from some state Z to state $\mathbf{W}_{X,Y}$.

For shortness, such second-order self-model states are sometimes called $\mathbf{H}_{\mathbf{W}}$-states, $\mathbf{M}_{\mathbf{W}}$-states, or $\mathbf{W}_{\mathbf{W}}$-states, whereas first-order self-model states can be called, for example, **W**-states or **T**-states. Following the different types of network characteristics used in the self-modeling network format, the following types of control can be distinguished.

- **Control by adaptive connectivity characteristics of first-order self-model states**
  - Adaptive connections of the causal pathways to the self-model states and their weights $\boldsymbol{\omega}$; for example:

    Choosing a mental model to be applied. For example, a decision to use a specific metaphor-based mental model as in the model in (Van Ments and Treur 2021b) or a decision to use a geometric mental model to support learning of an arithmetic mental model, as described in (Treur 2021a)
    Opening a communication channel from an information source to enable instructional learning of a mental model (decision to ask), as in the model in (Bhalwankar and Treur 2021b) and in the model described in (Treur 2021b)
    Opening an observation channel to enable observational learning of a mental model (decision to observe), as in the model described in (Treur 2021b)

  - Adaptive connections of the causal pathways from the self-model states to other states and their weights $\boldsymbol{\omega}$; for example:

    Modelling the effects of a chosen metaphor as in the model in (Van Ments and Treur 2021b)

- **Control by adaptive aggregation characteristics of first-order self-model states**
  - Adaptive choice of combination function; for example:

    For Hebbian learning of mental model connections a weighted average of $\mathbf{hebb}_{\boldsymbol{\mu}}(V_1, V_2, W)$ and $\mathbf{smin}_{\boldsymbol{\lambda}}(V_1, V_2)$, with adaptive weights $\boldsymbol{\gamma}_1$ and $\boldsymbol{\gamma}_2$.

  - Adaptive parameters of chosen combination functions; for example:

    Adaptive values for the persistence factor $\boldsymbol{\mu}$ of $\mathbf{hebb}_{\boldsymbol{\mu}}(..)$ as in the self-modeling network model for shared mental models described in (Van Ments et al. 2021) or for the scaling factor $\boldsymbol{\lambda}$ of $\mathbf{smin}_{\boldsymbol{\lambda}}(V_1, V_2)$.

- **Control by adaptive timing characteristics of first-order self-model states**
  - Adaptive adaptation speed (learning rate) $\boldsymbol{\eta}$; for example:

    Addressing the Plasticity Versus Stability conundrum (Sjöström et al. 2008) based on some context factors indicating when plasticity is needed fully and when plasticity should be limited or frozen.
    Accelerating adaptation speed upon increased stimulus exposure (Robinson et al. 2016), for example as applied in the example model in (Van Ments and Treur 2021b)

As discussed above, in a self-modeling network format, control of any of such network characteristics (for first-order self-models for adaptation of a mental model) is modeled by a second-order self-model. To illustrate this, based on the above distinctions, in Table 3 a summarized overview is given of several cases of applications of second-order self-models to control adaptation of mental models as also collected in (Treur and Van Ments 2022).

For case 2 depicted in Fig. 2 in row 2 of Table 4 it is indicated that for the learning of the flashback movie by Hebbian learning the adaptation speed is controlled; for case 3 depicted in Fig. 3, it is indicated in row 3 of Table 4 that for the learning by counterfactual thinking the exchange between the different mental models is controlled.

Table 4. Overview of example mental models: how they are controlled

		Control of adaptation speed	Control of persistence	Control of communication	Control of observation	Control of exchange
Multiple mental models	(Treur 2021a)					+
Flashbacks in PTSD	(Van Ments and Treur 2021a)	+				
Counterfactual thinking	(Bhalwankar and Treur 2021c)					+
Self-interpretation in therapy	(Treur and Glas 2021)					+
Metaphors for joint decision making	(Van Ments and Treur 2021b)	+				
Mental God-model and empathy	(Van Ments et al. 2022)	+				
Mental attachment model	(Hermans et al. 2021)	+				
Shared mental model	(Van Ments et al. 2021)		+			
Self-controlled learning	(Bhalwankar and Treur 2021b)			+		
Analysis and support	(Treur 2021c)	+				+
Bonding by homophily	(Treur 2021b)			+	+	

Moreover, in Table 5, a more complete overview is obtained of different types of control against different types of learning. Note that in most cells in Table 5 further references are included, but for those cells where no references are included, in general this means that these options are yet to be explored in detail. This contributes to a future research agenda.

Table 5. Overview of different types of controlled adaptation for example mental models as collected in (Treur and Van Ments 2022)

Learning by	Control of adaptation via connectivity	Control of adaptation via aggregation	Control of adaptation via timing
Observation and monitoring	Controlled learning by observation for a mental model **W**-state for bonding via a $\mathbf{W_W}$-state (Treur and Van Ments, 2022, Chap. 13); formation of a mental model of another person; see also (Hermans et al. 2021)	Hebbian mental model learning **W**-state persistence control via an $\mathbf{M_W}$-state (Treur and Van Ments 2022, Chap. 14); controlled forgetting of a mental model relation; see also (Van Ments et al. 2021)	Hebbian mental model learning **W**-state speed control via an $\mathbf{H_W}$-state: adaptation accelerates with increasing exposure (Treur and Van Ments 2022, Chap. 5, Chap. 7, Chap. 10, Chap. 11, Chap. 12); e.g., learning mental models for flashback experiences (Van Ments and Treur 2021b), analysis and support tasks (Treur 2021c), metaphors (Van Ments and Treur 2021c), mental God-model (Van Ments et al. 2022), self- and other-models (Hermans et al. 2021)
Excitability adaptation	Incoming connection for an adaptive mental model excitability **T**-state control via a $\mathbf{W_T}$-state (Treur and Van Ments 2022, Chap. 7); learning excitability (Debanne et al. 2019; Chandra and Barkai 2018) of a mental model's states; see also (Treur 2021c)	Excitability mental model learning **T**-state aggregation control, for example through adaptive (steepness $\boldsymbol{\sigma}$ and threshold $\boldsymbol{\tau}$) parameters of a logistic combination function used for the **T**-state represented by $\mathbf{S_T}$ - and $\mathbf{T_T}$ - states	Excitability mental model learning **T**-state speed control via an $\mathbf{H_T}$-state (Treur and Van Ments 2022, Chap. 7); learning excitability (Debanne et al. 2019; Chandra and Barkai 2018) of a mental model's states; see also (Treur 2021c)

(*continued*)

Table 5. (*continued*)

Learning by	Control of adaptation via connectivity	Control of adaptation via aggregation	Control of adaptation via timing
Communication	Learner-controlled instructional learning of a mental model **W**-state via a $\mathbf{W_W}$-state (Treur and Van Ments 2022, Chap. 9); opening a communication channel with the instructor by asking; see also (Bhalwankar and Treur 2021b) Controlled learning by communication for a mental model **W**-state for bonding via a $\mathbf{W_W}$-state (Treur and Van Ments 2022, Chap. 13); opening a communication channel with the other person by asking; see also (Treur 2021b)	Learner-controlled instructional learning of a mental model **W**-state via a $\mathbf{T_W}$-state for excitability (Debanne et al. 2019; Chandra and Barkai 2018) of the **W**-state (opening a communication channel with the instructor by more sensitive listening)	Learner-controlled instructional learning of a mental model **W**-state via an $\mathbf{H_W}$-state (controlling the timing of a communication channel with the instructor)
Other mental models	Controlled connection **W**-state for counterfactual activation of a mental model via a $\mathbf{W_W}$-state (Treur and Van Ments, 2022, Chap. 6); see also (Bhalwankar and Treur 2021c)	Controlled connection **W**-state for counterfactual activation of a mental model via a $\mathbf{T_W}$-state addressing excitability (Debanne et al. 2019; Chandra and Barkai, 2018) of the **W**-state	Controlled inter mental model exchange connection **W**-state via $\mathbf{H_W}$-state (Treur and Van Ments, 2022, Chap. 4); exchange from arithmetic mental model to geometric mental model; see also (Treur 2021a)

(*continued*)

Table 5. (*continued*)

Learning by	Control of adaptation via connectivity	Control of adaptation via aggregation	Control of adaptation via timing
Contextual factors	Controlled connection **W**-state for activation of mental model via a $\mathbf{W_W}$-state (Treur and Van Ments 2022, Chap. 6); activation of a mental model for possible future action; see also (Bhalwankar and Treur 2021c)	Controlled connection **W**-state for activation of mental model via a $\mathbf{T_W}$-state addressing excitability of the **W**-state based on contextual factors	Controlled adaptive mental model effect connection **W**-state via an $\mathbf{H_W}$-state (Treur and Van Ments 2022, Chap. 10); adapting the own choices based on the context given by the other person; see also (Van Ments and Treur 2021c)

7 Discussion

This paper addressed an analysis of how in mental and social processes, humans often apply specific mental models and learn and adapt them in a controlled manner. Part of it is based on (Bhalwankar et al. 2021). It was discussed how controlled adaptation relates to the Plasticity Versus Stability Conundrum in neuroscience (Sjöström et al. 2008). From the analysis, an informal three-level cognitive architecture for controlled adaptation was obtained. It was discussed from a self-modeling network viewpoint how this cognitive architecture can be modeled as a self-modeling network (Treur 2020). Making use of the specific network characteristics offered by the self-modeling network structure format, a large number of options for different types of adaptation of mental models and different types of control over adaptation of mental models were obtained and structured. Many but not all of these options were illustrated by a several realistic examples that were already formalized by self-modeling networks. The options that were not illustrated here in a formalized computational sense, provide interesting options for a future research agenda.

References

Abraham, W.C., Bear, M.F.: Metaplasticity: the plasticity of synaptic plasticity. Trends Neurosci. **19**(4), 126–130 (1996)

Bartholomew, K., Horowitz, L.M.: Attachment styles among young adults: a test of a four-category model. J. Pers. Soc. Psychol. **61**(2), 226–244 (1991)

Bhalwankar, R., Treur, J.: Modeling the development of internal mental models by an adaptive network model. In: Proceedings of the 11th Annual International Conference on Brain-Inspired Cognitive Architectures for AI, BICA*AI'20. Procedia Computer Science. Elsevier (2021a)

Bhalwankar, R., Treur, J.: A second-order adaptive network model for learner-controlled mental model learning processes. In: Benito, R.M., Cherifi, C., Cherifi, H., Moro, E., Rocha, L.M., Sales-Pardo, M. (eds.) COMPLEX NETWORKS 2020 2020. SCI, vol. 944, pp. 245–259. Springer, Cham (2021b). https://doi.org/10.1007/978-3-030-65351-4_20

Bhalwankar, R., Treur, J.: 'If only i would have done that…': a controlled adaptive network model for learning by counterfactual thinking. In: Maglogiannis, I., Macintyre, J., Iliadis, L. (eds.) Artificial Intelligence Applications and Innovations: 17th IFIP WG 12.5 International Conference, AIAI 2021, Hersonissos, Crete, Greece, June 25–27, 2021, Proceedings, pp. 3–16. Springer, Cham (2021c). https://doi.org/10.1007/978-3-030-79150-6_1

Bhalwankar, R., Van Ments, L., Treur, J.: Context-sensitive control of adaptation: self-modeling networks for human mental processes using mental models. In: Proceedings of the 3rd International Conference on Machine Learning and Intelligent Systems, MLIS 2021. Frontiers in Artificial Intelligence and Applications. IOS Press (2021)

Brogden, W.J.: Sensory preconditioning of human subjects. J. Exp. Psychol. **37**, 527–539 (1947)

Burtscher, M.J., Manser, T.: Team mental models and their potential to improve teamwork and safety: a review and implications for future research in healthcare. Saf. Sci. **50**(5), 1344–1354 (2012). https://doi.org/10.1016/j.ssci.2011.12.033

Chandra, N., Barkai, E.: A non-synaptic mechanism of complex learning: modulation of intrinsic neuronal excitability. Neurobiol. Learn. Mem. **154**, 30–36 (2018)

Craik, K.J.W.: The Nature of Explanation. University Press, Cambridge (1943)

Darling-Hammond, L., Austin, K., Cheung, M., Martin, D.: Thinking about thinking: metacognition. In: The Learning Classroom: Theory into Practice, pp. 157–172. Stanford University School of Education (2008)

Debanne, D., Inglebert, Y., Russier, M.: Plasticity of intrinsic neuronal excitability. Curr. Opin. Neurobiol. **54**, 73–82 (2019)

Du Plooy, M.C.: Visualisation as a metacognitive strategy in learning multiplicative concepts: a design research intervention. Ph.D. Thesis, University of Pretoria, Department of Mathematics Education (2016). https://repository.up.ac.za/handle/2263/51258

Hall, G.: Learning about associatively activated stimulus representations: Implications for acquired equivalence and perceptual learning. Anim. Learn. Behav. **24**(3), 233–255 (1996). https://doi.org/10.3758/BF03198973

Hebb, D.O.: The Organization of Behavior: A Neuropsychological Theory. Wiley (1949)

Hermans, A., Muhammad, S., Treur, J.: A second-order adaptive network model for attachment theory. In: Proceedings of the 21th International Conference on Computational Science, ICCS 2021. LNCS, vol. 12744, pp 462–475. Springer, Heidelberg (2021)

Hogan, K.E., Pressley, M.E.: Scaffolding Student Learning: Instructional Approaches and Issues. Brookline Books (1997)

Holland, J.H.: Adaptation in Natural and Artificial Systems. Ann Arbor, MI, University of Michigan Press (1975)

Hurley, S.: The shared circuits model (SCM): how control, mirroring, and simulation can enable imitation, deliberation, and mindreading. Behav. Brain Sci. **31**(1), 1–22 (2008)

Koedinger, K.R., Terao, A.: A cognitive task analysis of using pictures to support pre-algebraic reasoning. In: Gray, W.D., Schunn, C.D. (eds.) Proceedings of the 24th Annual Conference of the Cognitive Science Society, CogSci 2002, pp. 542–547. Lawrence Erlbaum Associates, Mahwah, NJ (2002)

Magerl, W., Hansen, N., Treede, R.D., Klein, T.: The human pain system exhibits higher-order plasticity (metaplasticity). Neurobiol. Learn. Mem. **154**, 112–120 (2018)

Mahdavi, M.: An overview: metacognition in education. Int. J. Multidisc. Curr. Res. **2**, 529–535 (2014)

March, J.G.: Exploration and exploitation in organizational learning. Organ. Sci. **2**, 71–87 (1991)

McPherson, M., Smith-Lovin, L., Cook, J.M.: Birds of a feather: homophily in social networks. Ann. Rev. Sociol. **27**(1), 415–444 (2001)

Piaget, J.: Origins of Intelligence in the Child (La Naissance de l'intelligence chez l'enfant). Routledge & Kegan Paul, London (1936)

Pintrich, P.R.: The role of goal orientation in self-regulated learning. In: Boekaerts, M., Pintrich, P., Zeidner, M. (eds.) Handbook of Self-regulation Research and Applications, pp. 451–502. Academic Press, Orlando, FL (2000)

Robinson, B.L., Harper, N.S., McAlpine, D.: Meta-adaptation in the auditory midbrain under cortical influence. Nat. Commun. **7**, e13442 (2016)

Seel, N.M.: Mental models in learning situations. In: Advances in Psychology, vol. 138, pp. 85–107. North-Holland, Amsterdam (2006)

Sjöström, P.J., Rancz, E.A., Roth, A., Hausser, M.: Dendritic excitability and synaptic plasticity. Physiol Rev **88**, 769–840 (2008)

Treur, J.: Network-Oriented Modeling for Adaptive Networks: Designing Higher-Order Adaptive Biological, Mental and Social Network Models. Springer Nature, Switzerland (2020)

Treur, J.: An adaptive network model covering metacognition to control adaptation for multiple mental models. Cogn. Syst. Res. **67**, 18–27 (2021a)

Treur, J.: Controlled social network adaptation: subjective elements in an objective social world. In: Yang, X.-S., Sherratt, S., Dey, N., Joshi, A. (eds.) Proceedings of Sixth International Congress on Information and Communication Technology. LNNS, vol. 235, pp. 263–274. Springer, Singapore (2021b). https://doi.org/10.1007/978-981-16-2377-6_26

Treur, J.: Self-modeling networks using adaptive internal mental models for cognitive analysis and support processes. In: Benito, R.M., Cherifi, C., Cherifi, H., Moro, E., Rocha, L.M., Sales-Pardo, M. (eds.) Complex Networks 2020 . SCI, vol. 944, pp. 260–274. Springer, Cham (2021c). https://doi.org/10.1007/978-3-030-65351-4_21

Treur, J., Glas, G.: A multi-level cognitive architecture for self-referencing, self-awareness and self-interpretation. Cogn. Syst. Res. **68**, 125–142 (2021)

Treur, J., Van Ments, L. (eds.): Mental Models and their Dynamics, Adaptation, and Control: a Self-modeling Network Modeling Approach. Springer Nature, Heidelberg (2022, in press)

Van Hoeck, N., Watson, P.D., Barbey, A.K.: Cognitive neuroscience of human counterfactual reasoning. Front. Hum. Neurosci. **9**, 420 (2015)

van Ments, L., Treur, J.: A Higher-order adaptive network model to simulate development of and recovery from PTSD. In: Paszynski, M., Kranzlmüller, D., Krzhizhanovskaya, V.V., Dongarra, J.J., Sloot, P.M.A. (eds.) ICCS 2021. LNCS, vol. 12743, pp. 154–166. Springer, Cham (2021a). https://doi.org/10.1007/978-3-030-77964-1_13

Van Ments, L., Treur, J.: Modeling adaptive cooperative and competitive metaphors as mental models for joint decision making. Cogn. Syst. Res. **69**, 67–82 (2021b)

Van Ments, L., Treur, J.: Reflections on dynamics, adaptation and control: a cognitive architecture for mental models. Cogn. Syst. Res. **70**, 1–9 (2021c)

van Ments, L., Treur, J., Klein, J., Roelofsma, P.: A second-order adaptive network model for shared mental models in hospital teamwork. In: Nguyen, N.T., Iliadis, L., Maglogiannis, I., Trawiński, B. (eds.) ICCCI 2021. LNCS (LNAI), vol. 12876, pp. 126–140. Springer, Cham (2021). https://doi.org/10.1007/978-3-030-88081-1_10

Van Ments, L., Treur, J., Roelofsma, P.H.M.P.: An adaptive network model for formation and use of a mental god-model and its effect on human empathy, Chap. 11. In: Treur, J., Van Ments, L. (eds.) Mental Models and their Dynamics, Adaptation, and Control: A Self-modeling Network Modeling Approach. Springer Nature, Heidelberg (2022, in press)

Wilson, R.C., Geana, A., White, J.M., Ludvig, E., Cohen, J.D.: Humans use directed and random exploration to solve the explore–exploit dilemma. J. Exp. Psychol. Gen. **143**(6), 2074–2081 (2014)

Interdisciplinary Approach in the Development of Geographic Information Technologies

D. V. Okunev, T. A. Salimova, Y. N. Sushkova, A. A. Yamashkina, and S. A. Yamashkin(✉)

National Research Mordovia State University, 68 Bolshevistskaya Street, 430005 Saransk, Russia

Abstract. The article is devoted to solving the scientific problem of the development of interdisciplinary research in the field of solving complex problems in the field of sustainable development of countries and regions on the basis of the introduction and effective use of new geoinformation technologies that ensure the collection, systematization and processing of data and knowledge about nature, economy and the population of the territories. The basis for the solution of the designated problem is determined by the need for the development of theory, methodology and methods for the development of interdisciplinary research for planning the sustainability of ecological-socio-economic systems (metageosystems) of territories. The central attention is paid to the problem of creating effective infrastructures of spatial data and geoportal systems as tools for solving the problem of integration, cataloging, dissemination and visualization of spatial data in order to support the adoption of managerial decisions in the field of optimizing the functioning of metageosystems, the development of the digital economy. It is shown that increasing the efficiency of diagnostics of spatio-temporal changes in metageosystems should be based on the introduction of intelligent systems used for the conjugate analysis of remote sensing data of the Earth, the results of monitoring environmental, social and economic processes and their interaction in the development of territories. While regional geoportals ensure the coordination of the search and rational use of natural resources, optimize territorial and sectoral planning and economic management, monitoring the ecological state of the region.

Keywords: Metageosystems · Geoportals · Spatial data infrastructures · Sustainable development

1 Introduction

Modern global and regional processes of changing the life support systems of the population put forward problems for science and practice, which must be solved fundamentally, on the basis of interdisciplinary research [1]. The generated knowledge reflects the current state of natural, social and production systems and their interaction. As a result of such interaction, complex metageosystems are formed, which represent the environment for the existence of society with environmental, economic, technological and social components.

R. Silhavy et al. (Eds.): CoMeSySo 2021, LNNS 231, pp. 648–658, 2021.
https://doi.org/10.1007/978-3-030-90321-3_53

The development of interdisciplinary research in the field of solving complex problems is associated with the development of special information systems that ensure the collection, systematization and processing of data and knowledge about the nature, economy and population of territories [2]. Such a system is built on a regional basis, reflecting the hierarchical organization of metageosystems and include a complex of information, the rational volume of which should provide a timely and scientifically grounded assessment of the state of local, regional and global problems. Ideally, a decision made at any level should be based on the necessary and sufficient amount of information for operational or strategic forecasting - to represent a certain link or a system of targeted long-term changes. Naturally, the amount of information, the principles of its processing and evaluation in each specific case should be rationally selected and scientifically grounded.

2 Methods

The geosystem approach to the analysis of the decisions made inevitably leads to the fact that each natural object enters the field of view of a certain subject of economic management. And, probably, for each object there are a finite number of possible, alternative decisions to be made. Decisions made, referred to a specific territory and a specific period of time, can be mapped and received maps of territory management. Any decision made on the basis of data processing and analysis can have various types of digital information: maps, tables, observation results, Earth remote sensing data. Therefore, one of the tasks of interdisciplinary research, when analyzing the problems of managing complex geosystems [3], is the practical use of spatial data infrastructures, geoportals and geographic information systems (GIS) as a basis for supporting management decisions. It should be borne in mind that geosystems differ in a certain territorial scope, with an increase in the size of the territory, the structure of decisions made also increases. Due to this, the need for a hierarchical systematization, typification and optimization of geoinformation materials as a basis for supporting decisions is clearly traced.

New directions and samples of research and practical activities in the field of analysis and assessment of the state of metageosystems are born in the development and operation of spatial data infrastructures. Geoinformation methods and technologies are associated with the prospects for integrating spatially defined and temporal data into integral information systems, and integrating the processes of creating and using maps. Its essence consists in information-cartographic modeling and research of natural, social and economic systems and their interaction based on digital databases, GIS technologies and special knowledge. Within the framework of this direction, an active search for effective computer methods of analysis and modeling is currently underway, operational and multivariate mapping is developing, new types of cartographic products and products are being created.

Elements of metageosystems are linked by flows of matter, energy and information. According to V.N. Solntsev, it is necessary to highlight the connections of relationships and interactions. In the author's interpretation, "relationship-relation is any form of correspondence of phenomena, which can be represented as a functional dependence of variables, which has either an unambiguous character (functional relationship in the narrow sense of the word) or probabilistic (correlation relationship). Communication-interaction is a process of mutual influence of objects (their properties), taking place in real space and time and expressed in a certain course (sequence) of events that occur with matter (and energy) in a certain area of space" [4, p. 27]. Disclosure of connections-relations and connections-interactions between components in the structure of metageosystems is the most important task of interdisciplinary research (Fig. 1).

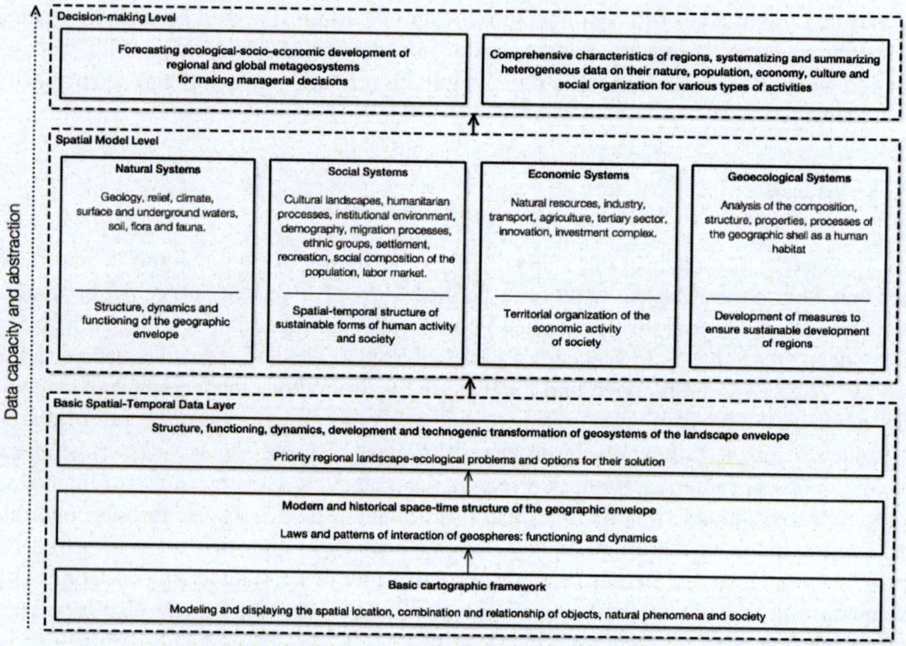

Fig. 1. General scheme of regional and global metageosystems

The problem of territorial development, especially actualized by developments within the framework of the global concept of sustainable development of society, occupies one of the leading places in the system of interdisciplinary research: the development of methods for the digital description of the processes of development of territorial systems, the search for criteria for quantifying its rates and determination of the management strategy [5].

Despite the introduction of the "human development index" into international practice, the indicator of per capita gross domestic product (per capita GDP) is still most often used to assess the level of development of countries [6]. To reduce the existing shortcomings of this indicator, which does not allow taking into account a number of nuances of the economic process taking place in the territory, per capita GDP began to be calculated taking into account the actual purchasing power of the national currency, which significantly changed the idea of the economic picture of the world. Recently, they have been trying to introduce coefficients into the calculations that take into account the level of the shadow sector of the economy, which even more "straightens out" economic realities. The national accounts of the industrially developed countries of the world keep records of the depreciation of the means of production, as well as the degree of use of natural resources and the negative impact of anthropogenic activities on the natural environment, which makes it possible to calculate the so-called "green GDP" and the "actual savings" indicator.

Taking into account the fact that sustainable, i.e. guaranteed in the longest term, the development of society is possible only under the condition of harmonious development of natural, social and production systems that determine:

- the complexity, all-round development of all structural constituent parts of the territorial natural-socio-economic system;
- uniformity of development of all spatial constituent parts of territorial systems, i.e., spatial homogeneity of ecological, socio-economic development;
- high place (high rating) of the managed (analyzed) institution in the world community in terms of the level of environmental, socio-economic development, calculated by a set of indicators.

When determining the strategy for managing the development of territorial systems, the development trends of the civilized world are the benchmark. The world strategy for the development of society (that is, the determination of the leading directions of the development of society, taking into account the strategic tasks of the development of world civilization) has both an active and a passive beginning. On the one hand, it is deliberately determined by specific entities (countries, corporations, individual scientists), and on the other hand, it is formed objectively under the influence of macroeconomic and geopolitical circumstances, including within the framework of world economic technological cycles, a special place among which is occupied by the cycles discovered by the Russian economist. N. Kondratyev and the American economist of Russian origin S. Kuznets. The activeness of the world strategy for the development of society is manifested in the dissemination of world experience to other countries.

Considering the great importance of knowledge of the trends in the dynamics of the world economy for forecasting the economic development of individual countries, including Russia that has returned to the market path, let us dwell in somewhat more detail on the issue of the cyclical development of the world economy. As you know, the idea of the periodicity of the capitalist economy going through successive stages of boom, crisis and depression was formed quite steadily by the middle of the 19th century. By the 1920s, most economists from the industrialized countries of the world had no doubts about the existence of 7–11-year cycles, which they already described quite well. By this time, there were reports of signs of the existence of cycles of other sizes, both smaller (about 3.5 years) and longer cycles. However, by that time no one had studied them in detail.

3 Results and Discussions

3.1 Territory Management Support System

The spatio-temporal heterogeneity of metageosystems, differences in the history of changes in the types of economic development of the territory determine regional differences in nature management and predetermine the development of certain spectra of environmental problems [7]. To implement an interdisciplinary approach based on modern geographic information systems, it is necessary:

- development of a program for the collection, processing and systematization of data on the natural resource potential, features of the economic development of landscapes, modern regional nature management, the correspondence of the territorial organization of the economy to the resource and landscape-ecological properties of the region, which opens ways to improve and optimize the territorial structure of nature management, landscape planning and forecasting development of ecological, socio-economic systems.
- creation of special problem-oriented infrastructures of spatial data for modeling ecological-socio-economic systems using cartographic methods, deep machine analysis technologies, as well as probabilistic and statistical methods for analyzing information in order to make management decisions based on the use of geoportal systems.
- development of regional geoportals based on spatial data infrastructures – information processing and visualization systems for optimizing information management processes in land use systems and cadastres, supporting managerial decision-making, monitoring technogenic systems, managing traffic flows and road traffic, analyzing the geoecological state of territories, forecasting natural and natural and man-made emergencies and mitigation of the consequences of natural disasters, dissemination of information about natural and cultural heritage, integration of complex scientific research and visualization of their results.

In the administrative management system, information on long-term processes of the spatio-temporal dynamics of economic objects, resources and development conditions is used mainly for forecasting purposes, medium- and long-term planning and corresponding types of control, i.e. focuses on the needs of strategic management [8]. In operational management, the use of information is limited to some types of spatial information averaged over the territory.

In the emerging economic management system, the role of complex information is growing sharply. The solution of strategic tasks determines more stringent requirements for the purposefulness, consistency and objectification in the spatio-temporal context of traditional forecasts, recommendations, reference data, assessments. Simultaneously with the increase in the economic independence of the primary links of production, the need for information is growing for the purposes of operational management, i.e. information should promptly respond to changes in resources, cover all aspects of management in specific territories. Operational information services should be carried out within the strict framework of the production regime, and the information supplied to consumers should have a form that allows it to be used for decision-making (quantitative characteristics of resources and conditions in economically significant ranges, alternative recommendations, etc.).

In addition to the development of technical means for obtaining and processing spatial information, an interdisciplinary basis is needed, including the solution of a wide range of issues – from the formulation of goals and the identification of objects of study, to the determination of the circle of information consumers of various territorial levels, types and content of output products, which is necessary for the design of systems databases, development of methods of applied GIS analysis for thematic processing of spatial information [9].

Among the key elements of the process of managing territorial systems should also include the problem of its theoretical and practical support. At the same time, it is necessary to distinguish two groups of support: ensuring the process of researching the problems to be solved and ensuring the practical actions themselves to regulate territorial systems [10]. The content of these groups is largely the same, but there are also differences. Information support is common, although in its pure form it is more manifested in research. At the stage of practical implementation of management actions, information support itself is complemented by the introduction of innovations, which represent the quintessence of information impact in the form of scientifically based proposals for improving the practical activities of society.

3.2 Goals and Benchmarks of Territory Management

A clear vision of the final result of the functioning of the territorial system, or, in other words, a clear idea of the goals, final and intermediate guidelines of society's activities in the controlled territory, is extremely important in the process of managing the territory. In detail, this issue is resolved in the process of studying a complex of problems occurring on the territory, at the stage of forecasting the development of metageosystems. However, a general, fundamental (one might even say, to a certain extent intuitive, but still more expert) idea of this should be already at the initial stage of the elementary management cycle when setting management tasks and preparing tasks for their solution (more precisely, for their elaboration, study).

The issue of determining the absolute values of mankind has long and consistently been of interest to both ordinary people and scientists, primarily philosophers and sociologists. It seems that many of the values of mankind in relation to the territorial issue can probably be grouped into three groups of final guidelines for the development of territories. Taking into account the hierarchy, determined taking into account the degree of practical significance of these values, they should be placed in the following order: in the first place should be the enduring values of humanity, that is, the interests of people, which are of absolute importance for ensuring human life. The second most important issue is the strategy for the development of world civilization. And, finally, the third group is formed by private values determined by the goals of functioning of the most governed territorial system.

The variety of enduring values of humanity can be summarized in three positions. The first category is made up of the conditions for the safety of human life, that is, everything that can be conditionally called the ecological-socio-economic conditions of human existence. The second group of conditions related to the enduring values of mankind should be considered the factors that ensure the stability of the material and spiritual activity of society, that is, its sustainable development. Specifically, the development of society is manifested in the development of the territory in which this society takes place. Therefore, speaking about the sustainable development of society, we are simultaneously talking about the sustainable development of the territory, and vice versa. The activity to ensure the sustainable development of society by the activity of manifestation is of an attacking nature.

The third group is the resultant one: actions to ensure the safety of human life, which in turn provide a basis for the implementation of activities to ensure sustainable development of society, ultimately ensure (must ensure) a high quality of life for the population. The quality of human life is the criterion by which the quality of all activities of society should be assessed, including, of course, the state. Considering the fact that the listed groups of factors that create the enduring values of mankind are always geographically determined, we can talk about the leading role of interdisciplinary research in the general process of cognition and regulation of this phenomenon. And since in all cases these actions are implemented in a complex, full-scale, it should be recognized that the theory of the foundations of ensuring the safety of human life, sustainable development of society and the quality of life of the population is an integral interdisciplinary scientific direction.

The enduring values of mankind, refracted in taking into account the real and predicted strategic directions of development of the Modern World, provide a basis for determining the goals of the functioning of a particular territorial system. At the same time, first of all, the interests of the majority of the society that has formed in a given territory should be taken into account, while respecting the legal rights of each individual and small institutions. The democratically developed values will become corporate for the entire society of this territory. These are the territorial-corporate values of society, or territorial-institutional values. At the national level, they form the basis of patriotism.

The definition of the goals of the functioning of metageosystems is carried out in two stages. The first refers directly to the management process, the second, although it is related to management, but enters directly into the research process, and through it - into the management process. In practice, management activities often end with the first stage, which is undoubtedly flawed. Preliminary management decisions taken at this time at the level of an expert assessment of the situation and even sometimes purely intuitively do not receive a sufficiently high level of information support, which is why they are far from always correct.

At the initial stage, territorial and institutional values are identified by the subject of management, based on the general attitude of the local society, which makes requirements for the safety of its life and relies on knowledge of the strategy of sustainable development of mankind at the present time and in the future. At the second stage, a purposeful study of the situation is carried out, projects of management decisions are specified and, after adjustment, are adopted in the final version. Goal-setting is accordingly also subdivided into two stages. The first is devoted to a general assessment of the prospects for the development of the territory. On the second, the diagnostics of territorial systems is carried out, as a result of which the tasks that need to be solved are identified and ranked. Then a tree of goals (or task clusters) is compiled, the algorithms of these tasks are carried out and the implementation of the compiled algorithms is carried out.

Determination of the driving forces of human behavior is of great practical importance. The main stimuli of people's social behavior: fear, conscience, benefit, are well known in practice (one can recall the existence of a popular rule for achieving efficiency in social management - the "carrot and stick" method) and, in general, have been well studied by macrosociology. Taking into account the non-specialized nature of the present work, we will note here only the main features of these concepts.

First of all, it should be said about the need for the integrated use of these incentives in real practice. Any bias in one direction or another is fraught with the loss of a certain amount of positive results of management, and in many cases, on the contrary, it is possible to obtain a negative effect from excessive enthusiasm for using one or another stimulus. Especially often in the practice of administrative-territorial management, a forceful method of solving arising problems is used, while often the most correct from the standpoint of achieving real results is to engage the interests of the masses, because in the first case only the consequences are most often removed, and in the second, the causes of negative phenomena are eliminated. The first is easier to execute and more effective politically, and therefore most often gets priority in choosing this method. The second is much more difficult to execute and at the same time is often not noticeable for the broad masses of the population (poorly distinguishable, and therefore poorly evaluated by "ordinary" people), due to which, as a rule, administrators are reluctant to execute.

Therefore, the relevance for society of all-round work on the development of moral principles of social behavior among the broad masses of the population becomes obvious.

3.3 Territorial Security Systems Against Man-Made Threats

The group of technogenic challenges includes all types of threats associated with violations of scientifically grounded rules and regulations for the operation of machinery and equipment, the implementation of production and technological processes, the construction and operation of engineering facilities. Violations of scientifically based rules and regulations for the operation of machinery and equipment are widespread in all spheres of production activity, primarily in industry, construction, agriculture and transport, with all legal forms of organization of production, as well as in everyday life. The prevention of these hazards at work is carried out within the framework of the rules of labor protection and safety measures by the relevant services available at all enterprises. This direction is the main one in school courses on the basics of life safety.

Just like the previous hazard, the engineering group of hazards includes violations of the rules for the implementation of production and technological processes and violations of the rules in the design, construction and operation of engineering facilities. However, they are purely industrial, since they include only hazards arising as a result of deviations from the norms for the implementation of technological processes leading to serious man-made accidents and disasters, including, for example, such as the accident at the Chernobyl nuclear plant.

3.4 Territorial Security Systems Against Natural Threats

The group of natural security threats includes threats of purely natural origin, i.e. natural disasters, and natural and anthropogenic hazards.

The first group of natural hazards is formed by natural disasters, among which the most important are earthquakes, seaquakes with tsunamis, volcanic eruptions, floods, hurricanes and tornadoes, forest fires and droughts, avalanches and mudflows. In some cases, a person has to deal with locally manifested natural phenomena, such as showers, hail, landslides, karst, linear erosion, erosion of banks by river flows, etc. Of course, many of the listed dangerous natural processes are also associated with human activities, often provoked by anthropogenic actions, however, the basis of these processes remains clearly pronounced natural, and not technogenic.

Natural hazards are specific to each location. One and the same state of the natural environment in some cases can lead to an emergency situation, while in others it will refer to the norm. The criteria for classifying the state of the natural environment as hazardous are periodically reviewed.

With a certain deviation from some conditionally fixed average value, the nature management regime within a particular locality turns from normal to extreme, which forms an emergency situation in a given territory, which negatively affects the life of the population living (located) on it, as well as creating ecologically unfavorable environment for the existence of the entire local biota. If the likelihood of emergencies in the territory is high, then the environmental risk of living or temporary staying there is correspondingly high. Due to the fact that natural-anthropogenic (primarily natural-technogenic) phenomena can manifest themselves over large territories and often have great power, including acting evolutionarily, the ecological safety of human life is one of the urgent tasks of society.

Environmental problems are extremely diverse, but the most important for humanity are such global and regional environmental phenomena as deforestation and desertification of steppes and savannas, degradation of chernozems by wind and water erosion, climate warming as a result of atmospheric pollution, ozone holes resulting from increased emissions into the atmosphere of powerful oxidants such as freons, pollution of the natural environment and residential areas with household and industrial waste, including chemically, physically and microbiologically hazardous.

Of great importance in the formation of a set of dangers for people are violations of the rules and regulations for the design, construction and operation of natural-man-made systems, that is, systems that are basically naturally functioning formations, but either created by human labor, or under strong and/or continuous anthropogenic impact. The most common in human practice and historically ancient are actions to regulate water flow, originally for irrigation purposes. It is therefore natural that violations of the rules for the design, construction and operation of hydraulic structures and irrigation and drainage systems have become classical. Both in ancient times and in our time, salinization of irrigated lands is usually common in areas with irrigation and drainage activities, and later times gave mankind many cases of overdrying of soils during the reclamation of waterlogged and swampy lands. The practice of exploitation of artesian basins gives many examples of exceeding the norms of groundwater withdrawal, as a result of which the dehydration of the production aquifers occurs. In this case, powerful depression funnels are formed, contributing to the drainage of surface waters, often polluted, which also leads to pollution of groundwater.

3.5 Territorial Systems of Natural and Man-Made Emergencies

Strictly speaking, emergencies can be caused by all, without exception, the hazards discussed above. However, in real practice, it is customary to designate by the term emergency only the consequences - potential and realized - of hazards emanating from nature and technogenic processes.

The consequences of natural and man-made hazards occurring in a certain territory form the corresponding territorial systems of natural and man-made emergencies. Depending on the severity of the manifestation, emergency situations are divided into local, local, territorial, regional, federal and transboundary. In the latter, depending on the types of collateral, private collaterals take place. For example, information and scientific-methodological support is a territorial system for monitoring and forecasting natural and man-made emergencies in the regions.

4 Conclusion

The most important challenges of modern science include the awareness of the need for a transition to a system of multipurpose use of renewable resources, the development of a spectrum of economically profitable activities, the popularization of natural, historical and cultural heritage, the organization of active participation of society in environmental protection, forecasting socio-economic and natural-man-made emergencies. The identified problems suggest the need for the development of theory, methodology and methods for the development of interdisciplinary research for planning the sustainability of ecological-socio-economic systems (metageosystems) of territories.

The project focuses on the problem of creating an effective architectural solution for spatial data infrastructures, as a tool for solving the problem of integration, cataloging, distribution and visualization of spatial data based on geoportal technologies in order to support management decisions in the field of optimizing the functioning of metageosystems, and developing the digital economy.

The project focuses on the problem of creating an effective architectural solution for spatial data infrastructures as a tool for solving the problem of integration, cataloging, dissemination and visualization of spatial data based on geoportal technologies in order to support managerial decision-making in the field of optimizing the functioning of metageosystems and developing the digital economy.

Improving the accuracy and efficiency of diagnostics of spatio-temporal changes in metageosystems is based on the introduction into practice of intelligent systems, deep machine learning technologies used for coupled analysis of Earth remote sensing (ERS) data, monitoring results of ecological, social and economic processes and their interaction in the development of territories.

The solution of a wide range of modern spatio-temporal tasks to optimize the interaction of natural, social production systems in the development of territorial planning schemes, the design of specially protected natural areas, the integrated use and protection of natural resources, general plans of settlements and forecasting natural and natural-man-made emergency ecological processes is associated with development of GIS technologies.

The second promising direction in solving regional environmental problems is determined by the development of IT technologies in the development of geoportals - systems that provide presentation and access to information resources about natural, social and economic objects of regions (countries). The introduction of geoportal systems contributes to the implementation of strategic goals in the development of regions, the creation of an effective tool for managing natural, social and economic resources. Regional geoportals ensure the coordination of the search and rational use of natural resources, optimize territorial and sectoral planning and economic management, and monitor the ecological state of the region.

Acknowledgments. The reported study was funded by RFBR, project number 20–37-70055.

References

1. Miklós, L., et al.: Landscape as a geosystem. In: Miklós, L., Kočická, E., Izakovičová, Z., Kočický, D., Špinerová, A. (eds.) Landscape as a Geosystem, pp. 11–42. Springer, Cham (2019)https://doi.org/10.1007/978-3-319-94024-3_2
2. Van Ree, C., Van Beukering, P., Boekestijn, J.: Geosystem services: a hidden link in ecosystem management. Ecosyst. Serv. **26**, 58–69 (2017)
3. Sochava, V.B.: Introduction to the Study of Geosystems. Nauka. Sibirskoe Otdelenie, Novosibirsk, USSR (1978)
4. Solntsev, V.N.: On the difficulties of introducing a systematic approach to physical geography. Geogr. Issues Syst. Res. Nat. **104**, 20–37 (1977)
5. Nikolaev, V.A.: Landscape Studies. Practical and Seminary Classes. Moscow University Publishing House, Moscow, Russia (2000)
6. Dunets, A.N., Vakhrushev, I.B., Sukhova, M.G., Sokolov, M.S., Utkina, K.M., Shichiyakh, R.A.: Selection of strategic priorities for sustainable development of tourism in a mountain region: concentration of tourist infrastructure or nature-oriented tourism. Entrep. Sustain. Issues **7**(2), 1217 (2019)
7. Yamashkin, S.A., Yamashkin, A.A., Zanozin, V.V., Radovanovic, M.M., Barmin, A.N.: Improving the efficiency of deep learning methods in remote sensing data analysis: geosystem approach. IEEE Access **8**, 179516–179529 (2020)
8. Cheng, G., Yang, C., Yao, X., Guo, L., Han, J.: When deep learning meets metric learning: Remote sensing image scene classification via learning discriminative CNNs. IEEE Trans. Geosci. Remote Sens. **56**(5), 2811–2821 (2018)
9. Lee, J.G., Kang, M.: Geospatial big data: challenges and opportunities. Big Data Res. **2**(2), 74–81 (2018)
10. Li, M., Guo, W., Duan, L., Zhu, X.: A case-based reasoning approach for task-driven spatial–temporally aware geospatial data discovery through geoportals. Int. J. Digit. Earth **10**(11), 1146–1165 (2017)

Algorithm for Group Planning of Movement on Uneven Surface for Ground Robots

Konstantin Zakharov(✉)

St. Petersburg Institute for Informatics and Automation of the Russian Academy of Sciences, St. Petersburg Federal Research Center of the Russian Academy of Sciences (SPC RAS), 39, 14th Line, St. Petersburg 199178, Russia

Abstract. This paper considers the problem of energy-efficient path planning for a group of ground robots in a three-dimensional environment. To solve it, an algorithm for planning the group movement of ground robots on a relief surface GPMUS (Group Planning of Movement on Uneven Surface) is proposed. This algorithm is based on the LRLHD-A* and ORCA-3D algorithms developed by the authors. The primary paths of the robots are built using the LRLHD-A* (Local Roughness Local Height Difference A*) algorithm, designed to plan an energy efficient path on uneven surfaces. To construct collision-free trajectories for a group of robots, which are based on the paths calculated by LRLHD-A*, it was proposed to use a modification of the ORCA (Optimal Reciprocal Collision Avoidance) algorithm - ORCA-3D. This algorithm uses the difference in height between adjacent vertices of the path to calculate the correct length of the 2D velocity vector for each robot, which provides more accurate positioning of the robots at each time step. When GPMUS was tested in the Gazebo simulator, it was found that the energy efficiency of the trajectory of the robots increased in the range from 1.3 to 6.3%, and the average deviation of the positions of the robots at each time step from the tops of the paths decreased in the range from 10 to 57%.

Keywords: Heightmap · Path planning · Path planning for multiple agents · Relief surface · Collision avoidance · Robot group

1 Introduction

Currently, there is an increased interest in solving problems related to autonomous navigation of robots based on data on the terrain in open space. The solution to this problem plays an important role in the use of robotic tools for agricultural purposes, for exploring hard-to-reach areas or mapping. The navigation system must provide an analysis of the surface on which the robot will move, checking for static obstacles and other impassable areas on it. In such cases, the greatest attention is paid to ensuring energy efficiency: robots with limited energy resources must serve large areas. In this case, energy efficiency is understood as an indicator that determines the energy consumption of the robot while overcoming the selected route. Thus, energy efficiency depends on the length of the path, terrain features and terrain parameters. Altitude data can be obtained

R. Silhavy et al. (Eds.): CoMeSySo 2021, LNNS 231, pp. 659–668, 2021.
https://doi.org/10.1007/978-3-030-90321-3_54

using a heightmap. Typically, a heightmap is presented as a grayscale image, in which the intensity of each pixel represents a specific value for the height of the relief. When planning the movement of robot group, in addition to the features of the landscape, it is also necessary to take into account the position of other objects in space at each time step.

Path planning algorithms for robot ground can be divided into five main categories: sample-based algorithms, node-based optimal algorithms, mathematical model-based algorithms, bioinspired algorithms, and combined algorithms.

Sample-based algorithms require some known information about the entire workspace, that is, a mathematical representation to describe the workspace. This type of algorithm usually represents the environment as a collection of nodes, cells, or other shapes. Sample-based algorithms include: algorithms for constructing rapidly exploring random trees (RRT - Rapidly-exploring random trees) [1], algorithms based on potential fields [2], algorithms for constructing probabilistic road maps (PRM - Probabilistic Road Maps) [3], algorithms for constructing a Voronoi diagram [4]. The main advantages of such algorithms can be attributed to their speed, however, the efficiency of algorithms based on sampling largely depends on the initial assumption, which is repeated at each stage. Collision detection and avoidance can be realized through these algorithms.

Using optimal node-based algorithms, you can explore a workspace that is represented as a decomposed graph. Also, these algorithms operate with information about the weight of nodes and arcs (edges); they calculate the cost by examining the nodes to find the best path. Analyzing from a search engine perspective, node-based optimal algorithms apply the same property they are examining to a set of nodes (cells) on a map where measurement and information processing procedures are already in progress. Algorithms of this kind can always find the optimal path according to a certain decomposition. This type of algorithms includes: Dijkstra's algorithm [5], algorithms of the A* family [6, 7], algorithms of the D* family [8]. Since nodes and arcs provide an incomplete structure of the workspace, these algorithms can achieve the best result, limited to a specific representation of the environment. These kinds of algorithms are single search methods. Dijkstra's algorithm has a high computational complexity, but it finds the route with the minimum cost on the graph. Algorithms A* and D* have lower complexity, which allows them to be used in real-time path planning.

Algorithms based on mathematical models include linear algorithms and optimal control. These algorithms simulate the environment (kinematic constraints) as well as the dynamic system, and then associate the cost function with all the kinematic and dynamic constraint boundaries, which are inequalities or equations to achieve the optimal solution. Algorithms based on mathematical models perceive the robot as a point and only take into account the limitations on acceleration and speed. They can simulate the environment as a time-varying system that is in sync with the current location of the robot. This type of algorithms includes: linear algorithms [9], optimal control algorithms [10]. Algorithms based on mathematical models allow one to fully describe the states and variables of the environment, when applied in three-dimensional environments with a large number of obstacles, these methods can be adapted, applying the necessary constraints depending on the environment. These algorithms, as a rule, have a more complex formulation, that is, high computational complexity.

Bio-inspired computing imitate the biological behavior of certain organisms to solve problems. This type of planning algorithms excludes the process of constructing complex models of the environment for finding a path close to optimal, based on stochastic approaches. It overcomes difficult situations in which algorithms based on mathematical models often fail or get stuck in a local minimum when solving an NP-hard problem with a large number of variables and nonlinear objective functions. This category of algorithms includes: genetic algorithms [11], ant colony algorithm [12], particle swarm method [13], memetic algorithms [14], Shuffled Frog Leaping Algorithm (SFLA) [15], algorithms based on neural networks [16]. Bio-inspired algorithms mimic natural behavior, they define a set of optimization rules as well as a model for information processing. The whole process is then performed based on rules and models through iterative optimization. Thus, the final results largely depend on the proposed rules and model. Often, as the environment becomes more complex, the algorithms require more computational resources to find a solution that meets expectations.

Algorithm fusion is a recent approach to improving their 3D path planning performance. Typically, algorithms tend to combine layer by layer in an effort to plan the optimal path. For example, artificial potential field algorithms usually tend to sink to local minima without navigation or other tricks. Probabilistic roadmaps also cannot generate a single optimal path by themselves. On the constructed roadmap, the optimal path is found using the A* algorithm. In [18], an algorithm is presented that integrates the construction of the visibility graph, the construction of the Voronoi diagram, and the potential field algorithm. When one examines the extension of the VVP algorithm to three-dimensional space, it shows an effective trade-off between the shortest and safest path. In [19], visibility graphs and Dijkstra's (or Geodesics) algorithm were combined to find the optimal solution to the path planning problem in three-dimensional space. The existing algorithms for group path planning are not intended for navigation on an uneven surface, since they do not take into account the geometric features of the landscape, which affects both the passability and the speed of movement of the robots. To solve this problem, the GPMUS algorithm was developed, which is based on the algorithms developed by the authors LRLHD-A* (Local Roughness Local Height Difference A*) [6] and ORCA-3D (Optimal Reciprocal Collision Avoidance 3D).

2 Path Planning for Group of Ground Robots in a 3D Environment

In this paper, we solve the problem of planning a path for robot group moving on an uneven surface, presented in the form of a height map. The initial data for the task of group planning of the movement of robots in a three-dimensional environment are: a height map presented in the form of an array of cells with a certain value of heights, the initial and final positions of robots, the boundary values of the parameters characterizing the terrain. To solve the presented problem, the GPMUS algorithm would be developed, which includes single traffic planning using LRLHD-A * and group traffic planning based on the ORCA-3D algorithm. Consider the stages of the GPMUS algorithm (Fig. 1).

In the Gazebo simulation environment, the heightmap can be represented as a set of cells, which forms a three-dimensional terrain model (Fig. 2).

At the initial stage, a connected graph is formed from the height map. The height map must be presented in the form of a connected graph for further planning of the

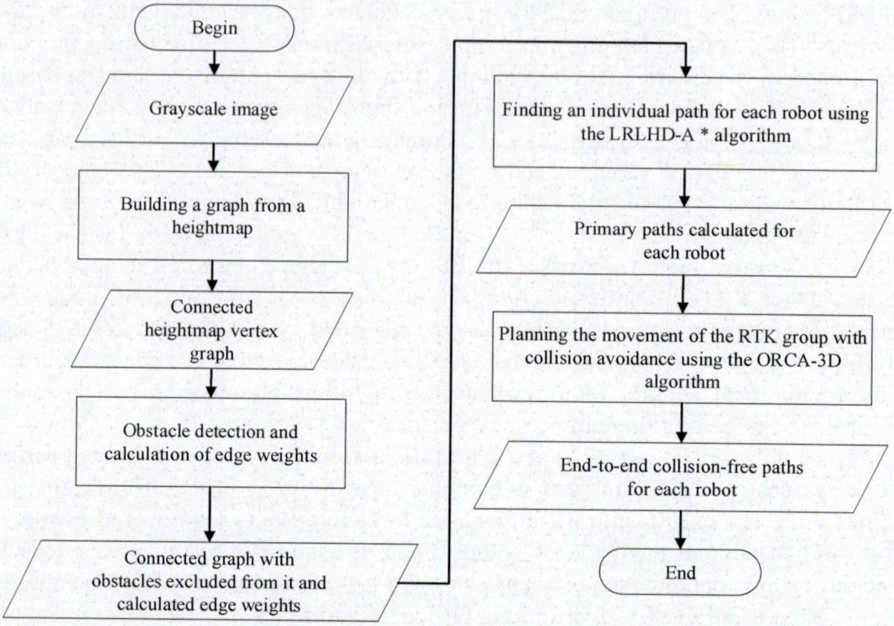

Fig. 1. GPMUS algorithm.

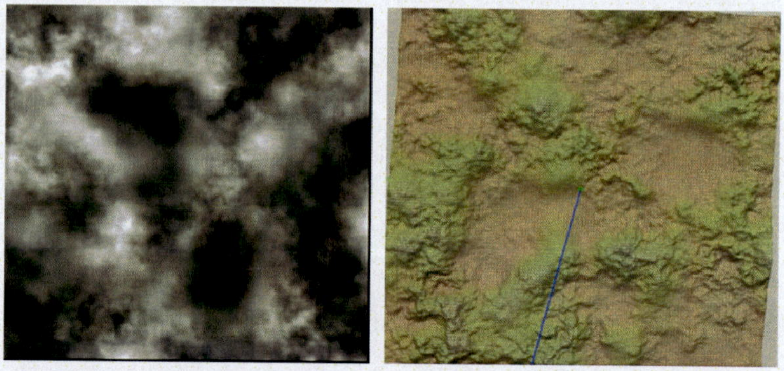

Fig. 2. Gazebo heightmap model generated from grayscale image.

path along it. To build a graph from a height map, each cell is assigned a list of eight neighboring vertices - two horizontally, two vertically and four diagonally. Further, the connectivity component will be used when finding obstacles and when building a route.

Figure 2 on the left is a grayscale image, and on the right is a Gazebo heightmap model built from the image. Each cell of the height map is a square with an area of 1 m^2 and a specific height value.

The next stage of the algorithm is to detect all areas impassable for the robot and calculate the weights of the edges. Before planning the path, it is necessary to identify all areas of the terrain along which the robot cannot move: holes, sharp uphill climbs and all uneven terrain, where the robot could overturn or get stuck. To do this, it is necessary to calculate the local roughness and the maximum local difference in heights for each vertex of the graph.

Local unevenness is the curvature of a surface caused by changes in relief. The roughness of the area $p(v)$ around the vertex v is expressed as the average angle of inclination of the surface for the nearest vertices $N_l(v)$ located at a distance less than l from the vertex v. The angle between the normals of the vertices is calculated by the formula $cr^{(u,v)} = acos\ (n(u)*n(v))$, and the average angle $a_{av}(v)$ is calculated by the formula (1), where $N_d(v)$ are all direct neighbors vertices v.

$$a_{av(v)} = \frac{\sum_{u \in N_d(v)} c_r^{(u,v)}}{\left|N_{d(v)}\right|}, \tag{1}$$

$$p(v) = \frac{a_{av(v)} + \sum_{u \in N_{l(v)}} a_{av(u)}}{\left|N_{l(v)}\right| + 1}. \tag{2}$$

To find irregularities in the relief, ravines, ditches, it is necessary to use an estimate of the local difference in heights. The local height difference $b(v)$ reflects the maximum height difference between the vertex v and each vertex from $N_l(v)$, and is calculated by the formula (3), where $v(z)$ and $u(z)$ are the values of the z-coordinates of the vertices v and u respectively.

$$b(v) = \max_{u \in N_l(v)} |v(z) - u(z)|. \tag{3}$$

Vertices for which the value of local roughness or local height difference exceeds a certain threshold are marked as obstacles and are excluded from the graph. The boundaries of the areas in which the obstacles are located are expanded by a distance comparable to the size of the robot, so that all points of the constructed path are sufficiently far from the areas dangerous for the movement of the robot. The gradual inflation algorithm smoothly reduces the weight of the vertices located at a distance from r_i to r_o from the area in which the obstacle is located. The parameter $r(v)$ reflects the degree of remoteness of the vertex v from the nearest obstacle and is calculated by the formula (4):

$$r(v) = c_{max} * \left(cos\left(\frac{(d_{lethal} - r_i)}{(r_o - r_i)} * \pi\right) + 1\right), \tag{4}$$

where c_{max} is the maximum total cost and d_{lethal} is the shortest distance to the obstacle.

For global path planning, the final edge weights must be calculated. The final weight $c(u, v)$ for each edge $e(u, v) \in E$ is calculated by the formula (5):

$$\begin{aligned} C^{(u,v)} &= \left(\left(k_p * max(p(u), p(v))\right) + 1\right) \\ &* ((k_b * \max(b(u), b(v))) + 1) * d^{(u,v)} + |r(u) - r(v)| \end{aligned} \tag{5}$$

where k_p and k_b are used to determine the effect of roughness p and local difference in heights b on the change in edge weight, and $d(u, v)$ is the Euclidean distance between vertices u and v.

After detecting obstacles and calculating the weights of the edges, the paths of the robots are calculated using the LRLHD-A* algorithm. The LRLHD-A* algorithm is designed to plan the path of the robot's movement on a three-dimensional map, which can be represented in the form of a graph. This algorithm is based on the A* algorithm, it uses the data on the local irregularity and the maximum local difference in heights of each vertex of the graph to detect obstacles and calculate the weights of the edges (Sect. 2.2). LRLHD-A* has a similar execution time to the original A* algorithm, however routes found by LRLHD-A* have a lower traversal cost. Application of this algorithm allows constructing safe initial routes of the robot's movement on an uneven surface, which will be necessary for the next stage of the GPMUS algorithm - group planning of movement.

The found routes are transferred to the modified ORCA (Optimal Reciprocal Collision Avoidance) algorithm [20], where collision-free trajectories are constructed to each point of the route. The input data for the ORCA-3D algorithm are: the number of agents (robots), their radius, movement speed, starting positions, orientation in space, and a set of waypoints. The original ORCA algorithm operates with positions and vectors of velocities of agents in two-dimensional space. The 3D and 2D velocity vectors of a robot moving on an inclined surface can be different, which can make it difficult to avoid potential collisions. To avoid such problems, the ORCA-3D algorithm adjusts the length of the velocity vector based on the height difference between adjacent vertices of the path. The length of the velocity vector is adjusted using the following formulas:

$$\begin{aligned} vel_{2D} &= \frac{dist_{2D}}{dist} * vel, \\ v_{vel} &= (x_{norm} * vel_{2D}, y_{norm} * vel_{2D}) \end{aligned} \tag{6}$$

where $dist_{2D}$, $dist$ is the distance between adjacent points of the path in two-dimensional and three-dimensional space, respectively, x_{norm}, y_{norm} are the coordinates of the direction of the normalized two-dimensional vector, vel is the speed of the robot in three-dimensional space, vel_{2D} is the speed of the robot in three-dimensional space. The operation of adjusting the length of the robot's velocity vector is performed for each segment of the path. This allows you to get more accurate data on the speeds and positions of robots when simulating their movement, which allows you to build safe paths for the movement of robots.

The output of the GPMUS algorithm will be energy efficient, slip-free trajectories for a group of robots. Next, we will consider the results of the experiments carried out.

3 Experiment Results

To test the developed solution, we used Gazebo robot simulator, into which a 256 * 256 vertices height map was imported, where the distance between adjacent vertices is 0.5 m. A group of robots consists of three Pioneer3ATs (Fig. 3).

Figure 3 shows a scene in Gazebo with a height map and three robots, for each of which a route is visualized. Experiments were carried out comparing the energy efficiency of paths built using the LRLHD-A* and A* algorithms (Fig. 4) to determine the appropriate algorithm for use in GPMUS.

From Fig. 4, which shows the dependence of the cost of the path on the distance between the initial and final vertices for the LRLHD-A* and A* algorithms, it can be

Fig. 3. Visualization of constructed paths in Gazebo.

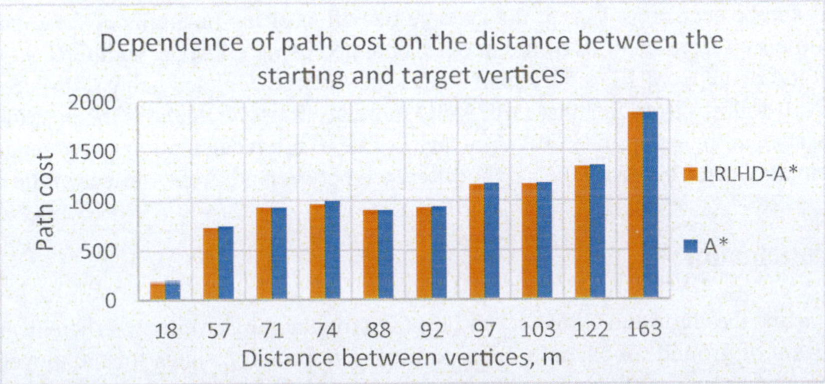

Fig. 4. Dependence of the path cost on the distance between the initial and target vertices for the LRLHD-A* and A* algorithms.

seen that the developed LRLHD-A* algorithm finds a path whose energy efficiency is 1.3–6.3% higher than that of the A* algorithm. It follows from this that the LRLHD-A* algorithm provides more efficient path planning on a raised surface.

For each combination of routes, GPMUS testing was carried out for group planning of movement using ORCA - without taking into account the height difference between adjacent vertices and ORCA-3D - taking into account the height difference and recalculating the length of the velocity vector. For the experiments, five different combinations of routes were selected for each of the robots calculated by the LRLHD-A*. During the movement, the positions of the robots were tracked in parallel streams and compared with their calculated positions. Figure 5 shows a graph with the average deviations of robots from routes calculated using GPMUS with ORCA and GPMUS with ORCA-3D.

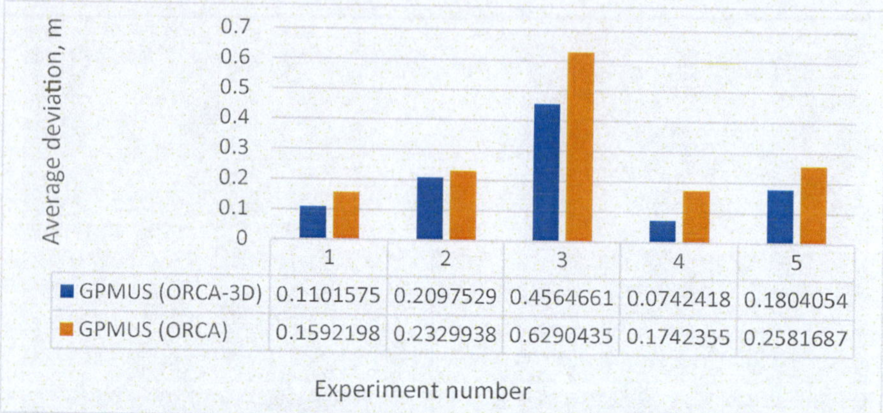

Fig. 5. Average deviation of robots from routes calculated by ORCA and ORCA-3D.

As can be seen from Fig. 5, the average deviation of the positions of the robots at each time step from the tops of the tracks calculated using GPMUS with ORCA-3D is less in the range from 10 to 57% than that of the tracks calculated using GPMUS with ORCA. It follows from this that ORCA-3D more accurately calculates the positions of robots at each time step than ORCA, which is due to the recalculation of the length of the velocity vector based on the height difference between adjacent vertices of the path.

4 Conclusion

In this work, the algorithm GPMUS was proposed for planning an energy-efficient, group movement of ground robots on a relief surface. The primary routes for the movement of robots on uneven surfaces are calculated using the LRLHD-A* energy-efficient path planning algorithm, which takes into account the terrain features. For group planning of movement, a modification of the ORCA algorithm is used for the movement of ground robots on an uneven surface - ORCA-3D. The correction of the length of the velocity vector of each robot is proposed, which depends on the height difference between the adjacent vertices of the path. The GPMUS algorithm was tested in the Gazebo simulator. The use of LRLHD-A* makes it possible to increase the energy efficiency of the robot's trajectory by 1.3–6.3% than that of A*. The use of ORCA-3D allowed at each time step to reduce the deviation of the positions of robots moving on an uneven surface in the range from 10 to 57% than using ORCA. This advantage allows the formation of safer routes for the movement of a group of ground robots in a three-dimensional environment. Further research will be aimed at increasing the speed of this algorithm and ensuring the smoothness and continuity of trajectories for robot group.

References

1. Yershova, A., Jaillet, L., Siméon, T., LaValle, S.M.: Dynamic-domain RRTs: efficient exploration by controlling the sampling domain. In: Proceedings of the 2005 IEEE International Conference on Robotics and Automation, pp. 3856–3861 (2005)
2. Sundar, S., Shiller, Z.: Optimal obstacle avoidance based on the Hamilton-Jacobi-Bellman equation. IEEE Trans. Robot. Autom. **13**(2), 305–310 (1997)
3. Gayle, R., Sud, A., Lin, M.C., Manocha, D.: Reactive deformation roadmaps: motion planning of multiple robots in dynamic environments. In: 2007 IEEE/RSJ International Conference on Intelligent Robots and Systems, pp. 3777–3783 (2007)
4. Kim, D., Kim, D.S.: Region-expansion for the Voronoi diagram of 3D spheres. Comput.-Aided Des. **38**(5), 417–430 (2006)
5. Verscheure, L., Peyrodie, L., Makni, N., Betrouni, N., Maouche, S., Vermandel, M.: Dijkstra's algorithm applied to 3D skeletonization of the brain vascular tree: evaluation and application to symbolic. In: 2010 Annual International Conference of the IEEE Engineering in Medicine and Biology, pp. 3081–3084 (2010)
6. Zakharov, K., Saveliev, A., Sivchenko, O.: Energy-efficient path planning algorithm on three-dimensional large-scale terrain maps for mobile robots. In: Ronzhin, A., Rigoll, G., Meshcheryakov, R. (eds.) International Conference on Interactive Collaborative Robotics. LNCS, pp. 319–330. Springer, Cham (2020). https://doi.org/10.1007/978-3-030-60337-3_31
7. Dergachev, S., Yakovlev, K., Prakapovich, R.: A combination of Theta*, ORCA and push and rotate for multi-agent navigation. In: Ronzhin, A., Rigoll, G., Meshcheryakov, R. (eds.) International Conference on Interactive Collaborative Robotics. LNCS, pp. 55–66. Springer, Cham (2020). https://doi.org/10.1007/978-3-030-60337-3_6
8. Koenig, S., Likhachev, M.: Fast replanning for navigation in unknown terrain. IEEE Trans. Robot. **21**(3), 354–363 (2005)
9. Yue, R., Xiao, J., Wang, S., Joseph, S.L.: Modeling and path planning of the city-climber robot part II: 3D path planning using mixed integer linear programming. In: 2009 IEEE International Conference on Robotics and Biomimetics (ROBIO), pp. 2391–2396 (2009)
10. Anderson, S.J., Peters, S.C., Pilutti, T.E., Iagnemma, K.: An optimal-control-based framework for trajectory planning, threat assessment, and semi-autonomous control of passenger vehicles in hazard avoidance scenarios. Int. J. Veh. Auton. Syst. **8**(2–4), 190–216 (2010)
11. Hacioĝlu, A.: Transonic airfoil design and optimisation by using vibrational genetic algorithm. Aircr. Eng. Aerosp. Technol. **75**, 350–357 (2003)
12. Duan, H., Yu, Y., Zhang, X., Shao, S.: Three-dimension path planning for UCAV using hybrid meta-heuristic ACO-DE algorithm. Simul. Modell. Pract. Theory **18**(8), 1104–1115 (2010)
13. Kennedy, J., Eberhart, R.: Particle swarm optimization. In: Proceedings of ICNN'95-International Conference on Neural Networks, vol. 4, pp. 1942–1948 (1995)
14. Moscato, P., Norman, M.G.: A memetic approach for the traveling salesman problem implementation of a computational ecology for combinatorial optimization on message-passing systems. Parallel Comput. Transput. Appl. **1**, 177–186 (1992)
15. Kundu, S., Parhi, D.R.: Modified shuffled frog leaping algorithm based 6DOF motion for underwater mobile robot. Procedia Technol. **10**, 295–303 (2013)
16. Yang, S.X., Luo, C.: A neural network approach to complete coverage path planning. IEEE Trans. Syst. Man Cybern. Part B (Cybern.) **34**(1), 718–724 (2004)
17. Yan, F., Liu, Y.S., Xiao, J.Z.: Path planning in complex 3D environments using a probabilistic roadmap method. Int. J. Autom. Comput. **10**(6), 525–533 (2013)
18. Masehian, E., Amin-Naseri, M.R.: A voronoi diagram-visibility graph-potential field compound algorithm for robot path planning. J. Robot. Syst. **21**(6), 275–300 (2004)

19. Schøler, F., la Cour-Harbo, A., Bisgaard, M.: Generating approximative minimum length paths in 3D for UAVs. In: 2012 IEEE Intelligent Vehicles Symposium, pp. 229–233 (2012)
20. Van Den Berg, J., Guy, S.J., Lin, M., Manocha, D.: Reciprocal n-body collision avoidance. In: Pradalier, C., Siegwart, R., Hirzinger, G. (eds.) Robotics Research, pp. 3–19. Springer, Berlin (2011). https://doi.org/10.1007/978-3-642-19457-3_1

Data Augmentation in Training Neural-Network Language Model for Ontology Population

Pavel Lomov, Marina Malozemova(✉), and Maxim Shishaev

Institute for Informatics and Mathematical Modeling, Kola Science Centre of the Russian Academy of Sciences, Apatity, Russia
{lomov,malozemova,shishaev}@iimm.ru

Abstract. This paper is a continuation of the research focused on solving the problem of ontology population using training on an automatically generated training set and the subsequent use of a neural-network language model for analyzing texts in order to discover new concepts to add to the ontology. The article is devoted to the text data augmentation - increasing the size of the training set by modification of its samples. Along with this, a solution to the problem of clarifying concepts (i.e. adjusting their boundaries in sentences), which were found during the automatic formation of the training set, is considered. A brief overview of existing approaches to text data augmentation, as well as approaches to extracting so-called nested named entities (nested NER), is presented. A procedure is proposed for clarifying the boundaries of the discovered concepts of the training set and its augmentation for subsequent training a neural-network language model in order to identify new concepts of ontology in the domain texts. The results of the experimental evaluation of the trained model and the main directions of further research are considered.

Keywords: Data augmentation · Neural network · Ontology population

1 Introduction

This work is a continuation of the study [1], aimed at automatic generation of a training set based on the analysis of domain texts and its use for training a neural-network model focused on solving one of the ontology training subtasks - the ontology population. The mentioned ontology learning is the process of deriving high-level concepts and relations as well as the occasional axioms from texts to form an ontology [2]. In turn ontology population involves adding new instances for the classes specified in existing ontology without changing its structure.

In the previous work, a technology was proposed that involves the analysis of an ontology to form a list of its concepts, the collection and analysis of texts

R. Silhavy et al. (Eds.): CoMeSySo 2021, LNNS 231, pp. 669–679, 2021.
https://doi.org/10.1007/978-3-030-90321-3_55

related to the domain of the ontology, with the formation of a training set of marked sentences. Further, this set was used to train a neural-network language model, focused on solving the named entity recognition (NER) problem. The model was subsequently used to extract new concepts from texts - candidates for addition to the ontology.

In view of the fact that the proposed technology is based on supervised learning, it is necessary to provide a sufficiently large volume of the training set for successful learning. One of the ways to increase it is data augmentation, which involves the automatic creation of new samples by some modification of the existing ones. This allows you to improve the efficiency of the trained model.

In this paper, we consider the problem of augmentation of the generated set to increase the efficiency of the language model trained on its basis with respect to the fullness and precision of detecting possible new elements of the ontology in the texts - classes and instances. In this case, it is important to ensure that the label still corresponds to its sample (text sentence) after modification of the latter during augmentation. Since the label includes the extracted concept, represented in the form of its boundaries (indices of its first and last tokens in the sentence), it is of particular importance to correctly define these boundaries for the correct definition of the part of the sentence that can be modified. This problem resembles the problem of nested Named Entity Recognition (Nested NER) [3], but in our case it has some specificity due to the extraction of concepts for ontology learning. In this work, after the stage of generating a training set, it is proposed to perform a stage for clarification of found concepts and then perform an augmentation stage - formation of additional variations of sentences in which clarified concepts occur.

The remaining part of the paper is structured as follows: Sect. 2 discusses existing approaches to text data augmentation, as well as approaches for solving the problem of nested named entities extraction, Sect. 3 presents the proposed augmentation procedure, Sect. 4 discusses the results of experimental evaluation of models trained on the initial and augmented set. The conclusion contains conclusions and directions for further work.

2 Related Work

Augmentation of text data means slight modification of initial samples (usually texts or sentences) without changing their meaning. In this case, the sample label, as a rule, remains unchanged.

Among the common types of augmentation techniques are the following:

- replacement of words in a sentence [4];
- rearrangement of words in a sentence or the sentences themselves within the texts [5];
- creating text noise, that is, adding minor errors to words, sentences or texts (changing text case, punctuation marks);
- generation of new sentences or texts based on changes in the structure of the initial ones [6].

Recently, it has become a very common practice to use models with BERT (Bidirectional Encoder Representations from Transformers) [7] architecture pre-trained on a large volume of texts. A key feature of this architecture is the ability to consider a single word in the context of the surrounding words during the training process. Such pre-training of BERT models on a large volume of texts allows to form the contextualized word embeddings for a certain natural language. These embeddings can be used to solve various NLP problems including performing augmentation.

For example, Sun and Jiang [8] propose an algorithm for correcting misspelled words by providing variants for their replacement using a BERT-based masked language model. In this case, augmentation is performed by concatenating the initial sentence and its variant containing masking tokens instead of misspelled words. According to the authors, such augmentation makes it possible to obtain variants for replacement consisting of more or fewer tokens than the word being replaced, as well as to "distract" the model from the erroneous words during the generation of variants for its replacement.

Wu et al. [9] propose a method of contextual augmentation using conditional BERT focused on the augmentation of labelled sentences. Conditional BERT is the result of fine-tuning of the initial BERT model using a dataset that additionally includes positive/negative labels. This allows the model to propose substitutions for masked tokens during augmentation, taking into account the label, and thereby ensure the correctness of the obtained samples.

Kang et al. [10] introduce an augmentation method using the so-called filtered BERT to solve the problem of de-identification (prevention of personal information disclosure) of protected health information (PHI) during secondary use of clinical documents. Filtered-BERT predicts the masked word, providing several possible variants for it and then filters them by comparing the cosine distance between their fastText vectors and the masked word vector. As a result, augmented sentences are formed with words that have passed through this filter.

The mentioned problem of clarifying the boundaries of the concept in a sentence is similar to nested named entity recognition. Early work used rule-based approaches to solve this problem. So, Zhang et al. [11] propose two such approaches to recognition of biomedical entities: a post-processing rule-based approach and an approach based on Hidden Markov Model (HMM). The first one assumes the use of patterns specially designed based on the GENIA corpus, which allow recognizing the longest entity names based on short (nested) ones. The HMM-based approach, in turn, involves the use of two pre-trained models: the first model recognizes short entities, and the second is used to subsequently expand these recognized short entities into long ones.

Finkel and Manning [3] introduce a discriminative constituency parser for nested NER. This parser is trained on sentences represented in the form of parse trees, which contain information about the components (tokens) of each named entity - "parent" and "grandparent", as well as their parts of speech.

Sohrab and Miwa [12] propose a simple neural-network model for nested NER. The model allows to find and classify all possible fragments of the input sequence, in which a potential nested entity is mentioned. The entities themselves are then recognized in these found fragments using the LSTM layer.

The next work [13] proposes a neural model for identifying nested entities by stacking so-called "flat" NER layers. A "flat" layer used to recognise a "flat" entity - the opposite of a nested entity. These layers, in turn, consist of LSTM layer, which captures a bidirectional contextual representation of a sequence, and CRF layer, which predicts a sequence of labels - BIO tags for this representation. Representation of each word of the input sequence in the embedding layer is the result of combining character embedding and word embedding. The number of "flat" layers depends on the level of entity nesting (e.g. New York - level 1, New York University - level 2). The entity recognition process stops if the current flat NER layer does not detect any entities.

Chen et al. [14] propose the regression operation to find nested named entities in a sentence. To execute it, a sentence is first transformed using a deep neural network into recurrent feature maps, i.e. into abstract representations that fix semantic dependencies between words. Each feature map defines the possible boundaries of an entity. Next, bounding boxes are generated from these feature maps, which are abstract representations of named entities. Each box includes information about the position of the entity (starting position and length) and its class label. During training, the regression operation predicts the offset value of the starting position and the offset value of the current box relative to the box of the true named entity. The predicted offsets allow the box to be "moved" correctly, thereby accurately identifying the entity.

Dadas and Protasiewicz [15] introduce an iterative algorithm for bidirectional nested NER. It involves training two neural-network models on one dataset to identify named entities in two directions: from external entity to nested one (from outside to inside) and from nested entity to external one (from inside to outside). Each word of the input sequence is represented as a concatenation of three vectors: the result of a character-based language model, a word embedding vector and a vector, each component of which indicates whether the word has been assigned to a certain category in previous iterations. At each iteration, the model generates new predictions based on the initial word sequence and the predictions made earlier. This process ends when new entities are no longer detected. The output predictions of both models are filtered using one of the some selection criteria (e.g. union, intersection, etc.) with the subsequent formation of the final set of detected entities.

Shibuya and Hovy [16] propose a decoding method that iteratively recognizes entities in an outside-to-inside way. It allows to identify nested entities in the range of each detected entity using the Viterbi algorithm [17].

The augmentation procedure proposed in this work also assumes the use of a pre-trained BERT model focused on solving the masked language modeling problem [7]. The main difference is the preparation of a masked sentence, taking into account the redefinition of the entity (concept) boundaries, performed as part of its "clarification".

3 Proposed Procedure for Concept Clarification and Training Set Augmentation

Augmentation involves some modification of the initial sentence. However, in this case, it is assumed the subsequent extraction of the concept for ontology population, therefore, the modification of the sentence should not affect it. Therefore, the correctness of defining the concept boundaries in the sentence acquires special relevance.

This circumstance forces us to reconsider the approach to identifying concepts during training set formation, proposed in the previous work [1]. It assumed a search in the texts of sentences containing the names of instances of classes already existing in the initial ontology. This allowed us to reduce the task of ontology population to named entity recognition (NER).

However, the analysis of the sets of labeled sentences obtained in this way revealed some features that must be taken into account in the context of extracting precisely ontological concepts. For example, when searching for sentences in the corpus of news texts containing an ontology instance named "Russia" which belong to the class "Country", in addition to sentences containing the corresponding token as the name of the country, sentences were also found that include combinations of this token with others: "government of Russia", "president of Russia", "company in Russia". These combinations of tokens can be interpreted both as ontology classes (e.g. "government", "president") and as instances of classes ("government of Russia", "president of Russia"). At the same time, they can implicitly define some relation (e.g. company "is-located-in" Russia) to the initial class "Country" and/or its instance "Russia", which can also considered for addition to the ontology. Thus, the identification of an ontological concepts in sentences during training set formation, in contrast to named entities in the NER, it makes sense to carry out taking into account some ontological context describing the position of the concept in the ontology.

Taking into account the ontological context in the formation of a training set, as well as the possibility of including it in the labels of sentences, is supposed to be considered in more detail in the continuation of the study. Within the framework of this work, for correct augmentation, it is necessary to clarify the boundaries of concepts in sentences, which were found in the texts at the first stage. For this, an additional step was introduced, which involves determining the position of the initial concepts in the dependency tree of the sentence and expanding it by tokens directly related to it by syntactic relations.

For example, the sentence "Международный аэропорт Шереметьево является крупнейшим в России." ("*Sheremetyevo International Airport is the largest one in Russia.*") has the following dependency tree as shown in Fig. 1. According to this tree the ontology concept "аэропорт" ("*Airport*") is related to tokens "Международный" ("*International*") and "Шереметьево" ("*Sheremetyevo*"). So the clarified concept will be "Международный аэропорт Шереметьево" ("*Sheremetyevo International Airport*").

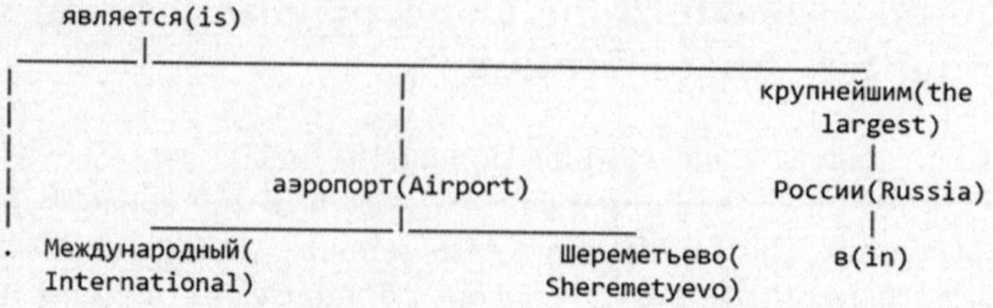

Fig. 1. Sentence dependency tree

After concept clarification, the initial sentence is augmented as follows:

1. Based on the analysis of the sentence dependency tree among the tokens of concept, the main token is determined (it is located at a higher level of the tree than the child tokens).
2. We identify the nearest tokens which are located to the left and right of the concept tokens and are not adjectives, stop words or prepositions. Using the dependency tree, we also determine the tokens dependent on them. Thus, we get the groups of the nearest left and right tokens.
3. In the groups obtained at step 2, we mark the main and child tokens as candidates for replacement.
4. We form all combinations of token indices marked for replacement.
5. Using the combinations of indexes formed in the previous step, we create modifications of the initial sentence, in which different combinations of tokens are replaced by a token-mask.
6. Using the pretrained BERT model, we obtain variants of tokens for replacement and substitute them instead of mask tokens in the modifications of the initial sentence.
7. For the received modified sentences, we redefine the boundaries of the ontological concepts, since the replacement of tokens leads to their change.

For example, Fig. 1 shows that the main token of the concept "Международный аэропорт Шереметьево" ("*Sheremetyevo International Airport*") is "аэропорт" ("*Airport*"). In the next step, we get only the group of the nearest right tokens ["является", "крупнейшим"] (["*is*", "*the largest*"]) and the group of the nearest left tokens is empty because entity tokens group is located at the beginning of the sentence. These two tokens from the nearest right group are candidates for replacement. Further, having formed combinations of the indices of these tokens, we create modifications of the initial sentence, where tokens "является" ("*is*") and "крупнейшим" ("*the largest*") are replaced with a mask tokens:

Международный аэропорт Шереметьево [mask] крупнейшим в России.
(Sheremetyevo International Airport [mask] the largest one in Russia.)
Международный аэропорт Шереметьево является [mask] в России.
(Sheremetyevo International Airport is [mask] one in Russia.)

Using the BERT model, we receive 16 possible tokens to replace mask ones. Finally, applying some of the proposed tokens we obtain the following modified sentences:

Международный аэропорт Шереметьево был крупнейшим в России. (Sheremetyevo International Airport was the largest one in Russia.)
Международный аэропорт Шереметьево является крупнейшим в России. (Sheremetyevo International Airport is the largest one in Russia.)
Международный аэропорт Шереметьево является единственным в России. (Sheremetyevo International Airport is the only one in Russia.)
Международный аэропорт Шереметьево является старейшим в России. (Sheremetyevo International Airport is the oldest one in Russia.)

To obtain sentence dependency tree, the Russian-language model from the spaCy framework [18] was used. The Russian-language RuBERT model from the deeppavlov [19] project was used as a language model for solving the language mask modeling problem and proposing a substitution tokens.

In order for the BERT model to provide substitute tokens that are relevant to the domain of the sentence, it is necessary to avoid replacing too many tokens in it. For this reason, in step 4, when generating combinations, parameters are used that define the maximum number of replaced tokens in a single combination. They were assigned the values 4 and 2 correspondingly. Along with this, in order to avoid frequent errors in the coordination of the token provided by the model and its environment (e.g. inconsistency in gender, number or case), combinations in which masked tokens were located one after another were not considered.

These parameters allow you to vary the grammatical and semantic correctness of the formed sentences and the volume of the final augmented set. The more combinations are considered, the more variants will be generated for each initial sentence. However, this increases the likelihood of grammatical and semantic errors in them.

After completing the presented augmentation procedure, the formed set can be used to train the language model.

4 Evaluation

In order to evaluate the effect of the proposed procedures for concept clarification and training set augmentation, an experiment was carried out, in which three models were trained and evaluated. The first model was obtained without using noted concept clarification and training set augmentation, the second - with the use of only concept clarification, the third - with the use of both procedures.

To form the initial training set, the corpus of Russian-language texts of the Internet edition Lenta.ru [20] was used. It contains about 800 thousand news texts on various topics (politics, economics, sports, etc.). To imitate a set of concepts of ontology being populated, an initial list of concepts typical for news texts (e.g. “company”, “police”, “action”, etc.) was formed manually.

As a result of the analysis of the text corpus using the initial list of concepts, a training set was formed. It included about 550 thousand marked samples - sentences with labels “concept and its category”. The first model was trained on this set.

Further, the proposed clarification and augmentation procedures were applied to the formed training set. As a result, two more sets were obtained: a set with clarified concepts and an augmented set with clarified concepts. The size of the last set has increased from 550 thousand to 2800 thousand samples. After that, two more models were trained on these sets.

The quality assessment of all trained models was performed on a test set. Its formation was carried out similarly to the formation of a training set but with the use of another part of the news text corpus. The size of the test set was 300 thousand samples.

The quality assessment was carried out in the following experiments:

Experiment 1. Detection of concepts from the test set that are "known" to models (concepts that were represent in the training sets):

- model trained without clarification and augmentation: precision = 0.002, fullness = 0.024;
- model trained with concept clarification: precision = 0.056, fullness = 0.694;
- model trained with clarification and augmentation: precision = 0.055, fullness = 0.593.

Experiment 2. Detection of concepts from the test set that are "unknown" to models (concepts that were not represent in the training sets):

- model trained without clarification and augmentation: precision = 0.0, fullness = 0.01;
- model trained with concept clarification: precision = 0.455, fullness = 0.424;
- model trained with clarification and augmentation: precision = 0.435, fullness = 0.321.

Separately, an expert assessment of the correctness of the concepts detected by the models that were not in the initial list (and, accordingly, in the test and training sets) was made. Thus, the percentage of those concepts that can be used for populating an ontology was estimated.

Experiment 3. Detection of concepts that are not represented in the initial list of concepts:

- model trained without clarification and augmentation: total new concepts - 39, percentage of correct concepts - 0.6;
- model trained with concept clarification: total new concepts - 3566, percentage of correct concepts - 0.82;
- model trained with clarification and augmentation: total new concepts - 3254, percentage of correct concepts - 0.85.

The second and third experiments are the most representative in relation to evaluating the effectiveness of using the obtained models for ontology learning. These experiments assess the ability of the models to find new concepts based on the contexts in which training set concepts were used.

The results of the experiments showed that the concept clarification procedure brings the main contribution to the increase in efficiency. This is probably due to the fact that the concepts from the initial list, representing the concepts of the initial ontology, are supplemented with the associated tokens from the sentences of analyzed texts during the clarification. This leads to the fact that these tokens are considered by the models as part of the concept, and not as its context during training. For example, the clarification of the concept "аэропорт" ("*Airport*") was "Международный аэропорт Шереметьево" ("*Sheremetyevo International Airport*") that allows the model to consider the context "... является крупнейшим в России" ("*... is the largest one in Russia*"), which is probably can be associated with more other concepts than the context "Международный ... Шереметьево является крупнейшим в России" ("*Sheremetyevo International ... is the largest in Russia*"), obtained without clarification. Thus, the clarification of concepts allows you to correct their context, which has a positive effect on the ability of the trained model to find new concepts.

The use of augmentation also slightly increased the efficiency in relation to the accuracy of detecting new concepts in the third experiment. However, this result was achieved after training the model on the augmented set of 2800 thousand samples, which took 5 times longer than training on unaugmented one.

5 Conclusion

Today ontology learning continues to be an urgent problem in the development of modern information systems focused on the representation of domain knowledge. The use of existing NLP and machine learning technologies has great potential in relation to automating the related tasks - from collecting, preprocessing and analyzing natural language texts to forming the initial structure of ontology concepts and its subsequent development.

This paper considers the extension of the previously proposed technology, focused on solving the problem of expanding the existing ontology with new instances of classes that are extracted from the domain texts using a trained neural-network language model. As additional steps, it is proposed to clarify the concepts during the formation of the training set. It involves analysis of the sentences containing the initial concepts of an ontology and expanding their boundaries by including some additional tokens associated with them. On the one hand, this made it possible to represent concepts of ontology in the form in which they were used in the analyzed texts, and on the other hand, to correct its context, which the model learns to recognize.

Another proposed step was sentence augmentation, which involved replacing some tokens that were not part of the clarified concept with a variant proposed by a pre-trained language transformer model capable of representing contextualised embeddings.

As a result of the experiments, it was found that the concept clarification has a positive effect on the detection of new concepts in the contexts that were represented in the sentences of the training set. In the future, it is planned to clarify concepts taking into account their ontological context, which describes their position in ontology. This will make it possible to interpret the tokens detected during clarification as another possible concept of ontology, which should also be represented in the training set. The latter circumstance may require repeating the stage of text analysis to include sentences with such concepts in the training set.

In addition, it is planned to consider the possibility of including the ontological context in sample label, which will allow trained models not only detect new concepts, but also to indicate their possible position in the ontology (subclass of some class, class instance, property carrier, property value, etc.).

With regard to the development of the proposed augmentation procedure, it is planned to consider the generating of new samples (text sentences) by changing the structure of the initial ones i.e. replacing/adding/removing their parts. In addition to the problem of defining modifiable parts and generating of substitutions, the problem of checking the semantic correctness of the result will become especially urgent. However, performing such augmentation will improve the quality of the trained model when using a smaller augmented set than with the current approach for augmentation.

Acknowledgements. The work is supported by the Ministry of Science and Higher Education of the Russian Federation (reg.n. 0226-2019-0036) and the Russian Foundation for Basic Research (RFBR), project number 20-07-00754.

References

1. Lomov, P., Malozemova, M., Shishaev, M.: Training and application of neural-network language model for ontology population. In: Silhavy, R., Silhavy, P., Prokopova, Z. (eds.) CoMeSySo 2020. AISC, vol. 1295, pp. 919–926. Springer, Cham (2020). https://doi.org/10.1007/978-3-030-63319-6_85
2. Wong, W., Liu, W., Bennamoun, M.: Ontology learning from text: a look back and into the future. ACM Comput. Surv. (CSUR) **44**, 1–36 (2011)
3. Finkel, J.R., Manning, C.D.: Nested named entity recognition. In: Proceedings of the 2009 Conference on Empirical Methods in Natural Language Processing, EMNLP 2009, USA, vol. 1, pp. 141–150. Association for Computational Linguistics, August 2009
4. Wang, W.Y., Yang, D.: That's so annoying!!!: A lexical and frame-semantic embedding based data augmentation approach to automatic categorization of annoying behaviors using #petpeeve tweets. In: Proceedings of the 2015 Conference on Empirical Methods in Natural Language Processing, Lisbon, Portugal, pp. 2557–2563. Association for Computational Linguistics, September 2015
5. Luque, F.M.: Atalaya at TASS 2019: data augmentation and robust embeddings for sentiment analysis. arXiv:1909.11241 [cs], September 2019
6. Coulombe, C.: Text data augmentation made simple by leveraging NLP cloud APIs, p. 33 (2018)
7. Devlin, J., Chang, M.-W., Lee, K., Toutanova, K.: BERT: pre-training of deep bidirectional transformers for language understanding. arXiv:1810.04805 [cs], October 2018

8. Sun, Y., Jiang, H.: Contextual text denoising with masked language model. In: Proceedings of the 5th Workshop on Noisy User-Generated Text (W-NUT 2019), Hong Kong, China, pp. 286–290. Association for Computational Linguistics, November 2019
9. Wu, X., Lv, S., Zang, L., Han, J., Hu, S., et al.: Conditional BERT contextual augmentation. In: Rodrigues, J.M.F. (ed.) ICCS 2019. LNCS, vol. 11539, pp. 84–95. Springer, Cham (2019). https://doi.org/10.1007/978-3-030-22747-0_7
10. Kang, M., Lee, K., Lee, Y.: Filtered BERT: similarity filter-based augmentation with bidirectional transfer learning for protected health information prediction in clinical documents. Appl. Sci. **11**, 3668 (2021)
11. Zhang, J., Shen, D., Zhou, G., Jian, S., Tan, C.-L.: Enhancing HMM-based biomedical named entity recognition by studying special phenomena. J. Biomed. Inform. **37**(6), 411–422 (2004)
12. Sohrab, M.G., Miwa, M.: Deep exhaustive model for nested named entity recognition. In: Proceedings of the 2018 Conference on Empirical Methods in Natural Language Processing, Brussels, Belgium, pp. 2843–2849. Association for Computational Linguistics, October 2018
13. Ju, M., Miwa, M., Ananiadou, S.: A neural layered model for nested named entity recognition. In: Proceedings of NAACL-HLT 2018, pp. 1446–1459 (2018)
14. Chen, Y., et al.: A boundary regression model for nested named entity recognition. arXiv:2011.14330 [cs], December 2020
15. Dadas, S., Protasiewicz, J.: A bidirectional iterative algorithm for nested named entity recognition. IEEE Access **8**, 135091–135102 (2020)
16. Shibuya, T., Hovy, E.: Nested named entity recognition via second-best sequence learning and decoding. Trans. Assoc. Comput. Linguist. **8**, 605–620 (2020)
17. Huang, Z., et al.: Iterative Viterbi A* algorithm for k-best sequential decoding. In: Proceedings of the Conference on 50th Annual Meeting of the Association for Computational Linguistics, ACL 2012, vol. 1, pp. 611–619, July 2012
18. Russian spaCy model. https://spacy.io/models/ru#ru_core_news_sm
19. Pre-trained embeddings: DeepPavlov 0.15.0 documentation. http://docs.deeppavlov.ai/en/master/features/pretrained_vectors.html#bert
20. News dataset from Lenta.Ru. https://kaggle.com/yutkin/corpus-of-russian-news-articles-from-lenta

Application of Visual Programming Methods to the Design of Neural Networks

E. O. Yamashkina[1], S. A. Yamashkin[2], Olga V. Platonova[1](✉), and S. M. Kovalenko[1]

[1] MIREA – Russian Technological University, 78 Vernadskogo Prospect, 119454 Moscow, Russia

[2] National Research Ogarev Mordovia State University, 68 Bolshevistskaya Street, 430005 Saransk, Russia

Abstract. In the modern world component development techniques are becoming increasingly popular. Solving the scientific problem of designing, training and using neural net-works to solve the problem of automated analysis of national and spatial-time data meets many open questions. This article addresses the challenge of developing an effective tool for designing neural networks through visual programming techniques. The designated module is a component of the repository of deep neuronetwork models created to address the scientific problem of the accumulation and systematization of models and machine learning algerits through the development for analysis and prediction. The Keras Open Network Library, in particular Layers API, is used as the basic algorithms for creating neural network architectures. Comprehensive studies have been carried out in the field of visual programming of neural networks. A graphical interface has been developed to configure neural network architectures. The presented approach of visual programming of neural networks simplifies the process of development, allows avoiding errors and creates more efficient and reliable systems.

Keywords: Visual programming · Spatial data · Neural networks · Spatial data infrastructure

1 Introduction

1.1 A Subsection Sample

The establishment, implementation and effective use of Digital Spatial Data Infrastructures (SDI) of regions aimed at rapid and long-term prediction of natural disasters is a cornerstone of the challenge of increasing connectivity between regions and the world. Diagnosis of the state of natural and social production systems. Due to the fact that data integrated into digital SDI are often abundant and have complex non-linear regularities, the core of a given class of systems now often have automated machine analysis methods and algorithms, it's possible to solve a complex of applied and project-oriented tasks – data classification and segmentation, anomaly detection, sign learning, data fusion.

R. Silhavy et al. (Eds.): CoMeSySo 2021, LNNS 231, pp. 680–689, 2021.
https://doi.org/10.1007/978-3-030-90321-3_56

The analysis may include digital landscape maps, Earth remote sensing (ER) data, aerosol materials, structured information on natural, social and economic objects with distributed geospatial capabilities organization [1]. However, the application of such analysis to the national economy ranges from improving the efficiency of agriculture through the deployment of point farming systems to assessing the impact of natural and natural disasters technological processes. Within this direction, a number of tasks should be identified for which the development and implementation of in-formation systems using artificial neural networks is relevant [2]: classification, clustering, pattern recognition, decision-making, data analysis and forecasting.

Solving the scientific problem of designing, training and using neural net-works to solve the problem of automated analysis of national and spatial-time data meets many open questions [3]. This article addresses the challenge of developing an effective tool for designing neural networks through visual programming techniques. The designated module is a component of the repository of deep neuronetwork models created to address the scientific problem of the accumulation and systematization of models and machine learning algerits through the development for analysis and prediction. Developing spatial processes to support management decision-making for sustainable development in the world's regions.

2 Materials and Methods

2.1 Analysis of Functional Requirements for Repository of Deep Neural Networks

Information and communication technologies, such as artificial intelligence, robotics, information security, cloud computing, virtual reality, big data and the Internet of things, have been continuously evolving over the past few years.

Artificial neural network (ANN) applications are increasingly being developed and implemented to address a number of problems for which the algorithm is not known. These include classification, clustering, pat-tern and speech recognition, decision-making, data analysis and forecasting.

The solution to the scientific problem of accumulating and systematizing models and algorithms of machine learning is possible through the repository of deep neuronetwork models. This class of systems are designed to analyse and predict the development of spatial processes to support management decision-making for the sustainable development of the world's regions.

Use Case Repository Diagram describes the main precedents of the system. Authorized access to the system is performed on the basis of authorization by entering authentication data (Fig. 1).

The functions of the repository of deep neural network models can be decomposed into the domain of neural network model management (case group M.), design tasks (group T.), and data (group D.). The role of the administrator of the system allows access to edit repository information, and the user has the rights to sue and read, including the use of the neural network recommendation step.

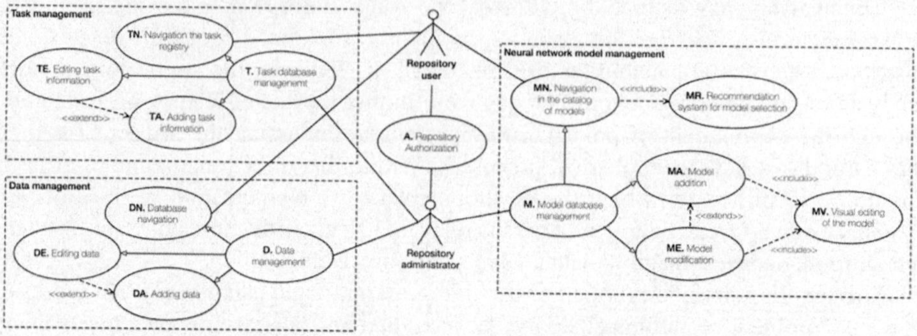

Fig. 1. Use case repository diagram of deep neuronetwork models.

This article focuses on the task of methodological and technological development of the MV use case. Visual editing of the model, which is included in the variety of uses associated with the creation and editing of deep neuronetwork models.

The study shows that using an artificial neural network to solve application problems is a time-consuming and, in some cases, time-consuming process. The use of graphic methods promises to dramatically increase the productivity of a programmer.

2.2 Neural Network Life Cycle

In order to understand the complexities of neural network design, it is necessary to isolate and describe several stages of its life cycle.

Stage 1 – Problem setting and choice of neural network architecture. Analysis of the problem with justification of application of artificial intelligence methods to solve it. Identification of the benefits that the neural network modelling method should provide. The architecture of the neural network is chosen for the specific problem.

Stage 2 – Determination of the quantity and quality of the inputs and outputs of the neural network. The quality of input and output variables of the neural network model is determined mainly by the task set out in the previous phase. Quantity composition may depend on the nature of input and output variables and the processes and properties they describe.

Stage 3 – establish a baseline sample of data. The initial sample of data can be derived from a practical or computational experiment. The larger the scale of the experiment, the more precise the neural network model can be obtained and used in practice.

Stage 4 – pre-processing and normalization of the reference sample. This step is determined by the chosen neural network architecture, the composition of input and output variables and, indeed, the actual data obtained from the experiment.

Stage 5 – splitting the initial sample into a training component and a test component. The existing reference sample is divided into a training and test sub-sample, taking into account the architecture of the neural network used. The training sub-sample is usually several times larger than the test sub-sample. The first is used strictly to adjust weights. The second is to test the correctness of the configured neural network model. In the case of a limited core sample, it can all be used for training, and the evaluation of the correctness of the network is tested during the actual operation phase.

Stage 6 – neural network structure determination. In addition to the previously defined composition of input and output variables, at this stage the number of hidden layers and neurons in each layer is given. For many neural network architectures, the choice of structure depends, among other things, on the amount of learning data.

Stage 7 – adjust neural network parameters and learning algorithm. At this stage the type and parameters of activation functions of hidden and output neurons are defined. For an iterative algorithm, the learning completion conditions, the speed coefficient, and the procedure for presenting examples of a learning sample are chosen.

Stage 8 – neural network learning. For precise algorithms, a single calculation of weighting factors is usually performed. For iterative, the repetition of learning eras. Only training sub-samples can be used for training.

Stage 9 – neural network contrast. If a neural network learning algorithm implied an iterative adjustment of weights throughout learning periods, it is not uncommon for some of the absolute weights to be so close to zero, that they have no practical influence on the results of calculations. How-ever, a considerable amount of computational resources is spent on such calculations, which significantly slows the learning process. In order to solve this problem, such weights are completely reset, that is, the appropriate connection between inputs and neurons or between two neurons is excluded from the network structure.

Stage 10 – neural network testing. The neural network is tested using sample samples. The results of the test assess the error of the neuronetwork model.

Stages 7–9 can be performed repeatedly in a loop until a configured neural network is obtained that can solve the problem with the required error level.

Stage 11 – practical use. The trained and tested neural network is used to practically solve the problem as new input data (values of input variables) become available. Results from the trained network can be obtained al-most instantaneously and interpreted by the user for subsequent decision-making.

Stage 12 – complete neural network training. The pairs of input and out-put vectors obtained during practical operation of the neural network can be used for further adjustment of weights. This is particularly important if the initial sample size was small. Some of the neural network architectures that use self-organization can not only change weighting factors in the course of pre-training, but can also change their structure.

It should be noted that the above sequence of steps is common to work with artificial neural networks. However, for a number of architectures there may be unique differences.

This structure makes it possible to see a clear approach to design, which can be implemented with the aid of block-based programming tools.

2.3 Modular Programming

The training of machine learning models can be very expensive, as it re-quires work on expensive hardware over a long period of time [1]. Thus, similar to imports of third party libraries to develop conventional systems, it is now common practice to reuse pre-trained machine learning models to develop machine learning applications. In particular, a recent research and consulting firm specializing in information technology markets,

Gartner has determined that the use of pre-trained machine learning models deployed as web-based services, is one of the main technological trends [2]. Popular frameworks such as Tensorflow and PyTorch provide high-level API support for easy integration of a set of common machine learning models [3].

Since there are a lot of solutions to avoid wasting valuable time inventing a bicycle, let's turn to the modular programming paradigm. A module is a logically completed piece of code with a specific function. For modular interaction, methods are used that do not allow to change parameters and functionality. The benefits of modular programming are obvious:

- acceleration of development;
- increased reliability
- simplification of testing • portability and fungibility.

As has already been said, in the field of neural networks, there are a great number of libraries, modules and patterns that can be implemented, combined and used for a wide range of solvable tasks. The visual design paradigm can be used to visualize the blocks being developed for effective application.

2.4 Visual Programming

Visual programming is the process of graphical representation of a program using a standard set of graphic elements [3].

In visual development, the number of forced control errors in the program is significantly reduced, hence the quality of the software product is improved.

Software visualization systems are generally classified as visual programming systems and programming visualization systems, as well as some programming systems if they use visual methods for presenting samples of input and output information.

A distinction should be made between a graphical programming language and visual development tools. A visual programming language is one that allows a programmer to create programs by manipulating graphic elements rather than printing text commands [4]. Visual Development Tools is a software development environment in which the most common soft-ware code blocks are represented as graphics objects [5]. In some cases, these concepts are linearly linked. For example, G is a visual programming language available through its native LabView development environment.

Visual programming languages may be classified according to different characteristics. For example, in the area of substantive application: training; used in industry; research. For the main modelled aspect: languages of the programme system structure description; Control Flow Description Languages Data Flow Description Languages Languages of behaviour (processes); languages to construct a graphical interface.

The following classification is also proposed:

Object-based languages where the visual programming environment provides graphical or symbolic elements that can be manipulated interactively according to certain rules. Example: Scratch.

Form editors that allow you to set user interface parts and configure their properties. Examples: Visual Basic, Delphi and C++ Builder by Borland, MS C#, MS

Access, C++ using wxSmith as part of the free cross-platform development environment Code::Blocks.

The languages of the schemes, based on the idea of «shapes and lines», where the figures (rectangles, ovals, etc.) are considered as subjects and are connected by lines (arrows, arcs, etc.), which are the relations. Examples: relationship editors in relational OBD, UML.

This type of visual programming tool typically solves the problems of building a user interface and simplifies application development by re-placing the writing method of a program with a design method. However, they do not affect the visual representation of program algorithms.

In designing visual programming languages, the basic approach is to draw graphs, that is, as a set of vertices (nodes) connected by arcs (edges).

Unlike the textual form of a record, in which objects (symbols and words) form a sequence, and each object is bound only to the left and right «neighbour», the graph form allows to visualize more complex relationships, because in it, each object can connect to several other objects. In this sense, the textual form is one-dimensional, while the graph form is two-dimensional. The ability to vary the geometry, shape and color of vertices, the appearance and thickness of arcs, and to change the mutual location of vertices without changing the topology of the graph significantly increases the expressive possibilities of the graph form representation of the program algorithm.

The creation of a visual (figurative) software development style is the main motivation for the development of graphometric programming technology.

GSP technology is a software algorithm design and coding technology based on the graphical representation of programs, with the aim of fully or partially automating software design, coding and testing processes [6].

This programming technology is based on two fundamental principles:

- visual, graphic presentation of program algorithms and other components of their specifications;
- the principle of structured procedural programming.

It is the implementation of this concept that is perfect for solving the problem of visual programming of neural networks.

Visual programming increases the visibility of submitted codes, significantly reduces the number of errors allowed during the design and coding phase of programs, and thereby speeds up the development process and increases the reliability of the codes of developed programs. Combined with a block approach, visual programming techniques will not only speed up development through simple graphic interchangeability of logical blocks, but will also ensure the simple portability of such programs [7–9].

3 Discussion of the Results of the Study

The practical use of deep neural networks integrated in the repository of neuronetwork models requires the decomposition of the ontology model of the repository into systems-generating domains: machine learning models, data and tasks (see Fig. 2). This makes it possible to create a platform solution for the consolidation, selection and efficient use of new methods and algorithms to solve problem-oriented problems related to the analysis of large spatial and spatial-time data [10].

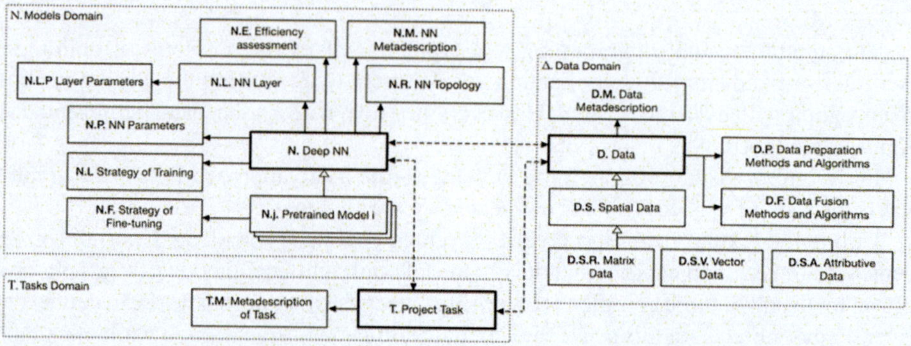

Fig. 2. Proposed ontological model repository of deep neural networks.

Machine learning models consolidated in the repository and designed using them always rely on sets of given different types. The domain of machine learning models is described by concepts defining deep neuronetwork models of different topology and methods, methods and algorithms of their learning. While machine learning is not limited to artificial neural networks, this study emphasizes them as a consequence of the fact that algorithms of this class are important for efficient analysis of large spatial data, aimed at identifying complex nonlinear dependencies in them.

The model of a deep artificial neural network is determined by its topology, including the use of layers of various types (drill, semi-socket, recurrence), activation and regulatory functions. The architecture of the model is also determined by the pattern of connection of layers, the format of representation of the input data of the model, and the characteristics of the output signal thereof. The deep neuronetwork model is also characterized by algorithms of initialization and optimization, a strategy of learning. It should also be noted that not only models of artificial neural networks but also their trained specimens, ready to solve design problems, are of practical interest. For each repository model, it is useful to compare a system category, a meta-description, a guide for use, a description of the project tasks to be performed, a description of the data to be analysed and the forms of their presentation [9, 11–13].

For the purpose of effective design of deep neurological models, a software package has been developed to address the identified visual programming methods. The created module is a component of the repository of deep neuronetwork models, which is created for solving the scientific problem of accumulation and C-stematization of models and algorithms of machine learning by means of development for analysis and forecasting.

In order to assess the convenience of visa programming of neural networks we turn to models with several inputs and outputs.

For example, if it is necessary to create a system for the automated distribution of geodata, the model being developed may have three inputs: data flow from IoT sensors, geolocation data and, for example, event tags (categorical information). In the result, the model should have two outputs: an estimation of the priority of the event from 0 to 1 (a scalar sigmoid output); a target division, which should process the received geoinformation (a SoftMax output by a set of departments). The result of the design of a neural network for the distribution of geodata is presented below (see Fig. 3).

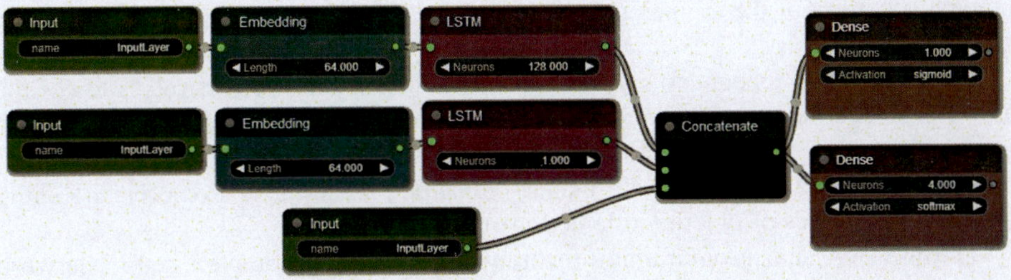

Fig. 3. Deep neuronetwork models through visual programming: a three-input, two-output model.

In addition to models with multiple inputs and outputs, manipulation of non-linear communication topologies in which the neuron layers are not sequentially connected is considered. An example of such an architecture is ResNet, which uses a deep residual learning structure. The ResNet model for CIFAR10 is presented below (see Fig. 4).

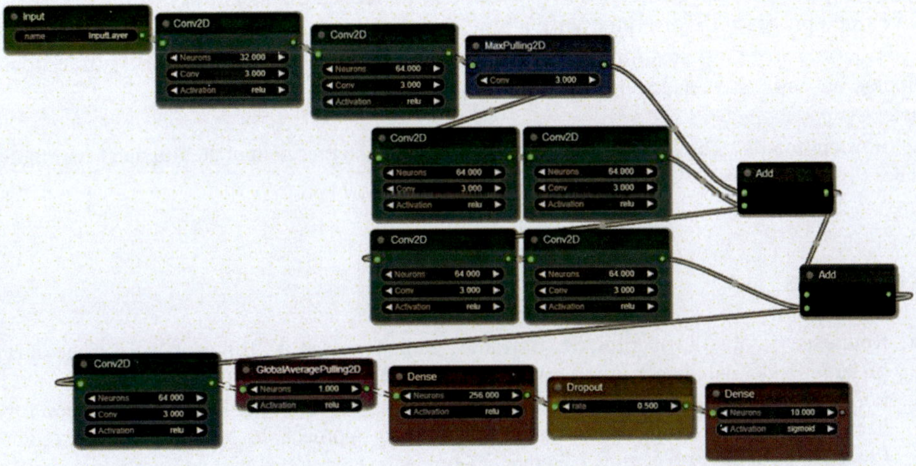

Fig. 4. Deep neuronetwork architecture ResNet.

The JavaScript programming language has been used to implement the graphical interface of the neuronetwork model design module using visual programming methods. The module has been tested in the task of forming basic models of neuronetwork architectures represented in the open network library Keras, in particular Layers API.

4 Conclusion

An important component of the repository of deep neuronetwork models with navigation functions, searching for a relevant model of machine learning to solve practiceoriented spatial data analysis tasks are a dynamic model visualization subsystem based on adaptive web interfaces in the form of dynamically curved graphs, with the possibility of interactive editing of the architecture and topology of the neuronetwork model via the web client.

Neural network repository graphics should be designed to effectively address the task of providing information on a specific neural network model (including systematic description, architectural class, format of analysed and output data, topology information, subjective and objective performance indicators, recommendations for finetuning the model, examples of practice-oriented use).

The use of visual programming techniques to describe complex neural network architectures greatly speeds up the development process and reduces errors by increasing visibility.

Together with the block approach, the visual programming methods will not only speed up the development of logical blocks by means of simple graphical interchangeability, but will also ensure the simple permeability of both whole programs and individual modules.

Visual programming increases the visibility of submitted codes, significantly reduces the number of errors allowed during the design and coding phase of programs, and thereby speeds up the development process and increases the reliability of the codes of developed programs. Combined with a block approach, visual programming techniques will not only speed up development through simple graphic interchangeability of logical blocks, but will also ensure the simple portability of such programs.

Acknowledgments. The study was supported by Grant of the President of the Russian Federation under Project no. MK-199.2021.1.6.

References

1. Roumelis, G., Vassilakopoulos, M., Corral, A., Manolopoulos, Y.: Efficient query processing on large spatial databases: a performance study. J. Syst. Softw. **132**, 165–185 (2016)
2. Solihin, M.I., Zekui, Z., Ang, C.K., Heltha, F., Rizon, M.: Machine learning calibration for near infrared spectroscopy data: a visual programming approach. In: Md. Zain, Z. et al. (eds.) Proceedings of the 11th National Technical Seminar on Unmanned System Technology 2019, pp. 577–590. Springer, Singapore (2021). https://doi.org/10.1007/978-981-15-5281-6_40
3. Aissaoui, O., Oughdir, L.: A learning style-based ontology matching to enhance learning resources recommendation. In: 2020 1st International Conference on Innovative Research in Applied Science, Engineering and Technology (IRASET), pp. 1–7 (2020)
4. Ahmed, F., Capretz, L.F., Bouktif, S., Campbell, P.R.J.: Soft skills and software development: a reflection from the software industry. Int. J. Inf. Process. Manage. **3**(4), 171–191 (2013)
5. Wang, Z., et al.: Plot2API: recommending graphic API from plot via semantic parsing guided neural network. In: 2021 IEEE International Conference on Software Analysis, Evolution and Reengineering (SANER), pp. 458–469 (2021)

6. Yamashkin, S., Yamashkina, E., Yamashkin, A.: Integration of knowledge, visualization and dissemination of spatio-temporal data through geoportal systems. In: E3S Web of Conferences, vol. 258, p. 03001 (2021)
7. Karpenko, A., Dobryakov, A.: Model for automated training systems. Overv. Sci. Educ. **7**, 1–63 (2011)
8. Haykin, S.: Neural networks: full course. Moscow, p. 113, 281–330 (2008)
9. San-Segundo, R., Gil-Martín, M., D'Haro-Enríquez, L.F., Pardo, J.M.: Classification of epileptic EEG recordings using signal transforms and convolutional neural networks. Comput. Biol. Med. **109**, 148–158 (2019)
10. Yamashkina, E.O., Kovalenko, S.M., Platonova, O.V.: Development of repository of deep neural networks for the analysis of geospatial data. In: IOP Conference Series: Materials Science and Engineering, vol. 1047, p. 012124 (2021)
11. Graves, A., Schmidhuber, J.: Framewise phoneme classification with bidirectional LSTM networks. In: Proceedings 2005 IEEE International Joint Conference on Neural Networks 2005, pp. 2047–2052. IEEE, Montreal, Que., Canada (2005)
12. Hamed, S.K., Ab Aziz, M.J.: A question answering system on holy Quran translation based on question expansion technique and neural network classification. J. Comput. Sci. **12**(3), 169–177 (2016)
13. Hafemann, L.G., Oliveira, L.S., Cavalin, P.: Forest species recognition using deep convolutional neural networks. In: 2014 22nd International Conference on Pattern Recognition, pp. 1103–1107. IEEE (2014)

Modeling the Competence of Experts

Alexander V. Solovyev(✉)

Federal Research Center "Computer Science and Control" of Russian Academy of Sciences, 44/2 Vavilova Street, Moscow 119333, Russian Federation
soloviev@isa.ru

Abstract. Rapid digitalization of economy intensified the need to conduct numerous researches. The most breakthrough discoveries currently lie at the intersection of different sciences, which raises the requirements towards the quality of expert appraisal for the evaluation of research potential and results. Experts appraising interdisciplinary projects that are run in organizations or under the grants from scientific foundations should have knowledge and expertise in different academic fields. This necessitates the search for competent experts who are able to appraise and evaluate interdisciplinary projects. The task of modeling the competence of experts is difficult primarily because it is not evident whether it is possible to assess the ability of an expert to adequately appraise a project based on the available information that is often non-structured, and how to do it. This article is dedicated to the development of concept for modeling the competence of experts. In the article we presented a methodological approach towards modeling the competence of experts who appraise scientific projects that are run in organizations or under the grants from scientific foundations. The suggested model of competence was tested and partially implemented as database information objects for future programmatic implementation. An array of documents on project appraisals was collected, the list of topics was defined and the list of keywords describing them was specified. We presented a prototype of software that allowed us making a preliminary modeling of subjects' breakdown by scientific fields, constructing matrices for graphs of correspondence between experts and subjects, producing of experts' profiles. Further we plan to elaborate procedures for matrices analyses for consistency and affiliation.

Keywords: Multi-criteria selection · Competence of experts · Modeling of competence

1 Literature and Problem Review

1.1 Introduction

The evident process of technological changes and economy digitalization associated with it has intensified the need to conduct researches that are more and more interdisciplinary. The most disruptive discoveries are made on interdisciplinary level, which raises the requirements towards the quality of expert appraisal of researches' potential and results. Experts involved into the appraisal of interdisciplinary scientific projects conducted by

R. Silhavy et al. (Eds.): CoMeSySo 2021, LNNS 231, pp. 690–697, 2021.
https://doi.org/10.1007/978-3-030-90321-3_57

both organizations and under grants of scientific foundations, should possess knowledge and expertise in different fields of science. Providing this, there is a need in competent experts who are able to perform appraisals of interdisciplinary scientific projects. While searching for the appropriate expert one should answer at least these questions: by what criteria is it possible to assess competence? How to avoid mistakes at the assessment of expert's competence? The problem of modeling the competence of experts is difficult primarily because it is not evident whether it is possible to assess the ability of an expert to adequately appraise a project based on available information that is often non-structured, and how to do it. This article is dedicated to the development of concept for modeling the competence of experts.

1.2 Review of Publications

Competence of an expert for the purposes of this article shall mean that the expert has knowledge, skills and experience that allow him making a reputable judgements about the subject of appraisal based on which those who conduct scientific researches or those who use their results can make the right decisions. The review of publications on this subject demonstrates that the task of modeling the competence of experts is quite timely.

The analyses of publications reveals that the main areas of application of recently developed competence models include: education [1], healthcare [2], businesses [3]. Competence models described in [4, 5] are the closest to the subject of this article. They are used in recruitment [4] and competence assessment in construction [5]. In open sources there are a lot of comparative studies on different competence models [6], their application in economy management, education, business processes of an organization. However, the application of these models for the search and assessment of experts is not covered. There is just an understanding that this should be done without any specific suggestions.

Another problem related to the modeling of competence is the modeling of trust. Under trust we shall mean the confidence that an expert expresses impartial, competent, skillful judgement at appraisal on a chosen subject, field or project. Trust modeling is also of great interest today judging by publication activity on this subject. Mainly this researches cover consumer trust and opinion, reputation of companies, banks, organizations or individual entrepreneurs and are targeted at economy, banking (credit sphere), insurance [7, 8]. Close to the subject of this research stand the models of personality confidence at social interaction [9]. Information available in the public domain is used as initial data to design the trust model both in these researches and in this article. Opinion model emerging under the influence of external factors (mass media etc.) is included into the trust model. The review of foreign publications on trust models [10–16] demonstrates that they cover more areas of application compared to domestic studies, namely e-commerce [10, 13, 15], healthcare [12], transportation [16]. There are different approaches towards modeling: probabilistic [12], graph [14], standardized models [15] used in e-commerce. Some of the models are implemented as software solutions [11].

In our country, decision-making theories are elaborated deeper, which makes it possible to apply already developed methods [17] to the solution of the problem of searching for experts by addressing the problem as a problem of multi-criteria selection in the space of signs of large dimension [18].

2 Modeling the Competence of Experts

2.1 Trust Model

As a part of modeling the competence of experts we suggested this formal trust model that includes the following blocks.

Opinion. Under opinion we shall mean an evaluative judgement expressed by an expert in his expert appraisal about a certain work done on a certain subject or under a certain project which was appraised by this expert. One of the main characteristic of the opinion is the attitude towards a scientific work expressed in the appraisal by the expert. It can be positive, negative or neutral. The opinion can also be complex containing both positive and negative judgements.

Block "Opinion" of the trust model is a structure that includes:

1. the text of an opinion
2. array of topics that defines subject domain (domains or projects)
3. an array of attitudes (positive, negative, neutral) on every judgement of an expert contained in the appraisal
4. an array of weighting coefficients (judgement value) generated through the analysis of expert appraisal (if possible, judgements are graded by value, otherwise all weighting coefficients are equal)
5. integral appraisal (possibly, a certain convolution of all provided attitudes with consideration of weighting coefficients of values)

Recommendation. Under recommendation we shall mean a judgement expressed by an expert in his expert appraisal about a certain work done on a certain subject or under a certain project which contains comments and suggestions on how to correct them. Recommendations can be explicit in the text of appraisal or can be obtained through analysis of the opinion text.

Block "Recommendation" of the trust model includes the following parts:

1. an array of topics that defines subject domain (domains or projects)
2. an array of N comments and suggestion on how to correct them
3. an array containing values of every comment (including verbal values, e.g. critical, very important, important, average, insignificant; if possible comments are graded by value based on the analysis of expert appraisal)
4. integral value of comments calculated by the value of comments

Reputation. Under the reputation we shall mean the level of expert's credibility based on the correlation analysis of the expressed opinions and recommendations against implementation results of a scientific project. Assessment of reputation is a separate and quite complicated task. Within the scope of this research we can only outline possible approaches to address it. The major assumption at the evaluation of reputation is as follows: if an expert acted in a certain way in the past he will act in the same way in the future.

One can evaluate reputation using the following approach:

1. arrays of successful and failed projects are composed with the use of retrospective subject analysis of information about the results of scientific projects in public domain (including that of available in the Internet)
2. analysis of expressed expert opinions and recommendations is done
3. correlation analysis of expressed experts' opinions and recommendations against implementation results of a scientific project is performed, it is defined whether the results correspond or not to the expressed opinion
4. integral level of expert credibility (reputation) is calculated, for example as the difference between the number of projects where the results correspond to the opinion and the projects where the results does not correspond to the opinion

Block "Reputation" of the trust model makes competence modeling process harder and requires further development. However, reputation modeling helps to provide more accurate evaluation of expert's competence.

2.2 Expert Competence Model

Competence of an expert was defined above and stipulates that the expert has knowledge, skills and experience that allow him making a reputable judgements about the subject of appraisal based on which those who conduct scientific researches or those who use their results can make the right decisions. The review of publications on this problem demonstrates that the problem of modeling the competence of experts is quite significant.

We shall use 2 types of indicators in expert competence model:

1. Indicators that characterize knowledge, experience and skills of an expert on a certain subject (subjects) that we shall call indicators of expert's profile
2. Indicators that allow evaluating the quality of practical application of knowledge, experience and skills of an expert that we shall call indicators of appraisal quality

Indicators of Expert's Profile. The values of expert's profile indicators can be calculated by the subject modeling method. Subject modeling is a generative probabilistic model that is effective for the subject identification in scientific publications [19].

Let us assume that there is an array of documents that characterize a certain expert appraisal by a certain expert available in the public domain. The set of documents is presented as a sampling of pairs document – word (d, w). Subjects are defined and keywords characterizing them are specified. An array of subjects defines a certain subject field. Probabilistic latent semantic analysis of the array of documents with the allocation of subjects in the array is performed. Experts and projects are determined from the array of documents. Bipartite graph of correlation between experts and subjects is generated. A profile is produced for every expert based on the generated graph with consideration of credibility level of the source from which the document was obtained.

The following indicators are suggested as expert's profile indicators:

1. The scale of projects (by budget size) where an expert took part (the budget or category depending on the sum e.g. by logarithmic scale)
2. A number of projects in every subject field of an expert
3. Experience in appraisal (in years) in a certain subject field
4. Professional qualification (education, position)
5. Age
6. A number of printed works in every subject field

Indicators of Appraisal Quality. To calculate the values of appraisal quality indicators we shall use opinion, reputation and recommendation blocks of the trust model defined in Sect. 2 above, and the assessment of appraisal dynamics (increase/decrease in the number of appraisals for a certain period of time).

The algorithm for the assessment of appraisal quality can be as follows:

1. Assessment of appraisal quality «through the eyes of other experts» is performed by the opinion block of the trust model
2. Assessment of comments value (criticism) of an appraisal is performed by the recommendation block of the trust model
3. Assessment of appraisal quality is performed by the comparison of expert assessment of a project and actual result of that project (success/failure)
4. Additional indicators that affect quality indicators are calculated with the use of values of credibility and trust indicators
5. Assessment of appraisal quality is performed by subject modeling method against the set of sources in public domain. This allows having up-to-date values of quality indicators and tracking their dynamics. Each indicator has a weighting coefficient defined by experts
6. Integral indicator of appraisal quality is calculated
7. The profile of competencies of an expert is presented in the form of a pie chart where subject areas are axes of reference where values of quality indicators are marked
8. Comparison of profiles of competencies can be performed automatically (by values of indicators) or graphically by overlapping the profiles of competencies. It is assumed that it is possible to compare the profiles of competencies for several subject areas simultaneously

2.3 Methodology for Modeling the Competence of Experts

The basis for modeling the competence of experts is the development of a model. Initial data used for analysis under methodological approach is: expert appraisals, sources in the public domain (including those in the Internet), description of a project (by the set of subjects) to which experts are selected.

Assumptions:

1. all information about experts is available in the public domain (meaning that unavailable or classified information is considered non-existent because it is not possible to analyze it)
2. this information is authentic
3. there is no fake information that is made available deliberately
4. the list of topics adequately characterizes the object of appraisal and considered subject areas
5. the list of researched open public sources fully covers the information array required for the decision-making
6. expert's past (reputation, competence) defines his behavior in future

Result: the list of experts' profiles in rank order by integral quality indicator of their work.

The main steps of methodological approach.

1. create the list of sources required to perform subject modeling and assessment of experts (sources in public domain, internet sources, expert appraisals, descriptions of scientific projects)
2. create the list of topics for every project for which the competence of experts is modelled
3. produce (or correct if they were produced earlier) the profiles of opinions, reputations and recommendations for every expert (Sect. 2)
4. produce (or correct if they were produced earlier) the profiles of competencies for every expert (Sect. 3)
5. perform the assessment of quality of work for every expert (Sect. 3). Quality indicator is corrected in accordance with the level of credibility calculated by Reputation block (Sect. 2)
6. perform an integral assessment of expert quality according to the competence model (Sect. 3)
7. define the list of organizations for which appraisal (appraisals) is performed
8. compare profiles of competencies of experts by choosing the maximum value of efficiency among alternative profiles. Comparison of profiles among each other is possible both automatically (by the values of indicators) including comparison for several subject areas simultaneously and graphically (manually, subject to justification of the choice of expert) by overlapping profiles of competencies. This step results in the list of profiles suitable for appraisal in descending order of quality assessment

3 Conclusion

In this article we described a methodological approach towards modeling the competence of experts. The suggested model was tested and partially implemented as database information objects for further programmatic implementation [20]. Also, we composed a test array of documents of projects appraisals of Russian Foundation for Basic Research (RFBR), created the lists of subjects and keywords that describe them. We presented a prototype of software that allowed us making a preliminary modeling of subject breakdown by scientific fields, constructing matrices for graphs of correspondence between experts and topics, producing of experts' profiles. Further we plan to elaborate procedures for matrices analyses for consistency and affiliation.

Providing that search space for information about experts is huge and the number of indicators is significant, the problem of expert selection is a problem of multi-criteria selection in the space of signs of large dimension. The problem of correct assessment and selection of the right profiles by the analysis of indicators (characteristics) in such conditions is quite complicated and requires additional researches. In our opinion, method [18] is the most suitable for these purposes. We suggest using it as the basis for further development and elaboration of methodology for addressing the problem of expert selection.

If implemented, the suggested methodological approach will allow increasing efficiency and quality of appraisal of interdisciplinary scientific projects.

References

1. Hong, J.-C., Chang, J.-O.: Constructing an educational policy acceptance model to explore attitudes toward policy implementation: using an industrial competency standards system in science and technology universities. J. Res. Educ. Sci. **63**(2), 251–284 (2018)
2. Violato, C., Gao, H., O'Brien, M.C., Grier, D., Shen, E.: How do physicians become medical experts? A test of three competing theories: distinct domains, independent influence and encapsulation models. Adv. Health Sci. Educ. **23**(2), 249–263 (2017). https://doi.org/10.1007/s10459-017-9784-z
3. Malekan, H.S., Adamiak, K., Afsarmanesh, H.A: Systematic approach for business service consolidation in virtual organizations. Serv. Oriented Comput. Appl. **12**(1), 41–57 (2018)
4. Jøranli, I.: Managing organisational knowledge through recruitment: searching and selecting embodied competencies. J. Knowl. Manage. **22**(1), 183–200 (2018)
5. Jazayeri, E., Liu, H., Dadi, G.B.: Creating a safety training and competence model for construction craft professionals. In: Construction Research Congress 2018: Safety and Disaster Management - Selected Papers from the Construction Research Congress, pp. 241–250 (2018)
6. Veshneva, I.V., Prokofyeva, L.P., Bol'shakova, A.A.: Evaluation of adequacy models of formation of competences built on the basis of status functions method [Ocenka adekvatnosti modeley formirovaniya kompetenciy, postroennih na osnove metoda statusnih funkciy]. In: Information Technologies in education (ITO-SARATOV-2017). Proceedings of the IXth All-Russian Scientific and Practical Conference, pp. 405–408. Publishing Center "Science", Saratov (2017)
7. Dankin, D.M.: Symmetrical trust – resource development strategy [Simmetrichnoye doveriye – resurs strategii razvitiya]. Secur. Eurasia **1**(53), 49–60 (2017)
8. Sinyaev, V.V.: Vector model of business reputation management in the market of insurance services [Vektornaya model upravleniya delovoy reputaciey na rynke strahovih uslug]. Educ. Sci. Sci. Pers. **1**, 105–108 (2018)
9. Gubanova, S.E.: Trust as a basis for harmonization of the interaction of business structures in the conditions of the digital economy [Doveriye kak osnova garmonizacii vzaimodeystviya biznes-struktur v usloviyah cifrovoy ekonomiki]. Bull. Volga Univ. **1**(2), 134–141 (2018)

10. Sfenrianto, S., Wijaya, T., Wang, G.: Assessing the buyer trust and satisfaction factors in the E-marketplace. J. Theor. Appl. Electron. Commer. Res. **13**(2), 43–57 (2018)
11. Bag, S., Azad, M.A., Hao, F.: A privacy-aware decentralized and personalized reputation system. Comput. Secur. **77**, 514–530 (2018)
12. Athanasiou, G., Anastassopoulos, G.C., Tiritidou, E., Lymberopoulos, D.: A trust model for ubiquitous healthcare environment on the basis of adaptable fuzzy-probabilistic inference system. IEEE J. Biomed. Health Inform. **22**(4), 1288–1298 (2018)
13. Dai, Y.N., Viken, G., Joo, E., Bente, G.: Risk assessment in E-commerce: how sellers' photos, reputation scores, and the stake of a transaction influence buyers' purchase behavior and information processing. Comput. Hum. Behav. **84**, 342–351 (2018)
14. Constantinov, C., Mocanu, M., Poteraş, C., Popa, B.: Using a graph database for evaluating and enhancing a social reputation engine. In: Proceedings of the 2018 19th International Carpathian Control Conference, ICCC 2018, pp. 518–523 (2018)
15. Zhang, Y., Jiang, H., Zhang, S., Wang, X., Ren, Y.: A C2C E-commerce trust model based on reputation. J. Theor. Appl. Inf. Technol. **96**(11), 3555–3566 (2018)
16. Sanchez-Paredes, A., Fernandez-Pena, F., Urrutia-Urrutia, P., Naranjo-Avalos, H.: A reputation-based shared transport system. A case study of E-collaboration in the City of Ambato. In: 5th International Conference on eDemocracy and eGovernment, ICEDEG 2018, pp. 269–274 (2018)
17. Petrovsky, A.B.: Decision theory [Theoriya prinyatiya resheniy]. Akademy. Moscow (2009). ISBN 978-5-7695-5093-5
18. Petrovsky, A.B., Lobanov, V.N.: Multi-criteria selection in the space of signs of large dimension: PAKS-M multimethod technology [Mnogokriterial'ny'j vy'bor v prostranstve priznakov bol'shoj razmernosti: mul'timetodnaya texnologiya PAKS-M]. Artif. Intell. Decis. Mak. **3**, 92–104 (2014)
19. Blei, D.M., Ng, A.Y., Jordan, M.I.: Latent Dirichlet allocation. J. Mach. Learn. Res. **3**, 993–1022 (2003)
20. Ivanova, A.A., Gladilin, S.A., Zhukovskiy, A.E., Pliskin, E.L.: Database for the administrative accounting of scientific publications [Baza dannih administrativnogo ucheta nauchnih publikaciy]. Proc. ISA RAS. Special Issue 83–89 (2018). https://doi.org/10.14357/20790279180509

Forecasting Stock Quotes by Generating a Differential Equation Based on the Statistical Data General Set

Roman I. Dzerjinsky(✉) and S. A. Nikulina

MIREA – Russian Technological University, 78, Vernadsky Pros., Moscow 119454, Russia

Abstract. The stock quotes values on the stock market are described using differential equations by analysing the data general set on the time series corresponding behavioural sections. For this, its main characteristics have been determined, the differential equation coefficients have been selected, which makes it possible to obtain a predictive estimate. Numerical experiments have been carried out.

Keywords: Stock quotes dynamics · Time series forecasting · Forecasting quality metrics

1 Introduction

Forecasting stock prices is one of the most difficult tasks in finance. Successful forecasting can bring significant profits in trading. The stock price is influenced by many factors: market conditions, macroeconomic conditions, economic events, investor preferences and companies management decisions.

Many traders base their analysis on the assumption that market price patterns will repeat themselves in the future. One of the most common and well-developed techniques used to make forecasts is time series analysis. This ideology application inevitably gives rise to the near-proportions problem, which can be identified in their structure using specialized algorithms for finding near-proportions. It is also interesting that this generates difficulties with the relations of near-periods [1–3].

Applying the near-periods ideology allows analysing different nature markets [4, 5].

The courtesy tribune for this report will also hear reports describing the oil consumption dynamics from 1965 to 2020 in the United States of America, world population growth, the prospects for the fight against terrorism, and some others.

This paper presents forecasting stock prices in a comparative analysis using three different models based on differential equations: the intensive growth model, the limited growth model and the geometric Brownian motion model.

As a result, we will be able to choose the highest quality model for predicting the stock price.

R. Silhavy et al. (Eds.): CoMeSySo 2021, LNNS 231, pp. 698–717, 2021.
https://doi.org/10.1007/978-3-030-90321-3_58

2 Near-Periods Definition

In general, data analysis and empirical data processing are often associated with difficulties in identifying fluctuations accompanying trends. To conduct fluctuation's a reliable analysis, it is necessary to extract the trend component, since the trend distorts the result obtained [2].

The trend characteristics are considered to be encoded through the anchor points. It is necessary to find these points in such a position that will ensure the trend exclusion. We will use 3 points: $y_{t-\Delta t}$, y_t, $y_{t+\Delta t}$, located at equal distances from the midpoint by argument.

In this formulation, the problem is reduced to the proportions' theory classical results, under which the segment is split, shown in Fig. 1.

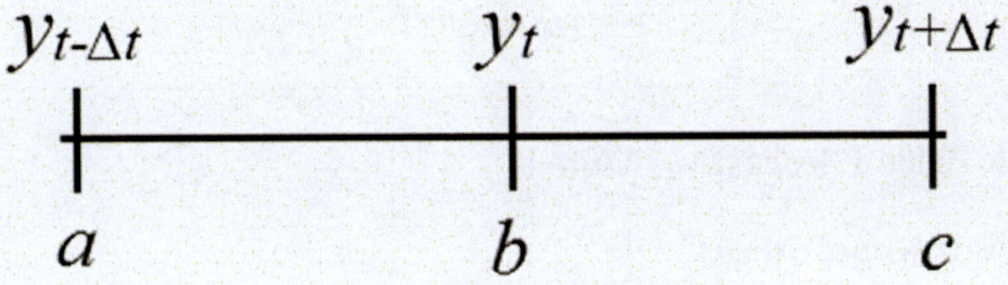

Fig. 1. Splitting a segment.

To exclude a trend, we will use a geometric proportion:

$$y_t = \sqrt{y_{t-\Delta t} * y_{t+\Delta t}}. \tag{1}$$

Then the system state is characterized by a dimensionless criterion:

$$P = \frac{y_{t-\Delta t} * y_{t+\Delta t}}{y_{t2}} = 1. \tag{2}$$

After taking the logarithm, we get:

$$\ln(P) = \ln\left(\frac{y_{t-\Delta t} * y_{t+\Delta t}}{y_{t2}}\right) = 0. \tag{3}$$

Converting empirical data to ln $(y_{t-\Delta t} * y_{t+\Delta t}/y_t^2) \sim t$ coordinates) results in the trend section's exclusion from the data.

Identify periods that are free from a priori assumptions, we use an approach that relies primarily on the function period fundamental characteristic property, which consists of repeating the function values through the independent variable change interval equal to the period:

$$f(t+\tau) - f(t) = 0. \tag{4}$$

The necessary generality is the identifying periodicities method by assessing the time series behaviour repeatability degree understudy at various time shifts.

As a rule, in real data, we have to deal with nonlinear fluctuations, and therefore pure periods are quite rare. As a result, it is only possible to identify the values that are closest to the periods. Such values are called near-periods [2, 6].

For the discrete case, if n is the function counts the total number $f(t)$, given by the experimental values, the following metric is introduced to determine the near-periods:

$$a(\tau) = \frac{1}{n-\tau}\sum_{t=0}^{n-\tau}|f(t+\tau) - f(t)|. \tag{5}$$

This function (5) is called the shift function or the Alter-Johnson function.

The near-periods system τ of the function $f(t)$ can be defined as shift function local minima a set:

$$\tau = \arg\min a(\tau). \tag{6}$$

3 Stock Price Forecasting Models

3.1 Intensive Growth Model

In the systems' development process, there are stages at which the reproduction rates do not decrease, but increase, that is, for Eq. (7), k increases. Such stages, as a rule, are of short duration, and therefore they are often dropped from the analysis.

$$\frac{dy}{dt} = ky, \tag{7}$$

where k is a function equal to y * a in the intensive growth case.

Then, as a result of substitution into Eq. (7), we obtain the equation:

$$\frac{dy}{dt} = ay^2. \tag{8}$$

As a result of solving this differential equation, we obtain:

$$y = -\frac{1}{at+C}. \tag{9}$$

Anamorphosis for this model type is:

$$\frac{1}{y} = -at - C. \tag{10}$$

Straightening is achieved in coordinates $1/y \sim t$.

This model determines the vertical asymptote position, at which the function value reaches an infinite value. Therefore, by the vertical asymptote position, an upper estimate can be made to determine the intensive growth process completion time.

3.2 Limited Growth Model

In the probability theory classical results, the integral distribution functions are represented by curves with a horizontal asymptote. In this case, the integral curves' structure itself can be asymmetric. The problem is the procedure for identifying both the distribution type curve and its parameters [2, 6, 7].

The only way to effectively identify the model parameters type is to construct empirical data in coordinates that give a linear relationship between the density function and the argument. In this case, the measurement results corresponding to this model are represented by linear dependences and the dependence slope change points determine the data homogeneity range.

For the Gompertz model, it has the following equation:

$$\frac{dy}{dt} = yAe^{-kt}. \tag{11}$$

As a result of the transformations, we get:

$$\ln(y) = -\frac{A}{k}e^{-kt} + \ln(C). \tag{12}$$

From the Gompertz model original equation, the following anamorphosis is obtained:

$$\ln\left(\frac{1}{y}\frac{dy}{dt}\right) = \ln(A) - kt. \tag{13}$$

Plotting the data corresponding to the Gompertz model gives a flattening in the coordinates $\ln(1/y \cdot dy/dt) \sim t$.

This defines the algorithm for determining the Gompertz equation parameters, as shown in Fig. 2. In this case, along the dependence linear section, we determine the parameter k, which is represented by the straight line inclination angle to the abscissa axis.

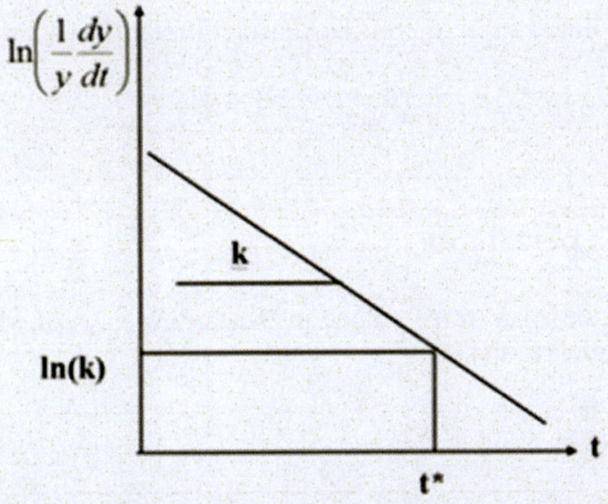

Fig. 2. Time determination to reach the inflexion point by the parameter k-value.

The Gompertz equation has an inflexion point at which:

$$\frac{d^2y}{dx^2} = 0. \tag{14}$$

At the inflexion point, we have Eqs. (15) and (16).

$$\frac{1}{y}\frac{dy}{dt} = k \tag{15}$$

$$\ln\left(\frac{1}{y}\frac{dy}{dt}\right) = \ln(k) \tag{16}$$

That is, we calculate the slope logarithm and plot the resulting value on the ordinate axis as a constant level line. This line intersection with the original straight line determines the time to reach the inflexion point on the curve, characterizing the resource total amount extracted by the given moment. This allows determining the inflexion point position, as shown in Fig. 2.

3.3 Model-Based on the Stochastic Differential Equation

One of the most widely used models for predicting stock prices is geometric Brownian motion, which satisfies the following equation:

$$\frac{dy}{dt} = \mu y dt + \sigma y \frac{dW}{dt}, \tag{17}$$

where

W - the Brownian motion,
μ - demolition parameter,
σ - volatility parameter [8, 9].

For an arbitrary initial value y_0, this stochastic differential equation has a solution:

$$y = y_0 e^{\left(\mu - \frac{\sigma^2}{2}\right)t + \sigma W}. \tag{18}$$

4 Numerical Experiments

As the initial data, the share price values in the trading days of Microsoft and IBM companies were taken [10, 11].

For Microsoft, the data were taken from 03.13.1986 to 05.15.2021, for IBM - from 01.02.1962 to 05.15.2021.

4.1 Microsoft Analysis

Let's build a chart where trading days are plotted on the abscissa, and the stock price in dollars at the trading day close on the ordinate. The graph is shown in Fig. 3.

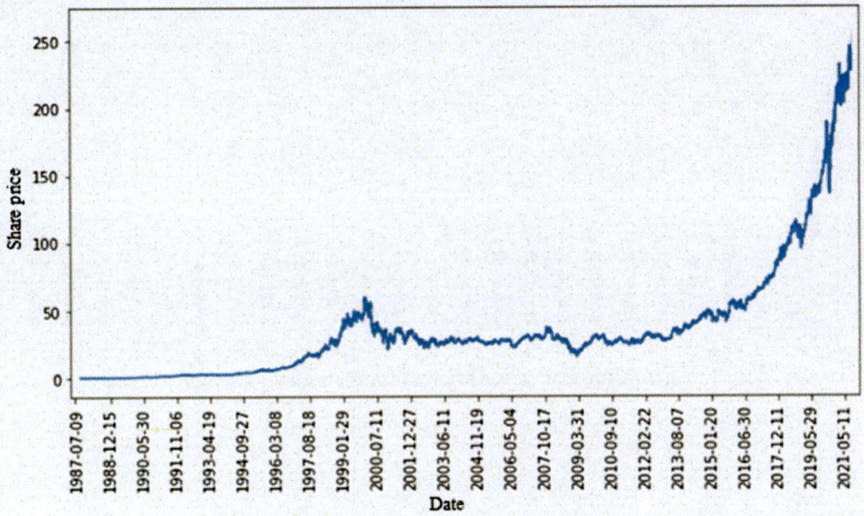

Fig. 3. Initial data for Microsoft.

To consider the initial data in more detail and identify the critical levels corresponding to the main boundaries' formation, it is necessary to construct the initial data in semi-logarithmic coordinates, that is, to log the values along the ordinate axis. The graph is shown in Fig. 4.

Also, using piecewise-linear approximation, select the trend lines in Fig. 4.

Figure 4 shows how the trend behaves and at what moments it changes its character, namely the inclination angle. Critical levels are the moments at which the process changes structure.

For a more detailed consideration, we will divide the graph into several parts.

In the intensive growth case, we have anamorphosis $1/y = -at - C$.

Let's build a graph in coordinates $1/y \sim$ t. The result is anamorphosis, which smooths out the graph in intense growth areas. These anamorphosis graphs are shown in Figs. 5, 6.

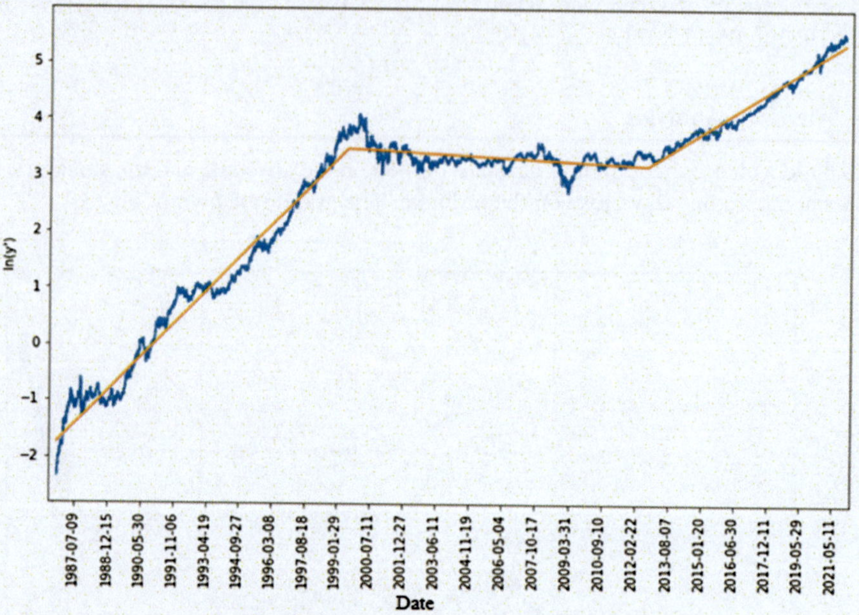

Fig. 4. Initial data for Microsoft on a semi-log scale.

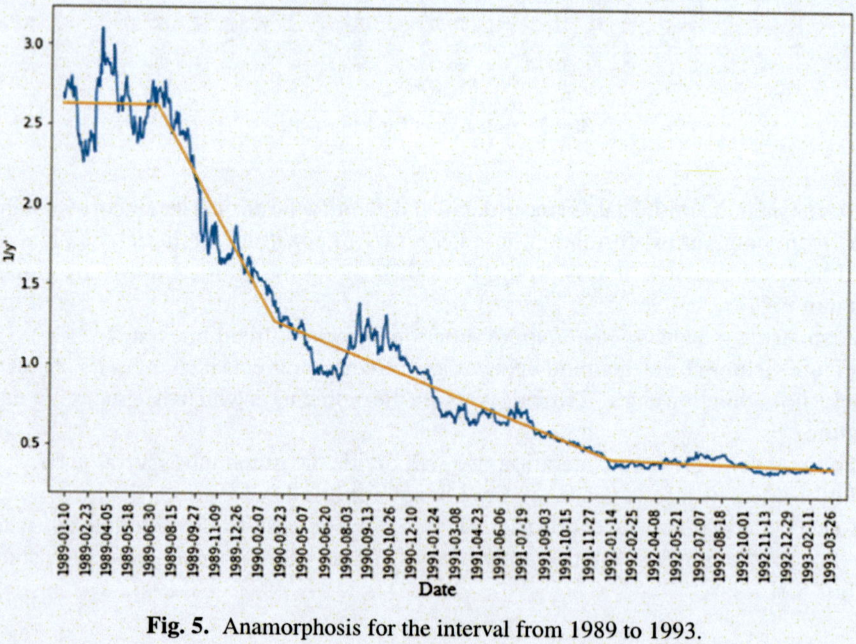

Fig. 5. Anamorphosis for the interval from 1989 to 1993.

Let's note some asymptotes, indicating the limits of intensive growth, on the initial data graph on a semi-log scale (Fig. 7). The growth limit is marked in red, the decline limit is shown in green.

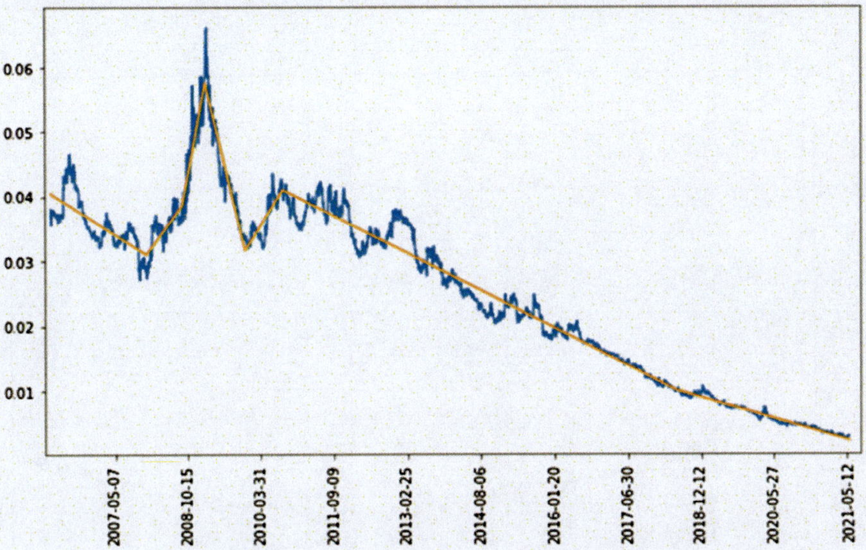

Fig. 6. Anamorphosis for the interval from 2006 to 2021.

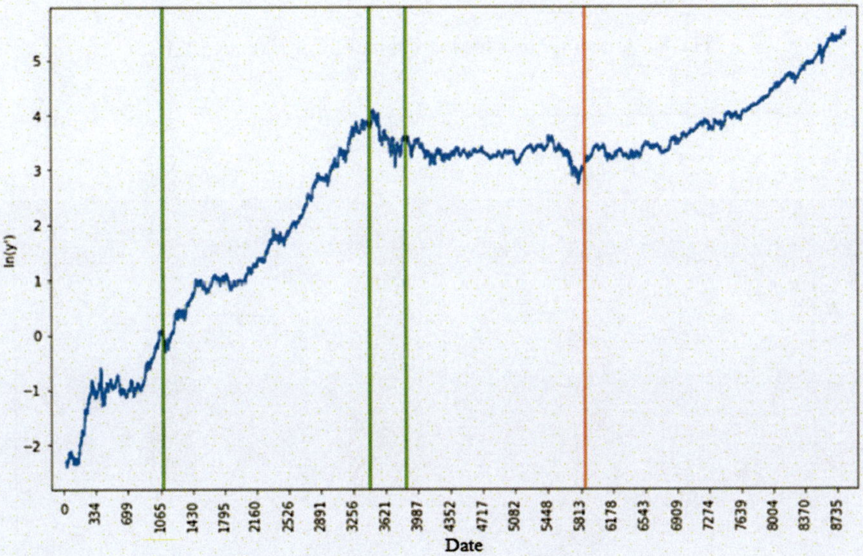

Fig. 7. Intensive growth and decline limits.

In the limited growth case, we have anamorphosis $\ln(1/y \cdot dy/dt) = \ln(A) - kt$.

Plotting the data corresponding to the Gompertz model gives a flattening in the coordinates $\ln(1/y \cdot dy/dt) \sim t$. These anamorphosis graphs are shown in Figs. 8, 9.

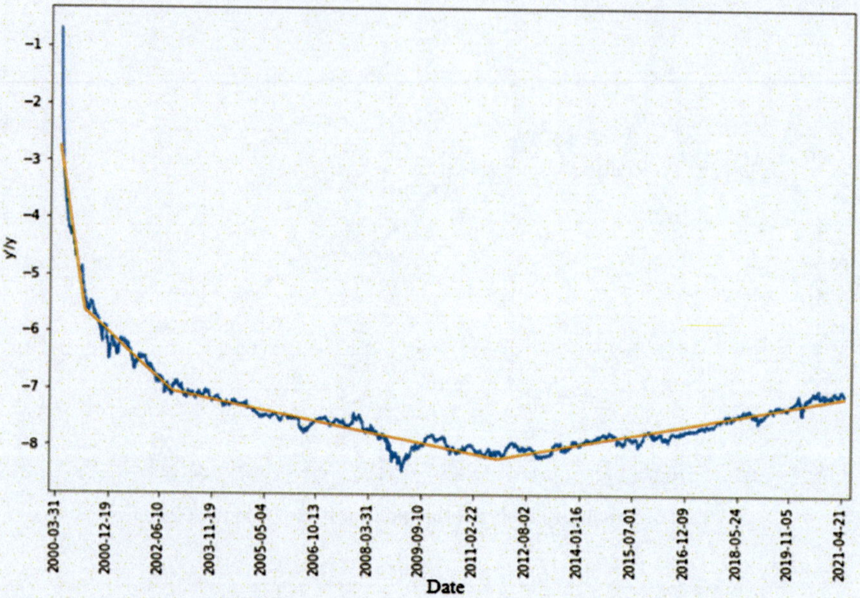

Fig. 8. Anamorphosis for the interval from 2000 to 2021.

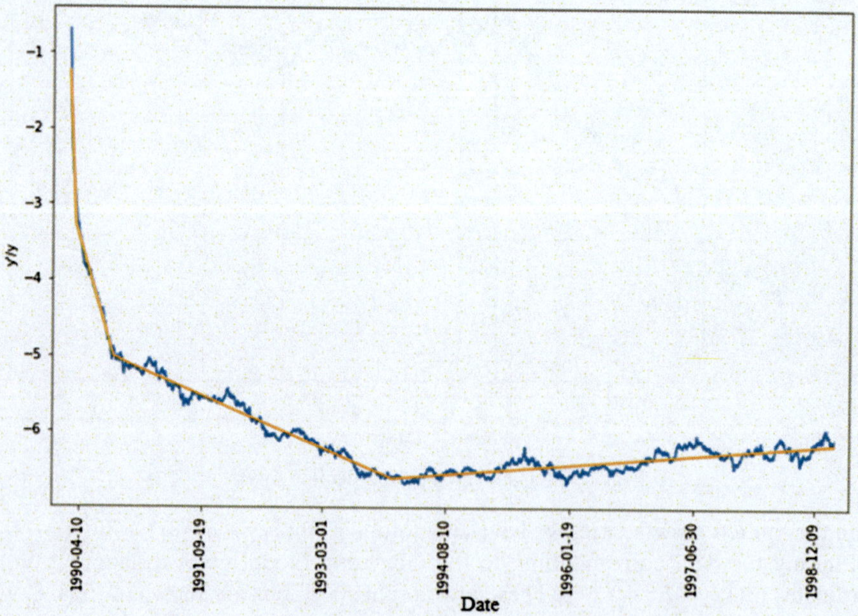

Fig. 9. Anamorphosis for the interval from 1990 to 1998.

Calculate the slope coefficient k equal to the trend line slope. Next, we calculate the k logarithm and find the intersection point with the trend line. We note the obtained values on the initial data graph on a semilogarithmic scale - these will be asymptotes, indicating the limited growth limit (Fig. 10).

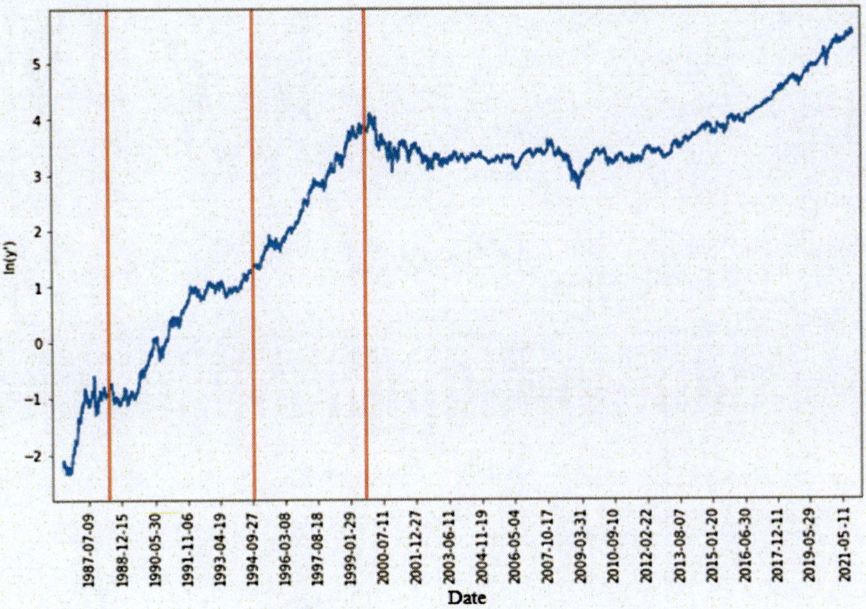

Fig. 10. Restricted growth limits.

4.2 IBM Analysis

Let's build a chart where trading days are plotted on the abscissa, and the stock price in dollars at the trading day close on the ordinate. The graph is shown in Fig. 11.

To consider the initial data in more detail and identify the critical levels corresponding to the main boundaries' formation, it is necessary to construct the initial data in semi-logarithmic coordinates, that is, to log the values along the ordinate axis. The graph is shown in Fig. 12.

Also, using piecewise-linear approximation, select the trend lines in Fig. 12.

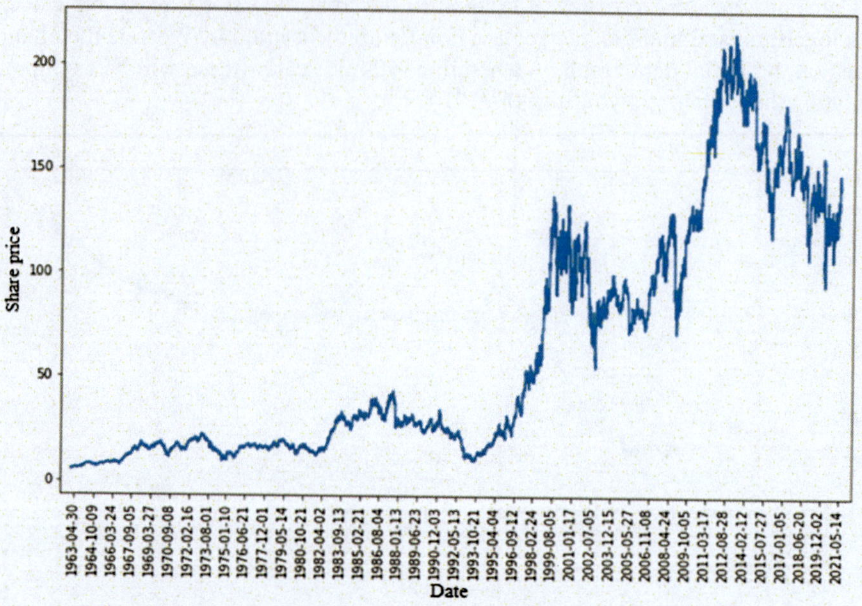

Fig. 11. Initial data for IBM.

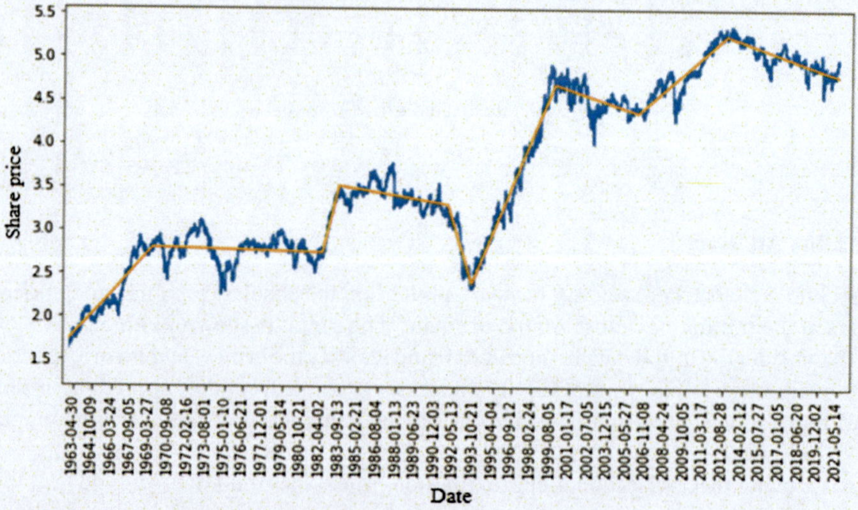

Fig. 12. Initial data for IBM on a semi-log scale.

In the intensive growth case, we have anamorphosis $1/y = -at - C$. These anamorphosis graphs are shown in Figs. 13, 14.

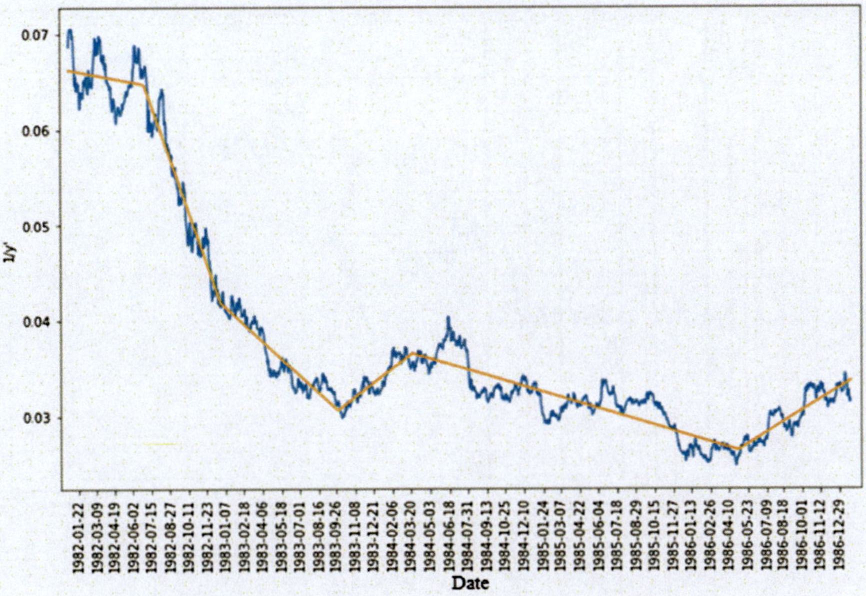

Fig. 13. Anamorphosis for the interval from 1982 to 1986.

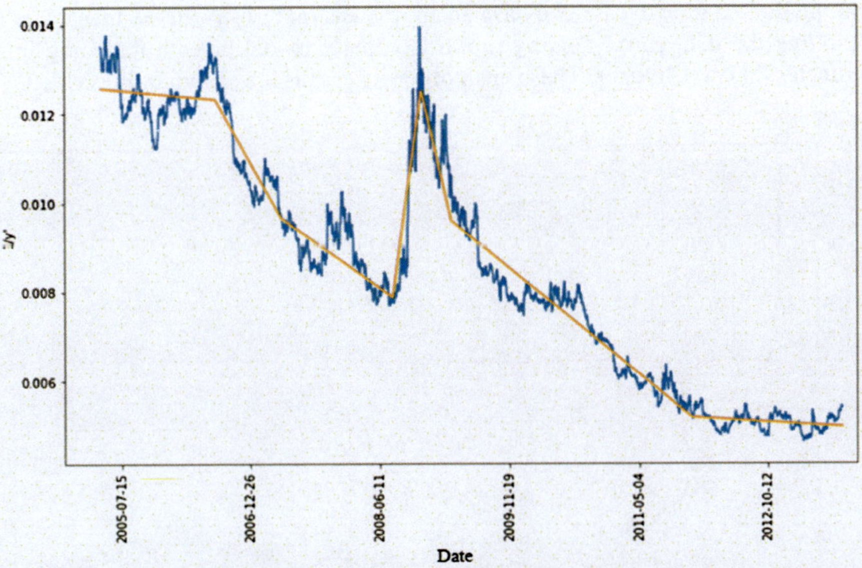

Fig. 14. Anamorphosis for the interval from 2005 to 2012.

Let's note some asymptotes, indicating the limits of intensive growth, on the initial data graph on a semi-log scale (Fig. 15). The growth limit is marked in red, the decline limit is shown in green.

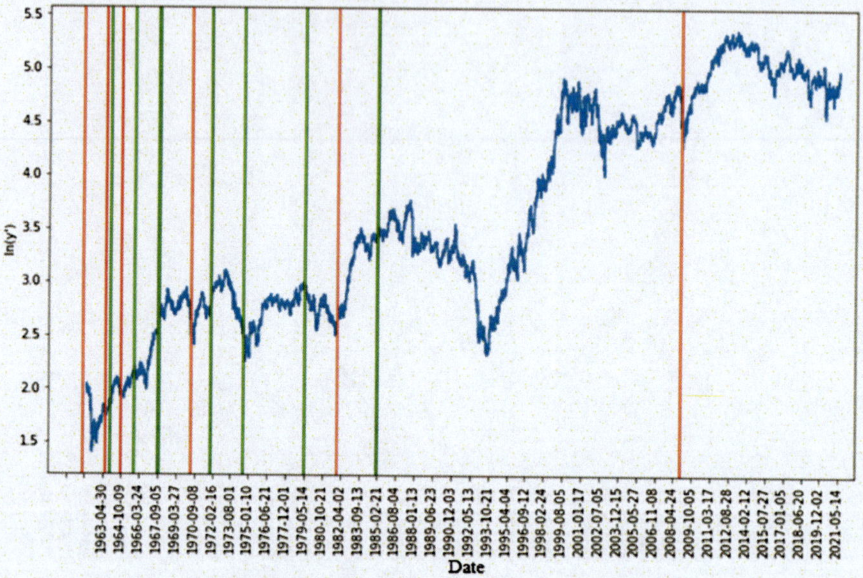

Fig. 15. Intensive growth and decline limits.

In the limited growth case, we have anamorphosis $\ln(1/y \cdot dy/dt) = \ln(A) - kt$.

Plotting the data corresponding to the Gompertz model gives a flattening in the coordinates $\ln(1/y \cdot dy/dt)$ t. These anamorphosis graphs are shown in Figs. 16, 17.

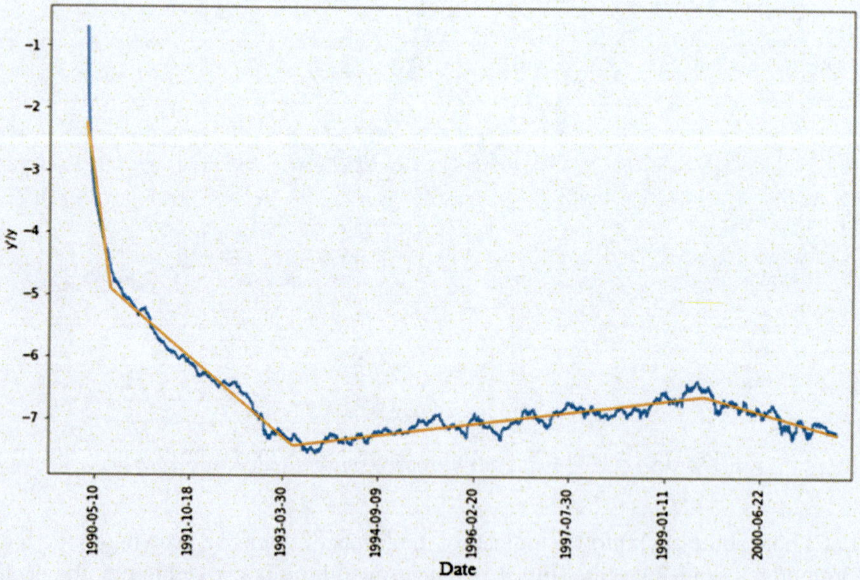

Fig. 16. Anamorphosis for the interval from 1990 to 2000.

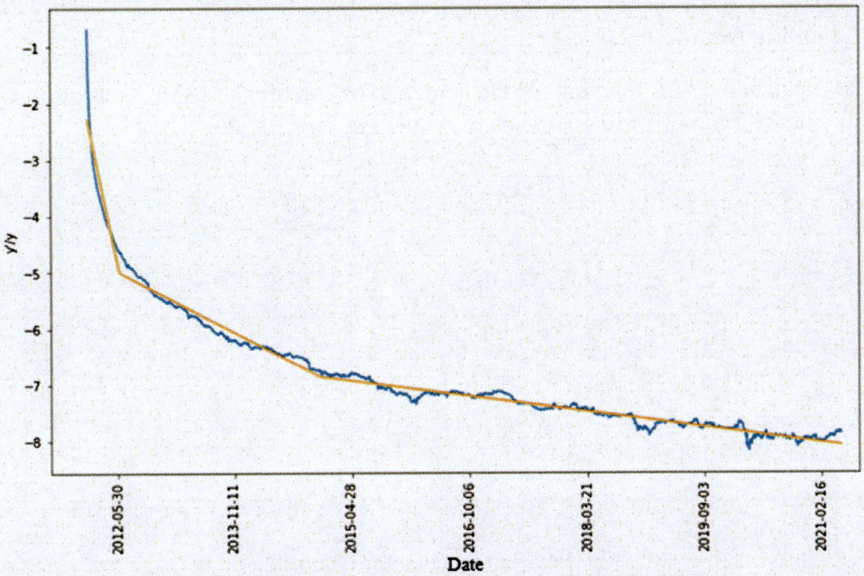

Fig. 17. Anamorphosis for the interval from 2012 to 2021.

Calculate the slope coefficient k equal to the trend line slope. Next, we calculate the k logarithm and find the intersection point with the trend line. We note the obtained values on the initial data graph on a semilogarithmic scale - these will be asymptotes, indicating the limited growth limit (Fig. 18).

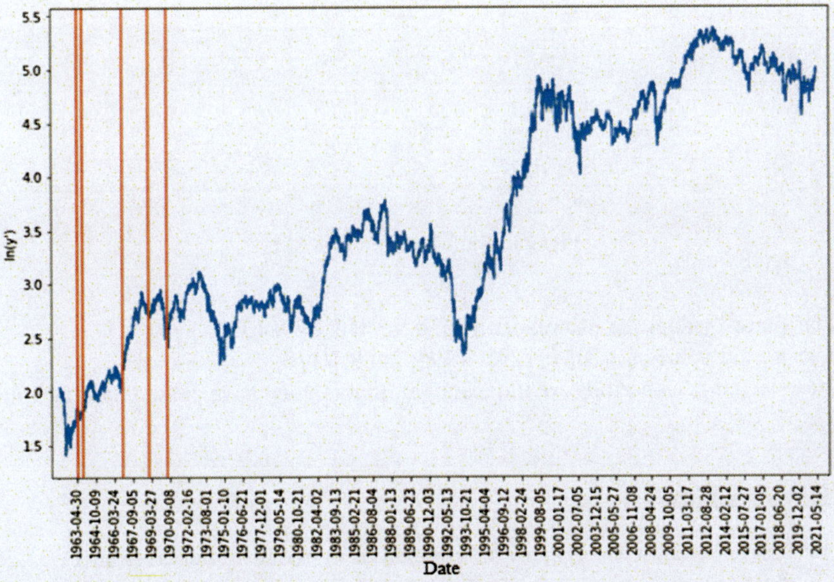

Fig. 18. Restricted growth limits.

4.3 Finding Near-Periods

Let's calculate the shift function for the quotes of Microsoft and IBM. As a result, after excluding the trend, we will get the charts shown in Figs. 19, 20.

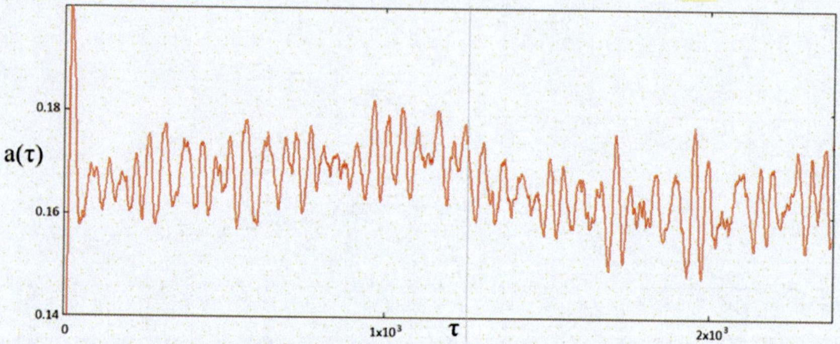

Fig. 19. Shift function for Microsoft.

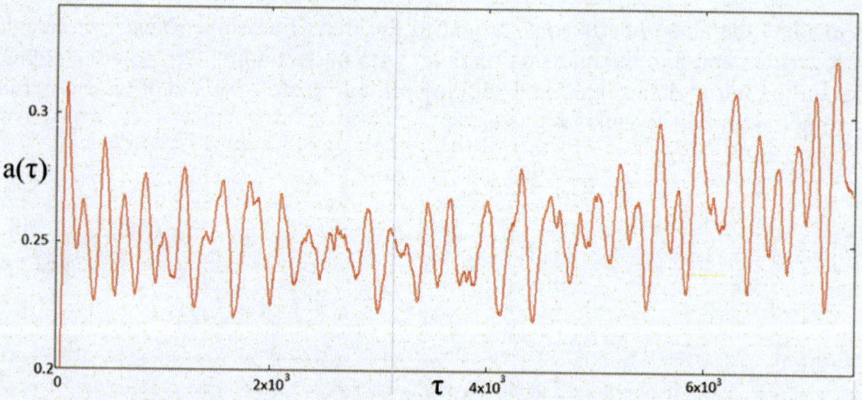

Fig. 20. Shift function for IBM.

Near-period values for Microsoft: 196, 428, 636, 679, 1550.

Near-period values for IBM: 1292, 1995, 5469, 7118.

The selected near-periods are independent, that is, they have passed the multiplicity check.

4.4 Making a Forecast

Let's perform a forecast assessment for companies on the test interval from 11.15.2020 to 04.26.2021.

The results are presented in Figs. 21, 22, 23, 24, 25 and 26.

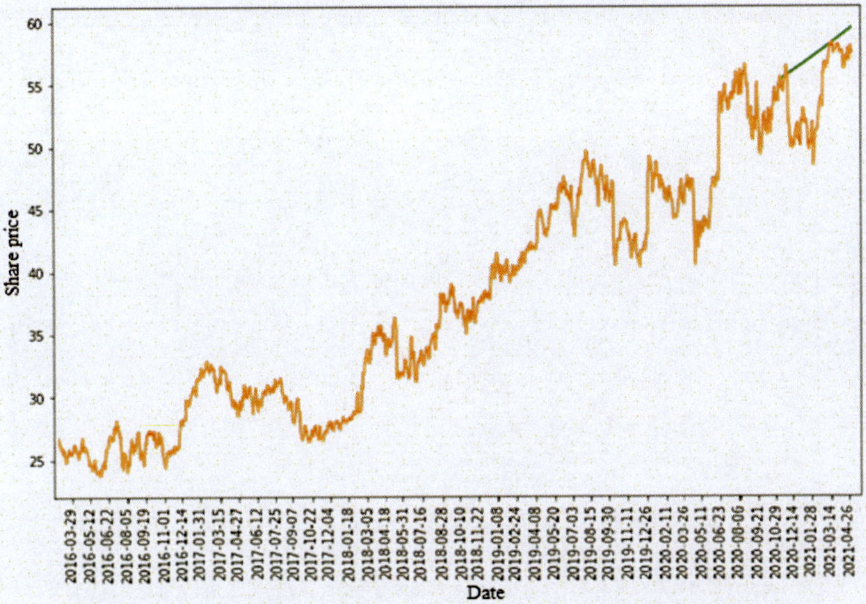

Fig. 21. The intensive growth model forecast assessment for Microsoft.

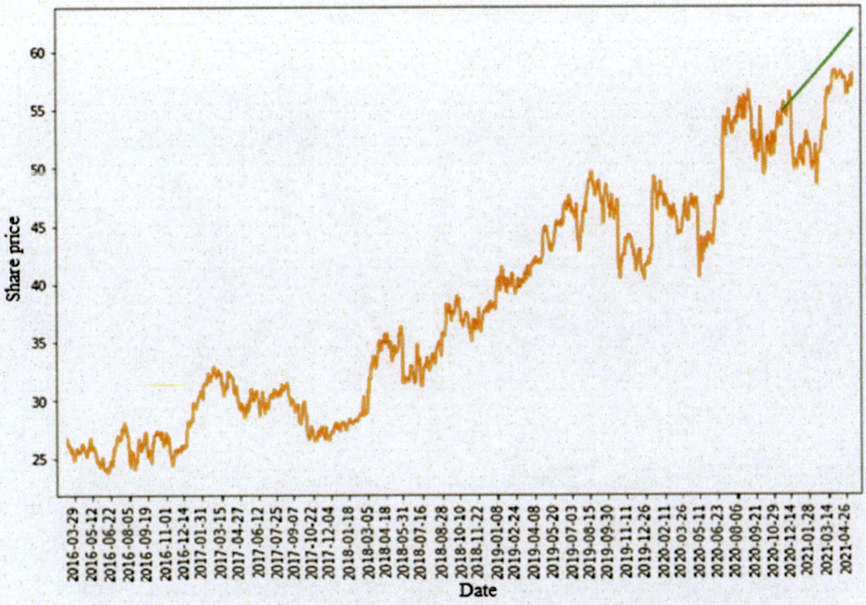

Fig. 22. The limited growth model forecast assessment for Microsoft.

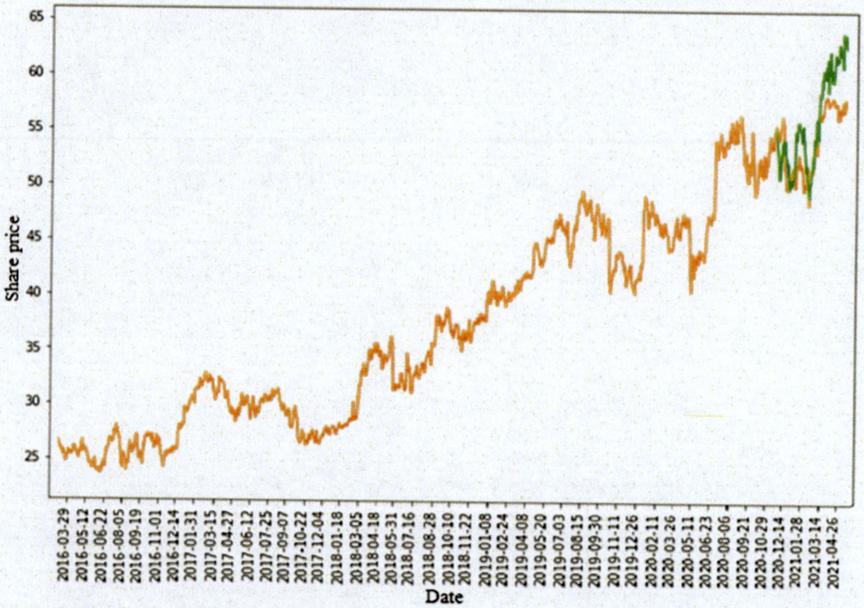

Fig. 23. The stochastic differential equation model forecast estimate for Microsoft.

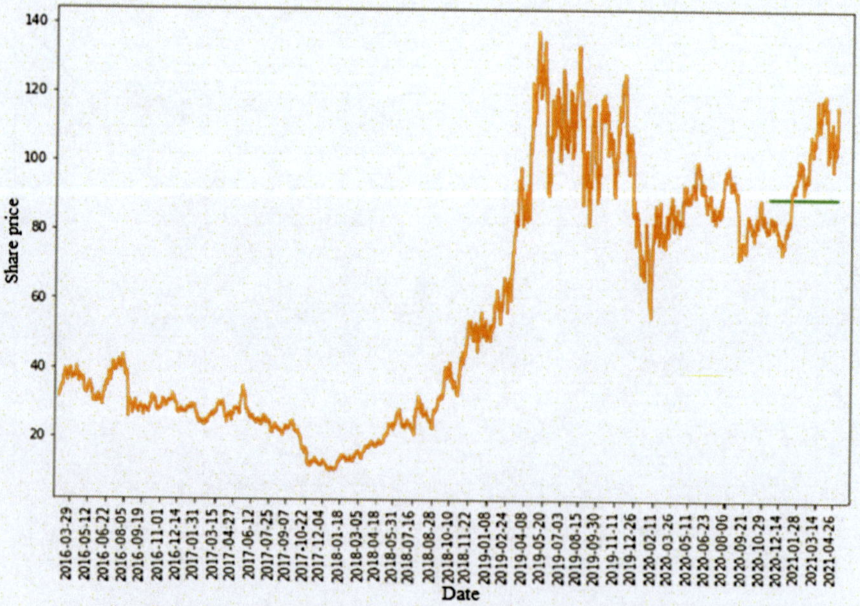

Fig. 24. The intensive growth model forecast assessment for IBM.

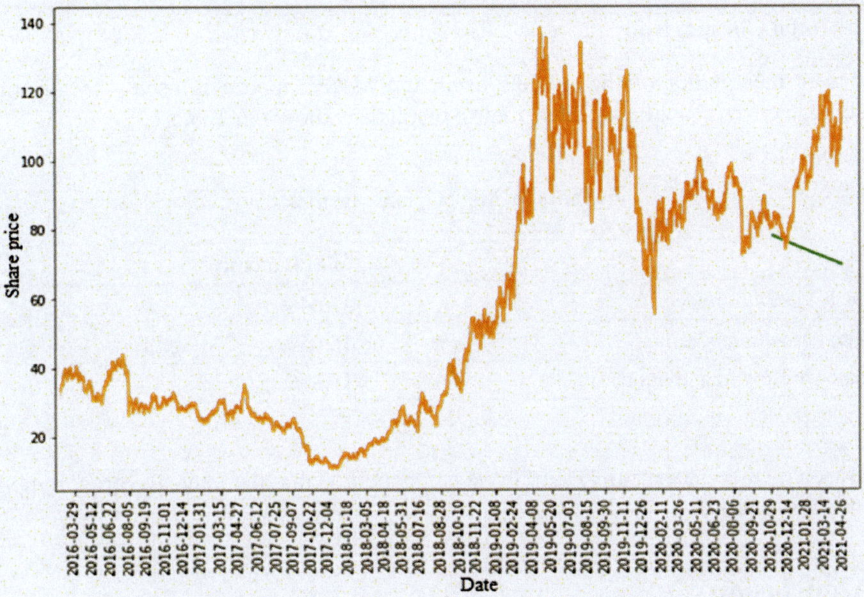

Fig. 25. The limited growth model forecast assessment for IBM.

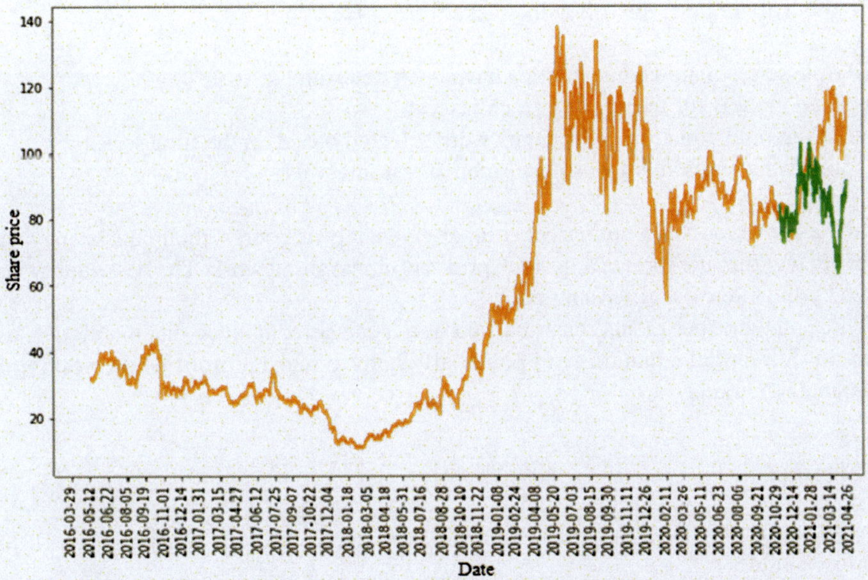

Fig. 26. The stochastic differential equation model forecast estimate for IBM.

4.5 Model Comparison

To assess the quality of the models used, the MAPE metric is used - the average percentage error. Calculating these metric results are shown in Table 1.

Table 1. Model quality metrics.

	MAPE metric
Intensive growth model	9.04%
Limited growth model	15.05%
Geometric Brownian motion	18.36%

Based on the values from Table 1, we can conclude that the intensive growth model has the best quality.

5 Conclusion

As this work a result, the subject area a study was carried out within the forecasting stock prices problem framework based on differential equations. The models used are also mathematically described and implemented in software: the intensive growth model, the limited growth model and the geometric Brownian motion model.

In the work course, the following tasks were solved:

- the time series main characteristics have been determined;
- selected models for forecasting stock prices;
- the differential equation coefficients were selected based on the time series;
- a forecasting assessment was performed at test intervals.

In the above models a numerical comparative analysis course, their consistency was established within the forecasting stock prices problem framework. The best quality was shown by the intensive growth model.

Thus, on the experiments' basis carried out, it can be concluded that the models use based on differential equations can potentially help to make a profit while working in the financial markets.

References

1. Veniaminov, S.S.: Revealing Latent Structure Laws in Processes and Signals, p. 216. URSS, Moscow (2014)
2. Kuzmin, V.I., Gadzaov, A.F.: Method of Construction of Models on Empirical Data. The Manual, p. 94. MGTU MIREA, Moscow (2012)
3. Serebrennikov, M.G., Pervozvansky, A.A.: Revealing of the Latent Periodicity, p. 244. Science, Moscow (1965)

4. Dzerjinsky, R.I., Pronina, E.N., Dzerzhinskaya, M.R.: The structural analysis of the world gold prices dynamics. In: Silhavy, R. (ed.) CSOC 2020. AISC, vol. 1225, pp. 352–365. Springer, Cham (2020). https://doi.org/10.1007/978-3-030-51971-1_29
5. Paramonov, A.A., Kuzmin, V.I., Dzerjinsky, R.I.: Analysis of almost-periodic and almost-proportional characteristics of a representative sample local minima time series. In: IOP Conference Series: Materials Science and Engineering, vol. 1047, p. 012045 (2021)
6. Kuzmin, V.I., Gadzaov, A.F.: Scientific and Technical Forecasting Models and Methods: Textbook. The Moscow State Institute of Radio Engineering, Electronics and Automation. The Technical University, p. 90. MIREA, Moscow (2016)
7. Kuzmin, V.I., Gadzaov, A.F.: Technical Analysis: Textbook. The Moscow State Institute of Radio Engineering, Electronics and Automation. The Technical University, p. 71. MIREA, MGUPI, Moscow (2015)
8. Ross Sheldon, M.: Variations on Brownian Motion. Introduction to Probability Models, 11th ed., pp. 612–614. Elsevier, Amsterdam (2014). ISBN 978-0-12-407948-9
9. Oksendal Bernt, K.: Stochastic Differential Equations: An Introduction with Applications, p. 326. Springer, Berlin (2002). ISBN 3-540-63720-6
10. Habr. https://habr.com/ru/. Accessed 22 May 2022
11. Investing. https://ru.investing.com/. Accessed 22 May 2021

Computer Program for Calculating Pipelines Destruction Probability

A. E. Alexandrov, T. B. Azhder, L. V. Bunina, S. S. Bikovsky(✉), and A. P. Titov

MIREA - Russian Technological University, 20, Stromynka, Moscow, Russia

Abstract. This article is devoted to probabilistic methods for diagnosing industrial facilities pipelines condition. To ensure the functioning of industrial equipment 24 h per day, constant diagnostics of their serviceability is required. The use of non-destructive testing methods, firstly, allows to reduce the costs of resources allocated for testing, and secondly, as a rule, it allows performing this control without interrupting the operation of the equipment. Thus, the topic raised in the article can be considered relevant. When preparing the material, the authors compiled mathematical models of the occurrence and development of typical pipeline defects (longitudinal and radial flaws, thickness degradation). Within the framework of the mathematical model, a new term is proposed for the probability of flaw survival. On the basis of modeling, a program for calculating the probability of pipelines destruction is presented. A significant part of the publication is the description of the modules that make up the program and the algorithm of its work. The developed program can be used to calculate the probabilities of the occurrence of leaks and large-scale destruction of cylindrical pressure vessels and pipelines under the influence of static and cyclic loads. The calculation can be performed with variable wall thicknesses and different temperature and stress fields. The calculation is performed by the Monte Carlo method. At the end of the article, the results of the application of the developed program for various models of the mechanical defects development in pipelines are presented.

Keywords: Probabilistic safety analysis · Fracture mechanics methods · Subcritical and critical flaw growth · Monte Carlo method

1 Introduction

In the operation of large electric power plants, one of the most important tasks is to ensure a given level of safety. An important place in this direction is occupied by the concept based on preliminary risk analysis and prediction of emergency situations [1]. To solve this kind of problems, modeling methods and computer programs are used. This work provides a universal program developed on the basis of a probabilistic model for pipeline calculating by the mechanism of ductile fracture, considered in this work [2]. Some instrumental means of obtaining primary data of pipeline condition can be found in that work [3, 4].

When writing the program, the task was to simulate the impact of the history of operation for a single flaw, including such events as:

R. Silhavy et al. (Eds.): CoMeSySo 2021, LNNS 231, pp. 718–733, 2021.
https://doi.org/10.1007/978-3-030-90321-3_59

- increase in flaw size due to cyclic loads;
- metal control with subsequent repair of detected defects;
- taking hydrotesting.

Due to the low probability of failure and the large amount of data, the Monte Carlo method was used to determine these probabilities.

If the reader of this material has additional questions related to the mathematical basis of the problem being solved, the authors recommend referring to the work [5–7].

2 Methods. Description of the Program Algorithm

2.1 Structure of the Developed Program

When describing the structure of a computer program, the following concepts were used:

Application – a computer program designed to perform a group of agreed functions or tasks for user benefit.

A module is a functionally complete piece of a program.

The computer program includes the following modules:

Module №1 "Calculation of defect size distributions parameters";
Module №2 "Import of stress condition";
Module № 3 "Formation of real flaws parameters distributions tables";
Module №4 "Calculation of cyclic flaw growth";
Module №5 "Probabilistic analysis of ductile fracture".

The Following Applications Were Formed on the Basis of the Developed Modules: *Application "Processing defects selection"*. Implements a graphical user interface for module №1. Calculates the parameters of several distribution laws for a user-provided sample of detected defects sizes and allows you to visually select the most suitable one.

Application "Project creation". Implements a graphical user interface for creating a project. A project is a text file that describes the initial calculation data, calculation settings and individual modules. The project is formed for one section and one operating mode.

Application "Calculation". Performs the calculation of the project by calling modules №2–5. The calculation is performed according to the calculation scheme and in accordance with the settings specified in the project. The calculation can be started both from the graphical user interface and from the command line.

The calculation scheme of the project in the application "Calculation":

1) Beginning of calculation.
2) Reading the source data described in the project.
3) Calculation of stresses using the "Import of stress condition" module.
4) Formation of the table with real flaws depths distribution density and the table with flaws semiaxes ratio distribution density the using the module "Generation of tables of real flaws parameters distributions". Depending on the options selected, undetected flaws can be added in this step and/or the repair simulated.

5) Flaw orientation cycle (axial, circumferential).
6) If the corresponding option is selected, then the tables of distributions of fracture parameters are corrected using the module "Calculation of fractures cyclic growth".
7) If the corresponding option is selected, then the probability of ductile fracture is calculated using the module "Probabilistic analysis of ductile fracture".
8) Recording results.
9) End of the cycle for fracture orientation.
10) End of calculation.

Using the applications described above, the calculation for the pipeline is carried out as follows:

1) Collect and prepare the initial data for the pipeline.
2) Calculation of force factors for all cross-sections of the pipeline and all operating modes of interest in the Astra or dPIPE program is performed.
3) If ready-made parameters for the distribution of detected defects are available, then proceed to the next step. Otherwise, the calculation of the distribution parameters of the detected defects is performed using the "Defect selection processing" application.
4) For each combination of section and operating mode, a separate project is formed. Since a project is a simple text file with special commands, the process of forming a set of projects can be automated.
5) Perform the calculation of the generated projects using the application: "calculation" To calculate multiple projects, it is recommended to use the command line launch. To do this, you can use a *.bat batch file or PowerShell script.
6) Perform processing of calculation results.

Thus, the enlarged algorithm of the application: "carrying out the calculation" is as follows:

1) Beginning of calculation.
2) Reading the source data described in the project.
3) Calculation of stresses using the "Import of stress condition" module.
4) Formation of a table with the density of distribution of the depths of real flaws and a table with the density of distribution of the ratio of the semiaxes of flaws using the module "Generation of tables of distributions of parameters of real flaws". Depending on the options selected, undetected flaws may be added in this step and/or a repair simulation may be performed.
5) Flaw orientation cycle (axial, circumferential).
6) If the corresponding option is selected, then the tables of distributions of fracture parameters are adjusted using the module "Calculation of cyclic growth of fractures".
7) If the corresponding option is selected, then the probability of brittle fracture is calculated using the "Probabilistic analysis of brittle fracture" module.
8) If the corresponding option is selected, then the probability of ductile fracture is calculated using the module "Probabilistic analysis of ductile fracture".
9) Recording results.

10) End of the cycle for fracture orientation.
11) End of calculation.

Next, we will consider each of the program modules in more detailed way.

2.2 Module "Calculation of Parameters of Defect Size Distributions"

The specified module calculates the distribution parameters for the provided sample of the detected defects sizes. The calculation is performed for the following distributions: normal, lognormal, exponential, two-parameter Weibull. In addition, the module calculates the parameters of the histogram of relative frequencies. There are no restrictions on the use of this module.

The input data of the module are: a sample of N elements $x1, x2, \ldots, xN$, each element describes the size of an individual detected defect in mm;

k is the number of columns (intervals) of the histogram of relative frequencies.

The output data of the module are: parameters of the distributions of the dimensions of the detected defects in mm; parameters of the histogram of relative frequencies (columns and their heights).

The calculation procedure for this module is as follows:

The parameters of the normal distribution are calculated:

$$\mu = \frac{\sum_{i=0}^{N-1} x_i}{N} \tag{1}$$

$$\sigma = \sqrt{\frac{\sum_{i=0}^{N-1} (x_i - \mu)^2}{N - 1}} \tag{2}$$

where μ is the mathematical expectation, mm; σ - standard deviation, mm.

The parameters of the lognormal distribution are calculated [5, 6]:

$$\mu = \frac{\sum_{i=0}^{N-1} \ln(x_i)}{N} \tag{3}$$

$$\sigma = \sqrt{\frac{\sum_{i=0}^{N-1} (\ln(x_i) - \mu)^2}{N - 1}} \tag{4}$$

where μ is the mathematical expectation of the logarithms of the values, mm; σ - standard deviation of logarithms of values, mm.

The parameters of the exponential distribution are calculated:

$$\lambda = \frac{\sum_{i=0}^{N-1} x_i}{N} \tag{5}$$

We calculate the parameters of the two-parameter Weibull distribution [5, 6]. We find the value of the shape factor k by solving the nonlinear equation:

$$\frac{\sum_{i=0}^{N-1} x_i^k * \ln(x_i)}{\sum_{i=0}^{N-1} x_i^k} - \frac{1}{N}\sum_{i=0}^{N-1} \ln(x_i) - \frac{1}{k} = 0 \tag{6}$$

The value of the scale factor λ is calculated by substituting the obtained value for k into the equation:

$$\lambda = \sqrt[k]{\frac{\sum_{i=0}^{N-1} x_i^k}{N}} \tag{7}$$

The parameters of the histogram of relative frequencies are calculated:

1) The minimum x_{min} and maximum x_{max} elements of the sample are found.
2) The interval $[x_{min}, x_{max}]$ is divided into k intervals with the same width Δx.
3) The number of hits of the sample elements in each interval m_i is counted.
4) The relative frequency of hitting the interval is $fr_i = m_i/N$, and $\sum_{i=0}^{k-1} fr_i = 1$.
5) The height of the i-th rectangle forming the histogram column will be equal to $h_i = fr_i/\Delta x$.

The resulting histogram of relative frequencies is an approximation of the theoretical distribution curve based on a sample from it.

End of the module.

2.3 Module "Formation of Real Fractures Parameters Distributions Tables"

The specified module is designed to perform the following operations:

- Formation of a table of the function *g* (*a*), approximating the density of distribution of the depths of real fractures in the interval $[a_{min}, s]$, depending on the options selected;
- Formation of a table of the function g (*ac_ratio*) approximating the distribution density of the ratio of the semiaxes of the flaws in the interval [0, 1];
- calculating the value of the density of defects in a pipeline element for real flaws.

The described operations are performed based on the data for the detected flaws and depending on the options selected. Available options:

1) only detected defects;
2) discovered defects with the addition of undetected defects;
3) detected defects with the addition of undetected defects and modeling of the repair.

The module has the following limitations of applicability:

The distribution of the ratio of the flaw semiaxes is assumed to be the same for detected and undetected defects. The addition of undetected flaws or repairs does not affect this distribution.

The value of the density of defects does not change when cutting the initial distribution of depths over the interval $[a_{min}, s]$.

The input data of the module are:

- The law and parameters of the distribution of the dimensions of the depths of the detected flaws, *mm*;

- The law and parameters of the distribution of the ratio of the semiaxes of flaws;
- ρ – mathematical expectation of the Poisson distribution law describing the density of detected defects in a pipeline element, $1/m$ or $1/m^2$;
- P_{func} – Table with defect detection function (probability of defect detection depending on its depth in mm);
- a_{min} – the minimum possible flaw depth (equal to the sensitivity of the test tool), mm;
- a_{rep} – repair value (detected defects with depth $a \geq a_{rep}$ are repaired), *mm*;
- s - pipe wall thickness, m;
- Δa – step with which the table of flaw depth distribution density is formed, mm;
- Δac_ratio – the step with which the distribution density table of the ratio of the semiaxes of the flaws is formed.

The output of the module is:

- Table of the function $g(a)$ approximating the density of distribution of the depths of real flaws (*in mm*) on the interval $[a_{min}, s]$;
- Table of the function $g(ac\_ratio)$ approximating the distribution density of the ratio of the semiaxes of the flaws in the interval [0, 1];
- ρ – mathematical expectation of Poisson distribution law describing the density of real defects in the pipeline element, $1/m$ or $1/m^2$.

The calculation procedure for this module is as follows:

Step 1. A table of probability density of the detected flaws depths is constructed. The table describes the piecewise constant function $g(a)$, which approximates the probability density function $f(a)$ of the fracture depths (see Fig. 1).

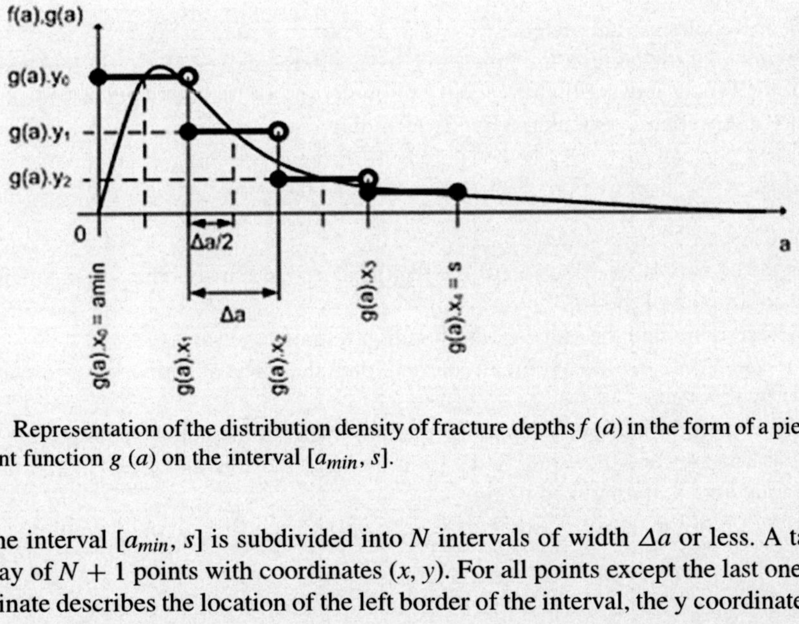

Fig. 1. Representation of the distribution density of fracture depths $f(a)$ in the form of a piecewise constant function $g(a)$ on the interval $[a_{min}, s]$.

The interval $[a_{min}, s]$ is subdivided into N intervals of width Δa or less. A table is an array of $N + 1$ points with coordinates (x, y). For all points except the last one, the x coordinate describes the location of the left border of the interval, the y coordinate is the value of the probability density of the specified law with the specified parameters for a point located in the middle of the interval to the right of the corresponding border. The last point is special: x – shows the location of the right border of the last interval, the y value for this border is equal to the y value of the penultimate border.

The area under the function g_{find} (a) is calculated by the method of left rectangles and the total area is normalized to one.

If the option "only detected defects" is selected, then $g\ (a) = g_{find}\ (a)$. The defect density of real flaws is equal to the original defect density, therefore, go to *step 5*.

Step 2. At this step, a table of the function g (a) is formed, which approximates the distribution density function of the depths of real flaws, and the value of the density of real flaws is calculated.

The points from the g_{find} (a) function table are iterated over. The value of the function describing the distribution density of real defects g (a) for the *i-th* boundary is calculated by the formula:

$$y_i = \frac{y_{ifind}}{P_{func}(x_i)} \tag{8}$$

where

y_i is the density of distribution of the depths of actual fractures in the *i-th* boundary;

y_{ifind} is the density of the distribution of the detected flaws depths in the *i-th* boundary;

x_i – coordinate of the i-th border, mm.

Since the P_{func} (a) function is specified in the table, the value for the required depth is calculated using linear interpolation.

The y-value of the last point is assigned the *y-value* at the penultimate point.

If the option "detected defects with the addition of undetected defects" is selected (that is, without simulation of repair), then go to *step 4*.

Step 3. Repairs are simulated.

The points of the table of the function g (a) are iterated over. If $xi \geq$ a_rem, then the value of the function describing the distribution density of real defects g (a) for the *i-th* boundary after repair is calculated by the formula:

$$y_i = y_i * \left(1 - P_{func}(x_i)\right) \tag{9}$$

where

y_i is the density of distribution of the depths of actual flaws in the *i-th* boundary;

x_i – coordinate of the *i-th* border, mm.

If $x_i <$ *a_rem*, then the value of the function remains the same.

If the value of *a_rem* is within an interval, then the interval is divided into two new intervals by this value.

The *y-value* of the last point is assigned the *y-value* at the penultimate point.

Step 4. The area under the function g (a) is calculated by the method of left rectangles and the total area is normalized to one.

Correction of the density of defects is carried out according to the formula:

$$\rho = k * \rho \tag{10}$$

$$k = \frac{S(g(a))}{1} \tag{11}$$

where

$S\ (g)$ is the area under the function $g\ (a)$ before normalization.

Step 5. A table of the probability density of the ratio of the semiaxes of the flaws is constructed. The table describes the piecewise constant function $g\ (ac\_ratio)$, which is approximated by the probability density function $f\ (g)$.

The table is formed similarly to the table of the function $g_{find}\ (a)$, by dividing the interval [0, 1] into N intervals with the width Δac_ratio or less.

The area under the function $g\ (ac\_ratio)$ is calculated by the method of left rectangles and the total area is normalized to one.

End of the module.

2.4 Module "Probabilistic Analysis of Ductile Fracture"

The module calculates the probability of ductile fracture from through flaws, the probability of ductile fracture from surface flaws, the probability of the formation of a through flaw, the probability of a stable through flaw, using Monte Carlo method for a pipeline element at a given operating mode [5–7]. The results for the element are obtained based on the calculation in the section of the pipe in the presence of one flaw.

Based on the lengths of stable through flaws, the probability of a stable through flaw in the element and the specified intervals of the nominal leak diameters, the module calculates the probability of stable leak formation for the intervals of the nominal diameters.

The module's limitations are as follows:

The random parameters are: flaw depth, ratio of flaw semiaxes.

The operating mode is set by tension and temperature [4]. It is assumed that the regime consists of one moment in time.

The tension and temperature are assumed to be uniform over the pipe wall thickness.

To calculate an axial or circumferential flaw, you must explicitly specify the orientation and choose a calculation method.

For the calculation of an axial flaw, it is necessary to use the circumferential stress, for the calculation of a circumferential flaw, the axial stress.

To calculate the circumferential flaw, you can use CAR method or moment method. For the calculation of an axial flaw, CAR or MRA method is used.

When calculating, it is necessary to explicitly indicate the type of steel (perlitic, austenitic).

The mathematical expectation of its distribution is taken as the value of the flow stress, which is determined according to the appropriate rules depending on the calculation method, type of steel, flaw orientation.

The maximum fracture half-length $c_{\max}$ is set by the user. If in the experiment the flaw half-length c exceeded c_{max}, then $c = c_{max}$ is taken.

The shape of the defects is limited by the ratio $a \leq c$.

The module uses the following input data:

- s – pipe wall thickness, *m*;
- r_a is the inner radius of the pipe, *m*;
- r is the average radius of the pipe, *m*;
- a_{leak} – depth above which the flaw is considered to be through, *m*;
- orientation – orientation of the flaw (*axial* – axial, *circum_ferential* – circumferential);
- *steel_type* – type of steel (*perlitic*, *austenitic*);
- σ_m is the membrane stress component in the section, *MPa*;
- σ_b – bending stress component in the section, *MPa*;
- T is the temperature in the section, °*C*;
- *Rp02* (*T*) – dependence of the value of the yield point on temperature in the form of a table for the interval (usually the interval [20, 350] is used), *MPa*;
- *Rm* (*T*) – dependence of the ultimate strength value on temperature in the form of a table for the interval (usually the interval [20, 350] is used), *MPa*;
- *E* (*T*) – dependence of the value of the elastic modulus of the material on temperature in the form of a table for the interval (usually the interval [20, 350] is used), *MPa*. It is used only for calculating an axial flaw by the *MRA* method;
- C_v – Charpy impact strength, *MPa*m*. It is used only for calculating an axial flaw by the *MRA* method;
- μ – Poisson's ratio;
- N is the number of experiments;
- *seed* – seed for the pseudo-random number generator;
- Table of the function g (a) approximating the density of distribution of the depths of actual flaws (in *mm*) on the interval [a_{min}; s];
- Table of the function g (*ac_ratio*) approximating the distribution density of the ratio of the semiaxes of the flaws in the interval [0; one];
- ρ – mathematical expectation of the Poisson distribution law describing the density of real defects in a pipeline element, $1/m$ or $1/m^2$;
- V is the estimated volume of a pipeline element in which the section is located (for the base metal – the area of the outer surface in m^2, for welded seams – the length of the seam along the outer surface in *m*);
- *intervals of nominal diameters of leaks.*

The output of the module is:

- *pf_leak* – the probability of ductile fracture from through flaws;
- *pf_se* is the probability of ductile fracture from surface flaws;
- *p_leak* is the probability of a through flaw formation;
- *p_stable_leak* – probability of a stable through fracture;
- the probability of a stable leak formation for the given intervals of nominal diameters.

The calculation procedure in this module is as follows:

Step 1. The boundaries of the distributions are determined.

Depth distribution limitations:

- a_{min} (the minimum possible depth, set equal to the sensitivity of the controls)
- $a_{max} = s$ (maximum possible depth)

Limitations for the distribution of the semiaxis ratio:

- $ac_{min} = 0$ (minimum possible)
- $ac_{max} = 1$ (constraint on the flaw shape)

The minimum boundary of the depth distribution can be optimized by excluding the zone in which destruction is impossible. For this, the *surface_fracture_a_min* value is calculated equal to *a_crit* for a surface flaw with a half-length c_{max}. It shows the depth of the surface flaw, less than which destruction is impossible. a_{min}_opt is used as the new minimum value, which is calculated using the formula:

$$a_{\min\_opt} = \min\left(a_{\min\_opt}, \mathit{surface\_fracture\_a}\_\min\right) \quad (12)$$

The original fracture depth distribution density table is cropped by the interval [a_{min}_opt; a_{max}].

After trimming the distribution density table, a normalization procedure is carried out to bring the area under the curve to one.

Since trimming the distribution of the semiaxes ratio does not give any practical effect, we will leave this distribution unchanged.

For each distribution, a cumulative distribution function (*cdf*) is plotted in tabular form over a given interval. The cumulative distribution function table is formed on the basis of the corresponding distribution density table.

Step 2. Initial values of event counters are set and an empty array *half_lengths* [] is created to store the half-lengths of stable through fractures.

Step 3. Calculate the values of *Rp02*, *Rm*, *E* (if the calculation is for an axial flaw by the *MRA* method), for the temperature T according to the corresponding tables of temperature dependences using linear interpolation. The value of the flow stress sf is calculated.

Step 4. A random value of the flaw depth a and the ratio of the semiaxes of the flaw ac_ratio is thrown out. To generate a random number, the corresponding tables of the cumulative distribution function are used.

Step 5. The fracture half-lengths are calculated using the formula:

$$c = \frac{a}{ac\_ratio} \quad (13)$$

Step 6. If the flaw half-length is greater than c_{max}, then $c = c_{max}$ is taken.

Step 7. If the flaw depth is less than a_{leak} (the depth after which the flaw is considered to be through), then the flaw is superficial. If the depth of the current flaw is greater than or equal to a_{leak}, then the flaw is through.

Step 8. The depth of the surface flaw is compared with the depth *a_crit* (c_{max}) (critical depth for a flaw with half-length c_{max}). If the flaw depth is less than *a_crit* (*cmax*), then the flaw is stable (increase the corresponding counter), otherwise the stability of the existing flaw is assessed.

Step 9. Depending on the chosen method and the orientation of the flaw, the critical depth is calculated.

Step 10. The depth of the existing flaw is compared with the critical depth. If $a \geq a\_crit$, then destruction occurs (the corresponding counter is incremented). Otherwise, the flaw is stable and no fracture occurs.

Step 11. Calculate the half-length of the through leak flaw *c_ls* in the following way:

$$l_{in} = 1.25 * 2c * \frac{c}{a} \tag{14}$$

$$l_{is} = \max(l_{is}, 3s) \tag{15}$$

$$c = 0.5 * l_{is} \tag{16}$$

where

l_{in} is the length of the through flaw on the inner surface of the pipe after the break of the bulkhead;

l_{ls} is the length of a rectangular leak flaw.

If $c_{ls} > c_{max}$, then we accept $c_{ls} = c_{max}$.

Step 12. Depending on the chosen method and orientation of the flaw, the critical half-length is calculated at $a = s$.

Step 13. c_ls of the flaw is compared with the critical half-length. If $c\_ls \geq c\_crit$, then destruction occurs (the corresponding counter increases). Otherwise, the flaw is stable (the corresponding counter increases) and no fracture occurs.

For a stable through fracture, its half-length (*c_ls*) is recorded in the *half_lengths* array. Based on this data, it will then be possible to calculate the probability of leaks of a certain size.

If there are still experiments left, then move on to the next one. Otherwise, we leave the loop.

Step 14. After looping through the experiments, we got the number of each event of interest. Dividing the number of events by the number of experiments, we get the frequency (probability) of each event.

Step 15. If we cut off the original depth distribution, now we need to take into account the probability of the depth falling into the selected interval of the depth distribution.

$$p = p * p_a \tag{17}$$

where

p is the probability of an event;

p_a is the probability of the depth hitting the interval $[a_{min}\_opt, a_{max}]$ for the distribution of depths in the interval $[a_{min}; s]$.

We apply this formula to the probability of each event.

Step 16. Up to this point, we have calculated the probabilities of events for a section in the presence of one flaw, but we need to determine the probabilities for a pipe-wire element, the number of flaws in which is described by the Poisson distribution. The number of flaws is normalized per unit size (area or length). We also know the estimated volume of the element. The probability of an event for an element is calculated by the formula:

$$p = p * \rho * V \tag{18}$$

where
ρ is the density of defects;
V is the estimated volume;
P is the probability of the event.
This formula is applied to the probability of each event.

Step 17. The probabilities of stable leaks are calculated in the specified intervals of the nominal diameters.

After all the experiments, we got an array *half_lengths* storing the *half-lengths* of all stable leaks. Each element of the array describes one leak.

For each leak, we calculate the opening area using the *B-K* method (Bartholome-Kastner) and calculate the nominal diameter of the leak using the formula:

$$D_y = \frac{A}{0.25 * \pi} \tag{19}$$

where A is the opening area, m^2.

After calculating the nominal diameter of all stable leaks, the number of hits in each of the specified intervals is counted. Dividing the number of hits in the interval by the number of elements in the half_lengths array, we get the probability of hitting the interval. Let us take into account the probability of a stable through flaw using the formula:

$$P_i = P_i * P_{stableleaks} \tag{20}$$

where P_i is the probability of a stable leak for the *i-th* interval of nominal diameters; *P_stable_leaks* is the probability of a stable through fracture (adjusted for steps 15 and 16).

The result is saved. End of the module.

3 Results. Examples of Calculations Using a Computer Program for Analyzing the Probability of Pipeline Destruction

In direct modeling of random flaws in the entire probability space, a problem arises – the events under study have a nonzero probability only for large defects. However, the probability of the appearance of a defect with large dimensions is very small, which leads to the need for a large number of experiments with a corresponding increase in the calculation time. To get around this problem, the ability to divide the probabilistic space into non-intersecting regions called cells was implemented (implementation of a stratified sample).

For each cell, separate Monte Carlo simulation is performed, and then the results of all cells are combined. To do this, each result obtained inside a cell must be multiplied by the probability of a flaw hitting the corresponding cell.

The use of cells can significantly reduce the variance of the results, reduce the number of experiments and the calculation time.

Below are the results of numerical experiments obtained for a different number of cells.

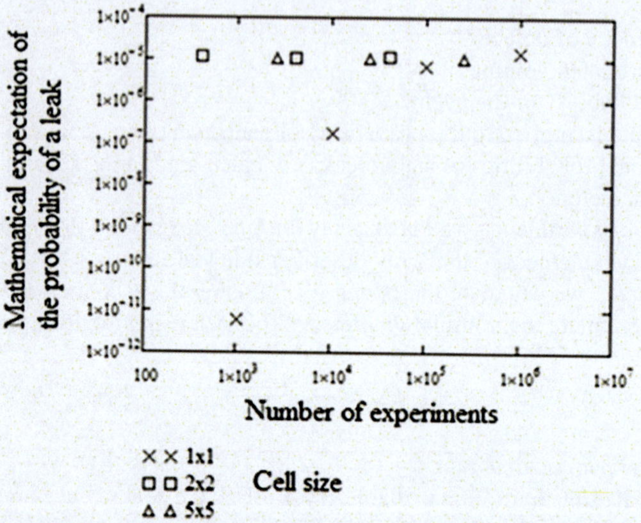

Fig. 2. Mathematical expectation depending on the number of experiments for a different number of cells.

The calculation results presented in Figs. 2 and 3 show that the use of even a small number of cells leads to a significant reduction in the number of experiments.

A similar graph for the standard deviation of the leak probability is shown in Fig. 3.

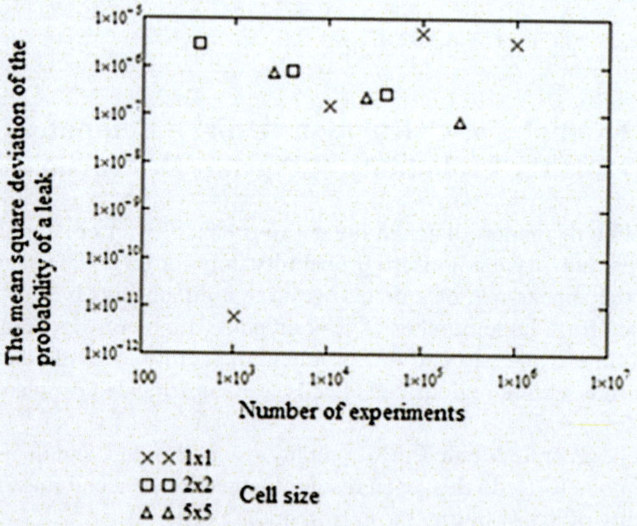

Fig. 3. Dependence of the standard deviation on the number of experiments with different cell sizes.

4 Discussions. Comparison of the Results Obtained with Different Models

Let us give a description of a new representation of the behavior of a flaw during operation, which also allows us to save the estimated time associated with the simulation of operational events.

Let us introduce a new concept for a specific flaw, which is declared in the modeling process and has given random dimensions: depth – a_0 and length – $2c_0$; the probability of the existence of a flaw or the probability of its "survival" to the current moment of time – $P_{life} = 1$. When carrying out metal inspection, the detected defects with a depth exceeding the repair value a_{rep} are repaired. The probability of detecting a defect is set as a function of its depth $P_{find}\ (a)$ Then the probability that a flaw will survive the metal inspection is described by the dependences: if $a > a_{rep}$, then $P_{life} = P_{life} * (1 - P_{life}(a))$, otherwise $P_{life} = P_{life}$.

Similarly, the likelihood of the existence of a flaw is corrected during hydraulic testing. If the dimensions of the flaw are such that they lead to the destruction of the structure, and there is no destruction during the hydrotesting process, then this means that there are no flaws of such dimensions in the structure. In this case, if the probability of structural failure is described by the dependence $P_f\ (a)$, then the probability that the flaw will survive hydrotesting is equal to: $P_{hydr} = P_{hydr} * (1 - P_f(a))$.

After simulating a given operating time, the design flaw is described by the following characteristics: $a \geq a_0$; $c \geq c_0$; $P_{life} \leq 1$. If we now calculate the probability of failure of the structure P_f, then the probability of failure, adjusted for the probability of fracture survival during operation, will be: $P_f = P_f * P_{life}$.

A comparison of the two different approaches to modeling operational events is shown in Fig. 4.

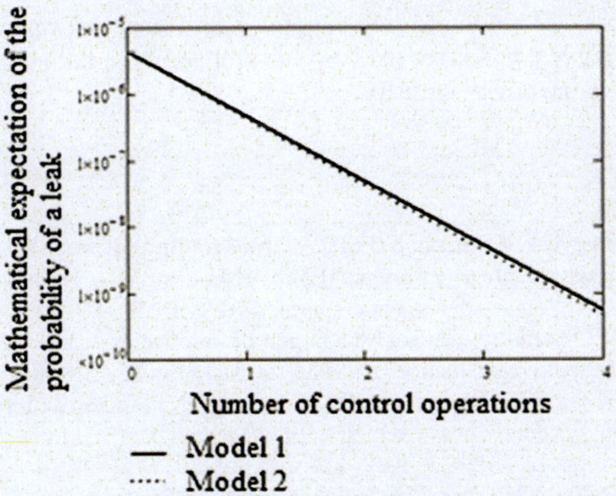

Fig. 4. Dependence of the standard deviation on the number of experiments with different cell sizes.

A possible option for further development of the program is to use the probability density function (Fig. 1) vtnjlbrb described in [8] to restore the probability density function.

5 Conclusion

The developed program can be used to calculate the reliability of pipelines under various operating conditions, including violation of normal operation modes, as well as emergency modes and prediction of accidents.

The current version of the program calculates the probability of destruction and formation of a leak for the base metal of straight pipeline sections, welded joints, bend made in the form of a bend, bent or welded bend. The given pipeline elements form independent objects for which the initial data are set, including operating modes, material characteristics and the results of metal control. Possible flaw paths are also specified for the selected structural element.

The initial data for the program are the results of calculating the temperature fields and stresses along the selected trajectories. This calculation is performed using external software. Based on information about temperature fields and stresses, the conditional probability of destruction is calculated for a given trajectory of the selected structural element. Using the obtained data of the conditional probability of destruction for each element, the calculation of the total probability of destruction of the investigated pipeline is carried out.

In order to reduce the computation time, the method was implemented in the program that allows one to divide the generated probabilistic space into separate non-intersecting regions – cells. This allows you to significantly save the estimated time of the experiments.

In addition, the developed program uses a new concept – the probability of flaw survival. On the basis of the introduced concept, a universal software has been created that allows modeling real processes to assess the probability of failure of high-reliability components used in the power industry.

References

1. Berens, A.P.: Probability of Detection (PoD) Analysis for the Advanced Retirement for Cause (RFC). Engine Structural Integrity Program (ENSIP) Nondestructive Evaluation (NDE) System Development, vol. 1, PoD Analysis. Finalreport. AFRL-ML-WP-TR-2001-4010, p. 88 (2000)
2. Berens, A.P.: NDE reliability data analysis in non-destructive evaluation and quality control: qualitative non-destructive evaluation. ASM Metals Data Book, ASM Int. **17**, 689 (1989)
3. Filinov, V.V., Mikaeva, S.A., Rodyukov, M.S., Filinova, A.V.: Modern architecture board information and control systems of heavy vehicles. Russ. Technol. J. **5**(3), 114–123. https://rtj.mirea.ru/upload/medialibrary/2c4/rtzh_3_2017_114_123.pdf (2017)
4. Getman, A.F., Kozin, Y.N.: Non-Destructive Control and Safety of Operation of Pressure Vessels and Pipelines, p. 288. Energoatomizdat, Moscow, Russia (1997)
5. Gandossi, L., Annis, C.: Probability of Detection Curves: Statistical Best-Practices. ENIQ report No. 41. European Commission Joint Research Centre Institute for Energy, EUR 24429 EN (2010)

6. Casella, G., Berger, R: Statistical Inference. Duxbury Press (1990)
7. Berens, A.P., Hovey, P.W.: Flaw Detection Reliability Criteria. vol. 1. Methods and Results, AD-A142 001, Dayton Univ OH Research Inst, Final Technical Report (1984)
8. Kryzhanovsky, A.D., Pastushkova, A.A.: Nonparametric method of reconstructing probability density according to the observations of a random variable. Russ. Technol. J. **6**(3), 31–38. https://rtj.mirea.ru/upload/medialibrary/a84/RTZH_3_2018_31_38.pdf (2018)

Harnessing Machine Learning and Big Data Analytics for Real-World Applications: A Comprehensive Survey

Soukaina Seddik(✉), Hayat Routaib, and Anass El Haddadi

Data Science and Competitive Intelligence Team, Applied Science Laboratory LSA, ENSAH, Abdelmalek Essaadi University, Al Hoceima, Morocco
seddik.soukaina@etu.uae.ac.ma, {hroutaib,a.elhaddadi}@uae.ac.ma

Abstract. Due to the exponential amount of data that is been generated every day, Big Data Analytics became a thriving research area in many domains especially computer science in all over the world. Several application areas used Big Data Analytics successfully such as social media, finance, healthcare, economy, etc. The humongous amount of data generated is challenging to analyse. As the velocity, the variety amount and the speed of data increase the uncertainty, which cause doubt and lead to a lack of confidence in the analysis process and the decisions made. Accordingly, numerous Machine Learning techniques have been developed to offer solutions for Big data Analytics challenges. Compared to traditional data techniques and platforms, machine learning delivers more accurate, faster, and more scalable results in big data analysis. In this paper, we provide a brief overview of previous researches in Big Data Analytics, machine learning, and we highlight the challenges of Big Data Learning especially the uncertainty, as well as the different machine learning algorithms and applications.

Keywords: Big data · Big data analytics (BDA) · Machine learning (ML) · Uncertainty

1 Introduction

The emergence of Big Data Analytics was in response to the surge of big data, which started in the 1990s. Businesses utilized large spreadsheets to look for trends and analyse numbers long before the term "big data" was invented. The vast amount of data generated in the late 1990s and early 2000s was nourished by new sources of data. In recent years, there has been an explosion of data due to the rapid rise in the number of smart devices in various sectors, and as a result, we are generating around 2.5 million data per day [1]. The Big Data concept is defined by Gartner [2]. Recently extensive research has been carried out on this vast amount of data generated by users referred to as "big data" and the need to extract useful information has become a necessity [3]. Therefore, big data analysis has become very important and inevitable as a result of

R. Silhavy et al. (Eds.): CoMeSySo 2021, LNNS 231, pp. 734–747, 2021.
https://doi.org/10.1007/978-3-030-90321-3_60

the adoption of Big Data Analytics (BDA) around the world in order to obtain many benefits from the data that is produced. This new digital resource can give the main motivational power to develop the smart products of tomorrow. As sources of data will continue to expand exponentially and as these sources are ubiquitous in nature it is hard to predict how the future might look like [4]. Big data analysis is challenging to perform using traditional data analytics [5] as they can lose effectiveness due to the five V's characteristics of big data: high volume, low veracity, high velocity, high variety, and high value. It is a daunting task to deal with big data with traditional data processing applications. Therefore, to uncover hidden data trends, patterns and associations, intelligent machine learning algorithms can be applied [6]. Machine learning has the power to extract information and find abstract patterns in a more accurate way than human experts. However, as the amount, variety, and speed of data increases, the uncertainty inherent within increases too, which leads to mistrust in the analyses process and the resulting decisions. In this paper, our goal is to present a comprehensive survey of all the related work in Big Data Analytics and machine learning, highlighting the most important challenges related to learning from huge amounts of data. The remainder of the paper is organized as follows. "Background" section presents background information on big data, Big Data Analytics, and machine learning. "Literature review" presents the surveyed works on Big Data Analytics and Machine Learning as well as a review on Machine Learning applications. "Challenges" section considers challenges of big data learning and focuses especially on the uncertainty challenge. Lastly, "conclusion and future works" section summarizes this paper and presents future directions of research.

2 Background

This section presents a background information reviews on the main features of Big Data, Big Data Analytics and Machine Learning.

2.1 Big Data

Researchers around the world refer to large data sets by the term big data [7]. The number of smart devices continue to grow and is expected to grow even more in the next few years. As the number of these smart devices increases, so does the amount of data generated, making traditional relational databases unable to handle such a huge amount of data sets (Big Data) that comes from various sources like social media, sensors, web machine data and transactional applications, etc. [8] (Fig. 1).

- **Volume:** represents the enormous amount of data generated and collected, across the world, in every second [9]. Since it applies to the scale and the size of a Dataset. Such massive volumes of data can introduce scalability but also uncertainty. Many existing data analysis techniques are not intended for massive databases and can miss the mark when attempting to analyse and understand large-scale data.

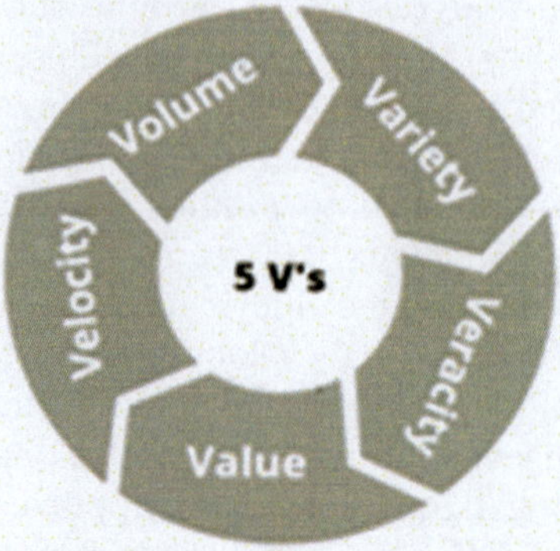

Fig. 1. 5V's model of big data

- **Variety:** refers to the different data formats [7]. The various forms of data in a dataset comprising structured data (the relational database, in which the information is stored according to a rigid and organized scheme). Semi-structured data (information that does not reside in a relational database but has organizational properties like CSV file, and excel file, etc.). Unstructured data (information that is not arranged according to a schema, preset data or model, hence cannot be stored in a traditional relational like images, videos, etc.). The data have very heterogeneous formats, which makes analysing effectively semi-structured and unstructured a challenging task.
- **Velocity:** refers to the high speed of generated data [10]. Several domains often rely on real-time operating systems enforcing strict timing standards on execution; therefore, optimized Big Data has to provide the right answer at the right time and. If the data failed to be delivered on time, many problems could be encountered.
- **Veracity:** refers to the inconsistencies and uncertainty in data [11]. Data veracity is categorized as bad, undefined, or good due to the fact that data can be incomplete, inconsistent, ambiguous, or noisy (for example False profiles on social networks, spelling mistakes, fraud ...). Data can be messy at times because of the multidimensional nature of data generated by diverse data types and sources, and it becomes difficult to establish quality and accuracy of Big Data Analytics.
- **Value:** refers to how useful the data in decision-making [3]. For example, companies such as Google, Facebook and Amazon have been able to leverage the value of Big Data and analyse it in the best possible way and they have all been enormously successful in retrieving useful information that has allowed them to make better business decisions. Thus, the value is the most important V of all 5V's.

The huge amount of data (big data) that is being generated continuously does not really have a use or significance unless it is processed and analysed in order to turn it into something valuable.

2.2 Big Data Analytics

Big Data Analysis, or Analytics (BDA) describes the process of analysing big data sets to uncover patterns, unknown correlations, market trends, user preferences and other valuable information that could not previously be analysed using traditional tools, which poses many challenges, make big data analysis much more difficult and complicated than normal size data analysis. Big Data Analytics (BDA) includes analysing data sets (within the form of text, audio, video, etc.), and reaching conclusions about the data they contain, regularly through specific frameworks, methods, and technics. BDA technologies are used on an industrial scale, in all business ventures since they empower organisations to make educated business decisions. Big Data Analytics is capable of handling various dominant research issues using computational intelligence [12]. All around the world, companies are exploiting the power of various data analysis techniques and using it to redraw their business models. As technology develops, new analysis software emerge, and as the Internet of Things (IoT) grows, the amount of data increases. Big data has evolved as a product of our increasing expansion and connection, and with it, new ways of extracting and analysing data have been developed. Various advanced strategies such as divide-and-conquer [13], incremental learning, parallelization [14], granular computing, sampling, feature selection, and instance selection can turn big problems into small ones and can be used to enhance processing efficiency, make better decisions, and reduce costs. As for data analysis techniques, they include spatial analysis, predictive modelling, association rule learning, network analysis and many, many more. There are five common and major big data analysis techniques:

- **Data fusion and data integration:** Various data sources are combined to provide more precise, valuable data. Data integration gives a unified perspective on these various sources. The fusion and integration process has global business and scientific applications. There are three common methods for data fusion: feature-level based, stage-based, semantic meaning-based [15]. By joining a number of techniques that analyse and fuse information from numerous sources and solutions, the information extracted is more productive and more accurate than if created through a single source of data.
- **Data mining:** It is a tool used commonly within big data analytics; data mining combines methods from statistics and machine learning, within database management in order to extract patterns from large data sets. In fact, efficient big data mining make it possible to improve the competitive advantage and to add value for various sectors [16].

- **Natural Language Processing (NLP):** Known as a sub-specialty of artificial intelligence, computational linguistics, and computer science, this data analysis tool uses algorithms to analyse, understand, and manipulate human (natural) language to fill the gap between human communication and computer understanding. The opportunity to benefit from emerging big data and NLP solutions is beneficial in various fields [17].
- **Statistics:** The way this technique works is that it collects, organises, and interprets data, in surveys and experiments. For instance, financial transaction files mostly consist of numeric data, which may handily be analysed using statistics however social behavior big data usually consists of nominal data such as phrases or words. Despite the fact that predicting the future events using nominal big data cannot generally be modelled, the analysis of non-metric fields for example attributes could distinguish patterns within the information [18].
- **Machine learning (ML):** ML is a fundamental component of Big Data Analytics [19]. Prominent in the field of artificial intelligence, machine learning is also used as a key tool to analyse data. Developed from computer science, it operates with computer algorithms to generate assumptions based on data. The predictions that it offers would be impossible for any human analysts to provide.

Accurate and efficient data analysis can reveal a wealth of information about the business, products and market.

2.3 Machine Learning

Machine learning is a research area that combines various branches of science majorly, computational linguistics, computer science, artificial intelligence, mathematics, information theory, etc. which makes it an interdisciplinary research area. The main goal of machine learning research is the development of efficient and fast learning algorithms that can make data based predictions [20]. A machine learning approach generally has two main phases: training phase and decision making phase. At the first phase (training phase), machine learning methods are employed to learn the system model with the help of the training dataset. At the second phase (decision making phase), the system uses the trained model to attain the estimated output for each new input. Machine learning algorithms are fundamentally characterized by three categories: *Supervised, Unsupervised* and *Reinforcement Learning* as shown in Table 1.

The majority of the traditional machines learning algorithms were developed for data sets that was completely able to fit into the memory. numerous big data predictive analytics problems emerges as the data continues to get bigger every second, a lot of intelligent learning methods are being applied to provide solutions to these problems.

Table 1. Machine learning algorithms

Supervised learning	Classification	Naive Bayes
		SVM
		K-Nearest Neighbour
	Regression	Decision Tree
		Linear Regression
		Logistic Regression
Unsupervised learning	Clustering	K-Means
		Mean Shift
		K-Medoids
	Dimensionality reduction	Principal Component Analysis (PCA)
		Feature Selection
		Linear Discriminant Analysis (LDA)
Reinforcement		Markov Decision Process
		Approximate Dynamic Programming
		Brute Force

- **Supervised Machine Learning** uses labelled training data. Each one the labelled training data comprises of an input value and a coveted target output value. The supervised learning algorithm analyses the training data and creates an inferential function, which can be used to map new values. It is used for forecasting or predicting [19].
- In **Unsupervised Machine Learning** technique, hidden insights are extracted from unlabelled data sets; it helps find all sorts of unknown patterns in data, for example, cluster analysis and association. Contrary to supervised learning, unsupervised learning refers to non-predictive goals [19].
- **Reinforcement Learning** enables a machine to learn its behavior from feedback received through interactions with the external environment. It is interested in how intelligent agents take action in an environment in order to maximize the idea of cumulative reward [21].

From a data processing point of view, both supervised and unsupervised learning techniques are preferred for data analysis and reinforcement techniques are preferred for decision making problems.

From a data-processing point of view, reinforcement techniques are preferred for decision-making problems and both unsupervised and supervised learning techniques preferred for data analysis.

3 Literature Review

This section displays the general findings of the review. First part presents the literature collection on Big Data Analytics and machine learning. Second part provides various application domains and sub-domains of machine learning applications as well as list of strengths, weaknesses and real word applications of different machine learning algorithms.

3.1 Big Data Analytics and Machine Learning

Hong-Ning Dai et al. [21], Focus on big data challenges in manufacturing data of MIot due to the heterogeneous types of data, the sheer volume and the velocity real time nature in manufacturing data. Moreover, it summarises the three phases in the life cycle of big data analytics: data acquisition, data pre-processing and storage, and data analytics and discusses the big data analytics enabling technologies in MIot (Table 2).

Table 2. Summary of various literature works

SN	Author	Year	Topic	Key findings
1	Hong-Ning Dai et al.	2019	Big data analytics for manufacturing internet of things: opportunities, challenges and enabling technologies	- An overview on big data analytics in manufacturing Internet of Things (MIoT) - The big data analytics enabling technologies of in MIoT
2	El-Sayed M. El-Alfy et al.	2020	A review of machine learning for big data analytics: bibliometric approach	- A review on machine learning for big data analytics following a bibliometric approach
3	Shweta Mittal et al.	2019	Big Data Analytics using Machine Learning Techniques	- Different challenges of machine learning with big data analytics and the different algorithms of machine learning
4	Rajesh Gupta et al.	2020	Machine Learning Models for Secure Data Analytics: A taxonomy and threat model	- ML based SDA architecture - The different existing ML models and the emerged challenges
5	Qolomany, B et al.	2019	Leveraging Machine Learning and Big Data for Smart Buildings: A Comprehensive Survey	- The techniques of machine learning and big data analytics that enables the delivery of smart services and infrastructures - The various issues that still need further research

Furthermore, Shweta Mittal et al. [22], Focused on the different challenges that could be faced if traditional machine learning technologies are used for Big Data Analytics and presented some of its possible solutions such as Parallel Processing, Map reduce jobs, Dimensionality Reduction techniques, Deep learning, GPUs and Online learning. And reviewed the different existing algorithms in machine learning. Jeffrey Hughes et al. [23], have complemented existing technical case studies with an account of the managerial and human elements of Big Data Analytics implementation. Effective deployment of persuasive practices potentially both embeds the benefits and mitigates the risks of BDA, sowing the seeds of many different forms of value. Other research is made by Changyong Lee [24] who developed a process based-on morphological matrix that provides a simple yet comprehensive view and enables the full spectrum of Data Analytics in Technological Forecasting (TF) to be examined. Specifically, the matrix consists of four dimensions and 12 factors: (1) awareness of TF contexts (objective, time horizon, field and level of analysis); (2) data collection and pre-processing (data source, data item and measure); (3) data analysis and validation (approach, methodology and performance evaluation); and (4) value creation (outcome and implication). He examined the practical implications of the process-focused morphological matrix, that Serves as a reference model for designing and developing data analytics in TF. Moreover, Benzidia et al. [25], developed a conceptual model to test a sample of data from 168 French hospitals using a partial least squares regression-based structural equation modeling method. The findings showed that the use of BDA-AI technologies has a significant effect on environmental process integration and green supply chain collaboration. Their results highlight the moderating role of green digital learning in the relationships between BDA-AI; green supply chain collaboration, and they help them in mobilizing BDA-AI technologies for supporting green supply processes and enhance environmental performance. These are an important findings that have not been highlighted in other previous literature. Furthermore, Surajit et al. [26] employed institutional theory and resource-based view theory to elucidate the way in which automotive firms configure tangible resources and workforce skills to drive technological enabling and improve sustainable manufacturing practices to develop circular economy capabilities. They tested the research hypothesis using primary data collected from 219 automotive and allied manufacturing companies operating in South Africa. The contribution of this work lies in the statistical validation of the theoretical framework, which provides insight regarding the role of institutional pressures on resources and their effects on the adoption of big data analytics-powered artificial intelligence, and how this affects sustainable manufacturing and circular economy capabilities under the moderating effects of organizational flexibility and industry dynamism.

Whereas, Qolomany, B et al. [27], focused on the main aspects of smart buildings with specific spotlight on the things that is being done and on the issues that require more research in ML and data analytics domains. Furthermore, the paper introduced a survey of the related works that discusses the use of big data analytics and machine learning especially for smart building services and infrastructure. Whereas, Rajesh Gupta et al. [28], Provided the architecture components of Secure Data Analytics (SDA) to classify normal or attack input data. Presented a comparative analysis of the existing machine learning models that learn from the patterns to make better decisions. Moreover, it discusses the different challenges such as reliability, accuracy, latency. Moreover, Hafyani. H. E [30], conducted a bibliometric analysis methodology of citation data in order to assess quantitatively and examine the exponential explosion and impact of literature and research performance, for the purpose of obtaining insights on the current research trends on machine learning for big data analytics.

The paper presented the scope and explained how the data was collected and the tools used in the collection. While conducting this survey, we summarized various ML algorithms with their real-world application in Table 3.

Table 3. Criteria and real-world application of ML algorithms

Learning algorithm	Strengths	Weaknesses	Real-world applications
Logistic Regression	• Study to noise • Output can be constructed as probability	• Categorical (binary) features are difficult to deal with • Only linear decision boundaries are handled, unless more features are created into polynomial terms	• Hotel Booking • Credit scoring • Medicine • Gaming • Spam classification
Naive-Bayes	• Manages noisy large datasets • Fits well in multi-class predictions	• It presumes that all properties contribute independently to the probability, which is almost impossible in real-life • If a categorical variable has a category in the test dataset, which was not present in the training data, then the category will be assigned zero probability and the model is will not be able to predict. (One of the solutions is the use of Laplace smoothing)	• Text Classification • Spam Filtering • Weather forecast • Recommendation System
Decision tree	• The feature is identified automatically (the first few nodes on which the tree is split are the most substantial variables within the dataset) • No effort to prepare data (not required to scale, normalize features) • Insensitive to outliers (since the segmentation occurs based on the scale of samples within the split ranges and it is not based on absolute values)	• Needs to be remodeled if new training set is added • Inclined to over processing, especially with data that has a lot of features (high variance). (One of the solutions to stop the tree from growing before it reaches one data point)	• Manufacturing (Fault diagnosis) • Finding prospective clients • Planning logistics and strategic management
Support Vector Machines (SVM)	• Performs well in complicated domains • It fits well with outliers (as it uses more relevant points for the separation it doesn't get influenced by them) • Works with non-linear classifications (using Kernel, for example Gaussian Kernels)	• Does not perform well for large noisy datasets • The choice of a good Kernel might be difficult	• Spam email filtering • Bioinformatics • Handwriting recognition • Face detection • Classification of images
k-Nearest Neighbors	• Study to noisy training data • It can detect non-linear or linear distributed data • It is non-parametric; thus, it does not make any assumption about the underlying data or its distribution	• Because it is non-parametric, it is in fact fast in training (vs. slow in querying) • Sensitivity to outliers can occur	• Finance (predicting whether the price of a particular stock will rise or fall, given past behavior) • Medicine (determine risk factors for prostate cancer, based on demographic and clinical variables)

4 Challenges

According to this related work study we noticed six main challenges that big data learning continues to face as shown in Fig. 2, focuses on the uncertainty as this challenge is present at all stages, and describes the different approaches used to deal with the uncertainty challenge.

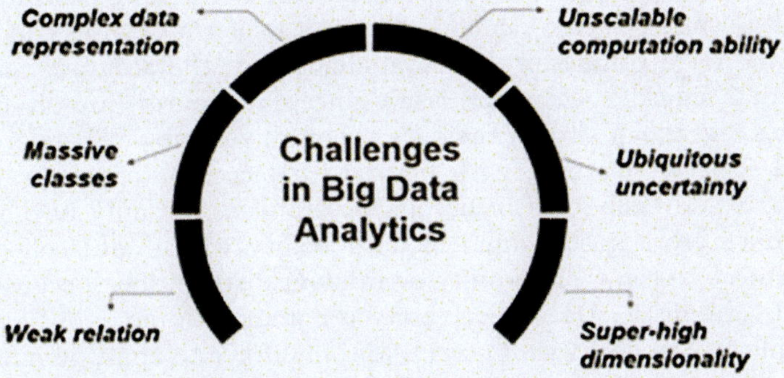

Fig. 2. Six main challenges in BDA

- **Complex data representation:** Deals with the great challenge of how to represent uniformly different types of features for multimodality big data [27].
- **Super-high dimensionality:** Big data in specific areas, specifically in life science computing or bio-informatics fields, is often highly dimensional. The problem is that scalability in current algorithms does not work well with high-dimensional data.
- **Massive classes:** In the era of big data, classification tasks with thousands of classes must be dealt with, for instance the large-scale recognition problem. The current classifiers appear to be eligible for the classification tasks, but their performance is seriously decreased.
- **Weak relation:** A mapping is more specific than a relation, and finding a relation is more challenging than finding a mapping when performing big data analytics. In most cases in a big data setting, all we need is to find a relation between two subsets of cases. This is because very often, it is impossible to find such exact mapping.
- **Unscalable computation ability:** The existing computational capacity is not scalable to the big data problem. Current learning algorithms find it hard to adapt to the new big data settings. We recognize a learning/mining algorithm as being effective for big data only if its complexity is quasi-linear or linear.
- **Ubiquitous uncertainty:** Uncertainty exists in every phase of big data learning [31]. For instance, big data contains a lot of noise, and most attribute values of a case in big data are missing. Obviously, some traditional learning

algorithms were not fit for processing data with 90% missing values; therefore, designing new learning algorithm to tackle the large-scale missing data is difficult.

4.1 Dealing with Uncertain Data

Previously, machine learning algorithms were usually fed with relatively accurate data from limited and quite known sources, so learning outcomes tend to be non-faulty; thus, veracity has never been an issue that causes concern. However, with the huge volume of data available today, the accuracy and trust of the source data quickly become an issue, since data sources are often from different origins and the quality of the data is not all verifiable. Therefore, veracity is a crucial issue for learning with big data to accentuate the significance of managing and addressing the incompleteness and uncertainty of data quality. Uncertain data are a special kind of data reality where data collections and readings subject to some probability or random distributions because they are no longer deterministic. Data uncertainty is common in numerous applications [29]. A simple way to deal with uncertainties in data is to apply summary statistics such as means and anomalies to abstract sample distributions. Another approach is to use the complete information brought by the probability distributions to build a decision tree, which is named distribution-based approach. Learning from these inaccurate data is a challenging task, since most existing machine learning algorithms cannot be applied directly. Taking classifier learning for example, handling incomplete data is a crucial issue, because incomplete data not only affects interpretations of the data or the models created from the data but may also impacts the prediction accuracy of learned classifiers. Here are two popular approaches that are frequently used to handle incomplete data:

- **Complete Case Analysis:** This method is the default in many statistical packages but must only be used when missing is under MCAR. All incomplete data points are just deleted from the dataset and only the complete data points used to build the mode. The approach reduces the volume of data and information available to the models and can bias the results. Tabachnick and Fidell [32] hypothesized that both the mechanisms and the patterns of missing values play a more important role than the proportion of missing data when using complete case analysis.
- **Imputation:** Imputation means replacing missing values in some way before the analysis [29]. Commonly mean or median imputation is used with mode imputation numerical and instances with the nominal instances. These simple imputation methods have been widely criticized, as they do not reflect the uncertainty in the data and may lead to bias in the analysis. However, multiple imputation, a more complex method, replaces missing values with a number of possible values that reflect the uncertainty even though the technique is more complex. A method is also required to combine the results of the analysis on multiple Datasets. For regression analysis, Rubin defined some rules for parameters estimation from multiple imputation analysis. For application to data mining, good methods may be required to synthesize the analysis.

5 Conclusion

Uncertainty exists in different types of data, such as finance or weather data, and even the most complex methods of data pre-processing cannot get rid of the underlying unpredictability. Furthermore, machine learning algorithms are not designed to deal with this kind of inaccurate data, therefore resulting in another set of unique and complex challenges for machine learning with Big Data Analytics.

In this paper, we present the survey on some of the challenges in Big Data Analytics. Knowing that the huge volume of data will continue to increase in an exponential, we will need more advanced and efficient technologies, better machine learning algorithms. Further studies will rely on finding possible solutions to better deal with data uncertainty using machine learning and, designing an efficient algorithm and generating a novel framework that will work on some specific data.

References

1. Petrov, C.: 25+ Impressive Big Data Statistics for 2020 (2020). https://techjury.net/blog/big-data-statistics/#gref
2. Beyer, M.A., Laney, D.: The importance of 'big data': a definition (2012)
3. Hashem, I.A.T., Ahmed, E., Ghani, N.A., Hamid, S.: Social media big data analytics: a survey. Comput. Hum. Behav. **101**, 417–428 (2019). https://doi.org/10.1016/j.chb.2018.08.039
4. Harerimana, G., Jang, B., Kim, J.W., Park, H.K.: Health big data analytics: a technology survey (2018). https://doi.org/10.1109/ACCESS.2018.2878254
5. Zhang, Y., Qiu, M., Tsai, C., Hassan, M.M., Alamri, A.: Health-CPS: healthcare cyber-physical system assisted by cloud and big data (2017). https://doi.org/10.1109/JSYST.2015.2460747
6. Mittal, S., Sangwan, O.P.: Big data analytics using machine learning techniques. In: 2019 9th International Conference on Cloud Computing, Data Science Engineering (Confluence), pp. 203–207 (2019). https://doi.org/10.1109/CONFLUENCE.2019.8776614
7. Oussous, A., Benjelloun, F.-Z., Lahcen, A.A., Belfkih, S.: Big data technologies: a survey (2018). https://doi.org/10.1016/j.jksuci.2017.06.001
8. El-Alfy, E.-S.M., Mohammed, S.A.: A review of machine learning for big data analytics: bibliometric approach. Technol. Anal. Strateg. Manag. **32**(8), 984–1005 (2020). https://doi.org/10.1080/09537325.2020.1732912
9. Manogaran, G., Thota, C., Lopez, D., Vijayakumar, V., Abbas, K.M., Sundarsekar, R.: Big data knowledge system in healthcare. In: Bhatt, C., Dey, N., Ashour, A.S. (eds.) Internet of Things and Big Data Technologies for Next Generation Healthcare. SBD, vol. 23, pp. 133–157. Springer, Cham (2017). https://doi.org/10.1007/978-3-319-49736-5_7
10. Liang, F., Yu, W., An, D., Yang, Q., Fu, X., Zhao, W.: A survey on big data market: pricing, trading and protection. IEEE Access **6**, 15132–15154 (2018). https://doi.org/10.1109/ACCESS.2018.2806881
11. Younas, M.: Research challenges of big data. Serv. Oriented Comput. Appl. **13**, 1863–2394 (2019). https://doi.org/10.1007/s11761-019-00265

12. Chen, L., Zhou, Y.: Quantile regression in big data: a divide and conquer based strategy. Comput. Stat. Data Anal. **144**. https://doi.org/10.1016/j.csda.2019.106
13. Alhegami, A.S., Alsaeedi, H.A.: A framework for incremental parallel mining of interesting association patterns for big data. Int. J. Comput. 106–117 (2020). https://doi.org/10.47839/ijc.19.1.1699. https://www.computing-online.net/computing/article/view/169916
14. Xie, P., Du, S., Teng, F., Yang, X., Liu, J., Li, T.: Urban big data fusion based on deep learning: an overview. Inf. Fusion **53**, 123–133 (2020). https://doi.org/10.1016/j.inffus.2019.06.016
15. Sowmya, R., Suneetha, K.R.: Data mining with big data. In: 2017 11th International Conference on Intelligent Systems and Control (ISCO), pp. 246–250 (2017). https://doi.org/10.1109/ISCO.2017.7855990
16. Alblawi, A.S., Alhamed, A.A.: Big data and learning analytics in higher education: demystifying variety, acquisition, storage, NLP and analytics. In: 2017 IEEE Conference on Big Data and Analytics (ICBDA), pp. 124–129 (2017). https://doi.org/10.1109/ICBDAA.2017.8284118
17. Choi, T.-M., Wallace, S.W., Wang, Y.: Big data analytics in operations management. Prod. Oper. Manag. 1868–1883 (2018). https://doi.org/10.1111/poms.12838. https://onlinelibrary.wiley.com/doi/pdf/10.1111/poms.12838
18. L'Heureux, A., Grolinger, K., Elyamany, H.F., Capretz, M.A.M.: Machine learning with big data: challenges and approaches. IEEE Access **5**, 7776–7797 (2017). https://doi.org/10.1109/ACCESS.2017.2696365
19. Annavarapu, C.S.R., Kumar, D.P., Amgoth, T.: Machine learning algorithms for wireless sensor networks: a survey. Inf. Fusion **49**, 1–25 (2019). https://doi.org/10.1016/j.inffus.2018.09.013
20. Liu, T., Tian, B., Ai, Y., Li, L., Cao, D., Wang, F.: Parallel reinforcement learning: a framework and case study. IEEE/CAA J. Autom. Sin. **5**(4), 827–835 (2018). https://doi.org/10.1109/JAS.2018.7511144
21. Dai, H.-N., Wang, H., Xu, G., Wan, J., Imran, M.: Big data analytics for manufacturing Internet of Things: opportunities, challenges and enabling technologies. Enterp. Inf. Syst. **14**, 1279–1303 (2020). https://doi.org/10.1080/17517575.2019.1633689
22. Mittal, S., Sangwan, O.P.: Big data analytics using machine learning techniques. In: 2019 9th International Conference on Cloud Computing, Data Science Engineering (Confluence), pp. 203–207 (2019). https://doi.org/10.1109/CONFLUENCE.2019.8776614
23. Hughes, J., Ball, K.: Sowing the seeds of value? Persuasive practices and the embedding of big data analytics. Technol. Forecast. Soc. Change (2020). https://doi.org/10.1016/j.techfore.2020.120300. www.sciencedirect.com/science/article/pii/S0040162520311264
24. Lee, C.: A review of data analytics in technological forecasting. Technol. Forecast. Soc. Change (2021). https://doi.org/10.1016/j.techfore.2021. www.sciencedirect.com/science/article/pii/S0040162521000780
25. Benzidia, S., Makaoui, N., Bentahar, O.: The impact of big data analytics and artificial intelligence on green supply chain process integration and hospital environmental performance. Technol. Forecast. Soc. Change (2021). https://doi.org/10.1016/j.techfore.2020.120557. www.sciencedirect.com/science/article/pii/S0040162520313834

26. Bag, S., Pretorius, J.H.C., Gupta, S., Dwivedi, Y.K.: Role of institutional pressures and resources in the adoption of big data analytics powered artificial intelligence, sustainable manufacturing practices and circular economy capabilities. Technol. Forecast. Soc. Change (2021). https://doi.org/10.1016/j.techfore.2020.120420. www.sciencedirect.com/science/article/pii/S0040162520312464
27. Qolomany, B., et al.: Leveraging machine learning and big data for smart buildings: a comprehensive survey. IEEE Access **7**, 90316–90356 (2019). https://doi.org/10.1109/ACCESS.2019.2926642
28. Tyagi, S., Kumar, N., Gupta, R., Tanwar, S.: Machine learning models for secure data analytics: a taxonomy and threat model. Comput. Commun. **153**, 406–440 (2020). https://doi.org/10.1016/j.comcom.2020.02.008
29. El Hafyani, H.: Big data series analytics in the context of environmental crowd sensing. In: 21st IEEE International Conference on Mobile Data Management (MDM), pp. 246–247 (2020). https://doi.org/10.1109/MDM48529.2020.00056
30. Marjani, M., et al.: Big IoT data analytics: architecture, opportunities, and open research challenges. IEEE Access **5**, 5247–5261 (2017). https://doi.org/10.1109/ACCESS.2017.2689040
31. Hariri, R.H., Fredericks, E.M., Bowers, K.M.: Uncertainty in big data analytics: survey, opportunities, and challenges. J. Big Data **6**(1), 1–16 (2019). https://doi.org/10.1186/s40537-019-0206-3
32. Fidell, L.S., Tabachnick, B.G.: Using Multivariate Statistics, 7th edn., p. 848. Pearson, Boston (2018)

Fine-Tuning GPT-3 for Russian Text Summarization

Nikolich Alexandr[1], Osliakova Irina[1](✉), Kudinova Tatyana[1], Kappusheva Inessa[1], and Puchkova Arina[2]

[1] Federal State Budget Educational Institution of Higher Education, «MIREA – Russian Technological University», Vernadsky Prospect, 78, Moscow, Russia
nikolich.a.d@edu.mirea.ru, {oslyakova,kudinova,kappusheva}@mirea.ru

[2] National Research University Higher School of Economics, Myasnitskaya Street, 20, Moscow, Russia
adpuchkova@edu.hse.ru

Abstract. Automatic summarization techniques aim to shorten and generalize information given in the text while preserving its core message and the most relevant ideas. This task can be approached and treated with a variety of methods, however, not many attempts have been made to produce solutions specifically for the Russian language despite existing localizations of the state-of-the-art models. In this paper, we aim to showcase ruGPT3 ability to summarize texts, fine-tuning it on the corpora of Russian news with their corresponding human-generated summaries. Additionally, we employ hyperparameter tuning so that the model's output becomes less random and more tied to the original text. We evaluate the resulting texts with a set of metrics, showing that our solution can surpass the state-of-the-art model's performance without additional changes in architecture or loss function. Despite being able to produce sensible summaries, our model still suffers from a number of flaws, namely, it is prone to altering Named Entities present in the original text (such as surnames, places, dates, etc.), deviating from facts stated in the given document, and repeating the information in the summary. #COMESYSO1120.

Keywords: Text summarization · rugpt3 · Russian language

1 Introduction

Automatic summarization seeks to present given information in a more compact form, determining the key messages of the text and eliminating unnecessary details and filler sentences. Such a task can be applied to various domains and informational sources. For illustration, numerous studies have been dedicated to summarizing news, lectures, and medical data, with some authors utilizing not only lexical but also acoustic and visual data [1].

In general, text summarization methods can be broadly divided into extractive and abstractive approaches. The extractive one aims to select the most relevant sentences or

R. Silhavy et al. (Eds.): CoMeSySo 2021, LNNS 231, pp. 748–757, 2021.
https://doi.org/10.1007/978-3-030-90321-3_61

tokens from the original document and comprise them into a summary. The abstractive approach, on the other hand, generates a summary 'from scratch' – it is conditioned on the source text but is capable of producing novel words, rearranging sentences, paraphrasing, replacing tokens with corresponding synonyms, and so forth. The advantages and disadvantages of this are obvious: generative models can ensure a higher level of linguistic fluency and provide more a coherent narrative imitating human speech, however, they can significantly alter the content of the original texts and distort the factual details.

Nonetheless, in current settings with the advances of sequence-to-sequence frameworks, abstractive models tend to outperform extractive ones. This holds true even in the cases of very limited supervision, with as little as 1000 training documents and respective summaries being provided to the model [2]. Most of this success can be attributed to the breakthrough of Transformer-based architectures [3]: due to their ability to 'understand' the logic of natural language, they can better decipher the relationship between sentences and the constituent words, thus, determining the general idea of the input texts.

Despite Transformers being widely used in recent works on text summarization in English, not many efforts have been made in order to apply them to the Russian language. To the best of our knowledge, current solutions employ Universal Transformer architecture [4] and BART-like models [5] to perform this task. Even so, BART is not exactly a generative model, and it is trained to reconstruct a document from its noisy version rather than put together text sequences on the fly. ruGTP3, however, is already capable of producing summaries and paraphrasing just as the original GPT3 model. We propose that with the additional fine-tuning on the contents of news articles and their corresponding summaries provided in Gazeta dataset [6] ruGPT3 can learn to deliver adequate summaries. Additionally, we experimented with hyperparameter-tuning in order to reduce the stochasticity of the text generation and penalize the repetition of the information. We calculate several metrics of text summarization quality alongside the human evaluation and conclude that our approach yields satisfactory results that can outperform the previous solutions. Our model, however, tends to produce a number of repeated errors that are yet to be fixed.

2 Related Work

First attempts of automatic text summarization utilized Recurrent Neural Networks [2]. However, the introduction of the self-attention mechanism Transformer architecture allowed to significantly improve the quality of many NLP tasks' solutions ensuring that the model learns the logic of the natural language and semantics of the words. Rush et al. [7] were the first to use these mechanisms for abstractive text summarization. Recent works in this field that employ Transformer-based architectures rely on several techniques. First, one can train a masked model or denoising autoencoders that can generate text as if restoring or filling the gaps in the original. Current solutions feature T5 network that was trained on corrupted texts (with deletions, replacement, or an excess of characters) with various spans of masking [8]. Another popular model used for summarization is BART [9] that was trained to restore the original texts after arbitrary noising functions were applied to them. Perhaps the most advanced model of text summarization, PEGASUS [2], follows the same logic: key sentences are being fully masked while the model is trained to restore them.

Another approach is to use purely generative models such as GPTs knowing that training on the large corpora makes them familiar with the task of summarization. For example, Radford et al. [10] show that a large Transformer trained on texts from the Internet is capable of generating summaries after the prompt "TL;DR" ('too long; didn't read' – this phrase usually precedes the summary of a post on the Internet forums). Analogously, ruGPTs can produce summaries after the prompt 'короче' ('briefly').

Several works employ GPTs for summarization of English texts [11], but to the best of our knowledge, no such experiments were done in regards to Russian. There is one work on fine-tuning ruGPT3 for a similar task of sentence simplification [12] that yields high values of the automatic quality metrics. The current state-of-the-art Russian summarization model, however, builds upon fine-tuning multilingual BART [6].

3 Dataset

Our model was trained on Gazeta dataset [6]. It contains 63435 articles published by gazeta.ru newsagency from June of 2010 up to March of 2020 with their respective URLs, dates of posting, titles, and summaries. After scraping the source, the author also cleaned the data by removing the summaries that are too short or too long and those that have too narrow or too wide n-gram overlap with the text.

The key advantage of this dataset is that it is one of the few datasets for summarization in Russian that provide actual summaries and not just headlines. While there is an abundance of data suitable for headline generation it is a separate task from text summarization since the summarization models aim not only to detect the main thought of the text but also to provide some necessary details.

For training, we split the dataset roughly in 90:10 proportion with 57665 examples belonging to a training set and another 5770 used for testing.

4 Model Description

For our experiments, we used a pretrained ruGPT3Small model (125M parameters, 2048 context length, 1024 sequence length) provided by Sber [13]. It was trained on 80B tokens for 3 epochs with total training time being around one week when using 32 GPUs.

The text-summary pairs were fed into the model with addition of special tokens in the following way: '<s>Text:{text} <|sep|> Summary:{summary} </s>' so that <|sep|> separates the source text from the target summary. During the model inference the input is also passed as 'Text:{text} < |sep|> Summary:' so that the model generates the summary as a continuation of the prompt.

5 Experiments and Hyperparameter Tuning

It is known that generative models such as ruGPT3 can deviate significantly from the source text, altering the details or even going onto a completely different subject. However, the randomness of the model's prediction can be somewhat controlled by hyperparameters such as temperature. The temperature of the GPT model ranges from 0 to 1, and generally with a lower temperature the model is more likely to choose tokens with a higher probability of occurrence (vice versa, the higher the temperature, the more 'creative' the model gets with its generation). For our task, we set the temperature to 0 so the model becomes more restricted to the contents of the source text.

Apart from that, model performance was further improved by tuning the beam search algorithm. The top-k method picks the most probable words, then selects top-k most probable words that follow the initial k words, and repeats the process generating numerous sequences (or beams), selecting the one with the highest compound likelihood as an output in the end. The top-p technique restricts the pool of word candidates to the minimum set of tokens whose cumulative probability of appearing exceeds p. Through experimenting, we set the top-p value at 3 and top-k value at 0.95. We also let the model generate 20 different beams (which is a comparatively large number) resulting in more coherent and sensible text generation. Additionally, we set early_stopping to 'True' meaning that the best output is chosen only when all beams reach the 'End Of Speech' special token (i.e. the end of the predicted sequence) allowing a better candidate selection.

Finally, we try to prevent a repetition of the information in the summaries, setting no_repeat_ngram_size to 3 so that no n-gram appears more than thrice in the generated sequence, and adjusting repetition_penalty to 2, penalizing the predicted probabilities of tokens that have already appeared in the generation. These hyperparameter settings were chosen empirically relying on the results of the evaluation metrics that are discussed below.

6 Evaluation Metrics

We evaluated the resulting summaries with different measures of the summarization quality. First, we compute the ROUGE score [14] that shows the n-gram overlap between the texts in terms of precision, recall, and F1 measure. We provide an F1 score for unigram and bigram overlap (ROUGE-1, ROUGE-2), and the longest common subsequence (ROUGE-L). Additionally, we calculate a BLUE score [15] – it also counts the unigram overlap between the texts, however, it penalizes words that appear in the summary more frequently than they do in the human-written text (this way, excessively repeating the words that are present in the text does not falsely increase precision).

We also employ BERTscore metric [16] that computes the pairwise cosine similarity between the tokens of the original and the generated summary using contextual embeddings. BERTscore may be a more adequate measure for our task since abstractive summarization techniques can reorganize sentences, replace words with their synonyms, and so forth. BERTscore makes use of the Transformers architecture that is able to grasp the context and semantics of words, thus, reporting higher scores if the general idea of the sentences is similar even if the word match is not exact. BERTscore also provides precision, recall, and F1 estimates.

7 Results

Table 1 displays scores achieved on the test section of Gazeta dataset. For comparison, we also included the scores of the current state-of-the-art model for summarization in Russian (mbart_ru_sum_gazeta) [] that was trained and evaluated on the same dataset that we used.

Table 1. Test scores on Gazeta dataset

	ruGPT3 score	mbart_ru_sum_gazeta
ROUGE-1 F1	11.4	**32.1**
ROUGE-2 F1	1.4	**14.2**
ROUGE-L F1	10.0	**27.9**
BLUE	23.1	**50.1**
BERTscore: precision	**0.87**	0.73
BERTscore: recall	**0.90**	0.73
BERTscore: F1	**0.89**	0.73

As can be seen, our model performs poorly when evaluating with metrics based on n-gram overlap, scoring approximately three times worse than mBART. Nonetheless, looking at BERTscore our solution shows significant improvements upon the previous results. As we have discussed above, BERTopic may be a better option for evaluating abstractive summarization, since such models can generate novel words, replace words with synonyms, rearrange the structure of the sentences, and so forth. Since the task is to detect the core ideas of the text and preserve its original message, BERTopic may provide a more reasonable measure of quality because it can gauge the meaning rather than merely score the exact matching of words.

On this basis, we conclude that our model delivers promising results in text summarization. Additionally, we validate a sample of generated summaries by hand, detecting several liabilities that can be further improved.

8 Analysis and Examples

In Table 2 we provide examples of good-quality model summarizations that come quite close to the human annotations.

However, manual validation also highlights some of the issues that our model faces. First of all, it tends to replace the Named Entities given in the text (i.e. surnames, companies' names, dates, places) with novel Named Entities. Perhaps, it is due to the fact that the model aims to paraphrase the text without taking into account that some

Table 2. Examples of good model performance

Human summary	ruGPT3 summary
Депутаты Госдумы одобрили в третьем чтении законопроект о пенсионном обеспечении граждан. Согласно документу, с 1 января 2020 года пенсии будут проиндексированы на 6,7%. Это позволит повысить размер пенсий, повысить их размер и размер страховых взносов	Работающие пенсионеры дождались индексации. Это следует из уже одобренных в первом чтении поправок в Конституцию. Так что работающие пенсионеры смогут получить замороженную с 2016 года прибавку к пенсии. В России их насчитывается более девяти миллионов. Эксперты называют это восстановлением справедливости
20 января состоится заседание Совета Госдумы, который рассмотрит вопрос о включении в повестку нижней палаты парламента на 21 января создание должности зампредседателя Совета безопасности РФ, которую займет бывший премьер-министр Дмитрий Медведев. Некоторые эксперты трактуют это как понижение, однако на деле Медведев фактически станет правой рукой Владимира Путина в сфере безопасности и вторым лицом одного из важнейших совещательных органов России	Дмитрий Медведев назначен на должность председателя совета безопасности при президенте РФ. Соответствующий указ был подписан президентом РФ Владимиром Путиным в ходе послания Федеральному собранию в декабре прошлого года. До этого Медведев занимал пост главы администрации президента
Владимир Путин позвонил премьер-министру Израиля Биньямину Нетаньяху и поздравил его с юбилеем. В ходе диалога они обсудили ситуацию в Сирии и некоторые актуальные двусторонние вопросы	Премьер-министр Израиля Биньямин Нетаньяху в ходе встречи с президентом России Владимиром Путиным заявил о готовности к диалогу с Россией по вопросу о транзите гуманитарных грузов через сирийскую территорию. Израильский премьер также выразил надежду на диалог с Москвой

information should not be changed. This can dramatically decrease the quality of the summarization because the model can correctly describe an event but mistakes essential details about who participated in it, or when did it happen. Examples of this behavior are shown in Table 3.

Table 3. Examples of named entities changes

Human summary	ruGPT3 summary
Роспотребнадзор дал новые советы россиянам в связи с распространением коронавируса COVID-19. Теперь ведомство рекомендует не пользоваться общественным транспортом в часы пик, а также воздержаться от посещения мест массовых скоплений. Кроме того, следует использовать одноразовые маски, меняя их каждые 2–3 часа, а также сократить тесные объятия и рукопожатия	Скандально известный вирус Эбола добрался и до России. Роспотребнадзор предостерег россиян от поездок за границу в связи с эпидемией лихорадки Западного полушария, которая может распространиться по всей стране в ближайшие несколько недель
В сети появились первые рецензии к сериалу «Ведьмак», премьера которого состоится на стриминговом сервисе Netflix 20 декабря. Кинокритики, получившие доступ к первым эпизодам шоу, отметили его превосходство над культовым сериалом «Игра престолов» от телеканала HBO	Американский телеканал NBC показал первый эпизод из культовой франшизы «ведьмаков». Главную роль в нем исполнит актер Бенедикт Камбербэтч, а главные роли в других частях шоу исполнят Джейсон Стэйтем и Эмма Уотсон
В США в продажу поступила книга Мэй Маск, матери миллиардера и основателя SpaceX Илона Маска. В книге она рассказала не только историю своего успеха, но и о том, как едва сводила концы с концами, пытаясь в одиночку прокормить троих детей	Популярная телеведущая и бывшая модель Мэй Макларен опубликовала в своем инстаграме новую книгу, в которой рассказала о своих отношениях с миллиардером и его семье. Женщина делится своими советами и рассказывает, какие моменты стали для нее переломными
В Сети появился тизер-трейлер фильма «Чернобыль: Бездна» с Данилой Козловским в главной роли. Лента расскажет о ликвидации последствий аварии на Чернобыльской АЭС и людях, которые ценой своей жизни предотвратили глобальную катастрофу	В российский прокат выходит первый в России сериал о Чернобыле, снятый режиссером Алексеем Козловым. Главные роли в нем исполнили актеры из Швеции, Дании, Великобритании и России

Sometimes the issue is more severe, so that model confuses the actors and their actions, and misses the point of a story completely. We hypothesize that this happens in a case when there is an abundance of Named Entities that confuse the model (examples can be seen in Table 4).

Table 4. Examples of the model missing the key idea of the text

Human summary	ruGPT3 summary
Австралиец Саймон Доранте-Дэй, уверенный, что он непризнанный внебрачный сын принца Чарльза и Камиллы Паркер-Боулз, решил потребовать признание этого факта через Верховный суд Австралии. Он пытается доказать, что наследник британского престола и его супруга — его настоящие родители. Накануне, 27 января, бельгийке Дельфин Боэль удалось добиться подтверждения отцовства бывшего короля Бельгии Альберта II — ей помог тест ДНК, который экс-монарх согласился пройти	Королева Великобритании Елизавета II и принц Чарльз подали иск против своего приемного отца, британского принца Уильяма и герцогини Эдинбургской. Они обвиняют его в незаконном лишении их прав на наследование престола
Телеведущий Джеймс Корден оправдался перед зрителями за появившийся в сети ролик, на котором видно, что в рамках «Караоке в машине» шоумен вместе с поп-исполнителем Джастином Бибером едут в автомобиле, который на самом деле везет грузовик. Комик заверил, что подобное скорее является исключением из правил ради безопасности	На канале CBS вышел новый выпуск популярного шоу «Карикатура», в котором ведущий Джеймсу Кэмерону задают вопросы, касающиеся его личной жизни, а также рассказывают о его отношениях с бывшей женой и друзьями
Выступая на конференции в Сан-Франциско, глава Apple Тим Кук осудил компании, занимающиеся сбором данных пользователей в качестве бизнес-модели. По словам менеджера, обмен персональной информации на доступ к технологиям — это «плохая сделка». В своей речи руководитель «яблочного» бренда также затронул другие темы, включая дружбу со Стивом Джобсом и слоган Apple «Think Different»	Apple хочет стать первой в мире компанией, продающей персональные данные пользователей iPhone и iPad. Корпорация планирует создать собственную социальную сеть, где пользователи смогут делиться своими данными с друзьями и родственниками

Another issue is the repetition of the information – the model successfully detects the key thought, but instead of providing other details just repeats it several times (examples can be seen in Table 5).

Table 5. Examples of the repetition of information

Human summary	ruGPT3 summary
Министр обороны США Марк Эспер заявил, что американские военные будут защищать сирийские месторождения нефти. Вашингтон намерен обороняться от «любых сил, которые будут ему угрожать», в том числе от России и сирийских военных	Американские военные продолжают защищать нефтехранилища на юго-западе Сирии от террористов «Исламского государства». В Пентагоне заявили о продолжении работы по защите нефтяных месторождений на юге Сирии
63-летний мужчина устроил стрельбу в центре Калининграда и убил супружескую пару, а также ранил их сына. Причем женщина была беременна и, как сообщают очевидцы, звала на помощь, но злоумышленник выстрелил в нее еще раз, добив ее. Сам же стрелявший пытался покончить собой и умер в реанимации. Предварительной версией конфликта стало место на местном Центральном рынке	Стрельбу в городе Калининграде открыл неизвестный, убивший троих человек. Об этом сообщили в полиции. Очевидцы сообщают о стрельбе, которая произошла в центре города

9 Conclusion

In this paper, we tested ruGPT3 ability to summarize text after fine-tuning and adjusting hyperparameters. We prove that our solution outperforms the current state-of-the-art model when using a BERT-based metric, however, it scores poorly in terms of n-gram overlap between the target summary and the generated one.

Our manual validation shows that the model still suffers from a number of flaws. Namely, it can change the important details of the story (surnames, dates, numbers, etc.), thus, severely decreasing the quality of summarization. Sometimes the model fails to grasp the key idea or confuses the narrative completely. Additionally, even with penalties for repetitions it still sometimes reiterates the information in the summary.

In future work, we will aim to fix these flaws. For example, one can penalize the model specifically for altering the Named Entities labeled during training. Alternatively, one can assemble a list of all words that are not given in the source text and penalize the model for using them. In theory, this will further restrict the model, and it will deviate much less from the original document.

Other solution would be multiple sampling. Since there can be various correct summaries of the same text, we can allow model to generate several examples and then select the best option. Additionally, one can employ dynamic modeling for generating a summary that is optimal in terms of semantics and n-gam overlap [17] (i.e. via the Bellman–Ford algorithm that computes the shortest paths in a matrix). Since ruGPT3 predicts the next token, one can accelerate the generation by dropping sequences that yield insufficient quality compared to others when adding the next word, thus, iteratively filtering the ones that score the highest according to chosen metrics. It also means that if two or more sequences score equally the model would prefer the 'less costly' option making the resulting summaries more compressed and less repetitive.

We also plan to train a larger model (ruGPT3medium) that can process much more complex text structure and contextual details.

References

1. Rezazadegan, D., et al.: Automatic speech summarisation: a scoping review. CoRR abs/2008.11897, pp. 1–21 (2020)
2. Zhang, J., Zhao, Y., Saleh, M., Liu, P.J.: PEGASUS: pre-training with extracted gap-sentences for abstractive summarization. CoRR abs/1912.08777, pp. 1–55 (2019)
3. Gavrilov, D., Kalaidin, P., Malykh, V.: Self-attentive Model for Headline Generation. In: Azzopardi, L., Stein, B., Fuhr, N., Mayr, P., Hauff, C., Hiemstra, D. (eds.) ECIR 2019. LNCS, vol. 11438, pp. 87–93. Springer, Cham (2019). https://doi.org/10.1007/978-3-030-15719-7_11
4. Vaswani, A., et al.: Attention is all you need. In: Guyon, I., et al. (eds.) Advances in Neural Information Processing Systems. vol. 30, pp. 5998–6008. Curran Associates, Inc., NY (2017)
5. Bukhtiyarov, A., Gusev, I.: Advances of Transformer-Based Models for News Headline Generation. In: Filchenkov, A., Kauttonen, J., Pivovarova, L. (eds.) AINL 2020. CCIS, vol. 1292, pp. 54–61. Springer, Cham (2020). https://doi.org/10.1007/978-3-030-59082-6_4
6. Gusev, I.: Dataset for Automatic Summarization of Russian News. In: Filchenkov, A., Kauttonen, J., Pivovarova, L. (eds.) AINL 2020. CCIS, vol. 1292, pp. 122–134. Springer, Cham (2020). https://doi.org/10.1007/978-3-030-59082-6_9
7. Rush, A.M., Chopra, S., Weston, J.: A neural attention model for abstractive sentence summarization. Empirical Methods in Natural Language Processing, pp. 379–389 (2015)
8. Raffel, C., et al.: Exploring the limits of transfer learning with a unified text-to-text transformer. J. Mach. Learn. Res. **21**, 1–67 (2020)
9. Lewis, M., et al.: Bart: denoising sequence-to-sequence pre-training for natural language generation, translation, and comprehension. CoRR abs/1910.13461, pp. 1–10 (2019)
10. Radford, A., Wu, J., Child, R., Luan, D., Amodei, D., Sutskever, I.: Language models are unsupervised multitask learners. OpenAI Blog **1**(8), 1–24 (2019)
11. Kieuvongngam, V., Tan, B., Niu, Y.: Automatic text summarization of covid-19 medical research articles using bert and gpt-2. CoRR abs/2006.01997, pp. 1–13 (2020)
12. Shatilov, A.A., Rey, A.I.: Sentence simplification with ruGPT3. In: Computational Linguistics and Intellectual Technologies: Proceedings of the International Conference "Dialogue 2021", pp. 1–13. Moscow (2021)
13. Russian GPT3 models: https://github.com/sberbank-ai/ru-gpts. Last accessed 21 July 2021
14. Lin, C.: ROUGE: a package for automatic evaluation of summaries. In: Text Summarization Branches Out: Proceedings of the ACL-04 Workshop, pp. 74–81. Association for Computational Linguistics, Barcelona (2004)
15. Papineni, K., Roukos, S., Ward, T., Zhu, W.J.: BLEU: a method for automatic evaluation of machine translation. In: Proceedings of the 40th Annual Meeting on Association for Computational Linguistics, pp. 311–318. Association for Computational Linguistics, Philadelphia (2002)
16. Zhang, T., Kishore, V., Wu, F., Weinberger, K.Q., Artzi, Y.: Bertscore: evaluating text generation with bert. CoRR abs/1904.09675, pp. 1–43 (2019)
17. Karpov, D.A., Struchenkov, V.I.: Dynamic programming in applied tasks which are allowing to reduce the options selection. Russ. Technol. J. **8**(4), 96–111 (2020)

Artificial Neural Networks for the Prediction of Mechanical Properties of Soils

Lusdali Castillo Delgado[1(✉)], Daniel Enrique Porta Maldonado[1], Juan J. Soria[2], and Leopoldo Choque Flores[1]

[1] Escuela Profesional de Ingeniería Civil, Universidad Peruana Unión, Lima 051, Perú
lusdalicastillo@upeu.edu.pe

[2] Escuela Profesional de Ingeniería de Sistemas, Universidad Peruana Unión, Lima 051, Perú

Abstract. In road projects it is important to obtain a correct value of the mechanical properties of the soils since these come to have a great influence on the pavement designs. In reference to this, it is known that conducting tests by traditional methods implies a high cost, time, and laboratory availability, in this context, using predictive models takes significance and importance to predict those values. The objective of the research was to predict mechanical properties of soils using software based on artificial neural network algorithms.

In this article a database of 289 values of granulometric tests, consistency limits, maximum dry density, optimum moisture content and CBR was compiled. The methodology corresponds to a quantitative approach, applied type, correlational level, and non-experimental-cross-sectional design.

In conclusion, 4 predictive models were obtained with the Neural Tools software, which are: the GRNN model for MDD, with an R^2 of 75% and an RMS of 0.09%, GRNN model for OMC, with an R^2 of 78% and an RMS of 1.67%, 2-node MLFN model for the CBR 95% MDD, with an R^2 of 79% and an RMS of 5.42%, 2-node MLFN model for the CBR100% MDD, with an R^2 of 82% and an RMS of 6.93%.

In addition, a comparison of values obtained in the soil laboratory vs ANN was made, where the results show a minimum variation of 0.002% in the MDD, 0.06% in the OMC, 0.03% in the CBR, 95% MDD and 0.04% in the CBR100% MDD.

Keywords: Artificial neural networks (ANN) · Predictive algorithms · Maximum dry density (MDD) · Optimal moisture content (OMC) · California support value (CBR)

1 Introduction

In road projects, the soil is an important factor, since it gives a stable behavior to the structure. That is why the soil study is carried out to determine the physical and mechanical properties of the soils, since these come to have an influence on the designs of the flexible or rigid pavement structures.

R. Silhavy et al. (Eds.): CoMeSySo 2021, LNNS 231, pp. 758–779, 2021.
https://doi.org/10.1007/978-3-030-90321-3_62

In large structures such as long-distance highways, soil studies are not detailed, since conducting this type of test by traditional methods implied a high cost, time and laboratory availability; This allowed to apply a methodology based on artificial neural networks (ANN) in order to predict the mechanical properties of soils, in that way this method can be efficient, accurate in the results, reducing time and cost in studies.

The objective of the study was to predict the mechanical properties of soils using software based on artificial neural network algorithms.

The neural models were obtained with Neural Tools software, these will obtain the mechanical properties values predictions of the soils in order to elaborate a first budget; making it clear that laboratory tests will not be replaced but will be applied in certain phases of a project, especially in the preliminary, profile and pre-feasibility phases [1].

The quantity of samples obtained for the soil study was determined according to the "Soil, Geology, Geotechnics and Pavement Highways Manual" or in the term of reference (TOR). Once the quantity of samples was obtained, the tests were carried out in the soil laboratory, in which the physical and mechanical properties of the soils were determined; These tests are the granulometry (percentage of gravel, percentage of sand, percentage of fines), consistency limits (liquid limit, plastic limit, plasticity index), modified proctor (maximum dry density, optimal moisture content) and the value of California support (CBR).

Finally, artificial neural models were developed, taking as input data the percentages of gravel, sand, fines, liquid limit, plastic limit, plasticity index, and the maximum dry density, the optimal moisture content to cbr95%mdd and cbr100%mdd were the results as an output data. These data were used to training, test and validate the neural network data, in which the determination factor (R^2) and Pearson's correlation coefficient (r) were obtained as a reliability indicator.

2 Methodology

In this research, the methodological procedure for predicting the values of the mechanical properties of soils using artificial neural networks (ANN) were described as shown in Fig. 1.

2.1 Methodological Procedure

The methodological procedure used in this research is as shown below.

Stage 1. A database of the Provias Nacional entity was compiled, of which 289 values of granulometric tests (percentage of gravel, percentage of sand, percentage of fines), consistency limits (liquid limit, plastic limit, plasticity index), proctor (maximum dry density, optimum moisture content) and California support value (CBR), was organized in a tabulation matrix generating 4 Excel files, which were cleaned according to the variables measurements.

Stage 2: A preliminary analysis was carried out with the database from the first stage, where the input and output data for the development of the neural network were identified.

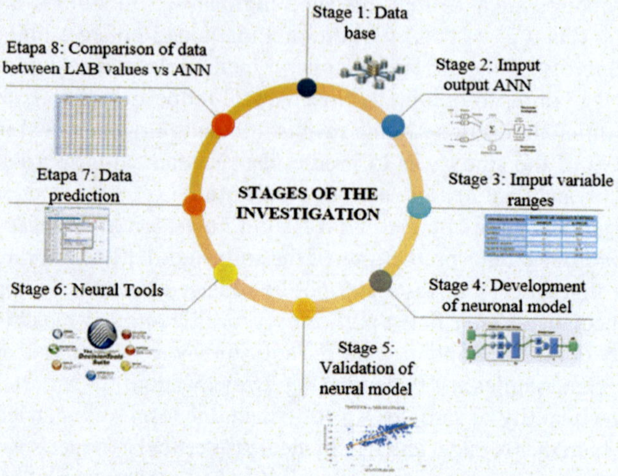

Fig. 1. Investigation stages.

Stage 3: Once all the data from the second stage have been obtained, the artificial neural model began to be developed in the Neural Tools software, where the training, test and data validation were made, in which the factor of determination (R^2) was evaluated between the values determined according to the artificial neural network (predicted) and those of the laboratory (the real data).

Stage 4: The neural network training was carried out, with the physical properties of the soil within certain ranges as shown in the following Table 1.

Table 1. Values that were used for the training of the artificial neural network.

Input variable ranges		
Variables	Lower	Upper
% Gravel	0.00	83.60
% Sand	7.80	64.60
% Fine	0.00	76.10
% Liquid	0.00	52.00
% Plastic limit	0.00	36.77
% Plasticity index	0.00	28.07

Stage 5: Table 2 is used for the validation of the artificial neuronal model. These values indicate how a model is qualified according to the coefficient of determination (R^2) [2].

Stage 6: With the Neural Tools software, a predictive model based on artificial neural networks (ANN) was obtained for the prediction of the mechanical properties of soils.

Table 2. Criterion of the parameter of the coefficient of determination (R^2).

Criterion	R^2
Excellent	>0.90
Good	0.70–0.89
Regular	0.40–0.69
Bad	0.20–0.39
Very bad	<0.19

Stage 7: Once the appropriate model was obtained, the prediction of the maximum dry density, optimum moisture content, CBR at 95% maximum dry density and CBR at 100% maximum dry density was made.

Stage 8: To finalize this research, a data comparison was made between the ANN values and the values of the laboratory tests.

2.2 Theoretical Bases

In recent years, artificial neural networks have been used to make predictions. Since, artificial intelligence is a branch of computer science whose purpose is to study the creation and design of machines capable of solving problems by themselves, based on human behavior. Expert systems, artificial neural networks, fuzzy systems, and genetic algorithms have been used to solve problems in the field of civil engineering [3].

Vila Zuñiga [4] aimed in his research to determine the linear behavior of the flexible pavement structure using artificial neural networks. His methodology was to use the Backvid and Michback software to perform the linear analysis of the structure and thus carry out the design of the neural model. In summary, an optimal neural network was built, composed of 5 layers, 13 input neurons, 25 neurons in the first two hidden layers, 5 in the last hidden layer and 5 neurons in the output layer.

Boza Capani and Merino Ortiz [5] aims in their research to determine the variation and correlation of the shear strength parameters of soils (friction angle and cohesion) from their physical properties (grain size, Atterberg limits, humidity and specific gravity) using artificial neural networks. The model was trained in which several simulations were carried out using the reverse propagation methodology (Feed – Forward Backprop). The results obtained by the statistical analysis of mean error are divided into: Training ($r = 0.93$), validation ($r = 0.99$), test ($r = 0.96$), obtaining an artificial neural network model equal to $r = 0.95$).

Paytán Ordoñez [6] mentions that in recent decades it has been shown that poor management of road infrastructure maintenance can result in costly expenses for the whole nation, reducing the useful life of the roads, increasing the operating costs of the vehicles and also increasing the travel time of the users. For this reason, his research aims to determine the value of the intentional regularity index of the PE-1S pathway using an artificial neural model of the value obtained by the laser profilometer. The results were compared with the IRI values obtained from measurements with the laser profilometer. The correlation coefficient obtained was r = 0.365, reflecting a weak relationship between the values and led them to conclude that Artificial Neural Networks did not perform well in this specific pathway.

Chávez Castro and Reyes Velasquez [7] in their research aimed to develop a computer vision system that allows the identification of superficial flaws in flexible pavement. As a result, they had 4 convolutional models, with a data set of 7200 images, being 1440 for training, 360 for validation and 77 for testing. It was 98.03% effective, 95.06% sensitive, and 95.63% specialty.

Tello Cifuentes et al. [8] carried out an evaluation of the deterioration of the roads in Colombia through manual inventories and visual inspections. The road condition assessment methods adopted by INVIAS are VIZIR and PCI, they determine the severity of damage to the flexible and rigid pavement; however, they can be tedious and require the expertise of a road condition assessor. This research presents an efficient methodology, with a reduction of time and cost using photometry and neural networks. The methodology was evaluated with real images of the pavement with three types of deterioration: longitudinal crack, crocodile skin and bump. As classifier, the multilayer neural network with configuration (12 12 3) was used, trained by the Levenberg Marquardt backpropagation algorithm. A precision of 96% was obtained in the classifier, a sensitivity of 93.33% and a kappa index of 0.936.

2.2.1 Artificial Neural Networks (RNN)

According to [9], he defines the artificial neural network as a computational model inspired by biological neurons, which can be considered as an information processing system with a distributed structure, made up of processing elements that are artificial neurons, they are interconnected by a large number of connections called synapses.

2.2.1.1 Classification of Artificial Neural Networks

Artificial neurons can be classified according to the values they can take [10]. Figure 2 shows the classification of artificial neural networks.

The artificial neural networks classification is according to the neural architecture, its learning, neural structure and its application [11].

2.2.1.2 Neural Models

According to Serrano [11] he affirms that every artificial neural model has four basic elements:

1. A set of connections, weights or synapses that determine the behavior of the neuron. These connections can be excitatory (have a positive sign), or inhibitory (negative connections).
2. An adder that is in charge of adding all the inputs multiplied by the respective synapses.
3. A nonlinear activation function to limit the amplitude of the neuron output.
4. An outer threshold that determines the threshold above which the neuron fires.

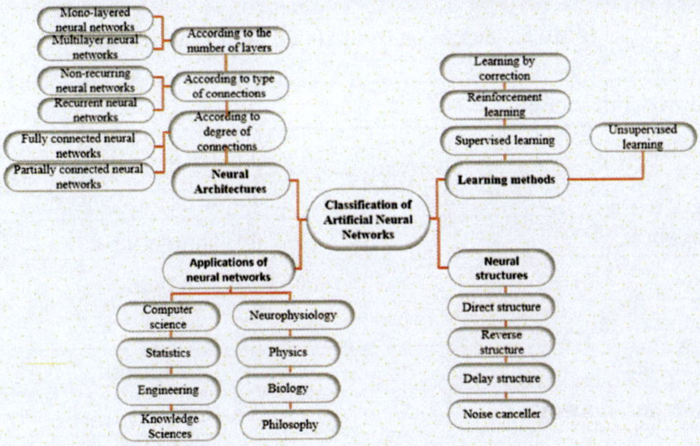

Fig. 2. Artificial neural networks classification.

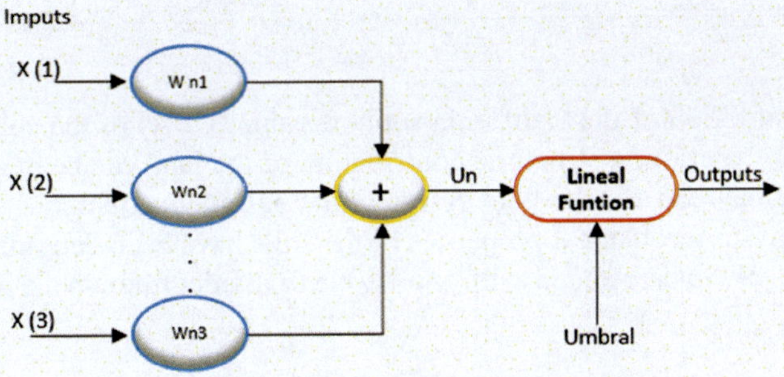

Fig. 3. Neural model schema.

Schematically, an artificial neuron would be represented by Fig. 3. Mathematically the operations to be carried out execute:

$$U_n = \sum\nolimits_{j=1}^{k} W_{nj}.x(j) \tag{1}$$

$$Output = \rho(U_n - threshold) \tag{2}$$

Where ρ is a non-linear function known as the activation function.

2.2.1.3 Selection of Artificial Neural Networks

The selection of a network is made based on the characteristics of the problem to be solved. Most of these can be classified into applications of prediction, classification, association, conceptualization, filtering and optimization [12]. Next, Table 3 shows the artificial neural network models.

2.2.2 Physical and Mechanical Properties of the Soil

In the soil studies, it is important to determine the physical – mechanical characteristics of the materials that are part of the foundation terrain and underlying soils, as well as

Table 3. Selection of artificial neural network.

Artificial neural network models	
1. Adaline y Madaline	11. DRS
2. ART	12. FLN
3. Back – propagation	13. Hamming
4. BAM	14. Hopfield
5. The Boltzman machine	15. LVQ
6. Brain-state-in a box	16. Perceptron
7. Cascade-correlation-networks	17. PNN
8. Counter – propagation	18. Recirculation
9. DBD	19. SOM
10. DNNA	20. SPR

the objective definition of the California support value (CBR) of the subgrade soil, in such a way that the existing structural contribution of the land can be quantified by the sectorization of the road and the final design of the pavement structure.

Once the physic-mechanical properties of the soils have been determined and if the results are correctly interpreted, it will be possible to predict future behaviors of a given soil [13].

2.2.2.1 Laboratory Test

The work made in the soil laboratory allowed evaluating the properties of the soils through physical and mechanical tests of the soil samples from each of the explorations that were carried out.

2.2.2.2 Soil Physical Properties

According to the manual of "soils, geology, geotechnics and pavements" in its section soils and pavement [14], it mentions that, in order to determine the physical properties of the soil, tests have to be carried out both in situ and in the laboratory, these are based in the Peruvian technical standards (NTP).

A. Natural soil moisture

According to the Peruvian technical standard NTP 339.127 and MTC-108/ASTM D-2216 [14, 15], it defines: moisture or moisture content as the ratio of the weight of water in a given mass of soil, to the weight of solid particles, expressed as a percentage. In other words, it indicates the amount of water that a sample contains. This is calculated with Eq. 3.

$$W = \frac{\text{Water weight}}{\text{Kiln Dry Soil Weight}} \times 100. \tag{3}$$

B. Soil granulometric analysis

According to the Peruvian technical standard NTP 339.128 and MTC-107/ASTM D-422 [14, 16] defines: the granulometric analysis consists of the particle se distribution of the soils. The classification of particles larger than 75 μm (retained on sieve No. 200) is carried out by sieving, while the determination of particles smaller than 75 μm is carried out by means of a sedimentation process based on Stokes' law using a suitable density meter.

To obtain the values of the percentages of gravel, sand, fines, the granulometric test was carried out by sieving as shown in Table 4.

Table 4. Classification of soils according to their particle size.

Type of material		Particle size
Gravel		75 mm–4.75 mm
Sand		Coarse sand: 4.75 mm–2.00 mm
		Medium sand: 2.00 mm–0.425 mm
		Fine sand: 0.425 mm–0.075 mm
Fine material	Limo	0.075 mm–0.005 mm
	Clay	Less than 0.005 mm

C. Consistency limits or Atterberg limits

According to NTP 339.129 [17] it mentions that Albert Atterberg originally defined six "consistency limits" for fine-grained soils: the upper limit of viscous flow, the liquid limit, the stickiness limit, the cohesion limit, the plastic limit and the shrinkage limit. In engineering, the liquid limit, the plastic limit and in some references the shrinkage limit are frequently used.

- Liquid limit (LL)

  According to the Peruvian technical standard NTP 339.129 and MTC-E110/ASTM D-4318 [14, 17] defines: it is the moisture content, expressed as a percentage, for which the soil is at the limit between the liquid and plastic. It is arbitrarily designated as the moisture content at which the separating groove of two halves of a soil paste closes along its bottom by a distance of 13 mm (1/2 in) when the cup is dropped 25 times from a height 1 cm at the rate of two falls per second.

- Plastic limit (PL)

  According to the Peruvian technical standard NTP 339.129 and MTC-E111/ASTM D-4318 [14, 17] defines: it is the moisture content, expressed as a percentage, for which the soil is at the limit between the liquid and plastic. It is arbitrarily designated as the moisture content at which the separating groove of two halves of a soil paste closes along its bottom by a distance of 13 mm (1/2 in) when the cup is dropped 25 times from a height 1cm at the rate of two falls per second.

- Plasticity index (PI)

  According to the Peruvian technical standard NTP 339.129 [17] defines: it is the range of moisture content over which a soil behaves plastically. Numerically it is the difference between the liquid limit and the plastic limit as shown in Eq. 4.

$$PI = LL - PL. \tag{4}$$

  Where:
  PI: Plasticity index
  LL: Liquid limit
  PL: Plastic limit.

2.2.2.3 Mechanical Properties of a Soil

A. Standard proctor test

According to the Peruvian technical standard NTP 339.142 [18], this test method covers the laboratory compaction procedures used to determine the relationships between the water content and the dry unit weight of soils (Compaction Curve) compacted in a 101.6 0 152.4 mm (4 or 6 in) diameter mold with a 24.4 N (5.5-lbf) piston that drops to a height of 305 mm (12 in) producing a compaction stress of (600 KN-m/m3 or 12,000 ft-lb/ft3).

The wet unit weight of the soil is calculated with the following Eq. 5:

$$\gamma = \frac{W}{V(m)}. \tag{5}$$

Where:
W: Compacted soil weight in the mold
V(m): Mold volume (943.3 cm3)

If the moisture content is known, the dry unit weight of the soil is calculated with the following Eq. 6:

$$\gamma d = \frac{\gamma}{1 + \frac{w(\%)}{100}}. \tag{6}$$

Where:
W (%): Percentage of moisture content

T values of the dry density of the soil can be plotted in relation to the moisture content to obtain the maximum dry unit weight and the optimum moisture content.

For each moisture content of the soil, a theoretical maximum unit weight is obtained, this occurs when there is no air in the empty spaces, this is when the degree of saturation is equal to 100%. Therefore, the maximum dry unit weight at a given moisture content with zero air voids can be given by the following Eq. 7:

$$\gamma cva = \frac{\gamma w}{w + \frac{1}{Gs}}. \tag{7}$$

Where:
γcva: Unit weight with zero air voids (gr/cm^3).
γw: Unit weight of water (gr/cm^3).
W: Moisture Content (%).
Gs: Specific gravity of soil solids.

B. Modified proctor test

According to NTP 339.141 and Mtc-e115/ASTM D-1557 [14, 19] defines: This test method covers laboratory compaction procedures used to determine the relationships between water content and dry unit weight of soils (Compaction Curve) compacted in a 101.6 or 152.4 mm (4 or 6 in.) diameter mold with a 44.5-N (10-lbf) piston dropping to a height of 457 mm (18 in.) Producing a compaction stress of (2,700 KN-m/m3 (56,000 ft-lb/ft3)).

The modified proctor arises due to the development of heavy rollers and the use in compaction in order to better represent field conditions.

The calculations performed in this test method are the same as those used in the standard proctor test method.

C. California Bearing Ratio CBR

According to NTP 339.145 and MTC E-132/ASTM D-1883 [14, 20] defines: this test method comprises the determination of the CBR (California Support Ratio) of subgrade of pavements, sub-base, base and materials granular specimens compacted in the laboratory. The test method is primarily to evaluate the strength of cohesive materials, with a maximum particle size of ¾ in. (19.0 mm), however it is not limited to this.

The standard PTS 339.145 [20] also mentions that this test is responsible for determining the CBR of a material with an optimal water content or a range of water content from compaction tests or a specified dry unit weight. The dry unit weight is generally presented as a percentage of the maximum dry unit weight of the compaction tests of the test methods PTS 339.141 or PTS 339.142.

The added compaction moisture can be calculated with the following Eq. 8:

$$\%moisture\ to\ add = \frac{H - h}{100 + h} \times 100 \tag{8}$$

Where:
H = Preset humidity
h: natural humidity

The soil sample, compacted with the required humidity according to the proctor test, is kept submerged for 4 days confined in a mold, with an overload equal to the weight of the pavement, this overload simulates the weight of the pavement, it is given by means of rings, in any case the total overload pressure should be less than 4.54 kg [14].

The expansion is calculated with the following Eq. 9:

$$\%Expansion = \frac{L2 - L1}{127} \times 100 \tag{9}$$

Donde:
Where:
L1: Final reading in mm.
L2: Final reading in mm.

3 Results and Discussions

The results and analysis according to the proposed methodology are shown below.

3.1 Data Presentation

To carry out this article, a database of Provias Nacional was compiled to carry out the training of the ANN model.

3.2 Data Analysis and Selection of Input Data

With the data collected from the Provias Nacional entity, the artificial neural network was trained with 6 input variables (percentage of gravel, percentage of sand, liquid limit, plastic limit and the plasticity index) and 4 output variables (maximum dry density, optimal moisture content and CBR).

3.3 Architecture and Training Stage of ANN

6 input variables and 4 output variables have been considered as shown in Fig. 4; with which we proceeded to build the architecture and carry out the training of the neural model.

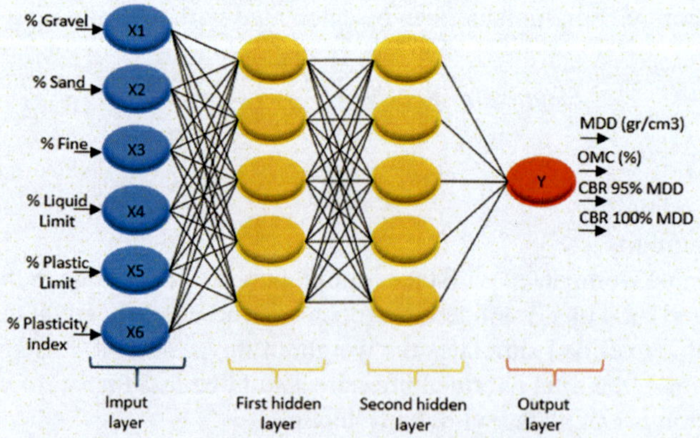

Fig. 4. Artificial neural network architecture.

3.4 Results of the Development of the Neural Network Model

The training was carried out until obtaining the appropriate regression coefficients. By default, when training the neural network is performed, it stops when it has achieved the best performance of the neural model.

3.5 Regression Analysis Was Performed

Scatter diagrams were made of the results of the output variables, obtained in the training using artificial neural networks vs the soil laboratory. The following results are shown below:

3.5.1 Regression Analysis of Maximum Density Training (MDD) with 80% of the Record

A scatter diagram was made with training prediction data at 80% of the record vs the data collected from Provias Nacional as shown in Fig. 5, depending on the way the points are located, it shows us how close they are to each other. With respect to the trend line, so that, as a result, we obtain a good coefficient of determination equal to $R^2 = 0.7487$.

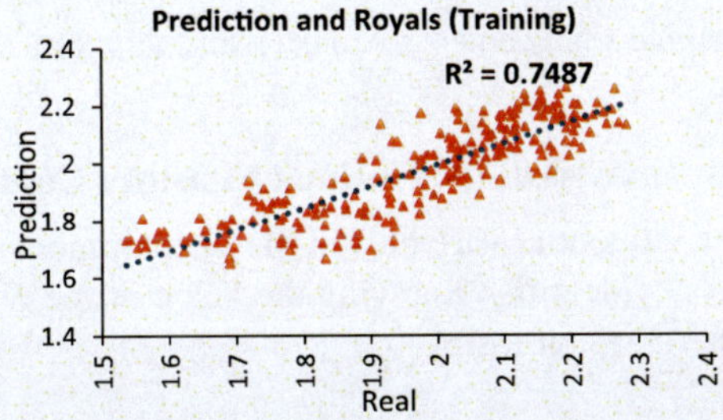

Fig. 5. Maximum dry density training results with 80% of the record.

3.5.2 Relative Impact Analysis of the Variable Maximum Dry Density (MDD)

Figure 6 shows the most influential variables in the development of the artificial neural network. In this way, it is evident that knowing the value of the physical properties it is possible to determine the MDD of a soil.

3.5.3 Regression Analysis of Optimal Moisture Content (OMC) Training with 80% of the Record

A scatter diagram was prepared with training prediction data at 80% of the record vs the data collected from Provias Nacional as shown in Fig. 7, according to the shape of the points, it shows how close they are to each other with respect to the trend line, as a result we obtain a good coefficient of determination equal to $R^2 = 0.7771$.

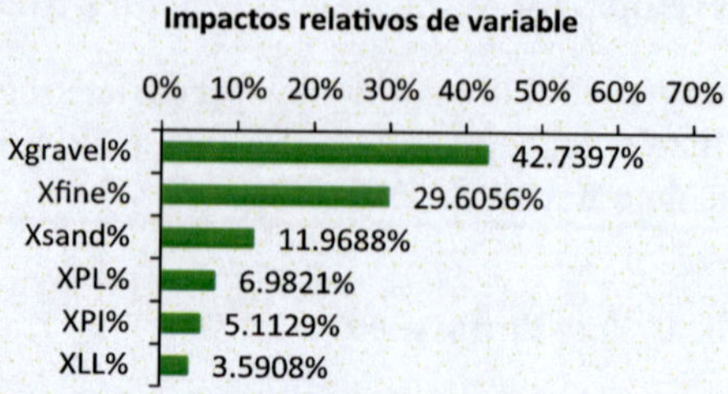

Fig. 6. Relative impact of the MDD variable.

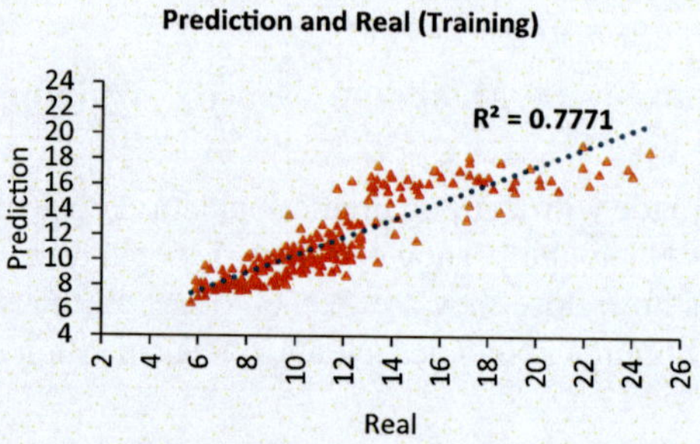

Fig. 7. Optimal moisture content training results with 80% of the record.

3.5.4 Relative Impact Analysis of the Optimal Moisture Content (OMC) Variable

Figure 8 shows the most influential variables in the development of the artificial neural network. In this way, it is evident that knowing the value of the physical properties it is possible to determine the OMC of a soil.

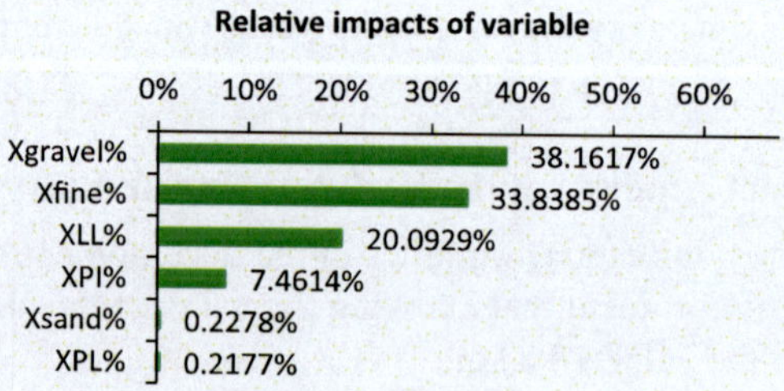

Fig. 8. Relative impacts of variables.

3.5.5 Regression Analysis of CBR Training 95% MDD with 80% of the Record

A scatter diagram was made with training prediction data at 80% of the record vs the data collected from Provias Nacional as shown in Fig. 9, according to the shape of the points, it shows us how close they are to each other. With respect to the trend line, therefore, as a result we obtain a good coefficient of determination equal to $R^2 = 0.789$.

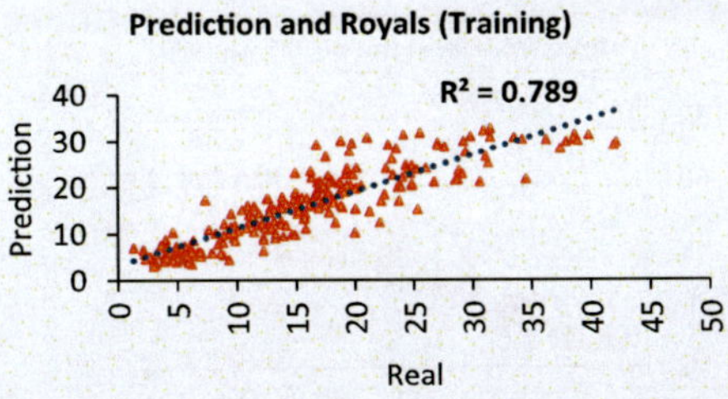

Fig. 9. Results of the training of the CBR 95% MDD with 80% of the record.

Over time, various investigations were recorded in search of the correct correlation between values.

3.5.6 Analysis of the Relative Impact of the Variable CBR 95% MDD

Figure 10 shows the most influential variables in the development of the artificial neural network. In this way, it is evident that once the value of the physical properties is known, it is possible to determine CBR 95% MDD of a soil.

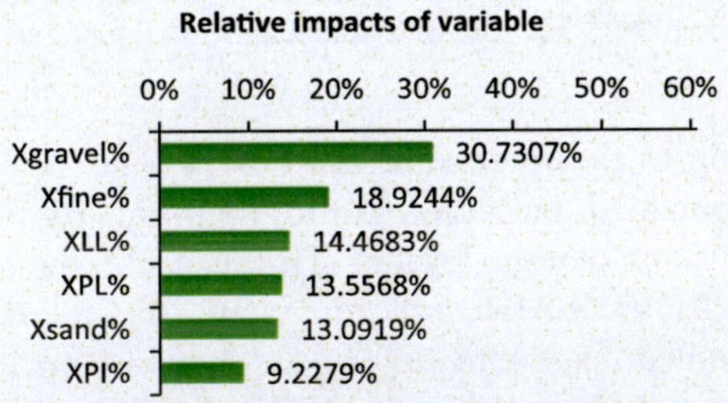

Fig. 10. Relative impacts of variable.

3.5.7 Regression Analysis of CBR Training 100% MDD with 80% of the Record

A scatter diagram was elaborated with training prediction data at 80% of the registry vs the data collected from the national provinces as shown in Fig. 11, according to the shape of the points shows us how close they are with respect to the trend line, therefore, as a result we obtain a good coefficient of determination equal to $R^2 = 0.811$.

3.5.8 Relative Impact Analysis of the CBR 100% MDD Variable

Figure 12 shows the most influential variables in the development of the artificial neural network. Thus, it is evident that knowing the value of the physical properties it is possible to determine the optimum CBR 100% MDS of a soil.

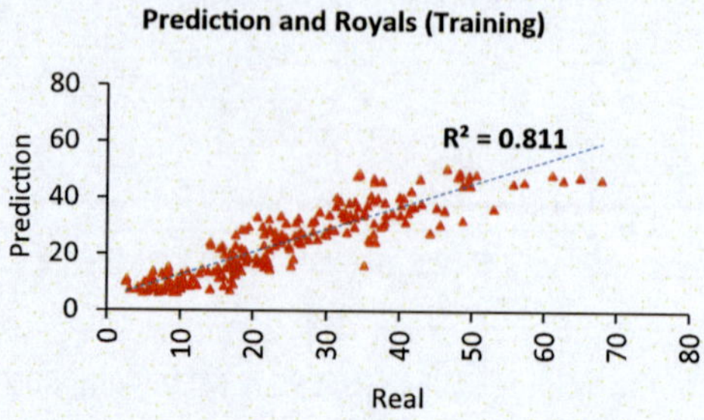

Fig. 11. Results of the training of the CBR 100% MDD with 80% of the record.

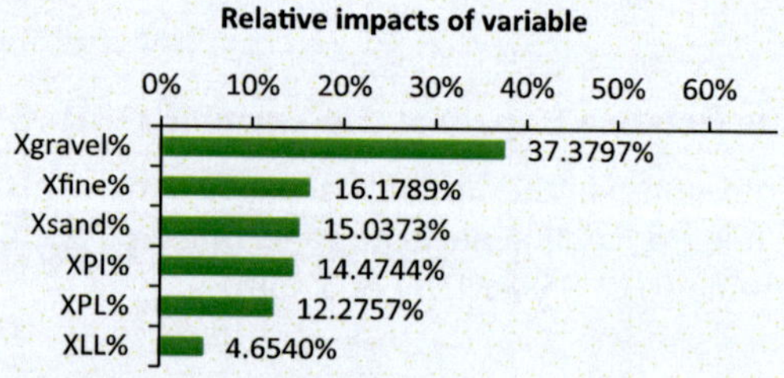

Fig. 12. Relative impacts of variable.

Discussion: According to the regression analysis in Figs. 5, 6, 7, 8, 9, 10, 11, and 12, a good model is obtained, but according to the normality test, there is no normal distribution of values, given that the level of significance in the MDD variables, OMC, CBR95% MDD and CBR100% MDD according to the test is less than 0.05 as shown in Table 9, so it is determined that the model is good, but with an unadjusted distribution. Likewise, a departure from some values is observed, this is due to the fact that at the time of the tests there was not a good quartering of samples, the equipment will not have been well calibrated and due to the variety of physical properties of the soil.

3.6 Search Results for the Best Neural Network

With the Neural Tools software, a model capable of adjusting to the type of data obtained from the Provias Nacional entity was sought, as shown in Tables 5, 6, 7, and 8. In addition, an analysis of the root means square error (RMS) to evaluate which is the appropriate model, once this was obtained it was used for new unknown values.

Table 5. Search for the best network – maximum dry density (gr/cm3).

Best network search – MDD (gr/cm3)			
Topology	%Error RMS (root mean square error)	Training time	Training stop reason
Linear prediction	0.09	00:00:00	Auto-stop
GRNN	0.09	00:00:00	Auto-stop
MLFN 2 nodes	0.10	02:00:00	Auto-stop

Table 6. Search for best network – optimum moisture content (%).

Best network search – OMC (%)			
Topology	%Error RMS (root mean square error)	Training time	Training stop reason
Linear prediction	18.36	00:00:00	Auto-stop
GRNN	1.67	00:00:00	Auto-stop
MLFN 2 nodes	1.91	02:00:00	Auto-stop

Table 7. Search for best network – CBR 95% MDD.

Best network search – CBR 95%MDD			
Topology	%Error RMS (root mean square error)	Training time	Training stop reason
Linear prediction	5.82	00:00:00	Auto-stop
GRNN	5.54	00:00:00	Auto-stop
MLFN 2 nodes	5.42	02:00:00	Auto-stop

Table 8. Search for best network – CBR 100% MDD.

Best network search – CBR 100%MDD			
Topology	%Error RMS (root mean square)	Training time	Training stop reason
Linear prediction	7.98	00:00:00	Auto-stop
GRNN	7.62	00:00:00	Auto-stop
MLFN 2 nodes	6.93	02:00:00	Auto-stop

3.7 Normality Test

The test was carried out to verify the normality of the database collected from Provias Nacional. The Kolmogorov-Smirnov method was used, since the amount of data exceeds 50 units.

Table 9. MDD Kolmogorov–Smirnov normality test summary.

Variable	Statistical	gl	Significance
LAB MDD	0.104	289	0.000
ANN MDD	0.087	289	0.000
Lab OMC	0.111	289	0.000
ANN OMC	0.147	289	0.000
LAB CBR95%MDD	0.076	289	0.000
ANN CBR95%MDD	0.073	289	0.001
LAB CBR100%MDD	0.069	289	0.002
ANN CBR100%MDD	0.087	289	0.000

Decision. The significance of the variables MDD, OMC, CBR95% and CBR100% are less than 0.05, so we reject the Ho and accept the Ha, that is, the data do not have a normal distribution, therefore, we apply non-parametric statistics.

3.8 Validation of Artificial Neural Models

With the artificial neural models already determined, we proceeded to make predictions with new data, in such a way that the results allowed us to evaluate the validation and performance of the neuronal model.

3.8.1 Maximum Dry Density (gr/cm3)

A scatter diagram was prepared with new values that were tested in the soil laboratory as shown in Fig. 13, using the GRNN algorithm for the prediction of the dependent variable, obtaining as a result equal to $R^2 = 0.7376$.

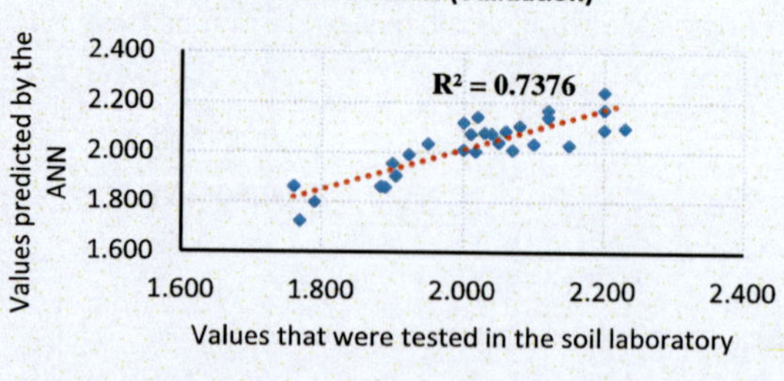

Fig. 13. Results of the prediction of the maximum dry density (gr/cm^3).

3.8.2 Optimal Moisture Content (%)

A scatter diagram was prepared with new values that were tested in the soil laboratory as shown in Fig. 14, using the GRNN algorithm for the prediction of the dependent variable, obtaining the result equal to $R^2 = 0.7815$.

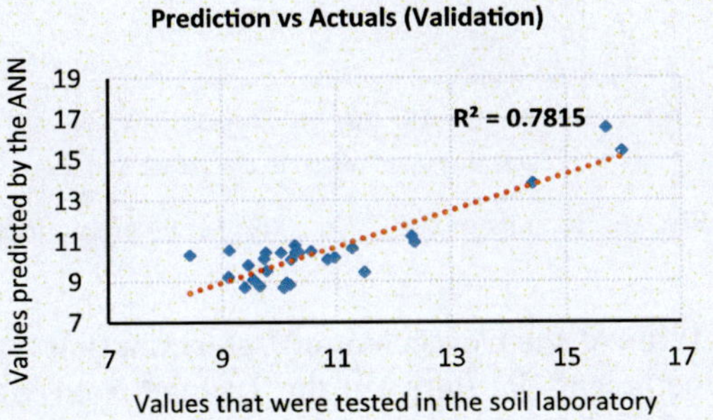

Fig. 14. Results of the prediction of the optimum moisture content (%).

3.8.3 CBR 95% MDD

A scatter diagram was prepared with new values that were tested in the soil laboratory as shown in Fig. 15, using the GRNN algorithm for the prediction of the dependent variable, obtaining as a result equal to $R^2 = 0.7935$.

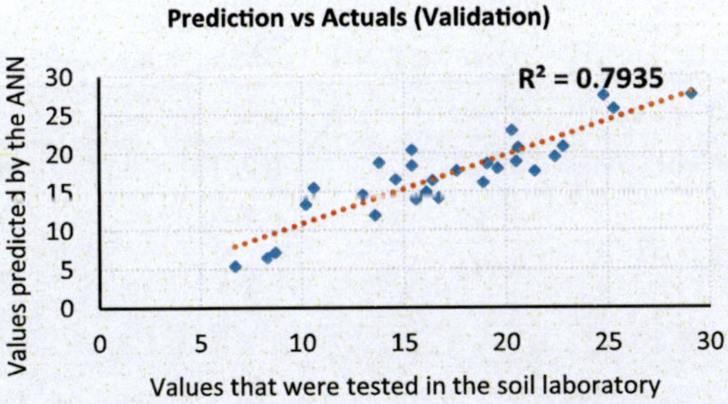

Fig. 15. Results of the CBR 95% MDD prediction.

3.8.4 CBR 100%MDD

A scatter diagram was elaborated with new values that were tested in the soil laboratory as shown in Fig. 16, using the GRNN algorithm for the prediction of the dependent variable, obtaining as a result equal to $R^2 = 0.8232$.

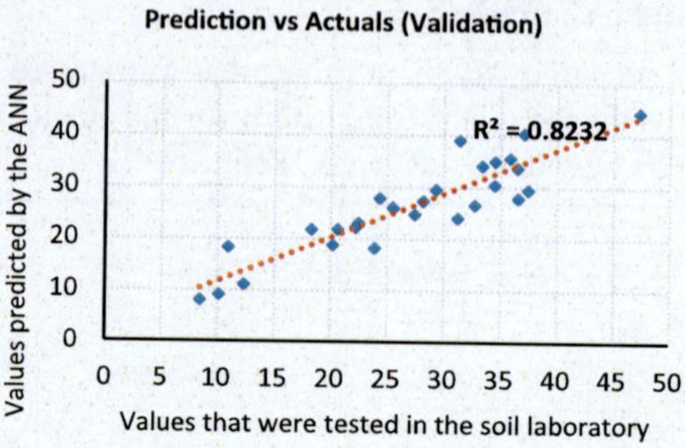

Fig. 16. Results of the CBR 100% MDD prediction.

3.9 Comparative Table of the Coefficient of Determination (R^2) and Pearson's Correlation Coefficient (R) Between the Trained Neuronal Model vs the Validation of the Model

In the following Tables 10 and 11 a comparison of the determination coefficient (R^2) and the Pearson correlation coefficient (r) between the training vs the validation of the model is shown. According to the statistical criteria with these results that we are observing, it indicates that the model is good.

Table 10. Comparative table between the trained model vs the validation of the model.

Regression analysis of the trained model vs model validation			
Variables	Training ANN	Model validation	Parameter criteria
Maximum dry density (gr/cm3)	0.7487	0.7376	Good
Optimal moisture content (%)	0.7771	0.7815	Good
CBR 95% MDD	0.789	0.7935	Good
CBR 100% MDD	0.811	0.8232	Good

Table 11. Comparative table between the trained model vs the validation of the model

Regression analysis of the trained model vs model validation			
Variables	Training ANN	Model validation	Parameter criteria
Maximum dry density (gr/cm3)	0.8653	0.7376	Good
Optimal moisture content (%)	0.8818	0.7815	Good
CBR 95% MDD	0.8789	0.7935	Good
CBR 100% MDD	0.8948	0.8232	Good

3.10 Comparative Table of the Dependent Variables Obtained Between the Predictions of the Neuronal Model vs the Soil Laboratory

Table 12 shows the results between the laboratory data vs the ANN data.

Table 12. ANN comparison table of soil laboratory data vs data predicted by ANN

Cuadro comparativo entre los resultados de laboratorio vs ANN								
SUCS	MDD (gr/cm3)		OMC (%)		CBR (95% MDD)		CBR (100% MDD)	
	LAB	ANN	LAB	ANN	LAB	ANN	LAB	ANN
SM	2.017	2.005	10.35	10.44	25.30	25.67	37.07	40.42
SP	1.905	1.907	10.29	10.44	24.80	27.38	36.47	33.78
SM	1.890	1.861	10.28	10.78	14.60	16.55	20.21	18.77
SM	1.884	1.864	10.03	10.41	17.60	17.72	34.49	30.54
CL	1.769	1.725	15.70	16.54	6.70	5.38	8.47	8.06
GM	2.020	2.143	9.66	8.78	29.10	27.49	34.53	35.05
CL	1.790	1.800	15.97	15.40	8.30	6.43	10.16	9.11
GC	2.030	2.079	10.97	10.16	15.60	13.94	22.21	22.17
GM	2.010	2.074	8.43	10.29	18.90	16.16	25.47	26.15
SM	1.923	1.992	10.85	10.06	22.40	19.51	25.52	26.00
GC	2.150	2.030	10.56	10.50	16.10	14.92	31.19	24.12
GC	2.230	2.099	9.55	9.09	20.50	18.86	37.43	29.58
GC	2.200	2.242	10.08	8.69	20.60	20.70	33.37	34.25
GC	1.950	2.035	11.28	10.63	13.00	14.54	18.30	21.78
GC	2.060	2.086	9.46	9.78	19.10	18.66	32.72	26.69
SC	1.900	1.956	12.31	11.22	10.20	13.33	10.95	18.28
GC	2.070	2.011	9.13	10.54	16.70	14.12	20.64	21.86
GC	2.100	2.035	9.76	10.42	10.60	15.47	22.46	23.05
GC	2.000	2.010	12.36	10.93	13.60	11.92	23.90	18.25
GC	2.200	2.093	9.78	9.52	15.40	18.38	24.33	27.92
GC	2.080	2.106	9.11	9.21	19.60	18.05	36.56	27.95
GC	2.050	2.042	9.73	10.13	16.40	16.43	27.41	24.84
GC	2.040	2.076	10.21	10.08	21.40	17.65	28.15	27.39
GP-GC	2.120	2.169	9.40	8.71	20.30	22.93	47.31	44.61
GC	2.200	2.171	10.13	8.94	15.40	20.39	31.38	39.17
GC	2.120	2.140	10.19	8.84	22.80	20.82	35.85	35.72
GC	2.000	2.120	11.49	9.44	13.80	18.79	29.27	29.63
CL	1.760	1.864	14.41	13.82	8.70	7.15	12.35	11.09

Discussion: The residual analysis shows that the minimum variation that exists between LAB value of soils vs ANN is equal to 0.002% in the maximum dry density, 0.061% in the optimal moisture content, 0.029% in the CBR95% MDD, 0.037% in the CBR100% MDD, this minimum variation occurred in a type of soil gravelly sand with little or no fines (SP). However, in gravelly clayey soils (GC) from the tests of CBR95% MDD, CBR100% MDD presents a significant variation of 4.991% and 8.609%.

4 Conclusion

A data cleaning had to be carried out, leaving 289 values to be evaluated and trained with the artificial neural network.

The GRNN and MLFN neural models works to predict unknown data, resulting in a coefficient of determination of 73.76% at the maximum dry density, 78.15% at the optimal moisture content, 79.35% at CBR, 95% MDD and 82.32%. in the CBR100% MDD.

Likewise, it is suggested to consider a high range of well-elaborated databases to develop complex models between the physical parameters vs the mechanical parameters of the soil.

Acknowledgements. To God for allowing this investigation to be carried out, to our advisor Leopoldo Choque Flores for the contribution of his professional knowledge, for the advice and review in this investigation. To our co-advisor Juan J. Soria, for sharing his knowledge in the field of artificial intelligence, for supporting us in the publication of our scientific article in an indexed journal. To our colleague, Eng. Axel Y. Montes for his knowledge provided in the field of civil engineering.

Lusdali Castillo Delgado.

I thank my mother J. Flor Delgado Fernández, my brothers Eduar Castillo Delgado, Edita Castillo Delgado and my friends for their constant support when I need it most.

Daniel Enrique Porta Maldonado.

I thank my parents Gloria Maldonado Flores and Jaime Porta Quintanilla for their emotional support and for their advice that allowed me to develop professionally.

References

1. Araujo, W., Ruiz, G.: Correlation equations of CBR with index properties of soils in the city of Piura, p. 69 (2016)
2. Pérez Eloarza, H.: Statistics for the Social, Behavioral and Health Sciences, 3a. edició. Cengage Learning Editores, S.A. de C.V., una Compañía de Cengage Learning, Inc. Corporativo Santa Fe, México (2008)
3. Alemán Morales, D.: Artificial Intelligence Techniques Applied to Civil Engineering Problems, vol. 6, pp. 164–175 (2017)
4. Vila Zuñiga, R.: Analysis of the linear behavior of the flexible pavement structure by means of neural networks on the Panamericana Norte highway. Universidad del centro del Perú (2017)
5. Boza Capani, M., Merino Ortiz, R.: Shear strength parameters of soils from their physical properties using artificial neural networks and triaxial equipment. UNH, Universidad Nacional de Huancavelica (2018)

6. Paytán Ordoñez, J.: Estimation of the International Evenness Index in Flexible Pavements Using Artificial Neural Networks. Universidad Nacional de Huancavelica (2018)
7. Chávez Castro, N.S., Reyes Velásquez, J.F.: Development of a Computer Vision System for the Identification of Surface Failures in Flexible Pavements in the City of Trujillo. Universidad Nacional de Trujillo (2019)
8. Tello Cifuentes, L., Aguirre Sánchez, M., Díaz Paz, J.P., Hernández, F.: Flexible pavement damage assessment using terrestrial photogrammetry and neural networks, vol. 24, no. 50 [Online] (2021). https://doi.org/10.22430/22565337.1686
9. Anderson, J.A.: An Introduction to Neural Networks. United States of America (1995)
10. Matich, D.J.: Neural Networks: Basic Concepts and Applications, p. 55 [Online]. ftp://decsai.ugr.es/pub/usuarios/castro/Material-Redes-Neuronales/Libros/matich-redesneuronales.pdf (2001)
11. Serrano, A.J., Soria, E., Martín, J.D.: Artificial neural networks (2010)
12. Basoqain Olabe, X.: Artificial neural networks and their applications. Vasco España
13. Crespo Villalaz, C.: Soil Mechanics and Foundations, p. 652. Editorial Limusa, México [Online]. https://stehven.files.wordpress.com/2015/06/mecanica-desuelos-y-cimentaciones-crespo-villalaz.pdf (2004)
14. D. G. de C. y F. MTC: 'Road Manual' Soil, Geology, Geotechnics and Pavements. Soils and pavements section R.D.N° 10-2014-MTC/14" (2014)
15. 339.127 NTP: Soils. Test method to determine the moisture content of a soil, 1a Edición. Lima, INACAL, 2019 (2019)
16. 339.128 NTP: *Soils.* Test method for particle size analysis, 1a Edición. Lima (2019)
17. 339.129 NTP: Soils. Test method to determine the liquid limit, plastic limit and plasticity index of soils, 1a Edición (2019)
18. 339.142 NTP: Soils. Test method for laboratory soil compaction using standard energy (600 KN-m/cm3 (12 400 ft-lbt/ft3)), 1a Edición. Lima (2019)
19. 339.141 NTP: Soils. Laboratory soil compaction test method using a modified energy (2 700 KN-m/m3 (56 000 ft-lbf/ft3), 1a Edición. Lima (2019)
20. 339.145 NTP: Soils. CBR (California Bearing Ratio) test method for compacted soils in the laboratory, 1a Edición. Lima (2019)

Knowledge-Based Automated Service Composition for Decision Support Systems Configuration

Nikolay Mustafin, Pavel Kopylov, and Andrew Ponomarev(✉)

St. Petersburg Federal Research Center of the Russian Academy of Sciences, St. Petersburg, Russian Federation
ponomarev@iias.spb.su

Abstract. The paper proposes a novel approach to reducing the complexity of building (service-oriented) decision support systems by applying knowledge-based automated service composition. The proposed approach is based on ontological representation of service capabilities, therefore, the paper discusses how these capabilities are encoded with a help of a modular ontology, including three components: DSS functional blocks ontology, service description ontology based on OWL-S, and application domain ontology. It is also discussed how these representations are used to form complex services using forward chaining process. The paper also presents an architecture of the configurable DSS, based on service composition. The proposed approach can be used for creating DSSs in a wide range of domains, especially, where quality conceptualizations in the form of ontologies already exist.

Keywords: Decision support systems · Service composition · Ontologies · Automated service composition · Service-oriented architecture

1 Introduction

Decision support systems (DSSs) play an important role in the information technology landscape, helping people to process and analyse information relevant to complex problems and improving the quality of decisions [1]. Traditionally, five classes of DSSs are distinguished [1]. In this paper, we focus mostly on model-driven and data-driven, because, on the one hand, today, most decision support systems are hybrid, offering features typically attributed to two or more types discussed in [1], on the other hand, features of data-driven and model-driven types are most common.

The design and development of a DSS is a complex effort, requiring both thorough domain analysis, data and knowledge engineering, building the most appropriate mathematical models, etc. At the same time, DSS is a software system, therefore, one need to program all the information gathering and transformation workflows, supporting the selected decision support procedures. The necessity of this technical (software programming) limits the flexibility of typical DSSs. The specifics of the decision making is that it

R. Silhavy et al. (Eds.): CoMeSySo 2021, LNNS 231, pp. 780–788, 2021.
https://doi.org/10.1007/978-3-030-90321-3_63

frequently relies on interactive and iterative workflows – decision-maker explores multiple approaches to the problem by experimenting and analyzing the stability and quality of the models and workflows. The need for severe technical (programming) work slows down the experimentation cycle, because it may force decision-maker to look for assistance with the technical side of work if he/she doesn't have certain skills. Therefore, the research aimed on simplifying this technical part of DSSs development is actual and may help to reduce the labor costs for the creation of a DSS and to shorten the decision-making cycle. It is also well-suited to modern dynamic environment requiring increasingly more speed from decision-makers managing commercial organizations and government agencies (see, e.g., [2]).

According to service-oriented architecture, a system can be constructed as a composition of relatively self-contained components (services), interacting via well-defined protocols and mechanisms (usually, supported by some middleware). There are multiple specific ways to implement service-oriented approach, starting from web services, to modern microservice approach [3]. Service-oriented architecture allows to solve several acute problems typical for software engineering field, and currently it is widely used for building various software systems, including DSSs [4–6]. With the service-oriented architecture, the creation of a DSS often can be reduced to building compositions of existing services that implement information processing functions so that the resulting composition provides necessary DSS functionality.

This paper continues our research on reducing the complexity of building (service-oriented) DSSs by applying knowledge-based automated service composition techniques. In our earlier work [7], we studied functional framework of different types of decision support systems, formulated a set of requirements for configurable service-oriented decision support systems and described their main components. In this paper, we propose descriptive ontological models, to describe both existing services of a DSS and the goal of the composition (a required service). We also propose a method for the composition of DSS services (creating composite processes that include invocation of several services) based on high-level machine-readable specifications composition method and an architecture of the configurable DSS.

The rest of the paper has the following structure. Section 2 describes related work in the field of automated service composition. Section 3 describes the ontological model to describe services and the composition goal. Section 4 presents the method of service composition. Finally, Sect. 5 describes the proposed architecture of such system.

2 Related Work

The idea to automate the process of service composition appeared almost simultaneously with the web services and service-oriented architecture (see, e.g., [8]). Most popular formal methods of automatic service composition are: HTN planning [9], colored Petri nets [10, 11], and various meta-heuristics (in particular, ant colonies, swarm intelligence, genetic algorithms) [12, 13].

An important aspect of service composition is how to describe the composition. Two main groups of languages have been proposed for that. First, there are a number of languages for describing composition, oriented mostly on "manual" way (BPEL, WS-CDL, etc.), allowing one to make a detailed specification of the composition. These languages are used mostly by software engineers. Second, there are languages designed primarily to support automatic service composition. A prominent example of such languages is OWL-S [14]. This language helps to link ontological model of service capabilities and the construction of a trajectory in the state space typical for automatic planning, allowing to apply automatic planning methods to construct compositions of services equipped with semantic metadata.

In the proposed approach, we also use OWL-S to model service capabilities and the goal of composition. We also adapt it to decision support systems taking into account the specifics of such systems (and models) when developing ways of flexible composition of services.

3 Ontological Models

The proposed approach to service composition is based on representing various parts of the problem in the form of ontologies. It allows to leverage ontological reasoning in order to infer some information that is not given explicitly (and/or, helps in matching service input/outputs by making use of a class taxonomy, contained in the ontology). This section introduces the proposed method of describing services using ontologies, as well as the proposed method of describing the goal of service composition.

3.1 Service Description Model

An ontological model, able to describe various aspects of a service-oriented DSS plays an important role in the proposed approach. This ontological model is constructed using the so-called modular ontologies approach [15–17], according to which the ontological model is divided into several (relatively) independent parts (modules), each of which is focused on solving certain problems (and, accordingly, describes the concepts and relationships relevant to these problems). These modules are interconnected and, as one modular ontology, meet the entire set of requirements for the describing elements of a DSS, enabling both service discovery and service composition algorithms.

The ontological model consists of the following modules: 1) DSS functional blocks ontology, 2) service description ontology based on OWL-S [14, 18], 3) application domain ontology, binding abstract OWL-S constructs to the information about the domain of the particular DSS.

The DSS functional blocks ontology is based on the typical functional framework of DSSs, which includes the following blocks: a) user interface and visualization components, b) data management, c) model management, d) solvers. Accordingly, the ontology introduces classes corresponding to the selected functional blocks, their inherent properties, as well as subcategories (for example, solvers are categorized depending on the supported classes of mathematical models and the algorithms used).

The service description ontology defines concepts and properties for describing both individual services and computational processes involving the execution of multiple services. Correspondingly, the cornerstone elements of this ontology are the service profile and the process specification. The service profile (*ServiceProfile*), allows one to describe the following three types of information, describing a service: a) input and output data (parameters), preconditions and effects, b) classification of the service by purpose within the application domain, c) additional characteristics associated with use of the service (cost, required resources, etc.). To describe the input and output data in the ontology, the ontology introduces *Parameter* concept, as well as the *hasInput* (input parameter of the service) and *hasOutput* (output parameter of the service) properties, linking the service profile and various types of parameters. To describe preconditions and effects, the ontology introduces *Condition* and *Result* concepts, as well as the properties *hasPrecondition* and *hasResult*, respectively. Classification of a service by purpose can be done using the *serviceCategory* property, allwing one to link the service profile with the concepts of the DSS functional blocks ontology. Additional (e.g., non-functional – time, cost, etc.) characteristics of a service can be described using the *serviceParameter* property.

Processes are divided into atomic (*AtomicProcess*) and composite (*CompositeProcess*). Actions performed by a single service during one interaction correspond to an atomic process. In other words, accessing a service and getting a result is an atomic process. A composite process is the execution of several atomic processes, "glued" by certain control constructs (*ControlConstruct*): sequential execution (*Sequence*), parallel execution (*Split*) and others. Similar to a service profile, each process is described by a set of input and output parameters, as well as preconditions and effects. Control constructs correspond to constructs of programming languages and allow one to define a composition of services, i.e., a complex (composite) service that can invokes several other services in order to accomplish user's task. In particular, the ultimate goal of composition is to create such specifications using available control constructs and a repository of basic services.

The application domain ontology defines domain concepts and properties. These entitles can also be used as inputs and/or outputs of services. In this way, the application domain ontology connects the service description ontology to the application area of the DSS. For example, the ontology tourism and infomobility domain includes such concepts as "Attraction", "Route", etc., as well as their properties. When creating a DSS in this domain the joint usage of this tourism and infomobility ontology and the service description ontology allows to describe services that search for and rank attractions, build routes and perform other tasks relevant to this DSS.

Let's consider an example of ontological description in the area of tourism and infomobility. A fragment of an application domain ontology is shown in Fig. 1. It introduces typical complex for any tourism application – "Point of interest", "Attraction" and "Route", as well as the notion of a user preference (in various forms). Figure 2 shows a simplified specification of a service that can retrieve an unordered collection (so-called, bag) of points of interest located in a certain geographical region. In this figures, we use prefix "ado" to designate application domain ontology, "geo" – geographical ontology, "owls" – OWL-S, and "owl" – for basic OWL entities.

In this example, service profile is identified by a universal unique identifier. This service is classified as a data provider (which is a subclass of data management services defined by the DSS functional blocks ontology), has a location (geo: Location) as an input and a BagOfAttractions as an output.

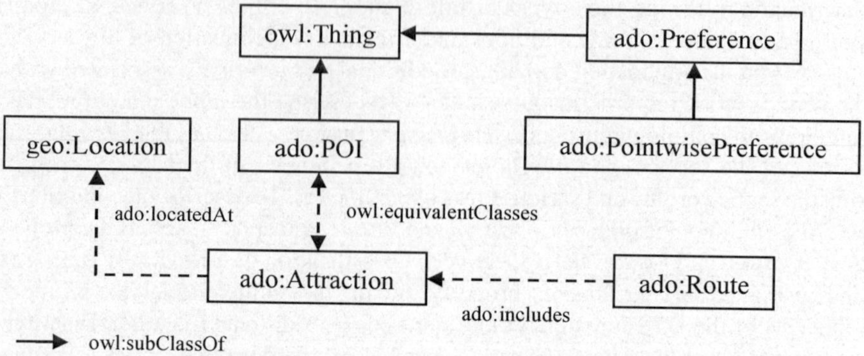

Fig. 1. Tourism and infomobility ontology (fragment)

```
<urn:uuid:6e8bc430-9c3a-11d9-9669-0800200c9a77>
   a owls:ServiceProfile ;
   owls:serviceName "Attractions locator." ;
   owls:serviceCategory _:cat1 ;
   owls:hasInput geo:Location ;
   owls:hasOutput :BagOfAttractions .

_:cat1
   owls:categoryName "Service role" ;
   owls:value "Data provider" .
```

Fig. 2. Service profile description (fragment)

3.2 Requirements Definition Model

The way of describing the requirements for the composition of services is closely related to the description of individual services and the processes they implement. Indeed, to specify the requirements to a missing service (that has to be built) one has to use three kinds of statements: 1) description of known information pieces that can be used as input for the composite service, 2) description of the desired (output) parameters, and 3) a description of quantitative constraints on the composite service (performance constraints, cost of execution, etc.). In practical implementation, the description of the available parameters can often be generated automatically (for example, based on the data model of the current project), but their specification is in any case the input data for the composition construction algorithm.

For the formal specification of these three types of statements, the same ontological structures are used as for describing the input and output data of a service (or process) and its characteristics: *hasInput*, *hasOuput*, *hasPrecondition*, *hasResult*, *serviceParameter*. The values for these properties are the concepts defined by the domain ontology. Thus, the description of the goals of the composition practically coincides with the description of the inputs and outputs of the composite process (while it is obviously devoid of the specification of the atomic processes composition, which is formed in an automated way to correspond to the given inputs and outputs).

4 Method for Automated Service Composition

The proposed service composition method allows to automatically create composite services (that include access to several other services in order to achieve the goal defined by the user) based on high-level machine-readable specifications. The developed method is based on the service ontology described above, the application domain ontology and the proposed model of describing the composition goals. The purpose of the created service is determined by the output parameter or several output parameters, the values of which are in demand by the decision maker. The proposed method consists in using a repository of services described using a service ontology to construct possible configurations of composite processes that allow obtaining the required output parameters. Composite processes are formed by choosing and building chains (or, in the general case, networks) for activating services. The resulting composite service configurations define the required input data (this is the input data of the services included in the composition). Input data sets are used to filter potential composite processes, and two options are allowed: automatic filtering, in which variants of composite processes for which some input data are missing, are excluded from consideration, and semi-automatic filtering, in which such options are also presented to the end user (possibly the corresponding types of initial data can be obtained manually or the task can be further defined by the end user).

To use the potential of logical inference when comparing the types of input and output data within the framework of constructing composite processes, the developed method assumes that part of the operations for the formation of such processes is carried out using an inference engine. In particular, inference is employed to check if an output concept of one service is basically the same as (or, compatible to) the input of another service. For example, one service may have an "*Attraction*" as an input, but another – "*Point of interest*" as an output. According to the considered application domain ontology these classes are equivalent, therefore, the respective services can be linked to form a service invocation chain (to solve some bigger problem).

The proposed method of service composition is an adaptation of the forward chaining search to the definition of the state as a set of available pieces of information about the problem. In particular, the composition engine starts with the active set of available parameters, initialized by the set of input parameters of the requirements definition model. Then, the procedure iterates though all the services that have all the input parameters in the active set of available parameters (the comparison is made via ontology inference, to account for domain-specific relationship between classes). The output parameters of all the services that can help be invoked are also added to the active set of available parameters. This process continues until either no new state can be obtained by invoking a service (no new pieces of information can be obtained), or each of the required parameters (given by the requirements definition model) belongs to the active set of available parameters. The procedure tracks which services should be invoked to obtain each parameter. If the invocation path has been successfully found, it is, first, pruned, by removing invocations that are not needed to obtain at least one of the required parameters, then, it is encoded into a composite process specification, using control constructs, supported by OWL-S.

5 System Architecture

The general architecture of the configurable DSS based on service composition is shown in Fig. 3.

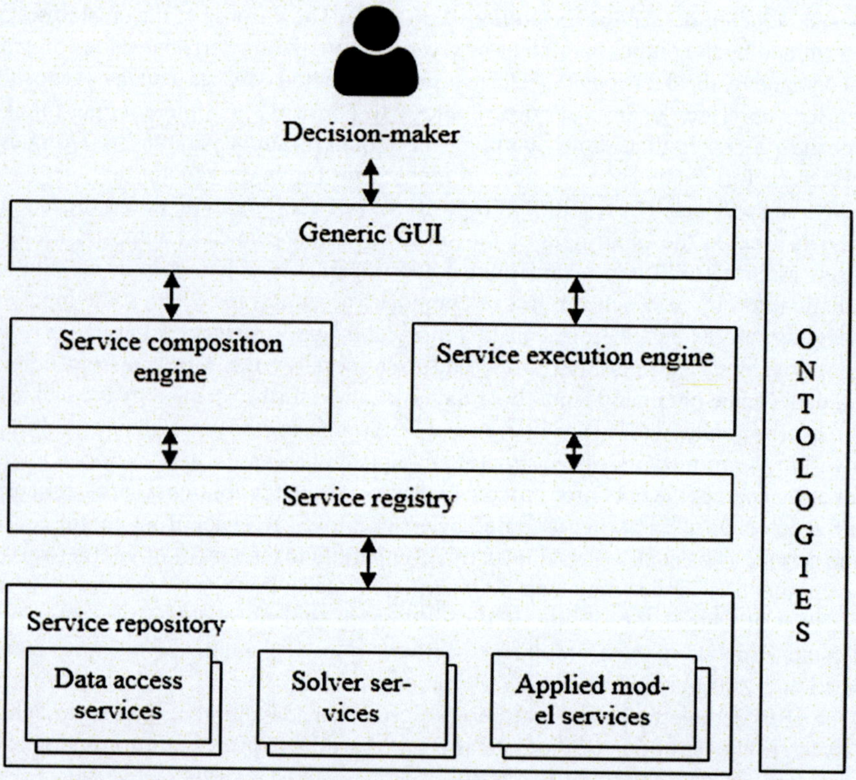

Fig. 3. Architecture of the DSS based on automatic service composition.

Foundational elements of the DSS are services. There are several types of services: data access services, solver services, services-applied models. To implement the capabilities of semantic search and subsequent automatic composition, services are provided with an ontological description using the ontological models, described in the previous section. Basic services constituting a service "palette" of the DSS are stored in the service repository. A special type of service is composite services formed from other services using certain composition operations (for example, linking the output parameters of one service and the input parameters of another) at the request of a decision maker in automatic or manual mode. All services of the DSS – basic services from the service repository and composite services – are registered in a special registry and are available for building composite services.

The core of the DSS is the infrastructure for building composite services, which can form a new composite service (or several options that are offered by the decision maker), based on the specification of the problem formulated by the decision maker using the ontological models and the descriptions of the existing services in the registry. To reduce the search space in the process of composition, service types are taken into account (based on the functional structure of DSS). Therefore, a typical composite service will include several data access services, a service-application model and a service solver, to which model-data binding and model-solver will be applied.

6 Conclusion

The paper proposes a novel approach to reducing the complexity of building (service-oriented) DSSs by applying knowledge-based automated service composition. The proposed approach heavily uses ontological representation of service capabilities, therefore, we discuss how these capabilities are encoded with a help of a modular ontology, including three components: DSS functional blocks ontology, service description ontology based on OWL-S, and application domain ontology. These representations are used to form complex services using forward chaining process.

The proposed approach is compatible (in a functional and architectural ways) with modern developments in the field of DSS (see, for example, [4, 6, 19]), and allows one to expand them by integrating different classes of DSS and providing the possibility of automatic composition of services in accordance with the tasks of the decision-maker.

Acknowledgement. The reported study was funded by RFBR, project number 19-07-00928. The study was also supported by Russian State Research, project 0073-2019-0005.

References

1. Power, D.J.: Decision Support Systems: Concepts and Resources for Managers. Praeger (2002)
2. Alemany, M.M.E., Boza, A., Ortiz, A., Fuertes-Miquel, V.S.: Configurable DSS for uncertainty management by Fuzzy sets. Procedia Comput. Sci. **83**, 1019–1024 (2016). https://doi.org/10.1016/j.procs.2016.04.217
3. Dragoni, N., et al.: Microservices: yesterday, today, and tomorrow. In: Mazzara, M., Meyer, B. (eds.) Present and Ulterior Software Engineering, pp. 195–216. Springer International Publishing, Cham (2017). https://doi.org/10.1007/978-3-319-67425-4_12
4. Dong, C.-S.J., Srinivasan, A.: Agent-enabled service-oriented decision support systems. Decis. Support Syst. **55**, 364–373 (2013). https://doi.org/10.1016/j.dss.2012.05.047
5. Demirkan, H., Delen, D.: Leveraging the capabilities of service-oriented decision support systems: putting analytics and big data in cloud. Decis. Support Syst. **55**, 412–421 (2013). https://doi.org/10.1016/j.dss.2012.05.048

6. Nada, A., Nasr, M., Salah, M.: Service oriented approach for decision support systems. In: 2014 IEEE 7th Joint International Information Technology and Artificial Intelligence Conference, pp. 409–413. ITAIC 2014 (2014). https://doi.org/10.1109/ITAIC.2014.7065081
7. Ponomarev, A., Mustafin, N.: Decision support systems configuration based on knowledge-driven automated service composition: requirements and conceptual model. Procedia Comput. Sci. **186**, 654–660 (2021). https://doi.org/10.1016/j.procs.2021.04.213
8. Peer, J.: A PDDL based tool for automatic web service composition. In: Ohlbach, H.J., Schaffert, S. (eds.) PPSWR 2004. LNCS, vol. 3208, pp. 149–163. Springer, Heidelberg (2004). https://doi.org/10.1007/978-3-540-30122-6_11
9. Georgievski, I., Nizamic, F., Lazovik, A., Aiello, M.: Cloud ready applications composed via HTN planning. In: 2017 IEEE 10th Conference on Service-Oriented Computing and Applications (SOCA), pp. 81–89. IEEE (2017). https://doi.org/10.1109/SOCA.2017.19
10. Kalamegam, P.: Usage of CPN models in web service compositions – a survey. In: 2017 International Conference on Technical Advancements in Computers and Communications (ICTACC), pp. 4–6. IEEE (2017). https://doi.org/10.1109/ICTACC.2017.12
11. Li, K., Li, W., Sun, X., Xia, Y.: A stochastic-petri-net-based model for ontology-based service composition. In: 2016 9th International Conference on Service Science (ICSS), pp. 108–112. IEEE (2016). https://doi.org/10.1109/ICSS.2016.40
12. Gil-Herrera, J., Botero, J.F.: A scalable metaheuristic for service function chain composition. In: 2017 IEEE 9th Latin-American Conference on Communications (LATINCOM), pp. 1–6. IEEE (2017). https://doi.org/10.1109/LATINCOM.2017.8240194
13. Dahan, F., El Hindi, K., Ghoneim, A.: An adapted ant-inspired algorithm for enhancing web service composition. Int. J. Semant. Web Inf. Syst. **13**, 181–197 (2017). https://doi.org/10.4018/IJSWIS.2017100109
14. OWL-S: Semantic markup for web services. https://www.w3.org/Submission/OWL-S/
15. Shimizu, C., Hirt, Q., Hitzler, P.: MODL: A Modular Ontology Design Library (2019)
16. LeClair, A., Khedri, R., Marinache, A.: Toward measuring knowledge loss due to ontology modularization. In: Proceedings of the 11th International Joint Conference on Knowledge Discovery, Knowledge Engineering and Knowledge Management, pp. 174–184. SCITEPRESS – Science and Technology Publications (2019). https://doi.org/10.5220/0008169301740184
17. Bouzidi, R., De Nicola, A., Nader, F., Chalal, R.: OntoGamif: a modular ontology for integrated gamification. Appl. Ontol. **14**, 215–249 (2019). https://doi.org/10.3233/AO-190212
18. Martin, D., et al.: bringing semantics to web services: the OWL-S approach. In: Cardoso, J., Sheth, A. (eds.) SWSWPC 2004. LNCS, vol. 3387, pp. 26–42. Springer, Heidelberg (2005). https://doi.org/10.1007/978-3-540-30581-1_4
19. Turban, E., Aronson, J.E., Liang, T.-P.: Decision Support Systems and Intelligent Systems. Prentice Hall (2004)

A Model of Continuous Integration and Deployment of Engineering Software

Golosovskiy Mikhail, Bogomolov Aleksey(✉), and Balandov Mikhail

St. Petersburg Federal Research Center, Russian Academy of Sciences, St. Petersburg 199178, Russia

Abstract. The article considers a model of continuous delivery of engineering software, including computer-aided design (CAD) systems. The main directions of engineering software development in the course of building projects are out-lined. The main phases and stages of building projects are highlighted. The necessity of developing engineering software at each of the listed stages and phases is shown. A model of forming requirements for the development of engineering software has been designed in order to indicate the need for development within the framework of building projects. The main problems affecting the reduction of the delivery time are considered. Possible methods and practices that will allow us to eliminate or reduce the influence of negative factors on the speed of software delivery are proposed. This algorithm includes tailored practices of DevOps framework for engineering software development life cycle.

Keywords: Computer-aided design · Engineering software · Developing engineering software · Software life cycle · DevOps

1 Introduction

In the modern world, much attention is paid to the speed of developing new versions of software. Engineering companies recognize that there is a need to enhance vendor software and develop their own software to perform critical tasks [1, 2]. Engineering software is understood as specialized software designed to perform tasks in the design, participating in the release of engineering design products. One of the types of engineering software is computer-aided design (CAD) systems. Examples of engineering software are Aveva E3D, Tekla Structures and Graphisoft Archicad [3, 4]. Due to the tight deadlines for the implementation of engineering projects and the high degree of uncertainty of software requirements at the initial stages of projects, there are strict restrictions on the time required for software completion [5]. In addi-tion, the software requirements are continuously updated during the project, which makes it necessary to quickly release new software versions that implement new or adjusted requirements. The main directions of development:

– Development of integration with external systems, especially in terms of changing the formats and forms of presentation of engineering data.

R. Silhavy et al. (Eds.): CoMeSySo 2021, LNNS 231, pp. 789–796, 2021.
https://doi.org/10.1007/978-3-030-90321-3_64

- Creation of modules for calculations based on available engineering data, taking into account industry specifics, requirements of customers and regulators.
- Development modules of reports and forms for project and procurement documentation.

At the same time, over the past 5–10 years, software engineering has been actively developing approaches to continuous integration and deployment of software-DevOps, aimed at accelerating the release of new software versions into the operation environment. DevOps is an abbreviation for Development and Operation [6–12]. A set of DevOps practices combined with flexible software development methodologies of the Agile family allows you to reduce the delivery time of new software versions and increase its reliability. According to the DORA State of DevOps 2019 [13] and Puppet State of DevOps 2021 reports [14], high-performance teams (Elite Performers) have 106 times higher delivery speed of new software versions (Lead Time) than low-performance teams (Low Performers), and the number of unsuccessful implementations (Change Failure Rate) is 7 times less. High team performance is ensured through the use of DevOps practices and techniques. In the development of engineering software, the following limitations are highlighted that prevent the implementation of DevOps practices [15–17]:

- All development and operation work is carried out in closed segments of the corporate network, which makes it difficult to create a large number of integration testing benches with external systems for parallel testing processes;
- Systems are often built on the "rich-client" principle, when most functions are performed on the client side. This complicates the deployment process.
- The introduction of a micro-service architecture is limited by a large number of vendors, the complete processing of which for a new architecture is not economically feasible.

2 Model Description

Building projects by the type of contracts can generally be divided into two categories: Engineering and Engineering Procurement and Construction (EPC). In Engineering only the release of project documentation is performed; EPC involves the execution of design, procurement and construction works in one contract. In general, building projects can be divided into the following phases and stages, presented in Table 1.

The stages may differ depending on the country, the industry and the type of contract. At each of the stages presented in Table 1, tools are selected and software is finalized or purchased to implement the requirements at this stage. The main part of these requirements is formed at the engineering phase, the main activities of which are related to the need to refine engineering software. These activities are common for each stage. The activity diagram in the UML notation with the main steps is shown in Fig. 1. It begins with the formation or clarification of the requirements for the entire construction project as a whole. Further, the requirements are divided into areas in which descriptions of procedures begin to be formed and comprise the primary requirements for engineering software. After that, the following activities begin: descriptions of procedures continue

Table 1. Stages and phases of building projects

Phase	Stage
Feasibility study	Feasibility study
Concept development	Concept development
Engineering phase	Conceptual design
	Preliminary front end engineering design (Pre-FEED)
	Front end engineering design (FEED)_
	FEED supervision & review
	Detailed engineering
	Design review
Procurement phase	Request for quote/request for proposal
	Purchasing (purchase order)
	Invoicing and payment
	Logistics and transport
	Receiving and incoming material quality check
Construction phase	Onsite and offsite fabrication
	Construction stage
	Erection and installation stage
	Pre-commissioning stage
	Commissioning stage
	Start-up stage

to be formed, followed by clarification of requirements for engineering software and activities for their implementation or purchase. The process is iterative. A decision regarding the purchase or development of software is made at each iteration on the basis of a technical and economic assessment. At the same time, there may often be cases when requirements are formed and refined in their details much faster than development teams have time to implement them. This is why queues accumulated, which additionally makes it necessary to reduce the delivery time (Lead Time).

To solve the problems described above, a model for ensuring continuous integration and deployment of engineering software is proposed (see Fig. 2). The model details the "Software developing" block in Fig. 1. The model focuses on the processes related to software delivery, deployment and operation, so the processes related to the practices of analytics, architecture development, coding, and unit testing are located on the scheme in one block – Development and bug fixing.

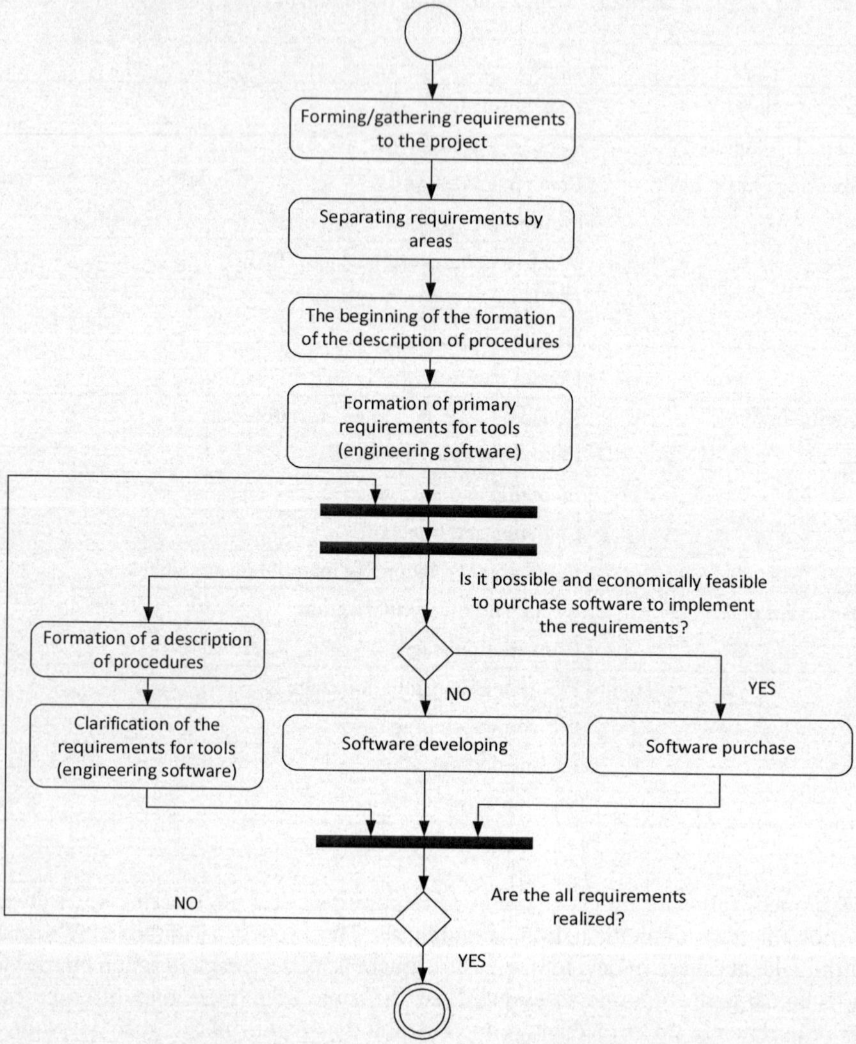

Fig. 1. Formation and implementation of requirements for the engineering software development

The emphasis in the model is on parallel execution with the processes of software development and static code analysis for information security vulnerabilities and the processes of developing and updating test scenarios and generating test data in order to obtain the necessary set of test data.

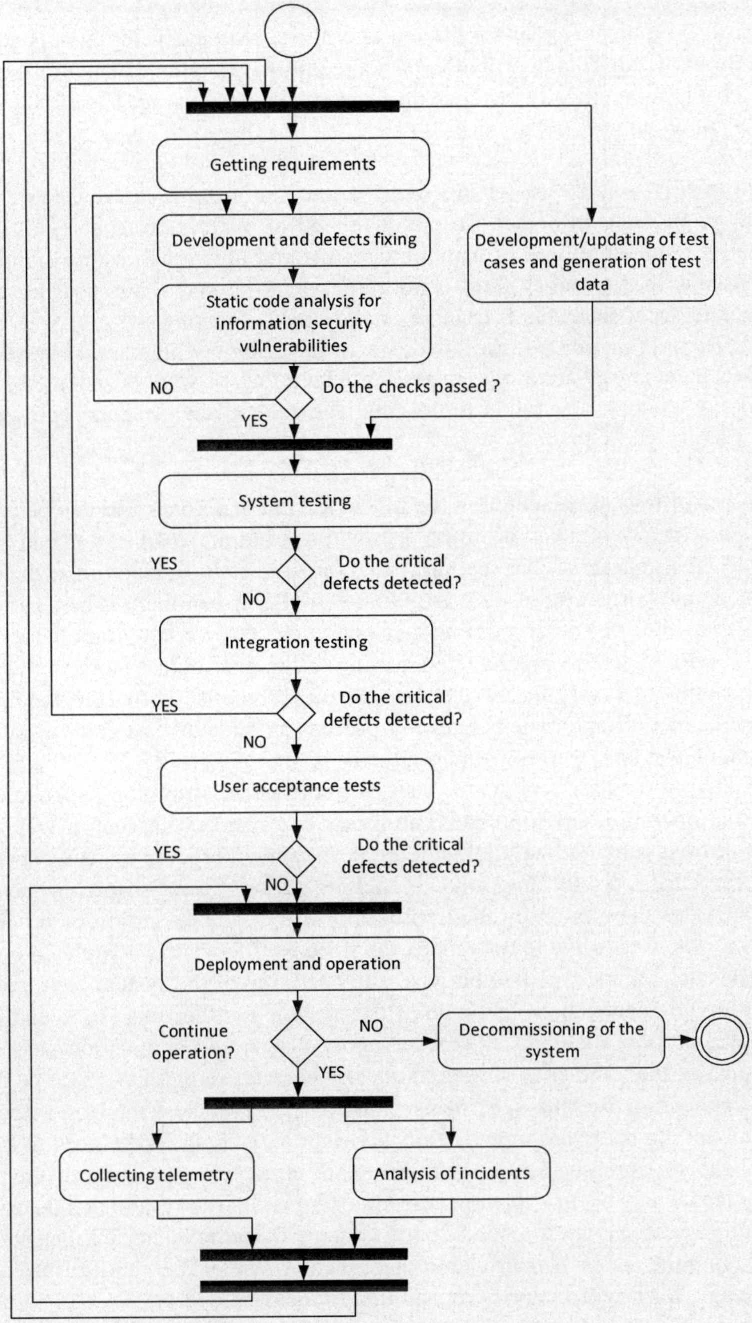

Fig. 2. A model of continuous integration and deployment of engineering software

The distribution package is assembled after the software development is completed and defects are fixed. This distribution package uploads into the distribution repository from which it is automatically installed on environments for testing. The model identifies three types of testing:

- System testing – software testing without integration with external systems. It is carried out in order to detect errors not related to integration as early as possible, in order to reduce the time of using integration and functional testing stands. Since integration with external systems is not required at this stage, it is possible to create a large number of such stands to increase the speed of testing.
- Integration and functional testing – testing of integration with external systems.
- User acceptance tests are a type of testing where the software is accepted by stakeholders, for example: customer representatives, related departments representatives, cyber security services.

The types of testing presented in the model are the main ones and can be expanded by adding new stands and variations on testing. For example, load testing can be added if necessary. An important clarification is that the collected distribution at the stage of development and correction of defects is transferred between the test benches without changes, up to the transfer to commercial operation. Only configuration files can be changed. If critical defects are detected, testing of the distribution stops and we return to the development stage and fix the defects. In accordance with DevOps practices, it is recommended to automate the testing process as much as possible, installing the distribution on test and operation environments.

After passing the user acceptance tests, the software distribution package is transferred to the operation environment. The model is organized in such a way that the operation process can go in parallel with the development process by using specialized nodes of the UML activity diagram – branching actions (wide stripes in the figure). After such nodes there is an implied possibility of parallel execution of processes in different threads. According to the model, the software lifecycle is completed only after the system is decommissioned. Otherwise, the operation of the system continues. During operation telemetry about the state of the system is collected. These can be both performance metrics, such as CPU and memory usage, and application performance metrics, such as the number of service failures when accessing the service or the time when the application executes specified scenarios. Based on these metrics, a decision is made about adding resources to operation environments, or about the need to optimize the code for performing certain operations. Also, during the operation of the system, various incidents may occur – an unplanned interruption of the system or a deterioration in the quality of the service provided by the system. These incidents are then sorted out in order to find the cause of their occurrence. If the cause of the incident is a software defect, the data for its correction is transmitted to the developers.

3 Conclusion

The model proposed in the article allows us to solve the main problems associat-ed with ensuring the continuous integration and deployment of engineering software. Problems with testing integration with related systems are reduced by using system testing environment. Problems that can arise when dealing with a rich client are solved by using automatic update modules and telemetry transmission modules. Problems with a large number of degrees in the test data are reduced by using data from the industrial operation environment.

As a further elaboration of the research topic, it is planned to conduct simulation modeling of the proposed model.

Acknowledgments. The study was supported by a grant from the President of the Russian Federation for state support of leading scientific schools of the Russian Federation (NSh-2553.2020.8).

References

1. Fillinger, S., Bonart, H., Welscher, W., Esche, E., Repke, J.-U.: Improving interoperability of engineering tools — data exchange in plant design. Chemie Ingenieur Technik **89**(11), 1454–1463 (2017). https://doi.org/10.1002/cite.201700032
2. Salehi, V., Wang, S.: Munich agile MBSE CONCEPT (MAGIC). Proc. Des. Soc. Int. Conf. Eng. Des. **1**(1), 3701–3710 (2019). https://doi.org/10.1017/dsi.2019.377
3. Beer, J., Depew, C.: The Role of Process Engineering in the Digital Transformation. Comput. Chem. Eng. (2021). https://doi.org/10.1016/j.compchemeng.2021.107423
4. Lupinetti, K., Cabiddu, D., Giannini, F., Monti, M.: CAD3A: a web-based application to visualize and semantically enhance CAD assembly models. In: 15th International Conference on Signal-Image Technology & Internet-Based Systems (SITIS), pp. 462–469 (2019). https://doi.org/10.1109/SITIS.2019.00080
5. Pan, W., Langosch, K., Bock, T.: Development of the process information modelling in the construction project: a case study of the ZERO-PLUS project. In: Proceedings of the 34rd ISARC. Taipei, Taiwan (2017). https://doi.org/10.22260/ISARC2017/0108
6. Forsgren, N., Humble, J., Kim, G.T.: Accelerate: The Science of Lean Software and DevOps: Building and Scaling High Performing Technology Organizations. IT Revolution Press (2018)
7. Eberhard, W.T.: A Practical Guide to Continuous Delivery, p. 288. Addison-Wesley Professional (2017)
8. Manifesto for Agile Software Development. http://agilemanifesto.org/. Accessed 28 July 2021
9. Golosovskiy, M.: Information and logical model of the software development process. Soft. Syst. Comput. Methods **1**, 59–68 (2015)
10. Golosovskiy, M.: Comparative analysis of software life cycle models with various ways of organizing work flows based on the results of simulation modeling. Cloud Sci. **4**(4), 676–690 (2017)
11. ISO/IEC 12207: Systems and software engineering — Software life cycle processes (2008)
12. Combemale, B., Wimmer, M.: Towards a model-based DevOps for cyber-physical systems. In: Bruel, J.-M., Mazzara, M., Meyer, B. (eds.) DEVOPS 2019. LNCS, vol. 12055, pp. 84–94. Springer, Cham (2020). https://doi.org/10.1007/978-3-030-39306-9_6
13. The 2019 Accelerate State of DevOps: Elite performance, productivity, and scaling. DORA GoogleCloud (2019)
14. 2021 State of DevOps Report. Puppet (2021)
15. Törngren, M., Sellgren, U.: Complexity challenges in development of cyber-physical systems. In: Lohstroh, M., Derler, P., Sirjani, M. (eds.) Principles of Modeling. LNCS, vol. 10760, pp. 478–503. Springer, Cham (2018). https://doi.org/10.1007/978-3-319-95246-8_27

16. Soldatov, E., Bogomolov, A.: Decision support models and algorithms for remote monitoring of the equipment state. CEUR Workshop Proceedings, 2843, (2021). http://ceurws.org/Vol-2843/shortpaper016.pdf
17. Larkin, E., Bogomolov, A., Privalov, A.: Data buffering in information-measuring system. In: 2nd International Ural Conference on Measurements (UralCon), pp.118–123 (2017). https://doi.org/10.1109/URALCON.2017.8120697

Malicious Emails Classification Based on Machine Learning

Eva Marková, Tomáš Bajtoš, Pavol Sokol(✉), Terézia Mézešová, and Patrik Pekarčík

Institute of Computer Science, Faculty of Science, Pavol Jozef Šafárik University in Košice, Jesenná 5, 040 01 Košice, Slovak Republic
{eva.markova,tomas.bajtos,pavol.sokol,terezia.mezesova, patrik.pekarcik}@upjs.sk

Abstract. An inherent part of everyday life and work on a computer is ownership and usage of the email address. This paper aims to analyze existing approaches to the classification of malicious emails into four subcategories: spam, scam, phishing, and emails that contain malware. First, we prepared a labelled data set. Next, we extracted 34 features from emails included in the data set, and then we found a set of more effective features to use. Finally, we have evaluated four supervised machine learning methods (Random Forest, Decision Tree, Support Vector Machines, k-Nearest Neighbors). According to our results, the Random Forest is the most suitable approach for email classification if accuracy is the most important value.

Keywords: Malicious email · Phishing · Spam · Machine learning

1 Introduction

It is well known that humans are the weakest link in the security of organizations. Attackers use current events worldwide and manipulate people's fears with social engineering techniques; as for many, this is an easier attack vector to execute than more technical exploits. The majority of security incidents nowadays involves some social engineering component, and organizations are trying to improve the awareness of their employees. However, identification of phishing emails is difficult also for aware individuals due to their diversity. The widespread phishing emails are an emerging threat to organizations now more than ever, with more of them adapting to epidemiological restrictions, and digital-first is becoming new normal [13]. Spam and phishing had many current events to tie to in 2020, especially in the first quarter [16]. Attackers often get credentials to various services or even bank details through malicious emails, which may contain fraudulent links. They also may require users to install some malicious software that can monitor their activity on the computer. Many email service providers can report an email as junk, but the users do not have a universal option to verify if an email is legitimate.

R. Silhavy et al. (Eds.): CoMeSySo 2021, LNNS 231, pp. 797–810, 2021.
https://doi.org/10.1007/978-3-030-90321-3_65

This paper is based on research [9], in which authors described how to implement an automated service for the users that would classify received emails and automatically reply with a tailored response based on the resulting classification. It is not enough today to distinguish just between legitimate and illegitimate emails. Even as users, we can observe various types of fraudulent emails. Each poses a different risk to the organization, and security teams need to adjust their approach when reacting to any security incidents stemming from such emails.

There are four main categories of illegitimate email that we considered: spam, scam, phishing and emails which contain malware. These labels come directly from the labels that Office 365 Quarantine functionality uses. Spam is any irrelevant or unsolicited message, mostly advertisement that is sent over the Internet. A scam is a popular form of fraud in which an attacker may convince a victim to pay a certain amount of money and promises a reward. Phishing contains links to malicious websites that appear legitimate or encourages one to click on a link or attempt to retrieve sensitive information. Finally, emails with malware are those that contain malicious code. It is complicated to find characteristics of emails that clearly distinguish these categories (e.g. in a scam, it is needed to look at the content of an email because it does not contain any suspicious links or any attachment).

Machine learning offers various classification methods that learn to assign a class label from already labelled examples. The training examples represent the problem well and contain multiple examples of each class. Each example is described by several features, and the model processes which features distinguish the most between the individual classes. It is interesting what features we should consider if we want to distinguish between categories of malicious emails. We chose to extract as many features as possible and discard those, which are irrelevant or worsen the accuracy results. We aimed to identify features that are suitable for different classification methods.

This paper compares and discusses the performance of four different classifier algorithms applied for the email classification problem. We evaluate the performance of a self-collected data set of 1000 emails. We compared Random Forest, Decision Tree, Support Vector Machines and k-Nearest Neighbors classifiers. The results show Random Forest achieves the best performance.

The main goal of this paper is to distinguish between malicious categories of emails and extract features, which would achieve the best results in accuracy and time spent for classification. We state the following research sub-goals:

- comparison of the performance of methods for classification of malicious emails, and
- find a set of attributes which would have the best results from as many as possible considered features.

This paper is organized into five sections. Section 2 focuses on the review of published research related to methods to detect and classify malicious emails. Section 3 focuses on the research methodology. Section 4 describes the processing of the emails. Section 5 contains results and discussion, and the last section contains the conclusion and our suggestions for the future research.

2 Related Works

Several approaches can detect and then classify malicious emails—for example, Natural Language Processing (NLP), blacklists, data mining algorithms, and others. In most existing solutions for detection, data mining algorithms are used because they turned out to be the most effective. Table 1 provides a comparison of the related works. We focused on the features that they extracted to classify emails and also the methods that they used. They mostly focused on comparing a few approaches to detect malicious emails and design and implement a system. First, we compared which features and methods the authors used in their papers. Then, as part of a more detailed analysis, we focused on what features authors categorized email. In most of the papers, they have focused on URL-based features because malicious emails often contain fraudulent links or lead to a phishing site, such as the number of URLs, the maximum number of dots in a URL, etc. Other types of attributes are header-based, such as sender domain, send time, domain address existence etc., which we can find in the email header. Body-based features are known as the most important. They include, for example, whether an email contains JavaScript, an HTML form or an attachment etc. Text attributes are essential to distinguish scam from other categories. Therefore, it is necessary to focus on the presence of some important words.

Table 1. Comparison of related works

Features	Methods				
	NLP	SVM	RF	DT	Other[a]
Textual	[12]	[2,3,21], [4,6]	[18,21], [1,3]	[1,21], [3,18]	[1,21], [2,4,14]
URL-based	[12]	[6,21]	[5,21]	[17,19,21]	[21]
Body-based	–	[6,21]	[5,21]	[17,19,21]	[21]
Header-based	–	[6]	–	[17,19]	–

[a]Other methods - k-NN, neural networks, Naive Bayes methods etc.

In [1,3,17,18,21], the authors tested and compared well-known classification techniques, e.g. Random Forest, Support Vector Machine and others. The best results were mostly achieved with Random Forest. Subasi et al. [18] used Decision Tree algorithms for spam email filtering. Form et al. [6] developed a system that handles data that are URL-based and content-based and also monitors features that can change the behaviour of the computer. Toolan and Carthy, in their research in [19] tried to find a suitable set of extracted features that would give the best results. Another system that was developed was called PFILFER [5], whereas the authors omitted headers. In [2], authors compared the performance of the regularized neural network on three spam data sets with four machine learning algorithms. Authors in [4] proposed a Support Vector Machine based on a naive Bayes filtering system. Saberi et al. [14] used a set of classifiers to detect scam emails, whereas features used in this work are all textual.

Related works focused mainly on the division of emails into two categories - phishing and non-phishing emails, which is easier to solve. We have considered more categories of malicious emails than only one - spam, scam, phishing and email that contains malware. Some works also considered searching for an effective set of features, that is a problem which we find interesting. We have extracted features from emails based on this literature.

3 Methodology

3.1 Data Set

For our research, it was necessary to create a data set that would meet the critical requirement of containing a vast number of examples from each class representing the variety in each category. It is also recommended that the number of examples used for training is larger than how many features are used to describe each example. We collected 1000 emails, 250 per category, from our organization's Office 365 quarantine. Office 365 itself marked these emails in various categories, and we subsequently checked and, if necessary, adjusted this classification. These labels were used for the training phase. A side objective of this paper was also to create a large data set of malicious emails that are composed of different categories and can be used for different research purposes. The data set was collected over 2.5 months between May and July 2020. Despite the fact, that the emails are from one organization and from a certain time frame, the emails are general, because spear-phishing emails were removed. The training sample and the learnt models, therefore, reflect current trends and are suitable for deployment. The data set was anonymized with several techniques - prefix-preservation of IP address, substitution with fake names of organizations or real-life persons, removal of DKIM signatures.

3.2 Selected Algorithms

We used the implementation of Random Forest (RF), Support Vector Machines (SVM), k-Nearest Neighbors (k-NN), and Decision Tree (DT) that is provided in the scikit-learn library [11]. We split the collected data set into two parts - the training sample, which was 80% of the collected emails (200 emails per category) and the testing sample, which was the remaining 20%. We ran each algorithm several times with different features, each time recording the time spent on learning the classification model and accuracy (ratio of correct predictions to a total number of examples) to express the performance of ML models.

RF is a computationally efficient technique that can quickly work with large data sets. It brings precise classification, a new way of determining the variable meaning and the ability to model complex interactions between variables. It is very flexible in performing several types of statistical data analysis, including regression, classification and others [8].

The principle of **SVM** [20] is the division of training examples into two classes, but the division into more classes than two is not excluded. The goal is to learn the classification rule from the training set to assign a specific class to all new subjects in the future. This approach has been well developed for binary classification. It divides the area into two classes and determines which object belongs to which class. Important is how the SVM works for $k > 2$, where k is the number of classes in the data set. One way to solve the problem is to divide it into a set of binary classification problems. For example, the k-th classifier constructs a hyperplane between the last and $k - 1$ other classes. Alternatively, $\frac{k(k-1)}{2}$ hyperplanes may be constructed, separating each class from the other. These methods are called "One-against-the-Rest" and "One-against-One".

The k**-NN** algorithm is a non-parametric classification technique which is effective in statistical recognition applications [7]. This algorithm is one of the simplest and effective methods for classification. It calculates the distance between the test sample and all existing samples in the training set. It creates a decision based on its k-nearest samples after applying the weight function. Creating predictions with a small data set also has disadvantages. The distance between each new email and all existing emails should be calculated before making a decision, which is computationally difficult for many emails.

In the structure of a **DT**, each inner node indicates a test on a feature, and each branch represents the result of the test data, and each leaf node identifies a particular class. The highest node in the tree is the root node. There are various algorithms for creating a Decision Tree, such as CART, ID3, C4.5 etc. These are greedy approaches in which the Decision Tree is constructed from top to bottom [15]. We chose the CART algorithm in the scikit-learn library.

3.3 Extracting Attributes from Email

The main objective of this paper is to find a set of features that would be appropriate for each of the used methods concerning the accuracy of the classification and learning time. In total, we have extracted eight information from headers, six body-based features, 13 information about links in email, one script-based feature and six textual features. Sixteen of these features are binary, and 18 are numeric - together, we have extracted 34 features from every email. Table 2 summarizes the features and also what category and type these features are.

There were 8 features extracted from emails' headers:

- **(1) diff_sender_reply _address** [6] - whether sender email address is different from reply address (return_path equals sender);
- **(2) not_work_time** [10] - sending time of email is not in normal work time which is 8AM to 5PM (UTC time zone);
- **(3) diff_domains** [10,17] - if there is difference between the sender's domain and the domain of embedded links;
- **(4) mime_format** [7] - version of MIME;
- **(5) num_receivers** [17] - number of receivers;

- **(6) domain_returnpath_sender** [6] - whether the domain name of return path is the same as the domain name of sender;
- **(7) text_email** [17] - whether Content-Type is "text/plain";
- **(8) html_email** [5,17,19,21] - whether Content-Type is "html".

There are 6 relevant body-based features which should be considered:

- **(9) contains_form** [17,19] - whether email contains HTML form;
- **(10) num_ characters_body** [19] - number of characters in body;
- **(11) num_words_body** [19] - number of words in body;
- **(12) richness_of_body** [19] - number of words divided by number of characters in body;
- **(13) email_size** [17] - size of email may indicate if there is an attachment or malicious code;
- **(14) num _attachments** [17] - number of attachments - if there is attachment, it may be malicious.

Malicious emails often contain fraudulent links and it can be a good indication that they are malicious emails. We have looked at 13 features related to URL links:

- **(15) num_URLs** [5,10,12,17] - number of URLs, because malicious emails often contain multiple URL links to fake websites;
- **(16) max_num_dots** [5,6,10,17,21] - maximum number of dots, because many attackers try to construct a legitimate-looking URL link by adding a few already known domains;
- **(17) num_IP_URLs** [5,6,10,17,19] - number of IP URLs are important, because some malicious websites may be hosted on computers that do not have DNS records;
- **(18) num_image_URLs** [17,19,21] - number of image URLs, which are found in the `<a href="">` tag;
- **(19) num_URLs_with_at** [6,10,17,19] - number of URLs with @ - if the URL contains @, the web browser will ignore the part before this symbol and retrieve the address that follows after;
- **(20) num_URLs_with_percent** [6] - number of URLs with % - an attacker can encode a part of a link with hexadecimal characters to hide a fraudulent URL;
- **(21) num_URLs_with_nonP** [17,19] - number of URLs with nonstandard port - we determine the number of URLs with a port different from 80 (http) and 443 (https);
- **(22) max_URL_length** [10] - maximum URL length - attackers often use long URL links to hide the dubious part;
- **(23) num_different** [5,6,10,17,19,21] - number of URLs, that are different for the `<a href="">` tag and the text after it - if there is a difference between the value in the `<a href="">` attribute and the text after it, we suspect it may be a malicious link;
- **(24) num_domains** [5,17,19,21] - maximum number of domains in URLs;

- **(25) num_URLs_with_dash** [10] - number of URLs with dash used in it;
- **(26) http_more_than_one** [10] - whether number of http/https in URL is greater than 1;
- **(27) shortened_URL** [5,10] - whether URL is using shortening service.
It is important to look if in an email is use of script language;
- **(28) contains_script** [5,10,17,19] - whether an email contains tag `<script>`.

Scam emails do not tend to contain any links, they are rather text-heavy. Therefore we need to look at the text features of the emails as well:

- **(29) body_words_spam** [6,10,19,21] - whether body contains one of these words: "delivery", "mail", "message", "sender", "your", "returning", "failed", "invoice", "images", "scanned";
- **(30) body_words_scam**, [6,10,19,21] - whether body contains one of these words: "payment", "urgent", "request", "attention", "important", "confidential", "immediate response", "transfer", "important update", "attn";
- **(31) url_words**, [5,6,10,21] - whether URL contains "click", "here", "login", "update";
- **(32) subject_words** [6,10,19,21] - whether subject contains "bank", "verify", "debit";
- **(33) is_a_reply** [10,19] - whether email is a reply;
- **(34) is_forwarded** [19] - whether email is forwarded.

Table 2. Features extracted from the emails and their type. NOTES: B - binary type, N - number type, SB - Script-based.

Feature	Category	Type	Feature	Category	Type
diff_sender_reply_address	Headers	B	num_image_URLs	URL	N
not_work_time	Headers	B	num_URLs_with_at	URL	N
diff_domains	Headers	B	num_URLs_with_percent	URL	N
mime_format	Headers	B	num_URLs_with_nonP	URL	N
num_receivers	Headers	N	max_URL_length	URL	N
domain_returnpath_sender	Headers	B	num_different	URL	N
text_email	Headers	B	num_domains	URL	N
html_email	Headers	B	num_URLs_with_dash	URL	N
num_attachments	Body-based	N	http_more_than_one	URL	N
contains_form	Body-based	B	shortened_URL	URL	B
num_characters_body	Body-based	N	contains_script	SB	B
num_words_body	Body-based	N	body_words_spam	Textual	B
richness_of_body	Body-based	N	body_words_scam	Textual	B
email_size	Body-based	N	url_words	Textual	B
num_IP_URLs	URL	N	subject_words	Textual	B
num_URLs	URL	N	is_a_reply	Textual	B
max_num_dots	URL	N	is_forwarded	Textual	B

3.4 Research Method

We searched for relevant attributes in two main phases, recorded the time spent on the learning, and evaluated the accuracy value for each run.

In the first phase, we extracted the 34 attributes that are presented in Table 2 and ran one round of learning for each chosen machine learning algorithm with all these 34 attributes. The recorded run time of the algorithms and the evaluated accuracy values became the benchmark for the next phase of the research. The second phase consisted of 34 parallel rounds of learning for each chosen algorithm. One parameter was removed in each of the 34 rounds, and the learning time and accuracy were recorded.

Our overall objective was to evaluate which attributes are meaningful for email classification and how each attribute affects the benchmark values for learning time and accuracy, i.e. if they improve or worsen. Using these results, we constructed sets of suitable attributes for each machine learning algorithm to get the best accuracy & learning time trade-off.

An observation of improved accuracy compared to the baseline means that we get better accuracy without this attribute. Therefore, it should not be in the final set of suitable features. The sets of suitable features were constructed by omitting the features that improved the accuracy metric and kept the features with worse or unchanged accuracy. In the final set of suitable features for any machine learning methods, we included features that achieve accuracy higher than the median value of maximum accuracy values achieved for each method.

In a third phase, we ran a round of learning for each algorithm with these sets:

- set of features with worse or unchanged accuracy observation - these are so-called; **improving and unchanging features** because when a feature from this set is skipped, the accuracy was worse than baseline;
- set of features with worse accuracy observation (i.e. **improving features**);
- set of features selected as most suitable for all machine learning methods (i.e. **chosen features**) - the whole set is listed in Sect. 4.2.

4 Results and Discussion

In this section, we present and discuss the second phase results, compare the results for the machine learning methods, and present the resulting sets of features. Table 3 provides an overview of achieved results of accuracy for each machine learning algorithm. The first row marked as "All features used" shows the first phase results, where the algorithms were run with all 34 features. Rows "Maximum accuracy" and "Minimum accuracy" show the respective observed maximum and minimum value from the 34 rounds of running the algorithm each time, with one feature removed in the second phase. The last three rows show the values after running a round of learning for each machine learning algorithm on the sets created in the third phase.

Table 3. Accuracy of methods

	RF	DT	SVM	k-NN
All features used	83.00%	65.50%	71.00%	66.50%
Maximum accuracy	84.50%	75.50%	71.50%	73.50%
Minimum accuracy	80.00%	61.00%	66.00%	63.50%
Average	82.22%	69.18%	69.79%	66.54%
Improving and unchanging features	82.00%	60.50%	70.50%	78.50%
Improving features	82.00%	51.50%	70.50%	76.50%
Chosen features	**81.00%**	**75.00%**	**66.00%**	**77.50%**

We can see that the best results regardless of features are achieved with the Random Forest algorithm. Using our set of chosen features with RF, we close the average of the accuracy values. However, Decision Tree and k-Nearest Neighbors reach much better results (75% and 77.5% respectively) than what the average for those methods is (69.18% and 66.54% respectively). k-NN had the most significant improvement when we used improving and stable features, but our chosen features accuracy is 1% worse than the best results. Support Vector Machines achieved its lowest accuracy with our set of selected features (66%). Its best accuracy (71.5%) was achieved when the attribute diff_sender_reply_address was skipped, and the other 33 were included. The sets were improving, and unchanging features or improving features achieved better than average accuracy (both 70.5%).

In Table 4 we can see an overall comparison of the accuracy with the chosen features against all features.

Table 4. Comparison of methods while using all 34 features and the chosen ones

Algorithm	Accuracy (all features)	Accuracy (chosen features)
RF	83.00%	81.00%
DT	65.50%	75.00%
SVM	71.00%	66.00%
k-NN	66.50%	76.00%

Concerning the learning time, the recorded values in seconds are provided in Table 5. We can see that the RF, even though the accuracy dropped from 83% with all features to 81% with just the final set of chosen features, the algorithm ran almost two times faster (drop from 0.2954 s to 0.1596 s). The cost of achieving higher accuracy with RF than with other algorithms is the time spent on learning. Table 5 shows that it is significantly slower on the sets of improving features and chosen features than the other algorithms.

SVM ran more than three times faster with our chosen features, but the accuracy is the lowest. However, SVM runs two times faster when compared to baseline while staying close to the average accuracy with the set of improving features.

k-NN is almost two times faster with the set of chosen features. It also achieves the second-highest accuracy. On the other hand, DT was the slowest with our selected features, but the difference between the quietest and fastest run is 0.0050 s.

Table 5. Learning time of methods in seconds

	RF	DT	SVM	k-NN
All features used	0.2954	0.0100	0.0957	0.0219
Maximum accuracy	0.2954	0.0110	0.6860	0.0269
Minimum accuracy	0.1574	0.0070	0.0577	0.0169
Average	0.1670	0.0082	0.1171	0.0186
Improving and unchanging features	0.1626	0.0120	0.0559	0.0219
Improving features	0.1521	0.0090	0.0479	0.0160
Chosen features	**0.1596**	**0.0120**	**0.0309**	**0.0170**

In Table 6 we show a comparison of accuracy, the learning time in seconds and which algorithm achieves this combination. We show only the accuracy for the higher algorithms than the median accuracy value, which resulted in 77%. If we need classification to be fast, we recommend using k-NN because the learning time is almost eight times faster than in every Random Forest's classification. On the other hand, if the classification needs to be precise, then Random Forest is recommended.

Table 6. Learning time and accuracy of algorithms

	Accuracy	Learning time	Algorithm
All features used	83.00%	0.2954	RF
Maximum accuracy	84.50%	0.2954	RF
Minimum accuracy	80.00%	0.1574	RF
Average	82.22%	0.1670	RF
Improving and unchanging features	82.00%	0.1626	RF
Improving features	82.00%	0.1521	RF
Improving features	78.50%	0.0219	k-NN
Chosen features	81.00%	0.1596	RF

4.1 Omitted Features

The results in the second phase showed that the attribute **html_email** when omitted, it worsened the accuracy regardless of the machine learning algorithm - it means it is suitable for the fraudulent email classification problem. We included it in the set of **chosen features**. Additionally, there are also features that when omitted the accuracy was unchanged or better with each algorithm, this suggests that they are not suitable at all. The attributes that behave this way are **num_URLs_with_at**, **num_URLs _with_percent**, **num_URLs_with_nonP** and they were not included in the **chosen features** set.

In Table 7, we present which features should be omitted for RF, DT, and *k*-NN. This table shows how is accuracy percentually improved when we omitted various features. In the case of the SVM, we have only two cases, in which SVM did have better results (diff_sender_reply_address and body_words_scam). The accuracy of RF when we used all features was 83.0%. RF had the best results (+1.5%) when we omitted num_characters_body. DT with all features had the accuracy of 65.5%. We achieved the highest accuracy (+10.0%) when we omitted num_different. Using all features we achieved with *k*-NN the accuracy of 66.5%. *k*-NN had the highest accuracy (+7.0%) when we omitted body_words_scam.

Table 7. Omitted features

Feature	RF	DT	*k*-NN	Feature	RF	DT	*k*-NN
num_URLs	—	1.5	—	domain_returnpath_sender	—	1.5	—
max_num_dots	1.0	1.0	—	text_email	—	2.0	—
num_IP_URLs	—	6.0	—	contains_form	—	1.5	—
num_URLs_with_at	0.5	3.0	—	num_characters_body	1.5	9.0	0.5
num_URLs_with_percent	0.5	2.0	—	num_words_body	—	3.0	0.5
num_URLs_with_nonP	0.5	7.5	—	richness_of_body	0.5	—	—
max_URL_length	—	5.0	—	num_domains	0.5	5.0	—
email_size	—	5.0	—	num_URLs_with_dash	0.5	5.0	—
num_different	—	10.0	—	http_more_than_one	—	5.0	—
num_attachments	—	4.0	—	shortened_URL	—	6.5	—
contains_script	—	2.5	—	body_words_spam	—	4.0	0.5
not_work_time	—	9.5	1.5	body_words_scam	—	7.5	7.0
diff_sender_reply_address	—	—	1.5	URL_words	—	2.0	1.0
domain_sender_links	—	6.5	—	subject_words	—	2.0	—
mime_format	—	8.0	—	is_a_reply	—	—	6.0
num_receivers	—	4.5	—	is_forwarded	—	5.5	10

4.2 Set of Chosen Features

When constructing the set of chosen features, we looked at the sets of remaining features for each algorithm and included them in the chosen features set if it was in the list of remaining features for at least three algorithms. The chosen features set includes the features:

- num_URLs;
- num_IP_URLs;
- max_URL_length;
- email_size;
- num_different;

- num_attachments;
- contains_script;
- diff_sender_reply_address;
- mime_format;
- num_receivers;
- domain_returnpath_sender;
- text_email,
- html_email;
- contains_form,
- richness_of_body;
- http_more_than_one,
- shortened_URL;
- subject_words; and
- is_reply.

5 Conclusion

The operational activities of security teams constantly confirm the growing trend of sophisticated fraudulent email messages. However, the art in the area of fraudulent email classification distinguishes between legitimate and malicious emails. This paper focuses on the next step in the analysis - classifying the malicious emails into different categories. The attribution of this work is the labelled data set of malicious emails and the effectiveness test of four data mining algorithms.

We analyzed the features from the reviewed literature on the four most commonly used classification methods. An interesting finding in this regard was that some features improve the classification results in all cases (e.g. number of URLs with the @ symbol), some worsen in all cases (e.g. whether Content-Type is "html"), and the impact of some depends on the classification method used. We have identified a set of features that improved either accuracy or time spent to classify. We launched the classification with all 34 proposed features and then omitted one feature in each round. After comparing the accuracy of each round, we chose 19 features that are suitable for each of the selected machine learning algorithms.

Furthermore, we compared the individual classification methods based on selected features. It turned out that the Random Forest is the best choice to classify malicious emails. Regardless of features, it always gave us an accuracy higher than 80%, but if we use our set of chosen features, time spent to classify is two times better than if we use all features. Furthermore, we selected features that bring the accuracy of other machine learning algorithms close to the one achieved by Random Forest while also keeping the time for learning low. As a result of this paper, we identified two open questions that are interesting problems for future work: looking into why ML methods, after removing all features that worsen accuracy, had worse results than before removing them and then which features better describe a particular category.

Acknowledgments. This research is funded by the Slovak Research and Development Agency project under contract No. APVV-17-0561, project No. VVGS-PF-2020-1436, and project No. VVGS-PF-2020-1427.

References

1. Abu-Nimeh, S., Nappa, D., Wang, X., Nair, S.: A comparison of machine learning techniques for phishing detection. In: Proceedings of the Anti-Phishing Working Groups 2nd Annual eCrime Researchers Summit, pp. 60–69. ACM (2007)
2. Barushka, A., Hájek, P.: Spam filtering using regularized neural networks with rectified linear units. In: Adorni, G., Cagnoni, S., Gori, M., Maratea, M. (eds.) AI*IA 2016 Advances in Artificial Intelligence. Conference of the Italian Association for Artificial Intelligence, pp. 65–75. Springer, Cham (2016). https://doi.org/10.1007/978-3-319-49130-1_6
3. Diale, M., Celik, T., Van Der Walt, C.: Unsupervised feature learning for spam email filtering. Comput. Electr. Eng. **74**, 89–104 (2019)
4. Feng, W., Sun, J., Zhang, L., Cao, C., Yang, Q.: A support vector machine based naive bayes algorithm for spam filtering. In: IEEE 35th International Performance Computing and Communications Conference (IPCCC), pp. 1–8. IEEE (2016)
5. Fette, I., Sadeh, N., Tomasic, A.: Learning to detect phishing emails. Technical report, Carnegie Mellon University (2006)
6. Form, L.M., Chiew, K.L., Tiong, W.K., et al.: Phishing email detection technique by using hybrid features. In: 2015 9th International Conference on IT in Asia (CITA), pp. 1–5. IEEE (2015)
7. Gascon, H., Ullrich, S., Stritter, B., Rieck, K.: Reading between the lines: content-agnostic detection of spear-phishing emails. In: Bailey, M., Holz, T., Stamatogiannakis, M., Ioannidis, S. (eds.) Research in Attacks, Intrusions, and Defenses. International Symposium on Research in Attacks, Intrusions, and Defenses, pp. 69–91. Springer, Cham (2018). https://doi.org/10.1007/978-3-030-00470-5_4
8. Genuer, R., Poggi, J.M., Tuleau-Malot, C., Villa-Vialaneix, N.: Random forests for big data. Big Data Res. **9**, 28–46 (2017)
9. Marková, E., Bajtoš, T., Sokol, P., Mézešová, T.: Classification of malicious emails. In: 2019 IEEE 15th International Scientific Conference on Informatics, pp. 279–284. IEEE (2019)
10. Niu, W., Zhang, X., Yang, G., Ma, Z., Zhuo, Z.: Phishing emails detection using CS-SVM. In: 2017 IEEE International Symposium on Parallel and Distributed Processing with Applications and 2017 IEEE International Conference on Ubiquitous Computing and Communications (ISPA/IUCC), pp. 1054–1059. IEEE (2017)
11. Pedregosa, F., et al.: Scikit-learn: machine learning in Python. J. Mach. Learn. Res. **12**, 2825–2830 (2011)
12. Peng, T., Harris, I., Sawa, Y.: Detecting phishing attacks using natural language processing and machine learning. In: 2018 IEEE 12th International Conference on Semantic Computing (ICSC), pp. 300–301. IEEE (2018)
13. Reiss, R.: After Q1 2020, digital transformation is no longer an option, March 2020. https://www.forbes.com/sites/robertreiss/2020/03/30/after-q1-2020-digital-transformation-is-no-longer-an-option/#5efcbf76584c
14. Saberi, A., Vahidi, M., Bidgoli, B.M.: Learn to detect phishing scams using learning and ensemble? methods. In: Proceedings of the 2007 IEEE/WIC/ACM International Conferences on Web Intelligence and Intelligent Agent Technology-Workshops, pp. 311–314. IEEE Computer Society (2007)

15. Sharma, A.K., Prajapat, S.K., Aslam, M.: A comparative study between Naïve bayes and neural network (MLP) classifier for spam email detection. Int. J. Comput. Appl. (2014)
16. Shcherbakova, T., Shcherbakova, T., Sidorina, T., Kulikova, T., K.P.o.J., *, N.: Spam and phishing in Q1 2020, May 2020. https://securelist.com/spam-and-phishing-in-q1-2020/97091/
17. Smadi, S., Aslam, N., Zhang, L., Alasem, R., Hossain, M.: Detection of phishing emails using data mining algorithms. In: 9th International Conference on Software, Knowledge, Information Management and Applications, pp. 1–8. IEEE (2015)
18. Subasi, A., Alzahrani, S., Aljuhani, A., Aljedani, M.: Comparison of decision tree algorithms for spam e-mail filtering. In: 2018 1st International Conference on Computer Applications & Information Security (ICCAIS), pp. 1–5. IEEE (2018)
19. Toolan, F., Carthy, J.: Feature selection for spam and phishing detection. In: 2010 eCrime Researchers Summit, pp. 1–12. IEEE (2010)
20. Weston, J., Watkins, C.: Multi-class support vector machines. Technical report, Citeseer (1998)
21. Yasin, A., Abuhasan, A.: An intelligent classification model for phishing email detection. arXiv preprint arXiv:1608.02196 (2016)

On Determining the Position of Expert Confidence Concentration Points When Modifying Membership Functions

Nickolay Barchev[1] and Vladimir Sudakov[2](✉)

[1] Plekhanov Russian University of Economics, Stremyanny Lane, 36, Moscow 117997, Russia
[2] Moscow Aviation Institute, Volokolamskoe Shosse, 4, Moscow 125993, Russia
sudakov@ws-dss.com

Abstract. The apparatus of fuzzy sets membership functions is a powerful tool widely used to express expert preferences when solving problems in various subject areas. Algorithmic ways to modify membership functions without additional involvement of experts are common modern means of obtaining new functions on the basis of the existing ones. This paper investigates methods of possible resolution of ambiguity in determining the position of the middle point of expert confidence concentration of triangular membership functions in cases of their modification with the application of algorithms that use the reference functions. The obtained membership functions of fuzzy sets can be used in modeling the behavior of agents under conditions of substantial uncertainty caused by COVID–19.

Keywords: Expert confidence · Membership function · Fuzzy set · Lexical variable · Linear transformations

1 Introduction

Modern practice of applying the apparatus of the theory of fuzzy sets in applied problems quite often involves the use of mechanisms for automating the construction of membership functions oriented to obtain new functions by applying to the existing ones some transformations usually called modifiers [1–6].

The use of such mechanisms allows to significantly reduce the amount of required expertise, thus reducing the cost of engaging appropriate specialists. In addition, it provides unification of the construction process of membership functions related to fuzzy sets, which are close to each other in one sense or another.

Substantial approach to work with membership functions modifiers implies their primary use in situations of significant linguistic relatedness of the relevant concepts formalized by the considered fuzzy sets. In particular, such a relatedness can be linguistically expressed by means of "very/not very" transitions and similar ones [2, 13]. In such cases the applied transformations should formalize the dynamics of changes in expert judgments regarding the degree of correspondence of certain values of the membership function definition area to some concept when approaching the areas in which the function is equal to one or zero, as well as when moving away from these areas [13].

R. Silhavy et al. (Eds.): CoMeSySo 2021, LNNS 231, pp. 811–817, 2021.
https://doi.org/10.1007/978-3-030-90321-3_66

A description of the corresponding dynamics can be performed quite easily using some "benchmark" (prototype), which is a pair of pre-selected known membership functions. One of the functions, meanwhile, plays the role of the initial (prototype), and the other plays the role of the resulting (image) [2, 13].

The dynamics of changes in expert judgments in the transition from the initial function to the resulting one at different sections of function changes is generally different, so the application of specific transformations in the methods that take into account this fact [13], can be performed separately for the corresponding sections of changes in the membership functions. The consequence of this is an ambiguous in general case determination of the positions of the boundary points separating the sections of the resulting accessory functions which have different dynamics of changes in expert judgments.

This circumstance presents a rather serious problem of not only technical, but also of fundamental nature, since the neighborhoods of such boundary points are, as a rule, the areas of expert confidence concentration. When using expert methods of constructing membership functions, significant, often abrupt changes in the behavior of functions occur precisely in these areas.

Thus, automating the construction of membership functions using the methods that apply the indicated approach implies the use of some ways to resolve the above ambiguity, which requires appropriate research. In particular, this is relevant for triangular membership functions, which often serve as a means of formalization of expert judgments [1–3, 12, 13].

2 Literature Review

Among the numerous methods for automating the construction of membership functions that take into account the linguistic relatedness of the concepts formalized by fuzzy sets, it is worth mentioning a group of methods that do not use any predefined reference functions. Such methods are often described in the literature [6–11], are common because of their relative simplicity, and reflect linguistic transitions of the "very/not very" type by means of one-time defined formal mathematical transformations applied to the values of the original membership function.

The use of tools of this kind has numerous disadvantages with almost the only advantage, which is simplicity. In particular, many of the corresponding transformations are exclusively heuristic, their application does not have a sufficiently weighty justification, and the results of application often seriously diverge from the actual expert evaluations. An additional side effect is the change in the general case of the type of the obtained membership function with respect to the original one, which is often undesirable.

In our opinion, the approach to automating the construction of membership functions using methods based on the above-mentioned practice of formalizing a particular linguistic transition using a previously defined (found or known) prototype (a pair of reference membership functions in some sense) is more substantively argued [2, 13].

Formalization of a linguistic transition on the basis of an existing prototype can also be performed in different ways – including those that imply obtaining, as a result of formalization, a single transformation to be subsequently applied to the original membership functions of one or another kind. For such automation methods, there is no considered problem of ambiguous determination of the positions of the boundary points separating sections of the resulting membership functions having different dynamics of changes in expert judgments.

Nevertheless, the choice of a particular transformation in one or another situation in such an approach in many cases is also insufficiently reasoned, which complicates the meaningful interpretation of the results obtained. In addition, the practice of applying the corresponding transformations shows that the obtained results of modifications are very sensitive to the choice of initial membership functions and often coincide with the actual expert evaluations only in some special cases [13]. An example of this type of modifications demonstrating the indicated behavior is the method considered in [2, 3] that uses fractional-linear transformations on a set of automorphic functions.

Earlier in [13], we proposed and justified a method for modifying the triangular membership functions, which is based on the above-mentioned principle of a separate application of different in general case transformations for function parts with different dynamics of changes in expert judgments. This method showed expert-acceptable results in most situations of its application and did not demonstrate any results that clearly contradicted the actual expert judgments.

In relation to the triangular membership functions, defined in their domains of definition by triples of divergent numbers:

$$\{x_i \in X; i = 1, 3\} : \mu(x_1) = \mu(x_3) = 0, \mu(x_2) = 1 \tag{1}$$

which are abscissas of expert confidence concentration points, the method involves finding a pair of linear transformations:

$$f_j(x) = a_j x + b_j; j = 1, 2 \tag{2}$$

each of which performs a transformation of one of the two fragments of the original membership function (prototype) into the corresponding fragment of the resulting function (image) [13].

Linear transformations (2) in this case are based on the existing prototype a pair of known membership functions, one of which, $\mu_A(x)$, initially is the original (prototype), and the other, $\mu_B(x)$ – the resulting (image). The calculation of the coefficients a_j is performed by means of relations reflecting the dynamics of changes in expert judgments, expressed by the prototype functions [13]:

$$a_1 = \frac{x_{B_2} - x_{B_1}}{x_{A_2} - x_{A_1}}; a_2 = \frac{x_{B_3} - x_{B_2}}{x_{A_3} - x_{A_2}} \tag{3}$$

$$\left\{x_{A_i} \in X; i = 1, 3\right\} : \mu_A\left(x_{A_1}\right) = \mu_A\left(x_{A_3}\right) = 0, \mu_A\left(x_{A_2}\right) = 1 \tag{4}$$

$$\left\{x_{B_i} \in X; i = 1, 3\right\} : \mu_B\left(x_{B_1}\right) = \mu_B\left(x_{B_3}\right) = 0, \mu_B\left(x_{B_2}\right) = 1 \tag{5}$$

The values of b_j are further found by substituting in (2) the found and previously known values, after which each of the obtained transformations (2) is applied to the corresponding parts of the definition areas of the modified functions.

The application of these transformations leads here in the general case to an ambiguous determination of the position of the expert confidence concentration point, which is the boundary point for the areas with different dynamics of changes in expert judgments (the average in the triple of points defining the desired triangular identity function). If it is necessary to preserve the form of the resulting function with respect to the original membership function, this ambiguity needs to be eliminated.

3 Methods

Modification of some original membership function $\mu_C(y)$ given by a triplet of divergent numbers:

$$\{y_{C_i} \in Y; i = 1, 3\} : \mu_C(y_{C_1}) = \mu_C(y_{C_3}) = 0, \mu_C(y_{C_2}) = 1 \tag{6}$$

using a pair of transformations of the form (2), found on the basis of the known prototype, allows us to obtain in general case two different values for the midpoint y_{D_2} of the resulting function $\mu_D(y)$ of triangular form:

$$y'_{D_2} = f_1(y_{C_2}) = a_1 y_{C_2} + b_1; y''_{D_2} = f_2(y_{C_2}) = a_2 y_{C_2} + b_2 \tag{7}$$

$$\{y_{D_i} \in Y; i = 1, 3\} : \mu_D(y_{D_1}) = \mu_D(y_{D_3}) = 0, \mu_D(y_{D_2}) = 1 \tag{8}$$

In this case, the positions of the extreme points characterizing the desired function are determined uniquely:

$$y_{D_1} = f_1(y_{C_1}) = a_1 y_{C_1} + b_1; y_{D_3} = f_2(y_{C_3}) = a_2 y_{C_3} + b_2 \tag{9}$$

As part of this study, several methods were heuristically reasoned and analyzed to determine the value of the midpoint y_{D_2} of the resulting function $\mu_D(y)$.

Assuming that both found values y'D2 and y"D2 fall inside the interval $[y_{D_1}; y_{D_3}]$ (otherwise, there are expert reasons to doubt the validity of the corresponding modification option for working with the considered membership functions), we can suggest the following methods to obtain the value of y_{D_2}.

Method A. Choosing from the available values y'_{D_2} and y''_{D_2} as the y_{D_2} value obtained using the transformation f_j that provides a more significant transformation of the corresponding fragments of the membership functions. This option responds to the researcher's subjective notion that such a transformation more realistically reflects the expert distinction between the concepts formalized by the fuzzy sets in question.

Since the area of values of the triangular membership functions under discussion is the segment [0; 1], the degree of transformation of the fragments of these functions is determined by the ratio of the lengths of the segments of the abscissa axis, on which the corresponding fragments are based, and is expressed by the coefficients a_j, calculated by means of expressions (3).

Thus, taking into account the possible both increasing and decreasing as a result of the transformation of the lengths of the segments that are the reference for the corresponding fragments of the resulting membership function:

$$y_{D_2} = \begin{cases} y'_{D_2}, max\left(a_1, \frac{1}{a_1}\right) > max\left(a_2, \frac{1}{a_2}\right) \\ y''_{D_2}, max\left(a_1, \frac{1}{a_1}\right) \leq max\left(a_2, \frac{1}{a_2}\right) \end{cases} \tag{10}$$

Method B. The opposite of the previous option is the selection from y'_{D_2} and y''_{D_2} the value obtained by the transformation f_j performing a less significant transformation. This approach expresses the researcher's subjective desire to minimize the expert distinction between the relevant concepts.

Taking into account the above reasoning:

$$y_{D_2} = \begin{cases} y'_{D_2}, max\left(a_1, \frac{1}{a_1}\right) < max\left(a_2, \frac{1}{a_2}\right) \\ y''_{D_2}, max\left(a_1, \frac{1}{a_1}\right) \geq max\left(a_2, \frac{1}{a_2}\right) \end{cases} \tag{11}$$

Method C. Formal averaging of the obtained values y'_{D_2} and y''_{D_2}, reflecting the situation of relative indifference of the researcher:

$$y_{D_2} = \frac{y'_{D_2} + y''_{D_2}}{2} \tag{12}$$

Method D. Determining the value of y_{D_2} as a weighted average of the values y'_{D_2} and y''_{D_2} using transformation coefficients of corresponding fragments of membership functions:

$$y_{D_2} = \frac{a_1 y'_{D_2} + a_2 y''_{D_2}}{a_1 + a_2} \tag{13}$$

The use of the above method seems appropriate when the researcher wishes to take into account the degree of transformation of function fragments, while maintaining a relatively neutral position.

Method E. An alternative method that does not use the found values y'_{D_2} and y''_{D_2} and involves directly obtaining the value of y_{D_2} from y_{C_2} by shifting the latter along the abscissa by a value Δ corresponding to the shift of the value of x_{B_2} (abscissa of the midpoint of the function-image of the chosen prototype) with respect to the value of x_{A_2} (abscissa of the midpoint of the function-image of the prototype):

$$\Delta = x_{B_2} - x_{A_2} \tag{14}$$

$$y_{D_2} = y_{C_2} + \Delta \tag{15}$$

The heuristic justification for the proposed one is the fact of the concentration of expert confidence in the vicinity of the central point of the triangular membership function, which, in the absence of a significant jump in the dynamics of changes in expert judgments, allows us to focus on the use of this option as one of the possible ones.

4 Conclusion

Experimental verification of the considered methods of determining the value of the mean point y_{D_2} of the modified membership function $\mu_D(y)$ demonstrates quite frequent obtaining of expert-acceptable results when using method B. The results of methods C and D are expectedly close to each other and on average less consistent with expert evaluations than the results obtained by method B.

Methods A and E show satisfactory behavior only for modifications corresponding to transitions involving not too significant qualitative changes.

The question of acceptable combinability of the pairs of reference functions (prototypes) selected for finding modifying transformations and the functions subjected to subsequent modifications remains debatable from the standpoint of expert evaluations at present.

The proposed methods can be used when modeling behavior in society. Usually a probabilistic approach is used for such problems, but it is difficult to obtain appropriate probabilistic estimates in practice. This is especially noticeable in the case of new epidemics, such as the COVID-19 pandemic. The lack of reliable statistical data makes it impossible to model epidemics with the necessary accuracy of prediction. Using the mathematical apparatus of fuzzy sets membership functions with points of expert confidence concentration will allow to use the potential of expert judgments in the field of COVID-19 spreading.

Acknowledgements. The reported study was funded by RFBR and CNPq, FASIE, DBT, DST, MOST, NSFC, SAMRC according to the research project No. 20-51-80002.

References

1. Borisov, A., Krumberg, O.: Theory of possibility for decision-making. Fuzzy Sets Syst. **9**(1), 13–23 (1983)
2. Merkuryeva, G., Savelyev, V., Borisov, A.: Synthesis of new technical decisions using fuzzy production systems. Cybern. Syst. **20**(1), 89–97 (1989)
3. Alexeyev, A., et al.: A linguistic approach to decision-making problems. Fuzzy Sets Syst. **22**(1–2), 25–41 (1987)
4. Sudakov, V.: Improving air transportation by using the fuzzy origin-destination matrix. Mathematics **9**(11), 1236 (2021)
5. Dombi, J., Csiszár, O.: Modifiers and membership functions in fuzzy sets. In: Explainable Neural Networks Based on Fuzzy Logic and Multi-criteria Decision Tools. SFSC, vol. 408, pp. 63–81. Springer, Cham (2021). https://doi.org/10.1007/978-3-030-72280-7_4
6. Atanassov, K., Vassilev, P.: Intuitionistic fuzzy sets and other fuzzy sets extensions representable by them. J. Intell. Fuzzy Syst. **38**(1), 525–530 (2020)
7. Bentkowska, U.: Fuzzy sets and their extensions. In: Interval-Valued Methods in Classifications and Decisions. SFSC, vol. 378, pp. 3–23. Springer, Cham (2020). https://doi.org/10.1007/978-3-030-12927-9_1
8. Bustince, H., et al.: A historical account of types of fuzzy sets and their relationships. IEEE Trans. Fuzzy Syst. **24**(1), 179–194 (2016)

9. Bentkowska, U., Bustince, H., Jurio, A., Pagola, M., Pękala, B.: Decision making with an interval-valued fuzzy preference relation and admissible orders. Appl. Soft Comput. **35**, 792–801 (2015)
10. Taneja, S., Suri, B., Narwal, H., Jain, A., Kathuria, A., Gupta, S.: A new approach for data classification using fuzzy logic. In: 6th International Conference – Cloud System and Big Data Engineering, pp. 22–27. Noida, India (2016)
11. Pękala, B.: Uncertainty Data in Interval-valued Fuzzy Set Theory. Properties, Algorithms and Applications. Studies in Fuzziness and Soft Computing, Springer, Cham (2019)
12. Kreinovich, V., Kosheleva, O., Shahbazova, S.N.: Why triangular and trapezoid membership functions: a simple explanation. In: Shahbazova, S.N., Sugeno, M., Kacprzyk, J. (eds.) Recent Developments in Fuzzy Logic and Fuzzy Sets. SFSC, vol. 391, pp. 25–31. Springer, Cham (2020). https://doi.org/10.1007/978-3-030-38893-5_2
13. Barchev, N., Sudakov, V.: On the question of modifying membership functions. In: Silhavy, R., Silhavy, P., Prokopova, Z. (eds.) CoMeSySo 2020. AISC, vol. 1295, pp. 657–662. Springer, Cham (2020). https://doi.org/10.1007/978-3-030-63319-6_61

Check for updates

Computational Time Evaluation of the Various Methods of Parameter Estimation for Pareto Distribution

Jakub Steinbach(✉) and Jan Vrba(✉)

Department of Computing and Control Engineering, University of Chemistry and Technology in Prague, Technická 1905/5, 166 28 Praha 6, Prague, Czech Republic
{jakub.steinbach,jan.vrba}@vscht.cz

Abstract. This article presents the evaluation of computational time of the methods of parameter estimation for Pareto distribution. The evaluation is done for various sample sizes. The evaluated methods are methods of moments and its four different modifications, uniformly minimum variance unbiased estimator and maximum likelihood estimator. The precision, in the sense of mean square error, of estimators is also evaluated for two different values of Pareto distribution parameters and for multiple sample sizes.

Keywords: Pareto distribution · Estimators · Computational time

1 Introduction

Pareto distribution (PD) [2] is a probability distribution introduced by Vilfredo Pareto [9]. The classical PD became a cornerstone for description of social phenomenons, i.e. distribution of wealth [14] or the distribution of human settlements [10]. The PD is a special case of generalized Pareto distribution (GPD) [4] that have been used in various economic models such as estimating marketing costs when introducing new products [1] or estimating potential hurricane damages for insurance purposes [5]. Nowadays, it is used in many fields, such as meteorology, i.e. to predict large waves generated by tropical cyclones [8] or in hydrology where it is used to estimate a frequency of floods [7]. Another area of applications is in geology, where is GPD used to estimate extreme occurrences in geomagnetic storms [11]. In some cases, such as monitoring bridge temperature gradients in civil engineering [16], GPD is used to estimate real-time changes of model parameters with sampling periods in the range of minutes. Despite its usefulness, the limitation of GPD in real-time application is given by computational time that is needed to estimate its parameters. This problem is even more significant for GPD that is extended to high-dimensional spaces [3].

We consider the classical PD to be a potentially suitable tool in many real-time applications, i.e. online novelty detection [12], as it features a number of computationally efficient methods to estimate its parameters, namely the

R. Silhavy et al. (Eds.): CoMeSySo 2021, LNNS 231, pp. 818–827, 2021.
https://doi.org/10.1007/978-3-030-90321-3_67

Method of Moments (MoM) and its modifications, the Maximum Likelihood Estimator (MLE) that comes with a closed solution or the Uniformly Minimum Variance Unbiased Estimator (UMVUE).

In this article we are going to investigate the computational time needed to estimate the scale and shape parameters of the Pareto Type I distribution. The computational time is evaluated for various sample sizes. The new samples are randomly generated for each estimation to work with various datasets. We also attested the precision of parameter estimation for all featured methods, namely MoM, MLE and UMVUE.

2 Materials and Methods

The probability density function of PD is given as follows

$$f(x) = \begin{cases} \frac{\alpha\gamma^{\alpha}}{x^{\alpha+1}} & x \geq \gamma \\ 0 & x < \gamma \end{cases} \tag{1}$$

with corresponding cumulative distribution function given as

$$F(x) = \begin{cases} 1 - \left(\frac{\gamma}{x}\right)^{\alpha} & x \geq \gamma \\ 0 & x < \gamma \end{cases} \tag{2}$$

where α is the shape parameter and γ is the scale parameter. This form is also know as the Type I Pareto distribution. The estimation of parameters α and γ is a crucial part of PD utilizing. For parameters estimation, we have evaluated following approaches and their modifications.

2.1 Method of Moments Estimator

Assume the r-th sample moment, that is given as follows

$$m_r = \frac{\sum_{i=1}^{n} x_i^r}{n} \tag{3}$$

where n is the number of samples and x_i is the value of i-th sample. For the PD, the raw moment is given as

$$r^n = \begin{cases} \frac{\alpha\gamma^n}{\alpha-n} & \alpha > n \\ \infty & \alpha \leq n \end{cases} \tag{4}$$

The Method of Moments (MOM) is based on the relation between sample moments, that are given by equation (3), and the dependence of the distribution moments on the parameters of PD (equation (4)) [13].

Since only two parameters need to be estimated, it is possible to use the first and second sample moments for the estimation, namely average mean and variance. Comparing both equations for the first and second moments gives following equations. Equation (5) relates first moments and equation (6) relates second moments.

$$\mu = \frac{\alpha\gamma}{\alpha - 1} \tag{5}$$

$$\mu^2 + \sigma^2 = \frac{\alpha\gamma^2}{\alpha - 2} \tag{6}$$

By solving equations (5) and (6), we get estimators for the scale and the shape parameters from the sample data. The estimation of scale and shape parameters is the given by following equations (7,8).

$$\hat{\alpha} = 1 + \sqrt{1 + \frac{\mu^2}{\sigma^2}} \tag{7}$$

$$\hat{\gamma} = \frac{\mu\sqrt{\sigma^2 + \mu^2}}{\sigma + \sqrt{\sigma^2 + \mu^2}} \tag{8}$$

2.2 Method of Moments Modification I

In this modification (MM1), the equation (8) is replaced by calculating the variance of the PD [6] that is given as follows.

$$\sigma^2 = \frac{\alpha\gamma^2}{(\alpha - 1)^2 \cdot (\alpha - 2)} \tag{9}$$

By combining equations (7) and (9), we get a new estimator for $\hat{\gamma}$.

$$\hat{\gamma} = \sqrt{(\sigma^2 + \mu^2)\left(\frac{\sqrt{\sigma^2 + \mu^2} - \sigma}{\sqrt{\sigma^2 + \mu^2} + \sigma}\right)} \tag{10}$$

2.3 Method of Moments Modification II

Another modification (MM2) is based on the calculation of the sample harmonic mean and the harmonic mean (h) for Pareto distribution [6]. The sample harmonic mean is given as follows.

$$h = \frac{\sum_{i=1}^{n} x_i^{-1}}{n} \tag{11}$$

The harmonic mean of PD is defined as

$$h = \gamma\left(1 + \frac{1}{\alpha}\right) \tag{12}$$

so combining equations (7) and (12) we get new estimators for both parameters.

$$\hat{\alpha} = \sqrt{1 + \frac{h}{\mu}} \tag{13}$$

$$\hat{\gamma} = \frac{h\sqrt{\mu + h}}{\sqrt{\mu} + \sqrt{\mu + h}} \tag{14}$$

2.4 Method of Moments Modification III

In the third modification (MM3), the formula for the second moment calculation is replaced by the mean of the first-order statistic of PD [6].

$$\frac{\alpha\gamma n}{\alpha n - 1} = x_{min} \tag{15}$$

Substituting μ with x_{min} in the equation (5) and solving the set of equations (5) and (15) gives the resulting estimators for the both parameters.

$$\hat{\alpha} = \frac{x_{min} - n\mu}{n(x_{min} - \mu)} \tag{16}$$

$$\hat{\gamma} = x_{min} - \frac{x_{min}}{n\hat{\alpha}} \tag{17}$$

2.5 Method of Moments Modification IV

In the last modification (MM4), the formula for the second moment calculation is replaced by expectation of the empirical CDF of the first-order statistic of PD [6].

$$\frac{1}{n+1} = 1 - \left(\frac{\gamma}{x_{min}}\right)^2 \tag{18}$$

Solving the set of equations (5) and (18) gives the equations for the estimators of both parameters.

$$\hat{\alpha} = \frac{(n-1)\mu}{n(\mu - x_{min})} \tag{19}$$

$$\hat{\gamma} = x_{min}\left(\frac{n}{n+1}\right)^{\frac{1}{\hat{\alpha}}} \tag{20}$$

2.6 Uniformly Minimum Variance Unbiased Estimators

Another evaluated method is the Uniformly Minimum Variance Unbiased Estimator (UMVUE) [15]. The estimators of PD parameters are given as

$$\hat{\alpha} = (1 - 2n^{-1})\ddot{\alpha} \tag{21}$$

$$\hat{\gamma} = (1 - (n-1)^{-1}\ddot{\alpha}^{-1})x_{min} \tag{22}$$

where $\ddot{\alpha}$ is a term defined as

$$\ddot{\alpha} = n\left(\sum_{i=1}^{n} \ln\left(\frac{x_i}{\hat{\gamma}}\right)\right)^{-1} \tag{23}$$

For more details and derivation see [15].

2.7 Maximum Likelihood Estimator

The Maximum Likelihood Estimator (MLE) uses the likelihood function as an objective function. By minimizing the function with respect to α, it is possible to obtain an estimator for this parameter.

$$\hat{\alpha} = n\left(\sum_{i=1}^{n} \ln\left(\frac{x_i}{\hat{\gamma}}\right)\right)^{-1} \tag{24}$$

It is not possible to get an estimator of γ the same way as the likelihood function is unbounded with respect to γ. However, since γ is defined as the lower bound of the sample x, the objective function can be optimized if γ is subject to the following constraint.

$$\hat{\gamma} = x_{min} \tag{25}$$

3 Results

To estimate the average and maximum computational time, totally 10,000 experiments for each presented estimation method (namely MoM, MM1, MM2, MM3, MM4, MLE and UMVUE) were performed. Each experiment consists of randomly generated data set that is represented by samples from PD with parameters $\alpha = 3$ and $\gamma = 5$. The computational time was measured for every single experiment with data sets that consisted of 50, 100, 500, 1000 and 5000 samples. All experiments were performed on a PC using Intel(R) Core(TM) i5-7400 4 cores CPU running at 3001 MHz with 32 GB RAM. The running operating system was Windows 10 Pro 64 bit version 10.0.19041 and the code was written in Python 3.8.3 using Numpy 1.19.5 and Scipy 1.4.1 libraries. The source code is available at www.github.com/vrbaj/pareto_estimators. The following Table 1

Table 1. The average and maximum computational time (in milliseconds) of PD parameters estimators

Method	$n = 50$		$n = 100$		$n = 500$		$n = 1000$		$n = 5000$	
	CT_{avg}	CT_{max}	CT_{avg}	CT_{max}	CT_{avg}	CT_{max}	CT_{avg}	CT_{max}	CT_{avg}	CT_{max}
umvue	23.55	272.20	28.26	255.00	66.97	336.80	114.60	447.90	490.64	1060.30
ml	19.70	383.80	24.52	170.00	63.18	270.20	111.32	615.10	488.04	948.00
mom	45.50	258.70	45.63	319.20	48.11	221.10	51.51	283.50	62.18	573.10
*mm*1	44.12	339.90	44.27	326.40	46.40	355.40	49.22	570.50	59.66	295.70
*mm*2	47.75	266.20	47.67	301.20	49.91	241.90	52.41	270.00	63.95	480.70
*mm*3	21.70	246.30	25.94	317.90	60.18	247.20	102.95	386.60	441.33	1745.20
*mm*4	21.28	258.60	25.56	271.80	60.09	293.70	103.16	371.10	441.39	909.50

contains the average time of computation CT_{avg} and the maximum time of computation CT_{max} in milliseconds. Parameter n represents the number of samples in the evaluated data set.

We also evaluated the precision of estimators for two different PD parameters settings to verify the published results. To replicate the results we made the estimation for multiple sample sizes, namely 20, 50, 100, 200, 500, 1000. For each sample size we performed 10.000 estimations and then evaluated the average mean square error (MSE). The results are in the following Tables 2, 3. Note that each estimation was made with different randomly generated data sets.

Table 2. Mean square error of PD parameters estimation ($\alpha = 2$, $\gamma = 1$)

n	True values		Method	MSE	
	α	γ		α	γ
20	2	1	*umvue*	$2.33 \cdot 10^{-1}$	$6.96 \cdot 10^{-4}$
			ml	$3.42 \cdot 10^{-1}$	$1.40 \cdot 10^{-3}$
			mom	1.73	$1.43 \cdot 10^{-1}$
			*mm*1	1.69	$1.70 \cdot 10^{-1}$
			*mm*2	$4.43 \cdot 10^{-1}$	$2.31 \cdot 10^{-2}$
			*mm*3	$3.16 \cdot 10^{-1}$	$6.78 \cdot 10^{-4}$
			*mm*4	$2.94 \cdot 10^{-1}$	$6.95 \cdot 10^{-4}$
50	2	1	*umvue*	$8.58 \cdot 10^{-2}$	$1.04 \cdot 10^{-4}$
			ml	$1.03 \cdot 10^{-1}$	$1.99 \cdot 10^{-4}$
			mom	$7.63 \cdot 10^{-1}$	$8.50 \cdot 10^{-2}$
			*mm*1	$7.50 \cdot 10^{-1}$	$1.21 \cdot 10^{-1}$
			*mm*2	$4.51 \cdot 10^{-1}$	$2.15 \cdot 10^{-2}$
			*mm*3	$1.14 \cdot 10^{-1}$	$1.11 \cdot 10^{-4}$
			*mm*4	$1.13 \cdot 10^{-1}$	$1.07 \cdot 10^{-4}$

(*continued*)

Table 2. (*continued*)

n	True values		Method	MSE	
	α	γ		α	γ
100	2	1	*umvue*	$4.11 \cdot 10^{-2}$	$2.58 \cdot 10^{-5}$
			ml	$4.32 \cdot 10^{-2}$	$5.30 \cdot 10^{-5}$
			mom	$4.71 \cdot 10^{-1}$	$7.19 \cdot 10^{-2}$
			*mm*1	$4.64 \cdot 10^{-1}$	$5.55 \cdot 10^{-2}$
			*mm*2	$4.54 \cdot 10^{-1}$	$2.14 \cdot 10^{-2}$
			*mm*3	$5.92 \cdot 10^{-2}$	$2.67 \cdot 10^{-5}$
			*mm*4	$5.78 \cdot 10^{-2}$	$2.54 \cdot 10^{-5}$
200	2	1	*umvue*	$2.02 \cdot 10^{-2}$	$6.57 \cdot 10^{-6}$
			ml	$2.16 \cdot 10^{-2}$	$1.24 \cdot 10^{-5}$
			mom	$3.12 \cdot 10^{-1}$	$6.42 \cdot 10^{-2}$
			*mm*1	$3.13 \cdot 10^{-1}$	$4.91 \cdot 10^{-2}$
			*mm*2	$4.56 \cdot 10^{-1}$	$2.13 \cdot 10^{-2}$
			*mm*3	$3.27 \cdot 10^{-2}$	$6.24 \cdot 10^{-6}$
			*mm*4	$3.22 \cdot 10^{-2}$	$6.36 \cdot 10^{-6}$
500	2	1	*umvue*	$8.01 \cdot 10^{-3}$	$1.03 \cdot 10^{-6}$
			ml	$8.21 \cdot 10^{-3}$	$2.01 \cdot 10^{-6}$
			mom	$1.99 \cdot 10^{-1}$	$3.00 \cdot 10^{-2}$
			*mm*1	$1.95 \cdot 10^{-1}$	$3.40 \cdot 10^{-2}$
			*mm*2	$4.57 \cdot 10^{-1}$	$2.12 \cdot 10^{-2}$
			*mm*3	$1.48 \cdot 10^{-2}$	$9.69 \cdot 10^{-7}$
			*mm*4	$1.50 \cdot 10^{-2}$	$1.01 \cdot 10^{-6}$
1000	2	1	*umvue*	$3.92 \cdot 10^{-3}$	$2.56 \cdot 10^{-7}$
			ml	$4.02 \cdot 10^{-3}$	$5.06 \cdot 10^{-7}$
			mom	$1.47 \cdot 10^{-1}$	$2.17 \cdot 10^{-2}$
			*mm*1	$1.47 \cdot 10^{-1}$	$2.40 \cdot 10^{-2}$
			*mm*2	$4.57 \cdot 10^{-1}$	$2.12 \cdot 10^{-2}$
			*mm*3	$8.55 \cdot 10^{-3}$	$2.48 \cdot 10^{-7}$
			*mm*4	$8.11 \cdot 10^{-3}$	$2.65 \cdot 10^{-7}$

Table 3. Mean square error of PD parameters estimation ($\alpha = 3$, $\gamma = 5$)

n	True values		Method	MSE	
	α	γ		α	γ
20	3	5	*umvue*	$5.28 \cdot 10^{-1}$	$7.87 \cdot 10^{-3}$
			ml	$7.70 \cdot 10^{-1}$	$1.53 \cdot 10^{-2}$
			mom	2.52	$3.94 \cdot 10^{-1}$
			mm1	2.48	$3.87 \cdot 10^{-1}$
			mm2	2.63	1.32
			mm3	$6.71 \cdot 10^{-1}$	$7.52 \cdot 10^{-3}$
			mm4	$6.72 \cdot 10^{-1}$	$7.24 \cdot 10^{-3}$
50	3	5	*umvue*	$1.93 \cdot 10^{-1}$	$1.14 \cdot 10^{-3}$
			ml	$2.24 \cdot 10^{-1}$	$2.20 \cdot 10^{-3}$
			mom	$9.30 \cdot 10^{-1}$	$1.88 \cdot 10^{-1}$
			mm1	$8.96 \cdot 10^{-1}$	$1.84 \cdot 10^{-1}$
			mm2	2.64	1.30
			mm3	$2.30 \cdot 10^{-1}$	$1.15 \cdot 10^{-3}$
			mm4	$2.24 \cdot 10^{-1}$	$1.12 \cdot 10^{-3}$
100	3	5	*umvue*	$9.35 \cdot 10^{-2}$	$2.80 \cdot 10^{-4}$
			ml	$9.96 \cdot 10^{-2}$	$5.76 \cdot 10^{-4}$
			mom	$4.82 \cdot 10^{-1}$	$1.17 \cdot 10^{-1}$
			mm1	$4.86 \cdot 10^{-1}$	$1.13 \cdot 10^{-1}$
			mm2	2.64	1.30
			mm3	$1.14 \cdot 10^{-1}$	$2.92 \cdot 10^{-4}$
			mm4	$1.11 \cdot 10^{-1}$	$2.82 \cdot 10^{-4}$
200	3	5	*umvue*	$4.50 \cdot 10^{-2}$	$7.18 \cdot 10^{-5}$
			ml	$4.79 \cdot 10^{-2}$	$1.40 \cdot 10^{-4}$
			mom	$2.73 \cdot 10^{-1}$	$7.94 \cdot 10^{-2}$
			mm1	$2.68 \cdot 10^{-1}$	$7.73 \cdot 10^{-2}$
			mm2	2.64	1.30
			mm3	$5.63 \cdot 10^{-2}$	$6.66 \cdot 10^{-5}$
			mm4	$5.61 \cdot 10^{-2}$	$6.80 \cdot 10^{-5}$
500	3	5	*umvue*	$1.80 \cdot 10^{-2}$	$1.05 \cdot 10^{-5}$
			ml	$1.80 \cdot 10^{-2}$	$2.34 \cdot 10^{-5}$
			mom	$1.37 \cdot 10^{-1}$	$4.89 \cdot 10^{-2}$
			mm1	$1.41 \cdot 10^{-1}$	$4.99 \cdot 10^{-2}$
			mm2	2.64	1.30
			mm3	$2.29 \cdot 10^{-2}$	$1.11 \cdot 10^{-5}$
			mm4	$2.30 \cdot 10^{-2}$	$1.08 \cdot 10^{-5}$

(*continued*)

Table 3. (*continued*)

n	True values		Method	MSE	
	α	γ		α	γ
1000	3	5	*umvue*	$9.09 \cdot 10^{-3}$	$2.68 \cdot 10^{-6}$
			ml	$9.06 \cdot 10^{-3}$	$5.58 \cdot 10^{-6}$
			mom	$8.82 \cdot 10^{-2}$	$3.76 \cdot 10^{-2}$
			*mm*1	$8.79 \cdot 10^{-2}$	$3.76 \cdot 10^{-2}$
			*mm*2	2.64	1.30
			*mm*3	$1.17 \cdot 10^{-2}$	$2.91 \cdot 10^{-6}$
			*mm*4	$1.17 \cdot 10^{-2}$	$2.80 \cdot 10^{-6}$

4 Conclusions

Based on the measured computational times, it is obvious that the Maximum Likelihood Estimator outperforms other estimators for small sample sizes (up to 100 samples), while the Method of Moments (especially modification $mm1$) is superior for larger sample sizes. For more details, see Table 1.

All results of estimators precision comparison are comparable with published data. Also the poor performance of $mm2$ estimator was verified. Especially for PD parameters setting $\alpha = 3$, $\gamma = 5$, it seems that $mm2$ estimator is producing unacceptable error.

Acknowledgements. This work was supported from the grant of Specific university research – grant IGA A1_ FCHI_2021_001.

References

1. Arkolakis, C.: Market penetration costs and the new consumers margin in international trade. J. Polit. Econ. **118**(6), 1151–1199 (2010)
2. Arnold, B.C.: Pareto distribution. Wiley StatsRef: Statistics Reference Online, pp. 1–10 (2014)
3. Clifton, D.A., Clifton, L., Hugueny, S., Tarassenko, L.: Extending the generalised pareto distribution for novelty detection in high-dimensional spaces. J. Sig. Process. Syst. **74**(3), 323–339 (2014)
4. Coles, S., Bawa, J., Trenner, L., Dorazio, P.: An introduction to statistical modeling of extreme values, vol. 208. Springer, London (2001)
5. Dey, A.K., Das, K.P.: Modeling extreme hurricane damage using the generalized pareto distribution. Am. J. Math. Manag. Sci. **35**(1), 55–66 (2016)
6. Hussain, S., Bhatti, S.H., Ahmad, T., Aftab, M., Tahir, M.: Parameter estimation of pareto distribution: some modified moment estimators. Maejo Int. J. Sci. Technol. **12**(1), 11–27 (2018)

7. Lamb, R.: Calibration of a conceptual rainfall-runoff model for flood frequency estimation by continuous simulation. Water Resources Res. **35**(10), 3103–3114 (1999)
8. MacAfee, A.W., Wong, S.W.: Extreme value analysis of tropical cyclone trapped-fetch waves. J. Appl. Meteorol. Climatol. **46**(10), 1501–1522 (2007)
9. Pareto, V.: Cours d'économie Politique, vol. 1. Librairie Droz, Switzerland (1964)
10. Reed, W.J.: On the rank-size distribution for human settlements. J. Reg. Sci. **42**(1), 1–17 (2002)
11. Thomson, A.W., Dawson, E.B., Reay, S.J.: Quantifying extreme behavior in geomagnetic activity. Space Weather **9**(10) (2011)
12. Vrba, J., Mareš, J.: Introduction to extreme seeking entropy. Entropy **22**(1), 93 (2020)
13. Wallis, J.R., Matalas, N.C., Slack, J.R.: Just a moment! Water Resource. Res. **10**(2), 211–219 (1974)
14. Wold, H.O., Whittle, P.: A model explaining the pareto distribution of wealth. Econ., J. the Econ. Soc. 591–595 (1957)
15. Woo, J.S., Kang, S.B.: Estimation for functions of two parameters in the pareto distribution. J. Korean Data Inf. Sci. Soc. **1**, 67–76 (1990)
16. Zhou, G.D., Yi, T.H., Chen, B., Zhang, H.: A generalized pareto distribution-based extreme value model of thermal gradients in a long-span bridge combining parameter updating. Adv. Struct. Eng. **20**(2), 202–213 (2017)

Knowledge and Datasets as a Resource for Improving Artificial Intelligence

Štefan Balogh(✉)

Faculty of Electrical Engineering, Slovak Technical University, Bratislava, Slovakia
stefan.balogh@stuba.sk

Abstract. This work discusses and analyzes options that could be helpful for the research and development of artificial intelligence. The work is based on analyzes and results of other authors as well as on the own experience. Preprocessing and selecting appropriate data is the key to success in using tools and algorithms for artificial intelligence. This means that the input data is largely a factor that affects the outcome and resulting success of the algorithms used in artificial intelligence. Data is essential for AI improvement, and this article gives our suggestion on how to improve that data, how to turn it into intelligent data that will provide more knowledge for AI algorithms. We describe our suggestions and experiences with the creation of datasets using ontology model and findings of what they should meet, and what are the problems associated with it.

Keywords: Artificial intelligence · Machine learning · Is intelligent datasets · Knowledge representation

1 Introduction

1.1 Artificial Intelligence and Role of Datasets

When using tools of artificial intelligence (AI) and machine learning (ML), one of the most important tasks is the correct selection and preprocessing of data that serves as input for AI and ML algorithms. We focus on analysis of data sources and data quality used in the security domain where artificial intelligence and machine learning algorithms are also most used in the detection of attacks and malware. The results of many works and summary articles point to the possibilities of using the given approaches in detection, but the authors consider the selection of suitable properties to be a very important aspect [1].

Alexander Wissner-Gross [2] in 2016 in a written response to the question posed by Edge: *"What do you consider the most interesting recent scientific news?"*
said that "perhaps the most important news of our day is that datasets—not algorithms—might be the key limiting factor to development of human-level artificial intelligence" also, the answer to question "*At the dawn of the field of artificial intelligence, in 1967, two of its founders famously anticipated that solving the problem of computer vision would take only a summer. What took the AI revolution so long?* "we get important answer

R. Silhavy et al. (Eds.): CoMeSySo 2021, LNNS 231, pp. 828–837, 2021.
https://doi.org/10.1007/978-3-030-90321-3_68

that many major AI breakthroughs have actually been constrained by the availability of high-quality training datasets, and not by algorithmic advances [2].

As the example, they mention some examples (see Table 1), like in 1994 when the achievement of human-level spontaneous speech recognition relied on a variant of a hidden Markov model algorithm initially published ten years earlier in 1984 but used a dataset of spoken *Wall Street Journal* articles and other texts made available only three years earlier in 1991. Also, in 1997, when IBM's Deep Blue defeated Garry Kasparov to become the world's top chess player, its core NegaScout planning algorithm was fourteen years old published in 1983, whereas its key dataset of 700,000 Grandmaster chess games was only six years old (published in 1991). More examples to prove his hypothesis can by find in Table 1 [2].

Table 1. Table show that the average elapsed time between key algorithm proposals and corresponding advances was about 18 years, whereas the average elapsed time between key dataset availabilities and corresponding advances was less than 3 years, or about 6 times faster [3].

Year	Breakthroughs in AI	Datasets (First Available)	Algorithms (First Proposed)
1994	Human-level spontaneous speech recognition	Spoken Wall Street Journal articles and other texts (1991)	Hidden Markov Model (1984)
1997	IBM Deep Blue defeated Garry Kasparov	700,000 Grandmaster chess games, aka "The Extended Book" (1991)	Negascout planning algorithm (1983)
2005	Google's Arabic- and Chinese-to-English translation	1.8 trillion tokens from Google Web and News pages (collected in 2005)	Statistical machine translation algorithm (1988)
2011	IBM Watson became the world Jeopardy! champion	8.6 million documents from Wikipedia, Wiktionary, Wikiquote, and Project Gutenberg (updated in 2010)	Mixture-of-Experts algorithm (1991)
2014	Google's GoogLeNet object classification at near-human performance	ImageNet corpus of 1.5 million labeled images and 1,000 object categories (2010)	Convolution neural network algorithm (1989)
2015	Google's Deepmind achieved human parity in playing 29 Atari games by learning general control from video	Arcade Learning Environment dataset of over 50 Atari games (2013)	Q-learning algorithm (1992)
Average No. of Years to Breakthrough:		3 years	18 years

So, at the end Alexander Wissner-Gross concluded that focusing on dataset rather than algorithm is a potentially simpler approach. "Although new algorithms receive much of the public credit for ending the last AI winter, the real news might be that prioritizing the cultivation of new datasets and research communities around them could be essential to extending the present AI summer." [3] Also Bowles [4] noted that efforts are out there to give machines "prebuilt knowledge" based on common sense, general knowledge, such as OpenCyc[1] or Off-The Shelf Knowledge, such as WordNet[2]. These computers will have a greater ability, based on their representations, to suggest medical diagnosis and treatments, analyze the impact of market trends or other advantages. To be able to obtain suitable data, usable as intelligent data, it is necessary to process the data using Taxonomies, Semantics and Ontology. Ontological (semantic) representation also brings other advantages: reasoning, deduplication, and integration.

1 https://www.ime.usp.br/~fr/opencyc/.

2 https://wordnet.princeton.edu/.

Semantic reasoning is another important part of AI. In addition to identifying potentially fraudulent transactions, determining users' intent based on their browser history and making product recommendations.

If we want to use such data, we need to have an up-to-date data source that would be in sufficient numbers for the needs of AI algorithms. Only large organizations like google, IBM and others can afford that. As a result, the possibilities of research can be significantly reduced and therefore it is necessary to look for a solution as to how to obtain the given data. One possibility is to create a system for sharing data in each domain. This idea can be supported, as data sharing is considered by many experts to be a key to differentiating knowledge. In the field of security, it is the key to early detection and prevention. Knowledge sharing is the key to innovation, and to improving security [5, 6]. Of course, there are many specifics in each domain that must be met in order for sharing to be realistic, in connection with the protection of sensitive data, which is especially important in the field of security.

The authors in the article [7] point out the keys four requirements for which it is necessary to meet for sharing. Another problem is the existence of various solutions, standards, and platforms for sharing today. A solution is needed that would be able to unify these platforms, as it does not seem realistic to integrate all platforms into one common solution.

The solution can take in the account some specification for security area as:

- Effective protection system can't by created because there are big field for attack. The offense only must find out about one vulnerability (or several) to compromise organization and the defense must protect against all of them. It's quite impossible to fix every vulnerability.
- Sharing knowledge about attack can help arrow your defense and successfully defeat attack.
- To achieve this, we need to have as soon as possible useful information about attack vector, Indicator of compromise (IoC), Techniques and Procedures (TTPs).
- We need integrate enough (as much as possible) information sources to get information's about attack in time, so we need open system.
- In [8] we can read "Much of the knowledge used by most of skilled analysts today is residing only in their heads. If we manage to model this knowledge and express it, not only more analysts would be able to consume this type of intelligence, but the analytics would be executed in a consistent way, contradicting the" confirmation bias" often referred to in an investigation".
- We need to sort out the variety of taxonomies. Organizations like MITRE (and others) started creating taxonomies and standards for security domain.
- There are multiple taxonomies that have a potential to become standard in security area like STIX, MAEC, CAPEC, but also taxonomies from other organizations are very popular, like OpenIOC from Mandiant company, Europol Incident – Europol class of incident taxonomy, MISP taxonomy and many more.
- To share security information and knowledge effectively we need to have one common language or a possibility to integrate different sources and formats.
- On a European or world level, we need to have a widely usable solution accepted by governments, commercial organizations, and the public.

To achieve this, we proposed Threat Intelligence Platform called ORBIS that is based on ontology models and built on open platforms and standards to integrate different sources and formats, analyze the behavior and intention of attackers, and share it within the world community using a common language. We have therefore developed our own proposal for sharing and obtaining data from various sources, which is designed to meet the requirements for the security domain. (secrecy of some information, sharing only data that brings benefits to other organizations and help increase protection and security).

The Science of Cyber-Security and Taxonomy Needs

In the report published by MITRE in 2010, titled "The Science of Cyber-Security" [9] addressed the question: What would be needed to make cybersecurity into a science?

The very first conclusion of the report was that the most important development would be the creation of a common language and a set of basic concepts about which the security community can develop a shared understanding. So, taxonomy is necessary, to create the cyber security science. Taxonomies according to Bowles [4] provide machines ordered representations, it represents the formal structure of classes or types of objects within a domain. Taxonomies can be stored "using a variety of different data structures as in a Relational Database, in a Draft Database, in a tool just for Taxonomies." but Bowles also stated, that "Taxonomies are a lightweight version." By adding Ontologies to a computer's representations, machines can "process the content of information instead of just presenting the information to humans." So that Artificial Intelligence can process such complexity and use Ontologies and the W3C recommends OWL, Ontology Web Language [4].

An Overview of Existing Ontologies Within the Cyber Security Domain

In the field of security, authors in article [8] performed taxonomy analysis and ontologies used in security domain. His conclusion was that "there is not any existing ontology readily available for use within cyber threat intelligence" or in other words there is no complete suitable model, only many partial models. The research on the use of ontologies for attack detection and detection automation produced several partial and specific ontologies. However, the creation of a unified ontological model for information sharing requires a standardized language. Existing standards for information sharing are built on XML and cannot be automatically transformed into an ontological model [10]. Authors in [11] have attempted to create a general ontological model (UCO). However, the ontology they proposed is only in a basic form, it is not yet directly usable and serves as a basic framework and as a bridge to other ontologies.

We tested the UCO solution in our work. Currently, in addition to the article [11] in the GitHub[3] for UCO ontology, was released version 2 and was added ontologies provided by the authors of [10]. To use the created ontology, we found that only basic classes are used in the CVE ontology and do not cover the knowledge that the CVE database provides. Therefore, we decided to create our own version of the CVE ontology that contained the required knowledge capture classes. The ontology is available in the GitHub[4]. Our effort is also to create a server for data sharing, which currently works in a limited version

3 https://github.com/Ebiquity/uco2.

4 https://github.com/rychtovy24/diplomova_praca/tree/master/Ontology.

and provides only data from the CVE and OVAL databases. We plan to gradually add more data. To store data, we try to use own ontological model. The model is still under development as the main goal is to find a model that will suitably store knowledge from the domain of security but should also provide up-to-date knowledge for finding suitable models for detecting attacks and malware. Research is leading us to this. The foreign model should meet the requirements for sharing, then for data deduplication, and the data should be usable for AI and ML tools. What data, in what format, how detailed it is necessary to store for the needs of AI is still a question of research. From our point of view, we tested the solution using static properties for detection in [12]. In the next part of article part 3.1, we also propose a solution for deduplication of records of malware activities in the system, which are monitored during dynamic analysis.

2 Design of Sharing Platform for Data Collection

2.1 Main Idea and Topology for Sharing Platform

This design is our suggestion that can be used for collecting data from security domain. The design can meet the requirement for sharing and for privacy of data. We use this design as a basic model for getting large datasets needed for AI and knowledge sharing. To get "intelligent data" we recommended to used ontology representation of data and we are also working on suitable ontology model. The proposal is to share not only the information about the incident, but also the techniques used in the attack. Used techniques can become an input for AI systems and we can look for attack patterns. To do this, it is necessary to create a uniform language for the description of individual techniques and procedures in the attack (network attacks, phishing campaigns, malware techniques).

There is a private and public part of the system, as can be seen in Fig. 1. In a private part we guarantee (roughly) the confidentiality of data and we know the identity of who was provided it. We assume that there will be more private blocks and it will group some departments or branches. E.g., Computer security incident response teams (CSIRT) can be one private block, Banks another block. They store only the knowledge they want to share with others on the server. Other information that can have value for the rest of the world (and it is possible to share it) is sent to a public server. But the public part will also provide the opposite mechanism, i.e., obtaining data from the public. Many experts or security analysts can share their knowledge about attacks or campaigns. It will then be possible for the private sector to retrieve knowledge from the public database (public blocks) and include this (but checked) knowledge in private servers and use it for analyzes.

The proposed system considers, so-called services, that will consume the data. The service will process and analyze the data using AI for detecting a security event or other activity.

Also, many research centers can use data obtained from public servers to create a new procedure or method for detecting a security event or other activity.

We want to make use of the idea offered by the open access initiative, but also consider the needs of private, closed security components.

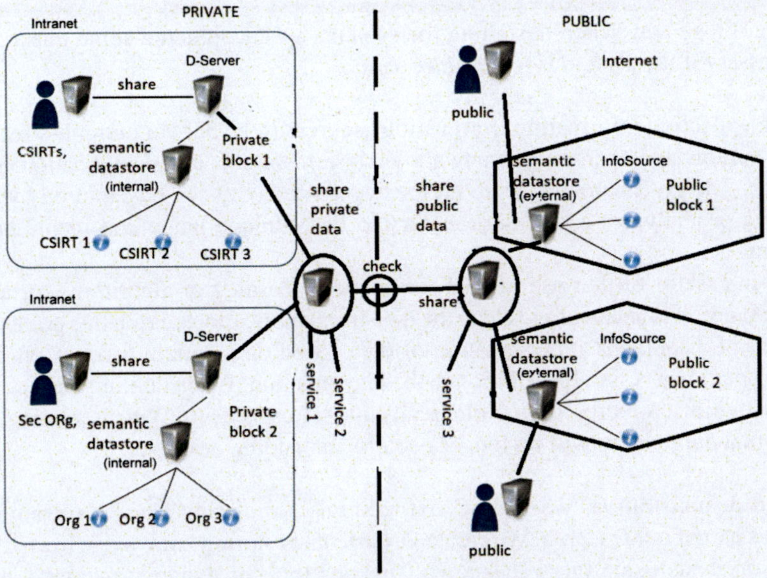

Fig. 1. Topology off sharing platform

So, there is a benefit for the public and private part, from the sharing of knowledge and can significantly contribute to the early capture of new attacks or campaigns, both for private blocks and for the public.

Another advantage of the solution is that many other organizations and research centers will be able to analyze data from public servers, which may lead to the creation of new, possibly more effective procedures and algorithms for detection or other security activity.

It can also be a situation where the research center uses data obtained from public servers to create a new procedure or method for detecting a security event or other activity and then a private block will order the service if it is interesting or necessary for it.

The idea and the goal of project is to use not raw data only to detect patterns. Semantic data from ontology model that are collected and shared from organizations and will be used for detection algorithm or detection patterns creations like attack details, attack chain, malware techniques, IoC, kill chain, technique and tactics used in attack[5], vulnerabilities used for attacks, used domain or IP for attack, etc. This information can be shared using the ORBIS system and ontology model (common language).

3 Testing the ORBIS System

3.1 Knowledge Collection from Existing Tools

To verify and try to identify the potential problems with data share used ontology design, we created our own system for collecting knowledge from various sources and tried to

5 https://attack.mitre.org/.

integrate it together. When designing the system, we encountered some questions that are necessary if the system is to be functional:

- When collecting information from various sources (e.g., network communication) or from various analyzers (e.g., malware analysis), we encounter duplication of information. The given knowledge about the event or activity can be captured by various sources or analyzers and it is necessary to have unique name and identifier in the system.
- Also, in case of further use of data for machine learning or algorithms for artificial intelligence, it is necessary to clean the data from duplications. From this point of view, the use of a semantic representation of data based on a certain form of ontological representation is a suitable solution to the phenomenon. When defining unique objects and activities correctly, the ontologically model can ensure (when creating unique IRIs) that duplication will no longer occur when adding new data.

For a demonstration, we add a brief example for clarification. An example was created as part of our research. We created a system for storing knowledge from dynamic malware analysis using the well-known Cuckoo[6] tool for dynamic malware analysis. We preserved knowledge in an ontological model. This was information about running processes, called API functions, used mutexes, etc.

Thus, if one malware triggers Process A and the other malware triggers Process A, we do not want to duplicate Process A in ontology, but we want both malwares to have a reference to one specific object that will represent Process A. However, the biggest problem is to devise a system for creating identifiers for specific Observable Objects. The amount of information we have about a given object varies significantly depending on the type of object, but also the program that performed the analysis.

The problem was caused, for example, by the PID of process. We could not include it in the creation of the identifier, because essentially every PID of the same process can be different in each analysis. After several tests and modifications, we finally chose the attributes: type, name, path, command line, key, mime type. However, for objects that do not have any of the attributes, this attribute is simply ignored. The identifier is formed from an MD5 hash of such a string. Of course, we are aware that this system is not perfect, but within our capabilities of analysis and verification of the functionality of the entire concept. Furthermore, we also needed to assign an identifier to individual actions. The Malware Action object basically has only a name and a possible reference to the input and output Observable Object. So, we take the already existing hash input or output object and at the end we attach the name of the action and create a new hash. The action created in this way is clearly identifiable for a given action with a specific object or objects. Now we can assign malware to actions that will be in a kind of common space and will not create duplicates, which will save us a lot of space. When we realize that one malware performs tens of thousands of actions, and really the basic ones like get-system-directory would be repeated to us to a large extent.

In the manual analysis of code, we found that some actions are performed very often with many repetitions. This is basically the case with our MEMZ test malware. So, the

[6] https://cuckoosandbox.org/.

malware performs actions that at first glance may not sound suspicious. However, if we calculate how many identical actions malware will do in a short time, this information can change the whole view of its behavior. In our case, the MEMZ malware scanned the list of already running processes and ran copies of them. In addition, it ran in the order of hundreds to thousands of applications such as a calculator, notepad, or browser. In addition to all this, he also duplicated himself. From these facts, we concluded that we need to capture such behavior.

So, we approached the alternative of creating another object that will serve as a step between the Malware Instance and the Malware Action. We called this object counter, and it has a single attribute "times executed". We created its identifier as a combination of an action hash and an instance hash. Thus, each instance has links to counter objects that hold information about the number of executions, and the counter also has a link to the Malware Instance. Thus, a counter is always unique for a particular pair of instance and action objects. The disadvantage of such a solution is that we increase the number of objects in the ontology, but we remain left with the advantage that the actions remain common to all malware, so newly imported instances can therefore conveniently connect to these actions using their own counter objects. In Fig. 2 is how the final structure of our ontology look like.

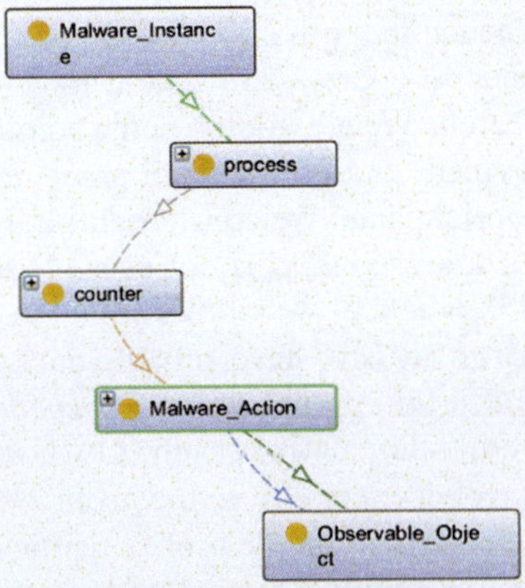

Fig. 2. Ontology model for store data from dynamic analysis of malware

We have repeatedly tested the import of data from our test malware as well as the analysis of harmless software to make sure that all the information and relationships between objects, that have been recorded in the MAEC report, are also recorded in our ontology. In other words, whether our method of converting JSON structured text to an OWL file did not result in a loss of information. All the information was included and thanks to the functionality of the ontological approach we saved a lot of memory. The MAEC report from the MEMZ malware analysis, created by Cuckoo, took up more than 8 megabytes, while our ontology had a disk size of approximately 280 kbyte after importing this data.

This example of the implementation of an ontological model points to the advantages of such a solution, but also to the problems that need to be solved in the given area. These are mainly the following:

- Creation of a procedure for a uniform way of defining identifiers for specific objects used in taxonomy and the standard. It is not enough to have a defined taxonomy, or a standard for sharing knowledge, it is necessary to define a way (procedure) for creating identifiers.
- In certain cases, it is necessary to create a methodology and method of recording the number of activities or actions.
- In certain cases, it is appropriate to address the sequence of events or find a way to capture them.
- For access information of this type, there should be special databases (ontological), which are linked together, and store specific information as in this case, the output of the dynamic analysis. The necessary data can be queried from the central information server. We suggest to use similar solution as is used by Takahashi et al. in [13].

4 Conclusion

In this work, we try to outline another possible direction of development in the field of AI and ML, where we consider the views and conclusions of other authors, as well as the conclusions of own research. We are adding some solutions for which we want to bridge barriers in the deployment and use of these procedures as a system design for sharing and collecting appropriate data. We consider this to be one of the key elements. Without a suitable source of knowledge, it is not possible to sensibly deploy and use systems for analysis and AI in practice. As follows from the overview of developments in this area, a substantial part of it is to have suitable data available so that the given algorithms can be tested and used appropriately in a specific area and domain. The analysis shows that there is still a lot of work, as the CVE ontology proposal shows us. It is certainly necessary to involve experts from the given areas where the main need is multidisciplinary of foreign experts in the field of semantic models and experts in the field of security. The parts of the solution created by us showed the urgency of solving duplications of knowledge, which can be a serious problem in such a system. Also, the data is not suitable for entry and must be used a complicated procedure to clean. That cleanup can only be performed by experts who know the data and can evaluate when it is a duplication (e.g., intelligence from various sources, sensors or honeypots, intelligence analysis of behavior from various dynamic malware analyzers (such as Cuckoo, Hybrid analysis[7], etc.) There is also an open area where it is necessary to search for suitable solutions. Our solution that uses hashing is not verified for other data types, such as network data.

But using ontology representation of data in right way we save up also some space, as confirmed ours testing environment. The MAEC report from the MEMZ malware analysis get by Cuckoo took up more than 8 megabytes, while our ontology had a disk size of approximately 280 kbyte after importing this data.

[7] https://www.hybrid-analysis.com.

In this work, we tried to outline a solution for the collection of intelligent data, which can be a great benefit for artificial intelligence. There is a need to create many solutions from the implementation of a common sharing system to the systems that will provide data for sharing. Also, it is necessary to work on appropriate ontology model for cyber security domain to store the data and knowledge in right way. In the work, we designed and tested a possible solution. We have also identified and pointed out some problems that need to be solved in the implementation of a comprehensive system for the collection of intelligent data. Our goal is to continue working on the system and gradually create individual components such as the necessary ontological models, integration solutions, transformations from other sources of ontological representations. Another important part is the development and modification of existing AI algorithms and techniques for the given types of datasets.

Acknowledgment. This work was supported by research grants APVV-19-0220.

References

1. Ucci, D., Aniello, L., Baldoni, R.: Survey of machine learning techniques for malware analysis. Comput. Secur. **81**, 123–147 (2019)
2. Wissner-Gross, A.: Datasets Over Algorithms. Edge. http://www.edge.org/response-detail/26587 (2016). Accessed 01 Aug 2021
3. Datasets Over Algorithms: https://www.kdnuggets.com/2016/05/datasets-over-algorithms.html (2016). Accessed 01 Aug 2021
4. Knight, M.: Taxonomy vs Ontology: Machine Learning Breakthroughs. https://www.dataversity.net/taxonomy-vs-ontology-machine-learning-breakthroughs/# (2017). Accessed 01 Aug 2021
5. LClaire, B., Patrick, J., Sébastien, M., Nicolas, P.: Data sharing is key to innovation in health care. https://www.technologyreview.com/2019/09/27/132847/data-sharing-is-key-to-innovation-in-health-care/ (2019). Accessed 01 Aug 2021
6. Baker, L.: Sharing Information is the Key. https://www.cyberwyoming.org/sharing-information-is-the-key/ (2019). Accessed 01 Aug 2021
7. Letouzé, E., Oliver, N.: Sharing is Caring Four Key Requirements for Sustainable Private Data Sharing and Use for Public Good. Data-pop alliance and Vodafone Institute for Society and Communications, London (2019)
8. Mavroeidis, V., Bromander, S.: Cyber threat intelligence model: an evaluation of taxonomies, sharing standards, and ontologies within cyber threat intelligence. In: 2017 European Intelligence and Security Informatics Conference (EISIC), pp. 91–98. IEEE (2017)
9. The MITRE Corporation: Science of Cyber-Security. https://fas.org/irp/agency/dod/jason/cyber.pdf (2010). Accessed 01 Aug 2021
10. Ulicny, B.E., Moskal, J.J., Kokar, M.M., Abe, K., Smith, J.K.: Inference and ontologies. In: Kott, A., Wang, C., Erbacher, R.F. (eds.) Cyber Defense and Situational Awareness. AIS, vol. 62, pp. 167–199. Springer, Cham (2014). https://doi.org/10.1007/978-3-319-11391-3_9
11. Syed, Z., Padia, A., Finin, T., Mathews, L., Joshi, A.: UCO: a unified cybersecurity ontology. In: Workshops at the Thirtieth AAAI Conference on Artificial Intelligence (2016)
12. Svec, P., Balogh, S., Homola, M.: Experimental Evaluation of Description Logic Concept Learning Algorithms for Static Malware Detection. In ICISSP, pp. 792–799 (2021)
13. Takahashi, T., Panta, B., Kadobayashi, Y., Nakao, K.: Web of cybersecurity: linking, locating, and discovering structured cybersecurity information. Int. J. Commun. Syst. **31**(3), e3470 (2018)

Knowledge Utilization and Code Reuse in Models Translation

Maxim Polenov(✉), Artem Kurmaleev, and Sergey Gushanskiy

Department of Computer Engineering, Southern Federal University, Taganrog, Russia
{mypolenov,kurmaleev,smgushanskiy}@sfedu.ru

Abstract. An approach of designing expert system for model's code reuse from existing knowledge is considered in this article. As a source of knowledge was taken apart a special module of multilanguage translation package – Multitranslator. This Multitranslator module itself allows to reuse models source codes in different modeling tools. After Multitranslator further development as a client-server application, we found some exceptional cases could cause unpredictable model translation results. All of the efforts to solve those cases lead us to add explicit software – an expert system based on CLIPS. It was suggested a way properly integrate it into translation part of Distributed Storage of Models complex and so took apart translation module of Multitranslator as a start to make sure that no exceptional cases can happen on this side. To properly use it the debugging interface was added that was sure to be made with dotNet due to Distributed Storage of Models compatibility issues. As basis of the expert system itself we chose message handling due to its OOP nature. The resulting version of expert system with such knowledge base shows expectable results and improves Distributed Storage of Models complex.

Keywords: Models reuse · Translation of models · Multitranslator package · Distributed storage of models · Expert system · CLIPS · dotNet

1 Introduction

The variety of existing modeling tools [1], their improvement and the development of the new packages leads to the task of existing models' translation due to the need of reuse and storage of a previously developed and debugged models of complex systems and their components for their further use by researchers in new tools. However, this process leads to significant additional time costs for each model source code both for learning a new modeling tool, language, and format of models and for translating existing models into the required format.

To solve this problem, we decided to develop tools for models' translation between the source and target modeling languages. As a result of solving this problem, a multi-language translation tool was developed and called Multitranslator (MT) [2]. The basis of the model conversion process is the translation module (TM) of the MT, which represent some set of rules for describing the grammars of the source modeling language in

R. Silhavy et al. (Eds.): CoMeSySo 2021, LNNS 231, pp. 838–845, 2021.
https://doi.org/10.1007/978-3-030-90321-3_69

the grammar description language and generating the output code of the transformation action models in the action description language. MT is implemented as a development environment for TM's, as well as a model conversion environment that uses already developed translation modules. Obviously, for most researchers, the function of translating models is mainly in demand since the development of TM requires both knowledge of the environment itself and the input and output languages for describing models, and the languages for describing the TM of the MT [3, 4]. In the process of working and using MT it was decided to create a version of the MT in the form of a client-server application based on the architecture of distributed systems [5], where MT itself would be on the server side and would translate models for remote researchers-clients, thereby performing a distributed implementation of an approach to models reuse [6]. It was also decided to additionally implement the function of a distributed storage (repository) of models in this client-server application, which will greatly simplify the access for researchers to the original and translated models.

Based on the described approach and combining the functions of translation and storage of the original and translated models the Distributed Storage of Models (DSM) was developed [7]. At the client side it was implemented simplified user interface to connect to the server, select a TM for conversion then select source models from local or server storage. Then upon completion of the translation procedure access the code of the translated model in the target modeling language.

The flow of the development of the DSM was shifted towards the automation of translation and the resolution of exceptional cases appearing during the translation of models. Such cases were caused by insufficient input data or the uncertainty of decisions when translating models and occurs when there are too many outcomes during parsing.

To solve this, we proposed to use an expert system (ES) [8] based on CLIPS [9]. The main element in the ES development [10] is the knowledge base (KB) [11]. To develop the KB, we have to go through a multi-stage process associated with the development of the ES itself [12, 13]. The main elements of the knowledge base are rules in the form of handlers-objects or simple rules-facts for the expert system. The main process of synthesizing the necessary knowledge for a KB is knowledge engineering.

At the current stage of the development to synthesize the basic rules of the knowledge base it is required to decompose the TM and so gather all the knowledge from it which will result in covering the same cases that MT does at current state. But first, let's discuss how we integrated the new element into Distributed Storage of Models.

2 Expert System Integration into Distributed Storage of Models

To properly add the expert system to the existing Distributed Storage of Models, we developed a debugging system [14] that interacted with CLIPS core. The ES core has sets of functions that control each step of user and CLIPS interaction, which is shown in Fig. 1. We will discuss it further in detail.

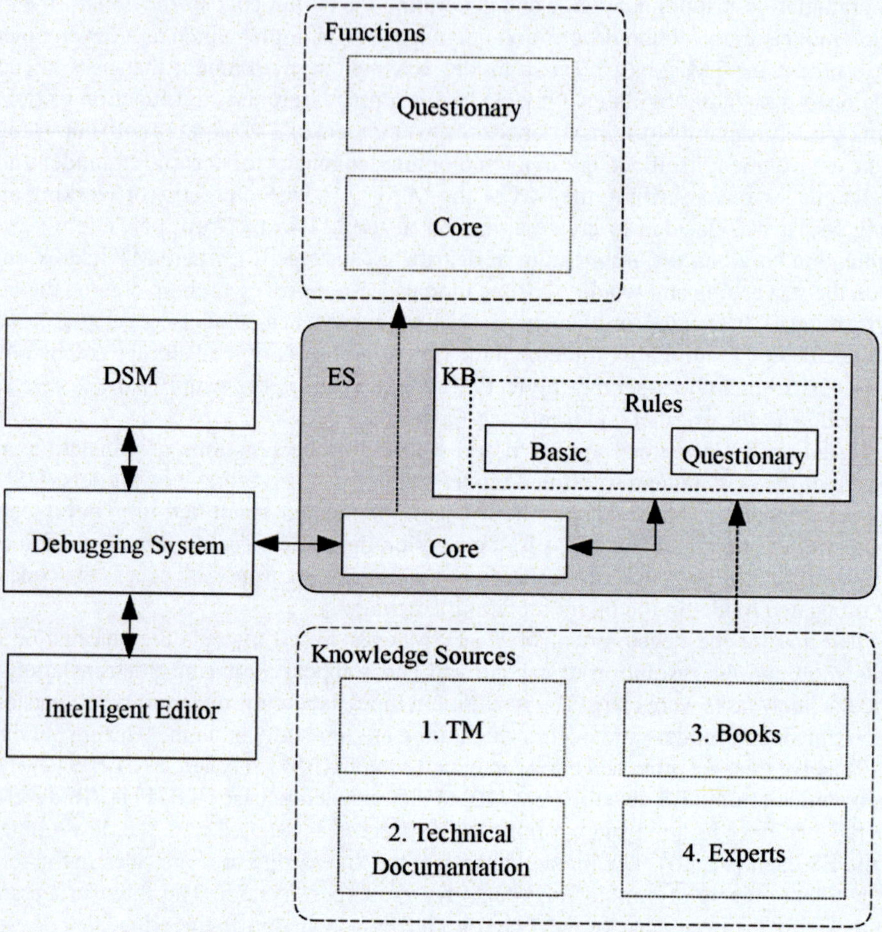

Fig. 1. The structure of CLIPS expert system integration

When working with the complex here, let's consider a full algorithm how the system shown in Fig. 1 works:

1. Information about exceptional case is transferred from the DSM to the debugging system.
2. Debugging system handles the response and sends it using dotNet [15] CLIPS library to the ES.
3. Expert system core uses its functions to process the data and either succeeds or offers a user to proceed as an expert. Next step happens only if user proceeds as an expert.
4. User is going through questionary to form the new rules of ES.
5. User is offered to either add the rules of the ES formed by questionary or edit/add them via intelligent editor.

After general explanation of the system flow now let's detail the ES inner architecture.

3 Expert System Architecture

The functions and rules interaction are complex so consider it here. ES functions are launched and controlled via debugging tools in proper order with required logics. Those functions interact with each other and with debugging tools for inputs and outputs with the flow is shown in Fig. 2.

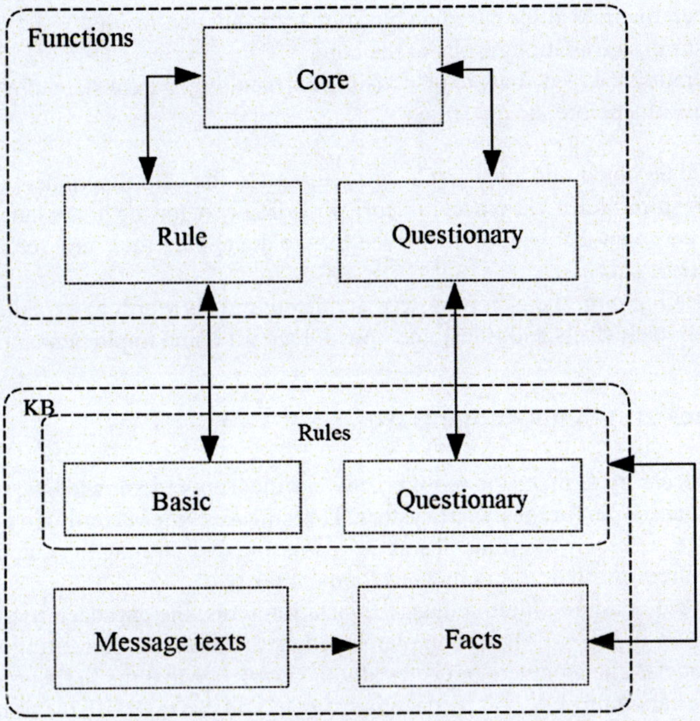

Fig. 2. Flow of the expert system inner elements

Let's consider each block to make a list of its consistency.

1. Core function are responsible for:
   - Generating rule names dynamically.
   - Translating results into required form.
   - Processing CLIPS inner format conversions.
2. Questionary functions behave to its name too and making next processes in ES possible:
   - Displaying current questionary state.
   - Processing questions throughout via states.

- Composing the questionary data with core functions of format conversions.
- Transmitting the questionary data to the validation functions.

3. Rule functions:

- Operating the facts for questionary if the state has to receive new rules from expert.
- Generating new rules based on questionary validation results.
- Returning generation results to the core.
- Operating the facts that are responsible for handling current state of KB to return the results before the questionary.

It should be noted that after results are returned to the core functions, they process them into required form via retrieving corresponding text for the given label identifier from rules and then the result is displayed by the debugging tools and returned to the DSM in case of success.

After dealing with the expert system architecture, it is worth to explain what was inside of KB itself (facts and rules) and how did we fill it and implemented the rules.

4 Synthesizing Knowledge Base

First, it is necessary to make the requirements for the acquired knowledge [13] of multilanguage translation for the expert system. This stage includes determining the source of knowledge, assessing the value of sources, assessing their availability and, as a result, choosing sources for their use in the development process.

While developing the multilanguage translation tools, the approach used for models' conversion in MT's TMs was implemented as the basis for converting the source codes of models. The module aggregates all the knowledge that the MT uses during the model's translation process, and by decomposing it, you can establish the foundation of the knowledge base. This source is the most reliable and accessible, since it has been repeatedly checked during the development of various models. This source of knowledge can be given the highest value and used as the basis for the suggested approach.

At the next stage, the official technical documentation of the source language for describing models and the output language in which the translation result is generated should be considered, since these sources are as reliable as the MT TM of the latest version. In terms of importance, such sources follow the TM and is not inferior to it in terms of availability and reliability.

After the official technical documentation, it is necessary to consider the main publications on the extended description of such documentation with practical examples that are less important and less reliable than technical documentation. In the case of the latest editions of books are moderately available due to the need to search for such in electronic libraries or purchases of electronic or printed versions, if the electronic versions of the latest editions are not available.

And the final stage is getting knowledge from experts. This source is less important than the previous ones, but after the completion of the previous steps, it may turn out to be the most significant for cases where the technical documentation, together with books describing the languages required for the TM, are insufficient to resolve all translation exceptions found. It should be noted here that the knowledge of the TM is basic and is not sufficient to resolve the exceptional cases of models' translation. This source is the least accessible and will require inviting outside experts and applying a knowledge management approach to collecting information from experts to make the information more reliable.

Thus, it is possible to list the stages and sources of knowledge acquisition during the development of the expert system as is shown in Fig. 1:

- Translation module of Multitranslator.
- Technical documentation of input/output modeling languages.
- Extended and additional publications and books about these languages.
- Experts of languages for describing translated models.

5 Expert System Implementation

Here is considered the fulfillment of the expert system functions with examples of these functions and debugging tools implementation.

To keep consequent order, let's start with examples of core functions:

```
(deffunction MAIN::find-message-by-id (?id)
   (bind ?fact
      (find-fact ((?f message-by-id))
                  (eq ?f:id ?id)))
   (if ?fact
      then
      (fact-slot-value (nth$ 1 ?fact) text)
      else
      ?id))
```

This function searches for the message by fact identifier and returns the text. This function is one of the main functions of entire virtual core component of ES since it's used for both the result output in the debugging interface and questionary.

For the next example here is one of the questionary functions:

```
(defmethod handle-state ((?state SYMBOL
(eq ?state interview))
  (?target SYMBOL (eq ?target gui))
                              (?message LEXEME)
                              (?relation-asserted SYMBOL)
                              (?response PRIMITIVE)
                              (?valid-answers MULTIFIELD)

                       (?display-answers MULTIFIELD))
  (assert (UI-state (display ?message)
                (relation-asserted ?relation-asserted)
                      (state ?state)
                      (response ?response)
                      (valid-answers ?valid-answers)
                 (display-answers ?display-answers)))
  (halt))
```

We have to notice here our naming simplification at structures since functions that don't return any data are called "methods" in CLIPS. This method handles questionary state when user that decides to take a role of an expert answers the questions to get reasonable rule generation.

The next example refers to the system of rule generation:

```
(deffunction create-name-of-rule (?uuid)
   (bind ?name (sym-cat rule- ?*rule-index*- ?uuid))
   (bind ?*rule-index* (+ ?*rule-index* 1))
   (return ?name))
```

This function takes part both in generation of templates for current rules and generation of new rule name depending on the input. The new rule name generation is more complex and takes input from questionary results, which will generate some basic rule name with UID as an input parameter for the function above. This still allows expert to change the rule name if required.

6 Conclusion

As the result of the study was proposed and implemented an approach for supporting the translation of models with intelligent tools to improve the functionality of the translation complex for model's source code reuse. The approach is based on the use of the DSM as a repository and tools of external models' translation for various modeling packages. Models' conversion implementation is based on the use of TMs (previously developed in the Multitranslator) with the support of the expert system. The formation of the necessary knowledge for the knowledge base of the ES is made based on the knowledge engineering.

The results obtained will make it possible to expand the functionality of the DSM when translating external models using the expert system and efficient processing of uncertainties arising during the translation process, which ultimately will lead to a reduction in the time spent on converting external models for complex systems modeling.

Acknowledgments. The reported study was funded by Russian Foundation for Basic Research (RFBR) according to the research project No. 19-07-00936.

References

1. Grigut, E., Kiriyakov, I., Senichenkov, Y.: Designing Application-dependent Tools for Modeling and Simulation on Basis of Universal Modeling Environment. In: Proceedings of the 3rd International Conference on Applications in Information Technology (ICAIT 2018), pp. 17–22. ACM (2018)
2. Chernukhin, Y., Polenov, M., Vemulapally, C., Solodovnik, E., Mantooth, A., Dougal, R.: Deploying modelica models into multiple simulation environments. In: Proceedings of 2005 IEEE International Behavioral Modeling and Simulation Workshop (BMAS 2005), pp. 134–139 (2005)
3. Chernukhin, Y., Guzik, V., Polenov, M.: Multilanguage Translation for Virtual Modeling Environments. Publishing house of Southern Scientific Center of Russian Academy of Sciences, Rostov-on-Don (2009). (in Russian)
4. Chernukhin, Y., Guzik, V., Polenov, M.: Multilanguage translation usage in toolkit of modeling systems. WIT Trans. Inform. Commun. Technol. **58**, 397–404 (2014)
5. Coulouris, G., Dollimore, J., Kindberg, T., Blair, G.: Distributed systems. Concepts and Design, 5th ed. Addison-Wesley (2012)
6. Robinson, S., Nance, R.E., Paul, R.J., et al.: Simulation model reuse: definitions, benefits and obstacles. Simul. Model. Pract. Theory **12**, 479–494 (2004)
7. Polenov, M., Guzik, V., Gushanskiy, S., Kurmaleev, A.: Development of the translation tools for distributed storage of models. In: Proceedings of 9th International Conference on Application of Information and Communication Technologies (AICT 2015), pp. 30–34. IEEE Press (2015)
8. Polenov, M., Gushanskiy, S., Kurmaleev, A.: Synthesis of Expert System for Distributed Storage of Models. In: Advances in Intelligent Systems and Computing, vol. 575, pp. 220–228. Springer (2017)
9. CLIPS. URL: http://www.clipsrules.net/
10. Waterman, D.A.: A Guide to Expert Systems. Addison-Wesley (1986)
11. Frost, R.: Introduction to Knowledge Base Systems. Macmillan Pub. Co. (1986)
12. Buchanan, B.G., Duda, R.D.: Principles of rule-based expert system. Adv. Comput. **22**, 163–216 (1983)
13. Durkin, J.: Expert Systems: Design and Development. Macmillan Coll Div. (1994)
14. Polenov, M., Kurmaleev, A., Gushanskiy, S., Madrigal, O.C.: CLIPS Utilization for automation of models' translation. In: Silhavy, R., Silhavy, P., Prokopova, Z. (eds.) CoMeSySo 2020. AISC, vol. 1295, pp. 788–796. Springer, Cham (2020). https://doi.org/10.1007/978-3-030-63319-6_73
15. dotNet. URL: https://docs.microsoft.com/en-us/dotnet/

Online Educational Platform as a Web Content Management System in the Organization of Student-Teacher Interaction

Alexey Sergeev[1], Natalia Kulikova[1], Elena Danilchuk[1](✉), and Natalia Borisova[2]

[1] Department of Informatics and Informatics Teaching Methods, Volgograd State Social-Pedagogical University, Volgograd, Russia

[2] Department of Computational Mathematics and Informatics Teaching Methods, Moscow State Regional University, Moscow, Russia

Abstract. The ubiquity of the Internet has led to a change in the ways of knowledge transfer and communication on the Web, which have become more intensive and focused on social interactions. Educational institutions have at their disposal various online platforms and tools for organizing online learning. But today it is becoming important to develop online educational platforms that take into account the specifics of educational institutions so that they meet their needs and requirements. The article analyzes the concept of "online educational platforms" and their features. The possibilities for educational institutions to organize the training process in the face of networking growth based on online platforms and Internet services are discussed. The capabilities of the developed online educational platform are presented, including various functions to create online learning courses, solve administrative tasks, and implement network interaction among participants of the learning process so that it meets the needs of teachers and students. All the tools of the developed online platform are easy to use by teachers of various academic disciplines and allow them to quickly adjust and supplement the online courses they develop themselves; they also enable to organize effective training and interaction between students and teachers, implement knowledge control, exchange pedagogical experience, and make it easier for students to create their own projects on the online platform.

Keywords: Online learning · Online educational platforms · Network technologies · CMS WordPress · Website development

1 Introduction

The Internet provides a teacher with a large number of tools that help to overcome the difficulties associated with the growing volume of the learning course contents and the insufficient number of hours allocated by the curriculum for its study. At the same time, when using network technologies and new models of education and training activities, teachers are required to involve all their personal resources, knowledge, competence

R. Silhavy et al. (Eds.): CoMeSySo 2021, LNNS 231, pp. 846–856, 2021.
https://doi.org/10.1007/978-3-030-90321-3_70

and creative skills. In that context, we will consider the Internet «as a special social and cultural space, the structure and dynamics of which are determined by network communities that unite people on the basis of common ideas and activities» [1]. The Internet provides diverse services which do not replace the traditional social environment but expand it, allowing to exchange information in online communities, discuss various problems, collaborate on web documents, and create own publications.

The influence of the Internet on the entire field of education is attributed to varying forms and intensity of communication on the Web, as well as social interactions, and methods of knowledge transfer. Another important factor is the influence of media content, especially the Internet, on the diversity of students' activities and communication. The implementation of pedagogical interaction on the Web using various telecommunications channels either in online or offline mode or in a blended form, involving mass-communication channels and social media is part of a multi-faceted pedagogical process, which is based on the goals and content of collaborative activities of the teacher and students. In order to facilitate the pedagogical interaction in the network, effective tools are needed to organize an online learning environment for students [2–5]. Special attention should be paid to involving students in online learning [6–8].

Today, educational institutions actively use various online platforms to organize the learning process, but not all online platforms meet their needs, which will require the search for alternative online platforms [9, 10] or the development of new ones [11, 12]. Next, we will analyze in more detail the very concept of "online educational platforms", the opportunities they offer for organizing the learning process to teach students in a network environment and present an online platform created to support the learning process at the Volgograd State Social and Pedagogical University.

2 Online Educational Platforms and Their Capabilities

Equipping educational institutions with computer and interactive facilities does not lead to a large increase in the quality of education without changing the approaches to learning and supplementing them with new forms and methods, as well as means of enhancing the cognitive activity of students. The search for new approaches to learning is driven by the spread of online educational platforms which ensure the constant emergence of new networking opportunities and the dynamic update of the technical base and content available on each such resource. The interest of the educational community in online educational platforms and the active use thereof imply an increase in the share of network interactions in education and the relevance of the analysis of this phenomenon and its possibilities [1, 13].

Practice shows that when networking on online learning platforms, there is a need to solve tasks such as: activating the individual intellectual processes of students with the arousal of their inner discourse; creating conditions for understanding and assimilation of new information by students; tailoring pedagogical interaction, with students evolving into subjects of influence; and organizing two-way communication with the exchange of information between participants in the learning process [14]. When planning classes in the networking environment, it is important for the teacher to:

- set goals and targets allowing for objective control;
- pre-design the learning process to subsequently reproduce it in pedagogical practice;
- choose the best forms, methods and means of training, depending on the selected information and communication technologies;
- organize in-process monitoring and feedback for quick and timely correction of the learning process.

It is important to note that in the new information reality, teachers need to be ready to use information and communication technologies competently, for which they should have a set of methodological recommendations for conducting classes using modern interactive pedagogical practices: group work; mutual learning; brainstorming; discussions and debates; applying and creating interactive educational presentations for public speeches; case studies; project activities, including long-term collective projects; as well as using Internet services.

Online educational platforms allow teachers to use their integrated innovative tools for the organization of the learning process. On their basis, it is possible not only to provide theoretical and practical training materials in an interactive and multimedia form but also to implement the interaction of all participants involved in the learning process, regardless of their location and time of the class.

Analyzing the pedagogical practice of teaching on the Internet, we can distinguish online educational platforms, virtual universities, and materials that are freely available on the Internet, for example, on various social network services. It is necessary to consider the concept of an online educational platform in more detail since there is no generally accepted definition of it at the moment.

A. A. Smirnova uses the term online educational platform to refer to software solutions that provide access to training content (mass open online courses, text-, audio- and video recordings, real-time broadcasts of lectures, tests, etc.) for users with any level of knowledge through the interface of the website [15].

We will understand by online educational platform a set of software solutions, Internet services and interactive technologies that provide students access to educational content (online courses, learning material in the form of text, audio, and video recordings, real-time broadcasts of lectures, various case studies, interactive tasks, test materials, etc.), as well as support teachers getting feedback from students in the process of interactive communication and ensuring control (sending reports on tasks, testing, evaluation, rating, etc.) implemented through the interface of the website [16].

As many studies show, today Russian universities mainly use a two-component information-educational environment, which combines the content they develop on their own with the available resources of international educational platforms, including foreign (Coursera, edX, Udacity, FutureLearn, Codecademy, KhanAcademy, Ted, etc.) and Russian educational platforms and resources (PostNauka, Lektorium, Open Education, Universarium, University Without Borders, Uniweb, etc.) [17, 18].

Today, Russia is actively analyzing the experience of foreign colleagues and developing its own online network training niche [15, 19].

Educational institutions actively use various online educational platforms and develop their own, which allows them to:

– create an extensive feedback system;
– add tools for effective diagnostics of students' development levels, tracking their intermediate results, and making timely adjustments;
– fill the platform with both educational information and electronic educational resources with varying degrees of interactivity, which the teacher can create independently or use resources on various educational sites and portals.

A great advantage of modern online platforms is that they provide the teacher with an opportunity to implement their curricular practices and individual style when creating their own virtual educational platform to achieve their teaching objectives.

Analyzing the results of the integration of foreign educational platforms into the national education system, we note that when actively developing and promoting national platforms, the use of foreign platforms with distance learning courses is acceptable provided that they are manageable, controllable, and accessible for moderation by the teacher and the academic community, including psychological protection of students [20]. Teachers who use online platforms in the modern training process need to continuously improve their expertise in the ability to design their contents for national education standards and evaluate the quality of educational resources. Online educational platforms enable teachers to select the necessary learning materials suitable for the subject by using online tools and introduce them to the learning process through the Internet.

3 Using an Online Educational Platform

Let us consider the possibilities of creating and using online platforms in teaching by the example of the e-learning portal of the Volgograd State Social and Pedagogical University (see Fig. 1).

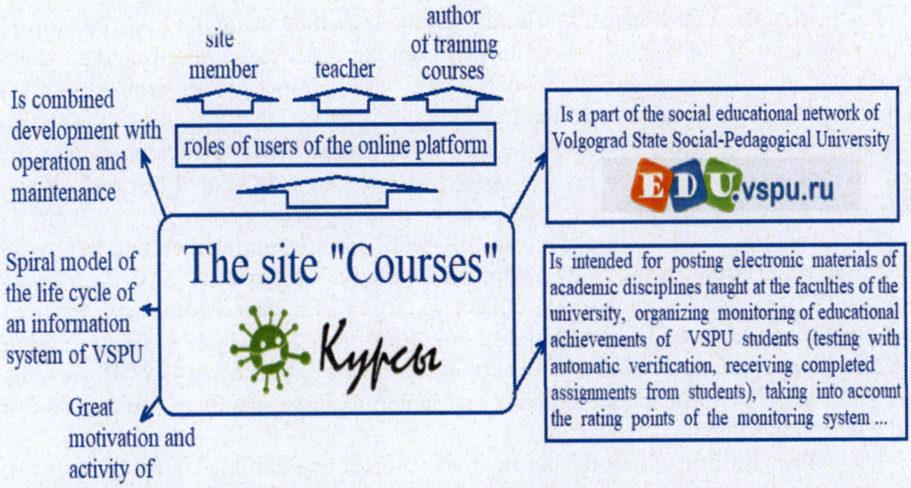

Fig. 1. Conceptual diagram of the online platform "Courses" of the e-learning portal

An important feature of this online platform is the ability to develop training courses by teachers themselves using convenient and intuitive tools, which is bringing huge changes to the traditional software development model. Teachers have the opportunity to continuously improve their online courses, adapt them to the changing needs and interests of students, not only post factum but also in the online mode. This increases the motivation of students to acquire new knowledge and their desire to improve their academic performance. Thus, the courses are constantly updated, and students get immediate communication with teachers due to advanced content delivery. Training courses of the online platform become part of the social network of the university, through which a high-quality student-teacher interaction is ensured. This feature allows Volgograd State Social-Pedagogical University to use a spiral model of the information system life cycle.

The online platform was created using the Content Management System (CMS) WordPress, which allows to provide and organize a joint process of creating, editing and managing content and the web content management system (website content).

The WordPress CMS enables to:

- provide teachers with the tools to create and publish their own content, allowing them to organize collaborative work;
- manage content: storage, version control, compliance with access mode, etc.;
- present information in a convenient form.

When designing an online platform based on the WordPress CMS, one can: «quickly» build «complex» websites; easily create content using built-in editors (no special knowledge in the field of web development is required); use feedback tools (comments, ratings, forums, etc.); differentiate access to content based on the built-in account system; categorize content, using built-in navigation and search systems; build multilingual websites; customize the appearance of URLs (Clean URL, SEO-friendly URL); follow web standards.

Features of the WordPress CMS implementation include using the LAMP platform (Linux, Apache, MySQL, PHP); installing by copying CMS files and setting the DBMS access parameters; using the admin dashboard (configuration, management, content creation); using design templates (themes), plugins, widgets; using caching to speed up access; regularly updating CMS software components (security, new features).

The online platform (see Fig. 2) is part of a social educational network of the Volgograd State Social and Pedagogical University (http://edu.vspu.ru).

The structure of the online platform of the e-learning portal includes five main blocks (see Fig. 3): electronic materials for the interactive whiteboard (1); online training courses (2), area with an open-access collection of guidance papers for the exchange of pedagogical experience (3), area with training documentation which facilitates design and development of curriculum and assessment materials by teachers (4), platform tools that make it easier for teachers to create test materials, integrate them into the course and control the students' knowledge (5).

The online platform allows the teacher to develop online training courses; use audio, video, photos, etc.; exchange various links, files and messages with students; embed video materials and various Internet services on the platform; create and upload original projects to the platform which students can do themselves; add tests and surveys; create

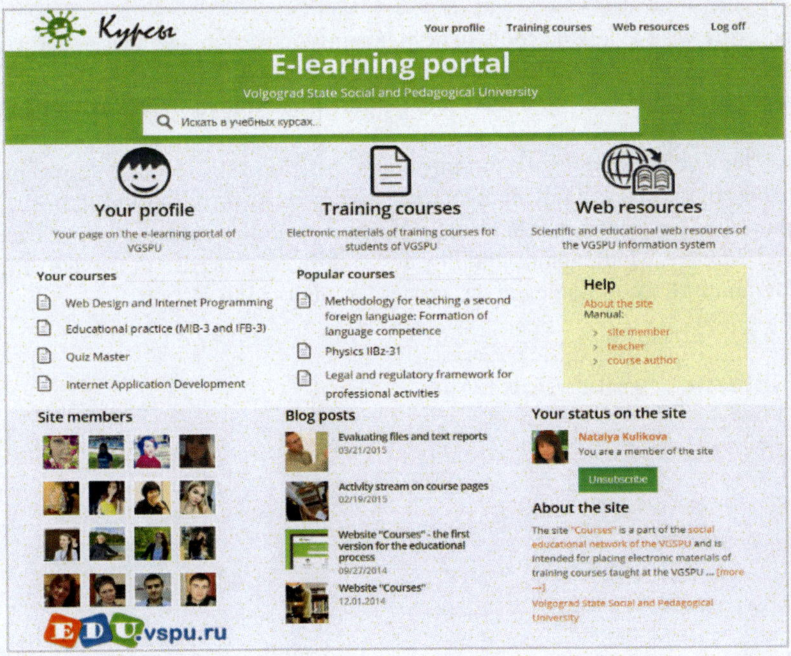

Fig. 2. Home page of the e-learning portal of the Volgograd State Social and Pedagogical University

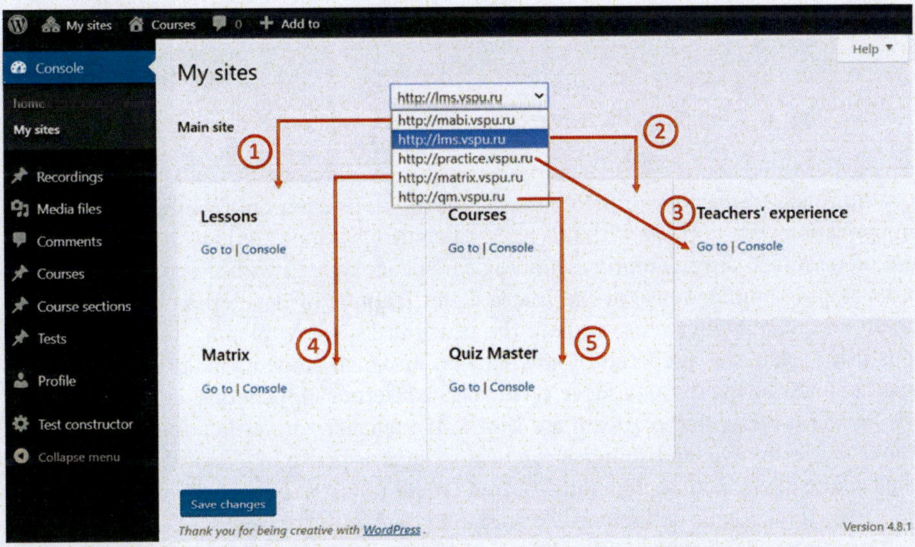

Fig. 3. Admin dashboard of the online platform of the e-learning portal

new platforms with available tools or use existing ones; manage access and rights of users and teachers; use monitoring tools (distribute students into groups, access files and text reports, automatically collect rating points, and upload student results in tabular form).

Let us look at an example of a training course developed by teachers shown in Fig. 4. Study of the course «Development of Internet applications» on the online platform (http://lms.vspu.ru/courses/razrabotka-internet-prilozheniy) allows students of the Faculty of Computer Science, Mathematics and Physics to develop expertise in the use of modern web programming technologies in the process of remote interaction.

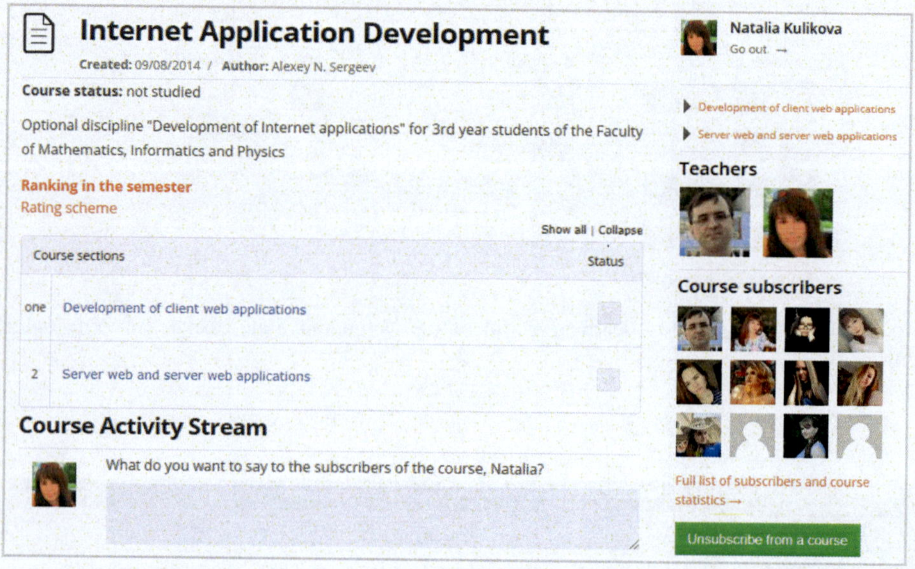

Fig. 4. Example of the online course «Internet Application Development»

Within the framework of the course, students receive not only methodological recommendations for working with this course but also the opportunities to study and analyze modern web programming technologies; get acquainted with the main approaches to developing applications for the Internet and features of developing client-side and server-side web applications; explore the possibilities of developing Internet applications using examples prepared by the teacher, in which students make changes to the program code during training; develop their own Internet applications, etc. All materials developed on the online platform are sent to the teacher for verification either in files or in a single archive, and students can also write a response in a text window or insert a link to the developed applications located on the Internet. For feedback, students can use personal messages on the portal or a shared comment area etc.

The results of the students' work are reflected in the rating table (http://lms.vspu.ru/courses/razrabotka-internet-prilozheniy/rating) on the platform, which can be used as a course navigator for jumping to sections, or downloaded to your device (see Fig. 5).

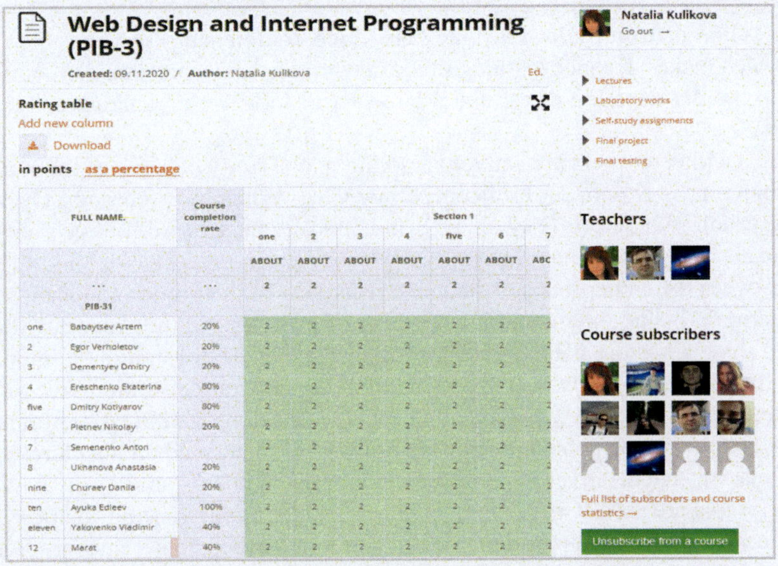

Fig. 5. Example of a training course rating table

Figure 6 shows the integration of the site Quiz Master (http://qm.vspu.ru) created on the online platform for monitoring knowledge and evaluating the competencies of students into the training course (http://qm.vspu.ru). The reference number of a test from the Quiz Master site is inserted into a special column for external evaluation in the training course section, where test materials are provided, and the students' test results are automatically reflected in the rating table.

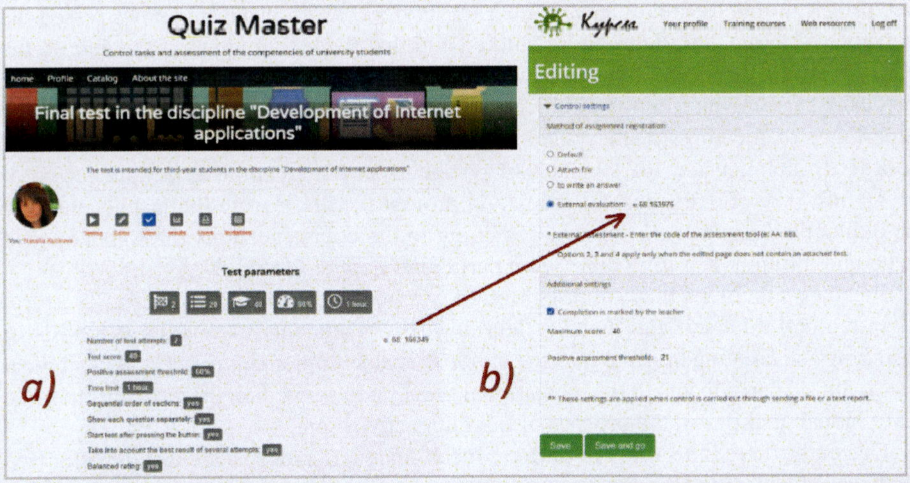

Fig. 6. Example of integration of tests from the site (a) Quiz Master into the course (b) «Internet Application Development»

The online platform allows to integrate video conferencing services, the optimal use of which makes distance learning in the networking environment more comfortable thanks to the participation effect and a large set of tools for organizing collaborative activities.

At the faculty, teachers are actively working on the study of services for organizing video conferences (Zoom.us, Proficonf, Draw Chat, Whereby, Webroom, OpenMeetings, BigBlueButton, etc.). They analyze and compare their possibilities for use in the educational process as an integrated distance learning technology based on virtual interaction between different groups of students, regardless of their time and location. Comparison criteria include the maximum number of participants, the broadcast time, the ability to share and store files, file demonstration, screen demonstration, joint management of the program window on a shared screen, the need for registration, free functions and features, etc. Particularly worth noting are OpenMeetings and BigBlueButton, open software solutions to support the learning process, implemented architecturally in the form of classic "client-server" systems, and comprising open-source software with all the features that are usually inherent in commercial video conferencing solutions. Users do not need to install the app on their device to join a video conference or a webinar. These applications are distributed under a free license and installed on your own (or third-party) webserver.

4 Conclusion

The analysis of the results of the survey proposed to students after completing the course showed that the study of web programming technologies through an online platform arouses great interest among students, with an especially positive response to training through practice: combining theoretical lecture material with its direct application in practice in the process of completing tasks for lectures, getting acquainted with new web programming technologies and Internet services. Most of the students stressed the relevance of the technologies being studied and that the technologies presented on the platform would be useful for them in their future professional activities. Many noted the desire to see even more interactive video resources on the platform, as well as the importance of meetings in the video conference format with the participation effect, in order to compensate for the lack of personal communication in remote forms of interaction. The effectiveness of quick and prompt feedback from the teacher through the online platform and the ease of switching to the course sections through a special menu and through the rating table were particularly appreciated.

It can be concluded that training through practice on the basis of an online platform for intramural, blended and distance learning allows students to develop the skills of designing and creating original Internet applications; to implement network pedagogical interaction of participants in the educational process; to show their creative abilities and professional qualities in the process of learning activity, etc.

Currently, there is a variety of online educational platforms that are available for free and have a fairly large amount of training resources at their disposal. To create their own educational platforms, teachers have a large selection of HTML editors, site designers, content management systems, in which they can embed and use Internet services etc.

These resources make it possible to create an extensive feedback system, use tools to diagnose students' competence levels, track their intermediate results and statistics of students' visits to the platform along with the dynamics of their work with materials available on it.

In our opinion, of a special value are the opportunities for teachers to improve the quality of education provided by modern online platforms: make timely adjustments to learning materials; fill it with a variety of educational information and multimedia interactive educational resources; use simple and convenient tools to ensure knowledge control.

Acknowledgment. Thanks to the reported study was founded by RFBR within the research project No. 19-29-14064 "Theoretical and methodological foundations and technological support of educational activities in online communities of school students".

References

1. Sergeev, A.N.: The educational social network in the structure of a University's web resources. Res. J. Appl. Sci. **9**(12), 909–912 (2014)
2. Allen, I.E., Jeff, S.: Digital Compass Learning: Distance Education En-rollment Report 2017. Babson survey research group (2017)
3. Andel, S.A., et al.: Do social features help in video-centric online learning platforms? A social presence perspective. Comput. Hum. Behav. **113**, 106505 (2020)
4. Chang, L.-J., et al.: EduXs: multilayer educational services platforms. Comput. Educ. **41**(1), 1–18 (2003)
5. Pirkkalainen, H., Pawlowski, J.M., Pappa, D.: Educators' open educational collaboration online: the dilemma of emotional ownership. Comput. Educ. **106**, 119–136 (2017)
6. Gaebel, M., et al.: E-Learning in European Higher Education Institutions. European University Association (2014)
7. Kim, D., et al.: Exploring student and teacher usage patterns associated with student attrition in an open educational resource-supported online learning platform. Comput. Educ. **156**, 103961 (2020)
8. Magalhães, P., et al.: Online vs traditional homework: a systematic review on the benefits to students' performance. Comput. Educ. **152**, 103869 (2020)
9. Koretz, D.M.: Measuring up. Harvard University Press (2008)
10. Yueqing, P.: Study of PE teaching evaluation system in general colleges [J]. J. Anhui Sports Sci. **4** (2010)
11. Dermaku, A., et al.: Educational and school management platform. IFAC PapersOnLine **49**(29), 138–143 (2016)
12. Glava, C.-C., Glava, A.-E.: On-line learning platforms as virtual classrooms. case study of initial primary teachers training at Babes-Bolyai University of Cluj-Napoca, Romania. Procedia Comput. Sci. **3**, 672–676 (2011)
13. Chuang, I., Andrew, H.: HarvardX and MITx: Four years of open online courses--fall 2012-summer 2016. Available at SSRN 2889436 (2016)
14. Kulikova, N.U., Danilchuk, E.V., Zhidkova, A.V.: Formation of readiness for future physics teachers by using interactive learning tools. AIP Conference Proceedings, vol. 1797, no. 1. AIP Publishing LLC (2017)

15. Smirnova, A.: Educational online platforms as a phenomenon of modern education: definition of the considered concept. Artif. Soc. **14**(1), 0 (2019). https://doi.org/10.18254/S207751800005274-0
16. Kulikova N.U.: Online educational platform as a factor in the study of interactive learning technologies in a network environment. World of Science. Pedagogy and psychology, [online] 4(8). Available at: https://mir-nauki.com/PDF/26PDMN420.pdf (in Russian) (2020)
17. Kiselev, B., Yakutenko, V.: An overview of massive open online course platforms: personalization and semantic web technologies and standards. Procedia Comput. Sci. **169**, 373–379 (2020)
18. Lobova, S.V., Ponkina, E.V.: Online courses: to accept impossible to ignore. Vysshee Obrazovanie v Rossii = Higher Educ. Russia **30**(1), 23–35 (2021). https://doi.org/10.31992/0869-3617-2021-30-1-23-35
19. Semenova, T., Vilkova, K., Shcheglova, I.: The MOOC market: prospects for Russia. Voprosy Obrazovaniya/Educ. Stud. Moscow **2**(2), 173–195 (2018)
20. Onosov, A.A., Bryzgalina, E.V., Savina, N.E., Tumanov, S.V.: Foreign learning platforms in Russian education system: potential capacity evaluation and risks forecasting. Higher Educ. Russia **27**(8–9), 135–151 (2018). https://doi.org/10.31992/0869-3617-2018-27-8-9-135-151

From the Real World to the Virtual and Vice Versa: Recognition of Man-Made Objects and Structures and Augmented Reality as Tools for Digitalization of Natural Science Education

N. O. Besshaposhnikov, A. G. Kushnirenko(✉), and A. G. Leonov

Federal Research Center "Scientific Research Institute for System Analysis of the Russian Academy of Sciences", 36-1 Nakhimovsky Prosp., Moscow 117218, Russian Federation
agk_@mail.ru

Abstract. The use of neural networks-based recognition algorithms for toy objects and structures and augmented reality generation algorithms opens up new methodological possibilities in the teaching of natural sciences. Artificial intelligence methods make it possible to transfer to the material world the processes of compiling and executing control programs for real or virtual robots and the processes of generating environments (game fields) along which robots move. This technical, at first glance, achievement opens up the possibility of developing and practically introducing a methodology for consciously (in the sense of L. Vygotsky) mastering by the preschoolers of the system of scientific concepts of programming. Children master this system of scientific concepts by working with tangible instructional kits and composing programs for managing real and virtual robots in the "PiktoMir" textless programming environment.

Keywords: PiktoMir · Pikto · Algorithmics · Textless · Preschooler · Augmented reality · Neural networks

1 Introduction

Today, few experts undertake to anticipate the needs of society in certain professions, competencies and skills for 10–20 years ahead. At the same time, most would agree that the education system should equip children with some universal competencies. Reading, writing and counting have long been considered such universal competencies; today, a growing number of experts rank as universal competences the ability to analyze and algorithmize plans for future activities, that is, the ability to program. In the 21st century, more and more scientists and educational practitioners are developing methods for introducing programming to preschoolers who still cannot read and write. To support such techniques, textless, pictogram programming environments are used, which make it possible to compose control programs for real toy robots and virtual screen robots [1]. In the past 10 years, the computing power of mass-produced personal computer devices has grown so much that it has made it possible to implement resource-intensive

R. Silhavy et al. (Eds.): CoMeSySo 2021, LNNS 231, pp. 857–869, 2021.
https://doi.org/10.1007/978-3-030-90321-3_71

artificial intelligence and augmented reality algorithms on such devices. This opens up new methodological opportunities for teaching programming to preschoolers and younger students. This article describes our experience in implementing and testing these new features [2].

2 Methodology

Teaching preschoolers to program, we will not only teach the child to compose simple control programs for real toy robots or virtual robots on the tablet screen, but also want the child to create in his imagination a model of the World in which the robot, the programmer, the program and the computer closely interact with each other in accordance with clear, clearly established "rules of the game" described in [2]: a) the robot executes the commands received from the computer, b) the computer issues the robot a sequence of commands specified by the program, c) this sequence is determined not only by the program, but also by the properties of the playing field on which the robot operates; d) a program is created by a person we call a programmer. Programming can be considered mastered if the child understands the "rules of the game" in which we immerse him in class, can tell and discuss what these rules are, that is, he or she is aware of the system of scientific concepts by which programming is based.

This awareness, according to L.S. Vygotsky [3], can arise only as a result of the child's immersion in various activities in which the child will play the roles of a programmer, robot, computer and participate in the interactions necessary to fulfill these roles. Pedagogical theory (stages of Piaget's development) and our practice [4] suggest that the development of a system of scientific ideas about programming in a child is possible at the age of 6+ years. According to our methodology, this development takes place in a one-year course, in which one adult once a week deals with a group of children at least 6 years old (the age of children in the graduating group of a Russian kindergarten), using the PiktoMir textless programming environment [5] developed in Russia (see Fig. 1). Three special training sets and two neural network algorithms implemented in the PictoMir environment play an important role in our methodology.

Instructional kits:

- a playing field composed of multi-colored mating square soft rugs with a side of 30–33 cm, (Fig. 4 and Table 1a);
- a real toy robot moving around the playing field obeying commands given in the form of sound signals heard by children, (Fig. 4);
- cubes with pictograms of robot's control commands on the faces and cubes with sets of 1 to 6 dots depicted on the faces, (Fig. 2 and Fig. 3).

Fig. 1. Typical lesson with the PiktoMir system.

Algorithms:

- recognition of the structure of the robot control program made up of cubes (Fig. 5) according to the rules of the "Pikto" language [6] and loading of this program into the textless program environment "PiktoMir";
- recognition of the combinatorial structure of the playing field from rugs (Fig. 8) and loading of this structure into "PiktoMir".

3 Results

The practices that we will now describe have been successful in teaching programming to thousands of Russian preschoolers [4]. First, children observe how adults control the robot by giving it audible commands using a sound remote control. Then each child is given the task of issuing to the robot a sequence of commands to solve some simple, verbally formulated problem by an adult. For example, guide the robot along a route from one corner cell of the field to another. Then the child is given the task to remember the sequence of commands that have just led the robot along a given route. The child solves this problem with the help of cubes, placing them with the necessary icons up in one row from left to right, and then in several rows one below the other in a sequence from left to right and from top to bottom.

The configuration of cubes laid out on the table (Fig. 2) is nothing more than a kind of program. By looking at the program and giving commands to the robot using the remote control, the child can execute the program one or more times. The execution of the program, drawn up by one child, can be entrusted to another child, if he is familiar with the rules for the creation and execution of the program.

Fig. 2. Program constructed with cubes.

After the steps described above have been completed by each child in the group, the child has already mastered, with one exception, the system of basic programming concepts. Children have already gained experience in controlling the robot via sound remote control, experience in executing programs. At this moment we introduce the computer as a device that takes over, automates, the boring for process of program execution. In our introductory lessons, the computer performs only two functions: memorizing the program shown to it in its memory, and commanding the robot according to the memorized program. The child shows the computer (photographs) a program made up of cubes and sees that the computer is trying to "understand" the program and displays on the screen how the computer understood the program (Fig. 3). If the child confirms that the program is understood correctly, the computer remembers the program in its memory and is ready to execute this program at the order of the child as many times as necessary.

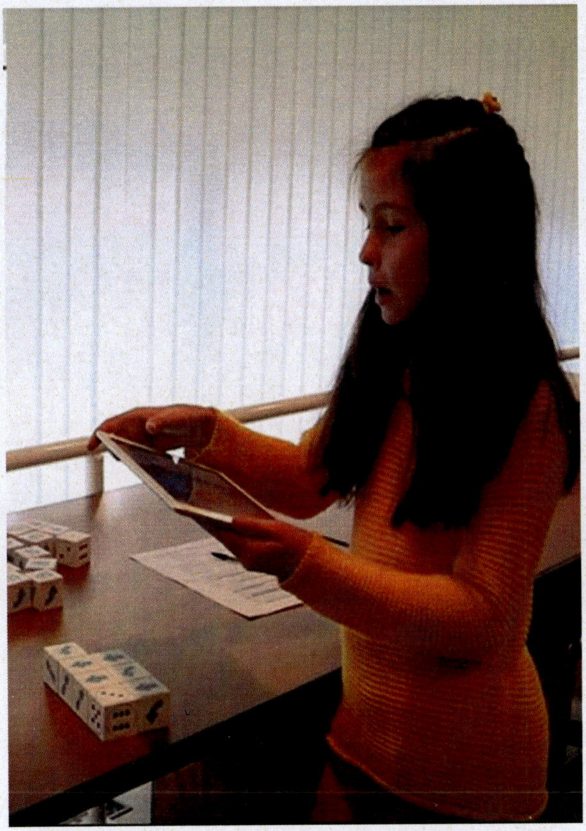

Fig. 3. Taking photo of a program.

When executing the program (see Fig. 2 and Fig. 4), the computer gives the robot sound commands identical to those that the robot received from the sound remote control. Thus, children observe (hear) the process of executing the program on the computer. It is important that the computer performs only those tasks that the children have learned to perform on their own by this time without resorting to the help of a computer. In this case, children can see how the computer replaces the Human, performs human functions. Since the child himself is able to recognize, memorize and execute programs laid out from cubes, the question of how exactly the computer performs these operations does not arise for children.

Fig. 4. Crowler the robot (Polzun).

4 Discussion

4.1 Recognition of a Program. Methodological Issues

The composition of a program from tangible objects is used in a number of robotic toys (Robot mouse [7], Primo Toys [8]). However, the programs compiled in these toys are too simple and can only support a few of programming lessons. For example, in Primo Toy system, the program element is a cube equipped with a human recognizable image on the front face and a hardware recognizable electrical connector on back face. To compose a program out of cubes, you have to place cubes in a grid of 20 sockets of special electrical device. This technique allows you to make only the simplest programs, for the structure of the program is rigidly fixed and only a small number of cubes of each type are included in robotic kit of this type.

The method we have implemented allows us to compile programs of a rather complex structure with nested blocks of control structures. The main advantage of our approach is that to designate control commands we use drawings that have an obvious meaning. We do not include text or specific labels [9] on our picto-grams (analogues of a QR code), which allows us to lower the age of using this technology. An additional advantage of our method is that the teacher can lay out a program from magnetic cards on a magnetic board, showing the process of drawing up a program to all children at the same time, and then "show" this program to his computer, which will then execute it, giving the robot sound commands that are audible and understandable by children.

4.2 Recognition of a Program. Technical Issues

Bringing the recognition algorithms of man-made objects to practical use in a real educational process turned out to be technically not an easy task and required considerable efforts to select the architecture of the neural networks used, to develop new approaches to the training of networks and to combine deterministic and heuristic algorithms.

The problem of generating in the computer memory a formal description of the program, starting from photographs of configuration of cubes placed on the table consists of two subproblems. Recognition of individual pictograms in the photograph and more difficult recognition of the two-dimensional structure that individual pictograms form. The software module built into "PiktoMir" recognizes individual pictograms and rows of pictograms in the image and, in the style of the Python language, endows this data with the structure of the "Pikto" program based on the horizontal and vertical indentation between the pictograms and the rows of pictograms. We managed to develop a software module that quickly and steadily solves this problem on a modern tablet for images obtained during manual shooting of a set of two or three dozen cubes laid out by a child with all kinds of spatial distortions and lighting variations.

Recognition of a program laid out of cubes consists of three main stages:

1. Finding icons using a neural network (iOs – CoreML Object Detection [10], Android – SSD MobileNet v2 [11, 12]).
2. Finding the combinatorial structure of the location of recognized pictograms using a neurotabulator [13].
3. Combining the program template in "PiktoMir" with recognition results.

The result of program recognition algorithm work is shown on Fig. 5.

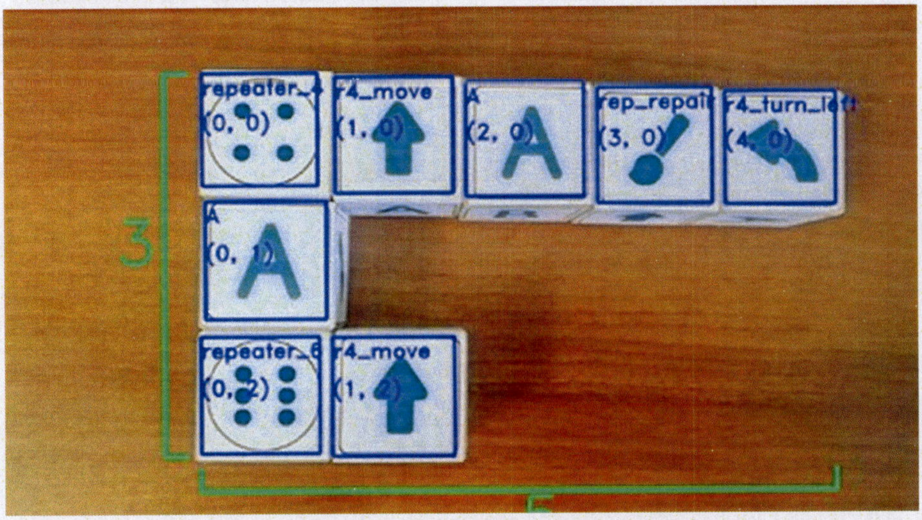

Fig. 5. Recognized program constructed with cubes.

4.3 Recognition of a Playing Field. Methodological Issues

The second problem we solved is recognition of the combinatorial structure of the playing field laid out on the floor along which toy robots move. The field is laid out of multi-colored articulated square rugs with digits printed on them. After the children, under the guidance of the teacher, assembled a real playing field from the rugs and discussed the task that the robot should perform, each child "photographs" the real field in order to quickly get its virtual copy in the memory of its personal tablet (Fig. 6).

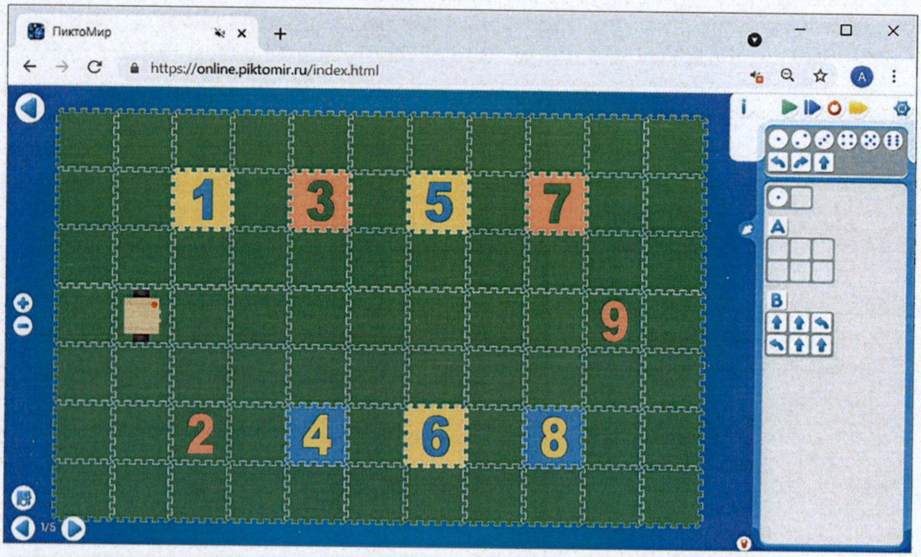

Fig. 6. The recognized playing field and the template of the control program appeared on the tablet screen of one of the children in the group.

This allows each child to predebug in the virtual environment his personal version of the robot control program (Fig. 7).

After individually debugging his or her program in a virtual environment, the child gets at his disposal a real robot in a real field and sees how a real robot in a real field from rugs under the control of a program compiled by a child performs actions synchronously with a virtual robot on the screen.

4.4 Recognition of a Playing Field. Technical Issues

The playing field recognition technology is based on the same principles; however, it has two additional preprocessing steps:

1. Finding key points – the corners of the field (Fig. 8).
2. Highlighting colored rugs (without numbers) using standard OpenCV library methods [14].

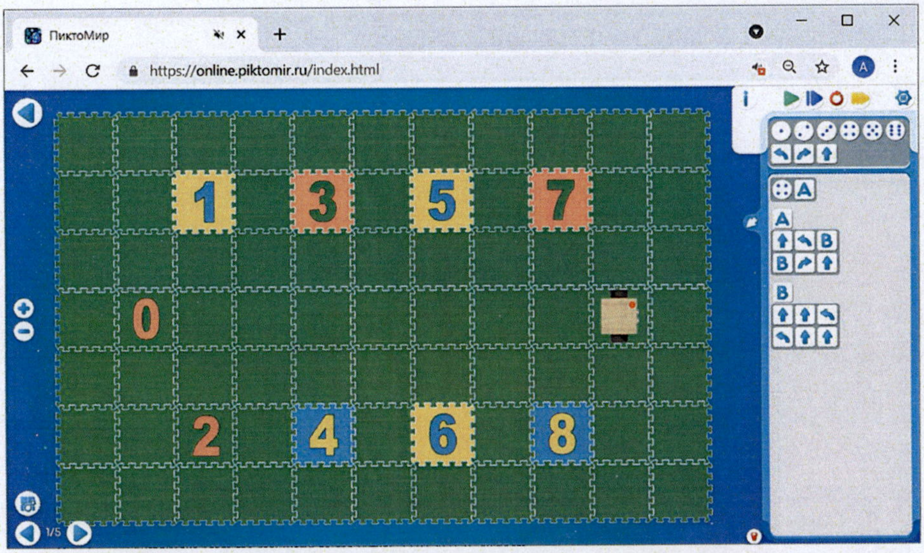

Fig. 7. Successfully debugged program on the child's tablet screen.

Fig. 8. Keypoints of a playing field.

4.5 Augmented Reality. Methodological Issues

We used augmented reality as an additional technique in our courses. The task was to enrich the image of a real playing field using augmented reality: a child, using a tablet camera, observes on the screen in augmented reality how a virtual robot moves across a playing field consisting of real rugs and some virtual objects. (see Table 1).

Table 1. Combination of real rugs and virtual object

	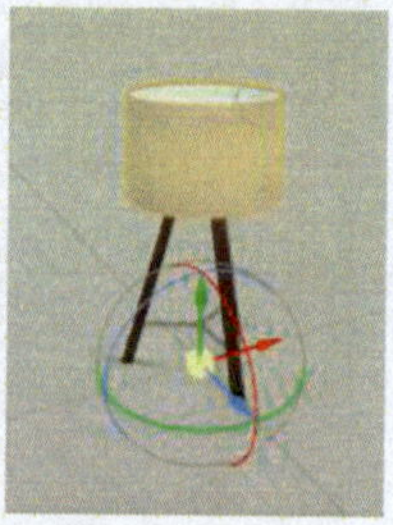	
a) The real playing field	b) Virtual object	c) The new AR playing field, combining real and virtual objects

Implemented two possible modes of operation with augmented reality:

- Total virtuality – both rugs and a robot are shown as "complemented"
- Real rugs and virtual Crawler Robot – "real" rugs are recognized, and the "augmented" robot moves on them (the good thing about this approach is that batteries of the virtual robot do not need maintenance and recharging ☺)

4.6 Augmented Reality. Technological Issues

The first mode of operation presents no difficulties in implementation: it is enough to use the standard set of functions of modern mobile augmented reality libraries ARKit [15] for iOs and ARCore [16] for Android.

The work in the second mode directly depends on the quality of the playing field recognition by its flat image obtained by the tablet. After recognition, the coordinates of the corners of the rugs on the flat picture received by the tablet at the time of the start of recognition should be embedded in the virtual three-dimensional AR space using augmented reality libraries. At this stage, additional difficulties arise: recognition of the field takes about one or two seconds, and during this time the tablet can move from the position it occupied at the beginning of the survey. Accordingly, the coordinates on the recognized image may be distorted. To fend off tablet movement, the projection of the center of the screen into AR space is recorded twice: during initial shooting and after recognition. Then the displacement vector in the augmented reality space is calculated, and all projected coordinates of the corners are displaced by this vector.

Then, using the coordinates of the corners, it is enough to calculate the centers of the field cells and move the "augmented" robot between them in accordance with the program compiled by the child (Fig. 9).

Fig. 9. AR robot moving on a playing field.

5 Conclusion

Our experience shows that after children have mastered a set of basic programming concepts in the course of role-playing games and consolidated their practice in the PiktoMir environment, the understanding of the classic basic program structures, namely, repetition, branching and subprogramming, occurs without any particular difficulties and by the end of the year each child in the group gains experience in independently compiling and debugging up to 120–150 programs. Moreover, in the last lessons of the course, children cope with the preparation of programs using cycles and auxiliary algorithms (subroutines).

And the augmented reality helps to decrease the cost of teaching aids kit excluding real robot from the kit. As a by-product of this solution, we also eliminate a bunch of real robot maintenance problems. Indeed, a virtual robot (Fig. 10) does not require recharging the battery and is not prone to failures due to articulation defects of real rugs.

Fig. 10. Crowler the AR robot

Acknowledgments. The work was carried out within the framework of the state assignment of Scientific Research Institute of System Analysis of the Russian Academy of Sciences (SRISA/NIISI RAS) on topic No. 0580-2021-0010, registration No. 121031300134-3).

References

1. Besshaposhnikov, N.O., Kushnirenko, A.G., Leonov. A.G.: PiktoMir: how and why do we teach textless programming for preschoolers, first graders and students of pedagogical universities. In: Proceedings of the 13th Central & Eastern European Software Engineering Conference in Russia (CEE-SECR '17), Article No. 21, pp. 1–7. ACM, New York, NY, USA (2017)
2. Betelin, V.B., Kushnirenko, A.G., Leonov, A.G.: Basic concepts of programming expounded for preschoolers. Inform. Primen. **14**(3), 55–61 (2020). https://doi.org/10.14357/19922264200308
3. Zavershneva, E, van der Veer, R.: Thinking and speech. In: Zavershneva, E, van der Veer, R. (eds.) Vygotsky's Notebooks. PCR, vol. 2, pp. 353–366. Springer, Singapore (2018). https://doi.org/10.1007/978-981-10-4625-4_21
4. Rogozhkina, I., Kushnirenko, A.: PiktoMir: Teaching programming concepts to preschoolers with anew tutorial environment. Procedia Soc. Behav. Sci. **28**, 601–605 (2011). https://doi.org/10.1016/j.sbspro.2011.11.114

5. Besshaposhnikov, N.O., Kushnirenko, A.G., Leonov, A.G., Raiko, M.V., Sobakinskikh, O.V.: Digital educational environment PiktoMir: experience of development and mass implementation of an annual programming course for preschoolers. Inf. Educ. (10), 28–40 (2020) (In Russ.). https://doi.org/10.32517/0234-0453-2020-35-10-28-40
6. Leonov, A.G., Besshaposhnikov, N.O.: Pictographic programming language "Pikto." Proc. Cybern. **4**(28), 173–190 (2017)
7. Learning resources: https://www.learningresources.com/code-gor-robot-mouse-classroom-set. Accessed 15 July 2021
8. Meet Cubetto: https://www.primotoys.com. Accessed 15 July 2021
9. Felix, H., Zekelman, A., Horn, M., Judd. F.: Strawbies: explorations in tangible programming. In: IDC 2015: Proceedings of the 14th International Conference on Interaction Design and Childre, pp. 410–413. ACM, New York, NY, USA (2015)
10. Turicreate Object Detecton: https://apple.github.io/turicreate/docs/userguide/object_detection/. Accessed 15 July 2021
11. Liu, W., et al.: SSD: single shot multibox detector. In: Leibe, B., Matas, J., Sebe, N., Welling, M. (eds.) ECCV 2016. LNCS, vol. 9905, pp. 21–37. Springer, Cham (2016). https://doi.org/10.1007/978-3-319-46448-0_2
12. Howard, A.G., et al.: Efficient Convolutional Neural Networks for Mobile Vision Applications. https://arxiv.org/abs/1704.04861. Accessed 15 July 2021
13. Besshaposhnikov, N.O., Leonov, A.G., Matyushin, M.A.: Problems of ordering objects on image by using neural networks and heuristic algorithms. Proc. Cybern. **4**(32), 136–142 (2018)
14. Rosebrock, A.: Open CV and Python Color Detection. https://www.pyimagesearch.com/2014/08/04/opencv-pythoncolor-detection/. Accessed 15 July 2021
15. Apple augmented reality: https://developer.apple.com/augmented-reality/arkit/. Accessed 15 July 2021
16. ARcore overview: https://developers.google.com/ar/discover. Accessed 15 July 2021

Integrated Approach to Automation of a Wide Range of Forestry Industry Enterprises Based on Universal Platform Solutions

A. I. Shabaev and E. A. Pitukhin(✉)

Petrozavodsk State University, 33, Lenin Street, Petrozavodsk 185910, Republic of Karelia, Russia
{ashabaev,eugene}@petrsu.ru

Abstract. The article proposes an integrated approach to the automation of forestry enterprises, which ensures the integrity and consistency of management processes and, as a consequence, increases the enterprise efficiency. The approach is based on universal digital platform solutions with modular structure. The novelty of the approach is the three-level organization of business processes according to the systemic principle, linking the external business environment, the integral parameters of the enterprise as a whole and the functioning of its production components. This approach to management allows solving not only tactical tasks of operational planning, but also strategic tasks of interaction with customers and suppliers through harmonization of their interests on the basis of mutually beneficial cooperation. A distinctive feature of the approach is its scientific sophistication, and developed mathematical models that take into account a large number of factors and limitations of technological and business processes, as well as methods and algorithms that enable optimal planning and control over production with increased speed. Examples of practical implementation of the approach are given in the form of production planning systems, successfully used by forestry industry enterprises with notable economic effect.

Keywords: Digitalization · Forestry industry · Mathematical models · Production planning software · Integrated approach

1 Introduction

Today, the concept of "industrial automation" has come to be used mainly in a narrow sense, namely, the automation of production processes. Under the influence of modern trends this concept has evolved into a broader concept "digitalization", when all business processes of an enterprise are handled by software: production, sales, procurement, advertising, human resources, etc. Thus, industrial automation has become a component of digitalization. At present many companies and organizations world-wide devote much attention to digitalization [1–3]. In addition to the Chief Information Officer, there even appeared the position of Chief Digitalization Officer.

Digitalization affects a wide range of areas of activity – marketing, sales, accounting, HR and many others, including, of course, manufacturing. It is believed that digitalization

R. Silhavy et al. (Eds.): CoMeSySo 2021, LNNS 231, pp. 870–883, 2021.
https://doi.org/10.1007/978-3-030-90321-3_72

is a fundamental necessity for industry [2]. Digitalization includes the collection and transmission of data in digital form using data sources and networks of various kinds (drones, mobile applications, IoT, etc.) between various organizations [1].

One of the main targets for companies is cost reduction. In particular, the lean manufacturing approach is an efficient way for identification and elimination of waste. Collecting and processing data in digital form can provide significant benefits – reduce staff time, avoid errors, increase visibility, etc. However, to "uncover" the full potential of the new data for reducing waste, companies need to and are beginning to pay more attention to analytics [1], artificial intelligence, and advanced process control (APC) [1, 2].

Thus, within the implementation of the import substitution concept, it is relevant to develop domestic automation systems with the functions of deep data analytics and advanced process control. Development of such advanced software systems in Russia will guarantee their legal independence from policies and decisions of foreign software development companies and significantly contribute to the development of industry and the digital economy in Russia.

2 Analysis of Solutions for the Automation of Forestry Enterprises

2.1 Large-Scale Enterprises in Forestry Industry

Forestry industry in Russia and other countries is characterized by the presence of a large number of geographically distributed multi-level holdings with a full cycle of raw materials processing and include dozens of large-scale production units – logging enterprises, sawmills, plywood mills, pulp-and-paper mills, and corrugated packaging mills.

For example, Segezha Group is the largest Russian vertically integrated holding with a full cycle of logging and deep wood processing, which includes more than 30 Russian and European enterprises of the forestry, woodworking and pulp and paper industries (PPI). The structure of the management company includes 5 business lines: paper and packaging, plywood and boards, woodworking, forest resources, energy.

Among the foreign forestry industry holdings, it is necessary to mention International Paper (USA) with a turnover of $21.7 billion, Kimberly-Clark (USA) with a turnover of $18.6, Oji Paper (Japan) with a turnover of $11.9 billion, and many others. More than 50 enterprises of UPM Kymmene (Finland) are located in 16 countries and grouped into 8 business lines: paper, pulp, forestry, sawmilling, labels, plywood, energy, as well as biofuels, biocomposites and biochemicals from wood raw materials. The holding's sales exceed €10 billion per year. Likewise, more than 30 enterprises of the Metsä Group (Finland) in different countries are grouped into 5 business lines: forestry; wood products; cellulose and sawmill products; paper and cardboard; sanitary paper. The holding's sales exceed €5 billion per year.

2.2 Expected Benefits of Digitalization in the Forestry Industry

The potential advantages of digitalization in the forestry industry include [1–3]:

- Increasing the predictability of production and its "transparency" for customers, including the ability to track the production and transportation of finished goods, as well as adjusting the purchase orders (within acceptable limits);
- While maintaining mass production, an increase in the individuality of products and the speed of response to changes, a more flexible pricing policy;
- Reduction of inventory and optimization of the use of equipment and resources, more flexible planning, taking into account the equipment condition and the market environment;
- A more flexible and informed business strategy based on modeling various scenarios, including both real and prospective work centers.

2.3 Barriers to Digitalization in the Forestry Industry

However, the forestry industry does not take full advantage of digitalization compared to the IT, aerospace and automotive industries [1].

The survey [3] shows that the reasons are mainly related to culture, organization, leadership and competencies of the personnel, i.e. lie within the companies, not in the external environment. Lack of clear vision and leadership from senior management has been cited as one of the three biggest problems in forestry industry companies, which significantly differentiates them from companies in other sectors.

Another most significant problem for forestry industry companies is the lack of a digital culture and insufficient skills of staff. Many forestry companies do not use a systematic approach to data analytics, but rely on skills of individual employees to a greater extent than in other industries. At the same time, many companies admit the insufficiency or even lack of data analytics skills of their staff. Only a third of companies have data analytics included in their HR functions, and 15% of companies have a dedicated data analysis department serving various departments.

The third most important problem is uncertainty about the economic benefits of digital investments. It is believed [3] that these reasons are interrelated.

Thus, the most significant obstacle for forestry companies on their way to digitalization is not the lack of "suitable" technologies, but mainly the need to transform personnel and culture, which requires long-term change programs and appropriate motivation at different levels of management, starting from the very top [3].

2.4 Problem Statement

As a result of the analysis of the situation, it can be concluded that, at the moment, forestry companies are mostly pessimistic about the return on investments in digitalization in general and in the advanced software systems, in particular. Therefore, many of the benefits of digitalization remain unused.

To overcome this conservatism, Russian forestry companies expect relatively inexpensive software systems with a proven economic effect, adapted both to Russian conditions in general and to the specific technological and business processes of these companies, in particular.

Foreign companies developing software for large-scale forestry enterprises quite often face challenges in promoting their systems in Russia due to unwillingness or impossibility of adapting their software to the specifics of Russian enterprises. The exceptions are Russian subsidiaries of foreign holdings (UPM Kymmene, Metsä Group, etc.), using corporate solutions, and ERP systems.

The variety of the product range and the production equipment complexity determine the significant labor intensity of production planning personnel, whereas the high productivity of equipment requires appropriate speed of decision-making. When solving problems of volume planning and scheduling, it is necessary to take into account not only the specifics of production, but also the external business environment.

Consequently, software and mathematical models should be based on an integrated holistic approach in order to solve the problems of not only planning, but also interaction with the external environment, for example, by harmonizing often conflicting interests of parties in the "customers – enterprise – suppliers" triad.

The need for a sufficiently accurate and complete consideration of the individual features of production technologies and the external environment of Russian enterprises necessitates developing and using in production planning domestic software systems based on complex mathematical models and effecient numerical methods.

Using this approach would yield the following economic benefits: increase the production volume and cost of goods sold, reduce inventory and work in progress, reduce unproductive equipment downtime, reduce costs in supply chains, select the most profitable orders into production, and other.

3 Related Work

Industrial leaders are digitalizing essential functions within their internal vertical operations processes, as well as with their horizontal partners along the value chain. At the end of this transformation process, successful industrial companies will become true digital enterprises, with physical products at the core, augmented by digital interfaces and data-based, innovative services. These digital enterprises will work together with customers and suppliers in industrial digital ecosystems. The forest, paper and packaging sector is no exception.

While the digital revolution has transformed many industries, its impact on forest products companies has been relatively limited, as the industry has lagged behind others in making investments in digital technology.

The problems of optimization and digital transformation driving into value chain in forestry industry are discussed in [1, 3, 4]. Their authors confirm that digital technology is becoming a fundamental necessity in the forestry.

The purpose of [5, 6] is to provide an overview of the wood-flow in forestry and to illustrate some problems that are modeled and solved using optimization models and techniques. Although the author admits impossibility to cover all applications arising in forestry, the chosen applications are claimed to give a flavor of the variety of problems arising in forestry.

Papers [5, 6] provide an overview of the different planning problems and reviews the past contributions in the field of forestry, with a focus on applications and problem descriptions by Canadian researchers to the field of forestry management.

Based on an extensive systematic literature review, the article [7] identifies general trends towards a smart wood supply chain and provides Industry 4.0 application examples, which are already in practical use or still in the stage of research and development.

In general, the theoretical and methodological basis for solving problems of planning and control of production processes are well-known methods of operations research and mathematical programming. Such methods are described in both domestic and Western works [8–10]. However, in practice classical optimization problems are rarely found in their pure form – applied tasks are significantly complicated by the need to account for the discrete and stochastic nature of production processes and products; the particular sequence of work, and many others.

Foreign applied research [11–13] is aimed at solving specific applied problems related to optimization of technological processes at forestry enterprises. The papers [14–16] also describe practical applications of optimization problems in the forestry industry. The solution of such special problems is based on the use of methods of integer optimization, dynamic and linear programming, branch & bound methods, heuristics and column generation, as well as others.

In Russia, the issues of optimization of wood processing and forestry companies management have also attracted much attention and, in particular, are presented in works [17–19]. Along with the integrated management of a forestry industry enterprise, more detailed models are being developed aimed at optimizing individual technological processes, such as the production of pulp, paper, corrugated packaging, etc. [20, 21], as well as auxiliary processes [22, 23].

Authors of [3] are seeing digitisation having an impact all the way from real-time information and monitoring of forestry and tree conditions through to highly customisable, on-demand manufacturing, with customers having real-time access to design, supply and demand systems. The relevance of harmonization of production and sales in forestry companies is confirmed also in [13, 24–26].

It is also necessary to note the evolutionary development of software at different stages of interaction with Customers:

- At the preliminary negotiations stage, a demo version with the maximum set of functions is required to demonstrate all significant scenarios of using the system;
- At the pre-project interaction stage, the Customer can narrow down the set of modules, but add a number of essential requirements that must be quickly implemented in a simplified version to prove the possibility of their implementation within the project;
- At the stage of implementing the system at a specific enterprise, a version appears with a number of unique settings and functions that further evolve during support and maintenance stage;

- In the process of interacting with customers, new or improved functions are added into the standard package and also distributed to supported customers using earlier versions.

To ensure widespread replication and rapid implementation of systems, as well as effective interaction with a wide range of customers at different stages of the project life cycle, it is necessary to minimize the time and resources on adapting systems to the specifics of technological and business processes of each customer. This requires maximum unification of development and implementation processes, which is possible within a single platform, including the use of "cloud" technologies.

4 Materials and Methods (Research Methodology)

The proposed integrated approach to automation of a wide range of forestry industry enterprises is based on the "many enterprises – one tool" concept.

The "many enterprises – one tool" concept is based on a systems approach that implements the "general-to-specific" principle. The concept allows describing each stage of a production process of a forestry enterprise by choosing appropriate procedure/operation models from the available set and tuning the parameters. Thus, despite the versatility of the tool, the model of functioning of each specific enterprise becomes unique, since unique the is set of models of the production process sections and their parameters induced by the specifics of the given enterprise.

In practice, this concept is based on the use of a binary matrix "processes – procedures", where the rows are formed by a formalized set of production processes of forestry enterprises, and the columns are the so-called "procedures/operations", which are, in the general case, dynamic mathematical models relating a variety of input parameters (material, energy, financial, time, human and information resources, and control algorithms) onto a set of output parameters (products, waste, costs) at a given moment of time.

The novelty of the approach is in the three-level organization of business processes according to the system principle, linking the external business environment, the integral parameters of the enterprise as a whole and the functioning of its production components. The external circuit of interaction connects the enterprise and the business environment. The internal circuit describes the production processes of the enterprise itself.

The idea of the unified approach is that when building a model of an enterprise it is necessary to rise to a higher level of abstraction, considering production as a set of technological processes formalized by mathematical models.

To account for the specifics of an enterprise, the task of choosing a specific procedure for the process is solved using binary information variables $I_{m,n} \in \{0, 1\}$. They assign to each technological process B_m, $m \in M$ an implementation in the form of the appropriate technological operation (procedure) A_n, $n \in N$, when $I_{m,n}$ are non-zero:

$$\left(I_{m,n} = 1\right) \wedge \left(I_{m,\tilde{n}} = 0\right), \forall \tilde{n} \neq n, \tilde{n} \in N.$$

Here M and N – and index sets of technological processes and procedures (operations) that they implement. The set of technological operations $A = \{A_n\}$ defines the set of processes of the enterprise $B = \{B_m\}$: $B = IA$, where $I = \{I_{m,n}\}$, $m = 1..M$, $n = 1..N$ – is the "processes – procedures" matrix.

In Fig. 1 shows a diagram of the implementation of a unified model of technological processes for the functioning of a forestry enterprise with a choice of procedures for their technological implementation, taking into account the specifics of a particular enterprise.

Figure 1 illustrates how each technological process $B_m = \sum_{n=1}^{N} I_{m,n}A_n$ of an enterprise from a set of resources $\{R_k\}, k \in K$ and a set of semi-finished goods $\{P_{m-1}\}_{|m=1...M-1}$ produces a set of intermediate products $\{P_m\} : B_m : \{R_k\} \cup \{P_{m-1}\} \rightarrow \{P_m\}$, $m = 1 \ldots M - 1$ and, at the final stage $m = M$ – the set of finished goods $\{P_M\}$.

The next step in the implementation of the integrated approach is to set a model of the production process of an enterprise in a matrix form. The production process connects all the operations of an enterprise producing intermediate or finished goods [27].

The production process of an enterprise with unified model based on unit processes (Fig. 1), is set by the three-dimensional matrix "Operation" – "Products" – "Operation number" $T = \{T_{n,m,o}\}$ using binary information variables $T_{n,m,o} \in \{0, 1\}$, $n \in N$, $m \in M$, $o \in O$, where O – is the index set. Dimension o – "Operation number" allows to order operations and processes in time, determine their duration $u_{n,o}$ and thereby implement any types of production structures: sequential, converging, converging-diverging, as well as with a reverse. In addition to the duration of the operation, its important characteristics are also the control value $v_{n,o}$ and the production unit $w_{n,o}$ that implements it.

In contrast to the classical matrix "operation – product" [27], the proposed approach allows taking into account the specifics of the technological processes of the enterprise, which makes it the most convenient for creating unified software in the design and implementation of information management systems.

Typical procurement, production and sales processes of a forestry enterprise have been formalized by many researchers, in particular, in [5, 22, 28] and others.

The external circuit of interaction (Fig. 2) shows the relations of the enterprise and its management system with the business environment in the form of resource suppliers and consumers of products, as well as external environment [27].

Resource suppliers are specified by the set $S = \{S_j\}, j \in J$, and product consumers – by the set $C = \{C_l\}, l \in L$, where J and L – are the index sets. Each supplier S_j and consumer C_l has a set of parameters that characterize the properties and features of their organizations and corresponding resources and products, the main of which are: the annual turnover; its brand, prestige and market reputation; nomenclature of available resources and products; unit price; the contract volume(s) and condition(s), the transactions history; allowable delay time; discounts and penalties for delays or quality and other factors. Taking into account these business environment properties in the proposed approach, which is the basis for the management information system of a forestry company, allows to harmonize the conflicting demands of production and sales within an enterprise.

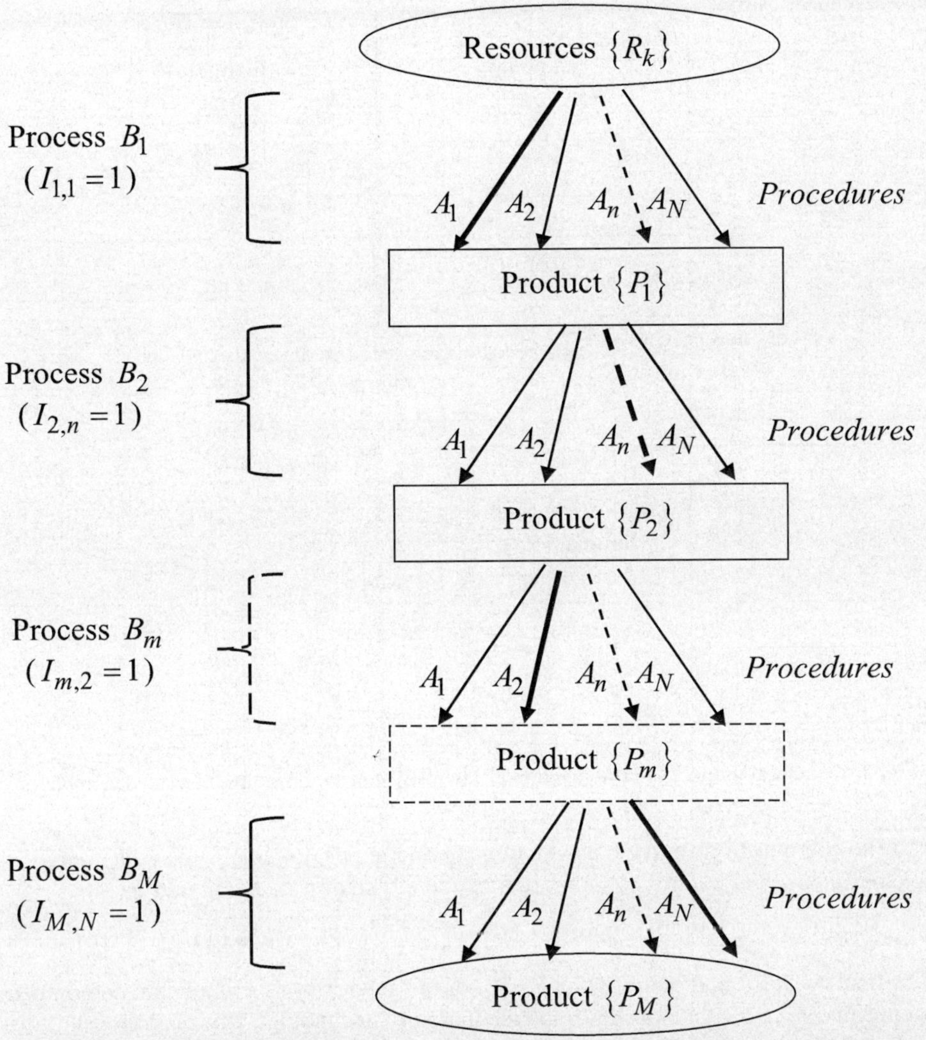

Fig. 1. An example of the implementation of a unified model of production processes of a forestry enterprise based on the "many enterprises – one tool" concept.

The external environment is specified by the set $F = \{F_d\}$, $d \in D$, where D – is the index set. The main factors of the external environment include: socio-political conditions of society and the state; financial and economic conditions; the impact of superior organizations and institutions; competition with domestic and foreign enterprises, etc.

The production processes of an enterprise are based on the set of production processes $\{B_m\}$, as a result of which resources $\{R_k\}$ are consumed in volumes V^R and goods $\{P_m\}$ are produced in volumes V^P.

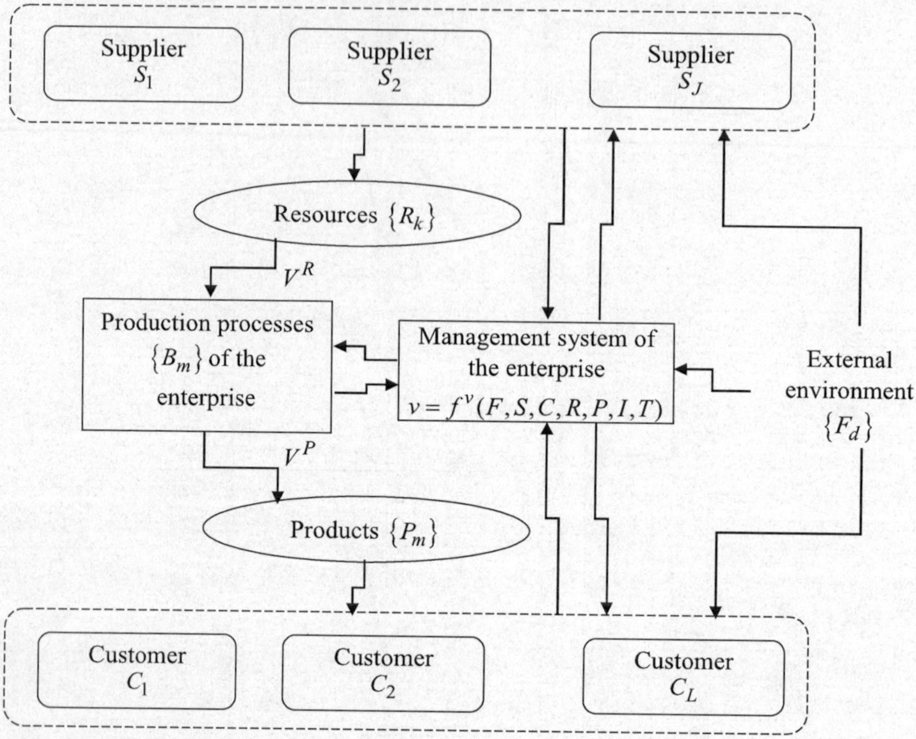

Fig. 2. Structural model of harmonization of interests of customers, production and suppliers.

The volumes of resource consumption V_n^R and goods output V_n^P for each operation A_n are functions of control $v_{n,o}$, duration $u_{n,o}$ and type of production unit $w_{n,o}$:

$$V_n^R = f_n^R(v_{n,o}, u_{n,o}, w_{n,o}),\ V_n^P = f_n^P(v_{n,o}, u_{n,o}, w_{n,o}).$$

The final costs V^R and output volumes V^P of the enterprise as a whole are determined taking into account the specific implementation of technological processes using the "processes – procedures" matrix I:

$$V^g = \{V_m^g(v_{n,o})\},\ V_m^g(v_{n,o}) = \sum_{n=1}^{N} I_{m,n} V_n^g,\ g \in \{R, P\},\ m \in M.$$

To estimate the production costs Z_m we define the value $Q_{n,m,o}^R$ of unit costs and resource consumption per product P_m for operation A_n per unit of time:

$$Q_{n,m,o}^R = f^R(F, S, C, R, P, I, T).$$

Then the production cost becomes a function of the enterprise production process T and unit resource consumptions Q^R taking into account the operation duration $u_{n,o}$:

$$Z_m = \sum_{n=1}^{N} \sum_{o=1}^{O} T_{n,m,o} Q_{n,m,o}^R u_{n,o}.$$

Hence, the total production expenses $X(v)$ will depend on the volume of consumed resources $V_m^R(v)$ and their costs, equivalent to the cost of production Z_m:

$$X(v) = \sum_{m=1}^{M} V_m^R(v) Z_m.$$

Revenue from sales of products $Y(\alpha, v)$ will depend on production volumes $V_m^P(v)$, production costs Z_m and profit margin α:

$$Y(\alpha, v) = \sum_{m=1}^{M} V_m^P(v) Z_m (1 + \alpha).$$

Here $v \in \Omega_v$ – is the vector of control actions, which includes not only the process control values $v_{n,o}$, but also takes into account the production structure and specifics, volume of resource consumption and goods output, as well as interaction with the business environment and the external environment:

$$v = f^v(F, S, C, R, P, I, T).$$

In a deterministic formulation of the problem of finding optimal control, one of the options for harmonizing the interaction of a forestry company with resource suppliers and product consumers is implemented through the objective function of maximizing profit subject to constraints on the interests of production and sales:

$$\{\alpha^*, v^*\} = \max_{\alpha > 0, v \in \Omega_v} \{E(\alpha, v)\},$$

where $E(\alpha, v) = (Y(\alpha, v) - X(v))/X(v)$ – is the enterprise's profit ratio.

In a stochastic formulation of the problem, a simulation block is added to the mathematical model, which enables setting not only fixed values of model parameters, but also estimates of their distribution densities as random variables based on retrospective data analysis. This makes possible to vary not only the production parameters combinations, but also external market conditions, financial environment, parameters of suppliers and customer demand. This formulation allows to first solve the expert problem "what if", and on its basis find the optimal values of parameters when solving the constructive problem "to be".

Thus, it is important in the production planning process not only to maximize profits, but also to find scenarios for optimal planning and management of production and sales. The proposed integrated approach to management based on the matrix implementation of the production processes model allows solving not only tactical tasks of operational planning, but also strategic tasks of interaction with customers and suppliers through the harmonization of their interests on the basis of mutually beneficial cooperation.

5 Results and Discussion

On the basis of the proposed approach, a suite of systems for optimal planning and management of complex production processes of industrial enterprises have been developed.

This is possible and efficient, because such systems typically use similar set of menus, forms, tables, dialog boxes and other components, as well as optimization algorithms [29].

The specialized library ("universal" solver") is another main distinctive feature of the platform (Kuznetsov et. al., 2008) developed based on 30 years of experience in customer projects for companies of Russia, and Finland. Algorithms for effective solution of complicated cutting problems, including linear and nonlinear optimization problems, high-dimension problems, problems with combined criteria, etc. are implemented based on the versatile solver. The library also includes a special module ("matrix constructor") for increasing efficiency of constructing, storing and using constraints matrix [30].

Opti-Sawmill [30, 31] is a decision-support system developed specifically for sawmills and woodworking mills as a tool for efficient planning, optimizing and managing production and sales. The use of the system allows to increase the yield of lumber by 1.1–1.6%. The system is installed at 10 sawmills in Russia (from North-West through Eastern Siberia to Far East) and at 2 sawmills in Germany including Ilim Nordic Timber GbmH (Wismar), which is the largest sawmill in world by processing volume.

Opti-Wood is a decision-support system for efficient planning of logging and wood transport operations, construction and maintenance of forest roads. It is designed for wood harvesting enterprises as a production optimization tool for planning and management. The system is used by a forestry company in Karelia. In [32, 33] we provide the mathematical model and solution approach to optimizing the profit measured as revenue from the sale of products, net of harvesting and transportation costs. During several case studies involving 198 sites and 14 machines harvesting up to 200,000 cubic meters from an available stock volume of about 300,000 cubic meters an increase in profit by 5% to 10% was observed.

Opti-Plywood [34] is a web-based service for plywood mills as a tool for optimizing production, starting from peeling and drying of veneer to pressing and cutting plywood. The system is in use at one of the largest softwood plywood mills in Russia (in Bratsk).

Opti-Paper [31] system for optimization of paper/cardboard production enables finding the optimal production plan for several paper/cardboard machines, taking into account the production technology, regardless of the number of customer orders and process equipment configuration. The system is in use at 3 pulp-and-paper mills in Russia.

Opti-Loading [31] is a web-based service for automatic solution of a complex of tasks related to the choice of vehicles and loading finished goods. The service enables to optimally distribute finished goods (rolls of paper/cardboard of various formats and diameters, as well as pallets and boxes or mixed load) to vehicles: containers, wagons, trucks. The system is in use at Mondi Syktyvkar (Russia's largest producer of paper good in), Kondopoga pulp-and-paper mill (Russia's largest producer of newsprint paper), 5 corrugated board producers, as well as TechnoNICOL corporation – one of the largest international producers of construction materials. The use of the service in many cases enables to increase by 4–5% the total load in a vehicle considering a wide range of constraints.

Opti-Corrugated [31] is the leading software tool in Russia for planning and managing the production of corrugated cardboard goods. The system unites commercial and production departments of an enterprise, ensures "just-in-time" execution of orders and production costs reduction. It utilizes advanced software development technologies together with complex mathematical models and optimization algorithms. The system ensures production of balanced amounts of rectangles of cardboard for boxes with minimal material losses, considering the production capacity, equipment re-tuning, order scheduling and priorities. The use of the systems enabled the customers to reduce material losses from 5–6% to 2–4%, and increase the production uptime by 3–4%, which saves several hundred thousand euro per year.

6 Conclusion

An important competitive advantage of production planning systems is their orientation on the future – the enterprise sets goals and the system computes the optimal "route" (according to different criteria) to achieve them.

Such systems must be characterized by a high degree of adequacy of mathematical models, high speed of calculations and flexibility of adaptation to the customer's enterprise, provide end-to-end coverage of production processes of an enterprise and/or a group of enterprises (if necessary).

The article describes an integrated approach to the automation of forestry enterprises, which improves the efficiency of the enterprise. This is achieved by means of the "many enterprises - one tool" concept, which allows building a mathematical model of production processes of an enterprise based on a matrix description. With the help of this concept, using universal modular platform solutions, unique systems can be created that take into account the specifics of the functioning of a particular forestry enterprise.

The platform for unification of development and implementation processes facilitates the development and maintenance of systems at installed at multiple customer sites.

The use of the systems enables Customers to reduce material losses, increase the profitability, and also reduce the production planning time, which leads to significant economic benefits that have been confirmed by installations at more than 50 large and medium-sized forestry enterprises in Russia and foreign countries.

The obtained results can be useful to automation specialists, to the management of forestry enterprises, industrial software developers, as well as to researchers in the field of improving the efficiency of industrial enterprises.

References

1. Forest Products: the Shift to Digital Accelerates. Natural Resources. https://www.accenture.com/us-en/insights/natural-resources/digital-forest-products-survey. Accessed Apr 2020
2. CEPI: What Digital Can do for the Paper Industry (2016). http://www.cepi.org/publication/paper-industry-40-what-digital-can-do-paper-industry
3. PricewaterhouseCoopers: Industry 4.0: Building the Digital Enterprise. Global Industry 4.0 Survey (2016). https://www.pwc.com/industry40
4. Pantsar, M., Schey, V., Mensink, M.: What digital can do for the paper industry (2018). https://www.pac.gr/bcm/uploads/paper-industry-4-0.pdf

5. D'Amours, S., Rönnqvist M., Weintraub A.: Using operational research for supply chain planning in the forest products industry. INFOR: Inf. Syst. Oper. Res. **46**(4), 265–281 (2017). DOI: https://doi.org/10.3138/infor.46.4.265
6. Rönnqvist, M.: Optimization in forestry. Math. Program. **97**, 267–284 (2003). https://doi.org/10.1007/s10107-003-0444-0
7. Müller, F., Jaeger, D., Hanewinkel, M.: Digitization in wood supply – a review on how industry 4.0 will change the forest value chain. Comput. Electron. Agricult. **162**, 206–218 (2019). https://www.elsevier.com/locate/compag
8. Bazaraa, M.S., Shetti, C.M.: Nonlinear Programming Theory and Algorithms. Mir, Moscow (1982)
9. Dreyfus Stuart, E., Averill, M.: Law, The art and Theory of Dynamic Programming. Academic Press, New York (1977)
10. Luenberger, D. G.: Linear and Nonlinear Programming. Addison Wesley, Boston (1984)
11. Gunn, E.A., Rai, A.K.: Modelling and decomposition for planning long-term forest harvesting in an integrated industry structure. Can. J. For. Res. **17**, 1507–1518 (1987). https://doi.org/10.1139/x87-233
12. Karlsson, J., Rönnqvist, M., Bergström, J.: An optimization model for annual harvest planning. Can. J. For. Res. **34**, 1747–1754 (2004). https://doi.org/10.1139/x04-043
13. Silva, E., Alvelos, F., Valrio de Carvalho, J. M.: An integer programming model for two- and three-stage two-dimensional cutting stock problems. Eur. J. Oper. Res. **205**(3), 699–708 (2010)
14. Dimitriadis, S., Kehris, E.: Cutting stock process optimisation in custom door and window manufacturing industry. Int. J. Decis. Sci. Risk Manage. **1**(1), 66–80 (2009)
15. Foronda Segundino, U., Honorio Carino, F.: A heuristic approach to the lumber allocation problem in hardwood dimension and furniture manufacturing. Eur. J. Oper. Res. **54**, 151–162 (1991)
16. Gramani, M.C.N., Franca, P.M.: The combined cutting stock and Lot-sizing problem in industrial processes. Eur. J. Oper. Res. **174**(1), 509–521 (2006)
17. Bugaev, Yu.: Taxonomy of models of timber cutting in wood processing technology. Interuniversity collection of articles Technology and equipment for woodworking production. SPb.: LTA im. CM. Kirov **15**, 1–3 (1986)
18. Voronin, A., Kuznetsov, V.: Mathematical Models and Methods of Planning and Management in the Pulp and Paper Industry. PetrSU Publishing House, Petrozavodsk (2000)
19. Petrovsky, B.: Automation of Technological Processes and Enterprises of the Forestry Industry. Academy, Moscow (2010)
20. Voronov, R.: Mathematical models and methods of automated paper production planning systems. PhD Thesis, Petrozavodsk (2004)
21. Soshkin, R.: On the use of mathematical methods to improve the efficiency of corrugated cardboard packaging production. Bull. St. Petersburg Forestry Acad. **181**, 165–173 (2007)
22. Voronin, A.: Features of formulating the tasks of planning and managing material flows of forestry enterprises. Logging, timber storage and timber transport: Interuniversity collection of articles SPb. Publishing house of SPbGLTA, SPb, pp. 29–32 (2002)
23. Voronin, A., Kuznetsov, A., Bulatov, A.: Analysis of tasks and methods for modeling information flows in forestry industry enterprises. Timber Industry Complex: Problems and Solutions: Collection of Scientific Works. KarNIILP, Petrozavodsk, pp. 16–20 (1998)
24. Galbraith, J.R.: Designing the customer centric organization: a guide to strategy structure and process. Jossey-Bass (2005)
25. Opacic L., Sowlati T., Mobini M.: Design and development of a simulation-based decision support tool to improve the production process at an engineered wood products mill. Int. J. Prod. Econ. **199**, 209–219 (2018). www.elsevier.com/locate/ijpe

26. Travis M., et al.: Conflicting paradigms in manufacturing and marketing decisions: The effects of situational awareness on team performance. Prod. Econ. **230**, 107–119 (2020). www.elsevier.com/locate/ijpe
27. Pervozvansky, A.: Mathematical Models in Production Management. Science, Moscow (1975)
28. Kuznetsov, V.: The use of applied optimization problems for improving the efficiency of pulp and paper production. Cellulose. Paper. Cardboard **5–6**, 28–31 (2003)
29. Shabaev, A., Kuznetsov, V., Kositsyn, D.: Optimal planning software platform development with cloud computing technology. In: Proceedings of the 12th Conference of Open Innovations Association FRUCT, pp. 129–135 (2012). https://www.fruct.org/publications/fruct12/files/Sha.pdf
30. Voronin, A., Kuznetsov, V., Arkhipov, I., Shabaev, A.: Software system for sawmill operation planning. In: Proceedings of the 12th Conference of Open Innovations Association FRUCT, pp. 165–171 (2012)
31. Shabaev, A., Voronin, A., Kuznecov, V., Arkhipov, I.: "Cloud" services for improving production efficiency of industrial enterprises. In: Proceedings of the Maintenance Performance Measurement and Management Conference, pp. 315–329 (2013)
32. Shabaev, A., Sokolov, A., Urban, A., Pyatin, D.: An approach to the optimal timber transport scheduling. In: E3S Web Conference, vol. 164, p. 03019 (2020). https://doi.org/10.1051/e3sconf/202016403019
33. Shabaev, A., Sokolov, A., Urban, A., Pyatin, D.: Optimal planning of wood harvesting and timber supply in Russian conditions. Forests **11**, 662 (2020). https://doi.org/10.3390/f11060662
34. Shabaev, A., Arhipov, I., Spirichev, M., Urban, A., Torozerov, M.: Development of planning system for plywood production using "Matrix Designer". In: Proceedings of the 14th Conference of Open Innovations Association FRUCT, pp. 140–147 (2014)
35. Morabito, R., Arenales, M.: Optimizing the cutting of stock plates in a furniture company. Int. J. Prod. Res. **38**(12), 2725–2742 (2000)

The Impact of Digitalization on the Bank Competitiveness

Vladislav Rutskiy[1,7], Skanda Moda Gururajarao[2], Nikita Chudopal[1], Nadezhda Kulakova[3], Natalia Bystrova[4], and Roman Tsarev[5,6](✉)

[1] Siberian Federal University, Krasnoyarsk, Russia
[2] SJCE, JSS Science and Technology University, Mysore, India
[3] Financial University Under the Government of the Russian Federation, Kaluga, Russia
[4] Minin State Pedagogical University of Nizhny Novgorod, Nizhny Novgorod, Russia
[5] V.A. Trapeznikov Institute of Control Sciences of Russian Academy of Sciences, Moscow, Russia
[6] International Academy of Science and Technologies, Moscow, Russia
[7] HSE University , Moscow, Russia

Abstract. The paper examines the bank competitiveness influenced by the digitalization of the economy. In the field of finance, digitalization implies the usage of electronic payments and e-commerce along with innovative technologies aimed at providing digital services. This paper aims to assess the digitalization level affecting the general competitiveness of banks and its components on the example of Russian banks. Authors use standardized input statistical indicators, comparison analysis and a regression and correlation analysis to examine this phenomenon. This approach lets authors evaluate the efficiency of banks in the course of digitalization and shows the attractiveness of bank's investment in digital tools and services.

Keywords: Digitalization · Fintech · Digital banking services · Development of the banking system · Competitiveness of bank · Impact of digitalization

1 Introduction

Since the banking sector is an integral part of the national economy, it affects financial and credit system and the potential for economic growth. Competition among banking institutions establishes market conditions in the sector through the use of more attractive banking tools and methods, as well as banking products for service consumers.

The digitalization process has become critical factor in the economies over the last few years. I has also appeared in banking operations through the active spread of electronic payments, e-commerce, and the introduction of innovative digital services. It expands the range of banking products, assists in customer engagement and thus shapes a competition policy and gives competitive advantages. Furthermore, the global spread of COVID-19 in 2020 necessitated the transition to a remote banking business model.

According to the abovementioned, analyzing and assessing the competitive capacity of banks influenced by digital transformation make it possible to identify both obstacles to the effective functioning of banks and the drivers of their effective performance.

R. Silhavy et al. (Eds.): CoMeSySo 2021, LNNS 231, pp. 884–894, 2021.
https://doi.org/10.1007/978-3-030-90321-3_73

2 Studies on Digitalization and Bank's Competitive Capacities

As for studies in the field of bank's competitiveness the Panzar-Rosse test [1] is the most common approach. It offers a means for assessing the bank's competitive position in the banking service market. Some researchers [2] actively apply the proposed toolkit, whereas others supplement it with factors and conditions for diagnosing the results obtained by the Panzar-Rosse test. Thus, Shaffer and Spierdijk and Bikker et al. [3] emphasized the need to consider the bank's operational life, its relative costs, strategy, banking product differentiation degree, and other determinants of competitiveness.

A considerable number of researchers and rating agencies assess the bank's competitive capacity relying on its financial indicators. The Banker, a periodical that is well-known among the banking sector experts, forms the Top 1000 World Banks rating based on the bank's Tier 1 capital indicators, total assets, their growth rate, profitability, return on assets, and volume of capital. Maochun and Zhixu [4] added a description of the bank's activity scale to the indicators list specified above. Pruteanu-Podpiera et al. [4] investigated the dependence of competitive capacity on banking efficiency.

Certain models for assessing the bank's competitive capacity address marketing performance. For instance, Qing [5] developed five indicators to assess the competitive capacity: uniqueness, strategic value, system integration, prevalence, and dynamics. Barbosa et al. [6] argue that banks offer not only suite of classic banking products but also financial services (including brokerage, insurance, and other) significantly predominant in terms of their competitiveness over banks that offer classic products only.

Krasovskis et al. [7] considered the analysis of the main characteristics peculiar to the competitive environment of the banking sector from a holistic perspective, namely, sector concentration, competition intensity, and financial indicators.

Financial indicators reflect both the results of competition (through a comparative assessment of the shaped assets, their profitability, acquired deposits, and existing loans) and the opportunities to finance innovative technologies for service delivery, updating the bank's business model and new products. However, the research studies do not explicitly attribute the digital transformation aspects in the banking sector to the competitive capacity assessment but determine them through the competitive capacity results in expanding the range and the customer base.

Mbama and Ezepue [8] explored the links among digital banking, customer satisfaction, loyalty, and financial indicators.

As demonstrated by Zhao et al. [9], the innovative Fintech implementation strategies strengthened the competitive capacity of China's banking sector. The authors specify the priorities for applying digital innovations: new business partners, new service concepts, organizational innovations, technological innovations, new customer interactions, and new revenue models.

Busby [10] compared different options for implementing the bank's basic digital technologies and observed the digitalization strategy as a significant factor contributing to customer loyalty in the current context.

Amin [11], Raza et al. [12], and Martins et al. [13] study how the quality of Internet banking services influences customer satisfaction and loyalty.

A wide range of research studies focus on mobile banking as a digital tool for banking development. A. Ali and H. Ali [14] used the mobile banking factor to prove the prerequisites for the increasing use of banking services. Other researchers [15] identify the relationship among customer trust, risk, volume of banking services, and mobile banking. Jun and Palacios [16] determined five criteria for customer satisfaction/dissatisfaction with mobile banking: convenient mobile devices, accuracy, various functions offered by mobile applications, ease of use, and constant improvement.

Dootson et al. [17] studied the use of social networks by banking institutions and revealed the customers' propensity to appreciate this digital tool in the case of maximizing the subjective value and benefits of banking products; as for banks, they also emphasized the improved dialogue with banking service consumers.

Koroleva and Kudryavtseva [18] and Stoica et al. [19] argue that digital banks are distinguished by higher efficiency, namely, return on assets, but leave undetermined the impact of digital transformation on the share of loans and deposits.

Thus, we could observe the digital transformation level as a factor of the bank's competitive capacity in research studies. Given the approaches and practices of assessing digital transformation and the bank's competitive capacity we could consider the relationship between the bank digitalization indicators and its competitive capacity.

3 Modelling the Relationship Between the Bank's Digitalization and Competitive Capacity

This paper aims to assess the impact of the level of digital transformation on the banks' competitive capacity in the broad terms and its certain components as illustrated by Russian banks.

The economic and mathematical modelling methods used at the preparatory stage of this research study allowed exploring the strength of relationhsip between objects and phenomena in the development of digital transformation processes, proposing approaches to improving methodological tools which could provide bank managers with reliable information about projected parameter values for certain types of banking activity, the bank's competitive positions and its strategic development.

The following methods have been applied to achieve the research study goal: standardization of input statistical indicators for twenty top Russian banks, which makes it possible to compare and rank them; a cluster analysis used to group banks by the level of their competitive capacity and digital transformation; a regression and correlation analysis performed to determine the strength and direction of the relationship between digital transformation and the bank's competitive capacity and its certain indicators.

While conducting the study, we proposed to evaluate the competitive capacity level of twenty top Russian banks by their financial indicators, namely, the Investment Security Coefficient, Scale of Current Operations Coefficient, Equity Share Coefficient, Use of Client Funds Coefficient, Liquidity Coefficient, Use of Attracted Assets Coefficient, Operational Efficiency Coefficient, Equity Use Coefficient. With this objective in view, the spatial data series were based on the banks' statistical data from the Financial Statements of Banks as of August 2020 and their linear transformation by MINIMAX in the range [0…1], where the minimum and maximum scaled values corresponded to 0 and 1, respectively (see Table 1).

The arithmetic average of eight indicators was used to determine the complex indicator illustrating the competitive capacity of each bank. The bank's rating among the banks in the research sample was established by the ranking method.

Table 1. Linear data transformation of the financial indicators designed to analyze the competitive capacity of Russian banks.

Bank	Investment security indicator	Scale of current operations indicator	Equity share indicator	Use of client funds indicator	Liquidity indicator	Use of attracted assets indicator	Operational efficiency indicator	Equity use indicator	Integral indicator
Tinkoff Bank	0,562	0,574	0,944	0,552	0,404	1,000	1,000	0,891	0,741
VTB Bank	0,227	0,240	0,605	0,864	0,648	0,114	0,113	0,161	0,371
Sberbank	0,478	0,453	0,987	0,355	0,273	0,461	0,481	0,471	0,495
Raiffeisenbank	0,626	0,442	0,820	0,985	1,000	0,447	0,552	0,462	0,667
Alfa Bank	0,454	0,382	0,892	0,481	0,422	0,902	0,998	1,000	0,691
Sovcombank	0,553	0,199	0,991	0,578	0,891	0,009	0,021	0,022	0,408
Otkritie Financial Corporation Bank	0,328	0,346	0,348	0,339	0,230	0,314	0,311	0,369	0,323
Home Credit & Finance Bank	1,000	1,000	0,929	0,038	0,000	0,000	0,000	0,000	0,371
AK BARS Bank	0,405	0,171	0,313	0,160	0,281	0,096	0,141	0,154	0,215
The Ural Bank for Reconstruction and Development	0,000	0,000	0,838	1,000	0,788	0,064	0,066	0,153	0,364
PromSvyaz Bank	0,146	0,212	0,669	0,170	0,061	0,256	0,232	0,368	0,264
Saint Petersburg Bank	0,362	0,265	1,000	0,479	0,458	0,245	0,286	0,324	0,427
Vostochniy Express Bank	0,333	0,358	0,732	0,469	0,326	0,035	0,034	0,045	0,292
Gazprombank	0,202	0,151	0,734	0,290	0,258	0,063	0,072	0,109	0,235
MTS Bank	0,598	0,560	0,632	0,123	0,086	0,025	0,027	0,026	0,260
Rosbank	0,260	0,271	0,902	0,374	0,264	0,148	0,148	0,198	0,321
Pochta Bank	0,585	0,632	0,994	0,408	0,268	0,335	0,323	0,283	0,479
ZENIT Bank	0,473	0,379	0,000	0,759	0,691	0,011	0,014	0,018	0,293
Uralsib	0,227	0,188	0,352	0,926	0,787	0,245	0,266	0,367	0,420
Russian Standard Bank	0,398	0,292	0,965	0,000	0,021	0,052	0,063	0,072	0,233

Similar analytical procedures were applied to consider the digital transformation indicators of banks in the sample (see Table 2 and Table 3).

Table 2. Linear data transformation of the digital transformation indicators designed to analyze the competitive capacity of Russian banks. Part 1.

Bank and indicators*	k_1	k_2	k_3	k_4	k_5	k_6	k_7	k_8	k_9	k_{10}
Tinkoff Bank	0,49	1,00	0,88	0,71	1,00	0,03	0,02	1,00	0,45	0,02
VTB Bank	0,30	0,77	0,51	0,08	0,58	0,01	0,00	0,13	0,05	0,04
Sberbank	1,00	0,91	1,00	0,56	0,88	0,01	0,04	0,13	0,28	0,25
Raiffeisenbank	0,71	0,97	0,47	1,00	1,00	0,13	0,04	0,18	0,65	0,00
Alfa Bank	0,48	1,00	0,71	0,24	1,00	0,01	0,02	0,02	0,12	0,06
Sovcombank	0,09	0,91	0,11	0,11	0,73	0,01	0,04	0,11	0,03	1,00
Otkritie Financial Corporation Bank	0,19	0,94	0,23	0,10	0,92	0,03	0,04	0,05	0,19	0,06
Home Credit & Finance Bank	0,37	0,97	0,85	0,58	0,96	1,00	0,02	0,17	0,10	0,02
AK BARS Bank	0,17	0,94	0,12	0,07	0,92	0,00	0,04	0,00	0,11	0,02
The Ural Bank for Reconstruction and Development	0,30	0,86	0,59	0,40	0,77	0,02	0,50	0,07	1,00	0,02
PromSvyaz Bank	0,19	0,97	0,37	0,17	0,96	0,05	0,02	0,31	0,26	0,02
Saint Petersburg Bank	0,05	0,83	0,22	0,10	0,96	0,01	0,50	0,07	0,19	0,06
Vostochniy Express Bank	0,01	0,17	0,80	0,03	0,27	0,01	0,00	0,34	0,20	0,00
Gazprombank	0,02	0,57	0,17	0,09	0,85	0,01	0,02	0,20	0,51	0,02
MTS Bank	0,00	0,00	0,17	0,01	0,00	0,02	0,04	0,05	0,31	1,00
Rosbank	0,02	0,66	0,22	0,06	0,85	0,01	0,04	0,04	0,04	0,06
Pochta Bank	0,09	0,51	0,80	0,49	0,85	0,01	1,00	0,08	0,15	0,02
ZENIT Bank	0,02	0,43	0,17	0,06	0,50	0,01	0,02	0,03	0,06	0,50
Uralsib	0,02	0,69	0,44	0,05	0,58	0,01	0,50	0,12	0,12	0,00
Russian Standard Bank	0,00	0,80	0,00	0,00	0,92	0,00	0,00	0,06	0,00	0,00

* k_1 – AppStore Reviews Indicator, k_2 – AppStore Rating Indicator, k_3 – GooglePlay Installs Indicator, k_4 – GooglePlay Reviews Indicator, k_5 – GooglePlay Rating Indicator, k_6 – Facebook Subscribers Indicator, k_7 – Facebook Replying Rate Indicator, k_8 – Average Monthly Posting Amount Indicator, k_9 – VK Subscribers Indicator, k_{10} – VK Replying Rate Indicator.

On the basis of the processed data it is possible to reveal the strength of the relationship between bank's digital tools and achievement of competitive advantages with the help of correlation and regression analysis. The relevant model shows the dependence of the bank's competitive capacity, namely integral financial indicator (y) on the digital transformation level, namely integral digital indicator (x): $y = b_1x + b_0$.

The arithmetic average according to Tables 1, 2 and 3 was used to determine the complex indicators illustrating the competitiveness and digitalization of each bank (see Table 4).

Figure 1 shows the graphical interpretation of the banks' integral indicators. Thus, a similar dynamic may be observed in most of the banks in the sample.

Table 3. Linear data transformation of the digital transformation indicators designed to analyze the competitive capacity of Russian banks. Part 2.

Bank and indicators*	k_{11}	k_{12}	k_{13}	k_{14}	k_{15}	k_{16}	k_{17}	k_{18}	k_{19}	k_{20}	k_{21}
Tinkoff Bank	1,00	0,96	0,59	1,00	1,00	1,00	1,00	1,00	1,00	0,39	0,73
VTB Bank	0,08	0,16	0,24	1,00	1,00	1,00	1,00	1,00	1,00	0,19	0,46
Sberbank	0,13	0,21	0,10	1,00	1,00	1,00	1,00	1,00	1,00	0,28	0,59
Raiffeisenbank	0,19	1,00	0,12	1,00	0,00	1,00	1,00	0,00	1,00	0,32	0,54
Alfa Bank	0,03	0,30	0,22	1,00	1,00	1,00	1,00	1,00	1,00	0,19	0,52
Sovcombank	0,11	0,05	0,47	1,00	1,00	1,00	1,00	1,00	1,00	0,24	0,50
Otkritie Financial Corporation Bank	0,03	0,61	0,27	1,00	0,00	0,00	1,00	1,00	0,00	0,33	0,35
Home Credit & Finance Bank	0,17	0,16	0,61	1,00	0,00	0,00	0,00	0,00	0,00	0,38	0,37
AK BARS Bank	0,00	0,38	0,16	1,00	1,00	0,00	1,00	1,00	0,00	0,20	0,36
The Ural Bank for Reconstruction and Development	0,14	0,49	0,14	1,00	0,00	0,00	1,00	1,00	1,00	0,20	0,48
PromSvyaz Bank	0,30	0,29	0,57	1,00	0,00	0,00	1,00	0,00	0,00	0,00	0,32
Saint Petersburg Bank	0,08	0,16	0,33	1,00	0,00	0,00	1,00	1,00	0,00	0,66	0,36
Vostochniy Express Bank	0,35	0,21	1,00	1,00	0,00	0,00	1,00	0,00	1,00	0,10	0,32
Gazprombank	0,14	0,44	0,22	1,00	0,00	0,00	1,00	0,00	0,00	0,13	0,27
MTS Bank	0,06	0,64	0,10	1,00	1,00	0,00	1,00	0,00	0,00	0,56	0,30
Rosbank	0,04	0,18	0,29	1,00	0,00	1,00	1,00	0,00	0,00	0,37	0,29
Pochta Bank	0,08	0,31	0,35	1,00	0,00	0,00	1,00	0,00	1,00	0,23	0,40
ZENIT Bank	0,03	0,23	0,27	1,00	0,00	0,00	1,00	1,00	0,00	1,00	0,32
Uralsib	0,12	0,11	0,00	1,00	0,00	1,00	1,00	1,00	1,00	0,26	0,40
Russian Standard Bank	0,07	0,00	0,33	1,00	0,00	0,00	1,00	0,00	1,00	0,51	0,28

* k_{11} – Average Monthly Posting Amount Indicator, k_{12} – Instagram Subscribers Indicator, k_{13} – Average Monthly Posting Amount Indicator, k_{14} – Faster Payments System Indicator, k_{15} – Self-employed Service Indicator, k_{16} – Online Accounting Indicator, k_{17} – Settlement Account Opening Indicator, k_{18} – E-banking Insurances, k_{19} – POS-smartphone Possibility Indicator, k_{20} – Indicator of Research and Development, Licensing, k_{21} – Integral Digital Indicator.

Table 4. Assessment of the competitive capacity and digital transformation of Russian banks

Bank	Integral digital indicator	Integral financial indicator
Tinkoff Bank	0,727	0,741
VTB Bank	0,457	0,371
Sberbank	0,589	0,495
Raiffeisenbank	0,540	0,667
Alfa Bank	0,520	0,691
Sovcombank	0,500	0,408
Otkritie Financial Corporation Bank	0,350	0,323
Home Credit & Finance Bank	0,368	0,371
AK BARS Bank	0,357	0,215
The Ural Bank for Reconstruction and Development	0,476	0,364
PromSvyaz Bank	0,324	0,264
Saint Petersburg Bank	0,361	0,427
Vostochniy Express Bank	0,325	0,292
Gazprombank	0,269	0,235
MTS Bank	0,298	0,260
Rosbank	0,294	0,321
Pochta Bank	0,398	0,479
ZENIT Bank	0,315	0,293
Uralsib	0,400	0,420
Russian Standard Bank	0,284	0,233

The relationship between the bank's competitive capacity by financial indicators (y) and the digital transformation level (x) was observed by the following regression equation: $y = -0{,}0583 + 1{,}10833x$.

The calculated value of the relationship strength (with the linear correlation coefficient equal to 0,859) indicates high level of relationship and the direct impact of digital transformation on the banks' competitive capacity.

The determination coefficient $R^2 = 0{,}7379$. In other words 73.79% of variation of the bank's competitive capacity is explained by the digital transformation level.

According to Fisher's criterion $F > F_{table}$, thus the determination coefficient is statistically significant, and the estimation of the regression equation is statistically reliable.

The regression dependence is illustrated in Fig. 2.

On the basis of the correlation-regression analysis results we could substantiate the hypothesis and prove strong relationship between the digital transformation of banks and their competitive capacity.

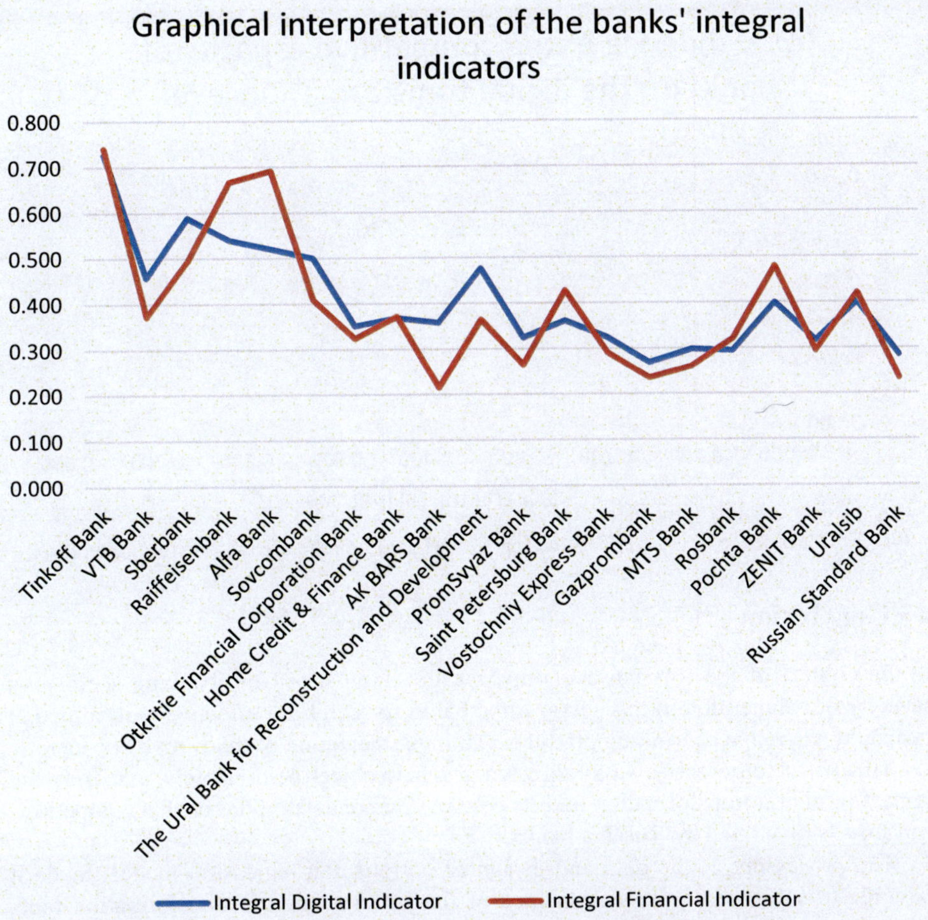

Fig. 1. Graphical interpretation of the banks' integral capacities' indicators.

The relationship between the banks' competitive position (evaluated by financial indicators of banks' activity) and the competitive level of digital transformation leads to the conclusion that the bank's digital technologies have the greatest and direct impact on the bank's competitive position. Especially the relationship between digitalization and indicator of personal deposits is observed. Also digitatlization shows a significant impact on the Investment Security Coefficient, Scale of Current Operations Coefficient, Equity Share Coefficient, Use of Client Funds Coefficient, Liquidity Coefficient, Use of Attracted Assets Coefficient, Operational Efficiency Coefficient and Equity Use Coefficient.

Thus the research findings can thoroughly demonstrate that digital transformation is a driving factor of the competitive capacity of Russian banks.

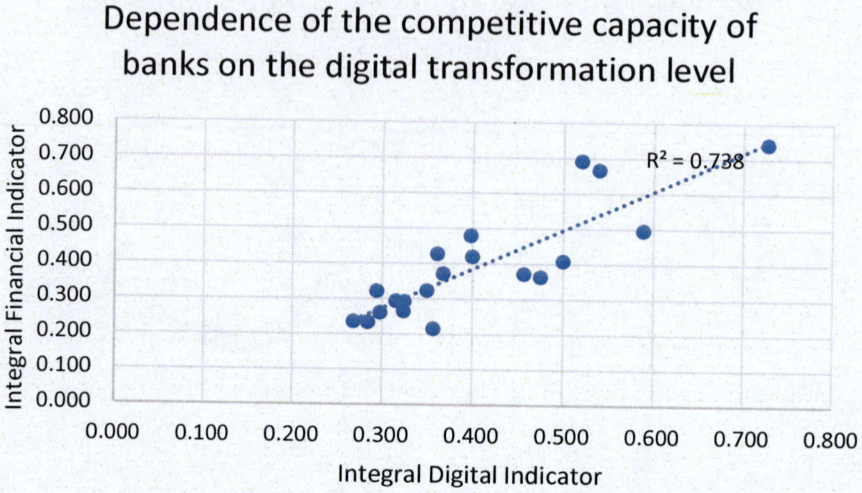

Fig. 2. Dependence of the competitive capacity of banks on the digital transformation level.

4 Conclusion

In the context of the Russian economy, the development of the banking sector and severety of competition in this sector force banks to search for effective tools aimed at managing the range of banking products and implementation of innovative services.

This paper represents a study directed at a new object field, namely, assessing the impact of digital transformation and its certain components on the bank's competitive capacity as illustrated by Russian banks.

The processing of the data and design of correlaction-regression models made it possible to substantiate the significance of the factor of digital transformation tools when assessing the competitive capacity of a bank in Russia.

The relationship between the level of bank's competitive capacity and the level of digital transformation was founded as close and direct one. Digital transformation showed impact on all significant indicators of the bank's activities which form its competitiveness.

In conclusion, it is worth noting that Russian banks should thoroughly aim to introduce innovative digital technologies in their activities to maintain a competitive position in the banking sector in Russia. The designed models are also capable of assessing the bank's efficiency and making advice for appropriate management decisions about investing in digital tools and services.

Further research studies could be aimed at developing a methodological approach to a comprehensive assessment of the bank's digital transformation tools in Russia and its effectiveness.

References

1. Panzar, J.C., Rosse, J.N.: Testing for monopoly equilibrium. J. Ind. Econ. **35**(4), 443–456 (1987). https://doi.org/10.2307/2098582

2. Shin, D.J., Kim, B.H.: Bank consolidation and competitiveness: empirical evidence from the Korean banking industry. J. Asian Econ. **24**, 41–50 (2013). https://doi.org/10.1016/j.asieco.2012.07.004
3. Bikker, J., Shaffer, S., Spierdijk, L.: Assessing competition with the panzar-rosse model: the role of scale, costs, and equilibrium. Rev. Econ. Stat. **94**(4), 1025–1044 (2009). https://doi.org/10.1162/REST_a_00210
4. Maochun, Z., Zhixu, S.: The research on the competitiveness of commercial banks by factor analysis. In: 6th International Conference on Information Management, Innovation Management and Industrial Engineering, pp. 367–370. Xi'an, China (2013). https://doi.org/10.1109/ICIII.2013.6703594
5. Qing, D.W.: The introduction of financial ethics business silver pound core competitiveness index system. Financ. Theory Pract. **3**, 27–30 (2009)
6. Barbosa, K., Rocha, B., Salazar, F.: Assessing competition in the banking industry: a multiproduct approach. J. Bank. Finance **50**, 340–362 (2015). https://doi.org/10.1016/j.jbankfin.2014.05.003
7. Krasovskis, D., Limanskis, A., Pancenko, E.: Measuring competitiveness of banks in Latvia. Copernican J. Financ. Account. **5**(2), 125–147 (2016). https://doi.org/10.12775/CJFA.2016.019
8. Mbama, C., Ezepue, P.O.: Digital banking, customer experience and bank financial performance: UK customers' perceptions. Int. J. Bank Market. **36**(2), 230–255 (2018)
9. Zhao, Q., Tsai, P.-H., Wang, J.-L.: Improving financial service innovation strategies for enhancing china's banking industry competitive advantage during the fintech revolution: a hybrid MCDM model. Sustainability **11**(5), 1419 (2019). https://doi.org/10.3390/su11051419
10. Busby, D.: Adopting the best approach for a digital banking solution: combine the benefits of the 'build', 'buy' or 'outsource' options. J. Digital Bank. **2**(1), 43–50 (2017). https://www.ingentaconnect.com/content/hsp/jdb001/2017/00000002/00000001/art00006
11. Amin, M.: Internet banking service quality and its implication on e-customer satisfaction and e-customer loyalty. Int. J. Bank Market. **34**(3), 280–306 (2016). https://doi.org/10.1108/IJBM-10-2014-0139
12. Raza, S.A., Jawaid, S.T., Hassan, A.: Internet banking and customer satisfaction in Pakistan. Qual. Res. Financ. Markets **7**(1), 24–36 (2015)
13. Martins, C., Oliveira, T., Popovic, A.: Understanding the internet banking adoption: a unified theory of acceptance and use of technology and perceived risk application. Int. J. Inf. Manage. **34**(1), 1–13 (2014). https://doi.org/10.1016/j.ijinfomgt.2013.06.002
14. Ali, A., Ali, H.: Customers' perception of m-banking adoption in the Kingdom of Bahrain: an empirical assessment of an extended tam model. Int. J. Managing Inf. Technol. **6**(1), 1–13 (2014). https://doi.org/10.5121/ijmit.2014.6401
15. Silva Bidarra, S.H., Muñoz-Leiva, F., Liébana-Cabanillas, F.: Analysis and modelling of the determinants of mobile banking acceptance. Int. J. Manage. Sci. Inf. Technol. (IJMSIT) **8**, 1–27 (2013). https://www.econstor.eu/handle/10419/97874
16. Jun, M., Palacios, S.: Examining the key dimensions of mobile banking service quality: an exploratory study. Int. J. Bank Market. **34**(3), 307–326 (2016). https://doi.org/10.1108/IJBM-01-2015-0015
17. Dootson, P., Beatson, A., Drennan, J.: Financial institutions using social media – do consumers perceive value? Int. J. Bank Market. **34**(1), 9–36 (2016). https://doi.org/10.1108/IJBM-06-2014-0079

18. Koroleva, E., Kudryavtseva, T.: Factors influencing digital bank performance. In: The 2018 International Conference on Digital Science, pp. 325–333 (2020). https://doi.org/10.1007/978-3-030-37737-3_29
19. Stoica, O., Mehdian, S., Sargu, A.: The impact of internet banking on the performance of Romanian banks: DEA and PCA approach. Proc. Econ. Financ. **20**, 610–622 (2015). https://doi.org/10.1016/S2212-5671(15)00115-X

Artificial Intelligence as a Disruptive Technology for Digital Marketing

Vladislav Rutskiy[1,7], Reza Mousavi[2], Nikita Chudopal[1], Younes El Amrani[3], Valentina Everstova[4], and Roman Tsarev[5,6](✉)

[1] Siberian Federal University, Krasnoyarsk, Russia
[2] PBXDom, Walnut, USA
[3] Abdelmalek Essaâdi University, Tangier, Morocco
[4] North-Eastern Federal University, Yakutsk, Russia
[5] V.A. Trapeznikov Institute of Control Sciences of Russian Academy of Sciences, Moscow, Russia
[6] International Academy of Science and Technologies, Moscow, Russia
[7] HSE University , Moscow, Russia

Abstract. The paper studies peculiarities of marketing techniques based on artificial intelligence used in relevant software products. The authors investigate trends in the global marketing industry and reveal problems of capturing customer attention and tailoring commercial offers to specific needs of the audience on the Internet. Competitive advantages of artificial intelligence for dealing with these problems are connected to the possibilities of this innovative technology for relevant data collection and data processing. Successful implementation of artificial intelligence into digital marketing is based on the agile principles of software products development in the marketing industry.

Keywords: Digital marketing · Artificial intelligence · Software development · Innovations · Agile

1 Introduction

At the moment in the era of the digital transformation of most processes of human activity and economic relations, advertising has begun to play a key role. Some companies and entrepreneurs are willing to spend up to half of their budget on just one ad. This state of affairs makes the very concept of advertising and marketing simply ineffective as long as people develop so-called "advertising blindness" because of omnipresent advertising.

An increasing number of companies are beginning to use artificial intelligence technology in the field of marketing to overcome this problem and gain competitive advantages in the market, both in terms of advertising quality and its costs. Only a few years ago the largest companies such as Netflix, Amazon, Google started to use it [1]. However, in recent years, artificial intelligence systems have penetrated deeply into the marketing industry, which has made this tool available for small and medium-sized businesses.

Currently, there is a variety of digital products and marketing tools powered by artificial intelligence technology. This paper explores the general principles of such

R. Silhavy et al. (Eds.): CoMeSySo 2021, LNNS 231, pp. 895–900, 2021.
https://doi.org/10.1007/978-3-030-90321-3_74

tools and assesses the capabilities of some marketing products developed using artificial intelligence technology and widely used nowadays.

2 Data Collection Mechanisms in Digital Marketing with the Help of Artificial Intelligence

The product and service recommendation system is possibly the first one and at the same time, a very well-known application of artificial intelligence in marketing. The recommendation systems range from the most simple, accessible to any person with some basic skills in working with online advertising products, to complex and intricate systems with several levels of identification of the needs of a potential customer.

At the moment, the application market is overflowing with offers of co-building bots that write review articles and positive reviews both on social networks and on regular websites. Such services are provided by many web studios. And the cost of such bots can range from $ 200 to $ 1500 on average, depending on the goals and complexity of the software development [2].

The system for identifying needs and building the psychological type (psychotype) of an Internet user as a potential client is an example of a more complicated implementation of artificial intelligence. Arthur Khachuyan, a well-known Russian expert in big data processing and a founder of Social Data Hub, distinguishes the following three levels of data identification concerning psychotype [3]:

1. The company data about its users (clients). The first surface level, which is characterized by the user's purchase history on the website, various transactions, traces of human interaction with the interface. All this allows the owner of a brand or online store to understand the primary needs of the buyer.
2. Open source user data about users ("digital footprint"). The huge set of personal data that a person spreads himself or herself in the digital space is an excellent source of information for forming the image of this person as a potential client.
3. User's environment on the Internet. Getting data about a person from his friends and acquaintances is a technique commonly used in marketing. An example is the analysis of the friends' profiles on social networks or accounts of acquaintances invited to various Internet resources through referral links and invitation programs.

Such a distribution of data identification (as a rule, 30% of the data appears from the analysis of the first level, 40% – from the second one, 30% – from the third one) allows identifying a set of goods and services of user's interest.

Improving the personalization of advertising is the next popular way to use of artificial intelligence in marketing [4]. There have been technologies of contextual and targeted advertising for about 15 years already. These allow a company to demonstrate the necessary content to the most relevant target audience. However, there is still a huge amount of "junk" advertising on the Internet, which does not correspond to the requests and needs of a potential buyer. Concerning this type of advertising, it is still difficult to form an effective set of offered goods and services that would be of interest to both the buying and selling parties. Therefore, in today's digital realities, an individual approach is needed that can help to introduce the use of artificial intelligence (AI) technology.

To implement this approach, personalized solutions are developed and implemented, which allow finding and sending individual proposals to clients [5]. For the most part, this works based on calculating and constructing the client's psychotype. Normally the following three models are used to determine it [6]:

1. "The content used" [7]. AI analyzes the words that the client uses in the digital space, the information he is looking for, as well as views of photo and video materials, especially if some kind of digital footprint has been left ("like", comment etc.).
2. "Interaction with the web interface" [8]. AI analyzes the keyboard buttons that people press when searching for something, the order in which they are pressed, and even the manner of printing. Some companies assess the psychotype of people by their keyboard handwriting.
3. "The feedback" [9]. The potential audience is divided into several fragments, and a certain set of goods and services are shown to them. A specific set is demonstrated to a specific segment of the audience. Thus, analyzing the reaction of each fragment of the audience, AI concludes on the positive perception of a particular set and the specific preferences of a certain psychotype.

Thus, such a personalized approach could generate the necessary response from the audience, as well as improve their user experience. This will help companies and entrepreneurs get more traffic and improve the conversion rate of marketing actions.

3 Data Processing in Digital Marketing with the Help of Artificial Intelligence

In addition to data search, a significant class of tasks for artificial intelligence is associated with data processing in marketing. For example, there are programs for improving work with influencers (opinion leaders). This became widespread among companies and spending on influencers exceeded $ 9 billion in 2020 [10]. Finding and selecting influencers could be time-consuming. Also, fraud in the face of "fake influencers" is highly widespread in this market which complicates the search for candidates for advertising. This is the field where the use of artificial intelligence is efficient allowing to select of the most relevant influencer in a matter of minutes, as well as to analyze his or her audience for the parameters of activity and authenticity.

The AI-based software products used by Socialbakers [11] recognize over 27 million micro and macro-influencers and provide data-backed influencer portraits, including interests, country, subscriber count, content, languages, brands mentions and actions on different social media channels.

AI search places the highest quality influencers first, reducing the priority of or completely removing bots and less preferred options [4].

However, influencers are not a panacea. Sometimes you have to create your unique content. Here another artificial intelligence technique can help, increasing the visibility of content through planning [12].

AI analyzes posting time, number of posts, comments and questions by time and day of the week. Based on several comparisons, it gives out the most appropriate time to publish content, which allows you to engage the largest possible audience. Also, such services provide recommendations for planning content based on audience behaviour. On top of historical data, AI adds layers of algorithms [13]: user activity and the time when users are online.

As a result, the advertiser gets a visual "heatmap" of the suggested hours and days of the week to post content that collects the most organic views [14]. The heatmap also includes missed views if a post is released outside the recommended time frame.

Another example of AI use in marketing is the automatic conduct of industry analysis. Knowing about audience intersections, discovering the most relevant content, and identifying the ups and downs in the industry allows a company to better position its brand to deliver exceptional content. For most brands, market research is part of their daily routine. Monitoring competitors in a wide data field helps to form a clear understanding of their advantages and disadvantages. With these insights, you can improve content performance and achieve important key performance indicators (KPIs) on social media [15]. This process can be time-consuming. It is sensible to include all available data in the form of easy-to-read comparisons for the analysis to be timely and efficient.

With a broader scope of the market, it is easier for marketers to determine the level of the company and set the best targets for competition. However, today specialists need even more information to play successfully in this field. The understanding of market trends in industries will become deeper in the process of using AI.

4 Specific Techniques in Digital Marketing Applicable for Artificial Intelligence Implementation

The successful implementation of artificial intelligence in digital marketing should be flexible, adapted to the changing tasks of the marketing departments of companies in the industry. At the same time, AI enables the integration of marketing, sales and supply chains. For example, AI-based software products can automate pricing, and thereby increase sales and profits without human intervention. To do this, it is necessary to allow AI to adjust sales proposals by setting a price for a product based on the behaviour of users and their needs, depending on the availability of the product and the demand for it [16]. This tactic can attract more new customers and encourage them to buy more.

In turn, machine learning allows a company to classify existing customers, predict their value to the company based on past behaviour and current intentions. After training, such data will be transmitted through identified signals from the customer base and converted into a unique message for those clients who can shop more and more often.

AI also allows a company to optimize work in Pay Per Click advertising (PPC) [17]. Typically, PPC campaigns are run either by their employees or by advertising agents, that is, people. AI can help to discover new advertising channels that competitors are not yet using, as well as optimize old channels [18]. With AI, advertisers can test more ad platforms and optimize targeting.

Perhaps the most popular use of AI in marketing is chatbot development [19]. Smart chatbots provide customer support in a wide variety of areas from fashion and entertainment to healthcare and insurance. Sometimes their style of service is more individual than human. Chatbots have access to thousands of data points associated with clients [20]. They can aggregate location-specific queries to identify patterns, find recurring problems, and predict what are the causes of the particular user problems. This makes chatbots more knowledgeable than any support person.

Various AI techniques provide ample opportunities for software development for the complex and rapidly changing tasks of the modern digital marketing industry listed above. For example, very high performance of the tasks could be achieved with the help of neural networks and the use of flexible (Agile) technologies for the development, implementation and refinement of the relevant marketing products [21].

5 Conclusions

Thus, the use of artificial intelligence technology gives breakthrough results in the digital marketing industry, expanding the possibilities of building real-time portraits of the target audience and adapting commercial offers to the specific needs of potential customers.

The techniques and mechanisms of digital marketing based on artificial intelligence are used for the following purposes mainly:

- Stimulating the growth of the company's profits.
- Using AI as a customer retention driver.
- More accurate targeting of the audience segments.
- Use of insights collected to analyze competitiveness in a particular industry.
- Creation, planning and smart management of content.
- Replacement of routine human labour with machine labour.

At the moment, it is clear that the future of digital marketing is largely related to artificial intelligence technology. The relevant software products could be developed on the principles of agile programming and used to save the advertising budgets and increase the effectiveness of marketing radically. Therefore, companies in various industries need to get prepared to accept this disruptive technology and use it to gain and secure their competitive edge in the market.

References

1. Artificial intelligence in marketing. https://polygant.net/ru/ai/iskusstvennyj-intellekt-v-marketinge/. Accessed 17 Feb 2021
2. Blog of the ArtisMedia company. https://artismedia.by/blog/iskusstvennyj-intellekt-v-marketinge/. Accessed 9 Feb 2021
3. Arthur Khachuyan: artificial intelligence in marketing. https://habr.com/ru/company/ua-hosting/blog/482816/. Accessed 16 Feb 2021
4. De Bruyn, A., Viswanathan, V., Beh, Y.S., Brock, J.K.-U., von Wangenheim, F.: Artificial intelligence and marketing: pitfalls and opportunities. J. Interact. Mark. **51**, 91–105 (2020)
5. Mustak, M., Salminen, J., Pléc, L., Wirtzd, J.: Artificial intelligence in marketing: topic modeling, scientometric analysis, and research agenda. J. Bus. Res. **124**, 389–404 (2021)

6. Artificial intelligence in digital marketing: how the work of a marketer is changing in 2020. https://vc.ru/marketing/107257-iskusstvennyy-intellekt-v-digital-marketinge-kak-menyaetsya-rabota-marketologa-v-2020-godu. Accessed 16 Feb 2021
7. Free Social Media Statistics., https://www.socialbakers.com/statistics. Accessed 16 Feb 2021
8. Feng, C.M., Park, A., Pitt, L., Kietzmann, J., Northey, G.: Artificial intelligence in marketing: a bibliographic perspective. Australas. Mark. J. **8**, 1–12 (2020)
9. Three major fields of artificial intelligence and their industrial applications. https://towardsdatascience.com/three-major-fields-of-artificial-intelligence-and-their-industrial-applications-8f67bf0c2b46. Accessed 16 Feb 2021
10. Artificial intelligence: applications in marketing, analytics and your life. https://lpgenerator.ru/blog/2017/04/11/iskusstvennyj-intellekt-primenenie-v-marketinge-analitike-i-vashej-zhizni/. Accessed 17 Feb 2021
11. Paschen, J., Paschen, U., Pala, E., Kietzmann, J.: Artificial intelligence (AI) and value co-creation in B2B sales: Activities, actors and resources. Australas. Mark. J. **2**, 1–9 (2020)
12. Arntz, M., Gregory, T., Zierahn, U.: The risk of automation for jobs in OECD countries: a comparative analysis (OECD Social, Employment and Migration Working Paper no 189), Paris: OECD (2016). https://doi.org/10.1787/5jlz9h56dvq7-en. Accessed 13 Feb 2021
13. Frey, C.B., Osborne, M.A.: The Future of Employment: How Susceptible are Jobs to Computerisation? University of Oxford, Oxford (2013)
14. Goos, M., Manning, A., Salomons, A.: Job polarization in Europe. Am. Econ. Rev. Papers Proc. **99**(2), 58–63 (2009)
15. Kenney, M., Zysman, J.: The rise of the platform economy. Issues Sci. Technol. **32**(3), 87–95 (2016)
16. Loureiro, A.L.D., Miguéis, V.L., da Silva, L.F.M.: Exploring the use of deep neural networks for sales forecasting in fashion retail. Decis. Support Syst. **114**, 81–93 (2018)
17. Rubio, F., Flores, M.J., Puerts, J.M.: Drawing a baseline in aesthetic quality assessment. Proc. SPIE **10696**, 361–364 (2017)
18. Kang, L., Ye, P., Li, Y., Doermann, D.: Convolutional neural networks for no-reference image quality assessment. Proc. IEEE Conf. Comput. Vis. Pattern Recognit. 1733–1740 (2014)
19. Talebi, H., Milanfar, P.: NIMA: neural image assessment. IEEE Trans Image Process **27**(8), 3998–4011 (2018)
20. Chollet, F.: Deep Learning with Python. Manning Publications, New York (2017)
21. Gruzenkin, D.V., et al.: Neural networks to solve modern artificial intelligence tasks. J. Phys: Conf. Ser. **1399** (2019). https://doi.org/10.1088/1742-6596/1399/3/033058

Driver's State Identification Based on the Vehicle Speed Analysis Taking into Account the Driving Context

Nikolay Shilov(✉)

SPC RAS, 39, 14 Line, St. Petersburg, Russia
nick@iias.spb.su

Abstract. Today, human drivers remain the main cause of traffic accidents. While not always being the main accident reasons, drowsiness and fatigue are often indirect reasons for accidents. As a result, it is still actual to classify the driver's state (drowsy or awake) while driving. The most popular approaches to driver's state classification today are the application of special medical equipment (which gives the highest results) or machine vision techniques. However, the former are nearly impossible for mass implementation due to the complexity and high costs, and the latter suffer from continuously varying lighting, which is the case for moving vehicles especially in dark conditions when the identification of the driver's state is even more important. Previous research has shown the fundamental possibility of identification of the driver's state based on an analysis of vehicle speed. However, its results were not very high. In this work, we are searching for a way to increase the quality of the classification by taking into account the driving context. The results show that (a) the classification of the vehicle's driving context can be performed based on the analysis of its speed with sufficiently high accuracy, and (b) that the preliminary classification of the vehicle's driving context can significantly increase the accuracy of the classification of the driver's state based on the analysis of the vehicle speed.

Keywords: Driver's state identification · Vehicle's speed · Driving context · CatBoost

1 Introduction

Despite the rapid development of self-driving vehicles, human drivers remain the main cause of traffic accidents [1, 2], with drowsiness and fatigue being often indirect reasons (20–30% according to [3]).

In the previous publication [4], a research effort was reported aimed at checking the hypothesis that the vehicle speed could be effectively used for the driver's state identification (in particular, the fatigue state). The motivation for that research effort was related to the fact that today most systems for driver state identification are either based on the application of rather complex medical equipment, which is nearly impossible for mass implementation, or computer vision techniques, which suffer from continuously

R. Silhavy et al. (Eds.): CoMeSySo 2021, LNNS 231, pp. 901–909, 2021.
https://doi.org/10.1007/978-3-030-90321-3_75

varying lighting due to the reflected or direct sun flares (when a car turns), lights of oncoming vehicles and road lights [5–9].

The carried out research showed the fundamental possibility of identification of the driver's state (drowsy or not) based on an analysis of vehicle speed. However, it was noted that the extension of the analyzed data could decrease the classification precision. As a result, it was decided to include additional data for the classification, namely, the driving context.

The vehicle's driving context is usually understood as the situation in which the vehicle is moving. For example, in [10], the following classes of the driving context are distinguished: driving on a highway, driving on city streets, driving on country roads, driving in a traffic jam, parking. Identification of the driver's state (in particular, identification of the drowsiness) is the most important for the first two classes (driving in the motorway and driving in city streets), which were analyzed in this study. Thus, the research questions in this section are as follows:

RQ1: What level of accuracy can be achieved for the classification of the vehicle's driving context based on the analysis of its speed?

RQ2: Will the preliminary classification of the vehicle's driving context improve the accuracy of the classification of the driver's state based on the analysis of the vehicle speed?

The paper is structured as follows. The next section describes the research approach including a description of the used dataset and the overall methodology used. Then, the experimentation results are presented and the answers to the research questions are given. The main results are summarized in the Conclusion.

2 Research Approach

2.1 Dataset

Unfortunately, still there are no datasets available for this task. This may be due to two factors. First, the data obtained on simulators cannot be used for this purpose, since the nature of the impact on the vehicle controls (primarily, the accelerator and brake pedals) on the simulator and in real conditions differs significantly (for example, see work [11]). Second, since drowsy driving is unsafe (and even illegal in some countries [12]), one cannot ask anyone to drive in a drowsy state, even for scientific purposes.

Thus, the dataset for identification of the driver's condition based on the speed analysis was collected (and continues to be collected at present) by the authors independently based on the speed records with the states of being drowsy and awake confirmed by the drivers among the users of the Drive Safely system personally known by the author [13] through interviewing.

As a result of the constant extension of the dataset, the accuracy of the previously built model [4] has significantly decreased and the identification of the vehicle's driving context for the subsequent separate context-dependent classification of the driver's state has become reasonable, given the relatively small amount of available data.

The dataset used in this research consists of 20 rides of one driver with a total duration of about 6500 min. The rides were recorded while driving different cars (equipped with both manual and automatic transmissions) with different external conditions. The speed measurements were recorded using the DriveSafely system running on an Android-based smartphone 5 times per second.

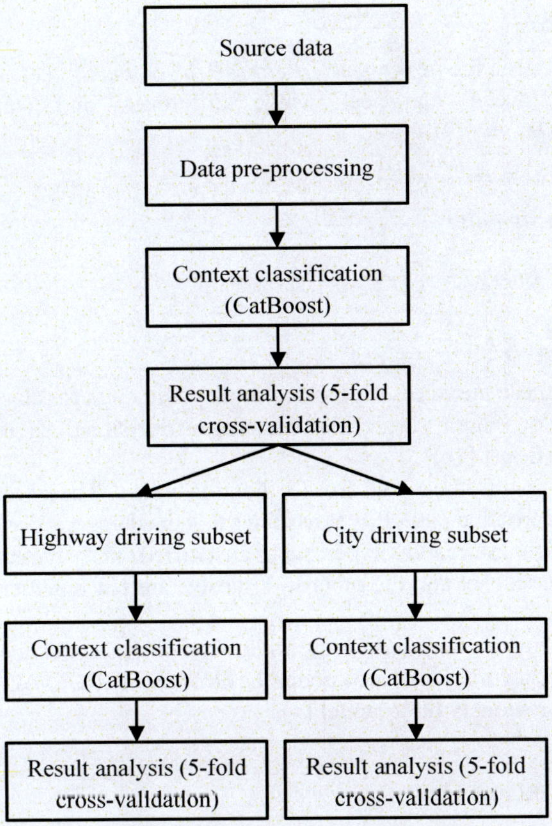

Fig. 1. Methodology of data analysis.

2.2 Methodology

The used methodology is shown in Fig. 1. First, the data is pre-processed. For this purpose, the data is filtered (the records with a speed less than 10 km/h are removed) and complemented with the calculated acceleration (the derivative of the speed) and jerk (the derivative of the acceleration) as well as their absolute values resulting in time series of five parameters (speed, acceleration, absolute value of acceleration, jerk, absolute value of jerk). Then, the time series are split into fragments with the duration of 60 s, and for each fragment, the following statistical measures are calculated for each of five parameters [14, 15]: Mean value: $Mean = \frac{\sum_{i=1}^{N} x_i}{N}$.

Standard deviation: $StdDev = \sqrt{\frac{\sum_{i=1}^{N} (x_i - Mean)^2}{N-1}}$.

Variance: $Variance = StdDev^2$.

The ratio of the standard deviation to the mean value: $StdDev\_Mean = \frac{StdDev}{Mean}$.

Skewness: $Skew = \frac{\sum_{i=1}^{N}(x_i - Mean)^3}{(N-1)StdDev^3}$.

Kurtosis: $Kurtosis = \frac{\sum_{i=1}^{N}(x_i - Mean)^4}{(N-1)StdDev^4}$.

Difference between the peak values: $PeakToPeak = max_{i=1}^{N}(x) - min_{i=1}^{N}(x)$.

The ratio between the peak value difference and standard deviation: $PeakToPeak\_StdDev = \frac{PeakToPeak}{StdDev}$.

Crest factor: $Crest = \frac{max_{i=1}^{N}(x)}{StdDev}$.

Impulse factor: $Impulse = \frac{max_{i=1}^{N}(x)}{Mean}$.

Energy factor: $Energy = \left(\frac{\sum_{i=1}^{N}\sqrt{|X_i|}}{N}\right)^2$.

Shape: $Shape = \frac{\sqrt{\frac{1}{N}\sum_{i=1}^{N}X_i^2}}{\frac{1}{N}\sum_{i=1}^{N}\sqrt{|X_i|}}$.

The values of these measures are used as input parameters for classification models.

The classification is done via gradient boosting on decision trees (implemented using Python library CatBoost [16]).

To analyze the results, the 5-fold cross-validation is used. The k-fold cross-validation procedure (in our particular case, k = 5) assumes that the dataset is divided into k subsets of approximately equal size, and k experiments are carried out so that each of the k subsets is used once as the test set and the other k–1 subsets are put together to form the train set.

The following indices are used for the evaluation of the model adequacy: precision, recall, F1-measure, accuracy, and area under the error curve (ROC-AUC). Table 1 shows the results of the previously built model [4].

3 Experimentation Results

3.1 Driving Context Classification

The classification of the vehicle's driving context has yielded the results presented in Table 2.

The analysis of the significance of the training parameters has led to the expected result. So, the most significant were the values of the mean speed and the coefficient of the energy of the speed (direct proportionality), as well as the ratio of the standard deviation to the mean value (in-verse proportionality). In other words, the movement on the highway, on average, is done at a higher speed and with a lower level of variability (Fig. 2, Fig. 3). This ob-servation basically confirms the adequacy of the built model.

An example of a trip with each point being classified by the constructed model is shown in Fig. 4.

Thus, the answer to the first research question (RQ1) is as follows: the context of the vehicle movement based on the analysis of its movement speed can be performed with sufficiently high accuracy.

Table 1. The results of the classification of the driver's state based on the analysis of the vehi-cle speed using the extended dataset.

Metric	Fold					Mean value
	1	2	3	4	5	
ROC AUC	0.576	0.575	0.592	0.522	0.585	0.570
Precision	0.512	0.513	0.548	0.514	0.500	0.518
Recall	0.525	0.500	0.425	0.450	0.450	0.470
F1-measure	0.519	0.506	0.479	0.480	0.474	0.491
Accuracy	0.552	0.552	0.570	0.547	0.535	0.551

Table 2. The results of the classification of the vehicle's driving context based on the analysis of its speed.

Metric	Fold					Mean value
	1	2	3	4	5	
ROC AUC	0.938	1.000	0.957	0.860	0.943	0.939
Precision	0.939	1.000	0.967	0.938	0.909	0.950
Recall	0.969	1.000	0.935	0.968	0.968	0.968
F1-measure	0.954	1.000	0.951	0.952	0.938	0.959
Accuracy	0.927	1.000	0.925	0.925	0.900	0.935

3.2 Driver's State Classification

In order to classify the driver's state separately based on the speed analysis for each context, the existing dataset was divided into two subsets depending on the context (city street driving and highway driving), and then the data was analyzed, the corresponding models were built and their evaluation was carried out separately for each context. However, in contrast to the process of building a model to determine the context described above, in these models, the speed and all statistical parameters calculated on its basis were excluded in order to reduce the dependence of the model on the congestion of the road network and road conditions, which directly affect the vehicle speed.

Tables 3 and 4 show the quality of the constructed models for driver's state classification when driving on city streets and on the highway, respectively.

It can be seen from the tables that the quality of the driver's state classification based on the analysis of the vehicle speed, taking into account the context, is significantly higher than without taking it into account (cf. Table 1).

So, the answer to the second research question (RQ2) is positive: the preliminary classification of the vehicle's driving context can significantly increase the accuracy of the classification of the driver's state based on the analysis of the vehicle's speed.

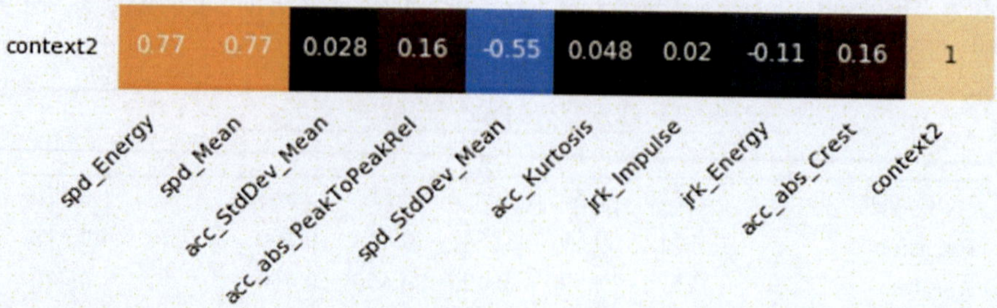

Fig. 2. Correlation of the most significant parameters with the value of the vehicle's driving context.

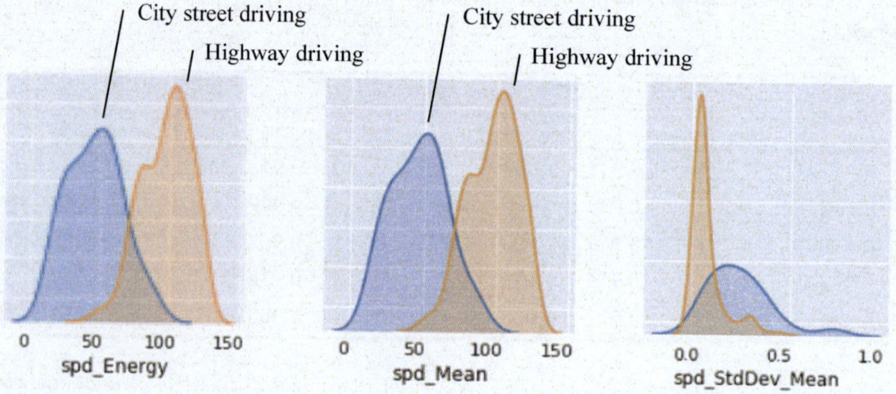

Fig. 3. Smoothed histograms of the number of observations of the values of the speed energy coefficient indicators (direct proportionality), the average speed and the ratio of the standard deviation to its mean value for various contexts of vehicle movement.

Table 3. The results of the classification of the driver's state based on the analysis of the vehi-cle speed when driving on city streets.

Metric	Fold					Mean value
	1	2	3	4	5	
ROC AUC	0.964	1.000	0.929	0.939	0.918	0.950
Precision	0.700	0.700	0.778	0.778	0.636	0.718
Recall	1.000	1.000	1.000	1.000	1.000	1.000
F1-measure	0.824	0.824	0.875	0.875	0.778	0.835
Accuracy	0.800	0.800	0.867	0.857	0.714	0.808

Table 4. The results of the classification of the driver's state based on the analysis of the vehi-cle speed when driving on city streets.

Metric	Fold					Mean value
	1	2	3	4	5	
ROC AUC	0.762	0.679	0.668	0.832	0.651	0.718
Precision	0.765	0.667	0.650	0.696	0.625	0.680
Recall	0.722	0.778	0.765	0.941	0.882	0.818
F1-measure	0.743	0.718	0.703	0.800	0.732	0.739
Accuracy	0.719	0.656	0.645	0.742	0.645	0.681

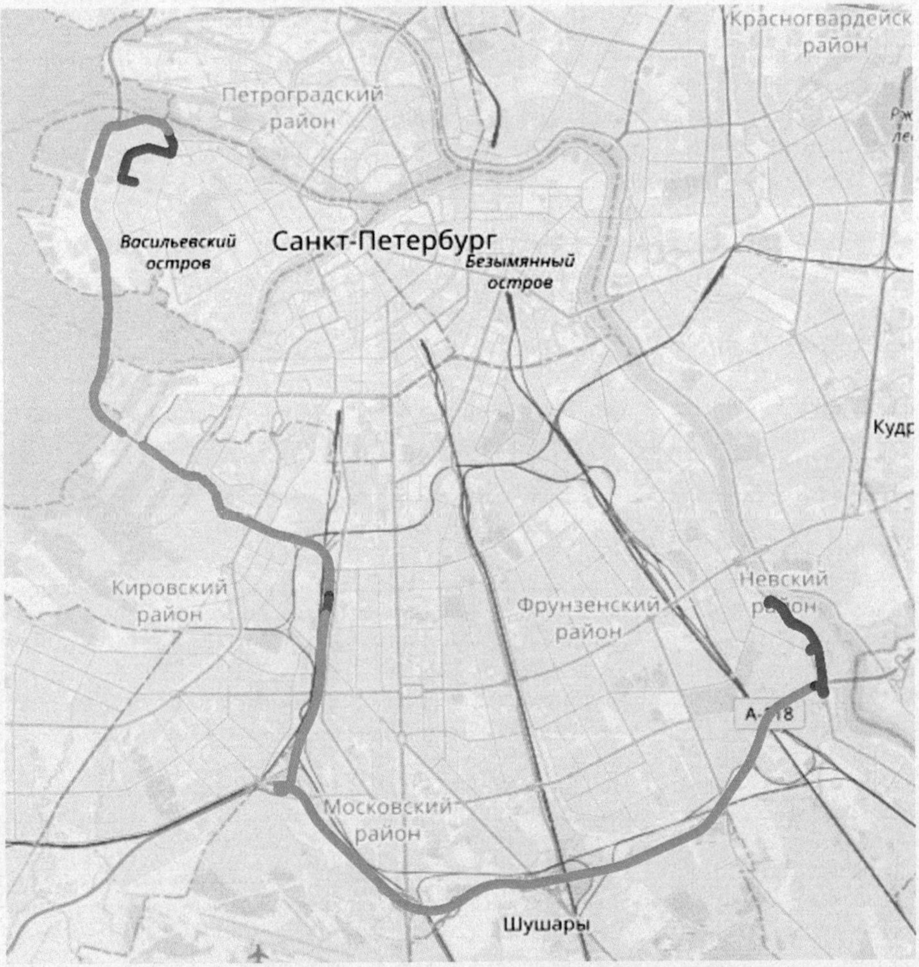

Fig. 4. An example of the results of the classification of the vehicle's driving context (lighter grey – driving on a highway, darker grey – driving on city streets).

Further research in this area is planned to be aimed at extending the dataset, publishing it in the public domain, as well as analyzing the possibility of using other machine learning models to solve the above classification problems.

4 Conclusion

The paper extends the previously published work aimed at driver's state identification based on the vehicle's speed analysis. In this publication, we analyze the driving context (namely, highway driving and city streets driving) in order to achieve a higher quality of the driver's state identification models. Two research questions are answered. We have shown that the vehicle's driving context (namely, highway driving and city streets driving) can be successfully classified based on the analysis of its speed with sufficiently high accuracy, and that the preliminary classification of the vehicle's driving context can significantly increase the accuracy of the classification of the driver's state based on the analysis of the vehicle speed.

The future work is planned to aim at extending the dataset and publishing it in the public domain. Besides, we are going to apply different machine learning models to solve classification problems.

Acknowledgments. The research is due to State Research, project number 0073-2019-0005.

References

1. Szumska, E., Frej, D., Grabski, P.: Analysis of the causes of vehicle accidents in Poland in 2009–2019. LOGI – Sci. J. Transp. Logist. **11**, 76–87 (2020). https://doi.org/10.2478/logi-2020-0017
2. Bucsuházy, K., Matuchová, E., Zůvala, R., Moravcová, P., Kostíková, M., Mikulec, R.: Human factors contributing to the road traffic accident occurrence. Transp. Res. Proc. **45**, 555–561 (2020). https://doi.org/10.1016/j.trpro.2020.03.057
3. Rolison, J.J., Regev, S., Moutari, S., Feeney, A.: What are the factors that contribute to road accidents? An assessment of law enforcement views, ordinary drivers' opinions, and road accident records. Accid. Anal. Prev. **115**, 11–24 (2018). https://doi.org/10.1016/j.aap.2018.02.025
4. Shilov, N., Kashevnik, A.: An effort to detect vehicle driver's drowsy state based on the speed analysis. In: 2021 29th Conference of Open Innovations Association (FRUCT), pp. 324–329. IEEE (2021). https://doi.org/10.23919/FRUCT52173.2021.9435466
5. Kashevnik, A., Lashkov, I., Gurtov, A.: Methodology and mobile application for driver behavior analysis and accident prevention. IEEE Trans. Intell. Transp. Syst. **21**, 2427–2436 (2020). https://doi.org/10.1109/TITS.2019.2918328
6. Luo, X., Hu, R., Fan, T.: The driver fatigue monitoring system based on face recognition technology. In: 2013 Fourth International Conference on Intelligent Control and Information Processing (ICICIP), pp. 384–388. IEEE (2013). https://doi.org/10.1109/ICICIP.2013.6568102
7. Masanovic, L., Vranjes, M., Dzakula, R., Lukac, Z.: Driver monitoring using the in-vehicle camera. In: 2019 Zooming Innovation in Consumer Technologies Conference (ZINC), pp. 33–38. IEEE (2019). https://doi.org/10.1109/ZINC.2019.8769377

8. Tipprasert, W., Charoenpong, T., Chianrabutra, C., Sukjamsri, C.: A method of driver's eyes closure and yawning detection for drowsiness analysis by infrared camera. In: 2019 First International Symposium on Instrumentation, Control, Artificial Intelligence, and Robotics (ICA-SYMP), pp. 61–64. IEEE (2019). https://doi.org/10.1109/ICA-SYMP.2019.8646001
9. Murugan, S., Selvaraj, J., Sahayadhas, A.: Driver hypovigilance detection for safe driving using infrared camera. In: 2020 International Conference on Inventive Computation Technologies (ICICT), pp. 413–418. IEEE (2020). https://doi.org/10.1109/ICICT48043.2020.9112568
10. Marina, L.A., Trasnea, B., Cocias, T., Vasilcoi, A., Moldoveanu, F., Grigorescu, S.M.: Deep Grid Net (DGN): a deep learning system for real-time driving context understanding. In: 2019 Third IEEE International Conference on Robotic Computing (IRC), pp. 399–402. IEEE (2019). https://doi.org/10.1109/IRC.2019.00073
11. Sahayadhas, A., Sundaraj, K., Murugappan, M.: Detecting driver drowsiness based on sensors: a review. Sensors **12**, 16937–16953 (2012). https://doi.org/10.3390/s121216937
12. Optalert: Drowsy driving laws, regulations and rules from around the world. https://www.optalert.com/drowsy-driving-laws-regulations-and-rules-from-around-the-world/. Accessed 17 Feb 2021
13. Kashevnik, A., Fedotov, A., Lashkov, I.: Dangerous situation prediction and driving statistics accumulation using smartphone. In: 2018 International Conference on Intelligent Systems (IS), pp. 521–527. IEEE (2018). https://doi.org/10.1109/IS.2018.8710548
14. Xia, Z., Xia, S., Wan, L., Cai, S.: Spectral regression based fault feature extraction for bearing accelerometer sensor signals. Sensors **12**, 13694–13719 (2012). https://doi.org/10.3390/s121013694
15. Caesarendra, W., Tjahjowidodo, T.: A review of feature extraction methods in vibration-based condition monitoring and its application for degradation trend estimation of low-speed slew bearing. Machines **5**, 21 (2017). https://doi.org/10.3390/machines5040021
16. Yandex: CatBoost - open-source gradient boosting library

Cloud-Based Architecture and Algorithms for Monitoring and Control of an Automated Greenhouse Complex

Olga Kachanova and Dmitriy Levonevskiy(✉)

St. Petersburg Institute for Informatics and Automation of the Russian Academy of Sciences, St. Petersburg Federal Research Center of the Russian Academy of Sciences (SPC RAS), 39, 14th Line, St. Petersburg 199178, Russia

Abstract. To grow crops in greenhouse complexes, it is necessary to maintain the microclimatic conditions required for these crops and to monitor them. Automation of these processes is an urgent task, and for its solution it is necessary to use both hardware and software tools. The paper proposes a cloud service architecture for managing automated vertical farms, which allows for reliable two-way data transfer between the greenhouse modules and the operator and can be used for growing crops both in vertical hydroponic farms and in single-level greenhouse complexes. Algorithms are described that make it possible to understand and evaluate the organizing principles of monitoring and control of growing plants. On the basis of the developed architecture, software has been implemented, which has been tested in the tasks of controlling the greenhouse complex and obtaining data on the microclimate. The use of wireless networks, the absence of restrictions on the type and location of modules, duplication of functionality and data replication makes such a system flexible and scalable.

Keywords: Cloud services · Software architecture · Climate monitoring algorithms · Control algorithms · Automated greenhouse complexes · Cyber-physical systems · Microclimate control

1 Introduction

To grow any crops in greenhouse complexes, it is required to maintain the microclimatic conditions required for these crops: ventilation, temperature regime, lighting and watering. It is necessary to monitor microclimatic conditions in the greenhouse, since even a short-term violation of them can lead to negative consequences, up to the death of plants and cause significant damage to the enterprise.

When automating greenhouse complexes, problems arise associated with the need to organize reliable two-way data transfer between the greenhouse modules and the operator. This necessitates the development of a cloud service architecture for managing robotic vertical farms, within which such protocols and data transfer algorithms will be implemented that would ensure the observability and controllability of greenhouse

R. Silhavy et al. (Eds.): CoMeSySo 2021, LNNS 231, pp. 910–921, 2021.
https://doi.org/10.1007/978-3-030-90321-3_76

complexes and take into account possible problems associated with interruptions in data transmission and equipment failure.

This work sets a goal of developing such an architecture using the example of automation of vertical hydroponic farms intended for the cultivation of microgreens, greenery and barren plants. Vertical greenhouses allow one to grow more products than in traditional greenhouses of a similar area, and also allow one to more flexibly scale the size of the greenhouse for production and the amount of the desired product. At the same time, the costs for the substrate and water are reduced due to their circulation in the hydraulic system of the greenhouse. Since vertical greenhouses are less demanding on environmental conditions, they can be used not only in rural environments, but also in urban areas.

2 Related Works

Let us analyze the current state of research and development in the field of automation of greenhouse complexes. The necessary conditions for automation are the availability of means for monitoring the microclimate of greenhouse complexes and the impact on it, which makes it possible to implement a control loop of a greenhouse complex for automated cultivation of crops in accordance with a given technology.

In [1], it is proposed to use a wireless sensor network to collect data from sensors evenly distributed over the greenhouse area. The Micaz network module is used, the main feature of which is low power consumption. The authors have developed a routing protocol used in the application to track parameters received from sensors. The developed system was tested in real conditions and showed work without failures and errors. The developed routing protocol is aimed at minimizing the power consumption of the sensor network, however, it is not resistant to collisions, is not suitable for transferring large amounts of data and is limited in scalability.

The authors of [2] also proposed a solution aimed at implementing a remote monitoring system for the state of the greenhouse microclimate. A bundle of MSP430 microcontroller with Ez430-RF2500 wireless sensor module for transmitting messages from temperature and humidity sensors is proposed. The network protocol SimpliciTI is used, the advantage of which is a low memory requirement and support for sleep modes. According to the results obtained by the authors, for the full operation of the system, 3.3 V lithium batteries are sufficient. However, this does not involve monitoring the level of illumination and parameters of air flow circulation in the room.

In [3], the authors propose to use a system based on the NetDuino3 microcontroller to monitor the level of humidity, illumination and temperature in the greenhouse. The advantages of such a solution are the ability to connect to Wi-Fi, the ability to use various sensors - soil moisture, room temperature and humidity, illumination and air flows entering the greenhouse. The data received from the sensors is transmitted to the mobile device.

In addition to microclimate monitoring systems, there is a separate type of systems for full-fledged automated microclimate control in closed greenhouse complexes. The work [4] presents an automatic control system for a greenhouse to reduce water consumption when watering plants. The system controls the parameters of the greenhouse microclimate around the clock, including soil moisture, and in case of deviation of any parameter, the microcontroller controls the operation of the executive devices to normalize the microclimate. The proposed automatic irrigation system combines drip irrigation and spray irrigation, which made it possible to achieve a water saving of 48.78% compared to manual irrigation.

In [5], the system includes 3 sensors - illumination, humidity and temperature. Such systems are used in greenhouses where different types of plants are grown, for which the authors described the rules according to which the necessary threshold values are set on various sensors to maintain healthy plant growth. The experiments carried out have proven the effectiveness of using fuzzy logic to automate the irrigation system by reducing the level of water consumption. With the help of the developed system, it was possible to maintain the temperature of the greenhouse within the normal range. System [6] differs from the one discussed above in that it additionally incorporates the ability to automatically control doors using an IR sensor. This prevents violations of the temperature regime inside the greenhouse.

In the article [7], the authors proposed an adaptive greenhouse microclimate control system, which uses a neural network to predict the climatic state of the greenhouse, and a parallel fuzzy logic approach to regulate the air flow rate and greenhouse temperature. The system is implemented in Matlab and the authors concluded that the coordinated control of the air flow rate and the temperature of the greenhouse makes it possible to achieve more efficient use of energy resources.

In modern scientific research, projects have also been proposed [8] that use ZigBee technology as a wireless sensor network, which has proven itself for use in agriculture as an inexpensive technology for providing wireless communication. In [8, 9], the authors propose to use this protocol for data collection and control of the greenhouse microclimate. The operation of ventilation and irrigation systems is based on an interval control algorithm. To reduce power consumption, the authors suggest putting the system nodes into sleep mode during their inactivity. The results of testing the proposed solution showed that the effective operation of such a system in terms of the share of lost packets is possible only with a small amount of interference between the sensors and the control module, as well as at short distances.

The authors of [10] proposed a web application for remote control of the greenhouse microclimate. ZigBee technology was also chosen as a protocol for communication with sensor devices. The experiments were carried out for a greenhouse of a small area with a small number of sensors, which is due to the above-mentioned disadvantages of this technology.

To transfer data between the controller and the control server in [11], the GSM-GPRS technology was used, which is more reliable and faster than ZigBee and radio frequency modules. For remote monitoring and data analysis from sensors installed in the greenhouse, the proposed system uses the concept of the Internet of Things, which is the most promising technology for connecting field devices to wireless Internet networks or remotely controlled devices [12, 13]. The ATmega328 microcontroller monitors the system and controls heaters/coolers, spray guns, lamps and water pumps to maintain the required microclimate for plant growth. The system was tested in a small vertical greenhouse. The experiments carried out showed the solution to work, however, during testing, the accuracy of monitoring the environmental parameters turned out to be low due to the complexity of the optimal positioning of the sensor elements within the framework of vertical greenhouse complexes, the design features of which lead to significant deviations in microclimate parameters at various points of such greenhouses.

A similar set of parameters in [14] was also controlled using an Atmega328 microcontroller. The main difference between this system and those discussed above is the use of a wired infrastructure, which complicates the use of such systems in operating greenhouse complexes, as well as in large greenhouses, despite the undoubted advantages in the speed and stability of communication between sensor and control modules.

A number of other works [15–19] also propose systems for monitoring and controlling the microclimate of greenhouse complexes, which are based on the use of fuzzy logic tools or other control models. The proposed solutions differ in the architecture of the software, the equipment and data transmission technologies used, as well as the set of sensors and actuators. Nevertheless, for them, as well as for the overwhelming majority of the solutions presented above, it is characteristic that they were developed with the aim of providing automated microclimate control in single-level greenhouses of closed ground. Taking into account the peculiarities of the complexes of this class, the focus of the researchers was not so much the tasks of increasing the number of controlled parameters as the tasks of minimizing energy consumption.

Modern vertical greenhouse complexes, studied in [20], have a number of features, such as a high density of crops, the location of support structures throughout the volume of the room and multilevel. These features increase the level of interference in the room and reduce the radius of stable signal transmission due to the presence of a greater amount of both plant matter and other interference. Such vertical structures have a serious impact on the structure of the microclimate in the premises, which leads to the formation of zones of local deviation of microclimate parameters. This requires an increase in the density of placement of sensor devices to maintain a high level of accuracy in tracking microclimate parameters.

The work [21] considers robotization of micro-climate monitoring processes in greenhouse complexes. The authors have proposed a model of a robotic intelligent control service for working in an automatic vertical greenhouse. The robot locally monitors environmental processes using various sensors and digital cameras. The collected information about the current state of the microclimate and plantings is transmitted to the user who manages the greenhouse. The disadvantage of this solution is low automation - the robot acts exclusively at the user's command and autonomously implements only individual low-level operations.

Thus, automation of microclimate monitoring processes in vertical greenhouse complexes is a promising and actively developing area. However, the existing solutions are designed either for manual use, or do not take into account the specific features of vertical greenhouses, which is why their use in this class of greenhouses without significant modernization is fraught with difficulties. At the same time, well-known developments do not offer sufficiently flexible architectural solutions for cloud services, which would allow building scalable greenhouse complexes with wireless modules.

This paper proposes a cloud architecture designed to control robotic vertical farms, and algorithms for automated monitoring and control of the growing process. Systems built on the basis of the proposed architecture allow monitoring the microclimate and the state of the equipment of vertical farms, provide reliable two-way data transfer between greenhouse modules, servers and operators, and provide the latter with the ability to remotely control the growing process. Such systems can be used for growing crops in vertical hydroponic installations as well as in single-level greenhouses in greenhouses. These capabilities are achieved due to the support of wireless data transmission, which allows collecting data on the microclimate of the greenhouse complex using robotic means, as well as using local servers that aggregate this data and transmit it to the cloud service using the database replication mechanism.

3 Materials and Methods

3.1 Service Architecture

To simplify the presentation, the article describes the operation of the system with one connected greenhouse complex. In general, the cloud architecture for this case is shown in Fig. 1. It consists of 6 main blocks: a cloud server, a local object server (LOS), a mobile application, a group control device, a device module, and a sensor module. Cloud and LOS have their own web applications: a cloud web client and a local web client available to a local device running in a greenhouse. Several small arms and light weapons can be connected to one cloud server. LOS is the main device of the greenhouse, which receives information about the status of sensors and actuators. The cloud service stores microclimate data collected from all LOS greenhouses. Replicating data from a local device to the cloud improves the reliability of the greenhouse complex and reduces the risk of data unavailability during network outages. In the absence of access to the cloud, the readings are recorded in the database of the local device, and when the connection is restored, the LOS synchronizes the readings with the database of the cloud service.

The proposed software system consists of 3 main parts: a cloud server, a local object server (LOS) and a mobile application. Each server has its own client application that allows monitoring based on the user's location, from the Internet or from the local network. The architecture is shown in Fig. 2. Several local objects can connect to one cloud server. LOS is the main device of the greenhouse, which receives information on the status of sensors and actuators.

The cloud service stores microclimate data collected from all local greenhouse facilities. Duplication of data from a local device to the cloud is necessary to improve the reliability of the greenhouse complex. If there is no access to the cloud due to the lack of a network, the readings are recorded in the database of the local device. When a connection with the cloud occurs, the LOS synchronizes the readings with the cloud database.

3.2 Monitoring and Control Algorithms

Monitoring means the functionality of the greenhouse, which is responsible for communicating information about the state of the microclimate and sensors to the user, both at the moment, and collecting information at a certain time specified by the user.

All monitoring functionality can be divided into 3 main parts:

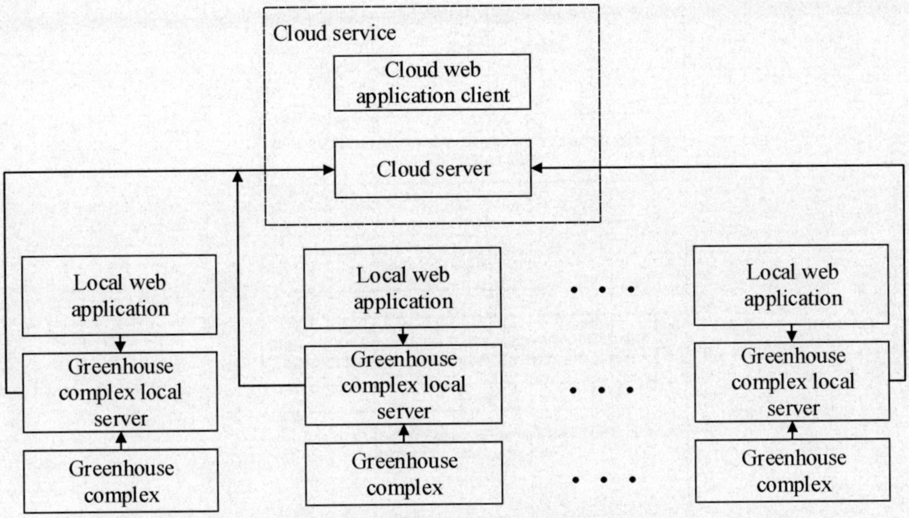

Fig. 1. General structure of the system.

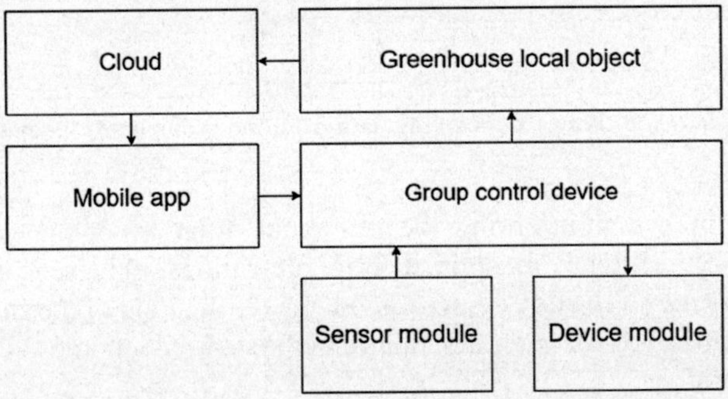

Fig. 2. Greenhouse complex architecture

- Receiving data from sensors;
- Generation of a report by analyzing data from sensors;
- Informing the user with notifications.

Consider an algorithm for visualizing data from sensors and devices. The group control device sends packets containing the readings of sensors and devices to the module for receiving data from the UUG on the local server. The data receiving module converts the data in the form of a new data object for each device and writes it to the Database. Then it sends a message to the data bus about the arrival of a new packet from the control device. A message from the data bus enters the server, which gives the latter a signal to update the user's page with the latest received packets. The user sees the changes in the web browser without reloading the page. The algorithm for visualizing data from sensors and devices is shown in Fig. 3. The screen displays data for the last hour.

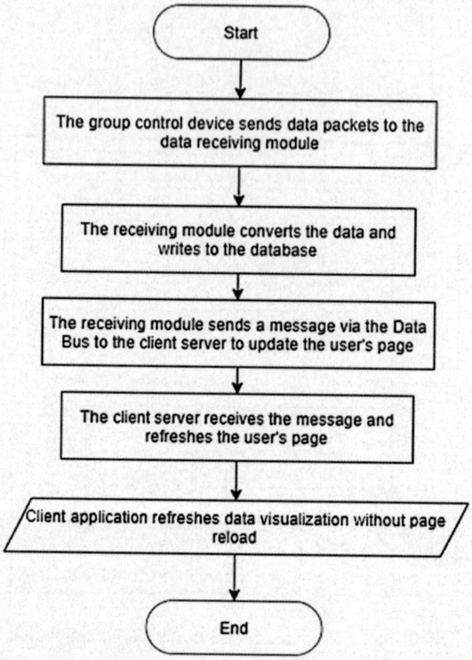

Fig. 3. Algorithm for receiving data from the group control device.

Consider the process of informing the user about violations or changes in the operation of the system. As already mentioned above, deviations of the sensor readings from those specified by the cultivation configuration, the termination of the supply of packets without prior approval by the user, or a failure in the internal modules of the system can be related to violations.

The data analysis module is responsible for detecting a discrepancy between the readings of sensors and devices. First, let's consider the work of the module in general. During initialization, the module requests and stores in its local memory active sensors and devices and container configurations to which these sensors are assigned. Next, the module requests data for a certain period of time, which can be set by the user on his own. By default, this period is 10 min, since during this time you can get the minimum data array for processing, taking into account the error and at the same time reduce the memory consumption for the operation of this module. After receiving all the data, the module compares the actual readings with the specified conditions and, in the event of a discrepancy between the values, sends a report containing the failure data, the time, the identifier of the sensor in which the failure occurred, and the board to which this sensor is attached, to the notification generation module. The latter sends a message to the server, and the server renders it to the user in the web application. In the absence of violations, the user is not informed. The algorithm of the data analysis module is shown in Fig. 4.

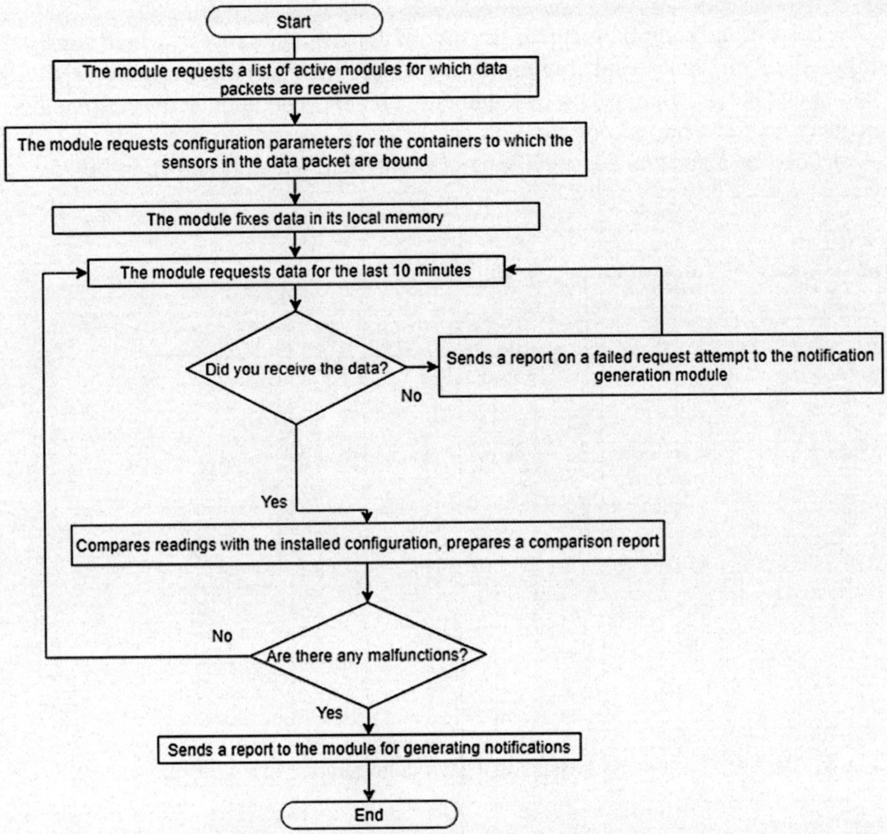

Fig. 4. Algorithm of the data analysis module.

Greenhouse complex management refers to the process of receiving commands from the user and ensuring that these commands are executed. For different user roles, different management functionality is implemented, which depends on the access allowed to them.

The functionality of control actions includes the following actions:

- Creation and filling of data about containers. Container is a description of a specific set of sensors and nodes that will present this system and settings for all control devices at a certain object. The user fills in the frequency channel number, group address, and indicates the number of boards with parameters of sensors and devices;
- Creation of data on cultivation technologies with reference to a specific container. The user fills in a set of certain parameters with their recommended values for cultivation, such as air humidity, air and nutrient solution temperature, level and intervals of illumination;
- Creation of data on growing configurations with reference to a specific container and growing technology. The user fills in a set of certain operating conditions of devices, depending on the readings of the sensors;
- Switching individual boards with sensors or modules on and off.

The basis of the control process is the task of the growing process, which consists of configurations, containers and growing technologies. First, the chief technologist fills the container and the growing technology, and only then the technologist of one greenhouse can concretize them into a configuration.

Consider the algorithm for specifying the container, which is shown in Fig. 5.

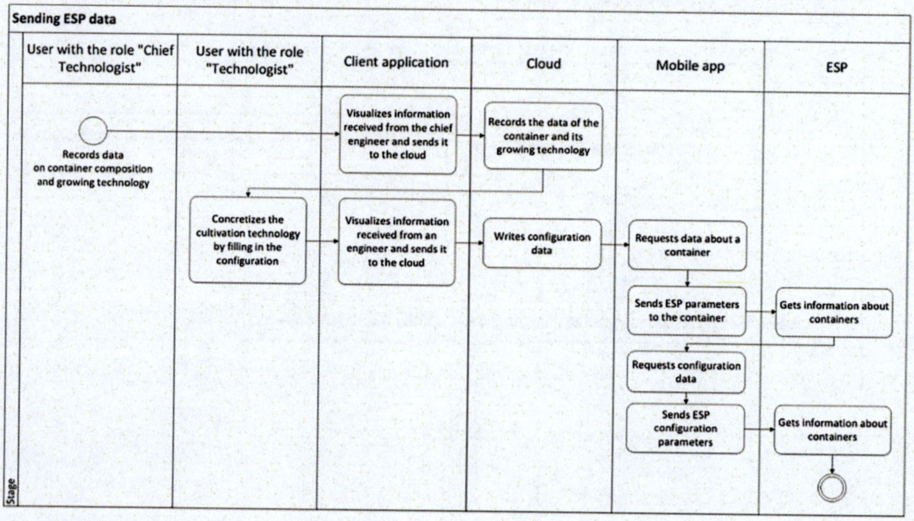

Fig. 5. The business process of sending a grow configuration to a group manager (ESP).

First, the user fills in the data for registering a new container: container name, frequency channel address, group address, composition and number of boards, with sensors and devices attached to them. After receiving data from the client, the server checks it and writes it to the database. After registering a container, you can create a growing technology for it, which will have the recommended data for growing in this container.

4 Results

In the course of the research, an automated microclimate control system was implemented using the above architecture of a cloud service in the laboratory of the St. Petersburg Federal Research Center of the Russian Academy of Sciences. Experiments were carried out to obtain data from the group control device (ESP) of the greenhouse microclimate readings: air temperature, air humidity and light level. The test results show that no more than two seconds pass from the moment the data packets are sent to their visualization on the user's screen, which can be considered an acceptable response time of the user interface [22].

Also, during testing, data were obtained on the load on the central processor of the virtual dedicated server of the hosting provider, on which the cloud service is located, the speed of reading and writing on the used drive and the throughput of the virtual dedicated server of the hosting provider. The results are shown in Figs. 6, 7 and 8.

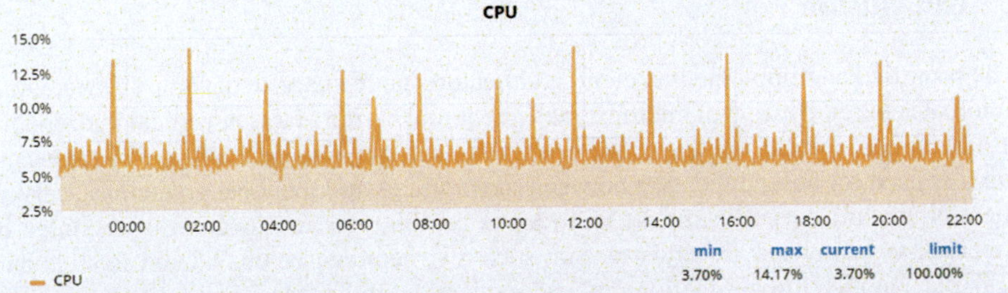

Fig. 6. The load on the central processor of the virtual dedicated server of the hosting provider.

The graph shows peaks of increase in the load on the central processor by 5–7% every 2 h. At this time, data is backed up.

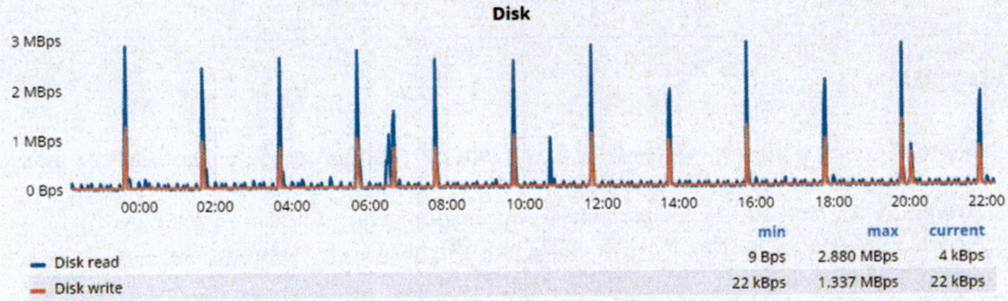

Fig. 7. Read and write speed on the used storage of the virtual dedicated server of the hosting provider.

On the graphs of the read and write speed on the used drive of the virtual dedicated server and throughput, you can also highlight the increase in the considered parameters during the backup.

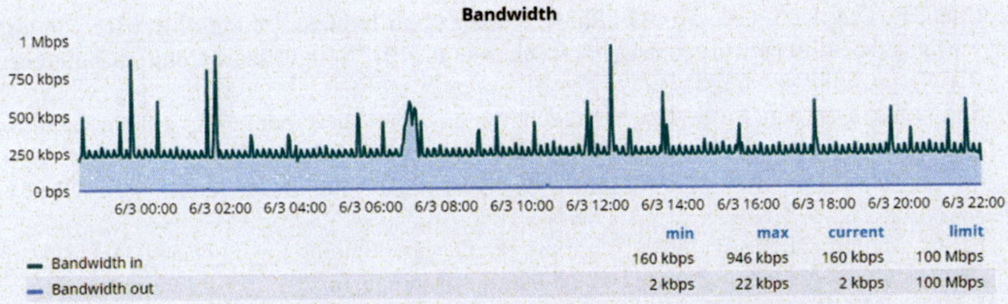

Fig. 8. Hosting provider's virtual dedicated server bandwidth.

5 Conclusion

The proposed and implemented cloud architecture can be used to develop software that automates the cultivation of microgreens, greens and barren plants in vertical hydroponic installations, as well as in horizontal greenhouses in greenhouses. Since a wireless network is used for communication between the modules and the local greenhouse server, the architecture does not impose restrictions on the type and location of modules in greenhouse complexes. For this reason, such a system can be considered flexible and scalable. Duplication of functionality and replication of databases between on-premises object servers and cloud services can improve data availability and system reliability.

In the future, the software developed on the basis of the proposed architecture is planned to be used to control a vertical greenhouse complex (multi-level hydroponic plant), which makes it possible to level the need for manual labor in technological production processes. This will expand the scope of agricultural robotization, as well as increase production efficiency through the integration of methods for point monitoring of the microclimate and assessment of plant growth dynamics.

References

1. Akkaş, M.A., Sokullu, R.: An IoT-based greenhouse monitoring system with Micaz motes. Proc. Comput. Sci. **113**, 603–608 (2017)
2. Aytekin, S.A., Levent, M.L.: Greenhouse Automation using Wireless System (2016)
3. Arvind, R.V., Raj, R.R., Raj, R.R., Prakash, N.K.: Industrial automation using wireless sensor networks. Indian J. Sci. Technol. **9**(11), 1–8 (2016)
4. Sivagami, A., Hareeshvare, U., Maheshwar, S., Venkatachalapathy, V.S.K.: Automated irrigation system for greenhouse monitoring. J. Inst. Eng. (India): Ser. A **99**(2), 183–191 (2018)
5. Cosman, S.I., Bilatiu, C.A., Marţiş, C.S.: Development of an automated system to monitor and control a greenhouse. In: 2019 15th International Conference on Engineering of Modern Electric Systems (EMES), pp. 1–4 (2019)
6. Harivardhagini, S.: LabVIEW based greenhouse automation. CVR J. Sci. Technol. **13**, 79–82 (2017)
7. Nicolosi, G., Volpe, R., Messineo, A.: An innovative adaptive control system to regulate microclimatic conditions in a greenhouse. Energies **10**(5), 722 (2017)
8. Chen, F., Qin, L., Li, X., Wu, G., Shi, C.: Design and implementation of ZigBee wireless sensor and control network system in greenhouse. In: 2017 36th Chinese Control Conference (CCC), pp. 8982–8986 (2017)
9. Shi, L., Li, Q., Qian, S.: Design of greenhouse environmental monitoring system based on Arduino and ZigBee. In: Wang, K., Wang, Y., Strandhagen, J.O., Yu, T. (eds.) IWAMA 2018. LNEE, vol. 484, pp. 576–582. Springer, Singapore (2019). https://doi.org/10.1007/978-981-13-2375-1_72
10. Li, Z., Wang, J., Higgs, R., Zhou, L., Yuan, W.: Design of an intelligent management system for agricultural greenhouses based on the internet of things. In: 2017 IEEE International Conference on Computational Science and Engineering (CSE) and IEEE International Conference on Embedded and Ubiquitous Computing (EUC), vol. 2, pp. 154–160 (2017)
11. Ullah, M.W., et al.: Internet of things based smart greenhouse: remote monitoring and automatic control. In: DEStech Transactions on Environment, Energy and Earth Sciences, (ICEEE) (2018)

12. Tangarife, H.I., Díaz, A.E.: Robotic applications in the automation of agricultural production under greenhouse: a review. In: 2017 IEEE 3rd Colombian Conference on Automatic Control (CCAC), pp. 1–6 (2017)
13. Diaz, P., Carrera, R.: IoT components for floriculture automation. In: 2019 IEEE CHILEAN Conference on Electrical, Electronics Engineering, Information and Communication Technologies (CHILECON), pp. 1–5 (2019)
14. Saha, T., et al.: Construction and development of an automated greenhouse system using Arduino Uno. Int. J. Inf. Eng. Electron. Bus. **9**(3), 1 (2017)
15. Azhar, M.R.A., Hamid, M., Irfan, M.H., Awais, M., Khan, U.S., Zeb, A.: Automated greenhouse system. In: 2019 2nd International Conference on Communication, Computing and Digital systems (C-CODE), pp. 215–219 (2019)
16. Basnet, B., Lee, I., Noh, M., Chun, H., Jaffari, A., Bang, J.: An smart greenhouse automation system applying moving average algorithm. Trans. Korean Inst. Electr. Eng. **65**(10), 1755–1760 (2016)
17. Siddiqui, M.F., Kanwal, N., Mehdi, H., Noor, A., Khan, M.A.: Automation and monitoring of greenhouse. In: 2017 International Conference on Information and Communication Technologies (ICICT), pp. 197–201 (2017)
18. Aydoğan, T., Ergün, S.: Controlling of a greenhouse by using intelligent agent system. Int. J. Inf. Electron. Eng. **7**(6), 1–5 (2017)
19. Azaza, M., Tanougast, C., Fabrizio, E., Mami, A.: Smart greenhouse fuzzy logic based control system enhanced with wireless data monitoring. ISA Trans. **61**, 297–307 (2016)
20. Gish, R.A., Karpenko, E.N.: Modernization and improvement of microclimate parameters control is the basis of the 5th generation greenhouses. Polythematic Netw. Electron. Sci. J. Kuban State Agrar. Univ. **123** (2016). (in Russ.)
21. Kim, H.-G., Park, D.-H., Chowdhury, O.R., Shin, C.-S., Cho, Y.-Y., Park, J.-W.: Location-based intelligent robot management service model using RGPSi with AoA for vertical farm. In: Jeong, H.Y., Obaidat, M.S., Yen, N.Y., Park, J.J.H. (eds.) Advances in Computer Science and its Applications. LNEE, vol. 279, pp. 309–314. Springer, Heidelberg (2014). https://doi.org/10.1007/978-3-642-41674-3_44
22. Nielsen, J.: Usability Engineering. Morgan Kaufmann, Burlington (1994)

A State of Power Based Deep Learning Model for State of Health Estimation of Lithium-Ion Batteries

Jisoo Shin, Inwhee Joe(✉), and Seokjoon Hong

Department of Computer Science, Hanyang University, Seoul 40763, Republic of Korea
iwjoe@hanyang.ac.kr

Abstract. Recently, research on eco-friendly fuels has been on the rise as various environmental problems have emerged. Many studies are being conducted on lithium-ion batteries among eco-friendly fuels for electric vehicles. Lithium-ion batteries are light and have high energy density, making them important in various fields. However, battery performance may deteriorate or fail due to various reasons. To solve these problems, many research are being conducted on battery care systems. For battery care systems, the values of SoH (State of Health) and SoC (State of Charge) are very important. In this paper, we propose a State of Power (SoP)-based deep learning model for predicting SoH performance of lithium- ion batteries, in that the battery life is determined according to the maximum power level. The proposed method is to calculate SoP values based on measurement from lithium-ion batteries and take them as input when using deep-learning models. Also, the formula for extracting SoP values from lithium-ion battery data is defined and the data extracted by the SoP values are applied to various deep learning models, such as the RNN, LSTM and GRU models. Our experiments are performed on four lithium-ion batteries 'B0005', 'B0006', 'B0007' and 'B0018' provided by NASA's open dataset 'Battery data set', and the verification of the experiment is made using root mean squared error (RMSE). The experimental results show that the proposed method in this paper achieves significantly low RMSE values for all deep learning models and all battery data.

Keywords: Battery management systems · Lithium-ion batteries · Neural networks

1 Introduction

Recently, many environmental problems have emerged, and eco-friendly transportation such as electric cars is drawing attention. The United States and China, one of the largest automobile markets, have set a goal of reducing carbon emissions by 12% and 18%, respectively, from 2017 to 2020 [1]. Electric cars can reduce carbon emissions because they use batteries rather than fossil fuels as fuel.

Lithium-ion batteries are very important not only in electric vehicles but also in various fields. Because lithium-ion batteries are light and have high energy density and

R. Silhavy et al. (Eds.): CoMeSySo 2021, LNNS 231, pp. 922–931, 2021.
https://doi.org/10.1007/978-3-030-90321-3_77

have very long cycles of charge and discharge [2]. However, changes in physical and chemical complexity in the course of use can lead to poor or faulty lithium-ion batteries [3]. Performance degradation can be directly attributed to accidents or cause economic losses. Therefore, many studies are being conducted on topics related to the Battery Management System (BMS).

The life cycle of a lithium-ion battery depends on the environment, such as temperature and humidity, or on the strength of its use. As a result, some problems occur while using the battery. So, BMS research is needed on predicting the performance of batteries that affect the remaining life of the battery, as even batteries of the same model can have different battery life in actual use. The BMS estimates State of Charge (SoC), State of Health (SoH), and State of Power (SoP) by measuring the voltage, current and temperature of the battery [4].

There are two types of technologies for battery monitoring. Model- based methods primarily analyze the physical and chemical principles of batteries. The Data-Driven method predicts the battery data in a short period of time for the same battery based on the data.

The purpose of this study is to accurately predict the lifetime of lithium-ion batteries using deep learning models. So, we propose a method to predict SoH using SoP based on supervised learning. Previously, a paper that mainly utilizes SoP in the Model-based method was published. In this paper, we conduct a study to make SoH predictions more accurate by utilizing SoP in the Data-Driven method.

The proposed method calculates SoP based on battery data, and predicts SoH using RNN, LSTM, and GRU methods. Furthermore, we obtain the reliability of the experimental results by comparing experimental data using the same deep learning technique without using the data from SoP. The composition of this paper is as follows. In Sect. 2 describes relevant research trends and Sect. 3 describes how to propose and how to experiment. In Sect. 4 compares and analyzes the results of experiments on the proposed method and existing research methods. In Sect. 5 draws conclusions on the method proposed in this paper.

2 Related Works

2.1 Model-Based Method

Ashwin et al. [5] developed an electrochemical battery model. The model was developed for battery aging according to periodically changing conditions. Using partial properties of the profile, Pola et al. [6] predicted SoC and discharge times for lithium-ion batteries using a prediction framework based on partial filtering. Xiaosong Hy et al. [7] predicted SoP, SoC, SoH using modified moving horizon estimation (mMHE) method.

Model-based methods have produced good results in battery state prediction. However, there are still some problems. The model proposed in [5] applies only to specific battery materials and operating environments. In addition, the model parameters depend on the physical characteristics of the battery. The method proposed in [6] is based on particle filters and Kalman filters. However, this method is vulnerable o noise and environmental interference, making it difficult to track dynamic characteristics. The method proposed in [7] has the limitation that the results change sensitively depending on the aging of the battery and the open circuit voltage (OCV)-SoC relationship.

For this reason, it is very difficult to establish an accurate model of SoH estimation in lithium-ion batteries, considering the complexity, noise, and environmental diversity of lithium-ion batteries.

2.2 Data-Driven Method

In the Data-Driven method, SoC prediction and SoH monitoring of lithium-ion batteries are implemented by analyzing historical data including current, voltage, capacitance, and impedance of batteries. Data-based methods are faster, more convenient, and less complex than model-based methods. Some experts and scholars use traditional statistical and mathematical optimization methods to predict the SoH of lithium-ion batteries.

Kernel-based technologies such as Support vector machine (SVM) and Relevance vector machine (RVM) are widely used. Weng et al. [8] proposed a model parameterization and adaptation framework that leverages the simple structure of the support vector regression (SVR) representation with support vector (SV) to allow real-time estimation of model parameters. Nuhic et al. [9] used SVMs to make diagnoses and predictions of system states to estimate SOH and RULs for lithium-ion batteries.

However, the kernel function must satisfy Mercer's conditions, is prone to local optimization, and has several limitations on the SVM, such as difficulty in determining loss functions and penalty factors.

Recently, research using neural network networks has been increasing. Ren et al. [10] proposed a lithium-ion battery RUL prediction method that combines autoencoders with deep neural networks. Liu et al. [11] proposed an adaptive recurrent neural network (RNN) for system dynamic state prediction with the pre-directing the RUL of lithium-ion batteries. In addition, Zhang et al. [12] proposed a method to predict RULs of lithium-ion batteries using long short-term memory (LSTM).

The above methods have achieved good results, but there are still problems that need to be improved. First, methods based on RNN and LSTM mainly use time-sliding windows as inputs. However, the characteristics of the sliding window have a different effect on the results. We need to find an appropriate way to identify this difference. Second, it has been demonstrated that key parameters have a significant impact on the performance of neural networks. We need to find a way to obtain the best key parameters. Third, to improve the performance of the model, it is necessary to dynamically update the model with state-of-the-art data.

3 Methodology

In this session, we describe the framework of the formulas used in theory and the methods we propose. We also describe the experimental method. Details are described below.

3.1 State of Power

State of power means the maximum power a battery can generate in a moment. (1) means the maximum power that a lithium-ion battery can provide at t time. In (1), Vt means the voltage measured in the battery at t seconds, and It means the current measured in the battery at t seconds. This is the SoP value when the two values are multiplied by t seconds.

$$SoPt = Vt * It \quad (1)$$

When using batteries, the value of SoP is very important for stable use without over-charging or over discharging. Therefore, the proposed method computes SoP from the given battery data and uses SoH as the predictive data.

Figure 1 is a graph of the changes in SoP over time over a cycle. The graph shows that the value of SoP gradually decreases as the number of cycles increases. It can also be seen that as the cycle increases, the time the battery can output decreases.

In summary, the more batteries are used, the less performance the battery will perform and the less time it will spend.

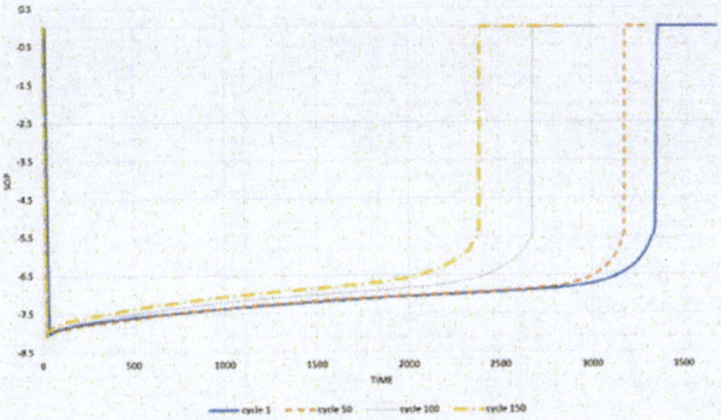

Fig. 1. Trend of SOP values in discharge cycle on B0005.

3.2 State of Health

State of health refers to the life of a battery. The initial state of the battery begins with a 100% SoH. The battery undergoes a continuous charging and discharging process, resulting in a gradual decrease in SoH over time. In this paper, SoH is calculated using (2). In (2), Qn indicates the capacitance of the battery in the nth cycle. Qint indicates the battery capacitance at its earliest state. SoH is an indicator of how much capacitance has decreased in the nth cycle for the initial capacitance.

$$SoH = Qn/Qint \quad (2)$$

$$Qn = In * t \quad (3)$$

$$Qn = SoP/Vt * t \quad (4)$$

In (3), In and t indicate current value in nth discharge cycle and using time. In discharge cycle, current always same. So, in can be converted to SoP/Vt. According to the (4), as the SoH decreases, the SoP output also decreases. SoH is arbitrarily determined by the method used. The critical value of SoH is set to 50% where the battery is of low importance, but the critical value of SoH is set to 90% where the battery is of high importance.

3.3 Battery Dataset

The battery dataset uses a lithium-ion battery dataset provided by NASA [13]. The battery dataset used a total of four datasets: B0005, B0006, B0007, and B0018.

3.4 Framework

The structure of the proposed method is shown in Fig. 2.

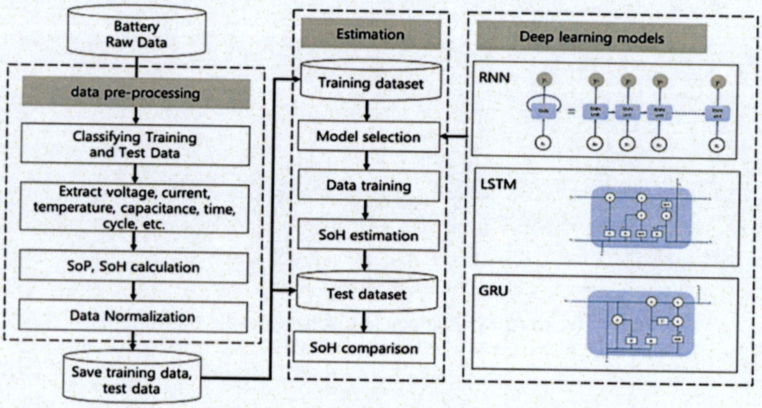

Fig. 2. An overall framework of the proposed method.

During pre-processing of battery data, it separates training data and data needed for testing. In this paper, we use the initial 70% of each battery data as training data and subsequently as test data for 30% data.

Battery data have a variety of data on charging and discharging. Therefore, we extract only the discharge data required for the experiment. The capacity value exists only in the discharge data.

Then, calculate SoP using (1) and SoH is calculated using (2). SoP refers to the output power of the battery and SoH means the remaining life of the battery. The reason for computing SoP and SoH is to compare the post-predictive results with the actual calculated results at the same time as the training.

The pre-processing data predict SoH with a total of three models: RNN, LSTM, and GRU. For deep learning model experiments, we use training data to learn for each model. Training data is 70% of the discharge data stored during pre-processing. Entering test data in the learned model yields the predicted results of SoH. Then, we compare the results of Eq. (4) with the SoH that came out as a result of the experiments.

In this paper, we compare how different the error of the estimation data varies for each model for the method that adds SoP, the proposed method, as training data. The stored data is used in the experiment.

3.5 Evaluation Method

In this paper, the evaluation method compares the results using root mean squared error (RMSE) method. The formula for RMSE is (5), as shown below.

$$\mathrm{RMSE} = \sqrt{\frac{1}{N}\sum\nolimits_{t=1}^{N}\left(\widehat{y_t} - y_t\right)^2} \tag{5}$$

In (3), $\widehat{y_t}$ means the value of SoH predicted at time t, and y_t means the value of SoH at the actual time t calculated according to (2). The RMSE values show the error between and without the proposed method. The smaller the RMSE value, the smaller the difference between the actual and predicted data values.

4 Performance Evaluation

In this session, shows experimental results on the proposed method. The environment in which the experiment was conducted and how the experiment was conducted are explained. And we can check the results from the experiment.

4.1 Experimental Setup

The experimental method was compared to a total of three models, and the parameter information for each model is shown in Table 1 below. When experiments were conducted for each hidden unit and time step, the result was chosen when the RMSE value was the lowest.

Table 1. Parameter information

Parameters	RNN	LSTM	GRU
Optimizer	RMSprop		
Learning rate	0.001		
Epoch	100		
Batch size	32		
Hidden Unit size	10, 15, 20, 32, 64		
Time step	10, 15, 20		

To ensure the reliability of the results, the parameters for each model are set equally as shown in Table 1.

4.2 Experimental Results

Table 2 shows a comparison of the RMSE values of the results using the proposed method and the results without the proposed method in this paper. The results were compared in the same parameters for each model. All results are rounded to five decimal places.

Table 2. RMSE results for each model-specific experiment

Heading level	B0005	B0006	B0007	B0018
Number of cycles	168	168	168	132
RNN with SoP	0.0170	0.0170	0.0176	0.0153
RNN	0.0191	0.0202	0.0182	0.0240
LSTM with SoP	0.0108	0.0143	0.0111	0.0114
LSTM	0.0116	0.0148	0.0125	0.0138
GRU with SoP	0.0079	0.0162	0.0080	0.0109
GRU	0.0084	0.0198	0.0090	0.0128

The results on Table 2 confirm that the method of predicting SoH using SoP proposed in this paper results in less error than the method without SoP. For B0005 and B0007, B0018 batteries, the best results in all experiments were shown to be GRU models with SoP. For B0006 batteries, the LSTM model with SoP was the best result. Table 2 shows the best results are the GRU models with SoP on B0007 batteries, with an RMSE value of 0.0080.

In this paper, we propose a method for predicting SoH using deep learning models by adding SoP to battery data. Table 2 shows that the method proposed in this paper performed better than when predicting SoH using conventional deep learning.

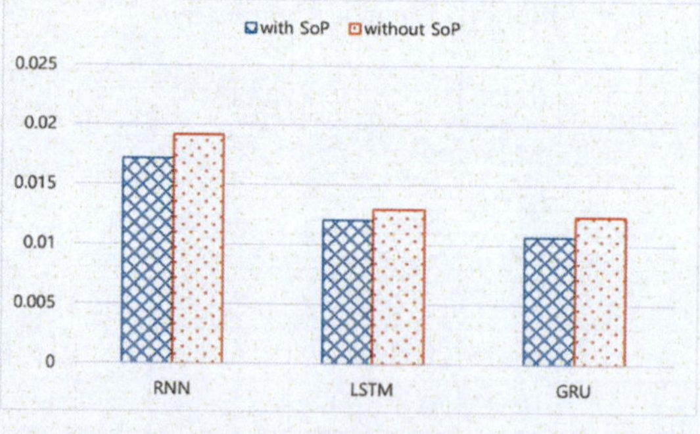

Fig. 3. RMSE results for each model-specific experiment.

Figure 3 shows an average error of B0005, B0006 and B0007 with the same cycle as the result of Table 2. Figure 3 shows that the use of the method proposed in this paper has low error values. Furthermore, we can identify deep learning methods that produce fewer errors in methods using RNN, SLTM, and GRU. Figure 3 shows that the method using GRU has the smallest mean of error.

Figure 4 is a graph of the results of predicting SoH using RNN for each battery. Of the battery data, the top 70% were used as learning data and the bottom 30% as test data. The green graph is a real SoH graph calculated using formulas. The blue graph is a graph using the method proposed in this paper, and the orange graph is a method without SoP. From all battery data, we can confirm that the blue graph approaches the green graph, which is the real SoH value.

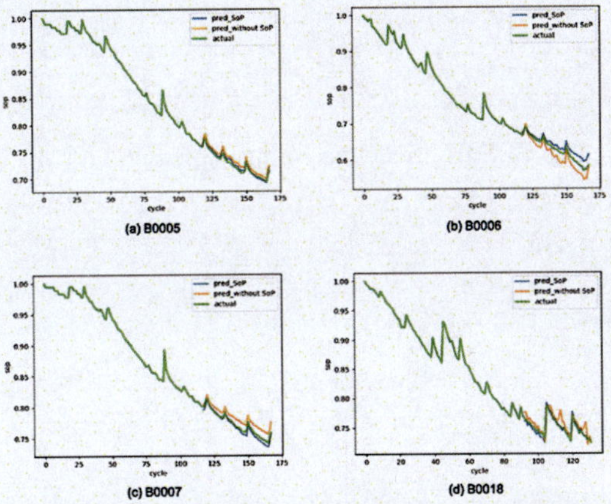

Fig. 4. SoH prediction comparison with RNN.

Figure 5 shows the results of predicting SoH using LSTM for each battery. We can confirm that LSTM performs better than RNN on most battery data. Furthermore, in the case of using LSTM, we can also confirm that the graph of the method that predicted SoH using SoP is closer to the real data graph.

Figure 6 shows the results of predicting SoH using a GRU for each battery. Figure 6 shows that the method using SoP is slightly closer to the real SOH value. Comparing the results, the best performance model for the rest of the battery except for the B0006 battery was GRU method, while the LSTM method performed the best for the B0006 battery.

Data from B0005, B0006 and B0007 are the same number of cycles and data, but the data from the B0006 battery shows that the initial and last values of the capacitance are large.

In contrast, the B0005 and B0007 battery data have a smaller width than the B0006 battery data in terms of the initial capacitance and the last capacitance. Based on these differences, we confirm that appropriate models can be selected for each battery data.

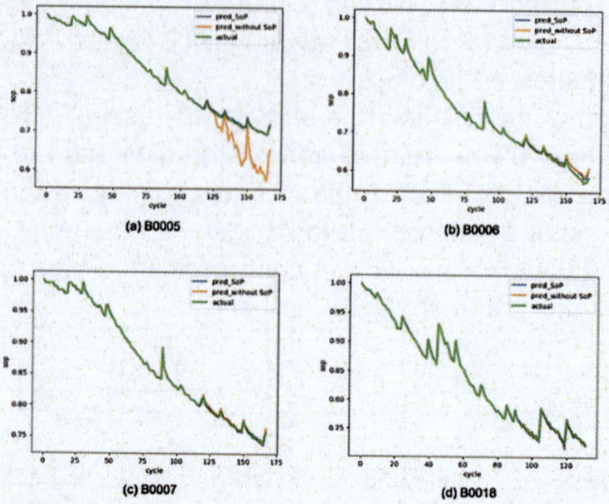

Fig. 5. SoH prediction comparison with LSTM.

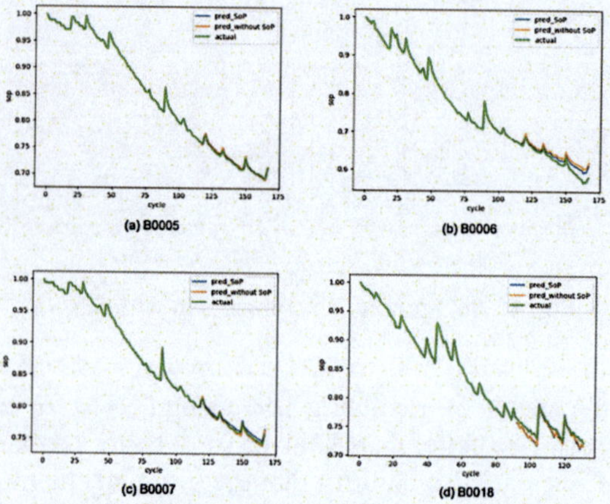

Fig. 6. SoH prediction comparison with GRU.

5 Conclusion

This paper proposes a state of power (SoP) based deep learning model as a way to predict the State of Health (SoH) of lithium-ion batteries, because the battery life relies heavily on the power level.

The proposed method uses the formula of SoP to derive the value of SoP for lithium-ion batteries and adds the derived value as input to the learning data for deep learning. In deep learning, the learning data included voltage, current, temperature, number of cycles, capacitance, etc. of lithium-ion batteries.

The experiments used NASA's "Battery data set", a public dataset, and for various validations, deep learning methods used three methods of RNN, LSTM, GRU, and lithium-ion batteries also used four data: B0005, B0006, B0007, and B0018. Verification of experiments uses root mean square error (RMSE) comparing the outcome and prediction values in deep learning.

The experimental results show that the method of adding SoP as learning data has a lower error rate than the method without adding it.

It is expected that the method of adding SoP as learning data will be able to produce more accurate results in predicting SoH, compared to later deep learning methods.

Acknowledgements. This research was supported by Basic Science Research Program through the National Research Foundation of Korea (NRF) funded by the Ministry of Education (NRF-2019R1I1A1A01058964).

References

1. Yang, Z.: Improving the conversions between the various passenger vehicle fuel economy/CO2 emission standards around the world. International Council on Clean Transportation (ICCT) (2017)
2. Hu, X., Zou, C., Zhang, C., Li, Y.: Technological developments in batteries: a survey of principal roles, types, and management needs. IEEE Power Energy Mag. **15**(5), 20–31 (2017)
3. Goebel, K., Saha, B., Saxena, A., Celaya, J.R., Christophersen, J.P.: Prognostics in battery health management. IEEE Instrum. Meas. Mag. **11**(4), 33–40 (2008)
4. Cuma, M.U., Koroglu, T.: A comprehensive review on estimation strategies used in hybrid and battery electric vehicles. Renew. Sustain. Energy Rev. **42**, 517–531 (2015)
5. Ashwin, T.R., Chung, Y.M., Wang, J.: Capacity fade modelling of lithium-ion battery under cyclic loading conditions. J. Power Sources **328**, 586–598 (2016)
6. Pola, D.A., et al.: Particle-filtering based discharge time prognosis for lithium-ion batteries with a statistical characterization of use profiles. J. IEEE Trans. Rel. **64**(2), 710–720 (2015)
7. Hu, X., Jiang, H., Feng, F., Liuc, B.: An enhanced multi-state estimation hierarchy for advanced lithium-ion battery management. Appl. Energy **257**, 110419 (2020)
8. Weng, C., Cui, Y., Sun, J., Peng, H.: On-board state of health monitoring of lithium-ion batteries using incremental capacity analysis with support vector regression. J. Power Sources **235**, 36–44 (2013)
9. Nuhic, A., Terzimehic, T., Soczka-Guth, T., Buchholz, M., Dietmayer, K.: Health diagnosis and remaining useful life prognostics of lithium-ion batteries using data-driven methods. J. Power Sources **239**, 680–688 (2013)
10. Ren, L., Zhao, L., Hong, S., Zhao, S., Wang, H., Zhang, L.: Remaining useful life prediction for lithium-ion battery: a deep learning approach. IEEE Access **6**, 50587–50598 (2018)
11. Liu, J., Saxena, A., Goebel, K., Saha, B., Wang, W.: An adaptive recurrent neural network for remaining useful life prediction of lithium-ion batteries. Department of Mechanical and Aerospace Engineering, Carleton University, Ottawa, ON, Canada, Technical Report ADA562707 (2010)
12. Zhang, Y., Xiong, R., He, H., Pecht, M.G.: Long short-term memory recurrent neural network for remaining useful life prediction of lithium-ion batteries. IEEE Trans. Veh. Technol. **67**(7), 5695–5705 (2018)
13. Saha, B., Goebel, K.: Battery data set. http://ti.arc.nasa.gov/project/prognostic-datarepository. NASA Ames, Moffett Field (2007)

An Adaptive Network Model for the Changes in Human Behaviour in Response to the Spread of COVID-19

Sharmayne Soh[1], Shihan Yu[2], and Jan Treur[3](✉)

[1] Lee Kong Chian School of Business, Singapore Management University, Singapore, Singapore
[2] Economic and Business, University of Amsterdam, Amsterdam, The Netherlands
shihan.yu@student.uva.nl
[3] Department of Computer Science, Social AI Group, Vrije Universiteit Amsterdam, Amsterdam, The Netherlands
j.treur@vu.nl

Abstract. The aim of the study reported here was to develop a model that can simulate the changes in human behaviour in response to the COVID-19 outbreak. To achieve this, a second-order adaptive social network model was designed integrating mental network models for each person. The model is based on adaptation principles such as the first-order Hebbian learning adaptation principle and the second-order 'adaptation accelerates with increasing exposure' adaptation principle.

1 Introduction

The Coronavirus Disease 2019 (COVID-19) pandemic has had widespread effects on the entire world. According to WHO, there have been over 194 million people confirmed of COVID-19, and over 4 million deaths as of today [11]. With the spread of the virus, we have seen significant changes to almost every aspect of our lives, including to our work, health, and even day-to-day routines. The virus is believed to be transmitted mainly by close contact, especially in indoor environments which have a higher risk of inadequate ventilation and higher rates of close contact [3]. As a result, we have seen various measures taken to control the spread of the virus, such as social distancing and mask-wearing.

Human behaviour plays a crucial role in preventing the spread of the virus. It is found that other than pharmaceutical interventions, changing human behaviour is essential to curb the transmission of the virus SARS-Cov-2 that causes COVID-19 [10]. In fact, behaviours such as social distancing, mask-wearing and staying at home are key contributors to slowing down the spread of the virus.

This paper focuses on building an adaptive network model to simulate the changes in behaviour of individuals in response to the spread of COVID-19. More specifically, this study explores the changes in their behaviour based on their perceptions of the external state and their own internal processes. With the aid of the model, we aim to have a better understanding of how our internal processes can affect our response to external stimuli in the COVID-19 context.

R. Silhavy et al. (Eds.): CoMeSySo 2021, LNNS 231, pp. 932–946, 2021.
https://doi.org/10.1007/978-3-030-90321-3_78

2 Background Literature

To model the behavioural changes in response to the COVID-19 outbreak, we have looked at existing literature related to this topic. Firstly, it is found that the COVID-19 virus is most commonly transmitted by close contact, particularly in indoor environments as they are more likely to have inadequate ventilation and higher rates of close contact amongst people [3], as described by WHO. With this in mind, the next step would be to find out the preventive methods to help curb the spread of the virus. It is found that the most essential and critical ways to prevent transmission of the virus, excluding non-pharmaceutical methods, are in fact largely related to human behaviour. Examples of such behaviours include social distancing, the use of face masks, hand-washing and frequent disinfecting [10]. In fact, we found that it was not just a one-way relation; not only does behavioural change impact the infection levels, the spread of COVID-19 in turn also causes human behaviour to change. To support this, a study conducted on the changes in human behaviour during the COVID-19 outbreak in Hong Kong shows that human behaviour is significantly affected by the pandemic, where a 52.0% fall was seen in the number of traveling passengers and 32.2% more time was spent at home [12].

Through further research, other relevant literature found was a deep-learning study conducted in the U.S. on human mobility and social behaviour in the COVID-19 context. This study used a long short-term memory neural network (LSTM) to model the effective reproduction number of the COVID-19 spread, in order to capture its temporal dependence on mobility and social behavioural variables. This LSTM explains the time lag between the change in mobility and social behaviour, as well as its effect on the COVID-19 spread. In addition, a sensitivity analysis showed that among the mobility parameters, being in enclosed areas had the largest effect on the effective reproduction numbers, unlike open areas or transit sites. In addition, all social behaviour parameters had relatively significant sensitivities, highlighting the importance of social interactions for the spread of the virus [2]. This supports the fact that human behaviour, in particular concerning social interactions, indeed have an impact on the spread of the COVID-19 virus.

3 The Self-modeling Network Modeling Approach Used

The network-oriented modeling approach [9] applied, uses nodes (also called states) with activation levels that vary over time and as network structure characteristics connection weights $\boldsymbol{\omega}$ for *connectivity*, combination functions $\mathbf{c}$ for *aggregation* and speed factors $\boldsymbol{\eta}$ for *timing* to specify a network model:

Connection weights: $\boldsymbol{\omega}_{X,Y}$ denotes the weight of the connection from node or state X to State Y; it usually ranges from -1 to 1.
Combination Functions: $\mathbf{c}_X(..)$ denotes a combination function used for state X. From the available library of combination functions, the following functions will be used in the introduced model: $\mathbf{alogistic}_{\sigma,\tau}$, $\mathbf{hebb}_{\mu}$, $\mathbf{stepmod}_{\rho,\delta}$; see Table 1.
Speed factors: $\boldsymbol{\eta}_X$ with values usually between 0 and 1 denotes the speed of change of state X with respect to the impact from other states.

Using the notion of *self-modeling network* (also called *reified network*) introduced in (Treur, 2020), every network characteristic can be made adaptive by adding a (self-model) state to the network that represents the value of this characteristic. This will be applied here to obtain a second-order adaptive network in which for some states X and Y:

(1) first-order self-model states $\mathbf{W}_{X,Y}$ are included in the network that represent the value of connection weight $\boldsymbol{\omega}_{X,Y}$, and
(2) second-order self-model states $\mathbf{H}_{\mathbf{W}_{X,Y}}$ are included in the network that represent the value of the speed factor of $\mathbf{W}_{X,Y}$ (learning rate).

Using the available dedicated software environment, simulation is based on the following equations based on the above network characteristics (where $X_1, \ldots, X_k$ are the states from which state Y gets incoming connections):

$$\mathbf{impact}_{X,Y}(t) = \boldsymbol{\omega}_{X,Y}X(t) \tag{1}$$

$$\mathbf{aggimpact}_Y(t) = \mathbf{c}_Y(\mathbf{impact}_{X_1,Y}(t), \ldots, \mathbf{impact}_{X_k,Y}(t)) = \mathbf{c}_Y(\boldsymbol{\omega}_{X_1,Y}X_1(t), \ldots, \boldsymbol{\omega}_{X_k,Y}X_k(t)) \tag{2}$$

$$\begin{aligned} Y(t+\Delta t) &= Y(t) + \boldsymbol{\eta}_Y\left[\mathbf{aggimpact}_Y(t) - Y(t)\right]\Delta t \\ &= Y(t) + \boldsymbol{\eta}_Y[\mathbf{c}_Y(\boldsymbol{\omega}_{X_1,Y}X_1(t), \ldots, \boldsymbol{\omega}_{X_k,Y}X_k(t)) - Y(t)]\Delta t \end{aligned} \tag{3}$$

Table 1. The combination functions used in the introduced network model

	Notation	Formula	Parameters
Steponce	$\mathbf{stepmod}_{\rho,\delta}(V)$	0 if t mod $\rho \leq \delta$, else 1	ρ repetition interval length, δ step time
Advanced logistic sum	$\mathbf{alogistic}_{\sigma,\tau}(V_1, \ldots, V_k)$	$[\frac{1}{1+e^{-\sigma(V_1+\cdots+V_k-\tau)}} - \frac{1}{1+e^{\sigma\tau})}](1+e^{-\sigma\tau})$	Steepness $\sigma > 0$ Excitability threshold τ
Hebbian learning	$\mathbf{hebb}_{\mu}(V_1, V_1, W)$	$V_1V_2\,(1-W) + \mu\, W$	V_1,V_2 activation levels of the connected states; W activation level of the self-model state for the connection weight μ persistence factor

4 The Adaptive Network Model

According to the study by Gentili and Cristea [4], the most widely explored preventive methods for COVID-19 are non-pharmaceutical measures. These include maintaining a social distance, mask-wearing, hand-washing, and frequent disinfecting. For our the scenarios addressed by our model, we have identified three of these behaviours that different individuals could exhibit, in order to study the different adaptations of their responses to the COVID-19 infection levels. The designed model provides a second-order adaptive network architecture to simulate behavioural adaptation of humans in the COVID-19 context. It covers both the social network of connections between persons and (part of) these persons' internal mental processes.

4.1 The Social Network Model

Based on data extracted from Hung et al. [6], we have selected three commonly exhibited behaviours in society to be used in our simulation scenario: whether an individual (i) keeps a 1.5 m distance from others, (ii) wears a mask, and/or (iii) stays at home. Table 5 (see the Appendix section) shows the different possible combinations in which eight hypothetical individuals could exhibit (some of) the aforementioned behaviours. The extent to which people exhibit risky behaviour will not only have an impact (through social contagion) on the behaviour of other individuals, but also on the infection levels, and vice versa. Figure 1 shows the connections between the eight individuals.

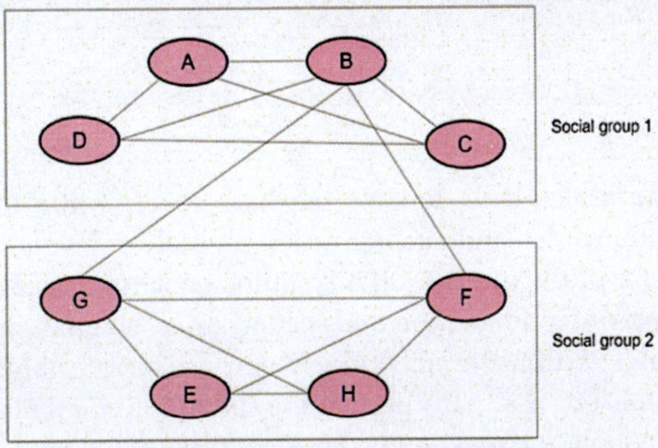

Fig. 1. The example social network addressed

We separated these individuals into two distinct social groups and Fig. 1 also shows links between these groups. For instance, Person A stays at home, wears a mask and observes a social distance of 1.5 m, which makes him/her unlikely to come into contact with Person H who fails to take any precautions. Since staying at home is the most efficient way to prevent contact between people, we therefore categorised Persons A to D in one social group and the rest in another social group. The possible interactions between Groups 1 and 2 include B with F and B with G, as they are the people that do not stay at home. These are hence the two most likely sites of contact between the two social communities.

4.2 The Adaptive Mental Network Model

In order to simulate both the behaviour and the underlying adaptation or learning of the individuals, our model also exploits the concept of network adaptation. For this, a second-order adaptive network architecture is used; the overall connectivity of the multi-level adaptive network model is expressed graphically in Fig. 2 (see Fig. 3 for a more detailed picture of the mental network model of a person).

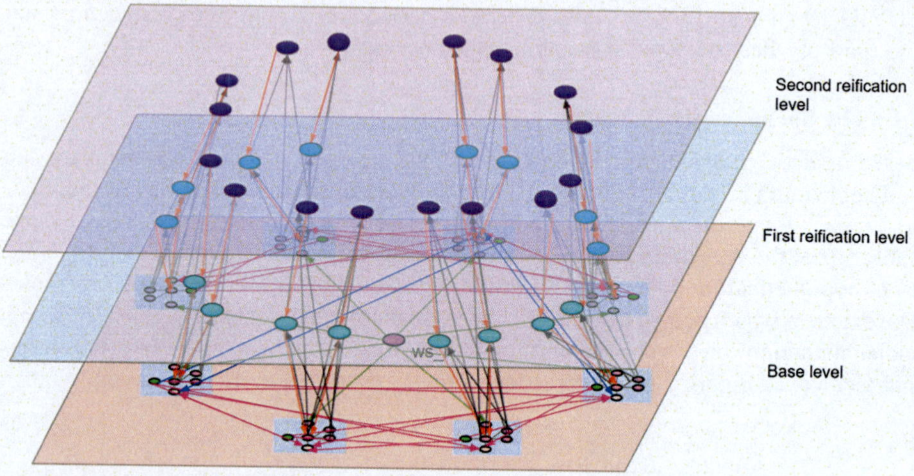

Fig. 2. The overall connectivity of the second-order adaptive network model

In this figure, the pink plane denotes the base level, while the blue and purple plane represent the first-order and second-order adaptation levels, respectively. Thus, the generic architecture of the multi-level adaptation process consists of three main levels: the base level, first-order adaptation and second-order adaptation [9]. The base level describes the dynamics within the network, which we have established earlier in the Social Network Model, but now also part of the individuals' mental processes is covered. The first-order adaptation represents the adaptation of the network itself by certain adaptation principles, which in this case concerns changes in behaviours of the individuals based on the behaviour learning. In Fig. 2, the first-order adaption level models Hebbian learning [5] and is represented by the blue plane. The second-order adaptation is the adaptation of these adaptation principles [1], which is reflected by the speed of change in the adaptation of such behaviours in our model; this is based on the second-order 'adaptation accelerates with increasing exposure' adaptation principle formulated in [8].

In Table 2 an overview of the different states can be found with their explanation. In Tables 3 and 4, an overview is provided of all types of connections used and their effects. Note that for the sake of transparency, not all connections are included in Fig. 2.

Base Level Processes

The base level interactions are graphically represented in the lower plane in Fig. 2; see also Table 2. Generally, since the world state ws_i represents information on the general infection level of the population, this information affects each person's sensory representation state $srs_{i,X}$ for the infection level. In turn, the world state ws_i will also be affected by the execution states es_X for the behaviour of our individuals A to H. As higher values of these states indicate 'bad' behaviour, such higher values have an increasing effect on the general infection level ws_i. Besides the world state for the infection level, other external stimuli that will influence the internal processes of each individual X concern the behaviours es_Y of the other individuals Y in the social network. This happens in each subgroup separately, but for interactions between the social groups, Person B from Group 1 has contact with Person F and G from Group 2.

Table 2. Description of the states

State	Description
ws_i	World state for the infection level
$srs_{i,X}$	Sensory representation state of Person X for the infection level i
$srs_{c,X}$	Sensory representation state of Person X for the general context c
ps_X	Preparation state of the behaviour of Person X
$srs_{e,X}$	Sensory representation state of Person X for the (predicted) effect e of the prepared behaviour
es_X	Executing state of the behaviour of Person X; note that a low value means that the person behaves in a safe manner by taking into account measures against infection and a high value means that the person does not take such measures into account.
sro_X	Sensory representation state of Person X for other people's behavior
$\mathbf{W}_{srs_{c,X},ps_X}$	First-order self-model state for the weight ω of the connection from $srs_{c,X}$ to ps_X
$\mathbf{W}_{srs_{e,X},ps_X}$	First-order self-model state for the weight ω of the connection from $srs_{e,X}$ to ps_X
$\mathbf{H}_{\mathbf{W}_{srs_{c,X},ps_X}}$	Second-order self-model state for the speed factor η of the first-order self-model state $\mathbf{W}_{srs_{c,X},ps_X}$
$\mathbf{H}_{\mathbf{W}_{srs_{e,X},ps_X}}$	Second-order self-model state for speed factor η of the first-order self-model state $\mathbf{W}_{srs_{e,X},ps_X}$

In our model, an example of a connection weight is $\boldsymbol{\omega}_{ws_i,srs_{i,A}}$, which represents the weight of the connection from the world state ws_i (infection level) to the sensory representation state $srs_{i,A}$ for the infection level perceived and represented by Person A. Another example is $\boldsymbol{\omega}_{srs_{i,X},ps_X}$, the weight for the connection from $srs_{i,X}$ to ps_X. As the connection from this sensory representation state for infection level to the preparation state of each person is assumed to be suppressing, the connection weights $\boldsymbol{\omega}_{srs_{i,X},ps_X}$ will be negative. In the model, we used −0.5 to represent this negative weight. Since each of the individuals' behaviours are risky to different extents, they would have varying impacts on the overall infection level. As such, we allocated different (initial) connection weights ranging from 0.4 to 0.9 to each individual, with staying at home having the highest weight, followed by wearing a mask, and lastly keeping a 1.5 m distance having the lowest weight. Additionally, as the connection between Social Group 1 and Group 2 is relatively weak, we assigned those with values ranging from 0.4 to 0.5.

The advanced logistic function, $\mathbf{alogistic}_{\sigma,\tau}(..)$, is used for all base level states with the exception of the sensory representation state $srs_{c,X}$ for the general context for each person. The $srs_{c,X}$ states will instead apply the **stepmod** function to model a repetitive type of context. All sixteen **W**-states use the Hebbian combination function denoted

Table 3. Description of the types of connections used at the base level

Connection type	Description for Person X
Intra-person links	
$srs_{c,X} \rightarrow ps_X$	Response link from the represented context c to preparation of behaviour
$srs_{i,X} \rightarrow ps_X$	Link for the (suppressing) effect of the represented infection level i on preparation of behaviour
$ps_X \rightarrow es_X$	Link for execution of the prepared behaviour
$ps_X \rightarrow srs_{e,X}$	Prediction link for the expected effect of the prepared behaviour
$srs_{e,X} \rightarrow ps_X$	Link for the effect of the predicted effect on the preparation of the behaviour
$sro_X \rightarrow ps_X$	Mirroring link for the effect of the (aggregated) represented behaviour of other people on the preparation of the own behaviour
Extra-person links	
$es_Y \rightarrow sro_X$	Link for aggregated representation by Person X of the observed behaviours of other Persons Y
$ws_i \rightarrow srs_{i,X}$	Link for perceiving and representing info on the overall infection level
$es_X \rightarrow ws_i$	Link for the effect of a person's behaviour on the overall infection level

by $\mathbf{hebb}_{\mu}(..)$, that models the first-order adaptation principle for Hebbian learning. The sixteen **H**-states also apply the combination function $\mathbf{alogistic}_{\sigma,\tau}(..)$.

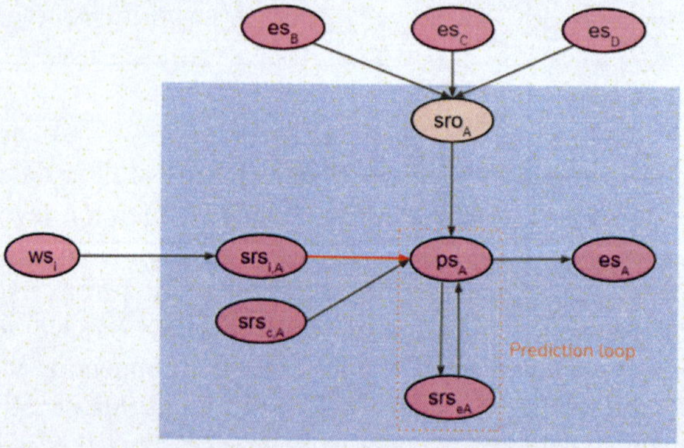

Fig. 3. The connectivity of the internal mental network model of Person A

To explain the internal mental network model of each individual in further detail, a pictorial representation of an individual's internal mental model can be found in Fig. 3, taking Person A as an example. We first start by looking at the overall infection level ws_i and the general context. As the individual perceives such external stimuli, his/her sensory representation states $srs_{i,X}$ and $srs_{c,X}$ for the infection level and the general context will be activated. Moreover, also the behaviours of other persons that are connected lead to an aggregated representation state sro_X of these behaviours. Following all these, the person's sensory representation states $srs_{i,X}$, $srs_{c,X}$, sro_X for the infection levels, general context, and other people's behaviour collectively affect his/her preparation state ps_X. Here the connection from sro_X to ps_X models in particular a *mirroring link*; e.g., [7].

Table 4. Description of the types of connections used for the self-model states

Connection type	Description for Person *X*
Connections for the first-order self-model states	
$srs_{c,X} \rightarrow \mathbf{W}_{srs_{c,X},ps_X}$	Upward link from $srs_{c,X}$ for Hebbian learning of the weight ω of the connection from $srs_{c,X}$ to ps_X
$ps_X \rightarrow \mathbf{W}_{srs_{c,X},ps_X}$	Upward link from ps_X for Hebbian learning of the weight ω of the connection from $srs_{c,X}$ to ps_X
$\mathbf{W}_{srs_{c,X},ps_X} \rightarrow \mathbf{W}_{srs_{c,X},ps_X}$	Self-link of $\mathbf{W}_{srs_{c,X},ps_X}$ for persistence of the learnt value of $\mathbf{W}_{srs_{c,X},ps_X}$
$srs_{e,X} \rightarrow \mathbf{W}_{srs_{e,X},ps_X}$	Upward link from $srs_{e,X}$ for Hebbian learning of the weight ω of the connection from $srs_{e,X}$ to ps_X
$ps_X \rightarrow \mathbf{W}_{srs_{e,X},ps_X}$	Upward link from ps_X for Hebbian learning of the weight ω of the connection from $srs_{e,X}$ to ps_X
$\mathbf{W}_{srs_{e,X},ps_X} \rightarrow \mathbf{W}_{srs_{e,X},ps_X}$	Self-link of $\mathbf{W}_{srs_{e,X},ps_X}$ for persistence of the learnt value of $\mathbf{W}_{srs_{e,X},ps_X}$
$\mathbf{W}_{srs_{c,X},ps_X} \rightarrow ps_X$	Downward link from $\mathbf{W}_{srs_{c,X},ps_X}$ for the effect on ps_X of the learnt weight ω of the connection from $srs_{c,X}$ to ps_X
$\mathbf{W}_{srs_{e,X},ps_X} \rightarrow ps_X$	Downward link from $\mathbf{W}_{srs_{e,X},ps_X}$ for the effect on ps_X of the learnt weight ω of the connection from $srs_{e,X}$ to ps_X
Connections for the second-order self-model states	
$srs_{c,X} \rightarrow \mathbf{HW}_{srs_{c,X},ps_X}$	Upward link from $srs_{c,X}$ for the second-order 'adaptation accelerates with increasing exposure' adaptation principle for the connection from $srs_{c,X}$ to ps_X
$ps_X \rightarrow \mathbf{HW}_{srs_{c,X},ps_X}$	Upward link from ps_X for the second-order 'adaptation accelerates with increasing exposure' adaptation principle for the connection from $srs_{c,X}$ to ps_X
$\mathbf{W}_{srs_{c,X},ps_X} \rightarrow \mathbf{HW}_{srs_{c,X},ps_X}$	Upward link from $\mathbf{W}_{srs_{c,X},ps_X}$ for the second-order 'adaptation accelerates with increasing exposure' adaptation principle for the connection from $srs_{c,X}$ to ps_X
$srs_{e,X} \rightarrow \mathbf{HW}_{srs_{e,X},ps_X}$	Upward link from $srs_{e,X}$ for the second-order 'adaptation accelerates with increasing exposure' adaptation principle for the connection from $srs_{e,X}$ to ps_X
$ps_X \rightarrow \mathbf{HW}_{srs_{e,X},ps_X}$	Upward link from ps_X for the second-order 'adaptation accelerates with increasing exposure' adaptation principle for the connection from $srs_{e,X}$ to ps_X
$\mathbf{W}_{srs_{e,X},ps_X} \rightarrow \mathbf{HW}_{srs_{e,X},ps_X}$	Upward link from $\mathbf{W}_{srs_{e,X},ps_X}$ for the second-order 'adaptation accelerates with increasing exposure' adaptation principle for the connection from $srs_{e,X}$ to ps_X
$\mathbf{HW}_{srs_{c,X},ps_X} \rightarrow \mathbf{W}_{srs_{c,X},ps_X}$	Downward link from $\mathbf{HW}_{srs_{c,X},ps_X}$ for the effect on $\mathbf{W}_{srs_{c,X},ps_X}$ of the adaptive speed factor η of the adaptation of the connection from $srs_{c,X}$ to ps_X
$\mathbf{HW}_{srs_{e,X},ps_X} \rightarrow \mathbf{W}_{srs_{e,X},ps_X}$	Downward link from $\mathbf{HW}_{srs_{e,X},ps_X}$ for the effect on $\mathbf{W}_{srs_{e,X},ps_X}$ of the adaptive speed factor η of the adaptation of the connection from $srs_{e,X}$ to ps_X

Moreover, another contribution to the activation of the preparation state ps_X comes from the predicted effect representation state $srs_{e,X}$, which is activated through a *prediction link* from ps_X to $srs_{e,X}$ which creates a prediction loop between the preparation state ps_X and the sensory representation $srs_{e,X}$ for (predicted) effect e. Eventually, the person's preparation to make a decision on his/her behaviour will lead to the executing state es_X. The execution of each person's behaviour would not only influence the sensory representation state sro_X of other individuals in their social group, but also influence the overall infection level ws_i. In accordance with what is listed in Table 2, for the considered example network extra-person connections exist from es_B to sro_G, es_B to sro_F, es_F to sro_B, and es_G to sro_B.

First-Order Adaptation

The first-order adaptation processes, as represented within the middle (blue) plane in Fig. 2, build on the base level by modelling the adaptation of the behaviours of the eight individuals. These adaptations take place by the (Hebbian) learning of the behaviours from the base level discussed above. As shown in Fig. 2, there are two first-order self-model states (at the first reification level) for each individual: $\mathbf{W}_{srs_{i,X},ps_X}$ represents the weight for the connection from $srs_{i,X}$ to ps_X, and $\mathbf{W}_{srs_{e,X},ps_X}$ represents the weight of the connection from $srs_{e,X}$ to ps_X. These weights are (partly) based on learning by a form of social contagion realized by internally mirroring others' behaviours. They determine the changes in the sensory representation state and preparation state of each individual that are used to generate her/his behaviour.

Second-Order Adaptation

Next, the second-order adaptation, as represented within the upper (purple) plane in Fig. 2, explains the adaptation (or control) of these adaptation principles, which is reflected by the speed of the change of the individuals' behaviours. Here, there are two (second-order) self-model states for each person: $\mathbf{H}_{\mathbf{W}_{srs_{i,X},ps_X}}$ represents the speed factor for $\mathbf{W}_{srs_{i,X},ps_X}$ and $\mathbf{H}_{\mathbf{W}_{srs_{e,X},ps_X}}$ represents the speed factor for $\mathbf{W}_{srs_{e,X},ps_X}$. These adaptive speed factors model the effect of (stimulus) exposure on the speed of change in the individuals' behaviours according to the second-order 'adaptation accelerates with increasing exposure' adaptation principle.

5 Simulation Results

In this section, some of the simulation results are discussed. The example scenario is based on the social network depicted in Fig. 1 and the more specific network models depicted in Fig. 2 and Fig. 3. More details about the network characteristics used can be found in the Appendix Sect. 7. The duration applied to our simulation is 800, and the interval iteration (time step) Δt was set to 0.25. Since all mental states are not assumed affected by any prior states, the initial values for all mental states (i.e., $srs_{e,X}$, es_X, ps_X, etc.) were set to 0. The initial values for the states that use Hebbian learning (all **W**-states) were set to 0.1, and the world state of the infection level was set to start with an initial value of 0.5.

Overall Results

As there was a total of 81 states, we categorised the states into each individual and generated a separate graph with simulation results for each person for a better view. Figures 4 and 5 display the results of the simulation for Person A to H. The curved blue line in the middle of each figure represents the world state of infection, which is the general infection level. The overall trend of the ws_i state suggests that the infection rate will initially decrease. However, as people learn more of the 'bad' behaviour over time, the infection rate will begin to increase again.

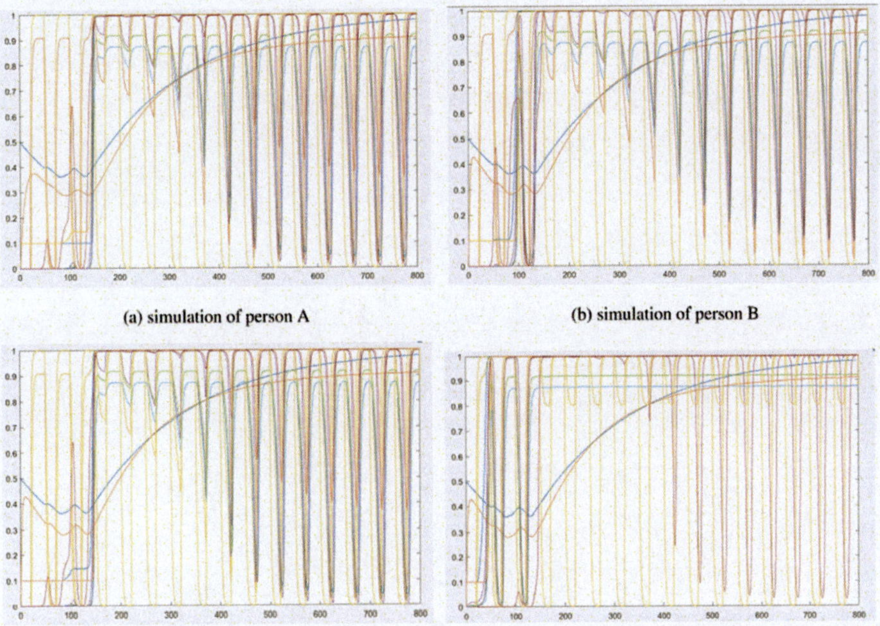

(a) simulation of person A

(b) simulation of person B

(c) simulation of person C

(d) simulation of person D

Fig. 4. Comparison of simulation results for social group 1

Results Social Group 1

Comparing the results in Fig. 4, Person A and Person C are less affected by the learning of bad behaviour, whereas Person B and Person D start to behave poorly from the second and first cycle respectively. The light blue line which represents the execution state es_X fluctuates considerably in the graphs for Persons A to C, but remains almost stable for Person D. While those who are more compliant with the social distancing behaviours are more likely to be cautious about the transmission of the virus, there could also be a tendency for their behaviour to become poorer after learning of the bad behaviours.

Results Social Group 2

As for the second social group in Fig. 5, the starting time of their learning is later than those in the first group. This could be attributed to the fact that since they already engage in two or three bad behaviours, their learning will take place relatively later than those who initially exhibit good social distancing behaviours. Despite this, even though the Group 2's starting time for learning is later, the sensory representation states $srs_{i,X}$ in

Group 2 have a steeper increasing trend compared to Group 1. This is represented by the dark red lines for each person in Figs. 4 and 5, which suggest that people in Social Group 2 require less time to respond to the information they receive from the external environment about the infection levels.

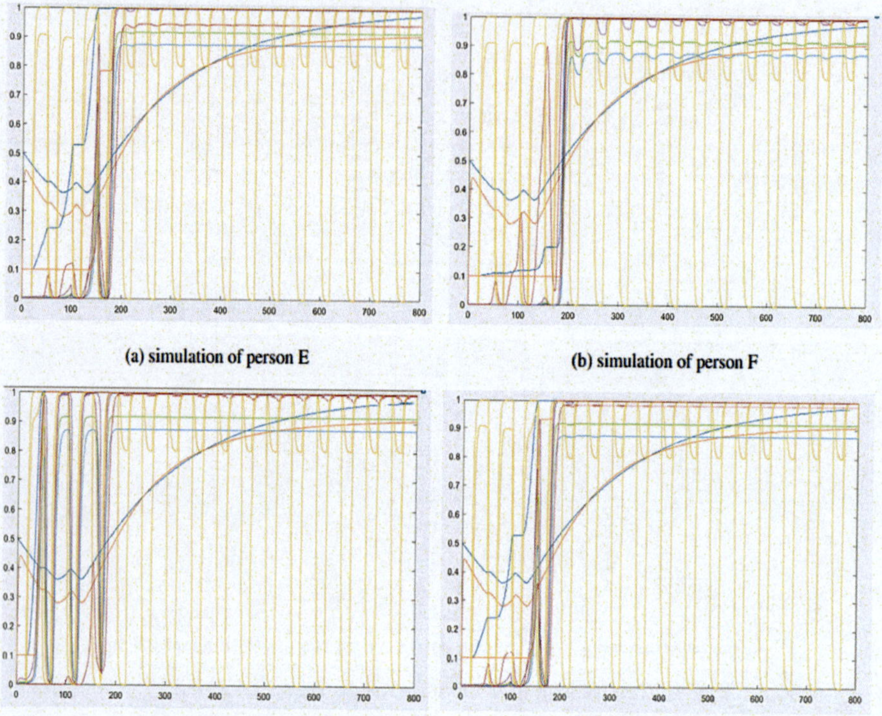

(a) simulation of person E

(b) simulation of person F

(c) simulation of person G

(d) simulation of person H

Fig. 5. Comparison of simulation results for social group 2

Comparison of the Execution States

Figure 6 shows a comparison of the execution states es_X of each individual. Generally, the execution states of every person reach the same end-point eventually. For Person B and D, the onset of their behavioural change are the earliest among the eight individuals, as represented by the two purple lines. This means that not wearing a mask and staying at home influence behavioural change more, subsequently.

Comparison of the W-States

Figure 7 depicts the comparison of the behavioural learning, as modelled by the **W**-states, for Persons A to H in two different scenarios. In the first scenario, where the Hebbian learning parameter (μ) was set to 1, Person B's, D's and H's learning begin considerably early, while Person E and F begin to learn relatively later.

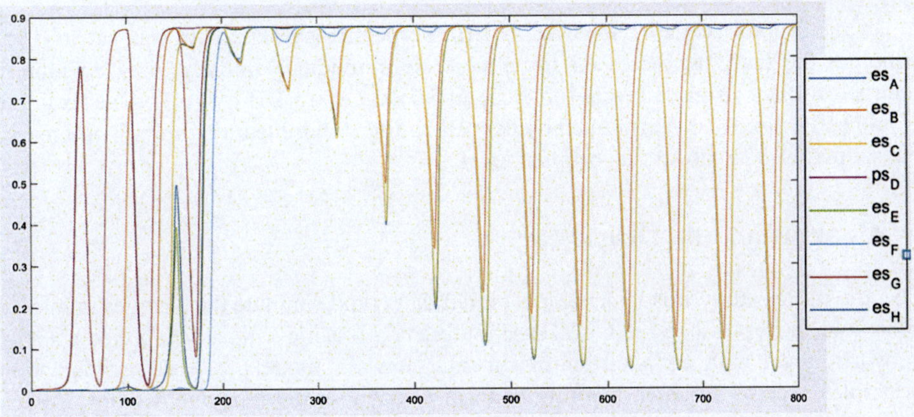

Fig. 6. Comparison of execution states

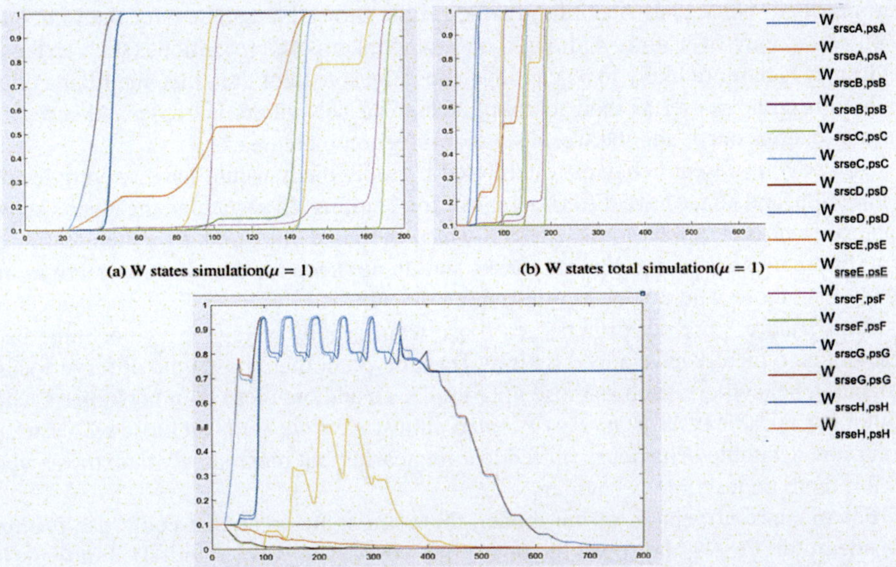

(a) W states simulation(μ = 1)

(b) W states total simulation(μ = 1)

(c) W states total simulation(μ = 0.95)

Fig. 7. Comparison of behavioural learning (**W**-states)

Also, for Person A, B, C, D, F and G, their learning process for $\mathbf{W}_{\text{srs}_{c,X},\text{ps}_X}$ and $\mathbf{W}_{\text{srs}_{e,X},\text{ps}_X}$ yield fairly similar results, except for a slight difference where their learning process for $\mathbf{W}_{\text{srs}_{c,X},\text{ps}_X}$ occurs more slowly than their prediction effect $\mathbf{W}_{\text{srs}_{e,X},\text{ps}_X}$. On the other hand, for Person E and H, the learning of $\mathbf{W}_{\text{srs}_{c,E/H},\text{ps}_{E/H}}$ and $\mathbf{W}_{\text{srs}_{e,E/H},\text{ps}_{E/H}}$ have a large difference in their graphs. In this first scenario where the persistence parameter μ for Hebbian function was set to 1, the learning for all individuals first increased and then stayed constant at 1 over time.

In the second scenario however, where the persistence factor μ was set to 0.95, although the learning increased for almost all individuals initially, they eventually returned to their original levels for those in Social Group 1 (Fig. 7(c)). One example would be Person A, as represented by the yellow line, whose learning initially increases but soon after decreases to its original level.

6 Conclusion and Discussion

The aim of this study was to develop a model that could simulate the changes in human behaviour in response to the COVID-19 outbreak. To achieve this, we applied a social network model with included internal mental network models, as well as adaptation principles such as Hebbian learning to form a second-order adaptive network model. The results agree to our initial expectations, and we have drawn three main findings from the results.

Firstly, individuals that exhibit any type of behaviour will eventually tend to start learning 'bad' behaviour. Accordingly, the results showed a tendency for the infection levels to increase over time. Although those who take more precautions (such as those in Group 1) are more likely to be cautious about the spread of the virus, their behaviour tends to become poorer as their learning of the bad behaviours increases. As a result, more infections occur and the world state of infection increases.

Secondly, different behaviours exhibited by individuals would have varying levels of infection, and hence have differing impacts on learning. In particular, the results show that those who do not wear masks generally have earlier starting points for their learning. In addition, those who do not wear masks and do not keep a 1.5 m distance learn more slowly than those who do not stay at home.

Furthermore, poor performers, i.e., those who initially engage in two or three bad behaviours, typically have a later starting learning point than those who observe social distancing behaviours, such as those in Group 1. To add on, these poor performers who exhibit undesirable behaviours also learnt at almost the same rate, but those who started with more desirable behaviours started learning earlier but more slowly than those who exhibit more undesirable behaviours.

For the current version of the model, there are some areas that could use further consideration. Firstly, the speed factor for the world state was set to 0.05, but there is still a fair bit of uncertainty in the speed of transmission of the virus. Therefore, further research could be conducted on the speed factor of ws_i in the simulation. Furthermore, as this social network model is a simplified version of reality, the current model could be improved to suit the complexities of social interactions in the real world, such as including more types of behaviours.

Appendix: More Details

In this section some more details of the simulation scenario are discussed. For the full specification, see URL https://www.researchgate.net/publication/353620668 under Linked Data. Table 5 shows the assumed combinations of behaviours considered in the simulation scenario.

Table 5. Combinations of behaviours for individuals A-H

Person	Behaviour
A	Wears a mask, Keeps 1.5 m distance, Stays home
B	Wears a mask, Keeps 1.5 m distance, Does not stay home
C	Wears a mask, Stays at home, but does not keep 1.5 m distance
D	Keeps 1.5 m distance, Stays at home, but does not wear a mask
E	Stays at home, but does not wear a mask, and does not keep 1.5 m distance
F	Keeps 1.5 m distance, but does not wear mask, and does not stay at home
G	Wears a mask, but does not keep 1.5 m distance and does not stay at home
H	Does not wear a mask, does not keep 1.5 m distance, does not stay at home

Generally, 50 was used as the repeated time duration $\boldsymbol{\rho}$ and $\boldsymbol{\delta}$ was 20 for the states with the stepmod function. For states applying the Hebbian function, the persistence factor $\boldsymbol{\mu}$ was valued at 1. As for the states applying the advanced logistic sum function, a steepness of 5 was used in general, with the exception of a few states with different threshold values 0.5–2. The speed factor represents the speed of the process to reach its corresponding state. For instance, $\boldsymbol{\eta}_{\mathrm{srs}_{i,X}}$ represents how fast Person X perceives and internalises the latest COVID-19-related news. Since news is often received at varying speeds by different people with different behaviours, their corresponding speed factors for their sensory representation states would also have differing values. Speed factors of 0.1, 0.4, 0.7 and 1 were allocated to the eight individuals who exhibit none to all three behaviours. Furthermore, because mental states will always change faster than actions, the preparation state ps_X would have a higher speed factor of 0.8, while the executing state es_X would have a slower speed of 0.3. The different speed factors for each state are shown in the Table 6.

Table 6. Speed factor for each state

ws_i	$\mathrm{srs}_{i,A}$	$\mathrm{srs}_{i,B\text{-}D}$	$\mathrm{srs}_{i,E\text{-}G}$	$\mathrm{srs}_{i,H}$	ps_X	es_X	Others
0.05	0.1	0.4	0.7	1	0.8	0.3	0.5

References

1. Abraham, W.C., Bear, M.F.: Metaplasticity: the plasticity of synaptic plasticity. Trends Neurosci. **19**(4), 126–130 (1996)
2. Bhouri, M.A., et al.: COVID-19 dynamics across the US: a deep learning study of human mobility and social behavior. Comput. Methods Appl. Mech. Eng. **382**, 113891 (2021)
3. Coronavirus disease (COVID-19): How is it transmitted? WHO. https://www.who.int/news-room/q-a-detail/coronavirus-disease-covid-19-how-is-it-transmitted. Accessed 13 Dec 2020

4. Gentili, C., Cristea, I.A.: Challenges and opportunities for human behaviour research in the coronavirus disease (COVID-19) pandemic. Front. Psychol. **11**, 1786 (2020)
5. Hebb, D.O.: The Organization of Behavior: A Neuropsychological Theory. Wiley, New York (1949)
6. Hung, M., et al.: Social network analysis of COVID-19 sentiments: application of artificial intelligence. J. Med. Internet Res. **22**(8), e22590 (2020)
7. Iacoboni, M.: Mirroring People: The New Science of How We Connect with Others. Farrar, Straus & Giroux, New York (2008)
8. Robinson, B.L., Harper, N.S., McAlpine, D.: Meta-adaptation in the auditory midbrain under cortical influence. Nat. Commun. **7**, 13442 (2016)
9. Treur, J.: Modeling Higher-order network adaptation by multilevel network reification. In: Treur, J. (eds.) Network-Oriented Modeling for Adaptive Networks: Designing Higher-Order Adaptive Biological, Mental and Social Network Models. SSDC, vol. 251, pp. 99–119. Springer, Cham (2020). https://doi.org/10.1007/978-3-030-31445-3_4
10. West, R., Michie, S., Rubin, G.J., et al.: Applying principles of behaviour change to reduce SARS-CoV-2 transmission. Nat. Hum. Behav. **4**, 451–459 (2020). https://doi.org/10.1038/s41562-020-0887-9
11. WHO Coronavirus (COVID-19) Dashboard. WHO. https://covid19.who.int/ (2021)
12. Zhang, N., et al.: Effects of human behavior changes during the coronavirus disease 2019 (COVID-19) pandemic on influenza spread in Hong Kong. Clin. Infect. Dis. **73**, e1142–e1150 (2020)

Early Prediction of Chronic Kidney Disease Using Data Mining Techniques

Hasin Shahed Shad[1], Zeeshan Jamal[1], S. M. Foysal Ahmed[1], Sifat Momen[1(✉)], and Nafees Mansoor[2]

[1] Department of Electrical and Computer Engineering, North South University, Bashundhara, Dhaka, Bangladesh
{hasin.shad,zeeshan.jamal,s.ahmed01,sifat.momen}@northsouth.edu

[2] Department of Computer Science and Engineering, University of Liberal Arts Bangladesh, Dhaka, Bangladesh
nafees.mansoor@ulab.edu.bd

Abstract. Chronic kidney disease (CKD) is a serious health condition that may result in multifaceted kidney related issues affecting kidney structure and function. CKD is identified as a global public health concern. Unfortunately, the symptoms of CKD do not get manifested during its early stage. In this work, we have used machine learning approaches to reliably detecting CKD. We have applied different machine learning classifiers to detect Chronic Kidney Disease using a publicly available data set. Empirical results indicate that the Random Forest offers better prediction in predicting chronic kidney disease.

Keywords: Chronic Kidney Disease (CKD) · CKD dataset · Data mining · Classification · Random forest classifier

1 Introduction

Chronic kidney disease (CKD) is a grievous health condition that may result in multifaceted kidney related problems affecting kidney structure and function [6]. This illness is referred to as 'chronic' because the damage to the kidneys exacerbates gradually over time [7].

CKD is currently acknowledged as a global public health problem. A study on discrepancies in Chronic Kidney Disease prevalence among males and females was conducted in 2016 [2]. According to their research, the total number of individuals with impaired kidney function was found to be 752.7 million, consisting of 417.0 million females and 335.7 million males [2]. Of this, around (2.5–11.2)% appertains to the adults residing in Asia, Europe, Australia, and North America [16].

CKD has been proven to be fatal when left untreated [3]. The consequences of chronic kidney disease are always extreme and dangerous. It has the potential to result in kidney failure and early cardiovascular diseases (CVD) [3]. If the illness is diagnosed during the later stage, the complications usually become

R. Silhavy et al. (Eds.): CoMeSySo 2021, LNNS 231, pp. 947–957, 2021.
https://doi.org/10.1007/978-3-030-90321-3_79

more severe [3]. Recent studies indicate that many adverse effects of CKD can be prevented or at least be delayed through early detection and treatment [11].

As there is a scarcity of nephrologists, they cannot exclusively manage all patients with CKD [8]. As a consequence, primary care physicians (PCP) are often responsible for managing chronic kidney disease [8]. Nonetheless, it has been observed that PCPs often struggle detecting chronic kidney diseases in the early stage [13]. As the illness develops slowly over time and does not manifest its symptoms during the early stage, it becomes difficult to diagnose CKD at an early stage. Hence, accurate and efficient techniques of detecting CKD is deemed important.

Recently, machine learning (ML) has been found to be very effective in detecting different kinds of diseases [1,5,9,10]. In this research, data mining techniques have been applied for detecting CKDs. Reliable prediction of CKD will be invaluable in detecting the disease at an earlier stage. The CKD dataset, that has been used for ML purposes, was acquired from the University of California, Irvine (UCI) Machine Learning Repository [15]. At the outset, the dataset was collected. Then it was pre-processed. The pre-processing step involved the conversion of categorical values to numeric values, the use of min-max scaler to normalize numerical attributes and replacing the missing values with the mean of the corresponding attribute. A total of five different classifiers were used in this work. Each classifier was evaluated after applying feature selection and stratified 10-fold cross validation using five different metrics- accuracy, precision, recall, f1-score and area under the ROC curve.

The remainder of the paper is structured as follows: The second section discusses the notable earlier works pertaining to chronic kidney diseases. Section 3 describes the methodology used in our research. The acquired findings are given and discussed in Sect. 4. Finally, Sect. 5 brings the paper to a conclusion.

2 Related Works

A number of research papers on the identification of chronic renal disease have been found in the literature. Recently, in a research conducted by Sinha and Sinha [14] to predict CKD used two machine learning algorithms including k-nearest neighbors and support vector machine. They collected the dataset from several medical labs, centers and hospitals. They achieved an accuracy of 78.75% and 73.75% using KNN and SVM respectively.

Revathy and colleagues used several classifiers for predicting CKD [12]. The accuracy from the decision tree, SVM, and Random forest were found to be 94.16%, 98.33%, and 99.16% respectively. In comparison to them, we obtained an accuracy of 95.9% using decision tree classifier, 93.40% using SVM, and over 99% using Random Forest classifier.

In [12], the researchers applied Neural Networks as well as K-Nearest Neighbor, Random Forest, and Support Vector Machine on the same data set. Using Random Forest classifier with 12 predictive features, they have achieved detection accuracy of 0.993.

Charleonnan et al. [4] conducted research on chronic kidney disease as well. They used four different machine learning classifiers for predicting CKD, namely the Decision Tree, K-nearest Neighbor (KNN), Logistic Regression (LR), and Support Vector Machine (SVM) classifier. They also applied Kfold cross-validation with $k = 5$. From 5-fold cross validation, the best accuracy they achieved was 98.3% using SVM. The mean accuracy of Decision Tree, KNN, and Logistic Regression was 94.8%, 98.1%, and 96.55%, respectively.

3 Methodology

Figure 1 explicates the methodology utilized in our research work. Initially the data set was acquired and pre-processed. Using 10-fold cross validation, the classifiers were evaluated against five different metrics (as mentioned earlier).

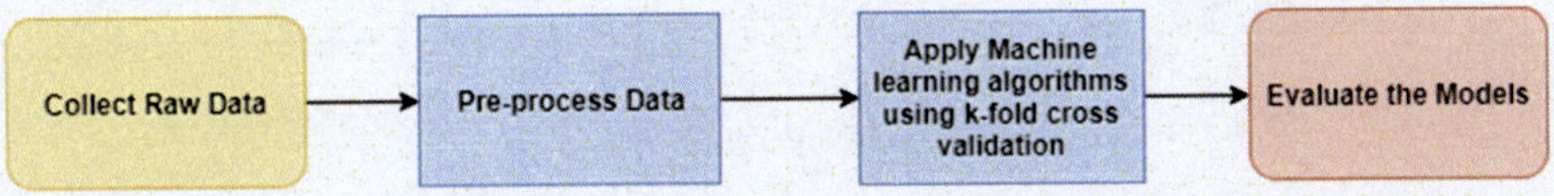

Fig. 1. Flow diagram of the work

The dataset contains a total of 400 instances and 25 features. There were some missing values in the dataset. Figure 2 illustrates the distribution of some of the attributes.

The relationship amongst the features for the data set is measured through a correlation heat map as shown in Fig. 3. Correlation heat map shows how closely the variables are related to each other. It ranges from +1 to −1. Closer to 0 indicates that the two variables do not have a linear dependency. They're significantly positively connected if the value is close to 1. On the contrary, if the value is closer to −1, the variables are strongly negatively correlated.

3.1 Preprocessing

To transform raw data into a format appropriate for data mining tasks, preprocessing is required. The raw data contains some missing entries. This was tackled by substituting the missing entries with the mean of the corresponding feature (Eq. 2). Also, the categorical data were all converted to numerical data. Feature scaling (Eq. 1) was applied on the numeric attributes so that all numeric values lie between 0 and 1.

$$X_{norm} = \frac{X - X_{min}}{X_{max} - X_{min}} \tag{1}$$

$$x = \frac{\sum x_i}{n} \tag{2}$$

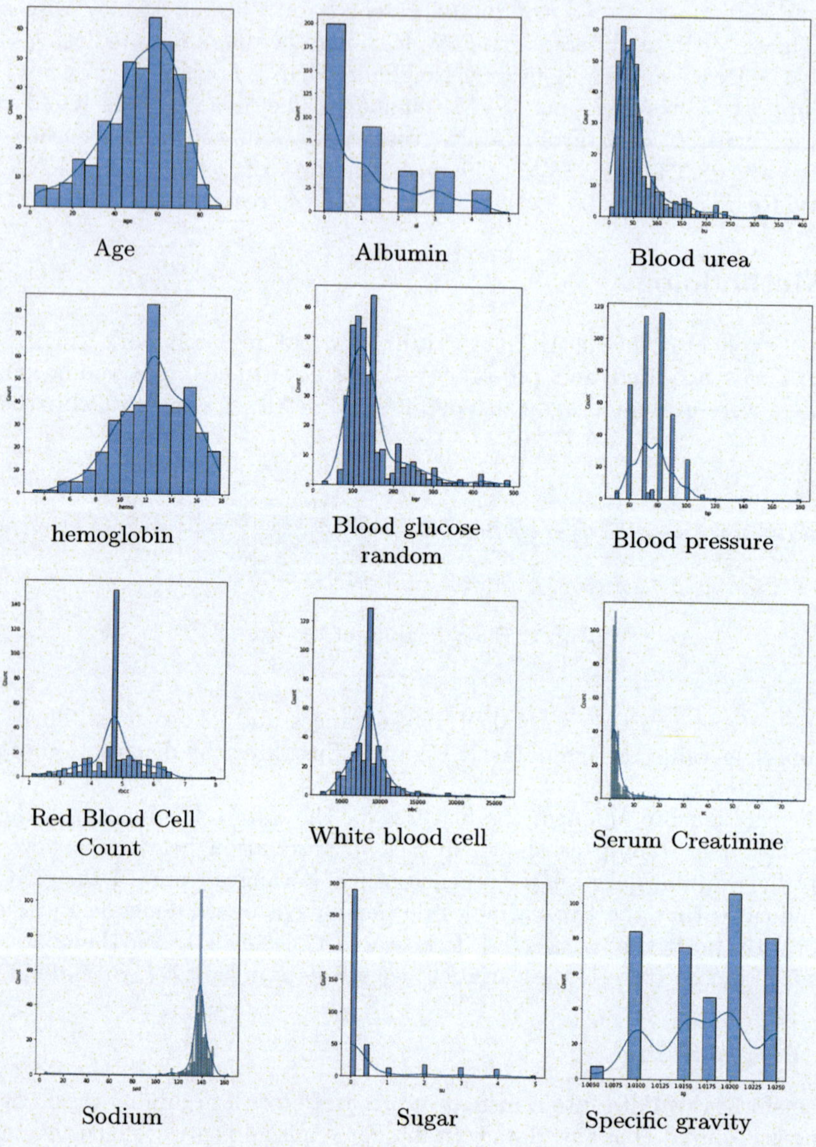

Fig. 2. Histogram representation of distributions according to the features

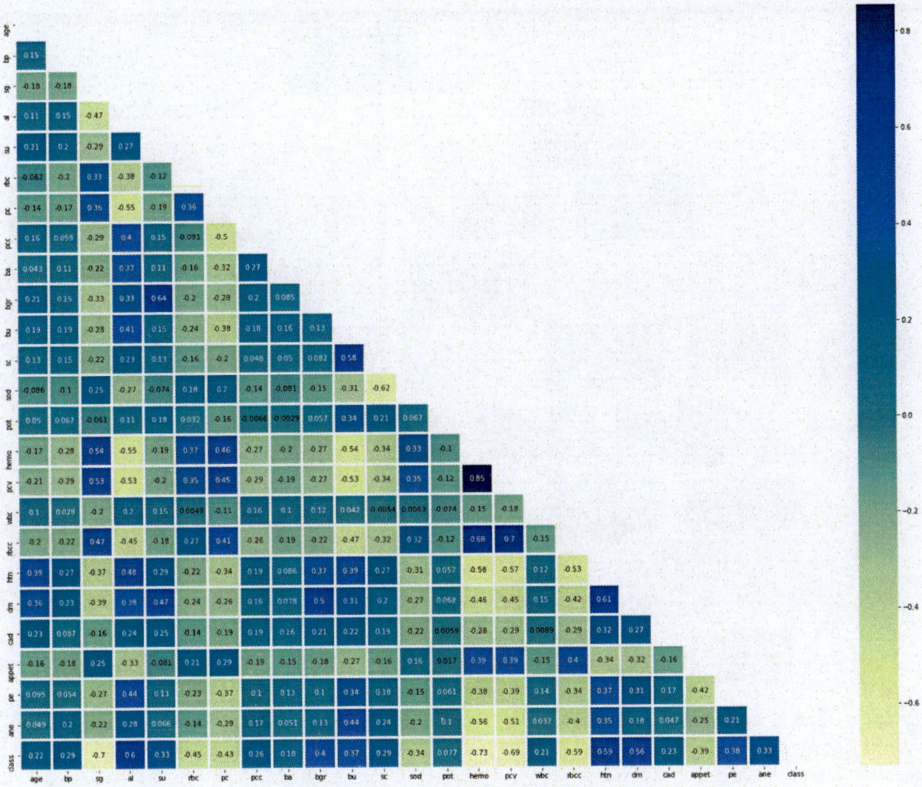

Fig. 3. Correlation heatmap

3.2 Machine Learning Algorithm Implementation

Five different classifiers were used in this work.

3.3 K-Nearest Neighbor

K-nearest neighbor (KNN) determines the distances of each instances from the query instance. Following this, the classifier identifies the k nearest instances. Each of the k instances vote for the output. Using the voting technique, the predicted result of the query instance is determined and returned. We used Euclidean distance (Eq. 3) metric as a means of computing the distance.

$$d(x, y) = \sqrt{\sum_{i=1}^{n}(x_i - y_i)^2} \tag{3}$$

For this dataset, we found that $k = 10$ provides the best result.

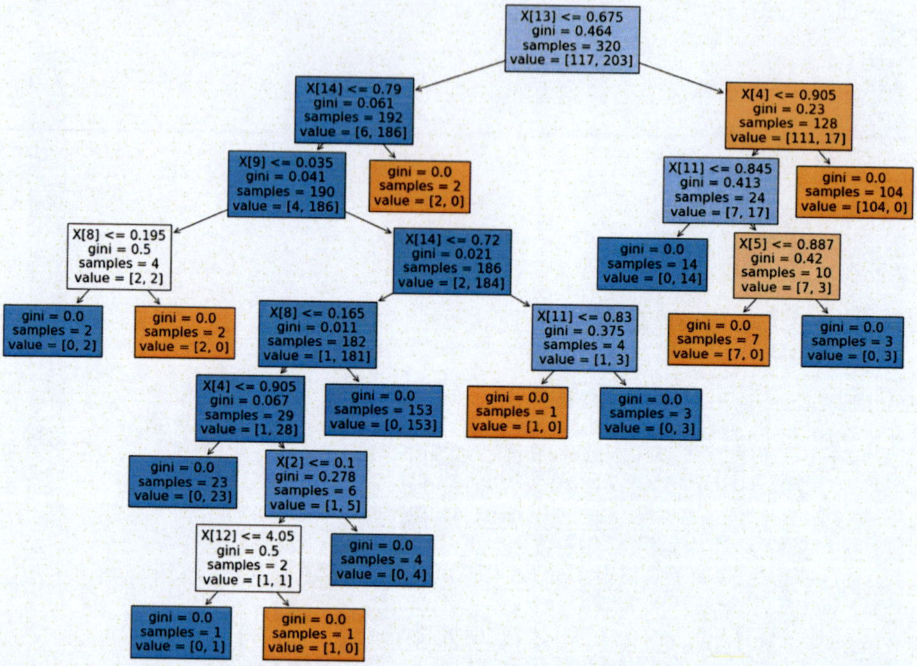

Fig. 4. Decision Tree diagram

3.4 Decision Tree Classifier

The main purpose of the Decision Tree is to develop a model that predicts classes or values of target variables using decision rules drawn from training data. This approach divides the data set into divisions by starting with the most appropriate attribute. The splitting process is repeated until all of the divisions are homogeneous. Gini (Eq. 4) has been used as the splitting criterion for our research work.

$$Gini_{index} = 1 - \sum_{i=1}^{c} p_i(t)^2 \tag{4}$$

where, $p_i(t)$ is the probability of class i belonging to node t.

The outcome of the Decision Tree for our dataset is shown in Fig. 4.

3.5 Support Vector Machine

A supervised machine learning algorithm called support vector machine that examines data for classification and regression problems. This algorithm is extensively used because it often provides better accuracy.

3.6 Logistic Regression

Logistic regression is familiar for solving binary classification problem. It can, be extended to solve multi-classification problem. Logistic regression uses the concept of linear regression and apply sigmoid function on the top of it.

3.7 Random Forest

Random Forest is an ensemble classifier that leverages L tree-structured base classifiers $h(X, \theta_n)$, n $= 1, 2, 3,$ . . . L, where X is the input data and a family of identical and dependent distributed random vectors is known as θ_n. Random Forest is a flexible model which often produces great results even without hyper parameter tuning. In this research, the highest accuracy was obtained using the random forest Classifier.

4 Experimental Results

After pre-processing the data, the classifiers described above were used to determine the best model for detecting chronic kidney disease. Kfold cross-validation approach has been followed to predict the best outcome. It is a useful strategy for evaluating different machine learning models. This is one of the popular and reliable methods where the entire dataset gets shuffled randomly and divided into k folds. Subsequently, one fold is treated as a test set while the rest of the $k - 1$ folds are treated as a training set. On the train set, the ML algorithm is applied, and it is tested on the test set. This is iterated k times, with each time test set consisting of a different fold. The performance of the classifier was assessed using five evaluation metrics which are described below.

4.1 Accuracy

The number of times the correct estimates made is referred to as accuracy. Accuracy can be calculated using the following equation.

$$accuracy = \frac{TP + TN}{TP + TN + FP + FN} \tag{5}$$

4.2 Recall

Recall can be used to measure how well the model detects true positives. A high recall is an indicator that the model has done well identifying true positives. On the contrary, if the recall value is low, it indicates the model encountered a lot of false negatives.

$$recall = \frac{TP}{TP + FN} \tag{6}$$

4.3 Precision

Precision (also known as positive predictive value) refers to how well the model predicts positive values out of all positive values predicted by the model.

$$precision = \frac{TP}{TP + FP} \quad (7)$$

4.4 Area Under the ROC Curve

The ROC curve depicts the relationship between the algorithm's sensitivity and specificity. The area under this graph is used as a performance metric as well.

4.5 F1-Score

The harmonic mean of precision and recall is the F1-score. It can be calculated using the Eq. 8.

$$F_1 = 2\frac{precision \times recall}{precision + recall} \quad (8)$$

Here,

TP = True Positives
TN = True Negatives
FP = False Positives
FN = False negatives.

Table 1 illustrates the prediction findings for different classifiers.

Table 1. Obtained results after applying the algorithms

Classifier	Accuracy	Precision	Recall	F1-Score	AUC
Decision Tree	0.959+/-0.5%	0.963+/-0.1%	0.960+/-0.5%	0.963+/-0.3%	0.943+/-0.3%
KNN	0.941+/- 0.2%	1.000+/- 0.0%	0.905+/- 0.1%	0.949+/-0.1%	0.981+/-0.1%
Logistic Regression	0.981+/- 0.1%	1.000+/- 0.0%	0.970+/- 0.1%	0.984+/-0.1%	0.995+/-0.1%
Random Forest	0.988+/- 0.2%	0.987 +/- 0.1%	0.995+/- 0.1%	0.993+/-0.0%	1.000+/-0.0%
SVM	0.934+/- 0.1%	1.000+/- 0.0%	0.896+/- 0.1%	0.944+/- 0.0%	0.995+/-0.0%

Figure 5 shows the comparison graph among all the classifiers after implementing with K-fold cross validation.

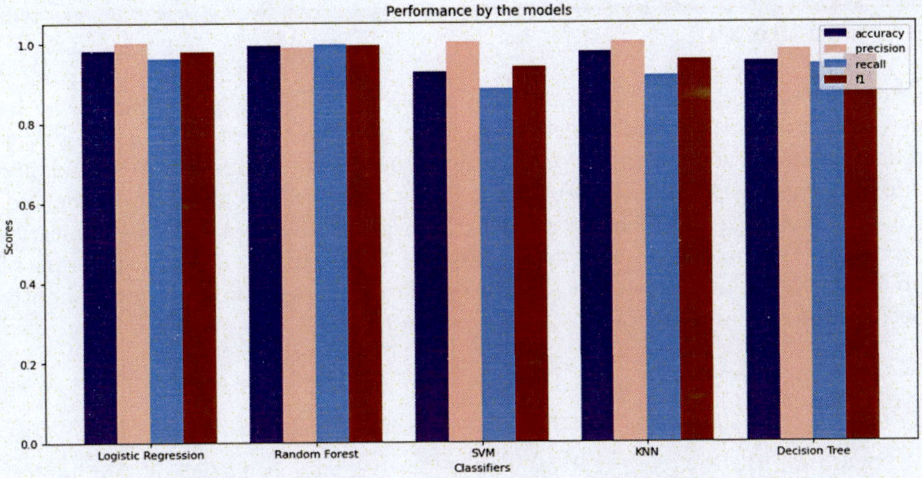

Fig. 5. Comparison graph among the classifiers

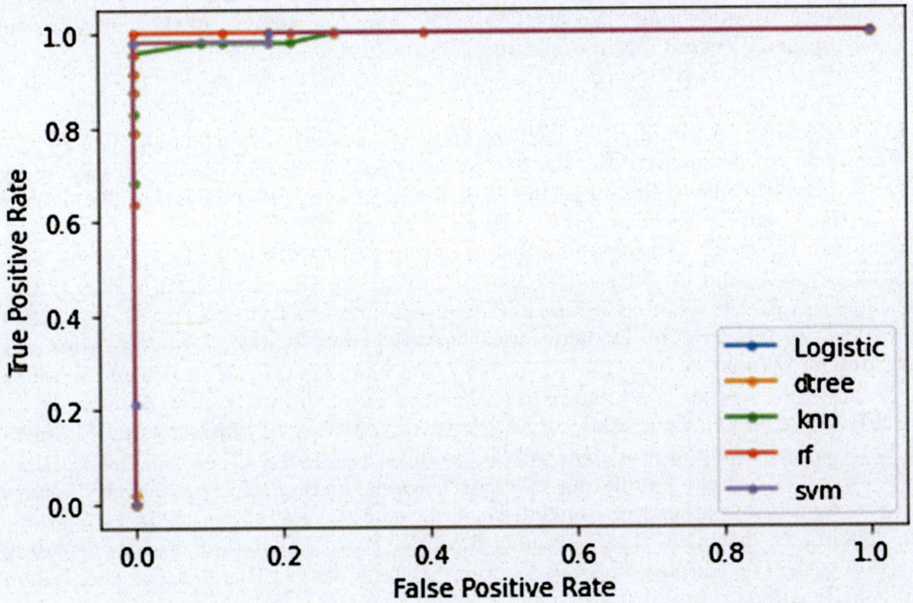

Fig. 6. ROC curve

Figure 6 represents the ROC curve for each classifier. The area under the ROC curve is also a metric for evaluating the performance of the classifiers.

5 Conclusion

Chronic kidney disease (CKD) is a prolonged condition which requires to be detected as quickly as possible. The major goal of this study was to find a reliable and accurate way of predicting chronic kidney disease. We used five different classifiers in this work. The results have been reliable and accurate. Among them, Random Forest classifier performed the best in several metrics (accuracy, f1-score, recall, and area under the ROC curve). However in terms of precision, KNN, support vector machine, and logistic regression performs slightly better that random forest.

References

1. Bhardwaj, R., Nambiar, A.R., Dutta, D.: A study of machine learning in healthcare. In: 2017 IEEE 41st Annual Computer Software and Applications Conference (COMPSAC), vol. 2, pp. 236–241. IEEE (2017)
2. Bikbov, B., Perico, N., Remuzzi, G., et al.: Disparities in chronic kidney disease prevalence among males and females in 195 countries: analysis of the global burden of disease 2016 study. Nephron **139**, 313–318 (2018)
3. Center for Disease Control and prevention. https://www.cdc.gov/kidneydisease/basics.html
4. Charleonnan, A., Fufaung, T., Niyomwong, T., Chokchueypattanakit, W., Suwannawach, S., Ninchawee, N.: Predictive analytics for chronic kidney disease using machine learning techniques. In: 2016 Management and Innovation Technology International Conference (MITicon), pp. MIT–80. IEEE (2016)
5. Fuhad, K., et al.: Deep learning based automatic malaria parasite detection from blood smear and its smartphone based application. Diagnostics **10**(5), 329 (2020)
6. Levey, A.S., Coresh, J.: Chronic kidney disease. Lancet **379**(9811), 165–180 (2012)
7. National Institute of Diabetes and Digestive and Kidney Diseases: Kina sea urchin regions in NZ (2017). https://www.niddk.nih.gov/health-information/kidney-disease/chronic-kidney-disease-ckd/what-is-chronic-kidney-disease
8. Plantinga, L.C., Tuot, D.S., Powe, N.R.: Awareness of chronic kidney disease among patients and providers. Adv. Chronic Kidney Dis. **17**(3), 225–236 (2010)
9. Pranto, B., et al.: Evaluating machine learning methods for predicting diabetes among female patients in Bangladesh. Information **11**(8), 374 (2020)
10. Pranto, B., Mehnaz, S.M., Momen, S., Huq, S.M.: Prediction of diabetes using cost sensitive learning and oversampling techniques on Bangladeshi and Indian female patients. In: 2020 5th International Conference on Information Technology Research (ICITR), pp. 1–6. IEEE (2020)
11. Remuzzi, G., Ruggenenti, P., Perico, N.: Chronic renal diseases: renoprotective benefits of renin-angiotensin system inhibition. Ann. Intern. Med. **136**(8), 604–615 (2002)
12. Revathy, S., Bharathi, B., Jeyanthi, P., Ramesh, M.: Chronic kidney disease prediction using machine learning models. Int. J. Eng. Adv. Technol. (IJEAT) **9**(1) (2019)
13. Salekin, A., Stankovic, J.: Detection of chronic kidney disease and selecting important predictive attributes. In: 2016 IEEE International Conference on Healthcare Informatics (ICHI), pp. 262–270. IEEE (2016)

14. Sinha, P., Sinha, P.: Comparative study of chronic kidney disease prediction using KNN and SVM. Int. J. Eng. Res. Technol. **4**(12), 608–12 (2015)
15. UCI Machine Learning Repository. http://archive.ics.uci.edu/ml
16. Zhang, Q.L., Rothenbacher, D.: Prevalence of chronic kidney disease in population-based studies: systematic review. BMC Public Health **8**(1), 1–13 (2008)

Utility Function Approach to Portfolio Selection Problem

Anna Andreevna Malakhova[1,2](✉), Elena Nikolaevna Sochneva[1], Svetlana Anatolyevna Yarkova[2], Albina Sergeevna Danilova[2], Elena Anatolyevna Yurieva[3], Alexey Yurievich Senchenko[3], Olga Valeryevna Starova[1], Dmitry Ivanovitch Kravtsov[1], and Dmitry Valeryevitch Zyablikov[1]

[1] Siberian Federal University, Krasnoyarsk, Russia
amalakhova@sfu-kras.ru
[2] Krasnoyarsk Institute of Railway Transport - Branch of the Irkutsk State Transport University, Krasnoyarsk, Russia
[3] Professor V.F. Voino-Yasenetsky Krasnoyarsk State Medical University, Krasnoyarsk, Russia

Abstract. In this paper portfolio selection problem is studied under premise that there are two basic approaches to determine an optimal portfolio: (1) mean-variance approach implemented into the Markowitz model; and (2) the concept of utility, which in regard to modern portfolio theory is defined in terms of portfolio risk and return. This paper focuses on these two approaches showing how they apply to single-period investment problem and demonstrating how they work together to produce strong and useful pricing relationships. The modification of the utility function based on the Markowitz model is proposed and an algorithm for investment portfolio optimization problem is developed. A modified utility maximization model is formulated in terms of portfolio variance and standard deviation. An illustrative experimental case-study is performed via 'Moscow Exchange' Trading Organizer. The investment portfolio is optimized on the basis of the modified utility maximization model in real time. It is shown that the introduced *k*-coefficient in the utility objective function can be considered as the coefficient of risk aversion. It is concluded that in contrast to the traditional Markowitz model, the proposed approach based on the utility function takes into account the investor's risk preferences and extends our understanding of the location of the efficient frontier in consideration of the parameters set for the investor.

Keywords: Utility function · Markowitz model · Investment portfolio · Mean-variance approach · Risk aversion

1 Introduction

The problem of portfolio optimization remains a widely discussed issue in investment analysis. With the securities market development the investment portfolio analysis has become one of the most topical issues nowadays as is demonstrated by many publications on this subject (see e.g. [1–3]). There are different formulations and models

R. Silhavy et al. (Eds.): CoMeSySo 2021, LNNS 231, pp. 958–972, 2021.
https://doi.org/10.1007/978-3-030-90321-3_80

developed to synthesize investment portfolios. The most popular approaches for the portfolio modelling include the Markowitz mean-variance approach, the Tobin extension of the Markowitz model, the Sharpe index model based on 'beta'-factor, the Capital asset pricing model, the Arbitrage pricing theory and their multiple modifications (see e.g. [1, 4, 5]).

The efficiency of the application of these models depends on the availability of the information for investor, diversity of the available investment assets and the level of development of securities market. However, the problem of determining the optimal portfolio on the efficient frontier remains an open question for all models [10, 11]. An investor needs a procedure for ranking random wealth levels. The present state of the art of investment portfolio management demonstrates that a utility function provides such a procedure [6, 7].

One strand of the literature is based on applying the exponential utility function with Arrow-Pratt risk-aversion index, which is constant for exponential utility independent of wealth [8]. The reason for choosing exponential utility is the assumption of its equivalency to quadratic utility and mean-variance analysis.

Another portfolio selection approach is based on the family of negative power utility functions, also called constant relative risk aversion [8]. It should be pointed out that such criteria depend on risk-aversion parameter and the initial wealth as well, so it is volatile and does not produce a unique value for portfolio selection.

One of the recent approaches is the use of utility functions with hyperbolic absolute risk aversion (HARA). One of the strong points of this approach is that such utility functions allow the derivation of a generalized two-funds separation theorem thus leading to sample capital market evaluation formulas, and to the generalization of the Sharpe ratio and the Treynor ratio as well [9].

Nevertheless, utility function approaches, though important, are rather subjective. The degree of investor's risk-aversion and the selection of utility function remain discussed questions. It is known that the most common used and efficient approaches for portfolio synthesis is the Markowitz model which presumes that the risk of the portfolio is measured by the variance of the portfolio return and depends on the correlation between the assets, measured by the covariance [5]. In this respect the Markowitz model can be reasonable for utility function construction in terms of mean-variance ratio optimization.

The aim of this paper is to formalize the utility function for portfolio selection problem based on the Markowitz model and to evaluate its efficiency relatively to the traditional mean-variance optimization approach.

The paper is organized as follows. The concept of utility, which is defined in terms of portfolio return and risk is formulated in the second section. The modification of the utility function based on the Markowitz model is proposed and an algorithm for investment portfolio optimization problem is developed. A modified utility maximization model is formulated in terms of portfolio variance and standard deviation. In the third section an illustrative experimental case-study is performed via 'Moscow Exchange' Trading Organizer. The investment portfolio is optimized on the basis of the modified utility maximization model. The results of model verification are discussed in the fourth section. The main findings of the research are summarized in the conclusion.

2 Methods

2.1 The Modified Utility Function Based on the Markowitz Model

In modern portfolio theory, the central place is occupied by the concept of utility, which is defined in terms of return and variance as follows [4]:

$$U = R_p - k\sigma_p^2 \tag{1}$$

where R_p is the expected return on the portfolio p, k is the coefficient that measures the degree of risk aversion of the investor, σ_p^2 is the variance of the portfolio p.

Such an objective function corresponds to the "expected value-variance" criterion used in decision-making theory as one of the criteria for evaluating a decision alternative [4].

As noted, variance and standard deviation are adequate measures of risk for symmetric distributions. Meanwhile, for distributions of portfolio returns that differ from normal, the utility function is formulated in a generalized form, taking into account the central moments of higher orders. In this case, the influence of the moments of higher orders on the value of the total utility is less significant than the influence of the moments of lower orders.

If skewness and kurtosis are important for an investor, then his utility function is formulated as follows:

$$U = R_p - k_1\sigma_p^2 - k_2 S_p^3 - k_3 S_p^4 \tag{2}$$

where, S_p^3, S_p^4 are the coefficients of skewness and kurtosis of the portfolio return p, k_1, k_2, k_3 are the coefficients of aversion to variance, skewness and excess, respectively. So, if a positive skewness of the distribution of returns ($S_p^3 > 0$) is preferable for an investor, then the contribution of asymmetry ($k_2 S_p^3$) increases the overall utility of the portfolio for a given investor, and, therefore, the coefficient of aversion to skewness should be set negative ($k_2 < 0$). Note that (1) is a special case of (2), which takes place for normal distributions ($S_p^3 = S_p^4 = 0$).

When choosing the parameter k, it is important to observe the investor's preferences and the requirement of non-negativity of the utility function: $U(k) \geq 0$. Thus, k is selected from the range: $k_{min} < k < k_{max}$, where k_{min} represents the lower bound of the values of k acceptable for the investor, depending on his preferences, k_{max} is the upper acceptable bound k, at which the value of the objective function is non-negative.

The prerequisites for the formulation of the modified Markowitz problem for portfolio selection are as follows [1, 10]:

1. Investors evaluate investment portfolios based on expected returns and standard deviations over the holding period.
2. Investors are never jaded, that is, other conditions being equal, they prefer a portfolio with a higher expected return.
3. Investors are rational, that is, other conditions being equal, they prefer a portfolio with less risk.

4. The portfolio consists only of risky assets, the risk measure of which is the standard deviation of return.
5. Investors assume a normal probability distribution of the expected return on assets. The portfolio return aggregates the returns on the securities included in it.
6. Private assets are infinitely divisible, that is, an investor can buy part of a share.
7. Taxes and transaction costs are immaterial.

The mean return of a portfolio of n assets is obtained as:

$$R_p = \sum\nolimits_{i=1}^{n} w_i \cdot \overline{R}_i, \tag{3}$$

where w_i is the weight of the asset i in the portfolio, $\overline{R}_i$ is the mean return on the asset i.

In respect that the risk of the portfolio is measured by the variance of the portfolio return and depends on the correlation between the assets measured by the covariance [Mark] the variance of a portfolio return is defined as [4]:

$$\sigma_p^2 = \sum\nolimits_{ij} w_i \cdot w_j \cdot Cov_{ij} \tag{4}$$

where w_i and w_j are the weights of the assets i and j in the portfolio, respectively, Cov_{ij} is the covariance between two assets i and j.

Hence, we modify the utility function (1) using (4) as follows:

$$U = \sum w_i \cdot \overline{R}_i - k \cdot \sum w_i \cdot w_j \cdot Cov_{ij} \tag{5}$$

The modified utility function can be formulated in terms of standard deviation as well. The initial prerequisites for the formulation of the utility function based on the Markowitz model will be prerequisites (1)–(7) and the following extra condition:
8. The distribution of returns is symmetric; the standard deviation is an adequate measure of risk.

In terms of standard deviation, the utility function (1) is assumed in the form:

$$U = \sum w_i \cdot \overline{R}_i - k \cdot \sqrt{\sum w_i \cdot w_j \cdot Cov_{ij}} \tag{6}$$

2.2 Utility Maximization Problem for Portfolio Selection

The main purpose of a utility function is to provide a systematic way to rank alternatives that captures the principle of risk aversion. The initial formulation of the Markowitz problem is to minimize (maximize) the variance (the mean return) of (on) the portfolio under fixed value R (σ^2) of the mean return (variance). That is [5]:

$$J_{Markowitz} = \sum\nolimits_{ij} w_i \cdot w_j \cdot Cov_{ij} \rightarrow min \tag{7}$$

subject to

$$\sum\nolimits_{i} w_i \cdot \overline{R}_i = R \tag{8}$$

$$\sum_i w_i = 1$$

$$w_i \geq 0,\ i = \overline{1, n}$$

.

or

$$J_{Markowitz} = \sum w_i \cdot \overline{R}_i \rightarrow max \tag{9}$$

subject to

$$\sum_{ij} w_i \cdot w_j \cdot Cov_{ij} = \sigma^2,\ i = \overline{1, n} \tag{10}$$

$$\sum_i w_i = 1,\ i = \overline{1, n}.$$

$$w_i \geq 0,\ i = \overline{1, n}.$$

The solution of the problem yields the optimal weight coefficients w_i^* for the assets in the portfolio.

Hence, we formulate the utility maximization problem based on the Markowitz model using (5):

$$J_{(R-k\sigma^2)} = \sum w_i \cdot \overline{R}_i - k \cdot \sum w_i \cdot w_j \cdot Cov_{ij} \rightarrow max \tag{11}$$

subject to

$$\sum_i w_i = 1,\ i = \overline{1, n} \tag{12}$$

$$w_i \geq 0,\ i = \overline{1, n}$$

$$\alpha_k \leq w_k \leq \beta_k,\ 0 \leq \alpha_k < 1,\ 0 < \beta_k \leq 1,\ k \in \overline{1, n},$$

where w_i is the weight of the asset i in the portfolio, $\overline{R}_i$ is the expected return on the asset i, k is the model parameter set in the range $k_{min} < k < k_{max}$; α_k and β_k are the boundaries of the interval of acceptable values of the asset weight k, for which there is a consideration of choosing certain weight values.

The solution to this problem yields weight coefficients $w^*_{(R-k\sigma^2)}$ for the assets in the portfolio.

In terms of standard deviation, we formulate the following utility maximization problem using (6):

$$J_{(R-k\sigma)} = \sum w_i \cdot \overline{R}_i - k \cdot \sqrt{\sum w_i \cdot w_j \cdot Cov_{ij}} \rightarrow max \tag{13}$$

subject to

$$\sum_{i} w_i = 1,\ i = \overline{1, n} \tag{14}$$

$$w_i \geq 0,\ i = \overline{1, n}.$$

$$\alpha_k \leq w_k \leq \beta_k,\ 0 \leq \alpha_k < 1,\ 0 < \beta_k \leq 1,\ k \in \overline{1, n}.$$

The solution to this problem produces weight coefficients $w^*_{(R-k\sigma)}$ for the assets in the portfolio.

The problems (11), (12) and (13), (14) are related to quadratic programming problems, since the objective function is quadratic and the constraints are linear [4]. This problem can be solved, for example, in MatLab and Excel software environment.

Solver Tool can be used to optimize portfolio in Excel. We will enter the portfolio utility into the target cell, and we will set the asset weights as the changing cells. In the constraints, we will set the sum of the asset weights, and constraints on the specific asset weights in the portfolio. To get an efficient set of portfolios, we will use the Visual Basic for Application add-in through recording a macros. The input parameters of the model are mean returns on the assets $\overline{R}_i$, obtained from series of moving stock returns R_{it}, variance-covariance matrix Cov_{ij}, assessing covariance between each pair of assets i and j, and risk aversion coefficient k acceptable for an investor.

2.3 Analysis of the *k*-Parameter in the Modified Utility Maximization Problem

In this section we estimate the influence of the coefficient k on the value of the objective utility function in problems (11), (12) and (13), (14). If k is set equal to zero, the objective function will reach the maximum possible value, which corresponds to the maximum expected return on the portfolio, and therefore to the one hundred percent share of investment in the asset with the maximum expected return (provided $\alpha_k = 0$, $\beta_k = 1$, $k = \overline{1, n}$). Obviously, a further increase of k-value will inevitably lead to a decrease in the objective function, since the desire to achieve a higher expected return will be constrained by the limitation of the second element of the objective utility function. Thus, we presume the dependence of the objective function on k as is conceptually shown in Fig. 1.

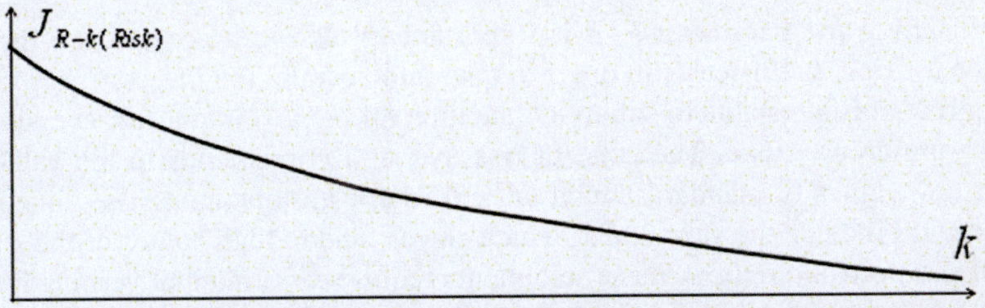

Fig. 1. Influence of the k coefficient on the value of the objective utility function.

To analyze the influence of the coefficient k on the parameters of the portfolio, let us trace how the shares of distribution in assets will change. It was found that zero k corresponds to a 100% investment in the most profitable asset. If we assume that as k increases, preference will continue to be given to the highest-yielding assets, we find that this would entail a sharp decrease in the objective function, since high-yielding assets entail a higher risk. Obviously, in order to avoid a sharp decline in the objective function, it will be necessary to reduce the portfolio risk, albeit at the expense of high returns. We conclude that with an increase in k, the share of investments will gradually redistribute from more profitable to less profitable assets, which will entail a decrease in the total return and risk of the portfolio. We forecast the dependance of the portfolio return on the k coefficient as is presented in Fig. 2. Thus, we consider the coefficient k as the degree of risk aversion of the investor.

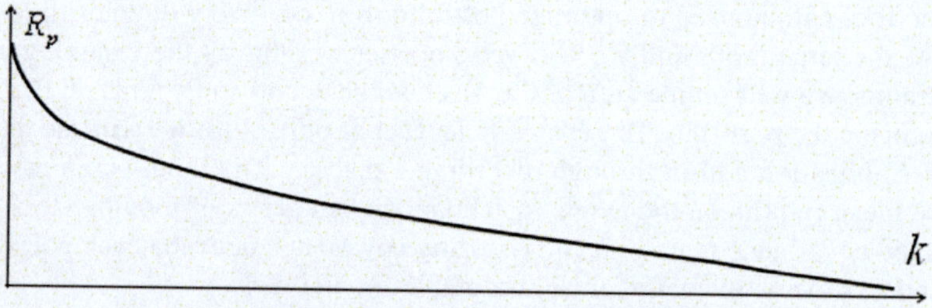

Fig. 2. Influence of the k coefficient on the portfolio return.

Note that it makes sense to consider the range of k values, $k \in [0; k']$, within which the utility, that is, the value of the objective function, is non-negative, that is $U(k) \geq 0$.

It is also advisable to distinguish the areas of values of k, in which the values of the portfolio parameters remain practically unchanged, and those areas for which there is a significant change in the values of the parameters of the portfolio. This emphasis is significant, since the changing values of return and risk describe the main part of the effective frontier of portfolios. Figure 3 shows the range of values of k and the corresponding sections of the efficient frontier.

It is notable that the major part of the efficient set corresponds to the interval $[k_1; k_2]$. Whereas on intervals $[0; k_1)$ and $(k_2; k']$ the parameters of the portfolio do not change substantially. Thus, the value of the parameter k should be selected from the interval $[k_1; k_2]$.

When setting the parameter k, it is important to take into account the investor's preference for risk. In this case, in order to take into account the investor's preferences regarding risk, it is advisable to subdivide the interval $[k_1; k_2]$ into a risk-aversion zone and a risk-preference zone. The zone of risk aversion corresponds to the values of k at which low values of standard deviation and return are obtained. The zone of risk preference will include the values of k, which correspond to high values of the standard deviation and portfolio return. These statements require experimental verification.

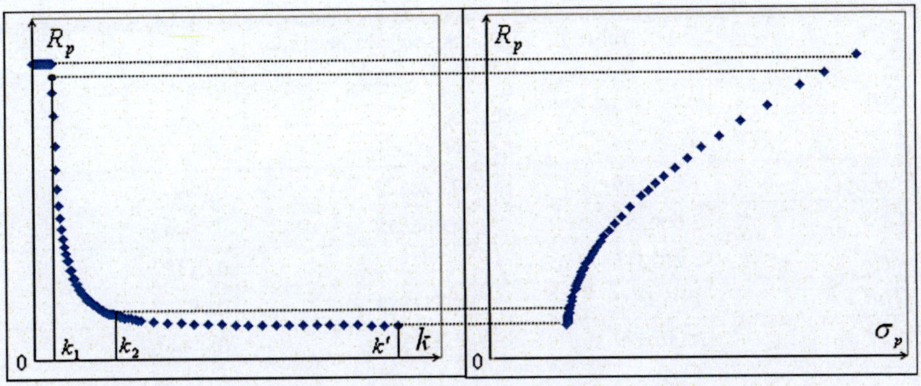

Fig. 3. Ranges of values of the k coefficient and sections of the efficient frontier.

3 Results

The 'Moscow Exchange' was chosen as the information basis for conducting this part of the experimental study as the oldest organizer and the leading trading platform for the securities market in Russia [12].

An important factor in the choice of the 'Moscow Exchange' was the open access to trading information; the prices that are formed on the 'Moscow Exchange' are a generally accepted guideline for investors making transactions with Russian stocks and depositary receipts. Trading information is broadcast to a huge number of consumers in Russia and abroad and is the basis for calculating the RTS and MICEX stock indices.

For further analysis and synthesis of the portfolio, diverse assets have been be selected that represent various sectors of the economy according to the chosen principle of branch diversification: electricity, oil and gas, financial sector, communications and telecommunications, metallurgical, air transportation, consumer sector. Preference is to be given to securities with good growth potential and liquidity.

Thus, six assets were selected that meet the listed characteristics of the following issuers: Public Joint stock company "ROSSETI" (ordinary share *RSTI*), Public Joint-Stock Company VTB Bank (ordinary share *VTBR*), Public Joint Stock Company "Mining and Metallurgical Company "NORILSK NICKEL" (ordinary share *GMKN*), Public Limited Liability Company Yandex N.V., shares of a foreign issuer (ordinary share *YNDX*), Public Joint Stock Company "Aeroflot-Russian Airlines" (ordinary share *AFLT*), Mobile TeleSystems Public Joint Stock Company (ordinary share *MTSS*) [12].

For the further analysis we selected one-year period. The price series of assets were transformed into series of moving returns with a smoothing window of one month. Monthly moving returns have been obtained for the one-year period, namely January, 01, 2020 – December 31, 2020. The initial parameters of the assets are presented in Table 1.

Variance-covariance matrix is given in Table 2.

Table 1. Initial statistic parameters

Asset	Asset parameters		
	$\overline{R}_i$	σ_i	σ_i^2
GMKN	1,023189	0,096822	0,009375
MTSS	1,006183	0,066658	0,004443
YNDX	1,062315	0,107592	0,011576
VTBR	1,058802	0,832589	0,693204
AFLT	1,002309	0,029359	0,000862
RSTI	1,035651	0,151427	0,022838

Table 2. Covariance matrix for the selected assets

Asset	Cov_{ij}					
	GMKN	MTSS	YNDX	VTBR	AFLT	RSTI
GMKN	0,009337	0,004135	0,002326	0,011534	0,000525	0,009864
MTSS	0,004135	0,004425	0,003937	0,009764	0,000862	0,007330
YNDX	0,002326	0,003937	0,011529	0,007151	0,000655	0,005283
VTBR	0,011534	0,009764	0,007151	0,690431	0,000995	0,024477
AFLT	0,000525	0,000862	0,000655	0,000995	0,000821	0,000875
RSTI	0,009864	0,007330	0,005283	0,024477	0,000875	0,022838

The portfolio optimization problem (7), (8) was solved using estimates of expected returns $\overline{R}_i$ from Table 1 and covariance matrix from Table 2. The solution of the $(w_i w_j Cov_{ij})$ - minimization problem yields the optimal weight coefficients w_i^* for portfolios from the efficient frontier, represented in Table 3.

In this part of the experiment, the portfolio is synthesized using the modified utility maximization models (11), (12) and (13), (14).

To solve utility maximization problems, the parameters *k* were set taking into account the nonnegativity of the objective function. For the corresponding k, weight distributions $w^*_{(R-k\sigma^2)}$, $w^*_{(R-k\sigma)}$, were obtained. It is notable that the obtained weight distributions $w^*_{(R-k\sigma^2)}$, and $w^*_{(R-k\sigma)}$ are close to each other: the difference in the minimum values of the weights does not exceed 0.4%, and the difference in the maximum values of the weights does not exceed 0.77%. The $w^*_{(R-k\sigma^2)}$ distribution is close to the distribution of the Markowitz problem: the difference in the minimum values of the weights does not exceed 0.02%, and the difference in the maximum values of the weights does not exceed 1.73%. The $w^*_{(R-k\sigma)}$ distribution is also close to the w^* distribution: the difference in the minimum values of the weights does not exceed 0.4%, and the difference in the maximum values of the weights does not exceed 1.73%.

Table 3. Portfolio parameters and weight distribution w_i^*

Portfolio return	Standard deviation	Weights of the assets					
		GMKN	MTSS	YNDX	VTBR	AFLT	RSTI
1,0410	0,008535	0,00000	0,00000	0,00000	0,00000	0,0173	0,9827
1,0400	0,006863	0,00000	0,00000	0,00000	0,00000	0,11831	0,88169
1,0390	0,005423	0,00000	0,00000	0,00000	0,00000	0,21941	0,78059
1,0380	0,004214	0,00000	0,00000	0,00000	0,00000	0,32052	0,67948
1,0370	0,003237	0,00000	0,00000	0,00000	0,00000	0,42172	0,57828
1,0360	0,002492	0,00000	0,00000	0,00000	0,00000	0,52273	0,47727
1,0350	0,001978	0,02683	0,00000	0,00000	0,00000	0,5577	0,41547
1,0340	0,001696	0,07572	0,00000	0,00000	0,00000	0,52285	0,39038
1,0330	0,001604	0,12021	0,02903	0,00000	0,00000	0,48914	0,36162
1,0320	0,001540	0,16475	0,04703	0,00000	0,00000	0,4554	0,33282
1,0310	0,001480	0,20921	0,06512	0,00000	0,00000	0,42172	0,30408
1,0300	0,001424	0,32608	0,10768	0,02489	0,00000	0,31901	0,22234
1,0290	0,001371	0,32484	0,09755	0,07761	0,00000	0,28996	0,21004
1,0280	0,001321	0,32362	0,08742	0,13033	0,00000	0,2609	0,19775
1,0270	0,001274	0,32236	0,07728	0,18305	0,00000	0,23185	0,18546
1,0260	0,001230	0,32117	0,06715	0,23576	0,00000	0,20277	0,17315
1,0250	0,001190	0,31996	0,05701	0,28847	0,00000	0,1737	0,16086
1,0240	0,001153	0,31864	0,04688	0,34122	0,00000	0,14468	0,14858
1,0230	0,001119	0,31741	0,03675	0,39394	0,00000	0,11563	0,13628
1,0220	0,001088	0,31616	0,02662	0,44666	0,00000	0,08657	0,12399
1,0210	0,001060	0,31492	0,01649	0,49937	0,00000	0,05752	0,1117
1,0200	0,001034	0,31369	0,00635	0,5521	0,00000	0,02846	0,0994
1,0190	0,001011	0,31094	0,00000	0,60461	0,00000	0,00000	0,08445
1,0180	0,000991	0,28597	0,00000	0,65878	0,00000	0,00000	0,05525

The dependence of the parameters of the portfolio and the objective utility function on the parameter k is analyzed. The dependences of the parameters of the portfolio R_p, σ_p, σ_p^2 and objective utility functions $(R_p - k\sigma_p^2)$, $(R_p - k\sigma_p)$ on the k coefficient are shown in Figs. 4, 5, 6, 7, 8 and 9.

According to the results of the experimental study, presented in Figs. 4, 5, 7, 8 the return and variance (standard deviation) curves may be seen as consisting of three parts: interval k, on which the risk and return values change insignificantly; interval k, at which portfolio risk and return change considerably; interval k, at which the parameters of the portfolio tend to a constant value. In this case, the objective function decreases from k,

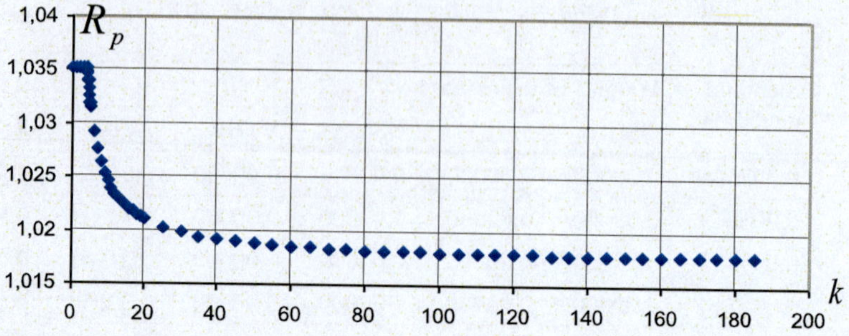

Fig. 4. Dependence of the portfolio return R_p on k coefficient for the $(R_p - k\sigma_p^2)$-maximization model

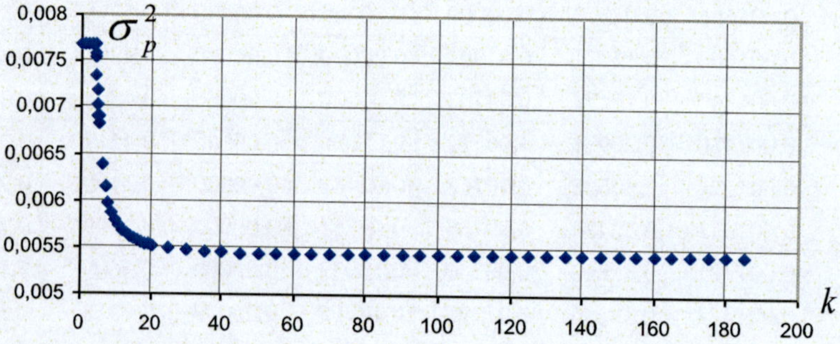

Fig. 5. Dependence of the portfolio variance σ_p^2 on k coefficient for the $(R_p - k\sigma_p^2)$-maximization model

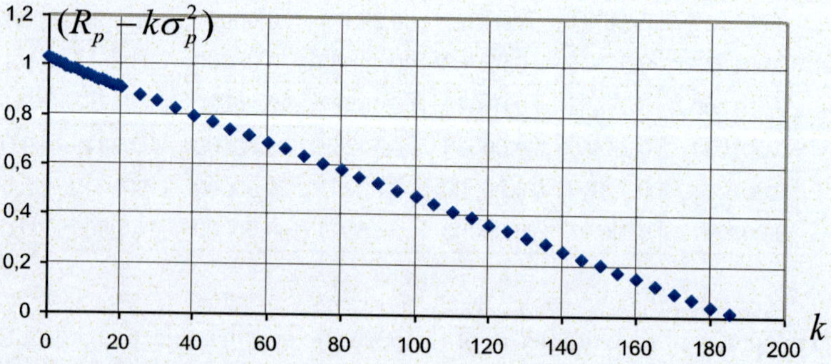

Fig. 6. Dependence of the objective $(R_p - k\sigma_p^2)$-utility function on k coefficient.

which confirms the assumptions made in Sect. 2. Thus, the coefficient k can be regarded as the coefficient of risk aversion.

The interval values of R_p, σ_p^2 and R_p, σ_p for different k-intervals of risk aversion coefficient for models $\left(R_p - k \cdot \sigma_p^2\right)$, $(R_p - k \cdot \sigma_p)$, are shown in Table 4 and Table 5, respectively.

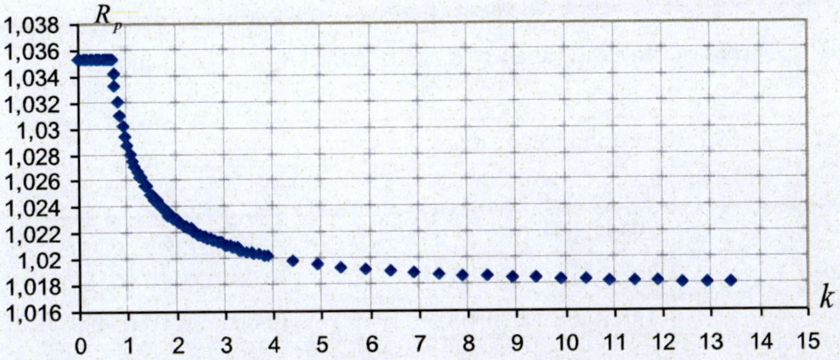

Fig. 7. Dependence of the portfolio return R_p on k coefficient for the $(R_p - k\sigma_p)$-maximization model

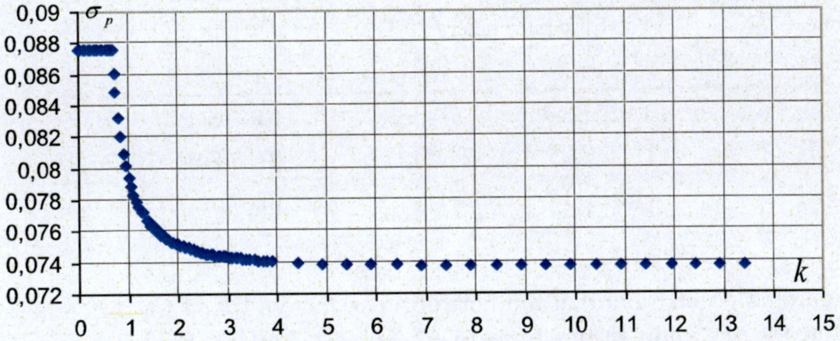

Fig. 8. Dependence of the portfolio risk σ_p on k coefficient for the $(R_p - k\sigma_p)$-maximization model

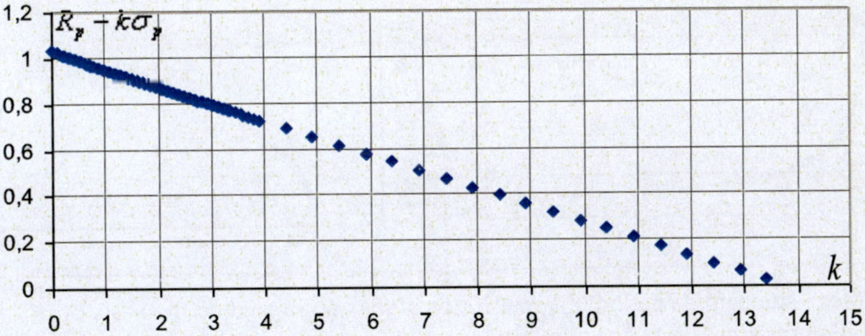

Fig. 9. Dependence of the objective $(R_p - k\sigma_p)$-utility function on k coefficient.

We recommend to set the coefficient k from the interval within which there is a significant change in the parameters of the portfolio. According to the results of the experimental study, these interval values for k are evaluated as [0.71; 3.1], [4.01; 20] for models $(R_p - k \cdot \sigma_p)$, $(R_p - k \cdot \sigma_p^2)$, respectively.

Table 4. Interval values of R_p and σ_p^2 for $\left(R_p - k \cdot \sigma_p^2\right)$-model

k-interval	Portfolio parameters	
	R_p	σ_p^2
[0; 4]	1.0353263	0,007675682
[4.01; 20]	[1.021131, 1.03495]	[0.005523, 0.007581]
[21; 185]	[1.017734, 1.020364]	[0.005428, 0.005488]

Table 5. Interval values of R_p and σ_p for $\left(R_p - k \cdot \sigma_p\right)$-model

k-interval	Portfolio parameters	
	R_p	σ_p
[0; 0.7]	1,0353263	0,087611
[0.71; 3.1]	[1.021095, 1.034242]	[0.074304, 0.087611]
[3.15; 13.4]	[1.018112, 1.020911]	[0.073697, 0.074244]

It should be pointed out that the portfolio parameters found for different k, corresponding to the maximum utility, form an efficient set similar to the Markowitz efficient frontier as shown in Fig. 10.

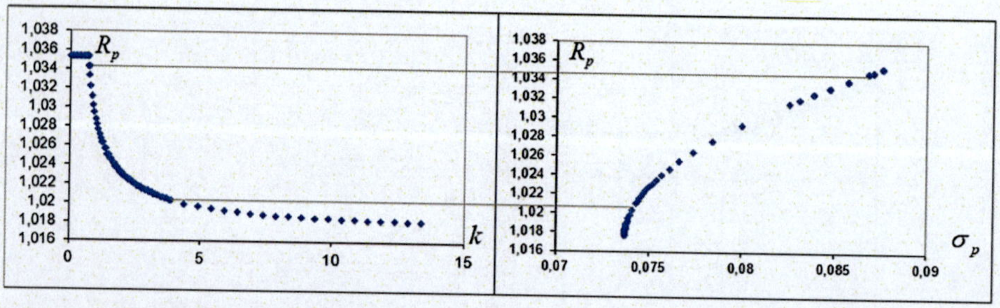

Fig. 10. Mean-standard deviation diagram for portfolios optimized on the basis of $(R_p - k\sigma_p)$-utility function for k-coefficient.

4 Discussions

It has been shown that an alternative to the risk-minimization approach is the objective utility function maximization. In Markowitz's problem, modified taking into account the utility function, the maximum value of this function corresponds to the trade-off between an increase in the expected return and a decrease in risk, specified by the value of the coefficient k. Unlike the risk minimization problem, the utility maximization problem does not require the introduction of an additional constraint on the expected portfolio return.

Another difference in the statement of the utility maximization problem is the introduced parameter k, which we consider as the degree of investor's aversion to risk: low values of k correspond to high values of portfolio return and risk; setting a higher value of k entails lower profitability and risk. It is advisable to set the parameter k from a range of values $k_{min} < k < k_{max}$, which makes it possible to take into account the investor's preferences regarding risk and to comply with the condition of non-negativity of the objective function for different measures of risk. We recommend choosing a value of k from the interval that corresponds to the main part of the effective set. For the case study under consideration these interval k-values have been determined as [0.7; 3.1], [4.01; 20] for models $(R_p - k \cdot \sigma_p)$, $(R_p - k \cdot \sigma_p^2)$, respectively.

It has been shown that the portfolio parameters found for different k, corresponding to the maximum utility, form an efficient set similar to the Markowitz efficient frontier. In contrast to the traditional Markowitz model, the proposed approach based on the utility function allows us to take into account the investor's risk preferences and extends our understanding of the location of the efficient frontier, taking into account the parameters set for the investor.

Note that the utility maximization problem is formulated depending on the accepted risk measure. For symmetric distributions, the problem is formulated as $(R_p - k\sigma_p^2)$ - or $(R_p - k\sigma_p)$-maximization. To take into account the left-sided deviations of asymmetric distributions, this approach should be developed for alternative risk measures taking into account non-normal distributions and substantial downside tail risk.

5 Conclusion

In this paper an alternative to the risk-minimization approach based on the objective utility function maximization is proposed.

The utility function is modified using the Markowitz model for portfolio risk assessment. It has been shown that the maximum value of this function corresponds to the trade-off between an increase in the expected return and a decrease in risk, specified by the value of the coefficient k. Unlike the risk minimization problem, the utility maximization problem does not require the introduction of an additional constraint on the expected portfolio return.

The results of experimental study demonstrate that the k coefficient can be considered as the coefficient of risk aversion: an increase in k leads to a decrease in the risk and return of the portfolio. The weight distributions obtained when solving the risk minimization problems are close to the distributions obtained when solving the problems of maximizing the objective utility function using the appropriate risk measures. It is recommended to set the coefficient k from the interval within which there is a significant change in the parameters of the portfolio.

To the best of our knowledge the portfolio parameters found for different k, corresponding to the maximum utility, form an efficient set similar to the Markowitz efficient frontier. In contrast to the traditional Markowitz model, the proposed approach based on the utility function allows us to take into account the investor's risk preferences and extends our understanding of the location of the efficient frontier, taking into account the parameters set for the investor.

It should be pointed out that the obtained results best apply to symmetric portfolio distributions. It remains to be settled if the excellency of utility maximization approach based on the Markowitz model holds for non-normal distributions setting, when asset and portfolio returns might exhibit substantial downside tail risk. This issue is left for further research.

References

1. Sharpe, W.F., Alexander, G.J., Beiley, J.V.: Investments. Prentice Hall International Inc., Upper Saddle River (1995)
2. Bodie, Z., Kane, A., Marcus, A.J.: Essentials of Investments. McGraw-Hill, New York (2001)
3. Fabozzi, F.J.: Quantitative strategies: multi-asset. J. Portfolio Manag. Multi-Asset Spec. Issue **46**(6), 1–3 (2020)
4. Luenberger, D.G.: Investment Science. Oxford University Press, New York (1998)
5. Markowitz, H.M.: Portfolio selection. J. Financ. **7**(1), 77–91 (1952)
6. Amilon, H.: Comparison of Mean-Variance and Exact Utility Maximization in Stock Portfolio Selection, Working Papers 2001:4, Lund University, Department of Economics (2001)
7. Sharpe, W.F.: Expected Utility Asset Allocation, Stanford University, September 2006, Revised June 2007. https://web.stanford.edu/~wfsharpe/art/euaarevised.pdf. Accessed 28 July 2021
8. Ziemba, W.T.: The Stochastic Programming Approach to Asset, Liability, and Wealth Management. The Research Foundations of The Association for Investment Management and Research, USA (2003)
9. Breuer, W., Gürtler, M.: Performance evaluation and preferences beyond mean-variance, swiss society for financial market research. Fin. Markets Portfolio Manag. **17**(2), 213–233 (2003)
10. Malakhova, A.A., et al.: Reward-to-variability ratio as a key performance indicator in financial manager efficiency assessment. In: Silhavy, R. (ed.) CSOC 2020. AISC, vol. 1225, pp. 598–613. Springer, Cham (2020). https://doi.org/10.1007/978-3-030-51971-1_49
11. Malakhova, A.A., et al.: Portfolio optimization model for asset allocation problem based on alternative risk measures. In: Silhavy, R. (ed.) CSOC 2021. LNNS, vol. 229, pp. 673–693. Springer, Cham (2021). https://doi.org/10.1007/978-3-030-77445-5_61
12. Moscow Exchange. https://www.moex.com/. Accessed 21 June 2021

Integrating Multilevel Adaptive Models to Develop Systematic, Transparent, and Participatory EIA Practice

Andreea Nita[1(✉)], Jan Treur[2], and Laurentiu Rozylowicz[1]

[1] Center for Environmental Research and Impact Studies, University of Bucharest, 1 Nicolae Balcescu, 010041 Bucharest, Romania
andreea.nita@cc.unibuc.ro, laurentiu.rozylowicz@g.unibuc.ro

[2] Social AI Group, Department of Computer Science, Vrije Universiteit Amsterdam, Amsterdam, The Netherlands
j.treur@vu.nl

Abstract. European Directive 85/337/EC introduced Environmental Impact Assessment (EIA) as a process governed by administrative rules with the aim of reducing environmental degradation and associated health problems generated by projects. Generally, the EIA process involves analyses and evaluations of the potential impacts that human activities may have on the environment by considering approaches such as the precautionary principle, prevention of conflicts, loss of natural resources, and environmental degradation. EIA influences decision-making at local, national, and transboundary levels, with the following overall objectives: potentially screen out environmentally harmful projects, predict significant adverse impacts, suggest measures to reduce or prevent major impacts, identify feasible alternatives, and engage communities or individuals potentially affected by the implementation of the project. Several issues obstructing the proper implementation of the EIA process are common in developing countries: low quality of assessment reports, lack of public participation, insufficient equipment, and trained staff, inadequate institutional framework, and low cooperation between policymakers, researchers, and stakeholders. The number of research studies focused on the investigation of EIA collaboration process through network analysis and multilevel adaptive models is worryingly limited, considering that the implementation of EIA procedures is deficient in most developing countries, and the contribution of science that envisages the collaboration between the actors involved to the process is minor at best. This paper aims to use Multilevel Network Reification to create Higher-order Adaptive Network Models. The results of multilevel network analysis will contribute to reshape impact assessment procedures and create opportunities for better communication and transparency between practitioners, researchers, policymakers, and other stakeholders. Therefore, integrating Multilevel Adaptive Models in EIA helps to raise the policy efficiency and define the dynamic interplay between EIA actors and diagnose the organizational structures that strongly influence this procedure. Thus, by using an adaptive computational network model, we aim to understand the roles of each actor and the connections established in different EIA networks. The findings will provide innovative information to find solutions and design a collaborative EIA

R. Silhavy et al. (Eds.): CoMeSySo 2021, LNNS 231, pp. 973–991, 2021.
https://doi.org/10.1007/978-3-030-90321-3_81

procedure to improve projects under evaluation considering the current threats to society and the environment.

Keywords: Environmental impact assessment · EIA actors · Cooperation · Multilevel adaptive model · Nonadaptive model

1 Introduction

European Directive 85/337/EC introduced Environmental Impact Assessment (EIA) as a process governed by administrative rules with the aim of reducing environmental degradation and associated health problems generated by projects. Generally, the EIA process involves analyses and evaluations of the potential impacts of human activities on the environment by considering approaches such as the precautionary principle, prevention of conflicts [1], loss of natural resources, and environmental degradation [2]. EIA influences decision-making at local, national and transboundary levels, with the following overall objectives: potentially screen out environmentally harmful projects, predict significant adverse impacts, suggest measures to reduce or prevent major impacts, identify feasible alternatives, and engage communities or individuals potentially affected by the implementation of the project. Developing countries tend to have weak EIA performance [3–5] and low scientific productivity in the EIA field [6], highlighting the necessity of improving the process and integrating it with planning and policy-making at multiple scales [7]. Moreover, poorly implemented EIA often led to environmental conflicts, lack of confidence in the system and infringement procedures against EU countries. EIA was introduced in European Union as a mandatory process for implementing projects with potential environmental impact [8]. In developing countries the procedure is mainly used to obtain environmental permits, not to improve the project (minimize the impact) or identify concrete mitigation measures, and most studies examine the environmental components in a vague/subjective manner without proper public participation or scientific oversight [6, 9].

Although the core EIA procedures are comparable worldwide, the quality of implementation varies from country to country and sometimes from project to project [10]. However, several issues obstructing the proper implementation of the EIA process are common in developing countries: low quality of assessment reports, lack of public participation, insufficient equipment and trained staff, inadequate institutional framework, low cooperation between policymakers, researchers, and stakeholders [4, 11–13]. High quality and streamlined EIA process [14] requires efforts from all involved EIA actors [15]. Identifying technical solutions to the problems of a single category of actors is not enough to develop sustainable projects, as this may neglect others' interests, generating conflicts [16]. Current approaches to question EIA process use case studies to discuss the quality of EIA reports, describe or compare stages or the whole procedure, review the performance of EIA within a specific domain of activities, analyse perceptions of EIA practitioners through surveys or evaluate public participation [4, 5]. The number of research studies focused on the investigation of EIA collaboration process through network analysis/multilevel adaptive models [17] is worryingly limited [10], the implementation of EIA procedures is deficient in most developing countries [6], and the contribution of science that envisages the collaboration between the actors involved to the process is minor at best.

Another issue affecting EIA worldwide is the gap between practice and research, and to our knowledge, no current approach captures the level of cooperation among EIA actors at the international level. It is necessary to demonstrate that interactions between practitioners, policymakers, and stakeholders can have a crucial impact on the quality and effectiveness of EIA procedures [18]. To fill this gap, the paper uses social network analysis (SNA) to investigate cooperation structures around EIA by using multilevel adaptive models and adaptive temporal-causal network models [19]. The analysis is performed for each stage of the EIA (i.e., screening, scoping, impact assessment & EIA reporting, public consultation, decision-making, monitoring), by analyzing the relationship between EIA actors. The analysis can help define best practice examples for decision-making and constructive participation of stakeholders [20] in the EIA process.

Given that actors cooperate, compete, and support one another [21], the quality of interactions makes a difference [22] in finding the key to success [23]. Furthermore, EU-funded research projects are increasingly asked to include policy and societal dimensions into their work plans and strengthen their dissemination and science-policy interface activities. In this regard, better communication and transparency between practitioners, researchers, policymakers, and other stakeholders have become a priority at the EU level. Therefore, integrating SNA in EIA helps define the dynamic interplay between EIA actors [24] to diagnose the organizational structures that strongly influence this procedure. The difficulties of the investigated issue come from the fact that EIA is a complex and collaborative process [25]; there is no single EIA actor powerful enough to coordinate the process and elaborate scientific research and professional practice at the same time. Thus, understanding the roles of each actor and the connections established in different EIA networks will provide innovative information to find solutions and design a collaborative EIA procedure to improve projects under evaluation, facts critical for countries as Romania.

In this paper we integrate a novel framework for increasing the effectiveness and accuracy of the EIA collaboration and overcome its low performance by exploring the interplay between actors and providing practical solutions for a proper collaboration in different scenarios for sustainable development. We analyze the impact of collaboration combinations of the EIA stakeholders involved by considering their tasks and dominance in an EIA project in obtaining the environmental permits. We use computationally modeling and simulation to answer the following research question: What is the most effective combination of the EIA actors, and if dominance can improve the EIA collaboration process.

2 Concepts and Methodology

2.1 EIA Network Modelling

This paper uses Social Network Analysis (SNA) [26], a well-developed scientific domain within network theory [27]. This methodological approach is widely used for resource management networks [28, 29] and for investigating patterns in environmental policy collaboration [7, 30]; however, to our best knowledge, it has not been used in a complex socio-ecological arena such as EIA.

In this regard, better communication and transparency between practitioners, policymakers and other stakeholders have become a priority at the European level. Therefore, integrating SNA in EIA helps to investigate the dynamic interplay between EIA actors in the context of social interactions and organizational structures that strongly influence the implementation of this procedure.

The difficulties of the investigated issue come from the fact that EIA is a complex and collaborative process [25]; there is no single EIA actor powerful enough to coordinate the process and elaborate scientific research or professional practice at the same time. Thus, creating the different scenarios and understanding the roles of each EIA actor and the connections established within an EIA network will provide innovative information for designing a collaborative EIA procedure and improve projects under evaluation.

Our findings of social network analysis applied to empirical EIA processes will help solving problems while leading to improved collaboration and knowledge exchange between practitioners, scientists, policymakers, and stakeholders.

2.2 Investigated EIA Structures

By mapping the models and accurately documenting EIA practice, we identify key individuals/organizations influencing the network's structure [31, 32]. Specifically, we reproduce here two scenarios in which the EIA network is influenced only by a few actors with low collaboration between Developer/Beneficiary and Public Authority and Stakeholders (Scenario 1) and the second one in which EIA decision-making process is influenced by all stakeholders (Scenario 2 - Level of cooperation according to EU EIA Directive).

In contrast to the non-adaptive network model [33], higher-order adaptation is addressed in the model introduced here [34, 35].

We further perform simulations of adaptive multilevel models in which we present a Homogeneous EIA model and a dominance scenario of all EIA actors that should take part to EIA.

For adaptive multilevel models for cross-sectional EIA Nets we compare the network of actors and collaboration process according to the EIA Directive (85/337/EEC) for each stage of the process (i.e., Screening, Scoping, EIA Reporting, EIA Report quality assessment, Public participation, Decision making and Follow-up and Monitoring)) with the resulted collaboration network from the actual EIA practice for the EIA permit and implementation of a project and/or implication of public in each EIA stage (Fig. 1), as follows:

1. **Screening** – The stage that establishes whether the environmental and social impacts of a proposed development project would be significant enough to develop and perform an EIA;
2. **Scoping** – The stage that determinates the limits/boundaries of the EIA, set the basis of the analyses that will be conducted at each stage, presenting the project alternatives;

3. **Impact assessment & EIA Reporting quality assessment** – The stage evaluating the socio-economic and environmental impacts of the planned project and its alternatives, and then identify the mitigation measures to reduce those impacts;
4. **EIA consulting (Public participation)** – The stage aiming to consult and include all stakeholders potentially affected by the proposed project;
5. **Decision-making** – The stage in which the competent authority grants or not the environmental permit for the implementation of the project;
6. **Information on decision** – The public/stakeholders are informed of the decision
7. **Follow-up and Monitoring** – The stage in which it is carried out monitors the implementation of the project to respect all the clauses and norms included in the environmental impact assessment report.

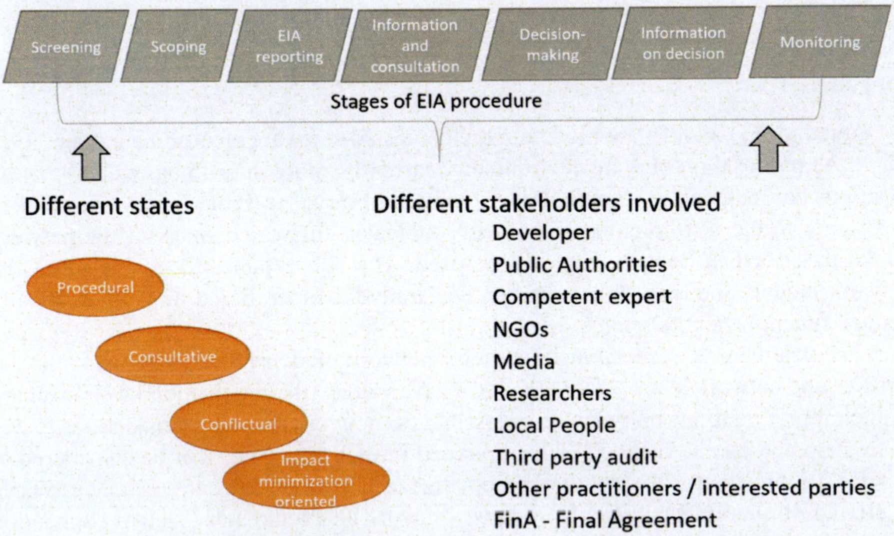

Fig. 1. EIA stages, states and stakeholders included in this study

3 The Network-Oriented Modeling Approach Used

In this section, the network-oriented modelling approach used here (see [35]) is briefly introduced. Following this approach, a temporal-causal network model is characterized by (here X and Y denote nodes of the network, also called *states* with activation values $X(t)$ and $Y(t)$ over time (t):

- **Connectivity characteristics:** Connections from a state X to a state Y and their weights $\boldsymbol{\omega}_{X,Y}$
- **Aggregation characteristics:** For any state Y, some combination function $\mathbf{c}_Y(..)$ (usually with some parameters) defines the aggregation that is applied to the impacts $\boldsymbol{\omega}_{X,Y}X(t)$ on Y from its incoming connections from states X

- **Timing characteristics:** Each state Y has a speed factor η_Y defining how fast it changes for a given impact.

The following difference (or differential) equations that are used for simulation purposes and also for analysis of temporal-causal networks incorporate these network characteristics $\omega_{X,Y}$, $\mathbf{c}_Y(..)$, η_Y in a standard numerical format:

$$Y(t + \Delta t) = Y(t) + \eta_Y[\mathbf{c}_Y(\omega_{X_1,Y} X_1(t), \ldots, \omega_{X_k,Y} X_k(t)) - Y(t)]\Delta t(1) \quad (1)$$

for any state Y and where X_1 to X_k are the states from which Y gets its incoming connections. Here the overall combination function $\mathbf{c}_Y(..)$ for state Y is the weighted average of available basic combination functions $\mathbf{c}_j(..)$ by specified weights $\gamma_{j,Y}$ (and parameters $\pi_{1,j,Y}$, $\pi_{2,j,Y}$ of $\mathbf{c}_j(..)$) for Y:

$$\mathbf{c}_Y(V_1, \ldots, V_k) = \frac{\gamma_{1,Y}\mathbf{c}_1(V_1, \ldots, V_k) + \cdots + \gamma_{m,Y}\mathbf{c}_m(V_1, \ldots, V_k)}{\gamma_{1,Y} + \cdots + \gamma_{m,Y}} \quad (2)$$

Equations (1) and (2) are hidden in the dedicated software environment; (see [35] - Ch 9). As part of this software environment, currently, more than 45 basic combination functions are included in a combination function library; see Table 1 for some of them used here; in the last column, it is already indicated for which states of the network model described in Sect. 4 they will be used. The above concepts enable to design network models and their dynamics in a declarative manner, based on mathematically defined functions and relations.

Note that there is a crucial distinction for network models between network characteristics and network states. Network states have values (their activation levels) and are explicit representations that may be accessible for network states by connections to and from them and can be handled or manipulated in that way. They can be considered to provide an informational view on the network; usually, the states are assumed to have a certain informational content. In contrast, network characteristics (such as connection weights and excitability thresholds) have values (their strengths) and determine (e.g., cognitive) processes and behaviour in an implicit, automatic manner. They can be considered to provide an embodiment view on the network. In principle, these characteristics by themselves are not directly accessible nor observable for network states; you can make connections between states, but you cannot make connections between network characteristics or between states and characteristics.

Table 1. Examples of basic combination functions from the library used in the current paper

	Notation	Specification	Parameters
Advanced logistic	**alogistic**$_{\sigma,\tau}$(V_1, …,V_k)	$[\frac{1}{1+e^{-\sigma(V_1+\cdots+V_k-\tau)}} - \frac{1}{1+e^{\sigma\tau}}](1+e^{-\sigma\tau})$	Steepness $\sigma>0$ Excitability threshold τ
Hebbian learning	**hebb**$_\mu$(V_1, V_2, W)	$V_1V_2(1-W)+\mu W$	Persistence factor $\mu>0$

As indicated above, 'network characteristics' and 'network states' are two distinct concepts for a network. Self-modeling is a way to relate these distinct concepts to each other in an interesting and useful way. A self-model is making the network characteristics (such as connection weights and excitability thresholds) explicit in the form of adding states (called self-model states) for these characteristics and also connections and other network characteristics for these additional states. Thus, the network gets an internal self-model of part of its structure: it explicitly represents information about its own structure. In this way, by iteration, different self-modeling levels can be created where network characteristics from one level relate to network states at the next level.

More specifically, adding a self-model for a temporal-causal base network is done in the way that for some of the states Y of the base network and some of the network structure characteristics for connectivity, aggregation and/or timing (i.e., some from $\boldsymbol{\omega}_{X,Y}$, $\boldsymbol{\gamma}_{j,Y}$, $\boldsymbol{\pi}_{i,j,Y}$, $\boldsymbol{\eta}_Y$), additional network states $\mathbf{W}_{X,Y}$, $\mathbf{C}_{j,Y}$, $\mathbf{P}_{i,j,Y}$, $\mathbf{H}_Y$ (*self-model states*, also called *reification states*) are introduced and connected to other states:

a) **Connectivity self-model**

- Self-model states $\mathbf{W}_{X,Y}$ are added representing connectivity characteristics, in particular connection weights $\boldsymbol{\omega}_{X,Y}$

b) **Aggregation self-model**

- Self-model states $\mathbf{C}_{j,Y}$ are added representing aggregation characteristics, in particular combination function weights $\boldsymbol{\gamma}_{j,Y}$
- Self-model states $\mathbf{P}_{i,j,Y}$ are added representing aggregation characteristics, in particular combination function parameters $\boldsymbol{\pi}_{i,j,Y}$

c) **Timing self-model**

- Self-model states $\mathbf{H}_Y$ are added representing timing characteristics, in particular speed factors $\boldsymbol{\eta}_Y$

The notations $\mathbf{W}_{X,Y}$, $\mathbf{C}_{j,Y}$, $\mathbf{P}_{i,j,Y}$, $\mathbf{H}_Y$ for the self-model states indicate the referencing relation with respect to the characteristics $\boldsymbol{\omega}_{X,Y}$, $\boldsymbol{\gamma}_{j,Y}$, $\boldsymbol{\pi}_{i,j,Y}$, $\boldsymbol{\eta}_Y$: here $\mathbf{W}$ refers to $\boldsymbol{\omega}$, $\mathbf{C}$ refers to $\boldsymbol{\gamma}$, $\mathbf{P}$ refers to $\boldsymbol{\pi}$, and $\mathbf{H}$ refers to $\boldsymbol{\eta}$, respectively. For the processing, these self-model states define the dynamics of any state Y in a canonical manner according to Eq. (1) and (2) whereby the values of $\boldsymbol{\omega}_{X,Y}$, $\boldsymbol{\gamma}_{j,Y}$, $\boldsymbol{\pi}_{i,j,Y}$, $\boldsymbol{\eta}_Y$ are replaced by the state values of $\mathbf{W}_{X,Y}$, $\mathbf{C}_{j,Y}$, $\mathbf{P}_{i,j,Y}$, $\mathbf{H}_Y$ at time t, respectively.

An example of an aggregation self-model state $\mathbf{P}_{i,j,Y}$ for a combination function parameter $\boldsymbol{\pi}_{i,j,Y}$ is for the excitability threshold $\boldsymbol{\tau}_Y$ of state Y, which is the second parameter of a logistic sum combination function; then $\mathbf{P}_{i,j,Y}$ is usually indicated by $\mathbf{T}_Y$, where $\mathbf{T}$ refers to $\boldsymbol{\tau}$. The network constructed by adding a self-model to a base network is called a *self-modeling network* or a *reified network* for this base network. This constructed network is also a temporal-causal network model itself, as has been shown (see [35], Ch. 10); for this reason, this construction can easily be applied iteratively to obtain multiple levels or orders of self-models, in which case the resulting network is called a *multi-order* or *higher-order self-modeling network*.

4 The Designed Network Models

Using the data discussed in Sect. 2, we designed a specific library to create multilevel adaptive models to understand how multiple EIA actors interact and identify organizations, initiatives, and key individuals that significantly influence [36] the general EIA process. These analyses will allow understanding of network connectivity, where there are critical network hubs (e.g., organizations/persons that bridge the network and increase the network's overall alignment and extend its reach and influence), and opportunities to strengthen connections.

The results of this statistical approach will be contrasted with the technical quality of the EIA reports and at each stage of the procedure to highlight best practice examples. In the next subsections, we present two nonadaptive models to compare the EU EIA Directive indications for the collaboration of stakeholders in the whole process (considering EIA stages mentioned in Sect. 2.2) and what happens most of the time when it comes to public consultation. Furthermore, we compute a model considering an adaptive scenario – Collaboration considering the Procedural aspects. Figure 2 shows the collaboration model considered for analysis (see Table 2 for the legend and Appendix Sect. 7 for a full specification of the adaptive model in role matrix format).

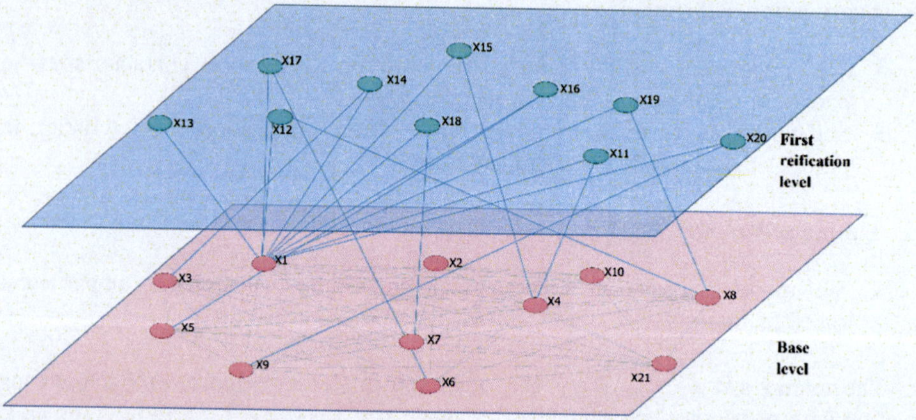

Fig. 2. Model in EIA process included in this study (Base level and First reification level)

5 Simulation Results

In this section, some of our simulation experiments are discussed.

5.1 Nonadaptive EIA Models

The simulation results presented in Fig. 3 show a proactive behavior of the actors' involvement in the EIA stages, in which everyone knows their role and intervenes in the stage in which they must, considering the EU Directive. This type of collaboration is less and less common in the EIA process because the public consultations are not made known and publicized in time so that the stages are passed quickly until the issuance of the environmental permit for the implementation of the proposed project. In this model, we can also see as variable the final agreement, which in Fig. 4 is missing due to lack of communication, existing conflicts, and various interests other than the quality of the environment.

Table 2. Overview of the states and their explanation.

State	Explanation	
D	Developer	Base Level
PA	Other Public Authorities	
CE	Competent experts	
NGO	Non-governmental organizations	
ME	Media	
R	Researchers	
LP	Local People	
TP	Third party audit	
OP	Other practitioners/interested parties	
FinA	Final Approval	
$W_{NGO,D}$	Self-model state for the weight of the connection from NGO to D	Self-Model Level
$W_{ME,D}$	Self-model state for the weight of the connection from ME to D	
$W_{D,PA}$	Self-model state for the weight of the connection from D to PA	
$W_{CE,PA}$	Self-model state for the weight of the connection from CE to PA	
$W_{NGO,PA}$	Self-model state for the weight of the connection from NGO to PA	
$W_{OP,PA}$	Self-model state for the weight of the connection from OP to PA	
$W_{R,PA}$	Self-model state for the weight of the connection from R to PA	
$W_{LP,PA}$	Self-model state for the weight of the connection from LP to PA	
$W_{ME,PA}$	Self-model state for the weight of the connection from ME to PA	
$W_{TP,PA}$	Self-model state for the weight of the connection from TP to PA	
ApEx	Approval Expected	Base Level

This lack of EIA stakeholder involvement can be explained by disinterest on the one hand but can be justified at the same time by the shorter duration of the whole process (all EIA stages) and also by reduced transparency and lack of provision of information by the public authorities and the developer (Figs. 3 and 4).

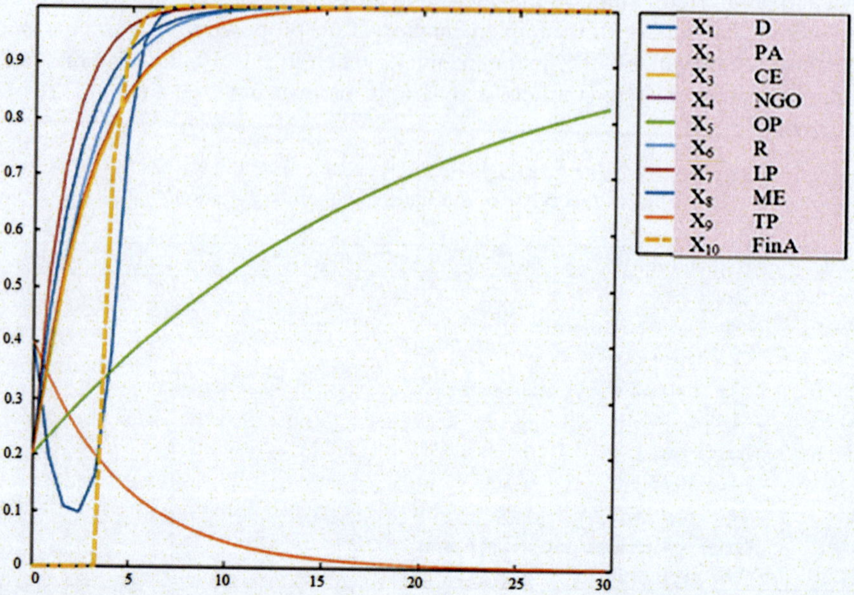

Fig. 3. Nonadaptive model - Scenario 1 – Developer/Beneficiary and Public Authority (decision making) influenced by all stakeholders (Legend in Table 2)

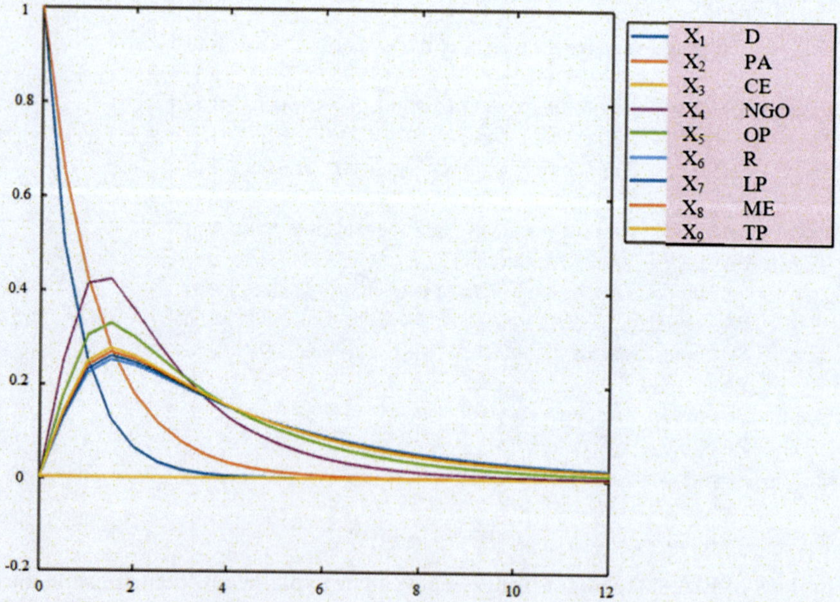

Fig. 4. Nonadaptive model - Scenario 2 – Low collaboration between Developer/Beneficiary and Public Authority and Stakeholders

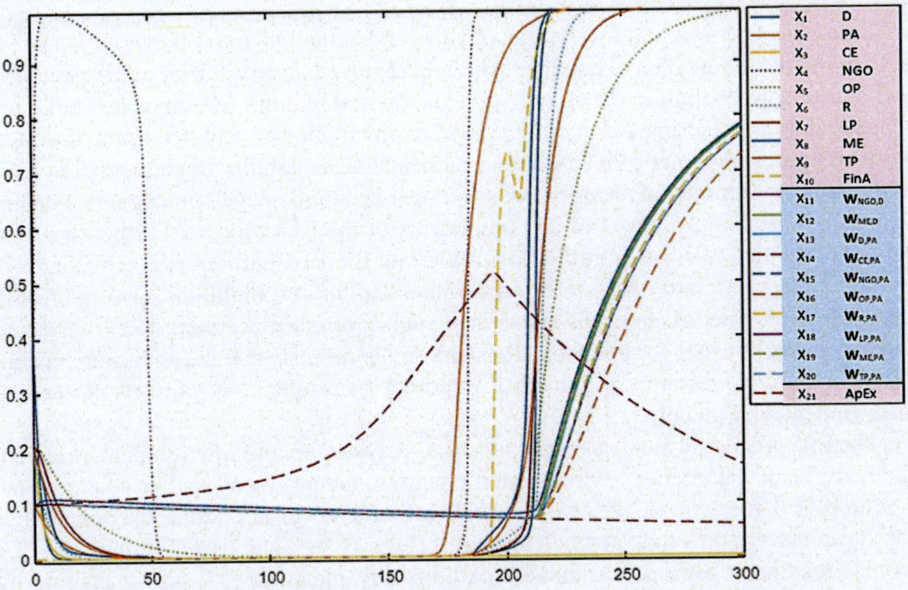

Fig. 5. Adaptive scenario – collaboration considering the procedural aspects

5.2 Actor Dominance Scenario

Figure 5 illustrates a computational EIA collaboration model where each actor influences the final agreement/decision-making process (authorize or not the implementation of the proposed project considering its negative impact on the environment). Here we highlight a dominance scenario in which simulations were performed by making each EIA stakeholder dominant within the group by increasing their connection weight.

6 Discussion and Future Work

The present research explored the patterns of collaboration between the stakeholders involved in the EIA process. Our results could be helpful and inform environmental policymakers by describing EIA collaboration patterns, weaknesses of the network, opportunities for greater involvement of key individuals and perspectives of building bridging communication hubs to avoid or reduce environmental conflicts.

From the economic and social point of view, understanding the collaboration process in EIA and the main types of conflicts will help remove barriers in EIA implementation and sustainable economic development [11]. Understanding how these networks are shaped while considering EU and transposed national policies will determine how network structures influence efficacy in environmental sustainability implementation [36]. Integrating environmental change with social and economic development is mandatory to ensure a positive evolution of all components of the EIA process. Furthermore, the paper contributes to **raising public awareness** on the EIA participatory. Our findings could be integrated into future policy recommendations and highlight opportunities to share best EIA practices between researchers, public authorities, interested groups, and experts. These avenues for outreach are based on the premise that the environment constitutes a common resource and that it is important to cooperate towards its protection, management, and planning.

The EIA procedure must enable connections between people, environment, research, and laws, leading to better environmental planning, management, and protection. Furthermore, it is fundamental for EIA stakeholders to change the traditional approaches based exclusively on single actors/institutions (Figs. 3, 4, 5 and 6) and rethink the EIA process leading it towards sustainable participatory planning [37] and managing by learning from past experiences, best practices and public attitudes.

Acknowledgment. AN and LR were supported by a grant of the Romanian National Authority for Scientific Research (https://uefiscdi.gov.ro), PN-III-P1-1.1-TE-2019-1039.

Appendix: Specification of the Adaptive Model by Role Matrices

In this section, the full specification of the model is provided in terms of role matrices, which is the standard format used for design of a model and which can also be used as input for the software environment. Each role matrix has three color blocks that match the colors in the graphical representation used earlier: pink for the base level, blue for the first-order self-model states and purple for the second-order self-model states.

Connectivity Characteristics Role Matrices

The first role matrix **mb** (see Table 3) represents the base connections between all the states, as presented in the graphical representation in Fig. 1. For example, row 4 with state NGO (also indicated by X_4) has five incoming connections from X_4, X_1, X_2, X_7, X_9, which are NGO itself, D, PA, LP, and ME. Role matrix **mcw** (see Table 4) shows the weights **ω** of the connections presented in role matrix **mb**. There is a difference in nonadaptive (green) and adaptive connections (peach-red). For example, in row 2 for state PA with ten incoming connections, there are two non-adaptive connection weights with vales (0.5 for the connection from X_1 and 1 for the connection from X_{21}, as can be seen in role matrix **mb**) and 8 adaptive connection weights. The latter 8 ones are represented by states X_{13} (i.e., $\mathbf{W}_{D,PA}$, a first-order self-model state), and similarly, the other self-model states X_{14} to X_{20}.

Table 3. Role matrix **mb** for base connectivity.

mb	base connectivity	1	2	3	4	5	6	7	8	9	10
X_1	D	X_1	X_2	X_3	X_4	X_5	X_6	X_7	X_8	X_9	
X_2	PA	X_2	X_1	X_3	X_4	X_5	X_6	X_7	X_8	X_9	X_{21}
X_3	CE	X_3	X_1								
X_4	NGO	X_4	X_1	X_2	X_7	X_9					
X_5	OP	X_5	X_1	X_2	X_6	X_7	X_8				
X_6	R	X_6	X_1	X_2							
X_7	LP	X_7	X_1	X_2	X_6	X_8					
X_8	ME	X_8	X_1	X_2	X_3	X_4	X_5	X_6	X_7		
X_9	TP	X_9	X_1	X_2							
X_{10}	FinA	X_1	X_2	X_3	X_4	X_5	X_6	X_7	X_8	X_9	X_{21}
X_{11}	$\mathbf{W}_{NGO,D}$	X_4	X_1	X_{11}							
X_{12}	$\mathbf{W}_{ME,D}$	X_8	X_1	X_{12}							
X_{13}	$\mathbf{W}_{D,PA}$	X_1	X_1	X_{13}							
X_{14}	$\mathbf{W}_{CE,PA}$	X_3	X_1	X_{14}							
X_{15}	$\mathbf{W}_{NGO,PA}$	X_4	X_1	X_{15}							
X_{16}	$\mathbf{W}_{OP,PA}$	X_5	X_1	X_{16}							
X_{17}	$\mathbf{W}_{R,PA}$	X_6	X_1	X_{17}							
X_{18}	$\mathbf{W}_{LP,PA}$	X_7	X_1	X_{18}							
X_{19}	$\mathbf{W}_{ME,PA}$	X_8	X_1	X_{19}							
X_{20}	$\mathbf{W}_{TP,PA}$	X_9	X_1	X_{20}							
X_{21}	ApEx	X_{21}	X_{10}								

Aggregation Characteristics Role Matrices

Role matrix **mcfw** (see Table 5) specifies which combination function is used for each state. For example, states X_1 to X_{10} and X_{21} use the combination function alogistic, which is indicated by a combination function weight $\boldsymbol{\gamma}$ of 1 in the first column. Role matrix **mcfp** (see Table 6) defines the exact parameters used for each state and function.

Table 4. Role matrix **mcw** for base connectivity.

mcw connection weights		**1**	**2**	**3**	**4**	**5**	**6**	**7**	**8**	**9**	**10**
X_1	D	1	0.5	0.5	X_{11}	0.2	0.5	0.2	X_{12}	0.2	
X_2	PA	X_{13}	0.5	X_{14}	X_{15}	X_{16}	X_{17}	X_{18}	X_{19}	X_{20}	1
X_3	CE	1	0.5								
X_4	NGO	1	1	1	1	1					
X_5	OP	1	0.4	0.4	0.4	0.4	0.4				
X_6	R	1	0.4	0.4							
X_7	LP	1	0.3	0.3	0.3	1					
X_8	ME	1	0.2	0.2	0.2	0.2	0.2	0.2	0.2		
X_9	TP	1	0.5	0.5							
X_{10}	FinA	1	1	1	1	1	1	1	1	1	1
X_{11}	$\mathbf{W}_{NGO,D}$	1	1	1							
X_{12}	$\mathbf{W}_{ME,D}$	1	1	1							
X_{13}	$\mathbf{W}_{D,PA}$	1	1	1							
X_{14}	$\mathbf{W}_{CE,PA}$	1	1	1							
X_{15}	$\mathbf{W}_{NGO,PA}$	1	1	1							
X_{16}	$\mathbf{W}_{OP,PA}$	1	1	1							
X_{17}	$\mathbf{W}_{R,PA}$	1	1	1							
X_{18}	$\mathbf{W}_{LP,PA}$	1	1	1							
X_{19}	$\mathbf{W}_{ME,PA}$	1	1	1							
X_{20}	$\mathbf{W}_{TP,PA}$	1	1	1							
X_{21}	ApEx	1	-1								

Table 5. Role matrix **mcfw** for combination function weights.

mcfw combination function weights		**1 alogistic**	**2 hebb**
X_1	D	1	
X_2	PA	1	
X_3	CE	1	
X_4	NGO	1	
X_5	OP	1	
X_6	R	1	
X_7	LP	1	
X_8	ME	1	
X_9	TP	1	
X_{10}	FinA	1	
X_{11}	$\mathbf{W}_{NGO,D}$		1
X_{12}	$\mathbf{W}_{ME,D}$		1
X_{13}	$\mathbf{W}_{D,PA}$		1
X_{14}	$\mathbf{W}_{CE,PA}$		1
X_{15}	$\mathbf{W}_{NGO,PA}$		1
X_{16}	$\mathbf{W}_{OP,PA}$		1
X_{17}	$\mathbf{W}_{R,PA}$		1
X_{18}	$\mathbf{W}_{LP,PA}$		1
X_{19}	$\mathbf{W}_{ME,PA}$		1
X_{20}	$\mathbf{W}_{TP,PA}$		1
X_{21}	ApEx	1	

Table 6. Role matrix **mcfp** for combination function weights.

mcfp combination function parameters		**1 alogistic**		**2 hebb**	
		1 σ	**2** τ	**1** μ	**2**
X_1	D	10	1		
X_2	PA	10	4.2		
X_3	CE	10	1.4		
X_4	NGO	10	0.7		
X_5	OP	10	0.7		
X_6	R	10	0.7		
X_7	LP	10	0.7		
X_8	ME	10	0.7		
X_9	TP	10	0.7		
X_{10}	FinA	30	7		
X_{11}	$\mathbf{W}_{NGO,D}$			0.9	
X_{12}	$\mathbf{W}_{ME,D}$			0.9	
X_{13}	$\mathbf{W}_{D,PA}$			0.9	
X_{14}	$\mathbf{W}_{CE,PA}$			0.9	
X_{15}	$\mathbf{W}_{NGO,PA}$			0.9	
X_{16}	$\mathbf{W}_{OP,PA}$			0.9	
X_{17}	$\mathbf{W}_{R,PA}$			0.9	
X_{18}	$\mathbf{W}_{LP,PA}$			0.9	
X_{19}	$\mathbf{W}_{ME,PA}$			0.9	
X_{20}	$\mathbf{W}_{TP,PA}$			0.9	
X_{21}	ApEx	18	0.2		

Timing Characteristics Role Matrix and Initial Values

The speed factors $\boldsymbol{\eta}_Y$ are specified in role matrix **ms**; see Table 7.

Table 7. Role matrix **ms** for speed factors.

ms speed factors		**1**
X_1	D	0.5
X_2	PA	0.1
X_3	CE	0.3
X_4	NGO	0.8
X_5	OP	0.05
X_6	R	0.2
X_7	LP	0.1
X_8	ME	0.4
X_9	TP	0.1
X_{10}	FinA	0.2
X_{11}	$\mathbf{W}_{NGO,D}$	0.02
X_{12}	$\mathbf{W}_{ME,D}$	0.02
X_{13}	$\mathbf{W}_{D,PA}$	0.02
X_{14}	$\mathbf{W}_{CE,PA}$	0.02
X_{15}	$\mathbf{W}_{NGO,PA}$	0.02
X_{16}	$\mathbf{W}_{OP,PA}$	0.02
X_{17}	$\mathbf{W}_{R,PA}$	0.02
X_{18}	$\mathbf{W}_{LP,PA}$	0.02
X_{19}	$\mathbf{W}_{ME,PA}$	0.02
X_{20}	$\mathbf{W}_{TP,PA}$	0.02
X_{21}	ApEx	0.01

References

1. Gavrilidis, A.-A., Nita, A., Niculae, M.-I.: Assessing the potential conflict occurrence due to metropolitan transportation planning: a proposed quantitative approach. Sustainability **12** (2020). https://doi.org/10.3390/su12020527
2. Toro, J., Requena, I., Duarte, O., Zamorano, M.: A qualitative method proposal to improve environmental impact assessment. Environ. Impact Assess. Rev. **43**, 9–20 (2013). https://doi.org/10.1016/j.eiar.2013.04.004
3. Kolhoff, A.J., Driessen, P.P.J., Runhaar, H.A.C.: Overcoming low EIA performance - a diagnostic tool for the deliberate development of EIA system capacities in low and middle income countries. Environ. Impact Assess. Rev. **68**, 98–108 (2018). https://doi.org/10.1016/j.eiar.2017.11.001

4. Dangi, M.B., Fernandez, D., Born, U.B., Belbase, S., Kaphle, R.: Evaluation of environmental impact assessment report preparation and public participation in landfill projects in Nepal. Habitat Int. **46**, 72–81 (2015). https://doi.org/10.1016/j.habitatint.2014.10.021
5. Hasan, M.A., Nahiduzzaman, K.M., Aldosary, A.S.: Public participation in EIA: a comparative study of the projects run by government and non-governmental organizations. Environ. Impact Assess. Rev. **72**, 12–24 (2018). https://doi.org/10.1016/j.eiar.2018.05.001
6. Niţă, A., Buttler, A., Rozylowicz, L., Pătru-Stupariu, I.: Perception and use of landscape concepts in the procedure of environmental impact assessment: case study—Switzerland and Romania. Land Policy **44**, 145–152 (2015). https://doi.org/10.1016/j.landusepol.2014.12.006
7. Berardo, R., Alcaniz, I., Hadden, J., Jasny, L.: Policy networks and environmental governance. In: Victor, J.N., Montgomery, A.H., Lubell, M. (eds.) The Oxford Handbook of Political Networks. Oxford University Press, Oxford (2016)
8. European Commission. Environmental Impact Assessment of Projects: Rulings of the Court of Justice, 72 (2013)
9. Khosravi, F., Jha-Thakur, U., Fischer, T.B.: Evaluation of the environmental impact assessment system in Iran. Environ. Impact Assess. Rev. **74**, 63–72 (2019). https://doi.org/10.1016/j.eiar.2018.10.005
10. Nita, A.: Empowering impact assessments knowledge and international research collaboration - a bibliometric analysis of Environmental Impact Assessment Review journal. Environ. Impact Assess. Rev. **78**, 106283 (2019). https://doi.org/10.1016/j.eiar.2019.106283
11. Bratman, E., Dias, C.B.: Development blind spots and environmental impact assessment: tensions between policy, law and practice in Brazil's Xingu river basin. Environ. Impact Assess. Rev. **70**, 1 (2018). https://doi.org/10.1016/j.eiar.2018.02.001
12. Li, W., Liu, J., Li, D.: Getting their voices heard: three cases of public participation in environmental protection in China. J. Environ. Manage. **98**, 65–72 (2012). https://doi.org/10.1016/j.jenvman.2011.12.019
13. Nita, A., Ciocanea, C.M., Manolache, S., Rozylowicz, L.: A network approach for understanding opportunities and barriers to effective public participation in the management of protected areas. Soc. Netw. Anal. Min. **8**(1), 1–11 (2018). https://doi.org/10.1007/s13278-018-0509-y
14. Jha-Thakur, U., Fischer, T.B.: 25years of the UK EIA system: strengths, weaknesses, opportunities and threats. Environ. Impact Assess. Rev. **61**, 19–26 (2016). https://doi.org/10.1016/j.eiar.2016.06.005
15. Simpson, N.P., Basta, C.: Sufficiently capable for effective participation in environmental impact assessment? Environ. Impact Assess. Rev. **70**, 57–70 (2018). https://doi.org/10.1016/j.eiar.2018.03.004
16. Hodgson, I.D., Redpath, S.M., Fischer, A., Young, J.: Who knows best? Understanding the use of research-based knowledge in conservation conflicts. J. Environ. Manage. **231**, 1065–1075 (2019). https://doi.org/10.1016/j.jenvman.2018.09.023
17. Treur, J.: Dynamic modeling based on a temporal–causal network modeling approach. Biol. Inspired Cogn. Archit. **16**, 131–168 (2016). https://doi.org/10.1016/j.bica.2016.02.002
18. Wood, G., Glasson, J., Becker, J.: EIA scoping in England and Wales: practitioner approaches, perspectives and constraints. Environ. Impact Assess. Rev. **26**, 221–241 (2006). https://doi.org/10.1016/j.eiar.2005.02.001
19. Commu, C., Theelen, M., Treur, J.: Modeling enabling learning of social interaction based on an adaptive temporal-causal network model. Neurocomputing **338**, 349–360 (2019). https://doi.org/10.1016/j.neucom.2018.05.127
20. Llopis-Albert, C., Palacios-Marques, D., Soto-Acosta, P.: Decision-making and stakeholders' constructive participation in environmental projects. J. Bus. Res. **68**, 1641–1644 (2015). https://doi.org/10.1016/j.jbusres.2015.02.010

21. Hossu, C.A., Ioja, I.C., Susskind, L.E., Badiu, D.L., Hersperger, A.M.: Factors driving collaboration in natural resource conflict management: evidence from Romania. Ambio **47**(7), 816–830 (2018). https://doi.org/10.1007/s13280-018-1016-0
22. Crossley, N., Bellotti, E., Edwards, G., Koskinen, J., Tranmer, M., Everett, M.: Social network analysis for ego-nets, p. 208 (2015)
23. Barabási, A.-L.: The Formula: The Science Behind Why People Succeed or Fail. Macmillan (2018)
24. Bodin, O.: Collaborative environmental governance: achieving collective action in social-ecological systems. Science **357**, eaan1114 (2017). https://doi.org/10.1126/science.aan1114
25. Armitage, D.R.: Collaborative environmental assessment in the Northwest Territories, Canada. Environ. Impact Assess. Rev. **25**, 239–258. (2005). https://doi.org/10.1016/j.eiar.2004.06.012
26. Borgatti, S.P., Everett, M.G.: Network analysis of 2-mode data. Soc. Netw. **19**, 243–269 (1997). https://doi.org/10.1016/S0378-8733(96)00301-2
27. Brandes, U., Erlebach, T.: Network Analysis - Methodological Foundations – Introduction, 1st edn. LNCS, vol. 3418 (2005). https://doi.org/10.1007/b106453
28. Bodin, O., et al.: Theorizing benefits and constraints in collaborative environmental governance: a transdisciplinary social-ecological network approach for empirical investigations. Ecol. Soc. **21**, 40 (2016). https://doi.org/10.5751/Es-08368-210140
29. Bodin, Ö., et al.: Improving network approaches to the study of complex social–ecological interdependencies. Nat. Sustain. **2**, 551–559 (2019). https://doi.org/10.1038/s41893-019-0308-0
30. Manolache, S., et al.: Governance networks around grasslands with contrasting management history. J. Environ. Manage. **273**, 111152 (2020). https://doi.org/10.1016/j.jenvman.2020.111152
31. Bonacich, P.: Power and centrality - a family of measures. Am. J. Sociol. **92**, 1170–1182 (1987). https://doi.org/10.1086/228631
32. Abbasi, A., Hossain, L., Leydesdorff, L.: Betweenness centrality as a driver of preferential attachment in the evolution of research collaboration networks. J. Informetr. **6**, 403–412 (2012). https://doi.org/10.1016/j.joi.2012.01.002
33. Gao, Z., Liu, R., Ullah, N.: A temporal-causal network model for age and gender difference in choice of emotion regulation strategies. In: Nguyen, N.T, Chbeir, R., Exposito, E., Aniorté, P., Trawiński, B. (eds.) ICCCI 2019. LNCS, vol. 11683, pp. 106–117. Springer, Cham (2019). https://doi.org/10.1007/978-3-030-28377-3_9
34. Ullah, N., Gao, Z., Liu, R., Treur, J.: A second-order adaptive temporal-causal network model for age and gender differences in evolving choice of emotion regulation strategies. J. Inf. Telecommun. **4**, 213–228 (2020). https://doi.org/10.1080/24751839.2020.1724738
35. Treur, J.: Network-Oriented Modeling for Adaptive Networks: Designing Higher-Order Adaptive Biological, Mental and Social Network Models. Studies in Systems, Decision and Control, pp. 3–24. Springer, Cham (2020). https://doi.org/10.1007/978-3-030-31445-3
36. Mitincu, C.-G., et al.: Licensing sustainability related aspects in Strategic Environmental Assessment. Evidence from Romania's urban areas. Land Policy **108** (2021). https://doi.org/10.1016/j.landusepol.2021.105572
37. Gavrilidis, A.A., Niţă, M.R., Onose, D.A., Badiu, D.L., Năstase, I.I.: Methodological framework for urban sprawl control through sustainable planning of urban green infrastructure. Ecol. Indic. **96**, 67–78 (2019). https://doi.org/10.1016/j.ecolind.2017.10.054

Modeling Context-Sensitive Metacognitive Control of Focusing on a Mental Model During a Mental Process

Gülay Canbaloğlu[1] and Jan Treur[2(✉)]

[1] Department of Computer Engineering, Koç University, Istanbul, Turkey
gcanbaloglu17@ku.edu

[2] Social AI Group, Department of Computer Science, Vrije Universiteit Amsterdam, Amsterdam, The Netherlands
j.treur@vu.nl

Abstract. Focusing on a proper mental model during mental processes is often crucial. Metacognition is used to control such focusing in a context-sensitive manner. In this paper, a second-order adaptive mental network model is introduced for this form of metacognitive control. The second-order adaptive network model obtained is illustrated by a case scenario concerning social interaction.

1 Introduction

Metacognition (Darling-Hammond et al. 2008; Shannon 2008; Mahdavi 2014; Flavell 1979; Koriat 2007; Pintrich 2000) is a form of cognition about cognition. In Koriat (2007) it is described as what people know about their own cognitive processes and how they put that knowledge to use in regulating their cognitive processing and behavior. A sometimes used closely related term is self-regulation and when the cognitive processes addressed by metacognition concern learning, the term self-regulated learning is used. For example, in Pintrich (2000), self-regulated learning is described as an active, constructive process whereby learners set goals for their learning and then attempt to monitor, regulate, and control their cognition, motivation, and behavior, guided and constrained by their goals and context.

Mental processes often use different mental models; e.g., (Gentner and Stevens 1983; Greca and Moreira 2000; Skemp 1971; Seel 2006). An important metacognitive control decision is the decision to focus on a specific mental model.

From a network-oriented modeling perspective, change of mental models is usually described by adaptive mental (or neural) network models, where some of the network characteristics such as connection weights or excitability thresholds change over time. If, in addition, metacognition is used to regulate or control the change process, this implies that the adaptation of the mental network is itself adaptive as well, which is called second-order adaptation. Thus, a network model for these processes has to address such complex structures and behaviour; e.g., (Bhalwankar and Treur 2021; Van Ments and Treur 2021; Treur and Van Ments 2022). Following this line in the current paper, using the modeling approach for higher-order adaptive networks from (Treur 2018, 2020a, b),

R. Silhavy et al. (Eds.): CoMeSySo 2021, LNNS 231, pp. 992–1009, 2021.
https://doi.org/10.1007/978-3-030-90321-3_82

a second-order adaptive mental network model is introduced for metacognitive control over the choice to focus on a specific mental model.

In this paper, first in Sect. 2 more background knowledge is discussed on metacognition and its role in controlling mental processes. In Sect. 3 the network-oriented modeling approach used is briefly explained. Next, in Sect. 4 the introduced second-order adaptive network model is described in some detail. In Sect. 5, it is shown how this model was used to perform simulations for the illustrative example scenario. Finally, Sect. 6 is a discussion and Sect. A is an Appendix with the full specification of the introduced adaptive network model.

2 Metacognitive Control of Mental Processes

Literature on metacognition, sometimes also called self-regulation, can be found, for example in (Darling-Hammond et al. 2008; Shannon 2008; Mahdavi 2014; Flavell 1979; Koriat 2007; Pintrich 2000). The focus is here on the role of metacognition in the context of learning. For example, in Pintrich (2000, pp. 452–453) the following assumptions for self-regulated learning are described by 'It is a process whereby learners set goals for their learning and monitor and control their cognition, motivation, and behavior guided by these goals', 'Learners can monitor, control, and regulate certain aspects of their own cognition, motivation, and behavior, and some elements of their environment.' and 'Self-regulatory activities are mediators between personal and contextual characteristics and actual performance'. In Koriat (2007, p. 290), metacognition is described by what people know about cognition and in particular their own cognitive processes, and how they use that in regulating their cognitive processes and behavior. Some of the assumptions mentioned there are:

- Self-controlled cognitive processes have measurable effects on behavior (Koriat 2007), pp. 292–293
- Feelings, such as the feeling of knowing are part of monitoring, and exert a causal role on the control of cognitive processing (Koriat 2007), p. 293, p. 314–315
- There is a causal relation from monitoring to control (Koriat 2007), p. 315

So, in both descriptions of Pintrich (2000) and Koriat (2007) on metacognition (as well as in most other literature on metacognition), monitoring and control of the own cognitive processes are central concepts (where Koriat also emphasizes the feeling or experiencing that comes together with monitoring). These processes work through a causal cycle where the own cognitive processes affect the metacognitive monitoring, this monitoring in turn affects the metacognitive control, and this control affects the own cognitive processes.

3 Higher-Order Adaptive Network Models

In this section, the network-oriented modeling approach used is briefly introduced. Following (Treur 2016, 2020b), a temporal-causal network model is characterised by (here X and Y denote nodes of the network, also called states):

- *Connectivity characteristics*
  Connections from a state X to a state Y and their weights $\omega_{X,Y}$
- *Aggregation characteristics*
  For any state Y, some combination function $\mathbf{c}_Y(..)$ defines the aggregation that is applied to the impacts $\omega_{X,Y}X(t)$ on Y from its incoming connections from states X
- *Timing characteristics*
  Each state Y has a speed factor η_Y defining how fast it changes for given causal impact.

The following difference (or related differential) equations that are used for simulation purposes and also for analysis of temporal-causal networks, incorporate these network characteristics $\omega_{X,Y}$, $\mathbf{c}_Y(..)$, η_Y in a standard numerical format:

$$Y(t + \Delta t) = Y(t) + \eta_Y[\mathbf{c}_Y(\omega_{X_1,Y}X_1(t), \ldots, \omega_{X_k,Y}X_k(t)) - Y(t)]\Delta t \quad (1)$$

for any state Y and where X_1 to X_k are the states from which Y gets its incoming connections. Within the software environment described in (Treur 2020b, Ch. 9), a large number of around 50 useful basic combination functions are included in a combination function library.

The above concepts enable to design network models and their dynamics in a declarative manner, based on mathematically defined functions and relations. Realistic network models are usually adaptive: often not only their states but also some of their network characteristics change over time. By using a *self-modeling network* (also called a *reified* network), a similar network-oriented conceptualisation can also be applied to *adaptive* networks to obtain a declarative description using mathematically defined functions and relations for them as well; see (Treur 2018, 2020a, b). This works through the addition of new states to the network (called *self-model states*) which represent (adaptive) network characteristics. In the graphical 3D-format as shown in Sect. 4, such additional states are depicted at a next level (called *self-model level* or *reification level*), where the original network is at the *base level*. As an example, the weight $\omega_{X,Y}$ of a connection from state X to state Y can be represented (at a next self-model level) by a self-model state named $\mathbf{W}_{X,Y}$. Similarly, all other network characteristics from $\omega_{X,Y}$, $\mathbf{c}_Y(..)$, η_Y can be made adaptive by including self-model states for them. For example, an adaptive speed factor η_Y can be represented by a self-model state named $\mathbf{H}_Y$ and an adaptive excitability threshold parameter τ_Y can be represented by a self-model state named $\mathbf{T}_Y$. Moreover, a persistence factor μ of a state Y of used for adaptation can be represented by a self-model state $\mathbf{M}_Y$.

As the outcome of such a process of network reification is also a temporal-causal network model itself, as has been shown in (Treur 2020b, Ch 10), this self-modeling network construction can easily be applied iteratively to obtain multiple orders of self-models at multiple (first-order, second-order, …) self-model levels. In the current paper, this multi-level self-modeling network perspective will be applied to obtain a second-order adaptive mental network model addressing metacognitive control of adaptation as needed to focus on a specific mental model.

4 A Mental Network Model for Metacognitive Control of Focusing on a Mental Model

In this section, the adaptive mental network model for metacognitive control of focusing on a specific mental model is introduced. This adaptive mental network model follows a cognitive architecture that has processes at three levels (see Fig. 1); e.g., (Van Ments and Treur 2021):

- The base level network for the mental models used
- The first-order self-model level for the adaptation of the mental models
- The second-order self-model level for control by adaptation of the first-order network for adaptation

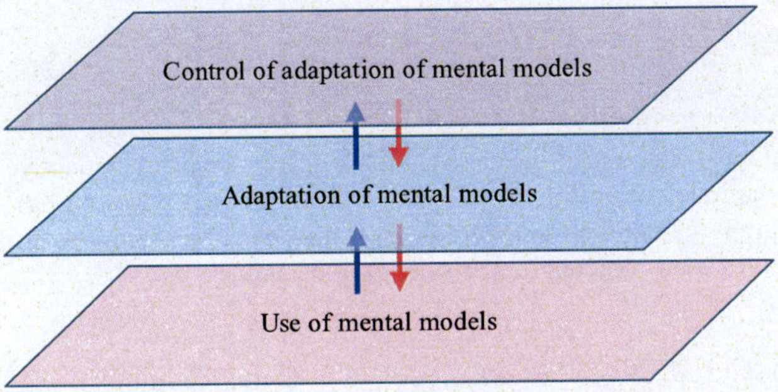

Fig. 1. Cognitive architecture for mental model handling with three levels of mental processing for mental models where each next level is modeled by relations one order higher than at the level below it

These three levels of mental processes have been modeled by a second-order adaptive self-modeling network model according to the modeling approach described in (Treur 2018, 2020a, b). The connectivity of this network model is depicted (according to three different views) in Figs. 2, 3 and 4. The states used are explained in Table 1. As a case study, in this second-order self-modeling network model, indirect and direct communication scenarios are modeled addressing the learning process of people trying to communicate with each other. Direct communication means the direct interaction between two people via asking a question and replying to it. On the other hand, indirect communication requires one or more intermediary people to interact with the target person. The scenario is constructed as a request for an appointment to focus on the understanding of the process. For indirect communication, in the example scenario person A asks the personal assistant PA of person B for an appointment. PA relays the request to B and receives the reply of B. Then PA transfers B's reply to A and gets A's notification receipt to let B know about this notification. In contrast, for direct communication, A and B interact with each other, therefore, there is no need for a personal assistant. A directly asks B for the appointment and can get the answer immediately.

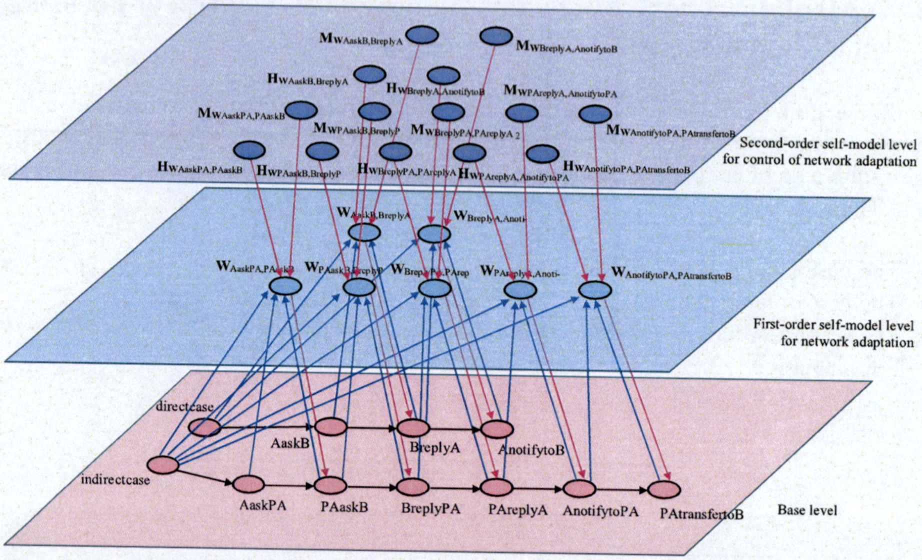

Fig. 2. Graphical representation of the connectivity of the second-order adaptive mental network model for metacognitive control of focusing on a mental model: view on base level and first-order self-modeling level and their interaction. Here the upward connections to the second-order self-model level have been depicted for the sake of transparency.

In the introduced adaptive network model, mental model connections are represented in an adaptive manner by first-order self-model states called **W**-states. Control of the adaptation uses second-order self-model states (called $\mathbf{H_W}$-states) for the adaptation speed of the **W**-states. A **W**-state can change when the corresponding $\mathbf{H_W}$-state representing the adaptation speed becomes nonzero; when this $\mathbf{H_W}$-state is 0, no change can take place. The $\mathbf{H_W}$-states can become nonzero in a context-sensitive manner because the context states have connections to them. This is how the context-sensitive metacognitive control over the change of a mental model takes place, thereby following the second-order adaptation principle 'Adaptation accelerates with increasing stimulus exposure' as formulated by Robinson et al. (2016).

The second-order self-model states called $\mathbf{M_W}$-states model the persistence of the mental model connections represented by the related **W**-states. If $\mathbf{M_W}$-states have activation level 1, there is full persistence, if they have level 0 there is no persistence at all. These second-order self-model states (together with the $\mathbf{H_W}$-states) model in a context-sensitive manner the 'Plasticity versus Stability Conundrum' described, for example, in Sjöström et al. (2008). They are used here to get rid of a mental model if the focus has to be shifted to another mental model and they also play a role in the Hebbian learning (and forgetting) that takes place during the use of a mental model for internal mental simulation (Hebb 1949).

Figure 2 depicts the view on the base level and first-order self-model level and the interactions between these two levels. It can be seen that from the context state for the indirect interaction case, by upward connections (in blue) the **W**-states for the mental model for the indirect case are activated (assuming that these **W**-states have a nonzero adaptation speed) and similarly for the direct case. Within the base level, the two context states also have connections to the first state of these mental models, so that these mental models are actually used by internal simulation.

Table 1. The states in the adaptive network model

Nr	State	Explanation	Level
X_2	PAaskB	Mental model state for personal assistant asks B for appointment for person A	
X_3	BreplyPA	Mental model state for person B replies to PA for appointment	
X_4	PAreplyA	Mental model state for personal assistant replies to person A for appointment	
X_5	AnotifytoPA	Mental model state for person A notifies receipt to personal assistant of B	Base level
X_6	PAtransfertoB	Mental model state for personal assistant transfers receipt of A to person B	
X_7	AaskB	Mental model state for person A directly asks person B for appointment	
X_8	BreplyA	Mental model state for person B directly replies to person A for appointment	
X_9	AnotifytoB	Mental model state for person A directly notifies receipt to person B	
X_{10}	indirect case	Context state for the indirect communication (option 1)	
X_{11}	direct case	Context state for the direct communication (option 2)	
X_{13}	$\mathbf{W}_{PAaskB,BreplyPA}$	Self-model state for speed factor for connection weight between X_2 and X_3	
X_{14}	$\mathbf{W}_{BreplyPA,PAreplyA}$	Self-model state for speed factor for connection weight between X_3 and X_4	First-order self-model level
X_{15}	$\mathbf{W}_{PAreplyA,AnotifytoPA}$	Self-model state for speed factor for connection weight between X_4 and X_5	
X_{16}	$\mathbf{W}_{AnotifytoPA,PAtransfertoB}$	Self-model state for speed factor for connection weight between X_5 and X_6	
X_{17}	$\mathbf{W}_{AaskB,BreplyA}$	Self-model state for speed factor for connection weight between X_7 and X_8	
X_{18}	$\mathbf{W}_{BreplyA,AnotifytoB}$	Self-model state for speed factor for connection weight between X_8 and X_9	
X_{20}	$\mathbf{H}_{\mathbf{W}_{PAaskB,BreplyPA}}$	Self-model state for speed factor for self-model state X_{13}	
X_{21}	$\mathbf{H}_{\mathbf{W}_{BreplyPA,PAreplyA}}$	Self-model state for speed factor for self-model state X_{14}	
X_{22}	$\mathbf{H}_{\mathbf{W}_{PAreplyA,AnotifytoPA}}$	Self-model state for speed factor for self-model state X_{15}	
X_{23}	$\mathbf{H}_{\mathbf{W}_{AnotifytoPA,PAtransfertoB}}$	Self-model state for speed factor for self-model state X_{16}	
X_{24}	$\mathbf{H}_{\mathbf{W}_{AaskB,BreplyPA}}$	Self-model state for speed factor for self-model state X_{17}	
X_{25}	$\mathbf{H}_{\mathbf{W}_{BreplyA,AnotifytoB}}$	Self-model state for speed factor for self-model state X_{18}	Second-order self-model level
X_{26}	$\mathbf{M}_{\mathbf{W}_{AaskPA,PAaskB}}$	Self-model state for persistence factor for self-model state X_{12}	
X_{27}	$\mathbf{M}_{\mathbf{W}_{PAaskB,BreplyPA}}$	Self-model state for persistence factor for self-model state X_{13}	
X_{28}	$\mathbf{M}_{\mathbf{W}_{BreplyPA,PAreplyA}}$	Self-model state for persistence factor for self-model state X_{14}	
X_{29}	$\mathbf{M}_{\mathbf{W}_{PAreplyA,AnotifytoPA}}$	Self-model state for persistence factor for self-model state X_{15}	
X_{30}	$\mathbf{M}_{\mathbf{W}_{AnotifytoPA,PAtransfertoB}}$	Self-model state for persistence factor for self-model state X_{16}	
X_{31}	$\mathbf{M}_{\mathbf{W}_{AaskB,BreplyPA}}$	Self-model state for persistence factor for self-model state X_{17}	
X_{32}	$\mathbf{M}_{\mathbf{W}_{BreplyA,AnotifytoB}}$	Self-model state for persistence factor for self-model state X_{18}	

Figure 3 shows the view on the interactions of the second-order self-model level with the other two levels for the $\mathbf{H_W}$-states. It can be seen here that by upward connections (in blue), the adaptation speed as represented by the $\mathbf{H_W}$-states depends on activation of the related mental model states at the base level and the related connection representation **W**-states at the first-order self-model level. This follows the second-order adaptation principle 'Adaptation accelerates with increasing stimulus exposure' (Robinson et al. 2016) mentioned above and makes that by the downward connections (in pink) from the $\mathbf{H_W}$-states to the **W**-states, these **W**-states indeed can change. All this depends on the context state that triggers activation of these specific base states and **W**-states for a mental model, which makes it a context-sensitive form of control of adaptation.

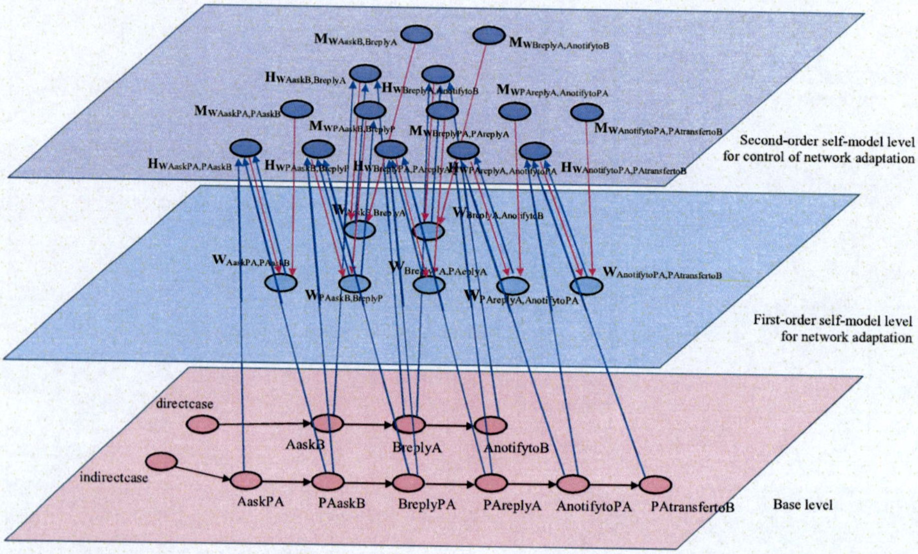

Fig. 3. Graphical representation of the connectivity of the second-order adaptive mental network model for metacognitive control of focusing on a mental model. View for the interactions of the second-order self-model level $\mathbf{H_W}$-states.

Figure 4 shows the view on the interactions of the second-order self-model level with the other levels in particular for the $\mathbf{M_W}$-states. The $\mathbf{M_W}$-states have values 1 when the related **W**-states needs to be fully persistent and low(er) values when these **W**-states should not be (fully) persistent. As mentioned above, this models in a context-sensitive manner the 'Plasticity versus Stability Conundrum' described, for example, in Sjöström et al. (2008). The context-dependence of this form of control is modeled specifically by the upward connections (in blue) from the two context states at the base level to their related $\mathbf{M_W}$-states.

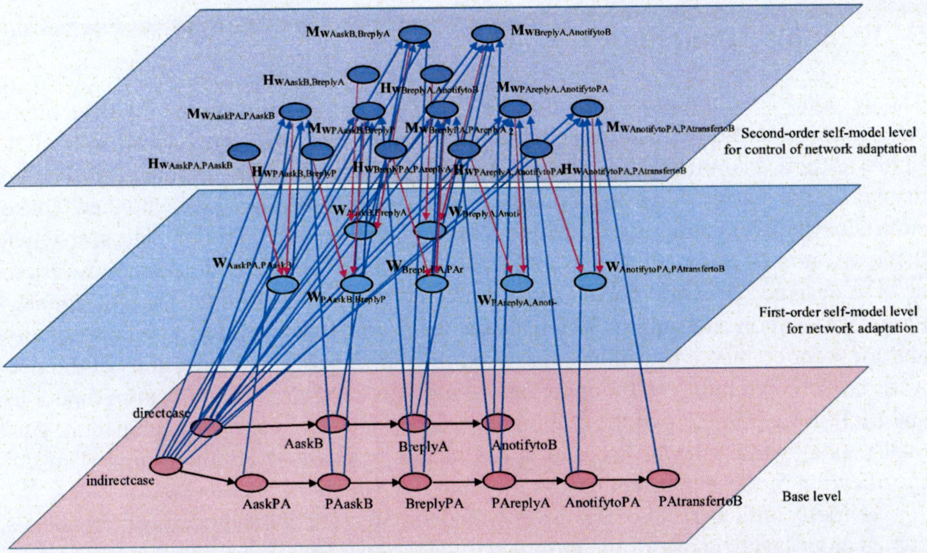

Fig. 4. Graphical representation of the connectivity of the second-order adaptive mental network model for metacognitive control of focusing on a mental model. View on the interactions of the second-order self-model level $\mathbf{M_W}$-states.

The combination functions from the library used in the introduced network model are defined as follows:

- The *advanced logistic sum combination function* $\mathbf{alogistic}_{\sigma,\tau}(V_1, \ldots, V_k)$ is defined by:

$$\mathbf{alogistic}_{\sigma,\tau}(V_1, \ldots, V_k) = \left[\frac{1}{1+e^{-\sigma(V_1+\cdots+V_k-\tau)}} - \frac{1}{1+e^{\sigma\tau}}\right](1+e^{-\sigma\tau}) \quad (2)$$

  where σ is a steepness parameter and $\tau_{\log}$ a threshold parameter and $V_1, \ldots, V_k$ are the impacts from the states from which the considered state Y gets incoming connections
- The *composed max-hebbian learning combination function* is defined by

$$\mathbf{maxhebb}_{\mu}(V_1, V_2, V_3, V_4, V_5) = \max(V_1 * V_2(1 - V_3) + \mu V_3, \max(V_4, V_5)) \quad (3)$$

  where μ is a persistence parameter, V_3 represents the weight of the connection, and V_1, V_2 are the activation levels of the connected states.
- The $\mathbf{steponce}_{\alpha,\beta}$(..) function is 1 between time points α and β and 0 else.

The full specification can be found in the Appendix Sect. A; see also the Linked Data at URL https://www.researchgate.net/publication/353667091.

5 Example Simulation Scenario

The introduced second-order self-modeling network model is illustrated for an example case concerning indirect and direct communication scenarios, as described in Sect. 4. In this section, a simulation example of the adaptive mental network model is discussed displaying the learning process of people trying to communicate with each other. Not so surprisingly, direct communication is much faster than indirect communication, so when that is an option it is not efficient to apply a mental model for indirection communication.

The designed network model can be characterized in terms of three aspects: connectivity, aggregation and timing. Two matrices for connectivity, **mb** for base connectivity and **mcw** for connection weights were used to characterize the connections of the network model. Two matrices for aggregation, **mcfw** for weight of combination functions and **mcfp** for parameters of the combination functions were used for aggregation. And finally, one matrix **ms** was used for speed factors to characterize the timing. The full specification of these matrices can be found in the Appendix Section A.

As briefly indicated in Sect. 4, the self-modeling feature of the modeling method puts higher-order levels above to the base level which includes all the simple non-adaptive states. In our model, base level states are the interaction states between people and two context states for switching between indirect and direct cases, beginning from X_1 to X_{11} inclusive.

In the base level, for the indirect case, there is a mental model of 6 states that can be seen as the steps of the communication, and for the mental model for the direct case there are 3. It indeed shows that indirect communication is a longer process.

At the first-order self-model level, there are 7 **W**-states for all connections of the two mental models shown at the base level both for the direct and indirect cases. This level makes the model adaptive in terms of the connections between base level states. For these **W**-states Hebbian learning is used to provide learning (and forgetting), and the **W**-states are changed to obtain focusing on them.

At the second-order self-model level, $\mathbf{H_W}$-states and $\mathbf{M_W}$-states are used to control the mentioned forms of adaptation itself in the first-order self-model level. In our model, there are 7 $\mathbf{H_W}$-states and 7 $\mathbf{M_W}$-states for all of the 7 **W**-states. The simulation outcomes depicted in the overall graphs in Fig. 5 and the partial graphs in Figs. 6, 7 and 8 show how the process runs. In each of the periods from 25 to 125 and from 200 to 300, the different mental model comes in focus. In both cases the mental model is not perfect as its connection weights are only 0.8. However, by using the mental model for internal mental simulation Hebbian learning (Hebb 1949) takes place by which the weights increase to 1 or close to one. The first context state finishes at time 125 and the next, other context state starts at time 200. In the meantime, the weights of the mental model that was first in focus drop to 0 (so, it is not in focus anymore) and instead, after time 200 the other mental model relating to the new context gets in focus.

The **steponce** function was used in our model to do the indirect-direct context case shift. In Figs. 5, 6, 7 and 8 the time interval 25–125 demonstrates the first case (indirect communication) and the time interval 200–300 demonstrates the second one (direct communication). As it can be seen, base states become higher in their level after the increase of the **W**-states because of the learning that takes place. Each different **W**-state affects its corresponding base states, thus, also in harmony with real life, within a mental model the represented communication steps do not trace the same route, they follow each other with a small time difference. Learning is a process, it does not happen all of a sudden.

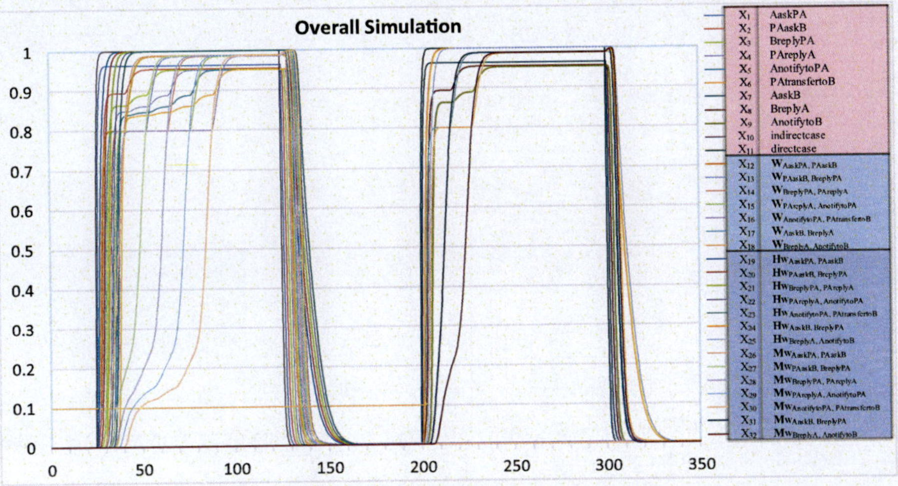

Fig. 5. Overall picture of all states in the simulation.

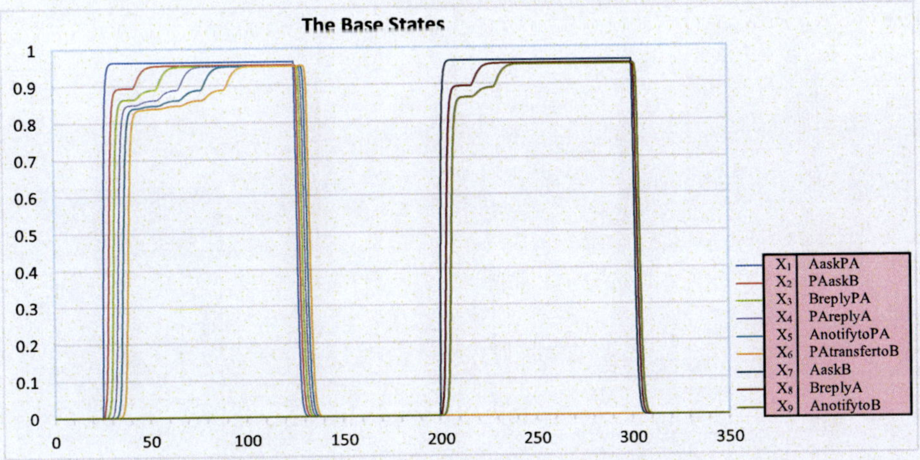

Fig. 6. Picture for the base states in the simulation.

The learning process of indirect communication only takes place in the first time interval (25–125) with the help of the context states based on the **steponce** function. All the base and adaptation states of the indirect case get increased, but between time 125 and time 150, all become 0 because the context state of this case becomes 0 after

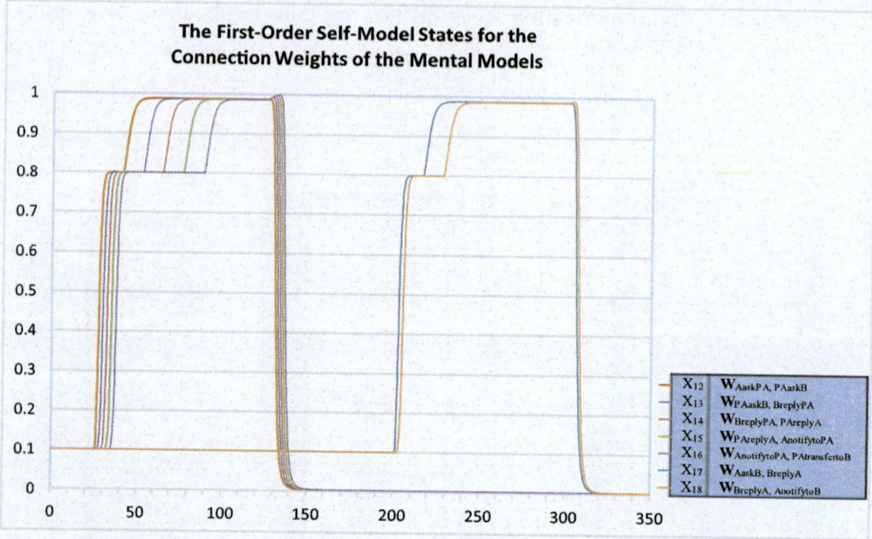

Fig. 7. Picture for the first-order self-model states representing the connection weights of the mental models in the simulation.

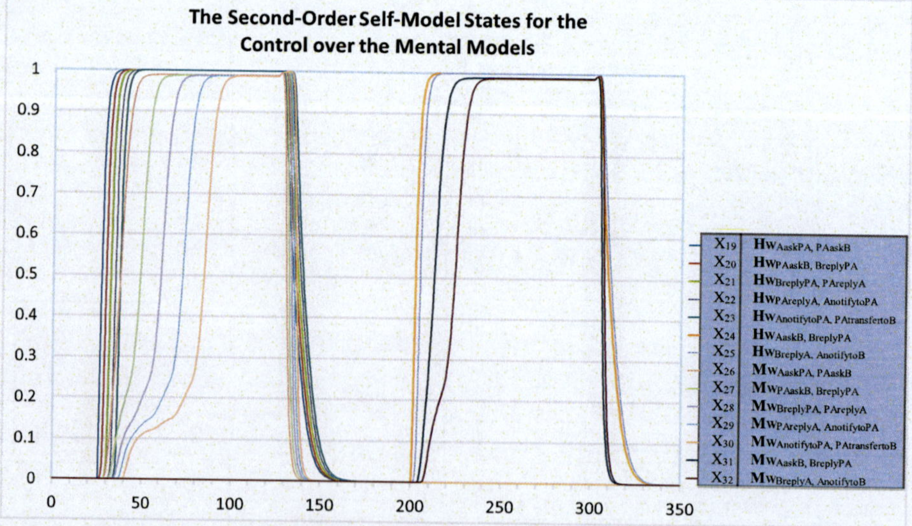

Fig. 8. Picture for the second-order self-model states in the simulation exerting context-sensitive control both over the focusing on a mental model and the learning of it.

time 125. During the gap between the cases, until the second case context state becomes nonzero, everything is zero except the **W**-states of the second case. A minimal nonzero weight is used for the initialization of the learning of direct communication. Then the second case starts and the learning process of this direct communication only happens in the second time interval from time 200 to about 300. Similar to the first case, adaptation states such as **W**-, $\mathbf{H}_{\mathbf{W}}$- and $\mathbf{M}_{\mathbf{W}}$-states increase hand in hand with the base states and that corresponds to the learning process.

6 Discussion

Within mental processes, often mental models are used for internal simulation to obtain predictions. These mental models are also often adapted. Adaptation processes can be described by adaptive mental (or neural) network models. If metacognition is used to regulate adaptation, for example during learning (Pintrich 2000), the adaptation of the mental network becomes itself adaptive as well, so then it involves second-order adaptation of the mental model.

In this paper, a second-order adaptive mental network model was introduced for metacognitive control of adaptation processes needed for focusing on a specific mental model and learning of this mental model. It was shown how a second-order self-modeling network model provides adequate means to model the different aspects that make the addressed topic complex: the network has a self-model about its own structure, it models mental models and adaptation for focusing on them, and it models context-sensitive metacognitive control of this adaptation. The model was applied to simulate an illustrative example scenario that explains what the model does. In further work other scenarios will be addressed as well, in particular for the use of mental models in teams in hospitals, building further, for example, on (Van Ments et al. 2021) and (Treur and Van Ments 2022).

A Appendix: Full Specification by Role Matrices

In Figs. 7, 8, 9, 10 and 11, the different role matrices are shown that provide a full specification of the network characteristics defining the network model in a standardised table format. Here in each role matrix, each state has its row where it is listed which are the impacts on it from that role.

A.1 Role Matrices for Connectivity Characteristics

The connectivity characteristics are specified by role matrices **mb** and **mcw** shown in Fig. 9 and Fig. 10. Role matrix **mb** lists the other states (at the same or lower level) from which the state gets its incoming connections, whereas in role matrix **mcw** the connection weights are listed for these connections.

mb	**base connectivity**	1	2	3	4	5	6
X_1	AaskPA	X_{10}					
X_2	PAaskB	X_1					
X_3	BreplyPA	X_2					
X_4	PAreplyA	X_3					
X_5	AnotifytoPA	X_4					
X_6	PAtransfertoB	X_5					
X_7	AaskB	X_{11}					
X_8	BreplyA	X_7					
X_9	AnotifytoB	X_8					
X_{10}	indirectcase	X_{10}					
X_{11}	directcase	X_{11}					
X_{12}	$\mathbf{W}_{AaskPA,PAaskB}$	X_1	X_2	X_{12}	X_{10}	X_{11}	
X_{13}	$\mathbf{W}_{PAaskB,BreplyPA}$	X_2	X_3	X_{13}	X_{10}	X_{11}	
X_{14}	$\mathbf{W}_{BreplyPA,PAreplyA}$	X_3	X_4	X_{14}	X_{10}	X_{11}	
X_{15}	$\mathbf{W}_{PAreplyA,AnotifytoPA}$	X_4	X_5	X_{15}	X_{10}	X_{11}	
X_{16}	$\mathbf{W}_{AnotifytoPA,PAtransfertoB}$	X_5	X_6	X_{16}	X_{10}	X_{11}	
X_{17}	$\mathbf{W}_{AaskB,BreplyA}$	X_7	X_8	X_{17}	X_{10}	X_{11}	
X_{18}	$\mathbf{W}_{BreplyA,AnotifytoB}$	X_8	X_9	X_{18}	X_{10}	X_{11}	
X_{19}	$\mathbf{H}_{WAaskPA,PAaskB}$	X_1	X_2	X_{12}	X_{19}		
X_{20}	$\mathbf{H}_{WPAaskB,BreplyPA}$	X_2	X_3	X_{13}	X_{20}		
X_{21}	$\mathbf{H}_{WBreplyPA,PAreplyA}$	X_3	X_4	X_{14}	X_{21}		
X_{22}	$\mathbf{H}_{WPAreplyA,AnotifytoPA}$	X_4	X_5	X_{15}	X_{22}		
X_{23}	$\mathbf{H}_{WAnotifytoPA,PAtransfertoB}$	X_5	X_6	X_{16}	X_{23}		
X_{24}	$\mathbf{H}_{WAaskB,BreplyPA}$	X_7	X_8	X_{17}	X_{24}		
X_{25}	$\mathbf{H}_{WBreplyA,AnotifytoB}$	X_8	X_9	X_{18}	X_{25}		
X_{26}	$\mathbf{M}_{WAaskPA,PAaskB}$	X_1	X_2	X_{12}	X_{26}	X_{10}	X_{11}
X_{27}	$\mathbf{M}_{WPAaskB,BreplyPA}$	X_2	X_3	X_{13}	X_{27}	X_{10}	X_{11}
X_{28}	$\mathbf{M}_{WBreplyPA,PAreplyA}$	X_3	X_4	X_{14}	X_{28}	X_{10}	X_{11}
X_{29}	$\mathbf{M}_{WPAreplyA,AnotifytoPA}$	X_4	X_5	X_{15}	X_{29}	X_{10}	X_{11}
X_{30}	$\mathbf{M}_{WAnotifytoPA,PAtransfertoB}$	X_5	X_6	X_{16}	X_{30}	X_{10}	X_{11}
X_{31}	$\mathbf{M}_{WAaskB,BreplyPA}$	X_7	X_8	X_{17}	X_{31}	X_{10}	X_{11}
X_{32}	$\mathbf{M}_{WBreplyA,AnotifytoB}$	X_8	X_9	X_{18}	X_{32}	X_{10}	X_{11}

Fig. 9. Role matrices for the connectivity: **mb** for base connectivity.

Nonadaptive connection weights are indicated in **mcw** by a number (in a green shaded cell), but adaptive connection weights are indicated by a reference to the (self-model) state representing the adaptive value (in a peach-red shaded cell). This can be seen for states X_2 to X_6 (with self-model states X_{12} to X_{16}) and states X_8 and X_9 (with self-model states X_{17} and X_{18}).

mcw	**connection weights**	1	2	3	4	5	6
X_1	AaskPA	1					
X_2	PAaskB	X_{12}					
X_3	BreplyPA	X_{13}					
X_4	PAreplyA	X_{14}					
X_5	AnotifytoPA	X_{15}					
X_6	PAtransfertoB	X_{16}					
X_7	AaskB	1					
X_8	BreplyA	X_{17}					
X_9	AnotifytoB	X_{18}					
X_{10}	indirectcase	1					
X_{11}	directcase	1					
X_{12}	$\mathbf{W}_{AaskPA,PAaskB}$	1	1	1	0.8	0	
X_{13}	$\mathbf{W}_{PAaskB,BreplyPA}$	1	1	1	0.8	0	
X_{14}	$\mathbf{W}_{BreplyPA,PAreplyA}$	1	1	1	0.8	0	
X_{15}	$\mathbf{W}_{PAreplyA,AnotifytoPA}$	1	1	1	0.8	0	
X_{16}	$\mathbf{W}_{AnotifytoPA,PAtransfertoB}$	1	1	1	0.8	0	
X_{17}	$\mathbf{W}_{AaskB,BreplyA}$	1	1	1	0	0.8	
X_{18}	$\mathbf{W}_{BreplyA,AnotifytoB}$	1	1	1	0	0.8	
X_{19}	$\mathbf{H}_{\mathbf{W}AaskPA,PAaskB}$	1	1	-0.1	0.6		
X_{20}	$\mathbf{H}_{\mathbf{W}PAaskB,BreplyPA}$	1	1	-0.1	0.6		
X_{21}	$\mathbf{H}_{\mathbf{W}BreplyPA,PAreplyA}$	1	1	-0.1	0.6		
X_{22}	$\mathbf{H}_{\mathbf{W}PAreplyA,AnotifytoPA}$	1	1	-0.1	0.6		
X_{23}	$\mathbf{H}_{\mathbf{W}AnotifytoPA,PAtransfertoB}$	1	1	-0.1	0.6		
X_{24}	$\mathbf{H}_{\mathbf{W}AaskB,BreplyPA}$	1	1	-0.1	0.6		
X_{25}	$\mathbf{H}_{\mathbf{W}BreplyA,AnotifytoB}$	1	1	-0.1	0.6		
X_{26}	$\mathbf{M}_{\mathbf{W}AaskPA,PAaskB}$	1	1	1	1	-1	-1
X_{27}	$\mathbf{M}_{\mathbf{W}PAaskB,BreplyPA}$	1	1	1	1	-1	-1
X_{28}	$\mathbf{M}_{\mathbf{W}BreplyPA,PAreplyA}$	1	1	1	1	-1	-1
X_{29}	$\mathbf{M}_{\mathbf{W}PAreplyA,AnotifytoPA}$	1	1	1	1	-1	-1
X_{30}	$\mathbf{M}_{\mathbf{W}AnotifytoPA,PAtransfertoB}$	1	1	1	1	-1	-1
X_{31}	$\mathbf{M}_{\mathbf{W}AaskB,BreplyPA}$	1	1	1	1	-1	-1
X_{32}	$\mathbf{M}_{\mathbf{W}BreplyA,AnotifytoB}$	1	1	1	1	-1	-1

Fig. 10. Role matrices for the connectivity: **mcw** for connection weights.

A.2 Role Matrices for Aggregation Characteristics

The network characteristics for aggregation are defined by the selection of combination functions from the library and values for their parameters. In role matrix **mcfw** it is specified by weights which state uses which combination function; see Fig. 11.

mcfw	**combination function weights**	1 **alo-gistic**	2 **step-once**	3 **max-hebb**
X_1	AaskPA	1		
X_2	PAaskB	1		
X_3	BreplyPA	1		
X_4	PAreplyA	1		
X_5	AnotifytoPA	1		
X_6	PAtransfertoB	1		
X_7	AaskB	1		
X_8	BreplyA	1		
X_9	AnotifytoB	1		
X_{10}	indirectcase		1	
X_{11}	directcase		1	
X_{12}	$\mathbf{W}_{AaskPA,PAaskB}$			1
X_{13}	$\mathbf{W}_{PAaskB,BreplyPA}$			1
X_{14}	$\mathbf{W}_{BreplyPA,PAreplyA}$			1
X_{15}	$\mathbf{W}_{PAreplyA,AnotifytoPA}$			1
X_{16}	$\mathbf{W}_{AnotifytoPA,PAtransfertoB}$			1
X_{17}	$\mathbf{W}_{AaskB,BreplyA}$			1
X_{18}	$\mathbf{W}_{BreplyA,AnotifytoB}$			1
X_{19}	$\mathbf{H}\mathbf{w}_{AaskPA,PAaskB}$	1		
X_{20}	$\mathbf{H}\mathbf{w}_{PAaskB,BreplyPA}$	1		
X_{21}	$\mathbf{H}\mathbf{w}_{BreplyPA,PAreplyA}$	1		
X_{22}	$\mathbf{H}\mathbf{w}_{PAreplyA,AnotifytoPA}$	1		
X_{23}	$\mathbf{H}\mathbf{w}_{AnotifytoPA,PAtransfertoB}$	1		
X_{24}	$\mathbf{H}\mathbf{w}_{AaskB,BreplyPA}$	1		
X_{25}	$\mathbf{H}\mathbf{w}_{BreplyA,AnotifytoB}$	1		
X_{26}	$\mathbf{M}\mathbf{w}_{AaskPA,PAaskB}$	1		
X_{27}	$\mathbf{M}\mathbf{w}_{PAaskB,BreplyPA}$	1		
X_{28}	$\mathbf{M}\mathbf{w}_{BreplyPA,PAreplyA}$	1		
X_{29}	$\mathbf{M}\mathbf{w}_{PAreplyA,AnotifytoPA}$	1		
X_{30}	$\mathbf{M}\mathbf{w}_{AnotifytoPA,PAtransfertoB}$	1		
X_{31}	$\mathbf{M}\mathbf{w}_{AaskB,BreplyPA}$	1		
X_{32}	$\mathbf{M}\mathbf{w}_{BreplyA,AnotifytoB}$	1		

Fig. 11. Role matrices for the aggregation characteristics: combination function weights.

In role matrix **mcfp** (see Fig. 12) it is indicated what the parameter values are for the chosen combination functions. Some of them are adaptive, as can be seen in the rows from X_{12} to X_{18} (see the indications for the persistence factors **μ** represented by the self-model states X_{26} to X_{32}).

mcfp	combination function parameters	1 alogistic		2 step-once		3 max-hebb	
		1	2	1	2	1	2
		σ	τ	α	β	μ	
X_1	AaskPA	5	0.3				
X_2	PAaskB	5	0.3				
X_3	BreplyPA	5	0.3				
X_4	PAreplyA	5	0.3				
X_5	AnotifytoPA	5	0.3				
X_6	PAtransfertoB	5	0.3				
X_7	AaskB	5	0.3				
X_8	BreplyA	5	0.3				
X_9	AnotifytoB	5	0.3				
X_{10}	indirectcase			25	125		
X_{11}	directcase			200	300		
X_{12}	$\mathbf{W}_{AaskPA,PAaskB}$					X_{26}	
X_{13}	$\mathbf{W}_{PAaskB,BreplyPA}$					X_{27}	
X_{14}	$\mathbf{W}_{BreplyPA,PAreplyA}$					X_{28}	
X_{15}	$\mathbf{W}_{PAreplyA,AnotifytoPA}$					X_{29}	
X_{16}	$\mathbf{W}_{AnotifytoPA,PAtransfertoB}$					X_{30}	
X_{17}	$\mathbf{W}_{AaskB,BreplyA}$					X_{31}	
X_{18}	$\mathbf{W}_{BreplyA,AnotifytoB}$					X_{32}	
X_{19}	$\mathbf{H}_{WAaskPA,PAaskB}$	5	0.3				
X_{20}	$\mathbf{H}_{WPAaskB,BreplyPA}$	5	0.3				
X_{21}	$\mathbf{H}_{WBreplyPA,PAreplyA}$	5	0.3				
X_{22}	$\mathbf{H}_{WPAreplyA,AnotifytoPA}$	5	0.3				
X_{23}	$\mathbf{H}_{WAnotifytoPA,PAtransfertoB}$	5	0.3				
X_{24}	$\mathbf{H}_{WAaskB,BreplyPA}$	5	0.3				
X_{25}	$\mathbf{H}_{WBreplyA,AnotifytoB}$	5	0.3				
X_{26}	$\mathbf{M}_{WAaskPA,PAaskB}$	5	2				
X_{27}	$\mathbf{M}_{WPAaskB,BreplyPA}$	5	2				
X_{28}	$\mathbf{M}_{WBreplyPA,PAreplyA}$	5	2				
X_{29}	$\mathbf{M}_{WPAreplyA,AnotifytoPA}$	5	2				
X_{30}	$\mathbf{M}_{WAnotifytoPA,PAtransfertoB}$	5	2				
X_{31}	$\mathbf{M}_{WAaskB,BreplyPA}$	5	2				
X_{32}	$\mathbf{M}_{WBreplyA,AnotifytoB}$	5	2				

Fig. 12. Role matrices for the aggregation characteristics: combination function parameters.

A.3 Role Matrices for Timing Characteristics

In Fig. 13, the role matrix **ms** for speed factors is shown, which lists all speed factors. Next to it the list of initial values can be found. Also here a number of them are adaptive: the speed factors of X_{12} to X_{18} are represented by self-model states X_{19} to X_{25}. In Fig. 13 also the initial values of all states used in the example simulation are shown.

ms	**speed factors**	1
X_1	AaskPA	1
X_2	PAaskB	1
X_3	BreplyPA	1
X_4	PAreplyA	1
X_5	AnotifytoPA	1
X_6	PAtransfertoB	1
X_7	AaskB	1
X_8	BreplyA	1
X_9	AnotifytoB	1
X_{10}	indirectcase	1
X_{11}	directcase	1
X_{12}	$\mathbf{W}_{AaskPA,PAaskB}$	X_{19}
X_{13}	$\mathbf{W}_{PAaskB,BreplyPA}$	X_{20}
X_{14}	$\mathbf{W}_{BreplyPA,PAreplyA}$	X_{21}
X_{15}	$\mathbf{W}_{PAreplyA,AnotifytoPA}$	X_{22}
X_{16}	$\mathbf{W}_{AnotifytoPA,PAtransfertoB}$	X_{23}
X_{17}	$\mathbf{W}_{AaskB,BreplyA}$	X_{24}
X_{18}	$\mathbf{W}_{BreplyA,AnotifytoB}$	X_{25}
X_{19}	$\mathbf{HW}_{AaskPA,PAaskB}$	0.5
X_{20}	$\mathbf{HW}_{PAaskB,BreplyPA}$	0.5
X_{21}	$\mathbf{HW}_{BreplyPA,PAreplyA}$	0.5
X_{22}	$\mathbf{HW}_{PAreplyA,AnotifytoPA}$	0.5
X_{23}	$\mathbf{HW}_{AnotifytoPA,PAtransfertoB}$	0.5
X_{24}	$\mathbf{HW}_{AaskB,BreplyPA}$	0.5
X_{25}	$\mathbf{HW}_{BreplyA,AnotifytoB}$	0.5
X_{26}	$\mathbf{MW}_{AaskPA,PAaskB}$	0.5
X_{27}	$\mathbf{MW}_{PAaskB,BreplyPA}$	0.5
X_{28}	$\mathbf{MW}_{BreplyPA,PAreplyA}$	0.5
X_{29}	$\mathbf{MW}_{PAreplyA,AnotifytoPA}$	0.5
X_{30}	$\mathbf{MW}_{AnotifytoPA,PAtransfertoB}$	0.5
X_{31}	$\mathbf{MW}_{AaskB,BreplyPA}$	0.5
X_{32}	$\mathbf{MW}_{BreplyA,AnotifytoB}$	0.5

iv	**initial values**	1
X_1	AaskPA	0
X_2	PAaskB	0
X_3	BreplyPA	0
X_4	PAreplyA	0
X_5	AnotifytoPA	0
X_6	PAtransfertoB	0
X_7	AaskB	0
X_8	BreplyA	0
X_9	AnotifytoB	0
X_{10}	indirectcase	0
X_{11}	directcase	0
X_{12}	$\mathbf{W}_{AaskPA,PAaskB}$	0.1
X_{13}	$\mathbf{W}_{PAaskB,BreplyPA}$	0.1
X_{14}	$\mathbf{W}_{BreplyPA,PAreplyA}$	0.1
X_{15}	$\mathbf{W}_{PAreplyA,AnotifytoPA}$	0.1
X_{16}	$\mathbf{W}_{AnotifytoPA,PAtransfertoB}$	0.1
X_{17}	$\mathbf{W}_{AaskB,BreplyA}$	0.1
X_{18}	$\mathbf{W}_{BreplyA,AnotifytoB}$	0.1
X_{19}	$\mathbf{HW}_{AaskPA,PAaskB}$	0
X_{20}	$\mathbf{HW}_{PAaskB,BreplyPA}$	0
X_{21}	$\mathbf{HW}_{BreplyPA,PAreplyA}$	0
X_{22}	$\mathbf{HW}_{PAreplyA,AnotifytoPA}$	0
X_{23}	$\mathbf{HW}_{AnotifytoPA,PAtransfertoB}$	0
X_{24}	$\mathbf{HW}_{AaskB,BreplyPA}$	0
X_{25}	$\mathbf{HW}_{BreplyA,AnotifytoB}$	0
X_{26}	$\mathbf{MW}_{AaskPA,PAaskB}$	0
X_{27}	$\mathbf{MW}_{PAaskB,BreplyPA}$	0
X_{28}	$\mathbf{MW}_{BreplyPA,PAreplyA}$	0
X_{29}	$\mathbf{MW}_{PAreplyA,AnotifytoPA}$	0
X_{30}	$\mathbf{MW}_{AnotifytoPA,PAtransfertoB}$	0
X_{31}	$\mathbf{MW}_{AaskB,BreplyPA}$	0
X_{32}	$\mathbf{MW}_{BreplyA,AnotifytoB}$	0

Fig. 13. Role matrices for the timing characteristics and initial values: **ms** for speed factors.

References

Bhalwankar, R., Treur, J.: Modeling learner-controlled mental model learning processes by a second-order adaptive network model. PLoS ONE (2021, in press)

Darling-Hammond, L., Austin, K., Cheung, M., Martin, D.: Thinking about thinking: metacognition (2008)

Flavell, J.H.: Metacognition and cognitive monitoring: a new area of cognitive–developmental inquiry. Am. Psychol. **34**(10), 906–911 (1979)

Gentner, D., Stevens, A.L.: Mental Models. Erlbaum, Hillsdale (1983)

Greca, I.M., Moreira, M.A.: Mental models, conceptual models, and modelling. Int. J. Sci. Educ. **22**(1), 1–11 (2000)

Hebb, D.O.: The Organization of Behavior: A Neuropsychological Theory. Wiley, Hoboken (1949)
Koriat, A.: Metacognition and consciousness. In: Zelavo, P.D., Moscovitch, M., Thompson, E. (eds.) Cambridge Handbook of Consciousness. Cambridge University Press, New York (2007)
Mahdavi, M.: An overview: metacognition in education. Int. J. Multidiscip. Curr. Res. **2**, 529–535 (2014)
Pintrich, P.R.: The role of goal orientation in self-regulated learning. In: M. Boekaerts, P. Pintrich, Zeidner, M. (eds.) Handbook of Self-Regulation Research and Applications, pp. 451–502. Academic Press, Orlando (2000)
Robinson, B.L., Harper, N.S., McAlpine, D.: Meta-adaptation in the auditory midbrain under cortical influence. Nat. Commun. **7**, 13442 (2016)
Seel, N.M.: Mental models in learning situations. In: Advances in Psychology, vol. 138, pp. 85–107. Amsterdam, North-Holland (2006)
Shannon, S.V.: Using Metacognitive Strategies and Learning Styles to Create Self-Directed Learners. Inst. Learn. Styles J. **1**, 14–28 (2008)
Sjöström, P.J., Rancz, E.A., Roth, A., Hausser, M.: dendritic excitability and synaptic plasticity. Physiol. Rev. **88**(769–840), 2008 (2008)
Skemp, R.R.: The Psychology of Learning Mathematics. Penguin Books, Harmondsworth (1971)
Treur, J.: Network-Oriented Modeling: Addressing Complexity of Cognitive, Affective and Social Interactions. Springer, Cham (2016). https://doi.org/10.1007/978-3-319-45213-5
Treur, J.: Multilevel network reification: representing higher order adaptivity in a network. In: Aiello, L., Cherifi, C., Cherifi, H., Lambiotte, R., Lió, P., Rocha, L. (eds.) COMPLEX NETWORKS 2018. SCI, vol. 812, pp. 635–651. Springer, Cham (2018). https://doi.org/10.1007/978-3-030-05411-3_51
Treur, J.: Modeling higher-order adaptivity of a network by multilevel network reification. Netw. Sci. **8**, S110–S144 (2020a)
Treur, J.: Network-oriented modeling for adaptive networks: Designing Higher-order Adaptive Biological, Mental and Social Network Models. Springer, Cham (2020b). https://doi.org/10.1007/978-3-030-31445-3
Treur, J.: An adaptive network model covering metacognition to control adaptation for multiple mental models. Cogn. Syst. Res. **67**, 18–27 (2021)
Treur, J., Van Ments, L. (eds.): Mental Models and their Dynamics, Adaptation and Control: A Self-Modeling Network Modeling Approach. Springer, Cham (2022)
Van Ments, L., Treur, J.: Reflections on dynamics, adaptation and control: a cognitive architecture for mental models. Cogn. Syst. Res. **70**, 1–9 (2021)
van Ments, L., Treur, J., Klein, J., Roelofsma, P.: A second-order adaptive network model for shared mental models in hospital teamwork. In: Nguyen, N.T., Iliadis, L., Maglogiannis, I., Trawiński, B. (eds.) ICCCI 2021. LNCS (LNAI), vol. 12876, pp. 126–140. Springer, Cham (2021). https://doi.org/10.1007/978-3-030-88081-1_10

Check for updates

The Problem of Statistical Instability of Samples of Biosystems Requires New Invariants

V. V. Eskov[2(✉)], V. A. Galkin[1], O. E. Filatova[1], M. A. Filatov[2], and V. M. Eskov[1]

[1] FGU "Federal Research Center Scientific Research Institute for System Research of the Russian Academy of Sciences", Separate Subdivision "FSC NIISI RAS" in Surgut, 34, Bazovaya Street, Surgut 628400, Russia

[2] BU VO KhMAO-Yugra "Surgut State University", 1, Lenina Street, Surgut 628400, Russia

Abstract. Traditional statistical methods require statistical stability of samples of any parameters $x_i(t)$ *of* biosystems. However, back in the middle of the 20th century, two scientists (one of the founders of information theory *W. Weaver* and biomechanist N.A. Bernstein) proposed to abandon statistics in the study of complex biosystems. Currently, the absence of statistical stability of any samples has been proven $x_i(t)$ for any biosystems, which means the completion of the further application of statistics (and the theory of dynamical systems) in the study of unstable biosystems. In this regard, serious problems arise in choosing an invariant in a rigorous, mathematical description of stationary regimes of unstable biosystems.

Keywords: Stochastics · Chaos · Uncertainty · Eskov-Zinchenko effect

1 Introduction

At present, in all modern science, the opinion is firmly rooted that deterministic theory (in particular, the theory of dynamical systems - TDS) and stochastics can quite adequately describe the state and behavior of complex biosystems. However, back in 1947, the outstanding biomechanist N.A. Bernstein proposed the "repetition without repetition" hypothesis. In it, he tried to introduce chaos in the behavior of biomechanical systems [1].

A year later, one of the founders of information theory *W. Weaver* proposed to abandon the further use of deterministic (including TDS) and stochastic science (DSS) in the study of systems of the third type (STT), i.e. living systems [2]. *W. Weaver* proposed to bring STT - living systems out of the bounds of DSS in general and to build a new third science for their description [2].

Following the logic of N.A. Bernstein and *W. Weaver,* we would have to prove the special properties of STT and build new models, concepts, and a theory that would go beyond the DSS. This third science (according to *K. Gödel* theorem) should contain new concepts and laws that would go beyond the DSS. Obviously, for such a construction (the third science), it was necessary to prove the special properties of biosystems (STT), which cannot be studied within the framework of the DSS.

R. Silhavy et al. (Eds.): CoMeSySo 2021, LNNS 231, pp. 1010–1022, 2021.
https://doi.org/10.1007/978-3-030-90321-3_83

At the turn of the 20th and 21st centuries, we began a detailed study of the behavior of the state vector $x = x(t) = (x_1, x_2, \ldots, x_m)^T$ of biosystems in the m-dimensional phase state space (PSS). At the same time, the hypotheses of N.A. Bernstein and W. *Weaver*, and the properties of the STT themselves can no longer be described within the framework of the DSS [3–9].

This communication provides arguments for this statement and poses the problem of choosing new invariants for studying stationary regimes of the STT or registering real changes in the dynamics of the behavior of the STT [9–14].

2 The Hypotheses of N. A. Bernstein, W. Weaver and Real Properties of STT

Note that N.A. Bernstein proved the existence of at least five different systems of regulation of movements (systems *A, B, C, D, E*). In his ideas, they could be chaotically included in the organization of any movements and chaotically turned off. What chaos could arise in this case, Bernstein did not present to us then [1].

Moreover, during these 74 years no one even tried to test this hypothesis. A year later *W. Weaver* strengthened Bernstein's thesis [2]. He gave a general classification of systems, in which biosystems – third type systems he generally took outside the DSS. He proposed to create a new, third science about STT (*Complexity*), which would not use TDS and all methods of stochastics (including the dynamic chaos of Lorentz). We have already proved this on numerous examples in the study of tremorograms (TMG), tapping grams (TPG), electromyograms (EMG), electroneurograms (ENG), cardiointervals (CI, RR), electroencephalograms (EEG), etc. [9–18].

W. Weaver tried to oppose *Complexity* (STT) to the entire DSS. The logic of his reasoning could be based on the following contradictions. In the TDS, the stationary mode of any system is estimated from the well-known equality for the vector $x(t)$*:* $dx/dt = 0$, $x_1 = const$. The final state of the system in the form $x(t_k)$ in the TDS can be repeated many times if we exactly repeat the initial parameters $x(t_0)$ and know the equation for the system [2–5].

These equations can be: differential, difference, integral, integral-differential and others. But they all have one basic property in common: $x(t_k)$ can be repeated many times and precisely. This means that the past of the system accurately determines the future. The arrow of time is at work, which spoke about many times *I.R. Prigogine* [19]. Recall that *I.R. Prigogine* declared the end of the deterministic approach to the study of biosystems, but he did not abandon stochastics and the theory of dynamic chaos [19].

In general, the deterministic approach requires strict certainty and the past always accurately determines the future ($x(t_k)$ can be repeated many times). This, however, is absent in stochastics, where initial state of the system $x(t_0)$ we must repeat the, but it isrepeat exactly $x(t_k)$ absolutely impossible toin stochastics. This is especially true for continuous random variables. Their $x(t_k)$ will never be exactly repeated and therefore in stochastic we work with samples $x(t_k)$. In this case, we repeat the experiments with the system many times and get a point cloud in the FPS.

However, for systems of the second type, in contrast to systems of the first type (according to's classification *W. Weaver*), we cannot repeat exactly and this is the very cloud $x(t_k)$ in the FPS. In stochastics, methods (and theories) have been developed by which it is possible to determine the invariability (stationary mode) for samples $x(t_k)$. In this case, the statistical distribution functions $f(x)$, their statistical means $<x>$, statistical variances D_x^*, spectral signal densities (SPS), autocorrelation (AC), etc. are found.

In this case, $f(x)$ and their characteristics can be used for (approximately, within the framework of statistical hypothesis testing, for example) checking the coincidence of samples. These two samples are obtained (systems of the second type are determined not at the point $x(t_k)$, as in the TDS) on different intervals Δt_1 and Δt_2, and they should be the same (approximately). For CBT, exact estimates come to an end and we do not have $dx/dt = 0$, because it is necessary to work with samples.

For CBT *W. Weaver* already introduces (due to what we have shown above) the first concept of *Complexity* (*Disorganized Complexity* - systems of the second type) [2]. This is where fundamental uncertainty appears. One point $x(t_k)$ cannot already describe systems of the second type (samples are needed). We went further 20 years ago and continued the logic of *W. Weaver*: if one point $x(t_k)$ cannot describe the systems of the second type, then one sample $x(t_k)$, which is obtained on the time interval Δt_1, cannot describe the STT. There is a fundamental *Complexity* and *Uncertainty* for all STTs (due to the uniqueness - the uniqueness of the sample $x(t_k)$ *itself*) [9–18].

However, this statement for the STT had to be proved. And we did it, aspointed out *W. Weaver* back in 1948. He wrote: "These new problems, moreover, cannot be handled with the statistical techniques so effective in describing behavior in problems of disorganized complexity. These new problems, and the future of the world depends on many of them, requires science to make a third great advance an advance that must be even greater than the nineteenth ~ century conquest of problems of simplicity or the twentieth ~ century victory over problems of disorganized complexity. Science must, over the next 50 years, learn to deal with these problems of organized complexity" [2].

At the turn of the 20th and 21st centuries, we proved the hypothesis of N.A. Bernstein on "repetition without repetition" and *W. Weaver*: STT is not an DSS object. In fact, by doing so, we go beyond the DSS, tk. already all existing models and theories cannot describe the statistical instability of a biosystem (STT). Any sample will be equivalent to one point in the CBT description. Any sample $x(t)$ *is* unique. The description of STT in the framework of the DSS is equivalent to history (we describe artifacts) [9–18].

As we have found, to prove these two hypotheses, it was enough to verify that the "repetition" of registration of samples of tremorograms (TMG), tapping gram (TPG), electromyogram (EMG) of any person on the planet Earth occurs without statistical "repetitions". If you register a sample any subject $x_i(t)$ for Δt_1 and for the next Δt_2 the second sample (TMG), then the frequency of their statistical coincidence is $p^*_{j,j+1} \leq 0.01$. For TPG and EMG, this frequency is slightly higher ($p_{j,j+1} \leq 0.1$).

In any case, it is arbitrarily impossible to repeat a sample of TMG, TPG or EMG twice in a row. The probability of such an event is much less than $\beta \geq 0.95$, which is accepted in statistics. Any sample $x(t_k)$ will be unique. For example, we present Table 1 for paired comparison of 15 samples of TMG, which were obtained from one subject in his unchanged (physiological and physical) state.

Table 1 we introduced Wilcoxon's criteria $p_{i,j} (i \neq j)$ when comparing the i-th and j-th samples of TMG. As a result, in Table 1 we have a very small number of k pairs of samples for which $p_{i,j} \geq 0.05$. In this case, these i-th and j-th samples can have one general population (they can statistically coincide). Here $k_I = 3$ (see Table 1).

Table 1. Matrix of paired comparison of samples of tremorograms (TMG) of the same person (no load, number of repetitions $n = 15$), Wilcoxon test was used (test of differences $p < 0.05$, number of matches $k_I = 3$).

	1	2	3	4	5	6	7	8	9	10	11	12	13	14	15
1		0.00	0.00	0.00	0.00	0.00	0.00	0.00	0.00	0.00	0.00	0.00	0.00	0.00	0.00
2	0.00		0.00	0.00	0.00	0.00	0.00	0.00	0.00	0.00	0.00	0.00	0.00	0.00	0.00
3	0.00	0.00		0.00	0.00	0.00	0.00	0.00	0.00	0.00	0.00	0.00	0.00	0.00	0.00
4	0.00	0.00	0.00		0.00	0.00	0.00	0.00	0.00	0.00	0.00	0.00	0.00	0.00	0.00
5	0.00	0.00	0.00	0.00		0.00	0.00	0.00	0.00	0.00	0.00	0.00	0.00	0.00	0.00
6	0.00	0.00	0.00	0.00	0.00		0.00	0.00	0.00	0.00	0.00	0.00	0.00	0.00	0.00
7	0.00	0.00	0.00	0.00	0.00	0.00		0.00	0.01	0.00	0.00	0.00	0.00	0.00	0.00
8	0.00	0.00	0.00	0.00	0.00	0.00	0.00		0.00	0.00	0.00	0.00	0.00	0.00	0.00
9	0.00	0.00	0.00	0.00	0.00	0.00	0.01	0.00		0.00	0.00	0.00	0.00	0.00	0.00
10	0.00	0.00	0.00	0.00	0.00	0.00	0.00	0.00	0.00		0.51	0.00	0.00	0.01	0.70
11	0.00	0.00	0.00	0.00	0.00	0.00	0.00	0.00	0.00	**0.51**		0.00	0.00	0.00	1.00
12	0.00	0.00	0.00	0.00	0.00	0.00	0.00	0.00	0.00	0.00	0.00		0.00	0.00	0.00
13	0.00	0.00	0.00	0.00	0.00	0.00	0.00	0.00	0.00	0.00	0.00	0.00		0.00	0.00
14	0.00	0.00	0.00	0.00	0.00	0.00	0.00	0.00	0.00	0.01	0.00	0.00	0.00		0.00
15	0.00	0.00	0.00	0.00	0.00	0.00	0.00	0.00	0.00	**0.70**	**1.00**	0.00	0.00	0.00	

Table 2 we present a matrix of paired comparisons of EMG samples of the same muscle of one subject (in its unchanged state). It is obvious that k_2 for EMG is slightly larger and this is a strict regularity: for TMG, k_1 is 2–3 times less than k_2 for EMG (and for TPG too). We emphasize that we have obtained several hundreds of such tables (matrices of paired comparisons of samples) for different parameters $x(t)$*of the* functions of the human body. Everywhere $k_I \leq 20\%$ and only for EEG we have some increase in k ($k \leq 30\%$), but these are also small values for statistics.

Recall that statistics require samples to be with a probability of β matched ≥ 0.95 (or higher), but in our case the opposite is true. In 85% of the samples, they do not statistically coincide, they are unique and cannot be studied within the framework of the DSS. We emphasize that we are not talking about the deterministic Lorentz chaos. It was on him that the four Nobel laureates (*M. Gell-Mann* [20], *IR Prigogine* [19], *R. Penrose* [21], V.L. Ginzburg [22]) in their publications. However, their expectations were not met.

V.L. Ginzburg singled out in a special way three "great" problems of physics, among which the third (it is the main) problem of reduction has now been precisely solved. Within the framework of the statistical instability of samples of any parameters $x_i(t)$ *of*

Table 2. Matrix of paired comparison of samples of electromyograms (EMG) of the same person (with a weak static load, $F_I = 50$ N), Wilcoxon's test was used (the criterion of differences $p < 0.05$, the number of matches $k_2 = 8$).

	1	2	3	4	5	6	7	8	9	10	11	12	13	14	15
1		0.00	0.00	0.00	0.00	0.00	0.46	0.00	0.00	0.00	0.00	0.05	0.00	0.00	0.00
2	0.00		0.00	0.00	0.00	0.00	0.00	0.00	0.00	0.00	0.00	0.00	0.00	0.00	0.00
3	0.00	0.00		0.08	0.00	0.00	0.02	0.00	0.00	0.00	0.00	0.13	0.00	0.00	0.00
4	0.00	0.00	**0.08**		0.00	0.00	0.00	0.00	0.00	0.00	0.00	0.10	0.00	0.00	0.00
5	0.00	0.00	0.00	0.00		0.00	0.00	0.00	0.00	0.00	0.00	0.05	0.22	0.00	0.00
6	0.00	0.00	0.00	0.00	0.00		0.00	0.00	0.77	0.00	0.00	0.00	0.00	0.00	0.00
7	**0.46**	0.00	0.02	0.00	0.00	0.00		0.00	0.00	0.00	0.00	0.20	0.00	0.00	0.00
8	0.00	0.00	0.00	0.00	0.00	0.00	0.00		0.00	0.00	0.00	0.00	0.00	0.00	0.00
9	0.00	0.00	0.00	0.00	0.00	**0.77**	0.00	0.00		0.00	0.00	0.00	0.00	0.00	0.00
10	0.00	0.00	0.00	0.00	0.00	0.00	0.00	0.00	0.00		0.46	0.00	0.00	0.00	0.00
11	0.00	0.00	0.00	0.00	0.00	0.00	0.00	0.00	0.00	**0.46**		0.00	0.00	0.00	0.00
12	0.05	0.00	**0.13**	**0.10**	0.05	0.00	**0.20**	0.00	0.00	0.00	0.00		0.00	0.00	0.00
13	0.00	0.00	0.00	0.00	**0.22**	0.00	0.00	0.00	0.00	0.00	0.00	0.00		0.00	0.00
14	0.00	0.00	0.00	0.00	0.00	0.00	0.00	0.00	0.00	0.00	0.00	0.00	0.00		0.00
15	0.00	0.00	0.00	0.00	0.00	0.00	0.00	0.00	0.00	0.00	0.00	0.00	0.00	0.00	

biosystems (this is called the Eskov-Zinchenko effect (EZE)), we cannot now use all DSS models and methods (and not only from physics) to describe and study complex biosystems [9–18].

We emphasize that tables of the form of Table 1 and Table 2, we obtained more than a thousand for different parameters of the functions of the human body, and everywhere one pattern is observed: any sample of TMG, TPG, EMG, CI, ENG, EEG is unique [9–18, 23–32]. On the next interval Δt_2, we get another sample, which with a frequency of 0.85 (and higher) does not coincide with the sample $x_i(t)$ in the previous interval Δt_1. The sample $x_i(t)$*is* unique. Statistics cannot be applied because it has a historical character (the sample is an artifact).

3 The Problem of Homogeneity of Samples $x_i(t)$ in Biocybernetics

The uniqueness of the samples $x_i(t)$ is manifested not only for one subject (in relation to his own samples $x_i(t)$). It turned out that 15 different subjects (if you register 5 min of CI for each of these subjects) also have 15 different CI samples. If we construct a matrix of paired comparisons of samples for 15 different subjects, then the number *k of* pairs for which the Newman-Keuls test (or any other criterion known in statistics) $p_{ij} \geq 0.05$ will also be very small.

Usually, k_g (for 15 different subjects) does not differ much from the number k_{CI} for one subject. As an example, we present two matrices of pairwise comparison of CI samples. Table 3, we have the number k_{KI} (for them $p_{ij} \geq 0.05$) is very small ($k_{KI} = 10$).

Table 3. Matrix of paired comparison of samples of cardiointervals (CI) of the same person (no load, number of repetitions of CI registration $n = 15$), Wilcoxon test was used (difference criterion $p < 0.05$, number of matches $k_{CI} = 10$).

	1	2	3	4	5	6	7	8	9	10	11	12	13	14	15
1		0.00	0.00	0.00	0.00	0.00	0.00	0.00	0.00	0.00	0.00	0.00	0.00	0.00	0.00
2	0.00		0.00	0.00	0.00	0.00	0.00	0.00	0.00	0.00	0.00	0.00	0.00	0.00 0.00	0.00
3	0.00	0.00		0.00	0.02	0.00	0.00	0.00	0.00	0.00	0.00	0.00	0.00	0.00	0.00
4	0.00	0.00	0.00		0.00	0.04	0.04	0.00	0.00	0.00	0.00	0.00	0.00	0.00	0.00
5	0.00	0.00	0.02	0.00		0.00	0.00	0.00	0.00	0.00	0.00	0.00	0.00	0.00	0.00
6	0.00	0.00	0.00	0.04	0.00		0.78	0.00	0.00	0.00	0.00	0.00	0.00	0.00	0.00
7	0.00	0.00	0.00	0.04	0.00	**0.78**		0.00	0.00	0.00	0.00	0.00	0.00	0.00	0.00
8	0.00	0.00	0.00	0.00	0.00	0.00	0.00		0.00	0.00	0.00	0.00 0.00	0.00	0.00	0.00
9	0.00	0.00	0.00	0.00	0.00	0.00	0.00	0.00		0.00	0.00	0.00	0.00	0.00	0.77
10	0.00	0.00	0.00	0.00	0.00	0.00	0.00	0.00	0.00		0.25	0.04	0.67	0.73	0.00
11	0.00	0.00	0.00	0.00	0.00	0.00	0.00	0.00	0.00	**0.25**		0.02	0.38	0.49	0.00
12	0.00	0.00	0.00	0.00	0.00	0.00	0.00	0.00	0.00	0.04	0.02		0.08	0.14	0.00
13	0.00	0.00	0.00	0.00	0.00	0.00	0.00	0.00	0.00	**0.67**	**0.38**	**0.08**		0.30	0.00
14	0.00	0.00	0.00	0.00	0.00	0.00	0.00	0.00	0.00	**0.73**	**0.49**	**0.14**	**0.30**		0.00
15	0.00	0.00	0.00	0.00	0.00	0.00	0.00	0.00	**0.77**	0.00	0.00	0.00	0.00	0.00	

This Table 3 is easy to compare with Table 4, where the matrix of paired comparisons of samples of 15 different subjects also contains a small number of $k_g = 15$. Obviously, both for one subject and for a group of 15 different subjects, we have the number of k samples (which may have a common general population, i.e. $p_{ij} \geq 0.05$) is very small. Usually these k_g and k_{ki} will be less than 15%, i.e. the number of statistically non-coinciding samples will be more than 85%. This is a very large value for statistics [23–31].

However, now (for a group of subjects) the situation has deteriorated sharply. The fact that the CI samples belong to different general populations indicates a loss of group homogeneity. These people cannot be united into a common group, because we do not know for what reasons their samples are statistically different (it may be that they have different diseases, but we could not identify them). The reasons may vary.

Table 4. Matrix of paired comparison of 15 samples of cardiointervals (RI) of a group of girls (no load), the Newman-Keuls test was used (difference test $p < 0.05$, number of matches $k_g = 15$).

	1 R: 1472.3	2 R: 2625.2	3 R: 2358.4	4 R: 2906.9	5 R: 1741.4	6 R: 1591.8	7 R: 3356.5	8 R: 1269.1	9 R: 2539.3	10 R: 2887.7	11 R: 1241.9	12 R: 564.67	13 R: 461.79	14 R: 349.71	15 R: 2765.9
1		0.00	0.00	0.00	0.57	1.00	0.00	1.00	0.00	0.00	1.00	0.00	0.00	0.00	0.00
2	0.00		0.62	0.38	0.00	0.00	0.00	0.00	1.00	0.70	0.00	0.00	0.00	0.00	1.00
3	0.00	**0.62**		0.00	0.00	0.00	0.00	0.00	1.00	0.00	0.00	0.00	0.00	0.00	0.00
4	0.00	**0.38**	0.00		0.00	0.00	0.00	0.00	0.02	1.00	0.00	0.00	0.00	0.00	1.00
5	**0.57**	0.00	0.00	0.00		1.00	0.00	0.00	0.00	0.00	0.00	0.00	0.00	0.00	0.00
6	**1.00**	0.00	0.00	0.00	**1.00**		0.00	0.09	0.00	0.00	0.03	0.00	0.00	0.00	0.00
7	0.00	0.00	0.00	0.00	0.00	0.00		0.00	0.00	0.00	0.00	0.00	0.00	0.00	0.00
8	**1.00**	0.00	0.00	0.00	0.00	**0.09**	0.00		0.00	0.00	1.00	0.00	0.00	0.00	0.00
9	0.00	**1.00**	**1.00**	0.02	0.00	0.00	0.00	0.00		0.03	0.00	0.00	0.00	0.00	1.00
10	0.00	**0.70**	0.00	**1.00**	0.00	0.00	0.00	0.00	0.03		0.00	0.00	0.00	0.00	1.00
11	**1.00**	0.00	0.00	0.00	0.00	0.03	0.00	**1.00**	0.00	0.00		0.00	0.00	0.00	0.00
12	0.00	0.00	0.00	0.00	0.00	0.00	0.00	0.00	0.00	0.00	0.00		1.00	1.00	0.00
13	0.00	0.00	0.00	0.00	0.00	0.00	0.00	0.00	0.00	0.00	0.00	**1.00**		1.00	0.00
14	0.00	0.00	0.00	0.00	0.00	0.00	0.00	0.00	0.00	0.00	0.00	**1.00**	**1.00**		0.00
15	0.00	**1.00**	0.00	**1.00**	0.00	0.00	0.00	0.00	**1.00**	**1.00**	0.00	0.00	0.00	0.00	

It is obvious that working with heterogeneous groups within the framework of stochastics is impossible in principle. You cannot work with unique samples either. As a result, we come to a real proof of's hypothesis *W. Weaver*: STT cannot be an object of DSS. Stochastics ends when STT-*complexity*. This is whatspoke about *W. Weaver* [2]. Then a natural question arises: how can the stationary (unchanging) state of an STT be mathematically identified from its real changes?

4 Is It Possible to Define Invariants for STT?

From a biological point of view, it is quite obvious that nothing happens to the biosystem, it is in a stationary (unchanged biologically) state. However, all its parameters are continuously and chaotically changing, the samples demonstrate a statistical "kaleidoscope" of changes. We cannot work with unique samples $x_i(t)$ for one subject. We cannot work with heterogeneous groups. At the same time, any group of subjects in biomedicine has been considered homogeneous until now. However, before us, no one checked the samples $x_i(t)$ of different people that biologists, doctors, psychologists, ecologists (etc.) combined into a group, no one checked the samples $x_i(t)$ of such people from the group (by statistical comparison their samples).

A priori, everyone believed that their samples $x_i(t)$ should belong to the same general population. But this was only an illusion. As the outstanding physicist of the 20th century E. Rutherford said, there is only one science - physics (the rest of the sciences are cooking). He expressed himself very accurately in the light of the EZE proved by us, i.e. lack of statistical stability of samples. In physics, if the samples do not coincide, then these are different systems, different states of the same system, etc.

Physics is an exact science and therefore it is definitely a science. In it, the past determines the future. This also applies to the dynamic chaos of Lorentz. Setting the initial parameters $x(t_0)$ and the equation for the system precisely determines the Lorentz attractor, in which we have the mixing property. In this case, the measure of the invariant and such an attractor will be predictable and checkable. The same cannot be said for the STT.

Statistical chaos STT has no forecast, because statistical functions, their numerical characteristics (statistical mean $<x>$, statistical variance D_x^* , spectral signal density (SSD), autocorrelation, etc.) change continuously and chaotically. The autocorrelation function does not tend to zero, Lyapunov constants continuously and chaotically (for each segment of these samples $x_i(t)$ change). This is the basic property of STT, which we have defined as the second type of uncertainty) [25–32].

This is another (special chaos EZE) chaos and then a natural question arises: is it possible to somehow register objectively the invariability of STTs or their variability? Are there invariants for STT? How can they be described mathematically? It turned out that there are answers to these questions, but they are within the framework of a new, third science, which spoke about *W. Weaver* 73 years ago [2].

To obtain these new invariants, we performed numerous repeated measurements in several separate groups of subjects. For example, we took 15 women (approximately the same age $<T> = 29$ years) and recorded the CI samples multiple times (225 times each). Each registration was 5 min, as required by the European Heart Association. For each subject from this group, 15 matrices were built (of the form of Table 3), in which we found the numbers k (for them $p_{ij} \geq 0.05$).

As a result, for each subject from such a group, we obtained a sample of values k_k, which contained 15 values of numbers k. In total, 15 such samples of numbers were obtained$k$ (for all 15 subjects). Having these 15 samples k_k for these 15 people, we calculated a matrix of paired comparisons of these 15 samples k_k for these 15 people.

The main task in this case was to check the statistical stability of these 15 samples k. As a result, such a matrix is presented in Table 5, where the number is located k_k. This number shows the number of pairs of samples k for which for the i-th and j-th pairs of samples k we have the Newman-Keuls *test* $p \geq 0.05$. It turned out that this number k_k *is* very large ($k_k \geq 95$).

Calculation of such matrices for numbers k showed two important conclusions. First, these 15 samples of values are k -statistically (very strong) the same. According to statistical criteria, $\beta \geq 0.95$, and in our Table 5 we have $k_k = 98$ (this is more than 0.95), then we can conclude that the numbers k can be invariants in assessing the immutability of biosystems. The entire group of these subjects can be considered homogeneous, because their samples of numbers k statistically very strongly coincide ($k_k \geq 0.95$). We assume that even one number k (for one matrix of paired comparisons of CI samples) can already be invariant, since it belongs to statistically matching samples k.

Table 5. Matrix of paired comparisons of samples of numbers k for 15 different subjects (before the load), represented by the Newman-Keuls test for each i-th and j-th samples k pair of comparisons (difference criterion $p < 0.05$, number of matches $k_k = 98$).

	1 R: 1472.3	2 R: 2625.2	3 R: 2358.4	4 R: 2906.9	5 R: 1741.4	6 R: 1591.8	7 R: 3356.5	8 R: 1269.1	9 R: 2539.3	10 R: 2887.7	11 R: 1241.9	12 R: 564.67	13 R: 461.79	14 R: 349.71	15 R: 2765.9
1		1.00	1.00	1.00	1.00	1.00	1.00	1.00	1.00	1.00	1.00	1.00	0.23	1.00	1.00
2	**1.00**		1.00	0.74	0.00	0.16	0.04	0.01	0.08	0.02	1.00	0.30	0.00	0.23	0.00
3	**1.00**	**1.00**		1.00	0.41	1.00	1.00	1.00	1.00	1.00	1.00	1.00	0.01	1.00	0.63
4	**1.00**	**0.74**	**1.00**		1.00	1.00	1.00	1.00	1.00	1.00	1.00	1.00	0.51	1.00	1.00
5	**1.00**	0.00	**0.41**	**1.00**		1.00	1.00	1.00	1.00	1.00	1.00	1.00	1.00	1.00	1.00
6	**1.00**	**0.16**	**1.00**	**1.00**	**1.00**		1.00	1.00	1.00	1.00	1.00	1.00	1.00	1.00	1.00
7	**1.00**	0.04	**1.00**	**1.00**	**1.00**	**1.00**		1.00	1.00	1.00	1.00	1.00	1.00	1.00	1.00
8	**1.00**	0.01	**1.00**	**1.00**	**1.00**	**1.00**	**1.00**		1.00	1.00	1.00	1.00	1.00	1.00	1.00
9	**1.00**	**0.08**	**1.00**	**1.00**	**1.00**	**1.00**	**1.00**	**1.00**		1.00	1.00	1.00	1.00	1.00	1.00
10	**1.00**	0.02	**1.00**	**1.00**	**1.00**	**1.00**	**1.00**	**1.00**	**1.00**		1.00	1.00	1.00	1.00	1.00
11	**1.00**	**1.00**	**1.00**	**1.00**	**1.00**	**1.00**	**1.00**	**1.00**	**1.00**	**1.00**		1.00	0.20	1.00	1.00
12	**1.00**	**0.30**	**1.00**	**1.00**	**1.00**	**1.00**	**1.00**	**1.00**	**1.00**	**1.00**	**1.00**		1.00	1.00	1.00
13	**0.23**	0.00	0.01	**0.51**	**1.00**	**1.00**	**1.00**	**1.00**	**1.00**	**1.00**	**0.20**	**1.00**		1.00	1.00
14	**1.00**	**0.23**	**1.00**	**1.00**	**1.00**	**1.00**	**1.00**	**1.00**	**1.00**	**1.00**	**1.00**	**1.00**	**1.00**		1.00
15	**1.00**	0.00	**0.63**	**1.00**	**1.00**	**1.00**	**1.00**	**1.00**	**1.00**	**1.00**	**1.00**	**1.00**	**1.00**	**1.00**	

At the same time, we can now talk about the homogeneity of this group, because samples of numbers k *are* statistically the same, all samples have some common properties in terms of parameters k. However, such checks require very large calculations and a lot of time to carry out repeated many times (we have 225 times for each) observations (experiments).

For the purity of our test, we gave these 15 subjects physical activity and the whole experiment was repeated (but after 30 squats). As a result, we got 15 more samples k before the load and 15 samples k after the load, which we compared in pairs. As a result, a special matrix (see Table 6) of paired comparisons of 15samples k before and 15samples k after loading was built. Table 6 we present the final result.

Obviously, matrix 6 showed a much smaller number of pairs for which $p_{ij} \geq 0.05$, i.e. when they are statistically the same. We emphasize that in this Table 6 there are much more pairs of samples k_{kk} than in Table 5 for k_k. Here all 225 pairs will be different and the total number of pairs is much greater than in the Table 5 (there were only 105 different pairs). However, the percentage of coincidences of samples k is small (38.6%).

Such a large difference in the samples k before and after the load proves that the parameter k can be an invariant for assessing the steady state of the CVS (and any STT, in general) and a new, unchanged state of the CVS of the same group of subjects. At the same time, any statistical characteristics cannot represent the state of STTs due to their statistical instability (EZE). The numbers do give a difference between the samples (before and after exercise) of CVS parameters (with a frequency of more than 60%).

Table 6. Matrix of paired comparisons of samples of numbers k for 15 different subjects (before and after exercise), represented by the Newman-Keuls test for each i-th and j-th samples k pair of comparisons (difference criterion $p < 0.05$, number of matches $k_{kk} = 87$).

	1 R: 164.12	2 R: 135.15	3 R: 122.53	4 R: 219.70	5 R: 289.48	6 R: 210.03	7 R: 163.05	8 R: 223.33	9 R: 252.88	10 R: 245.97	11 R: 232.10	12 R: 231.13	13 R: 332.62	14 R: 268.60	15 R: 291.80
1		1.00	1.00	1.00	0.02	1.00	1.00	1.00	0.86	1.00	1.00	1.00	0.00	0.20	0.02
2	**1.00**		1.00	1.00	0.00	1.00	1.00	0.91	0.05	0.10	0.41	0.45	0.00	0.01	0.00
3	**1.00**	**1.00**		0.40	0.00	0.96	1.00	0.28	0.01	0.02	0.12	0.13	0.00	0.00	0.00
4	**1.00**	**1.00**	**0.40**		1.00	1.00	1.00	1.00	1.00	1.00	1.00	1.00	0.08	1.00	1.00
5	0.02	0.00	0.00	**1.00**		1.00	0.02	1.00	1.00	1.00	1.00	1.00	1.00	1.00	1.00
6	**1.00**	**1.00**	**0.96**	**1.00**	**1.00**		1.00	1.00	1.00	1.00	1.00	1.00	0.03	1.00	1.00
7	**1.00**	**1.00**	**1.00**	**1.00**	0.02	**1.00**		1.00	0.78	1.00	1.00	1.00	0.00	0.18	0.01
8	**1.00**	**0.91**	**0.28**	**1.00**	**1.00**	**1.00**	**1.00**		1.00	1.00	1.00	1.00	0.12	1.00	1.00
9	**0.86**	0.05	0.01	**1.00**	**1.00**	**1.00**	**0.78**	**1.00**		1.00	1.00	1.00	1.00	1.00	1.00
10	**1.00**	**0.10**	0.02	**1.00**	**1.00**	**1.00**	**1.00**	**1.00**	**1.00**		1.00	1.00	1.00	1.00	1.00
11	**1.00**	**0.41**	**0.12**	**1.00**	**1.00**	**1.00**	**1.00**	**1.00**	**1.00**	**1.00**		1.00	0.29	1.00	1.00
12	**1.00**	**0.45**	**0.13**	**1.00**	**1.00**	**1.00**	**1.00**	**1.00**	**1.00**	**1.00**	**1.00**		0.26	1.00	1.00
13	0.00	0.00	0.00	**0.08**	**1.00**	0.03	0.00	**0.12**	**1.00**	**1.00**	**0.29**	**0.26**		1.00	1.00
14	**0.20**	0.01	0.00	**1.00**	**1.00**	**1.00**	**0.18**	**1.00**	**1.00**	**1.00**	**1.00**	**1.00**	**1.00**		1.00
15	0.02	0.00	0.00	**1.00**	**1.00**	**1.00**	0.01	**1.00**	**1.00**	**1.00**	**1.00**	**1.00**	**1.00**	**1.00**	

5 Discussion

At the turn of the 20th and 21st centuries, in biocybernetics and throughout biomedicine, the statistical instability of any samples of parameters of the functions of the human body was proved. This is called the Eskov-Zinchenko effect. This EZE proves the futility of further use of statistics in the study of complex biosystems - STT.

This EZE proves hypothesis *W. Weaver's* (STT is not an object of DSS) and N.A. Bernstein on "repetition without repetition." There is an urgent need to create a third (new) science, which, according to *K. Gödel* theorem, should completely go beyond the DSS. Such a new science requires the creation of new concepts, new models and a new mathematical apparatus.

In this process of creating a new science (new mathematics of statistically unstable biosystems), one of the main problems is associated with a new interpretation of the concept of rest and the concept of motion for the vector of the state of the biosystem. The old criteria ($dx/dt = 0$ or preservation of the statistical distribution function $f(x)$, its characteristics) can no longer work. They change continuously and chaotically from sample to sample. New invariants and new criteria are needed for special biosystems in which EZE is found [25–32].

Numerous experiments and observations show that the numbers k, which are obtained when calculating matrices of paired comparisons of samples of parameters $x_i(t)$*of* biosystems, can demonstrate the statistical stability of their samples. With numerous repeated registrations of CI samples from the same subject, we can obtain samples of the numbers themselves k (this is the number of pairs in the matrix of paired comparisons of samples for which $p_{ij} \geq 0.05$), which for one subject or for one (homogeneous) group remain approximately.

In the framework of stochastics, we have shown that the parameter k can characterize the homogeneity of the sample. Moreover, with a change in the physiological state of the subject, the numbers can also change k. This change in the number k *is* statistically significant. The parameter k can be a criterion for assessing the invariability of the state of the biosystem or a criterion for the homogeneity of a group. Traditional statistical characteristics can no longer objectively characterize TSS(they are unique) [28–32].

6 Conclusions

In the middle of the 20th century, two prominent scientists hypothesized about the inapplicability of traditional deterministic and stochastic science (DDS) in describing the behavior of living systems (STT). *W. Weaver* proposed to build a new (third) science of biosystems. In this he saw the main *Complexity* for the entire modern DSS.

At the turn of the 20th and 21st centuries (as planned by *W. Weaver*), we established the absence of statistical stability of any parameters of the functions of the human body. This was proof of EZE and proof that the past STT cannot determine (diagnose) the future state of the STT. The main principle of DSS is violated, any sample $x(t)$ for STT will be unique. This is proved by hundreds of matrices of paired comparisons of samples $x_i(t)$ from one subject, which were constructed by us for many parameters of various functions of the human body.

With a group of different subjects, everything is much more complicated, because samples of different subjects from one (supposedly homogeneous) group cannot belong to the general population. Any sample $x_i(t)$ in biomedicine is unique and the group cannot be homogeneous. The loss of stability and homogeneity of samples means the end of the use of DSS in the future. We need new invariants and a new science, new mathematics to describe biosystems.

One of the possible new invariants for STT is the use of numbers k - the number of pairs of samples in matrices of paired comparisons of samples from one subject. These numbers k remain stable, which we show in this communication. Obviously, for an STT, it is necessary to construct a new mathematics, in which there will be other concepts of stationarity $x(t)$ or real changes in the parameters of the biosystem state vector $x(t)$.

References

1. Bernstein, N.A.: The Coordination and Regulation of Movements. Pergamon Press, Oxford, New York (1967). p. 196
2. Weaver, W.: Science and complexity. Am. Sci. **36**, 536–544 (1948)
3. Eskov, V.M., Kulaev, S.V., Popov, Y., Filatova, O.E.: Computer technologies in stability measurements on stationary states in dynamic biological systems. Meas. Tech. **49**(1), 59–65 (2006). https://doi.org/10.1007/S11018-006-0063-2
4. Eskov, V.M., Eskov, V.V., Filatova, O.E.: Characteristic features of measurements and modeling for biosystems in phase spaces of states. Meas. Tech. **53**(12), 1404–1410 (2011). https://doi.org/10.1007/S11018-011-9673-4
5. Eskov, V.M., Gavrilenko, T.V., Vokhmina, Y.V., Zimin, M.I., Filatov, M.A.: Measurement of chaotic dynamics for two types of tapping as voluntary movements. Meas. Tech. **57**(6), 720–724 (2014). https://doi.org/10.1007/s11018-014-0525-x

6. Eskov, V.V., Gavrilenko, T.V., Eskov, V.M., Vokhmina, Y.V.: Phenomenon of statistical instability of the third type systems – complexity. Tech. Phys. **62**(11), 1611–1616 (2017). https://doi.org/10.1134/S106378421711007X
7. Betelin, V.B., Eskov, V.M., Galkin, V.A., Gavrilenko, T.V.: Stochastic volatility in the dynamics of complex homeostatic systems. Dokl. Math. **95**(1), 92–94 (2017). https://doi.org/10.1134/S1064562417010240
8. Zilov, V.G., Eskov, V.M., Khadartsev, A.A., Eskov, V.V.: Experimental verification of the Bernstein effect "repetition without repetition." Bull. Exp. Biol. Med. **163**(1), 1–5 (2017). https://doi.org/10.1007/s10517-017-3723-0
9. Eskov, V.M., Eskov, V.V., Vochmina, J.V., Gavrilenko, T.V.: The evolution of the chaotic dynamics of collective modes as a method for the behavioral description of living systems. Mosc. Univ. Phys. Bull. **71**(2), 143–154 (2016). https://doi.org/10.3103/S0027134916020053
10. Eskov, V.M., Eskov, V.V., Braginskii, M.Ya., Pashnin, A.S.: Determination of the degree of synergism of the human cardiorespiratory system under conditions of physical effort. Meas. Tech. **54**(7), 832–837 (2011). https://doi.org/10.1007/S11018-011-9812-Y
11. Eskov, V.M., Papshev, V.A., Eskov, V.V., Zharkov, D.A.: Measuring biomechanical parameters of human extremity tremor. Meas. Tech. **46**(1), 93–99 (2003). https://doi.org/10.1023/A:1023482026679
12. Eskov, V.M.: Models of hierarchical respiratory neuron networks. Neurocomputing **11**(2–4), 203–226 (1996). https://doi.org/10.1016/0925-2312(95)00048-8
13. Es'kov, V.M., Filatova, O.E., Ivashenko, V.P.: Computer identification of compartmental neuron circuits. Meas. Tech. **37**(8), 967–971 (1994). https://doi.org/10.1007/BF00977157
14. Zilov, V.G., Khadartsev, A.A., Eskov, V.V., Eskov, V.M.: Experimental study of statistical stability of cardiointerval samples. Bull. Exp. Biol. Med. **164**(2), 115–117 (2017). https://doi.org/10.1007/s10517-017-3937-1
15. Vokhmina, Y.V., Eskov, V.M., Gavrilenko, T.V., Filatova, O.E.: Measuring order parameters based on neural network technologies. Meas. Tech. **58**(4), 462–466 (2015). https://doi.org/10.1007/s11018-015-0735-x
16. Grigorenko, V.V., Eskov, V.M., Nazina, N.B., Egorov, A.A.: Information-analytical system of cardiographic information functional diagnostics. J. Phys. Conf. Ser. **1515**, 052027 (2020). https://doi.org/10.1088/1742-6596/1515/5/052027
17. Eskov, V.V., Filatova, D.Y., Ilyashenko, L.K., Vochmina, Y.V.: Classification of uncertainties in modeling of complex biological systems. Mosc. Univ. Phys. Bull. **74**(1), 57–63 (2019). https://doi.org/10.3103/S0027134919010089
18. Zilov, V.G., Khadartsev, A.A., Ilyashenko, L.K., Eskov, V.V., Minenko, I.A.: Experimental analysis of the chaotic dynamics of muscle biopotentials under various static loads. Bull. Exp. Biol. Med. **165**(4), 415–418 (2018). https://doi.org/10.1007/s10517-018-4183-x
19. Prigogine, I.R.: The End of Certainty: Time, Chaos, and the New Laws of Nature. Free Press (1996)
20. Gell-Mann, M.: Fundamental sources of unpredictability. Complexity **3**(1), 13–19 (1997)
21. Penrose, R.: The Emperor's New Mind: Concerning Computers, Mind and Laws of Physics. Oxford University Press, Oxford (1989)
22. Ginzburg, V.L.: What problems of physics and astrophysics seem now to be especially important and interesting (thirty years later, already on the verge of XXI century)? Phys. Uspekhi **42**, 353–373 (1999). https://doi.org/10.1070/PU1999v042n04ABEH000562
23. Zilov, V.G., Khadartsev, A.A., Eskov, V.V., Ilyashenko, L.K., Kitanina, K.Y.: Examination of statistical instability of electroencephalograms. Bull. Exp. Biol. Med. **168**(1), 5–9 (2019). https://doi.org/10.1007/s10517-019-04633-7
24. Zilov, V.G., Khadartsev, A.A., Eskov, V.M., Ilyashenko, L.K.: New effect in physiology of human nervous muscle system. Bull. Exp. Biol. Med. **167**(4), 419–423 (2019). https://doi.org/10.1007/s10517-019-04540-x

25. Filatova, O.E., Bashkatova, Yu.V., Shakirova, L.S., Filatov, M.A.: Neural network technologies in system synthesis. IOP Conf. Seri. Mater. Sci. Eng. **1047**, 012099 (2021). https://doi.org/10.1088/1757-899X/1047/1/012099
26. Grigorenko, N.B., Nazina, V.V., Filatov, M.A., Chempalova, L.S., Tretyakov, S.A.: New information technologies in the estimation of the third type systems. J. Phys. Conf. Ser. **1889**, 032003 (2021). https://doi.org/10.1088/1742-6596/1889/3/032003
27. Filatova, D., Bashkatova, Y., Filatov, M.A., Ilyashenko, L.K.: Parameter evaluation of cardiovascular system in schoolchildren under the conditions of latitudinal displacement. Hum. Ecol. **4**, 30–35 (2018)
28. Filatov, M.A., Ilyashenko, L.K., Kolosova, A.I., Makeeva, S.V.: Stochastic and chaotic analysis of students' attention parameters of different ecological zones. Hum. Ecol. **7**, 11–16 (2019)
29. Khadartseva, K.A., Filatov, M.A., Melnikova, E.G.: The problem of homogenous sampling of cardiovascular system parameters among migrants in the Russian North. Hum. Ecol. **7**, 27–31 (2020)
30. Eskov, V.M., Gudkov, A.B., Filatov, M.A., Eskov, V.V.: Principles of homeostatic regulation of functions in human ecology. Hum. Ecol. **10**, 41–49 (2019)
31. Filatova, O.E., Pyatin, V.F., Filatov, M.A., Shakirova, L.S.: Effect of low temperature on cardiointervals during physical training in men. Hum. Ecol. **1**, 17–21 (2021)
32. Filatov, M.A., Ilyashenko, L.K., Makeeva, S.V.: Psychophysiological parameters of students before and after translatitude travels. Hum. Ecol. **4**, 18–24 (2019)

Adaptive Data Envelopment Analysis Models of Ecosystems of Megalopolises

I. B. Mamai[1], B. I. Savelyev[1], and S. V. Pronichkin[2(✉)]

[1] Federal Center of Theoretical and Applied Sociology of Russian Academy of Sciences, Moscow, Russia

[2] Federal Research Center "Computer Science and Control" of Russian Academy of Sciences, Moscow, Russia

Abstract. Data envelopment analysis is an effective method for assessing the relative sustainability of ecosystems in megalopolises, which are characterized by a variety of environmental and socio-economic indicators. The existing models do not allow adapting the parameters of data envelopment analysis models to the preferences of the decision-maker and do not provide a consistent assessment of the effectiveness of complex socio-economic systems. The most important characteristics of complex socio-economic systems have to be calculated by indirect methods by solving a number of optimization problems. Adaptive data envelopment analysis models are developed to determine the sustainability of ecosystems in megalopolises on the basis of ecological and socio-economic indicators of anthropogenic load. In the proposed adaptive models, intermediate results at the previous stage of the functioning of the megalopolis ecosystem are only partially consumed at the next stage. A part of the input parameters of the functioning of the ecosystems of the megalopolis can be freely distributed. Additional input parameters are directly consumed at the next stage of the functioning of the megalopolis ecosystem. The optimal values of the parameters of the models are determined on the basis of the dialogue procedure of sequential cutting of multidimensional sets using affine subspaces. Indicators of sustainability of ecosystems of megalopolises, which characterize a low level of anthropogenic load, are calculated.

Keywords: Sustainability · Decision maker · Ecosystem · Megalopolis

1 Introduction

The sustainability of ecosystems of megalopolises is considered as the ability to function effectively and resist various inevitable negative impacts of exogenous and endogenous nature [1, 2]. The data envelopment analysis (DEA) is an effective method for assessing the relative sustainability of ecosystems in megalopolises, which are characterized by a variety of input and output indicators. In the paper [3], a minimax DEA model is proposed for calculating general indicators of the sustainability of socio-economic systems using a generalized model, in which structural parameters are added to obtain

R. Silhavy et al. (Eds.): CoMeSySo 2021, LNNS 231, pp. 1023–1028, 2021.
https://doi.org/10.1007/978-3-030-90321-3_84

the overall efficiency of a socio-economic system. Such approaches have been implemented for socio-economic systems with one input and several outputs. The authors of [4] improved this model by adding a search algorithm to reliably identify the values of the discriminating parameters for cases with one input and several outputs, including those obtained in the order scale. In the paper [5], mixed integer linear programming was used to develop a minimax sustainability model without taking into account the discriminating parameter. The authors of [6] proposed a mixed integer linear programming model, considering one input and several outputs. The developed models make it possible to classify indicators of the sustainability of ecosystems in megalopolises, but they have a number of limitations. Weights for inputs and outputs in some cases give inadequate results, while full ranking is not provided for each class, and thus additional information is required for decision-making.

2 Methods

In the papers [7, 8], minimax models of DEA are proposed to determine the most effective socio-economic system for the case, including several inputs and several outputs. The disadvantage of this approach is that the final estimate may give lower efficiency estimates for some minimax efficient systems compared to others that are considered ineffective according to minimax DEA models. In order to determine the single most effective socio-economic system, constraints are added on obtaining more than one effective object. Two models have been developed, one of which considers the ideal point as a reference object, and the other uses the anti-ideal point. The developed models provide a complete ranking of socio-economic systems using a unified mathematical programming model. These models do not provide a consistent assessment of the performance of complex socio-economic systems, since one of them provides a common set of weights for inputs and outputs, while the other allows you to choose weights. In the papers [9, 10], the technique of symmetric weight assignment was extended and four models of DEA were proposed. Existing approaches are applicable only when scaling inputs and outputs in commensurable indicators of the functioning of complex socio-economic systems. Such DEA models can be used when interval data is available. However, in practice, there is inconsistency in the structure of modeling, since at the first stage of existing approaches, a different set of weights is calculated for each socio-economic system, while the second stage gives a general set of weights. In the paper [11], an approach based on calculating distances was proposed by combining the principles of optimality of the main criterion and the ideal point. This approach can give more than one effective object, so the resolution of the combination of the principles of optimality is questionable. In the paper [12], a fuzzy model of total weight based on DEA is proposed, where the effectiveness of complex socio-economic systems is assessed using preference functions. In the paper [13], a model based on the DEA with a total weight is proposed, which is applicable for fuzzy data. Such a fuzzy DEA model leads to the determination of a general set of weights and can be reduced to an auxiliary fuzzy optimization problem with two goals. With regard to fuzzy DEA models, a set of weighting coefficients is determined for various boundaries and values of α-cuts. The efficiency values are then calculated as fuzzy sets. A minimax model of efficiency was developed taking into account intervals and/or inaccurate data [14]. However, the existing models do not allow adapting the parameters of the DEA models to the preferences of the decision maker (DM). Currently, DM cannot analyze the activities of the socio-economic system without using dialogue procedures [15, 16]. In the process of analyzing the activities of complex socio-economic systems, the efficiency coefficients are added with some weight coefficients, the result obtained is called the rating of the object, and the function of the activity coefficients is called the rating function [17]. This approach, perhaps, captivates many experts and analysts with its simplicity and clarity. However, the rating function strongly distorts the assessment of the performance of complex socio-economic systems. From the very beginning of the application of the DEA methodology to real problems, hidden effects and inaccuracies began to be revealed, since the calculations by the DEA model (measures of efficiency, scale elasticity, marginal coefficients, etc.) for some socio-economic systems did not coincide with the opinion of expert. The fact is that in the DEA methodology, the set of production possibilities is determined by real statistical data, this is the advantage of the approach. However, the correspondence function (functional dependence between the vector of costs and the vector of output) is not explicitly specified. Therefore, the most important characteristics of the activities of objects have to be calculated by indirect methods using the solution of a number of optimization problems, which significantly complicates the use of this approach to the full extent for experts and analysts.

3 Results

Adaptive DEA models have been developed to determine the sustainability of ecosystems of megalopolises based on environmental and socio-economic indicators of anthropogenic load. In the proposed adaptive models, intermediate results at the previous stage of functioning of the megalopolis ecosystem are only partially consumed at the next stage. Some of the input parameters of the functioning of the ecosystems of the megalopolis can be freely distributed. Additional input parameters are directly consumed at the next stage of the functioning of the megalopolis ecosystem.

Let there be n ecosystems of megalopolises $(i = 1, \ldots, n)$, that function for T periods $(t = 1, \ldots, T)$. Criteria characterizing the potential of the megalopolis ecosystem will be denoted by $X_{ji}^{(t)}$, $(j = 1, \ldots, m)$, and the results of functioning through $Y_{ri}^{(t)}$, $(r = 1, \ldots, z)$. Intermediate variables are denoted by $S_{di}^{(t-1,t)}$, $(t = 1, \ldots, T+1; d = 1, \ldots, q)$. The coefficients of the importance of criteria $X_{ji}^{(t)}$, $Y_{ri}^{(t)}$, $S_{di}^{(t-1,t)}$ denote by w_i, u_r, l_d. Then the overall potential and efficiency of the megalopolis ecosystem i during T periods can be presented in the following form.

$$\sum\nolimits_{t=1}^{T} \sum\nolimits_{r=1}^{z} u_r Y_{ri}^{(t)} + \sum\nolimits_{d=1}^{q} l_d S_{di}^{(T,T+1)} \tag{1}$$

$$\sum\nolimits_{t=1}^{T} \sum\nolimits_{j=1}^{m} w_j X_{ji}^{(t)} + \sum\nolimits_{d=1}^{q} l_d S_{di}^{(0,1)} \tag{2}$$

The optimal values of the coefficients w_i^*, u_r^*, l_d^* are determined on the basis of the dialogue procedure of sequential cutting of multidimensional sets using affine subspaces. The total anthropogenic load on the megalopolis ecosystem takes the following form:

$$AL_i = \frac{\sum_{t=1}^{T}\sum_{r=1}^{z} u_r^* Y_{ri}^{(t)} + \sum_{d=1}^{q} l_d^* S_{di}^{(T,T+1)}}{\sum_{t=1}^{T}\sum_{j=1}^{m} w_j^* X_{ji}^{(t)} + \sum_{d=1}^{q} l_d^* S_{di}^{(0,1)}} \tag{3}$$

Taking into account the restrictions on the value of the anthropogenic load of complex socio-economic systems of megalopolises, the adaptive model of the multicriteria analysis of sustainability can be presented in the following form:

$$SE = \max_i \frac{\sum_{t=1}^{T}\sum_{r=1}^{z} u_r Y_{ri}^{(t)} + \sum_{d=1}^{q} l_d S_{di}^{(T,T+1)}}{\sum_{t=1}^{T}\sum_{j=1}^{m} w_j X_{ji}^{(t)} + \sum_{d=1}^{q} l_d S_{di}^{(0,1)}} \tag{4}$$

$$\begin{cases} \frac{\sum_{t=1}^{T}\sum_{r=1}^{z} u_r Y_{ri}^{(t)} + \sum_{d=1}^{q} l_d S_{di}^{(T,T+1)}}{\sum_{t=1}^{T}\sum_{j=1}^{m} w_j X_{ji}^{(t)} + \sum_{d=1}^{q} l_d S_{di}^{(0,1)}} \le 1 \\ \frac{\sum_{r=1}^{z} u_r Y_{ri}^{(t)} + \sum_{d=1}^{q} l_d S_{di}^{(t,t+1)} + \sum_{d=1}^{q} l_d S_{di}^{(t)}}{\sum_{j=1}^{m} w_j X_{ji}^{(t)} + \sum_{d=1}^{q} l_d S_{di}^{(t-1,t)} + \sum_{d=1}^{q} l_d S_{di}^{(t)}} \le 1 \end{cases} \tag{5}$$

The developed model makes it possible to explore the structural relationship between the ecological and socio-economic characteristics of the anthropogenic load on the megalopolis ecosystem.

4 Discussions

Methods and tools are developed for constructing adaptive models for analyzing the functioning of complex socio-economic systems of megalopolises, in which there are no "strange" and inadequate results. Regardless of the purpose of use when assessing performance in terms of providing information about whether the systems being assessed are stable or not, existing DEA models have various hidden effects and imprecisions. First, these models need to be solved n times, to obtain sustainability estimates for all ecosystems. Existing DEA models do not define general weights for evaluating objects and allow each object to assign its own weights in order to maximize sustainability estimates, which leads to inaccurate results. The ranking of multi-feature objects according to performance indicators obtained on the basis of various sets of weights leads to the impossibility of a single assessment. Secondly, the existing models do not allow distinguishing a group of objects, since they consider all objects that have received a sustainability rating equal to one to be effective. Flexibility in defining weights leads to the definition of more than one effective object. The use of approaches such as cross-stability analysis and weight constraints either increase the computational load on the decision maker or distort the objective weighting scheme. The proposed models do not require subjective assessment to determine the weights for inputs and outputs, providing a common assessment for all objects. The choice of input and output weights of ecological and socio-economic indicators of the load on the ecosystem of a megalopolis is limited by a general assessment, which increases the discriminatory characteristics of the analysis of complex socio-economic systems. The use of common weights in adaptive models is more adequate for calculating indicators of the sustainability of ecosystems in megalopolises; there are no hidden effects and inaccuracies in the adaptive model. In addition, the proposed general-weight DEA models provide computational savings for determining ecosystem sustainability indicators in comparison with traditional DEA models. It is proposed to use the following indicators as criteria of the ecological load on the ecosystems of megalopolises: wastewater treatment; water consumption; water pollution; greenhouse gas emissions; power consumption; renewable energy; population density; household waste; waste recycling. The socio-economic load includes such aspects as: availability of health care; mortality of the population; health care costs; morbidity of the population; poverty of the population; business conditions; unemployment of the population; state debt; differentiation of incomes of the population; spending on science and education; crime; corruption; landscaping of territories; public transport. The following indicators are considered as the resulting indicators of the functioning of the megalopolis ecosystem: the comfort of the urban environment; safety; life span; GRP per capita; environmental friendliness. Indicators of sustainability of ecosystems of megalopolises are calculated, characterizing a low level of anthropogenic load for the following megalopolises: Los Angeles - 0.36; Mexico City - 0.37; Chicago - 0.4; Lima - 0.42; New York - 0.46; Buenos Aires - 0.33; Sao Paulo - 0.33; Rio de Janeiro - 0.43; Lagos - 0.6; London - 0.18; Paris - 0.16; Johannes Burg - 0.53; Istanbul - 0.39; Cairo - 0.55; Moscow - 0.41; Tehran - 0.57; Karachi - 0.87; Mumbai - 0.67; Delhi - 0.55; Dhaka - 0.86; Bangkok - 0.53; Jakarta - 0.61; Hong Kong - 0.55; Beijing - 0.44; Tianjin - 0.47; Manila - 0.65; Shanghai - 0.62; Seoul - 0.17; Osaka - 0.14; Tokyo - 0.15; Sydney - 0.27. The ranking of megalopolises according to the increasing level of anthropogenic load on ecosystems is as follows: Osaka > Tokyo > Paris > Seoul > London > Sydney > Buenos Aires = Sao Paulo > Los Angeles > Mexico City > Istanbul > Chicago > Moscow > Lima > Rio de Janeiro > Beijing > New York > Tianjin > Johannesburg = Bangkok > Cairo = Delhi = Hong Kong > Tehran > Lagos > Jakarta > Shanghai > Manila > Mumbai > Dhaka > Karachi.

5 Conclusion

The reasons for inaccuracies in the DEA models are identified, new approaches are developed for constructing adequate and correct models, and ways are found to improve the state of complex socio-economic systems of megalopolises in the multidimensional space of opportunities. The measure of efficiency calculated by the models of strong conditions of complementary slackness may not coincide with the measure of efficiency calculated by the existing models of the DEA methodology, which contradicts the theory of optimization. The reference set may contain ineffective socio-economic systems. It is advisable to calculate the most important indicators of the activity of complex socio-economic systems in the direct space of indicators; for this purpose, adaptive models are developed that implement algorithms for cutting multidimensional sets using affine subspaces. Adaptive models consist of several iterations that make it possible to effectively investigate the structure of multi-feature ecosystems of megalopolises.

Acknowledgments. The article is prepared with the financial support of the Russian Science Foundation, project № 19-78-10035.

References

1. Yang, Z., Yang, H., Wang, H.: Evaluating urban sustainability under different development pathways: a case study of the Beijing-Tianjin-Hebei region. Sustain. Cities Soc. **61**, 226–233 (2020)
2. Kates, R., Clark, W., Corell, R., Hall, J.: Sustainability science. Science **292**, 641–642 (2001)
3. Despotis, D., Sotiros, D., Koronakos, G.: A network DEA approach for series multi-stage processes. Omega **61**, 35–48 (2016)
4. Marbini, A., Toloo, M.: An extended multiple criteria data envelopment analysis model. Expert Syst. Appl. **73**, 201–219 (2017)
5. Dai, Q., Li, Y., Wu, D.: A DEA-based incentive approach for allocating common revenues or fixed costs. Eur. J. Oper. Res. **292**, 675–686 (2020)
6. Keshavarz, E., Toloo, M.: Efficiency status of a feasible solution in the multi-objective integer linear programming problems: a DEA methodology. Appl. Math. Model. **39**, 3236–3247 (2015)
7. Khezrimotlagh, D., Salleh, S., Mohsenpour, Z.: A new robust mixed integer-valued model in DEA. Appl. Math. Model. **37**, 9885–9897 (2013)
8. Toloo, M., Mensah, E.: Robust optimization with nonnegative decision variables: a DEA approach. Comput. Ind. Eng. **127**, 313–325 (2018)
9. Wang, Q., Wu, Z., Chen, X.: Decomposition weights and overall efficiency in a two-stage DEA model with shared resources. Comput. Ind. Eng. **136**, 135–148 (2019)
10. Foroughi, A., Tavassoli, M.: Discriminating extreme efficient decision-making units in DEA using random weight vectors. Comput. Ind. Eng. **128**, 305–312 (2018)
11. He, F., Xu, X., Zhu, L.: Interval efficiency improvement in DEA by using ideal points. Measurement **87**, 138–145 (2016)
12. Liu, S.: Restricting weight flexibility in fuzzy two-stage DEA. Comput. Ind. Eng. **74**, 149–160 (2014)
13. Marbini, A., Tavana, M., Beigi, Z.: A common-weights DEA model for centralized resource reduction and target setting. Comput. Ind. Eng. **79**, 195–203 (2015)
14. Amirkhan, M., Didehkhani, H., Hafezalkotob, A.: Mixed uncertainties in data envelopment analysis: a fuzzy-robust approach. Expert Syst. Appl. **103**, 218–237 (2018)
15. Tavana, M., Damghani, K., Hosseini, A.: A fuzzy multi-objective multi-period network DEA model for efficiency measurement in oil refineries. Comput. Ind. Eng. **135**, 143–155 (2019)
16. Peykani, P., Mohammadi, E., Malkhalifeh, M.: Fuzzy data envelopment analysis: an adjustable approach. Expert Syst. Appl. **136**, 439–452 (2019)
17. Bagherikahvarin, M., Smet, Y.: A ranking method based on DEA and PROMETHEE II. Measurement **89**, 333–342 (2016)

Time Detection of Malware Threads

Martin Strmiska(✉), Pavel Mesicek, Libor Pekař, and Roman Jašek

Tomas Bata University, Zlin, Czechia
strmiska@utb.cz

Abstract. Malware is an unwanted software that performs actions in computers or computer networks, which users might disagree with. One of the worst types of malware is ransomware that affects the victim's data by modifying, deleting, or blocking the access to them. Frequent malware attacks on organizations led to a change in malware detection from external identification (companies were dependent on other organizations or their products) to internal identification. Based on this, the time needed to detect ransomware (dwell time) has significantly decreased. Nowadays, internal detection prevails over the external one. The dwell time differs based on the continent. In the paper, the malware and ransomware descriptions with their variants are provided, and the concept of dwell time is described. Moreover, attention is not only paid to the reduction of dwell time within the recent years but also to how the most used vector attacks are connected.

Keywords: Malware · Detection · Dwell-time · Ransomware · Cybersecurity · Phishing · Remote Desktop Protocol

1 Introduction

A computer attack is any attempt to destroy, change, steal property, or gain unauthorized access to a property by using computer systems. One of the most problematic attacks is malware that spreads mainly due to software engineering. In this case, the attacker gets control over a computer or a computer network. Attackers manipulate users in order to infect their devices. It is often much easier to exploit users' weaknesses than it is to find a network to attack. For the identification of malware, it is essential to detect it just in time. This detection is called dwell time. Dwell time is calculated as the number of days the attacker is present in a victim's network before being detected.

A recent study (Fireeye, 2021) shows that 59% of the security incidents investigated by Mandiant last year were initially detected by the organizations themselves, which is an improvement of 12% from the prior year. Pervasive ransomware campaigns drove down the median dwell time as threat actors sought to capitalize on shifting trends in the workspace and a global crisis.

The purpose of this research is to summarize attacks which have been the most frequent over the last few years. In addition, the paper demonstrates how global dwell time has decreased and highlights the importance of internal notifications. The paper also shows a possible use of dwell time as a well-timed analysis of threads and prevention. Focusing on the ransomware field, we present important considerations regarding how

R. Silhavy et al. (Eds.): CoMeSySo 2021, LNNS 231, pp. 1029–1034, 2021.
https://doi.org/10.1007/978-3-030-90321-3_85

attacks are carried out, but we also want to point out that very often it is not the fault of the technology but poor-quality work of information technology (IT) department or social engineering.

The paper is organized as follows. The first part of the article describes malware, including its variants and a detailed description of ransomware and other terms for this work. The second part, the results section, explains the concept of dwell time, its reduction within the recent years and what attack vectors are most used and how they are related. The last paragraph of this article is a conclusion that briefly summarizes the whole paper.

2 Theoretical Background

2.1 Malware

Malware is malicious computer software that may perform an action that the user might disagree with, such as opening pop-ups, sending spam, or causing system breakdowns. Malware mainly spreads via a computer network. There are several types of malware such as computer viruses, computer worms, a Trojan horse, adware, spyware, ransomware, and others [2, 3, 17].

A computer virus can spread without the user's awareness. Its behavior is similar to any other's biological virus, and it may create its copies enabling further transmission. Those copies hide in files or programs [15].

A computer worm is a type of computer virus inside an infected file and enters a device or a computer network. Once inside the device, it distributes itself. Its goal is to gain information, make a freeway for a hacker, or cause harm (for example, making it impossible to use some of the programs) [15].

A Trojan horse is a virus pretending to be beneficial for the user; often, it offers a helpful feature, a free program, tool, or entertainment. In fact, it is used to open the back door to overtaking control over an attacked device, gain user's personal information and send them to hackers, download and launch another malware inside the attacked system [16].

An Adware is an unpleasant advertisement application popping up while visiting some websites. They are not threading but annoying and often offending [6, 18].

A Spyware is a virus collecting information about user's browsing history on the internet or personal data (for instance, credit card numbers, passwords, etc.) [6].

A Ransomware is a computer program that blocks a computer system or may encrypt data inside the system. Further, it demands ransom from the victim (user) in exchange for returning the stolen data. Some forms of a ransomware encrypt files on the storage device; others lock the system and try to make the victim pay with a denunciative message [3].

A Dwell time is detection time; in other words, it is the time between when an attack takes place and when it is discovered [6, 18].

A Remote Desktop Protocol (RDP) is a proprietary protocol developed by Microsoft company that offers to the user, who has a graphical user interface (GUI), connection to another computer via a computer network [3, 9, 13].

A Virtual Private Network (VPN) is an instrument for the safe interconnection of several computers when connecting to an unreliable public network [10, 11].

2.2 Ransomware as a Malware

Ransomware is a sophisticated malware program created by hackers with knowledge of computer systems. Devices may be infected when opening a virus-infected e-mail, web browsers, or visiting webs where the virus is. It also spreads similarly to computer worms or computer viruses. For example, WannaCry is the most common type of ransomware all around the globe. WannaCry has infected almost 125 000 organizations in more than 150 countries. It might appear on a mobile phone as well. Amongst ransomware that infects android phones is Doublelocker, Charger, Jisut, Lockerpin, Simplocker [3, 22].

Encrypting ransomware is a damaging program encrypting data on a hard drive based on which a ransom is demanded. The most used ones are Gpcode, Archiveus, Krotten, and Cryzip. The oldest was the AIDS trojan owned by Joseph Popp [21].

Non-encrypting ransomware is not as damaging as the encrypting ransomware; it only "blackmails" the user. In 2011, a new ransomware worm appeared. It pretended to be a Windows notification, informing Windows users about the urgency to restart their computer. The users had to call an international phone number and insert a six-digit code so they could reactivate the computer. However, even though the malware was claiming the call to be free; the opposite was true [3, 19, 20].

3 Results

Just as cyber-attacks are becoming more sophisticated, so are the methods by which companies and individuals can defend themselves. The time when companies depended only on external products and companies is long gone. Recently, they have understood that their business depends on cybersecurity, and they cannot afford financial losses. Especially with the threat of ransomware. It is not only about investing in products but also in people, who are the most vulnerable part of companies [5].

Improvements of internal departments in recent years have had a major impact on reducing global dwell time. The impact of external and internal notifications only started to be measured in 2015. From the beginning of the measurement, there is a clear evidence that internal departments detect a cyber-attack much earlier than external services. A significant difference is evident from the recent years, with the detection rate almost doubling between 2019 and 2020 [1].

The chart shows a significantly decreasing global dwell time and highlights the importance of internal notifications. In particular, thanks to internal departments, the current global dwell time is only 24 days.

As already mentioned, the global dwell time is 24 days, but this low figure is mainly due to America, where the dwell time is 17 days. In 2019, it was 60 days; this is the only world region that has reduced dwell time very significantly. Europe, the Middle East, and Africa had had a slight deterioration compared to 2019 where they had a dwell time of 54 days and in 2020 the dwell time was 66 days. The Asia-Pacific had an even more

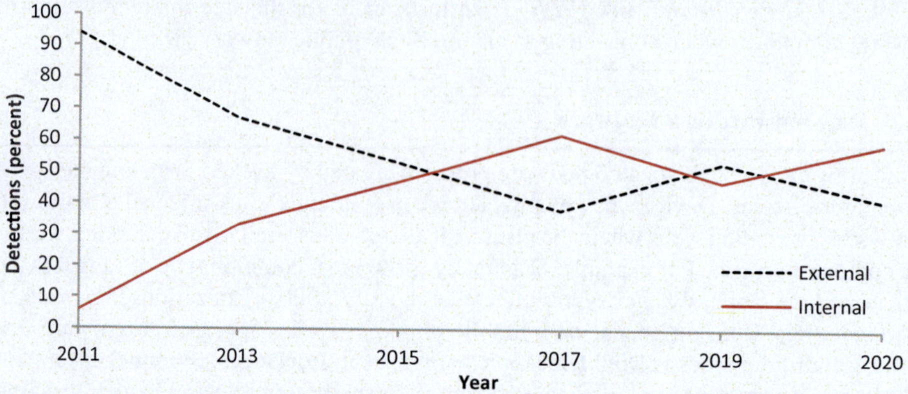

Fig. 1. Illustrates the results of a study dealing with the percentage of internal and external announcements varied over the years.

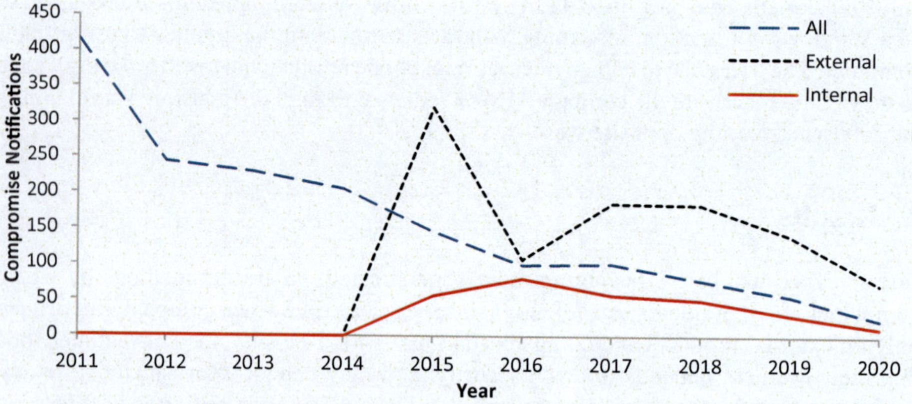

Fig. 2. Illustrates the results of reducing global dwell time between 2011–2020.

significant drop in values than Europe, the Middle East and Africa. From also 54 days in 2019, its dwell time dropped to 76 days in 2020.

There is another factor that significantly decreases the global dwell time and that is the threat of ransomware. According to Group-IB, the number of ransomware attacks grew by more than 150% in 2020 [7]. Other cyber-attacks that do not behave so conspicuously on the network are estimated to have a detection time of 45 days. Ransomware, on the other hand, is very aggressive and therefore has a detection time estimated to be 5 days. Ransomware can also be considered as one of the main reasons why internal notifications have such a low dwell time [4].

From the previous paragraph, it is obvious that companies need to focus on protection against ransomware. As for 2021, it is necessary to significantly improve the security policy. The most used vector to gain access was through Remote Desktop Protocol since more than half attacks were carried out through it. Phishing was responsible for 29% and exploitation of public-facing applications for 17%. Nevertheless, all factors are related [8].

RDP is often attacked because of short or dictionary passwords, disabled account locking, failure to use two-factor authentication, disabled network authentication, untrusted certificates, and use of known logins such as administrator, support printer, etc. However, none of these things are merely the fault of the RDP but also the poor-quality work of the Information Technology (IT) department or poor knowledge or improvidence of a company's workers. Because of these two factors, the statistic of the RDP is so high [9, 13].

What is also worth mentioning is that IT administrators rely strongly on the Virtual Private Network, which is an important security system that should be set up well. The first mistake is that users commonly use the same password for the VPN as for the domain. Another mistake is that attackers can easily guess passwords since the VPN is accessible from all over the world. Vulnerabilities have also already been proven for VPNs, especially with Palo Alto, FortiNet and Pulse Secure [9–12, 14].

4 Conclusion

Malware has become more advanced and greater in numbers over the last years since it also attacks computers, phones, and other smart devices. It shows that internal notifications involved in global dwell time attacks are prevailing. In 2020, internal notifications accounted for almost 60% of all announcements. Our results show how ransomware has changed global dwell time in recent years. We would like to point out that we did not have access to the hardware from the project when we published this paper. For one of our many future extensions, we plan to send targeted phishing e-mails that we craft ourselves. These e-mails would contain a link that will not be malicious in real, but simply report clickthrough rates to us, the researchers. Subsequently, we would like to send out a questionnaire asking why the questioned people will have decided or have not decided to click on the link and what exactly they have found suspicious.

Acknowledgement. The paper was supported by Internal Grant Agency of Tomas Bata University in Zlín under the project number IGA/CebiaTech/2021/006.

References

1. FireEye, M-Trends (2020). https://content.fireeye.com/m-trends/rpt-m-trends-2020. Accessed 6 July 2021
2. Shetty, N., Praveen, R.: A survey paper on malware detection techniques. Int. J. Adv. Trends Comput. Sci. Eng. **10**, 558–563 (2021)
3. Malware | Detection, Prevention, Protection & Removal | Avast. https://www.avast.com/c-topic-malware. Accessed 13 June 2021
4. Kovacs, E.: Breaches Detected Faster, But Ransomware Surge a Major Factor. https://www.securityweek.com/breaches-detected-faster-ransomware-surge-major-factor-fireeye. Accessed 13 Apr 2021

5. Why Cybersecurity should be part of Internal Communication Strategies. https://kingeclient.com/why-cybersecurity-should-be-part-of-internal-communication-strategies. Accessed 14 May 2019
6. Resource Center - Spyware and Adware. Hamilton College. https://my.hamilton.edu/offices/lits/rc/spyware-and-adware. Accessed 23 Mar 2018
7. Leyden, J.: Ransomware attacks more than doubled last year as cybercrime operations scale up during coronavirus pandemic. The Daily Swig | Cybersecurity News and Views. https://portswigger.net/daily-swig/ransomware-attacks-more-than-doubled-last-year-as-cybercrime-operations-scale-up-during-coronavirus-pandemic. Accessed 30 June 2021
8. Group-IB, Ransomware uncovered: Attackers' latest methods. https://go.group-ib.com/rs/689-LRE-818/images/Group-IB_Ransomware_Uncovered_whitepaper_eng.pdf
9. Haller, M.: Proč zabezpečení RDP pomocí VPN hackery nezastaví (2020). https://martinhaller.cz/bezpecnost/proc-zabezpeceni-rdp-pomoci-vpn-hackery-nezastavi. Accessed 24 Feb 2020
10. Tsai, O., Chang, M.: Attacking SSL VPN - Part 3: The Golden Pulse Secure SSL VPN RCE Chain, with Twitter as Case Study! https://devco.re/blog/2019/09/02/attacking-ssl-vpn-part-3-the-golden-Pulse-Secure-ssl-vpn-rce-chain-with-Twitter-as-case-study. Accessed 2 Sep 2019
11. Tsai, O., Chang, M.: Attacking SSL VPN - Part 2: Breaking the Fortigate SSL VPN. https://devco.re/blog/2019/08/09/attacking-ssl-vpn-part-2-breaking-the-Fortigate-ssl-vpn. Accessed 9 Sep 2019
12. Tsai, O., Chang, M.: Attacking SSL VPN - Part 1: PreAuth RCE on Palo Alto GlobalProtect, with Uber as Case Study! https://devco.re/blog/2019/07/17/attacking-ssl-vpn-part-1-PreAuth-RCE-on-Palo-Alto-GlobalProtect-with-Uber-as-case-study. Accessed 17 July 2019
13. Securing Remote Desktop (RDP) for System Administrators. https://security.berkeley.edu/education-awareness/securing-remote-desktop-rdp-system-administrators. Accessed 8 Aug 2021
14. Lakshmanan, R.: Pulse Secure VPNs Get New Urgent Update for Poorly Patched Critical Flaw. https://thehackernews.com/2021/08/pulse-secure-vpns-get-new-urgent-update.html. Accessed 9 Aug 2021
15. Aycock, J.: Computer Viruses and Malware. Springer, Berlin, pp. 14–15 (2006)
16. Fortinet, What Is a Trojan Horse? Trojan Virus and Malware Explained | Fortinet. https://www.fortinet.com/resources/cyberglossary/trojan-horse-virus. Accessed 9 Aug 2021
17. Souri, A., Hosseini, R.: A state-of-the-art survey of malware detection approaches using data mining techniques. Human-centric Comput. Inf. Sci. **8**, 1–22 (2018)
18. Zavrak, S.: Adware: a review. Int. J. Comput. Sci. Inf. Technol. **6**, 5599–5604 (2015)
19. Brewer, R.: Ransomware attacks: detection, prevention and cure. Netw. Secur. **9**, 59 (2016)
20. Forbes, Another Ransomware Campaign Threatens to Expose Victims' Data. https://www.forbes.com/sites/leemathews/2020/01/23/another-ransomware-campaign-threatens-to-expose-victims-data/#271fbefd770a. Accessed 9 Aug 2021
21. eTrust, eTrust Spyware Encyclopedia - AIDS Information Trojan. http://www3.ca.com/securityadvisor/pest/pest.aspx?id=175. Accessed 9 Aug 2021
22. BBC, Cyber-attack: Europol says it was unprecedented in scale - BBC News. https://www.bbc.com/news/world-europe-39907965. Accessed 9 Aug 2021

Analysis of the Effectiveness of National Research Support Systems Based on Neural Network Models

V. L. Rozaliev[1], Y. A. Orlova[1], and S. V. Pronichkin[2](✉)

[1] Volgograd State Technical University, Volgograd, Russia
[2] Federal Research Center "Computer Science and Control" of Russian Academy of Sciences, Moscow, Russia

Abstract. Analysis of the effectiveness of national research support systems includes processes related to the creation, sharing and application of knowledge. A key feature and limitation in the implementation of the analytical functions of research support systems is that only publication activity indicators are used as an efficiency criterion. Mathematical and software support has been developed for the systems analysis of the effectiveness of research support systems. The formation of a database for the analysis of factors affecting the effectiveness of the national research support system has been carried out. To analyze the input and output indicators of the effectiveness of the national research support system, a neural network approach was chosen. Distribution of factors over several neural structures, depending on the functional, process or object component of the analysis, has been performed. The most pressing scientific and technical problems form the first layer of the neural network. For the scientific study of ways to solve scientific and technical problems, four main functional blocks of interdisciplinary research have been identified, forming layers of a neural network that cover scientific directions and areas of using scientific results.

Keywords: Science foundation · Neural network · Efficiency · Scientific knowledge

1 Introduction

The development and adjustment of the system of priorities expressing in a concentrated form the position of the scientific community and the needs of management structures of various levels in the scientific study of options for managerial decisions allows to form an effective strategy for the development of the national research support system and to introduce a clear logic in support of priority projects [1, 2]. Science policy management emphasizes the strategic processes for the development, implementation and use of various research support tools [3, 4]. This aspect focuses on strategy and supporting mechanisms for individual research groups. Strategy, coordination and integration are the main management approaches aimed at updating various research support tools. Effective mechanisms such as reward and recognition, as well as efficient allocation

R. Silhavy et al. (Eds.): CoMeSySo 2021, LNNS 231, pp. 1035–1040, 2021.
https://doi.org/10.1007/978-3-030-90321-3_86

of resources, are critical for the sustainable development of research support systems [5, 6]. Continuous improvement and performance evaluation are also key factors in updating the research support systems policy [7]. Knowledge management is viewed as managing the cycle of creating scientific knowledge within the framework of initiative projects of scientific foundations [8, 9]. This governance aspect describes how the flow of knowledge from creation, diffusion and accumulation affects the achievement of the goals of national research support systems.

2 Methods

Research into the effectiveness of national research support systems includes processes related to the creation, sharing and application of knowledge. A systems analysis of the Russian Foundation for Basic Research and Russian Science Foundation scientific funds as actors of the national research support system will create a basis for the effective development of various tools for supporting scientific projects in Russia. A key feature and limitation in the implementation of the analytical functions of research support systems is that only publication activity indicators are used as an efficiency criterion [10, 11]. In order to take into account the noted features, mathematical and software support has been developed for the systems analysis of the effectiveness of scientific research support. The formation of a database for the analysis of factors affecting efficiency was carried out. To analyze the input and output indicators of the effectiveness of the national system for supporting scientific research, a neural network approach was chosen [12, 13]. The neural network approach does not require a priori specification of the equation and, in comparison with other methods, it is not required to choose the boundaries of independent variables, initial values, and convergence criteria. Distribution of factors over several neural structures was performed, depending on the functional, process or object component of the analysis. The most pressing scientific and technical problems form the first layer of the neural network. For the scientific study of ways to solve scientific and technical problems, four main functional blocks of interdisciplinary research have been identified, forming layers of a neural network, which cover scientific directions and areas of using scientific results.

3 Results

In the developed neural network model, the level of support for scientific research is determined by three factors X_1, X_2 and X_3 for neural layers RI_1, RI_2 and RI_3 respectively. Values for RI_1, RI_2 and RI_3 were calculated using an artificial neural network. A neural network can be considered as a nonlinear regression model with identity mapping, and logistic functions can approximate any continuous function arbitrarily well, provided that a sufficient number of elements appear at the middle levels [14, 15]. The following variables are used.

Science foundation organizational infrastructure: expert mechanisms; organizational culture; structure of the scientific fund; personnel management; leadership; competence.

Science foundation policy management: strategy development mechanisms; mechanisms for implementing the strategy; resource management mechanisms; mechanisms for ensuring sustainability; mechanisms for continuous improvement.

Science foundation knowledge management: formation of internal knowledge; acquisition of external knowledge; organizational learning; exchange and use of knowledge; accumulation of knowledge.

With the involvement of experts from the Russian Science Foundation and the Russian Foundation for Basic Research, the following most urgent problems of scientific and technical policy were identified as the most important in the near future: solving current and long-term problems of social development; increasing the manageability and social efficiency of the regions; ensuring national security; preservation and strengthening of intellectual sovereignty; observance of Russian interests in a changing world.

The selected most pressing state problems form the first layer of the neural network. For the scientific study of ways to solve state problems, four main functional blocks of interdisciplinary research have been identified, forming layers of a neural network, which cover scientific directions and areas of use of scientific results.

1. Study of patterns and long-term trends of socio-economic and political development of Russian society and state.

Scientific directions: social and economic geography; sociology and demography; economic theory; social psychology; ethnology and religion studies; socio-political conflictology.

The area of use of the results: the development of conceptual and political-institutional foundations of state-building.

2. Expert and analytical support for decisions of public authorities on key issues of national development.

Scientific directions: socio-political and economic history of modern and contemporary times; national and international security; international law; theory of management and social organization; research of political ideologies; complex socio-economic and political forecasting.

The area of using the results: working out a long-term goal-setting of national political and economic development, determining the place of Russia in the world.

3. Preservation and reproduction of national culture and education, strengthening the ideological foundations of the development of the state and society, its further modernization and adaptation to global transformations.

Scientific directions: philology and literary criticism; national and international security; domestic and world history of various periods; history of philosophy and social thought; cultural studies and archeology.

The area of using the results: the development of the education system and patriotic education, the sustainability of national and cultural identity and the development of civil society, the development of the state system of values.

4. Formation in Russia of its own, based on scientific knowledge and logic of vision (discourse) of universal history.

Scientific directions: historical sciences.

The area of use of the results: the preservation of Russia among the historiographic superpowers, in which the entire temporal and spatial volume of the history of mankind is studied, and the Russian language is among the developed languages, in which a scientific description of traditional and modern societies is possible.

The reliability of the developed neural network model was assessed using cross-validation. The modeling accuracy was compared with the accuracy of multiple linear regression methods and nonlinear approximation by the method of least squares [16]. To substantiate the connections in the developed neural network model, three hypotheses were tested.

Hypothesis No. 1. The organizational infrastructure of scientific foundations has a significant and positive effect on the effectiveness of the national research support system $X_1 \rightarrow Y$.

Hypothesis No. 2. Policy management of scientific foundations has a significant and positive effect on the effectiveness of the national research support system $X_2 \rightarrow Y$.

Hypothesis No. 3. Knowledge management of scientific foundations has a significant and positive effect on the effectiveness of the national research support system $X_3 \rightarrow Y$.

4 Discussions

To train the neural network, a tenfold cross-validation was carried out using data from a survey of experts from the Russian Foundation for Basic Research and the Russian Science Foundation. After completing the training of the neural network model, it became possible to predict the effectiveness of the national research support system for combinations of different values of the input data, which makes it possible to develop specific management decisions to level the dysfunctions of the national research support system. All hypotheses were statistically confirmed. On the basis of the study, a model of relations was built. Established cause-and-effect relationships between three inputs: organizational infrastructure of the scientific foundation (X_1), policy management (X_2) and knowledge management (X_3), and the main result: the effectiveness of the national research support system (Y). Then, using a neural network, the coefficients are calculated $RI_1 = 0, 47$, $RI_2 = 0, 26$ and $RI_3 = 0, 27$. A neural network with feedforward and back propagation of an error was used. The transfer functions on the hidden layers and on the output layers are hyperbolic sigmoid $f(x) = \frac{1}{1+e^{-RI_i X_i}}$ and the linear operator $\sum_{i=1}^{N} RI_i X_i$, respectively. Depending on the needs of the decision-maker, the critical factor can be considered as the main result, and other factors contributing to this result, as input data for neural network modeling.

5 Conclusion

The key factors for supporting scientific research are highlighted and divided into three main aspects: the organizational infrastructure of the scientific foundation, the management of the foundation's policy and the management of scientific knowledge. These three aspects were identified in relation to the development of management of scientific production, that is, from the organizational aspect concerning the cultural, structural and human resources of scientific foundations, to the management aspect with the deployment of the policy of financing scientific research, as well as fundamental support of science from the point of view of knowledge management perspective. As input parameters of the neural network model, three internal factors associated with the organization of support for scientific research are identified. The organizational infrastructure includes management issues related to the shared values, beliefs, and principles of the science foundation. Organizational culture and values, as well as the structural dimension and competence of human resources, are important elements of this internal factor. Breakthrough scientific results come from ideas that are cultivated by the infrastructure of scientific foundations. An effective organizational structure can facilitate the internal flow of information, which is a key source of effective research foundations. The regulatory impacts of scientific and technical policy have been determined, contributing to the strengthening of the positive and mitigation of negative factors in the functioning of the national research support system. Based on the dualism of external and internal incentives for the development of science, state scientific foundations are able to significantly influence the solution of interrelated problems of sustainable development of the national research support system in Russia.

Acknowledgments. The article is prepared with the financial support of the Russian Science Foundation, project № 18-78-10047-P.

References

1. Faria, A., Oliveira, M., Borini, F.: Pubic funding for innovation: the importance of individual resources of the entrepreneur and the relational resources of the firm. Technol. Soc. **59**, 159–167 (2019)
2. Chan, K., Lau, T.: Assessing technology incubator programs in the science park: the good, the bad and the ugly. Technovation **25**, 1215–1228 (2005)
3. Plank, J., Doblinger, C.: The firm-level innovation impact of public R&D funding: evidence from the German renewable energy sector. Energy Policy **113**, 430–438 (2017)
4. Wonglimpiyarat, J.: Challenges and dynamics of FinTech crowd funding: an innovation system approach. J. High Technol. Manag. Res. **29**, 98–108 (2018)
5. Frigolett, H., Pyka, A., Leoneti, A.: On the performance and strategy of innovation systems: a multicriteria group decision analysis approach. Technol. Soc. **67**, 132–144 (2021)
6. Grupp, H., Mogee, M.: Indicators for national science and technology policy: how robust are composite indicators? Res. Pol. **33**, 1373–1384 (2004)
7. Marvel, M., Wolfe, M., Kuratko, D.: Escaping the knowledge corridor: how founder human capital and founder coachability impacts product innovation in new ventures. J Bus. Ventur. **35**, 60–72 (2020)
8. Wang, Y., Li, J., Furman, J.: Firm performance and state innovation funding: evidence from China's Innofund program. Res. Pol. **46**, 1142–1161 (2017)
9. Meski, O., Belkadi, F., Furet, B.: Towards a knowledge-based framework for digital chain monitoring within the industry 4.0 paradigm. Procedia CIRP. **84**, 118–123 (2019)
10. Abuhay, T., Kovalchuk, S., Lees, M.: Analysis of publication activity of computational science society in 2001–2017 using topic modelling and graph theory. J. Comput. Sci. **26**, 193–204 (2018)
11. Saarela, M., Karkkainen, T.: Can we automate expert-based journal rankings? analysis of the Finnish publication indicator. J. Informetr. **14**, 101–114 (2020)

12. Wang, X.: Research on inversion of ecosystem dynamics model parameters based on improved Neural Network algorithm. Microprocess. Microsyst. **80**, 605–612 (2020)
13. Young, W., Millie, D., Fahnenstiel, G.: Modeling net ecosystem metabolism with an artificial neural network and Bayesian belief network. Environ. Model. Softw. **26**, 1199–1210 (2011)
14. Hanin, B.: Universal function approximation by deep neural nets with bounded width and ReLU activations. Mathematics **7**, 992–1004 (2019)
15. Knaggard, A., Slunge, D., Sahlin, U.: Researchers' approaches to stakeholders: interaction or transfer of knowledge? Environ. Sci. Policy. **97**, 25–35 (2019)
16. Findlay, C., Zheng, L.: Estimating ecosystem risks using cross-validated multiple regression and cross-validated holographic neural networks. Ecol. Modell. **119**, 57–72 (1999)
17. Paruelo, J., Tomasel, F.: Prediction of functional characteristics of ecosystems: a comparison of artificial neural networks and regression models. Ecol. Modell. **98**, 173–186 (1997)

Designing Smart Healthcare Systems Using Fuzzy Relations

Shady Aly[1](✉), Jan Tyrychtr[2], and Ivan Vrana[2]

[1] Faculty of Engineering at Helwan, Helwan University, Helwan, Egypt
shady_ali@h-eng.helwan.edu.eg

[2] Faculty of Economics and Management, Department of Information Engineering, Czech University of Life Sciences in Prague, Prague, Czechia

Abstract. The investment in the clinic and hospitals environments while becoming necessary very attractive, it must be justifiable with obvious level of reliability based on logical and systematically understandable steps. In this paper, we present a simple, systematic, and objective fuzzy relation-based approach for the design of a smart healthcare system. The proposed approach establishes fuzzy relations among three model building blocks: the smart enabling technologies, the healthcare system smartness features, and the healthcare system operational objectives that are desirable to realize. The max-min composition operator is utilized for combining the aforementioned relations for attaining the target relation among the enabling technologies and the operational objectives. A priority of the smart enabling technology is computed based on the total impact relation of each enabling technology on all the operational objectives. Then, the design of the smart healthcare system is reached by adopting the enabling technologies in order of priority.

Keywords: Smart healthcare system · Smart environment (SE) · Smart enabling technologies (SETs) · Fuzzy logics (FL) · Fuzzy relations (FR)

1 Introduction

Smart hospitals, clinics and other healthcare systems are fertile applications of smart enabling technologies (SETs) which have been proven beneficial in many applications. In fact, that fast pace of development continues rapidly in the realm of smart sensors, actuators and microcontrollers opening unlimited number of opportunities for new applications and improvement in the healthcare arena.

Currently there exist so many options and brands for various sensing technologies, like motion, position, physiological and biomedical, and various physical measurement sensors (e.g., positioning sensors: RFID [1], motion sensors [1, 2], pressure sensor [3], physiological sensors: Electrocardiography (ECG) [1, 4] and EMG (Electromyography) [1, 5], etc.). Additionally, other sophisticated smart systems include Smart watch [6], Eye tracker [7], Computer brain interface (CBI) [8]). However, in spite of the fact that the rapid and continual emergence of SETs and availability of so many alternatives for each

R. Silhavy et al. (Eds.): CoMeSySo 2021, LNNS 231, pp. 1041–1052, 2021.
https://doi.org/10.1007/978-3-030-90321-3_87

technology type, actually this makes the problem of assessing and justifying investment in these technologies more and more difficult. Currently there is no structured, objective, and systematic means for differentiating among these wide spectra of available SETs, nor exist an approach that considers the costs of the technologies and the available budget for the project of developing smart hospital. In this article, we present a new approach to develop smart healthcare systems systematically and objectively, including hospitals, clinics, medical centers, etc. We shall employ fuzzy relations to bridge the gap in modeling and understanding the relation among SETs, and the operational objectives of a healthcare facility.

The article is organized as follows. Next section, we review the basic concepts of smart enabling technologies and smartness features, and we review the related literature. In Sect. 3, we introduce the major building blocks of the proposed model. Section 4 presents the methodology and in Sect. 5 we provide some illustrative example.

2 Smart Environment, Technologies and Smartness Features

A smart environment would typically consist of the following basic building blocks: Environmental objects: equipments, devices, instruments, furniture's, etc., Sensors: precepts, measures, monitor activities and statuses of devices, objects, users or resources, and send related readings or messages to a control system, Actuators: receive control signals from the microcontroller to execute real-time physical actions, Control system: a computer control or microcontroller that applies a set of programmed rules based on the information received from sensors and feedbacks from actuators, Middleware: integrating and interoperating heterogeneous devices by an intermediary software layer called middleware, Communication networks: enables communication between the control system, sensors, actuators. This includes wireless networks: LAN, Wi-Fi, Bluetooth, etc., and finally the Interfaces: enable users to communicate with system.

Table 1 shows an example set of relevant smart enabling technologies (SETs), studied based on the current literature.

The following are some of smartness features because of significant application areas of SETS in smart healthcare fields:

1. Monitoring health parameters: Various health parameters and indicators can be monitored such as contact and remote thermometry and physiological sensors like Electrocardiography (ECG) and EMG (Electromyography) [4, 5], blood pressure, etc.
2. Hospital assets tracking: Inside the hospitals, it's important to know exactly where all the valuable assets are placed. In addition to this, hospitals should quickly access to location of different assets to ensure availability in case of emergency.
3. Automated patient check-in: Patients and visitors perform check-in before arriving physically on site.
4. Location aware access to the nearest doctor: Real-time location-based technology can be employed to know the place of the nearest doctor.
5. Indoor navigation & optimized patient flow: It might be complex to understand hospitals' routing to go through. Indoor navigation guides the person inside the hospital using indoor navigation technologies.

Table 1. List of the smart enabling technologies (SETs) and their potential application.

Tech. No.	Technology concept	Measure/senses/support	Application and deployment in smart hospital
T1	Motion sensor (e.g., PIR [1, 2])	Motion/direction	Detecting motions and presence
T2	Outdoor positioning system	GPS tracker [9]	Tracking patients outdoor
T3	Indoor Positioning (e.g., RFID) [1]	Position & proximity/ distance	Detecting and identifying people
T4	Indoor navigation system [10, 11]	Bluetooth Low Energy (BLE) beacon	Guiding patients and visitors to the target destination in the hospital building
T5	Video camera [12]	Real time videos	Real time monitoring
T6	Speech/Voice recognition engine [13]	Senses voice and speech	Provide for natural, multimodal interaction and control
T7	Head Mounted Display (HMD) [14]	Supporting technologies	virtual reality for displays and visualization, simulation and training
T8	Smart watch [6]	GPS, accelerator, thermometer, etc	Tracking physiological parameters, indoor and positioning
T9	EMG (Electromyography) [1, 5]	Detect muscle malfunctions/injuries	Monitors motor health and detect movement disorder for patients and staff
T10	Electrocardiography (ECG) [1, 4]	Detect cardiac abnormalities	Monitors cardiac parameters for patient and staff
T11	Contact thermometry (Skin temperature) [1]	Measures body temperature	Monitors temperature of patients and hospital staff
T12	Activity tracker wearables [15]	Tracking physical activities	Monitor physiological health parameters like heart beat rate, blood pressure, etc
T13	Computer brain interface (CBI) [8]	Use brain signals for controlling devices	Enabling disable to communicate

(*continued*)

Table 1. (*continued*)

Tech. No.	Technology concept	Measure/senses/support	Application and deployment in smart hospital
T14	Eye tracking system [7]	Measures eyes movement	Uses measures of eyes movements to control devices or monitor behavior
T15	Smoke detector [16]	Sense smoke	Detect smoke at hospital rooms
T16	Fall detector [17]	Detect fall or acceleration	Monitor patent or staff to detect fall or measure their acceleration
T17	Bed pressure sensor (Medical) [3]	Patient wakeup sensor	Monitor patient sleeping and wakeup to ensure safety
T18	Wireless network router (e.g., WiFi, ZigBee, Bluetooth) [1, 5]	Wireless data communication	Necessary to enable internet access, data communication and indoor/outdoor positioning
T19	Control software (e.g., Home Gateway, etc.) [17]	control of sensors and actuators	Connect, control and coordinate various sensors, actuators

In fact, the literature on smart healthcare systems development is continuing rapidly due to the accelerated progress and continual emergence of new SETs supporting technologies including IoT (Internet-of-Things), Cloud computing, etc. Khan and Chattopadhyay [18] presented a smart health monitoring system that uses biomedical sensors to check patient's condition and report it through the internet to inform the concerned doctors either through an LCD monitor or through a developed smart phone mobile application. Bansal and Gandhi [19] emphasized the role of utilizing IoT together with Big Data technologies in transmitting ECG sensors data from patient to the doctors and to other concerned persons for effective and efficient smart health monitoring. Almazroa et al. [20], proposed a smart sensing mobile application that is able to collect data coming from the body temperature sensor through the network. Kamruzzaman [21] proposed architecture of a smart healthcare system using artificial intelligence as a tool for optimizing patient care plans. Rajakumari et al. [22] presented convolution neural network based smart hospital and patient scheduling system, to assist patients comfortably while planning to visit the clinic and without any heavy burden to consult doctor for their needs. Zhao et al. [23] presented a smart IoT platform in hospital that can relieve postoperative pain. Ahmid et al. [24] proposed an intelligent health monitoring system based on based on IoT and agents.

3 Smart Clinic Design: Model Building Blocks

The proposed approach for designing a smart healthcare system environment is conceptually based on understanding the relationships among three model building blocks:

- The **smart enabling technologies (SETs)**: sensors, actuators, microcontrollers and other basic software and hardware constituents that shall be installed in the hospital or healthcare facility.
- The **healthcare system smartness features (HSFs)**: the intelligent characteristics, aspects or attributes that will be introduced by installing the SETs.
- The **healthcare system operational objectives (HOOs):** the objectives of the hospital management operations and activities that are significant to achieve.

The logic for the three building blocks is based on the following intuitive statement. The implementation of the Smart enabling technologies (SETS) achieve specific intended set of smartness attributes which in their turn can positively impact the realization of a set of hospital operational objectives.

Next section, we introduce the methodology for designing smart healthcare using fuzzy relations.

4 Designing Healthcare Using Fuzzy Relations

Fuzzy relations have been widely used in several fields like in fuzzy control, clustering, and uncertainty reasoning. The concept of fuzzy set theory was introduced in 1965 by Zadeh [25]. A fuzzy relation is defined also on the Cartesian product of two universal sets. Let, $X, Y \subseteq R$ be universal sets, then:

$$R = X \times Y = \{(x, y), \mu_R(x, y) | x \in X, y \in Y\} \tag{1}$$

$$R : X \times Y \rightarrow [0, 1] \tag{2}$$

R is called a fuzzy relation in $X \times Y \subseteq R$. The value 1 means complete membership of pair *(x,y)* in relation *R*, whereas, the value 0 means complete non-membership, and any value in between means intermediate degree of membership in R. A fuzzy relation *R* is usually presented in the form of a two-dimensional $m \times n$ matrix:

$$R = \begin{matrix} x_1 \\ x_2 \\ \dots \\ x_m \end{matrix} \begin{bmatrix} \mu_R(x_1, y_1) & \dots\dots & \mu_R(x_1, y_n) \\ \dots & \dots\dots & \dots \\ \dots & \dots\dots & \dots \\ \mu_R(x_m, y_1) & \dots\dots & \mu_R(x_m, y_n) \end{bmatrix} \tag{3}$$

where $\mu_R(x_i, y_j)$ are membership functions for pair(x_i, y_j)intherelation. This means that if the relation Q is named “Greater than or equal”, then $\mu_Q(x_1, y_1)$ is the membership grade of the relation between the pair of elements (x_1, y_1), which can take on values from 0 to 1, depending on the viewpoint or judgement of the decision analysts.

In this research, the assessed fuzzy relation between the model building blocks is: “**Positively impact**”. That is, we are interested in quantifying the positive impact influence among these building block, in order to adopt the enabling technologies (SETs) with highest positive impact. There are two fuzzy relations among the three building blocks of the proposed model: the smart enabling technologies (SETs), Healthcare system Smart features (HSFs) and the Healthcare system operational objectives (HOOs). The main goal is to compose the two relations using the max-min composition operation, to arrive at the direct positive impact relation between the available SETs and the HOOs, so that this can represent a necessary guide towards their prioritization, adoption, and implementation. The construction of fuzzy relation matrices that quantify relations among the mentioned three sets of the model building blocks (i.e., the SETs, smart features, and the operational objectives) requires a unified numerical scale to assign relation membership a linguistic value with a corresponding numerical value for each element or relation pair. This scale will express the degree of impact among the elements of the positive impact relation. Therefore, in order to simplify assignment process, the decision analysts can assign a linguistic term with corresponding membership values for each relation pair (see Table 2).

The membership values of the fuzzy relation “Positively impact” is graded from 0 up to one. Negative and neutral impacts are considered as zero and expressed linguistically as “Never”. We are only interested in knowing the positive impact relations for the smart enabling technologies on the operational objectives. Using the above membership, the decision analyst can quantify the relation membership value between each two elements of any two related sets, as defined in linguistic scale Table 2.

Table 2. Fuzzy relations linguistic and numerical membership values scale

Linguistic values of fuzzy relation “positively impact” (PI)	Abbreviation	Membership value (μ)
“Never”	N	0
“Very slightly”	V	0.2
“Slightly”	S	0.4
“Moderately”	M	0.6
“Highly”	H	0.8
“Extremely”	E	1

In the proposed model of the methodology, we have three sets that constitute the model building blocks. Consequently, we can investigate 2 impact relations among these three sets. They are as follows:

- Relation 1: Smart enabling technologies (SET) – Healthcare system smartness features (HSF).
- Relation 2: Smart Features (HSF) – Healthcare system operational objectives (HOO).

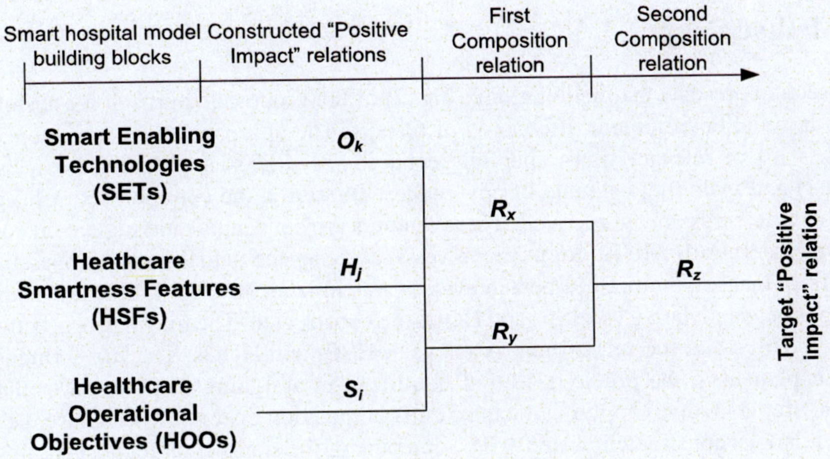

Fig. 1. Compositions of fuzzy relations to attain the SETs – HOOs composed fuzzy relation.

The Max-Min composition operator [25, 26] is used to combine two fuzzy relations between three sets of elements in order to obtain derived composed relations between the two indirectly related sets. The operator is formally defined as follows:

Let X, Y, Z be universal sets and let R and Q be relations that relate elements from X to Y and from Y to Z, respectively, i.e.

$$R = \{((x, y), \mu_R(x, y))\}\ x \in X, y \in Y, R \subset X \times Y \tag{4}$$

$$Q = \{((y, z), \mu_Q(y, z))\}\ y \in Y, z \in Z, Q \subset Y \times Z \tag{5}$$

Then, S will be a relation that relates elements in X that R contains to the elements in Z that Q

$$S = R^\circ Q = \{((x, z), \mu_s(x, z))\}, x \in X, z \in Z, S \subset X \times Z \tag{6}$$

Max-Min composition is then defined as

$$\mu_s(x, z) = \max_{y \in Y}(\min(\mu_R(x, y), \mu_Q(y, z))) \tag{7}$$

Example: Fuzzy relations and composition operator:

The Max-Min composition is similar to multiplication of matrices, but instead of multiplication and then addition, we apply the Min then Max operators. As an example, Let R and Q be as follows:

$$R = \begin{bmatrix} 0.4 & 0.6 \\ 0 & 1 \end{bmatrix}, Q = \begin{bmatrix} 0.5 & 0.25 & 0.4 \\ 0.7 & 1 & 0.8 \end{bmatrix}, \text{ then } S = R^{\circ}Q = \begin{bmatrix} 0.6 & 0.6 & 0.6 \\ 0.7 & 1 & 0.8 \end{bmatrix}$$

Next section introduces the methodology upon which the proposed smart hospital environment is based.

5 Methodology

This section presents the methodology (Fig. 2) of the proposed approach for designing smart hospital environment. It consists of 6 steps. The approach starts with **step 1,** in which a list of relevant smart enabling technologies (SETs) is prepared (as a list in Table 1) and including estimate of cost ranges. **In step 2**, the common set of hospital smartness features are enumerated. These features are conceptual and logical outcomes of implementing the SETs. **Step 3** involves specifying the actual hospital operational objectives that are required to be achieved or maximized. **In step 4**, the fuzzy impact relations between the SETs, HSFs and HOOs are constructed. The first matrix is between SETs & HSFs and the second matrix between HSFs and HOOs. The fuzzy linguistic scale explained in the previous section is utilized in assigning fuzzy impact relation values. **Step 5** involves applying the max-min composition operator to compose the two fuzzy relations constructed in step 4. The outcome of the step 5 is target positive impact relation between the SETs and the HOOs, which is the result of applying the composition operator to the aforementioned two fuzzy relations (i.e., fuzzy relation SETs – HSFs & fuzzy relations HSFs and HOOs). **In step 6**, the priority of each SETs is computed by summing its row in the composed fuzzy matrix (SETs – HOOs).

These priorities are utilized eventually for specifying the important smart enabling technologies, and hence design the smart healthcare system based on these priorities.

Next section, the example will illustrate the application of the proposed approach.

6 Illustrative Case Example

A clinic (Fig. 2) is currently investigating the potential of improving its operational performance through investing in SETs to develop smart hospital environment. Eleven smartness features were considered: F1: Location aware asset tracking, F2: Patient/Staff location tracking, F3: Patient/Staff Physiological activity monitoring and tracking, F4: Location-based services (patient, visitor, nurse and physician), F5: Automated patient check-in (no wait), F6: Indoor navigation, F7: Context aware data and analytics, F8: Smart energy management, F9: Smart access and security, F10: Proactive safety response and F11: Multimodal/Natural interaction.

The clinic's management has specified seven operational objectives which are desired to improve. They are: O1: Speed up patient check-in, O2: Optimize patient flow, O3: Enhance information support, O4: Speed up diagnosis, O5: Improve safety and security, O6: Minimize energy consumption, O7: Track people and assets. Now, implementing the proposed methodology, fuzzy relations max-min composition operator is utilized to compose the two constructed fuzzy relations (Fig. 3). This step yields the most valuable outcome of the proposed approach which is the overall priority ranking of each SET (Table 3) below.

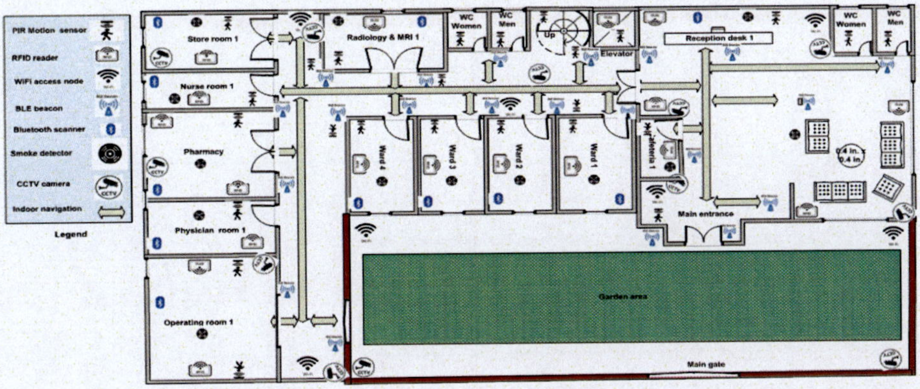

Fig. 2. Illustrative case example: Designing a smart clinic.

HSF

SET	F_1	F_2	F_3	F_4	F_5	F_6	F_7	F_8	F_9	F_{10}	F_{11}
T_1	0.8	1	0.6	1	0.8	0.8	0.8	1	1	1	1
T_2	1	1	1	1	0.6	0.2	0.4	0	0.4	1	0.8
T_3	1	1	0.8	1	0.8	1	0.8	0.6	0.8	1	1
T_4	0.6	0.6	0.2	1	1	1	0.8	0.4	0.4	0.8	0.6
T_5	0.8	0.8	0.8	0.4	0.6	0.6	0.4	0.2	1	1	0.8
T_6	0	0	0	0.6	0.8	0.8	0.4	0.8	1	0.6	0.4
T_7	0	0	0	0	0	0.8	0.6	0	0	0.4	0
T_8	0.2	1	1	0.8	0.4	1	0.8	0.4	0.4	0.8	0.6
T_9	0	0	1	0	0	0	1	0	0	0.6	1
T_{10}	0	0	1	0	0	0	1	0	0	0.8	1
T_{11}	0	0	1	0	0	0	1	0	0	1	1
T_{12}	0	0	1	0	0	0	1	0	0	1	1
T_{13}	0	0	1	0	0	0	0.4	0	0	1	1
T_{14}	1	0	0	1	0	0	0.6	0	0	1	1
T_{15}	0	0	1	0	0	1	0.8	0	0	0	1
T_{16}	0	0	1	0.4	0	1	0.8	0	0	0.8	1
T_{17}	1	1	1	1	1	1	1	1	1	1	1
T_{18}	1	1	1	1	1	1	1	1	1	1	1
T_{19}	1	1	1	1	1	1	1	1	1	1	1

○

OO

SF	O_1	O_2	O_3	O_4	O_5	O_6	O_7
F_1	0.8	0.8	0.8	0	1	0.8	1
F_2	0.8	0.8	1	0.4	1	1	1
F_3	0	0.2	1	0.8	1	1	1
F_4	0.8	0.8	1	0.4	1	1	1
F_5	1	0.4	1	0	0.8	0	0
F_6	0	1	1	0	1	0.4	0.8
F_7	0.8	0.6	1	1	1	0.8	0.8
F_8	0	0	0	0	0.4	1	1
F_9	0.8	0.6	0.8	0	1	0	1
F_{10}	0.8	0.8	0.8	0.6	0.8	0.8	1
F_{11}	0.6	0.4	0.8	0.6	0.6	0.6	0.2

=

Result

HOO

SET	O_1	O_2	O_3	O_4	O_5	O_6	O_7
T_1	0.8	0.8	1	0.8	1	1	1
T_2	0.8	0.8	1	0.8	1	1	1
T_3	0.8	1	1	0.8	1	1	1
T_4	1	1	1	0.8	1	1	1
T_5	0.8	0.8	0.8	0.8	1	0.8	1
T_6	0.8	0.8	0.8	0.6	1	0.8	1
T_7	0.6	0.8	0.8	0.6	0.8	0.6	0.8
T_8	0.8	1	1	0.8	1	1	1
T_9	0.8	0.6	1	1	1	1	1
T_{10}	0.8	0.8	1	1	1	1	1
T_{11}	0.8	0.8	1	1	1	1	1
T_{12}	0.8	0.8	1	1	1	1	1
T_{13}	0.8	0.8	1	0.8	1	1	1
T_{14}	0.8	0.8	1	0.6	1	1	1
T_{15}	0.8	1	1	0.8	1	1	1
T_{16}	0.8	1	1	0.8	1	1	1
T_{17}	1	1	1	1	1	1	1
T_{18}	1	1	1	1	1	1	1
T_{19}	1	1	1	1	1	1	1

Fig. 3. Composition of the two fuzzy relation matrices using the max-min composition operator.

Table 3. The overall priority of the SETs.

SET	Operational objectives							Σ	Rank
	O1	O2	O3	O4	O5	O6	O7		
T1	0.8	0.8	0.6	1	1	1	1	6.2	3
T2	0.8	0.8	1	1	1	1	1	6.6	2
T3	0.8	1	0.8	1	1	1	1	6.6	2
T4	1	1	0.2	1	1	1	1	6.2	3
T5	0.8	0.8	0.8	0.4	1	0.8	1	5.6	5
T6	0.8	0.8	0	0.6	1	0.8	1	5	7
T7	0.8	0.8	0	0	0.8	0.8	0.8	4	8
T8	0.8	1	1	0.8	1	1	1	6.6	2
T9	0.8	0.6	1	0	1	1	1	5.4	6
T10	0.8	0.8	1	0	1	1	1	5.6	5
T11	0.8	0.8	1	0	1	1	1	5.6	5
T12	0.8	0.8	1	0	1	1	1	5.6	5
T13	0.8	0.8	1	0	1	1	1	5.6	5
T14	0.8	0.8	0	1	1	1	1	5.6	5
T15	0.8	1	1	0	1	1	1	5.8	4
T16	0.8	1	1	0.4	1	1	1	6.2	3
T17	1	1	1	1	1	1	1	7	1
T18	1	1	1	1	1	1	1	7	1
T19	1	1	1	1	1	1	1	7	1

7 Conclusion

In this article, we have proposed a simple, structured, systematic, and objective approach for designing smart healthcare facilities in general. The proposed approach, in the lack of similar methodologies, contribute significantly to improving the practice of developing smart healthcare environment, while talking into account the available budget and improvement priorities. The proposed methodology is simple, practical, and computationally efficient. It helps organize and simplify the design problem space to those technologies with high priorities, and at the same time satisfy the budget constraints. We have devoted more focus on the logical steps to attain the priorities of the smart technologies alternatives and used a logical deployment scenario of the smart technologies in the smart clinic design example.

The most significant notion conveyed by this article is that the design of smart healthcare environment should be based on the impact relation analysis of the enabling technologies and their priorities and should be naturally limited by the available budget of the development project.

Acknowledgements. This work was conducted within the project Ambient intelligence in decision-making problems in uncertainty conditions (2019B0008) funded through the IGA foundation of the Faculty of Economics and Management, Czech University of Life Sciences in Prague.

References

1. Rashidi, P., Mihailidis, A.: A survey on ambient-assisted living tools for older adults. IEEE J. Biomed. Health Inform. **17**(3), 579–590 (2013)
2. Chan, M., Est'eve, D., Escriba, C., Campo, E.: A review of smart homes—present state and future challenges. Comput. Methods Prog. Biomed. **91**(1), 55–81 (2008)
3. Monroy, E.B., Rodríguez, A.P., Estevez, M.E., Quero, J.M.: Fuzzy monitoring of in-bed postural changes for the prevention of pressure ulcers using inertial sensors attached to clothing. J. Biomed. Inform. **107**, 103476 (2020)
4. Riyadi, M.A., Iskandar, I.A., Rizal, A.: Development of FPGA-based three-lead electrocardiography. In: International Seminar on Intelligent Technology and Its Applications (ISITIA), pp. 67–72 (2016)
5. Han, H., Ma, X., Oyama, K.: Towards detecting and predicting fall events in elderly care using bidirectional electromyographic sensor network. In: 15th International Conference on Computer and Information Science (ICIS), IEEE/ACIS (2016)
6. Reeder, B., David, A.: Health at hand: a systematic review of smart watch uses for health and wellness. J. Biomed. Inform. **63**, 269–276 (2016)
7. Bissoli, A., Lavino-Junior, D., Sime, M., Encarnação, L., Bastos-Filho, T.: A Human-Machine Interface Based on Eye Tracking for Controlling and Monitoring a Smart Home Using the Internet of Things Sensors **19**, 859 (2019). https://doi.org/10.3390/s19040859
8. Vourvopoulos, A.B., Badia, S.B.: Usability and cost-effectiveness in brain-computer interaction: is it user throughput or technology related? In: AH 2016: Proceedings of the 7th Augmented Human International Conference 2016, (19), pp. 1–8 (2016). https://doi.org/10.1145/2875194.2875244.
9. He, Z.M., Peng, L., Han, H.Y., Lu, H., Wang, Z.F., Zhao, P.: Research on Indoor and Outdoor Comprehensive Positioning Technology Based on Multi-Source Information Assistance, Procedia Computer Science **166**, 361–365 (2020)
10. Mohsin, N., Payandeh, S., Ho, D., Gelinas, J.P.: Study of activity tracking through bluetooth low energy-based network. Hindawi J. Sensors, 21 (2019)
11. AL-Madani, B., Orujov, F., Maskeliunas, R., Damasevicius, R., Venckauskas, A.: Fuzzy logic type-2 based wireless indoor localization system for navigation of visually impaired people in buildings, MDPI. Sensors, **19**(9), 2114 (2019). https://doi.org/10.3390/s19092114
12. Rougier, C., Meunier, J., St-Arnaud, A., Rousseau, J.: Monocular 3D head tracking to detect falls of elderly people. In: 28th Annual International Conference of the IEEE Engineering in Medicine and Biology Society, EMBS 2006 (2006)
13. Bhavsar, M., Kosaraju, P., Anan-thakrishnan, G., Subray Shet, G., Anand, S.: Dynamic improvements in a cloud based speech recognition engine by incorporating trending data. In: 4th IEEE International Conference on Mobile Cloud Computing, Services, and Engineering (MobileCloud), pp. 60–66 (2016)
14. Ehsan, A., Jayfus, D., Peter, K.: Augmented reality goggles with an integrated tracking system for navigation in neurosurgery. In: Proceedings of IEEE Virtual Reality, pp. 123–124, Orange County, CA (2012)
15. Shin, G.D.: Investigating the impact of daily life context on physical activity in terms of steps information generated by wearable activity tracker. Int. J. Med. Inform. **141**, 104222 (2020)

16. Festag, S.: Analysis of the effectiveness of the smoke alarm obligation – experiences from practice. Fire Safety J. 103263 (2020)
17. Gomez, C., Paradells, J.: Wireless home automation networks: a survey of architectures and technologies. IEEE Commun. Mag. **48**(6), 92–101 (2010)
18. Skubic, M., Alexander, G., Popescu, M., Rantz, M., Keller, J.: A smart home application to eldercare: current status and lessons learned. Technol. Health Care, **17**(3), 183–201 (2009)
19. Khan, T., Chattopadhyay, M.K.: Smart health monitoring system. In: 2017 International Conference on Information, Communication, Instrumentation and Control (ICICIC) (2017)
20. Bansal, M., Gandhi, B.: IoT & Big Data in Smart Healthcare (ECG Monitoring). In: 2019 International Conference on Machine Learning, Big Data, Cloud and Parallel Computing (COMITCon) (2019)
21. Almazroa, A., Alsalman, F., Alsehaibani, J., Alkhateeb, N., AlSugeir, S.: Easy clinic: smart sensing application in healthcare. In: 2nd International Conference on Computer Applications & Information Security (ICCAIS) (2019)
22. Kamruzzaman, M.M.: Architecture of smart health care system using artificial intelligence. In: 2020 IEEE International Conference on Multimedia & Expo Workshops (ICMEW) (2020)
23. Rajakumari, K., Madhunisha, M.: Intelligent and convolutional-neural-network based smart hospital and patient scheduling system. In: 2020 International Conference on Computer Communication and Informatics (ICCCI) (2020)
24. Zhao, Y., Ge, S., Feng, Y.: Smart IoT data platform in hospital and postoperative analgesic effects of orthopedic patients. Procedia Comput. Sci. (2019)
25. Ahmid, M., Kazar, O., Benharzallah, S., Kahloul, L., Merizig, A.: An intelligent and secure health monitoring system based on agent. In: IEEE International Conference on Informatics, IoT, and Enabling Technologies (ICIoT) (2020)
26. Zadeh, L.A.: Fuzzy sets. Inf. Control **8**, 338–353 (1965)
27. Fuller, R.: Introduction to Neuro-Fuzzy Systems, Advances in Soft Computing Series, Springer-Verlag, Berlin/Heildelberg (2000)

Author Index

R. Silhavy et al. (Eds.): CoMeSySo 2021, LNNS 231, pp. 1053–1056, 2021.
https://doi.org/10.1007/978-3-030-90321-3